International Conference on Applied Chemistry and Industrial Catalysis (ACIC 2019)

Clean Production and High Value Utilization

IOP Conference Series: Materials Science and Engineering
Volume 729

Shenzhen, China
15 - 17 November 2019

ISBN: 978-1-7138-2588-3
ISSN: 1757-8981

Printed from e-media with permission by:

Curran Associates, Inc.
57 Morehouse Lane
Red Hook, NY 12571

Some format issues inherent in the e-media version may also appear in this print version.

Institute of Physics
Dirac House, Temple Back
Bristol BS1 6BE UK

Phone: 44 1 17 929 7481
Fax: 44 1 17 920 0979

techtracking@iop.org

Additional copies of this publication are available from:

Curran Associates, Inc.
57 Morehouse Lane
Red Hook, NY 12571 USA
Phone: 845-758-0400
Fax: 845-758-2633
Email: curran@proceedings.com
Web: www.proceedings.com

TABLE OF CONTENTS

IOP Conf. Series: Materials Science and Engineering **729** (2020) 011001 doi:10.1088/1757-899X/729/1/011001

Preface

2019 International Conference on Applied Chemistry and Industrial Catalysis (ACIC 2019) is a significant scientific event, which was recently held in Shenzhen, China during November 15-17, 2019. This conference was held by several top universities in China, such as Xinjiang University and Zhengzhou University, and some provincial key laboratories. The theme of this conference is "Clean Production and High Value Utilization". It brought together experts and scholars from the Chinese government, top universities, business associations, research centers and high-tech enterprises to discuss how to reduce environmental footprint at source and produce high value-added end products in chemical manufacturing. This conference is committed to building an enabling platform for the cooperation among the Chinese government, Chemical industry and scholars. The ultimate goal is to build a bridge between R&D results and the Chemical industry.

We were honored to have Prof. Xianyong Wei from XinJiang University/China University of Mining and Technology, as our Conference Chairman, who is the High-level academic leader of XinJiang University. The Conference Vice-Chairmen consist of Prof. JianSheng Zhang, the Executive vice President of Qinghua University, Prof. Jianglong Yu from Liaoning University of Science and Technology, and Prof. Sheng Han, the Director of science and technology from Shanghai Institute of Technology. The conference program mainly included keynote speeches. We were thrilled to have excellent speakers Prof. Xianyong Wei from Xinjiang University/ China University of Mining and Technology, China, and Prof. Linbing Sun from Nanjing Tech University, China to present marvelous speeches on Applied Chemistry and Industrial Catalysis, whose insightful talks gained continuous applause on site.

We would like to express special gratitude to members of the conference committee and organizers of the conference. We would also like to thank all distinguished speakers, Chairman, and participants for their great supports. Finally, we want to thank the authors for submitting their papers, which were compiled to the Proceedings of ACIC 2019 and edited to meet the requirements of publication standard. We hope that all the articles in this conference can provide valuable summary and reference for researchers in this field.

The Organizing Committee of ACIC 2019

IOP Conf. Series: Materials Science and Engineering **729** (2020) 011001 doi:10.1088/1757-899X/729/1/011001

Committee member

Conference Chairman

Prof.Xianyong Wei, XinJiang University/China University of Mining and Technology

Conference Vice-Chairman

Prof. JianSheng Zhang, Qinghua University/Taiyuan University of Technology
Prof.Jianglong Yu, Liaoning University of Science and Technology
Prof. Sheng Han, Shanghai Institute of Technology

Special Guest

Prof. Richard Axelbaum, Washington University in St. Louis
Prof. Yifan Han, East China University of Science and Technology/Zhengzhou University, China
Prof. Haoquan Hu, Dalian University of Technology, China
Prof. Qingixa(Chad) Liu, Universit of Alberta
Prof. Zhongwen Liu, Shanxi Normmal University, China
Prof. Jieshan Qiu, Beijing University of Chemical Technology, China
Prof. Guangwen Xu, Shenyang University of Chemical Technology, China
Prof.Hongbo Zeng, Universit of Alberta

Academic Committee

Prof. Jin Bai, Shanxi Institute of Coal Chemistry, Chinese Academy of Sciences, China
Prof. Jingpei Cao, China University of Mining and Technology, China
Prof. Zhongxiu Chen, Zhejiang Gongshang Universtiy, China
Prof. Hongxing Dai, Beijing University of Technology, China
Prof. Yiwei Fei, Air Force Logistics University, China
Prof. Xueli Huang, XinJiang University , China
Prof. Xun Hu, University of Jinan, China
Prof. Min Ji, Dalian University of Technology, China
Prof. Ruifeng Li, Taiyuan University of Technology, China
Prof. Xuanke LI, Wuhan University of Science and Technology, China
Prof. Xian Li, XinJiang University, China
Prof. Younian Liu, Central South University, China
Prof. Lei Liu,Taiyuan University of Technology, China
Prof. Fengyun Ma, XinJiang University, China
Prof. Jun Shen, Taiyuan University of Technology , China
Prof. Hengfu Shui, Anhui University of Technology, China
Prof. Linbing Sun, Nanjing Tech University , China
Prof. Shujiang Tu,Jiangsu Normal University , China
Prof. Baojun Wang, Taiyuan University of Technology, China
Prof. Guanghua Wang, Wuhan University of Science and Technology, China
Prof. Jiancheng Wang, Taiyuan University of Technology, China
Prof. Weize Wu, Beijing University of Chemical Technology, China
Prof. Dulin Yin, Hunan Normal University , China
Prof. Jianglong Yu, Liaoning University of Science and Technology, China
Prof. Yaxin Zhang, Xinjiang University , China
Prof. Tiansheng Zhao, Ningxia University, China
Prof. Anning Zhou, Xi'an University of Science and Technology, China
Prof. Jianbin Zhou, Nanjing Forestry University, China
Prof. Shixue Zhou, Shandong University of Science and Technology, China
Prof. Zhimin Zong, China University of Mining and Technology, China

Organizing Committee

Hongcun Bai, Ningxia University, China
Subing Fan, Ningxia University , China
Prof. Xing Fan, Xinjiang University , China
A.Prof. Gang Feng, Nanchang University , China
Prof. PingQiang Gao, Yulin University , China
Prof. Jianqiang Hu, Air Force Logistics University , China
Prof. Lijun Jin, Dalian University of Technology , China
A.Prof. Yuhong Kang, Yulin University , China
Prof. Xiaoyong Lai, Ningxia University , China
Prof. Zhiping Lei, Anhui University of Technology, China
Prof. Peng Liang, Shandong University of Science and Technology, China
Prof. Chunli Liu, Zaozhuang University, China
Prof. Yi Liu, Xuzhou Medical University, China
Ziwu Liu, China University of Mining and Technology, China
Prof. Zhongyi Liu, Zhengzhou University, China
Prof. Yajun Ma, Yulin University, China
A.Prof. Qingxiang Ma, Ningxia University, China
Prof. Guoyu Ren, Ningxia University, China
Prof. Shibiao Ren, Anhui University of Technology, China
Prof. Sergei Manzhos, Université du Québec
Wenzhong Shen, Shanxi Institute of Coal Chemistry, Chinese Academy of Sciences, China
Prof. Yaguang Sun, Shenyang University of Chemical Technology, China
Prof. Jichang Wang, University of Windsor
Prof. Kangjun Wang, Shenyang University of Chemical, Technology, China
Prof. Xiaodong Wang, North China Electric Power University, China
Prof. Yi Wang, Huazhong University of Science and Technology, China
Prof. Guanghui Wang, Wuhan University of Science and Technology, China
A.Prof. Yugao Wang, Taiyuan University of Technology, China
Prof. Zhicai Wang, Anhui University of Technology, China
Prof. Shiyong Wu, East China University of Science and Technology, China
Prof. Lin Wu, Wuhan University of Science and Technology, China
Prof. Long Xu, Northwest University, China
Prof. Biao Yan, Yulin University, China
Prof. Long Yan, Yulin University, China
Prof. Xueqiong Yin, Hainan University, China
Prof. Xinhua Yuan, Jiangsu University, China
Jianli Zhang, Ningxia University, China
Prof. Riguang Zhang, Taiyuan University of Technology, China
Prof. Shu Zhang, Nanjing Forestry University, China
Prof. Wei Zhao, China University of Mining and Technology, China
Prof. Yunpeng Zhao, China University of Mining and Technology, China
Prof. Zhifang Zhang, Yulin University, China
Prof.Tao Zhu, China University of Mining and Technology, (Beijing), China

IOP Conf. Series: Materials Science and Engineering **729** (2020) 011002 doi:10.1088/1757-899X/729/1/011002

Peer review statement

All papers published in this volume of *IOP Conference Series: Materials Science and Engineering* have been peer reviewed through processes administered by the proceedings Editors. Reviews were conducted by expert referees to the professional and scientific standards expected of a proceedings journal published by IOP Publishing.

IOP Conf. Series: Materials Science and Engineering **729** (2020) 012001 doi:10.1088/1757-899X/729/1/012001

The Performance of Enhancement MOSFET—The trade-off on Transfer and Output Characteristics

Wentao Zhang

Fengqi Road No.580, Hangzhou No.14 High School, Hangzhou, Zhejiang Province, 310006, China

*Corresponding author's e-mail: angela@cas-harbour.org

Abstract. MOSFET (Metal-Oxide-Semiconductor Field-Effect Transistor) is commonly used in electronic devices, and one of its major applications is power switching. The main research topic of the paper is to introduce the principles of functioning of n-channel enhancement MOSFET, discuss the trade-off in MOSFET when one factor is changed in order to improve one aspect of its performance and the method to improve the condition of trade-off. The author analyzes the trade-off problems when improving one of the transfer characteristics and output characteristics of MOSFET, respectively by decreasing the thickness of dielectric layer and changing the length and width of the channel. The author demonstrates the results will occur when decreasing the thickness of dielectric layer and length of channel, and increasing the width of channel; at last author introduces the method to relieve the problems of trade-off by applying third generation semiconductor.

1. Introduction

Currently, the size of the transistor needs to be decreased in order to place more of them into an integrated circuit. An ideal transistor needs to have an excellent performance on its transfer characteristics and output characteristics, high resistance to high voltage and operating temperature. When engineers try to improve one aspect of the devices, there is always a trade-off, meaning that engineers have to make a compromise among the trade-off of all factors.

The first part of the paper analyzes the functioning process of MOSFET from the aspect of its structural design. Then, the paper discusses the transfer and output characteristics of devices, the trade-off for both characteristics when people try to achieve one parameter to an expected magnitude and the side effect occurring respectively when the dielectric layer gets thinner, channel length is decreased, and channel width is increased. At last, the paper introduces the method to improve the trade-off problems by applying third generations semiconductor. The main methods applied in the research are qualitatively analyzing the formula to predict the theoretical change of one factor and comparing the functions using graphs to illustrate the change under different conditions.

2. The Operation of MOSFET

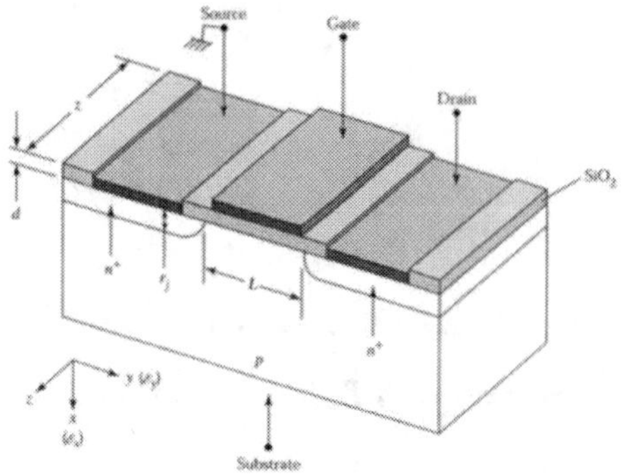

Figure 1. Perspective view of n-channel enhancement MOSFET [1]

When n-type semiconductor joints with p-type semiconductor, since n-type material contains a large concentration of electrons and few holes (vice verse for p-type material), carrier diffusion occurs, which electrons diffuse to p-side while holes diffuse to n-side and holes diffuse to the opposite direction, leaving p-side fixed negative acceptor ions and n-side positive donor ions near the junction. Thus, space charge region is formed at the junction, and an electric field from n-side to p-side is formed. At the same time, due to this inner electric field, few electrons in p-side drift to n-side and few holes in n-side drift to the p-side, which have the opposite direction with diffusion, causing the space charge region to narrow.

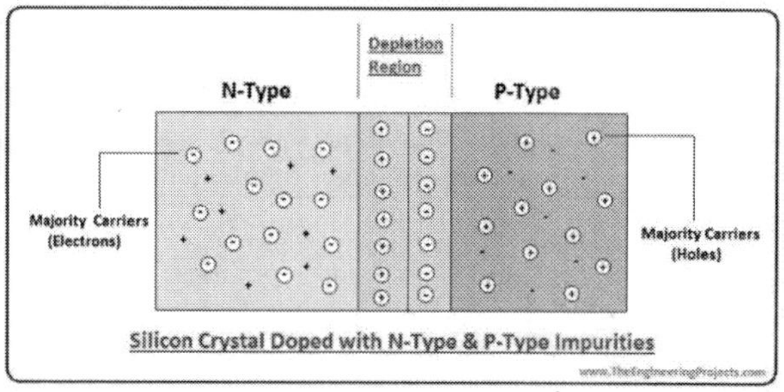

Figure 2. A closer look at pn-junction [2]

As the process of diffusion and drift reaches balance, the final space charge region is formed, which is also called depletion region, and this region is pn-junction. Pn-junction is commonly used in diode. Under the forward bias, the resistance is low; under the reverse bias, the resistance is extremely high. In MOSFET, two pn-junction are formed between source and substrate and drain and substrate (as the depletion region shown in Fig. 2). When voltage is applied at source and drain, no current flow because there will reverse bias always exist regardless of the direction of otter electric.

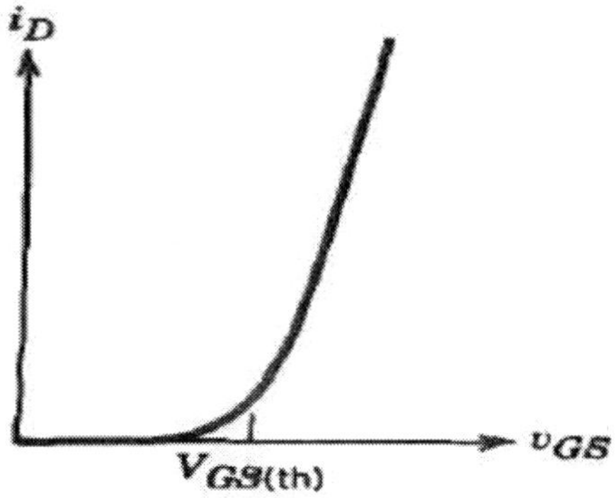

Figure 3. The graph of transfer characteristics of n-channel enhancement MOSFET [3]

However, when there is gate voltage applied, which generates an electric field pointing downwards, the electrons in source and drain will be attracted and flow to the gate. Because of the block of dielectric layer, electrons will concentrate on the channel which is between n+ source and n+ drain. Thus, the current is able to flow through the channel. The gate voltage can control the amount of electrons flows from the source to the drain. This represents the transfer characteristics of MOSFET. The gate voltage should first reach threshold voltage so that transistor has current passing through can leaves off-state; otherwise, even if the gate voltage increases, the drain current is still nearly zero (shown in Fig. 3 as a horizontal line). After threshold voltage is reached, as gate voltage increases, the thickness of channel increases since more electrons are attracted, so the amount of electrons flowing increases; when gate voltage decreases, the thickness of channel decrease, so the amount of electrons flowing decreases. This property of MOSFET is widely used in power switching and resistance control.

Drain voltage also has an effect on channel, and this represents the output characteristic of MOSFET. Under the condition that gate voltage is applied, when a small drain voltage is applied, electrons flows from the source to the drain through the channel. The channel acts as a resistor, and the drain current is proportional to the drain voltage. This can be presented on the graph as linear region (shown in Fig. 4). When the drain voltage increases and reaches saturation voltage, the thickness of the channel near the drain is reduced to zero. This is called the pinch-off point (the red curve in Fig. 4 presents the locus of Vsat vs. Isat). Beyond the pinch-off point, the drain current remains the same. This is the saturation region, where drain current is constant no matter how much the drain voltage increases. In Fig. 4, the saturation region is presented as horizontal lines.

Generally speaking, the keypoints of how n-channel enhancement MOSFET works are that under the control of gate, electrons move from source to drain, and thus current is formed. And the movable carries in MOSFET are attributed to the properties of n-type and p-type semiconductors.

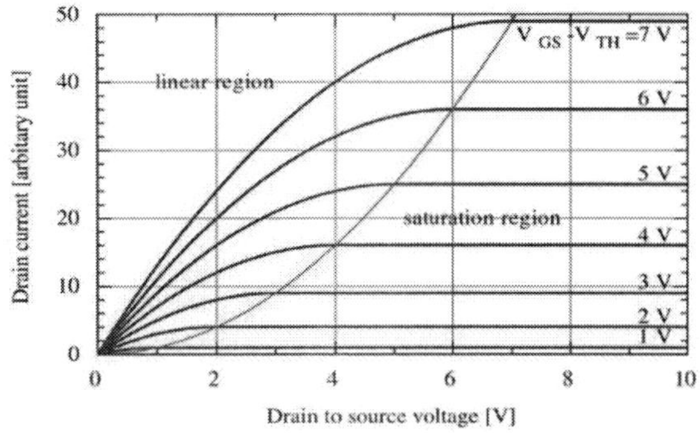

Figure 4. The graph of output characteristics of n-channel enhancement MOSFET [4]

IOP Conf. Series: Materials Science and Engineering **729** (2020) 012001 doi:10.1088/1757-899X/729/1/012001

3. Trade-off in MOSFET Devices

3.1. Trade-off for Transfer Characteristics

Nowadays, the transistors are made in the dimension of nanometers, which means each integrated circuit can contain more transistors. Moore' s law points out that the number of transistors placed in an integrated circuit doubles every two years. But problems occur when the size of transistor keeps on decreasing: the dielectric layer in MOSFET will be too thin, which has a negative effect on gate control, and other problems like heat dissipation, leakage current will appear. The limitation to the size of transistor leads to a trade-off problem.

An ideal transistor as a power switching has following characteristics: when power switching is at on-state, it should be switched on immediately and have no power dissipation, meaning the resistance should be zero; when it is at off-state, there' s no leakage current, meaning the resistance should be infinitely high. Let's first analyze from the aspect of transfer characteristic of an enhancement MOSFET:

As can be seen in Fig. 3, when the VG hasn' t reached VT, the curve is horizontal at where Idrain= 0, this means the power switch can' t be turned to on-state immediately, and there is energy dissipated to turn MOSFET to on-state. Thus, VT should be as small as possible. Below is the formula calculating VT:

$$V_T = \frac{\sqrt{2\varepsilon q N_A (2\psi)}}{C} + 2\psi$$

Where C, the capacity between the gate and semiconductor, is equal to:

$$C = \epsilon \frac{A}{d}$$

When d decrease, C will increase; according to the formula calculating threshold voltage, VT will decrease as a whole. Thus, the horizontal part of the curve in Fig. 3 will be shorter, so the switching frequency can be improved.

However, one problem is that breakdown voltage the layer can stand will be small. Under a relatively big bias, the tunneling current will conduct through dielectric layer, meaning that carriers will move through the layer. According to percolation theory, defects will accumulate in the layer. As this high gate voltage applies for longer time, the higher electric field at gate will give higher energy for carriers. When there will be enough defects to form a chain between the gate and the semiconductor, the breakdown path occurs, and the breakdown of dielectric layer happens. Fig. 5 shows the voltage at gate oxide vs. time to dielectric breakdown for different thickness: for the same thickness, as Vox increases, tDB decreases; for a same value of Vox, the thinner layer can sustain for a longer time. Dielectric breakdown is quite damageable for MOSFET devices it will affect the switching performance of MOSFET and increase the energy wasted. In order to avoid it, dielectric layer in MOSFET must have a certain thickness.

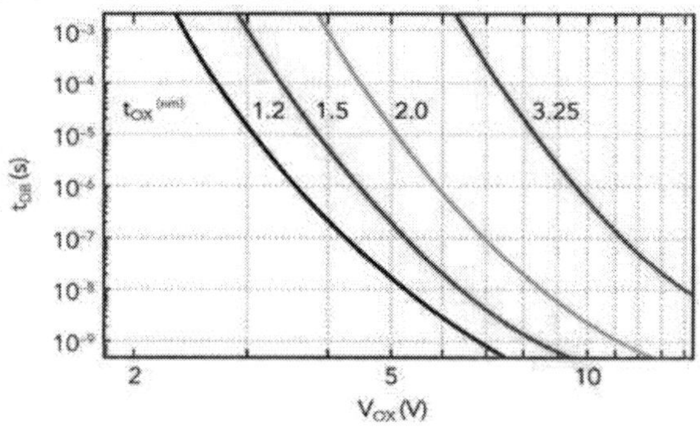

Figure 5. Graph of voltage at gate oxide vs. time to dielectric breakdown for different thickness [5]

The problem of trade-off is obvious: in order to improve MOSFET' s switching frequency, which means having a good transfer performance, the thickness of dielectric layer should be decreased; however, if the layer is too thin, MOSFET can' t sustain a high gate voltage, which may lead to dielectric breakdown. Therefore, when designing a MOSFET, an engineer must make a compromise between high transfer frequency and high gate voltage sustainability.

3.2. Trade-off for Output Characteristics

For the output characteristic of enhancement MOSFET, the ideal situation is that the resistance should be infinitely small so that there won' t be little joule heating, which is the unintended waste of energy. To be presented in graph, the slope at linear region in Fig. 4 should be almost vertical. In order to decrease the resistance, let' s first analyze the formula below calculating resistance:

$$R = \rho \frac{L}{A}$$

Where ρ can be considered as the resistivity of MOSFET as a whole, which is a constant not considering the effect of temperature; L is the length, which can be presented as the distance between drain and source; A is the cross section area, which can be presented as the width of the channel. Thus, there are two ways to reduce the resistance: increasing A or decreasing L. However, when L decreases too much, short channel effect happens, which will lead to the problems like drain-induced barrier lowering (DIBL), surface scattering and so on. In this part, only DIBL and surface scattering will be detailedly discussed.

The first problem is DIBL. In the normal condition, there is a potential barrier to stop electrons flowing from source to drain, and gate voltage controls the height of the barrier to determine the drain current. However, when the source gets too close to the drain, the field of drain and source will cross each other to form a special depletion region called punch-through, and this will lower the barrier height, which means carriers can flow from source to drain with less energy supplied. Thus, there will be a stronger leakage current. Fig. 6 depicts the relationship between drain current ID and gate voltage VG for different channel lengths L at different drain voltage; for example, for L=1.5 μ m and 3 μ m, as VG approaches to negative axis, ID is relatively higher than for L=1.5 μ m. That means MOSFET device is less likely to be at "off-state" when the channel length is short.

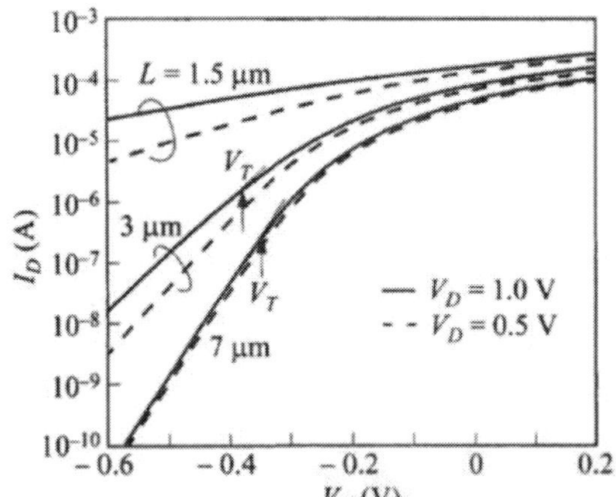

Figure 6. The relationship between ID and VG for different channel lengths at VD=1.0V and 0.5V [6]

More information can be obtained from the graph. Below is the equation of subthreshold swing:

$$S = \frac{dV_g}{dlogI_D}$$

Subthreshold swing represents how MOSFET' s performance of the frequency to switch between on and off-state, and the smaller S is, the faster the switching is and vise versa. According to Fig. 6, the inverse of slope for L=1.5 μ m is bigger than that for L=3 μ m. Thus, the switching performance of MOSFET is worse when the channel length is small because there is no clear boundary for on and off-state.

The another problem of short-channel effect is surface scattering. When carriers flow in the channel, they are moving under the effect of drain-to-source electric field and gate electric field. Due to this effect, carriers continuous collide with the lattice structure of semiconductor, so they are moving in a Z-path just like Fig. 7 shows.

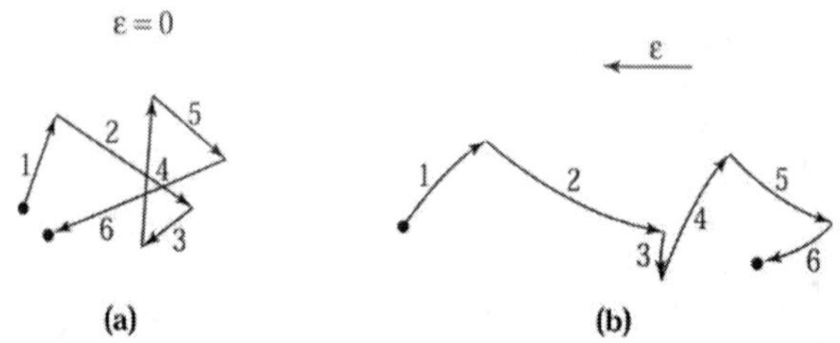

Figure 7. The moving path of an electron. (a) Random thermal motion; (b) Affected by electric field [7]

When the channel is shortened, drain-to-source field will be strengthened. In order to effectively control the state of MOSFET, a stronger gate voltage is needed. Thus, the collision against the lattice gets fiercer and more frequent, which means the electron mobility μ decreases. According to the equation:

$$\sigma=n\mu e,\ \ \rho=1/\sigma$$

where n is electron density, q is the quantity of charges, as μ decreases, σ decreases, and ρ , the resistivity, increases. Thus, the current-voltage relationship is impacted, and the output characteristics of MOSFET actually can' t reach the people' s original expectation, meaning that the performance is not increasing proportionally as the channel length decreases.

According to the formula calculating the resistance, another way to decrease the resistance is to increase the width of the channel, but a new consideration occurs: when developing better devices, engineers not only need to pay attention to the performance of the devices, but also the cost. The increase in width will surely increase the total size of the transistor, and for the same size of an integrated circuit, there will be fewer components can be placed in it. Facing the trend that people' s electronic devices are getting smaller while have a better function, increasing the width is not a wise choice just in order to improve the current-voltage relationship but ignore the economic benefits.

Thus, the trade-off for output characteristics of enhancement MOSFET is that when people try to decrease the output performance, people need to consider the short-channel effect, which will bring the problems of DIBL, increasing subthreshold swing and surface scattering. And these problems will respectively strengthen the leakage current, make the boundary of on and off-state unclear, worsen transfer performance and inversely decrease the output performance. What' s more, increasing the channel width is not economic efficiency way though the trade-off problem can be relieved.

3.3. Method to Relieve the trade-off

Silicon, the first generation semiconductor, is the commonest semiconductor material right now. Silicon, a tetrels (+4), is often dopped with elements of triels (+3) and pentad (+5) to create holes and electrons. However, compared to the third generation semiconductor, the first generation is not good enough though. According to Fig. 8, many aspects of silicon are weaker than GaN and SiC, which are two major types of third generation semiconductor materials. Both GaN and SiC generally have the same performance energy gap, and electron velocity better than Si: higher energy gap represents stronger ability to tie valance electrons; higher electron velocity represents a better current-voltage relationship. Respectively, MOSFET will have a better switching ability and output characteristic. Based on their own properties, engineers can apply either GaN or SiC in different conditions. For example, when the transistors need to work under a high temperature, SiC is a better choice for its high thermal conductivity and melting point; when the transistors need to work under a high electric field, GaN is a better choice for its high breakdown voltage.

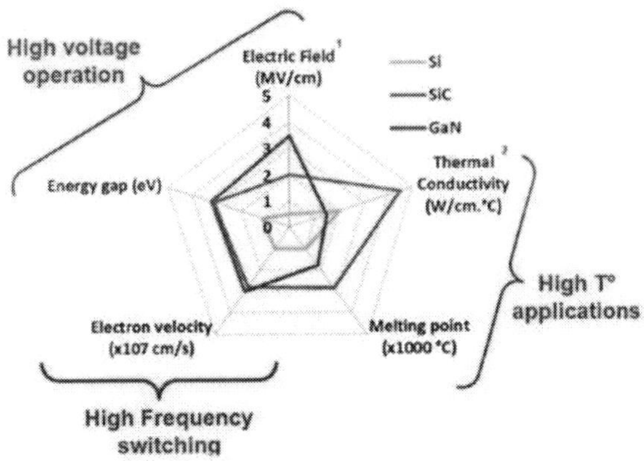

Figure 8. Radar diagram comparing five properties for Si, SiC and GaN [8]

However, widely applying the third generation semiconductor materials is not practical, because the growth and manufacture of the raw materials is very hard, especially for SiC.

4. Discuss about the Improvement

Generally speaking, under the trend that transistors are made in a smaller scale, trade-off problems are inevitable by changing the structure of basic n-channel enhancement MOSFET, meaning that the change in a parameter will negatively affect another. In order to improve the transfer characteristics of MOSFET, the dielectric layer should be thinner; due to the reduce of thickness, the resistance of continuously high gate voltage is weakened, and dielectric breakdown will happen and damage the device.

The first method to decrease the resistance is to shorten the channel length, but as the length decreases to a certain point, short-channel effect occurs, which brings the problems of DIBL, higher subthreshold swing and surface scattering. These problems will respectively strengthen the leakage current, make the boundary of on and off-state unclear, worsen transfer performance and inversely decrease the output performance. Another method is to increase the channel width. Though this approach can relieve the trade-off, but the consideration of economic benefits will rule out this choice.

To solve the problem of trade-off, applying third generation semiconductor materials is a method from the aspect of material. GaN and SiC have a relatively prominent performance on high voltage operation, frequency switching and high temperature application. However, the lack of raw materials is a problem still unsolved.

The trade-off in transfer and output characteristics is inevitable. As the research of relieving the trade-off keeps on going, a more practical way to face this issue is to specialized design the transistors based on their different applications.

5. Conclusion

From the study of n-channel enhancement MOSFET, the author illustrates the working principles of MOSFET, trade-off problems and method to improve the performance. However, several shortages still exist during the research process: all the illustrations are based on theories and only focus on main parameters that will significantly affect the performance of MOSFET when discussing the trade-off. Thus, some possible factors are ignored in the research, which means the study in the paper is should be brought into experimental condition to test whether other factors may affect MOSFET. In the later study, the main direction should base on observation, and quantitatively describe the phenomena. For example, when studying the relationship between the thickness of dielectric layer and breakdown voltage, control variate should be applied, recording how the time of breakdown varies with the gate voltage for a certain thickness. Generally speaking, the practice to demonstrate the points made in the paper is significant for the following research.

References

[1] Sze, M, S & Lee, K. M. (2012) Semiconductor Devices: Physics and Technology, 3rd ed. John Wiley & Sons, Inc., New York. pp.181.
[2] The Engineer Projects. Introduction to PN Junction https://www.theengineeringprojects.com/2018/05/introduction-to-pn-junction.html/.
[3] Power electronic A to Z. IGBT-An introduction http://www.completepowerelectronics.com/igbt-basics/.
[4] Modular Circuit. H-Bridge Driver. http://www.modularcircuits.com/blog/articles/h-bridge-secrets/h-bridge_drivers/#comment-171340/.
[5] EDN NETWORK. The Backstory of Backdriving. https://www.edn.com/design/test-and-measurement/4385741/The-backstory-on-backdriving/.
[6] Sze, M. S & Lee, K. M. (2012) Semiconductor Devices: Physics and Technology, 3rd ed. John Wiley & Sons, Inc., New York. pp. 199.
[7] Sze, M. S & Lee, K. M. (2012) Semiconductor Devices: Physics and Technology, 3rd ed. John Wiley & Sons, Inc., New York. pp. 44.
[8] Bigorra, J., Borrego, C., Fontanilles, J., and Giró, J., Innovative Electrical and Electronic Architecture for Vehicles With Dual Voltage Power Networks. In-Vehicle Application, SAE Technical Paper 2000-01-0452, 2000, doi:10.4271/2000-01-0452.e

IOP Conf. Series: Materials Science and Engineering **729** (2020) 012002 doi:10.1088/1757-899X/729/1/012002

The Difficulty and Solution of Mg-doped in GaN

Guoting Cheng

College of Physics, Qingdao University, Qingdao, Shandong, 266071, China

Abstract. GaN could be the representative third generation semiconductors, it becomes a significant material in the optoelectronic and microelectronic devices. However, the production of atomic H dramatically increases the resistance of Mg-doped GaN Large amount of Mg-H complex are produced during the MOCVD process, dramatically reducing the performance of Mg-doped GaN. Low-Energy Electron Beam Irradiation (LEEBI) could eliminate surface H^+ and thus reduce the resistance. Furthermore, Rapid Thermal Annealing (RTA) could decompose Mg-H complex, increasing the concentration of hole. By studying and improving every method, the electrical properties and optical properties are still not enough. Finally, the research points out the future direction of improving the performance of Mg-doped GaN.

1. Introduction

GaN is a direct bandgap semiconductor with the binary compound of III/V elements and an extremely stable as well as hard compound with the high melting point. Third generation semiconductors, represented by GaN, are of great advancement due to the special structure characteristic. Because Mg and Ga have similar electronic structural properties, Mg has been used in p-type doping species widely to produce Mg-doped GaN. However, the production of atomic H dramatically increases the resistance of Mg-doped GaN, and intrinsic difficulties due to atomic H seriously limit more application of Mg-doped GaN. Consequently, due to the high resistance of the material, the energy consumption is dramatically increased, and the energy efficiency is to be pretty low. The quality of optoelectronic devices based on GaN, such as LEDs and LDs, are seriously limited, handicapping the development of this kind of devices. According to E. Yamaguchi's study, the n-type doping technology is relatively simple with Si standing as acceptor, so the drift mobility could excess $300cm^2/(V \cdot s)$. However, Mg has larger covalent radius, so Mg-H will be produced to make Mg become passive, so the resistance is extremely high[1]. Therefore, in this paper, the difficulty and Solutions of Mg-doped GaN will be discussed in detail, as well as the future direction of development of Mg-doped GaN will be proposed in the end.

2. Current situation of GaN

GaN can be doped with silicon to n-type or with magnesium to p-type. With p-type doping, the performance of GaN could be dramatically improved. Because the deep nature of Mg acceptor has a pretty high doping levels, which is about 1020cm-3, Mg-doped p-type GaN has a significant value for the family of devices based on nitride, including photodetectors, light emitting diodes (LEDs), lasers and solar cells. Because of the high thermal conductivity and wide band gap, compared with Si and GaAs, Mg-doped GaN devices can operate at higher temperatures above 200°C, which means that it could tolerate higher energy density, so Mg-doped GaN has higher reliability. Furthermore, the efficiency of GaN could be dramatically increased due to the decreasing of resistance, with more wide band gap and dielectric breakdown electric field. Also, the devices are allowed to operate at high speed with a fast electron saturation speed and high carrier mobility.

3. The mechanism introduction and limitation analysis

3.1. The development and limitation of growing Mg-doped GaN by MOCVD

The growth of Mg-doped GaN is depended on MOCVD. In 1968, metalorganic chemical vapor deposition (MOCVD) was proposed in America. In the process of MOCVD, the ultrapure gases are injected into the reaction chamber, followed by precisely dosed to deposit an extremely thin layer of doping atoms onto a semiconductor wafer. The condition of reaction to happen on the surface for crystalline growth required compound semiconductors and the epitaxy of materials, with metalorganics and organic compounds and hydrides containing the necessary chemical elements. For example, gallium nitride could be produced in a reactor on heated sapphire substrates, then ammonia (NH_3), Trimethylgallium (TMGa) and biscyclopentadienyl-magnesium (Cp_2Mg) are reacted as growth precursors. The Mg-doped layers were produced in a H_2 ambient, followed by a post-growth annealing step which was applied to activate the dopants. However, when doping Mg in the epitaxial layer of GaN, because the covalent radium of Mg is larger than Ga and Mg is reacted as deep acceptor, the Mg acceptor in the epitaxial layer will be passivation because of the production of Mg-H, and the atomic H come from H_2 reacted as carrier gas and the decomposition of reactant NH_3. Therefore, the hole mobility of p-type doping is much low via the conventional metal organic chemical vapor deposition (MOCVD), with high concentrations of Mg which are about $10^{19}cm^{-3}$ and the high remaining hydrogen concentration. The remaining hydrogen atoms can be decreased by thermal activation, but under room temperature (RT), there is still considerable quantities of atomic hydrogen with concentration exceeds $10^{18}cm^{-3}$ remaining in the films. The consequence of it is that the resistivity of GaN could reach to $10^8\Omega$, which extremely limits the conduction of material, the reliabilities of devices such as lasers and light-emitting diodes (LEDs) are reducing, because the diffusion of atomic H is hardly driven by current injection.

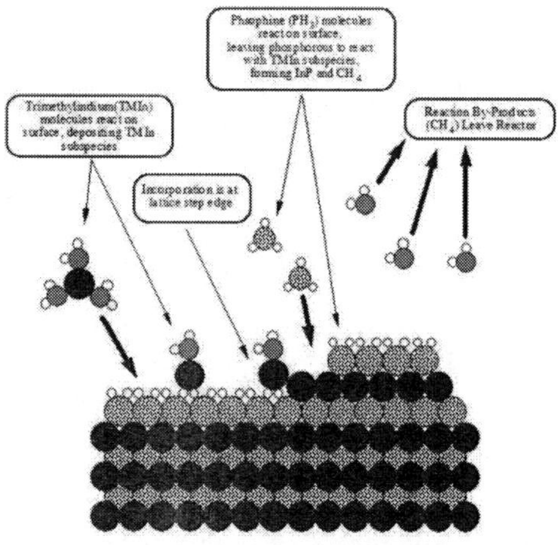

Figure 1. Illustration of MOCVD process.

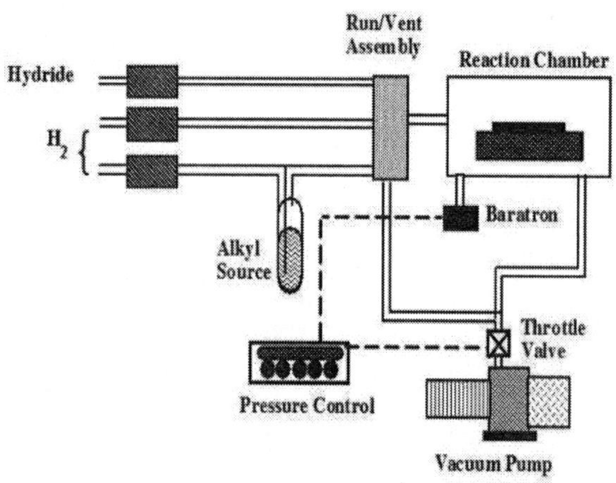

Figure 2. MOCVD apparatus

Accounting for that there is a 2.8 eV photoluminescence band in metalorganic chemical vapor deposition grown GaN, Kaufmann proposed a viewpoint of donor–acceptor (D–A) pair recombination character. He claims that blue band is produced by the recombination of dependent acceptor Mg_{Ga} and deep donor energy level, the deep donor level refers to $Mg_{Ga}V_n$, which is recombined by acceptor Mg_{Ga} and nearby vacancy of nitrogen V_N.V_N is the defection produced in the process of MOCVD with high pressure of nitrogen balance, the concentration of V_N reaches the highest point in the 1000 °C temperature[2].

Young pointed out that the increase of p-type doping is relative with the density of dislocation, dislocation becomes the migration path of hydrogen atom[3].

With the purpose of increasing the conductivity of hole and obtaining low resistance p-type GaN, researchers invented many new doping techniques. T. Yamamoto proposed co-doping method. Both of n-type and p-type dopant are doped at the same time, then the donor level is reduced and the acceptor level is increased, which will produce higher hole concentration. Meanwhile, the hole mobility is going to be increased, because short-range dipole scattering substitutes long-range Coulomb scattering of isolated impurity charges and the intensity of scattering is weakened[4].

3.2. The development and limitation Low-Energy Electron Beam Irradiation (LEEBI)

When the single-crystalline GaN is prepared to produce, a sapphire substrate is used to work as heteroepitaxial growth. However, because of huge difference in the thermal expansion coefficient between sapphire and GaN, as well as the huge lattice mismatch and the high-quality epitaxial growth must operate on a flat surface free from cracks, which is much difficult. Therefore, low-energy electron beam irradiation is applied before the growth of GaN by MOVCD. After handling with LEEBI, crystalline quality, the film uniformity, luminescence and electrical properties of GaN films are greatly improved, and resistance of p-type doped GaN is dramatically reduced. The hall effect measurement of Mg-doped GaN which had been treated with Low-Energy Electron Beam Irradiation at room temperature (RT) showed that the hole concentration is $2*10^{16}cm^{-3}$, the resistivity is 35Ω·cm and the hole mobility is $8cm^2/V$ approximately[5].

J. A. Van Vechten did research about the mechanism of this treatment, proposing that a large quantity of electron-hole pairs are produced in the GaN by the low energy electron beam irradiation, and there pairs are tend to assemble to the area where acceptors and donors $H+$-Mg^- complex are aggregated, therefore, the interstitial ion H^+ is neutralized by electron forming neutron H, and some of the neutron diffuse to the surface of GaN, then it forms H_2 with other neutron H to escape outside, leaving much activated acceptor[6]. The resistance of Mg-doped GaN could be reduced by this process. However, the resistance of Mg-doped GaN is far depended on the penetration depth of LEEBI, only on the area with very thin surface of GaN could present p-type characteristic distinctively.

3.3. The development and limitation of Rapid Thermal Annealing (RTA)

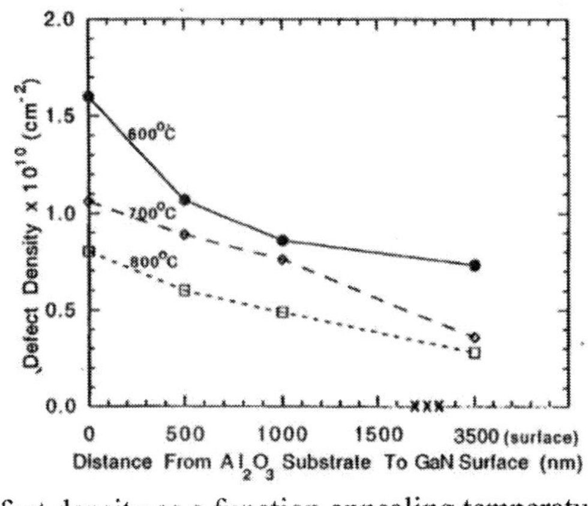

Figure 3. Defect density as a function annealing temperature and distance vary from the sapphire substrate.

Despite the hydrogen removing was first achieved with a low energy electron beam irradiation, subsequent work showed that a thermal anneal of 700°C for 20 minutes could still activate Mg acceptors. The contact of metal/GaN and film properties are influenced by different thermal annealing temperature and time. Therefore, different thermal annealing is used based on the difference of necessary.

Recently, there is a method of using Rapid Thermal Annealing (RTA) at temperatures about 1150°C to activate Mg-doped GaN in a light emitting diode (LED) structure. Compared with conventional thermal annealing, RTA works in a higher temperature. The high temperature step could improve the samples photoluminescence. If an excellent conductivity neither n-type or p-type, a rapid thermal annealing temperature of more than 1000°C is required to achieve.

The high-temperature rapid thermal anneals effectively improve the morphology of surface and photoluminescence of GaN growing on sapphire. Particularly, GaN samples rms roughness is reduced when annealed in an N_2 ambient at 1100°C for 15s, the average value of roughness, originally 3.9nm, could reach to 1.1nm[7]. Therefore, the band-edge photoluminescence emission also increases correlated with the improvements of surface morphology. Furthermore, the electron mobility is also reduced by thermal annealing, partly because the two bands conduction mechanism or an increasing of compensation ration.

The thermal annealing of GaN film growth usually occupies in the growth room, and the ohmic contact and rectifying contact in the production of GaN devices occupy in the special annealing furnace. With the purpose of protecting the surface of GaN film surface, the overpressure method is usually adopted. Hydrogen is usually adopted as carrier gas to transport NH_3 to growth room, providing overpressure of nitrogen, so the purpose of protecting surface could be achieved. However, this method could bring two-sided problems. The first one is that the volume of growth room is large normally, which is lead to require long process to warming and cooling. The another one is that NH_3 is inevitably decomposed worked as annealing gas atmosphere, and H could disperse to material, causing hydrogen passivation.

M. W. Cole has researched the impact of rapid thermal annealing temperature to the mass of film crystal. The annealing occupied in nitrogen atmosphere, lasting for 1 minute and the temperature range is 600~ 800°Cobserved by TEM, the quality of GaN film surface is dramatically enhanced, and the crystallographic defect of GaN film surface is reduced approximately 25% ~ 30% compared with the interface of substrate and buffer layer. The annealing temperature higher, the less flat defect, dislocation mostly, extend to surface. Furthermore, Mg-H complex still be found in the Mg-doped GaN after rapid thermal annealing, because H^+ are not be totally neutralized after annealing, so Mg-h could reproduce once temperature fall to the room temperature[8]. Theoretically, less defect could increase the mobility of carrier, but researchers hasn't observed any increasing of carrier mobility after annealing, partly because the negative influence of impurities including H and C cannot be removed by thermal annealing.

With purpose of obtaining better quality devices, scholars have studied the passive effect of Mg-H complex, the mechanism of acceptor stimulation, the relationship between the condition of thermal annealing and the concentration of external surface of GaN, the crystal quality, the luminescence character. Meanwhile, they have also proposed some invented doping method. The research of Kim indicated that when annealing temperature is 800°C, deep energy band originated from Mg-N-H complex concentrates on 0.22 eV, and more hydrogen atom will come out from complex with high annealing temperature[9]. K S Ahn applied Two-step Rapid Annealing Process, which means that GaN firstly is annealed under low temperature of 600°C for 5 min, then it is annealed under high temperature of 900°C for 1 min. The electric quality, crystal quality and surface smoothness of GaN are apparently improved[10]. Waki I invented that plating Ni which thickness of 1.5 nm with chemical vapor deposition (CVD), followed by annealing in the atmosphere of N_2. Observed by SIMS, atomic hydrogen is effectively ejected and the absorption of atomic hydrogen is apparently improved by Ni[11].

4. Result analysis

One of the most important reason of handicapping the development of GaN is that p-type GaN with high quality is hard to obtained, which far limited further the application of GaN devices. Although scientists have deeply studied the concrete mechanism of p-type GaN and invented several doping methods, the electric quality, photoluminescence and crystal characteristic of p-type GaN still remain in poor condition. In the future, more attempt should focus on activating the dopant of GaN, achieving higher concentration of hole carrier and lower resistivity and finding more proper acceptor dopant. The fact is that, there are only 1%~2% of Mg atoms in GaN could be acted as acceptor, so the amount of Mg atoms should be added to two orders of magnitude higher than hole concentration. However, with such huge amount, the surface of GaN will be rough, reducing the performance of materials. Therefore, only to find more active acceptor could reduce the amount of acceptor dopant.

5. Conclusion

In summary, this paper contains only a theoretical analysis of the development and limitation of MOCVD, LEEBI and RTA technology, lacking concrete experimental operation. Therefore, the proposition is based on theoretical deduction. In practical operations, the results will be influenced by several factors, including operation accuracy and material quality. In the subsequent research, the theory and practical operation will combine together, and the theory will be modified based on experiment results, also, the operation will be improved with the development of theoretical knowledge. With the deeper cognition of doping mechanism and improvement of doping techniques, the technique of p-type doping tends to be more improved, so meet more high-quality requirements. Therefore, the performance of optoelectronic and microelectronic devices will be significantly increased, embracing a bright future of microelectronic world.

References

[1] Yamaguchi, E., Junnar, M. R. (1998) Effects of nitrogen vacancy on optical properties of nitride semiconductors. J. Cryst. Growth., 189:570

[2] Kaufmann, U., Kunzer, M., Maier, M., Obloh, H., Ramakrishman, A., Santic, B., Schlotter, P. (1998) Nature of the 2.8 eV photoluminescence band in Mg doped GaN. J. Appl. Phys. Lett., 72(11): 1326~1328.

[3] Youn, D., Laehab, M., Hao, M., et al. (1999) Investigation on the p-type activation mechanism in Mg-doped GaN films grown by metalorganic chemical vapor deposition. Jpn. J. Appl. Phys., 38(2): 631~634.

[4] Yamaoto, T., Katayama-Yoshida, H. (1998) Electronic structures of p-type GaN co-doped with Be or Mg as the acceptors and Si or O as the donor condopants. J. Cryst. Growth., 189: 532.

[5] Amano, H., Kito, M., Hiramatsu, K., Akasaki, I. (1998) P-Type Conduction in Mg-Doped GaN Treated with Low-Energy Electron Beam Irradiation (LEEBI). J. Journal of Applied Physics, pp. 2112~2114.

[6] Van Vechten, J. A., Zook, J. D., Horning, R. D., et al. (1992) Defeating compensation in wide gap semiconductors by growing in H that is removed by low temperature de-ionizing radiation. Jpn. J. Appl. Phys., 31(11): 3662~3664.

[7] Zolper, J. C., Crawford Hagerott, M., Howard, A. J., Ramer, J., Hersee, S. D. (1996) Morphology and photoluminescence improvements from high temperature rapid thermal annealing of GaN. J. Appl. Phys. Lett., 68: 200.

[8] Cole, M. W., Ren, F., Pearton, S. J. (1997) Post growth rapid thermal annealing of GaN The relationship between annealing temperature, GaN crystal quality, and contact- GaN interfacial structure. J. Applied Physics Letters, 71: 3004.

[9] Kim, K., Harrison, J. G. (2003) Critical Mg doping on the blue-light emission in p-type GaN thin film grown by meta-organic chemical-vapor deposition. J. Vac. Sci. Technol., A21:134~139.

[10] Ahn, K. S., Kim, D. J., Moon, Y. T. (2001) Rapid thermal annealing method of p-GaN thin film growth. J. Vac. Sci. Technol, B19(1):215~217.

[11] Waki, I., Fujioka, H., Oshima, M. (2001) Doping Mg on UHCVD add Ni research. J. Appl. Phys. Lett., 78(4): 2899~2900

IOP Conf. Series: Materials Science and Engineering **729** (2020) 012003 doi:10.1088/1757-899X/729/1/012003

Suggestions on Efficiency Droop of GaN-based LEDs

Jixiang Ding

College of material science and engineering, Sichuan University, Chengdu, Sichuan, 610065, China

angela@cas-harbour.org

Abstract. InGaN/GaN-based light-emitting diodes (LEDs) are widely used in modern society and industry among different areas. However, InGaN/GaN LEDs suffer from an efficiency droop issue: The internal efficiency decreases during high current injection. The efficiency droop significantly affects the development of GaN-based LEDs devices in efficiency and light-output areas. Therefore, the improvement of the droop phenomenon has become a significant topic. This paper introduces several possible mechanisms of droop phenomenon based on different hypotheses including Auger Recombination, Carrier Delocalization and Electron Leakage. Furthermore, some proposals to mitigate efficiency droop, including semipolar LEDs, electron blocking layer(EBL), quaternary alloy and chip design will be discussed and analyzed. Also, it will provide some suggestions for the further optimization of droop phenomenon in each proposal.

1. Introduction

In 1993, Nakamura et al. demonstrated InGaN/GaN light-emitting diodes (LEDs) [1] for the first time. After that, LED materials and devices ushered in a period of rapid development, so now GaN-based LED is entering the lighting application field. Despite significant advances in GaN-based LED, higher efficiencies and light-output are two areas that researchers are primarily working to enhance. In the field of efficiency, GaN-based LED is seriously plagued, which is called efficiency droop. Usually, the LED reaches the highest efficiency during low injection current, when the current increases, the efficiency drops drastically. Therefore, in order to further develop high-power and high-efficiency LED devices, careful consideration must be given to the efficiency droop. This paper will list and analyze some striking theories of droop mechanisms, including Auger Recombination, Carrier Delocalization and Electron Leakage. Despite the controversy of the basics behind droop phenomenon, the attempts of diminishing efficiency droop never stop. The next section of this paper introduces some solutions of efficiency droop including semipolar LED, electron blocking layer(EBL), quaternary alloy and chip design based on different models of droop phenomenon basics.

2. Overview of GaN-based LED and efficiency droop

Due to the promising advantages over traditional light sources, including the largely extended lifetime and greatly improved power efficiency, the commercialization of GaN-based LED lighting applications has become very rapid. Potentially, the use of blue and green InGaN/GaN LEDs on large TV screens or micro full colour panels should be promising because of their excellent brightness. As flashlights, GaN-based LEDs consume less power and provide longer lighting time, as a result, helping to improve the environment and save costs. Furthermore, GaN-based LEDs are ideal sources for data transmission, thanks to the two characteristics of LEDs: the capability of being modulated at

IOP Conf. Series: Materials Science and Engineering **729** (2020) 012003 doi:10.1088/1757-899X/729/1/012003

high speed and solid-state source. In addition, with the further development of the optical application of LEDs, more optimizations, such as increased chip and operation power, will help GaN-based LEDs to achieve a higher output power of LEDs in single unit.

As I mentioned, efficiency droop describes the phenomenon that the internal efficiency decreases significantly with the gradual increase of the current. Furthermore, even under short pulse, low duty cycle and constant temperature injection, it has been proven that the well-known efficiency droop phenomenon strongly affects GaN-based LEDs over a wide range of wavelength spectrum. Therefore, the efficiency droop can have a significant negative impact on the GaN-based LED applications and limit their basic functionality.

With the deepening of the understanding of droop phenomenon, different theories have been put forward to explain the droop phenomenon. Among them, auger recombination, carrier delocalization and electron leakage have gradually become mainstream.

The augur recombination can be discussed in the ABC model. Due to the SRH linear dependence and the radiation recombination square dependence, the radiation recombination becomes a controlling recombination when LEDs are working at higher currents. With this in mind, the augur recombination is likely to cause the droop phenomenon.

Another theory, the carrier delocalization, is demonstrated in both logic and experiments. The SRH model describes a typical non-radiative electron-hole recombination at crystal defects. The unified IQE droop formula is that if other carrier loss mechanisms are neglected (Ileak=0, C=0), the equation will be transformed into. In addition, the SRH carrier lifetime is equal to 1/2A. The carrier lifetime of SRH decreases as carrier density increases, and the continuous linear rise of parameter A leads to efficiency droop phenomenon. In addition, some experiments also support this theory, which will be discussed in more detail below.

According to the electron leakage theory, the droop phenomenon is mainly caused by the asymmetric distribution of holes and electrons. The increasing current will continuously weaken the electron trap of quantum wells and thus result in droop phenomenon.

3. Droop mechanisms

3.1. Auger recombination

On account of the simplicity and flexibility of ABC model, researchers prefer taking it to briefly model the electron- hole recombination mechanism in LEDs[2].

$$\mathrm{R} = \mathrm{An} + Bn^2 + Cn^3$$

In the above formula, four variables, n, A, B, C in this equation, respectively represent the quantum wells (QWs) carrier density and Shockley - Read - Hall (SRH), radiative, nonradiative Auger-like recombination parameters. The radiative recombination turns into the governing recombination when LEDs work at higher current, attributed by the both SRH linear dependence and the radiative recombination square dependence. In addition, the domination of Auger contributions in recombination could be successfully predicted when LEDs are at a high current level by the ABC model.

Several researchers, based on some measurements and simulations, argued that the primary reason of the efficiency decrease with growing current and the basic nonradiative approach for carriers at LED working currents, is the Auger recombination. Shen et al. firstly published the measured result of InGaN Auger coefficient(ranging from 1.4*10−30 to 2.0*10−30 cm6 s−1), which they claimed was large enough to diminish the internal quantum efficiency of the standard-operating-currents LEDs.

IOP Conf. Series: Materials Science and Engineering **729** (2020) 012003 doi:10.1088/1757-899X/729/1/012003

Following the first published experimental measured result of Auger coefficient[3], some other groups also published their experimental reports of Auger coefficient respectively within later few years and demonstrated the relevance between Augur coefficient and droop phenomenon .

After that, however, some theoretical groups calculated the nitride Auger coefficient directly, and in terms of these low theoretical value reports, several different variations and their theoretical reports to fortify the Auger recombination theory, including interband Auger recombination and indirect Auger recombination, have also been proposed.

However, the probability of the interband Auger recombination commonly drastically reduces with changing band gap, therefore it seems won' t literally become the unique reason for the efficiency droop across a wide wavelength range. In addition, all yielded values of theoretical calculation of the indirect Auger recombination coefficient vary from 1 × 10−31cm6 · s−1 to 2 × 10−31cm6 · s −1[4], which is still not high enough to cover the discrepancy between experimental records and theoretical yielded values. Instead of the Band to band Auger recombination, it is likely that Phonon or defect-assisted Auger recombination are probable reasons for droop phenomenon according to the comparison of carrier density and temperature dependent electroluminescence and photoluminescence measurements.

Method of calculation	Author	Room temperature Auger coefficient ($cm^6 \cdot s^{-1}$)
Direct Auger recombination	Hader *et al.* (2008)	3.5×10^{-34}
Indirect Auger recombination	Kioupakis *et al.* (2011)	$0.5\times10^{-31}-2.0\times10^{-31}$
Interband Auger recombination	Delaney *et al.* (2009)	2.0×10^{-30}
	Bertazzi *et al.* (2010)	1.0×10^{-32}
Experimental reports of Auger coefficient or *C* coefficient	Brendel *et al.* (2011)	1.8×10^{-31}
	Shen *et al.* (2007)	$1.4\times10^{-30}-2.0\times10^{-30}$
	Zhang *et al.* (2009)	1.5×10^{-30}
	David and Grundmann (2010)	1.0×10^{-29}
	Dai *et al.* (2011)	8.0×10^{-29}
	Ryu *et al.* (2009)	$1.0\times10^{-27}-1.0\times10^{-24}$

Figure 1. Theoretical and experimental reports of the Auger and C coefficients in the GaInN/GaN material system[5]

3.2. Carrier delocalization

Typical non-radiative electron–hole recombination at crystal defects is described by the SRH model. In the model, the primary unified IQE droop equation(1) could be transformed into equation (2),under the condition that other carrier loss mechanisms are neglected (I_{leak}=0, C=0). As a result, theoretical, if the SRH carrier lifetime falls down as carrier density grows, the continuous linear rise of parameter A will result in efficiency droop phenomenon.

$$\eta_{IQE} = \frac{qV_{QW}Bn^2}{(I_{QW}+aI_{QW}^m)} \quad (1)$$

$$\eta_{IQE} = \frac{Bn^2}{(An + Bn^2)} \tag{2}$$

Primary unified IQE droop equation(1) and transformed IQE droop equation(2) with the electron charge q, carrier recombination current inside the QWs I_{QW}, the active volume V_{QW} of all QWs, the QW carrier density n, the SRH parameter A, the radiative coefficient B.

The carrier delocalization is observed due to the non-uniform distribution of indium in the InGaN quantum wells. The random distribution of gallium and indium atoms results in indium and gallium-rich areas on crystal lattice sites. Usually, these indium- rich areas where carriers prefer to stay are with lower bandgap energy, while gallium-rich areas are with higher bandgap energy and tendency of repelling carriers. As a result, normally, this particular carriers localization effectively prevents the carriers diffusion and further taking advantage of dislocation and defects to recombine non-radiatively.With high injection current, however, quantum wells regions could not bear continuous increasing carriers and end up in a filled-up indium-clusters.Then more spilling out carriers will diffuse into quantum wells and recombine at defects non-radiatively and reduce SRH lifetime, then further cause droop phenomenon .

Although the theory seems reasonable and tenable, it might still be disputable. Firstly, the existence of indium-clusters turned into contentious later. Wang et al. [6] compared the external quantum efficiency and injection current of two light-emitting diode samples ,and they concluded that although when at a low injection current ,efficiency droop is dominated by the carriers delocalization, while at a higher injection current, efficiency droop is mainly on account of the carrier leakage. Shen et al.[7] argued against Carrier delocalization theory. They argued the decrease would be apparent with a larger threading dislocation density if dislocations were indeed in charge of the efficiency decrease, while the actual observation didn’ t literally reflect to the expectation.

Nevertheless, despite controversy, since the mechanism of low non-radiative recombination rate in InGaN quantum wells remains nearly mysterious, the carrier delocalization is still counted as reasons of efficiency droop.

3.3. Electron leakage

Another commonly discussed potential mechanism of efficiency droop is electron leakage. Theoretically, under a driving current, considering the large mobility difference between electrons and holes in GaN, the electron diffusion-drift distance between n-GaN and p-GaN is far longer than that of holes, resulting in the asymmetric distribution of holes and electrons. Hence, the increasing current will continue to weaken the electron trap of quantum wells. Finally, the recombination region moves from MQWs to the p-side of LED, which then prevents holes from reaching the active region. The conversion from kinetic energy to thermal energy leads to the efficiency droop phenomenon. The experimental evidence of electron leakage have been reported extensively.

The basic cause of electron leakage might be complicated. Some studies have shown that the electron leakage comes from the enhanced active region caused by sheet charges at hetero interfaces. The sheet charges might be further caused by severe polarization mismatch between different layers in the active region.

4. Suggestion

Although the mechanism of efficiency droop is still in dispute. However, based on different theoretical assumptions, some researchers have been working on the optimization of efficiency droop. In this part, several solutions have been listed, analyzed and reviewed.

4.1. Semipolar LEDs

Conventional LEDs ("polar" LEDs) typically grow on the c-plane substrates and are usually accompanied with large built-in electric fields caused by spontaneous polarization. Then, since the m/a-plane and the intermediate plane were applied to the second generation of bulk-GaN-substrate-based LED, the semipolar GaN-Based LED was invented and became a remarkable breakthrough, contributing to the reduction of efficiency droop.

The basic mechanism for overcoming droop for semipolar blue LEDs has been thoroughly studied. As previously discussed, it is believed that the reduction of the effective active region volume results in dense carriers in the active region, which then leads to severe Auger recombination and electron leakage, and finally causes droop phenomenon. Then, since polarization electric field, which mainly causes reduction of the effective active region volume, is significantly reduced in semipolar LEDs, a great decrease of efficiency droop could be observed. In addition, the SQW active region could eliminate carrier non-uniformity, mitigate the carrier density and electron leakage, and then assist LEDs in obtaining high efficiency. Furthermore, some observed phenomenon like narrow spectral width and a small blue shift with increased current density could vividly demonstrated that InGaN QWs grown on the semipolar (20-2-1) plane has relatively no potential fluctuation, therefore, the minimum band fill for localization degree and localization state of low carriers could be proved. [8]

4.2. Electron blocking layer (EBL)

Many efforts have been made to improve the efficiency of the the GaN emitters in the development of GaN-based LEDs, and EBL is nearly the most commonly used structures to prohibit electron leakage.

The primary goal of a thin AlGaN cap as an EBL commonly grown over the MQWs of InGaN-based LEDs is to maintain electron overflow in the active region and prevent the InGaN active region from being damaged by the subsequent high-temperature growth of p-type layers.

The EBL structure is still under development to improve the efficiency of LEDs. For instance, to overcome the inability of AlGaN layers in protecting the MQWs region from the high temperatures growth of the p-type GaN contact layers, Hansen et al. develop a Mg-doped AlGaN EBL grown at low temperatures directly above the final InGaN well of the InGaN QWs. Then R. C. Tu et al.[9] demonstrate that the growth method at low temperature can reduce the electron leakage current and improve the external efficiency[10].

Another example, in order to solve a severe issue that EBL acts as a barrier for holes impeding the holes transportation into the active region, alternatives of the Mg doped AlGaN EBL are proposed to enhance the hole injection with the electrons blocking capability. Later, graded AlGaN electron blocking layer (GEBL) LEDs were invented to optimize the efficiency droop.

In spite of the widespread use of EBL, due to the effect of EBL on the efficiency droop in InGaN/GaN MQWs LEDs, the EBL does not seem to be suitable for every situation. However, since they came to different conclusion and they mainly discussed the influence of different current magnitudes on the function of EBL, it seems that more researches in EBL field are required to provide specific answer.

4.3. Quaternary alloy

Recently, some researchers proposed the polarization matching theory to explain efficiency droop. To prevent polarization charge from enhancing the carrier' s non-capture and carrier escaping from the active region, quaternary alloys have been proposed by controlling interfacial polarization charges and bandgap to mitigate electron leakage from the active region and to reduce efficiency droop. Kim et al. demonstrated theoretically polarization-matched GaInN/AlGaInN MQWs LEDs could exhibit high efficiency at large injection current and the quaternary alloys $Al_xGa_{1-x-y}In_yN$ of an appropriate bandgap and lattice constant. After that, Schubert et al.[11] reported their LEDs with polarization-matched AlGaInN barriers. The quaternary alloys enhances the illumination performance at high currents and reduces the forward voltages, which increases external quantum by 18.5% and reduces droop phenomenon.

Despite the encouraging features of quaternary InAlGaN in reducing droop phenomenon, it seems uneasy to utilize the common method of metal-organic-chemical-vapor deposition (MOCVD) to grow quaternary InAlGaN owning to tremendous discrepancies in the optimal growth temperatures of nitrides and the difficulty in balancing growth temperature and the diffusion length of Al adatoms. Hence, it is necessary to further study the optimization of quaternary InAlGaN growing methods.

4.4. Chip design

Lots of efforts have been put into the chip design of the GaN-based LEDs to improve LEDs light extraction efficiency. With the advent of conventional chip (CC) design and flip-chip (FC) approach, the current spreading problem and a tradeoff between current spreading and light extraction have been solved. Then as the droop phenomenon becomes a hot topic, the relationship between chip design and efficiency droop has been studied intensively. The picture of the relationship seems clear that magnifying the device chip areas could lead to the significant reduction of efficiency droop due to the current decrease. Hence, large-sized LED chip is usually employed.

In another way, efficiency droop might be related to overheating of the LED chip at high current density caused by the current crowding effect. The current crowding effect is mainly attributed to the non-uniform carrier mobility between holes and electrons in GaN-based materials, leading to carriers crowding near the electrode and then the efficiency droop. In order to mitigate the overheat problem, some researchers attempted to develop microchips due to the small chip area, high voltage and low operation current. Yen et al. reported that their multi-microchip design uses self-rectifier in alternating-current LEDs. Wang et al.[6] studied the application of microchips on high-voltage LEDs and reported significant enhancement and reduction in luminous efficiency and droop phenomenon due to more uniform current distribution in LEDs. In the future, it is expected to conduct more researches to reduce the overheating problem.

5. Conclusion

In this paper, the well-known efficiency droop phenomenon is discussed in detail. The theoretical discussion and experimental evidence of the basis of efficiency droop are elaborated, including Auger recombination, carrier delocalization and electron leakage. In the future, more researches will be conducted on the correlation between different possible mechanisms causing efficiency droop. Since these mechanisms are often interrelated and cross-linked, some recent studies showed that efficiency droop might be caused by several different factors.

Several proposed solutions to efficiency droop problem are listed, and the basic structure, functions and technical problems that need to be paid attention to in the future, such as semipolar LEDs, electron blocking layer(EBL), quaternary alloy and chip design, are investigated. Despite these extraordinary achievements, some issues still need more thorough investigations, such as the applicable electrical current range of the EBL, the quaternary in AlGaN growth methods and more details on microchip improvements.

References

[1] H. Search, C. Journals, A. Contact, M. Iopscience, and I. P. Address. P-GaN / N-InGaN / N-GaN Double-Heterostructure Blue-Light-Emitting Diodes, vol. 8.

[2] O. J. Glembocki, J. D. Caldwell, J. A. Mittereder, J. P. Calame, S. C. Binari, and R. E. Stahlbush.(2009) Measurement of Local Temperatures Using μ-Raman of SiC and AlGaN-GaN/SiC Power and RF Devices, Mater. Sci. Forum, vol. 600 - 603, pp. 1111-1114.

[3] Y. C. Shen, G. O. Mueller, S. Watanabe, N. F. Gardner, A. Munkholm, and M. R. Krames.(2007) Auger recombination in InGaN measured by photoluminescence, Appl. Phys. Lett., vol. 91, no. 14.

IOP Conf. Series: Materials Science and Engineering **729** (2020) 012003 doi:10.1088/1757-899X/729/1/012003

[4] Kioupakis E. , Rinke P. , Delaney K. T. and Van de Walle C. G.(2011) Indirect Auger recombination as a cause of effi ciency droop in nitride light- emitting diodes, Applied Physics Letters 98 , 161107.

[5] D. S. Meyaard, G. B. Lin, J. Cho, and E. F. Schubert, *Efficiency droop in gallium indium nitride (GaInN)/gallium nitride (GaN) LEDs*. Woodhead Publishing Limited, 2013.

[6] Wang, J. et al. (2010) Understanding efficiency droop effect in InGaN/GaN multiple-quantum-well blue light-emitting diodes with different degree of carrier localization, Applied Physics Letters, 97(20), pp. 11-14. doi: 10.1063/1.3520139.

[7] Y. C. Shen, G. O. Mueller, S. Watanabe, N. F. Gardner, A. Munkholm, and M. R. Krames(2007)Auger recombination in InGaN measured by photoluminescence, Appl. Phys. Lett., vol. 91, no. 14.

[8] M. Hansen, J. Piprek, P. M. Pattison, J. S. Speck, S. Nakamura, and S. P. DenBaars. (2002) Higher efficiency InGaN laser diodes with an improved quantum well capping configuration, Appl. Phys. Lett., vol. 81, pp. 4275-4277.

[9] R. C. Tu et al.(2003) Improvement of near-ultraviolet InGaN-GaN light-emitting diodes with an AlGaN electron-blocking layer grown at low temperature, IEEE Photonics Technol. Lett., vol. 15, no. 10, pp. 1342-1344.

[10] M. H. Kim et al. (2007) Origin of efficiency droop in GaN-based light-emitting diodes, Appl. Phys. Lett., vol. 91, no. 18, pp. 1-4.

[11] M. F. Schubert et al.(2008) Polarization-matched GaInNAlGaInN multi-quantum-well light-emitting diodes with reduced efficiency droop, Appl. Phys. Lett., vol. 93, no. 4, pp. 2006-2009.

GaN-based LEDs using Homo-Epitaxial technology and the Progress and Challenges of HVPE method

Shuhang Tan

School of electrical and power technology engineering, China University of Mining and Technology, Xuzhou, Jiangsu, 221116, China

angela@cas-harbour.org

Abstract. GaN and other group III-nitrides are promising materials for optoelectronic devices applications. Currently, sapphire is the most widely used substrate material for GaN LEDs due largely to economic considerations. During the past decade or so, free-standing GaN substrates fabricated via HVPE have gained technological importance and circumventing the shortcomings of sapphire substrates in terms of thermoconductivity and lattice mismatch. This technology, however, does have its own set of challenges. This review summarizes the current state of freestanding GaN substrate technology, the challenges it is facing, and outlook.

1. Introduction

The appearance of blue LEDs based on GaN gave people the chance to create white LEDs, which are more economical, environmental-friendly and have higher efficiency, so they attract a lot of attention and become popular very soon. To make a LED chip, it needs monocrystal substrates as the base and this is the key to decide the quality of a LED. Although using homo-epitaxial technology is the best choice, the growth of GaN monocrystal faces some difficulties, so people tend to let GaN grow on other materials, which is called hetero-epitaxial, and sapphire is the most successful material to make LEDs in business nowadays.

Sapphire has some advantages over other materials. Considering all possible monocrystal substrates materials, we can see that SiC seems to be the one suitable material, due to lower lattice mismatch and smaller difference in the coefficient of thermal expansion. However, it needs a very complex procedure to process properly and the expense is high. Therefore, it cannot be a good choice to use for large-scale production. If the LEDs want to be industrialized, the process to produce substrates should be simple and the expense ought not to be too high. That's why though the mismatch between sapphire and GaN is up to $10^8 cm^{-2}$[1], it is widely used around the world. **Table 1**. shows some physical properties of sapphire.

Table 1:The Physical Properties of Sapphire[2]

Physical Properties	Numerical Value	Scope of Application
The Coefficient of Thermal Expansion / K^{-1}	9.03×10^{-6}(//c axis)	20~1000°C
Thermal Conductivity /[$W/(m \cdot K)$]	23(//c axis)	296K
Energy Gap / eV	8.1~8.6	300K
Tensile Elongation / GPa	190	300K
Poisson's Ratio	0.25~0.30	300K

Though the sapphire-substrate LED is produced and sold worldwide, it also has some limitations. The one is the large lattice mismatch between foreign substrates and GaN, this will affect the continuum of lattice. Even on the most suitable film of sapphire, the lattice mismatch is still as high as 13.3%, and this will cause the dislocation density up to $10^{10}cm^{-2}$[2]. Meanwhile, the mismatch of thermal is also as large as -25.5% [2], which may break the chip in some extreme situations. Besides, low light extraction efficiency (LEE) is also a big problem.

When LEDs are connected with batteries, the carriers, electronics or holes, are recombined in the quantum confinement and emit photons, then LEDs emit light. However, since chips may absorb and reflect some photons, a large bunch of photons cannot arrive at people's eyes. According to the calculation, the conventional sapphire-substrate LEDs' LEE is only around 29%[3].

Considering all of the defects mentioned above, people are focusing on looking for other better substrates and free-standing GaN substrate was considered again. Using homo-epitaxial technology can reduce the mismatch between lattice and improve the thermoconductivity. Meanwhile, it can improve the quality of the chip and increase the internal quantum efficiency. What's more, GaN substrate can fabricate vertical-structure chip, which can enlarge the current expansion scale. Therefore, free-standing GaN substrate is the most valuable research direction for people to devote to.

This passage uses meta-analysis method to analyze the potential technologies which can be used to fabricate GaN substrates and thinks the free-standing technology is the best one to improve the quality of GaN-based LEDs. HVPE is used most to produce free-standing GaN, therefore, the passages focus on discussing this technology and want to make people understand this technology better.

2. Literature review

Many studies have devoted to finding effective ways to improve the quality of LEDs' substrates, such as new substrates. However, hetero-epitaxial has its own defect, the high lattice mismatch. Thus, more and more groups deter to homo-epitaxial technology, although making GaN substrates has some obstacles that need to be overcome.

There have been many institutions studying the GaN fabricating process all over the world and some discoveries and improvements have been achieved[2]. The Dam's[3] group found that the shape and size of the reactor influence the rate of reaction a lot. A slice change of the angle of quartz nozzle can have a big effect on the growth rate. Safvi's[4] group found that the relationship between the growth rate of GaN and temperature is similar to MOCVD technology through contrast tests. The increasement of growth rate is proportional to the partial pressure of GaCl. The partial pressure of NH_3 has little effect on the growth rate but affects the quality of the GaN layer. Zellweger's[5] group found that using N_2/H_2 instead of purify N_2 as the carrying gas can improve the quality of GaN crystals. Meanwhile, some companies in United States and China have been devoted to achieving the pyriform artificial GaN crystals with HVPE then people can attain GaN crystals by using linear saw. The following GaN can continue to grow in the GaN crystals attained before, thus, the dislocation density can drop step by step and people can get perfect material ultimately.

The previous researches mainly focus on the effects of variables during manufacturing process but do not analyze the most popular way to produce the free-standing GaN. Therefore, this passage discusses the HVPE technology thoroughly using meta-analysis method and wants to let people understand the development of free-standing GaN substrates better.

3. Free-standing GaN substrates and its improvement

Traditionally, free-standing GaN substrate means LED chips that they can remove the extra substrate and grow on large GaN monocrystal directly. However, because of the high temperature and pressure required, producing large enough GaN monocrystals as the substrates needs to overcome lots of technology problems. There are still some efforts need to do to arrive at the industrialized goal.

Nowadays, the free-standing GaN substrate is fabricated with another substitutable technology. The widest used fabrication method is that getting GaN thick single crystal layer via hetero-epitaxial technology at first, then stripping the GaN layer from the foreign substrates and doing following processes to make the substrates be a suitable substrate. Though GaN substrates made by this way cannot be as perfect as ideal large GaN substrates, it still makes an impressive improvement in the quality and efficiency of optoelectronic devices.

In theory, homo-epitaxial is the best way to produce GaN-based chips, but there are still some obstacles that need to be overcome before devoting chips into market. The main challenge is the high cost, due to the difficulty to produce large size bulk of single GaN crystals. People have made large efforts to find useful methods and researchers have made 2 inches GaN single crystals successfully and the dislocation is reduced to $5*10^3 cm^{-2}$[6] since now. However, this method achieved success just in the laboratory and cannot be industrialized. Therefore, the common way to produce GaN crystals is to strip the thick GaN layer which grows on hetero-epitaxial substrates, such as sapphire. The following passage will mainly discuss how to produce the large size of GaN single crystals and the sequence process technology.

4. The main methods of making GaN single crystals

There have been some applied methods to produce free-standing GaN substrates, such as ammonothermal crystal growth, Na-flux method, high pressure growth method and HVPE. Table 2 shows the comparisons of some important growing qualities of each method. Among these methods, HVPE is the critical choice, because it costs less and has developed maturely. The following passage will discuss this method.

Table 2: Comparison of Common Methods of Producing GaN Single Thick Crystal Film

Growing Methods	Ammonothermal crystal Growth	Na-flux Method	High Pressure Growth Method	Na-flux Method
Growing Conditions T_g/°C	1000~1050	400~500	400~500	600~800
Time / h	<10	170	100~200	100
Dislocation Density / cm^{-2}	1×10^5	5×10^4	10~100	2×10^5
p(N_2) / MP_a	Ordinary Pressure	200~300	1~2	5~10
FWHM(0002) / arcsec	100~145	15~50	20~30	55

4.1. The development of HVPE

The hydride vapor phase epitaxy (HVPE) method is a very old method for semiconductor's growth. It has also been the most popular technology for producing free-standing GaN substrates, since it is cheap and the process is simple. Meanwhile, GaN has a very high growth rate over 800 μm/h, and it can grow evenly on a large scale.

Its manufacturing-system includes furnace stack, quartz reactor, gallium boat and the equipment for gas to flow. The following are the related chemical equations[2].

$$2HCl(g) + 2Ga(l) == 2GaCl(g) + H_2(g)$$

$$GaCl(g) + NH_3(g) == GaN(g) + HCl(g) + H_{2(g)}$$

HCl flows into the gallium boat mixing with N_2 and reacts with Ga in it to create GaCl; then GaCl and NH_3 go into the equipment which is above the substrates to create GaN. HVPE growth reactions usually occurs in the environment where temperature is over 1000℃ and under the normal air pressure. Currently, most companies aim to attain the GaN crystals. Usually, they let GaN thick layer formed on the foreign substrates at first, such as sapphire or GaAs, and then they remove the foreign substrates using physical or chemical methods, like using laser or etching the substrates.

There are many institutions studying the GaN fabricating process all over the world and some discoveries and improvements have been achieved[3]. The Dam's group found that the shape and size of the reactor influence the rate of reaction a lot. A slice change of the angle of quartz nozzle can have a big effect on the growth rate. Safvi's group found that the relationship between the growth rate of GaN and temperature is similar to MOCVD technology through contrast tests. The increasement of growth rate is proportional to the partial pressure of GaCl. The partial pressure of NH_3 has little effect on the growth rate but affects the quality of the GaN layer. Zellweger's group found that using N_2/H_2 instead of purify N_2 as the carrying gas can improve the quality of GaN crystals. Meanwhile, some companies in United States and China have been devoted into achieving the pyriform artificial GaN crystals with HVPE then people can attain GaN crystals by using linear saw. The following GaN can continue to grow in the GaN crystals attained before, thus, the dislocation density can drop step by step and people can get perfect material ultimately.

4.2. The challenge of HVPE technology
Although this technology has many advantages and has been widely used, there are also some problems remained. The main problem is the vice production NH_4Cl appears during the growth process. It will form some corrosive powder below 340℃ and this powder will pollute the pyriform GaN crystals and prevent the current from flowing, so that it will take longer time for GaN to grow and waste a lot of money. People need to find a way to keep the powder away from the reaction areas. One of the possible solutions is to use vertical HVPE reaction chamber, where pyriform GaN crystals are hang on the top of the chamber, so that powder cannot drop on crystals' surface. This is mentioned by Aixtron Company and used commercially.

4.3. Process of removing foreign substrates
To get native thick GaN crystals, researchers need to strip the GaN substrates from the foreign substrates. The way chosen to strip the foreign substrates is dependent on the chosen foreign substrates. Chemical etching is often used when GaAs is regarded as the substrates and laser lift-off is applied on sapphire-based GaN.

Although this method seems to be perfect, there is great challenge during removing process. Taking the laser lift-off process as the example, as impulse UV lasers irradiating through sapphire substrates, the temperature of the film between GaN and sapphire increases soon. Followed by thermal decomposition of nearby GaN, GaN can separate from sapphire substrate. To apply this technology, the energy of laser should be a little higher than the energy band of GaN (3.4eV)[7]. GaN cannot be stripped totally with low energy while it may be broken if the energy is too high. It is very easy for GaN to break during removing process and because of the limitation of the energy density of lasers, researchers have to find other ways to overcome the challenge.

Some possible solutions have been found. The first one is that the adopted lasers should have moderate energy density. The second way is that researchers can start to scan the substrates from the bounder, so that the gas formed during the decomposition process can emit from the bounder which can reduce the air pressure above the GaN layers. Also, people can stick temporary substrates to GaN crystals, such as Si. The temporary film can support the GaN during the stripping process so GaN substrates keep integrity. Besides these methods, there is also a technology called self-separation. The State Key Laboratory in Peking University has studied it and already had some discoveries [7]. The self-separation technologies include using chemical pre-treatment and pre-thermal treatment on sapphire surface they found have obtained 2-inch GaN crystals successfully. This is also a potential technology.

5. Future outlooks
Free-standing GaN substrates can bring a lot of benefits such as simplifying the fabricating-process sharply. Meanwhile, due to the perfect thermal stability and chemical stability, there is still room for people to explore to find more efficiency epitaxial technology. Plus, there is no lattice mismatch and the dislocation density has been reduced[8]. It has a bright future in LEDs with much higher power.

Homo-epitaxial technology also has big potential in producing vertical structure chips, which can increase the photon emitting efficiency and enlarge the LEDs' lighting area. What's more, it can provide high quality non-polarization films, which may reduce the inter-quantum efficiency and can multiple the lighting efficiency. Seoul Semiconductor Company has released their production, nPola, and its lighting efficiency has reached 500lm/W. This is much higher than LEDs which are available in the market.

6. Conclusion

GaN has several advantages in producing optoelectronics devices and some of them have been used in business all over the world, especially LEDs. However, with the fast development of optoelectronics business, people are asking for high power devices with much higher efficiency and quality but the current technology cannot satisfy people's demands.

Homo-epitaxial technology is mentioned to solve this problem. Free-standing substrates can eliminate the defects like lattice mismatch, and it has high thermoconductivity. Also, it can reduce the dislocation density. Lots of methods have been explored to produce thick bulk GaN crystals and HVPE is a practical one. Since it has a high grown rate and can reduce the cost, this technology has been used in business successfully. However, it still cannot solve the dislocation problem totally and it is easy to break the GaN substrates during stripping process, and this will increase the expense and refine its development.

Although there are many obstacles need to be overcome, the advantages of free-standing GaN have attracted a lot of interests of companies and research groups worldwide. It can multiply the efficiency of inter-quantum and increase the lighting ability sharply. Moreover, it can apply for non-polar or semi-polar substrates and produce vertical structure chips. Thus, free-standing GaN substrates are promising materials for producing high power LEDs or other solid light devices. With the development of technology, the cost of GaN can be reduced and more and more GaN-based LEDs will be applied in business.

References:

[1] Tian, M., Zhu, Y. (2012) The Lighting Technology of White LEDs. M. Science Press, 2012: 162-165.
[2] Li, G. (2014) Blue/White LED epitaxial materials and chips on new substrates. M. Science Press, 2014: 4-9, 18-22, 41, 195-216.
[3] Dam, C. E. C., Grzegorczyka, A. P., Hagemana, P. R., et al. (2004) The effect of HVPE reactor geometry on GaN growth rate-experiments versus simulations. Journal of Crystal Growth., 271: 192-199.
[4] Safvi, S. A., Perkins, N. R., Horton, M. N., et al. (1997) Effect of reactor geometry and growth parameter on the uniformity and material properties. Journal of Crystal Growth., 182(3): 233-237
[5] Zellweger, C., Wagner, V., Py, M. A, et al. (2002) AlGaN/GaN HEMTs growth on SiC substrates. Institute of Physics Conference Series., 166: 455-459
[6] Li, J., Wang, J. (2016) The Technologies and applications of LEDs based on III-nitrides materials. M. Science Press, 2016: 6-18, 33-37.
[7] Tong-Jun, Y. (2015) GaN substrate and GaN homo-epitaxy for LEDs: Progress and challenges. J. Chin. Phys. B., 24(6).
[8] Li, G., Wang, W., Yang, W., et al. (2016) GaN-based light-emitting diodes on various substrates: a critical review. J. Reports on progress in physics, Physical Society (Great Britain), 79(5): 56501.

IOP Conf. Series: Materials Science and Engineering **729** (2020) 012005 doi:10.1088/1757-899X/729/1/012005

A Comparison between Si and SiC MOSFETs

Jiawei Wang

School of material science and engineering, Northwestern Polytechnical University, Xi'an, Shaanxi, 710072, China

angela@cas-harbour.org

Abstract. The Metal-Oxide-Semiconductor Field-Effect Transistor (MOSFET) has developed quickly and become one of the most important devices for a wide range of applications. There is an increasing need for power MOSFET devices with low power consumption and high energy efficiency. Silicon (Si) and Silicon Carbide (SiC) are two kinds of materials used in power MOSFET devices, which have their own advantages of performance for each use. This paper makes a comparison of the on-resistance and high-temperature performance between Si and SiC MOSFETs. The analysis of the differences between the two will mainly come out from the perspective of material properties, and the conclusion of what is the ideal material for power MOSFET devices will be finally drawn.

1. Introduction

Power electronic systems are of great significance in many areas, as they can proceed and control the flow of electric power[1]. One of the most important devices in power electronic systems is the metal-oxide-semiconductor (MOS) capacitor. The metal-oxide-semiconductor field-effect transistor (MOSFET) is composed of an MOS capacitor and two p-n junctions placed closely to the MOS capacitor[2]. MOSFET was firstly demonstrated in the 1960s and has developed quickly to become an important device for advanced integrated circuits, such as microprocessors and semiconductor memories. A lot of researchers have contributed to developing power MOSFET devices with many unique and unprecedented features, including low-power consumption and high switching speed and good efficiency[3].

Silicon (Si) and silicon carbide (SiC) are two kinds of important semiconductors, which are widely used by power MOSFET devices, thanks to their low-power loss and high-power efficiency[4,5,6]. Figure 1 shows an overview of Si and SiC based on physical material parameters which fundamentally affect the performance of any power switch in a power system[7]. However, based on these different properties of materials, few researchers have made the comparison of their properties of devices. This paper will mainly focus on the comparison of resistance and high-temperature performance between Si and SiC MOSFETs from the perspective of material.

IOP Conf. Series: Materials Science and Engineering **729** (2020) 012005 doi:10.1088/1757-899X/729/1/012005

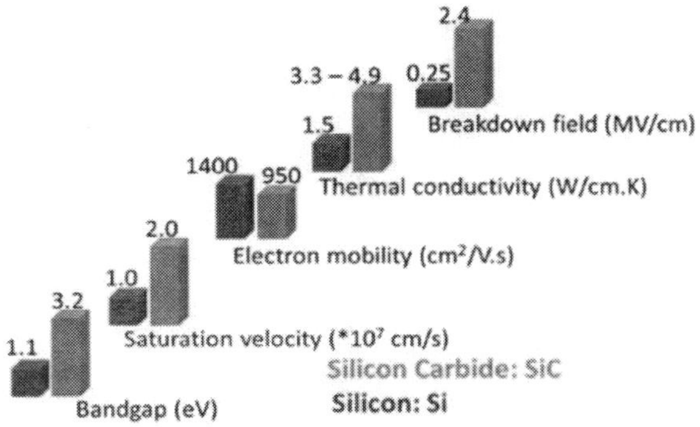

Figure 1. An overview of Si and SiC based physical material parameters that fundamentally affect the performance of any power switch in a power system

2. On-resistance comparison

2.1. The role of on-resistance in MOSFET

On-resistance is the total resistance between drain and source in on-state. It is an important parameter to determine the highest allowed current. For the same kind of MOSFET devices, lower on-resistance means lower power loss which could be essential for this industry. Figure 2 shows the composition of the total on-resistance of the VD-MOSFET and it could be calculated by the equation:

$$R_{on} = J_c(R_{N^+\cdot sp} + R_{CH\cdot sp} + R_{A\cdot sp} + R_{J\cdot sp} + R_{D\cdot sp} + R_{S\cdot sp})$$

where $R_{N^+\cdot sp}$ is the source diffusion resistance, $R_{CH\cdot sp}$ is the channel resistance, $R_{A\cdot sp}$ is the accumulating layer resistance, $R_{J\cdot sp}$ is the JEFT region resistance, $R_{D\cdot sp}$ is the drift-layer resistance and $R_{S\cdot sp}$ is the substrate resistance.

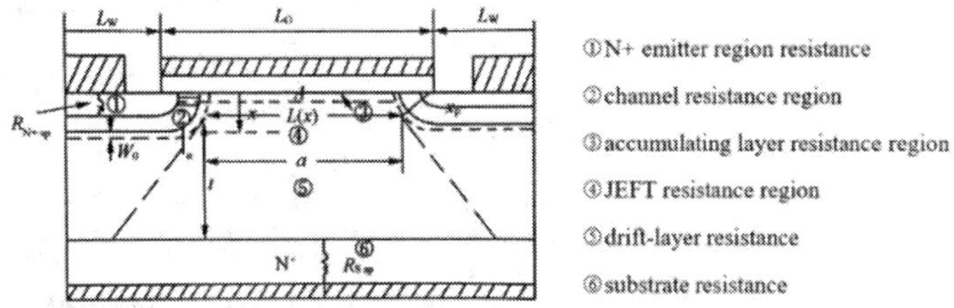

Figure 2. The VD-MOSFET cellular on-resistance schematic diagram

In the idealized case, it is assumed that the structure can support the blocking voltage without degrading due to the edge termination or local electric field enhancement. Additionally, all the resistance in device structure could be seen as parasitic resistance with zero resistance, except the drift region. In such a case, the drift region resistance is called ideal on-resistance. An equation was derived for the resistance per unit area for the drift region:

$$R_{on-sp,ideal} = \frac{4BV_{PP}^2}{\varepsilon_S \mu E_C^3}$$

Where BV_{pp} is the parallel-plane breakdown voltage, ε_S is the dielectric constant for the semiconductor, μ is the mobility, and E_c is the critical electric field for breakdown.

2.2. Comparison of on-resistance between Si and SiC MOSFET

The relationship between ideal on-resistance and breakdown voltage based on the equation above may be more directly shown by Figure3 which plots the minimum specific on-resistance against the breakdown voltage for Si and SiC unipolar devices[8].

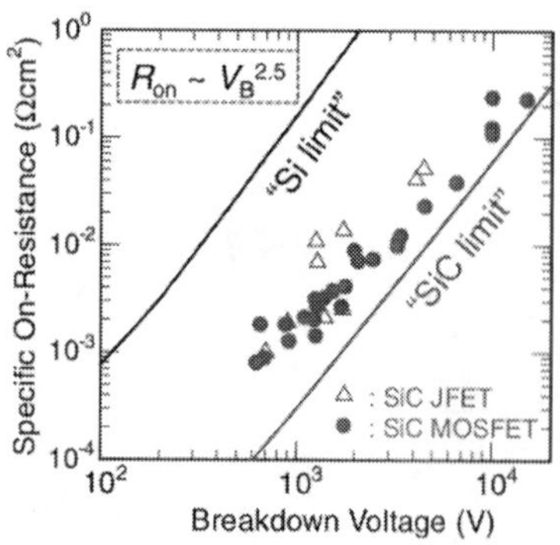

Figure 3. Minimum specific on-resistance for Si and SiC unipolar devices (Si limit and SiC limit) versus the blocking voltage

For Si power devices, it is assumed that the mobility is constant due to low doping concentration in the drift region. For the case of an N-type drift region that is utilized for n-channel power MOSFETs, the ideal specific on-resistance is given by[9]

$$R_{on-sp,ideal}(n-channel) = 5.93 \times 10^{-9} BV_{PP}^{2.5}$$

The resistance of high voltage power Si MOSFET is close to the ideal on-resistance in drift region. The equation above gives the 2.5th ergs increasing relationship between ideal on-resistance of Si and breakdown voltage, which means the on-state voltage drop is more than 10V when on-current density is 100A/cm^2. As a result, a large power loss will happen inside the device. Although the power loss can be drained by decreasing on-current density, such measure is implausible because it will increase the chip area and finally increase the cost of the chip.

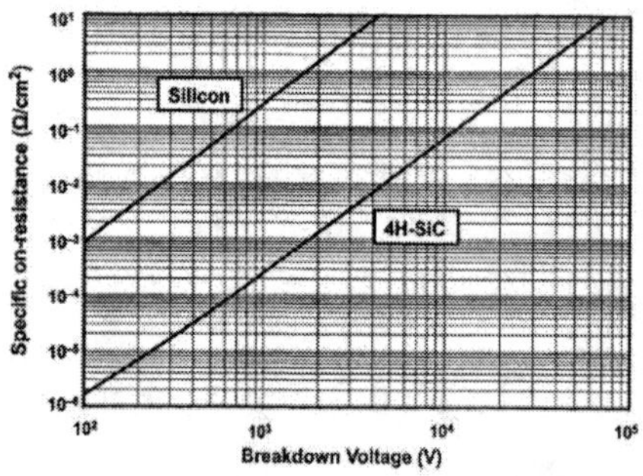

Figure 4. Ideal specific on-resistance for Si and 4H-SiC power devices

The ideal specific on-resistance for the drift region in 4H-SiC devices is compared with that for silicon in Figure4 with breakdown voltages ranging from 100V to 100,000V. It shows that there is a significant reduction of specific on-resistance of drift regions between Si and 4H-SiC devices. The ratio of the specific on-resistance for Si to that for 4H-SiC increases from 527 at a breakdown voltage of 100V to 1280 for breakdown voltages above 40,000V.

Table 1. The doping concentration, electron mobility, energy barrier width and on-resistivity versus the breakdown voltage

	Breakdown voltage/ V	Doping concentration/ cm^{-2}	Electron mobility/ cm^2/(V·s)	Energy barrier width/ μm	On-resistivity/ Ω·cm^2
6H-SiC	200	1.60×10^{17}	267	1.16	1.69×10^{-5}
	1000	1.81×10^{16}	435	7.69	6.11×10^{-4}

	3000	4.08×10^{15}	523	28.02	8.20×10^{-3}
	5000	2.04×10^{15}	530	51.12	2.95×10^{-2}
3C-SiC	200	7.85×10^{-15}	348	1.65	3.77×10^{-5}
	1000	8.87×10^{15}	731	10.97	1.06×10^{-3}
	3000	2.01×10^{15}	1000	39.97	1.24×10^{-2}
	5000	1.00×10^{5}	1000	73.14	4.54×10^{-2}
Si	200	1.72×10^{15}	1336	12.48	3.35×10^{-3}
	1000	2.01×10^{14}	1356	81.59	1.88×10^{-1}
	3000	4.65×10^{13}	1360	293.94	2.92×10^{0}
	5000	2.35×10^{3}	1360	533.44	1.05×10^{1}

Table 2. Ratio of Si to SiC for on-resistance $(R^{Si}_{on\cdot sp}/R^{SiC}_{on\cdot sp})$

Breakdown voltage/V	50V	200V	1000 V	5000 V
Si / 6H-SiC	92.9	198.2	305.7	355.9
Si / 3C-SiC	49.3	88.8	177.4	229.8

Table 1 shows the ideal breakdown voltage, doping concentration, electron mobility, energy barrier width and on-resistivity for 6H-SiC, 3C-SiC, and Si MOSFET[10]. Table 1 shows that for a given breakdown voltage, the on-resistance of SiC MOSFET will at least 2 orders of magnitude lower than in their Si counterparts for the same blocking voltage, and SiC exhibits more excellent conduction characteristic under higher breakdown voltage. This is particularly important because lower on-resistance contributes to lower on-state loss. The on-resistance ratio of Si to SiC is shown in Table 2.

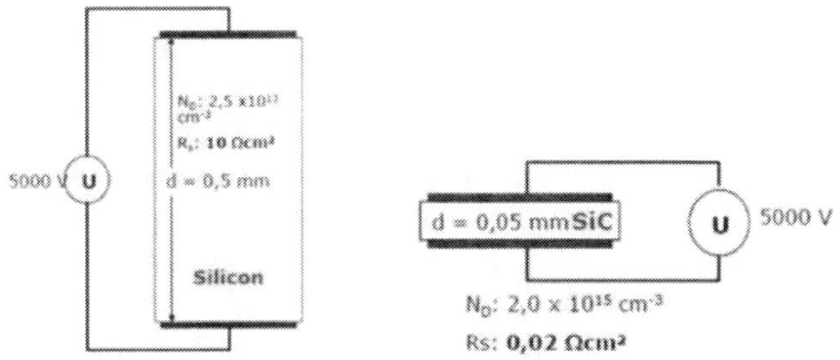

Figure 5. Si MOSFET and SiC MOSFET under the voltage of 5000V

As can be seen from figure 5, if a withstand voltage of 5000V is to be obtained, the Si power device with a substrate material doped with 2.5×1013/cm^3 requires a drift layer thickness of 0.5mm and an area resistance of 10Ω/cm^2, while the SiC MOSFET with a drift layer doped with a concentration of 2.0×1015/cm^3 only requires a thickness of 0.05mm and an area resistance of 0.02 Ω/cm^2. SiC power devices allow the use of thinner drift regions to maintain higher blocking voltages, significantly reducing forward voltage drop and conduction losses.

2.3. Conclusion

From the analysis above, it can be sure that SiC can be used to make the device which has lower area of specific on-resistance.

In conclusion, the on-resistance of SiC power MOSFET is much lower than Si power MOSFET at the same breakdown voltage. This is significant for MOSFET's application.

3. High temperature characteristics comparison

3.1. The role of high temperature characteristics in MOSFET

As is known to all, the efficiency of a MOSFET device will never be 100% because it will unavoidably lose energy mainly by heat generation when operating. In addition, the heat energy will make

IOP Conf. Series: Materials Science and Engineering **729** (2020) 012005 doi:10.1088/1757-899X/729/1/012005

temperature rise up gradually and the device will work in a higher temperature, thus some properties of the device will be influenced, especially the electrical characteristics.

Firstly, the threshold voltage will be influenced. The threshold voltage for a power MOSFET structure is given by

$$V_{TH} = \frac{\sqrt{4\varepsilon_S kTN_A \ln(\frac{N_A}{n_i})}}{C_{OX}} + \frac{2kT}{q}\ln(\frac{N_A}{n_i}) - \frac{Q_{OX}}{C_{OX}}$$

The equation shows that the threshold voltage for a given power MOSFET structure decreases with increasing temperature because the concentration of intrinsic carrier increases. Figure6 shows the change of the threshold voltage, in which a P-base doping concentration is $1\times10^{17}cm^{-3}$. It shows that the threshold voltage in 200°C decreases about 50% of it in -50°C.

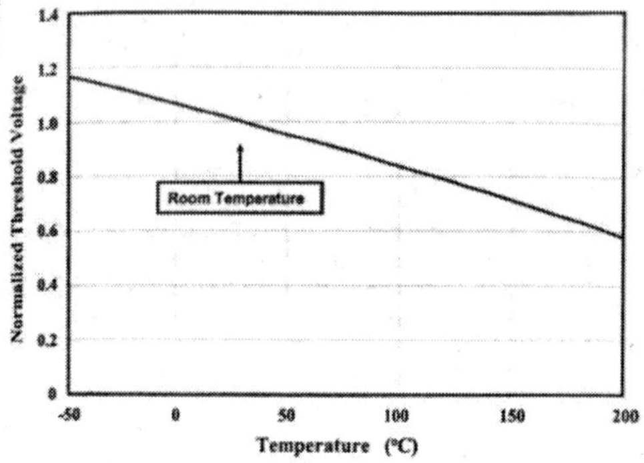

Figure 6. Impact of increasing temperature on the threshold voltage of the power MOSFET

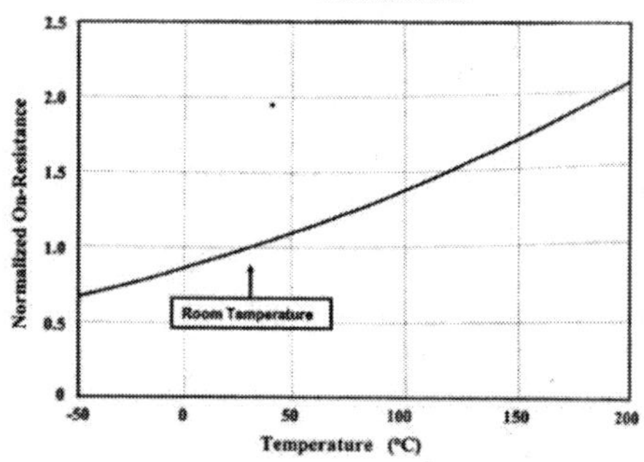

Figure 7. Impact of increasing temperature on the on-resistance of the power MOSFET

In addition, the on-resistance increases with the increasing temperature due to the reduction of the mobility in the drift region and the inversion layer. Figure 7 shows the change in the on-resistance with temperature for the case of 30 V power VD-MOSFET structure with a gate width of 10μm and a cell pitch of 18μm. It shows that the on-resistance increases about 70% at 150°C and about 100% at 200°C. Thus, a positive feedback between the device resistance and its junction temperature will occur when such device is conducting significant current in reality, which causes further increases in both. As a result, increasing on-resistance in high temperature environment will bring out a huge on-state power loss[11].

Therefore, superior high-temperature operating capability is very important. A comparison of high temperature characteristics for SiC and Si MOSFET devices will be discussed in the following section.

3.2. Comparison of high temperature characteristics between Si and SiC MOSFET

R.Kibushi conducted numerical simulation to investigate thermal properties of Si and SiC power MOSFET[12]. Figure 8 and 9 show the temperature of Si and SiC power MOSFET with different drain voltage. As can be seen from these figures, the temperature of SiC power MOSFET is much lower than Si device, despite the same structure and operating condition.

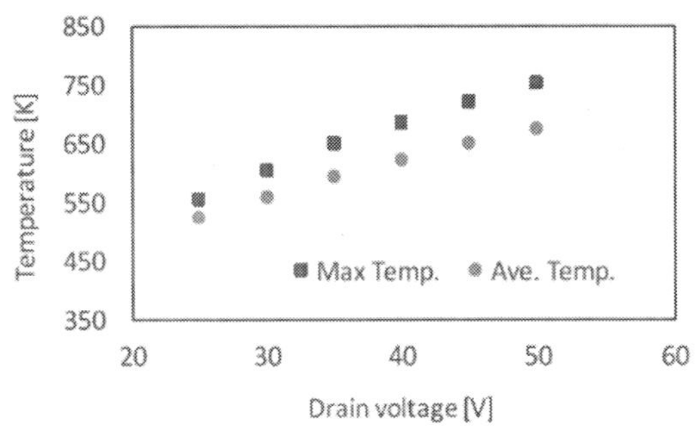

Figure 8. Temperature of Si power MOSFET

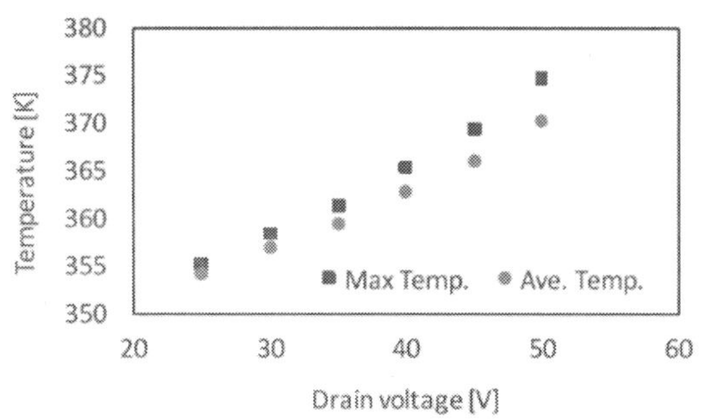

Figure 9. Temperature of SiC power MOSFET

Figure 10 shows the temperature dependence of thermal of Si and SiC. As is shown in the figure, thermal conductivity of both devices decreases when temperature increases. However, the thermal conductivity of SiC device is always much higher (about 2 times higher) than that of Si device. This can be one reason why SiC power MOSFET can be used in higher temperature environment and it can be explained by the strong chemical bonding between Si and C atoms.

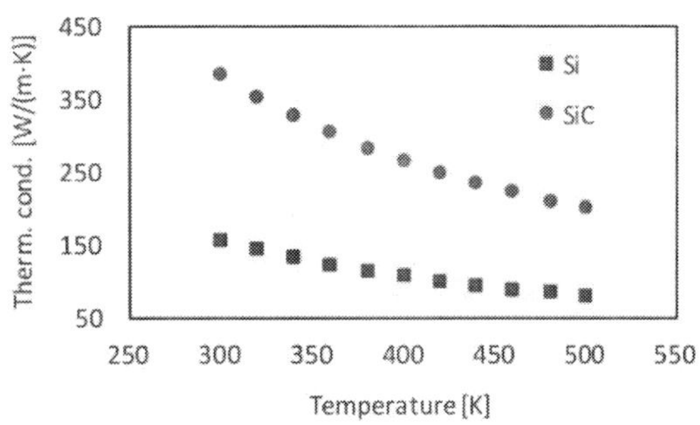

Figure 10. Temperature dependence of thermal conductivity

Figure 11 shows the heat generation density around hot pot in Si and SiC MOSFET device. Obviously, SiC generates much lower heat than Si, though the Heat generation density of Si MOSFET decreases with increase in drain voltage and that of SiC MOSFET increases. Therefore, the SiC power MOSFET produce much lower heat and thus the temperature will rise much more slowly. As a result, SiC power MOSFET, at the same time, will lose less energy in the operating process. Additionally, the different tendency can be explained by the electrical current through power MOSFET shown in figure 12. Although drain current of Si MOSFET decreases with increase in drain voltage, it is already saturated in calculation condition and it decreases due to high heat generation and high temperature of the device. However, in SiC power MOSFET, the drain current increases with the increase in drain voltage because of high saturation electron velocity of SiC. Consequently, the two types of power MOSFETs have different tendency of heat generation.

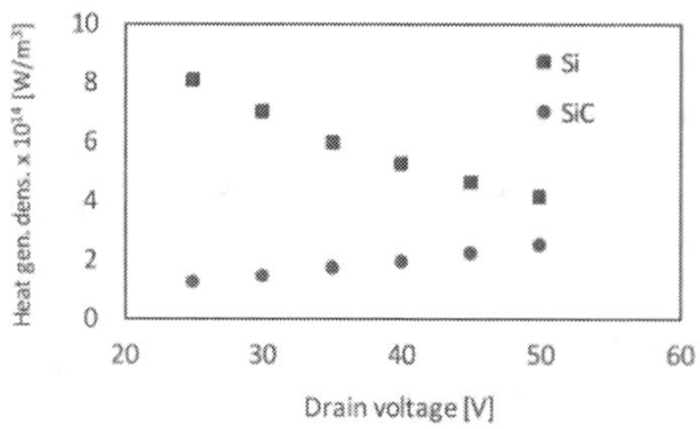

Figure 11. Heat generation density around hot spot

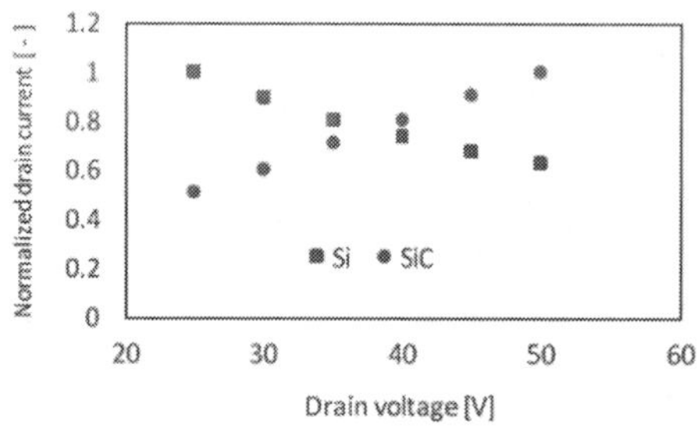

Figure 12. Normalized drain current

Moreover, Shenai proved that the junction temperature of Si power MOSFET is limited by 200°C, while SiC power MOSFET can still operate in a high temperature of 600°C.

3.3. Brief Summary
Since SiC has higher thermal conductivity, the temperature will rise amid lower speed. Moreover, while Si power MOSFET is in the condition of saturated current and calculation voltage, drain current of SiC power MOSFET still rises with the increase of applied drain voltage.

Consequently, SiC power MOSFET has a more superior high-temperature operating capability than Si power MOSFET's.

4. Discussion
Apparently, improving the performance of Si MOSFET devices has been limited by the physical properties of the material itself. Nowadays, Si power device technology is almost mature and it is hard to get innovative breakthroughs. Therefore, SiC becomes an ideal replacing material for power MOSFET due to its excellent physical properties. SiC power MOSFET has not been widely used in electric industry, due to the cost and low efficiency of electron drift mobilities in the channel of MOSFETs, which is caused by the extremely high density of interface states close to conduction band edge[13,14]. SiC is an ideal replacing material for power MOSFET, which is full of opportunities and challenges.

5. Conclusion
In this research, it makes the comparison of characteristics of on-resistance and high-temperature between Si and SiC power MOSFET devices, and analyzes the result. It can't be ignored that the experiments above still have lots of defects. For example, in the electro-thermal analysis, if applied voltage becomes much higher and saturation current condition appears, the temperature of SiC power MOSFET may rise sharply. Therefore, in future work, the condition of higher voltage and smaller scale will be examined. Moreover, it is expected to use more professional analyzing methods to examine the material structure, and to explore ways to improve some physical properties of MOSFET.

In conclusion, SiC power MOSFET has lower on-resistance and better performance amid high temperature. More researches about SiC, as an ideal replacing material for power MOSFET, is waiting to be conducted.

Reference

[1] Flack, T. J., Pushpakaran, B. N., Bayne, S. B. (2016). GaN technology for power electronic applications: a review. J. Journal of Electronic Materials., 45(6). 2673-2682.
[2] Nicollian, E. H., Brews, J. R., Nicollian, E. H. (1987). MOS (metal oxide semiconductor) physics and technology. New York et al.: Wiley.
[3] Sze, S. M. (2008). Semiconductor devices: physics and technology. John wiley & sons.
[4] Borgeest, K. (2015). Tested once, forever right? Influence of aging and temperature on susceptibility and emissions. In: 2015 IEEE International Symposium on Electromagnetic Compatibility (EMC). Dresden. pp. 271-276.
[5] Yin, L., Chen, C. P., Kapusta, C., Ghandi, R. (2015). Electronic packaging of SiC MOSFET-based devices for reliable high temperature operation. In: 2015 IEEE International Symposium on Circuits and Systems (ISCAS). Lisbon. pp. 1170-1173.
[6] Avenas, Y., Dupont, L., Baker, N., Zara, H., & Barruel, F. (2015). Condition monitoring: A decade of proposed techniques. IEEE Industrial Electronics Magazine, 9(4): 22-36.
[7] Nawaz, M., Ilves, K. (2016). Replacing Si to SiC: Opportunities and challenges. In: European Solid-State Device Research Conference (ESSDERC). Lausanne. pp. 472-475.
[8] Kimoto, T. (2015). Material science and device physics in SiC technology for high-voltage power devices. Japanese Journal of Applied Physics, 54(4): 040103.
[9] Baliga, B. J. (2010). Fundamentals of power semiconductor devices. Springer Science & Business Media.
[10] Zhu, R., Chow, T. P. (1998). A comparative study of the quasi-saturation in the high voltage vertical DMOS for different cell Geometries. In: Proceedings of the 10th International Symposium on Power Semiconductor Devices and ICs. ISPSD'98 (IEEE Cat. No. 98CH36212). Kyoto. pp. 343-346.
[11] Zhou, W., Zhong, X., & Sheng, K. (2013). High temperature stability and the performance degradation of SiC MOSFETs. J. IEEE Transactions on Power Electronics, 29(5): 2329-2337.
[12] Kibushi, R., Hatakeyama, T., Yuki, K., Unno, N., & Ishizuka, M. (2017). Comparison of thermal properties between Si and SiC power MOSFET using electro-thermal analysis. In: 2017 International Conference on Electronics Packaging (ICEP). Yamagata. pp. 188-192.
[13] Pensl, G., Bassler, M., Ciobanu, F., Afanas' ev, V., Yano, H., Kimoto, T., Matsunami, H. (2000). Traps at the SiC/SiO 2-interface. MRS Online Proceedings Library Archive., 640.
[14] Schörner, R., Friedrichs, P., Peters, D., Stephani, D. (1999). Significantly improved performance of MOSFET's on silicon carbide using the 15 R-SiC polytype. J. IEEE Electron Device Letters, 20(5): 241-244.

Research and application of intelligent control system for variable speed operation of pumping unit

Song Yang[1], Xuemin Bai[1], Zhixue Xia[1], Shengli Zhang[1], ShiTong Wu[1] and yun Li[1]

[1] Engineering technology research institute of Huabei oilfield company, Renqiu, Hebei, 062550, China

*ysyj_1220@163.com

Abstract: At present, the pumping machine lifting technology still has problems such as low efficiency, high energy consumption and short pumping cycle. In order to further improve the level of artificial lifting technology of pumping wells, achieve quality and efficiency, save energy and consume reduction, huabei oilfield has developed an intelligent control system for variable speed operation of pumping unit, which takes “closed-loop control technology”, “flexible control technology” and “variable speed operation technology” as a core, and sets digital supporting technology such as “mobile intelligent management and control”, “big data analysis” and “OPRS intelligent expert” as a whole. The system successfully applied more than 600 wells in Huabei Oilfield. Through large-scale on-site application, intelligent control system for variable speed operation of pumping unit realizes the most economical production with minimum energy consumption, achieves the goal of energy saving and consumption reduction, and enhances the intelligent production management level of the oilfield.

1. Introduction

In China's oilfield production, mechanical oil production wells account for 98% of all oil wells, and the beam pumping unit wells account for 96% of all mechanical oil production wells. At the same time, the major oilfields have been in the middle and late stages of development. With the deepening of the inferiority of resources, the working conditions of oil wells have become more and more complex, and the proportion of low-yield, low-efficiency wells and producing fluctuating wells has increased year by year. Crude oil has become more difficult to mine, and the stable production situation is grim.

A new technical means is urgently required to further improve the production management level of pumping wells, to achieve energy conservation, efficiency increasing, production increasing, and consumption reducing. To this end, a new type of intelligent control system for pumping wells was developed to realize the automation of the production process, the visualization of production orders, the scientizaion of analysis and decision-making of the intelligent production of pumping wells, and the oilfield quality improvement and sustainable development was promoted.

2. overall ideas and structure of the system

In the construction of the intelligent control system of the pumping well, the three technical objectives of the system are first established: (1) The pumping well can actively adjust the operating system according to the change of the oil supply in the oil well, and realize the attack-platoon relationship from passive regulation to active coordination, and the control mode from extensive to fine; (2) The driving mode is from rigid to flexible, reducing pole stress and motor energy consumption, improving system efficiency and prolonging pump inspection cycle; (3) "Internet + Big Data" technology is used to explore a new means of pumping well control to achieve digital production decision-making functions. Based on these three technical ideas, the intelligent control system for variable speed operation of pumping wells was developed.

In the framework of the intelligent control system for variable speed operation of pumping wells, the operating mode of the pumping well includes two types: power frequency operation mode and frequency conversion operation mode. Among them, the frequency conversion operation mode is the core of the intelligent control system, it integrates closed-loop control technology, flexible control technology, variable speed operation control technology and mobile intelligent management and control technology, and completes the three major technical objectives of supply and discharge coordination, energy saving and cost reducing and digital management. As shown in Figure 1.

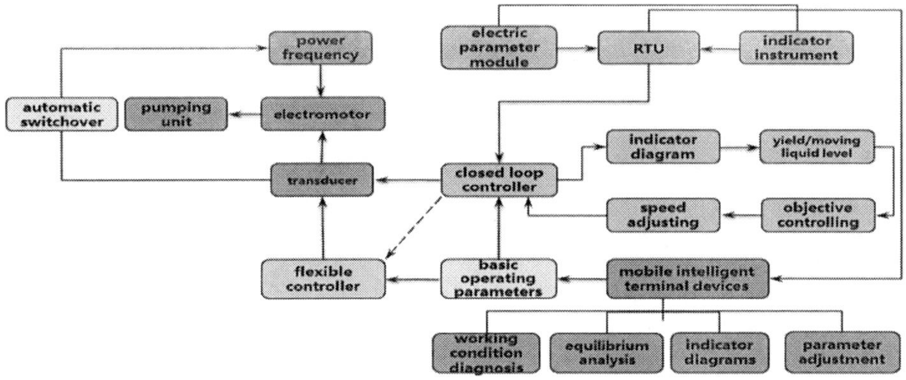

Figure 1 Frame diagram of intelligent control system for variable speed operation of pumping unit

3. Key technology of intelligent control system

3.1. Closed loop control technology of pumping unit well

The closed-loop control technology uses the oil production volume or the dynamic liquid surface depth of the pumping well as the control basis for adjusting pumping work system of the pumping well. First, the intelligent terminal equipment (mobile phone, tablet, etc.) sends data reading commands to the closed-loop controller, after receiving the above instruction, the closed loop controller starts to communicate with the well field RTU to obtain the current ground indicator diagram data and the running stroke of the pumping well, the closed loop controller calculates the current pumping efficiency of the pumping well based on the ground indicator diagram data, and the data is sent to the intelligent terminal equipment for display. The on-site technicians set the target pump efficiency on the intelligent terminal equipment according to the current pump efficiency of the pumping well. The intelligent terminal equipment calculates the target rate according to the closed-loop coordination calculation formula, and sends the target rate to the closed-loop controller in the form of an instruction, after receiving the instruction, the closed-loop controller controls the variable frequency controller to adjust the operating frequency of the system to achieve the purpose of changing the rate. If the adjustment does not achieve the desired pumping efficiency, repeat the above operation. As shown in Figure 2.

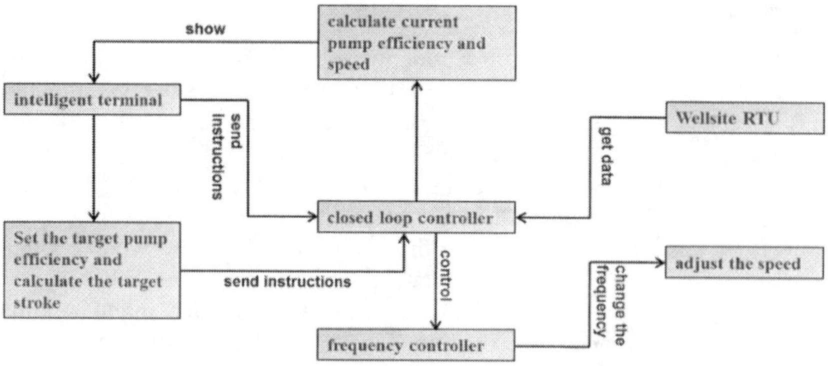

Figure2 Closed-loop control structure diagram of pumping well

3.2. *flexible control technology of pumping unit well*

The flexible control technology is based on the principle of "current monitoring-frequency control-load verification method". It consists of four parts: flexible controller, frequency converter, temperature-sensing flow switch and flexible control panel, the flexible controller is the backbone of the whole system, it is responsible for the analysis and processing of the data and the output of the results; the inverter is the direct controller of the motor to achieve variable speed and constant power operation; the temperature-sensitive flow switch utilizes the temperature change of the wellhead over-current pipeline, and it is responsible for measuring the amount of liquid per rate of the pumping well, and mark the rotational position of the pumping unit crank by monitoring the current. The basic principle of realizing the flexible operation of the pumping well is: wired connection flexible controller, frequency converter, temperature-sensing flow switch. the flexible control panel is used to set the basic operating parameters of the pumping well, and the flexible controller monitors the crank position data according to the current, uses the built-in flexible control algorithm performs analysis and calculation, and outputs the analysis and calculation result to the frequency conversion controller. The frequency converter outputs a specific frequency at a specific time according to the received analysis processing result, thereby controlling the motor to output different powers at different stages, and realizing the on-demand output of the oil motor power, the fluctuation of the pumping unit load is reduced, the ground drive and the fatigue of the pole are reduced by adjusting the motor speed and the speed distribution during the plunger pumping process, it achieves the purpose of improving working conditions and extending the pump cycle.

It can be seen from Fig.3, under the normal power frequency operation, the peak power is 87.67 KW, the peak power after the flexible optimization is 49.83 KW, the change trend of the power curve becomes slow, and there is a phenomenon of negative work before the flexible optimization, the impact of negative work on the system is eliminated under the flexible operation, " peak clipping and valley filling" are achieved. As shown in Figure 3.

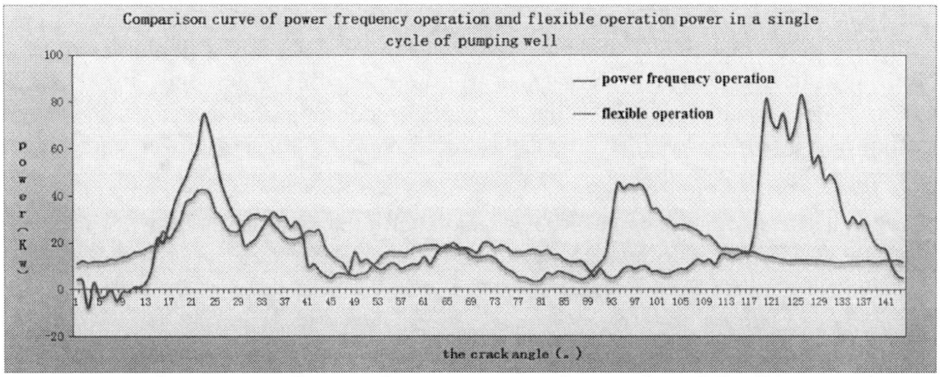

Figure 3 Comparison curve of power frequency operation mode and flexible operation mode

3.3. *Pumping well shifting operation control technology*

When the motor runs at a constant speed, the operation law of the polished rod of the pumping well is a mandatory quasi-sinusoidal operation law, it does not consider the changing factors of the oil well working condition. But if the motor changes speed, it can adjust the running speed of polished rod by changing the motor speed when there is no optimization space for regulating the pumping time of oil well, and then changes the running ratio of the up-and-down stroke, further optimizes the working condition of the well without changing the operating rate of the well, and achieves the effect of “excellent superiority”. The mathematical model is established to provide the basis for automatic learning, and the value of the frequency of each operating phase and the distribution time of the corresponding frequency are determined. After the frequency variation curve is determined, the motor output frequency is in strict accordance with this curve, and the oil well distributes and runs according to this frequency, the optimal operating condition of the pumping unit is achieved.

3.4. *the mobile intelligent control technology of pumping well*

Using the "Internet + APP Control" technology, the auxiliary decision functions such as production calculation, working condition diagnosis and parameter optimization are transplanted from the background PC end to the front end. Through establishing universal data communication interface among the data acquisition sensor, the industrial control RTU and the intelligent terminal device, and deploying the corresponding APP application on the intelligent terminal equipment (mobile phone, tablet, etc.), the well field autonomous acquisition and analysis processing of pumping unit well electric parameters and load displacement data are realized, and the analysis and processing results are visually displayed on the intelligent terminal equipment. Through the connection of various business models based on big data analysis and application, OPRS expert system, knowledge base and other systems, the on-site staff can fully understand the real-time working conditions of the oil well without being restricted by the geographical space. The production and management efficiency are greatly improved.

4. Field test and application

The technology successfully applied more than 600 wells in Huabei Oilfield from 2015 to 2018. As tested by the relevant energy conservation department in Huabei Oilfield, the average energy saving rate of single wells reached 17.14% and the system efficiency increased by 5.38 percentage points.

Taking the XL10-47 well and the Y63-140x well in Huabei Oilfield No. 1 Oil Plant as an example, the application results of the variable speed operation intelligent control system are further evaluated from the detailed parameters of the machine mining system. Before the implementation of the variable speed intelligent control technology, the XL10-47 well and the Y63-140x wells were all operated in the power frequency mode, and then converted to the inverter mode operation. The results of energy saving and production increasing before and after implementation are shown in Table 1 and Table 2.

It can be seen from Table 1 that after applying this technology, the daily output of XL 10-47 well increased by 0.11t, the pump efficiency increased from 29.35% to 63.11%, the maximum load decreased by 19.1kN, and the daily power consumption decreased 36.43 KW•h, the system efficiency increased by 8.6 percentage points, and the power consumption per ton of liquid per 100m reduced by 0.27 KW•h. It can be seen from Table 2 that after applying this technology, the maximum load of Y63-140x well decreased by 10.62kN, the daily power consumption decreased by 150.5kW•h, the system efficiency increased by 2.94%, and the liquid consumption per ton of liquid per 100m The electricity dropped by 0.12kW•h.

Table 1 running effect comparison table before and after implementing intelligent control technology of warible speed operation of XXX well

Operation mode	Pump diameter (mm)	Pump deep (m)	Stroke (m)	Speed (Min^{-1})	Liquid production capacity (t/d)	Pump efficiency (%)	Degree of balance	Maximum load(kN)	Active power (kW)	Day-power consumption (kW•h)	System efficiency (%)	Per ton of liquid per 100m electricity consumption (kW•h)
Power frequency operation	38	1963	4.8	3.75	8.25	29.35	1.12	86.9	6.29	151.03	27.06	1.02
Frequency conversion operation	38	1963	4.8	1.51	8.36	63.11	1.21	67.8	4.78	114.6	35.66	0.75
Difference comperison					0.11	33.76	0.09	-19.1	-1.51	-36.43	8.6	-0.27

Table 2 running effect comparison table before and after implementing intelligent control technology of warible speed operation of XX well

Operation mode	Pump diameter mm	Pump deep (m)	Stroke (m)	Speed (Min^{-1})	Maximum load (kN)	Active power (kW)	Day-power consumption (kW•h)	System efficiency (%)	Per ton of liquid per 100m electricity consumption(kW•h)
Power frequency operation	38	1946	4.8	4	88.34	11.75	282	20.81	1.30
Flexible operation	38	1946	4.8	2	77.72	5.48	131.52	23.75	1.18
Difference comperison					-10.62	-6.27	-150.5	2.94	-0.12

5. Conclusion

(1) Field application shows that the intelligent control system for the variable speed operation of the pumping unit realizes the most economical output with minimum energy consumption, achieving the purpose of improving quality and efficiency, saving energy and reducing consumption.

(2) The on-site application of the intelligent control system for the variable speed operation of the pumping unit greatly reduces the workload of on-site management, improves the work efficiency of the on-site management personnel, and further improves the intelligent control level of production.

Acknowledgments

The auther is very grateful to engineer Bai for providing ideas of the article, to engineer Xia for his support in the design process of "Integration of Oil Recovery ,Well Completion and Drilling", and to engineer Zhang for his help in the funding of the article. Thank you all!

References

[1] Hancheng Liu. Calculation method of reasonable submergence when swabbing gas mixture crude oil[J]. Oil Drilling & Production Technology, 1995, 17 (5): 90 - 93.

[2] Yuguang Geng, Yunhui Zhai. Supply and discharge coordination diagram and application of rod pump pumping oil[J]. Oil Drilling & Production Technology, 1995, 17 (6) : 78 - 81, 107.

[3] Junhai Li, Tianzhong Wu. Determination method of reasonable submergence of rod pump[J]. Journal of Jianghan Petroleum Institute, 2001, 23 (2): 36 – 37.
[4] Qi Zhang. Principles and Design of Oil Recovery Engineering[M]. Dongying: China University of Petroleum Press,2000: 103 - 117.
[5] Jinxia Fan, Jingli Duan, Xiaognag Jiang,etc. General Solution about the Crack Reasons of Top Cage of Plunger for Domestic and Overseas[J]. Oil field equipment, 2011, 40 (6): 26 - 29.
[6] Jianhua Rao, Zhaohong Liu, Dongping Li, etc. Dynamic numerical simulation and predictionof rod pumping system in directional well[J]. China petroleum machinery, 2004, 32 (8): 4 – 6.
[7] Lea J F. Boundary condition used with dynamic models of beam Pump Perfor-mance. AmocoProduction Co.Southwesten Petroleum short Course [J].1988, 120-131.
[8] Jennings J W, Lame R E. A Method for design Fiberglass sucker rod string Using APRIL[J].SPE18188, 1988, 44-78.
[9] Gibbs S G. A.general method for predicting rod pumping system performance [J].SPE685050.1977, 1-8.
[10] Zhenwen Dong, Shengjie Dong, Dongfeng Lu, etc. Simulation models for optimization design of suction parameters for rod pumping system in directional wells[J]. Acta petrolei sinica, 2008, 29 (1): 120 – 123.
[11] Zhenwen Ni, Jun Qi, Jianhua Hao. Motor rotate speed control technology on the beam-pumping unit[J]. Oil field equipment, 2001, 30 (5): 46 – 49.

IOP Conf. Series: Materials Science and Engineering **729** (2020) 012007 doi:10.1088/1757-899X/729/1/012007

The Green Building Development Status in Guangzhou

HE Junfei, HUANG Jiayu*, XIE Zeqiong, XIAO Zhenqian, FENG Chao

Guangzhou Institute of Energy Testing,Guangzhou 511447, China

Corresponding author e-mail: ziyuan2007@126.com

Abstract. The paper mainly introduces the green building development status in Guangzhou. As one of the well developing areas of green building in China, Guangzhou's green building projects are gradually increasing. However, the development is still in its infancy. There are certain problems and limitations of the green building development.

1. Introduction

Since the 18th century, the industrial revolution has opened the era of industrialization of human society, and the impact of human activities on the climate and environment is growing. With the rapid development of economy, energy consumption is also growing rapidly, causing serious problems of resources and environment. At the Copenhagen Conference in 2009, China pledged to significantly reduce carbon intensity and actively develop a green low-carbon economy. In November of the same year, China announced the goal of reducing carbon intensity: by 2020, carbon dioxide emissions per unit GDP of China will be reduced by 40%-45% compared with 2005.[1]With the increasingly prominent issues of the environment, the green low-carbon research has become an important topic at home and abroad. Many places in China have proposed concepts e.g. the carbon emission rights, the carbon benefits, and the green buildings, which helps to seek construction and life style that can reduce the impact to the environmental.

2. The concept of the green building

Architecture is the main place for life, work and entertainment. The whole life of the architecture includes the beginning of planning and design, the construction, the operation, the decoration and the final demolition. Except from the planning and designing, the other stages of an architecture are accompanied by resources utilization, energy input, and discharge of waste water and waste gas. [2]In the 1990s, the concept of green building was introduced to China, which attracted great attention. The relevant departments issued many relevant management methods and normative documents. [3]Green building refers to the maximizing resource conservation during the entire life cycle of the architecture, including energy saving, land saving, water saving, material saving, etc. It can protect the environment and reduce the pollution. Moreover, it provides people with healthy, comfortable and efficient use of space, and the nature harmonious symbiotic buildings. [4]

Due to the difficulty of data acquisition in China and the lack of a unified and standardized database and calculation method, the research on carbon emissions from building life cycle in China is still in its infancy. Although there is a big gap with developed countries, in recent years, Chinese scholars have made some attempts in obtaining basic data of carbon emissions and calculating carbon emissions, which laid a good foundation for the study of carbon emissions in the whole life cycle of buildings.

3. The green building development status in Guangzhou

On July 17, 2015, the Guangdong Provincial Development and Reform Commission officially released the "Guangdong Province Carbon GSP Pilot Work Implementation Plan" and the "Guangdong Province Carbon GSP Pilot Construction Guide". It encourages different regions to select different industry sectors, explore different Construction methods and actively carry out trials. In order to cooperate with the construction of carbon-saving work in Guangzhou, the Guangzhou Municipal People's Government officially released the "Notice of the General Office of the Guangzhou Municipal People's Government on Printing and Distributing the Implementation of the Carbon Trading Mechanism and the Carbon GSP Pilot Project in Guangzhou's Construction and Transportation Sector". In the notice, it is pointed out that residents can obtain corresponding carbon credits or carbon coins when they purchase (rent) green buildings, which encourages more residents to choose green buildings when they choose to buy or rent houses.

On November 11, 2013, the Guangdong Provincial People's Government officially issued the "Guangdong Province Green Building Action Implementation Plan 《". The plan determined that it is in line with the actual green building development technology route in Guangdong Province. Moreover, it pointed out that it is necessary to effectively increase the proportion of green buildings in new buildings in Guangdong Province. The plan emphasizes that the new large-scale public buildings, the government-invested new public buildings, and the new affordable housing in Guangzhou and Shenzhen must fully implement green building standards. Therefore, Guangzhou's work in promoting green buildings needs to play a demonstration role throughout the province. On October 16, 2014, the Guangzhou Municipal People's Government issued the "Guangzhou Green Building Action Implementation Plan". The plan clarifies that Guangzhou will further increase the proportion of new green buildings in the city. By the end of 2020, green buildings will account for more than 40% of the proportion of new buildings.

As of August 2016, some projects have obtained the green building logo in Guangzhou. In these project, public buildings and residential buildings accounted for 46.2% and 53.8% of the total construction area respectively. Moreover, there is only one industrial building.

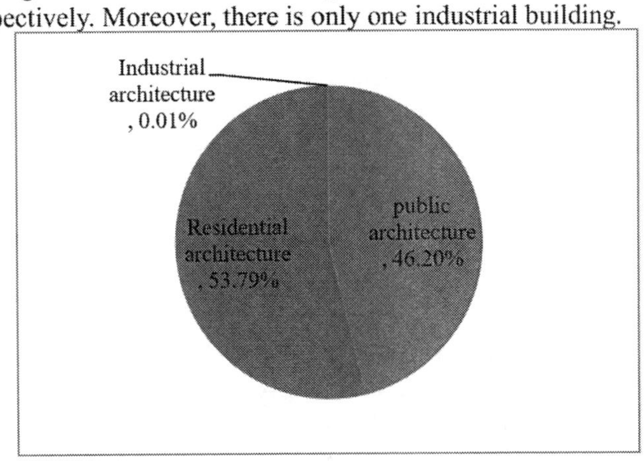

Fig. 1 Proportion of Different Types of Green Buildings in Guangzhou

There are two main types of green building logo, namely "design evaluation logo" and "operation evaluation logo". Among them, design evaluation marks are the evaluation marks for residential buildings and public buildings in the planning, design and construction stages, which should be carried out after the completion of construction drawings and the validity period of the marks is two years; operation evaluation marks are the evaluation marks for residential buildings and public buildings that have been completed and put into use, and should be put into use. One year later, the validity period of the logo is 3 years. [5] From the type of sign, the building area of the design mark project accounts for 96.0% of the total building area, while the operation mark project is a public building, and its building area accounts for 4.0% of the total building area.

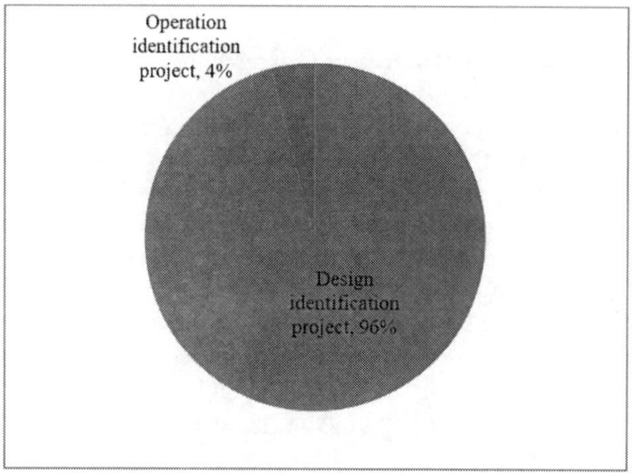

Fig. 2 Proportion of Green Buildings with Different Sign Types in Guangzhou

According to GB/T 50378, green buildings are classified into three grades: one star, two star and three star. As of August 2016, the number of green buildings rated as one star of the national standard in Guangzhou was the largest, accounting for 65.3% of the total, followed by two stars of the national standard, 19.3% and three stars of the national standard, 12.4%.

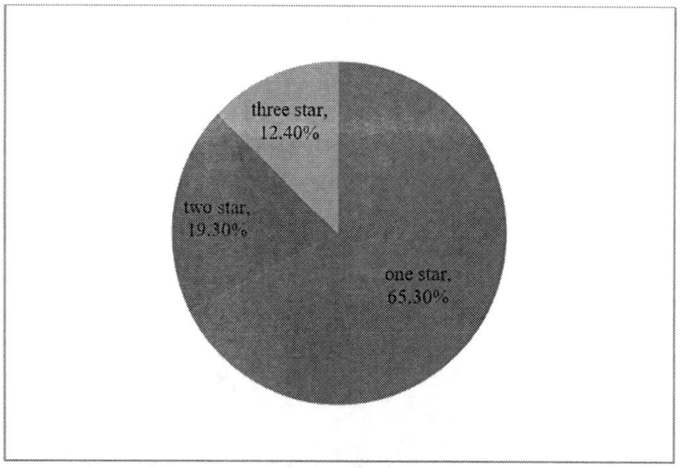

Fig. 3 Proportion of Green Buildings of Different Star Types in Guangzhou

Guangzhou has made great efforts in promoting green building work and actively responded to the requirements of green building development in Guangdong Province. However, as the incentive policy system for the whole process of green building has not been perfected, the construction cost of green buildings is higher than that of conventional buildings. Additionally, the public's awareness of green buildings is not very high, which has caused difficulty for Guangzhou to promote green buildings.

4. Conclusion

Under the background of vigorously promoting sustainable development, Guangzhou's green buildings have achieved great developments[6]. However, due to the fact that Guangzhou's green building development is still in its infancy, there are still some problems and limitations. As one of the well development of green building in China, it is of great significance to carry out research on Guangzhou green buildings as a target. It will greatly promote the development of green buildings in China.

Acknowledgements
Fund Project: Project funded by Guangzhou Municipal Bureau of Quality and Technical Supervision (No. 2017kj25).

References

[1] Li BY, Fan RX 2010. *China Rubber* 9 7-11(in Chinse).
[2] Ye ZD 2012. *Modern Urban Research* 9 42-48(in Chinse).
[3] Tian HF, Zhang H, Sun DM, Liang Y, Wang YW 2012. *Architecture Science* 28(in Chinse).
[4] Fu W 2014. *Detailed explanation of common terms used in construction engineering* (Beijing: China Electric Power Press).
[5] Li HY, Han JH, Cheng ZJ, Zeng J 2014. *Assessment standard for green building* (Beijing: China Standards Press).
[6] Zhu JH, Xu JT, Liang JJ 2018. *South Architecture* 6 73-80(in Chinse).

IOP Conf. Series: Materials Science and Engineering **729** (2020) 012008 doi:10.1088/1757-899X/729/1/012008

A Case Study on Energy Audit of the Public Institution

HUANG Jiayu*, HE Junfei, XIE Zeqiong, XIAO Zhenqian, FENG Chao

Guangzhou Institute of Energy Testing,Guangzhou 511447, China

Corresponding author e-mail: ziyuan2007@126.com

Abstract. In 2016, China issued the "Interim Measures for the Administration of Energy Auditing of Public Institutions". Energy audits of public institutions have received increasing attention and have been widely used. Through the energy audit, a public institution found that the energy management, the energy measurement and the energy statistics system have not been perfected. The transformer load rate is low. Moreover, there are high energy consumption and backward equipment existed. Therefore, the energy conservation proposals are put forward.

1. Introduction

With the rapid development of the economy, energy consumption has grown rapidly, causing serious resource and environmental problems. At the 2009 Copenhagen conference, China promised to significantly reduce carbon intensity and actively develop a low-carbon economy. [1] In order to strengthen the energy conservation management of public institutions and standardize the energy audit work of public institutions, in 2016, China issued the "Interim Measures for the Administration of Energy Auditing of Public Institutions". Energy audits of public institutions have received increasing attention and have been widely used. [2]

2. Analysis of energy performance of public institutions

One public agency has about 3,200 employees and a building area of 40,000 m^2. Its energy consumption includes electricity, natural gas, gasoline, diesel and tap water. All energy sources are outsourced. The energy and resources used by public institutions are mainly electricity, natural gas, steam/diesel, and water. The power is used in various power supply areas after being transformed, mainly for power equipment, lighting, office and living areas, etc. Natural gas is mainly used for canteens and water heaters. Tap water is mainly used for office, districts, canteens, swimming pools, fire and green. Gasoline and diesel are mainly used in vehicles. The consumption of energy resources is shown in Table 1.

As can be seen from Table 1, the power consumption is the largest, accounting for 70.46%, followed by natural gas consumption, accounting for 12.50%, gasoline accounting for 15.36%, and diesel accounting for 1.68%.

Table 1 2017 energy and energy consumption

Energy type	Quantity		Standard coefficient		Standard coal	
	Value	**Unit**	**Value**	**Unit**	**Value/tce**	**%**
Electricity	100.22	10^4kW·h	1.229	tce/1e4kW·h	123.17	70.46%
Natural gas	16434	m^3	1.3300	kgce/m^3	21.86	12.50%
Gasoline	18.25	t	1.4714	tce/t	26.85	15.36%
Diesel	2.01	t	1.4571	tce/t	2.93	1.68%
Water	32000	m^3	---	---	---	---
Total	---	---	---	---	174.81	100%

3. Comprehensive energy consumption comparison

In 2016, the total consumption power was 1,307,700kWh, natural gas was 17,641m^3, gasoline was 20.85t, diesel was 3.58t, tap water was 81,053m^3, and the total energy consumption was 220.07tce. In 2017, total electricity consumption was 1,022,000 kWh, natural gas was 16,434m^3, gasoline was 18.25t, diesel was 2.01t, tap water was 32000m^3 and total combined energy consumption was 174.81tce. The total combined energy consumption in 2017 is 11.89% less than that of 2016.

Table 2 Comprehensive energy consumption

Items	Unit	2016	2017	Growth rate /%
Electricity consumption	10^4kW·h	130.77	100.22	-23.36%
Natural gas consumption	m^3	17641	16434	-6.84%
Gasoline consumption	t	20.85	18.25	-12.47%
Diesel consumption	t	3.58	2.01	-43.85%
Water consumption	m^3	81053	32000	-60.52%
Total consumption	tce	220.07	174.81	-20.57%
Comprehensive energy consumption/unit of building area	kgce/m^2	68.77	54.63	-20.57%
Per capita comprehensive energy consumption	kgce/ capita	5.50	4.37	-20.57%

Table 3 Comparison electricity consumption between 2016 and 2017

month	2016	2017
1	9.11	6.68
2	5.54	6.64
3	7.04	5.62
4	7.96	6.32
5	8.75	7.09
6	11.78	11.52
7	17.88	12.38
8	17.91	13.15
9	14.68	11.95
10	12.06	8.59
11	11.12	5.97
12	6.94	4.31
Total	130.77	100.22

4. Energy saving measures

In 2017, the public institution carried out a number of energy-saving projects such as energy-saving renovation of LED lamps, procurement of energy-saving products, renovation of canteen gas stoves, and renovation of aged water supply pipelines. The comprehensive energy consumption was reduced by 11.89% compared with 2016. The energy-saving effect was remarkable.

5. Conclusion and suggestion

Public institution energy audits help public institutions to use energy efficiency and save on fiscal expenditures. Through this audit, it is found that the energy management, energy measurement and energy statistics systems of a public institution have not been perfected. Therefore, there is room for improvement and improvement. For example, the transformer load rate is low, and there are high energy-consuming backward equipment.

For the problems found, the following energy saving suggestions are proposed:

(1) Energy conservation management. Firstly the management system of energy-using units should be improved. Secondly, the established energy management institutions and energy management systems should be sorted out. Thirdly, energy audits of public institutions shouled be conducted regularly. Fourthly, energy consumption indicators should be formulated. Fifthly, energy-saving and water-saving publicity should be strengthened.

(2) Energy measurement and energy statistics. Establish a sound energy measurement and energy statistics system and strengthen analysis.

(3) Energy saving system recommendations. Eliminate high-energy-consumption equipment, purchase new high-efficiency energy-saving products, and conduct detailed single-phase testing of key energy-using equipment.

Acknowledgements

Fund Project: Project funded by Guangzhou Municipal Bureau of Quality and Technical Supervision (No. 2017kj25).

References

[1] Li BY, Fan RX 2010. *China Rubber* 9 7-11(in Chinse).
[2] Liu D, Li AG 2011. *Journal of Xi'an University of Science and Technology* 31 493-499(in Chinse)

IOP Conf. Series: Materials Science and Engineering **729** (2020) 012009 doi:10.1088/1757-899X/729/1/012009

An AGC dynamic control method based on DQN algorithm

Su Jianjun[1], Ma Changhui[1], Li Shan[1], Li Xiaoyu[2], Zhang Bing[3], Liu Wenxue[1]

[1] State Grid Shandong Electric Power Research Institute, Jinan 250003, China

[2] Key Laboratory of Power System Intelligent Dispatch and Control of Ministry of Education (Shandong University), Jinan, China

[3] State Grid Shandong Electric Power Company, Jinan 250001, China

*Corresponding author's e-mail: 626211631@qq.com

Abstract. In order to cope with the incompatibility of traditional AGC control under CPS control performance standard, this paper proposes a hierarchical control framework based on DQN algorithm. It effectively solves the problem of dimensionality disaster of Q learning by using neural network instead of state action pair, and speed up the convergence. Through simulation verification, this method can effectively improve the system CPS control performance index and effectively reduce the system power generation cost.

1. Introduction

Automatic generation control (AGC) is one of the important aspects of energy management system (EMS). It is mainly divided into two processes: 1) tracking of total power command,2) allocation of total power command. The CPS pass rate is usually used as an important criterion for evaluating AGC control performance[1].

The traditional load frequency control is the narrow AGC uses the negative feedback control, and the proportional integral link can always eliminate the error and reach the steady state. However, due to the nonlinear link of the thermal power system, the dynamic performance based on linear programming control is not satisfactory. At the same time, the proportional integral gain coefficient of the secondary control loop is not only related to the system structural parameters, but also related to the load change. It needs to be adjusted in real time with the change of the system load to meet the frequency change requirements of the load change[2-3].

The AGC system under the CPS standard can be regarded as "uncertain stochastic system". The mathematical model is modeled by Gauss-Markov stochastic process, and the power dynamic allocation problem can be understood as a discrete time Markov decision process[4]. Traditional analytical methods rely on the accuracy of grid structure, parameters and operational measurement data. Complex iterative algorithms often have weak robustness. The reinforcement learning (RL) algorithm with Markov decision process (MDP) as the basis of strict mathematics only needs to respond to the evaluation information of current control effects, with higher control real-time and robustness, and conventional control methods. Gradual learning and optimization ability that is not available. In recent years, a large number of studies have emerged to widely apply RL algorithms to the decision-making of power system operation and planning[5-6]. However, these studies are generally based on the more mature Q learning algorithm, which discretizes the continuous state of the grid state, and solves the maximum future reward expectation by iterative Berman optimal equation. It is difficult to avoid the phenomenon of "dimension disaster" when facing complex problems.

The Deep Reinforcement Learning (DRL) algorithm, which combines deep learning (DL) and reinforcement learning, replaces the "state-action" pair in traditional Q learning with deep neural network (DNN), which can directly form the mapping relationship between continuous state quantity and reward expectation. Through the organic combination with deep learning, not only the reinforcement learning algorithm has the ability to process complex continuous state quantities, but also the deep learning algorithm can pre-learn the deep neural network parameters to effectively improve the convergence speed of the reinforcement learning algorithm.

2. Introduction of deep reinforcement learning

2.1. Deep learning

The neural network is a special nonlinear function structure with high versatility. Its hierarchical structure is very suitable for solving the parameters by using the gradient descent method chain-guided rule back propagation. The basic principle of the deep learning algorithm is to solve the maximum likelihood estimation of the neural network parameters, that is, the optimization problem with the maximum likelihood function as the objective function. Each training of the neural network corresponds to an iterative process of the optimization problem, and the iterative process. The initial values are usually randomly generated by the program. The maximum likelihood estimation uses the probability model and is a kind of statistic-based phylogenetic tree reconstruction method. That is to say, if the training set is large enough, the noise generated by the training data due to measurement error or the like is regarded as a normal distribution $\sim N(0,\ \sigma^2)$ satisfying the mean value of 0. For a given input quantity and neural network parameters, the output is subject to the exponential family. Distribution, and the goal of deep learning is to predict the mean E(y) of the output.

The most essential feature of deep learning is that neural networks have the ability to transform features. The output of a layer in the middle of the network can be thought of as another expression of data, which is considered to be a feature learned through the network. The implementation of traditional algorithms usually requires artificially specifying some feature quantities, then inputting variables and then transforming them according to the specified feature quantities. The deep learning algorithm is difficult to determine each layer of nerves even after the network training is completed and put into practical use. What characteristics are corresponding to the feature transformation of the network. This is one of the characteristics of today's big data analysis. It does not require experts in various fields to carry out detailed principles and feature analysis. Instead, it uses automatic data summarization to form corresponding feature transformations. This is why neural networks are highly generalized. One of the reasons for sex. However, on the other hand, this is the shortcoming of the deep learning algorithm, and the comprehensibility is very poor. It is difficult for us to interpret the trained neural network in the traditional sense. The ability of this kind of feature transformation of neural networks has been widely used in methods such as migration learning and reinforcement learning.

In summary, deep learning is not a "smart" algorithm, but a type of regression analysis problem in the statistical sense of big data. It is an algorithm for predicting the mean value of the probability distribution of the output based on the training data fitting function. The training process is essentially a solution to the optimization problem.

2.2. Q-Learning

Reinforcement learning (RL) is one of the methods for studying multi-stage decision problems based on Markov decision process. In the field of control theory, some experts also call reinforcement learning "approximate dynamic programming", which is different from dynamic programming. Reinforcement learning does not require the state transition matrix of the corresponding Markov decision process, reward function and all state sets. As a known condition, only the action set and the initial state of the system need to be known. The goal of reinforcement learning is to get the agent to get the most cumulative reward in the process of interacting with the environment.

The Q-learning algorithm has been proposed as early as 1989. The basic idea is to form a Q-table as shown in the figure after a large number of interactions with the environment after determining the discrete state set and action set of the multi-stage decision problem. The maximum future reward expectation of taking different actions in a given state is used as a criterion for evaluating the effect of the control action.

During the iterative process of the algorithm, the control unit selects a control action to act on the environment according to the current state s, and the environment feeds back to the control unit reward and punishment signal R and the new state s' to which the environment is transferred under the action of the control action a. The maximum future reward expectation is the reward value R of action a under the current state s, plus the maximum future reward expectation of the discount factor that can be obtained by the new state s' to which the environment is transferred under the action of this control action a.

$$Q(s,a)=E\left[R(s,s',a)+\gamma\max_{a'\in A}Q(s',a')\right] \tag{1}$$

Where: $R(s,s',a)$ is the reward value R obtained by taking action a under state s, γ is a discount factor, and $\max_{a'\in A}Q(s',a')$ is the maximum value among the row of Q values corresponding to the new state s'.

The iterative solution formula for the Q value is:

$$Q^{k+1}(s,a)=Q^{k}(s,a)+\alpha\left[R(s,s',a)+\gamma\max_{a\in A}Q(s',a')-Q(s,a)\right] \tag{2}$$

When all the Q values in the Q table converge, we can select the optimal control action for each discrete state of the system by querying the Q table to form a "state-action" pair. Because the algorithm always chooses the control action that is most likely to get a bigger reward, Q-learning is generally considered a greedy algorithm that pursues only the greatest rewards without taking into account the risks.

$$a^{*}(s)=\underset{a\in A}{\operatorname{argmax}}\,Q(s,a) \tag{3}$$

The Q-learning algorithm is essentially a statistical method. After ensuring that each discrete case in the Q table has experienced enough times and iterating to convergence, the algorithm for selecting the optimal control action according to the Q table is only It can cope with the state that has experienced enough times, and can't do anything about the state of insufficient or even unknown experience. It is difficult to avoid the problem of “dimensional disaster” in the face of complex problems.

2.3. Deep Q Network

In order to solve the shortcoming of the Q-learning algorithm, the Q value is no longer stored in a tabular form, but a function approximation (FA) method is used to construct a mapping relationship between states, actions and Q values. . However, at that time, the academic community generally believed that large-scale nonlinear approximation functions for value function theory could not guarantee the convergence of the algorithm, instead of using large nonlinear approximation functions, Q-learning algorithm can only be used to solve some simple problems, lacking practical value. The research has entered the bottleneck. Until 2013, Google's deep mind team proposed the DQN algorithm, using the large-scale nonlinear function structure of deep neural network (DNN) as an approximation function, and guaranteed the convergence of the algorithm through special processing, which broke the recognition of people. know.

The DQN algorithm, which introduces neural network as an approximation function, faces severe convergence problems of neural network parameters. Deep learning requires training samples to be independent of each other, and there is a correlation between the state before and after the multi-stage decision-making problem of reinforcement learning. At the same time, in deep learning, each training (iterative) parameter is advanced toward the same target value, and its training process target distribution is fixed. In reinforcement learning, especially in the initial stage of strengthening learning to explore the state space, the new data added by each training may trigger the change of the convergence target of the neural network parameters. The reinforcement learning training process belongs to the optimization problem of the variable objective function, which makes It is difficult to converge neural network parameters. In order to solve the problem of convergence, the DQN algorithm uses experience replay and target Q network, which is a method of establishing a memory and a dual neural network using time difference update. Experience replay establishes a memory and uses storage-random sampling to break the correlation between data. The target Q network introduces an auxiliary neural network whose structure is identical to the main neural network, but the parameters do not participate in the iterative operation. Its role is to form a parameter time difference update with the main neural network, and solve the convergence problem by reducing the correlation between the two neural networks.

By using a neural network instead of a Q-table, the DQN algorithm no longer divides the system state into a number of independent points like the Q-learning algorithm, but treats each state experienced during the interaction with the environment as a continuous state. A point on the domain that makes the DQN algorithm somewhat capable of coping with new states that have not been experienced in the training process. At the same time, the combination with the neural network enables deep reinforcement learning to be pre-learned by means of deep learning algorithms to speed up the convergence.

3. Design of AGC dynamic control method based on DQN algorithm

3.1. Hierarchical control framework

The closed-loop control of AGC dynamic optimization process under CPS standard is divided into two processes: CPS control and CPS instruction allocation. The two-layer control framework of this paper corresponds to these two processes.

The first layer of the framework is CPS control, and the AGC total scheduling instruction is calculated according to the system frequency deviation. The second layer of the framework allocates CPS commands. The distribution factor is calculated according to the characteristics of the AGC total dispatching command and the remaining capacity of each unit of the system. The distribution power of each unit is the product of the total CPS command and the distribution factor. The control purpose is to realize the economic system. run. In the initial stage of training, the two parts are separately trained. When training the CPS control part, the instruction allocation is implemented using the PROP algorithm based on the adjustable capacity ratio allocation; when training the CPS instruction allocation part, the set load variation is directly used as the AGC total scheduling. instruction. After the two parts are separately trained to converge, the two parts are combined for random disturbance simulation to further train to convergence and use the control effect of the CPS standard test algorithm.

3.2. Design of reward function

The power system frequency should be kept within the range of 50 ± 0.2 Hz during normal operation, so the segmentation reward function for CPS control is:

$$R = \begin{cases} -\mu_1 * |\Delta f|, |\Delta f| < 0.2 \\ -\mu_2 * |\Delta f|, |\Delta f| > 0.2 \end{cases} \tag{4}$$

μ_1, μ_2 are the segment weights corresponding to the bonus function. This paper is set to 1 and 5.

The purpose of CPS instruction allocation is to realize the minimum cost operation of the system considering constraints such as unit capacity limitation and climbing speed. The unit capacity constraint and the climbing speed are controlled by the hard constraint participation algorithm, and only the system power generation cost is used as a reward. Function, no need to consider the weight factor.

$$R=-\sum_{i=1}^{n} C_i \Delta P_{Gi} \tag{5}$$

3.3. *Flow of DQN algorithm*

The unified control process of the above two-layer control framework is as follows:

1) Initialize algorithm parameters such as Q-value neural network, initial state s, learning speed, and discount factor.

2) Select the control action a according to the current state, and perform additional processing considering hard constraints such as unit capacity and climbing speed.

3) Applying the processed control action a to the power system simulation environment, obtaining the feedback system new state s', and calculating the reward R corresponding to the control action according to the bonus function formula.

4) Update the Q value neural network parameters according to the reward R of the control action.

5) Update the system state to s' and increase the number of iterations by one. When the current iteration number is greater than the maximum number of iterations, the algorithm ends; otherwise, jump to step 2).

4. Examples

Refer to the AGC unit data of a provincial power grid, and divide 77 units into 10 groups according to the characteristics of the unit. The information of the AGC unit is detailed in the literature[7]. The AGC control period takes 2s. The system model and its controller are built on the Python language environment. The CPS control and CPS instruction allocation part of the algorithm parameters are shown in the table.

Table 1. parameters of DQN algorithm

	Learning rate	**Reward decay**	**ε greedy**	**τ replace**	**Memory size**	**Batch size**
CPS control	0.02	0.9	0.8	0.02	500	32
CPS distribution	0.005	0.9	0.9	0.01	1500	64

In view of the random exploration of the state space by the depth enhancement algorithm, this paper runs the CPS controller part and the CPS dynamic instruction allocation part of the training process 50 times each time, taking the best-performing neural network parameters as the initial parameters of the final AGC dynamic control training part.

In order to verify the effectiveness of the proposed algorithm, the traditional PI-PROP control and DQN algorithm are used to solve the AGC dynamic scheduling problem respectively. The CPS index and system power generation cost are shown in the table.

Table 2. CPS indicator comparison

	CPS1/%	**CPS2/%**	**CPS/%**	**Costs/w¥**
PI-PROP	156.47	100	84.52	179.36
DQN	183.42	100	100	146.38

From the table, the CPS index of the proposed algorithm is significantly better than the traditional PI-PROP algorithm, and the total cost of system power generation is significantly reduced.

5. Conclusion

Aiming at the problem that the dynamic performance of the existing AGC unit dynamic optimization scheduling is poor, and the CPS assessment cost is not considered, the DQN hierarchical control algorithm is proposed. The simulation results show that the algorithm can effectively improve the CPS index and reduce the total cost of system power generation.

The success of the deep reinforcement learning algorithm represented by Alpha-go shows its potential and possibility beyond the traditional classical algorithms in the face of complex scenes, surpassing the existing level of human capabilities. Although the deep reinforcement learning algorithm is not yet perfect, it is undoubtedly one of the algorithms closest to artificial intelligence.

References

[1] Ai X , Liao S , Wen J . An AGC control performance index for main tie-lines in interconnected power grids[J]. Automation of Electric Power Systems, 2013, 37(21):111-117.

[2] LI H, YAN Z. Application of Q-learning Approach with Prior Knowledge to Non-linear AGC System [J][J]. Automation of Electric Power Systems, 2008, 23.

[3] Li H, Yan Z. Application of reinforcement learning method in a hydro-thermal hybrid automatic generation control system[J]. Dianli Xitong Zidonghua(Automation of Electric Power Systems), 2010, 34(9): 39-43.

[4] Y , Liu R , Li W . Comparison of CPS and its assessment between North America and China[J]. Dianli Xitong Zidonghua/Automation of Electric Power Systems, 2012, 36(15):63-72.

[5] Yu T, Liang H, Zhou B. Smart power generation control for microgrids islanded operation based on R (λ) learning[J]. Power System Protection and Control, 2012, 40(13): 7-13.

[6] Zhang X, Yu T, Tang J. Optimal CPS command dispatch based on hierarchically correlated equilibrium reinforcement learning[J]. Autom Elect Power Syst, 2015, 39(8): 80-86.

[7] Yu T, Wang Y, Ye W, et al. Multi-objective dynamic optimal dispatch method for CPS order of interconnected power grids using improved hierarchical reinforcement learning[C]//Zhongguo Dianji Gongcheng Xuebao(Proceedings of the Chinese Society of Electrical Engineering). Chinese Society for Electrical Engineering, 2011, 31(19): 90-96.

IOP Conf. Series: Materials Science and Engineering **729** (2020) 012010 doi:10.1088/1757-899X/729/1/012010

Distributed Energy Storage Optimization Configuration of Active Distribution Network

Jiang Zhe[1], Wang Liang[2], Huang Xuxiang[3], Han Xueshan[3], Yang Dong[1], Zhou Ning[1]

[1] State Grid Shandong Electric Power Research Institute, Jinan 250003, China

[2] State Grid Shandong Electric Power Company, Jinan 250001, China

[3] Key Laboratory of Power System Intelligent Dispatch and Control of Ministry of Education (Shandong University), Jinan, China

*Corresponding author's e-mail: hxxsdu@163.com

Abstract. With the high proportion of renewable energy access to the distribution network, the volatility and randomness of renewable energy output will have a serious impact on the distribution network, especially at the peak of renewable power output, the reverse trend phenomenon is obvious. The active distribution network displays the "power" characteristics externally, which makes the voltage problem and transmission congestion problem in the distribution network very serious. This paper establishes a distributed energy storage optimization configuration model that takes into account the voltage quality, and considers the energy storage operation strategy optimization in the energy storage site selection and volume model, which makes the energy storage configuration more reasonable, improves the voltage quality and enhances the renewable energy consumption capacity, improve the economics of system operation.

1. Introduction

In the future, renewable energy, especially photovoltaics[1-2], will exist in a distributed form in the distribution network, making the distribution network from the passive acceptance of electrical energy to a certain initiative. After a high proportion of renewable energy is connected to the distribution network, the volatility and randomness of the renewable energy output will have a serious impact on the distribution network[4], especially at the peak of the renewable power output, the reverse trend phenomenon is obvious. The source distribution network displays the "power" characteristics externally, which makes the voltage problem and transmission congestion problem in the distribution network very serious. These will become the key issues restricting the consumption of renewable energy in the context of the significant increase in renewable energy penetration in the future.

The energy storage system can improve the ability of the system to flexibly adjust. In the context of the high-ratio renewable power supply being connected in a distributed form, how the energy storage is configured and in what form is the key to cope with the high-ratio renewable energy grid connection. This paper considers the problem of renewable energy consumption from the distribution network level. Firstly, it analyzes the influence of reverse tidal current on the voltage of renewable energy grid-connected points, and establishes a distributed energy storage optimization configuration model considering voltage quality. Considering the optimization of energy storage operation strategy in the volume model, the energy storage configuration is more reasonable, the voltage quality is improved, the renewable energy consumption capacity is improved, and the system operation economy is improved.

2. Distributed energy storage optimization configuration model

The first paragraph after a heading is not indented (Bodytext style). Energy storage system as a flexible resource can improve the system's regulation ability and assist the system to absorb renewable energy. The distributed energy storage system existing in the active distribution network can not only improve the system's active power regulation capacity through charge and discharge, but also change the power output adjusts the voltage.

2.1. Distributed energy storage value analysis

After a high proportion of renewable energy is connected in a distributed form, the voltage of the grid-connected point is increased due to the reverse tidal current at the peak of the renewable energy output. The reactive power adjustment capability is limited by the power circle, resulting in the necessity of limiting the renewable The energy source contributes to limiting the voltage. Ignore the voltage drop horizontal component. When there is only renewable energy and no energy storage, the voltage of each node can be expressed as

$$\mathbf{U}_{n\text{-}1} - \mathbf{U}_n = \mathbf{U}_{diag}(\mathbf{R}_n\mathbf{P} + \mathbf{X}_n\mathbf{Q}) = \mathbf{U}_{diag}\mathbf{R}_n\begin{bmatrix} P_1 \\ \cdots \\ P_n - P_v \end{bmatrix} + \mathbf{U}_{diag}\mathbf{X}_n\begin{bmatrix} Q_1 \\ \cdots \\ Q_n - Q_v \end{bmatrix} \tag{1}$$

In the formula, $\mathbf{U}_{n\text{-}1} = [U_0,...,U_{n-1}]^T$ 、 $\mathbf{U}_n = [U_1,...,U_n]^T$ denotes the first and last node voltage column vectors of each branch respectively. For the convenience of analysis, a radial network without branches is considered here; $\mathbf{U}_{diag}$ is a diagonal matrix, the diagonal elements are $1/U_n$; $\mathbf{R}_n$ 、 $\mathbf{X}_n$ is a diagonal matrix, The corner elements are the resistance and reactance of each branch respectively; $\mathbf{P}$ 、 $\mathbf{Q}$ is the active power and reactive power column vector of each branch end respectively, wherein it is assumed that node n has distributed photovoltaic, and its output power is opposite to the original power flow direction.

$$\mathbf{U}_{n\text{-}1} - \mathbf{U}_n = \mathbf{U}_{diag}\mathbf{R}_n\begin{bmatrix} P_1 \\ \cdots \\ P_n - P_v + P_{ch} - P_{dis} \end{bmatrix} + \mathbf{U}_{diag}\mathbf{X}_n\begin{bmatrix} Q_1 \\ \cdots \\ Q_n - Q_v + Q_{ch} - Q_{dis} \end{bmatrix} \tag{2}$$

It can be seen from formula (2) that the role of energy storage in improving the capacity of renewable energy absorption, on the one hand, storage energy stores electricity at the peak of renewable power output, and discharges electricity when it is low, to reduce the amount of abandoned wind and light. On the other hand, the energy storage can reduce the amount of terminal voltage increase due to the reverse power, thereby increasing the capacity of renewable energy absorption.

2.2. Distributed energy storage site selection and capacity optimization model

Cost is an important trade-off indicator in energy storage planning. From the perspective of distribution network, this paper considers energy storage investment cost, power loss cost, electricity purchase cost to main network (sales revenue), and establishes energy storage site selection and capacity optimization. The comprehensive model of the strategy: determine the energy storage capacity of each node by site selection and volume, and perform the time series production simulation according to the actual load curve and the renewable energy prediction curve, with the line loss cost and the purchase cost to the main network (the sales revenue) the minimum The goal is to optimize the energy storage and discharge strategy to promote the recovery of renewable energy while ensuring voltage quality.

$$\min C = C_{investment} + C_{operation} \tag{3}$$

$$C_{investment} = \frac{\delta(1+\delta)^{N_{ess}}}{(1+\delta)^{N_{ess}} - 1}\sum_{1}^{k}(C_{pcs} \cdot P_{pcs,k} + C_{esu} \cdot E_{esu,k}) \tag{4}$$

$$C_{operation} = \sum_{d=1}^{365} \sum_{t=1}^{24} \Delta t \cdot (C_{ev}^{t} \cdot P_{change}^{t} + C_{ep} \cdot \sum_{l=1}^{Line} P_{loss,l}^{t}) \tag{5}$$

The total cost C includes: annualized investment cost $C_{investment}$, including investment cost of power capacity and energy capacity, $P_{pcs,k}$ 、 $E_{esu,k}$ is the power capacity and energy capacity of the energy storage, respectively, C_{pcs} 、 C_{esu} is its unit cost coefficient, and N_{ess} is used for energy storage device years, δ represents the discount rate; annualized operating cost $C_{operation}$, including the cost of purchasing electricity to the main network (sales revenue) and line loss costs, Δt is the time interval (this paper takes 1h), P_{change}^{t} is the network root node and the main The network exchanges power, C_{ev}^{t} is the system electricity price at that time, $P_{loss,l}^{t}$ is the line loss power, C_{ep} is the unit loss power cost (yuan), and $Line$ is the total number of lines.

Including constraints:

1) each node installs a capacity constraint:

$$0 \le P_{pcs,k} \le P_{pcs,\max} \tag{6}$$

$$0 \le E_{esu,k} \le E_{esu,\max} \tag{7}$$

In the formula, $P_{pcs,\max}$ 、 $E_{esu,\max}$ is the maximum energy capacity and energy capacity of the energy storage allowed for each node.

2) power equation constraint:

This paper utilizes the accurate AC flow equation DistFlow model applicable in radial distribution networks.

$$P_{ij}^{t} - r_{ij} \frac{(P_{ij}^{t})^2 + (Q_{ij}^{t})^2}{(U_i^{t})^2} = \sum_{k \in \omega(j)} P_{jk}^{t} + (P_{j,l}^{t} - P_{j,v}^{t} + P_{j,ch}^{t}) \tag{8}$$

$$Q_{ij}^{t} - x_{ij} \frac{(P_{ij}^{t})^2 + (Q_{ij}^{t})^2}{(U_i^{t})^2} = \sum_{k \in \omega(j)} Q_{jk}^{t} + (Q_{j,l}^{t} - Q_{j,v}^{t} + Q_{j,ch}^{t}) \tag{9}$$

$$(U_j^{t})^2 = (U_i^{t})^2 - 2(r_{ij} P_{ij}^{t} + x_{ij} Q_{ij}^{t}) + (r_{ij}^2 + x_{ij}^2) \cdot \frac{(P_{ij}^{t})^2 + (Q_{ij}^{t})^2}{(U_i^{t})^2} \tag{10}$$

Each variable superscript t represents a time; P_{ij}^{t} 、 Q_{ij}^{t} 、 r_{ij} 、 x_{ij} represents the active power, reactive power, resistance, and reactance between the node and the node; $\omega(j)$ represents a set of nodes connected to the node but not on the path of the node j to the root node; $P_{j,l}^{t}$ 、 $Q_{j,l}^{t}$ respectively represents a node Active and reactive load; $P_{j,v}^{t}$ 、 $Q_{j,v}^{t}$ respectively represents the active and reactive power of the renewable power supply on the node; $P_{j,ch}^{t}$ 、 $Q_{j,ch}^{t}$ is the active and reactive power absorbed by the node energy storage device, and negative if the power is emitted; U_i^{t} 、 U_j^{t} is i and j Node voltage.

3) line constraint:

$$P_{loss,l}^{t} = r_{ij} \frac{(P_{ij}^{t})^2 + (Q_{ij}^{t})^2}{(U_i^{t})^2}, l \in i,j \tag{11}$$

$$(P_{ij}^{t})^2 + (Q_{ij}^{t})^2 \le (S_{ij})^2 \tag{12}$$

Equation (11) is the line loss constraint, nodes i 、 j are the first and last nodes of line l; Equation (12) is the line capacity constraint, and S_{ij} is the branch transmission capacity.

4) voltage constraint:

$U_{\max}$ 、 $U_{\min}$ is the voltage upper limit and lower limit, and the root node ($i=1$) voltage is the reference value.

$$U_{\min} \le U_i^{t} \le U_{\max}, i \ne 1 \tag{13}$$

5) renewable power supply constraints:

$$(P^t_{i,v})^2+(Q^t_{i,v})^2\le(S_{i,v})^2 \tag{14}$$

$$0\le P^t_{i,v}\le P^t_{i,vf}, -S_{i,v}\le Q^t_{i,v}\le S_{i,v} \tag{15}$$

Equation (14) is the renewable power power circle constraint, $S_{i,v}$ is the apparent power; Equation (15) is the active and reactive power constraints, and $P^t_{i,vf}$ is the predicted value of the renewable power output.

6) distributed energy storage system constraints:

$$-x_i\cdot P_{pcs,i}\le P^t_{i,ch}\le x_i\cdot P_{pcs,i} \tag{16}$$

$$(P^t_{i,ch})^2+(Q^t_{i,ch})^2\le(P_{pcs,i})^2 \tag{17}$$

$$E^t_{ess,i}=E^{t-1}_{ess,i}+\eta\cdot\Delta t\cdot P^t_{i,ch},\quad \eta=\begin{cases}\eta_{ch}, & P^t_{i,ch}\ge 0\\ 1/\eta_{dis}, & P^t_{i,ch}\le 0\end{cases} \tag{18}$$

$$x_i\cdot\alpha_{low}\cdot E_{esu,i}\le E^t_{ess,i}\le x_i\cdot\alpha_{high}\cdot E_{esu,i} \tag{19}$$

$$E^0_{ess,i}=E^{24}_{ess,i} \tag{20}$$

Equation (16), Equation (17) is the energy storage power constraint, $P_{pcs,i}$ is the energy capacity of the energy storage device; Equation (18)- Equation (20) is the energy storage energy constraint, and $E_{esu,i}$ is the energy storage. The energy capacity of the device, η is the energy conversion efficiency coefficient, η_{ch}、η_{dis} is the charging and discharging efficiency coefficient, respectively, and α_{low}、α_{high} is the upper and lower limit coefficients of the energy storage state. Equation (20) is to ensure the continuity of the state of charge storage, $E^0_{ess,i}$ is the initial value, and $E^{24}_{ess,i}$ is the energy state of the last time energy storage.

In summary, the distributed energy storage location constant volume and operation strategy optimization model is a Mixed-Integer NonLinear Programming (MINLP) problem involving 0,1 quantities, which is difficult to solve and linearized for calculation.

3. Example

This chapter adopts the IEEE33 node distribution network example. The wiring diagram is shown in Figure 1. There are a total of 32 branches in the tie line. The data of each node and the resistance reactance of each branch are shown in the appendix. According to the actual data, distributed photovoltaics account for a large proportion of distributed power sources[8]. For comparison analysis, it is assumed that nodes 13 to 18 each have distributed photovoltaics with a capacity of 1 MW. The maximum network load is $3715+j2300$ kvar and the reference voltage is 12.66 kV.

The unit power capacity cost of the energy storage system is 4,000 yuan/kW, the unit energy capacity cost is 3,500 yuan/kWh, the service life is 20 years (full charge and full discharge), the discount rate is 10%, and the energy storage efficiency of the energy storage device is charged. The discharge efficiency is 0.8, the installation node is allowed to be 2-33 nodes, the maximum allowable installation power capacity of each node is 300kW, the maximum allowable installation energy capacity is 1000kWh, and the maximum number of installed energy storage devices is 10.

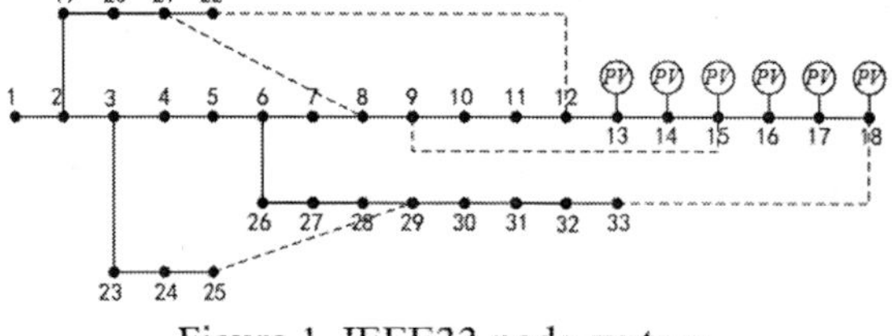

Figure 1. IEEE33 node system

The domestic industrial electricity time-sharing electricity price period is divided into: 8:00-12:00, 17:00-21:00 is the peak period; 12:00-17:00, 21:00-24:00 is the flat period; 0:00-8:00 is the valley time. According to this division mode, there will be a small peak load during the flat period, that is, the time division of the time-of-use electricity price is not completely divided according to the load level. Therefore, if the storage energy is charged and discharged according to the time-sharing electricity price, there may be a "peak". The phenomenon of "adding peaks", so the time-of-use electricity price data used in this paper comes from the literature [89]. Assume that the daily time-of-use electricity prices are consistent throughout the year, as shown in Table 1.

Table 1. Electricity price data

t/h	Electricity price data /yuan/kWh
0:00~8:00	0.4
8:00~12:00	1.0
12:00~15:00	0.6
15:00~21:00	1.0
21:00~24:00	0.6

The energy storage capacity and energy capacity of each node obtained by the distributed energy storage optimization configuration model are shown in Table 2.

Table 2. Electricity price data

Node	Power Capacity /kW	Energy capacity /kWh
24	238	573
25	295	718
29	300	720
30	300	720
31	257	720

To compare and analyze the impact of the energy storage configuration on the operating cost of the distribution network, consider the total annual operating cost and the cost of each part in the scenario of configuring energy storage and unconfigured energy storage, and obtain the results as shown in Table 3.

Table 3. Annual operating cost of the system in two scenarios

Cost/104	Unconfigured ESS	Configured ESS
Investment cost	0	365.6
Electricity purchase cost	1563.2	1128.9
Line loss	603.4	492.7
total cost	2166.6	1987.2

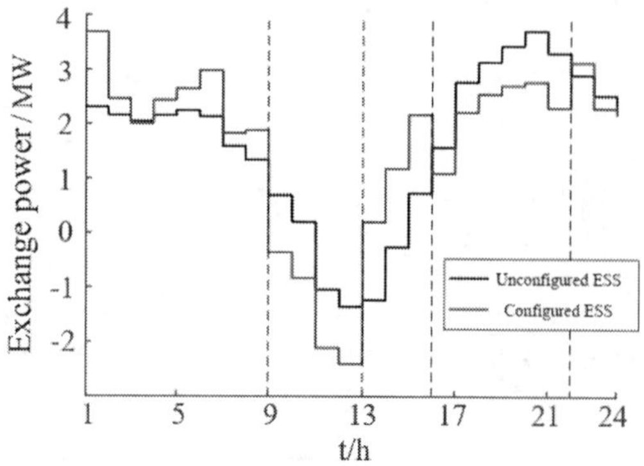

Figure 2. Exchange power with the main network in a typical scenario

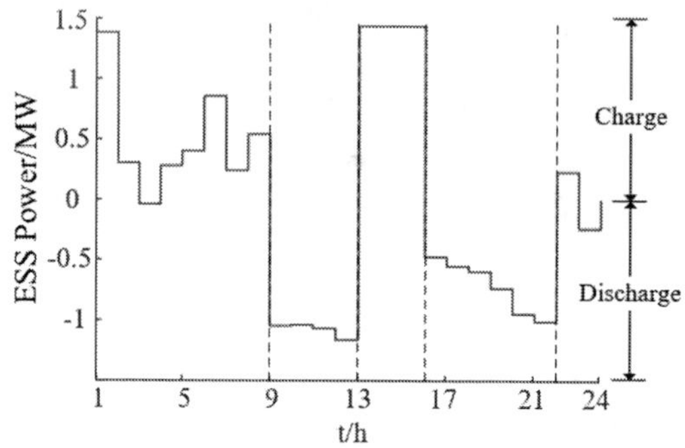

Figure 3. Charge and discharge power of a typical daily energy storage system

For the analysis, take the active power exchange between the root node and the main network in a scenario of 24 hours in a typical day. Figure 2 shows the total active power of the energy storage device. Figure 3 shows the time sharing. During the low price period (0:00-8:00) and the price period (12:00-15:00, 21:00-24:00), the energy storage device absorbs electric energy, and the distribution network purchases electricity from the main network (selling electricity) Reduced); during the peak hours of time-of-use tariffs (8:00-12:00, 15:00-21:00), the energy storage device releases electricity, especially at 9:00-12:00, the PV output is large, so At this time, the network sells electricity to the main network. During the rest of the period, due to energy storage and storage of electrical energy, the distribution network purchases electricity from the main network. In summary, after the energy storage device is configured, the system greatly reduces the cost of purchasing electricity. After calculation, the energy storage device transfers 3.12 GWh per year, accounting for 11.98% of the total load.

At the same time, the energy storage device can generate reactive power, so that the reactive power flow of the branches 24, 25, 26-30 is reduced, and the line loss cost is also reduced.

4. Conclusion

This paper analyzes the example, the high-proportion distributed photovoltaic access distribution network, its active output peak time, the reverse power flow phenomenon is serious, resulting in the voltage increase of the photovoltaic access point, limiting the outward transfer of photovoltaic active power, and distributed storage. The energy device not only can directly reduce the light rejection rate by charging and discharging active power, but also can reduce the line voltage drop by adjusting the reactive power, maintain the voltage of the photovoltaic access point, facilitate the active power delivery, and improve the system operation economy.

References

[1] Energy Research Institute of the National Development and Reform Commission. China 2050 High Proportion of Renewable Energy Development Scenarios and Paths [R]. Beijing: Energy Research Institute of the National Development and Reform Commission, (2015).

[2] Schellekens G, Battaglini A, Lilliestam J, et al. 100% Renewable electricity: a roadmap to 2050 for Europe and North Africa[R]. London, UK: Price water house Coopers, (2010).

[3] Baran M E,Wu F F.Network reconfiguration in distribution systems for loss reduction and load balancing[J]. IEEE Transactions on Power Delivery, 1989, 4(2):1401-1407.

[4] Jia Wenzhao, Kang Chongqing, Li Dan, et al. Evaluation method of wind power capacity based on wind power forecasting[J]. *Power System Technology*,(36(8):69-75,2012).

[5] Franco J F,Rider M J,Lavorato M,et al.A mixed-integer LP model for the optimal allocation of voltage regulators and capacitors in radial distribution systems[J]. International Journal of Electrical Power & Energy Systems, 2013, 48(48):123-130.

[6] Xu Y,Dong Z,Zhang R,et al. Multi-timescale Coordinated Voltage/var Control of High Renewable-penetrated Distribution Systems[J]. IEEE Transactions on Power Systems, 2017.

[7] Sortomme E,Hindi M M,Macpherson S D J,et al. Coordinated Charging of Plug-In Hybrid Electric Vehicles to Minimize Distribution System Losses[J]. IEEE Transactions on Smart Grid, 2011, 2(1):198-205.

[8] Haoran Ji,Chengshan Wang,Peng Li,Guanyu Song,Hao Yu,Jianzhong Wu. Quantified analysis method for operational flexibility of active distribution networks with high penetration of distributed generators[J]. Applied Energy,2019,239.

IOP Conf. Series: Materials Science and Engineering **729** (2020) 012011 doi:10.1088/1757-899X/729/1/012011

Improve dielectric properties of Lamellar Calcium Niobate ceramics by spark plasma sintering

Xinxin Liu[1, 2] Xiong Xie[1, 3] Yunling Zhai[1, 2] Liang Wang[1, 2*]

[1]Chongqing Key Laboratory of Multi-Scale Manufacturing Technology, Chongqing Institute of Green and Intelligent Technology, Chinese Academy of Sciences, Chongqing, 400714, P.R. China

[2] University of Chinese Academy of Sciences, Beijing, 100049, P.R. China

[3]State Key Laboratory of Mechanical Transmission, College of Materials science and engineering, Chongqing University, Chongqing, 400044, P.R. China

[*]Corresponding author's e-mail: wangliang@cigit.ac.cn

Abstract. The structure and dielectric properties of $HCa_2Nb_3O_{10}\cdot 1.5H_2O$ (HCNO) and its exfoliated products (e-HCNO) ceramics were investigated. Where two-dimensional e-HCNO sheets were obtained by ion-exchange and exfoliating consequently. Spark plasma sintering (SPS) was introduced for products processing. After which, the dielectric constant of e-HCNO is 1300 and the loss is 0.002 at 2 M Hz. Compared to ordinary solid phase sintering, the dielectric constant of ceramics was found to be directly improved by the SPS processing.

1. Introduction

Energy storage and conversion have always been a long-lasting technology focus, especially, for the production, storage and full use of electrical energy. Capacitors, as an indispensable item for power distribution and storage, have been widely used in electronic industry, high voltage field, communication field, new energy-powered vehicles and so on [1]. To satisfy the growing demand of the electronics industry, multilayer ceramic capacitors (MLCC) composed of a plurality of dielectric and electrode inner layers are being actively sought. However, the dielectric properties of conventional titanium oxides are continuously reduced when the thickness is reduced to less than 1 μm, which makes it not fit for downsized MLCC. Therefore, thin film materials with excellent dielectric properties and constant thickness are keys for triggering progress. Niobate materials such as $KCa_2Nb_3O_{10}$ and $KSr_2Nb_3O_{10}$, are belonging to the Dion-Jacobson family [2], the film combined niobate nanosheets that exfoliated from their parent bulk phase through Van der Waals force [3] exhibits a high dielectric constant even at a thickness of less than 100 nm. The dielectric films made from titanate and niobate nanosheets were also investigated by Sasaki and co-workers [4]. While HCNO (synthesized from KCNO by ion exchange) sintered by furnace exhibits an almost constant ε_r (~23.5) and dielectric loss (tanθ = 0.009) in a wide frequency range [5]. This phenomenon can be explained by the constant dielectric properties of films made from $Ca_2Nb_3O_{10}$ nanosheets [6]. Moreover, the crystal structure of exfoliated nanosheets is usually well-defined and almost free of defects such as oxygen vacancies [7].

Sintering is the most critical procedure to the final quality of ceramics. In addition to conventional sintering (CS), hot pressing (HP), two-step sintering (TSS), and spark plasma sintering (SPS) are used to synthesize ceramic materials. Among them, SPS is widely used due to its fast heating rate and short sintering time, which is beneficial to gain uniform and dense ceramics with good performance [8]. However, the dielectric properties of SPS sintered HCNO ceramics are rarely systematically studied or reported so far.

Here, we synthesized HCNO and e-HCNO nanosheets by combining with methods of ion-exchange, liquid phase stripping and SPS sintering. Their synthesis conditions were explored and samples were characterized using XRD, EDS and XPS. And the dielectric properties of HCNO and e-HCNO nanosheets obtained by liquid phase stripping were systematically studied.

2. Experimental

K_2CO_3, $CaCO_3$, and Nb_2O_5 were mixed at a molar ratio of 1.1: 4: 3 and ground for 30 min. Then, the ground mixture was placed in a Muffle furnace and calcined at 1,200 °C for 12 h in air. White $KCa_2Nb_3O_{10}$ (KCNO) powder was obtained after cooling down the mixture to room temperature. The identification of KCNO powder was performed using XRD and XPS analysis. Then, $HCa_2Nb_3O_{10}\cdot 1.5H_2O$(HCNO) was obtained by ion-exchange, 0.4 g of KCNO powder and 40 mL of nitric acid (1 mol/L) were put in a sealed beaker, the suspension was then stirred at 250-300 rmp with a magnetic stirrer for 72 hours. Next, the suspension was filtered and rinsed with water to completely remove the scrap. The residue was dried at 30 °C overnight (~12 h), and was confirmed as HCNO by XRD analysis. Then, for powder stripping, TBAOH/H^+ (H^+ refers to the protons of HCNO powder, i.e. $[H^+]$ = [HCNO]) suspension was made by mixing HCNO powders (0.1 g) with 25 mL TBAOH solution (7.3 mmol/L). Both stripped and unstripped HCNO powders were placed in a graphite mold and sintered at 800 °C for 5 minutes in a vacuum of 7 Pa (SPS 625). After sintered, samples were annealed for 3 hours. The exfoliated powder was especially sintered at 850 °C for comparison.

The crystalline phases of the KCNO and HCNO powders were characterized by X-ray diffraction (XRD, Rigaku Dmax-2000 diffractometer) using Cu Kα radiation from 20°-80° (2-theta). The morphology and compositions of the products were characterized by scanning electron microscopy (SEM, JSM-7800F) with energy dispersive X-ray spectroscopy (EDS) and X-ray Photoelectron Spectroscopy (XPS, Thermo escalab 250Xi). The dielectric properties of sintered samples were measured by Agilent E4980A LCR meter in a wide frequency range of 20-2 M Hz.

3. Results and discussion

The XRD patterns of the synthesized compound are shown in Fig. 1. The synthesized product were confirmed to be KCNO and HCNO as expected, and all the samples are in their single perovskite phases since no obvious impurity peaks was observed.

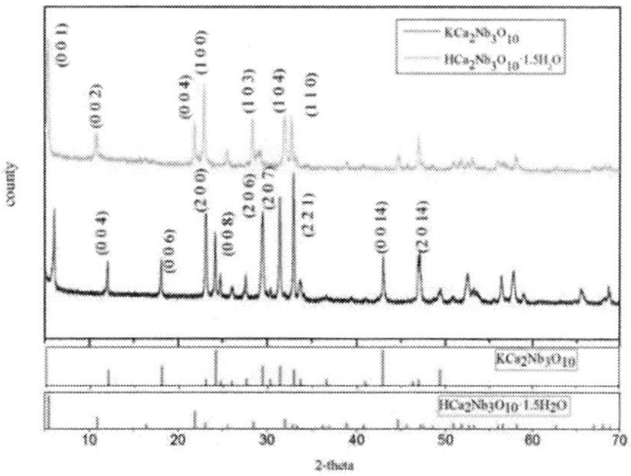

Fig. 1 XRD patterns of the KCNO and HCNO ceramics, 2θ = 5°-70°

IOP Conf. Series: Materials Science and Engineering **729** (2020) 012011 doi:10.1088/1757-899X/729/1/012011

We could evaluate the surface chemical composition and the oxidation state of individual element in the powders by XPS shown in Fig. 2 where the positions of peaks were corrected using the C 1s peak at 285 eV as a reference. The XPS survey spectra of KCNO in a wide energy range are shown in Fig. 2a.

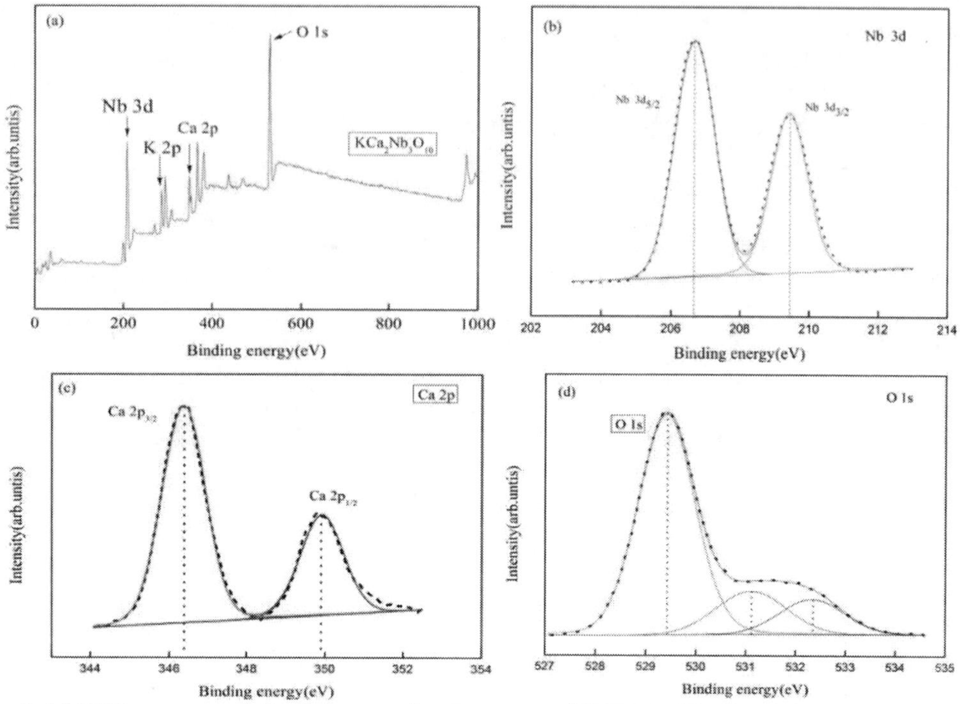

Fig. 2 (a) XPS survey and (b-d) the core level spectra of KCNO: (b) Nb 3d, (c) Ca 2p, (d) O 1s

Fig. 2b reveals the typical Nb 3d core level. The 3d peaks are split into two components, $3d_{3/2}$, $3d_{5/2}$, respectively, as a result of spin-orbit coupling with ratio of 2:3 [9]. The peaks of Nb 3d centred at 206.8 eV and 209.5 eV can be assigned to $3d_{5/2}$ and $3d_{3/2}$ states respectively, since they are identical to the binding energy of Nb^{5+} in Nb_2O_5 [10]. It indicates that niobium in prepared samples exists as Nb^{5+} ion. As can be seen from Fig. 2c, the characteristic peaks of Ca 2p located at about 346.4 eV and 349.9 eV, are assigned to the Ca $2p_{3/2}$ and Ca $2p_{1/2}$ binding energies, respectively. At the same time, The O 1s XPS spectrum of the KCNO sample is displayed in Fig. 3d. It can be deconvoluted to three components. The major peak of lower binding energy (529.4 eV) was ascribed to the lattice oxygen in Nb-O bond; while the minor peak located at 531.2 eV [11] and 532.4 eV was assigned to the hydroxyl OH- group and absorbed oxygen in the surface, respectively [12] .

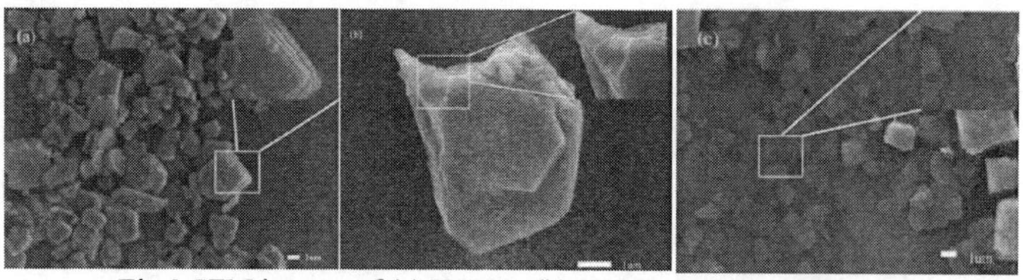

Fig.3 SEM images of (a) KCNO, (b) HCNO and (c) e-HCNO

Fig. 3 shows the SEM images of synthetic (a) KCNO, (b) HCNO and post-stripped HCNO samples. In Fig (a) and (b), particles with a size of 2-8 μm are observed, and the images clearly show that the compound exhibits a layered structure. The HCNO sample retained the morphology of its precursor (KCNO). However, compared with KCNO, the interval between the CNO layers in the HCNO remained unchanged, but the HCNO sample after peeling was significant distinction. To confirm whether all K ions were extracted and replaced during the ion exchange process, the XRD analysis was performed and shown in Fig. 1.

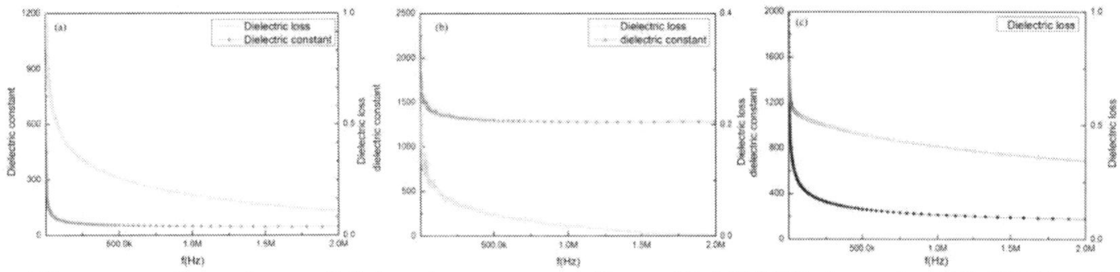

Fig.4 Frequency dependence of dielectric constant and loss of (a) HCNO sintered at 800 °C (b) e-CNO sintered at 800 °C and (c) e-CNO sintered at 850 °C from 100 Hz to 2 M Hz.

Fig. 4 shows the dielectric constant spectrum and dielectric loss spectrum of samples. From Fig. 4a, 4b, and 4c, the HCNO and e-HCNO samples sintered at 800 °C, and the e-HCNO samples at 850 °C, respectively, we can see that both the dielectric constant and loss of all samples decrease with increasing frequency. The dielectric properties of e-HCNO are superior to HCNO after 800 °C sintering, which may be caused by the decrease of H^+ in the e-HCNO sample during stripping. Because the smaller H^+ that intercalated between the CNO layers can cause a small interlayer distance. It was reported that the dielectric constant of films made of CNO nanosheets [6] keep constant, which could explain the behaviour of e-HCNO dielectric constant in a wide frequency range. Comparing with Fig. 4b and 4c, it is found that the dielectric constant of 800 °C is higher than that of 850 °C while the loss is lower.

4. Conclusion

In this paper, the dielectric properties of plasma spark sintered calcium niobate ceramics are studied. The HCNO precursor was synthesized by substituting K ions with H ions. And CNO nanosheets were obtained by removing a large amount of H^+ between the HCNO perovskite layers by liquid phase exfoliating. The samples were analysed by XRD, SEM and EDS. It shows that there is almost no structural change in ion exchange, and both samples exhibit a layered perovskite structure. This layered structure could be also observed by SEM. The dielectric constant and loss factor are measured over a frequency range of 20-2M Hz. The e-HCNO sample obtained by liquid phase exfoliated, then sintering at 800 °C for 5 min, exhibits a higher relative dielectric constant (ε_r = 1300) and lower dielectric loss (tanθ = 0.002).

Acknowledgments

This work was supported by the National Natural Science Foundation of China (No.21875245) and the Program of Hundreds Talents of Chinese Academy of Science.

References

[1] Chauhan, A., Patel, S., Vaish, R., Bowen, C.R., (2015) Anti-Ferroelectric Ceramics for High Energy Density Capacitors. Materials, 8: 8009-8031.
[2] Jacobson, A.J., Johnson, J.W., Lewandowski, J.T., (1985) Interlayer chemistry between thick transition-metal oxide layers—synthesis and intercalation reactions of $K[Ca_2Na_{n-3}Nb_nO_{3n+1}]$ (3 less-than-or-equal-to n less-than-or-equal-to 7). Inorganic Chemistry, 24: 3727-3729.
[3] Geim, A.K., Grigorieva, I.V., (2013) Van der Waals Heterostructures. Nature, 499: 419-425.

[4] Osada, M., Sasaki, T., (2012) Two-dimensional dielectric nanosheets: novel nanoelectronics from nanocrystal building blocks. Advanced Materials, 24: 210-228.
[5] Yim, H., Yoo, S.Y., Nahm, S., Hwang, S.J., Yoon, S.J., Choi, J.W., (2013) Synthesis and dielectric properties of $HCa_2Nb_3O_{10}$ layered structure ceramics. In: 8th Asian Meeting on Electroceramics. Malaysia. pp. S611-S614.
[6] Osada, M., Akatsuka, K., Ebina, Y., Funakubo, H., Ono, K., Takada, K., Sasaki, T., (2010) Robust high-kappa response in molecularly thin perovskite nanosheets. ACS Nano, 4: 5225-5232.
[7] Wang, Y., Sun, C.H., Yan, X.X., Xiu, F.X., Wang, L.Z., Smith, S.C., Wang, K.L., Lu, G.Q., Zou, J., (2011) Lattice Distortion Oriented Angular Self-Assembly of Monolayer Titania Sheets. Journal of the American Chemical Society, 133: 695-697.
[8] Guillon, O., Gonzalez-Julian, J., Dargatz, B., Kessel, T., Schierning, G., Rathel, J., Herrmann, M., (2014) Field-Assisted Sintering Technology/ Spark Plasma Sintering: Mechanisms, Materials, and Technology Developments. Advanced Engineering Materials, 16: 830-849.
[9] Kim, Y., Schlegl, H., Kim, K., Irvine, J.T.S., Kim, J.H., (2014) X-ray photoelectron spectroscopy of Sm-doped layered perovskite for intermediate temperature-operating solid oxide fuel cell. Applied Surface Science, 288: 695-701.
[10] Zhang, W.B., Wu, W.D., Wang, X.M., Cheng, X.L., Yan, D.W., Shen, C.L., Peng, L.P., Wang, Y.Y., Bai, L., (2014) The investigation of NbO_2 and Nb_2O_5 electronic structure by XPS, UPS and first principles methods. Surface and Interface Anaylsis, 45: 1206-1210.
[11] Brunckova, H., Kanuchova, M., Kolev, H., Mudra, E., Medvecky, L., (2018) XPS Characterization of $SmNbO_4$ and $SmTaO_4$ Precursors Prepared by Sol-Gel Method. In: 15th International Symposium on Novel and Nano Materials, Lisbon, Portugal. pp. 1-5.
[12] Ramos-Moore, E., Ferrari, P., Diaz-Droguett, D.E., Lederman, D., Evans, J.T., (2012) Raman and x-ray photoelectron spectroscopy study of ferroelectric switching in $Pb(Nb,Zr,Ti)O_3$ thin films. Journal of Applied Physics, 111.

Antibacterial performance of Ag-doped TiO_2/wood surface under visible light irradiation and its superior mould-resistance

Fu'an Zhang[1]

[1]Zhejiang A&F University, Hangzhou, China

E-mail: wlshzj105@162.com, Tel: +86-571-63732718

Abstract. The TiO_2/wood (TW) based on fast-growing poplar wood was in situ synthesized through one-step solvothermal method and then Ag-doped TiO_2/wood (ATW)was also prepared via the solvothermal method. The morphologies, crystal structure, and chemical characteristics of the composites were carried out by SEM, XRD, XPS and FTIR. The results of XPS and FTIR showed that the nanoparticles (TiO_2 and Ag) were chemically bonded to the wood surface through the combination of hydrogen groups. Spherical-like Ag nanoparticles (Ag-NPs) was about 66 nm diameter. What's more, the TW and ATW showed excellent antibacterial against Gram positive and negative bacterium under visible light irradiation and mould-resistance properties, which illustrated a significance and practical value in high added-value of wood.

1. Introduction

As one of four building materials (plastic, cement, steel, wood), wood is the only sustainable one and wood possesses the advantages of high specific gravity, low production cost, beautiful structure and texture, good thermal insulation performance, small energy consumption, renewable and recycle, wide range of application, no poison, environment friendly, and so on [1, 2]. However, wood is easily attacked by microorganisms like bacteria and mould [3, 4]. There are multiple of researches in antibacterial species such as various natural and inorganic templates [5]. With a high specific surface area and a fraction of surface atoms, metal nanoparticles such as Ag, Cu, and Nd, have been explored widely due to their unique physicochemical performances such as optical or electronic properties, catalytic activity, and antibacterial activity [6, 7]. Among these metal nanoparticles, Ag-NPs are well known because of inhibitory and bactericidal effects. Resistance to antibacterial agents by pathogenic bacteria has developed in recent years, which is also a major sustainable development issue [8]. It can be expected that the high specific surface area and high fraction of surface atoms of Ag-NPs will lead to high antibacterial activity as compared with bulk silver metal. Unfortunately, it is limited due to high-cost for its applications especially as large-area antibacterial coatings. Therefore, the combination the advantages of Ag with economic nanoparticles of metal oxides such as TiO_2, ZnO, and WO_3 could design an economical antimicrobial coating [9, 10]. Among these nanoparticles, TiO_2, a wide band gap and nontoxic semiconductor metal oxides, has attracted considerable research interest owing to its unique optical, electronic, and atalytic properties. The nanostructured synthesis of TiO_2 consisted of the nano-particles, nano-rods or nanotubes, which were applied to various fields such as catalysts, dye sensitized solar cells, pigments, and self-cleaning surfaces [11, 12]. However, there were only several researches on the antimicrobial activity of TiO_2. Furthermore, the synergy of Ag-NPs with the antimould and antibacterial performances of TiO_2 on surface of wood under visible light irradiation have not been studied.

In this study, the prepared Ag-doped TiO_2 on wood surface were evaluated against two types of bacterias. Furthermore, mould-resistance activity was evaluated against three different kinds of moulds.

2. Experiments

2.1 Materials

The size of wood, which was sampled from the sap wood sections of poplar wood (Populusussuriensis Kom) in Hangzhou City, Zhejiang Province, is 10mm (length) × 10mm (width) × 10mm (thickness). The material was marked as OW. All the reagents were provided by Shanghai Boyle Chemical Co., Ltd. All the reagents were all nalytically pure grade and could be used directly without secondary purification.

2.2 Preparation of TiO_2/wood (TW) and Ag-doped TiO_2/Wood (ATW) by solvothermal method

First, wood was ultrasonically rinsing in ethanol, acetone and deionized water for 30 min, respectively, and dried in a vacuum oven at 45 °C for 48h. The humidity of the wood was measured by drying the wood until the weight was constant. The resulting humidity was 7%. The wood was immersed in of tetrabutyl titanate [$Ti(OC_4H_9)_4$] (4 mL) in 50 mL absolute ethanol and then hydrochloric acid (1 mL) was dropped dropwise into the solution with stirring for 30 min After that, the solution with the wood was transferred into a 100 mL Teflon container in an oven at 95 °C for 5h. Finally, the specimen was taken out, cleaned with acetone and ethanol for 30 min, respectively. Then dried in a vacuum oven at 45 °C for 48 h. The prepared specimen was marked TW.

The TW was placed in an Teflon container with 4 mL tetrabutyl titanate solution, 50 mL acetic acid and1.0 g of $AgNO_3$. The solvothermal reaction was carried out at 160 °C for 3 h. After the solvothermal process, the specimens were washed by deionized water and then dried at 60 °C. The prepared specimen was marked ATW.

2.3 Characterizations

The *Characterizations* of the samples were shown in Text S1

2.4 The tests of antibacterial and mould-resistance performances

2.4.1 Antibacterial performance

According to the bacterial inhibition ring method (a gar plate diffusion test/CEN/TC248WG 13) and the reduction of bacterial growth test (ENISO 20743:2007 Transfer Method), antimicrobial performance were carried out carefully[13]. Antibacterial properties of the specimens were evaluated against was evaluated against two different types of bacteria--- *Escherichia coli* (*E. coli*) and *Staphylococcus aureus* (*S. aureus*, ATCC 25922, Gram-positive bacterium). In detail, the nutrient broth and nutrient agarin 1L distilled water at pH 7.2 as well as the empty Petri plates were mixed and autoclaved. The agar medium was then cast into the Petri plates and cooled in laminar airflow. Approximately 105 colony-forming units of each bacterium were inoculated on plates, and then specimens were grown onto the agar plates. All the plates were incubated at 37°C for 24h with a visible lamp (λm (mean)= 525 nm, 18 K lux), following which the zone of inhibition was measured. Three parallel specimens were tested. The number of living bacteria was counted. Remove (R %) is considered as antimicrobial efficiency, which is as follows:

$$R (\%) = (B - M)/B \times 100\% \quad (1)$$

B is the average bacteria counts on blank specimen. In contrast, M is the average bacteria counts on modified specimen.

2.4.2 Mould resistance test

Mould-resistance performance of OW, TW, and ATW was evaluated against three different kinds of mould------*Aspergillus niger* (*A. niger*), *Aspergillus terreus* (*A. terreus*), and *Paecilomyces variotii* (*P.*

variotii). The specimens were exposed in an environment chamber to allow fungal growth under controlled conditions (95% relative humidity, 25 °C). Mould resistance test was condcted referenced to AWPA Standard EN24-06. Samples were tested after three weeks. Mould growth on each specimen was visually rated using the following criteria (complete description of visual rating scheme given in the AWPA Standard) (Table S1). The lower the infection value means the better the efficiency of treatment.

3. Results and discussion

Fig. 1 showed the SEM images and EDS spectra of the specimens. Fig. 1a exhibited that the OW showed typical wood structure and smooth surface. After a solvothermal process, the surface showed a coat of TiO_2 microparticles, (Fig. 1b). Then the TW was covered with a compact coat of Ag-NPs through the solvothermal synthesis to form the ATW shown in Fig. 1c and d. Further, the mean diameter of Ag nanoparticles was about 66 nm learned from Fig. 1d. Fig. 1e displayed that there were only C, Ti, Au, and O elements in TW. The element of Au derived from the coating layer used for SEM observation. The C and O elements were from the wood substrate and Ti element derived from TiO_2. From the EDS spectrum of ATW in Fig. 1f, obvious peak of Ag element was presented except for the signals of C, Ti, Au, and O element, indicating there was Ag loaded onto the surface of TW.

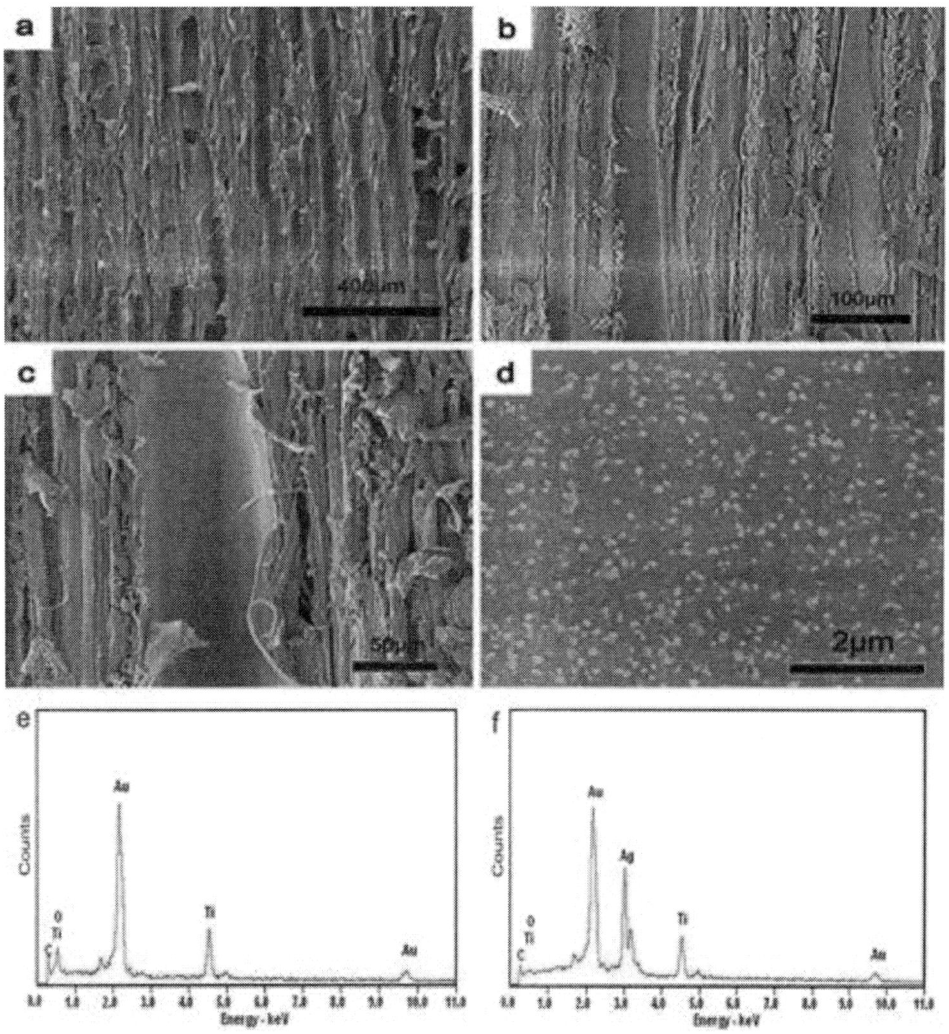

Fig. 1 SEM images and EDS spectra of OW (a), TW (b) and ATW (c and d); EDS of TW (e) and ATW (f)

Fig. S1 presented the XRD patterns of ATW and TW. The peaks at 22° and 16° were referred to the typical planes (002) and (101) in wood cellulose [14], while sharp peaks at 2θ of centered at 25.2°, 38.0°, 47.8°, 54.2°, 62.5°, 68.8°, and 74.9°, which could be ascribed to (101), (004), (200), (211), (204), (116) and (215) planes, which can be attributed to TiO_2 in the anatase form (PDF card 21-1272, JCPDS) [15]. Furthermore, for ATW, new sharp peaks at 2θ of 38.1°, 44.2°, 64.4°, and 77.4° could be ascribed to (111), (200), (220) and (311) planes of cubic phase of Ag (PDF card 04-0783, JCPDS) [16]. These results confirmed that the TiO_2-doped (TW) and Ag-doped TiO_2 (ATW) on the surface of wood were successfully synthesized.

In Fig. 2a, the fully spectra showed that the Ti, O, and C elements in both of the ATW and TW while a new binding energy of around 370 eV assigned to Ag on the ATW. The C element could be attributed to the wood. These results were well consistent with that of XRD and EDS. In Fig. 2b, the Ti 2p XPS peak of the ATW could be fitted into two components Ti^{4+} species (459.3eV) and Ti^{3+} species(458.9eV). In addition, in the Ti 2p XPS spectra of the TW, the two peaks at 459.0eV and 464.8 eV were attributed to the Ti $2p_{3/2}$ andTi$2p_{1/2}$, respectively. The existence of Ti^{4+} was illustrated due to the gap of 5.7eV between the two peaks. However, the red shift of the peak of Ti2p in the ATW compared with that of the TW confirmed a lower electron density of the Ti atom in the ATW, illustrating a strong interaction between metallic Ag and TiO_2 (the insert in Fig. 3b). From Fig. 2c, O1s XPS spectra with three peaks at binding energies of 530.4eV, 532.9 eV, and 533.2 eV, attributing to O (Ti-O), O(O-H), and adsorbed O (O2 or C-O). As shown in Fig. 2d, the Ag3d XPS spectra with two peaks at 367.6 eV and 373.6 eV were corresponded to the Ag$3d_{5/2}$ and Ag $3d_{3/2}$ peaks, respectively. The gap of 6.0 eV between the two peaks was also existing form of metallic Ag instead of Ag^+. Moreover, the peaks of Ag3d showed the blue shift to the lower position, illustrated that the electrons migrated to metallic Ag from the TiO_2 particles in the interface of ATW heterostructures.

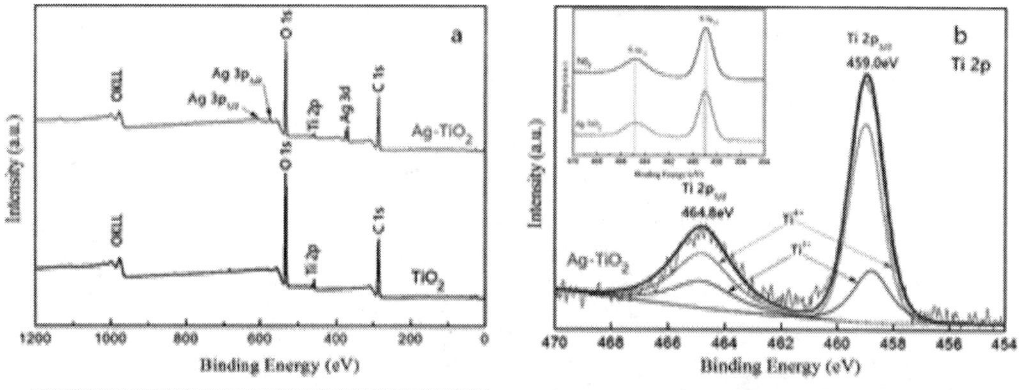

Fig. 2 XPS spectra of TW and ATW: (a) survey spectra; (b) Ti 2p; (c) O1s; (d) Ag 3d.

Fig. S2 showed the FTIR spectra of OW, TW and ATW. For OW, the absorption peak at the 3400 cm^{-1} derived from the wood surface due to the stretching vibrations of –OH. The intensity of hydroxyl absorption peak at 3420-3400 cm^{-1} band of TW was weakened, whil the hydroxyl absorption peak intensity of ATW prepared by solvothermal synthesis significantly reduced, indicating that the Ag- TiO_2 may be combined with the surface of the wood through hydroxyl bond to a composite. The absorption peaks at 1423 cm^{-1}, 1261 cm^{-1} and 1114 cm^{-1} were the deformation vibration of Ti and the stretching vibration of Ti-OC [17]. The absorption peaks at 619 cm^{-1} and 597 cm^{-1} belonged to O-Ti [18]. However, both of the TW and ATW showed strong interaction of titanium dioxide at the 700-600 cm^{-1} [19], which further showed that titanium dioxide is generated on the wood surface by the solvothermal method.

Antibacterial and mould-resistance properties analysis: More importantly, the both of TW and ATW exhibited superior antimicrobial and mould-resistance activities in the dark shown in Fig. 3. The R% of TW reached 94.9% and 96.4% toward *E. coli* and *S. aureus*, respectively. Simultaneously, the R% of ATW reached 99.0% and 90.5% toward E. coli and S. aureus, respectively. The ATW exhibited better antibacterial performance than TW as for gram-negative bacterium while the TW showed better antibacterial property than ATW for gram-positive bacterium. The rating of OW reached four toward *A. niger*, *A. terreus*, and *P. variotii* (Table 1). However, both of the TW and ATW illustrated grade zero toward *A. niger*, *A. terreus*, and *P. variotii*, further indicating that TW and ATW were better mould-resistance ability. As well known, silver was proved to possess antibacterial performance [20]. Ag-NPs in the ATW played a metal antibacterial agent role and a kind of doping metal in TW. It was reported that tetravalent Ti ions in TiO_2 would be substituted by monovalent Ag ions, resulting in various defects like oxygen vacancies[21], which might be benefit for the strengthen of antibacterial property of the ATW. Hence, both of the TiO_2/wood and Ag-doped TiO_2/wood possessed superior antibacterial and mould-resistance properties and Ag-doped TiO_2/wood was better than TiO_2/wood in antibacterial performance as for gram-negative bacterium.

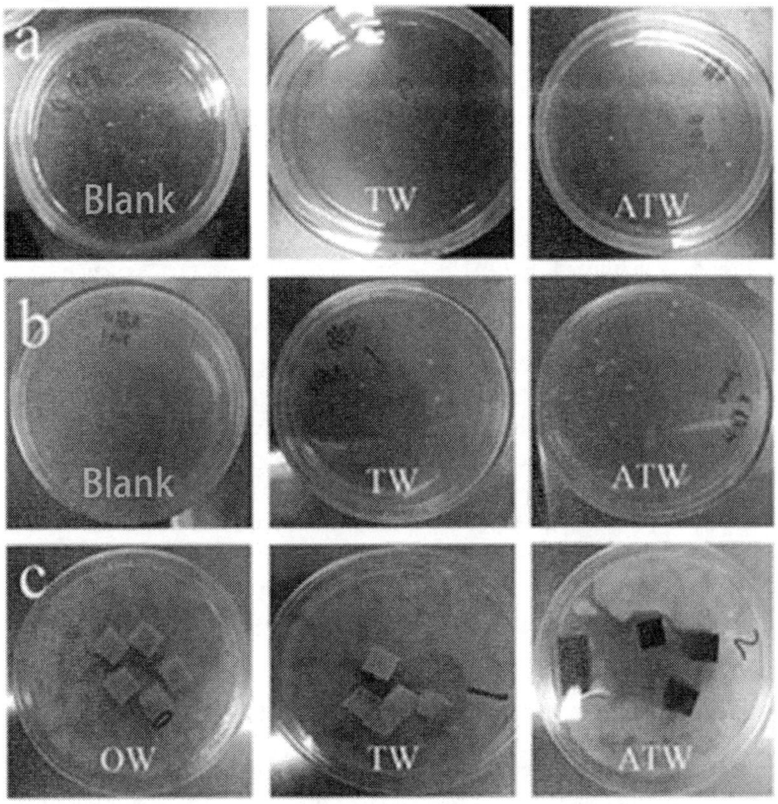

Fig. 3. Antibacterial activity of TW and ATW toward *E. coli* (a) and *S. aureus* (b); Mould-resistance performance of OW, TW, and ATW toward *A. niger*, *A. terreus*, *and P. variotii*, respectively.

Table 1 Antibacterial rate and mould-resistance Rating

Specimen	R (%) against *E. coli*	R (%) against *S. aureus*	Rating
OW			4
TW	94.9	96.4	0
ATW	99.0	90.5	0

4. Conclusion

In summary, the Ag-doped TiO_2 on wood surface was fabricated successfully through the solvothermal synthesis method. Furthermore, TiO_2 and Ag nanoparticles were connected on wood surface in situ by hydrogen bond. The results exhibited that the average particle size of Ag-NPs was about 66 nm and they were well dispersion. The antibacterial tests against *S. aureus*, *E. coli* and mould-resistance tests against *A. niger*, *A. terreus*, and *P. variotii* showed that Ag-doped TiO_2/wood had good antibacterial and mould-resistance performance, which illustrated that the Ag-doped TiO_2/wood could be applied in fields such as for the low-cost materials for antibacterial floor, nano-antibacterial furniture. Furthermore, the methods may be regarded as a candidate for the profitable antibacterial biomass-based composite materials in the home decoration, public places, as well as work office.

References

[1] Chen, C., Zhang, Y., Li, Y., Dai, J., Song, J., Yao, Y., Gong, Y., Kierzewski, I., Xie, J. and Hu, L. (2017) All-wood, low tortuosity, aqueous, biodegradable supercapacitors with ultra-high capacitance. Energy & Environmental Science, 10(2): 538-545.
[2] Wang, Q., Li, J. and Winandy, J. (2004) Chemical mechanism of fire retardance of boric acid on wood. Wood Science and Technology, 38(5).
[3] Aviat, F., Gerhards, C., Rodriguez-Jerez, J.-j., Michel, V., Bayon, I.L., Ismail, R. and Federighi, M. (2016) Microbial Safety of Wood in Contact with Food: A Review. Comprehensive Reviews in Food Science and Food Safety, 15(3): 491-505.
[4] Gerullis, S., Pfuch, A., Spange, S., Kettner, F., Plaschkies, K., Küzün, B., Kosmachev, P.V., Volokitin, G.G. and Grünler, B. (2017) Thin antimicrobial silver, copper or zinc containing SiOx films on wood polymer composites (WPC) applied by atmospheric pressure plasma chemical vapour deposition (APCVD) and sol–gel technology. European Journal of Wood and Wood Products, 76(1): 229-241.
[5] Shahid ul, I. and Mohammad, F. (2015) Natural Colorants in the Presence of Anchors So-Called Mordants as Promising Coloring and Antimicrobial Agents for Textile Materials. ACS Sustainable Chemistry & Engineering, 3(10): 2361-2375.
[6] Abbasi, E., Milani, M., Fekri Aval, S., Kouhi, M., Akbarzadeh, A., Tayefi Nasrabadi, H., Nikasa, P., Joo, S.W., Hanifehpour, Y., Nejati-Koshki, K. and Samiei, M. (2016) Silver nanoparticles: Synthesis methods, bio-applications and properties. Crit Rev Microbiol, 42(2): 173-180.
[7] Palza, H. (2015) Antimicrobial polymers with metal nanoparticles. Int J Mol Sci, 16(1): 2099-2116.
[8] Ahmed, S., Ahmad, M., Swami, B.L. and Ikram, S. (2016) A review on plants extract mediated synthesis of silver nanoparticles for antimicrobial applications: A green expertise. J Adv Res, 7(1): 17-28.
[9] Elumalai, K. and Velmurugan, S. (2015) Green synthesis, characterization and antimicrobial activities of zinc oxide nanoparticles from the leaf extract of Azadirachta indica (L.). Applied Surface Science, 345: 329-336.
[10] Podporska-Carroll, J., Panaitescu, E., Quilty, B., Wang, L., Menon, L. and Pillai, S.C. (2015) Antimicrobial properties of highly efficient photocatalytic TiO_2 nanotubes. Applied Catalysis B: Environmental, 176-177: 70-75.
[11] Tang, J., Wu, Y., McFarland, E.W. and Stucky, G.D. (2004) Synthesis and photocatalytic properties of highly crystalline and ordered mesoporous TiO_2 thin films. Chem Commun (Camb), (14): 1670-1671.
[12] Yoo, K., Choi, H. and Dionysiou, D.D. (2004) Ionic liquid assisted preparation of nanostructured

TiO_2 particles. Chem Commun (Camb), (17): 2000-2001.
[13] Zhang, Q., Sun, C., Zhao, Y., Zhou, S., Hu, X. and Chen, P. (2010) Low Ag-Doped Titanium Dioxide Nanosheet Films with Outstanding Antimicrobial Property. Environmental Science & Technology, 44(21): 8270-8275.
[14] Yao, Q., Jin, C., Zheng, H., Ma, Z. and Sun, Q. (2015) Superhydrophobicity, Microwave Absorbing Property of $NiFe_2O_4$/Wood Hybrids under Harsh Conditions. Journal of Nanomaterials, 2015: 1-8.
[15] Gao, L., Zhan, X., Lu, Y., Li, J. and Sun, Q. (2015) pH-dependent structure and wettability of TiO_2-based wood surface. Materials Letters, 142: 217-220.
[16] Yao, Q., Fan, B., Xiong, Y., Wang, C., Wang, H., Jin, C. and Sun, Q. (2017) Stress sensitive electricity based on Ag/cellulose nanofiber aerogel for self-reporting. Carbohydr Polym, 168: 265-273.
[17] Colomer, M.T. and Velasco, M.J. (2007) Rutile-type dense ceramics fabricated by pressureless sintering of $Ti_{1-x}Ru_xO_2$ powders prepared by sol–gel. Journal of the European Ceramic Society, 27(6): 2369-2376.
[18] Li, J., Yu, H., Sun, Q., Liu, Y., Cui, Y. and Lu, Y. (2010) Growth of TiO_2 coating on wood surface using controlled hydrothermal method at low temperatures. Applied Surface Science, 256(16): 5046-5050.
[19] Tseng, T.K., Lin, Y.S., Chen, Y.J. and Chu, H. (2010) A review of photocatalysts prepared by sol-gel method for VOCs removal. Int J Mol Sci, 11(6): 2336-2361.
[20] Sambhy, V., MacBride, M.M., Peterson, B.R. and Sen, A. (2006) Silver Bromide Nanoparticle/Polymer Composites:? Dual Action Tunable Antimicrobial Materials. Journal of the American Chemical Society, 128(30): 9798.
[21] Es-Souni, M., Fischer-Brandies, H. and Es-Souni, M. (2008) Versatile Nanocomposite Coatings with Tunable Cell Adhesion and Bactericidity. Advanced Functional Materials, 18(20): 3179-3188.

IOP Conf. Series: Materials Science and Engineering **729** (2020) 012013 doi:10.1088/1757-899X/729/1/012013

Research on fire prevention and extinguishing technology in underhand working face of easy spontaneous combustion coal seam

Xueli Liu[1*], Jijun You[2]

[1]School of Civil and Environmental Engineering, Anhui Xinhua University, Hefei 230031, China

[2]Huainan Coal Mining in Western Investment Management Ltd, Er dos 017000, China

Email: lxlpgy@126.com; youjijun@163.com TEL:18255105636

Abstract: The management of fire prevention and extinguishing in goaf becomes a major problem due to short spontaneous combustion period, poor roof caving conditions, large air leakage in goaf and difficulties for grouting. Oxygen consumption rate and the productivity of carbon monoxide are measured through the XK-VI spontaneous combustion test-bed, while prediction index gas for spontaneous combustion of the coal is obtained. Through the engineering practice test, the range of the gob "triple-zone" under the condition of nitrogen injection is acquired. Take 113101 working face of Bojianghaizi coal mine as an example. The method for detecting the fire prevention effect is to inject nitrogen into the goaf, supplemented by measures such as forced anchoring in the intake and return wind lane, constructed of the sandbags separation wall, and embedded the beam tube through the goaf. The volume fraction of carbon monoxide in the return corner and the R_2 value are used as predictors of fire prevention. Engineering practice test shows that there is no danger of spontaneous combustion in the working face when the volume fraction of carbon monoxide in the return corner is less than $200*10^{-6}$ and the value of R_2 is less than 0.2. The process has reference significance for the fire prevention management of the Bojianghaizi Mine and other similar conditions.

1. Introduction

The spontaneous combustion of the coal is one of the main natural disasters in China's mines, while residual coal in gobs is one of the prime reasons of the mine fire[1]. With the development of mine fire prevention technology, nitrogen injection has become one of the main technologies to prevent spontaneous combustion of residual coal in gobs[2,3]. At present, the research on fire extinguishing technology of nitrogen injection is mainly based on engineering practice field test and numerical simulation. Then it can confirm the technological parameters and nitrogen injecting volume, which is beneficial to study the oxygen concentration and the change of oxidation heating zone in the goaf after nitrogen injection[4-8]. Li Zongxiang et al[9]carried out simulation research based on heterogeneous air leakage seepage equation, gas seepage-diffusion equation and porous media seepage comprehensive equation. Yang Shengqiang et al[10,11]used fluent software to simulate the air leakage field in the goaf, which proved that spontaneous combustion in the goaf is a process of oxygen-heat microcirculation. Shi Guoqing et al[12,13]obtained the distribution of seepage velocity and oxygen concentration by numerical simulation in the goaf. These research results have effectively promoted the development of fire-fighting techniques of nitrogen. Due to the short spontaneous combustion period of 113101 working face in Bojiang Haizi coal mine and the influence of unfavorable factors such as residual coal, roof caving and underhand stopes of working face, the hidden danger of spontaneous combustion in the goaf is more serious. By means of laboratory experiments and engineering practice, the author has confirmed the oxygen consumption rate, carbon monoxide production rate, spontaneous ignition period, fire prevention prediction index, the range of spontaneous combustion "triple-zone" and the technological parameters of nitrogen injection in the goaf. The spontaneous combustion in the goaf is eliminated through the injection of nitrogen, supplemented by forced anchoring in the intake and return wind lanes, constructed sandbag separation wall in the goaf.

2. General situation of mine

Bojianghaizi coal mine is located in the northwestern part of Ordos. The overall topographical trend is high in the south and low in the north. The main coal-bearing strata are the Jurassic Middle and Lower Yan'an Formation.The main coal seam in the mine is non-coking and long flame coal with low metamorphism. The 113101 working face is the first mining face in mine, which the length is 2600 meters, the width is 200 meters, and the thickness of coal seam is from 3.8 to 6.9 meters, with an average of 4.8 meters. It belongs to spontaneous combustion coal seam. The elevation of working face is from 850m to 799.2m, which is a sublevel mining face. There are three reasons of fire hazard in the goaf during coal extraction. First, there is amount of residual coal in the intake and return wind lanes, which provides the necessary conditions for spontaneous combustion; Second, the roof caving conditions of the intake and return wind lanes are poor, which leads to increasing air leakage in the goaf, and provides oxygen for spontaneous combustion. At last, the effect of grouting for fire prevention is poor in the underhand stopes , and the coal falling from the upper roof of the goaf can not be covered by slurry[14,15]. Therefore, it is urgent to confirm the characteristic parameters and the "triple-zone"of spontaneous combustion in the 3-1 working face. It can also provide reference for fire prevention management in the following working face.

3. Combustion simulation experiment and prediction index determination

3.1. Analysis of characteristic parameters of coal combustion

In order to accurately grasp the rule of spontaneous combustion, it is necessary to confirm the change of index gas in different stages of spontaneous combustion, especially in the exothermic oxidation stage. Using the XK-VI experimental equipment developed by Xi'an University of Science and Technology, the spontaneous combustion simulation experiment was carried out by collecting the 3-1 coal samples. The experimental conditions are shown in Table1. The corresponding relationship between the rate of oxygen consumption , the rate of carbon monoxide production and temperature in the 3-1 coal seam at different stages was confirmed.

Table 1. Experimental conditions for spontaneous combustion of the coal samples.

Coal sample	Average particle size	Height	Quality	Volume	Density	Bulk density	Void fraction	Volume of blast	Initial temperature
	d_{50}/(mm)	/(cm)	/(kg)	/(cm^3)	/(g/cm^3)	/(g/cm^3)		/(m^3/h)	/(T)
3-1	2.71	180	1828.8 1	203472 0	1.40	0.8988	0.358	0.1～ 1.6	32.9

According to the experimental data, the oxygen consumption rate is calculated, and the relationship between the rate of oxygen consumption and temperature is obtained, as shown in Figure 1. It can be seen that the rate of oxygen consumption in the oxidation process is basically exponential with temperature.The increase of this coefficient indicates that the possibility of spontaneous combustion of coal is increasing.The curve of oxygen consumption can provide key technical parameters for the fire

IOP Conf. Series: Materials Science and Engineering **729** (2020) 012013 doi:10.1088/1757-899X/729/1/012013

area treatment, which is of great significance to the safe of the fire zone.

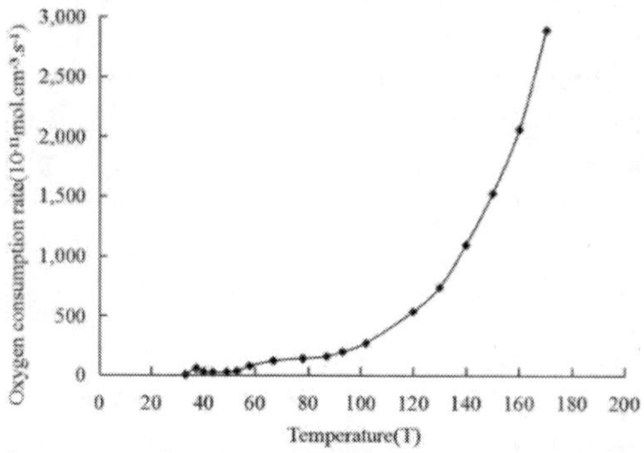

Figure 1. the curve of oxygen consumption rate with temperature.

Carbon monoxide and carbon dioxide are the products from exothermic oxidation of coal. According to the programmed heat experiment, the amount of them produced at different temperature are measured. On the basis of the calculation, the curves of carbon monoxide and carbon dioxide production rate are obtained, as shown in figure 2. The relationship between the volume fraction of ethylene and ethane and temperature is shown in Figure 3. With the increase of molecular velocity, more carbon monoxide and carbon dioxide will be produced. Since the coal sample is in an oxygen-enriched state, the growth rate of the carbon dioxide is much larger than that of the carbon monoxide. Furthermore, the rates of carbon monoxide and carbon dioxide production are exponentially related to the temperature.

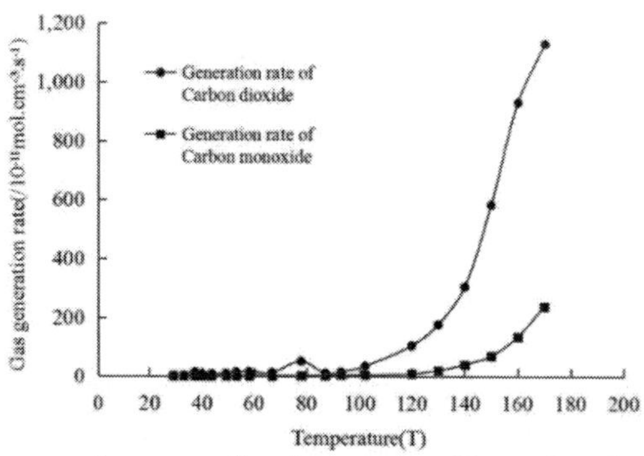

Figure 2. The curves of generation rate of carbon monoxide and carbon dioxide with temperature

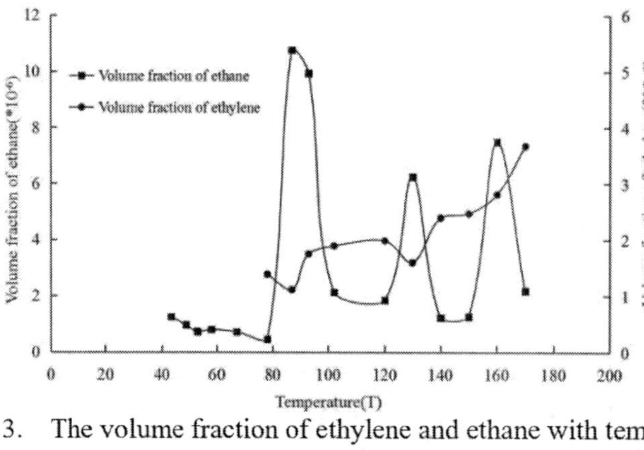

Figure 3. The volume fraction of ethylene and ethane with temperature

Under the experimental conditions, the spontaneous combustion period of the 3-1 is 29 days. According to the actual temperature of the mine is 25°C, the spontaneous combustion period is 31days[16].

3.2. Analysis of fire prediction index

The release of gases in the process of exothermic oxidation are different due to different conditions. So the indicator gas must have sensitivity, regularity and measurability. In the process of spontaneous combustion, the index gases can be divided into oxidation gases and pyrolysis gases. Oxidation gases are mainly carbon monoxide and carbon dioxide; and pyrolysis gases are mainly methane, ethane, propane, ethylene and acetylene. Although the measurable temperature range of the oxidation gases is relatively wide and the productions is large, it is greatly affected by the wind and the environment. However, the measurable temperature range of the pyrolysis gases is relatively narrow, the amount of production is relatively small, and it is easily diluted by the wind, which is greatly affected by factors such as detection means. The oxidation gases is compared with the oxidation gases, and the ratio can be used as a characterization parameter for confirming the degree of spontaneous combustion. The advantage is that the influence of air volume and environmental factors can be excluded. R_2 (the ratio of the increase of the volume fraction of carbon monoxide to the decrease of the volume fraction oxygen) can be used as an indicator of the speed of oxidation. Through the experiment of temperature increasing, it was found that the carbon monoxide appeared in the first and in the whole process. When the coal temperature reached 78.4°C, ethylene began to appear. Therefore, the appearance of ethylene can be used as the temperature of the residual coal in the goaf reaches about 78.4°C.

4. Classification of gobs spontaneous combustion "triple-zone"

Due to the short spontaneous combustion period of the 3-1 coal seam,there is hidden danger of spontaneous combustion in goaf. Therefore, it is of great significance to accurately confirm the range of gobs spontaneous combustion “triple-zone”. Combined with the relationship between the rate of oxygen consumption, the rate of carbon monoxide and the temperature in the spontaneous combustion test. The volume fraction of oxygen and carbon monoxide in goaf are measured by bundles buried in the intake and return wind lanes. Then, under the condition of nitrogen injection, the variation law of the oxygen volume fraction on the intake and return sides with the distance of the measuring point buried can be obtained respectively, as shown in Figures 4 and 5.

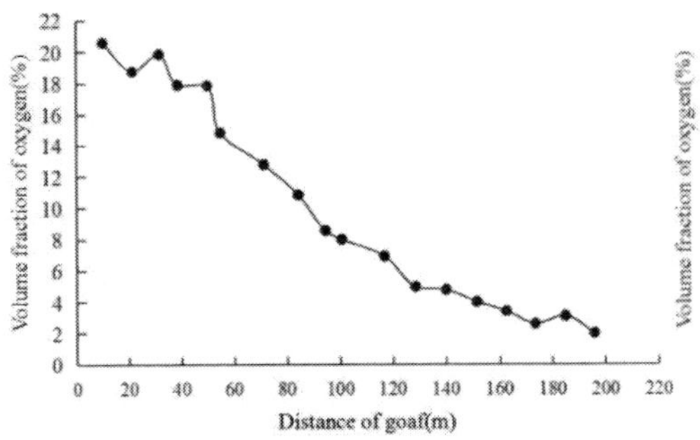

Figure 4. Volume fraction of oxygen on the intake side.

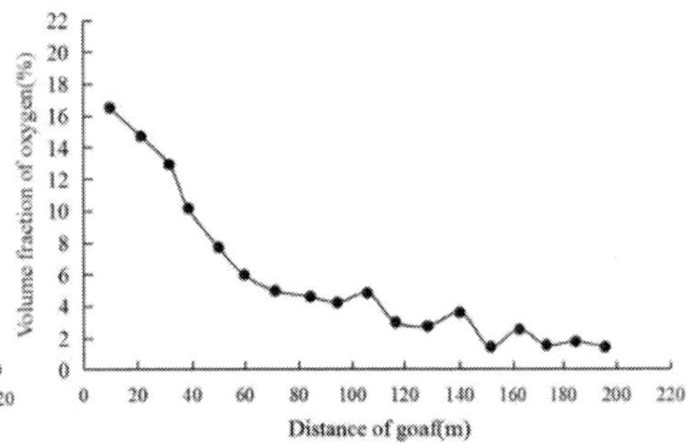

Figure 5. Volume fraction of oxygen on the return air side.

From Figure 4, it can be seen that the air leakage on the intake side is more serious, and the volume fraction of oxygen is smaller in the range of 0 to 55m. It is almost linearly decreased between 55m and 130m, and the volume fraction of oxygen is between 5% and 15%. It tends to be flat after 130m, which drops below 5%. From Figure 5, it can be seen that the volume fraction of oxygen are less than 15% at 20m and 5% at 70m on the return side respectively. With the depth of the measuring point buried in the goaf continues to increase, the measuring point enters the asphyxiation zone, and the volume fraction of oxygen gradually decreases and tends to be stable. Therefore, assuming that the thickness of floating coal is enough to accumulate heat, the potential spontaneous combustion hazard is most likely in the range of 55m-130m on the intake side and 20m-70m on the return side of the working face, which is the key area of fire prevention management.The range of the "triple-zone" under the conditions of nitrogen injection is shown in Table2[17].

Table 2. The range of "triple-zone" in 113101 working face.

	Scattered tropical	Oxidation temperate zone	asphyxiation zone
Intake air side goaf(m)	0~55	55~130	>130
Return air side goaf(m)	0~20	20~70	>70
Oxygen concentrations（%）	>15	5~15	<5

5. Analysis of fire prevention and extinguishing measures and effect

5.1. Fire prevention measures of the nitrogen injection

According to the range of "triple-zone" , the nitrogen release port of nitrogen injection pipe is set in the exothermic oxidation zone on the intake side, which is 0.3m higher than the coal seam floor and protected. The burial of the nitrogen injection pipe adopts a cyclic step method. The first nitrogen injection pipe is buried in the gobs from 55m to 130m, then the second nitrogen injection pipe is buried. When the first nitrogen release port enters the suffocation zone, the nitrogen injection is stopped, then the second nitrogen injection pipe is activated to inject nitrogen. The nitrogen release port is always kept in the oxidation heating zone until the end of the mining face.

5.2. The measures of mandatory retreating anchors (cables) and sandbags

Reducing the air leakage can minimize the spontaneous combustion of the residual coal in the goaf. The air leakage mainly comes from the intake and return wind lanes. Especially when the roof conditions of the two lanes are poor, the air leakage increases rapidly. Therefore, the mandatory retreating anchors (cables) are taken to create conditions for the roof collapse. If it is adopted, the two lanes still cannot be slammed in time, then the sandbag partition wall should be constructed at the end of the two sides of the working face.

5.3. Analysis of fire prevention effect

Based on the analysis of the pre-buried bundle tube sampling, the relationship between the volume fraction of carbon monoxide and oxygen and the distance of buried is obtained, as shown in Figure 6. The relationship between the volume fraction of carbon monoxide and the volume fraction of oxygen in the corner of the stope during the whole mining is shown in Figure 7. The variation of R_2 value in the corner of return side is shown in Figure 8.

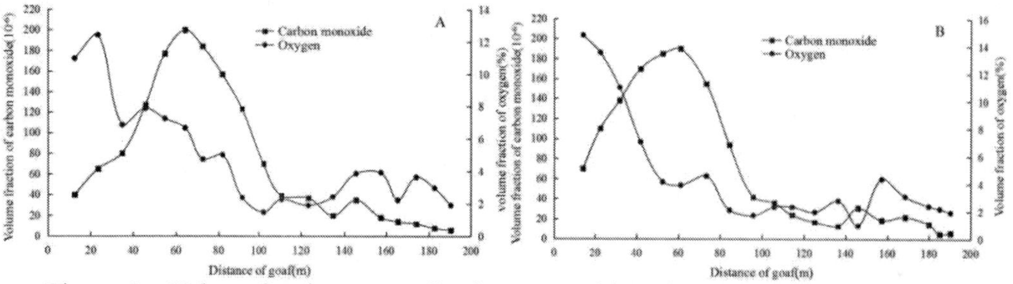

Figure 6. Volume fraction curve of carbon monoxide and oxygen on the return side.

From Figure 6, it can be seen that with the depth of the measuring point buried in the goaf increasing, the volume fraction of carbon monoxide increases at first and then decreases gradually,

IOP Conf. Series: Materials Science and Engineering **729** (2020) 012013 doi:10.1088/1757-899X/729/1/012013

while the volume fraction of oxygen decreases all the time.It is because of the measuring buried point enters thc oxidation zone from the heat-dissipating zone, the thermal storage is conducive to the rapid oxidation of the residual coal. It produces a large amount of carbon monoxide. As the working face continues to mining, the measuring point slowly enters into the asphyxiation zone from the oxidation zone. After the fraction of oxygen volume is reduced to 5%, the oxidation rate of the residual coal gradually decreases and eventually stabilizes. From the volume fraction curve of carbon monoxide and oxygen , the volume fraction of carbon monoxide starts to rise rapidly at 25m, and reaches its peak at about 65m. At this time, the volume fraction of oxygen is between 5% and 15%, and then begins to gradually decrease.The results show that the range of oxidation heating zone is located in the 25-65m, which is the key area for fire prevention and monitoring. It basically coincides with the range of the gobs spontaneous combustion "triple-zone", which is 20-70m. It also verified the accuracy of the "triple-zone".

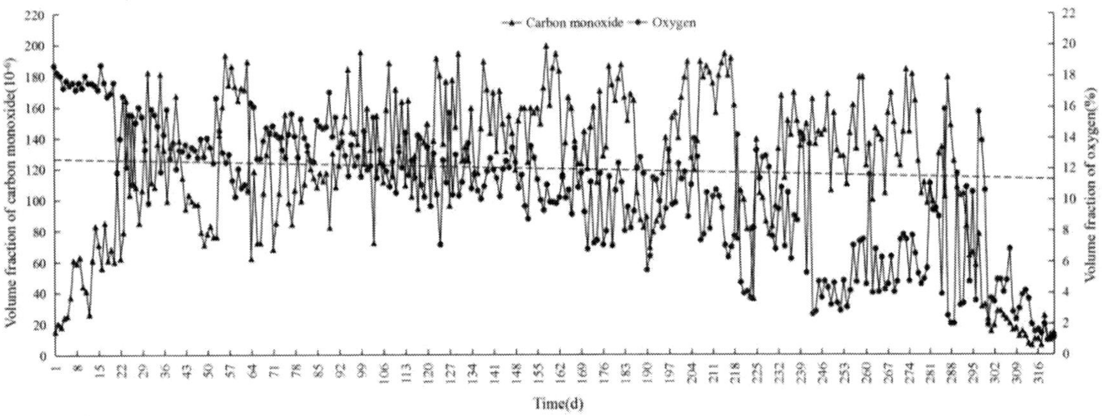

Figure 7. Volume fraction curve of Carbon monoxide and oxygen in the corner of return side.

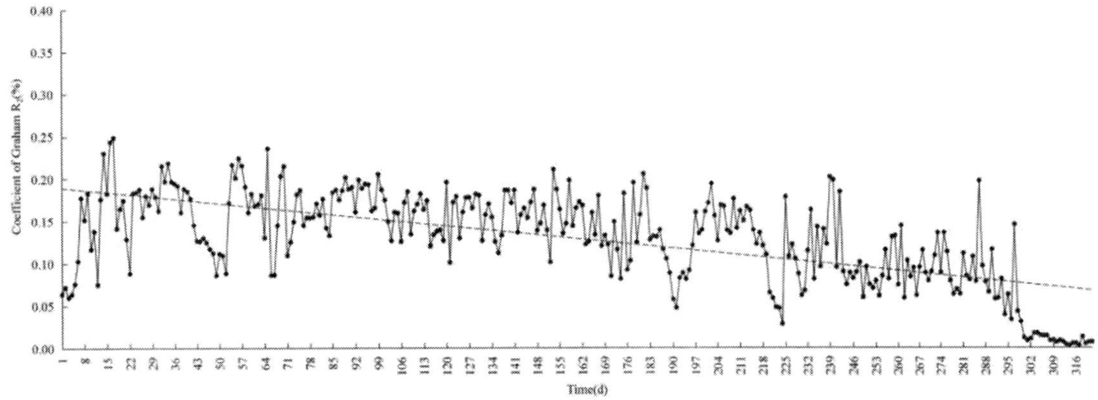

Figure 8. Curve of R_2 value in the corner of return side.

From Figure 7, it can be seen that the volume fraction of carbon monoxide in the corner of return side decreases gradually during the whole mining. It is low at the beginning and after stopping of mining. Because when the working face is just beginning to mining, the residual coal is in the range of the heat dissipation zone, the heat of oxidation of the coal is difficult to accumulate.Then the oxidation rate is slow, and the amount of carbon monoxide generated is small. When mining is stopped, the sandbags partition wall at the corner of the intake and return sides is blocked in time. The pressure difference between the intake and return wind lanes is reduced, so the air leakage is reduced, and the oxidation speed of the residual coal is delayed to a certain extent. The volume fraction of carbon monoxide in the working face is between $69*10^{-6}$ and $200*10^{-6}$ under normal circumstances.

From Figure 8, it can be seen that the the change trend of R_2 value in return corner is consistent with the volume fraction of carbon monoxide, both of which show a gradual downward trend. R_2 value ranges from 0.06 to 0.2. There have been several cases where the R_2 value is greater than 0.2 during the initial mining of the working face.

At present, the working face has been safely harvested, and there is no abnormality of the volume fraction of carbon monoxide during the whole process. There is no sign of spontaneous combustion in the goaf. Based on the statistical analysis of gas test data during mining and plotting of curves, it is found that when the volume fraction of carbon monoxide in the corner of return side is less than $200*10^{-6}$ and the R_2 value is less than 0.2, there will be no hidden danger of spontaneous combustion occurs in the goaf.

6. Conclusion

(1) The most effective fire prevention measures for the working face with easy spontaneous combustion are the nitrogen injection, forced bolt withdrawal (cable) in the return wind lane and constructed sandbags partition wall.

(2) Through engineering practice test, the range of the 3-1 coal spontaneous combustion "triple-zone" under the condition of nitrogen injection is obtained.

(3) The variation curves of the volume fraction of carbon monoxide and R_2 value in the corner of return side can be used as the prediction index of the fire prevention. Ethane and ethylene can be regarded as the temperature index gases with the temperature of residual coal reaching 43.5°C and 78.4°C respectively.

(4) There will be no potential spontaneous combustion in the goaf when the volume fraction of carbon monoxide in the return corner of the working face is less than $200*10^{-6}$ and R_2 value is less than 0.2.

Fund projects:

Study on transition metal oxide synergistic flame retardant polyurethane in different fire scenes(2018M642541); Preparation of layered compound catalytic synergistic flame retardant polyurethane thermal insulation material and its flame retardant and toxic reduction mechanism studying(KJ2018A0593); Industrial ventilation and dust control(2016jpkcx17); Preparation and properties of new intumescent flame retardant low density polyethylene(2017zr007).

References

[1] Wang, X.SH., ZHang, G.SH. Prevention and Control of Mine Fire[M]. Xuzhou, China University of Mining and Technology Press,1990:20-27.

[2] Liu, Y.L., Peng, B.SH., Meng, W.J., et al. Goaf Nitrogen Injection Fire-fighting Technology of Shallow Buried Depth Coal Seam[J].Safety in Coal Mines,2017, 48(5):73-76.

[3] Wu, Y.G., Wu J.M., ZHang D.P., et al. Distribution law of gas and change rule of"three zones"in the goaf of fully mechanized top coal caving working face under the continuous nitrogen injection[J].Journal of China Coal Society,2011,36(6):964 -967.

[4] Gao, K., Liu, J., Liu, Y.J. Innovated nitrogen-injection project and numerical simulation in gob at steep inclined fully-mechanized mining face[J].Journal of Safety and Environment,2014,14(5):49-53.

[5] Pan, R.K., CHeng, Y.P., Yu, M.G., et al. New technological partition for "three zones" spontaneous coal combustion in goaf[J]. International Journal of Mining Science and Technology, 2013, 23(4): 489-493.

[6] ZHao, W.B., Wang, J.F., Min, Y.T., et al. Division on spontaneous combustion zone and judgement on influence zone of goaf under the condition of nitrogen injection[J].Coal Science and Technology,2016,44(3):60-64.

[7] Zhu, H.Q., Liu, X.K. Fire prevention effects and construction parameters of continuous nitrogen

injection in gob area[J].Journal of Liaoning Technical University: Natural Science Edition,2011,30(5):706-712.

[8] Deng, J., Ren, L.F., Ai, SH.W., et al. Influence of nitrogen injection on oxygenated spontaneous combustion zone in gob of fully mechanized caving face[J].Journal of XI'AN Uniwersity of Science and Technology,2017,37(4):468-469.

[9] Li, Z.X. Numerical simulation and parameter determination of nitrogen injection process in fire prevention and extinguishing in fully mechanized longwall top-coal goaf[J].China Safety Science Journal,2003,13(5):57-61.

[10] Yang, SH.Q., Xu, Q., Huang, J., et al. The"three zones"microcirculation theory of goaf spontaneous combustion and numerical simulation of the air leakage flow field[J].Journal of China University of Mining and Technology,2009,38(11):769-773.

[11] Guo, H.,Yuan, L. An integrated approach to study of strata behaviour and gas flow dynamics and its application[J].International Journal of Coal Science and Technology, 2015, 2(1): 12-21.

[12] Shi, G.Q., Wang D.M., ZHong, X.X., et al. CFD simulation of oxidation zones distribution in goaf of fully-mechanized caving face[J].Journal of Mining and Safety Engineering,2010,27(4):568-571.

[13] Shi, G.Q., Hu, F.K., Wang D.M., et al. Unsteady simulation of distribution of three zones for spontaneous combustion in goaf areas[J].Journal of China University of Mining and Technology,2014, 43(2):189-194.

[14] Ti, ZH.Y., Qin, H.Y., Qiao, N., et al. Simulation of slurry diffusion involved in fire prevention in large dip angle fully mechanized caving gob by grouting[J]. China Safety Science Journal,2016, 26(1):51-57.

[15] Si, L.L., Ju, G.G., Wang, M., et al. Study on fire prevention technology with grout consolidation at mining terminal line of coal mining face in steep inclined seam[J].Coal Science and Technology,2016,44(03):80-85.

[16] Liu, W.Y.,Wang, W.G., ZHang, Z.T.,et al.Experimental report on spontaneous combustion of coal in Bojiang Haizi Coal Mine[R].Technology Energy College of Xi'an University of Science,2014,2-6.

[17] Luo, X.R., Tang, G.CH., Li, Y.W., et al. CFD model based design of both gas drainage and nitrogen injection against fire in goaf[J].China Safety Science Journal,2016,26(6):70-71.

IOP Conf. Series: Materials Science and Engineering **729** (2020) 012014 doi:10.1088/1757-899X/729/1/012014

Research progress of modified macroporous adsorption resin in adsorption separation

Tao Sun

School of Chemistry and Chemical Engineering, North Minzu University, Yinchuan, China

Corresponding author's e-mail: 2633219607@qq.com

Abstract: Macroporous adsorption resin (MAR) is widely used in molecular adsorption and material separation due to its unique porous nature, surface activity and easy modification. Although there may be multiple synergistic effects in the adsorption process, its adsorption capacity is weak, and it is difficult to meet the problem of separation and purification of complex samples. The functional modification of macroporous adsorption resin is an ideal way to improve its separation efficiency and selectivity. In this paper, recent advances in the preparation of modified macroporous adsorption resins and their application in adsorption separation are reviewed, which provides a reference for further research.

1. Introduction

Macroporous adsorption resin (MAR) is a new type of high molecular polymer adsorption functional material that has been rapidly developed in recent years,because its high degree of cross-linking, spatial network structure, porosity, large specific surface area, etc. It has important application value in the field of adsorption and separation of natural products and sewage treatment.However, there are many kinds of commercialized MARs. For different sample components, the workload is huge when predicting and screening MARs with the best separation efficiency. In addition, the preparation process of MARs is special, the structure is complex, and the pore structure parameters vary widely. Sexuality makes it difficult to carry out theoretical research on its adsorption. However, the research on the structure-activity relationship between MARs and target molecules and its separation law is weak. It is difficult to realize the design and preparation of MARs, which makes the selectivity and specificity of MARs not strong, and the adsorption separation effect is not high.

Therefore, in order to obtain MARs with higher adsorption capacity and higher selectivity, designing resin structure for different target components, functional modification of its surface and structure, and improving adsorption performance are the hotspots and focuses of people. In this paper, recent advances in the application of functional group modified MARs to adsorb and remove different compounds have been reviewed.

2. MARs surface graft modification

The commonly used MARs are non-polar resins, and the adsorption driving force is mainly hydrophobic, and the selectivity is low. In order to improve its adsorption performance and selectivity, a specific functional group is often introduced into the MARs to achieve separation of the target components. However, the direct substitution of the benzene ring can result in less functional groups grafting, which is not conducive to modification. Therefore, Friedel-Crafts catalyzed[1] and Blanc[2] chloromethylation are one of the main methods for functional modification of MARs substrates.

IOP Conf. Series: Materials Science and Engineering **729** (2020) 012014 doi:10.1088/1757-899X/729/1/012014

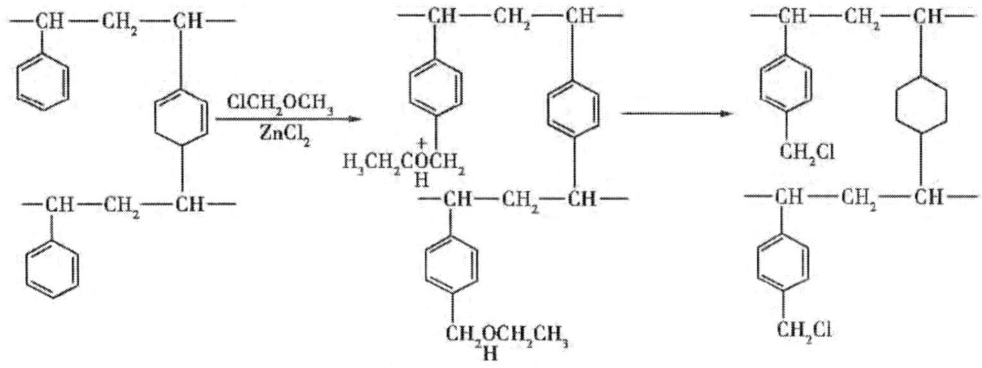

Figure 1 Schematic diagram of the preparation process of chloromethylation

2.1. Amino modified resin
Lou et al[3,4] grafted amino groups by reacting chlorine spheres with chloromethylated D101 [5] with excess triethylenetetramine. Ye et al[6] modified the grafted amino group with chlorosphere and hexamethylenetetramine. Li Aimin et al[7] used chlorine with a chlorine content of 18.5% to 19.0%, Wang Dexing[8] with chlorine containing 17.5% chlorine, Zhang et al[9] with a cross-linking degree of 8% and a chlorine content of 3.8 mmol/g. The ball is the substrate, Xie Xingyu et al[10] reacted with toluene to prepare a toluene ultrahigh crosslinked macroporous resin with a chlorine content of 6.38%, and then chemically treated with 2-aminopyridine. Modified, grafted amino. Ma et al[11] grafted glycidyl methacrylate onto chloromethylated polystyrene resin by surface-initiated atom transfer radical polymerization followed by 5-aminosalicylic acid and 2-(4-imidazolyl) Ethylamine is modified to graft the amino group. Xiao et al[12] modified chloromethylated polystyrene resin by 2-aminopyridine. Xu et al[13] prepared a composite functional resin by further introducing nitro and amine groups into the chloromethylated polystyrene skeleton by optimizing the conditions of the Friedel-Crafts reaction.

2.2. Phenolic hydroxyl modified resin
Huang[14] and Hu[15] prepared a phenolic hydroxyl group modified by macroporous cross-linked chloromethylated styrene-divinylbenzene copolymer by Friedel-Crafts post-crosslinking reaction and esterification reaction. Supercrosslinked polymer adsorbent. He et al[16] prepared a sorbent containing multiple phenolic hydroxyl groups from a chloromethylated polystyrene resin by Friedel-Crafts reaction. Wei et al[17] grafted phenol onto the surface of non-porous gel-type chloromethylated polystyrene by Friedel-Crafts post reaction. Qi Fengpei et al[18] used chloromethylated cross-linked polystyrene microspheres as starting materials to react with p-hydroxybenzaldehyde to prepare hydroxylated modified polystyrene microspheres.

2.3. Other group graft modified resin
Long Yu et al[19] chloromethyl polystyrene microspheres were bonded to naphthalene and anthracene with strong conjugation effect by Friedel-Crafts reaction, and two new types of macroporous super high crosses were synthesized by post-crosslinking reaction. Copolystyrene resin. Cai et al[20] synthesized a novel bifunctional polymeric resin by introducing sulfonic acid groups into the surface of the resin during the cross-linking of chloromethyl low-crosslinking macroporous polystyrene resin. Li et al[21] synthesized two carbonyl-modified XAD-4 resins by chemically modifying Amberlite XAD-4 with carbon tetrachloride and benzoyl chloride. Li[22] chloromethyl macroporous polystyrene resin was used as the substrate, and a novel formaldehyde carbonyl modified self-crosslinking polystyrene resin was synthesized by Friedel-Crafts reaction. Huang Xiaojuan et al[23] used a macroporous ultra-high crosslinked polystyrene resin for nitration reaction to obtain a nitrification

IOP Conf. Series: Materials Science and Engineering **729** (2020) 012014 doi:10.1088/1757-899X/729/1/012014

modified macroporous ultra-high crosslinked polystyrene resin.

In summary, the basic path of functional graft modification of polystyrene resin is to first activate the polystyrene resin substrate, generally using chloromethylation, acylation or bromomethylation; The functional group required for the branch is a modified resin which prepares a series of functional groups.

3. Application of modified MARs

Xu et al[13] used chloromethyl polystyrene resin as substrate to further introduce nitro and amine groups into the framework by Friedel-Crafts reaction to prepare phenol adsorption behavior. The results showed that the modified resin was phenol. The adsorption capacity of the compound was 412.9 mg / g, which was 16.4 to 47.7% higher than that of a commercially available resin having a similar physical structure. Wang et al[24] grafted four groups of 2-pyrrolidone, 2-imidazolidinone, trimellitic anhydride and sulfamic acid onto the surface of chloromethyl polystyrene resin by Friedel-Crafts reaction, and 3-amino in sewage. The adsorption behavior of acetanilide was studied. As a result, the adsorption capacity of the modified resin on 3-aminoacetanilide in aqueous solution was improved compared with that before modification, and the modified resin of grafted 2-pyrrolidone was effective from aqueous solution. 3-aminoacetanilide was removed. Li et al[22] have polyphenolic groups and hydrophobic structures for licorice in licorice, and introduced amino groups on chloromethyl polystyrene resin, which has a adsorption capacity of 45.5 mg/g, which is significantly higher than that of licorice. Adsorbent before (15.2 mg/g) without modification. Ye et al[6] used chloromethylated styrene-co-divinylbenzene copolymer as substrate, modified by hexamethylenetetramine by amination reaction, and obtained a new adsorbent with amino group. Continuous modification of formylphenylboronic acid to obtain modified resin with phenylboronic acid group, the adsorption of rebaudioside A and stevioside on chloromethyl, amino and phenylboronic acid resin in chamomile was studied. It is indicated that the resin having an amino group and a phenylboronic acid group preferably adsorbs stevioside instead of rebaudioside A. Yan et al[25] a porous polymer antibacterial resin containing an N-halamine functional group and subjected to an antibacterial test. As a result, the modified resin Escherichia coli (E. coli) and Staphylococcus aureus (S. aureus) have Strong antibacterial effect. Liu et al [5] used agaric catechin gallate (EGCG) and caffeine (CAF) in discarded tea to introduce amino and aniline onto chloromethyl polystyrene resin, and aniline added more to EGCG. Hydrogen bond and π-π interaction reduce the hydrophobic effect on CAF, and the adsorption performance is significantly different, thus effectively removing CAF impurities in EGCG; introducing phenol, increasing the adsorption active sites of EGCG and CAF, and improving separation ability [26] .

The non-polar polystyrene type MARs and the adsorbate are mainly π-π interaction and hydrophobic interaction. The introduction of an amino group easily forms a hydrogen bond or an electrostatic interaction with the hydroxyl group on the adsorbate, and the hydrophilicity of the surface of the resin is also improved, which facilitates the entry and adsorption of the hydrophilic component; the introduction of the aniline and the nitrogen atom is more easily associated with the hydrogen-containing bond donor. The organic matter forms a hydrogen bond; the large π bond of the benzene ring can participate as a hydrogen bond acceptor to form a hydrogen bond; at the same time, the pore structure and distribution state also change, the molecular sieve principle changes, and the separation efficiency and adsorption performance also change. Therefore, the introduction of different functional groups on the surface of the polystyrene resin substrate according to different target components is an effective way to improve adsorption performance and selectivity.

4. Prospecive Future

At present, the research on the modification method of polystyrene resin is mostly used in the treatment of environmental wastewater. The natural products are not much involved, but the graft modification method still has great reference value. The polarity, hydrophilicity, pore size distribution and adsorption of MARs changed after graft modification. These factors together led to the change of adsorption efficiency of target components, which provided a way of thinking about the separation of MARs.

With the gradual deepening of natural product research, efficient separation and purification technology is an urgent problem to be solved. Conventional MARs are difficult to meet when obtaining active ingredients with low content, high added value, and difficulty in obtaining. Therefore, the design and preparation of specific MARs for the target components is a new opportunity for the separation and purification of specific active components in natural products.

References

[1] Gao B,Liu Q,Jiang L.(2008)Studies on performing chloromethylation reaction for polystyrene by micellar catalysis in aqueous surfactant solutions[J].Chemical Engineering and Processing: Process Intensification, 47(5):852-858.

[2] Ma Ping, Tian Lili.(2011)Blanc Chlorine Methylation Polystyrene Preparation and Conditions Screening [J].Guangdong Chemical Industry,38（11）: 209-210

[3] Lou Song,Chen Zhenbin,Liu Yongfeng,et al.(2012)Synthesis of Functional Adsorption Resin and Its Adsorption Properties in Purification of Flavonoids from Hippophae rhamnoides L. Leaves[J].Ind Eng Chem Res,51(6):2682-2696.

[4] Lou Song,Chen Zhenbin,Liu Yongfeng,et al.(2011)New Way to Analyze the Adsorption Behavior of Flavonoids on Macroporous Adsorption Resins Functionalized with Chloromethyl and Amino Groups[J].Langmuir,27(15):9314-9326.

[5] Liu Yongfeng, Bai Qingqing, Lou Song, et al.(2012) Adsorption Characteristics of (−)-Epigallocatechin Gallate and Caffeine in the Extract of Waste Tea on Macroporous Adsorption Resins Functionalized with Chloromethyl, Amino, and Phenylamino Groups [J].J Agric Food Chem,60(6):1555-1566.

[6] Fayin Ye,Ruijin Yang,Xiao Hua,et al.(2014)Adsorption characteristics of rebaudioside A and stevioside on cross-linked poly(styrene-co-divinylbenzene) macroporous resins functionalized with chloromethyl, amino and phenylboronic acid groups [J].Food Chemistry,159:38-46.

[7] Jinnan Wang,Li Xu,Aimin Li,et al.(2010)Adsorption Study of Gallic Acid by Hyper Cross-linked Resin Modified by Amino Function Groups [J].Chemical Journal of Chinese Universities,31(1):193-198.

[8] Dexing Wang.(2016)Synthesis of Amino-modified Ultra-high Crosslinking Resin and Its Adsorption of Dimethyl Phthalate [J].Resource Information and Engineering,31(04):97-98.

[9] Zhang W M,Dua Q,Pan B C,et al.(2009)Adsorption equilibrium and heat of phenol onto aminated polymeric resins from aqueous solution[J].Colloids and Surfaces A:Physicochemical and Engineering Aspects,346(1/2/3):34-38.

[10] XIE Xing-yu,WANG Bin,QIN Hai-jiao,et al.(2016)Synthesis of Toluene Hypercross-Linked Macroporous Resins Modified by Aminopyridine and Their Adsorptive Properties [J].Chemistry and Bioengineering,1:26-29.

[11] Ma Jingchong,Shen Jiwei,Wang Chaozhan,et al.(2018)Preparation of dual-function chelating resin with high capacity and adjustable adsorption selectivity to variety of heavy metal ions [J].Journal of the Taiwan Institute of Chemical Engineers,91:532-538.

[12] Guqing Xiao,Ruimin Wen.(2019)Comparative adsorption of glyphosate from aqueous solution by 2-aminopyridine modified polystyrene resin, D301 resin and 330 resin: Influencing factors, salinity resistance and mechanism [J].Fluid Phase Equilibria,411:1-6.

[13] Xu Chao,Jiang Long,Qin Xiaoli,et al.(2109)Enhancement mechanism behind the different adsorptive behaviors of nitro/amine modified hypercrosslinked resins towards phenols [J].Journal of the Taiwan Institute of Chemical Engineers,102:340-348.

[14] Jianhan Huang,Xiaomei Wang,Xing Deng.(2009)Synthesis, characterization, and adsorption properties of phenolic hydroxyl group modified hyper-cross-linked polymeric adsorbent [J].Journal of Colloid and Interface Science,337:19-23

[15] Huanxiao Hu,Xiaomei Wang,Shengyong Li,et al.(2012)Bisphenol-A modified hyper-cross-linked polystyrene resin for salicylic acid removal from aqueous solution: Adsorption equilibrium, kinetics and breakthrough studies [J].Journal of Colloid and Interface Science,372:108-112.
[16] Chunlian He,Kelong Huang,Jianhan Huang.(2010)Surface modification on a hyper-cross-linked polymeric adsorbent by multiple phenolic hydroxyl groups to be used as a specific adsorbent for adsorptive removal of p-nitroaniline from aqueous solution [J].Journal of Colloid and Interface Science,342:462-466.
[17] Wei Kuang,You-NianLiu,Jianhan Huang.(2107)Phenol-modified hyper-cross-linked resins with almost all micro/mesopores and their adsorption to aniline [J].Journal of Colloid and Interface Science,487:31-37.
[18] Qi Fengpei, Hu Yifan, Ou Panlin,et al.(2019)Study on the Adsorption Properties of Modified Polystyrene Microspheres Containing Polyhydroxy Compounds for Pb2+ [J].Journal of Hunan City University(Natural Science),4:65-59.
[19] LONG Yu LIAO Hui-lin FU Yu-li,et al.(2018)Synthesis of Polystyreme Nenaphthalene Resin and Its Adsorption Property for Theophylline [J].Fine Chemical Intermediates,4:47-50
[20] Cai Jianguo,Li Aimin,Shi Hongyan,et al.(2005)Adsorption characteristics of aniline and 4-methylaniline onto bifunctional polymeric adsorbent modified by sulfonic groups [J].Journal of Hazardous Materials,124:173-180.
[21] Chengyong Li,Maowen Xu,Xiucheng Sun,et al.(2013)Chemical modification of Amberlite XAD-4 by carbonyl groups for phenol adsorption from wastewater [J].Chemical Engineering Journal, 229:20-26.
[22] Xiaoting Li，Yongfeng Liu，Duolong Di，et al.(2016)A formaldehyde carbonyl groups-modified self-crosslinked polystyrene resin: Synthesis, adsorption and separation properties [J].Colloids and Surfaces A: Physicochemical and Engineering Aspects,500:1-9.
[23] HUANG Xiao-juan,LONG Yu,FU Yu-li,et al.(2017)Study on Nitrification and Adsorption Properties of Macroporous Hypercross-linked Polystyrene Resin [J].Fine Chemical Intermediates, 47(06):37-42
[24] Tao Wang,Chunhui Shena,Nan Wang,et al.(2019)Adsorption of 3-Aminoacetanilide from aqueous solution by chemically modified hyper-crosslinked resins: Adsorption equilibrium, thermodynamics and selectivity [J].Colloids and Surfaces A: Physicochemical and Engineering Aspects,575:346-351.
[25] Shulan Yan,Xiufang Yan,Xuexiang Shao,et al.(2015)Porous polymeric antimicrobial resin containing N-halamine functional groups [J].Reactive and Functional Polymers,96:71-77.
[26] Xiaofeng Zhang,Yi Xu,Qing Zhang,et al.(2016)Simultaneous separation and purification of (−)-epigallocatechin gallate and caffeine from tea extract by size exclusion effect on modified porous adsorption material [J].Journal of Chromatography B,1031:29-36.

Surface/Deep Defects manipulated Fluorescence Properties and LED application of Copper Nanoclusters

Lin Ai[1,2], Ji Li[2], Min Tang[2], Hsiaohsien Chen[2] and Hong Meng[1*]

[1]School of Advanced Materials, Peking University Shenzhen Graduate School, Guangdong Shenzhen 518055, CHN

[2]Shenzhen China Star Optoelectronics Technology Co., Ltd. Guangdong Shenzhen 518132, CHN

[*]Corresponding author's e-mail: chaofenzi2015@163.com

Abstract. The contribution of metal defects to the emission enhancement of Cu nanoclusters has been proposed. But the insufficient understanding of the defects luminescence mechanism limits its further effective adjustment and control to the emission color. In this work, the surface metal defects of the assembled fluorescent materials are deliberately engineered. Metal defects have effects on luminescent energy levels, which are relative to the dynamic process of exciton radiation transition and regulate the photophysical properties of the materials. Further studies demonstrate the surface metal defects possess shallow energy level and multi sublevels, broaden the emission colors. Moreover, the continuous alteration of emission color is achieved by simply adjusting the relative ratio of the surface and deep metal defects. Based on the full-color emission of copper clusters, the white light emitting diodes (WLEDs) with different color temperatures (warm, cold, and pure white) are prepared for illumination application.

1. Introduction

As new generation fluorescent nanomaterials, ultra-small copper nanoclusters have attracted enormous interests in the applications of illumination and display on account of their non-poisonous, large amount of storage in earth, simple synthetic route, and relative low cost[1-4]. Recent 3 years, thanks to the self-assembly induced emission enhancement (SAIEE), the fluorescent quantum yields (QYs) of the copper clusters are greatly improved to 37%, comparing to previous 1%~10%. However, as the fluorescence mechanism is still an open topic, the confused mechanism makes it difficult to systematically regulate the emission color. In addition, the methods to control the emission color are finite and complicated, always involving varying size of the metal core by utilizing the quantum size effect[5], or heterogeneous metal doping to influence the ligand to metal charge transfer or ligand to metal-metal charge transfer (LMCT/LMMCT) process[6]. The scarce regulatory methods and narrow adjustable luminous bands greatly limit the associated academic researches and practical applications comparing to the competitive alternatives of organic phosphors and quantum dots (QDs).

Lately, the contribution of the metal defects to the regulation of the fluorescent properties has been investigated by our group[7], the defects-related Cu(I) rich structures influence the LMMCT process and facilitate the radiative relaxation of excitons, which drastically heightens the emission intensity. Besides, the introduction of the defect energy level shows a bright prospect for tailoring the emitting colors. As for the ambiguous fluorescence mechanism, further control and exploration are in progress.

The defects-related emission mechanism in QDs field is very well depicted to regulate emission intensity and color, and it is inspiring to observe similar behavior for quantum-sized copper clusters. It is comparatively difficult to construct atom missing and dangling bonds as the ultrasmall clusters composed of only several to several tens atoms. Surprisingly, through deliberately altering the capping ligands, the partly modified coordination environment of Cu atoms alters the surface properties of Cu NCs. The induced surface defects possess shallow defects energy level and multi sublevels, which is differing from the previous reported deep defects emission, making the emission color show significant bathochromic-shift. Remarkably, the emission color experience continuous variation by intentionally adjusting the relative ratio of the surface and deep metal defects.

In this work, solely altering the capping ligands of the copper nanoclusters, we can engineer the deep and surface defects-rich materials by renovating the coordination condition of Cu atoms. Further studies corroborate the metal defects make correlated radiative transition energy levels, which arouses the additional electron relaxation process from the intrinsic LMMCT-determined triplet state to the metal defects-related intermediate state. This process interferes the luminescence properties of copper clusters, making a substantial change in the emitting color. Additionally, the relative proportion of the two kinds of defects can be adjusted, resulting in the continuously changed emitting colors. Based on the obtained manifold color-emitting of fluorescent powders, the white light emitting diodes (WLED) by composition of pure copper clusters are prepared with tunable color temperature.

2. Materials and methods

2.1. Chemicals and reagents

Copper(II) chloride ($CuCl_2$, 99.0%, A. R.) was purchased from Alfa Aesar. Hexanethiol, hexanedithiol, and mercapto-1-hexanol were purchased from TCI. Alcohol, acetone and chloroform were all commercially available products and used as received without further purification.

2.2. Instrument and characterization

Transmission electron microscopy (TEM) was conducted using a Hitachi H-800 electron microscope. Photoluminescence (PL) spectroscopy was performed using a Shimadzu RF-5301 PC spectrophotometer. The spectra of the LEDs were measured under ambient conditions at room temperature by combining a Spectrascan PR-650 spectrophotometer with an integrating sphere and a computer-controlled direct-current power. The color of the light was identified using the CIE (Commission Internationale de L'Eclairage 1931) calorimeter system. Any color could be described by the chromaticity (x, y) coordinates on the CIE diagram.

2.3. Preparation of copper clusters fluorescent nanomaterials

$CuCl_2$ 100 mg was dissolved in 20 mL alcohol at room temperature. Different kinds of thiols capping ligands (10 mmol), which acted as a ligand cum reductant, were added and stirred at room temperature for 10 min. After reaction finished, 1 mL reaction suspension of the products was washed and precipitated by adding 2 mL chloroform and 4 mL acetone. Then, the products were centrifuged by 4500 rpm for 8 min. The purification process was repeated twice. Finally, the products were redispersed in 2 mL chloroform.

3. Results and discussion

Cu NCs were prepared in alcohol as the solvent, using different kinds of thiols as the capping ligand at 30 °C. Interestingly, we get three emission colors, corresponding to blue, yellow and red (Figure 1d). To investigate such fluorescent materials thoroughly, transmission electron microscopy (TEM) shows that the fluorescent materials are as-assemblies of Cu NCs (Figure 1a-c). High magnification TEM image confirms that the assemblies are composed of ultrasmall Cu NCs with the average diameter of <2 nm (Figure 2).

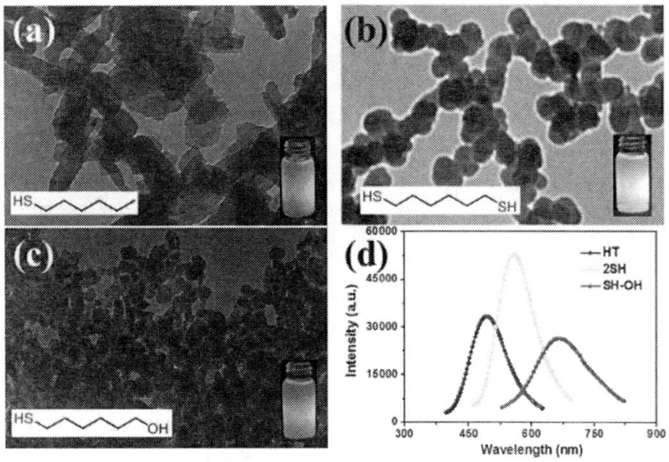

Figure 1. TEM images for blue emission (a), yellow emission (b) and red emission (c) assemblies of copper clusters, the corresponding photo images and capping ligands structures are shown inset. The fluorescence spectrum of assemblies.

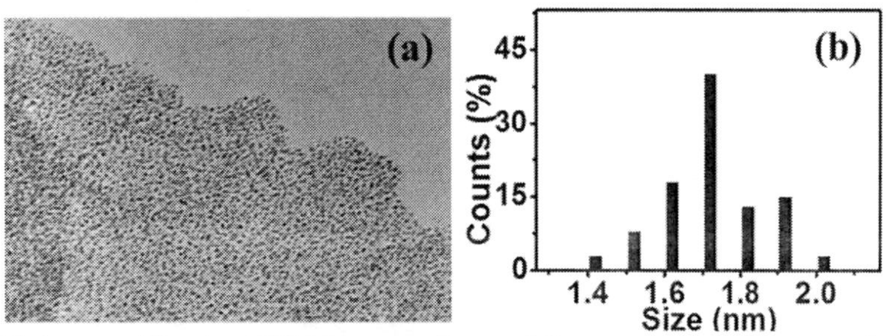

Figure 2. (a) High magnification TEM images of assemblies composed of copper clusters. (b) The corresponding size calculated by 200 spots of Cu clusters in (a).

The fluorescence mechanism of tremendously disparate emitting colors is explored. Obviously, the employed capping ligands structures are key points. The capping ligands can alter the coordinate micro environment of copper source. The over reducing power of ligands makes imperfect clusters, thus facilitating to form metal defects. So, the hexanedithiol ligands easily makes deep defect. In contrast, the OH group in mercapto-1-hexanol can adsorb on the copper atom, it prefers to form surface defects at the interface where the ligands contact with the copper core. In consideration of previous work and theoretical calculation[7], besides the contribution to the relaxation of excited electrons via a radiative pathway as of the increased Cu(I), the metal defects also can generate additional energy level and narrow the emission band gap, inducing bathochromic-shift of the emission color (Figure 3).

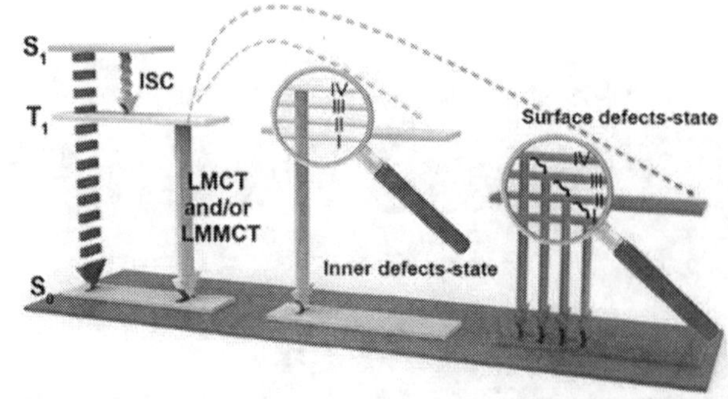

Figure 3. Energy level of materials with blue, yellow and red emitting, where S0, T1, and S1 represent ground state, excited triplet state, and singlet state. I, II, III, and IV represent the sublevels of defect state. ISC represents the intersystem crossing process. The dotted arrow represents the luminous excitons relax to the defect state from the intrinsic triplet state.

Next the deep and surface defects should be proved further. As the surface defects are generally more sensitive to the environment, when raising the reaction temperature, due to the instability of the defects, the higher reaction temperature will gradually remove the defects, the defects luminescence will gradually quench out. The fluorescence of the metal triplet states completely replaces the one from metal defects, making the blue fluorescence appear. When reaction is conducted at high temperature of 120 °C, the red emitting materials need only 20 min while the yellow one needs nearly 1 h for the restoration of the intrinsic emitting color. That is the effect of temperature on shallow defects is greater, so the light color changes more rapidly (Figure 4).

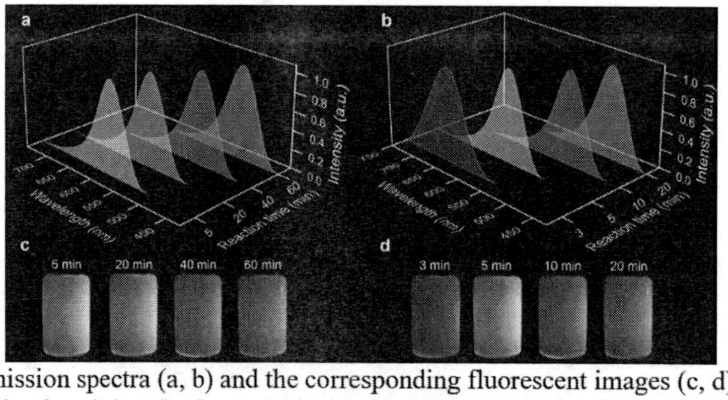

Figure 4. PL emission spectra (a, b) and the corresponding fluorescent images (c, d) of the yellow-emitting (a, c) and red-emitting (b, d) assemblies prepared at 120 °C at different reaction time. 365 nm excitation is adopted.

Here, the continuous adjustment of emission colors is also achieved in this work by adjusting the two kinds of defects. Found that just simply alter the varieties and proportion of capping ligands, the continuous changes of emission color from yellow to red are present. From Figure 5, various proportions between hexanedithiol and mercapto-1-hexanol bring gradient changing of the fluorescent colors. When the ratio of the mercapto-1-hexanol rises, it is easier to form surface defects relatively, consequentially, more red shift of the emission color appears, demonstrating the construction of defects has reached a certain degree of controllability.

IOP Conf. Series: Materials Science and Engineering **729** (2020) 012015 doi:10.1088/1757-899X/729/1/012015

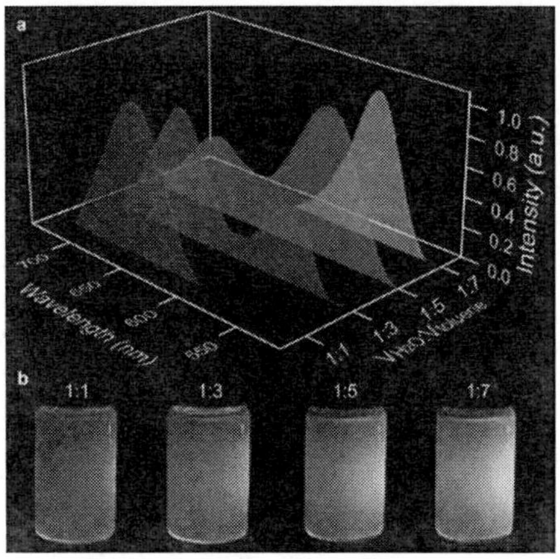

Figure 5. PL emission spectra (a) and the corresponding fluorescent images (b) of the assemblies prepared in hexanedithiol: mercapto-1-hexanol = 1:1, 1:3, 1:5 and 1:7 as capping ligands with 365 nm excitation.

As the simple synthesis, no fluorescence resonance energy transfer (FRET) and excellent stability, the obtained fluorescent powders are suitable for preparing WLED with mixed corresponding emission color. As of the well manipulative emission colors, these powders are mixed by the ratio of 2.5/0.5/1.2/1, 2/0.7/1.3/1.2 and 1.7/0.6/1.5/1.3, the high-performance WLED are constructed with cold (9587 K), natural (5122 K) and warm (3214 K) color temperature for illumination application (Figure 6).

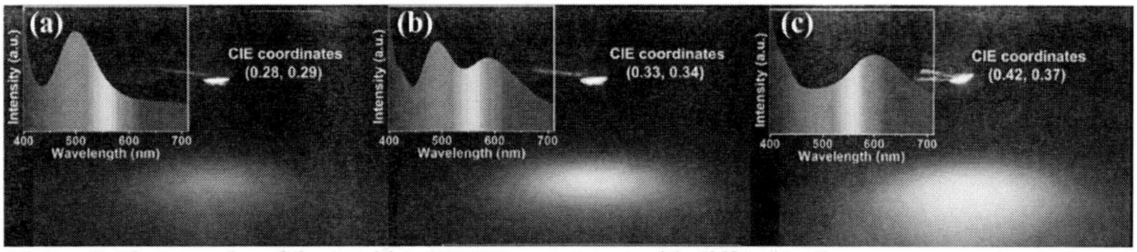

Figure 6. Photographs of the working WLED prototypes with different color temperature (a–c). Insets in (a–c) are the corresponding emission spectra of WLEDs.

4. Conclusion

In summary, the emission color is well manipulated by intentionally inducing the metal defects. The defects energy levels can influence the radiation transition photodynamic process, which is greatly related to the emission property. The surface/deep defects of metal clusters are constructed by means of utilizing different reducing power of capping ligands. Simultaneously, through controlling the defects, the continuous change of emission color is realized, providing a new method for the adjustment of emission color, which facilitates to devise high performance WLED architectures.

Acknowledgments

This work was supported by the National key research and development program of China (2016YFB0401701), NSFC (51425303, 11674123), the 973 Program of China (2014CB643503), and the Special Project from MOST of China.

References

[1] Shang, L., Dong, S., Nienhausa, G.U. (2011) Ultra-small fluorescent metal nanoclusters: Synthesis and biological applications. Nano Today, 6: 401−418.

[2] Zhang, L., Wang, E. (2014) Metal nanoclusters: New fluorescent probes for sensors and bioimaging. Nano Today, 9:132−157.

[3] Wu, Z., Liu, J., Gao, Y., et al (2015) Assembly-Induced Enhancement of Cu Nanoclusters Luminescence with Mechanochromic Property. J. Am. Chem. Soc., 137:12906−12913.

[4] Niesen, B., Rand, B. P. (2014) Thin Film Metal Nanocluster Light-Emitting Devices. Adv. Mater., 26:1446–1449.

[5] Gan, Z., Lin, Y., Luo, L., et al (2016) Fluorescent Gold Nanoclusters with Interlocked Staples and a Fully Thiolate-Bound Kernel. Angew. Chem. Int. Ed., 55:11567–11571.

[6] Kang, X., Wang, S., Song, Y., et al (2016) Bimetallic Au_2Cu_6 Nanoclusters: Strong Luminescence Induced by the Aggregation of Copper(I) Complexes with Gold(0) Species. Angew. Chem. Int. Ed., 55:3611–3614.

[7] Wu, Z., Liu, H., Li, T., et al (2017) Contribution of Metal Defects in the Assembly Induced Emission of Cu Nanoclusters. J. Am. Chem. Soc., 139:4318−4321.

IOP Conf. Series: Materials Science and Engineering **729** (2020) 012016 doi:10.1088/1757-899X/729/1/012016

Research on High Efficiency IO Control Technology of Seismic Data Based on Data Segmentation

Liting Liu[1], Xi Zhang[2]*, Yu Chen[1]

[1] BGP Southwest Geophysical Company, CNPC, Huayang road, Chengdu, Sichuan, 610213, China

[2] BGP Research & Development Center Chengdu Branch Center, CNPC, Chengdu, Sichuan, 610213, China

*Corresponding author's e-mail: zhangqian_wt@cnpc.com.cn

Abstract. With the development of geophysical prospecting technology, seismic operation processing is terabytes of data in actual production. A seismic operation processing time can be as long as ten days. In order to complete these processing tasks in as short a time as possible, efficient use of node resources is needed. Therefore, it is necessary to segment and decouple the input data in order to improve the computational efficiency when carrying out such large amount of data, intensive calculation and long running time operations. IO control system can decompose such jobs into several sub-jobs for parallel processing, so as to shorten the running time, improve the efficiency of seismic data processing and enhance the computing performance of cluster system.

1. Introduction

With the development of geophysical prospecting technology, seismic operation processing is terabytes of data in actual production. A seismic operation processing time can be as long as ten days. In order to complete these processing tasks in as short a time as possible, efficient use of node resources is needed. Therefore, it is necessary to segment and decouple the input data in order to improve the computational efficiency when carrying out such large amount of data, intensive calculation and long running time operations.

2. Data separable job

Efficient IO control of seismic data is a necessary step in the process of seismic data processing. Job scheduling can allocate available computing resources for submitted jobs and deal with abnormal situations when jobs are running. Job management through the platform can improve the efficiency of computer utilization and the computing performance of cluster system.

Separable jobs refer to jobs that can be divided into a number of sub-jobs, which are independent of each other and have no data dependence and operation dependence. The submission, scheduling, IO control part of this kind of job will implement job parsing, data segmentation, job distribution start-up, data collection, state feedback and other functions. For the operation of data separable jobs, the main control program is needed to complete the operation of data separable jobs. The sub-jobs are sent to the sub-nodes in the form of common job requests, which realizes the decomposition of jobs and parallel batch submission.

2.1. The structure framework of data separable job

Data separable job processing framework receives job requests from server to start job IO control process according to job scheduling system. Job IO control process parses job files, dynamically loads function modules to complete the processing of corresponding jobs, and returns the mode of job execution status to process jobs. Specifically, job scheduler sends jobs to the master node, and other idle computing nodes IP are sent to the master node. The master node starts the master IO control, analyses the data, calculates the number of sub-jobs and the data range of each sub-job according to the segmentation strategy described in the job. The master node partitions the job file. The master node distributes sub-jobs to other idle nodes to run sub-IO control system and perform sub-job processing. The sub-node starts IO to control process to read the data from the master node and send the calculation results back to the master node. The master node merges the temporary data files of the child nodes into the final output data files of the job.

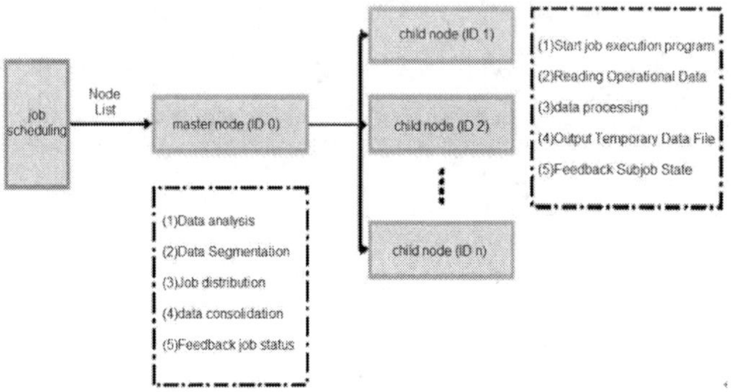

Figure 1. The structure framework of data separable job

2.2. IO control of data separable job

Data separable job IO control program is the underlying core program that describes, parses and submits operations for processing jobs. Data separable job IO control software provides a control execution software system that can complete job parsing, job IO control and job runtime information output function. It helps to improve the monitoring ability of job operation, enhance job runtime information output ability and exception handling ability.

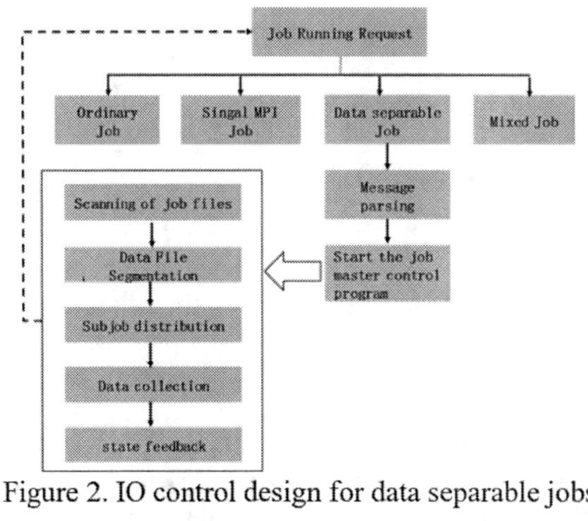

Figure 2. IO control design for data separable jobs

The main technologies of data separable job IO control system implementation process are as follows:

(1) Job Parsing Technology: Parsing input and output data file names from job files.

(2) Data Segmentation Technology: According to different segmentation strategies, the file is divided.

(3) Job Distribution Startup Technology: The distribution of sub-jobs is initiated through job scheduling system. Create job run requests and distribute them to each node. The job type in the job run request is set to the normal job type. When a node receives a job execution request, it calls the execution function according to the common job type to start the job execution process.

(4) Dynamic Data Collection Technology

After the sub-job process runs, it feeds back the completion status of job execution to the primary node. According to the temporary file strategy of sub-job data, the master node merges the temporary files output by sub-job into one data file. At the same time, the temporary data file of the sub-job is deleted and the completion status of the job is reported to the server.

(5) State Feedback Technology

In addition to reporting the current status to the server, the sub-job process also reports the current status to the primary node. The master node reports the status of job completion to the server after completing data merging.

3. IO Control System for Data Separable Job

IO control program is an important part of modern geophysical prospecting processing software. It is the core of processing system. IO control system organically links all processing modules in series to form processing flow and complete seismic data processing tasks. The main task of IO control program is to analyze and submit the operation of work flow.

In the process of seismic data processing, the processors may use hundreds of functional modules. These processing function modules need to be integrated into the workflow, run directly on the computing nodes or run in batches. At this time, we need a job IO control system program to describe, translate and control the execution of the jobs defined by the processors, and finally get the seismic data processed by the functional modules. Job IO control system can solve the problems of weak IO control, poor exception handling ability and inconvenience for job management and scheduling in pipeline mode. By strengthening job operation control, information output ability and monitoring ability of job operation are enhanced, and exception handling ability of job operation is improved.

3.1. Structure framework of job IO control program

Operational IO control system consists of four parts (as shown in Figure 3): (1) Job interactive editing software, which mainly completes job editing and submission; (2) Job server and scheduler, complete job status management and scheduling; (3) Calculating the node residence program, completing the resource acquisition and starting the job control process; (4)Job IO control program, realizing the specific operation of the job. The communication platform is responsible for the communication between the above four parts, which is realized by sending messages, in order to control the operation of complex operations and complete seismic data processing.

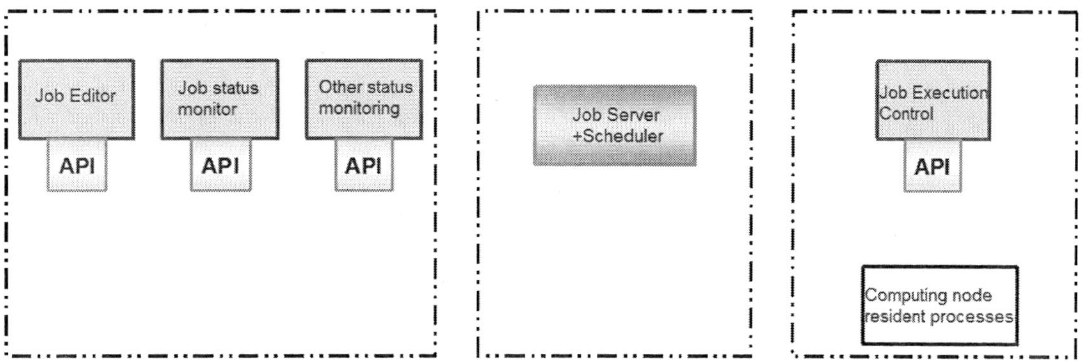

Figure 3. Composition of operational IO control system

IOP Conf. Series: Materials Science and Engineering **729** (2020) 012016 doi:10.1088/1757-899X/729/1/012016

Job IO control system is the underlying core program that describes, parses and submits operations for processing jobs. Job IO control system software provides a control execution software system that can complete job parsing, job IO control and job runtime information output function. It can improve the monitoring ability of job operation, enhance job runtime information output ability and exception handling ability. The overall framework of job IO control is shown in Figure 4.

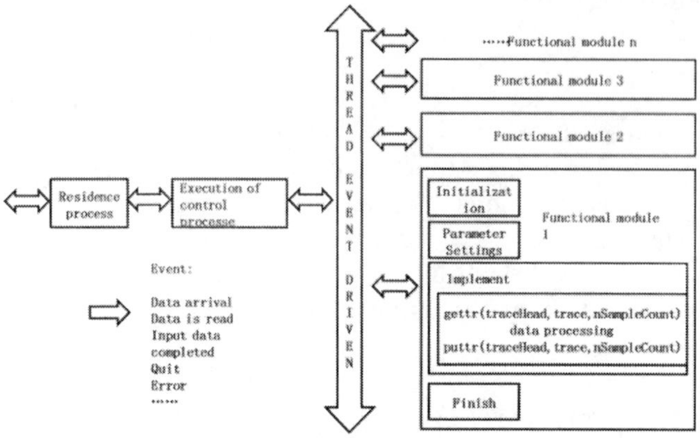

Figure 4. Overall structure framework of job IO control

3.2. Implementation of job IO control program

Node resident process receives job requests from server to start job IO control process. Job IO control process parses job files, dynamically loads function modules to complete the processing of corresponding jobs, and returns job execution status. Five modules are designed to complete the function of IO control program, including functional module base class, functional module dynamic loading, job parser, data adapter and IO control main program module.

Function module base class provides a unified programming interface for all functional modules, provides thread time-driven, and default processing of various events. At the same time, it provides functions that other functional modules will call (data processing main cycle, function module thread processing, debugging output, etc.).

Job (function module) has its inherent characteristics compared with conventional programming: the name of the executor (dynamic link library) of the function module is known only at runtime; the topological connection mode of the module is known only at runtime; and the parameters of the function module are known only at runtime. The system uses dynamic loading technology and parameters of dynamic configuration function module to solve the above problems.

Job parser can parse all parameters of Xml job file. Job parser is called by IO control program to analyze job steps, function modules, topological connections and parameter information.

Data adapter is mainly used for matching data formats of functional modules, so that seismic data can be read and calculated normally.

IO controls the main program to complete job parsing, module dynamic loading, topological connection, parameter setting, function module thread start, multi-thread running job, waiting for exit of the job process.

At the same time, each part is linked through the operation file, and the coupling degree between them is small, which is easy to maintain. At the same time, the operation IO control system software also has better robustness.

Dynamic loading job IO control technology system has strong IO control power. It can realize complete parameter analysis information output and job operation information output, provide effective exception handling mechanism, and meet the processing requirements of various complex jobs.

3.3. Cooperative operation of modular jobs

The cooperative operation technology of modular jobs driven by seismic data is very suitable for software systems centered on job processing. In seismic data processing, some processing operations are usually taken as a whole, and the input seismic data are output to the next processing step after the same operation. The smallest separable unit of input data is a seismic channel. According to the relationship between seismic traces and operational function modules, input data can be input to the input of functional modules in the way of one seismic traces in turn. In the form of data flow, a topological connection relationship among all functional modules in a job is established. For a functional module, as long as the data (channel) arrives, it can be calculated, which is conducive to improving the efficiency of seismic data processing.

The data between functional modules is mainly realized by input and output port classes, which manage the creation of shared memory. The subsequent function module fetches data from the output port buffer of the previous module through the input port, and the data processed by the operation of this function module is written into the output port buffer of this function module.

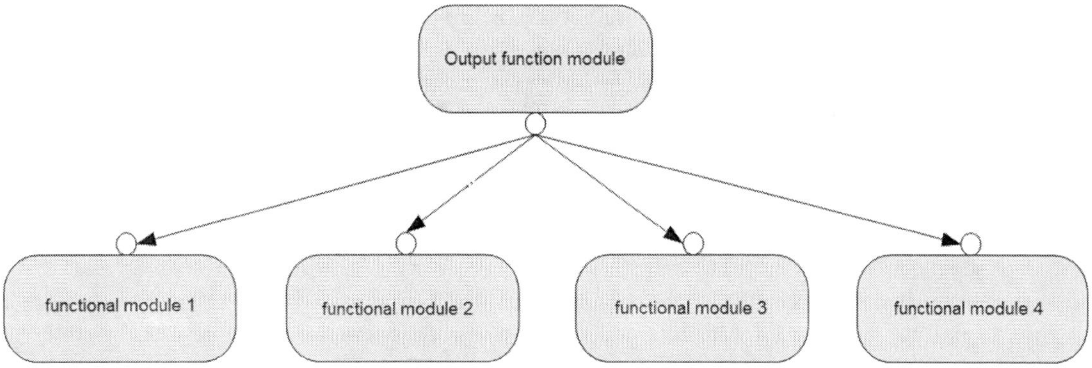

Figure 5. Output port connect to follow-up function modules

The input port reads data from the associated output port. Because each function module reads data at different speeds, the position of each subsequent function module reads is different. But when and only when the data is read by all subsequent functional modules, it can be written to cover. The length of data that can be read by all subsequent functional modules is the data from the reading position of the subsequent functional modules to the starting position of the output port. The read-write (ring buffer) of shared memory is a ring buffer simulated by using a block of memory pointed by a pointer, so when reading and writing data, if the end boundary is involved, it needs to be read or written twice.

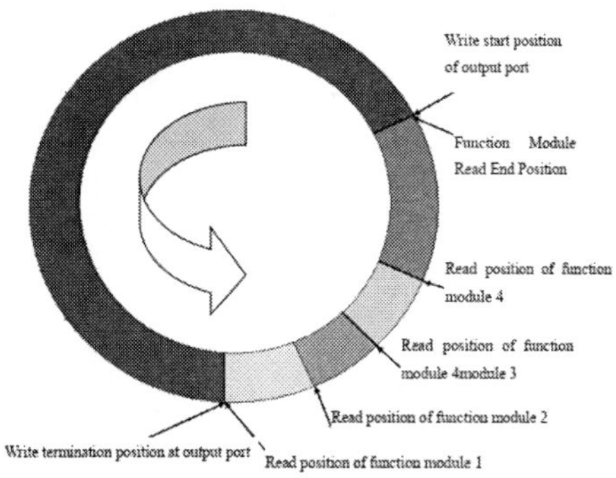

Figure 6. Shared memory design diagram

IOP Conf. Series: Materials Science and Engineering **729** (2020) 012016 doi:10.1088/1757-899X/729/1/012016

When the input port reads the data, it notifies the associated output port that the data has been read. When the data is read by all the related input ports, the output port updates the write pointer position, and the data buffer can be covered by the new data.

When the output port writes new data, it notifies the associated input port that new data is readable; when the input port has no data port to read, the output port updates the write pointer position, and the data buffer can be covered by the new data.

When the output port writes new data, it notifies the associated input port that new data is readable; when the input port does not write data, it sets up the data and maintains the read status information of the associated input port data.

Data separable job IO control system forms the cooperative operation technology of modular job, realizes the further research of the cooperative operation program of modular job based on shared memory, expands the type of data interaction, improves the operation efficiency, and enables it to provide more high-quality service for the efficient operation of modular job.

4. Conclusion

When the input port reads the data, it notifies the associated output port that the data has been read. When the data is read by all the related input ports, the output port updates the write pointer position, and the data buffer can be covered by the new data.

When the output port writes new data, it notifies the associated input port that new data is readable; when the input port has no data port to read, the output port updates the write pointer position, and the data buffer can be covered by the new data.

When the output port writes new data, it notifies the associated input port that new data is readable; when the input port does not write data, it sets up the data and maintains the read status information of the associated input port data.

Data separable job IO control system forms the cooperative operation technology of modular job, realizes the further research of the cooperative operation program of modular job based on shared memory, expands the type of data interaction, improves the operation efficiency, and enables it to provide more high-quality service for the efficient operation of modular job.

Acknowledgments

The paper was supported by the national project numbered 2016ZX05023-004 and we also appreciate the contributions of our colleagues.

References

[1] S. Ghemawat, H. Gobioff, S. Leung. The Google file system[C]. Acm Sigops Operating Systems Review. ACM, 2003:29-43

[2] M. Burrows. The Chubby lock service for loosely-coupled distributed systems[C]. 7th Symposium on Operating Systems Design and Implementation (OSDI '06), November 6-8, Seattle, WA, USA. 2006:335-350

[3] F. Chang, J. Dean, S. Ghemawat, et al. Bigtable: A Distributed Storage System for Structured Data[J]. ACM Transactions on Computer Systems, June 2008, Volume 26 Issue 2

[4] J. Dean, S. Ghemawat. MapReduce: Simplified Data Processing on Large Cluster[C]. Communications of the ACM, January 2008, Volume 51 Issue 1, Pages 107-113

[5] M. L. Massie, B. N. Chun, D. E. Culler. The ganglia distributed monitoring system: design, implementation, 2004, 30(7): 817-840

IOP Conf. Series: Materials Science and Engineering **729** (2020) 012017 doi:10.1088/1757-899X/729/1/012017

Analysis of Noise Characteristics in Microseismic Monitoring

Liting Liu[1]*, Liang Kang[1], Xi Zhang[2], Yong Tang[1], Zilong Li[1]

[1] BGP Southwest Geophysical Company, CNPC, Huayang road, Chengdu, Sichuan, 610213, China

[2] BGP Research&Development Center Chengdu Branch Center, CNPC, Chengdu, Sichuan, 610213, China

*Corresponding author's e-mail: liulit_wt@cnpc.com.cn

Abstract. The effective signal energy of microseismic data is weak, and the received data will contain a lot of noise because of its special source excitation mode. At the same time, because the wave will inevitably lose energy in the process of propagation, the signal on the section is weak amplitude, weak continuity, even the effective signal on the whole line is submerged by noise, and the signal-to-noise ratio of the data is very low. In this paper, the common noise in different monitoring methods is analyzed, and the common processing methods in production are summarized.

1. Introduction

In the early stage of unconventional reservoir development, different wells need different observation methods. The commonly used micro-seismic monitoring measures are deep well observation, shallow well observation and surface observation. Deep well observation has a high S/N ratio because the geophones are close to the fracturing zone, and can record extremely small events during the fracturing process. However, in the early stage of unconventional oil and gas reservoir exploitation and development, it is not conducive to a wide range of promotion. Because of the sparse well location, surface observation is to lay hundreds of geophones on the surface of the fractured area to improve the signal-to-noise ratio and reliability of positioning data through the advantage of quantity. In shallow observation, the geophones are placed in 4-6 wells at depths of about 300 meters. To a certain extent, this method can avoid the low velocity zone near the surface, which is beneficial to the reception of microseismic signals, but the cost of drilling makes this monitoring method more expensive. However, the characteristics of noise and microseismic signals are quite different in different observation modes under different fracturing modes. Through the spectrum analysis of the data, we find that in most cases, the frequency of the effective signal is significantly different from that of the noise, and the ideal effect can be achieved only by filtering processing.

2. Noise characteristic analysis and filtering effect of different monitoring modes in hydraulic fracturing environment

Because of the excitation mode of the special source, the energy of the effective signal is very small, and the received data will contain a lot of noise. At the same time, because the wave will inevitably lose energy in the process of propagation, the signal on the section is weak amplitude, weak continuity, even the effective signal on the whole line is submerged by noise, and the S/N ratio of the data is very low. In this section, we expose the noise characteristics and filtering effect under different monitoring modes by analyzing the data under different observation modes.

2.1. Noise characteristics and processing methods of microseismic deep well monitoring in hydraulic fracturing environment

The event S/N ratio of microseismic deep well monitoring is the highest relative to surface and shallow well monitoring. Generally speaking, the obvious event waveform can be seen before filtering. A data file with relatively clean background, obvious event energy and no obvious random noise is selected to analyze its background noise and event signal. It can be found that the frequency of events is mainly concentrated in about 100Hz, and the noise is mainly low-frequency noise below 20Hz and some random noise, as shown in Figure 1.

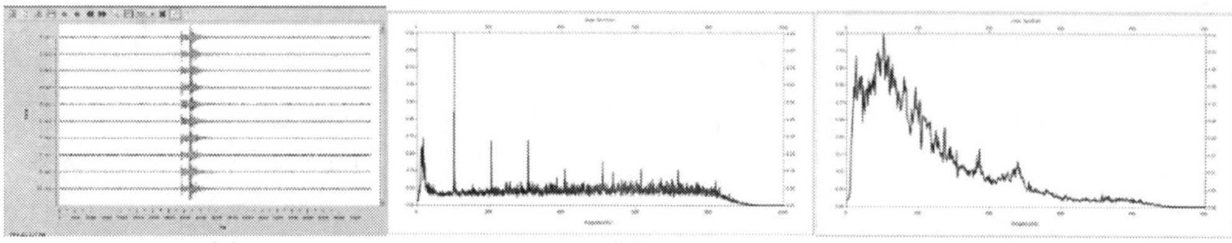

(a) (b) (c)

Figure 1. Spectrum analysis of background noise and events in deep well monitoring. (a) Raw data. (b) Spectrum Analysis of the noise. (c) Spectrum Analysis of the event.

(1) Strong low frequency interference

There are often strong low frequency interference signals in micro-seismic deep well monitoring, as shown in Fig. 2. The frequency of the interference signal is stable, and the interference signal is about 50HZ from the beginning to the end of the interfered channel. This interference mainly comes from AC and can be eliminated by bandpass filtering. From the filtered data, it can be seen that the event signal can be distinguished more clearly.

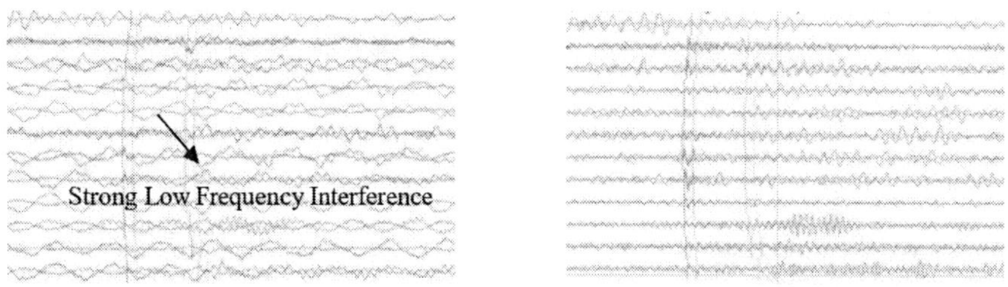

(a) (b)

Figure 2. Contrast diagram before and after filtering processing for the strong low frequency interference. (a) Before filtering. (b) After filtering.

(2) Strong high frequency interference

At present, microseismic data acquisition often uses geophone sensitive to high-frequency information, which makes very weak high-frequency signal stronger. Therefore, regardless of the quality of field construction, there will be a lot of high-frequency noise. As can be seen from Fig. 3, the high frequency noise of deep well monitoring is stable with time, and its energy relative event is weak. This kind of high frequency noise belongs to high frequency environmental noise. The noise source is mainly the thermal noise of the instrument and various relatively stable high frequency interference sources in nature. The noise can be effectively removed by band-pass filtering.

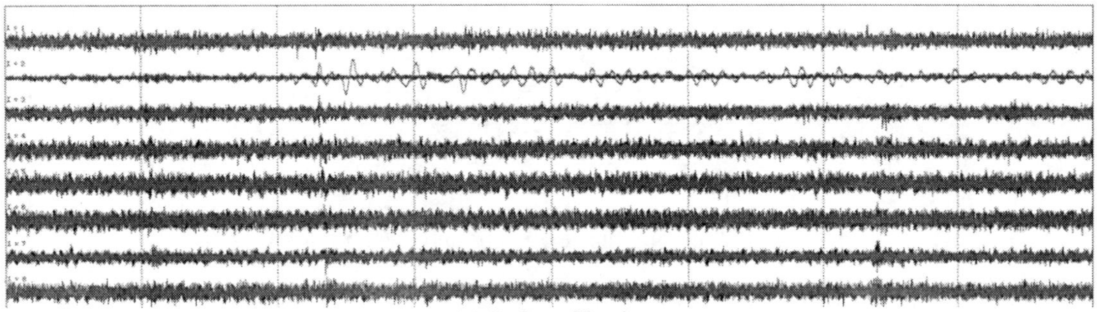

(a) Before filtering

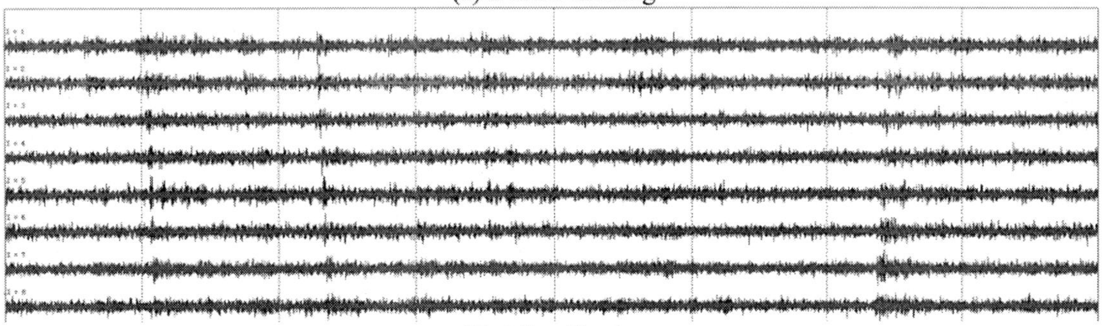

(b) After filtering

Figure 3. Contrast diagram before and after filtering processing for the strong high frequency interference

(3) Strong linear interference

Because the strong linear interference noise produced by wellbore operation, this kind of noise has obvious linear characteristics in waveform record, and multiple waves are produced by the reflection or refraction of the formation. This kind of noise is difficult to carry out spectrum analysis, so it can not be removed by band-pass filtering.

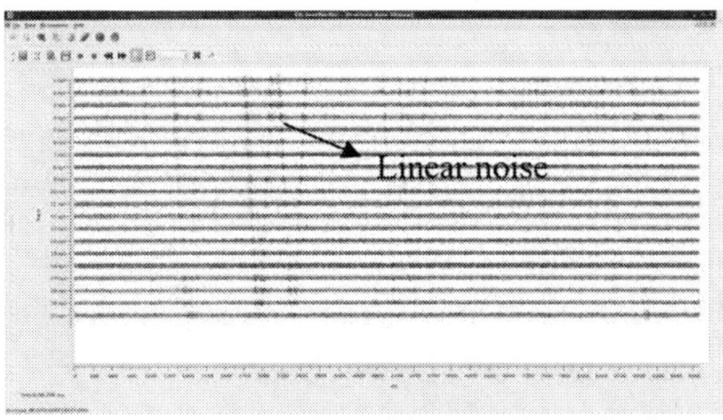

Figure 4. Linear noise of deep well monitoring

(4) Random interference

Microseismic deep well monitoring detectors generally have a deeper settling depth than conventional geophones, and the environment is very complicated, and the detector is more difficult to fix. Therefore, there will be poor coupling of geophoneor or local vibration, which is the main cause of random noise monitoring in deep wells. As shown in Fig. 5, this kind of random noise only occurs in individual channels, and the vibration waveforms are not correlated with each other. This shows that the vibration does not propagate with the formation, but comes from the geophone itself. This kind of noise only occurs occasionally, and the chance of simultaneous occurrence with the event signal is not very much, so it will not have a great impact on location without special processing.

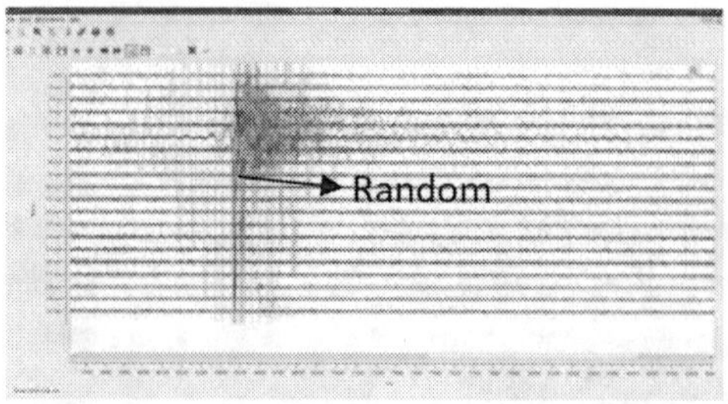

Figure 5. Random interference of deep well monitoring

2.2. Noise characteristics and processing methods of microseismic surface monitoring in hydraulic fracturing environment

Surface survey makes the noise interference of ground microseismic monitoring very complex, which is mainly the environmental noise collected during the acquisition process. The signal-to-noise ratio of surface microseismic is much lower than that of deep well monitoring data. The low signal-to-noise ratio is due to the weak effective signal, which is relative to noise. In order to effectively present the event signal of microseismic ground monitoring, noise must be effectively suppressed. There are two basic types of serface monitoring noise: one is periodic noise, the other is accidental linear noise.

(1) Strong DC component

In surface monitoring of microseismic, severe strong low-frequency noise occurs usually, as shown in Fig. 6. Its amplitude is usually 1-2 orders of magnitude larger than the microseismic event signal, and its frequency is concentrated below 5HZ. Because of the frequency difference between this kind of signal and microseismic event signal, this kind of noise can be effectively eliminated by band-pass filtering and single-frequency noise elimination technology. From the data after filtering, it can be seen that the strong low frequency noise has been basically removed, and the original masked event signal can be highlighted.

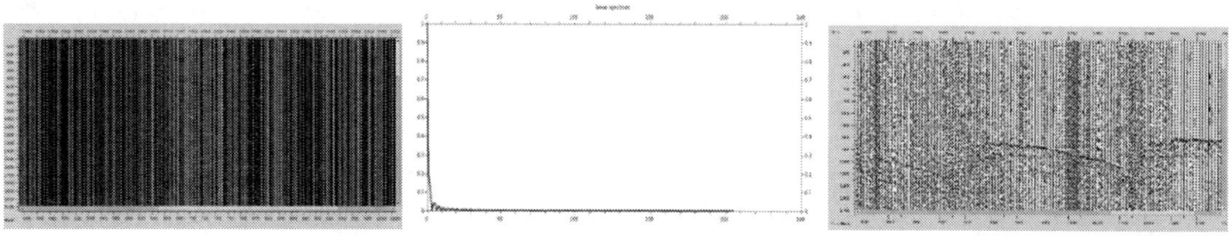

(a) (b) (c)

Figure 6. Strong low-frequency noise of surface monitoring. (a)Raw data.(b) Spectrum Analysis of the raw data.(c) The signal is highlighted after filtering

(2) Periodic noise

Geophones are widely distributed in surface observation, and there are periodic persistent disturbances of a certain frequency in local traces due to mechanical vibration.

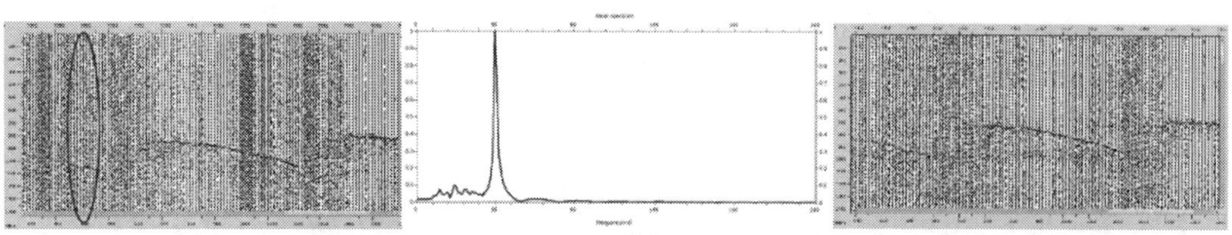

(a) (b) (c)

Figure 7. Periodic noise of surface monitoring. .(a)Raw data. The area marked in the figure is periodic noise. (b) Spectrum Analysis of the noise data.(c)Noise signal weakened after filtering.

(3) Linear interference

In the process of fracturing monitoring, strong linear noise occurs due to fixed large-scale vibration. As shown in Figure 8, the event signal of monitoring data is clear, the waveform and continuity are good is continuous, and the energy of noise signal is similar to that of effective event signal. The adaptive FK filtering method can suppress the noise and enhance the effective events, but it also makes the waveform changes in the filtering process. If the parameters are not selected properly, the travel time information of the waveform itself will be destroyed.

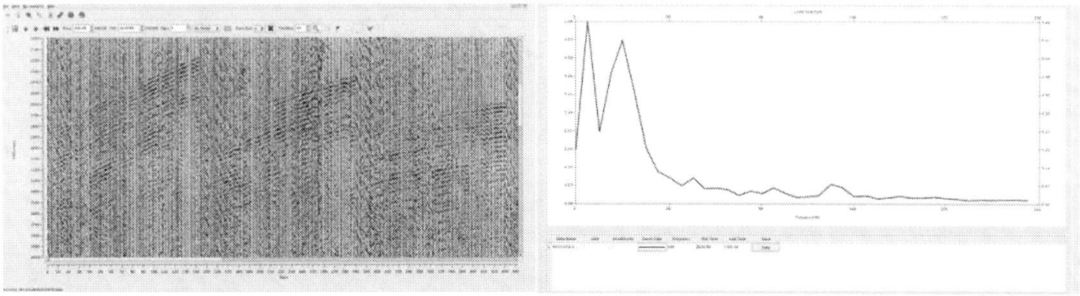

(a) (b)

Figure 8. Strong Linear Noise in Ground Monitoring.(a)Raw data.(b) Spectrum Analysis.

(4) Mechanical Vibration in Fracturing Construction

In the process of fracturing construction, such as pumping, strong mechanical vibration is induced, which is transmitted along the surface or air to the geophone and causes obvious interference. As shown in Fig. 9, the spectrum analysis shows that the main frequency is mainly around 100Hz. This kind of noise has strong energy and irregular occurrence. So, it usually does not need special treatment.

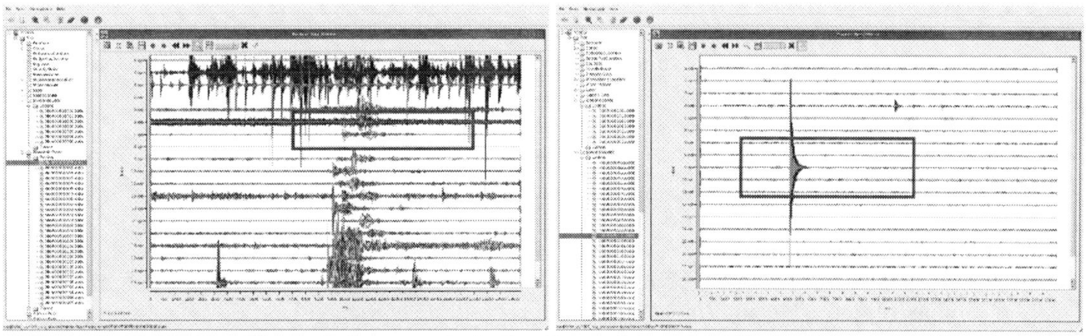

Figure 9. Long period vibration noise and random noise

2.3. Noise characteristics and processing methods of microseismic shallow well monitoring in hydraulic fracturing environment

The noise of shallow observation is mainly strong low frequency and random noise. It is difficult to see the effective waveform in the general gain when displaying the image. Because the amplitude of the strong low-frequency noise is about 20 times that of the signal, and the energy of the effective signal is relatively suppressed. This kind of noise can be removed by de-direct current component, while random noise can be effectively removed by band-pass filter, as shown in Figure 10.

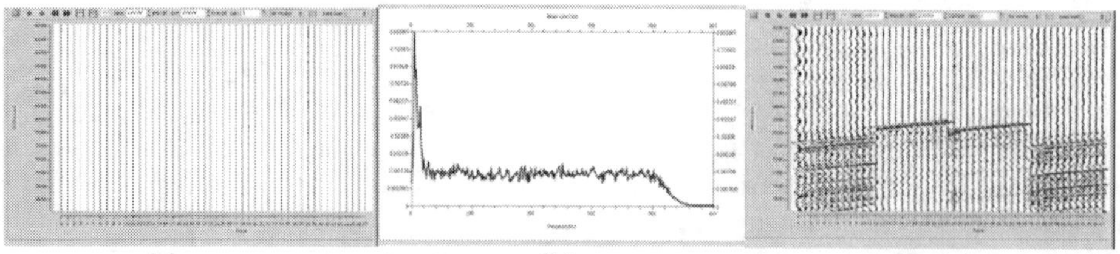

(a) (b) (c)

Figure 10. Comparison of pre-denoising and post-denoising of shallow observation data. (a)Raw data. (b) Spectrum analysis (c) Raw data are processed by removing low-frequency noise and random noise

In addition, shallow observation is more widely distributed than deep well observation. As micro-seismic surface observation, there exists periodic continuous interference of a certain frequency in local traces due to mechanical vibration. As can be seen from Fig. 11, the frequency of this periodic noise is mainly concentrated in about 50HZ, which can also be removed by predictive filtering.

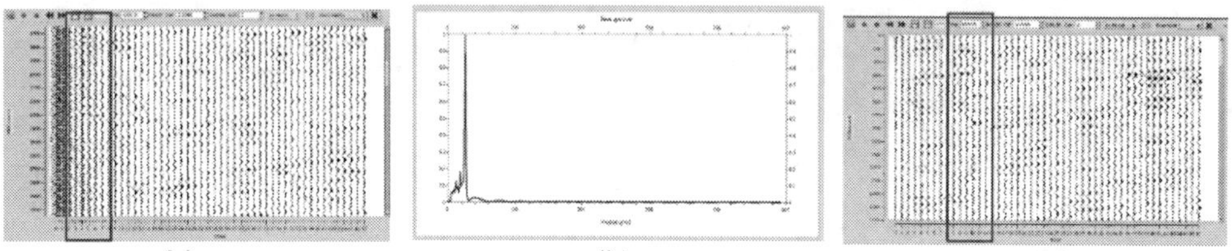

(a) (b) (c)

Figure 11. Elimination of periodic noise in shallow observation.(a)Raw data. The area marked in the figure is periodic noise. (b) Spectrum Analysis of the noise data.(c) Noise signal weakened after filtering.

3. Conclusion

By analyzing the monitoring data of deep well observation, surface observation and shallow wells observation, we get the noise characteristics under different monitoring conditions. According to its different characteristic information, the corresponding denoising processing has achieved good results.

Acknowledgments

The paper was supported by the national project numbered 2016ZX05023-004 and we also appreciate the contributions of our colleagues.

References

[1] Chen Yin, Yalin Li, and Furong Wu. (2018) The feasibility analysis of joint microseismic monitoring in Sichuan, China. https://doi.org/10.1190/segam2018-2997911.1

[2] Y. Tang,H. Zhang,J. Li,C. Yin and F. W. 2018.Focal mechanism determination for induced seismicity using the neighbourhood algorithm.Geophysical Journal International.Volume 214 :1715-1731

[3] Huang, W., Wang, R., Gong, X., & Chen, Y.2018.Iterative deblending of simultaneous-source seismic data with structuring median constraint.IEEE Geoscience and Remote Sensing Letters.Volum15.No.1:58-62

[4] Chen, X., Wang, R., Huang, W., & Jiang, Y..2018.Clustering based stress inversion from focal mechanisms in microseismic monitoring of hydrofracturing.Geophysical Journal International.Volume 215.No.3:1887-1899

[5] Huang, W., & Wang, R..2018.Random noise attenuation by planar mathematical morphological filtering.Volume 83.01:11-25

IOP Conf. Series: Materials Science and Engineering **729** (2020) 012018 doi:10.1088/1757-899X/729/1/012018

Density functional theory study on bond dissociation energy of polystyrene trimer model compound

Jinbao Huang, Hanxian Meng*, Xiaocai Cheng*, Guiying Pan, Xunming Cai and Jiangtao Liu

School of Mechatronics Engineering, Guizhou Minzu University, Guiyang, Guizhou Province, 550025, China.

*Corresponding author's tel: 15585138870

*Corresponding author's e-mail: 1658856263@qq.com (H. Meng), 2364115733@qq.com (X Cheng).

Abstract: The bond dissociation energy (BDE) of syndiotactic polystyrene(s-PS) was calculated by using four density functional theory methods, which were M062X, B3LYP, B3P86, BHandHLYP with different LANL2DZ, 6-31G(d), 6-31G(d, p), and 6-311++G(d, p) basis sets. The influence of these computational methods on the accuracy of the BDE of s-PS was analyzed. It was found that the BDE had a higher accuracy at B3P86/6-31G (d, p) level than other methods, and B3P86/6-31G (d, p) spent more less time to calculate the bond dissociation energy of s-PS, and the average of the C-C bonds' BDE values calculated by B3P86/6-31G (d, p) is about 8.3 kJ/mol lower than the experimental value (310.0 kJ/mol), which is derived from the i-bond database. Besides, the bond dissociation energy values of the C-C bonds computed by M062X/LANL2DZ, B3LYP/6-31G(d), M062X/6-31G(d), B3LYP/LANL2DZ and BHandHLYP/6-311++G(d, p) methods were as follows: 330.8 kJ/mol, 345.3 kJ/mol , 283.2 kJ/mol, 267.6 kJ/mol, 267.4 kJ/mol. The calculation conclusion is drawn that different calculation methods with the same basis set have an obvious influence on the result of BDE values' calculation. Finally, based on the calculation of bond dissociation energy of polystyrene, three possible reaction paths of polystyrene pyrolysis are designed to analyze the influence of bond dissociation energy on polystyrene pyrolysis.

1. Introduction

It is generally believed that polystyrene has three isomeric structures: syndiotactic, isotactic and atactic, which are composed of C-C bonds, C-H bonds and C-$C_{aromatic}$ bonds. The reaction mechanism of free radicals is considered as the main mechanism in the pyrolysis process of polystyrene [1-4]. And the BDE value is a necessary parameter for the chemical properties of radical reactions to represent the strength of chemical bonds that is the energy required while the chemical bonds are breaking. The larger the BDE is that the less likely the chemical bond is to break [5, 6]. Polystyrene is a kind of high poly compound, which is polymerized by small molecules, its bond dissociation energy values are very important for the pyrolysis of polystyrene. The calculation of polystyrene's BDE value is helpful to understand the pyrolysis mechanism of polystyrene. The BDE of s-PS is mainly calculated and studied in this paper because that it is the most stable structure for polystyrene molecules [7]. We use density functional theory methods, which are MO62X/LANL2DZ, MO62X/6-31G(d), B3LYP/6-31G(d), B3LYP/LANL2DZ, BHandHLYP/6-311 ++G(d, p), B3P86/6-31G(d, p) to calculate polystyrene trimer model compound's C-H bonds, C-C bonds and C–$C_{aromatic}$ bonds, and analyze the polystyrene trimer model compounds' BDE to compare the influence of the BDE values by these methods. The purpose is to find a more accurate method to calculate the BDE values of polystyrene, and to explore the influence of the bond dissociation energy values on the pyrolysis mechanism of polystyrene trimer model compounds.

2. Computational Method

The bond dissociation energy values were calculated by using Gaussian 09 program in this paper [8]. Four density functional theory methods, including M062X, B3LYP, B3P86 and BHandHLYP, and five different accuracy's basis sets, such as LANL2DZ, 6-31G (d), 6-31G (d, p), 6-31++G (d, p) were selected to calculate and study the BDE of s-PS. The bond dissociation energy (BDE) of syndiotactic polystyrene(s-PS) was calculated by using the combinaton of the same method with different basis sets, different methods with the same basis set and different methods with different basis sets. M062X was a good hybrid functional in density functional theory method, which played a great role in the calculation of chemical bonds main groups [18]. B3LYP and B3P86 are all hybrid functional of Becke three-parameter, B3LYP uses the non-local correlation of LYP expression, B3LYP can relatively calculate molecular configuration, energy and frequency accurately. The non-local correlation of B3P86 is provided by Perdew 86 [9]. BHandHLYP is also mixed density functional. Some studies have shown that BHandHLYP is an unexceptionable method in all functional tests and it is one of the applicable methods to describe hydrogen bonding systems. Different density functional methods use different correlative exchange function. As a result, the calculation results were different [10, 11]. The calculation formula of bond dissociation energy for each major bond splitting reaction is [12]:

$$E_B(R\text{-}X) = E(R) + E_{ZP}(R) + E(X) + E_{ZP}(X) - E(RX) - E_{ZP}(RX)$$

In this formula, E_B represents the bond dissociation energy when zero correction is taken into account, E is the molecular energy, and E_{ZP} is the zero correction energy.

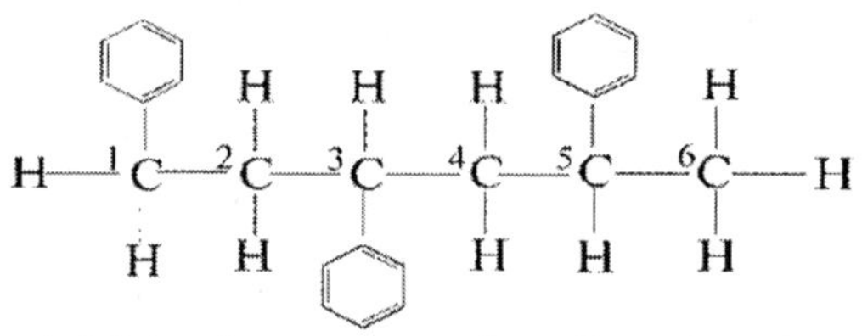

Figure 1. Molecular structure of syndiotactic polystyrene.

3. Results and Discussion

3.1. The energy dissociation of the C-C bonds

Figure 1 shows the structure of s-PS, which consists of main-chain C-C bonds and three alternating benzene rings. The experimental values of BDE of C-C bond, C-H bond and $C\text{-}C_{aromatic}$ of s-PS are 310.0 kJ/mol, 351.5 kJ/mol and 415.1 kJ/mol respectively. Table 1 lists the values of the BDE of the C-C bonds calculated by different methods. The average value of the BDE obtained by M062X/ 6-31G (d) is 345.3 kJ/mol, and the average values of the BDE calculated by M062X/LANL2DZ is 330.8 kJ/mol, its deviations with the experimental value are respectively 35.3 kJ/mol and 20.8 kJ/mol. The computational results of M062X/6-31G (d) and M062X/ LANL2DZ method have a large deviation with the experimental value, and the error of M062X/6-31G (d) is larger than that of M062X/LANL2DZ method. The average BDE value of C-C bonds of syndiotactic polystyrene at B3LYP/6-31G(d) and B3LYP/LANL2DZ level are respectively 283.2 kJ/mol and 267.6 kJ/mol, and its errors with the experimental values are respectively 26.2 kJ/mol and 42.4 kJ/mol. By comparing the results calculated by the combination of the above methods and basis groups, it is concluded that LANL2DZ basis set is relatively accuracy than the basis set 6-31G (d) when M062X method is used for C-C bonds calculation. The LANL2DZ basis set is not as accurate as the 6-31G (d) when the method B3LYP is used to calculate the BDE of C-C bonds. For LANL2DZ basis set, the values obtained by M062X method are more accurate than the result obtained by B3LYP method. The average values calculated by BHandHLYP/6-311++G (d, p) is 267.4 kJ/mol, it is about 42.6 kJ/mol lower than the experimental value. The calculated result by B3P86/6-31G (d, p) method is 301.7 kJ/mol, its calculation time is shorter than above five methods, and the error value is 8.3 kJ/mol, which is the most similar to the experimental value.

From the above discussion, the C-C bonds' BDE values are calculated by using the combination of the same method with different basis sets, and the combinations of the same set with different methods. The results obtained by M062X method, B3LYP method and BHandHLYP/6-311++G (d, p) have a great error with the experimental values. B3P86/6-31G (d, p) method is the closest to the experimental value and relatively accurate.

Table 1. The values of C-C bond dissociation energy (kJ/mol).

methods / C-C bonds	M062X/6-31G(d)	M062X/LANL2DZ	B3LYP/6-31G(d)	B3LYP/LANL2DZ	B3P86/6-31G(d, p)	BHandHLYP/6-311++G(d, p)
C_1-C_2	347.1	332.4	288.5	272.1	306.5	270.3
C_2-C_3	348.2	334.4	278.9	264.9	298.2	264.6
C_3-C_4	345.7	332.5	277.6	264.0	297.6	264.2
C_4-C_5	340.3	326.7	276.7	261.6	294.7	262.7
C_5-C_6	345.2	327.9	294.1	275.2	311.4	275.1
Average value	345.3	330.8	283.2	267.6	301.7	267.4

3.2. The energy dissociation of C-$C_{aromatic}$ bonds

The following table 2 marks the required values of C-$C_{aromatic}$ bonds calculated by M062X/6-31G (d), M062X/LANL2DZ, B3LYP/6-31G (d), B3LYP/LANL2DZ and B3P86/6-31G (d, p) methods. The calculated results of C-$C_{aromatic}$ are respectively 435.8 kJ/mol and 433.0 kJ/mol by using M062X/6-31G (d) and M062X/LANL2DZ methods, and the errors between them with the experimental values are respectively as high as 20.7 kJ/mol and 17.9 kJ/mol. This calculation result indicates that M062X/6-31G (d) method isn't too different from the result of the M062X/LANL2DZ, but all the calculation values are more higher than experimental value. The average BDE values of C-$C_{aromatic}$ bonds calculated by B3LYP/6-31G (d) and B3LYP/LANL2DZ method are 388.7 kJ/mol and 384.2 kJ/mol respectively, and calculated results are respectively about 26.4 kJ/mol and 30.9 kJ/mol lower than the experimental value. The average BDE values at BHandHLYP/6-311++G (d, p) level has a highest error to experimental value, which is 38.3 kJ/mol. A relatively accurate calculation results can aquired by B3P86/6-31G (d, p), the average value is 406.4 kJ/mol, which is the most similar to experimental value, it is only 8.7 kJ/mol lower than experimental value.

As for M062X and B3LYP methods, the results obtained by M062X/ 6-31G (d) and M062X/LANL2DZ methods are semblable, the result obtained by B3LYP/ 6-31G(d) are similar to the result of B3LYP/LANL2DZ method, But the C-$C_{aromatic}$ bonds BDE values obtained by these four methods are very different from the experimental values. The average values obtained by BHandHLYP/6-311++G (d, p) has a largest deviation. The results show that those computing methods of M062X/6-31G (d), M062X/LANL2DZ, B3LYP/6-31G (d), B3LYP/LANL2D and BHandHLYP/6-311++G (d, p) are not suitable for C-$C_{aromatic}$ bonds dissociation energy's calculation. B3P86/6-31G (d, p) is the most accurate calculation way, which is more suitable for the calculation of polystyrene C-$C_{aromatic}$ bond dissociation energy.

Table 2. The values of C-$C_{aromatic}$ bond dissociation energy (kJ/mol).

methods / C-$C_{aromatic}$ bonds	M062X/6-31G(d)	M062X/LANL2DZ	B3LYP/6-31G(d)	B3LYP/LANL2DZ	B3P86/6-31G(d, p)	BHandHLYP/6-311++G(d, p)
C_1-$C_{aromatic}$	443.6	438.6	399.9	392.1	415.8	384.6
C_3-$C_{aromatic}$	429.1	428.8	384.1	382.2	403.6	374.3
C_5-$C_{aromatic}$	434.6	431.5	382.0	378.3	399.7	371.6
Average value	435.8	433.0	388.7	384.2	406.4	376.8

3.3. The energy dissociation of the C-H bonds

Table 3 shows the values of the bond dissociation energy of C-H bonds. Through calculation, the average value calculated by M062X/ 6-31G (d) was 397.6 kJ/mol, and the error with the experimental value is 46.1 kJ/mol. The C-H bonds's average BDE values calculated by M062X/LANL2DZ method is 397.0 kJ/mol, its error with the experimental value is 45.5 kJ/mol. The average bond dissociation energy value calculated by B3LYP/6-31G (d) method was 385.4 kJ/mol, its error of BDE value with experimental value as 33.9 kJ/mol. The value calculated by B3LYP/LANL2DZ method is 383.9 kJ/mol, it is about 32.4 kJ/mol higher than experimental value. The average bond dissociation energy of C-H bonds calculated by B3P86/6-31G (d, p) method is 398.6 kJ/mol, the error with the experimental value is 47.1 kJ/mol. The average dissociation energy of C-H bonds obtained by BHandHLYP/6-311++G (d, p) method was 313.7 kJ/mol, and the error with the experimental value is 37.8 kJ/mol.

By comparing the results obtained by each method to experimental value. It is concluded that the values calculated by M062X/6-31G (d) and M062X/LANL2DZ methods are basically the same for C-H bonds'BDE, the values obtained by B3LYP/6-31G (d) and B3LYP/LANL2DZ are not significantly different. But all of them are quite different from the experimental values. The result computed by M062X/6-31G (d) is more accurate than B3LYP/6-31G (d) method. Comparing M062X/LANL2DZ with B3LYP/LANL2DZ, the M062X/LANL2DZ method is not applicable to calculate C-H bonds' BDE. The results obtained by the combination of these methods with the same basis set show that different methods with the same basis set have a great influence on the calculation of C-H bonds'BDE. The values calculated by BHandHLYP/6-311++G (d, p) is about 37.8 lower than experimental value. The values obtained by B3P86/6-31G (d, p) method is different from for calculating C-C bonds and C-$C_{aromatic}$ bonds, its error is relatively large. In summary, in the calculation of C-H bonds dissociation energy, the selection of 6-31G (d) basis set and LANL2DZ basis set has a little influence on the calculation results. The error of B3LYP method is smaller than that all of other methods, but the result is not ideal to calculate the C-H bond dissociation energy.

Table 3. The values of bond dissociation energy of C-H bonds (kJ/mol).

methods / C-C bonds	M062X/6-31G(d)	M062X/LANL2DZ	B3LYP/6-31G(d)	B3LYP/LANL2DZ	B3P86/6-31G(d, p)	BHandHLYP/6-311++G(d, p)
C_1-H	375.9	374.6	354.1	352.8	369.1	348.0
C_2-H	414.8	413.4	403.0	402.4	416.3	395.8
C_3-H	391.0	393.1	385.1	381.7	396.2	395.1
C_4-H	412.8	412.0	402.2	402.4	416.2	408.1
C_5-H	364.4	364.6	348.9	347.5	362.0	384.3
C_6-H	426.7	424.5	419.0	416.5	431.5	346.1
Average value	397.6	397.0	385.4	383.9	398.6	313.7

3.4. Analysis on the formation mechanism of polystyrene pyrolysis products

Liu, X.X., et al. [13] concluded that styrene, ethylbenzene, toluene, alpha-methyl styrene products and so on are the main products of polystyrene pyrolysis reaction. This part mainly proposes three possible reaction paths to the formation mechanism of styrene dimer, styrene, ethylbenzene, alpha-methyl styrene products during the pyrolysis of polystyrene. According to the density functional theory method in Gaussian 09 (B3P86/6-31G (d, p)), reaction kinetics parameters and thermodynamic parameters are calculated and the formation mechanism of main products is studied in the pyrolysis reaction process. Figure 2 describes three possible reaction paths in the pyrolysis process of polystyrene, it can reflect the evolution process of each product generated by PS pyrolysis. Figure 3 depicts the energy barrier diagram of each reaction stage in the three reaction paths, which can well show the kinetic principles of the reaction process.

In the reaction of path**1**, the styrene monomer **2** and free radical **3** are generated by breaking of the **C_4-C_5** bond of polystyrene free radical **1** under the transition state **TS_1**. In this process, there is 82.6 kJ/mol heat absorbed, and the reaction energy barrier is 162.3 kJ/mol. The free radicals **3** generate styrene monomer **2** and a free radical **4** by **TS_2** after the **C_2-C_3** bond's breaking. It absorbs 82.6 kJ/mol quantity of heat, and the reaction energy barrier is 118.6 kJ/mol in this process. Radical **4** get one hydrogen to form ethylbenzene **5**, which releases 367.2 kJ/mol energy. Or generates styrene monomer **2** by dehydrogenation, and absorbs 218.5 kJ/mol energy.

In the reaction of Path**2**, the reaction mechanism of styrene dimer generated by polystyrene pyrolysis is mainly **1, 3** hydrogen transfer within polystyrene molecules. The **C_1**'s H of the free radical **1** is transferred to **C_3** to form free radical **6** through the transition state **TS_3**, it absorbs 27.1 kJ/mol heats, and the reaction energy barrier is 165.2 kJ/mol during the process, which is so low. The **C_2-C_3** bond of the free radical **6** is broken and the styrene dimer **7** and free radical **4** were generated through the transitional state **TS_4**. This process absorbs heat of 27.1 kJ/mol, and the reaction energy barrier is 165.6 kJ/mol. Free radical **4** releases 367.3 kJ/mol heat by getting a hydrogen to form ethylbenzene **5**, or by dehydrogenating to make styrene monomer **2**, which absorbs 218.4 kJ/mol heat. The reaction of this path not only produces styrene dimer, but also styrene monomer and ethylbenzene. This path makes the reaction reactive fully.

In the reaction path Path**3**, the main reaction mechanism of polystyrene pyrolysis to produce alpha-methyl styrene is **1, 2** hydrogen transfer in polystyrene molecules. The **C_1**'s H of the free radical **1** is transferred to the position of **C_2** by transition state **TS_5** to form radical **8**. In this process, 47.3 kJ/mol is absorbed, and the energy barrier of this reaction is 182.7 kJ/mol. The **C_3-C_4** bond of Free radical **8** is broken, and the product **9** and free radical **10** are formed under the function of the transition state **TS_6**. This process absorbed 47.3 kJ/mol of heat, and the energy barrier of this reaction is 180.8 kJ/mol. Radical **10** generates alpha-methyl styrene monomer **11** by removing one hydrogen, which absorbs 146.4 kJ/mol of heat.

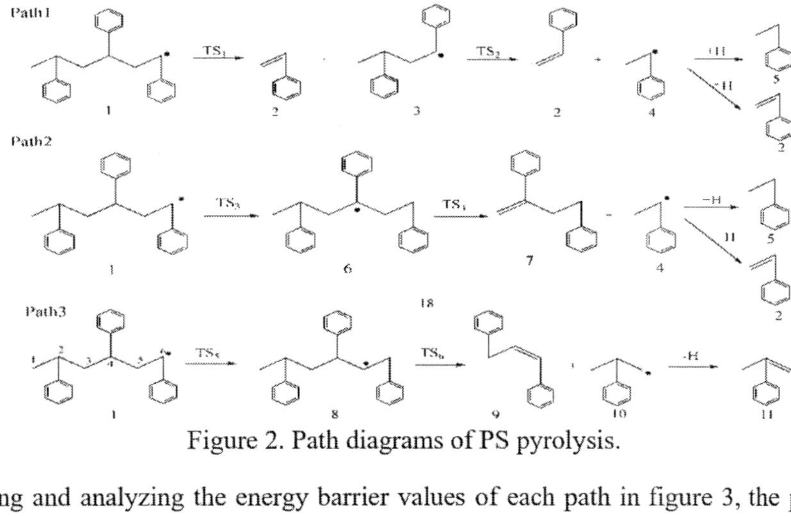

Figure 2. Path diagrams of PS pyrolysis.

By comparing and analyzing the energy barrier values of each path in figure 3, the path1 has the lowest reaction rate control step energy barrier to generate styrene monomer. It can draw a conclusion that the reaction is easier to carry out, so a large number of styrene monomers can be generated, it is consistent with the findings [16]. In addition, in the three reaction paths, the pyrolysis reaction of polystyrene mainly involves the fracture of C-C bonds on the main chain of polystyrene molecule, free radical reaction, dehydrogenation reaction and hydrogen transfer. In the calculation of the bond dissociation energy of C-C, C-Caromatic and C-H bonds, the values of C-C and C-H bonds are generally lower than the C-Caromatic bonds. The reaction is not easy to form benzene rings, which is consistent with the calculation results [13].

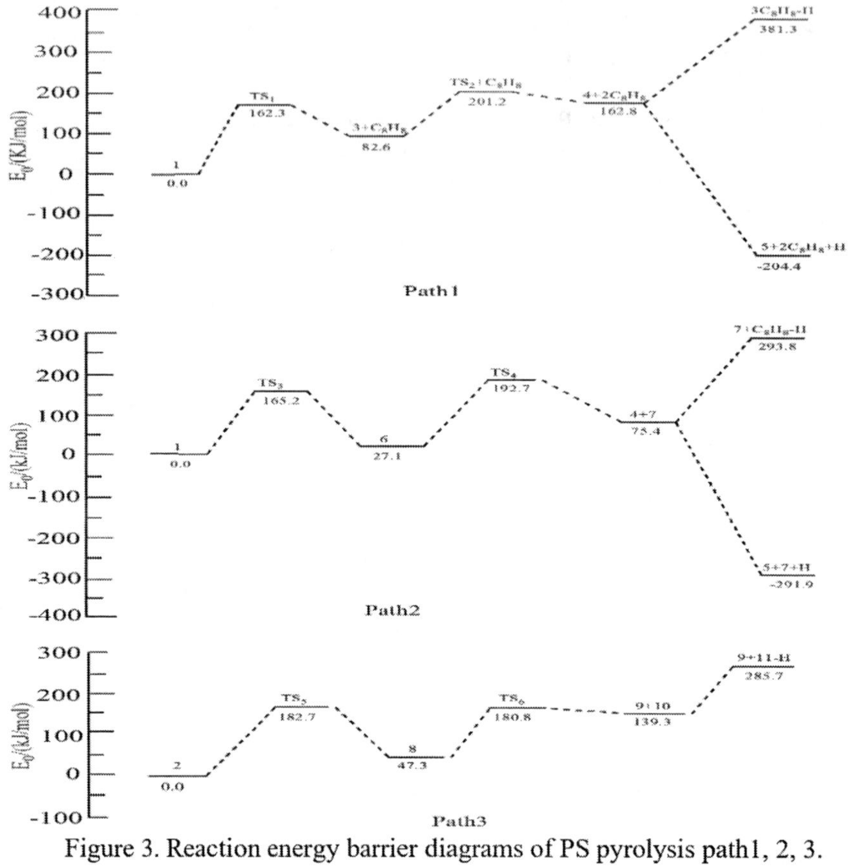

Figure 3. Reaction energy barrier diagrams of PS pyrolysis path1, 2, 3.

4. Conclusion

Bond dissociation energy is an important parameter for radical reaction. It's calculated by using the combination of the same density functional method with different base sets, different methods with the same basis set. According to the calculative results, the following conclusions are drawn:

1. The methods of M062X/6-31G (d), M062X/LANL2DZ, B3LYP/6-31G (d), B3LYP/LANL2DZ and BHandHLYP/6-311++G (d, p) has a large error in polystyrene BDEs' calculation. And the combination of different basis sets with the same method has little influence on the calculation results, while the same basis set with different methods has certain influence on the calculation results.

2. The method of B3P86/6-31G (d, p) has a large error in C-H bonds' calculation. But generally speaking, it is relatively accurate for calculating the BDEs of polystyrene, and the calculation time is shorter than other methods. It is feasible to utilize in the calculation of t polystyrene's BDEs.

3. The pyrolysis reaction of polystyrene mainly involves the fracture of C-C bonds, free radical reaction, dehydrogenation reaction and hydrogen. In the calculation of BDEs, the values of C-C and C-H bonds are generally lower than the $C\text{-}C_{aromatic}$ bonds. The results show that the smaller the bond

dissociation energy, the more likely the bond is to break, the easier the reaction is to carry out.

Acknowledgments
This work was supported by the National Natural Science Foundation of China (No. 51266002), the Innovation Group Major Research Project of Department of Educationin in Guizhou Province (No. KY [2016]028) and the Open Research Funds of Key Laboratory of Jiangxi Province for Persistent Pollutants Control and Resources Recycle (Nanchang Hangkong University) (No. ES201880055).

References

[1] Liu, B. (2016) Research on flame retardant properties and flame retardant mechanism of polystyrene foam insulation board [D]. Chang An University.
[2] Zeng, W.R., Zhou Y.J., Huo, R., Yao, B., et al. (2006) Non-isothermal kinetic integral-differential analysis of polystyrene pyrolysis reaction mechanism and kinetic parameters [J]. Polymer materials science and engineering, (05):162-165.
[3] Wang, L.H. (1989) Comprehensive utilization of waste polystyrene foam [J]. Modern chemical industry, (02):62-63. (In Chinese)
[4] Zhou, J.B. (2016) Experimental study on pyrolysis of typical common vinyl plastics under the condition of secondary reaction minimization [D]. Hua Zhong University of science and technology.
[5] Wang, H.J., Zhao, Y., Wang, C., et al. (2009) Theoretical study on the pyrolysis process of lignin dimer model compounds [J]. Journal of Chemical, 67(9): 893.
[6] Zhang, F.P., Cheng, X.L., Liu, Z.J., et al. (2005) Density functional theory research on dissociation energy and pyrolysis mechanism of propyl nitrate bond [J]. Journal of high pressure physics, 19(2): 189-192. (In Chinese)
[7] Cheng, X.C., Huang, J.B., Pan, G.Y., et al. (2019) Theoretical study on thermal degradation mechanism of polystyrene [J]. Journal of fuel chemistry, 47(07):884-896. (In Chinese with English abstract)
[8] Frisch, M.J., Trucks, G.W. (2009) Schlegel H B, et al. Gaussian 09 Rev. D.01.Pittsburgh, PA: Gaussian, Inc.
[9] Zhao, Y., Truhlar, D.G. (2008) Density Function with Broad Applicability in Chemistry [J]. Accounts of Chemical Research, 41(2):157-67.
[10] Jiang, H. (2010) Study on ground state properties of diatomic and triatomic molecules of transition elements [D]. Qingdao University.
[11] Yu, W.B. (2009) Theoretical study on the structure and properties of amino acids and oligopeptides [D]. University of science and technology of China. (In Chinese)
[12] Masayoshi, N., Takashi. K., Kenji, K., et al. (2005) Second hyperpolarizabilities of polycyclic aromatic hydrocarbons involving phenalenyl radical units [J]. Chemical Physics Letters, 418(1).
[13] Huang, J.B., Wu, S.B., Cheng, H., et al. (2015) Theoretical study on the bond dissociation energy of lignin compound [J].Journal of fuel chemistry, 43(04): 429-436.
[14] Liu, X.X., Yin, D. L., Huang, W. Z. (2009) Pyrolysis and catalytic pyrolysis of waste polystyrene [J], Petrochemical industry, 18(32):189-1992. (In Chinese)

IOP Conf. Series: Materials Science and Engineering **729** (2020) 012019 doi:10.1088/1757-899X/729/1/012019

Application of VSP well drive data in the prediction of complex carbonate reservoirs

Li Qin [1]*, Kun Luo [1], Lili Cheng[1]

BGP Geophysical Company, CNPC

Abstract. The qixia formation carbonate rocks in the northwest of sichuan are of great buried depth, thin reservoir thickness and strong heterogeneity. Therefore, VSP data are used to control the seismic data on the surface, VSP well drive data lay a good foundation for interpretation. Then, through waveform clustering analysis, favorable facies zones were developed. Secondly, natural faults are predicted based on maximum likelihood attribute technique. Finally, the reservoir quantitative prediction is carried out by using phase control inversion technique. The drilling well location is optimized based on reservoir and fracture prediction information, and the drilling effect is good, which provides strong technical support for subsequent exploration and development in this area.

1. Introduction

The research area is located in the northwest of sichuan basin, and qixia formation has great potential for oil and gas exploration and development. For example, well A was drilled in qixia formation and obtained a high-yielding gas flow of 876,000 m^3 per day [1]. Qixia formation in this area has large buried depth, thin reservoir thickness and strong heterogeneity. Due to the limitation of resolution of conventional seismic data, it is very difficult to identify reservoirs. Therefore, the well flooding research of seismic data VSP with improved resolution was carried out, and reservoir prediction was carried out on this basis, which improved interpretation accuracy and achieved good drilling results.

Qixia formation is carbonate platform deposit in the basin. In the carbonate deep slope environment during the first habitat period, a set of thick deep limestone and biogenic (clastic) micritic limestone was mainly accumulated. Except for the coral reef limestone with partial deep water deposition, there was no large biological beach. Its reservoir performance was poor, but it had good oil generation ability. During qi-ii period, affected by the decline of sea level and the reactivation of the ancient fault in longmenshan, the study area is generally a carbonate gentle slope sedimentary environment, which is conducive to the development of high-energy beach reservoirs.

Qixia formation is relatively stable in the area. The thickness of the first habitat member is about 50 ~ 60m, and the lithology is argillaceous limestone and micritic limestone. The second qi formation is about 60 The application of seismic exploration technology has developed from geological structure to formation lithology. In order to meet the needs of fine structural interpretation and reservoir prediction in the study area, and to improve the accuracy of lithology interpretation, the research on 3d seismic data processing technology of VSP well drive in this area was carried out 70m thick and is lithologically composed of fine-mesocrystalline dolomite, micritic biogenic (clastic) limestone and cloud-like "leopard spot" limestone. The reservoir is mainly controlled by sedimentary facies and dolomization. Reservoir thickness (phi >2%) lateral variation is large, in 1.25 ~ 13.5m, reservoir velocity in 5200 ~ 5800m/s, mainly gas layer and poor gas layer.

IOP Conf. Series: Materials Science and Engineering **729** (2020) 012019 doi:10.1088/1757-899X/729/1/012019

2. Method

2.1. VSP well control treatment

The application of seismic exploration technology has developed from geological structure to formation lithology. In order to meet the needs of fine structural interpretation and reservoir prediction in the study area, and to improve the accuracy of lithology interpretation, the research on 3D seismic data processing technology of VSP well drive in this area was carried out.

VSP well flooding is to fully analyze, combine and constrain high-quality well point data and low-resolution ground seismic data, conduct quantitative analysis and optimize processing parameters, and improve data resolution without reducing signal to noise ratio. The key parameters of VSP well flooding include true amplitude compensation factor, seismic wavelet extraction of target layer, quality factor Q, etc.

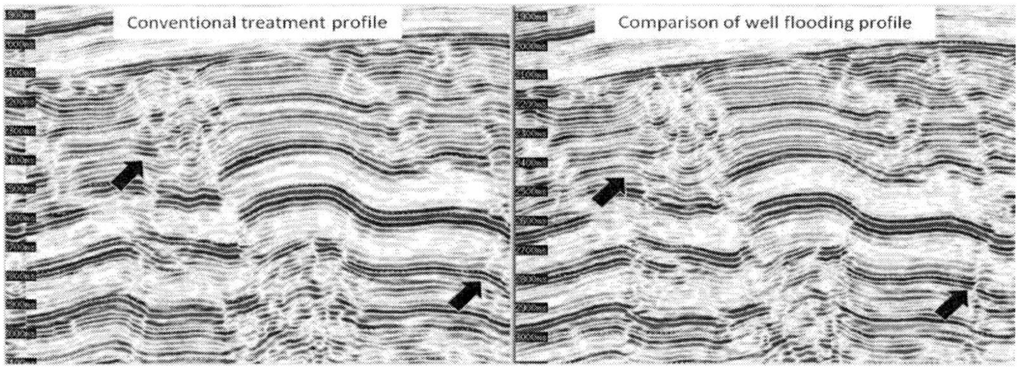

Figure 1. Comparison of well flooding profile and Conventional treatment profile.

From the perspective of the imaging section (as shown in figure 1), the frequency of the well-flooding profile can be increased by about 8 ~ 10Hz compared with the conventional one, and the diffraction return position is reasonable. The breakpoint is clear and crisp.

2.2. VSP logging horizon fine calibration technique

Seismic exploration and drilling in the working area are relatively low. Therefore, in order to ensure the accuracy and reliability of the geological horizon. The seismic reflection horizon of the target layer is precisely calibrated by combining the acoustic synthesis record and the migration time profile. It can be seen from FIG. 3 that the characteristics of the target layer wave formation relation, waveform, phase and time difference in the offset time profile are in good agreement with the acoustic synthesis records. Therefore, the reliability of horizon calibration is high.

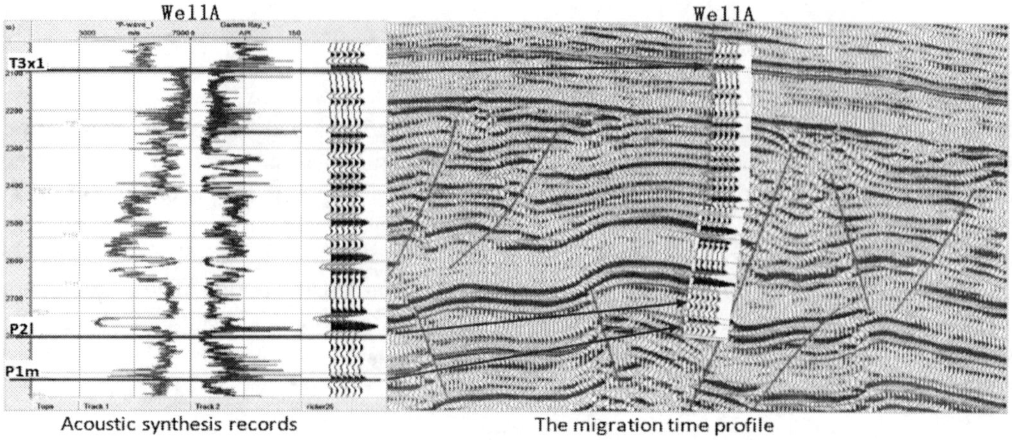

Figure 2. Fine calibration diagram of layers.

2.3. *Maximum likelihood attribute fracture prediction technique*

In seismic exploration, coherence, curvature and maximum likelihood are used to detect faults, fractures and geological boundaries. The prediction effect of the maximum likelihood attribute is good. Maximum likelihood attribute is a property proposed by Hale [2]. In the study of section extraction and fault distance estimation in 2012, which is mainly used to enhance the seismic imaging effect of faults. This method highlights discontinuity, improves lateral resolution, and makes the planar characterization of faults clearer.

2.4. *Seismic phase analysis technique*

There are mainly three seismic phases reflected from qixia formation seismic profile .You can see from figure 3. Phase one is parallel or subparallel reflection. The second is messy or dome-like reflection; The third is the broad and slow trough reflection. Seismic facies three-dimensional space using neural network identification technology, seismic facies to qixia group division, can be seen from the figure 4 earthquake phase plane distribution, south of the study area are mainly composed of parallel or the parallel reflection, mainly rolling troughs reflection in northern area, in the central area, high structure parts for clutter or mounded reflection.

Based on the analysis of rock facies and well logging facies, three points are known. Firstly, the low-energy facies zone under open platform tide shows parallel or subparallel reflection in seismic profile. Secondly, the high energy phase zones in the open platform show chaotic or dome-like reflections in the seismic profile. The third point is that the slope facies zone in the seismic profile is a broad moderate trough.. According to the geological characteristics, the chaotic or dome-like and broad and moderate trough seismic facies are favorable areas for reservoir development.

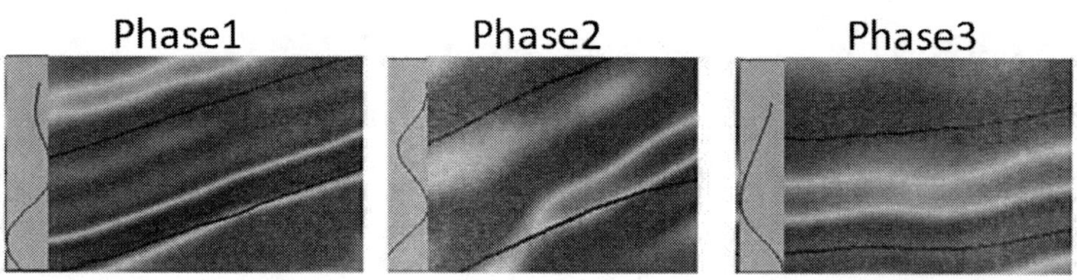

Figure 3. Typical seismic phase characteristic map.

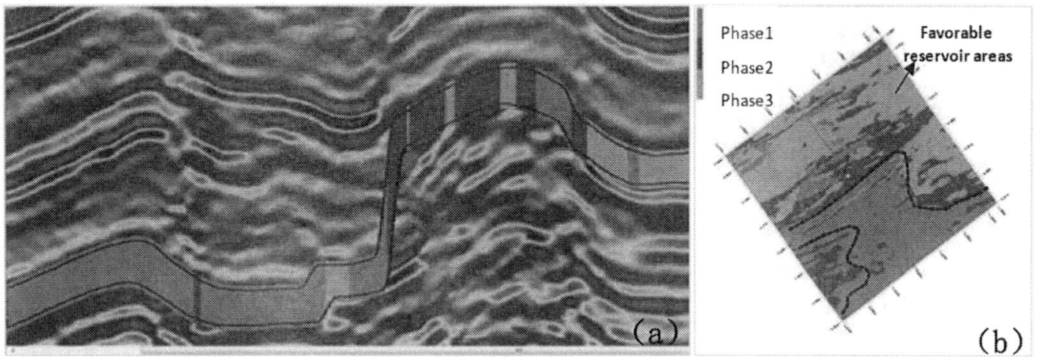

Figure 4.seismic facies classification profile (a) and seismic facies plane distribution map (b).

2.5. Phase-controlled inversion technique

The vertical and horizontal heterogeneity of qixia formation reservoir is very strong, and the conventional stratified difference method is not applicable to the carbonate platform reservoirs with massive and dam-like distribution, especially in the few Wells and low exploration areas, the contradiction is more prominent. Therefore, the seismic facies boundary information is integrated into seismic inversion, and the initial reservoir parameter model is established in the facies division zone. The inversion results are more consistent with the geological knowledge of the study area, which is conducive to improving the prediction accuracy.

3. Effect analysis

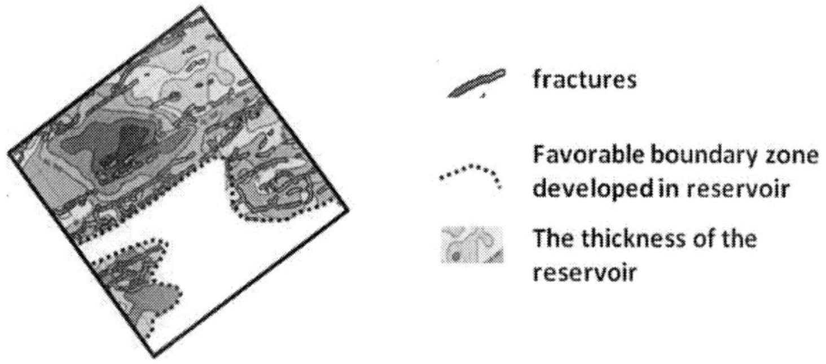

Figure 5. Composite map of fracture, seismic facies and reservoir thickness prediction results.

You can see from figure5, fractures are relatively developed in the research area. Qixia formation Reservoir in the research area is characterized by thickness in the northwest, which is more than 30 meters. According to this rule, two drilling Wells are proposed within the zone.

4. Conclusion

Using VSP well drive data to carry out reservoir prediction of qixia formation in northwest sichuan, the following understanding is obtained:

Well flooding of seismic data using VSP logging data can improve the fidelity and resolution of seismic data and facilitate the subsequent fine interpretation of structures and reservoirs.

By scanning the entire data volume to calculate the similarity between the data sample points, the maximum likelihood attribute significantly improves the accuracy of fracture zone characterization, and provides a relatively clear description of the fracture zone interior, which also serves as an indicator for fracture density development zones.

Based on the seismic data of VSP well drive, the maximum likelihood attribute fracture prediction, seismic phase analysis and phase control inversion were carried out, and the reservoir of qixia formation was qualitatively and quantitatively predicted, and the drilling well location was optimized, and good results were obtained. The process of this method has certain application value.

Acknowledgements
This article was supported by Scientifice research project of CNPC No. 2017F-14. Meanwhile, the author will appreciate BGP for permission to publish this article.

Reference

[1] ShenPing,Zhang Jian, Song Jiarong,elal.Significance of new breakthrough in and favorable tartgets of gas exploration in the Middle Permian system,Sichuan Basin【J】.Natural Gas Industry,2015,35(7):1-9.
[2] Hale Dave.Fault surface and fault throws from 3D seismic images[C]//Seg Technical Program Expanded Abstracts, 2012:205-220.DOI:10,1190/SEGAM2012-0734.1

IOP Conf. Series: Materials Science and Engineering **729** (2020) 012020 doi:10.1088/1757-899X/729/1/012020

Spectrophotometric Determination of Tiopronin using 2,2'-Bipyridyl as Chromogenic Reagent

Changqing Tu and Xinrong Wen*

College of Chemistry and Environment, Jiaying University, Meizhou, Guangdong 514015, P. R. China

Corresponding author's e-mail: 198601012@jyu.edu.cn

Abstract. The hydrosulfuryl(-SH) in tiopronin molecule has reducibility, and it can reduce Fe^{3+} to Fe^{2+}. We use 2,2'-bipyridyl as chromogenic reagent of Fe^{2+}, Fe^{2+} react with 2,2'-bipyridyl to form the orange red complex which the maximum absorption is at 522 nm. By measuring the absorbance of the orange red complex, the content of tiopronin can be determinated. A new method for the spectrophotometric determination of tiopronin has been established. The content of tiopronin is determinated using this method, and the results are satisfactory.

1. Introduction

Tiopronin(TPN, Figure 1) is a glycine derivative containing reductive sulfhydryl group and widely used in the clinical treatment of liver diseases. So far, the Chinese Pharmacopoeia used the titration[1] method to determinate the content of TPN. In addition, spectrophotometry[2,3], flow injection analysis[4], LC-UV[5], chemiluminescence[6], Electrochemical analysis[7] have been used to determinate the content of TPN.

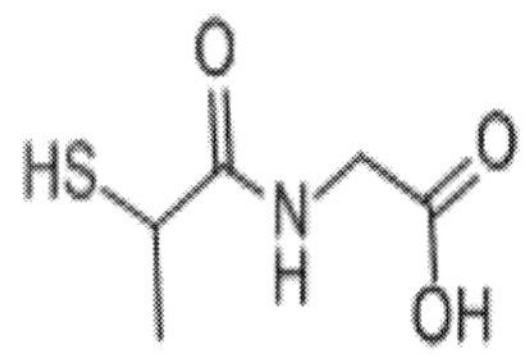

Figure 1. The molecular structure of tiopronin

The hydrosulfuryl(-SH) in TPN molecule has reducibility, and Fe^{3+}can be reduced to Fe^{2+} by hydrosulfuryl(-SH), Fe^{2+} react with the chromogenic reagent of 2,2'-bipyridyl to form the orange red complex which the maximum absorption is at 522 nm. By measuring the absorbance of the orange red complex, the content of TPN can be determinated. A new method for the spectrophotometric determination of TPN has been established. The results show that in 0.0004000-0.004000 mg/mL, the relationship between absorbance of the orange red complex and mass concentration of TPN is A=0.0057+44.652C (mg/mL). The content of TPN is determinated using this method, and the results are in good agreement with the standard method.

2. Experimental

2.1. Equipment and reagents

UV-2401 UV-visible spectrophotometer; 723S spectrophotometer.

TPN standard solution: 0.1000 mg/mL, stored at 4°C in dark place; Fe^{3+} solution: 02000 mg/mL; 2,2'-bipyridyl(BPY) solution: 0.02000 mg/mL. Buffer solutions of different pH was Prepared as references 8.

Analytical reagent grade reagents are used. Bidistilled water is used.

2.2. Method

Fe^{3+} solution (1.80 mL), bidistilled water (10.00 mL) and a given amounts of TPN solution or TPN sample solution are added into a 25 mL comparison tube, mixed well. Aftering this mixture reacted for 10 min at room temperature, 0.40 mL of BPY solution and 4.00 mL pH=5.0 buffer solution are added. The solution is diluted to 25 mL and shaked up, then the absorbance is measured at 522 nm against the reagent blank (prepared in the same way).
Figure 2. Absorption spectrum

3. Results and discussion

3.1. The optimal conditions

3.1.1. Absorption spectrum

The absorption spectrum of the orange red complex of Fe(II)-Byp is shown in Figure.1. It show that the the orange red complex of Fe(II)-Byp has an absorption peak at 522 nm. So, 522 nm is selected.

3.1.2. Effects of the reaction temperature and reaction time

According to the experimental method, Fe^{3+} solution (1.50 mL), bidistilled water (10.00 mL), BPY soluton(1.00 mL) and TPN solution (1.00 mL) are added. The effect of the reaction temperature is studied. The results show that the absorbance is not affected by the reaction temperature. So, room temperature is selected.

Keep other conditions unchanged and the reaction temperature is room temperature, the effect of the reaction time indicate that the absorbance remains constant when the reaction time is 5~80 min. So, 10 min is chosen.

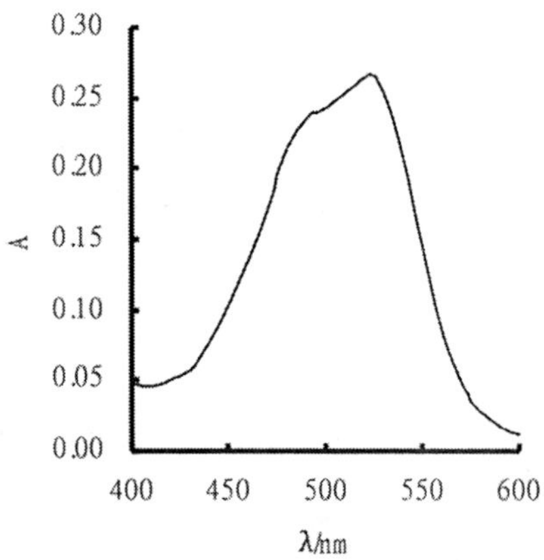

Figure 2. Absorption spectrum

3.1.3. pH buffer solution

The experimental results from figure 3 show that the absorbance achieves maximum and no longer change when pH buffer solution is 4.8~5.6. Therefore, pH 5.0 buffer solution is used.

3.1.4. The amount of pH 5.0 buffer solution

The effect of the dosage of pH buffer solution on absorbance is investigated (Figure 4). The results show that the absorbance run up to maximum and basically unchanged when the amount of pH buffer solution is 3.00 mL～8.00 mL. Hence, 4.00 mL pH 5.0 buffer solution is ensured.

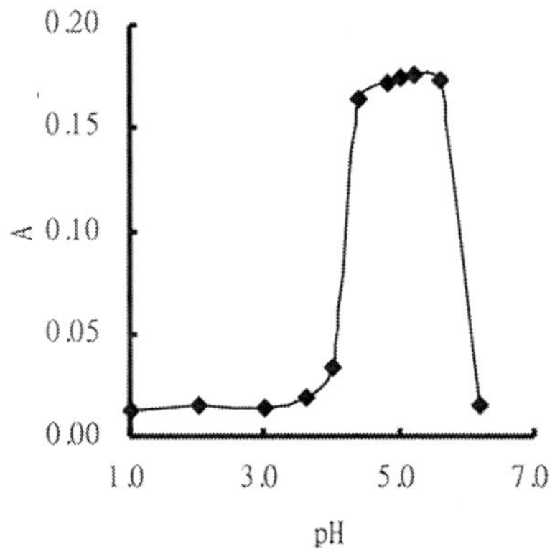

Figure 3 Effect of the pH buffer amount of solution Fe^{3+}:1.50mL; TPN:1.00mL; BPY:1.00 mL; reaction time: 5 min.

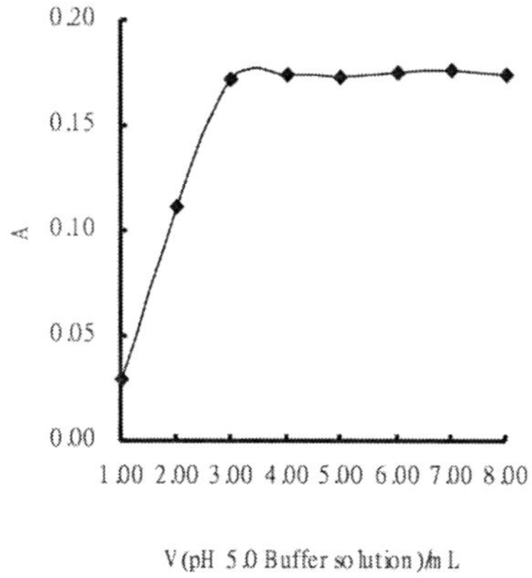

Figure 4 Effect of the pH 5.0 buffer solution Fe^{3+}:1.50mL;TPN:1.00mL;BPY:1.00 mL; reaction time: 5 min.

3.1.5. The amount of BPY solution

Figure 5 show that in the range of 0.20～1.80 mL, the absorbance is basically constant, or the absorbance is not affected by the amount of the BPY solution. Thus, 0.40 mL is used.

3.1.6. The amount of Fe^{3+} solution

It can be seen form Figure 6 that the absorbance is basic stability when the amount of Fe^{3+} solution is 1.60～2.60 mL. So, 1.80 mL Fe^{3+} solution is selected.

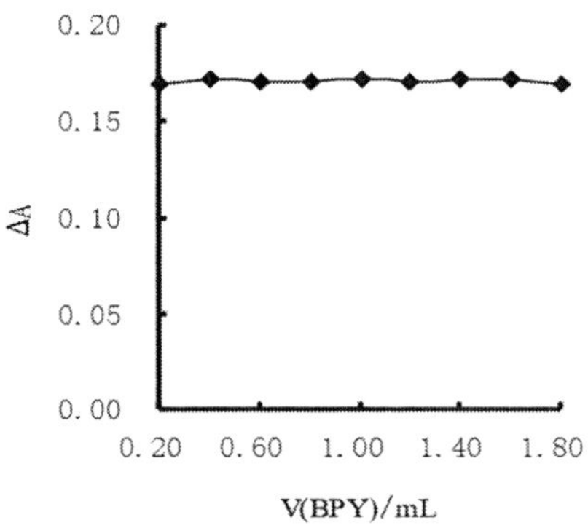

Figure5 Effect of the amount of BPYFe^{3+}:1.50mL;TPN:1.00mL;pH 5.0 buffer solution :4.00 mL; reaction time: 5 min.

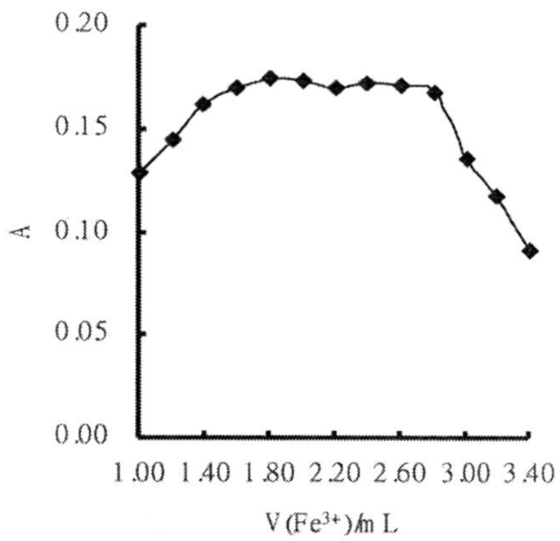

Figure 6 Effect of the amount of Fe^{3+}TPN:1.00mL; BPY:0.40 mL;pH 5.0 buffer solution :4.00 mL; reaction time: 5 min.

3.2. Calibration curve

According to the experimental method, take the mass concentration of tiopronin as the transverse coordinate and the absorbance as the longitudinal coordinate, the standard curve is drew(figure 7). In the range of 0.0004000-0.004000 mg/mL, the standard curve equation is A=0.0057+44.652C (mg/mL) with linear correlation coefficient is 0.9993.

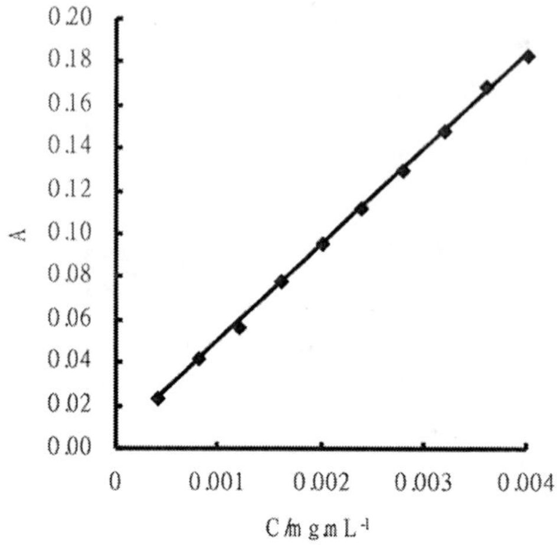

Figure 7 Calibration curve Fe^{3+}:1.80mL; BPY:0.40 mL; pH 5.0 buffer solution :4.00 mL; reaction time: 5 min.

3.3. Determination the content of tiopronin

After removing the sugar coating, round and blended,12 tablets of tiopronin tablet are weighed 3.1635g. 2.2505 g powder of tiopronin is weighed precisely and dissolved, then is transferred into a 250 mL volumetric flask and is diluted to 250.0 mL and mixed well, this is tiopronin sample solution.

The solution is preserved from light at 4°C.

According to the experimental method,the content of tiopronin is determined by tiopronin sample solution. Meanwhile, the standard addition recovery experiment and standard method experiment are performed. The results can be seen in Table 1.

Table 1. The content of tiopronin in tiopronin tablet n = 5

Sample	By proposed method ($mg \cdot tablet^{-1}$)	RSD (%)	By standard method[1] ($mg \cdot tablet^{-1}$)	Added ($\mu g \cdot mL^{-1}$)	Recovered ($\mu g \cdot mL^{-1}$)	Recovery (%)
Tiopronin tablet	91.40	0.7	90.57	0.8000 1.200	0.7928 1.218	99.1 101.5

It can be seen that the content of tiopronin is 91.40 $mg \cdot tablet^{-1}$ by proposed method, and the result is 90.57 $mg \cdot tablet^{-1}$ by standard method. The recovery yields are 99.1%~101.5%.

4. Conclusion

A new method for the determination of tiopronin has been established. The content of tiopronin is 91.40 $mg \cdot tablet^{-1}$ by this method, and the result is 90.57 $mg \cdot tablet^{-1}$ by standard method. It can be seen that the results are consistent. Obviously, this method has certain significance and application prospect.

References

[1] China Food and Drug Anministration. (2002) National Drug Standard WS1-(X-054)-2002 Tiopronin Tablet[S]. Standards Press of China: Beijing. (In Chinese)

[2] LIU L T, LI J, LI Q M. (2012) A novel method for the determination of tiopronin by using potassium ferricyanide as spectroscopic probe reagent in pharmaceutical and urine samples[J]. J.Anal.Chem., 67(1):41-46.

[3] ZHANG Y P, LI Q M, ZHAO X H, et al. (2007) The Study of indirect determination of tiopronin by extraction flotation copper(II) with an ammonium sulfate-water-n-propyl alcohol system[J]. J. Chin. Chem. Soc., 54(3), 743-748.

[4] GARCIA M S, PEDRENO C S, ALBERO M I, et al. (1993) Determination of penicillamine or tiopronin in pharmaceutical preparations by flow injection analysis[J]. J.Pharm.Biomed. Anal., 11(8):633-638.

[5] KRZYSZTOF K, EDWARD B. (2007) Simultaneous determination of tiopronin and D-penicillamine in human urine by liquid chromatography with ultraviolet detection[J]. Anal. Chim. Acta., 590(1):132-137.

[6] Lu J Z, LAU C, YAGISAWA S, et al. (2003) A simple and sensitive chemiluminescence method for the determination of tiopronin for a pharmaceutical formulation[J]. J. Pharm. Biome. Anal., 33(5): 1033-1038.

[7] SIANGPROH W, WANGFUENGKANAGUL N, CHAILAPAKUL O. (2003) Electrochemical oxidation of tiopronin at diamond film electrodes and its determination by amperometric flow injection analysis[J]. 499(1-2):183-189.

[8] CHANG W B, LI K A. (1981) Brief handbook of analytical chemistry[M]. Beijing university press: Beijing, 262. (In Chinese)

Study on Optimization Application of Iron Ion Stabilizer in the Polymer Injection Process of Offshore Oilfield

Shanshan Zhu[1], Shijie Zhu[1]*, Hong Zhang[2], Changlong Liu[2], Leiting Shi[1]*, Zhongbin Ye[1]

[1] State Key Laboratory of Oil & Gas Reservoir and Exploitation Engineering, Southwest Petroleum University, Si Chuan, Chengdu 610500, China;

[2] Tianjin Branch of CNOOC Ltd. Tianjing, 300000, China.

*Corresponding author Shijie Zhu's e-mail: 289045557@qq.com

*Corresponding author Leiting Shi's e-mail: slting@swpu.edu.cn

Abstract. A large amount of Fe^{2+}/Fe^{3+} ions are produced by corrosion and fouling in the high-pressure polymer injection process of offshore platforms, which will affect the viscosity of the polymer solution, and is difficult to effectively control by the addition of stabilizers in the curing tank. In order to solve this problem, this paper has carried out related research and the results show that: 1) Fe^{2+} mainly generates free radicals (·OH) by oxidation catalysis, to destroy long chains of polymer molecules, causing irreversible fragment breakage, resulting in a significant decrease in polymer viscosity; 2) Fe^{3+} mainly binds to hydroxyl groups, resulting in flocculation and viscosity reduction of polymers; 3) Stabilizers with complexation can effectively control Fe^{3+} to form water-soluble complexes, while excess stabilizers increase salt content and reduce polymer viscosity by electrostatic shielding effect. In order to effectively control the influence of Fe^{2+}/Fe^{3+} and the stabilizer on the polymer solution in the production process, an appropriate amount of complex stabilizer should be added to the diluent inlet of the mixture of the polymer mother liquor and the diluent.

1. Introduction

With the deepening of the development of oilfields and the application of polymer flooding technology, it is difficult to satisfy the needs of water for formulating polymer solutions simply by relying on fresh water/clean water. At the same time, environmental protection requires that the sewage generated in the oil field be re-injected into the oil layer/stratum. Most oilfields use a polymer injection process of "Preparing polymer mother liquor with clean water and then diluting and injecting the oil production wastewater" to meet the needs of oilfield production. Although the sewage is treated by ground equipment or process (such as filtration, exposure to oxygen, etc.), the salinity, bacterial content, dissolved oxygen content, reducing ions content, etc. of the treated sewage after treatment will still cause the viscosity of the polymer solution to decrease [1]. The effect of several metal ions on the viscosity of partially hydrolyzed polyacrylamide (HPAM) solution in the water preparation from large to small [2] is $Fe^{2+} > Fe^{3+} > Mg^{2+} > Ca^{2+} > Sr^{2+} > Ba^{2+}$, where Fe^{2+} is the factor of a sharp decrease in the viscosity of the polymer solution, causing the viscosity of the polymer solution reaching the oil layer is only about 30% of the designed injection viscosity [3-5], and it is difficult to satisfy the performance requirements of the reservoir for the displacement agent. In order to maintain the viscosity of the polymer, different types of treatment agents and treatment techniques have been developed [6, 7], which has greatly improved the viscosity of the polymer solution prepared by the sewage.

In the engineering application, the polymer injection process always has corrosion scale/fouling corrosion, even if it is treated by related techniques (such as oxygen exposure, adding water freshener and stabilizer), but a certain amount of iron ions (Fe^{2+} and Fe^{3+}), still reduce the viscosity of the polymer solution after reaching the bottom of the well. Although the mechanism of the influence of Fe^{2+} and Fe^{3+} on the viscosity of the polymer solution is relatively clear, the influence of Fe^{2+} and Fe^{3+} on the viscosity is a function of time, and the degree of influence on the viscosity is different during the process. Iron ion stabilizers have different effects and even poor effects [8-10]. Therefore, the mechanism and process of Fe^{2+} and Fe^{3+} with polymer solution are studied in detail, and the mechanism of its effect on solution viscosity is further analyzed, which provides guidance and ideas for the screening and optimization of iron ion stabilizer.

2. Experimental conditions and methods

2.1. Experimental materials and instruments

Iron source: analytically pure $FeCl_2$, analytically pure $Fe_2(SO4)_3$.

Polymer: Hydrophobically associating polymer (AP-P4), produced by Sichuan Guangya Company, with relative molecular mass of 6.6×10^6, solid content of 88.9% and hydrolysis degree of 21.7%.

Stabilizer: PA-TL stabilizer, the main component is sodium nitrilotriacetate

Instruments: BROOKFIELD DV-III cloth viscometer, beaker, suspended agitator (IKA), FEI Quanta450 environmental scanning electron microscope

2.2. Experimental method

(1) Effect of ion concentration (Fe^{2+} and Fe^{3+}) on viscosity of polymer solution with concentration of 1750mg/L: ① $FeCl_2$ and $Fe_2(SO_4)_3$ were prepared with Fe^{2+} and Fe^{3+} mother liquors with ion concentration of 5mg/L, respectively, and diluted to Fe^{2+}(Fe^{3+}) solution according to proportion of 0.5mg/L (0.5mg/L), 1.0mg/L (1.0mg/L), 1.5mg/L (5mg/L), 3.0mg/L (10mg/L); ② A polymer solution with a concentration of 1750mg/L was prepared by the above solution. After stirring and dissolving, the solution viscosity was measured by a cloth viscometer at a shear rate of 7.34 s^{-1}.

(2) Effect of time of action of iron ion and ferrous ion on viscosity of polymer diluent: In order to reduce the experimental error, 5.0mg/L (100mg/L), 10 mg/L (300 mg/L)of higher Fe^{2+}(Fe^{3+}) content was prepared in 1750mg/L polymer solution, the solution viscosity was measured every 8 min.

(3) Effect of pure stabilizer concentration on polymer solution: polymer mother liquor prepared by using distilled water, the ratio of the actual stabilizer to polymer solution concentration is 1:40, the amount of stabilizer added and the concentration of polymer mother liquor ratios are 1/20, 1/25, 1/30, 1/40, 1/50, 1/61.5, 1/99, 1/124, 1/165.6, 1/249, 1/499, 1/999 .

(4) Explore the control effect of stabilizers on iron ions and ferrous ions: Select a solution of Fe^{2+} and Fe^{3+} ions with concentrations of 0.5mg/L and 1.5mg/L respectively as solvent to prepare 5000mg/L polymer mother liquor, and determine the solution every 8min. Viscosity, after the viscosity is stable, the actual application ratio of 1:40 and the 1:99 ratio of the double reduction are studied separately to study the viscosity change.

(5) Environmental scanning electron microscopy was used to analyze the microscopic morphology of polymer solution under different conditions: ① the morphology of 5000mg/L polymer mother liquor prepared by solvent solution with Fe^{2+}(Fe^{3+}) ion concentration of 0.5mg/L and 1.5mg/;L②step (4) separately added stabilizer ratios 1/40 and 1/99 micro-morphology; ③ step (4) prepared in the system solution method. The scanning electron microscope can be used as follows: 1) Sample preparation: take a drop of the polymer solution from the configured polymer solution and place it on the metal sample bench and fix it. Quickly solidify the solution with liquid nitrogen to freeze the solution, and lock the structural form of polymer solution; then the water molecules are sublimed in a freeze vacuum dryer to maintain the original structure and form of the polymer. 2) Coating gold film: A special device is used to

evaporate gold with a small resistivity and cover the surface of the sample. After the gold film is applied on sample, there is not only a path for incident electrons is provided, but also the charge phenomenon of charge accumulation is eliminated, which can also improve the secondary electron emission rate, increase the signal-to-noise ratio, and obtain images with rich details and high resolution. In addition, after the sample is coated, it can also improve the mechanical strength of the sample surface and enhance the bombardment ability of the electron beam. 3) The microstructure of the polymer solution was observed by scanning electron microscopy.

3. Results

3.1. Effect of Fe^{2+}/Fe^{3+} on viscosity of polymer solution

The long-term production and development of the oilfield has caused the composition of the liquid water to become complex, and the process corrosion and scaling phenomenon has increased. Fe^{2+} and Fe^{3+} will inevitably appear in the injection polymerization process, which will have certain influence on the preparation of the polymer solution. In order to avoid the influence of other ions in the aqueous solution on the polymer, different concentrations of Fe^{2+} and Fe^{3+} were prepared by using distilled water to study the effect of the polymer on the viscosity of the polymer solution.

3.1.1. Effect of concentration on viscosity of polymer solution

The effect of different ion concentration on the viscosity of polymer solution with a concentration of 1750mg/L is shown in the following Figure 1:

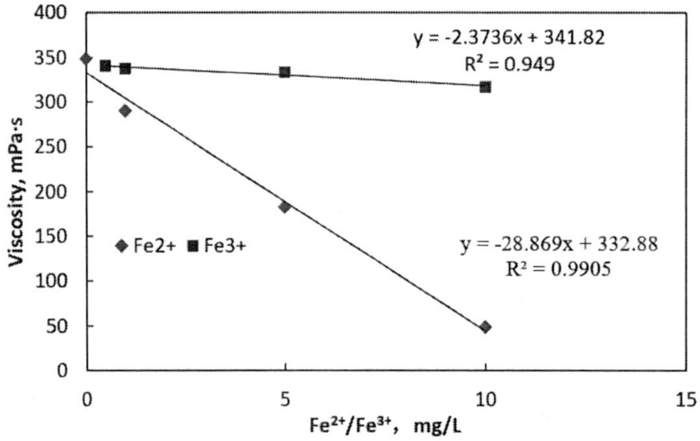

Fig.1 Effect of different Fe^{2+}/Fe^{3+} concentration on polymer solution

It can be seen from Figure 1 that the polymer solution in the Fe^{2+} solution increases linearly with the increase of Fe^{2+} content, and the fitting degree is as high as 99.05%. According to the fitting function, for each increase of 1mg/L Fe^{2+}, the viscosity of the solution decreased by 28.9mPa·s. It can be seen that the effect of Fe^{2+} concentration on the polymer solution is very significant; while in Fe^{3+} solution, the content of Fe^{3+} is increased by 1mg. The viscosity of the solution was only reduced by 2 mPa·s. Therefore, compared with Fe^{3+}, Fe^{2+} has a 14-fold increase in the influence on the solution, which indicates that the effect of Fe^{2+} with oxidation catalysis on the polymer solution is much greater than that of Fe^{3+}. The damage of Fe^{2+} in the solution is huge. .

3.1.2. Effect of action time on viscosity of polymer solution

The effects of different action times of Fe^{2+} and Fe^{3+} on the viscosity of the solution were further studied. For the Fe^{2+}, three ion concentrations of 1mg/L, 5mg/L and 10mg/L were selected, and Fe^{3+} was selected as 10mg/L and 100mg/L to prepare the polymer solution. 1750mg / L) see Figure 2 and Figure 3:

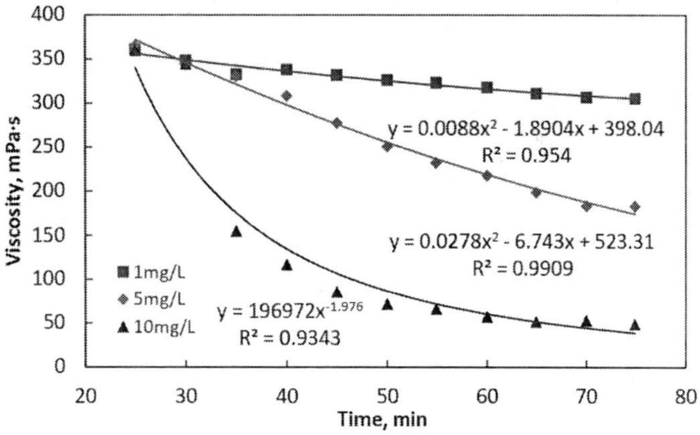

Fig.2 Effect of Fe^{2+} on viscosity of solution at different time

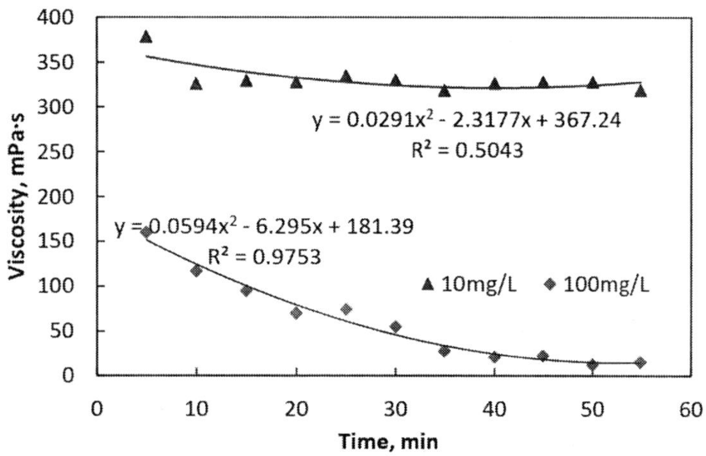

Fig.3 Effect of Fe^{3+} on viscosity of solution at different time

It can be seen from Figure 2 that different concentrations of Fe^{2+} exhibit different influence characteristics, and have a better linear decreasing relationship under lower concentration conditions (<5mg/L); the concentration decreases at 10mg/L. This indicates that Fe^{2+} will destroy the polymer in a short period of contact, resulting in a large viscosity loss. It can be seen from Figure 3 that the influence of Fe^{3+} is relatively small, and the concentration is enlarged to 100 mg/L, which is equivalent to the viscosity loss of Fe^{2+} concentration of 5 mg/L. It can also be seen from the figure that as the action time of Fe^{2+} and polymer solution is prolonged, the viscosity loss of the polymer solution gradually increases, and finally stabilizes, which means that the Fe^{2+} of the polymer solution is prepared, the earlier the treatment is better. .

3.1.3. Microscopic morphology analysis of ion-pair polymer solution

The surface morphology of the polymer solution was further analyzed by environmental scanning electron microscopy. The influence of Fe^{2+}/Fe^{3+} on the polymer solution was analyzed, as shown in Figure 4:

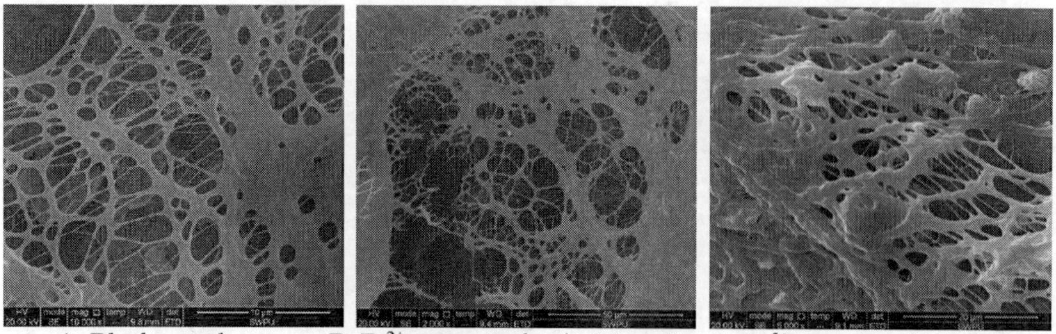

A. Blank sample B. Fe^{2+} concentration is 0.5 mg/L C. Fe^{3+} concentration is 0.5mg/L

Fig.4 Micro morphology of polymer parent under different conditions

It can be seen from Figure 4 that: 1) the blank polymer mother liquor forms a compact spatial network structure under the association; 2) the microscopic topography of the polymer solution with Fe^{2+} concentration of 0.5 mg/L, There are obvious fracture chains in some regional structures. On the more concentrated chain, the polymer is denser than the blank, because the catalytic oxidation causes the molecular chain to break. 3) The Fe^{3+} concentration is 0.5mg/L. Without destroying the polymer molecular chain, only the polymer molecular chain is aggregated by complexation, which shows a phenomenon of concentrated chain bunch aggregation.

The reason that Fe^{2+} ions have a great influence on the viscosity of the polymer solution is that Fe^{2+} ions have strong reducibility and are prone to redox reactions, resulting in the formation of free radicals. Active radicals easily cause the polymer skeleton to break, making the polymer The molecular weight is reduced, which ultimately results in a substantial loss of polymer viscosity. In addition, the oxidation of Fe^{2+} ions to Fe^{3+} ions also has a viscosity-reducing effect on the polymer solution. Therefore, Fe^{2+} ions can significantly reduce the viscosity of polymer solutions by catalytic oxidation [10].

3.1.4. Mechanism of action of Fe^{2+} and Fe^{3+} on polymers

The mechanism analysis of the influence of iron ions and ferrous ions on the performance of polymer solution: Fe^{2+} is different from Fe^{3+}, which is a reducing substance, which is easy to undergo redox reaction to form Fe^{3+}; in solution, O2 oxidizes Fe^{2+} to generate free radicals (· OH) is extremely unstable and quickly captures electrons from adjacent molecules (long chains of polyacrylamide molecules), leaving itself in a stable state, causing the polymer to break long chains, showing a viscosity rapid decline phenomenon of the polymer shown in Fig. 2, the mechanism is as shown in formula (1)-(8).

$$O_2+H_2O+2e \rightarrow \cdot OH^-+HO_2^- \quad (1)$$

$$\cdot HO_2^-+H_2O+2e \rightarrow 3OH^- \quad (2)$$

$$HO_2^- + H^+ \xrightarrow{Fe^{2+}} 2\cdot OH \quad (3)$$

$$\cdot OH+PH \rightarrow P\cdot +H_2O(P{:}HPAM) \quad (4)$$

$$P\cdot +O_2 \rightarrow P\text{-}O\text{-}O\cdot \quad (5)$$

$$P\text{-}O\text{-}O\cdot +PH \rightarrow P\text{-}O\text{-}O\text{-}H+P\cdot \quad (6)$$

$$P\text{-}O\text{-}O\text{-}H+Fe^{2+} \rightarrow P\text{-}O\cdot +Fe^{3+}+OH^- \quad (7)$$

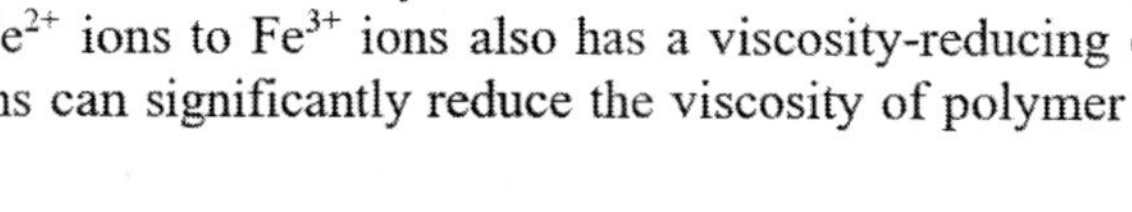

(8)

The damage of Fe^{2+} to the polymer can be manifested by the physical form (spatial geometric factor) of the polymer molecule in the aqueous solution and the change of the chemical structure. It mainly means that the "body shape becomes thinner" and the viscosity of the polymer solution becomes smaller; the polymer in the aqueous solution Molecular degradation, its molecular weight becomes smaller, the main mechanism refers to "chain becomes shorter" [5]. On the other hand, Fe^{2+} is oxidized to Fe^{3+}, and Fe^{3+} is easily bonded to a carboxyl group (-COOH) bond in the polymer molecule to form a coordination bond compound (complex), which destroys the structural characteristics of the polymer molecule and further leads to polymer viscosity. reduce.

Therefore, the effect of Fe^{3+} on Fe^{2+} on polymer molecules lacks rapid and strong oxidative degradation, and the viscosity reduction rate and extent are significantly weaker than Fe^{2+}, which confirms the experimental phenomena in Fig. 2 and Fig. 3.

3.2. Effect of stabilizer on viscosity of polymer solution

In view of the serious influence of Fe^{2+}/Fe^{3+} on the viscosity of polymer solution, the mine uses iron ion stabilizer to control the addition in the polymer preparation process, but the effect is not obvious. Therefore, the effect of the iron ion stabilizer on the polymer solution and the solution containing Fe^{2+}/Fe^{3+} polymer was investigated under distilled water conditions.

3.2.1. Effect of stabilizer on polymer solution

The effect of stabilizer content on the apparent viscosity of the polymer solution is shown in Figure 5:

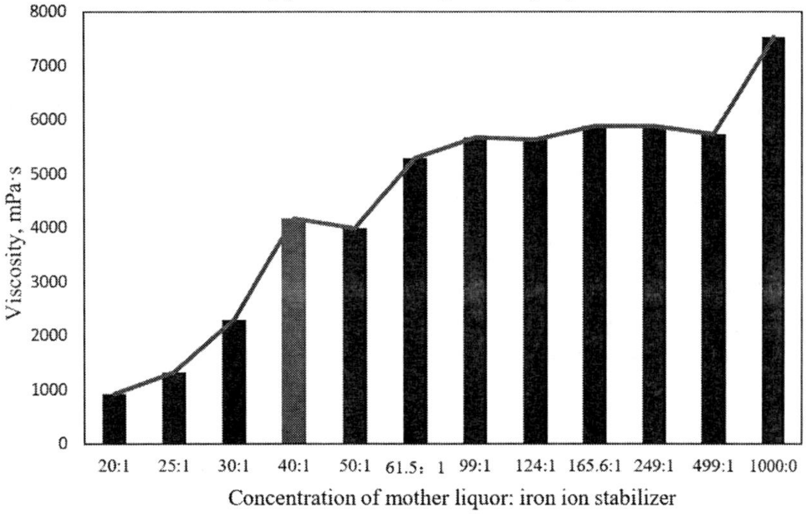

Fig.5 Effect of stabilizers with different content on polymer solution

It can be seen from Figure 5 that as the specific gravity of the stabilizer increases, the viscosity of the polymer mother liquor gradually decreases; in the range of ratio 61.5:1-499:1, the stabilizer has less damage to the polymer mother liquor, and relatively stable (viscosity loss is around 20%); in the range of ratio 40:1 and 50:1, the loss of mother liquor viscosity is further increased to 40%; and the ratio further increase, the viscosity loss increases rapidly. Since the polymer solution is prepared by using distilled water, the influence of other substances on the polymer is excluded in the aqueous solution, and only the presence of the stabilizer affects the viscosity of the polymer solution.

In order to more clearly understand the effect of the stabilizer on the solution, the analysis was compared by scanning electron microscopy, as shown in Figure 6:

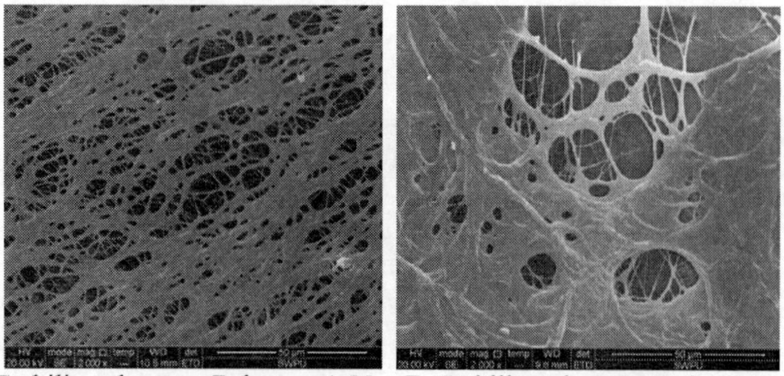

A. Stabilizer dosage : Polymer=1:99 B. Stabilizer dosage : Polymer=1:40

Fig. 6 Effect of different stabilizer dosage on the micro-morphology of polymer solution

Comparing the blank morphology of the blank sample of Figure 4 with the different amount of stabilizer in the polymer matrix of Figure 6, it can be found that the spatial structure of the polymer molecular chain in the blank sample is stretched; and the polymer solution space of the 1/99 stabilizer is added. The structure is deteriorated, the entanglement between the chains is weakened, and the viscosity of the polymer solution is decreased. As the amount of stabilizer increases (1/40), the spatial structure of the polymer solution is forced to deteriorate further, and part of the network structure "collapses". The influence of the stabilizer mainly composed of sodium nitrilotriacetate on the polymer is chemically analyzed as the polymer prepared by distilled water. Through the electrostatic shielding effect of Na+ salt brought by the stabilizer on the polymer, the viscosity of the polymer solution decreases [11].

3.2.2. Analysis of the Effect of Stabilizer on Fe^{2+}/Fe^{3+}

The polymer mother liquor was prepared by dissolving Fe^{2+} water. After being dissolved for a certain period of time, a stabilizer was added to determine the viscosity change of the solution, wherein the time of 1/99 addition was 72 min, and the time of 1/40 addition was 84 min. The results are shown in Figure 7:

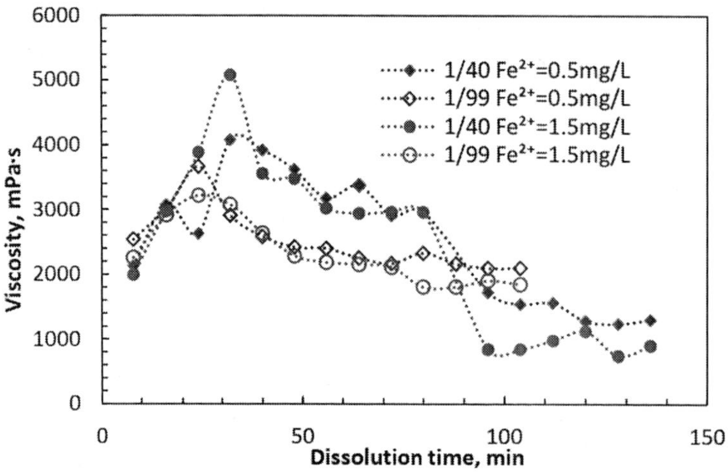

Fig.7 Stabilizer Controls the Viscosity Change of Fe^{2+} Solution

It can be seen from Figure 7 that: 1) With the stirring time, the viscosity of the solution is gradually decreased by the influence of Fe^{2+}. After adding the stabilizer, the viscosity of the polymer mother liquor is not recovered, indicating that the damage caused by Fe^{2+} to the polymer solution is irreversible; 2) Comparing different amounts of stabilizers, the stabilizer with a low adding ratio of

1/99 has little effect on the viscosity of the polymer mother liquor, while the stabilizer with a higher adding ratio of 1/40 ratio controls the Fe^{2+}, and the excess stabilizer Further damage to the polymer solution, so the viscosity of the solution is further reduced.

The experimental results of the influence of Fe^{3+} and stabilizer on the viscosity of polymer solution are shown in Figure 8:

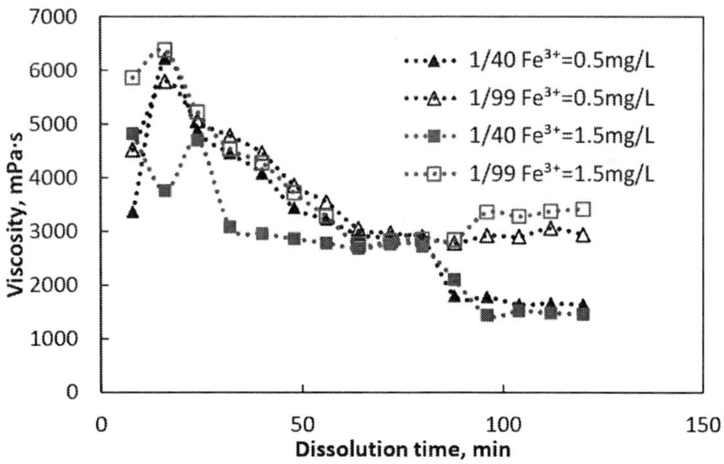

Fig.8 Stabilizer Controls the Viscosity Change of Fe^{3+} Solution

It can be seen from Figure 8 that the 1/40 ratio addition of the stabilizer further deepens the damage to the viscosity of the polymer solution, and the experimental phenomenon of the 1/99 ratio addition of the stabilizer shows the opposite state. The analysis considers that the stabilizer complexes Fe^{3+}, which reduces the adverse effect of Fe^{3+} on the polymer solution; while 0.5mg/L and 1.5mg/L appear opposite, the analysis considers that 1.5mg/L matches the amount of stabilizer added. The complexation between each other; while 0.5mg/L is complexed by the stabilizer, the excess stabilizer replaces Fe^{3+} to damage the polymer solution, so 1.5mg/L shows a good viscosity recovery phenomenon.

Based on the macroscopic experimental results of the effect of stabilizer on the performance of polymer solution, in order to further analyze the control of Fe^{2+}/Fe^{3+} in the polymer solution by stabilizer, the microscopic morphology analysis by environmental scanning electron microscopy is shown in Figure 9 and Figure 10:

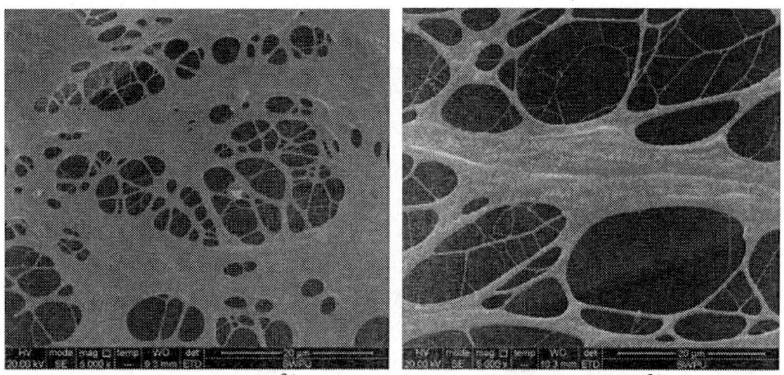

A. 0.5mg/LFe^{2+} B. 0.5mg/LFe^{2+}

Fig.9 Microscopic characteristics of stabilizer dosage of 1/99 in polymers with different Fe $^{2+}$ content

It can be seen from Figure 9 that the destruction of the polymer solution by Fe^{2+} with a concentration of 0.5 mg/L is irreversible, and the addition of the stabilizer cannot repair the broken polymer molecular chain, but merely complexes the remaining Fe^{2+} and Fe^{3+} to form a stable. The structural complex is present in the polymer solution.

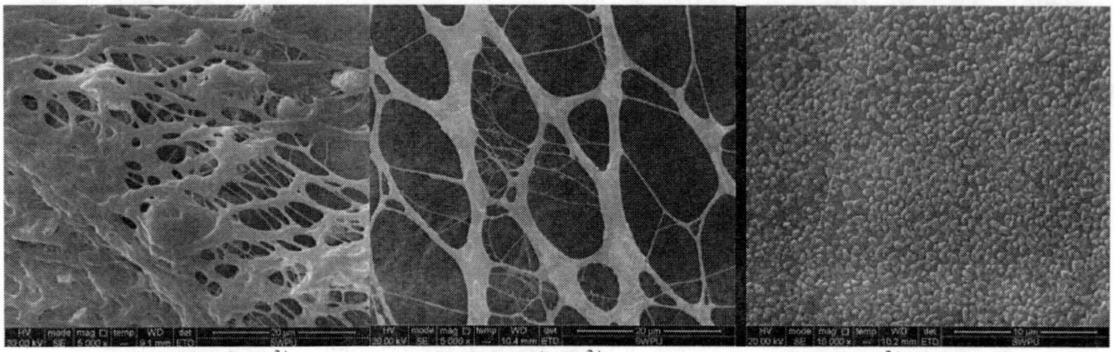

A. 0.5mg/LFe^{3+} B. 0.5mg/LFe^{3+} c. 0.5mg/LFe^{3+}(Enlarged sample)
Fig.10 Microscopic characteristics of stabilizer dosage of 1/99 in polymers with different Fe^{3+} content

It can be seen from Figure 10 that for the Fe^{3+} polymer solution with a concentration of 0.5 mg/L, before the addition of the stabilizer (left) (middle), the flocculation of the polymer due to the presence of Fe^{3+} is alleviated, and Fe^{3+} is stabilized. The complexation of the agent is stable, forming the water-soluble complex microparticles shown in the middle figure, and further magnifying the microscopic ratio of the solution (as shown in the right of Figure 10), it can be seen that many tiny granular particle complexes are formed in the solution, It shows that an appropriate amount of stabilizer can effectively complexing Fe^{3+} to form a complex to achieve the purpose of protecting the polymer solution.

Therefore, the effective control of Fe^{2+} and Fe^{3+} caused by process corrosion and scaling needs to meet the following two points: 1) The added stabilizer should be mainly complex metal cations, and does not have oxidation, otherwise it will further damage polymer solution in the oxidation of Fe^{2+}; 2) Stabilizer should be added in a high-pressure process line that is prone to corrosion. It is added to the diluted aqueous solution and mixed with the polymer mother liquor to dilute the injection layer. Premature addition will affect the polymer mother liquor.

4. Conclusion

(1) The higher the content of Fe^{2+}/Fe^{3+} in the polymer solution, the longer the contact time with the polymer, the greater the viscosity loss to the polymer ($Fe^{2+}>Fe^{3+}$); the oxidative degradation of Fe^{2+} is irreversible However, the damage to the polymer by Fe^{3+} is reversible.

(2) Excessive stabilizer will increase the sodium salt content of the solution, and causing the electrostatic shielding to reduce the viscosity of the polymer solution. The higher the content, the greater the damage caused; the proper amount of stabilizer can effectively control the corresponding amount of Fe^{3+} to form water-soluble complex.

(3) For the prevention and treatment of Fe^{2+}/Fe^{3+} generated by corrosion scale in the process, an appropriate amount of complex stabilizer should be added to the dilution water pipe before the high-pressure process pipeline poly/water mixing.

Acknowledgments
The authors acknowledge the financial support from the "13th Five-Year" National Science and Technology Major Project 25 task 3 "offshore oilfield chemical flooding technology"(No. 2016ZX05025-003)

Reference

[1] Jinwei Bao. Laboratory Study on Polymer Solution Prepared from Oil Production Sewage. Dong ying: China University of Petroleum (East China), 2007.
[2] Haiying Wang. Study on Influencing Factors of Viscosity of HPAM Solution and Selection of Complexing Agent. Daqing: Northeast Petroleum University, 2014.
[3] Xiaoming Yin, Yigeng Zhu, Yali Zhu. Effect of Sulfur Ion on Polymer Viscosity and Its Treatment. Oil and gas field surface engineering, 2010, 29(4): 54-55.
[4] Xinsheng Xue, Yongjun Guo, Rusen Feng, et al. Factors Affecting Viscosity Aging of Associated Polymer Solution in Bohai SZ36-1 Oilfield Reservoir, Oil Drilling & Production Technology, 2007, 29(5):58-61.
[5] Xin Lv. The Research on Influence Factor of the Preparation of Polymer Solution with Produced Water. Journal of Southwest Petroleum University (Natural Science Edition), 2010, 32(3): 162-166.
[6] Yuting Wang. Study on Influencing Factors of Diluted Polymer Solution Containing Polymeric Sewage and Optimum Treatment Process. Harbin: Harbin Industrial University, 2015.
[7] Shide Feng. Study on Viscosity Control Factors of Polymer Solution in Daqing Oilfield. Daqing: Northeast Petroleum University, 2013.
[8] Fang Wang. Removal of Ferrous Ions and Sulfides from Polymerized Sewage and Study on Viscosity Reduction Mechanism. Shandong: Shandong University, 2016.
[9] Xiangliang Li, Fang Wang, Shiping Xu, et al. Study on the Optimum Removal Method of Fe~ (2+) in Polymerization Wastewater from Oilfield. Industrial water treatment, 2015 ,35 (09): 73-75+109.
[10] Pengfei Yuan. Formulation of weak gel flooding system after block H1 flooding. Daqing: Northeast Petroleum University, 2015.
[11] Leiting Shi, Shijie Zhu, Zhongbin Ye, et al. Effect of salinity on aggregation behavior of hydrophobic associating polymers. Speciality Petrochemicals, 2017, 34(5): 9-14.

IOP Conf. Series: Materials Science and Engineering **729** (2020) 012022 doi:10.1088/1757-899X/729/1/012022

CdSe quantum dots (QDs) were synthesized by a hot-bubbling strategy

Renbao Wang, Qiang Sun, Zhenfeng Xu

Department Electronic and Electrical Engineering, Hefei University, Hefei 230601, PR China

ABSTRACT: A hot-bubbling strategy was studied to synthesize CdSe quantum dots. The synthetic parameters including temperature and growth time, were studied to show great influence on the optical properties. At the different reaction time CdSe quantum dots could emit fluorescence of different color, ranging from green to red, under the ultraviolet lamp which had irradiation of 365nm. The size of CdSe quantum dots was adjusted by the temperature and reaction time. CdSe quantum dots with sphalerite structure were prepared by this method.

1. Introduction

There were many disadvantages in the synthesis of CdSe quantum dots by hot-injection method. The high cost of materials, the high temperature of synthesis and the small scale of synthesis led to poor application prospects [1-5]. In this paper, a new method that was hot-bubbling was studied for preparing CdSe quantum dots. Compared with the hot-Injection method, this method had many advantages. Because of the presence of gas reactants, the heat transfer was better. The synthesis temperature was lower, which made the synthesis route be easy to control. And the method also reduced the difficulty of experimental operation. It was the large-scale synthesis which had better application prospects.

In this method, $NaBH_4$ replaced TOP which reduced the cost. $NaBH_4$, selenium powder and water were used to generate H_2Se, which was flowed into hot cadmium oleic acid solution with inert gas. CdSe quantum dots were successfully synthesized. The effects of reaction temperature and reaction time of CdSe quantum dots were studied.

2. Experimental

2.1. Materials

Cadmium oxide (CdO, Aldrich), trioctylphosphine oxide (TOPO, Aldrich), oleic acid (OA, Aldrich), selenium power (ODE, Aldrich) and $NaBH_4$(Aldrich) were used as received to prepare CdSe quantum dots.

2.2. Synthesis of CdSe QDs

The mixture of CdO(500mg), TOPO(200mg), OA(2.8ml) and ODE(16ml) were added into the flask with four necks of 80ml. The Schlenk-Line device was vacuum-pumped repeatedly. The system was heated to 200-250°C under the nitrogen. The mixture was fully dissolved and the clarified solution was obtained. Then, the mixture of $NaBH_4$(200mg) and Selenium powder(100mg) were added to the flask with three necks of 50ml. After injecting water slowly, H_2Se gas was generated, which entered the flask with four necks with nitrogen, CdSe quantum dots were grown at 210°C. In the reaction process, a small amount of reaction products were extracted at different time intervals and stored in toluene. The remaining reaction liquid was heated until the reaction was finished.

3. Results and Discussion

3.1. Influence of growth time

A small amount of samples was extracted separately and dispersed in toluene during the growth of CdSe quantum dots. The luminescent photographs of the samples were obtained under fluorescent and ultraviolet light irradiation. It was showed in the Figure 1. It could be seen from the figure that samples with different reaction time could emit fluorescence of different colos, ranging from green to red, under the ultraviolet lamp which had irradiation of 365nm. Similar to the hot-Injection method, it was intuitively confirmed that hot-bubbling method could successfully synthesize quantum dots materials with different light-emitting bands. It also showed that the size of quantum dots could be controlled by the reaction time in a certain range.

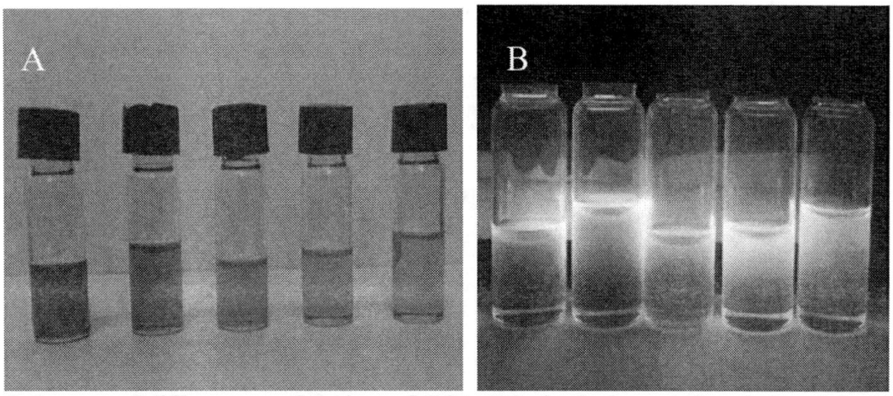

Figure 1. Images of different reaction time of CdSe QDs in the lamp (A) and ultraviolet light (B)

The properties of CdSe quantum dots were effected by the reaction time. The size of CdSe quantum dots was adjusted by the reaction time to obtain quantum dots with different light-emitting bands. Some samples were taken out and dispersed in toluene at different reaction time. After purification, CdSe quantum dots were characterized by TEM showed in Figure 2.

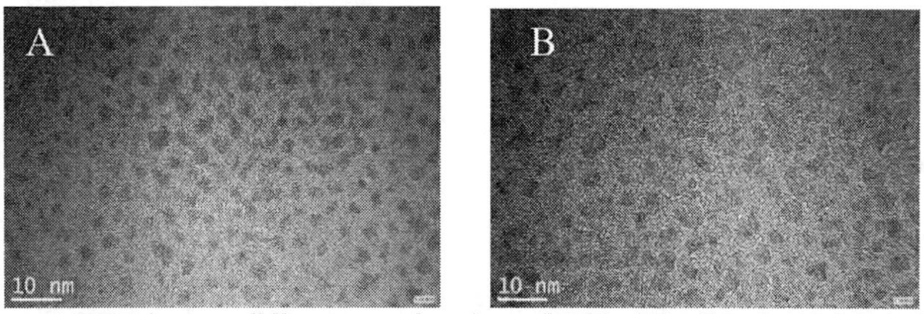

Figure 2 TEM images different reaction time of CdSe QDs (30min(A),80min(B))

3.2. Influence of different temperature

The luminescent properties of CdSe were studied on the basis of reaction temperature, which obtained quantum dots with different luminescent bands. The small amount of reaction liquid of CdSe quantum dots, at different reaction temperature, was extracted. The CdSe quantum dots were dispersed in toluene, and then characterized by XRD and TEM.

The XRD spectra of QDs was showed in Figure 3. Three main diffraction peaks of CdSe nanocrystals synthesized by this method were at 25.35°、42°、49.69°. By comparing with the library of spectra, these three peaks could be classified as crystal planes of (111), (220), (311) respectively. It showed that CdSe quantum dots with sphalerite structure were prepared by this method [6,7].

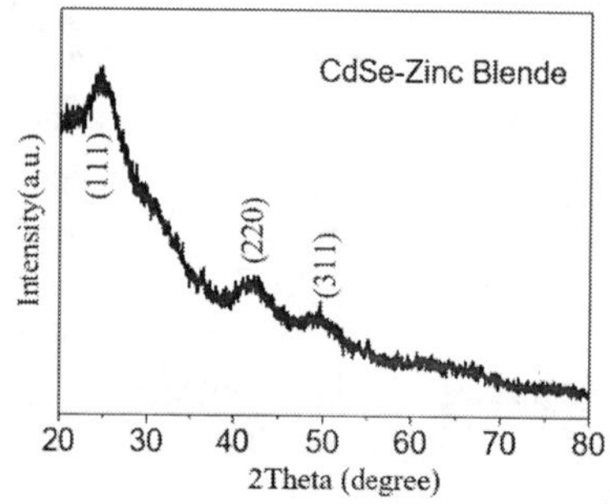

Figure 3 XRD spectra of zinc blende structure CdSe

TEM diagrams of CdSe quantum dots at 230°C and 280°Cwere showed in the Figure 4 respectively. It could be seen from the figure that the particles were well dispersed, but the size distribution was not uniform. The size of CdSe nanocrystals synthesized at 230°Cwas estimated by the empirical formula. The size of CdSe nanocrystals was about 2.0 nm. The size of particles was basically the same as TEM showed in Figure 4.

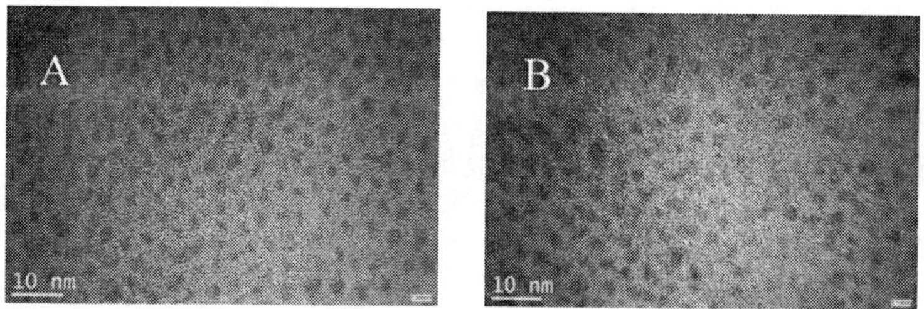

Figure 4. TEM images different temperature(230°C (A), 280°C(B))

4. Conclusions

In summary, the hot-bubbling strategy was studied for the large-scale synthesis of CdSe QDs. The optical properties of the QDs could be controlled by the reaction temperature and time. It was intuitively confirmed that hot-bubbling method can successfully synthesize quantum dots materials with different light-emitting bands.

Acknowledgements

The work was supported by the Talent Fund Project of Hefei University (18-19RC41) and PhD Foundation Research Project of Anhui University of Technology (ZY527).

References

[1] Yordanov G.G., Yoshimura H., Dushkin C.D. (2008) Phosphine-free synthesis of metal chalcogenide quantum dots by means of in situ-generated hydrogen chalcogenides. Colloid. Polym. Sci., 286: 813-817.

[2] Mandal S., Rautaray D., Sanyal A., Sastry M. (2004) Synthesis and assembly of CdS nanoparticles in keggin ion colloidal particles as templates. J. Phys. Chem. B, 108: 7126-7131.

[3] Peng X.G. (2002) Green chemical approaches toward high-quality semiconductor nanocrystals.Chem-Eur. J., 8: 334-339.

[4] Biju V., Itoh T., Anas A. (2008) Semiconductor quantum dots and metal nanoparticles: syntheses, optical properties, and biological applications. Anal Bioanal Chem, 391: 2469-2495.
[5] Wang R. , Wu X. , Xu K. , Zhou W. , Shang Y. , Tang H. , Chen H., Z. (2018) Highly Efficient Inverted Structural Quantum Dot Solar Cells. Adv. Mater., 30: 319-326.
[6] Zhang W., Wang C. Zhang L., Zhang X., Liu X., Tang K. (2000) Room Temperature Synthesis of Cubic Nanocrystalline CdSe in Aqueous Solution. J. Solid State Chem., 151: 241-244.
[7] Wang W.Z., Geng Y., Yan P. (1999) Synthesis and characterization of MSe (M=Zn, Cd) nanorods by a new solvothermal method. Inorg. Chem. Commun., 2: 83-85.

Chemical Fixed Point Sealing Technology in Oil Well

Lili Zhang[1]* Haijun Yan[1] Huijun Hao[2] Gang Xie[1] Aijun Wei[1] Lili Wei[1] Lei Sun[3] Hongfeng Wang[4]

[1]Engineering Technology Research Institute of Huabei Oilfield Company, Renqiu, Hebei 062552, China

[2]CNPC Engineering Technology R & D Company Limited, Beijing 102200, China

[3]International Engineering Company, Bohai Drilling and Exploration Engineering Company, Ltd., Tianjin 300280, China.

[4]Oil and Gas Test Company, Bohai Drilling and Exploration Engineering Company, Ltd., Tianjin 300280, China.

*Corresponding author's e-mail: cyy_zhangll@petrochina.com.cn

Abstract. When most oil fields have entered high water cut stage, the distribution of oil and water is more and more complex. Profile control and water shutoff is an important technology for water control and oil stabilization. Some types of water invasion in special reservoirs such as bottom water coning or edge water propelling in horizontal wells, water invasion in micro facture and big pore carry more challenges for profile control and water shutoff technology. Beyond that this technology is also affected by the plugging agent injection's uncertainty and migration randomness. On the basis of earlier research, this paper introduces the plugging agent formulas, technology and field application of three types of technologies which are as follows, casing channeling treatment technology, water invasion plugging technology in interlayer and water invasion in horizontal well. The three types of fixed point sealing technology can achieve the accurate injection of shutoff agent and good results had obtained in site test.

1. Introduction

Profile control and water shutoff is an important technology on water control and oil stabilization that has been commonly used in many oil fields. The water and oil relation becomes more and more complex in high water cut oil field. Water invasion in special reservoir such as bottom water coning or edge water injection in horizontal, water invasion in interlayer of thick oil layer, water invasion in fracture and big pore which have brought challenges for water control and water shutoff technology.

Water and oil wells have been operated repeatedly for profile control and water shutoff under the conditions that the reservoir is undetected and residual oil is unpredicted. The injection' uncertainty and randomness of migration of plugging agent both restrain the effect of profile control and water shutoff. Fixed point sealing technology is on the initial stage. Mechanical water plugging technique has complicated process and is limited by the type of well.[1~2] The chemical sealing in annulus is proposed by Du Yong and Zhou Zhaochuan, etc, it won't work without the packer.[3~4] Despite of the bigger agent injection radius, the commonly used general injection technique tend to contaminate the oil reservoir. Such kind of plugging agent as oil well cement is easier to cause accident. People in our country have studied the factor of plugging agent's placement on profile control and water shutoff.

Zhao Xiutai simulates the process of profile control and water shutoff in big pore reservoir using visual areal sand-packed model.[5] But there are not many reports about field application. In this paper, three types of fixed point sealing technology are studied, the result shows that they can achieve the accurate injection of shutoff agent and obtain good results in site test.

2. Cyclic plugging to channel outside casing

Factors that leading to water channeling outside casing are numerous. Pressure rises too rapidly in the process of injecting cement plugging agent to work continuously because of its property. The key techniques in this system include synthesis, modification, matching of material, chemical reaction rate control and execution program. The applied temperature scope is from 40°Cto 120°C. Through running in retrievable bridge plug inserted with a tube and perforation etc, the plugging agent cyclic channel is formed. The agent can be placed at desired position to plug water.

2.1. Thermosetting resin

Thermosetting resin is mainly composed of phenol formaldehyde resin and epoxy-modified phenolic resin. To ensure the controllability of reaction rate and curing strength, unsaturated resin, curing agent etc are added in different proportion according to reservoirs temperature. Before the injection to formation the agent is viscous liquid, and it will be solidified finally with hardly any permeability. The thermosetting resin has good chemical stability, mechanical property and high curing strength, it overcomes the disadvantages of curing shrinkage of cement slurry.

Viscosity: 50-300mPas, reaction time: 3~8h, tensile strength≥15MPa, compressive strength≥25MPa, shear strength≥12MPa, the dissolved rates in the acid, alkali and salt are≤0.5%,≤0.45%,≤0.3% respectively.

Figure 1. Sample

2.2. Application example

Yong 11 well is one development well of FU Shan oilfield in Hainan. 167A layer of Yong 11 well was perforated in June 10th 2015, by June 15th the total liquid yield is 76.1m3, oil is 54.45m3, from 16th to 18th the water yield is 10~20m3, no oil is extracted. There is water layer above and below 167A layer because of the unqualified cement quality. Water is channeled to the wellbore through perforation of 167A layer with the rising production pressure difference. To recover the capacity of 167A layer, cyclic plugging to channel outside casing technology is used.

Focusing on water channeling below 167A layer, layer 2155~2155.5m is perforated. Running in retrievable bridge plug inserted with a tube, the plugging agent is injected to water channel with cyclic plugging techniques. Focusing on water channel above 167A layer, the 167A layer is protected by sand, layer 2125~2155.3m is perforated, the plugging agent is injected to the above water channel (perforation is avoided of casing coupling). After completion, swabbing drainage from November 19th to 26th, the daily oil production is from 4.5m^3 to 18.95m^3, the daily water production is from 12.89m^3 to 0.76m^3.

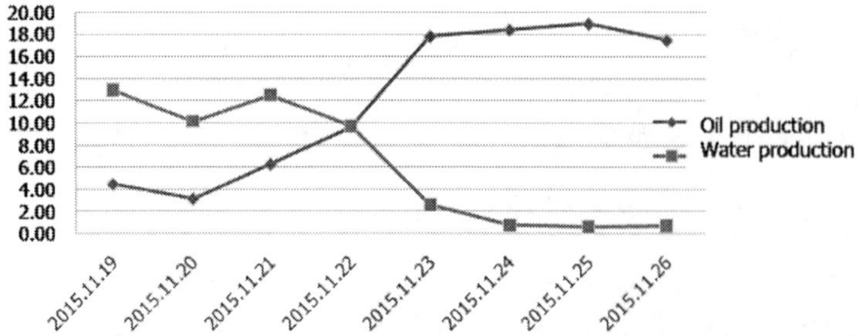

Figure2. Well testing for acquiring production after plugging completion

3. Selective water shutoff in thick oil formation

Most oil fields have entered the high water cut stage, the heterogeneity of formation is stronger. The conventional water plugging and injection profile modification technology is not incompetent now, especially for those complicated water-yielding in thick oil formation. The commonly used plugging agent can shut-off the water and oil layer at the same time. The selective plugging agent of emulsified thickened oil proposed by Gao yujun has low strength.[6] The oil- soluble material the main agent is processed into water-based emulsion at certain condition. Accumulation and adhesion of the main agents and other assistant agents will happen after being injected into formation which forming plug to water layer. Because of the oil solubility of main agent it can form temporary plug to oil layer and the plug will be eliminated soon. Fix-position plugging will be achieved at a lower speed of injection rate and pressure at which the plugging agent can be entered to water layer selectively.

3.1. Property of plugging agents

The plugging agents brown emulsion has good compatibility with water before injection to the formation and the oil solution ratio is 86.3%.Core flow experiment shows that plugging rate is over 93% and breakthrough pressure increases by about 10 times.

Figure 3. Plugging agent before and after injecting into formation

3.2. Application example

The main target layer of development is Chang2, Chang213 which belongs to lithologic-structural reservoir. The buried depth is 1100~1300m, average effective porosity is 18.2%, air permeability is 33.8mD, and the sedimentary type is distributary channel on delta plain. Nature fracture and high permeability channel cause the rising of water cut. Based on the analysis of reservoir characteristic, selective plugging agent is used in layer. This technology has been applied in 8 wells, as Dong 76-18, Dong 78-19 etc successively.

4. Fix-point plugging with modified cyanogen coagulation

Bottom water coning or edge water propelling in horizontal wells usually cause water flooding in partial or whole. The technique of water plugging in horizontal well is in the early stage. Drawing lessons from dam cracks and waterproof, fix-point plugging with modified cyanogen coagulation is proposed. It had been applied in 3 wells,among them the L horizontal with screen pipe completion used this technology for the first time.

4.1. Formula system

Cyanogen coagulation(contain –NCO performed polymer) is a chemical grouting material with polyurethane and it has bad inter solubility with water and oil. Isocyanate group can react intensely with water and polyurethane is produced in addition polymerization. The reaction equation is described as follows.

To realize the reaction rate controllability and inject ability, cyanogen coagulation is modified. On the basis of theory of compact packing of particle size distribution, hollow glass bead is doped in cyanogens coagulation. The hollow glass bead has low density (0.4~0.6g/ml), high strength (12~25MPa) and can disperse uniformly in slurry surface.

Point contact replacing surface contact makes the bubble volume and reaction rate controllable and the uniform solidification is formed finally.

Table1.Formulas at different temperatures

Temperatures °C	Solvent	Assistant	Dosage %	Glass bead %	Initial setting time (h)	Curing time (h)	Expansion multiple
25-50	Organic ester/ketone	Amine/dibutyltin Dilaurate	0.05~0.3	0.05~0.5	≥2	18~24	2
50-70		——	——		≥2	12~18	2~3
70-90		Benzoyl chloride	0.1~0.4		1~2	9~13	3

4.2. Bearing capacity of plugging agent

Test result shows that compressive strength is ≥12MPa. The agent is integrity after being compressed. Pressure test shows that cyanogen coagulation can also react in high pressure.

Figure4. The cured plugging agent

4.3. Bridging doped with fiber and process pipe with single-flow piston

Ultra fine fiber with rough surface and plasticity is injected into fracture and pore throat. They occupy the space and absorb water to saturation (the density it up to 0.9g/cm3), then the network structure is formed. The network structure is combined with the following slurry. Fluidity test shows that the viscous force increases significantly with the addition of 0.2~0.5% ultra fine fiber. Plugging pipe string is composed of several single-flow pistons, oil sealed isolating liquid, plugging agent slurry, cleaning liquid and substitution fluid. The piston is controlled by the wellhead pressure then the plugging agent can be injected to the right place in isolated space. After the injection, the oil pipe will be lift up and the well is closed for plugging agent's consolidation, at last the plugging effect is by tested by well flushing circulation.

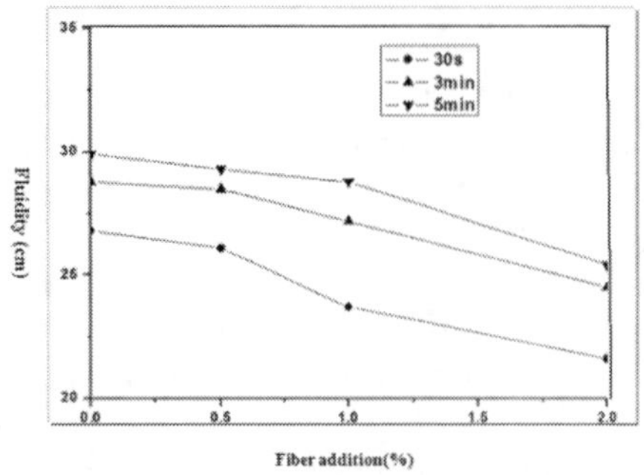

Figure5. Fiber's addition to the fluidity of slurry

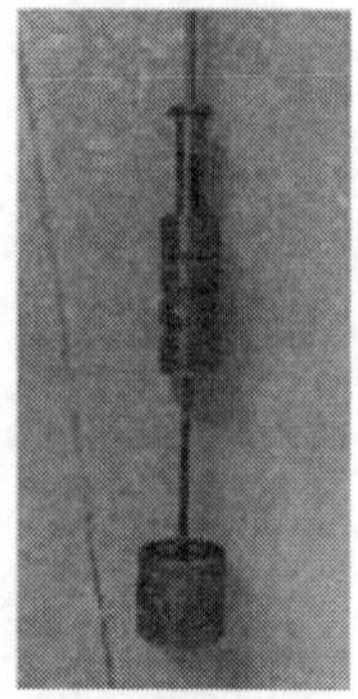

Figure6. The single-flow piston and operation pipe

4.4. Field test

F71H4 well is one L type horizontal well of Jinchen slope at Fanzhuang block in south of Qin Shui basin, Shan Xi. The daily water production is up to 90m^3 and no combustible gas is shown. The analysis shows that there is large fracture in the depth of 1709m which communicating the channel of roof and floor water of limestone cave and with no clear information of karst development. Coal-bed methane well has such characteristics of low temperature, lost circulation, collapse and strong sensitivity. Well completion and production were carried on when drilling on large fracture, the fracture of toe is not only the water invasion point but also the objective plugging layer. The adaptability of plugging agent must be considered to make sure it will not contaminate the oil layer. The depth of artificial well bottom is detected after the operation. At the same time circulation at 0.2m3/min should be formed, the plugging agent should be located within 50m of toe. The daily gas production is 1100m3 after recovery and the daily production of water is 20~30m^3.

Fixed -point sealing technology with modified cyanogen coagulation has been used in L horizontal well with screen pipe completion for the first time.

5. Conclusion

Cyclic plugging to channel outside casing, selective water shutoff in thick oil formation and fix-point plugging with modified cyanogen coagulation can achieve the accurate injection of shutoff agent and also have good results in site test.

References

[1] WEI Falin, et al. Developing trend and current situation of water plugging technology for slotted pipe horizontal wells[J].Oil Drilling & Production Technology, 2007, 29(1):40-43

[2] GE Honggiang, GOU Jingfeng, LEI Qiling, et al. Chemicals for water shutoff in horizontal wells[J].Oilfield Chemistry, 2009,26(4):387-390.

[3] DU Yong, Research and application of ACP positioning water control technology in edge-bottom water reservoir horizontal well [J].Drilling & Production Technology, 2015,38(5):44-46.

[4] ZHOU Zhaochuan, CHEN Liqin, GAO Shang, et al. Application of annulus chemical plugging for horizontal well with CESP screen pipe in Bohai Oilfield [J]. Fault-Block Oil & Gas Field, 2013, 20(3):400-402.

[5] ZHAO Xiutai, FU Minjie, BAI Baojun, et al. Visualized simulation of fine profile control and water plugging in reservoirs with secondary porous channels[J]. Oil Drilling & Production Technology, 2013, 35（5）: 84-87.

[6] GAO Yujun, MA Chunbao, BAI Baojun, et al. Selective water plugging in production wells by viscous crude oil admixtured with emulsificator[J]. Oilfield Chemistry, 1997,14（3）: 224-229.

[7] High Polymer Teaching and Research Department of Chemical Engineering, Tianjin University. Development and application of urethane precondensate grouting[J]. Chemistry, 1978(3):11-13.
[8] LIU Shangping, HAO Jian, Zhang Shunjun, et al. Modified urethane precondensates OPPT and WTPT as gelling materials for water shutoff/profile modification agents [J]. Oilfield Chemistry, 1999, 16（4）: 314-317.
[9] LIU Xuepeng, ZHANG Pingchang, FENG Minghui, Research and application of composite hollow microbead low density cement slurry ,Oil Drilling & Production Technology, 2014, 36（6）: 39-41.
[10] BAO Guiquan, Study on Some Focal Technical Problems of Drilling Engineering for Coal Bed Methane [J].Exploration Engineering, 2007, 12: 4~8.

IOP Conf. Series: Materials Science and Engineering **729** (2020) 012024 doi:10.1088/1757-899X/729/1/012024

Laboratory Study on Active Aging Oil Plugging Agent

Shasha Liu[1*], Yanru Zhang[2], Haijun Yan[1], Gang Xie[1], Cheng Lu[3], Jizhe Jing[1], Aijun Wei[1], Zhonghai Qin[1], Lei Shi[1]

[1] Engineering Technology Research Institute, Huabei Oilfield Company, Renqiu, Hebei, 062552, China

[2] Underground Gas Store Management Agency, Huabei Oilfield Company, Langfang, Hebei, 065000, China

[3] Erlian Branch, Huabei Oilfield Company, Xilinhot, Inner Mongolia, 026000, China

[*]Corresponding author's e-mail: cyy_liuss@petrochina.com.cn

Abstract: Aging oil refers to the stable emulsion of crude oil formed by emulsification that cannot be treated by conventional surfactants due to the influence of bacteria, chemical agents, self asphaltene and colloid in the process of crude oil exploitation, transportation and processing. There is a large amount of aging oil stored in the joint station of the production plant in Huabei oilfield, which not only causes certain harm to the gathering and transportation system, but also brings heavy environmental burden to the oilfield. In this study, on the basis of physical property analysis of aging oil samples, the properties of aging oil and emulsifying resin plugging agent were compared. And the water plugging principle of emulsion resin was used for reference. The system of active aging oil plugging agent is constructed by experiment. Core displacement experimental evaluation shows that active aging oil core water plugging rate was 98%, the breakthrough pressure gradient was 26 $MPa.m^{-1}$, and it has good plugging effect.

1. Introduction

Various stimulation measures in the middle and late stage of oilfield development make the composition of crude oil produced more and more complex. Aging oil is a very stable emulsion with complex components and difficulty in demulsification and dehydration. It will enter into other treatment links with the purified oil and sewage after separation, seriously damaging the oil-water separation process, affecting the quality of crude oil dehydration and the effect of oily water treatment. Current technical means cannot achieve ideal governance effect. [1-3]

After the intermediate layer of suspended oil sludge in the sewage tank and water injection tank of a jizhong oil production plant, the electric field of the electric dewatering device is unstable, and the water content of crude oil exported exceeds the standard. At the same time, a large amount of aging oil flocculent was found in sewage tank and water injection tank, which was difficult to recover and brought heavy environmental burden to the oilfield. [4-6]

This study analyzed the suspended sludge aging oil samples in sedimentation tank by experiment. And we got to know about its content and function of components. The component characteristics of aging oil and emulsion resin plugging agent were compared. On the basis of reference to the plugging principle of emulsion resin water, the experiment system of active aging oil plugging agent was constructed. Core displacement experiment confirmed that the plugging agent has good sealing performance.

2. Aging oil sample analysis

2.1. Experimental methods

Aging oil samples were taken from suspended oil sludge in settling tanks. And water content analysis of samples was mainly conducted by GB/T/8929-2006 which was "water content determination of crude oil by distillation method". The free sewage of aged oil was filtered. The polymer content was analyzed by ultraviolet spectrophotometer. The aged oil was dissolved by petroleum ether and extracted by water. The water phase was observed under a microscope of 40 times. And the particle size was analyzed by laser particle size analyzer.

2.2. Composition analysis of aging oil samples

The moisture content was 48.4%. The colloidal asphaltene content was 61.5%. The wax content was 31.9%. The polymer content in oily sewage from sewage tank was 74.75mg/L. The polymer content in oily sewage from air-floating inlet was 65.06mg/L, which was relatively high. When the particle size was a number, the median particle size is 6.1976 micron and the average particle size is 6.9637 micron. When the particle size reference was volume, the median particle size is 14.5113 microns and the average particle size is 31.8446 microns.

2.3. Dehydration effect of conventional surfactant

The effect of demulsification and dehydration of suspended oil sludge was evaluated by using surfactants such as conventional demulsifier, paraffin remover, oil displacement agent and microbial surfactant. The on-site sewage and suspended oil sludge were stirred and mixed according to the mass ratio of 100:5. Biological decomposing agents and biological enzymes were added. And the experimental phenomena were observed after 2 days of standing at 40°C. Conventional surfactants were ineffective in dehydrating aged oils, while microorganisms had the opposite emulsifying effect on suspended sludge. [7-9]

2.4. Experimental results analysis

Suspended oil sludge is flocculated at room temperature and disappears after heating. This phenomenon indicates that flocculent adsorb waxy crude oil and collect and suspend in aqueous phase at low temperature. It is difficult to penetrate into the wax crystals to play a role at normal temperature. Suspended particles and impurities in the produced liquid also enhance the strength of oil film at oil-water interface to a certain extent, which makes it difficult to demulsify oil-water.

The colloidal asphaltene content was 61.5%, and the waxy content was 31.9% in the aging oil. The properties were similar to those of main components of emulsified resin. After mixing with water the aging oil formed a stable oil-in-water emulsion, and the viscosity reached the highest at the phase transition point of the emulsion. Rapid demulsification of the emulsion could be achieved by adding inorganic salts or alkalis. And the oil droplets stick together and formed clumps that block the water outlet.

3. Construction experiment of active aging oil plugging agent system

3.1. Experimental materials and instruments

Tween, Span, OP, C01, quaternary amine cationic surfactant, nonionic surfactant, etc. Aging oil emulsification apparatus, ultrasonic apparatus, constant temperature water bath apparatus, measuring cylinder with plug, etc.

IOP Conf. Series: Materials Science and Engineering **729** (2020) 012024 doi:10.1088/1757-899X/729/1/012024

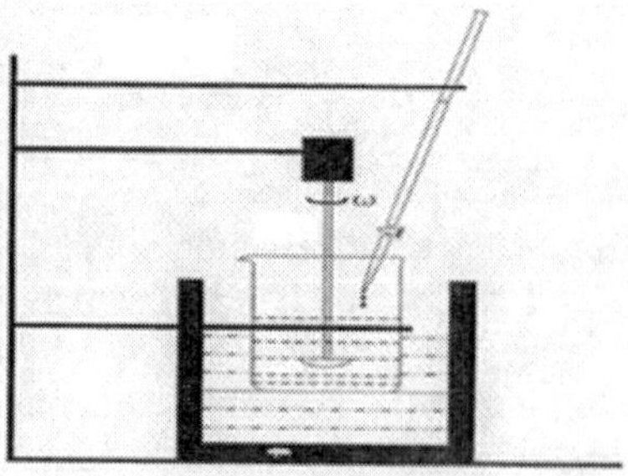

Figure 1. Emulsifying apparatus for aging oil

3.2. Experimental method

The moisture content of aging oil was determined according to GB/T/8929-2006 which was "water content determination of crude oil by distillation method". Aging oil that had been aged for a long time usually contained high moisture content. During the distillation experiment, the higher water content of the sample will affect the results. Therefore, electromagnetic stirring was carried out when the sample is heated to prevent local temperature form exploding. [10] Rheological properties of aged oil were measured by a rotary viscometer, and the shear rate was $100s^{-1}$.

3.3. Results and discussion

3.3.1. Determination of viscosity-temperature curve and water phase transition point

The aging oil was pretreated by physical ultrasonic heating dewatering method. When the dehydrated temperature of crude oil was higher than 40°C, the viscosity was significantly reduced. And the viscosity was 144mPa·s at 50°C.

The aging oil gradually formed a stable oil-in-water emulsion after mixing with water, and its viscosity increased. When the water content exceeded 52%, the emulsion viscosity decreased rapidly and changed phase.

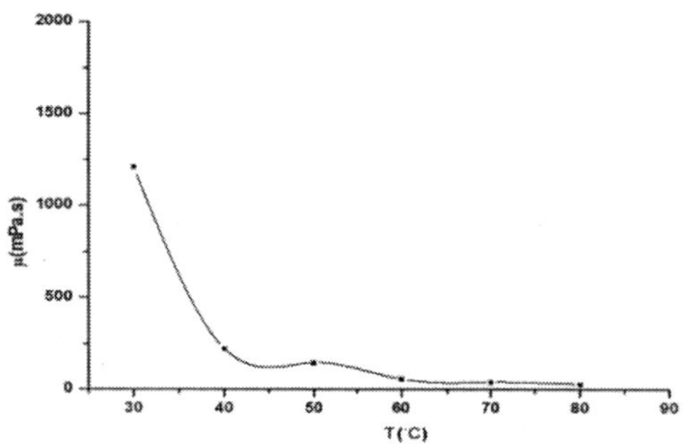

Figure 2. Viscosity-temperature diagram

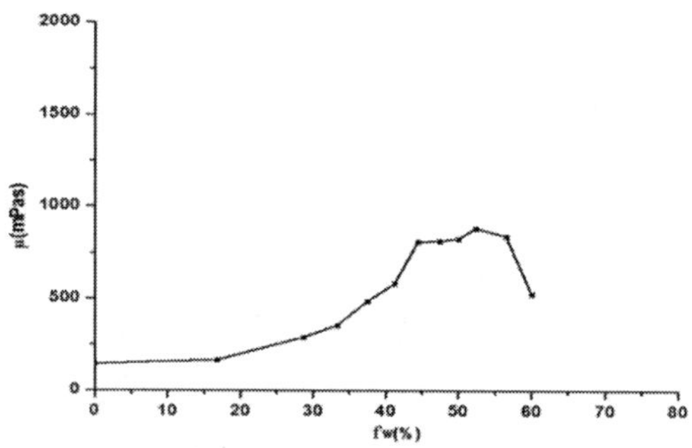

Figure 3. Emulsion water curve

3.3.2. Emulsifier screening for dehydrated aging oil

Three kinds of nonionic and two kinds of cationic surfactants were preliminarily screened. Test temperature was 50°C. Firstly, the dehydrated aging oil with a moisture content of 52% was placed in a constant temperature water bath at 50°C for 0.5hr. Emulsifier was added to the oil sample and stirred for 5min. The emulsification and viscosity of emulsion was tested by adding different kinds of emulsifiers. The results showed that the nonionic surfactant SP had good emulsifying and viscous effect.

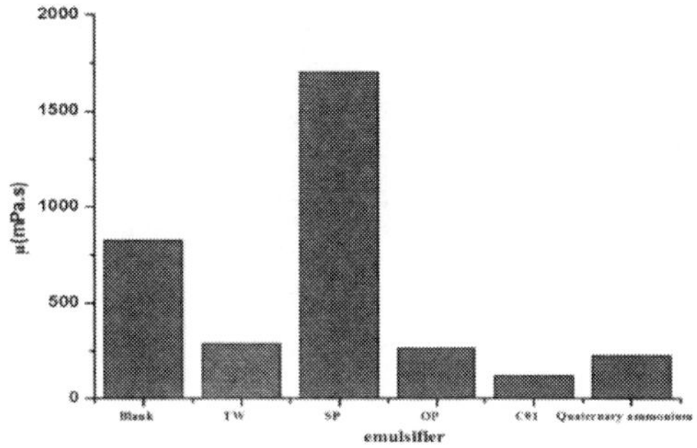

Figure 4. Emulsifying effect of different emulsifiers

3.3.3. Optimization of emulsifier concentration

After the addition of emulsifier, the water-in-oil capacity of aging oil was significantly increased. And its phase transition point was increased to 60%, as shown in figure 5. As the concentration of emulsifier increased, the viscosity of emulsion increased continuously. And the viscosity reached 1800 to 3000mpa.s when the concentration was 2% to 5%, as shown in figure 6.

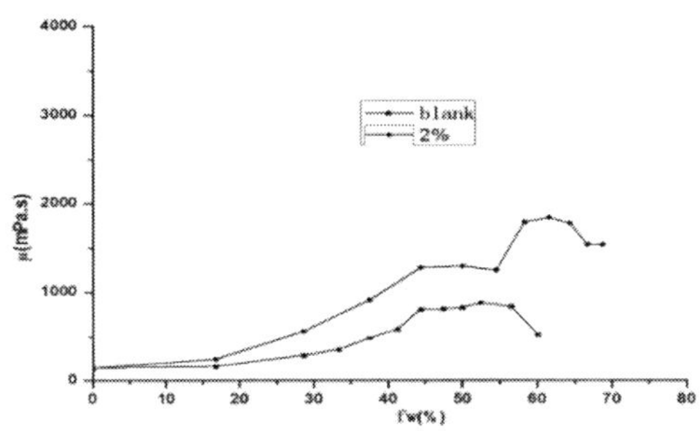

Figure 5. Phase point change after adding emulsifier

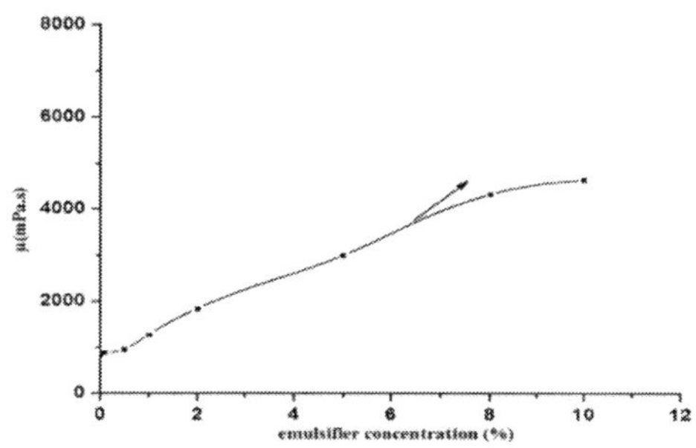

Figure 6. Viscosity of crude oil emulsion varies with the content of emulsifier

3.3.4. Demulsification experiment of active aging oil

As the emulsion is a thermodynamic unstable system, the phenomenon of partial or total separation of oil and water under certain conditions is called demulsification. First under the effect of external force, zeta potential gradually reduced as the diffusion layer gradually thinning. Droplet charge repulsion was more and more small. The droplet slowly approached, or even started to stick together and flocculate. And then the water completely flowed out between the droplets. The droplets merged and coalesced to eventually demulsify.

Active aging oil was in uniform and stable at room temperature and indoor environment. Adding some common electrolytes, such as sodium chloride, magnesium chloride, calcium chloride, aluminum nitrate, etc., can break the emulsion by reducing the electronegativity on the surface of oil beads and changing the hydrophilic and hydrophilic balance of emulsifiers, forming clumps and blocking the water outlet passage.

3.3.5. Screening of demulsifiers

According to the industry standard [11], the strength of the gel is compared according to the depth of the standard needle vertically falling into the gel. The deeper the standard needle falls, the stronger the gel. The concentration of demulsifier is optimal when the depth is the minimum. The experimental results showed that when the concentration of sodium hydroxide solution was 2% to 3%, and the concentration of calcium chloride solution was 3% to 5%, the depth of the standard needle falling into the gel was 6.5mm. And the optimal dosage of demulsifier was obtained as follows: 2% to 3% sodium hydroxide solution and 3% to 5% calcium chloride solution.

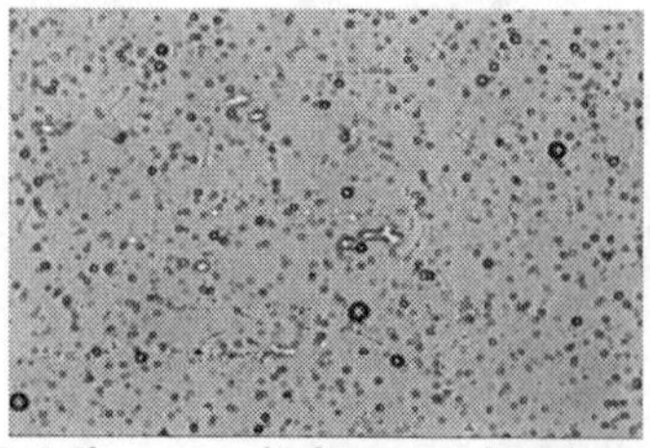

Figure 7. Microscopic image of active aged oil

4. Core displacement experiment

4.1. Instruments and materials of displacement experiment

Horizontal flow pump, hand pump, vacuum device, core gripper, intermediate container, pressure gauge, aging oil plugging agent, active aging oil and demulsifier plugging agent, clean water, artificial columnar cementing core with a permeability of 200 to 2000×10^{-3} μm^2 (diameter was 2.49cm).

4.2. Steps of displacement experiment

The single tube core displacement experiment was conducted to evaluate the breakthrough pressure, water plugging rate and erosion resistance of the core with permeability of 200 to 2000mD by two plugging agents, aging oil and active aging oil demulsification system.

Firstly, the core was flooded with water at a velocity of 2ml/min and reached the saturation state. Then, plugging agent 2PV was injected at a velocity of 0.5ml/min, and finally water flooding at 200PV. The pressure data during the experiment were recorded. The breakthrough pressure gradient and the permeability before and after core plugging were calculated, as shown in table 1.

Table 1. Single core displacement data

Core no.	Blocking adjustment type	Core permeability/mD		Water plugging rate/%	Core length/cm	Breakthrough pressure gradient/ (MPa•m^{-1})
		Before plugging	After plugging			
1#	Aging oil	1534	89	94	8.0	10
2#	Active aging oil	1538	21	98	8.0	26

The core of #2 has a higher breakthrough pressure gradient and water erosion resistance after plugging with the active aging oil system. While the core of #1 has a lower breakthrough pressure gradient after plugging with the aging oil, and the permeability increases significantly after 90PV water flooding. Therefore, the active aging oil plugging system has good injectivity and stability.

5. Cognition and conclusion

The core water shutoff rate of the active aging oil system was 98%, and the breakthrough pressure gradient was 26MPa.m^{-1}. It has good injection and water erosion resistance. The construction experiment of active aging oil plugging agent system proves that it not only has a good plugging effect, but also realizes the reuse of aging oil, reducing the pressure on the environment and the crude oil collecting and transporting system.

References:

[1] Yang Xiaoli, Lu Wanzhen. (1998) Advances in stabilization and destabilization of water in crude oil emulsions. Oilfield Chemistry, 15(1): 87-96.

[2] Liao Kejian, Sun Daquan, Yan Feng. (1999) A comprehensive evaluation of high viscous crude from no.1 Shuguang zone of Liaohe oil. Journal of Petrochemical Universities, 12(4): 5-6.
[3] Wu Bencheng, Zhu Jianhua, Liu Hongyan. (2004) Electric-chemical deaquation technique development for the extra-heavy crude oil in Liaohe oilfield. Journal of Petrochemical Universities, 17(1): 46-47.
[4] Hao Qingying. (2009) Technology for recovery of aging oil and application in Daqing oilfield. Chemical Industry, 27(3): 32-35.
[5] Xue Qiang, Wang Jian. (2008) Technology for recovery of aging oil. Inner Mongolia Petrochemical Industry, 2:19-22.
[6] Lu Baochun. (2009) Technology for recovery of aging oil. Daqing Petroleum College, Daqing.
[7] Feng Jinyu, Tang Haiyan. (2018) Research progress of microbial treatment of aging oil. Contemporary Chemical Industry, 8(47): 1691-1694.
[8] Li Huimin, Zhang Yanping, Xiao Fengpu. (2016) Test of bioremediation method for aging oil sludge treatment in Xinjiang oilfield. Oil-gas Field Surface Engineering, 35(5):62-65.
[9] Meng Xiangjuan, Chen Defei, Cao Xianping. (2016) Progress on aged oil treatment by biological technology. Chemical Industry and Engineering Progress, 35(8): 2406-2411.
[10] GB/T/8929-2006, water content determination of crude oil by distillation method.
[11] Zhang Xiaoying, Gao Qilin. (2011) GB/T4509-2010, bitumen needle penetration method. China Standard Publishing House, Beijing.

The Design of Car Head-up Display System

Min Ying

Guangzhou Civil Aviation College, Guangzhou, Guangdong, 510500, China

2402981185@qq.com

Abstract. In order to meet the requirements of driving safety and provide consumers with safer travel, this paper designs a car head-up display system. The system consists of two parts: circuit system and optical system. The circuit system is connected to the car through CANBus interface to realize data communication. Through the circuit system, the data of car driving can be obtained, and then the data can be displayed in front of the driver through the optical system, so that we can know the speed of driving, navigation and other related data without lowering our head. The test results show that the circuit system and optical system can run normally and steadily.

1. Introduction

Heads-up displays, which first appeared on fighter jets, displayed important information on a clear piece of glass in front of the plane so that pilots could focus without looking down at the instrument frequently. The technology was introduced in cars a few years ago, but few mass-produced cars come with the standard technology, which has poor display and high costs. With the continuous development of optical technology and electronic technology, head-up display plays an increasingly important role in car mounted products. The head-up display device enables drivers not to need to look down to check the display and data of the instrument, but to always keep the head up, reduce the fast change of the external environment between the head down and the head up, and reduce the delay and discomfort caused by the constant adjustment of eye focus, so as to improve driving safety.

This paper designs a head-up display system, including optical system and circuit system. The optical system uses highlighted TFT as the display light source to project the driving information to the distance in front of the car. NXP vehicle gauge platform is adopted in the circuit system, RGB888 display interface is used to connect to the display system, and CANBus is used to connect to the car to realize real-time display of car body data.

2. The structure of display system

2.1. The optical system

The optical system mainly realizes the magnified virtual image of the front of the car, which is composed of three parts: LCD, Mirror, Combiner. Refer to figure 1 for system schematic diagram.

a) LCD: In the optical system as a plane light source, in order to ensure visibility under strong light and interface permeability, LCD needs to meet the following conditions: LCD surface brightness needs to reach 30000cd/m^2, The contrast is required to reach 1200, RGB888 interface is connected to the CPU unit of the circuit system.

b) Mirror: In order to ensure that the image is far enough, the object distance should also be large enough. The use of plane or curved mirror can effectively lengthen the object distance.

c) Combiner: A concave mirror is used to magnify and extend the plane light source. Image quality can be adjusted by adjusting the Angle of Mirror and the curvature of the X and Y axes of Combiner.

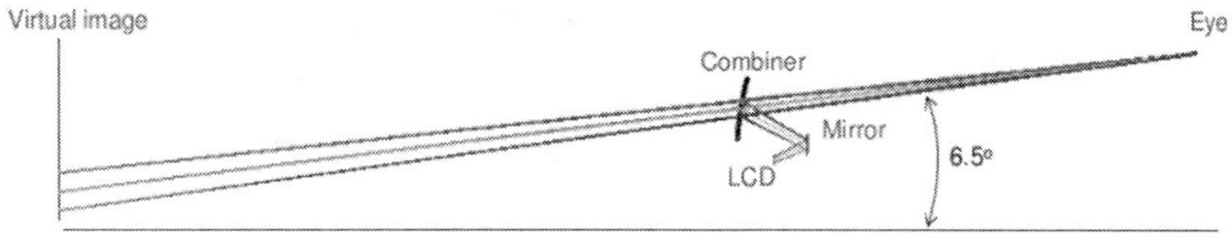

Figure 1. Schematic diagram of optical system.

The design parameters related to the optical system are shown in Table 1.

Table 1. Optical system parameter setting.

Item	Design parameters
TFT Size	1.8" 480×240
The virtual image size	> 7"
Virtual image distance	>1.6m
Down look angle	< 6.5 degree
Eye relif	800mm
Combiner	116×102mm / R475(V) & R435(H)
Mirror	110×36mm / Plate Plane

With optical simulation, Grid Distortion max -1.4939% meets the requirement of less than 2%. Simulation figure refers to figure 2.

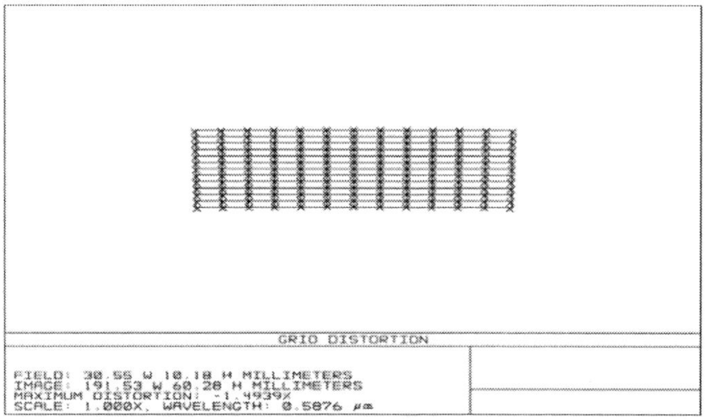

Figure 2. Optical simulation-Grid Distortion.

TV Distortion Max is 1%, which meets the requirement of less than 2%. The actual UI simulation diagram is shown in figure 3.

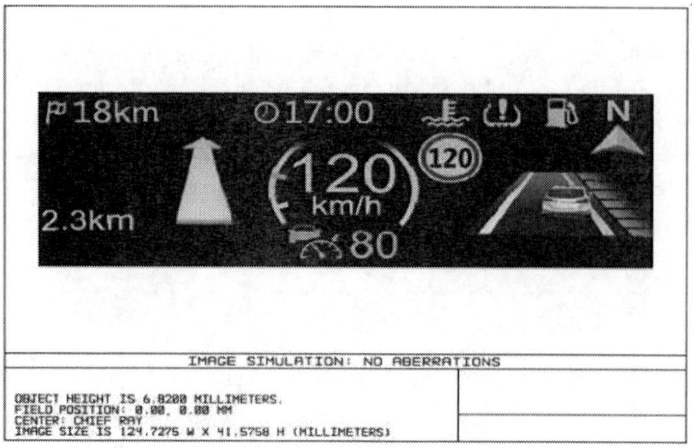

Figure 3. Optical simulation-TV Distortion.

MTF 0.42@2lp/mm, Simulation figure refers to figure 4.

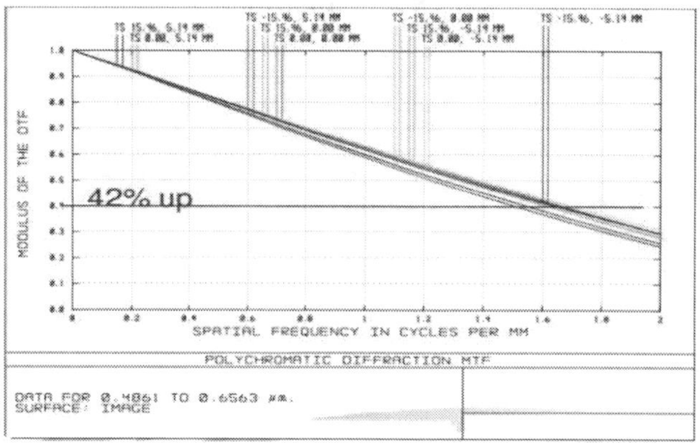

Figure 4. Optical simulation- MTF.

According to the optical parameters in table 1, after simulation, the actual imaging size is 8 inches, and the virtual image is 1.99 meters away from the human eye.

2.2. The circuit system

The circuit system is mainly responsible for collecting information of CAN transceiver unit, including vehicle driving information, vehicle alarm information, navigation information, entertainment information and function setting information, which is presented to the driver in front through the display unit (optical system). Responsible for collecting the data of the automatic photosensitive unit and realizing the automatic adjustment of backlight with ambient light to ensure safer and more comfortable driving.

The circuit system is composed of five parts, as follows:

a) Display unit (optical system interface)
b) CAN transceiver unit
c) Automatic photosensitive unit
d) Hardware interface unit
e) Power supply unit

Circuit system diagram refers figure 5.

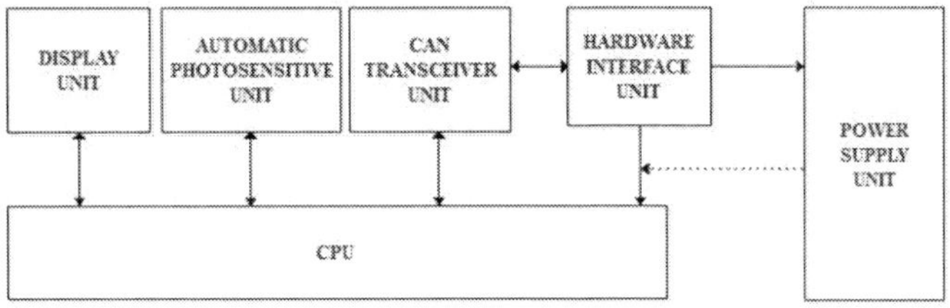

Figure 5. Circuit system block diagram.

2.2.1. CPU unit

CPU uses NXP- LPC4357 based on Arm ® architecture ®-M4/M0 kernel of high-performance microcontroller, integrated with asymmetric dual-core architecture of high performance and flexibility, as well as the serial connection, senior timer, simulation and high-speed connection, including high speed USB, CAN 2.0, 10/100 Ethernet, graphics LCD controller. Detailed block diagram refers to figure 6.

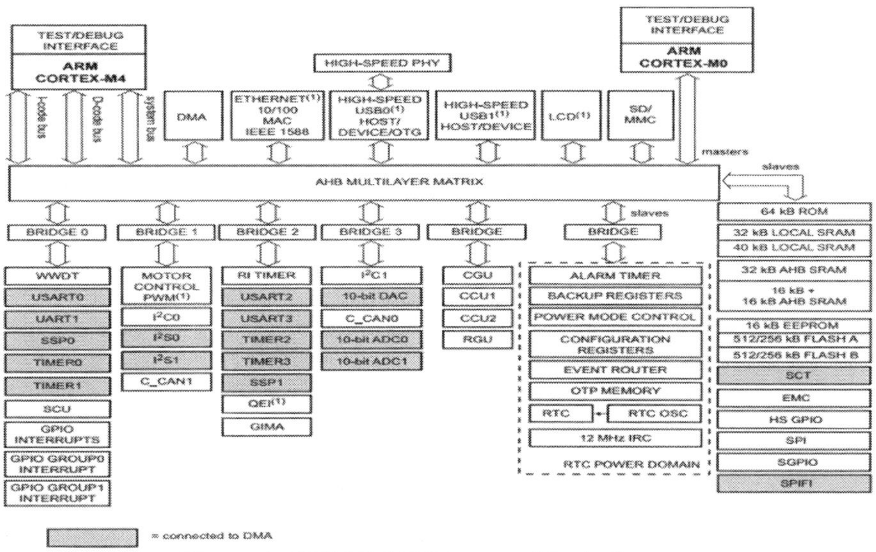

Figure 6. CPU LPC4357 system block diagram.

2.2.2. CAN transceiver unit

CAN transceiver module adopts NXP CAN transceiver chip TJA1042T/3, the main features are as follows:

a) Comply with ISO 11898-2 standard, suitable for 12V and 24V systems, with extremely low electromagnetic radiation characteristics and super anti-electromagnetic interference performance.

b) Standby mode with very low current, which can be awakened by the host bus. The power-off transceiver disconnects from the bus.

c) Protect the bus pin in the transient environment of the vehicle. Overheating protection.

As shown in figure 7, common mode inductance ACT45B-510-2P is used in the figure, mainly to suppress Electro Magnetic Compatibility (EMC) problems.

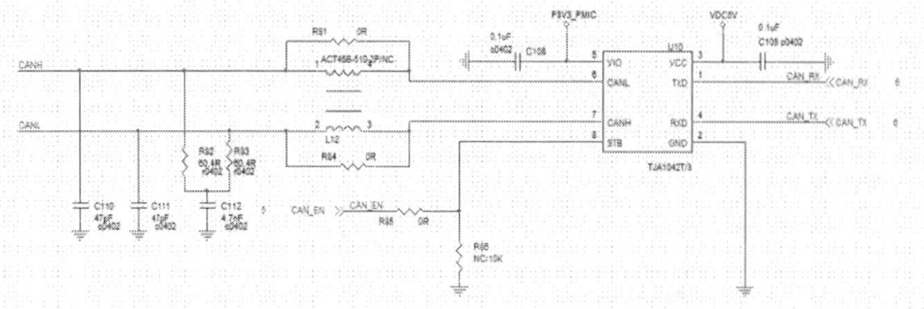

Figure 7. CAN transceiver diagram.

2.2.3. *Automatic photosensitive unit*

The light intensity of the environment will be collected by the automatic photosensitive unit in real time, and then fed back to the CPU unit through the I^2C interface to achieve the purpose of adaptive ambient light. The relevant process is as follows, refers to figure 8.

a) The optical sensor module accepts the current ambient brightness.

b) The optical sensor module sends brightness to CPU through I^2C interface.

c) after anti-shake, filtering and other relevant algorithms, the ambient brightness data received by CPU will finally output appropriate PWM to adjust the brightness of TFT backlight.

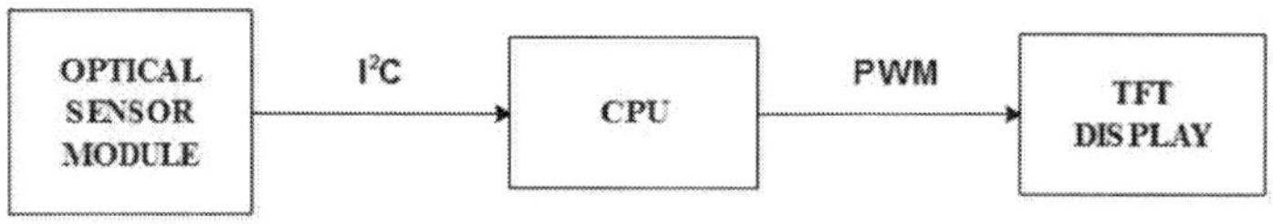

Figure 8. Automatic photosensitive unit flow chart.

The automatic photosensitive unit uses the chip ISL29023IROZ as the receiver and communicates with the CPU through I^2C. The circuit diagram refers to figure 9.

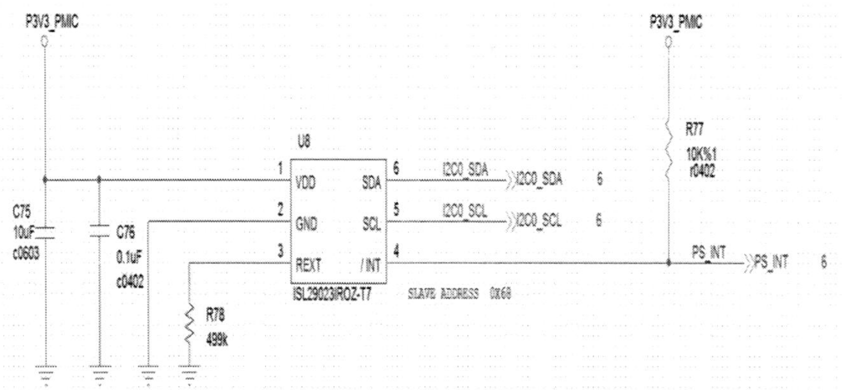

Figure 9. Automatic photosensitive unit circuit diagram.

2.2.4. *Hardware interface unit*

The hardware interface unit is responsible for connecting with the vehicle signal, and the specific distribution of the signal is shown in figure 10 below.

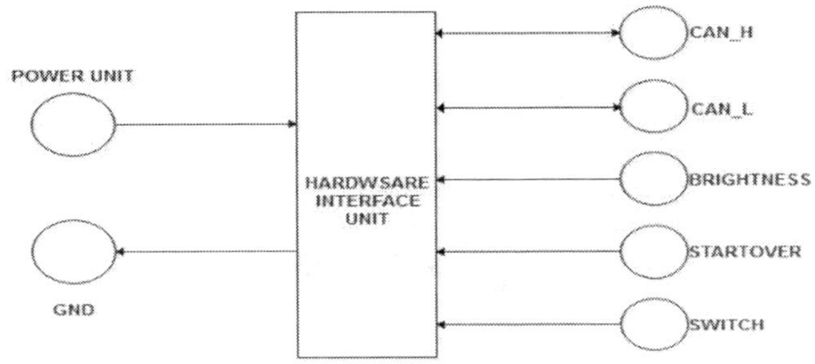

Figure 10. Hardware interface unit signal allocation diagram.

2.2.5. Power unit

The power supply unit is mainly responsible for supplying power to each module. There are three types of power supply: 12V,5 V, 3.3V. Power tree is shown in figure 11. The total power consumption of the system is 6W.

a) 5V power supply: transfer 12V to 5V CAN transceiver through DC-DC IC MPQ4423GQ-AEC1.

b) 3.3v power supply: transfer 5V to 3.3v for system CPU and IO port through LDO LM317AEMP.

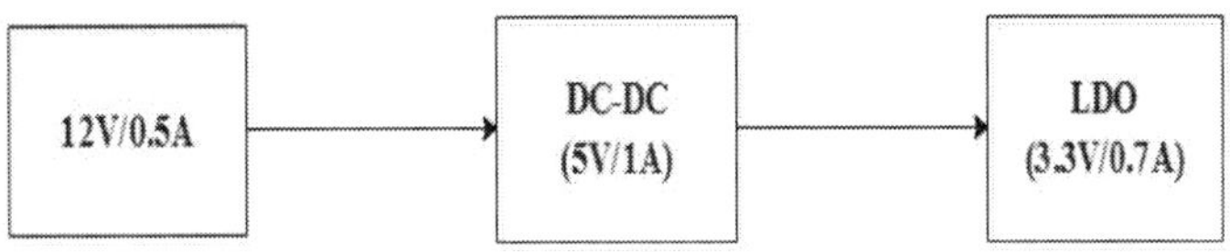

Figure 11. Power tree.

3. System software design

The system software architecture consists of three layers: Linux Kernel layer, Framework layer, Application layer.

a) Linux Kernel layer: load peripheral drivers, responsible for mounting and communication of peripherals, and obtain relevant signals effectively.

b) Framework layer: get Kernel data and carry out relevant processing to provide API interface for Application layer.

c) Application layer: which realizes relevant functions of custom APP by calling API interface of Framework layer.

Refer to figure 12 for the system software architecture diagram.

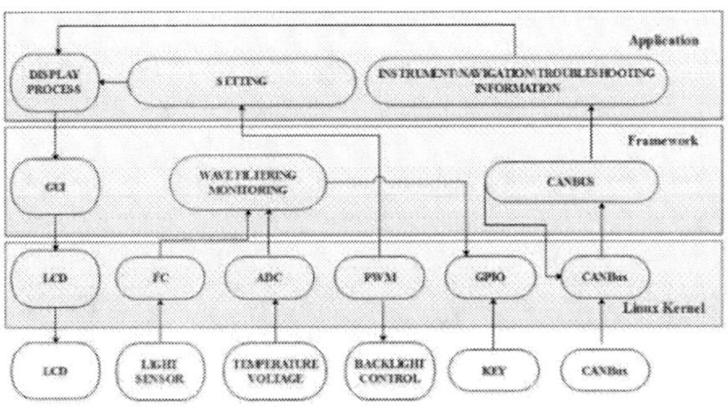

Figure 12. System software architecture diagram.

IOP Conf. Series: Materials Science and Engineering **729** (2020) 012025 doi:10.1088/1757-899X/729/1/012025

4. Physical confirmation

According to the simulation result of optical system and the design of circuit system, the production of optical material and circuit system is carried out. The actual effect diagram is shown in figure 13. According to the measurement, the actual imaging size is 8.1 inches, and the virtual image is 2.1 meters away from the human eye. All functions can operate normally and steadily.

Figure 13. Confirmation drawing of physical results.

5. Conclusion

The car head-up display system displays the driving data and pictures of the car in front of the driver through the optical system and circuit system, and it can safely drive without lowering the head. The design scheme of the system has been verified to be stable and reliable by production. It is applicable to all types of vehicles and can meet the relevant requirements of vehicle manufacturers.

References

[1] Shen, C.L., Hou, L.M. (2018) Development and trend of optical inspection technology for vehicle head - up display. Journal of lighting engineering.,5:64-67.

[2] Li, C.L. (1982) Optical design of head-up displays. The fire control technology.,1:31-41.

[3] Liu, C.H., Wei, J.M. (2014) New BMW 7 series head-up display HUD technology analysis. Vehicle electrical.,7:29-32.

[4] Yang, F., Lin, S.H., Wu, C.J. (2013) Application of head-up display system in automobile., 11:24-26.

[5] Hu, Y. (2012) Research on vehicle head-up display system. Wuhan university of technology, Wuhan.

[6] Cao, B., Wang, J.M., Luo, W.D. (2015) Research on header display system based on CAN Bus. Computer measurement and control., 7:2497-2500.

[7] Chu, Y.T. (2017) Application of head-up display technology based on automobile safety. Hebei agricultural machinery., 9: 56-56.

Finite element analysis of damage region of CBM well based on underbalance reconstruction technology

Xitao Bao [1,2], Jiang Yang[1], Yifei Yan[2], Xiangzhen Yan[2]

[1] College of Petroleum Engineering, China University of Petroleum, Qingdao 266580, People's Republic of China;

[2] Oil and Gas CAE Technology research Centre, China University of Petroleum, Qingdao 266580, People's Republic of China;

baoxitao@126.com

Abstract. The reserves of coalbed methane (CBM) in China are very high. But the characteristics of "three low" (low pressure, low permeability, low porosity), thin formation and shallow burial depth for the CBM reservoirs may cause problems of long production cycle and slow increase in production. For above problems, a new reconstruction method for coal bed methane formation named underbalance reconstruction was introduced. By this method, a jet generator was lowered to perforated interval. Then, with the nitrogen injected and blowout, reconstruction results of coalbed reservoir can be achieved. Considered the process of underbalance reconstruction, coalbed reservoir properties and mechanical properties of coalbed, mechanical model of coalbed methane well was built and analyzed by finite element method. Afterwards, damage area of different nitrogen injecting cycle can be characterized. And the effect of underbalance reconstruction for coal bed methane formation can be obtained. Then, parameters of reconstruction technology can be optimized and optimization scheme can be supplied for the underbalance reconstruction for coal bed methane formation.

1. Introduction

In recent years, natural environmental stress increases with the rapid development of economy. As an efficient clean energy and an unconventional resource with high interest (Supartono 2013, Shan et al 2005), development of coalbed methane (CBM) receives more and more attention. Commonly, hydraulic fracturing is the main technical mean for the development of coalbed methane (Shan et al 2005, Wu et al 2006, Luo et al 2008, Chen 2008). By hydraulic fracturing method, the connectivity between cleats and well can be improved. And the gas channel and flow conductivity in gas channel can be improved at the same time. Then, the pressure of CBM producing layer is released. And the pressure of gas in the producing layer is reduced at the same time. After that, CBM adsorbed in reservoir becomes free state and flows into wellbore through the gas channel that has been improved by hydraulic fracturing. The aim of increasing production of CBM well can be achieved finally. However, in CBM reservoir, there are characteristics of cleats development, shallow burial depth, low temperature and et al. These characteristics may cause many problems in process of hydraulic fracturing. The flowing time of fracturing fluid in wellbore is short. And this leads to the short crosslinking time of crosslinked fracturing fluid and linear gel fluid in wellbore. The activity of tackifier in fracturing fluid may be low. Gel breaking incompletely, filter cake and residue of fracturing fluid would produce formation contamination. In order to avoid the above problems in the process of fracturing, from the perspective of underbalanced drilling and using the principle of underbalanced drilling (Xian et al 2008, Ma et al 2000, Salehi et al 2010, Zhao et al 2001), a new reconstruction method named underbalance reconstruction was used to reconstruct CBM formation. After drilling and completion, a jet generator was lowered to slotted screen located in perforated interval. Then, five cycles of nitrogen injection were performed. Based on the constitutive model of coalbed, mechanical parameters of coalbed and pump injection parameters, the stress state and ranges of plastic zones around wellbore under five cycles of nitrogen injection can be analyzed. Thus, the effect of underbalanced reconstruction can be demonstrated. According to the analysis results, optimization scheme can be provided for underbalanced reconstruction while applied on site operation.

2. Damage characteristics of coalbed surrounding CBM well

The drilling of borehole will disturb the initial in-situ stress field and redistribute the stress distribution of coalbed surrounding CBM well. And at the same time, there is stress concentration of far field stress on the borehole wall coalbed that can cause the deformation damage or even failure of the coalbed around the wellbore.

There are a large number of joints and fractures in the coalbed mass around the wellbore, which are potential damage development sources. The strength of the weak joint surface is far less than the tensile strength of the coalbed. Due to the influence of high in-situ stress and drilling construction, when the coalbed itself does not break, it can only cause the fracture of the weak joint and crack, and the fracture length of these weak surfaces is very short without connection.

In the damage range of coalbed surrounding coalbed, the damage does not mean the deterioration of the coalbed mass caused by single fracture growth, but refers to a coalbed mass environment formed by a large number of fracture growth that complies with the statistical law. There is a regional characteristic for the damage coalbed surrounding coalbed. And with only a few fractures, this regional characteristic cannot be produced. The generation of these fractures is caused by the original defects or weak plane with fracture initiation and propagation due to the action of the in-situ stress. When the number of such fractures in the coalbed mass reaches a certain extent, the properties of coalbed mass will change obviously. Thus, the surrounding coalbed failure damage zone is formed.

Bui elastoplastic damage constitutive model can be used to analyze the damage in the coalbed mass around the wellbore. In the one-dimensional case, the damage evolution equation is given as follows:

$$D=0 \quad (\varepsilon \le \varepsilon_a) \tag{1}$$

$$D=\frac{\lambda}{E}\left(\frac{\varepsilon}{\varepsilon_a}\right) \quad (\varepsilon \ge \varepsilon_a) \tag{2}$$

where λ is reduced modulus, E is elasticity modulus, ε is the strain and ε_a is the strain under the condition of peak strength. In the triaxial case, let the principal strain be ε_1, ε_2 and ε_3, and equivalent strain can be given as follows:

$$\bar{\varepsilon}=\frac{\sqrt{2}}{3}\sqrt{(\varepsilon_1-\varepsilon_2)^2+(\varepsilon_2-\varepsilon_3)^2+(\varepsilon_3-\varepsilon_1)^2}\,. \tag{3}$$

Replace the strain of ε in one-dimensional case with the equivalent strain of and the damage evolution equation in triaxial case can be obtained as follows:

$$D=\frac{\lambda}{E}\left(\frac{\bar{\varepsilon}}{\varepsilon_a}-1\right). \tag{4}$$

By the damage evolution equation of coalbed surrounded wellbore, the stress near wellbore formation can be calculated and the stress distribution can be obtained thereby.

3. Probability statistics of mechanical properties of coalbed

Before the numerical simulation of coalbed damage in the near wellbore area by using the damage model, probability statistics of mechanical properties of coalbed surrounded wellbore were carried out based on logging and mechanical test data of CBM Wells in block W of China, as shown in table 1. And the probability distribution and the fitting curve correspondingly for coalbed surrounded wellbore can be shown in Figures.1-4.

Table 1 Statistics of coalbed mechanical parameters of W block in China

Parameters	Elasticity modulus/GPa	Poisson's ratio	Tensile strength /MPa	Shear strength /MPa
Minimum	0.241	0.162	0.054	0.065
Maximum	5.703	0.431	0.651	0.693
Average	2.981	0.292	0.361	0.354
Standard deviation	1.126	0.063	0.142	0.147
The lower limit of 95% confidence interval for the mean	2.751	0.276	0.312	0.309
The upper limit of 95% confidence interval for the mean	3.202	0.303	0.411	0.398
Coefficient of variation	0.379	0.218	0.392	0.415

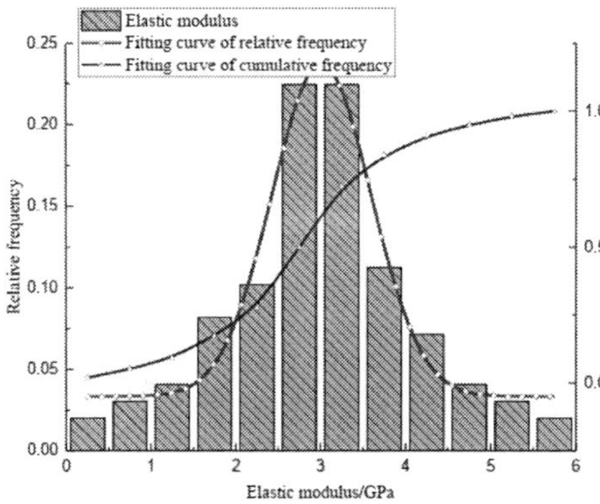

Figure 1 Probability distribution and fitting curve of elastic modulus of coalbed

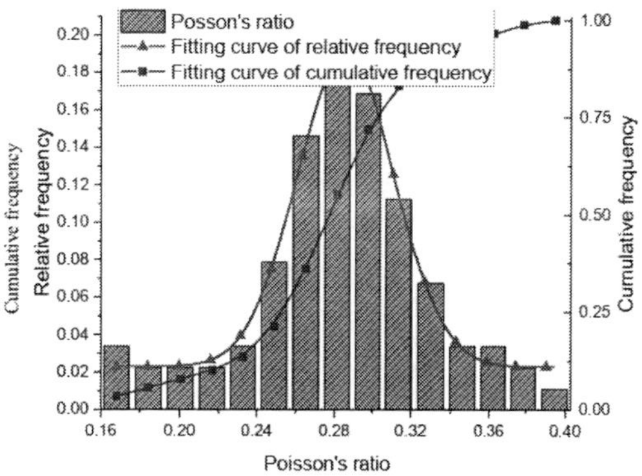

Figure 2 Probability distribution and fitting curve of poisson's ratio of coalbed

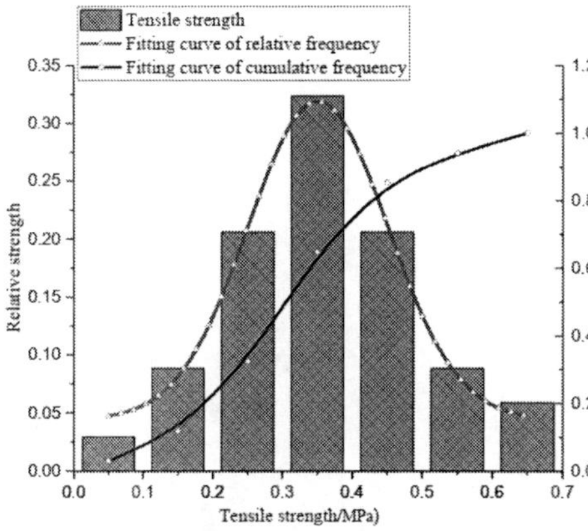

Figure 3 Probability distribution and fitting curve of tensile strength of coalbed

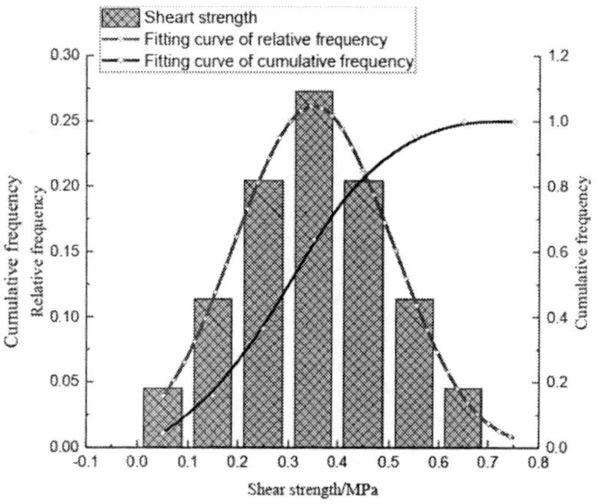

Figure 4 Probability distribution and fitting curve of shear strength of coalbed

IOP Conf. Series: Materials Science and Engineering **729** (2020) 012026 doi:10.1088/1757-899X/729/1/012026

4. Analysis of stress damage zone of coalbed surrounded wellbore under unbalanced transformation

Based on the principle of underbalanced drilling (Xian et al 2008, Ma et al 2000, Salehi et al 2010, Zhao et al 2001), after the completion of CBM drilling, slotted screen with certain specifications is lowered to perforated interval. Then, nitrogen is injected with certain pump injection pressure. And coalbed surrounded wellbore is damaged. After that, the CBM is discharged and there is an underbalanced environment for coalbed surrounded wellbore. And exfoliated pulverized coal particles in the damage zone flows into wellbore through screens and goes upwards with air current recirculation at last. So far, the first cycle of unbalanced transformation has been finished. After five cycles of unbalanced transformation, the transformation of CBM formation will be achieved finally. For the damage of coalbed surrounded wellbore, A finite element model with parameters: wellbore diameter with 0.1524m, screen fractured width with 12mm, number of circumference screens with 6, density of coalbed with 1400kg/m^3, shear strength of coalbed with 2.6MPa and pump injection pressure with 14MPa is built up. Take five cycles of unbalanced transformation and colored stress patterns applied to analyze the radial extension of the damage of coalbed surrounded wellbore can be obtained as follows:

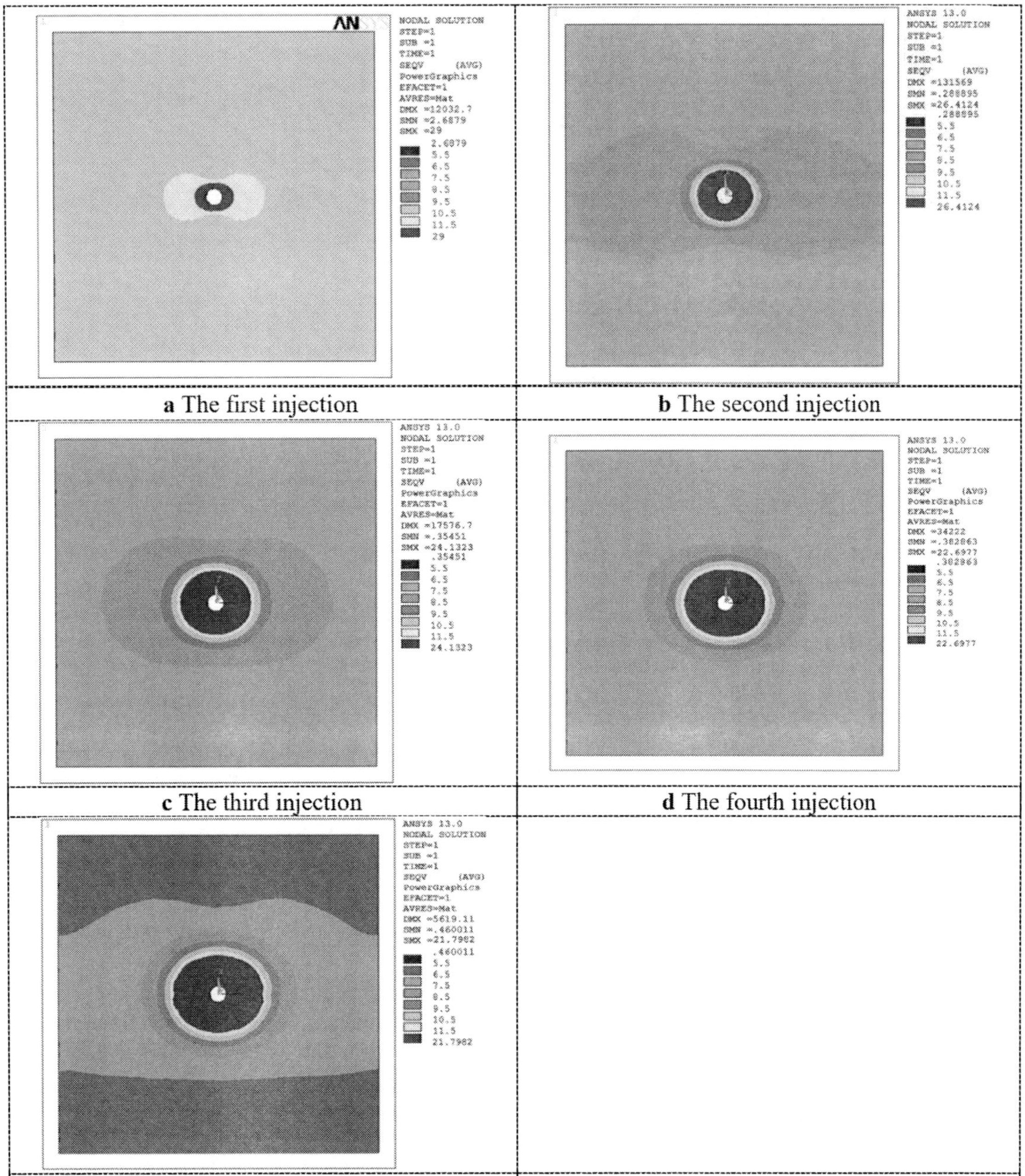

a The first injection
b The second injection
c The third injection
d The fourth injection
e The fifth injection

Figure 5. Stress distribution of coalbed surrounded wellbore with pump injection of 14MPa

As the shear strength of 2.6 MPa and the pump injection pressure of 14MPa, the relationship of shear stress and the distance from damage zone edge to the axle center of wellbore with five cycles of injection can be obtained in Figure 6. As shown in Figure 6, the stress decreases with the increase of distance in the single injection process. Around the wellbore, the stress of the next injection is less than that of the last injection. But the action region of the next pump injection pressure is wider than condition of the last pump injection pressure.

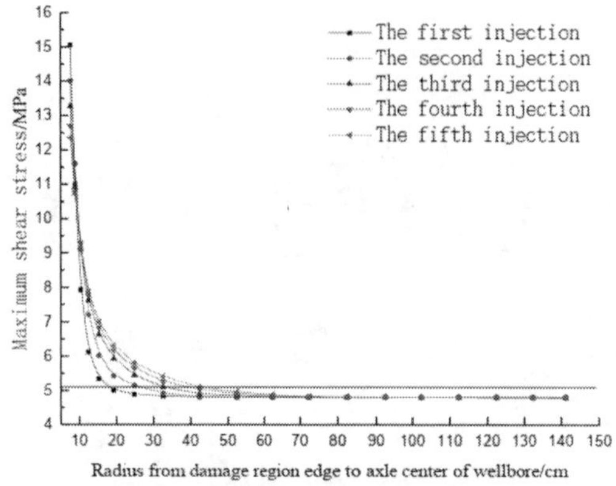

Figure 6. Relationship between axial radius and shear stress with 5 injection cycles

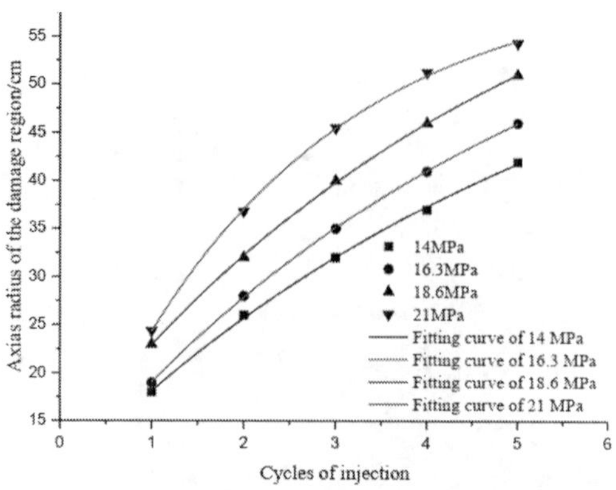

Figure 7. Axial radius of the damage region under different cycles with different pump injection pressure

As shown in Figure 7, as the shear strength of 2.6 MPa and after the first injection, damage region with the radius of 18cm was formed. After the second injection, the radius of the damage region increased 8cm. And the radius of the damage region increased 6cm further after third injection. The radius of the damage region increased 5cm after the fourth injection and 5cm after the final injection. Increase the injection pressure to 16.3 MPa, 18.6 MPa and 21 MPa correspondingly, the extension of the damage region can be seen Figure 7. And The extension of the damage region through the whole process of injection conforms to exponential decay law. The fitting formula can be seen in Table 2.

Table 2. Fitting formula of axial radius of the damage region under different pump injection pressure

Pump injection pressure	Fitting formula	Correlation coefficients	Reduced Chi-Sqr
14MPa	R=-55.81572*exp(-p/5.67629)+ 64.93243	0.99923	0.13596
16.3MPa	R=-60.34962*exp(-p/5.12622)+ 68.70652	0.99992	0.019
18.6MPa	R=-61.85799*exp(-p/4.91001)+ 73.38633	0.99981	0.04705
21MPa	R=-56.5705*exp(-p/2.37077)+ 61.39579	0.99959	0.11906

Under five cycles of pump injection, the damage region increases during five cycles of pump injection. And the increasing degree conforms to exponential attenuation law. The total axial radius of the damage region increases with the increasing of pump injection pressure. And the degree of attenuation increases with the increasing of pump injection pressure.

5. Conclusions

(1) Based on the principle of underbalance reconstruction, properties of coalbed reservoir and damage characteristics of coalbed surrounding CBM well, a finite element model was built to analyze the extension of the damage region for the CBM well.

(2) With the increasing of pump injection pressure, the damage region was enlarged. And the enlarged degree conforms to exponential attenuation law.

(3) As the effect of pump injection pressure on the damage region, the underbalance reconstruction technology can be optimized to maximize CBM well productivity.

Acknowledgements
This research was supported by National Natural Science Foundation of China (51804330), the Fundamental Research Funds for the Central Universities (Grant No. 18CX02154A) and Major Project of the National Science and Technology of China (Grant No. 2016ZX05017-003).

References
[1] Supartono Y.C., Lee J. (2013) Well test simulation as a validation method for new CBM reservoir model. International Society of Offshore and Polar Engineers pp 50-3.
[2] Shan X.J., Zhang S.C., Li A.Q., et al (2005) Analyzing the fracture extended law of hydraulic fracturing in coalbed gas wells. Natural Gas Industry 2:130-2.
[3] Wu X.D., Xi C.F., Wang G.Q. (2006) The mathematic model research of complicated fractures system in coalbed methane well. Natural Gas Industry 12:124-6.
[4] Luo T.Y., Zhao J.Z., Wang J.H., et al (2008) Forming mechanism of complicated fractures in hydraulic fracturing Fult-Block Oil & Gass Field 3:46-9.
[5] Chen M., Jin Y., Zhang G. Q. (2008) Petroleum engineering rock mechanics. Science Press Beijing pp.30-9.
[6] Xian B.A., Sun P., Wang Y.B., et al (2008) Research and application of under-balanced drilling technology in drilling horizontal CBM well. China coalbed methane 1:5-8.
[7] Ma Z.J., Xiao R.D. (2000) Recent underbalanced drilling technology. Driling & production technology 3:1-4.
[8] Salehi S., Hareland G., Nygaard R. (2010) Numerical simulations of wellbore stability in under-balanced-drilling wells. Journal of Petroleum Science and Engineering 3-4:229-35.
[9] Zhao J.Z., Sun R.H., Yan X.L. (2002) Applications of under-balanced drilling technology in development of low permeability reservoirs. Petroleum Drilling Techniques, 1:21-2.

IOP Conf. Series: Materials Science and Engineering **729** (2020) 012027 doi:10.1088/1757-899X/729/1/012027

Analysis of Disaster Gas Accumulation Characteristics in Local Area of Upper Corner Based on Entropy Variable Dissipative Structure Model

Yin-shang Wei Pei-yao Liu* Yu-quan Jia

School of Safety Science and Engineering, Xi'an University of Science and Technology, Xi'an 710054, China

TEL: 17792577195

Email: 947638750@qq.com; 563471248@qq.com;

Abstract: This paper takes the angle CO of the 1406 working face of Fengjiata Coal Mine as main research object, by combining field measurement, theoretical research and numerical simulation, to explore essence of energy dissipation during gas accumulation in the upper lag zone and seek effective governance solutions. The result shows that the irreversible process such as eddy current loss caused by the spatial morphology of the upper sag angle is an important cause of gas accumulation. Thus, a flexible inflatable airbag for mining to change local ventilation mode is proposed and designed. The numerical simulation method is used to verify the venting of the mine inflatable airbag, which helps significantly eliminate the eddy current phenomenon in the lag zone.

1. Preface

In recent years, with the continuous improvement of mechanized mechanization level in northern Shaanxi mining area, the large-volume shifting device installed at the tail of the machine causes the tail of the working face to be in a state of partial bracket lag for a long time. The problem of gas gathering and exceeding limits of the working face returning to the wind and the corner is becoming more and more serious, which now is a major hidden danger that seriously restricts coal mine safety production.

Research and practice show that: accumulation of disaster gas in upper corner is the result of material, energy and information exchange between the local ventilation system of the upper corner and surrounding environment. However, currently there is limited research about energy dissipation characteristics of the gas accumulation process in the upper corner. Therefore, based on the equivalence and unity of information entropy and thermodynamic entropy, this paper conducts entropy analysis of the concentration of disaster gas accumulation in the lag zone of the upper corner, combing field tests and numerical simulations, to exploring the characteristics of disaster gas accumulation in the upper sag angle region from the perspective of energy dissipation, Finally, it is proposed to use the mine flexible inflatable airbag to deflate and drain, introducing " negative entropy flow" to control limitation exceeding of gas in the upper corner.

2. Analysis of dissipative structure characteristics of the upper corner ventilation system

The theory of dissipative structure was firstly proposed by Belgian unbalanced statistical physicist Prigogine at an international conference [1]. Dissipative structure refers to the open system that is far from equilibrium and the dynamic exchange of substances in the surrounding environment、 energy and information, when the state of surrounding environment changes to a certain threshold, the induced system produces random fluctuations that are mutated from the previous chaotic state, and transition to a relatively stable orderly state in time and space. The process of dissipating energy and information must be accompanied by the generation of entropy and dissipative structures must constantly absorb negative entropy from the outside to maintain their order, as its formation must meet four conditions [2] [3]:

(1) The system must be open to introduce negative entropy flow by exchanging matter, energy and information with the outside world;

(2) The system must be in a state of being far from equilibrium;

(3) There must be nonlinear interactions between the various elements within the system;

(4) The system implements the existing order through random rise.

Using information entropy to characterize the uneven accumulation of disaster gas in the local area of the upper corner, that is, the local area of the upper corner of the reaction is continuously subjected to material and energy exchange processes with the surrounding environment. Macroscopically, when external environment is interrupted due to periodic pressure, maintenance, and speed change, this will keep the upper corner ventilation system away from equilibrium. From the point of view of molecular motion theory, the molecules in the material system moving from order to chaos is a process of entropy increase; due to the elimination of disturbances, according to the system principle, the molecules in the system are exchanged with surrounding environment for substances, energy and forces due to different factors such as wind flow field, gas source and density. This exchange for the upper corner ventilation system, will transit equilibrium state to a new orderly steady state in time and space, that is, the air volume and flow field are dynamically stable. Due to constant mass transfer, disaster gas molecules changes in entropy within the transient steady state system. From the perspective of the principle of entropy, entropy is a state. "Negative entropy" is used to characterize the exchange of systems with the environment, and uses the exchanges absorbed from the environment to maintain their orderly state, resisting the irreversible entropy increase process of itself. The increase in negative entropy means a way to promote better synergy within the system. Thus the power of development towards ordering increases, finally making the local disaster gas accumulation phenomenon of the upper corner an orderly near equilibrium state.

Based on the points above, the upper corner local ventilation system is used as a subsystem of the mine ventilation system, maintaining its stability through dynamic material and energy exchange with the surrounding environment, fluctuating a system that is far from a steady state from an unstable state to a new state, meeting the conditions to form a dissipative structure.

3. Analysis of the entropy value of disaster gas concentration in the local area of the upper corner

3.1. The unified relationship between thermodynamic entropy and information entropy [4]-[5]

3.1.1. Information Entropy Deriving Thermodynamic Entropy

The s in information theory is called Shannon entropy, which represents the average amount of information of the source. If $p(x_i)$ represents the probability of any microscopic state of the macroscopic state of the system, which is $\sum_i^n p(x_i)=1$ .The S of the source X can be seen as a function of the n-dimensional probability vector.which is:

$$E\{J(x)\} = H(x) = -\sum_{i=1}^{n} p(x_i)\log_a p(x_i) \tag{2.1}$$

The microscopic state is equal probability according to the balance of the system, and the information entropy is the maximum value when the equal probability is known , which is $p(x_i) = 1/\Omega$ substitute (2.1) get $H(X) = -K\ln\frac{1}{\Omega} = K\ln\Omega = S$.

Information entropy is the uncertainty that describes the falling of a single particle at any energy level, the significance of statistical entropy is the degree of disorder of the microscopic states of N particles distributed at all energy levels. This shows that, the information entropy is consistent with the essence of Boltzmann statistical entropy, and there is an equivalence and unity relationship between the two.

3.1.2. Thermodynamic Entropy Deriving Information Entropy

The upper corner is relative to the entire mining area ventilation system, is a local ventilation system, think of the upper corner as an isolated system. Set the microscopic state number Ω of the local area system of the corner to be:

$$\Omega = \frac{N!}{N_1!N_2!\cdots N_m!} \tag{2.2}$$

Boltzmann statistical entropy is:

$$S = K\ln\Omega \tag{2.3}$$

Substituting (2.2) into (2.3):

$$S = K\ln\Omega = K\ln\frac{N!}{N_1!N_2!\cdots N_m!} \tag{2.4}$$

According to the Sterling formula $\ln N! = N\ln N - N$, Substituting (1.3):

$$S = K\ln\frac{N!}{\prod_{i=1}^{m} N_i!} = KN(-\sum_{i=1}^{m}\frac{N_i}{N}\ln\frac{N_i}{N}) \tag{2.5}$$

Hypothesis $\frac{N_i}{N} = p_i$, then the relationship between thermodynamic entropy and information entropy is:

$$S = KN(-\sum_{i=1}^{m} p_i \ln p_i) = KNH \tag{2.6}$$

3.2. Information entropy of disaster gas accumulation in the upper corner area and its calculation

3.2.1. Information entropy of disaster gas accumulation in the upper corner area

In information theory, information entropy represents the average degree of uncertainty of the source, which is a measure of information disorder. Let X denote a finite number of random statesX_1, X_2, ……, Xn, $p(X_i)$ representative $X_i(i=1,2,...,n)$ probability of occurrence, Then the information entropy H(X) of the system is defined as [6]:

IOP Conf. Series: Materials Science and Engineering **729** (2020) 012027 doi:10.1088/1757-899X/729/1/012027

$$H = -\sum_{i=1}^{n} p(x_i) \ln p(x_i) \tag{2.7}$$

among them: $0 \le p(X_i) \ge 1$ and $\sum_{i=1}^{n} p(x_i) = 1$.

3.2.2. Entropy analysis

It can be seen from the above formula that the calculation of information entropy is mainly to determine a specific probability distribution or probability density function. However, in the actual mining process, due to the complex mining environment and numerous interference factors, it is impossible to determine the probability distribution function that can express the change in the concentration of the gas in the upper corner. Literature [7] proposes to obtain the entropy estimate directly based on the rank estimation method of the surveyed observation data. Details as follows:

The samples are $x_1, x_2, \ldots, x_N$, arranged in a new sequence $x_{(1)}, x_{(2)}, \ldots, x_{(n)}$ from small to large, and the rank is defined as nk:

$$r_k = \int_{-\infty}^{x_k} p(x)dx = \int_{-\infty}^{x_k} dp(x)dx = p(x_{(k)}) \quad 0 \le r_k \le 1 \tag{2.8}$$

Where is the probability distribution function of x, where:

$$\overline{p}(x_{(k)}) = \overline{r} = \frac{k}{n+1} \tag{2.9}$$

Then the information entropy calculation formula is:

$$H(x) = -\int_{-\infty}^{+\infty} p(x) \ln p(x) dx = -\int_{-\infty}^{+\infty} \ln \frac{dP(x)}{dx} dP(x) \tag{2.10}$$

The entropy estimate based on the discretized form of the monitored data is:

$$\overline{H}(x) = -\sum_{k=1}^{n} \ln \frac{\Delta P(x_{(k)})}{x_{(k)}} \Delta P(x_{(k)}) = -\sum_{k=1}^{n} \ln \frac{\overline{r_{k+1}} - \overline{r_k}}{x_{(k)}} (\overline{r_{k+1}} - \overline{r_k}) \tag{2.11}$$

4. Analysis of disaster gas accumulation characteristics in the local system of the upper corner

Combined with the upper corner angle to test the change trend of gas concentration, the grid size of the measuring point is in the form of non-uniform grid, and the measuring point is located at the intersection of the grid. As shown in Figure 1, in Fengjiata 50104 working face, the corner space y=1.6m height plane, with the intersection of the return air duct and the goaf as the starting point, a total of 26 measuring points are arranged. Among them, 7 measuring points are arranged along the direction of the return airway, and the measuring point positions are 0.2m, 0.7m, 1.4m, 2.4m from the junction, and 8 measuring points are arranged along the working surface, and the measuring points are respectively at a distance of 0.3.m, 0.8m, 1.6m, 2.6m.

Finally, the sample mean value set of the upper corner CO concentration detection data is D={Xi丨(=1, 2, , 78)}, using the Meshgrid function in MATLAB software and the Griddata function to perform plane difference processing on the data [8], then use the mesh function to draw the measured concentration distribution map of CO gas in different horizontal planes, as shown in the figure.

IOP Conf. Series: Materials Science and Engineering **729** (2020) 012027 doi:10.1088/1757-899X/729/1/012027

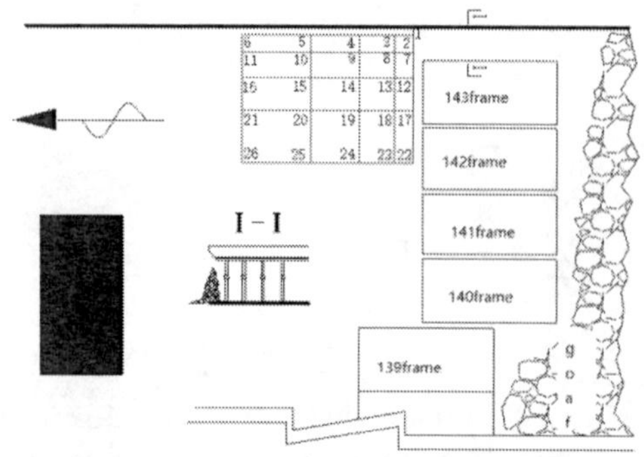

Fig.1 Schematic diagram of the arrangement of the angle measuring points on the working surface at y=1.6m

According to the CO concentration detection data sample of the upper corner, the information entropy of the upper corner gas concentration at y=1.6 m is calculated by using Equation 2.11, as shown in Table 1.

Table 1 y = 1.6m at the upper corner of the measuring point CO concentration statistical value and CO concentration information entropy calculated value

Measuring point number	Concentration mean /ppm	Information entropy	Measuring point number	Concentration mean /ppm	Information entropy
Measuring point 1	24.81	79.0594	Measuring point 14	20.36	76.5591
Measuring point 2	25.94	80.6659	Measuring point 15	18.06	76.5687
Measuring point 3	25.06	80.0184	Measuring point 16	16.72	69.8153
Measuring point 4	20.56	78.4855	Measuring point 17	9.89	76.7809
Measuring point 5	20.75	76.8278	Measuring point 18	18.39	71.9047
Measuring point 6	16.69	75.6577	Measuring point 19	10.67	71.3352
Measuring point 7	17.72	76.1856	Measuring point 20	9.86	71.3324
Measuring point 8	17.61	75.8823	Measuring point 21	9.83	68.0593
Measuring point 9	22.78	79.2725	Measuring point 22	7.61	67.9763
Measuring point 10	17.61	76.8995	Measuring point 23	7.44	66.2461
Measuring point 11	15.00	75.2192	Measuring point 24	6.33	62.0977
Measuring point 12	18.72	77.0312	Measuring point 25	5.44	57.4349
Measuring point 13	21.44	78.2494	Measuring point 26	2.83	51.4724

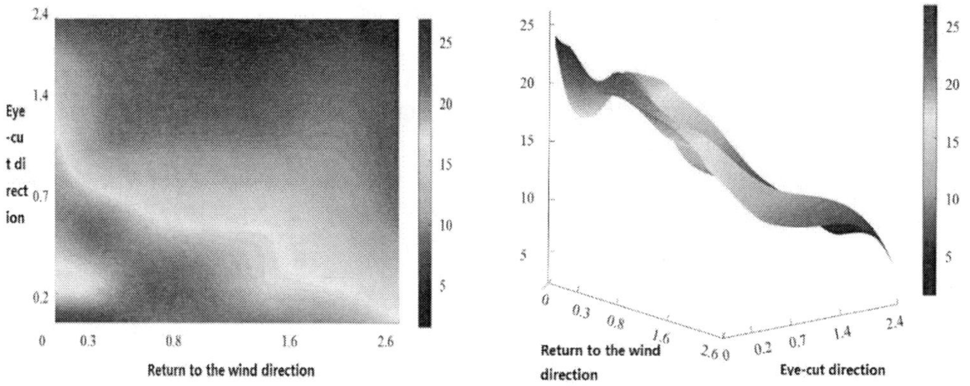

Fig. 2 Distribution of CO concentration in the working plane at y=1.6m

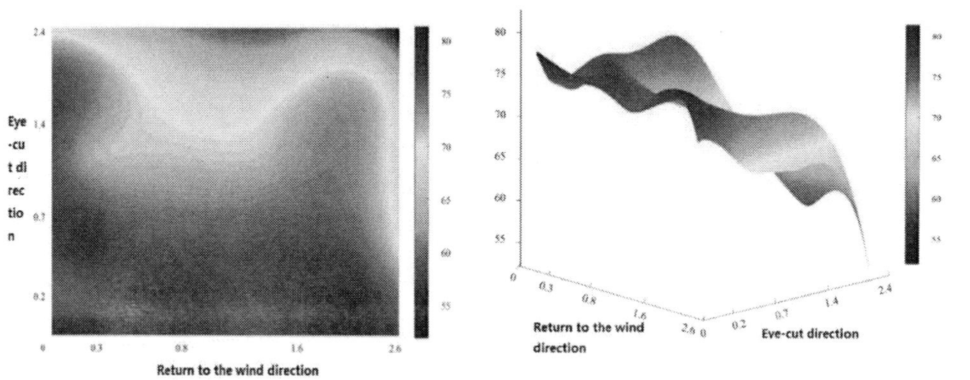

Fig.3 Isoline contour map of the CO concentration of the corners on the working surface at y=1.6m

It can be seen from Fig. 2 that the information entropy of the CO angle of the upper corner and the CO concentration of the upper corner are generally in the plane of the direction of the working surface, and the overall fan-shaped change trend gradually decreases from the upper corner to the main wind region, and in the direction of the return airway, the higher variation law of the isolation coal wall is obtained, in the dissipative structure formed by the upper corner region, the gushing and diffusion process of harmful gases such as CO is an irreversible process of increasing entropy flow; In the direction of the cut-eye, there is a slight air leakage due to the goaf near the rear of the single hydraulic support, it is shown that there is a small area of air leakage near 0.2 to 0.7 m in the direction of the cutting eye, resulting in a relatively low concentration of CO gas in this range. Information entropy is a basic quantity used to measure the amount of information. As can be seen from Figure 3, there is also a small range of entropy relative reduction in the corresponding position of the upper corner. According to the theory of dissipative structure, external leakage introduces “negative entropy”, it is characterized by the exchange of substances, energy and information between the ventilation system in the upper corner area and the surrounding environment, forming the dissipation of energy and information; It can be seen from Fig. 2 and Fig. 3 that the trend of information entropy between CO concentration and CO concentration is basically consistent. It can be seen that information entropy can well characterize the cohesion characteristics of CO gas in the upper corner, the overall trend of the fan shape gradually decreases from the upper corner to the main airflow region, it indicates that the system tends to increase the order degree, and the direction of total entropy reduction, that is, the essence of CO gas accumulation in the local system of the upper corner is the entropy reduction process of the dissipative structure.

IOP Conf. Series: Materials Science and Engineering **729** (2020) 012027 doi:10.1088/1757-899X/729/1/012027

Numerical Simulation

(I) Model and boundary condition settings

The Fengjiata Coal Mine uses inclined shafts to develop. Main layout of the panel is to transport the main road, and the coal mining face is arranged in strips. The coal conveyor face scraper conveyor has high power, and a large-volume shifting device is installed at the tail, so that the tail of the working face is in a state of partial bracket lag for a long time. According to the actual arrangement of four fully mechanized mining supports and four single hydraulic supports in the return air corner of the 1406 working face of Fengjiata Coal Mine, the physical model of the local lag of the return wind angle is shown in Fig. 4. Among them, in order to facilitate the study, the influence of the working surface tail scraper transport equipment on the flow field is neglected, the gas emission source is only the rear goaf, and the hysteresis area is simplified to a space area of 6m×1.5m×3m.

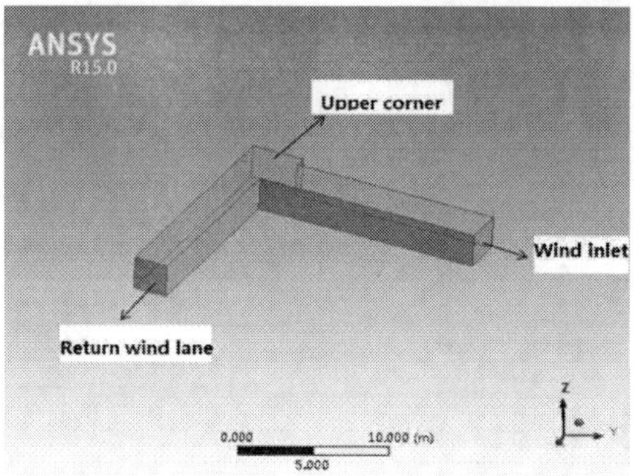

Figure 4 1406 working face return air corner local lag physical model

According to the actual supply air, the inlet boundary condition is set to the speed inlet inlet1 with an average wind speed of 1.5 m/s, according to the field measurement and analysis, the air leakage at the single hydraulic support of 0.2~0.7m along the direction of the cutting eye is significant, it is set to a speed inlet inlet2 with an average air leakage wind speed of 0.22 m/s, and a return air outlet port is set as a free outlet outflow. As shown in Fig. 5, the cloud volume fraction distribution cloud map and the flow field velocity distribution cloud map of the upper corner angle at y=1.6 m are respectively taken.

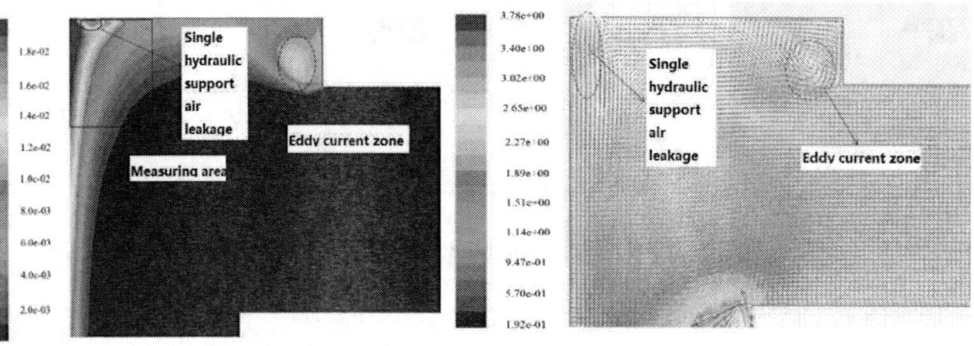

Figure 5 a: y = 1.6m at the upper corner of the CO volume fraction distribution plan
b:y=1.6m at the upper corner of the flow field velocity distribution plan

As shown in Figure 5 (b), due to the partial lag of the upper corner, the special environmental boundary causes the wind flow flowing from the working surface into the lag zone to form a local eddy current zone inside the upper corner lag zone, the position of the upper corner is in a windless or breezy state. When the average wind speed of the working face is 1.5m/s, the average wind speed in the lag zone of the upper corner is <0.3m/s. As shown in Figure 5 (a), due to the presence of eddy currents in the lag zone and lower wind speeds, the disaster gas that causes the rear goaf and the isolated coal pillar to overflow flows in this area, and it is difficult to merge into the return air lane with the main wind flow, the occurrence of disaster gas overruns often occurs at the upper corner; As shown in Fig. 5 (a, b), there is a wind leakage speed at a single hydraulic support of 0.2 to 0.7 m in the direction of the cutting eye, in the case where the air leakage of the single hydraulic support occurs, in the case where the average air leakage wind speed is 0.42 m/s, the CO volume fraction shows a small range of decline in the affected area, however, due to the intricate environment and various influencing factors, the degree of decline is different from the actual measurement on site, but the overall trend is in line with the actual situation on site.

(II) Numerical simulation analysis of the over-limit of gas in the upper corner of the mine with flexible inflatable airbag

The mine flexible inflatable airbag is a multifunctional mining equipment that integrates gas detection and alarm, temporary sealing and pressure air drainage. The airbag is made of wear-resistant and anti-static flame-retardant rubber material, and has the characteristics of softness, good sealing effect, simple operation and convenient movement. As shown in FIG. 6, the airbag is mainly composed of a main structure such as an airbag main body, a base, an adjustable dense grille vent, and a gas detector; the pressure air line is connected to the gas quantity detector through a sensor; the gas detector is connected to the gas alarm controller and the deflation valve through sensors, and two adjustable dense grid pressure relief ports are respectively placed at the front end of the airbag and the middle of the right side of the airbag. Mainly installed between the working face support and the isolated coal pillar, and the gas is expanded by the mine pressure air pipe, the control gas supply is between 30m3/min and 50m3/min, when the upper gas angle disaster gas is detected, automatically open 2 dense grille pressure relief ports to deflate, control the amount of venting from 10m3 / min to 20m3 / min, dilute the accumulated gas until the gas concentration will reach the safe range.

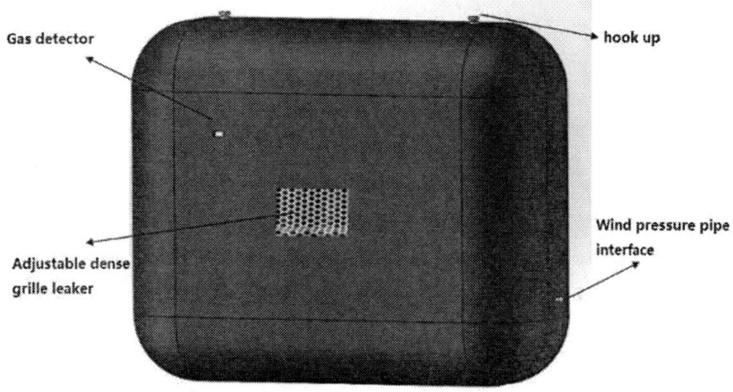

Figure 6 Schematic diagram of a flexible inflatable airbag for mining

In the numerical simulation, other boundary conditions remain unchanged, and only the mine flexible inflatable airbag vent is added as the speed inlet inlet3, the average draft wind speed at which the speed inlet is set is 0.42 m/s. As shown in Fig. 7, the cloud volume fraction distribution cloud map and the flow field velocity distribution cloud map of the upper corner of y=1.6m are respectively taken.

IOP Conf. Series: Materials Science and Engineering **729** (2020) 012027 doi:10.1088/1757-899X/729/1/012027

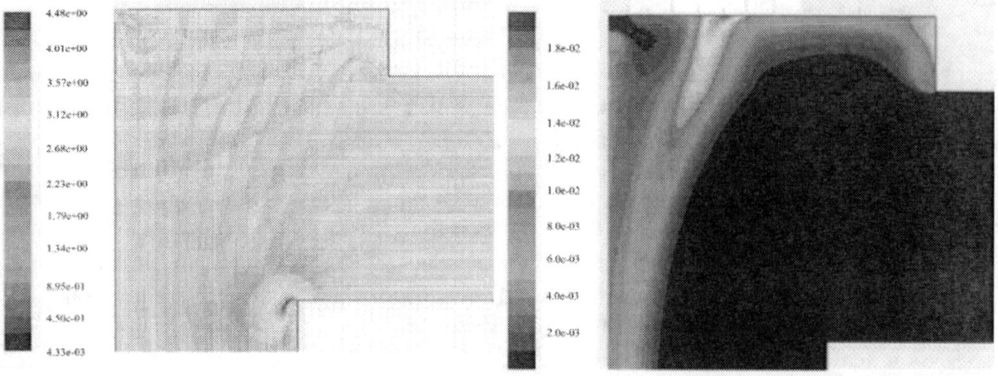

Fig. 7a Cloud flow velocity distribution diagram of the upper corner at y=1.6m with airbag
b When there is a balloon, y=1.6m at the upper corner of the CO volume fraction distribution cloud map

When passing through the front and left ends of the airbag, the airflow of the dense grille vents to the upper corner space is 20m3/min, that is, when the average wind speed around the airbag shown in Fig. 7a is 1.5 m/s, the concentration of the disaster gas at the upper corner can be effectively reduced; as can be seen by comparing Figures 5b and 7b, the air pressure of the airbag can significantly increase the wind speed in the lag zone of the upper sag angle, it acts as a disturbance to the flow field in the lag space of the upper corner, and can obviously eliminate the eddy current existing in the lag space of the upper corner, significantly rectified the phenomenon of overcapacity of the gas in the upper corner.

5. Conclusion

Introducing information theory and comparing the field test results with the entropy value analysis, it is concluded that the information entropy can characterize the gas accumulation characteristics of the upper horn disaster. From the perspective of energy dissipation, through field measurement and numerical simulation, the characteristics of the gas accumulation in the upper corner of the fault are analyzed. It is concluded that essence of the gas accumulation in the local system of the upper corner is the entropy reduction process of the dissipative structure. The paper proposes and designs a flexible inflatable airbag for mining to introduce the "negative entropy" through the pressure-drainage method to control the CO accumulation problem of the upper corner. Practicality and effectiveness of CO gas accumulation in Fengjiata 1406 working face by airbag treatment were verified by numerical simulation.

Reference

[1] Shannon C E. A mathematical theory of communication[J]. Bell Labs Technical Journal, 1948, 27(4):379-423.
[2] ZHAO Yi-xin, JIANG Yao-dong, TIAN Su-peng. Investigation on the characteristics of energy dissipation in the preparation process of coal bumps[J]. Journal of China Coal Society, 2010, 35(12):1979-1983.
Wu Xue-juan. Negentropy of Dissipative Structure System and Its Realizing Process[J]. Chinese Journal of Systems Science, 1995(2):76-79.
[3] QIN Si-qing. Prmary discyssion on form ation mechanism of dissipative structure in instability process of rock mass[J]. Chinese Journal of Rock Mechanics and Engineering, 2000, 19(3):265-265.
[4] ZHAO Wen-tong, ZHAO Pei-zhang. The rmodynamic entropy and information entropy[J]. Journal of Henan Normal University(Natural Science Edition, 1988(1):44-50.
[5] LI He-ling. The relation between information entropy, Boltzmann entropy and Clausius entropy[J].College Physics, 2004, 23(12):37-40.

[6] ZHANG Yan-bo, LIANG Peng, LIU Xiang-xin, et al. Experimental study on precursor of rock burst based on acoustic emission signal dominat-frequency and entropy[J]. Chinese Journal of Rock Mechanics and Engineering, 2015(s1):2959-2967.
[7] WANG Zhong-yu, XIA Xin-tao, ZHU Jian-min. Non-statistical theory of measurement uncertainty[M]. Defense press, 2000.
[8] Yu Ming gao, Chang Xu hua, Jia Hai lin, et al. Analysis of spontaneous combustion "three-zone" in goaf based on Matlab[J]. Journal of China Coal Society, 2010(4):600-604.

IOP Conf. Series: Materials Science and Engineering **729** (2020) 012028 doi:10.1088/1757-899X/729/1/012028

Study on Yield Mechanical Properties of High Availability Materials before and after Aging for Intelligent Measurement Equipment

Angang Zheng[1], Xiangyu Yuan[1], Yifan Wang[2], Huaiying Shang[1], Yan Liu[1], Wulei Zhang[3]

[1]China Electric Power Research Institute, Beijing, China

[2]North China Electric Power University, Beijing, China

[3]State Grid Henan Electric Power Co., Ltd., Henan, China

shanghuaiyingip@163.com

Abstract. One of the key technical research points in the field of smart grid measurement is to make a breakthrough in the detection and evaluation of high availability materials in intelligent measurement equipment. The comprehensive cross-action of multi-environmental stress brings a lot of data, how to find the key parameters in it, and how to analyze the complex environment into the characteristic combination of multiple single environmental factors, and the mathematical model of nonlinear evolution and service life of material aging based on the characteristic expression of the combination, all of the above are the key points and difficulties of the research. On the basis of the service life model, the research on differential protection measures and the study of life evaluation technology for smart grid measurement equipment can enable the equipment to realize wide-area application and high availability under more harsh conditions, which is one of the key points to improve the quality control in NQI system.

1. Introduction

The research of detection and evaluation of high availability materials of intelligent measurement equipment is an important link of the National Quality Infrastructure system. On the one hand, it can provide basic technical support and technical reserve for the wide-area reliable application of equipment in the field of measurement of smart grid, on the other hand, it can provide the corresponding technical standards, testing technology, evaluation technology, life control and so on for the quality control of equipment in the field of measurement, so as to make a positive push for the continuous improvement and leading development of NQI system in the field of smart grid measurement. [1-3]

Relying on the experimental base to design the aging test, the key index test data of the material in the typical environment of high cold, high dry heat, high altitude, high salt spray and high humidity heat are studied, and the nonlinear evolution and service life model of material aging under the comprehensive cross action of multi-environmental stress are abstracted, and the anti-aging and service life evaluation technology of the measurement equipment is studied.

According to the test data in extreme environment, the test design and testing technology optimization supplement or revise the corresponding technical standards, and based on the service life evaluation technology of the measurement equipment, the quality standard system covering the quality control of the measurement equipment is constructed, so as to achieve the NQI four elements of high availability to support the smart grid measurement equipment technology reserves. [4-6]

2. Test Overview

2.1. Test materials subsection

This test is for the aging test of two materials used in the metering box: polycarbonate resin and acrylbutylene-styrene resin (PC-ABS) and glass fiber-reinforced unsaturated polyester mold plastic (SMC) under GB/T 23641.

(1)Polycarbonate resin and a butyl triamcine-styrene restyrene (PC-ABS) is a combination of polycarbonate resin (PC) and acrylonitril dipen-styrene (ABS) plastic, overcoming the original shortcomings of the two raw materials, to play their own advantages, that is, to enhance ABS heat resistance, impact and tensile strength, It also reduces the melt viscosity of polycarbonate, improves the stress cracking condition of THE PC, has good heat resistance, anti-impact, flame retardant properties, and has good mechanical properties in the ordinary temperature.

(2)Glass fiber enhanced unsaturated polyester mold plastic (SMC) has good strength, corrosion resistance, impact resistance, arc and smooth flat and other advantages, is widely used to replace metal materials.

2.2. Sample design

This test piece refers to the national standard "GBT 1043.1-2008 plastic simple beam impact performance determination" and the national standard "GBT 9341-2008 plastic bending performance determination" related content production, test type and size as shown in the figure.

2.3. Pilot scheme

Samples are collected from the metering box equidistant collection, each metering box collects about 28 samples, respectively, refers to the relevant standards described in the previous section for impact test and bending test, and records the impact performance before aging, pre-aging bending performance, post-aging impact performance and aging after bending performance.

2.3.1. Sample grouping

(1) Impact Group. A total of 3200 samples participated in the impact test, according to whether the aging is divided into 2 groups, each group is subdivided according to the intercepted box source, in order to avoid the accident of the test, each 7 for a group to test related performance, each group to remove the maximum and minimum value, the remaining 5 spline as the team's record value.

Impact strength a_{cU} is calculated as:.

$$a_{cU} = \frac{E_c}{h \cdot b} \times 10^3 \qquad (1)$$

In the formula: $a_{cU}$for the impact strength of the specimen, E_c for the repaired specimen damage absorption of energy, h for the pattern thickness, b for the pattern width.

(2) Bending group. A total of 3200 samples participated in the impact test, according to whether the aging is divided into 2 groups, each group is subdivided according to the intercepted box source, in order to avoid the accident of the test, each 7 for a group to test related performance, each group to remove the maximum and minimum value, the remaining 5 spline as the team's record value.

Impact strength a_{cU} is calculated as:

$$\sigma_b = \frac{3PL}{2bh^2} \qquad (2)$$

In the formula: σ_b for the specimen bending strength, P for the specimen destruction load, L for the span, b， h are the width and thickness of the specimen.

2.4. Test equipment

(1) Material aging equipment

The equipment used in the aging of the test parts is the Q-sun xenon lamp aging test machine produced by Q-Lab, the model is xe-3-H, according to GB/T 16422.2-2014, the aging environment in this test is set to: wavelength 290 to 800 nm, choose 550 W/m2 Temperature 65℃, humidity 65%, each spray 18 min, no water 102 min, cycle 500 hours.

Figure 1. Material aging equipment.

(2) Impact equipment

Test impact using Chengde Jinhe Instrument Manufacturing Co., Ltd. production of XCJD-5 type micro-controlled simple beam impact test machine, model XCJD-5, in line with the GBT 1043.1-2008 plastic simple beam impact performance determination Part 1: non-instrumental impact test, BS EN ISO179-1 : 2001 Plastics-Kry of Charpy impact properties -Part 1: Non-instrumented impact test, GBT 21189-2007 Plastic simple girders, cantilevers Beam and stretch impact test with the hammer impact test machine test, JB/T 8762-1998 plastic simple beam impact test machine and other standards, impact speed 2.9 m/s, swing hammer angle 150 degrees, the strike center distance of 230 mm, the test used in the 5 J swing hammer torque PL =2.6795 N·m .

(3) Bending equipment

The test piece bending is made by Chengde Jinhe Instrument Manufacturing Co., Ltd. of the WDW-2 type microcomputer-controlled electronic universal test machine, which can produce a maximum test force of 2 kN, test force accuracy of 0.1%, test force measurement range of 2%-100% Fs, displacement measurement accuracy of 0.1%, speed range 0.01 mm/min- 500 mm/min; speed accuracy of up to 1%, resulting in a maximum compression travel of 600 mm, the loading speed of 2 mm/min was used in this test.

Figure 2. Impact equipment.

Figure 3. Bending equipment.

3. Test results and analysis

3.1. Experimental physical phenomena and characteristics

IOP Conf. Series: Materials Science and Engineering **729** (2020) 012028 doi:10.1088/1757-899X/729/1/012028

In the course of the impact test, due to polycarbonate resin and a butyl trilene - styrene resin relative to glass fiber-enhanced unsaturated polyester mold plastic has better toughness, the spline in the absorption of pendulum fall kinetic energy, the production of about 20 degrees of small plastic deformation; The sample breaks quickly and only a small part is connected.

In the process of bending test, glass fiber enhanced unsaturated polyester mold plastic due to higher strength, load increase early, the appearance of the sample no significant change, no significant change in the loading process, with the increase in deflection, the sample produces local fracture, the end of the test. Polycarbonate resin and a butyl triene - styrene resin in the test process to produce a more obvious bending, reached the specified deflection after the load disappeared, the sample shape recovered, did not break.

3.2. Related data analysis

Calculate the impact strength before and after the sample aging based on the absorption energy of the sample shock process collected by the impact test equipment, as well as the pre-measured sample size and span, and calculate the bending strength before and after the sample aging according to the maximum load of the sample collected by the equipment and the sample size and span. The average impact strength and average bending strength of each batch of samples are calculated according to the results. [7-9]

3.2.1. Impact. After the aging of the two materials, the impact strength of most of the impact strength has a certain degree of impact intensity decay. Among them, polycarbonate resin and a butyl-butylene-styrene resin (PC-ABS) impact strength decreased by an average of 2.36%, and glass fiber-enhanced unsaturated polyester mold plastic (SMC) impact strength decreased by an average of 8.23%. [10-11]

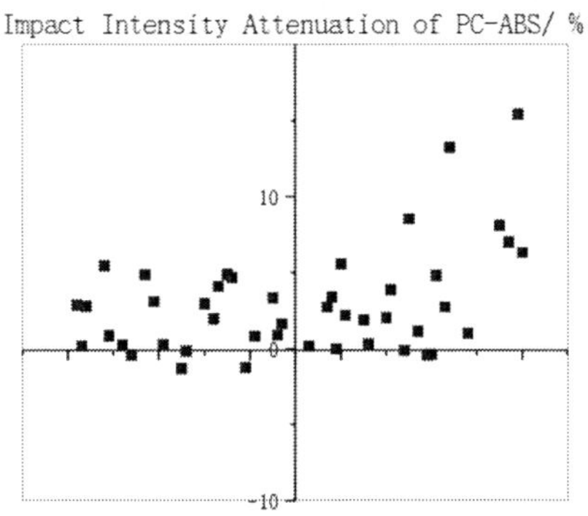

Figure 4. Impact Intensity Attenuation of PC-ABS

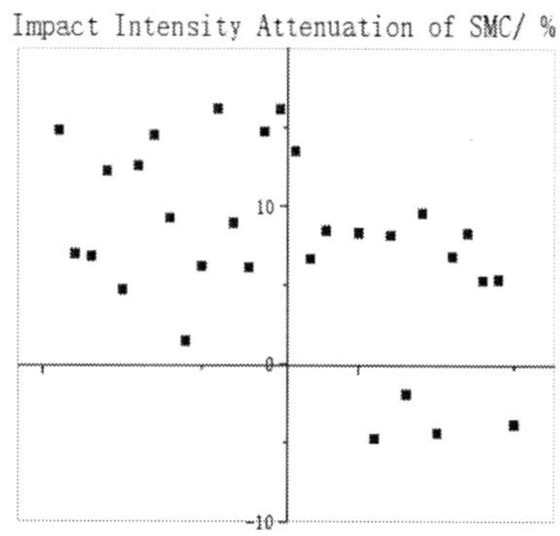

Figure 5. Impact Intensity Attenuation of SMC

3.2.2. Bending. After UV aging, the fiberglass-enhanced unsaturated polyester mold plastic (SMC) has a more obvious bending strength attenuation, the average bending strength attenuation of 8.25%, polycarbonate resin and acrylbutene-styrene resin (PC-ABS) bending strength attenuation is not obvious, about half of the samples have a small performance enhancement, Approximately 52.5% of the sample review, the average bending strength of the sample increased by 0.3%. [11-13]

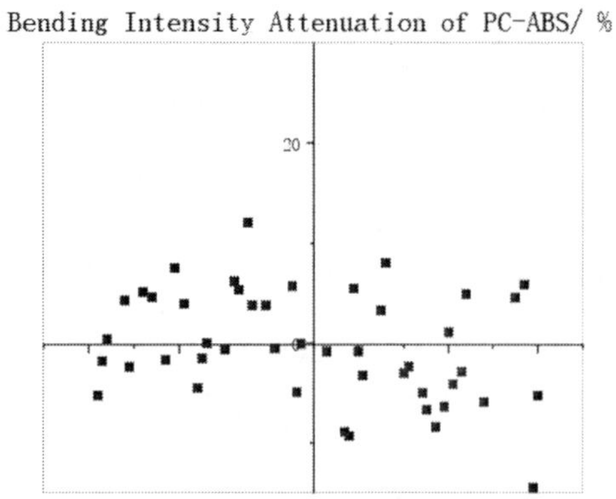

Figure 6. Bending Intensity Attenuation of PC-ABS.

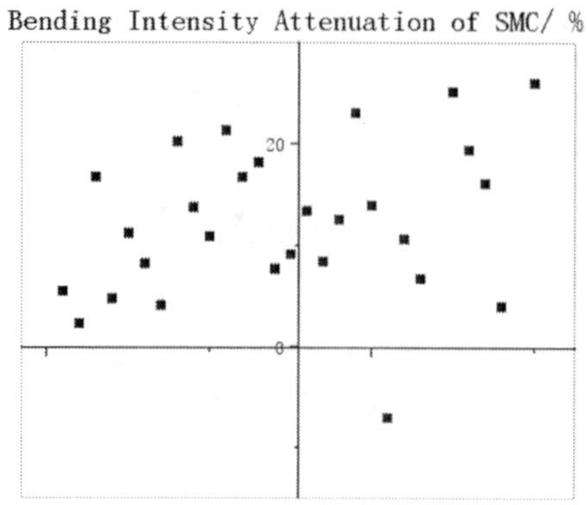

Figure 7. Bending Intensity Attenuation of SMC.

By using Pearson, Spearman and Kendall, the correlation between the impact performance and bending performance before and after aging of the two materials was analyzed, and the impact strength and bending intensity of the two materials were not correlated. That is, high availability materials do not have the correlation of aging resistance in different directions. [14-15]

Table 1. Relevance Description Calculation.

	Correlation coefficient	p
Pearson	0.06748	0.72312
Spearman	-0.00156	0.99348
Kendall	0.0069	0.95732

4. Conclusion

（1）After the test set UV aging process, polycarbonate resin and abutene-styrene resin (PC-ABS) can still maintain good bending strength. In contrast, UV aging has a more obvious impact on the material performance of fiberglass-enhanced unsaturated polyester mold plastic (SMC), which reduces the impact strength and bending strength of the material, and the elastic mode has a certain increase, indicating that the external aging environment will reduce the material's extreme deformation rate. It is recommended to pay attention to the adverse effects of the external environment when designing glass fiber-reinforced unsaturated polyester mold plastics that are exposed to the external environment for a long time.

（2）After the experimental application of the UV aging process, there is no definite relationship between the bending strength and impact strength of the two materials. That is, when the impact strength performance of the two materials with aging significantly reduced, it can not be concluded that its bending strength performance will be reduced at the same time, the opposite is the same. Whether this correlation will change with the change of aging time and the change of aging process has yet to be further experimentally verified.

Acknowledgments

This work is supported by Innovation Fund Project of China Electric Power Research Institute. (Research on NQI Top-level Design and Measurement Key Technologies in Some Areas of Smart Grid)

References

[1] Sousa, J. M. , Correia, J. R. , & Cabral-Fonseca, S. . (2016). Durability of glass fibre reinforced polymer pultruded profiles: comparison between quv accelerated exposure and natural weathering in a mediterranean climate. Experimental Techniques, 40: 207-219.
[2] Xian, G. , Li, H. , & Su, X. . (2012). Water absorption and hygrothermal ageing of ultraviolet cured glass-fiber reinforced acrylate composites. Polymer Composites, 33: 1120–1128.
[3] Kaushal, S. , Tankala, K. , Rao, R. M. V. G. K. , & Kishore. (1991). Some hygrothermal effects on the mechanical behaviour and fractography of glass-epoxy composites with modified interface. Journal of Materials Science, 26: 6293-6299.
[4] Cabral-Fonseca, S. , Correia, J. R. , Rodrigues, M. P. , & Branco, F. A. . (2012). Artificial accelerated ageing of gfrp pultruded profiles made of polyester and vinylester resins: characterisation of physical–chemical and mechanical damage. Strain, 48: 162-173.
[5] Jiang, B. K. , Xu, J. , & Li, Y. H. . (2014). Flapwise vibration analysis of a rotating composite beam under hygrothermal environment. Composite Structures, 117, 201-211.
[6] Cecchi, A. , & Di Marco, R. . (2004). Glass fiber reinforced polymer pultruded members: constitutive model and stability analysis. Journal of Engineering Mechanics, 130: 961-970.
[7] Ibrahim, S. , Polyzois, D. , & Hassan, S. K. . (2000). Development of glass fiber reinforced plastic poles for transmission and distribution lines. Canadian Journal of Civil Engineering, 27: 850-858.
[8] Zhang, X. D. , Feng H. , & Jie, G. X. . (2010). Aging Behavior of Polycarbonate in Typical Atmospheric Environment of China. China Building Material Science and Technology ,S1: 108-111.
[9] Zhao, Y. Y. , Gao, J. G. , & Zhao, Y. X. . (2012). Advances in aging mechanism and research methods of polycarbonate. Aging and application of synthetic materials, 41: 48-54.
[10] Wang, X. J. . (2016). Analysis of aging performance and molecular simulation of dieniene rubber and polycarbonate materials. (Doctoral dissertation).
[11] Huang, Y. J. , Ye, L. .(2017). New advances in the study of aging law, life prediction and prevention of polymer materials under complex conditions. Chinese Polymer Bulletin, 10: 52-63.
[12] Xu, Y. J. , Zhao, Y. , Tang, B. J. , Wang, X. , & Duan, Y. X. . (2013). Effect of ULTRAViolet radiation on the performance of indoor carbon fiber-reinforced epoxy-based composite materials. Journal of Composites, 30: 63-69.
[13] Guo, M. Y. , & Li, H. B. , et al. (2008). Effects of UV aging on the performance and structure of aramid/epoxy composites. Fiber Composites, 25: 35-38.
[14] Yu, S. J. , Wei, D. L. , et al. (2016). Study on uv-accelerated aging performance of Pvc/modified swords and wood-plastic composites. Plastics in China, 30: 54-59.
[15] Shokrieh, M. M. , & Bayat, A. . (2007). Effects of ultraviolet radiation on mechanical properties of glass/polyester composites. Journal of Composite Materials, 41: 2443-2455.

IOP Conf. Series: Materials Science and Engineering **729** (2020) 012029 doi:10.1088/1757-899X/729/1/012029

Application of Point-by-Point Vector Synthesis in Walkaway VSP Logging Data Processing

Fei Liu[1*], Li Qin[1], Jian Tang[1], Lili Cheng[2]

[1] BGP Southwest Geophysical Company, CNPC, Chengdu, Sichuan, 610213, China

[2] Southwest Research Institute of BGP, CNPC, Chengdu, Sichuan, 610213, China

[*]Corresponding author's e-mail: 414923573@qq.com

Abstract. At present, it is necessary to use Walkaway VSP logging imaging data to drill horizontal shale gas wells in exploration areas lacking three-dimensional seismic data.Walkaway VSP logging imaging profile can delineate the formation structure around wells, obtain detailed information of dip, inclination and fault around wells, and guide the trajectory design of horizontal wells. But VSP logging is richer and more complex than conventional seismic data. The precondition of up-going wave imaging is to separate up-going P-wave or converted S-wave from complex wave field. Nowadays, most vector synthesis techniques can not completely separate up-going P-wave from up-going converted wave, which affects the accuracy of final imaging results. In order to achieve the best wave field separation effect of VSP logging data, the point-by-point vector synthesis technique is adopted. Firstly, the directional rotation synthesis of horizontal components is carried out according to the maximum energy criterion. Then the rotational synthesis of horizontal and vertical components uses ray tracing based on velocity model to estimate the polarization angle of time-varying upward reflected P-wave or converted wave. The method of "velocity fine-tuning scanning" is applied to search for the best synthesis angle, and the point-by-point vector synthesis is carried out by calculating the time-varying polarization angle. Finally, the best separation effect data are selected as the final vector synthesis result. This technique solves the problem that conventional vector synthesis cannot completely separate up-going P-wave from up-going converted S-wave. Upward P-waves acquired after separation has clear characteristics, more continuous phase axis and good fidelity processing effect, which lays a solid foundation for fine imaging of target formation in the study area and has become an indispensable key supporting technology for optimizing the trajectory of horizontal wells in shale gas development in Sichuan Basin.

1. Introduction

Vertical seismic profile (VSP) is a kind of borehole seismic observation technology. Compared with ground earthquakes, VSP data have high signal-to-noise ratio, high resolution and obvious kinematic and dynamic characteristics of waves [1]. VSP technology provides the most direct correspondence between underground stratum structure and ground survey parameters. This technology can provide precise time-depth conversion and velocity model for processing and interpretation of surface seismic data, improve the accuracy of surface seismic imaging, and solve the imaging problem of small structures near wellbore that surface seismic technology cannot solve. VSP technology realizes high resolution imaging of strata around wellbore, makes lithologic characteristics research and reservoir evaluation results more reliable, thus providing reliable location and depth of drilling target geology

IOP Conf. Series: Materials Science and Engineering **729** (2020) 012029 doi:10.1088/1757-899X/729/1/012029

body, and avoiding target missing in drilling. However, VSP logging data are more abundant and complex than conventional seismic data, and downward direct wave, downward multiple wave, downward converted wave, upward reflection wave, upward converted wave and upward multiple wave can usually be observed. In order to achieve fine imaging and accurately depict the details of formation structure, it is necessary to separate up-going P-wave or converted S-wave from the complex wave field of VSP logging, so as to ensure the final imaging accuracy. In order to separate the wave field more reliably, cleanly and objectively, it is necessary to synthesize the three-component data of Walkaway VSP by vector rotation. Vector rotation synthesis of Walkaway VSP three-component wave field is usually achieved by projecting the observed wave field to the polarization direction of a particular waveform. However, at present, most vector synthesis techniques only rotate the same seismic trace at a fixed angle, often fail to achieve ideal synthesis results. In order to achieve the best wave-field separation effect of VSP logging data, a method of using ray tracing to estimate the polarization angle of up-going wave and point-by-point vector synthesis based on velocity fine-tuning scanning is developed. Combined with multi-channel filtering technology, the up-going P-wave or converted S-wave with clean wave field are finally obtained, which lays a solid foundation for high-precision imaging of VSP logging data. This technology plays an important supporting role in the optimization design of horizontal well trajectory for deep shale gas development in Sichuan Province.

2. Technical principle

2.1. Principle introduction

Vector wavefield rotation synthesis mainly achieves wave field separation by projecting the observed wavefield to the polarization direction of a particular wavefield (e.g. direct P-wave, up-going P-wave and converted wave).The polarization directions of these specific waves are different when they reach the receiver at different observation times.Estimation of polarization angle of up-going P-wave field and converted wave field at different time is the difficulty of this technique.The technology includes three parts: directional rotation synthesis of horizontal component, point-by-point vector synthesis of horizontal component and vertical component, and subsequent multi-channel filtering wave field separation.The horizontal component directional rotation synthesis uses a single direction for the whole channel synthesis, and does not need to be processed at each sampling point.Horizontal and vertical components are synthesized by point-by-point vector synthesis, which is based on the time-varying polarization angle.The final wave field separation uses the sequence from strong to weak to separate the required wave fields one by one.

2.2. Directional rotation synthesis of horizontal components

The observation direction of horizontal component in Walkaway VSP well is not necessarily in the ray plane of shooting point and receiving point, so it is necessary to synthesize horizontal component by directional rotation. According to Fig. 1, it is assumed that the projection of the ray plane in the horizontal direction is in the direction of Hp, which is at an angle of θ_1 with the direction of the horizontal component X.

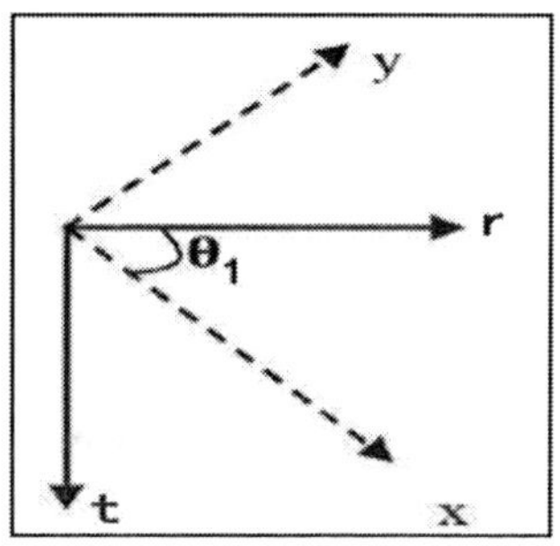

Figure 1.Horizontal component redirection diagram (top view)

Then, the directional rotation synthesis of horizontal components of X and Y components can be expressed as

$$\begin{cases} r = x\cos\theta_1 + y\sin\theta_1 \\ t = x\sin\theta_1 - y\cos\theta_1 \end{cases} \tag{1}$$

In the above formula, X and Y represent the two horizontal components measured by VSP, and r and t represent the two components polarized in the plane of Hp and in the plane perpendicular to Hp respectively after directional rotation. The key of formula (1) is to obtain the polarization angle $\theta 1$. Because the polarization direction of P-wave is the same as its propagation direction, the polarization angle $\theta 1$ can be estimated according to the maximum energy near the first arrival of P-wave. The calculation formula is as follows:

$$\theta_1 = \frac{1}{2}\arctan\frac{2\sum_{i=1}^{N} x_i y_i}{\sum_{i=1}^{N}(x_i^2 - y_i^2)} \tag{2}$$

In the formula (2), X_i and Y_i represent the sampling values of two horizontal components within a time window including the first arrival of P-wave, and N represents the sampling points in that time window.

2.3. Point-by-point vector synthesis of horizontal and vertical components

Figure 2 shows a schematic diagram of up-going wave propagation direction.$\theta 2$ represents the polarization angle of the up-going wave to the receiving point (the angle between the wave propagation direction and the vertical direction).

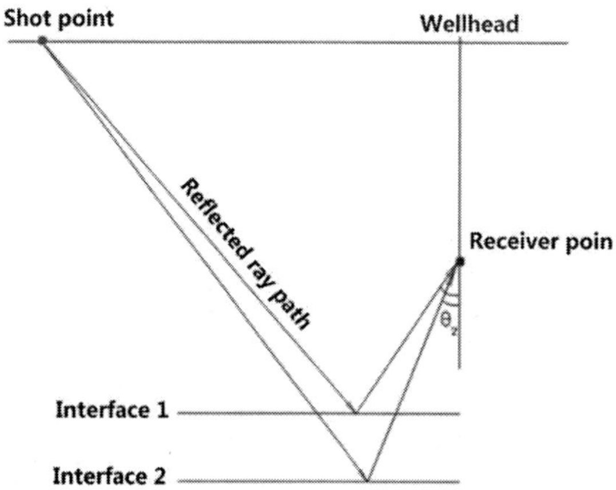

Figure 2. Schematic diagram of up-going wave propagation direction of VSP

If $\theta 2$ is known, the horizontal component and the vertical component can be projected into the coordinate system of the propagation direction, so that theup-going P-wave and S-wavecan be rotated into two new components. The calculation formula is as follows:

$$\begin{cases} R = z\cos\theta_2 + r\sin\theta_2 \\ T = z\sin\theta_2 - r\cos\theta_2 \end{cases} \tag{3}$$

In the above expressions, R and T represent two components of polarization in the direction of up-going P-wave propagation and perpendicular to the direction of up-going P-wave propagation, Z represents the vertical component measured by VSP, and r represents the horizontal component of polarization in the ray plane after horizontal directional rotation.For the same receiving point, the up-going waves received at different times come from the reflective interface of different depths. The direction of propagation of these up-going waves to the receiving point is different, that is, the polarization direction is different.Therefore, in order to synthesize the required wave onto the same component record, it is necessary to synthesize the horizontal and vertical components point by point with time-varying vector rotation [2]. Therefore, the determination of polarization angle $\theta 2$ is very important.

Assuming that the velocity model is known, ray tracing can be carried out and the angle of $\theta 2$ can be calculated according to the ray path.However, velocity models can only be estimated by various methods, such as direct wave estimation based on zero offset VSP, but absolute true values cannot be obtained. Therefore, the angle of $\theta 2$ calculated from ray tracing results also has a large error.In this case, in order to obtain better rotation effect, a rotation synthesis technology based on "velocity fine-tuning scanning" is proposed.That is to say, in a certain range of velocity values, several groups of velocity parameters are divided into equal intervals to calculate the corresponding $\theta 2$ angles, and then the corresponding rotation synthesis results are calculated. Finally, the group of data with the best separation effect is selected as the final result.This step improves the effect of vector synthesis.Through the above rotational synthesis steps, the up-going P-wave and the up-going converted shear wave with similar apparent velocities can be separated into different components. However, there are still down-going and other interference waves. Because the apparent velocities of these waves are quite different from those of up-going P-wave and converted wave (the direction of apparent velocity is opposite), they can be filtered out successively by traditional multi-channel filter or median filter, and finally the separated up-going P-wave and converted wave fields can be obtained.

3. Application Effect of Point-by-Point Vector Synthesis Technology

3.1. Directional rotation synthesis of horizontal components

Fig. 3 is the original X and Y horizontal component records of two shoots in Walkaway VSP logging of a well in Sichuan Basin. It can be seen that the horizontal component records are chaotic and the continuity of the up-going and down-going traveling wave coaxial is poor, which indicates that the down-hole geophone has no orientation in the acquisition process, and the direction of its X and Y components is randomly distributed.According to the maximum energy criterion, the two horizontal components are rotated to the horizontal plane and perpendicular to the ray plane, and two new components are synthesized, which are recorded as r and t components respectively (Fig. 4).After directional rotation synthesis, we can see clearly and continuously down-going and up-going traveling waveform events in the horizontal component records of r and t.

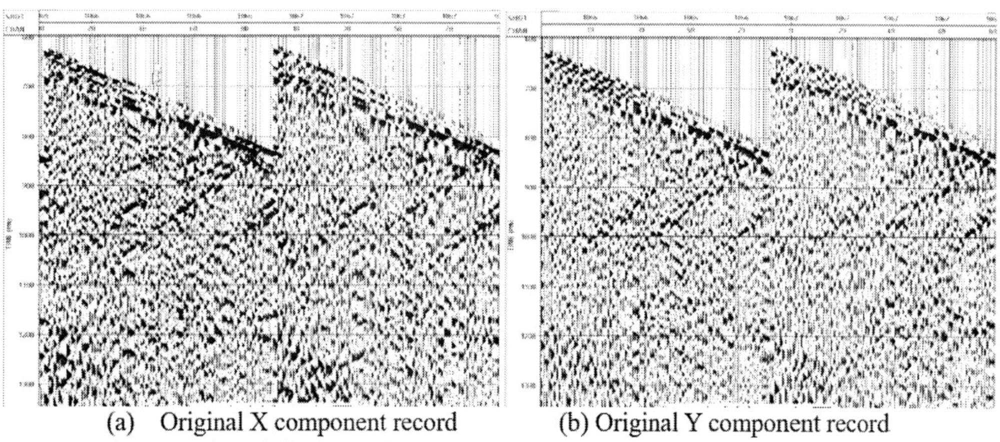

(a) Original X component record (b) Original Y component record

Figure 3.The original X, Y horizontal component records of walkaway VSP

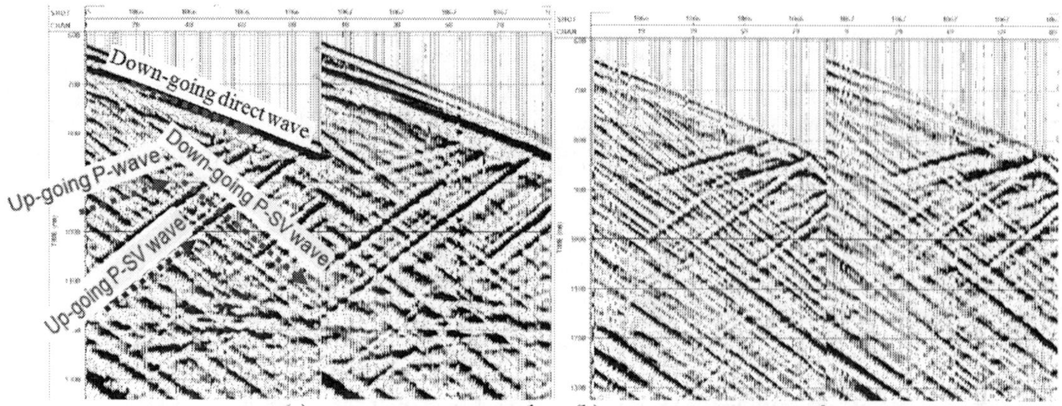

(a) r component record (b) t component record

Figure 4. Horizontal component records of r and t after horizontal directional rotation synthesis

3.2. The point-by-point vector synthesis of horizontal and vertical components

Before point-by-point vector synthesis, the records of the vertical component are shown in figure 5, and the records of the horizontal component are shown in figure 4 (a).In order to compare the application effects, the traditional vector rotation synthesis method was first tested, and the results were shown in figure 6.The traditional method is to determine the polarization angle of P-wave based on the first arrival of direct wave, and then rotate the whole seismic trace with the same polarization angle.From the synthetic results, the up-going P-wave and the up-going converted S-wave obtained by traditional methods still exist in both radial and tangential components, which are not completely separated, indicating that the separation effect is not good. However, from the implementation effect, the target function of this traditional method has been achieved, because in the two records after the traditional vector rotation synthesis, the energy near the first arrival is obviously a strong one and a weak one, which is the goal of this method. This shows that the poor separation effect of the traditional vector rotation synthesis method for P-wave and converted S-wave is the defect of the method itself. It is not caused by improper handling of parameters. The main reason is that the traditional method uses the same polarization angle to synthesize the whole seismic trace.

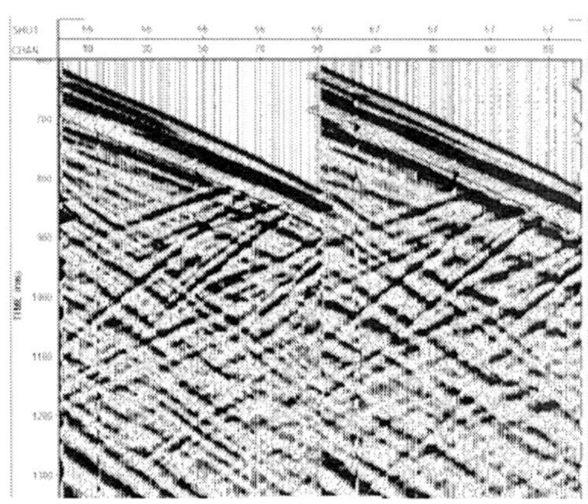

Figure 5. Vertical component (Z component) recording before point-by-point vector composition

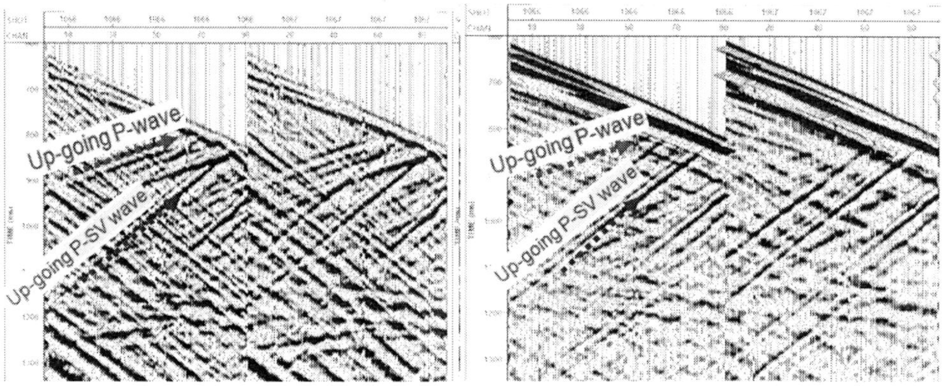

(a) T component record (b) R component record

Figure 6. Radial R and tangential T component recording after vector composition by traditional method

Figure 7 shows the result of synthesis using point-by-point vector synthesis technique. After point-by-point time-varying vector synthesis, the radial R component is mainly composed of up-going P-wave and down-going converted S-wave, while the tangential T component is dominated by down-going P-wave and up-going converted S-wave. It can be seen that the up-going P-wave and the up-going converted S-wave are synthesized into two components respectively, and no longer mix and exist in one component at the same time. Point-by-point time-varying vector synthesis has achieved better results than the traditional vector rotation synthesis method, which is well prepared for the next wave field separation.

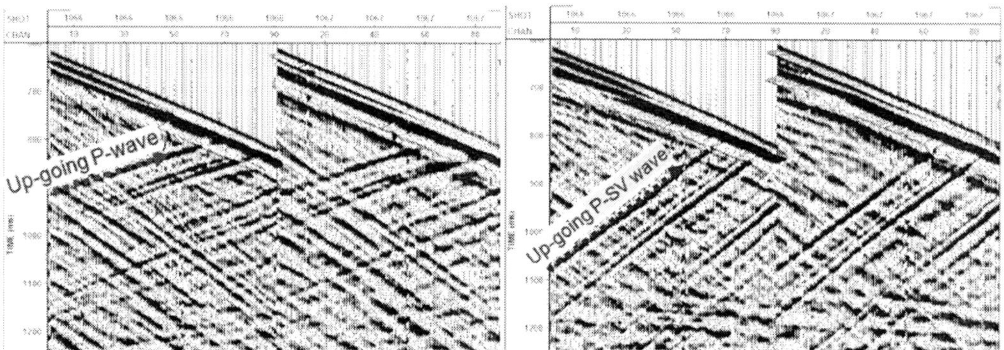

(a) R component record (b) T component record

Figure 7. Radial R and tangential T component recording after point-by-point vector composition

Figure 8 shows the radial R component recordings synthesized by point-by-point vectors, in which the up-going wave is mainly reflected P-wave and the down-going wave is mainly the interference wave. Yellow line indicates the direction of down-going P-wave coaxiality to be removed, blue line indicates the direction of down-going converted S-wave coaxiality to be removed, and the red line indicates the direction of up-going P-wave coaxiality to be extracted. Using two-step median filter, down-going P-wave and down-going converted S-wave can be removed separately, and the useful up-going P-wave records can be obtained. Generally, the up-going P-wave group obtained by median filtering technology has clear characteristics, more continuous coaxial, good fidelity processing effect, and is more conducive to later imaging [3].

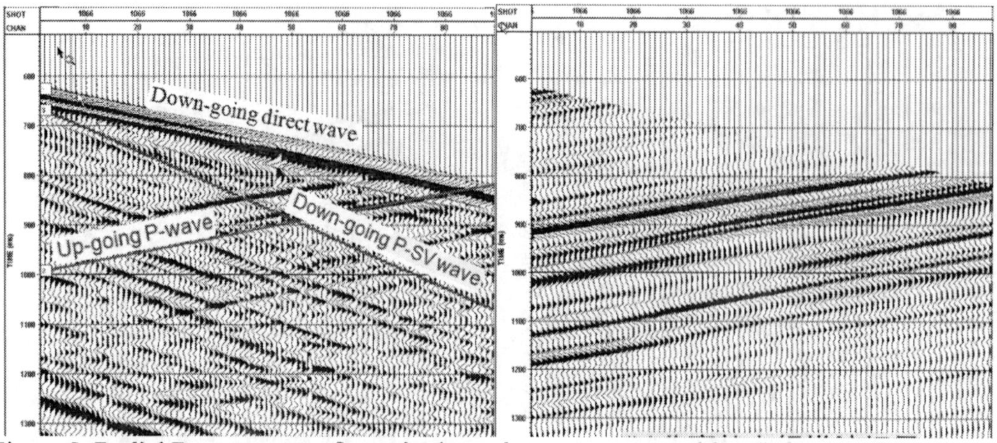

Figure 8. Radial R component after point-by-point vector composition (left upper figure) and Up-going P-Wave recording after wave field separation (right upper figure)

4. Conclusions

(1) The imaging effect of VSP logging depends on the separation accuracy of up-going P-wave and up-going converted S-wave. If P-wave and S-wave are not completely separated, the accuracy of subsequent imaging results will be affected. Point-by-point vector synthesis technology uses ray tracing to estimate the polarization angle of up-going P-wave and velocity fine-tuning scanning method to synthesize. It solves the problem that conventional vector synthesis cannot completely separate the up-going P-wave from the up-going converted S-wave. The up-going P-wave group obtained by this technology has clear characteristics, more continuous coaxial, and good fidelity processing effect. This technology has laid a solid foundation for fine imaging of target formation in the study area, and has become an indispensable key supporting technology for optimizing the trajectory of horizontal wells in shale gas development in Sichuan Basin. Point-by-point vector synthesis technology has been widely used in Chang Ning, Yong Chuan and Mo Xi blocks of Sichuan Province, with good application effect and broad application prospects.

(2) Point-by-point vector synthesis method and the most commonly used Walkaway VSP logging imaging methods (including VSP-CDP conversion imaging method, Kirchhoff VSP migration method, etc.) are sensitive to velocity model. The accuracy of velocity model will ultimately affect the imaging results, and how to build a high-precision velocity model with anisotropy needs further study. It is suggested that the research on the method of "joint anisotropic velocity model establishment between VSP data and surface seismic data" should be further strengthened in the future.

Acknowledgments

The paper was supported by the national project numbered 2016ZX05062-001 and we also appreciate the contributions of our colleagues.

References

[1] Guangming Zhu. (1998) Vertical seismic profile method. Beijing Petroleum Industry Publishing, Beijing.

[2] Wei Zhang, Liting Liu, Chunhua Chen, Fei Liu, Hong Mi. (2018) Application of Walk-away VSP technology in shale gas horizontal well design. In: CPS/SEG Beijing 2018 International Geophysical Conference & Exposition Electronic papers. Beijing. pp. 1319-1322.

[3] Libin Cao, Furong Wu, Zhirong Li, Chuan Deng, Kun Luo. (2009) Application effect of mountain Walk-away VSP technology. Oil Geophysical Prospecting, 44:292-297.

Study on Thermal Storage Performance of Industrial Paraffin and Fatty Acid Binary Mixture

Quanying Yan, Qingzhu Fan*, Jing Zhang.

College of environmental and energy engineering, Beijing University of Civil Engineering and Architecture, Beijing, 100044, China.

*Corresponding author's e-mail: whitehate@163.com

Abstract. Taking the industrial paraffin and fatty acid binary mixture as the research object, the phase change temperature, latent heat of phase change and supercooling degree were studied by differential scanning calorimeter, aiming at reducing the application cost of simple fatty acid as phase change material , broadening the range of phase transition and increasing the amount of latent heat value of phase change. The results show that the paraffin-tannic acid and paraffin-lauric acid binary mixture reduces the phase transition temperature and increases the latent heat of phase change compared with the single fatty acid, which is suitable for the energy storage of the envelope structure. The phase transition temperature of paraffin-myristic acid, paraffin-stearic acid, paraffin-palmitic acid mixture is lower than 40 °C. Although it is lower than the phase transition temperature of single fatty acid under certain mixing ratio conditions, which is not suitable for application. In the energy storage of the structure, it can be used in other energy-saving systems such as phase change heat exchangers for waste heat recovery. Paraffin-tannic acid and paraffin-lauric acid binary mixtures are less subcooled and are ideal for energy storage in envelope structures.

1. Introduction

The unique superiority of phase change material energy storage technology in the field of energy storage has made phase change materials more and more attention as a new energy storage material. The phase change material energy storage technology absorbs/releases a large amount of heat through an approximately isothermal phase change process to achieve thermal energy conversion while ensuring thermal comfort.

The application of phase change material energy storage technology in the field of energy conservation has been relatively mature, and experts and scholars at home and abroad have conducted in-depth research on this. Sharma[1] tested the phase transition parameters and thermal storage properties of paraffin, stearic acid and acetamide. Keles[2] prepared a composite phase change material using lauric acid-myristic acid as a phase change material, and applied it to the thermal storage system of a vertical coaxial casing, and studied its heat storage performance and application feasibility. Sari[3] mixed lauric acid, stearic acid, myristic acid and palmitic acid to test the phase transition temperature and latent heat of phase change of the binary complex fatty acid. Hou Jingpeng et al[4] tested the temperature control effect of paraffin phase change temperature control concrete and the thermal conductivity of paraffin phase change temperature control mortar. The research results show that the temperature control effect of paraffin phase change temperature control concrete is obvious, and the paraffin phase change temperature control The thermal conductivity of the mortar is significantly improved. Wu Qisheng et al[5] prepared a binary organic eutectic (C citric acid CA and cetyl alcohol H) / expanded graphite

IOP Conf. Series: Materials Science and Engineering **729** (2020) 012030 doi:10.1088/1757-899X/729/1/012030

composite phase change energy storage material, and studied its structure and thermal properties, found that expanded graphite to CA-H The physical adsorption effect reaches 80%, and the composite phase change energy storage material has better thermal stability. Yan Quanying et al[6] studied the phase transition temperature and phase change latent heat of two binary mixtures consisting of 48# paraffin and liquid paraffin, tannic acid and stearic acid, respectively, using a differential scanning calorimeter. The thermal stability of the mixture is good, and the thermal stability of the fatty acid mixture is superior to the paraffin mixture. Li Lisha[7] studied the stability of paraffin-like and fatty acid phase change materials, and found that fatty acids have better heat storage properties and thermal stability than paraffin, and are more suitable for building energy conservation. Zhang Lin[8] formulated a variety of fatty acid phase change materials and selected fatty acid composite phase change materials suitable for wall applications. After repeated cycles experiments, the fatty acid phase change materials prepared have good stability. Zhang Yi[9] formulated a dibasic fatty acid composite phase change material of tannic acid-lauric acid, and tested its thermal properties by differential scanning calorimetry (DSC), using porous diatomaceous earth to adsorb fatty acid phase change materials, and The heat storage performance and thermal conductivity of the diatomaceous earth/fatty acid composite phase change material were tested. Huang Xue[10] selected 10 kinds of ternary eutectic fatty acid composite phase change materials with five kinds of single fatty acids, such as tannic acid, lauric acid, myristic acid, stearic acid and palmitic acid, and studied their thermal properties by DSC. A fatty acid composite phase change material with excellent performance.

In this study, binary mixtures of low-cost industrial paraffin and high-priced fatty acids were used to study the range and regularity of the phase transition temperature and latent heat of phase change. Paraffin and fatty acid are used as low-temperature energy storage phase change materials. According to the phase change temperature that can be achieved by the formulated mixture, it can be applied to building envelopes such as walls and floors, and can also be used for phase change energy storage heat transfer of waste heat recovery heating systems.

2. Meterial and Methods

2.1. Materials and instruments

Industrial paraffin, citric acid, lauric acid, palmitic acid, stearic acid, nutmeg, Sinopharm Chemical Reagent Co., Ltd. The heat storage performance data of the experimental paraffin and fatty acid phase change materials are shown in Table 1. Differential Scanning Calorimeter (DSC, Model 200PC) is made by Germany's NETZSCH; Electronic Balance (AR1530/C) is made by House International Trading Co., Ltd.

Table1. Thermal storage performance data of experimental paraffin and fatty acid phase change materials

Phase change material	Phase transition temperature/°C	Phase change latent heat/($J \cdot g^{-1}$)
Industrial paraffin (P)	66.0	180.2
Tannic acid (CA)	30.5	151.1
Lauric acid (LA)	42.8	222.3
Palmitic acid (PA)	61.0	241.0
Stearic acid (SA)	67.4	239.9
Myristic acid (MA)	52.7	217.6

2.2. Preparation of samples

The industrial paraffin and five kinds of fatty acids were weighed according to the preset mass ratio, and the electronic balance (AR1530/C type, House International Trading Co., Ltd.) was weighed and placed in a beaker, and the beaker was placed on an alcohol lamp to heat it. . The mixture of paraffin and fatty acid is completely melted, stirred with a glass rod until the material is evenly mixed, and naturally cooled

IOP Conf. Series: Materials Science and Engineering **729** (2020) 012030 doi:10.1088/1757-899X/729/1/012030

and solidified in the air. This process was repeated 3 times, and finally 46 kinds of paraffin and fatty acid binary mixture samples were separately prepared.

2.3. Test of samples

Forty-six samples were tested by differential scanning calorimeter (DSC, Model 200PC, Nexus, Germany) and their test results were analyzed.

3. Results

Organize and summarize the measured data and draw the following graphics. Among them, the paraffin-tannic acid mixture phase transition temperature and phase change latent heat curve are shown in Figure 1. The phase transition temperature and phase transition latent heat curve of the paraffin-palmitic acid mixture are shown in Figure 2.

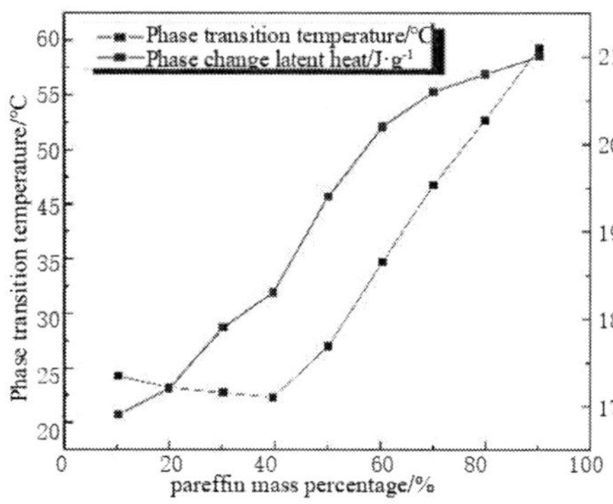

Figure 1. Phase transition temperature and phase change latent heat curve of paraffin-decanoic acid mixture.

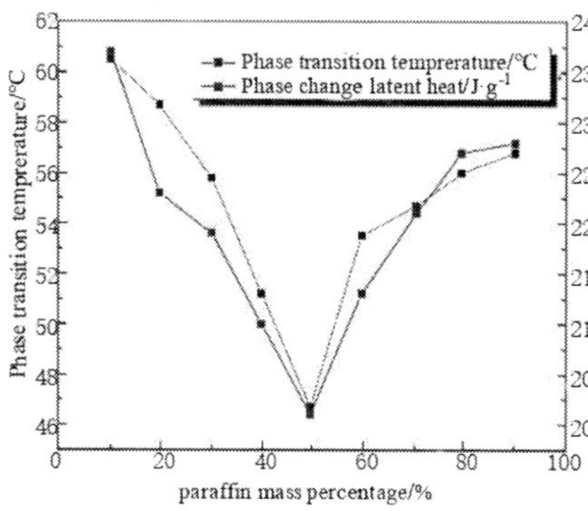

Figure 2. Phase transition temperature and phase change latent heat curve of paraffin-palmitic acid mixture.

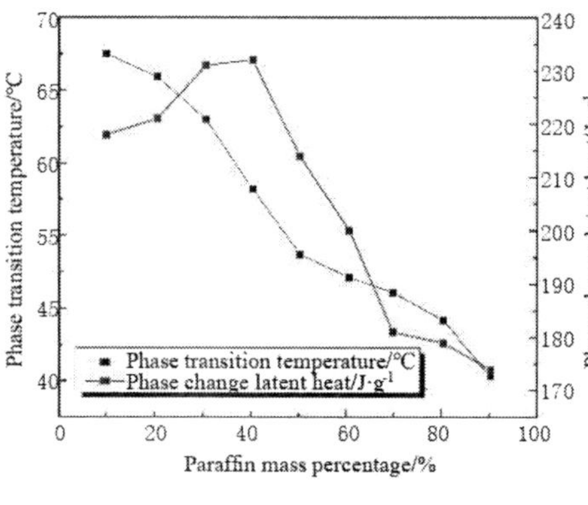

Figure 3. Phase transition temperature and phase change latent heat curve of paraffin-stearic acid mixture.

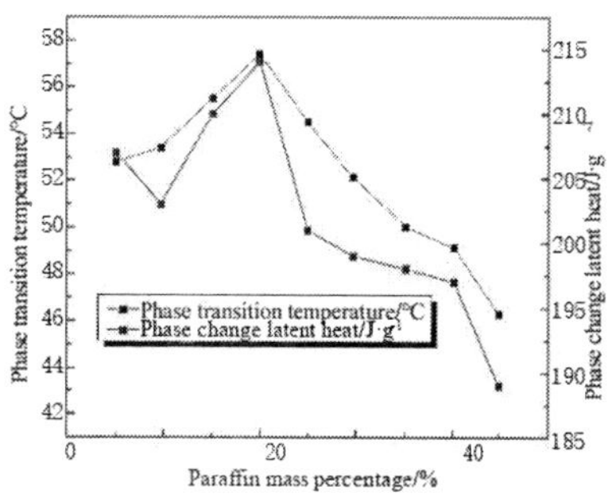

Figure 4. Phase transition temperature and phase change latent heat curve of paraffin-myristic acid mixture.

The phase transition temperature and phase transition latent heat curve of the paraffin-stearic acid mixture are shown in Figure 3. The phase transition temperature and phase change latent heat curve of the paraffin-myristic acid mixture are shown in Figure 4. The phase transition temperature and phase transition latent heat curve of the paraffin-lauric acid mixture are shown in Figure 5. The subcooling curve of paraffin and 5 fatty acids under different mixing ratios is shown in Figure 6.

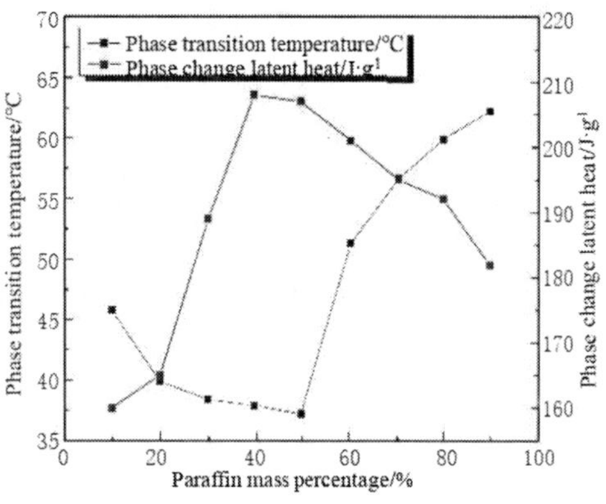

Figure 5. Phase transition temperature and phase change latent heat curve of paraffin-lauric acid mixture

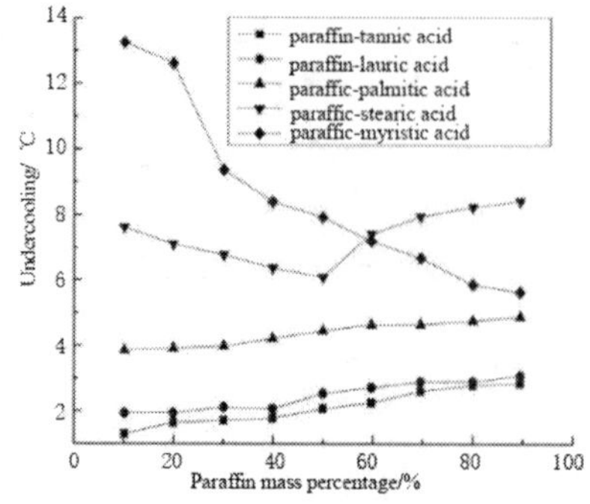

Figure 6. Undercooling curve of paraffin and five fatty acids under different mixing ratios

4. Discussion

4.1. Analysis of phase transition temperature and latent heat of phase change of paraffin-citric acid mixture

It can be seen from Figure 1 that the phase transition temperature of the binary mixture of industrial paraffin and tannic acid fluctuates between 27.3 and 60.0 °C. As the amount of paraffin increases, the phase of the binary mixture of paraffin and tannic acid The temperature is lowered first, then the situation is raised. Under the condition of 40% paraffin (wt, mass fraction, the same below), the mixture reaches a low in-melt state, the phase transition temperature reaches a minimum value; the paraffin amount is <40%. Under the condition that the phase transition temperature of the mixture fluctuates between 27.3 and 29.7 °C, it is suitable for the energy storage of passive phase change wall; the phase transition temperature of the mixture is between 50% and 60% of paraffin. Between 30 and 40 °C, it is suitable for use in active phase change wall and phase change floor. Under the condition of paraffin dosage >60%, the phase transition temperature of the mixture increases rapidly and is higher than 40 °C, which is not suitable for the envelope structure. The low-temperature energy storage application can be used as the energy storage material for the phase change heat recovery of waste heat recovery; in addition, the latent heat of phase change of paraffin-citric acid mixture increases with the increase of paraffin consumption , Latent heat between 168.6 ~ 209.2J / g change, capric acid itself is large relative to the latent heat. As the amount of paraffin increases, the latent heat value is greater than the phase change latent heat of the two single components.

4.2. Analysis of phase transition temperature and latent heat of phase change of paraffin-palmitic acid mixture.

It can be seen from Figure 2 that the phase transition temperature of the paraffin-palmitic acid binary mixture decreases rapidly with the increase of the amount of paraffin, and then rapidly increases the situation. The mass ratio of paraffin to palmitic acid is 5:5. The phase transition temperature and the latent heat of phase change are the minimum values, the phase transition temperature is 46.7 ° C, the latent heat of phase change is 201.9 J / g, which is lower than the melting point of palmitic acid and paraffin, and has high latent heat. The paraffin-palmitic acid phase transition temperature is high and is suitable for use in waste heat recovery phase heat exchangers.

4.3. Phase transition temperature and latent heat of phase change of paraffin-stearic acid mixture

It can be seen from Figure 3 that as the amount of paraffin increases, the phase transition temperature of paraffin-stearic acid is always reduced, and the minimum is 45.3 ° C, which is lower than the melting point of paraffin and stearic acid single substance; The latent heat of phase change of stearic acid mixture increases slowly and then decreases rapidly with the increase of paraffin dosage. The minimum value of latent heat value is 174.8J / g, which is lower than that of paraffin and stearic acid. Change the latent heat.

4.4. Phase transition temperature and latent heat of phase change of paraffin-myristic acid mixture

It can be seen from Figure 4 that the phase transition temperature of paraffin-myristic acid increases slowly with the increase of paraffin dosage. Under the condition of 40% paraffin, the phase transition temperature of paraffin-myristic acid is 57.38°. C, the latent heat of phase change is 213.97J/g. Under the condition of paraffin dosage >40%, the phase transition temperature decreases rapidly. Under the condition of paraffin dosage of 60%-90%, the phase transition temperature is lower than paraffin and myristic acid. The melting point of any single substance can achieve the purpose of lowering the phase transition temperature by mixing; however, the latent heat of phase change has no obvious change law, and the whole fluctuates between the latent heat values of the two substances. By mixing, the latent heat of phase change of paraffin-myristic acid was not significantly improved.

4.5. Phase change temperature and latent heat of phase change of paraffin-lauric acid mixture

It can be seen from Figure 5 that as the amount of paraffin increases, the phase transition temperature of the paraffin-lauric acid mixture decreases first and then increases. The latent heat of phase change increases first and then decreases, and the amount of paraffin is 20%-50. Under the condition of %, the phase transition temperature varies between 30 and 40 °C. Under the condition of 50% paraffin, the phase transition temperature of the paraffin-lauric acid mixture reaches the minimum value of 37.1 °C, and the latent heat of phase change is 207.1 J / g, very close to the maximum. This combination ratio is suitable for use in the field of energy storage for active enclosures.

4.6. Undercooling analysis

In the process of phase transition from liquid to solid, the temperature at which the phase transition actually occurs is lower than the theoretical transition temperature. This temperature difference is called subcooling. A large number of studies have shown that organic phase change materials have the advantage of less subcooling than inorganic phase change materials, but the specific values of subcooling of different kinds of organic phase change materials are large and small, the phase transition temperature is suitable, and the latent heat is better. Large and subcooled phase change materials are more suitable for phase change energy storage applications. The subcooling curve of paraffin and 5 fatty acids under different mixing ratios is shown in Figure 6. It can be seen from the figure that the binary mixture of paraffin-myristic acid and paraffin-stearic acid has a higher degree of subcooling, and the subcooling ratio of the paraffin-nonanoic acid binary mixture and the paraffin-lauric acid binary mixture is higher. The subcooling degree of small, paraffin-myristic acid gradually decreased with the increase of paraffin dosage. The subcooling degree of the mixture under different mixing ratios varied greatly from 13.1 °C to 5.6 °C; - The degree of subcooling of stearic acid mixture decreases first and then increases with the increase of paraffin amount, and the degree of supercooling fluctuates between 6 and 9 °C; paraffin-antimonic acid binary mixture, paraffin-lauric acid binary The supercooling degree of the mixture and the paraffin-palmitic acid binary mixture increased slowly with the increase of the amount of paraffin in the mixture; the subcooling degree of the paraffin-antimonic acid binary mixture and the paraffin-lauric acid binary mixture were all compared. Small, between 1 and 3.5 ° C, the degree of subcooling varies little with the mixing ratio, and the smaller degree of subcooling provides a broader space for the application of the binary mixture.

5. Conclusions

(1) The phase transition temperature of the paraffin-niobate binary mixture decreases slowly and then increases rapidly with the increase of paraffin dosage. Under the condition of paraffin dosage less than 60%, the phase transition temperature of the mixture is 20.0. Between 40.0 °C fluctuation, the latent heat of phase change is greater than the latent heat of phase change of single component, broadening the range of phase transition temperature, and increasing the latent heat of phase change. It can be used for active or passive enclosure under different mix ratio conditions. Structure energy storage field.

(2) The phase transition temperature of paraffin-lauric acid binary mixture decreases first and then increases with the increase of paraffin dosage. The latent heat of phase change increases first and then decreases, and the mass ratio of paraffin to lauric acid increases. Under the condition of 5:5, the phase transition temperature is the lowest, 37.1 °C, and the latent heat of phase change is 207.2 J / g, which is close to the maximum value, which is suitable for energy storage in the active wall or floor area.

(3) The phase transition temperature of the paraffin-myristate binary mixture, the paraffin-stearic acid binary mixture, the paraffin-palmitic acid binary mixture, although in some mixing ratios, is higher due to the higher phase transition temperature of the industrial paraffin itself. Under the condition, the phase transition temperature of the single fatty acid is lower than that of the single fatty acid, but it is still above 40 °C. It is not suitable for the active wall energy storage field. Some mix ratios can be applied to the active wall or floor energy storage field, or phase change storage. Can be in the field of heat exchangers.

(4) The subcooling degree of the binary mixture formed by paraffin wax and different kinds of fatty acids is different, the binary mixture of paraffin-myristate and the paraffin-stearic acid binary mixture has a large degree of supercooling, and the paraffin-nonanoic acid binary mixture The degree of subcooling is at least between 1.0 and 3.5 °C.

References

[1] Sharma A,Sharma S D,Buddhi D.(2002)Accelerated thermal cy-cle test of acetamide, stearic acid and paraffin wax for solar thermal latent heat storage applications. Energy Convers Manag.43(14):1923-1930.

[2] Keles S, Kaygusuz K, Sarl A.(2005) Lauric and myristic acids eutectic mixtureas phase change material for low-temperature heating applications. International Journal of Energy Reseach. 29:857- 870.

[3] Sari A, Sari H,Onal A. (2004)Thermal properties and thermal reliability of eu-tectic mixtures of some fatty acids as latent heat storage materials. Energy Conversion and Management.45:365- 376.

[4] Hou JP, Shi w, Zhang X. (2012)Study on Thermal Properties of Paraffin Phase Change Temperature Control Concrete.jouranl of building materials. 15(6):767-770.

[5] Wu QS, Chou Y, Li SP, et al. (2014)Preparation and Study of Citrate-Cetyl Alcohol/Expanded Graphite Composite Phase Change Energy Storage Materials. jouranl of building materials. 17(1):84-88.

[6] Yan QY, Wang LJ, Yu D, et al.(2015) Phase change material properties for walls and floors. jouranl of building materials. jouranl of building materials.18(2):302-306

[7] Li LS.(2011) Preparation of fatty acid mixture and experimental study on shaped phase change wall.Beijing: Beijing University of Civil Engineering and Architecture.

[8] Zhang L. (2010)Preparation of fatty acid mixture and experimental study on shaped phase change wall[D]. Beijing: Beijing University of Civil Engineering and Architecture.

[9] Zhang Y, Zhang JY, Huang B, et al. (2012)Thermal Conductivity Test of Fatty Acid Phase Change Materials and Numerical Simulation of Phase Change Heat Transfer Process.journal of functional materials.43(14):1950-1954.

[10] Huang X. (2015)Preparation and Thermal Properties of Ternary Eutectic Fatty Acid Composite Phase Change Materials. Guangzhou:Guangdong University of Technology.

IOP Conf. Series: Materials Science and Engineering **729** (2020) 012031 doi:10.1088/1757-899X/729/1/012031

Experimental and theoretical research on the thermal properties of carbonate mixtures

Quanying Yan[1], Chenyu Wang[1] *, Chao Liu[1], Jing Zhang[1]

[1]Beijing Municipality Key Lab of Heating, Gas Supply, Ventilating and Air Conditioning Engineering, Beijing University of civil Engineering and Architecture, Beijing Engineering Research Center of Sustainable Energy and Buildings, Beijing, 100044, China

*Corresponding author's e-mail: wangchenyuc@163.com

Abstract. In this paper, Na_2CO_3, Li_2CO_3 and K_2CO_3 were selected to prepare carbonate mixtures according to different proportions. The thermal property and stability of the samples were tested by differential scanning calorimeter (DSC). The phase change temperature and latent heat of carbonate mixtures were calculated by theoretical formula, and compared with the experimental data to verify whether the theoretical formula are applicable in the field of high temperature molten salt. The results show that the melting point of carbonates which melted together is about 400°C. And the melting point of ternary carbonates is lower than the single component. The initial crystallization points of some samples are higher than the melting points. The differences between the melting point and the primary crystallization temperature are 5~20°C. The phase change latent heat of the carbonate mixtures is lower than the single component's. The latent heat of melting is more than the latent heat of solidifying. The DSC testing curves of the carbonate mixtures are almost coincident. The stability of mixed carbonate is very good. This experiment will provide reference and basis for its application in solar thermal power generation.

1. Introduction

High temperature heat transfer and heat storage technology of solar energy is a core technology in solar thermal power generation system. It is very important to select a suitable high-temperature heat storage medium. In recent years, molten salts have been a good choice as high temperature heat transfer and heat storage medium of solar energy, because they have wide temperature range, low viscosity, high liquidity, low vapor pressure, low pressure resistance to the pipeline, large relative density, high specific heat capacity, good heat storage capacity and low cost, etc..

The American Oak Ridge National Laboratory studied the physical and chemical properties of various molten salts and their compatibility with the container materials [1]. American Sandia National Laboratory has been engaged in studying the thermal properties of mixed molten salt and developing new low melting point mixed molten salts [2]. Araki et al studied the thermal conductivity of molten carbonate [3]. Kourkova et al fitted the specific heat curve of Li_2CO_3 at 303.15~563.15K [4]. Ma Zhongfang and Wu Yuting prepared hundreds of mixed molten salts, especially a number of molten salts with low melting point to reduce the cost of molten salts [5]. The team of Professor Ding Jing of Zhongshan University has made a deep research on the aspect of preparation of new molten

IOP Conf. Series: Materials Science and Engineering **729** (2020) 012031 doi:10.1088/1757-899X/729/1/012031

salt, reducing the melting point and increasing the boiling point [6]. Ren Nan et al prepared and researched ternary mixed carbonates with different proportions [7].

We selected Na_2CO_3, Li_2CO_3 and K_2CO_3 to prepare 36 kinds of mixed molten salts according to different proportions. Thermal properties of the samples were studied by differential scanning calorimeter (DSC). The melting point, phase change latent heat and the initial crystallization points were studied and carried out cyclic experiment. The 36 kinds of mixed carbonates were optimized according to the standard of low melting point, high latent heat and wide using temperature.

2. Experimental methods

2.1. Experimental materials and experimental equipment

The carbonates used in the experiment were sodium carbonate(Na_2CO_3), lithium carbonate(K_2CO_3) and potassium carbonate(Li_2CO_3), which produced by Sinopharm Group Chemical Reagent Co., Ltd. level are analytically pure. The main instruments used in the experiment included vacuum drying box, electronic balance, muffle furnace and differential scanning calorimeter (DSC).

2.2. Experimental processes

The three kinds of carbonates were placed in the vacuum drying box for 24 hours at 120°C, and then mixed according to different quality ratio to prepare 36 kinds of samples (showed in Table 1). Put them into the muffle furnace to melt,the temperature of the muffle furnace was set to 600°C, and heating time was 2 hours. Then the sample was taken out and solidified in the air. And then ground into a powder in a mortar. There were 6 kinds of samples not to melt in the muffle furnace, and they were number 08, 15, 31, 34, 35 and 36.

Differential scanning calorimeter (DSC) was used to test the mixed carbonate with different proportions, the phase change temperature and phase change latent heat of the samples during the heating process and the crystallization point of the sample during cooling were determined. The proportions of the different components are shown in Table 1.

Table 1. Mass ratio of ternary carbonate（Na_2CO_3-Li_2CO_3-K_2CO_3）

NO.	Na_2CO_3	Li_2CO_3	K_2CO_3	NO.	Na_2CO_3	Li_2CO_3	K_2CO_3	NO.	Na_2CO_3	Li_2CO_3	K_2CO_3
01	10%	10%	80%	13	20%	50%	30%	25	40%	40%	20%
02	10%	20%	70%	14	20%	60%	20%	26	40%	50%	10%
03	10%	30%	60%	15	20%	70%	10%	27	50%	10%	40%
04	10%	40%	50%	16	30%	10%	60%	28	50%	20%	30%
05	10%	50%	40%	17	30%	20%	50%	29	50%	30%	20%
06	10%	60%	30%	18	30%	30%	40%	30	50%	40%	10%
07	10%	70%	20%	19	30%	40%	30%	31	60%	10%	30%
08	10%	80%	10%	20	30%	50%	20%	32	60%	20%	20%
09	20%	10%	70%	21	30%	60%	10%	33	60%	30%	10%
10	20%	20%	60%	22	40%	10%	50%	34	70%	10%	20%
11	20%	30%	50%	23	40%	20%	40%	35	70%	20%	10%
12	20%	40%	40%	24	40%	30%	30%	36	80%	10%	10%

3. Experimental results and analysis

3.1. Test and analysis of phase change temperature

In the solar thermal power generation system, the lower the melting point of the mixed molten salt which used as heat transfer and heat storage medium is, the lower the operation cost is, the better the economy. Simultaneously, the lower the freezing point of the mixed molten salt is, the safer the system is. The melting point and the initial crystallization temperature of the mixed molten salts are very important for the optimization of mixed molten salts. The melting starting point and the ending point of each sample are shown in Table 2.

Table 2. Melting point and temperature range of 21 kinds of mixed carbonates

Number	Starting point (°C)	Ending point (°C)	Number	Starting point (°C)	Ending point (°C)
01	444.6	466.8	19	401.6	410.4
02	428.9	471.7	20	396.8	405.9
07	400.9	408.1	21	393.8	403.6
09	428.5	447.0	22	408.8	423.3
10	424.6	445.0	24	397.0	405.8
12	399.6	407.4	25	399.7	406.2
13	398.4	406.4	27	403.1	412.4
14	402.4	409.3	28	398.8	406.0
16	428.2	437.7	29	393.9	402.3
17	403.1	412.4	30	397.4	406.3
18	398.6	407.1			

The DSC curves of the 36 kinds of samples are shown in Figure 1. The melting point of most mixed molten salts is concentrated at 400°C. In all 36 kinds of samples, there were 9 kinds of samples didn't form regular melting peak in the melting process, and they are number 03, 04, 05, 06, 11, 23, 26, 32 and 33. Figure 2 shows an irregular DSC curve of sodium carbonate, lithium carbonate, and potassium carbonate at a mass ratio of 1: 3: 6. The curve has two melting peak, and the melting temperature range is about 100°C. This not only increases the cost of system operation, but also makes it easy to separate part of the group when the temperature changes, increasing the probability of pipeline plugging, and it is detrimental for the solar power generation system.

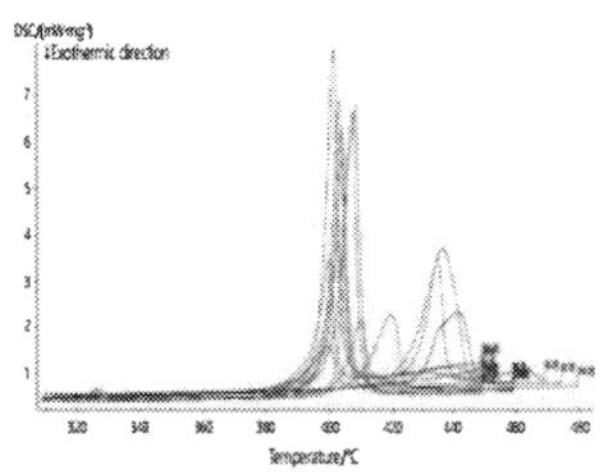

Figure 1. DSC curves of 30 kinds of mixed carbonate salts

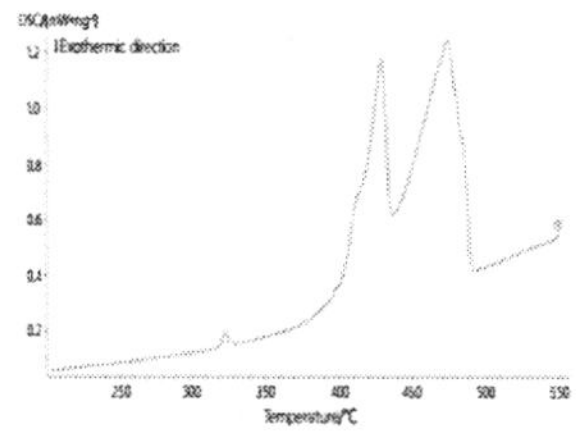

Figure 2. Abnormal melting peaks of ternary carbonate (sample 03)

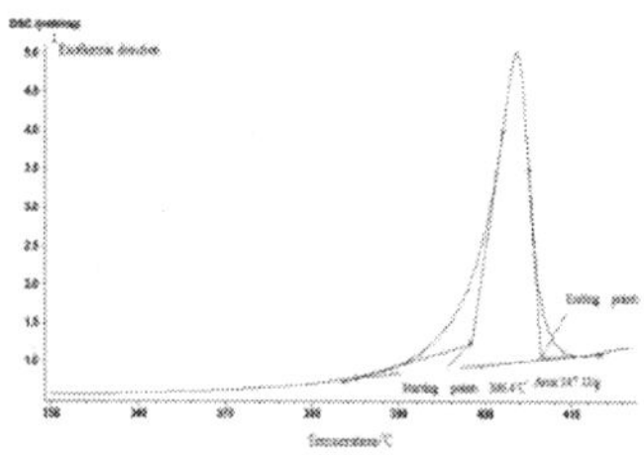

Figure 3. DSC curve of sample 13

According to the experimental results, 15 kinds of ratio with low melting point and regular melting peak regular were preliminarily preferred. DSC curve of sample number 13 is shown in Figure 3. From the figure, we can see that it has typical melting peak. It's phase change latent heat is 147.1J/g, the starting melting temperature is 398.4°C, and the ending melting temperature is 406.4°C, the melting temperature range is 8°C. The mixed molten salt with the ratio can be completely melted in a small temperature range. It is more suitable for solar power generation and is conducive to the initial operation of the power generation system.

In all 9 kinds of samples with irregular melting peaks, 6 kinds of samples have a same characteristic that one of the components accounted for a small mass ratio, and only 10%, such as sample 03, 04, 05, 06, 26 and 33. They were not eutectic in the heating process, maybe because minor component reduce the interaction with other components and can't melt together at the same temperature.

3.2. Test and analysis of initial crystallization point

When the mixed molten salt changed from a molten state to a solid state, the precipitation of each component was in order. So we can obtain the primary crystallization temperature (initial crystallization point) by analyzing the DSC curve of the cooling process. The heating curve is compared with the cooling curve.

The DSC curves of sample 07 and 30 appeared two exothermic peaks, the components of mixed molten salts were precipitated in two phases, and it can increase the risk of pipeline clogging. They are not suitable to be used as thermal storage medium. Figure 4 is the heating and cooling curve of sample 07. From the figure, we can see that its melting point is 400.9°C, the first crystallization temperature is 407.9°C, and the second crystallization temperature is 389.9°C. The melting points and the primary crystallization temperatures are shown in Table 3.

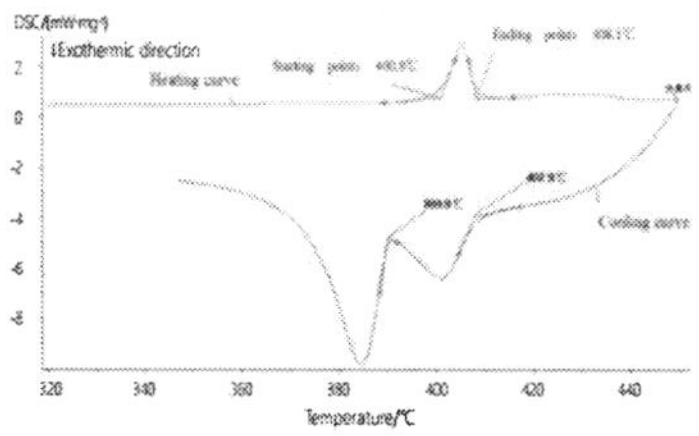

Figure 4. Heating and cooling curves of No. 07

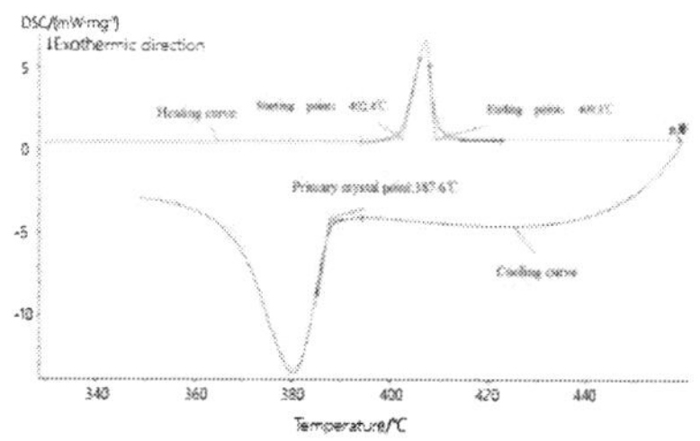

Figure 5. Heating and cooling curve of sample 14

Table 3. Melting points and primary crystallization temperatures of mixed carbonates

Number	Melting point (°C)	Primary crystallization temperature (°C)	Number	Melting point (°C)	Primary crystallization temperature (°C)
01	444.6	432.7	19	401.6	385.5
02	428.9	413.6	20	396.8	385.7
09	428.5	420.7	21	390.8	381.9
10	424.6	404.8	22	408.8	390.2
12	399.6	389.5	24	397.0	384.0
13	398.4	389.1	25	399.7	385.2
14	402.4	387.6	27	403.1	381.6
16	428.2	413.4	28	398.8	378.8
17	403.1	398.2	29	393.9	384.6
18	398.6	385.7			

From the table, we can see that the primary crystallization temperatures of samples are lower than the melting points. The differences between the melting point and the primary crystallization temperature are 5~20°C. Figure 5 is the heating and cooling curve of sample 14. From the figure, we can see that the starting melting point of the sample is 402.4°C, the ending melting point is 409.3°C. When the sample is cooled, the components precipitate at 387.6°C。

3.3. The stability of mixed carbonate

The sample of No.18 was selected to study the stability. The sample was heated to melt and cooled to solidify continuously for 5 times. And the phase change temperature and latent heat after melting and solidifying each time were tested. The results of repeated melting are shown in the Figures 7 and Table 5. The results of repeated solidifying are shown in the Figures 8 and Table 6.

IOP Conf. Series: Materials Science and Engineering **729** (2020) 012031 doi:10.1088/1757-899X/729/1/012031

Table 4. The repeated melting experimental results of mixed carbonate（sample 18）

Experimental order(times)	Starting point（°C）	Ending point（°C）	Latent heat of melting（J/g）
1	398.6	407.1	263.7
2	397.2	406.2	261.4
3	397.2	406.2	262.6
4	397.3	406.1	263.2
5	397.4	406.3	263.4
Average value	397.54	406.38	262.86

Table 5. The repeated solidifying experimental results of mixed carbonate（sample 18）

Experimental order(times)	Primary crystallization temperature（°C）	Latent heat of solidification（J/g）
1	385.7	244.4
2	384.5	243.9
3	384.5	241.5
4	384.8	243.9
5	384.4	242.6
Average value	384. 8	243.3

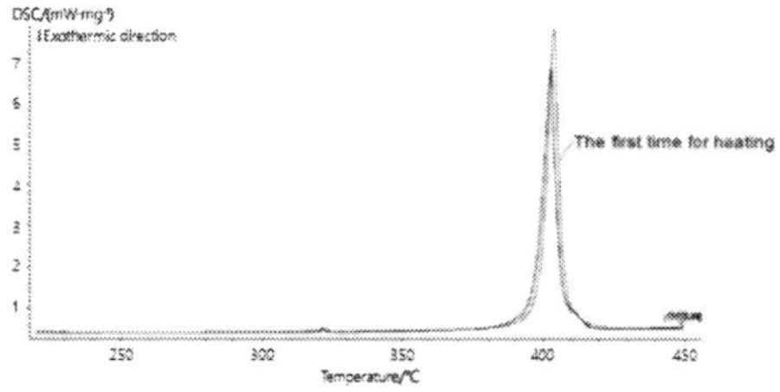

Figure 6. DSC comparison curves for repeated heating of mixed carbonate

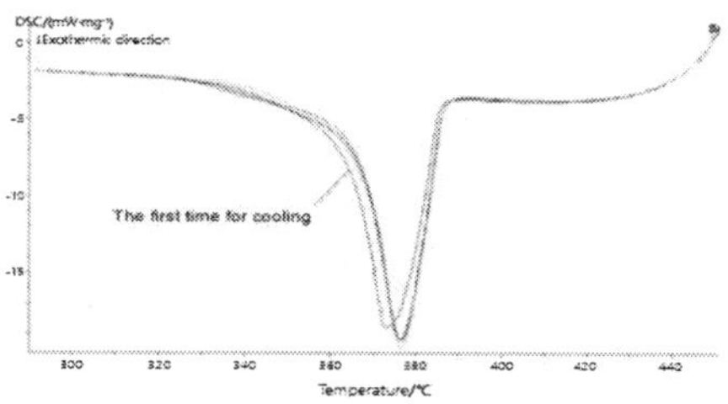

Figure 7. DSC comparison curves for repeated cooling of mixed carbonate

From the figures and tables, we can see that the stability of mixed carbonates is very good. The DSC testing curves are almost coincident. The DSC testing curve in the first time has a little deviation compared with others. The reason is that the sample contact incompletely with the bottom of the crucible in the first time. The phase change temperature and latent heat in the heating and cooling process are almost equal. The biggest differences in the phase change temperatures are about 1°C (relative difference 0.25%) and in the latent heat are 3J/g (relative difference 1.2%).

4. Theoretical calculation of mixed carbonate

It is assumed that several inorganic salts are completely immiscible in the solid state and they are completely miscible in the liquid state. Zhang Yinping deduced the theoretical formulas for calculating the phase change temperature and latent heat of eutectic system according to the second law of thermodynamics and phase equilibrium theory.

$$T_{\mathrm{m}} = \left[\frac{1}{T_i} - \frac{R\ln X_i}{H_i}\right]^{-1} \quad H_{\mathrm{m}} = T_m \sum_{i=1}^{n} \frac{X_i H_i}{T_i} \quad (\mathrm{i=A,\ B})$$

T_m -The melting point of the binary mixture (K); H_m -The melting latent heat of the binary mixture (J/mol); T_{i} -The melting point of single component (A or B) (K); H_{i} -The melting latent heat of single component (A or B) (J/mol); X_i -The mole percentage of A or B in the mixture;R-Gas constant(8.315J/mol.K).

According to the above formulas, several binary mixed chloride salts prepared in this experiment were calculated theoretically and compared with the experimental values. The results are shown in Table 8. Basic Calculation parameters of this two kinds of carbonates are shown in Table 7.

Table 6. Basic calculation parameters of three kinds of carbonates

Type	Melting point（°C）	Melting latent heat（J/g）	Molecular weight
Na_2CO_3	851	138	105.99
Li_2CO_3	723	506	73.89
K_2CO_3	891	197	138.21

Table 7. Theoretical values and experimental values of phase change temperature and latent heat of ternary mixed carbonate

Sample Number	Phase change temperature （°C）		Phase change latent heat（J/g）	
	Experimental value	Theoretical value	Experimental value	Theoretical value
13	398.4	467.76	157.4	244.11
14	402.4	397.63	173.3	243.73
18	398.6	539.20	263.7	208.91
19	401.6	474.29	257.4	217.54
20	396.8	402.68	120.6	219.59
22	408.8	612.94	93.96	163.81
24	397.0	481.20	250.2	190.56
25	399.7	408.00	134.1	195.14
28	398.8	488.54	178.7	163.14

Results in the Table 8 shows that the theoretical calculation and the experimental value have a large difference. So the above assumptions do not hold and the above formulas are not suitable for the calculation of ternary mixed carbonate.

5. Conclusion

(1) In all the ternary carbonate mixtures prepared by Na_2CO_3, Li_2CO_3 and K_2CO_3, some samples are not eutectic, and some samples' DSC curves have not regular melting peak. For other samples, the melting points of the mixtures are lower than that of the single component and between 393°C and 445°C. From starting to ending, the melting temperature range are between 7°C and 43°C.

(2) The primary crystallization temperature of the ternary carbonate mixtures are less than the melting points. The differences between the melting point and the primary crystallization temperature are 5~20°C.

(3) The phase change temperature and latent heat in the heating and cooling process are almost equal. The biggest differences in the phase change temperatures are about 1°C and in the latent heats are about 3J/g. The stability of the mixtures is good.

(4) The theoretical formula for calculating phase change temperature and latent heat of the eutectic is not suitable for the theoretical calculation of ternary mixed carbonate.

Acknowledgments

The project was supported by Science and Technology Key Program of Beijing Municipal Education Commission (KZ201710016011) and Beijing Scholar Program.

IOP Conf. Series: Materials Science and Engineering **729** (2020) 012031 doi:10.1088/1757-899X/729/1/012031

References

[1] Williams D.F., Toth L.M. and Clano K.T. (2006) Assessment of candidate molten salt coolants for the advanced high-temperature reactor [R]. ORNL/TM-2006/12, USA.

[2] Bradshaw R.W. and Siegel N.P. (2008) Molten Nitrate Salt Development for Thermal Energy Storage in Parabolic Trough Solar Power Systems. In: Proceedings of ES2008 Energy Sustainability 2008. Jacksonville, Florida USA. 2008: 55~67.

[3] Araki N., Matsuura M. and Makino A, et al. (1988) Measurement of thermophysical properties of molten salts：Mixtures of alkaline carbonate salts[J]. International Journal of Thermophysics, 9（6）：1071-1080.

[4] Kourkova L.，Sadovska G. (2007) Heat capacity, enthalpy and entropy of Li_2CO_3 at 303.15~563.15K [J]. Thermochimica Acta, 452(80-81):153-154.

[5] Wu Y.T., Ren N., and Wang T, et al. (2011) Experimental study on optimized composition of mixed carbonate salt for sensible heat storage in solar thermal power plant [J]. Solar Energy，85（9）：1957-1966.

[6] Liao M., Wei X.L. and Ding J. et al. (2010) Prepartion and experimental investigation for LNK carbonate molten salts [J]. Acta Energiae Solaris Sinica, 31(7): 863-867.

[7] Ren N. (2011) Study on the preparation and thermal properties of mixed carbonates and low melting point molten salts [D], master's thesis of Beijing University of Technology. 2011:1-18.

IOP Conf. Series: Materials Science and Engineering **729** (2020) 012032 doi:10.1088/1757-899X/729/1/012032

Preparation of a gel-type microbial-missile for biodegradation of oil pollution

Jianqiang Shi[1], Shaojun Zhang[1] and Mingyu Wang[1*]

[1]College of Naval Architecture and Marine Engineering, Shandong Jiaotong University, Weihai, Shandong, 264310, China

*Corresponding author's e-mail: 13792750903@163.com

Abstract. The removal of oil spill pollution is a global problem. In order to solve the problem of low biodegradation efficiency, a series of microbial-missile materials with high floatability, hydrophobicity and biological affinity were prepared to enhance the degradation of oil spill. Hydrophobic macroporous materials with polar groups inside and hydrophobic aliphatic polyester outside were prepared by physical and chemical methods using SA/PVA as substrate for gel-type microbial-missile. The experimental results show that the saturated adsorption rate of macroporous gel-type microbial-missile to oil particles is as high as 18.17 g/g, the contact angle is 149.300°, the average density is 0.5883 g/cm^3, the average mechanical strength is 2.52 mN, the penetration rate of 40 s is 82.72%, and the degradation rate of crude oil in 7 days is 88.95%.

1. Introduction

The pollution caused by the entry of petroleum is a worldwide serious marine pollution [1]. At present, the main methods of oil spill cleaning include physical method, chemical method and biological method [2]. Petroleum hydrocarbon-degrading bacteria has the advantage of clearing up oil film more thoroughly than physical and chemical methods, and is considered as the fundamental way to solve the pollution of petroleum hydrocarbons [3,4]. However, due to the specific gravity of petroleum hydrocarbon-degrading bacteria, most of them are in sedimentation state after being added to water, which hinders the effective contact between petroleum hydrocarbon-degrading bacteria and oil slick [5]. Therefore, a floating, hydrophobic and biocompatible immobilized carrier is likely to achieve a qualitative leap in bioremediation of marine oil spill pollution.

Sodium alginate (SA) is a by-product of extracting iodine and mannitol from kelp or Sargassum of brown algae, containing a large amount of -COO- [6,7]. SA can rapidly form gel under extremely mild conditions. When Ca^{2+}, Sr^{2+} and other cations exist, ion exchange reaction between Na^+ and two valence cations on G unit and G unit stacking form a cross-linked network structure, thus forming hydrogels[8]. The conditions for forming alginate gel are mild, which can avoid the inactivation of sensitive substances such as sensitive drugs, proteins, cells and enzymes [9]. In this study, SA was used as a substrate to explore an oil-affinity carrier that can track and disperse oil pollution, referred to as "microbial-missile".

2. Experimental materials and methods

2.1. Reagents and materials

Reagents such as dichloromethane, *n*-hexane and methyl trichlorosilane (MTS) were purchased from China Pharmaceutical Group. Agar, peptone, beef extract, glucose and yeast powder were purchased from Beijing Oberstar Company. All strains come from China Microbial Species Preservation Center. Crude oil, diesel oil and heavy oil are presented by COSCO.

2.2. Preparation of gel-type microbial-missile and immobilized microbial

SA and PVA were fed according to the experimental design in order to prepare the carrier material. Under the condition of 85°C water bath heating, the mixture was stirred by a rotor agitator, and silyl coupling agent was added to the mixture. When foam was observed in the mixture, the heating was stopped, and when the temperature dropped to 40°C, the mixture was mixed with the bacterial culture medium in equal volume and stirred for 5 min. The mixed liquid droplets were added to 2% calcium chloride saturated boric acid solution by sterile syringe, and the spheres were formed under the cross-linking effect. After 24 h of cross-linking, the carrier was washed 2~3 times with normal saline and stored in 4°C refrigerator for reserve.

Bacterial consortium enrichment solution (including *Bacillus sp.*, *Alpha sp.* and *Pseudomonas sp.*) with good degrading effect on crude oil were obtained by incubation in Zobell 2216E medium at 24°C for 3 days at constant temperature. The prepared carrier was weighed at 5.0 g and added to it. After shaking for 3 hours, the culture medium was filtered out and the gel-type microbial-missile was obtained.

2.3. Characterization test of samples

Contact angle measuring instrument (Easy-Drop, RUSS company, Germany) measures surface contact angle. Scanning electron microscopy (JCM-7000, JEOL, Japan) was used to observe the internal appearance of the carrier. Ultraviolet spectrophotometer (UV-1800, Shimadzu, Japan) Measure residual oil concentration. Adsorption performance test: including material oil absorption ability, oil retention ability test [10]. Oil absorption ratio is used to characterize the oil absorption capacity of materials. The oil absorption capacity is measured by mass difference ratio method. As seen formulas (1) and (2), the mass of the sample before adsorbing oil was recorded as m_1. The mass of the sample after absorbing oil was recorded as m_2. After 15 min, the sample was weighed again, and the mass of the sample was recorded as m_3.

$$Gs = \frac{(m_2 - m_1)}{m_1} \quad (1)$$

$$q = \frac{m_3 - m_1}{m_2 - m_1} \times 100\% \quad (2)$$

3. Results and analysis

3.1. Reaction principle of preparing immobilized microbial community carrier with PVA/SA as substrate

The reaction principle of preparing hydrogel microbial-missile carrier material with SA/PVA as substrate is shown in figure. 1. The SA molecular chain contains a large number of hydroxyl and alkyl groups. $CaCl_2$ is used as crosslinking agent. The calcium alginate polymer is obtained by crosslinking reaction between the two. The prepared gel carrier with SA/PVA as the substrate not only has the advantages of biocompatibility, modification, and complete degradation, but also has the characteristics of three-dimensional network structure and interstitial water combination. Therefore, the adsorption and degradation of spilled oil by floating macroporous gel-type microbial-missile is a promising alternative to solve the problem of marine oil pollution.

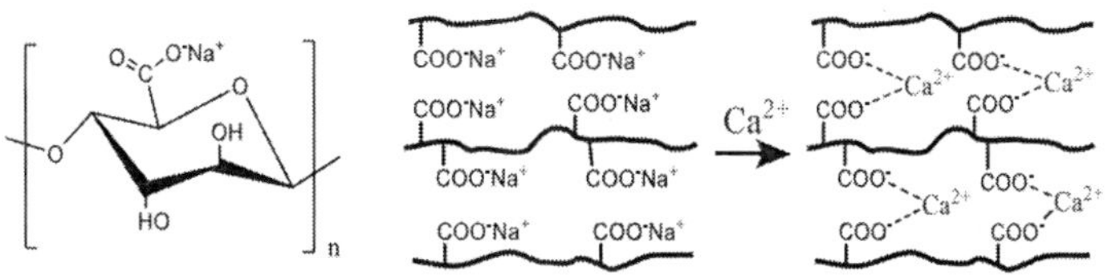

Figure 1 Principle of hydroxyl cross-linking reaction for preparation of microbial-missile using PVA/SA as substrate

3.2. Adsorption performance test

According to formula (1) and (2), the final saturated oil absorption rate Gs (g/g) and the final sustained release oil retention rate q(%) are calculated(Table 1). The adsorptive capacity of the prepared carrier for crude oil and diesel oil is quite different, which indicates that the variety of oil is related to the adsorptive capacity.

Table 1 adsorption test results of macroporous gel-type microbial-missile for different oil products

No.	Type	m_1(g)	m_2(g)	m_3(g)	q_a(g/g)	q_k (%)
1	Heavy oil	0.150 0	1.864 4	1.791 4	12.43	96.10
2	Crude oil	0.150 0	1.154 8	1.055 3	7.70	91.38
3	Diesel oil	0.150 0	0.654 1	0.527 9	4.36	80.71

3.3. Static contact angle measurement

The hydrophobicity of macroporous gel-type microbial-missile is characterized by static contact angle. The test results are shown in figure 2. The average contact angle of macroporous gel-type microbial-missile is 149.300°, which is between 90° and 150°, and is nearly 150°, showing hydrophobicity. Hydrophobic carriers contact and adsorb with weak polar oil slick under the impetus of ocean wind and current, which provides carbon source for the survival of bacteria in carriers, and is conducive to improving biodegradation efficiency.

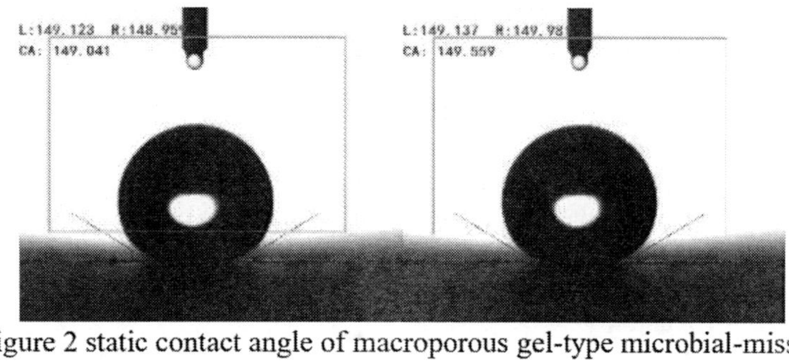

Figure 2 static contact angle of macroporous gel-type microbial-missile

3.4. Density, mechanical strength and mass transfer performance of macroporous gel-type microbial-missile

The density of carrier material is measured by drainage method. The mechanical strength of the carrier is measured by front pressing method. The mechanical strength of each carrier is indicated by pressure gauge. The mass transfer performance of the particles was tested by color infiltration method. The results are shown in Table 2, Table 3 and Table 4 respectively.

The average density of the carrier is 0.5883 g/cm^3, which is less than the density of the water under the standard pressure. The carrier has good floating performance on the water surface.

Table 2 Density determination results of macroporous gel-type microbial-missiles

No.	1	2	3	4	5	6	Average
m(g)	1.0201	1.0365	1.0157	1.0368	1.0229	1.0268	1.0265
v(cm^3)	1.81	1.69	1.77	1.73	1.87	1.62	1.7483
ρ (g/cm^3)	0.563	0.613	0.574	0.599	0.547	0.634	0.5883

The average mechanical strength of the carrier is 2.52 mN, which provides an effective protective barrier for bacteria in the marine environment.

Table 3 Mechanical strength determination results of macroporous gel-type microbial-missile

No.	1	2	3	Average
M_i (mN)	38.0	36.7	38.7	37.8
F_i (mN)	2.53	2.45	2.58	2.52

The permeability of carrier increased fastest in 5~20 s, reached osmotic equilibrium in 20 s. The permeability of carrier reached 82.72% in 40 s. It shows that the carrier has good mass transfer performance, low diffusion resistance of substrates and products, and good fluidity of bacteria and secretions in the carrier. It is conducive to immobilization of bacteria to be nutritious and dissolved oxygen, timely discharge of metabolites, thus conducive to the growth and reproduction of bacteria, and improve degradation efficiency.

Table 4 Mass transfer performance determination results of macroporous gel-type microbial-missile

Time(s)	5	10	15	20	25	30	35	40
Penetration distance (mm)	0.55	1.06	1.35	1.76	1.78	1.79	1.81	1.82
Carrier radius (mm)	2.20	2.20	2.20	2.20	2.20	2.20	2.20	2.20
mass transfer rate (%)	25.0	48.18	61.36	80.0	80.91	81.36	82.27	82.72

3.5. Biocompatibility of gel-type microbial- missile

The SEM image (figure. 3) show that the hydrocarbon-degrading bacteria are distributed between the pore of the material and in a good growth state, which shows that the material has good biological affinity. Large pore network structure with coral-like protuberance inside the carrier is not only conducive to the floatation of the carrier, but also conducive to capture dissolved oxygen at the oil-water interface. The three-dimensional mesh structure of large pore can reduce the surface tension between oil/water/gas/solid phases, reduce the rolling angle and thus increase the static contact angle of the surface. The carrier shows oil-affinity, which is conducive to the carrier contacting with oil spill under the impetus of wind, wave and flow.

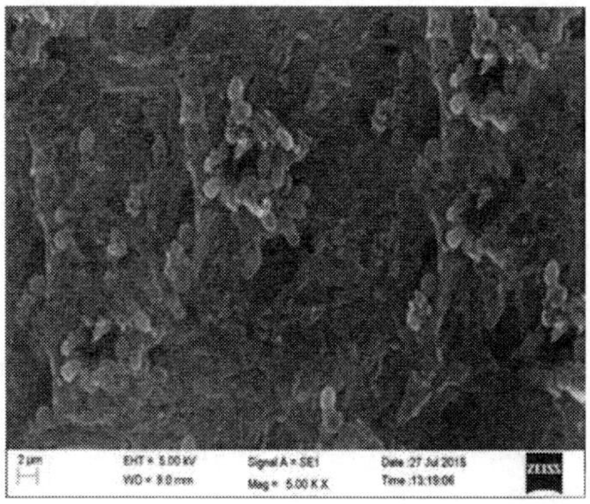

Figure 3. SEM image of macroporous gel-type immobilized bacterial consortium

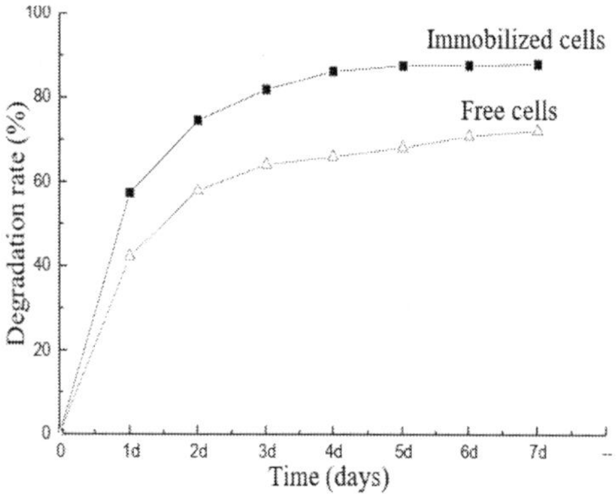

Figure 4. Crude oil degradation rate of immobilized and free bacterial consortium

3.6. Comparisons of degradation rate between gel-type microbial- missile and free-form bacterial consortium

The degradation rates of immobilized and free bacteria were compared. The results were shown in figure. 4. The degradation rate of crude oil by immobilized and free bacteria was 57.36% and 42.21% respectively in 24 h. The degradation rate of immobilized bacteria was 1.36 times that of free bacteria. The degradation rate of immobilized and free bacteria reached 88.95% and 72.22% respectively in 7 days. The results showed that the immobilized bacterial consortium showed higher degradation rate than the free consortium. Combining with uncertainties such as wind, wave, current and temperature, the preparation of immobilized natural macromolecule carriers is of great practical value.

4. Conclusion

A kind of floating gel-type microbial-missile carrier was prepared. Through the preparation, characterization and application of the material, the material with inner polar group and outer non-polar lipophilic group was obtained. It can automatically track and absorb oil spill and affinity contact degradation within a short distance, with obvious advantages. The contact angle is 149.300°, and the average density is 0.5883 g/cm^3. It has good floater performance and the average mechanical strength is 2.52 mN, which can resist the impact of wind and waves. The mass transfer equilibrium was achieved in 20 s and the permeability reached 82.72% in 40 s. Compared with the free bacterial consortium, the degradation rate of gel-type microbial-missile reached 88.95% in 7 days. It shows that the gel-type microbial-missile material improves the degradation effect of degrading bacteria.

Acknowledgments

This work was funded by National Natural Science Foundation of China (51609131); Shandong Provincial Natural Science Foundation, China (ZR2017MEE031); Shandong Province Key Research and Development Program（2019GHY112018）; Shandong Provincial Higher Educational Science and Technology Foundation, China (KJ2018BBA015); Weihai Science and Technology Development Plan (2017GNS18). Shandong Jiaotong University Doctoral Research Initiation Fund, and Shandong Jiaotong University "Climbing" Research Innovation Team Program（SDJTUC1802）.

References

[1] John, D. U., John C. M. M., Marcello, *et al.* (2019) Mycoremediation of petroleum contaminated soils: progress, prospects and perspectives. Environ. Sci. Proc. Impacts, 21(9):1446-1458.
[2] Alsadat, H. E., Hosein, A., Asghar, N. A., *et al.* (2018) Data on assessment of physical, chemical and biological characteristics of effluent from wastewater treatment in Torbat Heydarieh, Iran. Data in brief, 19: 1287-1290.
[3] Wang, C. Y., Liu, X., Guo, J., *et al.* (2018) Biodegradation of marine oil spill residues using

aboriginal bacterial consortium based on Penglai 19-3 oil spill accident, China. Ecotox. Environ. Safe., 159:20-27.

[4] Kadri, T., Rouissi, T., Brar, S. K., *et al.* (2017) Biodegradation of polycyclic aromatic hydrocarbons (PAHs) by fungal enzymes: A review. J. Environ. Sci., 29(1):52-74.

[5] Hou, L. G., Li, J., Chen, G. G., *et al.* (2019) Application of Microorganism Immobilization Technology in Wastewater Treatment. Tech. Water Tr., 45(01):1-5.

[6] Xiong, X. J., Xu, H., Zhang, B. P., *et al.* (2019) Floc structure and membrane fouling affected by sodium alginate interaction with Al species as model organic pollutants. J. Environ.Sci., 82(08):1-13.

[7] Yu, W. T., Zhang, D. M., Liu, X. D., *et al.* (2019) Amphiphilic sodium alginate-vinyl acetate microparticles for drug delivery J. Oceanol. Limn., 37(03):855-862.

[8] Ji, X. A., Ou, K. K., Dong, X., *et al.* (2019) Effect of waterborne polyurethane on structure and properties of polyvinyl alcohol-sodium alginate hydrogels. J.Mat. Sci. Eng., 37(03):436-441.

[9] Yao, W. H., Yu, F., Ma, J. (2018) Preparation of Alginate Composite Gel and Its Application in Water Treatment. Prog. Chem., 30(11):1722-1733.

[10] Zhang, S. J., Wang, M. Y., Liu, B. B., *et al.* (2017) Preparation and Application of Core-Shell Magnetic Particles. J. Shandong Jiaotong Univ., 25(04):54-59.

IOP Conf. Series: Materials Science and Engineering **729** (2020) 012033 doi:10.1088/1757-899X/729/1/012033

The research of Oil Producing Evaluation on Northern-2 Daqing Blocks and Solution Scheme

Yanyan Zhang[1]

[1] The Second Oilfield Technical Work Area of the Third Oil Production Plant of Daqing Oilfield Co.Ltd.Daqing, Daqing,HeiLongjiang, ZIP 163000, China

*Corresponding author's e-mail: 93003413@qq.com

Abstract. Among current oil in place in Daqing Oilfield, a large part of which is in oil reservoir of the pilot area on Northern-2 Blocks of Daqing oilfield, whose natural flow potential is very low. A Theoretical Analysis of producing state for the pilot area in west of Northern District of Daqing oilfield is studied to find affecting factors on producing state in reservoir and some renovations are presented, in order to provide theory evidence for effective development in reservoir. Based on evaluation of data points of producing state from 261 wells, the pilot area on Northern-2 District of Daqing oilfield is studied to establish mathematical model of injection-production ratio and water-oil ratio. The paper finds affecting factors on producing state in reservoir and ascertain major limitations on technical policy, in order to provide main measures for effective development in reservoir. The study on the effect of producing state in reservoir of the pilot area in west of Northern District of Daqing oilfield, especially carrying on the theoretical analysis to form a set of reasonable limitation on technical policy, have important directive significance on reservoir development.

1. Introduction

With the statistical analysis of the water drive injection profile on Northern-2 Blocks of Daqing oilfield, it has measured about 102 water well injection profiles totally in 2018. As classified by the effective thickness of the reservoir, from the sheet about the hygroscope producing rate condition by counted(sheet 1), we may find that the rate of the water well injection profile producing between 2017 and 2018 remained steady, and the numbers of the producing reservoirs which effective thickness not lower than 2m accounts for 82.8%, the rate of the producing sand stone thickness accounts for 80.3%, the rate of the producing effective thickness accounts for 79.2%. Form the hermetic well interpreting data,it shows that the rate of the break through from the reservoirs which effective thickness not lower than 2m accounts for 100%.From that, we can find that the reservoir which effective thickness not lower than 2m has a lower producing.[1]

Table 1. The condition of the water drive hygroscope producing rate from north area

The number of the wells in 2017	The number of the wells in 2018	The level of the effective thickness	Hygroscope in 2017%			Hygroscopein 2018%			Difference(%)		
			Layers number	Sand stone	Effective	Layers number	Sand stone	Effective	Layers number	Sand stone	Effective
131	102	>=2.0	81.8	78.9	78.9	82.8	80.3	79.2	1.0	1.4	0.2
		1.0-2.0	64.2	71.8	71.2	67.1	75.4	73.7	2.9	3.6	2.5
		0.5-1.0	55.6	57.5	55.7	56.0	58.9	56.7	0.4	1.5	1.0

<0.5	40.9	46.2	41.5	38.9	44.3	39.1	-2.0	-1.9	-2.4
other	30.4	34.4	55.3	31.5	35.6	51.3	1.1	1.2	1.5
total	42.2	54.1	65.3	42.6	55.1	65.8	0.4	1.0	0.4

2. The analysis of the factor which impact the isotope injection profile producing degree

2.1. The difference of the connection

The difference of the connection of the profit well is an important factor of impact the injection profile. In the water flooding in west of the north two, the reservoir which effective thickness higher than 2m appeared a condition that it was not hygroscope,and most of them. Because the connection was not very well with the oil well which around it. Most of them are belong to the second type connection, they have developed the interstream and out side, or they have exists the part which sand stone go bad and located in the middle of the connected oil-water wells. Such as the implantation well in north well #1, the whole well shoot sand stone thickness is 10.8m, and the effective thickness is 9.3m. In April 20,2018,the isotope injection profile showed that the level 6-8 has not hygroscope, the shoot sand stone thickness is 3.7m, and the effective thickness is 3.5m.The north well #2 has developed the lump sand in level 6-8, and with the north well #22, the north well #18 and the north well #20, they were both belong to the second type connection, and both developed the interstream, and they have exists the part which sand stone go bad. The average sand stone thickness which connected with these wells is 0.8m, and the average effective thickness which has connected is 0.5m. Because of these wells belong to the high inject and low product,they have caused the difference producing in level 6-8 in north well #9.[2-3]

2.2. The relationship between injection and production is not consummate

The relationship between injection and production is not consummate may impact the producing degree of the reservoir,although the water well developed well, there has a little oil wells to connect around it, may caused the difference of the thick reservoir producing.

For example, the injection well in north well #31,according to the isotope injection profile, the north slanting well #31 has not produced in level PII2, the shoot sand stone thickness of it is 2.8m,and the effective thickness is 2.6m.According to analysis,the injection well has connected 7 production wells around it, but it was only connected with the north slanting well #045 in level PII2, and the north slanting well #045 has developed the outer level in this level, and the shoot sand stone thickness is 1m,belong to the second type connection.Because of the little wells has connected, the level PII2 has not produced.

2.3 The reservoir pollution

When the injection well was under pit operating, the kill mud may enter the formation and make it jam, or we may take some measures such as acidify not very well to make the formation rock texture be damaged or caused the bridge over both can caused the information jam and impact the reservoir producing degree. For example, the injection well in north well #28 had a heavy repair reset operation in April, 2013. Before the operation, the relative water intake rate in level SII7+8 is 23.6%, and the level SII10+1(1) is 22.9%. But after it, they were both not hygroscope(in sheet 2), and the reservoir producing became not very well. From the well test data, we can find that before the operation, the skin coefficient of that well is -3.30, but now is 1.64, so we think after the well operation, the reservoir may be polluted and then caused the reservoir producing go bad.[4-6]

Table 2. The vary condition after the well operation of the north well #28

Stratigraphic position	Sand stone thickness (m)	Effective thickness (m)	The relative water intake rate before the operation (%)	The relative water intake rate after the operation (%)
SII7+8	6.4	5.2	23.6	0
SII10+11(1)	2.6	2.3	22.9	0

3. Analysis of reasonable injection-production ratio and water-oil ratio

Injection-production ratio for Water-Flooding Oilfield can be shown by the following exponential equation[7-8]

$$Lg\ (W_i\text{-}F) = C + DN_p$$

Formula: W_i——cumulative water injection, 104m^3;

N_p——cumulative oil production, 104m^3;

C、D、F—— fitting coefficient.

Relation of injection-production ratio:

- $$IPR = \frac{\frac{dW_i}{dt}}{\frac{dN_p}{dt}\left(\frac{\beta_o}{\gamma_o} + WOR\right)}$$

Formula: IPR—— injection-production ratio, dimensionless ;

WOR——water-oil ratio, dimensionless;

β_o——oil volume factor, dimensionless;

γ_o——the density of dead oil, dimensionless.

The calculation formula of injection-production ratio in sandstone oilfield is derived.

- $$IPR = \frac{WOR^H}{\frac{\beta_o}{\gamma_o} + WOR}$$

Formula: G、H——comprehensive coefficient

In west of 2nd Northern District of Daqing oilfield, oil volume factor was β_o=1.15 and the density of oil was γ_o=0.816.According to dynamic Datas of west of 2nd Northern District of Daqing oilfield, draw curves of $IPR \bullet (\frac{\beta_o}{\gamma_o} + WOR)$ and WOR. In logarithmic coordinate system, $IPR \bullet (\frac{\beta_o}{\gamma_o} + WOR)$ has a linear relationship with WOR

The values of G and H were drawn by linear regression as shown in Figure 1.

IOP Conf. Series: Materials Science and Engineering **729** (2020) 012033 doi:10.1088/1757-899X/729/1/012033

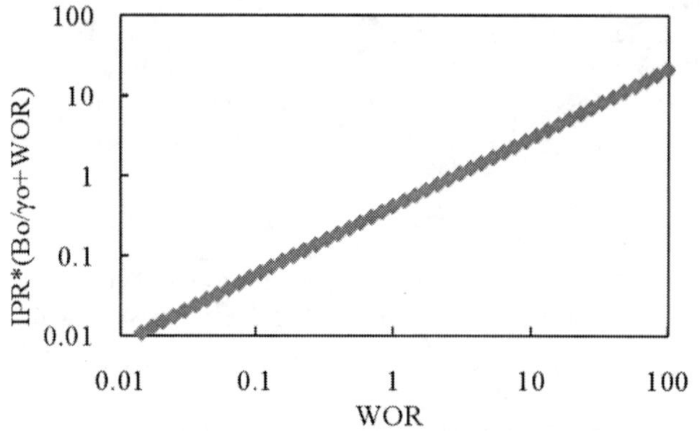

Figure 1. Comprehensive Coefficient Nomogram.

According to Fig.1 Comprehensive Coefficient Nomogram, G was 0.39 and H was 0.843.

From G and H , Relation Diagram of injection-production ratio and water-oil ratio was drawn as shown in Fig 2.

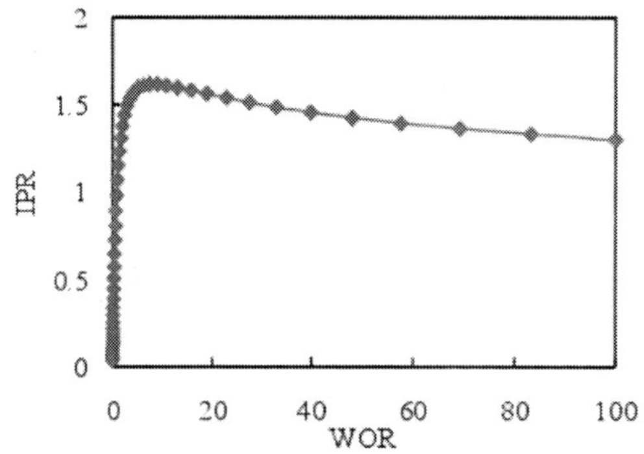

Figure 2. Relation Diagram of injection-production ratio and water-oil ratio

According to Fig.2, when the current water content was 94.15%, water-oil ratio was 14.1765 and reasonable injection-production ratio was 1.61.

4. The method to improve the injection profile

4.1. Improve the data accuracy of the injection profile

When we measure the injection profile, we should choice different isotope by its diameter in different geology condition, and make up the well clean-up system to make sure the injection well pit clean and remove the impact of the absorb pollution.

4.2. Improve the relationship between injection and production

When the reservoir connected condition not very well due to the condition that no product, we should perforations adding the correspondent position of the well, and improve the relationship between injection and production.

4.3. Resolve the plane contradict problem

If the well had the product problem made the reservoir hygroscope go bad, we should resolve it by above ground operation on time,and improve the hygroscope condition.

4.4. Avoid the reservoir pollution
In order to avoid the mud make inroads on formation,and the bottom bridge over caused by the wrong measures or operations,we should adopt the operation such as no killing the well and no beaning up when we work for well.

4.5. Improve the quality of the injection water
We must guarantee the water quality well, and do better in water quality treatment in all respects. Make sure the water quality in sewage plant exits,main line, metering plant and well head "four points"both reach the standard,and then reduce the reservoir pollution to a minimum degree.

5. Conclusion
Based on evaluation of data points of producing state from 261 wells, the pilot area on Northern-2 Blocks of Daqing oilfield is studied to establish mathematical model of injection-production ratio and water-oil ratio. In the reservoirs which higher than 2m and have producing difference,the inaccuracy factors of the injection profile data accounts for a huge rate. When the current water content was 94.15%, water-oil ratio was 14.1765 and reasonable injection-production ratio was 1.61.The paper finds affecting factors on producing state in reservoir and ascertain major limitations on technical policy, in order to provide main measures for effective development in reservoir. The study on the effect of producing state in reservoir of the pilot area on Northern-2 Blocks of Daqing oilfield, especially carrying on the theoretical analysis to form a set of reasonable limitation on technical policy.

Acknowledgments
The Third Oil Production Plant of Daqing

References
[1] Huo Jin, Shi Guoxin.(2010)The Horizontal Well Production Technique with Thin Bedded Bottom Water Reservoir in Luliang Oil Field .SPE131890.
[2] Zeito G.A.(2015)Interbedding of Shale Breaks and Reservoir Heterogeneities, JPT. 1065,10. P1223-1228
[3] Ye Jigen,Li Junshi.(2010)Water Flooding Regime Study of High Dip Reservoir with Narrow Oil zone .SPE131998.
[4] A.W.Cense,S.Berg.(2011)Direct Visualization of Designer Water Flooding in Model Experiments .SPE144936.
[5] Pradeep Cherukupalli.(2010)Analysis and Flow Modeling of Single Well Micro Pilot to Evaluate the Performance of Chemical EOR Agents .SPE136767.
[6] Chris Reddick,BP.（2007）Field of the Future Program:Planning for Success.SPE106859.
[7] Economides,M.J,Nolte,K.G.Reservoir Stimulation.(2002)3rd ed.Old Japan,NJ:Prentice Halllnc.P1-7.
[8] S.Kuzmina,Rosneft,K.K.Butula.(2009)Reservoir Pressure Depletion and Water Flooding Influencing Hydraulic Fracture Orientation in Low-Permeability Oilfields.SPE120749.

Thermal Properties of Poly(L-lactide) Containing Succinic Dihydrazide Derivative

Lisha Zhao, Yanhua Cai*

Chongqing Key Laboratory of Environmental Materials & Remediation Technologies, Chongqing University of Arts and Sciences, Yongchuan-402160, P.R. China

*Corresponding author's e-mail: mci651@163.com

Abstract: The slow crystallization speed is one of the most important factors restricting the development of Poly(L-lactide) (PLLA). In this paper, a succinic dihydrazide derivative BPASD was prepared to study its influence on the PLLA's non-isothermal crystallization behavior and thermal decomposition behavior. A comparative study on the melt-crystallization of the PLLA/BPASD and pure PLLA showed that the BPASD could promote the melt-crystallization of PLLA in cooling, and a larger amount of BPASD exhibited the better nucleation effect for PLLA. However, the cold-crystallization peak moved to the lower temperature with increasing of BPASD loading. Additionally, the cooling rate and the heating rate were two improtant factors to the crystallization process of PLLA. Thermal stability measurements showed that the pure PLLA and all PLLA/BPASD samples have only one thermal decomposition profile, but the presence of BPASD reduced the thermal stability of PLLA.

1. Introduction

Among biodegradable polyester, Poly(L-lactide) (PLLA) has played a leading role in research and development or production field. Meantime, its noteworthy advantages such as the excellent compatibility and biodegradability, easy processability lead to a wide application in packaging materials [1], 3D printing [2], etc. Passador et al [3] prepared the PLA composites containing carbon black to be suitable for use as the antistatic packaging for electronic components. The introduction of the carbon black could increase the degradable temperature of PLA, meantime, the addition of the carbon black leaded to a decrease of the resistance of PLA/carbon composites.

However, PLLA itself suffers from some inherent defects including slow crystallization speed and low heat distortion temperature [4], which seriously restricts its development and application. Aiming at the defect of slow crystallization rate, the blending with a crystallization accelerator is often an important method to promote the crystallization performance of the pure PLLA. Talc [5], zinc phenylphosphonate [6], orotic acid [7], *N, N'*-bis(1H-benzotriazole) adipic acid acethydrazide [8], and WS_2 [9] could serve as efficient nucleating agents for promoting the PLLA's crystallization of. For instance, with addition of only 0.8 wt% *N, N'*-bis(benzoyl) suberic acid dihydrazide and upon the crystallization of 115 °C, the crystallization half-time decreased from 26.5 min to 1.4 min comparing with the pure PLLA [10].

In this work, an organic compound, *N, N'*-succinic bis(4-phenylbutyric acid) dihydrazide (designated here as BPASD), was synthesized based on the succinic dihydrazide and 4-phenylbutyric acid to be used as a functional additive in PLLA matrix, and the melt-crystallization and cold-crystallization behavior and thermal decomposition behavior of PLLA/BPASD were studied. This work might be helpful to overcome some flaws of PLLA.

2. Experimental section

2.1 PLLA and BPASD

Nature Works LLC supplied the 2002D PLLA. The BPASD (The detailed compound structure is shown in Fig.1) was synthesized in our lab, and the characterization data of the molecular structure are as follows: Infrared spectroscopy (KBr) υ: 3446.5, 3207.8, 3051.0, 2934.9, 2854.9, 1686.4, 1592.4, 1487.7, 1450.9, 1416.3, 1433.2, 1201.4, 1151.9, 1028.1, 911.9, 696.1, 669.3 cm^{-1}; 1H nuclear magnetic resonance (DMSO, 400 MHz) δ: ppm; 9.74 (s, 1H, NH), 9.68 (s, 1H, NH), 7.16~7.30 (m, 5H, Ar), 2.56~2.60 (t, 2H, CH2), 2.37 (s, 2H, CH2), 2.11~2.14 (t, 2H, CH2), 1.78~1.82 (t, 2H, CH2).

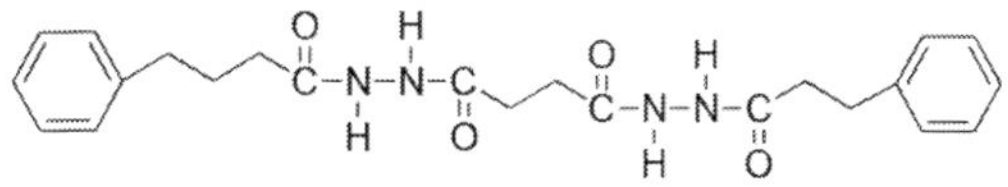

Figure 1. The structure of BPASD

2.2 Preparation of PLLA/BPASD

The melting blend of PLLA and BPASD was carried out using a counter-rotating mixer, and the set of the blending parameter and detailed operational process were the same as our previous work [11].

2.3 Characterization and testing

^{1}H nuclear magnetic resonance spectrometer and infrared spectrometer were used to obtain the structure date of BPASD. The melt-crystallization and cold-crystallization behaviors of the pure PLLA and PLLA/BPASD were tested through DSC under nitrogen of 50 mL/min. The thermal decomposition process of the PLLA/BPASD and pure PLLA was observed using TGA from 40 °C to 650 °C under flowing air of 60 mL/min.

3. Results and discussion

3.1 Non-isothermal crystallization behavior

The melt-crystallization is very important to practical manufacturing, thus, this study firstly focuses on the melt-crystallization process. Figure.2 displays the melt-isothermal crystallization behavior of the pure PLLA and PLLA/BPASD samples in cooling of 1 °C/min. For the pure PLLA, there is almost not the melt-crystallization in cooling DSC curve, indicating that the pure PLLA has very poor crystallization ability. However, all PLLA/BPASD samples exhibit the obvious melt-crystallization peaks in DSC curves, meaning that the

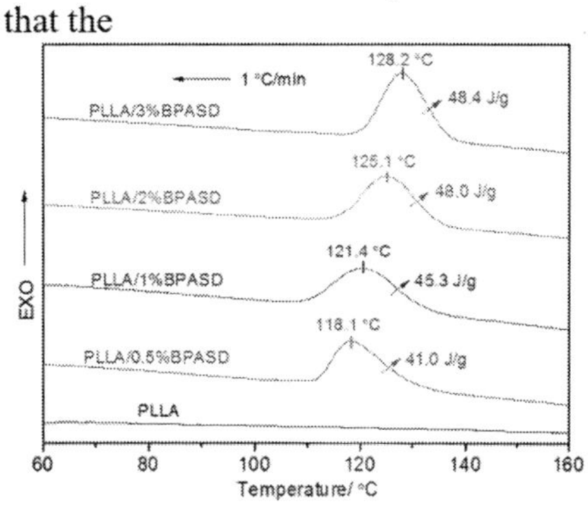

Figure 2. Melt-crystallization behavior of the pure PLLA and PLLA/BPASD in cooling of 1 °C/min

BPASD promotes the crystallization of PLLA during process. Furthermore, with increasing of BPASD content, the melt-crystallization peak shifts toward the higher temperature, implying that a larger amount of BPASD possesses the better promoting role for the PLLA's crystallization, because a larger amount of BPASD often has the higher nucleation density. 3 wt% BPASD can make the melt-crystallization peak temperature and the melt-crystallization enthalpy increase to the maximum values of 128.2 °C and 48.4 J/g comparing with the neat PLLA, respectively.

However, the negative role of an increase of the cooling rate on the melt-crystallization process is obvious (See Fig.3), that is, the melt-crystallization peak becomes wider and shifts to the lower temperature with increasing of cooling rate to 3 °C/min, which indicates that a higher cooling rate weakens the PLLA/BPASD's crystallization. This effect may depend on the competitive relationship between the cooling rate and BPASD. An increase of cooling rate causes the PLLA molecular chain to have no sufficient time to form a regular structure, although the presence of BPASD can promote the crystallization of PLLA.

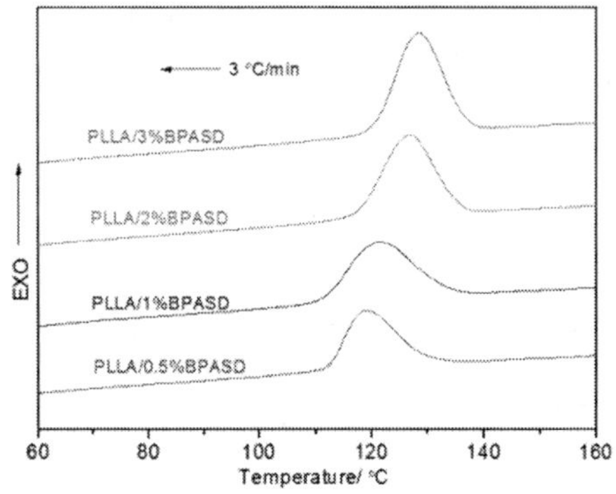

Figure 3. Melt-crystallization process of PLLA/BPASD in cooling of 3 °C/min

For the cold-crystallization process, PLLA itself has fast nuclear rate *via* homogeneous nucleation of PLLA itself in the low temperature region, possibly exhibiting the different crystallization behaviors. Figure.4 is the cold-crystallization of PLLA/BPASD sample in heating of 1 °C/min. As shown in Fig.4, except PLLA/1%BPASD sample, an increase of BPASD content makes the cold-crystallization peak move to the lower temperature, and the cold-crystallization enthalpy of other PLLA/BPASD samples increases, which differs from the result of the melt-crystallization. Additionally, it is found from Fig.4 that the PLLA containing the larger amount of BPASD content has the higher melting temperature, showing a better perfection of crystals formed in heating.

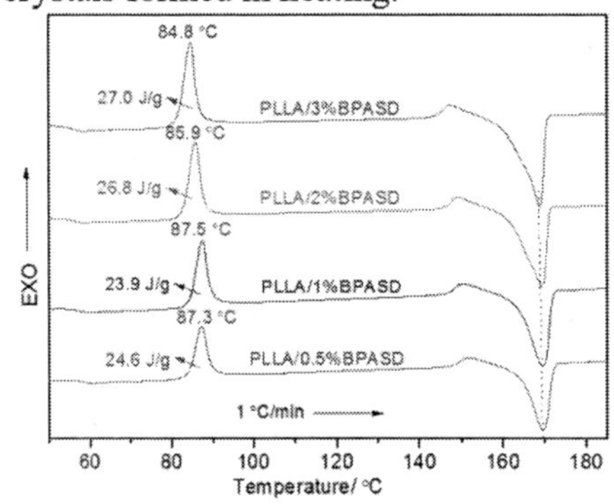

Figure.4 Cold-crystallization process of PLLA/BPASD in heating of 1 °C/min

Similarly, the heating rate also significantly affects the cold-crystallization of BPASD-nucleated PLLA. As shown in Fig.5, for a given PLLA/BPASD sample, the shift to the higher temperature occurs with increasing of heating rate owing to the thermal inertia. But the effect of BPASD content on the cold-crystallization process of PLLA at a given heating rate is consistent with the aforementioned result of Fig.4.

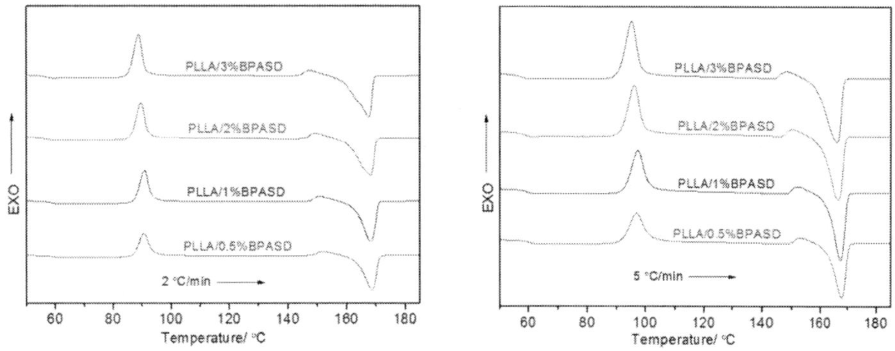

Figure.5 Cold-crystallization of PLLA/BPASD at a different heating rate

3.2 Thermal decomposition behavior

The heating decomposition behaviors of the pure PLLA and all PLLA/BPASD were further studied using TGA (See Fig.6). All PLLA/BPASD samples like the neat PLLA have only one thermal decomposition stage, and the thermal weightlessness occurs in the temperature range from 275 °C to 375 °C, showing that the BPASD cannot change the PLLA's thermal decomposition profile. However, the BPASD reduces the PLLA's onset thermal decomposition temperature (T_d), and the corresponding onset thermal decomposition temperature of the PLLA, PLLA/0.5%BPASD, PLLA/1%BPASD, PLLA/2%BPASD, PLLA/3%BPASD are 341.3 °C, 333.2 °C, 327.0 °C, 334.7 °C and 334.3 °C, respectively. Although the onset thermal decomposition temperature of any PLLA/BPASD slightly decreases, the onset thermal decomposition temperatures of all PLLA/BPASD samples are above 300 °C, which still satisfies the daily use requirement.

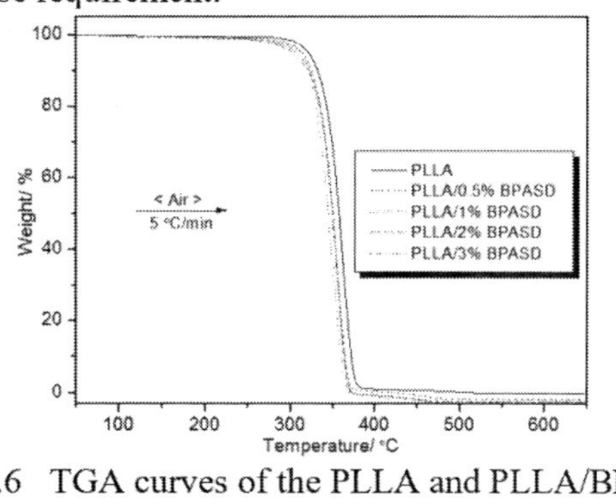

Figure.6 TGA curves of the PLLA and PLLA/BPASD

4. Conclusions

In this work, BPASD as an organic nucleating agent could obviously accelerate the PLLA's crystallization in cooling, and the melt-crystallization peak, with increasing of BPASD content, shifted to the higher temperature, as well as the 3 wt% BPASD had the best crystallization promoting role. For the cold-crystallization process, an increase of BPASD content caused the cold-crystallization peak to move to the lower temperature, exhibiting the opposite result with the melt-crystallization. The increasing of cooling rate could weaken the crystallization ability of PLLA/BPASD, but a higher heating rate caused the cold-crystallization peak to shift to the higher temperature. The BPASD could not change the PLLA's thermal decomposition profile, however, and reduced the PLLA's thermal stability.

Acknowledgements

This work was supported by Foundation of Chongqing Municipal Science and Technology Commission (project number cstc2017shmsA20021), Scientific and Technological Research Program of Chongqing Municipal Education Commission (project number KJQN201801319), and Foundation of Yongchuan District (project number Ycstc, 2018cc0801).

References

[1] Kamthai, S., Magaraphan, R. (2019) Development of an active polylactic aicd (PLA) packaging film by adding bleached bagasse carboxymethyl cellulose (CMCB) for mango storage life extension. Packag. Technol. Sci., 32(2): 103-116.

[2] Teixeira, B.N., Aprile, P., Mendonca, R.H., Kelly, D.J., Thire, R.M.D.M. (2019) Evaluation of bone marrow stem cell response to PLA scaffolds manufactured by 3D printing and coated with polydopamine and type I collagen. J. Biomed. Mater. B., 107(1): 37-49.

[3] da Silva, T.F., Menezes. F., Montagna, L.S., Lemes, A.P., Passador, F.R. (2019) Preparation and characterization of antistatic packaging for electronic components based on poly (lactic acid)/carbon black composites. J. Appl. Polym. Sci., 136(13): 47273.

[4] Fan, Y.Q., Yu, Z.Y., Cai, Y.H., Yan, S.F., Chen, X.S., Yin, J.B. (2013) Crystallization behavior and crystallite morphology controlling of poly (L-lactic acid) by adding N, N'-bis(benzoyl) sebacic acid dihydrazide. Polym. Int., 62(4): 647-657.

[5] Cipriano, T.F., da Silva, A.L.N., da Silva, A.H.M.D.F.T., de Sousa, A.M.F., da Silva, G.M., Rocha, M.G. (2014) Thermal, rheological and morphological properties of poly(lactic acid)(PLA) and talc composities. Polimeros, 24(3): 276-282.

[6] Pan, P.P., Liang, Z.C., Cao, A., Inoue, Y. (2009) Layered metal phosphonate reinforced Poly(L-lactide) composites with a highly enhanced crystallization rate. ACS Appl. Mater. Inter., 1(2): 402-411.

[7] Qiu, Z., Li, Z. (2011) Effect of orotic acid on the crystallization kinetics and morphology of biodegradable poly(L-lactide) as an efficient nucleating agent. Ind. Eng. Chem. Res., 50: 12299-12303.

[8] Cai, Y.H., Zhao, L.S., Zhang, Y.H. (2015) Role of *N, N'*-bis(1H-benzotriazole) adipic acid acethydrazide in crystallization nucleating effect and melting behavior of poly (L-lactic acid). J. Polym. Res., 22: 246.

[9] Naffakh, M, Marco, C, Ellis, G. (2015) Non-isothermal cold-crystallization behavior and kinetics of poly (L-lactic acid)/WS2 inorganic nanotube nanocomposites. Polymers, 7(11): 2175-2189.

[10] Cai, Y.H., Yan, S.F., Yin, J.B., Fan, Y.Q., Chen, X.S. (2011) Crystallization behavior of biodegradable poly (L-lactic acid) filled with a powerful nucleating agent-*N, N'*-bis(benzoyl) suberic acid dihydrazide. J. Appl. Polym. Sci., 121(3): 1408-1416.

[11] Zhao, L.S., Cai, Y.H. (2018) Investigating the physical properties of poly (L-lactic acid) modified using a aromatics succinic dihydrazide derivative. Polym. Sci. Ser. A., 60(6): 777-787.

IOP Conf. Series: Materials Science and Engineering **729** (2020) 012035 doi:10.1088/1757-899X/729/1/012035

Evaluation and Optimization of Corrosion Inhibitor System

Liu Yun[1*], Du Qingzhen[1], Xiong Wei[1], Gao Yanfang[1] , Shang Fang[1], Zhang Wei[1] and Li Gangliang[2]

[1]Engineering Technology Research Institute, Petrochina Huabei Oilfield Company, Renqiu, Heibei, ZIP 062552, China

[2] Engineering Technology Research Institute, Petrochina Huabei Oilfield Company, Renqiu, Heibei, ZIP 062552, China

[*]Corresponding author's e-mail: yjy_liuyun@petrochina.com.cn

Abstract: By combining two methods of weightlessness and electrochemistry, the corrosion inhibitors have been selected and optimized. The orthogonal experiment has been carried out to optimize the inhibitor's performance. The range analysis results indicated that corrosion inhibition performance of optimized corrosion inhibitor (HZ-1) has been improved remarkably. The corrosion inhibition rate can reach 94.66% with lower corrosion current density. The optimum concentration of HZ-1 is 90 mg/L and working temperature is 80～100°C. Its site application shows that the inhibitor also has remarkable corrosion inhibition effect, the concentration of iron ions is decreases, and the period of pumping wells has been prolonged significantly.

1. Introduction

With the oilfield entering the high water cut development stage, oil well corrosion is becoming seriously. According to statistics, there are more than 30% oil wells having corrosion phenomena in Huabei Oilfield. Among them, the pump inspection cycles of 825 wells are less than one year, accounting for more than 50% of the number of production wells. Because of the synergistic effect of chloride ion, hydrogen sulfide, free carbon dioxide, bacteria and other corrosion factors in the fluid of oil wells, it is very difficult to control the corrosion of oil wells[1]. There are variety of corrosion inhibitors in the market which make the traditional screening experiments have a lot of problem such as large workload, long cycle and lacking pertinence. It is really an urgent problem to effectively improve the study level for the typical problems existing in oil well corrosion. The research of new type corrosion inhibitor has been developed according to the liquid characteristics of oil wells in Huabei Oilfield.

2. Experiment methods

2.1. Experiment materials and instruments

Experiment materials: anhydrous ethanol; petroleum ether; sodium chloride; magnesium chloride; sodium sulphate; calcium chloride; sodium bicarbonate; steels N80(72mm×12mm×2mm);

Experiment instruments:

CT-2 potentiostat (Huazhong university of science and technology);

Electric blast drying oven (Shanghai experimental instrument company);

Multi-parameters water quality tester (Shanghai experimental instrument company).

2.2. Experiment design

2.2.1. Simulated fluid

According to the standard of produced fluid, the fluid has been prepared to simulate characters and environment in oil wells. The salinity of simulated fluid is between 5000～70000 mg/L.

Table1. Main components of corrosion solution

Component	NaCl	$MgCl_2 \cdot 6H_2O$	Na_2SO_4	$CaCl_2$	$NaHCO_3$
ρ，g/L	50	2	6	4	4

2.2.2. Electrochemical evaluation experiment

The electrochemical specimens have been prepared according to the standard GB/13671-92. The N80 steel specimens have been sealed by high temperature resistant resin with a surface area of 1 cm^2 at the exposed end. The reference electrode, auxiliary electrode and working electrode of CT-2 potentiostat respectively are saturated calomel electrode, platinum electrode and N80 working electrode.

Add chemicals into the simulated fluid, and the ion concentration of the solution is 80mg/L. Then potential scanning has been carried out at 70°C using CT-2 potentiostat. The scanning is ranged from -250 mV to 250 mV relative to the natural corrosion potential, and the scanning rate is 0.5mV/s. The polarization curves of the solution are obtained.

2.2.3. Weightlessness evaluation experiment

According to the standard of SY/T5273-2000, the corrosion rate and inhibiting rate under different corrosion conditions have been evaluated. The corrosion experiments have been carried out in wide-mouth bottles of 500ml. The experiments are during seven days. After the corrosion experiments the loss weights of steel sheets have been measured, and the corrosion rate and corrosion inhibition rate have been calculated according to the following formula.

$$r_{corr} = \frac{8.76 \times 10^4 \times (m_0 - m_1)}{st\rho} \times 100\% \tag{1}$$

r_{corr}—corrosion rate，mm/a；

m_0—test specimens quality before experiment, g；

m_i—test specimens quality after experiment, g；

s—corrosion surface area of specimens,cm^2；

t—corrosion time, h；

ρ—density of specimen material, g/cm^3；

$$\eta_i = \frac{\Delta m_0 - \Delta m_i}{\Delta m_0} \times 100\% \tag{2}$$

η_i—corrosion inhibiting rate,%；

Δm_0—loss quality in blank sample，g；

Δm_i—loss quality in adding chemical sample, g；

2.2.4. Ion analysis

The concentration of total iron and ferrous ions has been measured by the field water quality rapid tester.

3. Experiment results and discussions

3.1. Primary selection of different inhibitors

Imidazoline corrosion inhibitors, as a new and broad-spectrum corrosion inhibitor, have good corrosion inhibition effect on carbon steel in fluid containing CO_2[2]. They also have the characteristics of good thermal stability and low toxicity[3]. They are widely used in oil field at home and abroad.

Firstly, four different types of imidazoline corrosion inhibitors have been screened according to the properties of the produced fluid in Huabei oil field. As the evaluation result shows in Figure 1, HS-2 cationic corrosion inhibitor has the best performance in four inhibitors. The corrosion potential has the maximal negative displacement in the polarization curves. But HS-2 also needs more optimization to attend satisfied performance in field application.

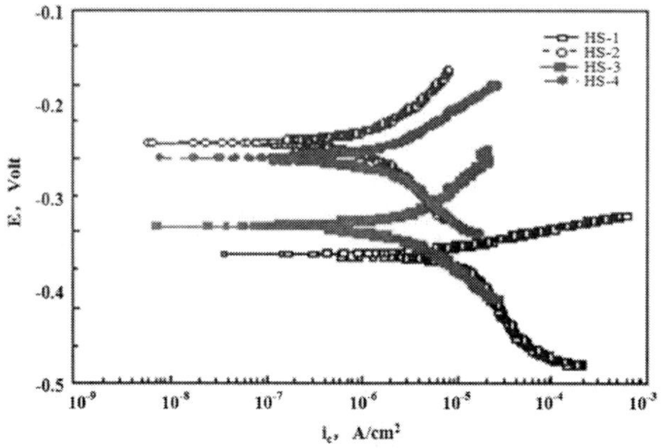

Figure 1. The electrochemical evaluation results of inhibitors

3.2. Formulation system optimization

According to the field requirements, the selected HS-2 needs to be optimized. Therefore, FJ-1, FJ-2 and FJ-3 additives have been added together which have been selected according to the corrosion characteristics of oil wells and polarization curves. The experiment design has been carried out according to $L_9(3^4)$ orthogonal table as shown in Table 2. The inhibition efficiency has been evaluated by static weightlessness method, the concentrations of the corrosion inhibitors are 80mg/L, and the temperature is 70 °C.

The range analysis has been carried out on the weightlessness experiment results. The analysis indicates that the order of chemicals impacting the inhibitor's performance is HS-2, FJ-1, FJ-3, FJ-2 from high to low. Nine groups of experiments have been carried out. The optimal level is group 1 named HZ-1, and the ratio of HS-2, FJ-1, FJ-2 and FJ-3 is 20:20:2:2. The performance of the initial corrosion inhibitor has been significantly improved, and the corrosion inhibiting rate reaches 94.66% with no pitting corrosion.

Table 2. The result of orthogonal experiment

	HS-2	FJ-1	FJ-2	FJ-3	η_i, %
1	20	20	2	2	94.66
2	20	10	1	1	90.25
3	20	5	0.5	0.5	89.03
4	10	20	1	0.5	87.43
5	10	10	0.5	2	85.50
6	10	5	2	1	85.00
7	5	20	0.5	1	84.21
8	5	10	2	0.5	82.11
9	5	5	1	2	83.23
K_1	273.94	266.3	262.89	263.39	
K_2	257.93	257.86	260.91	259.46	
K_3	248.55	257.26	258.74	258.57	
k_1	91.31	88.77	87.63	87.80	
k_2	85.98	85.95	86.97	86.49	
k_3	82.85	85.75	86.25	86.19	
R	8.46	3.02	1.37	1.61	

3.3. Performance evaluation of optimization formulation system

3.3.1. Evaluation of electrochemical performance

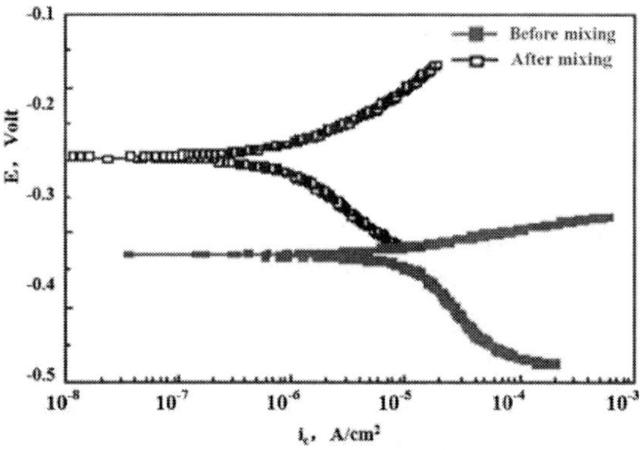

Figure 2.The polarization curves before and after mixing

By comparing two polarization curves before and after optimization, the current density decreases obviously after optimization, which indicates that additives have good synergistic effect. It forms a continuous and compact protective film on the metal surface. The FJ-1 shows weak alkaline and therefore it plays the role of dissolving free carbon dioxide in water and adjusting pH value[4]. At the same time, it also contains a lone pair of electrons, which is profit the molecular structure of imidazole rings to fill into the empty d orbit of iron, thus forming a stable multicentre adsorption structure[5]. This structure is favourable to form an adsorption film on the metal surface by directional alignment of corrosion inhibitor molecules, which are closely wrapped around the metal surface like a network[6]. The additive FJ-2 effectively controls the growth of bacteria and algae in production fluid, so the corrosion inhibition efficiency is grown better after using. The FJ-3 forms stable complex compounds with various metal ions, and dissolves some oxides on the metal surface, which help the main agent to form protective film on the metal surface.

3.3.2. The influence of concentration

Through the weightlessness method find the optimum concentration of optimization inhibitor in different blocks. The experiment time is 7 days, the temperature is 70 °C, and the concentration respectively are 60 mg/L, 70 mg/L, 80 mg/L, 90 mg/L and 100 mg/L.

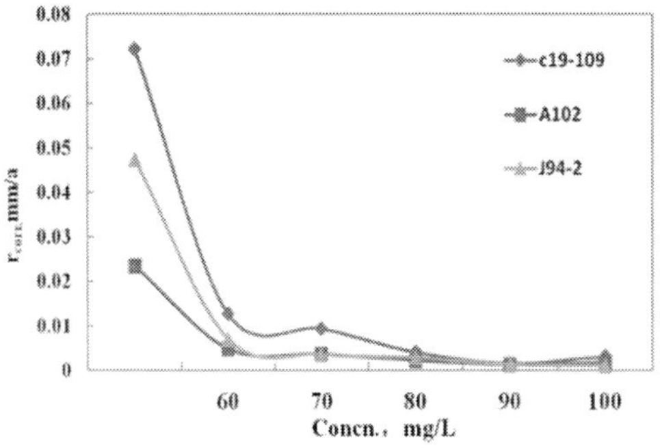

Figure 3. The influence of different concentrations at corrosion rate

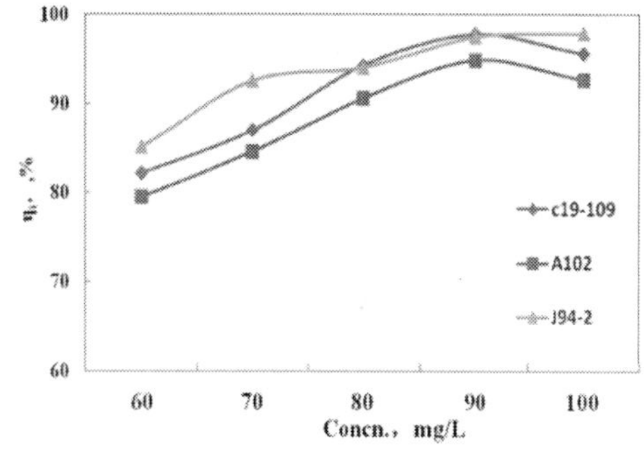

Figure 4. The influence of different concentrations at corrosion inhibition rate

Because imidazoline corrosion inhibitors belong to adsorption type, the adsorption capacity on the metal surface increases with the concentration of inhibitor. The inhibitor molecules are aligned on the metal surface to form a more complete protective film. It shows the inhibition rate reaches 97.8% when the inhibitor concentration is 90 mg/L, and a complete protective film formed on the metal surface with strong anti-corrosion ability.

3.3.3. The influence of temperature

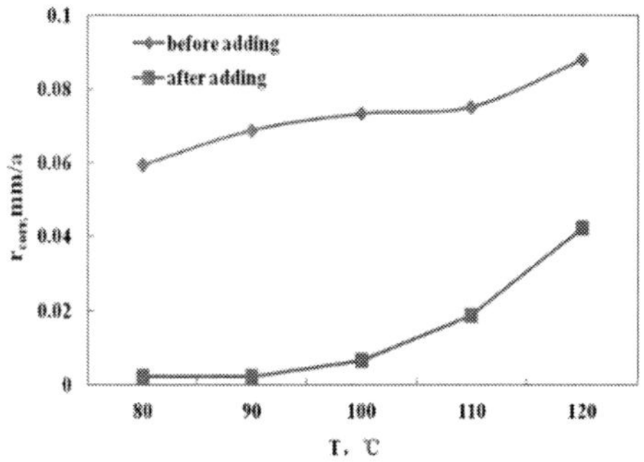

Figure 5. The influence of different temperatures at corrosion rate

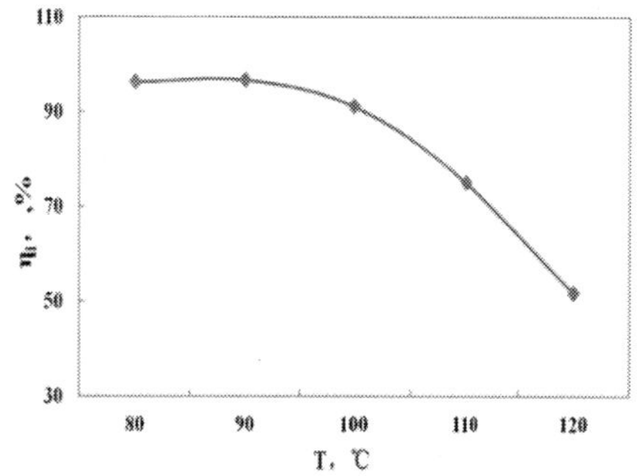

Figure 6. The influence of different temperatures at corrosion inhibition rate

Because Huabei Oilfield belongs to a complexity fault reservoirs with different reservoir temperatures in each block, it is necessary to study the influence of temperature on corrosion inhibition. The static corrosion evaluation experiment has been carried out at five different temperatures from 80°C to 120°C using producing liquid from C19-109 oil well, the concentration of HZ-1 is 90 mg/L, and the experiment lasted 7 days.

The corrosion rate decreases slowly when the temperature is range from 80°C to 100°C. But when the temperature is above 100°C the corrosion inhibition rate decreases rapidly. The corrosion inhibition rate decreases to only 51.9% when the temperature is 120°C. The surfaces of steel sheets are darkness and lots of the corrosion productions have been accumulated at the bottom of bottle. The result shows that the HZ-1 performance is relatively stable when the temperature at range of 80°C to 100°C. But it is easy to break the molecular bond at high temperature which leads to lack integrity of the adsorption films on the metal surface. Therefore, the formulation HZ-1 is suitable for oil wells below 100°C.

4. Field Application

Well c19-22 has been selected to evaluate the application effect. The production fluid is 31m^3/d, and the water cut is 97%. It has been found a serious corrosion in April 2010. The 144th rod has been severely corroded and broken. The average pump inspecting period was only 193 days.

Through the analysis of the production liquid from well c19-22, the free CO_2 concentration is 26.7 mg/L, the bacterial content of SRB is 12/ml, the salinity is 18641 mg/L, and the chloride ion concentration is 10599 mg/L. According to the corrosion characteristics of rods and pipes in site, the optimized HZ-1 formula has been selected to use. Through indoor evaluation the corrosion inhibition rate of HZ-1 has reached 86.7%. Thus it has been selected for corrosion treatment in well c19-22, and the adding chemical period is 7 days. In order to investigate the long-term corrosion treatment effect, a half-year monitoring is carried out as shown in Figure7.

The field monitoring indicates that the ion indexes have a significant downward trend after using 3 days. The total iron concentration has decreased from 18.80 mg/L to 0.70 mg/L and remained at about 1 mg/L for a long term. The ferrous ion has also decreased from 18.50 mg/L to 0.63 mg/L. The period of pump inspection has been prolonged from 193 days to 966 days after application. The corrosion situation is grown better and the corrosion trend has been restrained significantly.

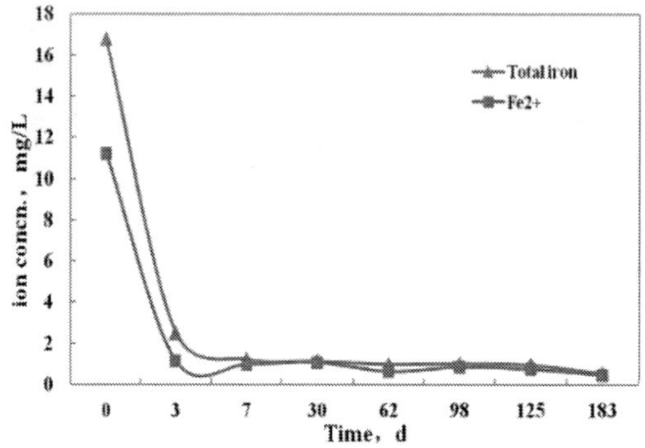

Figure 7. The long term ion monitoring results

5. Conclusion

According to the characteristics of oil wells produced fluid in Huabei oil field, different types of corrosion inhibitors have been selected, optimized and evaluated by electrochemical method and weightlessness method. The range analysis shows that the formulation HZ-1 has good corrosion inhibition performance and its inhibition rate is above 90%. The well c19-22 application indicates that optimized HZ-1 has satisfied corrosion inhibition effect. After application the concentration of iron ions decrease significantly, the period of pump inspection has been prolonged significantly, and the corrosion situation on site has been inhibited.

Reference

[1] Ward C D, Millians D E. (1991) Prediction of CO_2 corrosion of carbon steel. Corrosion, 47(12): 976.Another reference.

[2] Sloan J J, Basta N T. (1995) Remediation of acid soils by using alkaline biosolids. Environ Qual, 24: 1097-1103.

[3] Jovancicevic V. (1999)Inhibition of carbon dioxide corrosion of mild steel by imidazolines and their precursors. NACE Corrosion.55(5): 449.

[4] Shen T D, Schwarz R B (2001)Bulk ferromagnetic glasses in the Fe-Ni-P-B system. Acta Materialia, 49 (5): 837-847.

[5] Chaiebe E., Bouyanzer A., Hammouti B..(2009)Limonene as green inhibitor for corrosion in hydrochloric scidsolutions, Acta Physic-Chimica Sinica, 25(7): 1254-1258.

[6] Shangguan Changhuai, Wang Xia, Du Lei,Yin Mingxue, Li Ke.(2011),Synthesis and performance evaluation of an imidazoline inhibitor ,Corrosion and Protection, 32(7): 528-537.

IOP Conf. Series: Materials Science and Engineering **729** (2020) 012036 doi:10.1088/1757-899X/729/1/012036

The research of relative permeability curve for microbial flooding in Baolige oilfield

Shasha Liu[1*], Yaxiu Fu[2], Yi Gu[1], Rui Wang[1], Guan Wang[1], Lisha Duan[1], Suzhen Guo[1], Lili Wei[1]

[1] Engineering Technology Research Institute, Huabei Oilfield Company, Renqiu, Hebei, 062552, China

[2] Erlian Branch, Huabei Oilfield Company, Xilinhot, Inner Mongolia, 026000, China

*Corresponding author's e-mail: cyy_liuss@petrochina.com.cn

Abstract: To adequately simulate environment of Baolige oilfield reservoir, research the seepage variation of microbial in the reservoir, we selected the current microbial flooding system, nature core, dehydration crude oil and injection water of Baolige oilfield. And the evaluation experiments of microbial flooding system and the relative permeability experiments of water flooding and microbial flooding were carried out. Results showed that the flooding system had obvious effect to reduce the emulsification, viscosity and IFT. Comparing to the water flooding, the two-phase seepage area expanded, oil phase permeability endpoint shifted to the right, and residual oil saturation decreased when using microbial flooding. The water phase permeability decreased and the oil phase permeability increased. The relative permeability in the isotonic point changed little, but shifted to the right. Microbial flooding anhydrous recovery period was longer, and the anhydrous recovery rate was higher. In the condition of the same injection, microbial flooding had higher degree of reserve recovery and the displacement pressure was lower. The experiment results fully reflected the two-phase flow in the microbial flooding process, and provided important parameters and basic basis for the prediction and effect evaluation of microbial flooding.

1. Introduction

MEOR technology is a technology that achieves enhanced oil recovery by directly or indirectly acting on reservoir and reservoir fluid by microorganisms and their metabolites. Compared with other EOR technologies, MEOR technology has the advantages of low cost, simple process, wide application and environment friendly. At present, this technology is being paid more and more attention by oilfield production units. Multiple oilfields in our country have conducted related field experiment [1-3]. Microbial laboratory experiments are mostly based on microbial laboratory evaluation. Emulsification and viscosity reduction tests are carried out to evaluate the effects of microbial and crude oil [4-6]. In the aspect of physical model experiment, the simulated core is generally used to carry out microbial flooding test [7-10]. Some scholars observed the migration, adsorption and action mechanism of microorganisms in the model with visual microscopic model [11, 12]. Some results have been obtained in the preliminary experiment on the mechanism and effect evaluation of microbial oil displacement. However, due to the limitation of experimental conditions, the effects of microorganisms in reservoir environment cannot be fully explained. Therefore, nature core, dehydration crude oil and injection water of Baolige oilfield were applied in this paper. First, the performance of microbial flooding system was evaluated in laboratory. Secondly, the experiment of microbial flooding relative permeability was carried out under reservoir temperature by means of unsteady state method.

2. Experimental materials

2.1. Microbial flooding system
The microbial flooding species were H, z-2 and HB3. The activated medium (LB) included peptone, yeast extract and sodium chloride. The nutrition formula included glucose, urea, ammonium chloride, yeast paste and potassium dihydrogen phosphate.

2.2. Oil and water samples
The viscosity of dehydrated crude oil in Baolige oilfield was 36.606mpa.s at the experimental temperature of 58°C. The viscosity of injected water was 0.495mpa.s at the experimental temperature of 58°C. And the salinity of injected water was 6137mg/l.

2.3. Experimental instruments and cores
The main equipments included 2PB2020 advection pump, Ts-2112b shaker, Ty-3 core gripper, HAAKE MARS viscometer, TX550 interface tensiometer and DGT2006 thermostat. The cores sampling depth of Baolige oilfield was 1504 to 1559 meters. And the core was in the fourth member of the alshan formation of cretaceous.

2.4. Experimental conditions
To simulate the reservoir environment of Baolige oilfield, the experiment adopted constant temperature displacement at 58°C and constant temperature culture at 58°C.

3. Experimental methods

3.1. Preparation of microbial flooding system
H, Z-2 and HB3 strains were first activated by LB medium. The activated bacterial solution was prepared as microbial flooding bacteria solution in a ratio of 1:1:1. Then, the produced reservoir fluid was used to prepare micro-flooding nutrient solution according to the nutrient formula of 1.1. And the activated bacterial solution was added into the nutrient solution at the concentration of 1.0% to prepare the microbial flooding system.

3.2. Microbial flooding system evaluation experiment
The dehydrate crude oil from Baolige oilfield was added into the microbial flooding system at a ratio of 50g/l. It was placed in a constant temperature shaking table and cultured at 58°C and 200r/min for 5 days. At the same time, blank samples were made for comparison. The emulsification, viscosity reduction and interfacial tension reduction of crude oil were analyzed.

3.3. Relative permeability experiment

3.3.1. Core preparation
First, the core needed to be washed, dried and polished. Then we measured the length and width of the core and weighed it. Gas permeability experiment was carried out. The differential pressure and gas flow at both ends of the core were recorded. And the permeability of the core was calculated. After 3 to 4 hours of core pumping, the vacuum pump was turned off. And then the core was saturated with formation water. The core wet weight was then weighted and the porosity was calculated. The core parameters were shown in table 1.

Table 1. Core base data

Core number	Length (cm)	Diameter (cm)	Porosity (%)	Gas permeability (μm^2)
1# Water flooding	5.04	2.52	17.5	0.814
2# Microbial flooding	5.04	2.52	17.8	0.757

3.3.2. Relative permeability test procedure
The relative permeability curve was measured according to the operating procedures of industry standard sy-5354-1999. And the unsteady state method was adopted in this experiment. Experimental procedures were as follows.

- The core was dried and vacuumed to saturate the simulate formation water. The simulated water was produced by formation. The water type was sodium bicarbonate type with a salinity of 6199.2mg/l. The water phase permeability (kw) was measured. The simulated oil was prepared from white oil and dehydrated crude oil produced by Baolige oilfield. The viscosity of simulated oil was 10.5mPa.S at 58°C. Firstly, the core was saturated with simulated oil. The irreducible water was established to be 22.4%. Then the oil phase permeability (ko) was measured.
- The simulate oil was flooded by crude oil at least 2 times of the volume. The core was aged for 2 to 3 days. Then the crude oil permeability (kco) was measured.
- Water flooding experiment was carried out to determine the oil and water phase permeability curve.
- The core was removed, cleaned and send to be washed again. After washing the oil, the core was retested for permeability.
- The core was dried and vacuumed to saturate the simulate formation water. The water phase permeability (kw) was measured. Firstly, the core was saturated with simulated oil. The irreducible water was established to be 21.9%. Then the oil phase permeability (ko) was measured.
- The simulate oil was flooded by crude oil at least 2 times of the volume. The core was aged for 2 days. Then the crude oil permeability (kco) was measured.
- The core saturated with oil was injected into nutrient solution with concentration of 0.81% and bacterial strain with concentration of 1%. After water breakthrough, the injection was stopped and the core was cultured in a constant temperature box at 58°C for 5 days. Then, water flooding was carried out to measure the oil and water phase permeability curve.

4. Results and discussion

4.1. Indoor evaluation results of microbial flooding system

4.1.1. Emulsification effect
The experimental results showed that the flooding system has good emulsifying and dispersing ability. The oil and water could be completely miscible after emulsification. And the emulsification grade was above grade 4. There was no obvious diving line between oil and water, and no stratification after 2 hours standing.

4.1.2. Viscosity reduction effect
The viscosity was determined by MARS viscometer. The results showed that compared with the blank sample, the viscosity reduction effect of the flooding system was good. The viscosity of crude oil decreased from 217mPa.s to 117.83mPa.s. And the viscosity reduction rate was up to 45.7%.

4.1.3. Ability to reduce interfacial tension
The interfacial tension was determined by the interfacial tensionmeter. The interfacial tension of blank sample and microbe was measured. After the action of microbial flooding system, the interfacial tension decreased from 4.32mN/m to 0.92mN/m. And the interfacial tension reduction rate reached

79.2%.

4.2. Effects of different flooding modes on water content

In the water flooding process, 20.5PV was injected. The flooding time of anhydrous oil recovery period was 70s, and the recovery degree was 8.8%. After water breakthrough, the recovery degree increased by 8.8. The flooding time was 990s when water content reached 100%. The cumulative flooding time was 5013s, and the final recovery degree was 17.6%. In the microbial flooding process, 20.3PV was injected. The flooding time of anhydrous oil recovery period was 346s, and the recovery degree was 11.4%. After water breakthrough, the recovery degree increased by 9.6. The flooding time was 1157s when water content reached 100%. The cumulative flooding time was 2807s, and the final recovery degree was 21%. Microbial flooding process showed obvious longer without water production period, and anhydrous recovery was 2.6% higher than that of water flooding process. And the ultimate recovery increased by 3.4%.

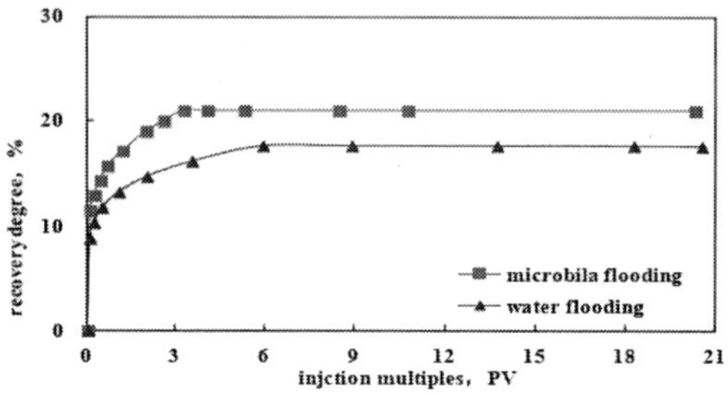

Figure 1. Relationship curve between production degree and injection PV number

4.3. Influence of different flooding modes on recovery degree

The comparison of water content curves between water flooding and microbial flooding showed that both flooding modes had relatively large anhydrous recovery rate as shown in figure 3. After water phase breakthrough, water content of the two flooding modes rose rapidly with similar changing rules, and they all presented a typical s-shape. After the analysis of experimental data, the microbial flooding curve reflected the strong concavity. At the beginning of the flooding, the water cut rise rate of microbial flooding rose slower than water flooding after water breakthrough. In the late stage of microbial flooding, the water cut rose faster than the water flooding. And the microbial flooding was more uniform. The ultimate recovery was 2.4% higher than that of water flooding. It showed that the oil- water interfacial tension was lower in the process of microbial flooding and the oil-water viscosity ratio could be effectively reduced by reducing the crude oil viscosity.

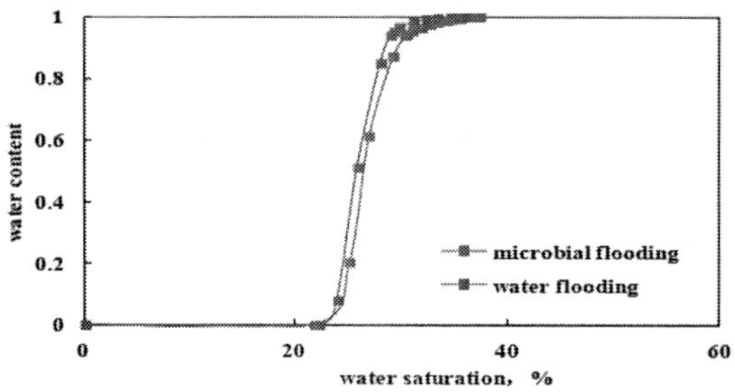

Figure 2. Relationship curve between water content and water saturation

4.4. Effects of different flooding modes on relative permeability curve

The comparison of relative permeability curve between water flooding and microbial flooding was shown in figure 4. As can be seen from the figure, microbial flooding has the following characteristics.

IOP Conf. Series: Materials Science and Engineering **729** (2020) 012036 doi:10.1088/1757-899X/729/1/012036

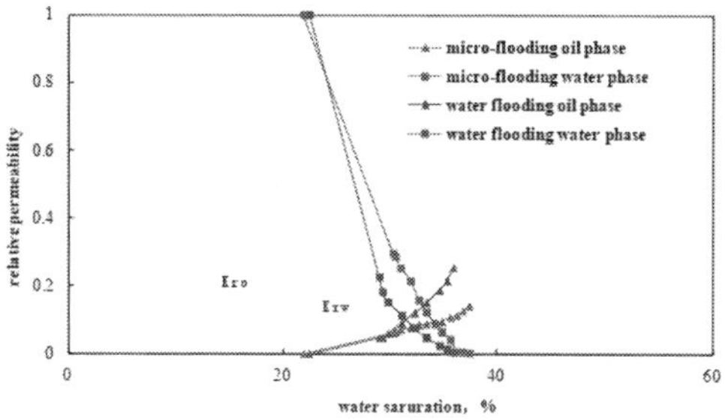

Figure 3. Relative permeability curve

The water phase curve of microbial flooding decreased, and the oil phase curve increased. In the incubator culture process, the microorganisms interacted with the crude oil. And the crude oil viscosity could be effectively reduced by self-degradation and interaction of metabolites such as surfactants. Therefore, at the early stage of flooding, low viscosity oil was easier to be driven out. And the percolation ability of the oil phase increased significantly. With the increase of water saturation, the oil phase gradually changed from continuous phase to non-continuous flow. And the flow resistance of the water phase increased, leading to the decrease of the percolation ability of the oil phase. In the process of flooding, leading to the decrease of the seepage capacity of the water phase. According to the analysis, the first was that the oil-water viscosity ratio decreased due to the decrease of crude oil viscosity. And the oil phase seepage capacity increased, which affected the water phase seepage. Second, with the analysis of the isotonic point, the water-wet degree of the core increased with the progress of microbial flooding, which ultimately leaded to the continuous decrease of the water phase permeability.

The isotonic point of microbial flooding was shifted to the right. The relative permeability value at the isotonic point of microbial flooding and water flooding was basically consistent, which indicated that the two-phase permeability was similar under the two flooding modes. And microbial flooding had little influence on capillary pressure. At the same time, the water saturation at the isotonic point of microbial flooding and water flooding was not significantly different, which was less than 40%. It indicated that the strong oil-wet state of the core has not changed. Although microbial flooding has a tendency to change the core into water wet, but it did not change the wettability of the core.

The two-phase co-permeation zone increased. And the end-point value of oil phase moved to the right. It indicated that the flooding capability of microbial flooding was enhanced. And the residual oil saturation was reduced. Then the final recovery of microbial flooding was improved. On the whole, it was considered that by improving the fluidity ratio of flooding and no flooding phase, microbial flooding could reduce the occurrence of viscous fingering and make the leading edge advance evenly. Thus, the water flood swept volume was increased.

4.5 Flooding pressure analysis of different flooding modes

The flooding pressure under two flooding models was analyzed. When the pressure was stable, the flooding pressure of water flooding fluctuated in 0.72 to 0.78MPa, and the flooding pressure of microbial flooding changed in the range of 0.44 to 0.55MPa. The flooding pressure of microbial flooding was obviously smaller than that of water flooding. It was included that the interaction of microorganisms and their metabolites with crude oil reduced the viscosity and interfacial tension of crude oil. Thus, the flow resistance was reduced and the flooding pressure of microbial flooding was reduced.

5. Conclusion

- The interaction of microorganisms and their metabolites with crude oil could effectively reduce the viscosity of crude oil and the interfacial tension between oil and water.

• Compared with water flooding, the relative permeability curve of microbial flooding has the following characteristics. The water phase curve decreased, and the oil phase curve increased. The two-phase co-infiltration zone was enlarged and the end - point value of oil phase was shifted to the right. The isotonic point shifted to the right, but the relative permeability value did not change much.

• The period of anhydrous recovery was longer and the anhydrous recovery was higher of microbial flooding. With the same PV number injected by water flooding, the recovery degree of microbial flooding was higher.

• The flooding pressure of microbial flooding was significantly lower than that of water flooding. The flow resistance of microbial flooding was reduced.

References:

[1] Sun Sangdun. (2014) Field practice and analysis of MEOR in Shengli oilfield. Journal of Oil and Gas Technology, 36(2): 149-152.

[2] Wu Xiaolin, Le Jianjun, Wang Rui. (2013) Progress in pilot tests of microbial enhanced oil recovery in Daqing oilfield. Microbial China, 40(8):1478-1486.

[3] Huang Shiwei, Zhang Yanshan, Huo Jin. (2006) Field experiments on microbial enhanced oil recovery of heavy oil in Xinjiang oil field. Xinjiang Geology, 24(1):84-87.

[4] Li Qing, Wu Gang, Xie Gang. (2011) Research on evaluation of microbe oil recovery bacterium of low-temperature inspissated pool. Oil Drilling & Production Technology, 33(2): 114-116.

[5] Pornsunthorntawee O, Arttaweeporn N, Paisanjit S. (2009) Laboratory study on activating indigenous microorganisms to enhance oil recovery in Shengli Oilfield. J Petrol Sci Eng, 66(1):42-46.

[6] Song Z Y, Guo L Y, Qu Y Y. (2008) Microbial surfactant-enhanced mineral oil recovery under laboratory conditions. Colloid Surface B: Biointerf, 63(1): 73-82.

[7] Liu Baolei, Dong Hanping, Yu Li. (2011) Displacement characteristics analysis of bacterial microscopic profile control in reservoir. Journal of Shenzhen University Science and Engineering, 28(4): 356-359.

[8] Yu Li, Wu Guoqing. (2000) The mechanisms study of profile modifying and pluging in porous media for microbe in porous media. Journal of Chongqing University (Natural Science Edition), 23: 142-145.

[9] Yang Peng, Huang Lixin, Yu Li. (2006) Study on the capacity of microbial migration in the low permeability oil reservoir. Petroleum Geology and Recovery Efficiency, 13(2), 85-87.

[10] Han Jianhua, Li Zhanxing, Ju Dengfeng. (2003) The research on physical analog testing for improving microbial oil displacement efficiency. Drilling & Production Technology, 26(6): 86-88.

[11] Ma Jiye, Guo Xingxue, Lei Guanglun. (2008) Microscopic mechanism studies on microbial enhanced oil recovery under high temperature high pressure reservoir conditions. Oilfield Chemistry, 25(4): 369-373.

[12] Saikrishna Maudgalya, Roy M. Knapp, Michael J. Mclnemey. (2007) Microbial Enhanced-Oil-Recovery Technologies: A Review of the Past, Present, and Future. In: 2007 SPE Production and Operations Symposium. Oklahoma City, Oklahoma, U.S.A. SPE 106978.

IOP Conf. Series: Materials Science and Engineering **729** (2020) 012037 doi:10.1088/1757-899X/729/1/012037

Preparation and Properties of Silane Molecular Sieve Supported Composite Polyether Ionic Liquids Catalysts

Zhongyi Cui, Liying Guo*

School of Petrochemical Engineering, Shenyang University of Technology, Liaoyang 111003, China

*Corresponding author's e-mail: lyguo1981@163.com

Abstract: In this work, three composite polyether imidazole ionic liquids (PIILs) supported on molecular sieve (MCM-22) with terminal hydroxyl, carboxyl and amino groups were synthesized with the help of silane coupling agent 3-chloropropyltriethoxysilane (CPTES). They were used for cycloaddition of CO_2 and propylene oxide (PO) in a fixed-bed reactor. The results showed that MCM-22-CPTES-HOOC-[PECH-MIM]Cl/[$ZnBr_2$] was the best catalytic properties in this reaction. The conversion rate and selectivity were 92.7% and 82.7%.

1. Introduction

As a greenhouse gas, carbon dioxide is an abundant C1 resource. If carbon dioxide can be converted into chemical products, waste can be turned into treasures and new ideas for greenhouse gas treatment can be provided[1-2]. However, carbon dioxide is chemically stable, so the choice of catalyst is especially important for the conversion of carbon dioxide[3-5]. Ionic liquids are used in the cycloaddition reaction of carbon dioxide with alkylene oxide to synthesize propylene carbonate(PC) due to their excellent catalytic properties[6-7]. However, traditional ionic liquid catalysts are difficult to separate and recover products because of their high viscous liquid state, and are limited to batch reactions in autoclave, which limits the industrial application of ionic liquids (IL)[8]. The above problems can be effectively solved by immobilizing ionic liquids onto the carrier to realize the immobilization of IL.

In this work, we reported three kinds of composite polyether imidazole ionic liquids (PIILs): MCM-22-CPTES-HO-[PECH-MIM]Cl/[$ZnBr_2$], MCM-22-CPTES-HOOC-[PECH-MIM]Cl/[$ZnBr_2$] and MCM-22-CPTES-H_2N-[PECH-MIM]Cl/[$ZnBr_2$] immobilized on molecular sieve (MCM-22) supporter covalently with the help of silane coupling agent 3-chloropropyltriethoxysilane (CPTES), which exhibited good chemical and thermal stability, and were applied in the preparation of PC by cycloaddition of CO_2 and PO as a novel heterogeneous catalyst. In addition, flexible automation, reliability process and low mechanical consumption were easily achieved in industrial continuous operation.

2. Experimental

2.1. Reagents and Instruments

Three PIILs, i.e., HO-[PECH-MIM]Cl, HOOC-[PECH-MIM]Cl and H_2N-[PECH-MIM]Cl, were prepared according to our published procedures. Zinc bromide (98.0%), toluene (99.5%), acetonitrile (99.0%), propylene oxide (99.5%). Above all reagents and materials (analytic grade) were purchased from Sinopharm Chemical Reagent Co., Ltd. (China)3-chloropropyltriethoxysilane (CPTES, 95%) was purchased from Aladdin Chemical Co. CO_2 gas (99.95%) was offered by Petrochina Liaoyang Yifang Petrochemical Company (China). The MCM-22 was provided from Petrochina Fushun Petrochemical Company (China). All materials were used without further purification.

Instruments used in this study include MAGNA-IR750 Fourier Transform Infrared Spectrometer (FT-IR); TM3000 Scanning Electron Microscope (SEM); D/max-2400 Automatic X-ray Diffractometer (XRD);PARR4523 Catalytic Device (PARR, USA); 1790F Gas Chromatograph (GC) (Agilent Technologies, Inc.); D08-8C Carbon Dioxide Flowmeter; DF-101S Magnetic Stirrer; SFX-2L Rotary Evaporator; DZF-6050 Vacuum Drying box; 2-XZ-4 rotary vane vacuum pump. The catalytic experimental fixed bed reactor (Tianjin Pengxiang Technology Co., Ltd, China) was applied in the process of catalysis.

2.2. Preparation of MCM-22-CPTES-PIILs/($ZnBr_2$)$_2$

2.2.1. Synthesis of MCM-22-CPTES. CPTES (3.5 mL) was added a flask where existed a mixture of MCM-22 (3.5 g) and toluene (50 mL) solvent slowly to prepare MCM-22-CPTES. The mixture was refluxed at 110°C for 24 h, after that the flask was cooled to room temperature. The solvent was removed by distillation, and the solid product was washed three times with ethyl acetate and then dried at 80 °C for 12 h under vacuum to afford the MCM-22-CPTES (4.2g).

2.2.2. Synthesis of PIILs/($ZnBr_2$)$_2$. Three PIILs, i.e., HO-[PECH-MIM]Cl, HOOC-[PECH-MIM]Cl and H_2N-[PECH-MIM]Cl, were prepared according to our published procedures. Epichlorohydrin (ECH) was polymerized to form poly-epichlorohydrin (PECH) which had –OH groups, used chloroacetic acid and ammonia to reacted with PECH to achieve –COOH and –NH_2 groups respectively. Then three types of PECH with –OH, -COOH and –NH_2 reacted with imidazole to obtain three PIILs. Then, a given amount of $ZnBr_2$ was added into three PIILs, respectively, and resulting mixtures were stirred at reflux for 24 h. As the reaction proceeded, the mixtures became more viscous with a darker color change. When the reactions were finished, three reddish brown liquids were obtained, namely three composite PIILs, i.e., HO-[PECH-MIM]Cl/[$ZnBr_2$], HOOC-[PECH-MIM]Cl/[$ZnBr_2$], and H_2N-[PECH-MIM]Cl/[$ZnBr_2$].

2.2.3. Synthesis of MCM-22-CPTES-PIILs/($ZnBr_2$)$_2$. Then PIILs/($ZnBr_2$)$_2$ and MCM-22-CPTES (4.2g) were poured into the flask with acetonitrile solvent (50 ml), the intermixture was refluxed at 80°Cfor another 24 h, the acetonitrile solvent was removed by rotary evaporation to obtain the immobilized catalyst ,i.e., MCM-22-CPTES-PIILs/($ZnBr_2$)$_2$. Among the synthesis procedure of MCM-22-CPTES-HO-[PECH- MIM]Cl/[$ZnBr_2$] is illustrated in Scheme 1.

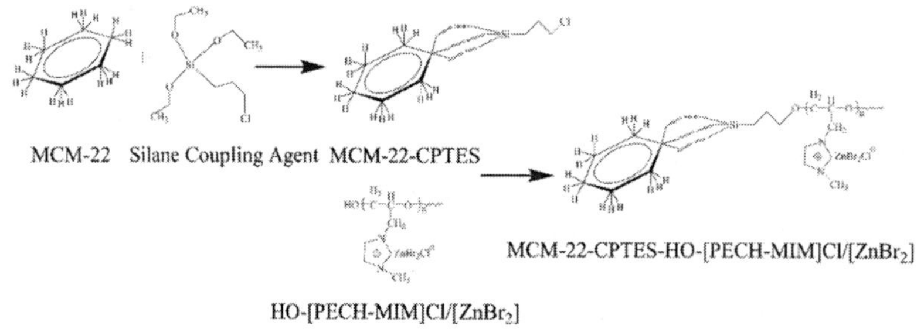

Scheme 1. Synthesis of the immobilized catalyst MCM-22-CPTES-HO-[PECH-MIM]Cl/[$ZnBr_2$]

2.2.4. Typical procedure for the synthesis of PC from PO and CO_2. A small amount of quartz sand were filled in the front and rear ends of the catalyst container, and 7.0 g of the immobilized catalyst was placed in the middle of the container of the fixed bed catalytic reactor. CO_2 was output from a high-pressure cylinder while PO was output through a pump into fixed bed reactor. PO was heated in

IOP Conf. Series: Materials Science and Engineering **729** (2020) 012037 doi:10.1088/1757-899X/729/1/012037

the preheater into the packed column to react with CO_2 and there was no additional solvent would be added during the reaction. The product from the reactor was collected every 5 h after cooling, and the purity of the product propylene carbonate (PC) was tested, and the yield was calculated.

3. Results and discussion

3.1. Characterization of MCM-22-CPTES-PIILs/($ZnBr_2$)$_2$

3.1.1. FT-IR. FT-IR studies on MCM-22-CPTES and its immobilized catalysts were carried out (Figure 1).The siloxy groups of the molecular sieves in the spectrum were reflected in the four curves of 1210 cm^{-1} and 797 cm^{-1}. Compared with the infrared spectrum (a) of the silane molecular sieve as it is, the spectrum (b, c, d) of the catalyst is greatly changed. For immobilized PIILs/($ZnBr_2$)$_2$ catalysts, there displayed 3145 cm^{-1} and 2938 cm^{-1} characteristic peaks assigned to stretching vibrations of CH_3 and CH_2. The hydroxyl peak broadening in 3360 cm^{-1} as water molecules adsorbed on catalysts surface. These functional groups of PIILs/($ZnBr_2$)$_2$ were removed during the preparation of immobilized catalysts, so the terminal peaks were not existed. All vibration peaks indicated the successful immobilization of PIILs/($ZnBr_2$)$_2$.

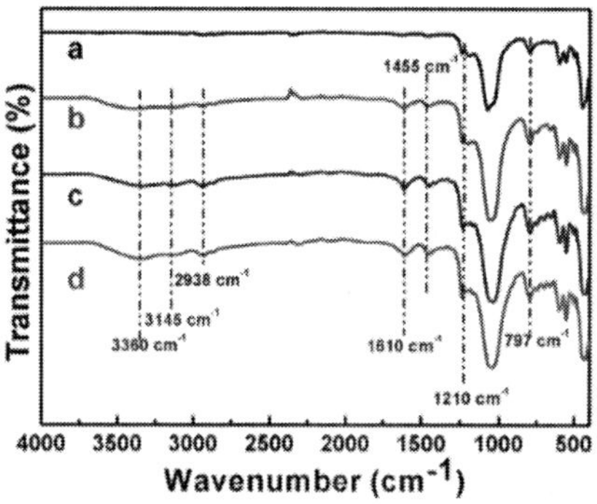

Figure 1. FT-IR spectra of the silane molecular sieve MCM-22-CPTES and immobilized catalysts (a) MCM-22-CPTES; (b) MCM-22-CPTES-HO-[PECH-MIM]Cl/[$ZnBr_2$]; (c) MCM-22-CPTES-HOOC-[PECH-MIM]Cl/[$ZnBr_2$]; (d) MCM-22-CPTES-H_2N-[PECH-MIM]Cl/[$ZnBr_2$]

3.1.2. XRD. The XRD patterns of MCM-22-CPTES and its immobilized catalysts were shown in Figure 2.It can be seen from the XRD spectrum that the diffraction peaks of the three catalysts were largely different from the crystallinity of the silane molecular sieve. In pattern(a), the high intensive peaks at 7.14°, 7.85°, 9.59°, 12.8°, 19.8°, 22.0°, 24.9°and 25.9° are the characteristic diffraction peaks of MCM-22, indicating a good crystallinity of synthesized MCM-22-CPTES.

After the three composite PIILs were immobilized, the XRD spectrum changed greatly. The strong and sharp diffraction peak of the original molecular sieve became a very weak peak, and some peaks were not observed. This indicated that during the PIILs immobilization process, the hydrogen on the terminal group was condensed with the chlorine at the end of the silane molecular sieve, and the released strong acid HCl gas destroys the crystallinity of the molecular sieve. The XRD analysis resulted further demonstrate the successful preparation of the silane molecular sieve supported PIILs catalyst.

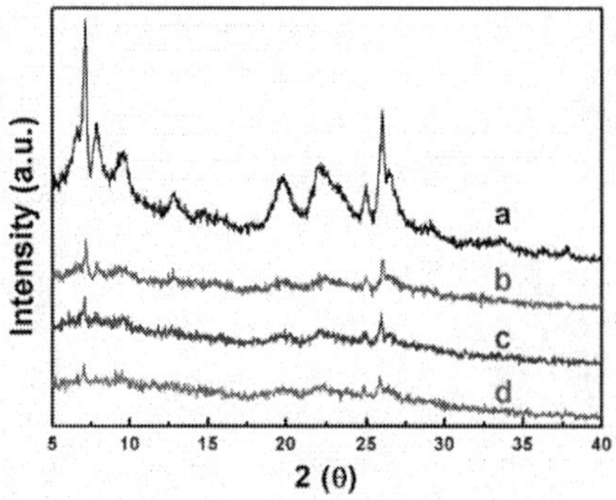

Figure 2. XRD of the silane molecular sieve MCM-22-CPTES and immobilized catalysts (a) MCM-22-CPTES; (b) MCM-22-CPTES-HO-[PECH-MIM]Cl/[$ZnBr_2$]; (c) MCM-22-CPTES-HOOC-[PECH-MIM]Cl/[$ZnBr_2$]; (d) MCM-22-CPTES-H_2N-[PECH-MIM]Cl/[$ZnBr_2$]

3.1.3. SEM. SEM images of MCM-22-CPTES and its immobilized catalysts were provided in Figure 3.

Compared with the original MCM-22-CPTES(a) of silane molecular sieve, the surface of the immobilized catalyst was covered with a thin but less pronounced material. This means that only a small amount of complex PIILs was immobilized on the surface of the molecular sieve, and most of it was retained in the pores. SEM results further suggested the successful immobilization of PIILs/$(ZnBr_2)_2$ catalysts on MCM-22-CPTES and the solid state of obtained immobilized catalysts.

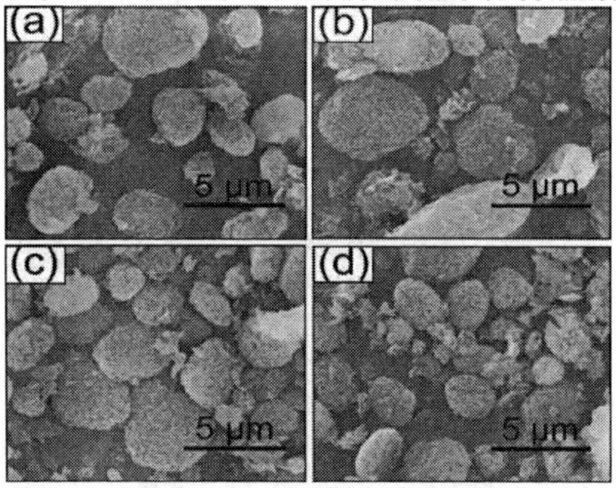

Figure 3. SEM of the silane molecular sieve MCM-22-CPTES and immobilized catalysts (a) MCM-22-CPTES; (b) MCM-22-CPTES-HO-[PECH-MIM]Cl/[$ZnBr_2$]; (c) MCM-22-CPTES-HOOC-[PECH-MIM]Cl/[$ZnBr_2$]; (d) MCM-22-CPTES-H_2N-[PECH-MIM]Cl/[$ZnBr_2$]

3.2. Effect of Different Catalyst on Catalytic Performance

The catalytic performance of the above-mentioned three immobilized catalysts for synthesizing PC with CO_2 and PO was investigated and the results were shown in Table 1.

Table.1 Effects of different catalysts on the catalytic performance

Catalyst	conversion rate (%)	selectivity(%)	yield (%)
MCM-22-CPTES-HO-[PECH-MIM]Cl/[$ZnBr_2$]	91.3	81.8	74.7
MCM-22-CPTES-HOOC-[PECH-MIM]Cl/[$ZnBr_2$]	93.2	83.6	77.9
MCM-22-CPTES-H_2N-[PECH-MIM]Cl/[$ZnBr_2$]	92.8	82.7	76.7

[1] Reaction condition: catalyst, 2.5% (wt%); temperature, 120 °C; pressure,3.0 MPa; time, 5 h.

It can be seen from the results in the table that the type of the composite PILs still affected the cycloaddition reaction. Among which MCM-22-CPTES-[PECH-MIM]Cl/[$ZnBr_2$] has better catalytic activity than the other two immobilized catalysts, because there was still a small portion of HOOC-[PECH-MIM]Cl/[$ZnBr_2$] physical adsorbed on the surface of MCM-22-CPTES still had acidity which was good for ring opening[9-10]. The conversion rate and selectivity were 92.7% and 82.7%.

4. Conclusions

In conclusion, three composite PILs that MCM-22 was immobilized with CPTES catalysts were prepared successfully and used in the cycloaddition reaction of CO_2 and PO. Among the three catalysts investigated, MCM-22-CPTES-HOOC-[PECH-MIM]Cl/[$ZnBr_2$] exhibited excellent catalytic performance in fixed bed reactor. MCM-22-CPTES-HOOC-[PECH-MIM]Cl/[$ZnBr_2$] exhibited the highest catalytic activity in the synthesis of PC under the optimized condition: the reaction temperature of 120 °C, the pressure of 3MPa and the time of 5h, the conversion rate was 93.2% and selectivity was 83.6%.

Acknowledgements

This work was supported by National Natural Science Foundation of China (21706163) and Department of Education Liaoning province (LQGD2017020).

References

[1] Rulev Y A, Zalina G, Maleev V I, et al. (2005) Robust bifunctional aluminium–salen catalysts for the preparation of cyclic carbonates from carbon dioxide and epoxides[J]. Beilstein Journal of Organic Chemistry., 11: 1614-1623.
[2] Yang L H, Wang H M. (2014) Recent advances in carbon dioxide capture, fixation, and activation by using N-heterocyclic carbenes[J]. ChemSusChem., 7: 962-998.
[3] Liu X F, He L N. (2018) High value utilization of carbon dioxide[J]. Science., 70: 14-19.
[4] Chen H P, Liang Y H, Wang B. (2009) Research progress on chemical utilization of carbon dioxide and catalytic system [J]. Chemical Industry Progress., 28: 271-278.
[5] Chen S M, Liu Y, Guo J P, et al. (2015) A multi-component polyoxometalate and its catalytic performance for CO_2 cycloaddition reactions[J]. Dalton Transactions., 44: 1-3.
[6] Peng J J, Deng Y Q. (2001) Catalytic Synthesis of Propylene Carbonate by Room Temperature Ionic Liquid[J]. Chinese Journal of Catalysis., 6: 598-600.
[7] Li F W, Xiao L F, Xia C G. (2005) Highly Efficient Catalytic Synthesis of Cyclic Carbonates by Zinc Bromide-Ionic Liquid Composite Catalytic System[J]. Chemical Journal of Chinese Universities., 2: 343-345.
[8] Chen C, Ma Y, Zheng D N, et al. (2017) Insight into the role of weak interaction played in the fixation of CO_2 catalyzed by the amino-functionalized imidazolium-based ionic liquids[J]. Journal of CO_2 Utilization., 18: 156-163.
[9] Guo L Y, Deng L L, Jin X C, et al. (2017) Composite ionic liquids immobilized on MCM-22 as efficient catalysts for the cycloaddition reaction with CO_2 and propylene oxide[J]. Catalysis Letters., 147: 2290-2297.
[10] Sang Y, Liu H X, He S C, et al. (2013) Catalytic performance of hierarchical H-ZSM-5/MCM-41 for methanol dehydration to dimethyl ether[J]. Journal of Energy Chemistry., 22: 769-777.

IOP Conf. Series: Materials Science and Engineering **729** (2020) 012038 doi:10.1088/1757-899X/729/1/012038

Study on Electromagnetic Wave Absorption performance of Dendritic-like Co

Fuli Yang[1], Xingzhe Hou[1], Lirui Wang[2], Youjiang Li[2], Xuecong Zhang[2], Miao Yu[2*]

[1] Electric power research institute, Chongqing Electric Power Corporation, State Grid, Chongqing 401123, China

[2] College of Optoelectronic Engineering, Chongqing University, Chongqing 400044, China

* Corresponding author's e-mail: yumiao@cqu.edu.cn

Abstract. Dendritic-like Co particles consisting of leaf-shaped nano-units were successfully synthesized by chemical reduction method. The absorbing particles have abundant multi-stage branches. Since the crystal structure, size and special morphology have an important influence on the microwave absorption performance, the X-ray diffractometer, scanning electron microscope, vibration sample magnetic measurement and vector network analysis are used to characterize the dendritic Co. Absorbing mechanism was analyzed systematically. The nano-sized, multi-stage branches, porous structures and rough surfaces of the prepared particles all contribute to the penetration, scattering and attenuation of EM waves. Hence the composite, in the test band, especially in the S-band (2-4 GHz), presents excellent microwave absorption properties.

1. Introduction

In recent years, microwave technology has been widely used in transportation, satellite communications, ship radar, power grid and other fields. The latest 5G network communication technology has been promoted in the Internet of Vehicles and Automated Driving, Remote Surgery, Smart Grid and other industries. However, the national defense safety problems and ecological environment caused by electromagnetic radiation are becoming more and more serious. Therefore, it is a key and urgent task to develop electromagnetic absorbing materials that can reliably and effectively reduce radiation[1, 2].

Magnetic metal powder has the advantages of higher snooker's limiting frequency、 high saturation magnetization、 high relative permeability at radar wave frequency, etc[3]. In addition, it can rely on dielectric loss and magnetic loss to consume electromagnetic waves that has a wide range of applications in the field of electromagnetic protection and stealth. Inevitably, electromagnetic waves will arouse magnetization of magnetic metal materials, resulting in eddy current effect and decreased absorption efficiency[4]. Nano-type structure (such as flower, sea urchin, dendritic) is a hyper-branch structure formed by nanoparticles with different shapes and sizes, which can effectively weaken the skin effect and eddy current effect[1, 5, 6]. What's more, this structure introduces quantum effect and small size effect, etc. This nano-type structure has shown superior performance in the field of microwave absorption and has attracted wide attention of many researchers. Sun et al. prepared dendritic α-Fe by hydrothermal and redox method. When the thickness of sample was 1.5 mm, the minimum reflection loss RL was -22.5 dB[7]. Yu et al. prepared flower-like carbonyl iron powder by in-situ reduction. The experimental results showed that the flower-like carbonyl iron powder had excellent electromagnetic absorbing properity in the X (8-12 GHz) band, compared with the spherical carbonyl iron powder. Minimum reflection loss increased by 46%, and effective reflection loss (RL < -20dB) band broadened by 57%[3]. It can be seen that the absorbing performance of the nano-typed absorbing material was greatly improved. However, in order to realize engineering applications, it is still a challenge to prepare high-performance electromagnetic absorbing materials. In particular, most of the electromagnetic waves in the civilian field are concentrated in the S-band (2-4 GHz). Thus, the development of the absorbing materials in the S-band is important to electromagnetic protection.

In this paper, dendritic Co structure self-assembled by nano-leaf cells was synthesized by chemical reduction method. These leaf units with a thickness of about several nanometers can effectively suppress eddy currents. The shape is highly anisotropic and helps to exceed the limits of Snoek. The prepared dendritic Co has a rich grain boundary and can induce strong intrinsic electric dipole polarization. Due to the large number of first, second, third or even higher branches, multiple interfacial polarizations can be formed, and a continuous network is easily formed between the dendritic particles. All of these factors contribute to favorable microwave absorption properties.

2. Experiment

2.1. Synthesis of dendritic Co particles

The specific preparation steps of the dendritic Co were as follows: First, $CoCl2 \cdot 6H2O$ (14.28 g) was weighed and dissolved in 300 ml of absolute ethanol, and mechanically stirred until the $CoCl2 \cdot 6H2O$ powder was completely dissolved to form a transparent solution. Then, NaOH (12 g) powder was added into the above-mentioned solution of CoCl2 and mechanically stirred for 20 min to obtain a homogeneous mixed solution. Then, 30 ml of ethylenediamine (C2H8N2) and 30 ml of hydrazine hydrate 80% (N2H4H2O) were slowly added dropwise, accompanying by mechanical stirring for 10 min. The whole reaction process lasted for 1 h. A black solid appeared on the bottom of the flask after the reaction was completed. The black precipitate was separated by a magnet and washed repeatedly with deionized water and absolute ethanol. Finally, the separated black solid was transferred to a vacuum drying oven at a temperature of 80° C. The final product was obtained by drying for 12 h.

2.2. Testing and characterization

To be able to analyze the chemical composition of the product, crystal structure was examined by X-ray diffraction XRD (copper Kα radiation source). The morphology and size of the product were also observed by field emission scanning electron microscopy (SEM).

For the purpose of studying the electromagnetic absorbing properties of the as-prepared product in the 2~18 GHz band, ring-shaped samples having an outer diameter of 7 mm, an inner diameter of 3.04 mm, and a thickness of 2.78mm were pressed by paraffin-bonded dendritic cobalt. The mass fraction ratio of particles to paraffin wax was 1:2 (sample 1) and 1:1 (sample 2), respectively. The test technique used coaxial reflection/transmission technology.

2.3. Microwave Absorption performance calculation

For a single-layer absorbing material with a metal backing, the usual reflection loss (RL) can be calculated from the following equation according to the transmission line theory[8, 9]:

$$Z_{in} = \sqrt{\frac{\mu_0 \mu_r}{\varepsilon_0 \varepsilon_r}} \tan h(j(\frac{2\pi}{c}) fd\sqrt{\mu_r \varepsilon_r}) \quad (1)$$

$$RL = 20\log\left|(Z_{in} - Z_0)/Z_{in} + Z_0\right| \quad (2)$$

Where the complex permittivity ε_r and permeability μ_r are measured at a given thickness and frequency. Z_0 is the impedance of free space, Z_{in} is the input impedance between free space and the material interface, c is the speed of light, f is the frequency of the microwave, and d is the thickness of absorbing material.

3. Results and discussion

3.1. Characterization of the samples

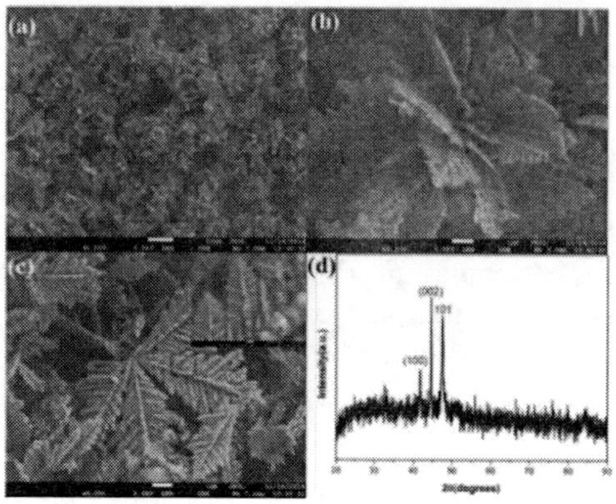

Figure 1. (a-c) SEM image of dendritic cobalt; (d) XRD pattern of dendritic cobalt

The SEM images showed that the prepared product has a typical layered dendritic structure. The entire structure had a radius of about 5 microns and radiated from the center by a plurality of leaf-like elements. There was a large space between the leaf-like unit layers, which facilitated the scattering of incident electromagnetic waves. The single leaf-like structure was composed of trunk, secondary branches and tertiary branches. The lateral branches were evenly distributed on both sides of the trunk at a certain angle. The length of the trunk was about 1~5 microns, while the length of the lateral stem was 50 nm-1um. Since the order of adding sodium hydroxide was adjusted based on the previous experimental scheme[10], the particles showed many small pores and the surface of the structure was rougher than before. The XRD pattern of the dendritic cobalt was shown in figure. 1(d), and the diffraction peaks of these samples were located at 2θ=41.54, 44.54, and 47.36, corresponding to the (100), (002) and (101) planes. It had a good match with the close-packed (hcp) Co crystals (space group P63/mmc1 (194); a = 0.2503 nm, c = 0.4060 nm; JCPDS No 05-0727)[10].

3.2. Microwave absorption

As could be seen from figure 2, the permittivity (ε) of the two samples increased as the content of dendritic cobalt particles increased. This reason was that the distance between the particles became shortened when the content of the dendritic cobalt was increased. This would undoubtedly enhanced charge transporting ability, resulting in a stronger space charge polarization, thereby increasing the value of permittivity. The ε' of sample 1 remained substantially constant at 4.5, and ε" value was almost constant in 2-12 GHz, while there was large fluctuation in 12-18 GHz. However, sample 2, whether it was the real part (ε') or the imaginary part (ε"), had large fluctuations throughout the whole test frequency range. This complex nonlinear vibration behavior was caused by space charge polarization, dipole polarization and interfacial polarization, which was usually related to the special structure of dendritic cobalt particles and their distribution in paraffin[11, 12]. At the same time, with the increase of frequency, the electric dipole polarization gradually become the dominant factor of metal-insulation system, and the electron polarization caused by the non-coincidence of the equivalent centers might also have a great influence on the dielectric response[13]. The relationship between the complex permeability and frequency was shown in figure 2(c,d). The complex permeability of Sample 2 was substantially higher than that of Sample 1 in the frequency range of 2-18 GHz. The real permeability (μ') of sample 1 increased first with frequency and then fluctuated around 1.2, while the overall trend of the real part (μ') of permeability of sample 2 increased with frequency. The whole ascent process was accompanied by fluctuations and fluctuated greatly in the range of 14-18 GHz. As for the imaginary part of magnetic permeability, the imaginary part (μ") of permeability of sample 1 had a distinct peak in the 6-8 GHz band, and the sample 2 had multiple peaks.

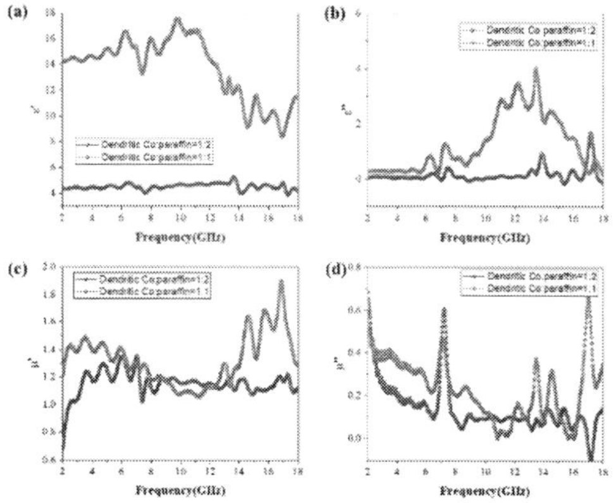

Figure 2. (a) and (b) are the curves of the real and imaginary parts of permittivity of dendritic cobalt with frequency; (c) and (d) are the curves of the real and imaginary parts of the dendritic cobalt permeability with frequency.

Figure 3 showd the reflection loss (RL). Sample 1 had multiple absorption peaks located at 7 GHz, 14 GHz, 17 GHz, and minimum reflection loss (RL) was -16 dB. Sample 2 had a minimum reflection loss of -16.3 dB at 8.9 GHz and effective bandwidth (RL < -10 dB, electromagnetic wave absorption of 90%) at 7.5-10.1 GHz. As the thickness increased, the absorption peak gradually shifted to low frequency, and multiple absorption peaks appeared. In the thickness from 2mm to 5mm, the absorption band of sample 2 covered 2-10GHz, which means that it has good absorption at low and medium frequencies.

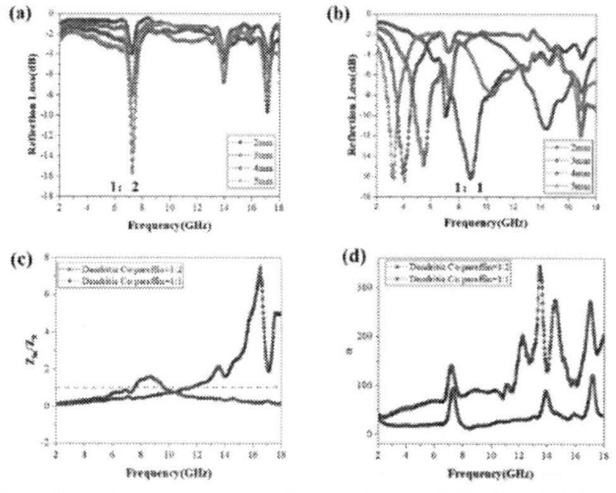

Figure 3. (a), (b) are the reflection loss curves of samples with dendritic Co and paraffin mass fraction ratios of 1:2 and 1:1 at different thicknesses; (c) normalized impedance matching; (d) loss factor

It was well known that the good absorption characteristics of electromagnetic absorbing materials need to meet two conditions: impedance matching and loss factor α. The loss factor α is a kind of characterization of the ability of absorbing material to absorb electromagnetic waves, which is used to describe the integral attenuation ability based on the following equation:[14]

$$\alpha=\frac{\sqrt{2}\pi f}{c}*\left[\left(\mu''\varepsilon''-\mu'\varepsilon'\right)+\left(\left(\mu''\varepsilon''-\mu'\varepsilon'\right)^2+\left(\mu'\varepsilon''+\mu''\varepsilon'\right)^2\right)^{1/2}\right]^{1/2} \quad (3)$$

The higher the loss capability, the stronger the ability to absorb electromagnetic waves. From figure 3(c), when the thickness was 2 mm, the two samples had good impedance matching near 6 GHz and 9 GHz, respectively, and the frequency range was the best absorption position at the current thickness. In addition, the loss factor of sample 2 was significantly higher than that of sample 1, which also showed that sample 2 had a better effect of absorbing electromagnetic waves. This might be because the content of the dendritic cobalt absorbing particles in the sample 2 was higher than that in the sample 1, and the dense absorbing particles could easily form a continuous network, and the electromagnetic waves entering the inside of the material could be quickly dissipated as a current.

In order to better understand the loss mechanism of dendritic cobalt particles, the dielectric loss tangent (tan δ_ε =ε''/ε') and magnetic loss tangent (tan δ_μ =μ''/μ') of sample 2 were calculated. In the range of 2-10 GHz and 16.2-18 GHz, the magnetic loss tangent of the dendritic cobalt was higher than the tangent value of the electrical loss, which indicated that the loss mechanism in these two frequency bands was mainly magnetic loss, and the dielectric loss was dominant in other frequency bands. Generally, main mechanisms causing magnetic loss are hysteresis loss, domain wall resonance loss, eddy current loss, natural resonance and exchange resonance loss. Under weak external magnetic field, the hysteresis loss is very weak and can be ignored[14, 15]. Magnetic domain wall resonance occurs in the low frequency range of 1~100MHz for multi-domain wall material[17]. Thus, the loss of electromagnetic waves is mainly caused by eddy current loss, natural resonance and exchange resonance loss. The contribution of the eddy current loss can be judged according to the change of the value of the formula$C_0 = \mu''(\mu')^{-2}f^{-1} = 2\pi\mu_0\sigma$ with the frequency[18, 19]. If the magnetic loss is only from eddy current losses, then the value of C0 should remain constant as the frequency changes[20]. As shown in figure 4 (b), the C0 value of the dendritic cobalt particles decreases with increasing frequency, and there is significant fluctuation, so the influence of the eddy current on the magnetic permeability is negligible. A peak appears in C0 curve in the 6-8 GHz band, mainly due to the appearance of the ferromagnetic resonance of nanomaterials, so the magnetic loss is contributed by natural resonance. Multiple formants appear in the frequency band above 12 GHz. According to Aharoni's theory, the frequency of natural resonance should be lower than the exchange resonance[21]. Therefore, it can be inferred that the peak appearing after 12 GHz is an absorption peak caused by exchange resonance.

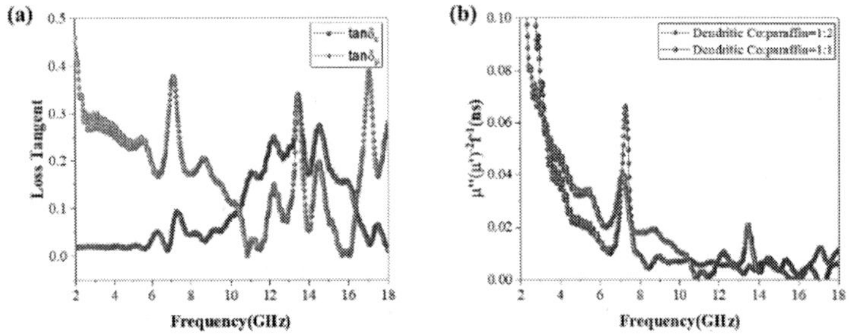

Figure 4. (a).Dielectric loss angle and magnetic loss angle;(b) the value of $\mu''(\mu')^{-2}f^{-1}$ in 2-18GHz

4. Conclusion

In this paper, the absorbing particles of dendritic cobalt were successfully prepared by chemical reduction method. The structure is a super-branched structure formed by self-assembly of nano-leaf cells. On the one hand, the nano-leaf unit effectively overcomes the skin effect and eddy current effect of the magnetic metal material, and improves the high-frequency magnetic permeability; on the other hand, the rich secondary structure is favorable for wave propagation and scattering; Rough surface, porous structure and large specific surface area are beneficial to the interface polarization, thus enhancing the loss of electromagnetic waves. The experimental results show that the absorption band of sample 2 covers 2-10 GHz when the thickness is changed from 2 mm to 5 mm, which means that it has a good absorption effect at low and medium frequencies. Therefore, the dendritic cobalt absorbing particles prepared in this paper have a good application prospect in electromagnetic protection.

Acknowledgments

The financial support from Electric power research institute of Chongqing Electric Power Corporation of State Grid (52202317001D) is highly acknowledged.

References

[1] Lv H, Liang X, Ji G, et al. Porous Three-Dimensional Flower-like Co/CoO and Its Excellent Electromagnetic Absorption Properties[J]. ACS Applied Materials & Interfaces. 2015, 7(18): 9776-9783.
[2] Saini L, Patra M K, Dhaka M K, et al. Ni/graphitic carbon core–shell nanostructure-based light weight elastomeric composites for Ku-band microwave absorption applications[J]. CrystEngComm. 2018, 20(32): 4630-4640.
[3] Yu M, Yang P, Fu J, et al. Flower-like carbonyl iron powder modified by nanoflakes: Preparation and microwave absorption properties[J]. Applied Physics Letters. 2015, 106(16): 161904.
[4] Wang Z, Bi H, Wang P, et al. Magnetic and microwave absorption properties of self-assemblies composed of core-shell cobalt-cobalt oxide nanocrystals[J]. Phys Chem Chem Phys. 2015, 17(5): 3796-3801.
[5] Yu Z, Yao Z, Zhang N, et al. Electric field-induced synthesis of dendritic nanostructured α-Fe for electromagnetic absorption application[J]. Journal of Materials Chemistry A. 2013, 1(14): 4571.
[6] Facile synthesis of urchin-like ZnO hollow spheres with enhanced electromagnetic wave absorption properties[J].
[7] Sun G, Dong B, Cao M, et al. Hierarchical Dendrite-Like Magnetic Materials of Fe3O4, γ-Fe2O3, and Fe with High Performance of Microwave Absorption[J]. Chemistry of Materials. 2011, 23(6): 1587-1593.
[8] Zhao B, Shao G, Fan B, et al. Preparation and electromagnetic wave absorption properties of novel dendrite-like NiCu alloy composite[J]. RSC Adv. 2015, 5(53): 42587-42590.
[9] Meng F, Wei W, Chen X, et al. Design of porous C@Fe3O4 hybrid nanotubes with excellent microwave absorption[J]. Phys. Chem. Chem. Phys. 2016, 18(4): 2510-2516.
[10] Yu M, Wang L, Yang P, et al. Preparation and high-performance microwave absorption of hierarchical dendrite-like Co superstructures self-assembly of nanoflakes[J]. Nanotechnology. 2017, 28(48): 485703.
[11] Zhao B, Shao G, Fan B, et al. Synthesis of flower-like CuS hollow microspheres based on nanoflakes self-assembly and their microwave absorption properties[J]. J. Mater. Chem. A. 2015, 3(19): 10345-10352.
[12] Liu T, Pang Y, Zhu M, et al. Microporous Co@CoO nanoparticles with superior microwave absorption properties[J]. Nanoscale. 2014, 6(4): 2447.
[13] Ma F, Qin Y, Li Y. Enhanced microwave performance of cobalt nanoflakes with strong shape anisotropy[J]. Applied Physics Letters. 2010, 96(20): 202507.
[14] Xie P, Li H, He B, et al. Bio-gel derived nickel/carbon nanocomposites with enhanced microwave absorption[J]. Journal of Materials Chemistry C. 2018, 6(32): 8812-8822.
[15] Fu L, Jiang J, Xu C, et al. Synthesis of hexagonal Fe microflakes with excellent microwave absorption performance[J]. CrystEngComm. 2012, 14(20): 6827.
[16] Zhang X F, Dong X L, Huang H, et al. Microwave absorption properties of the carbon-coated nickel nanocapsules[J]. Applied Physics Letters. 2006, 89(5): 53115.
[17] Spontaneous and Induced Magnetisation in Ferromagnetic Bodies[J].
[18] Murali K P, Sharma H, Markondeya Raj P, et al. Structure-magnetic property correlations in

IOP Conf. Series: Materials Science and Engineering **729** (2020) 012038 doi:10.1088/1757-899X/729/1/012038

nickel-polymer nanocomposites[J]. Journal of Materials Science: Materials in Electronics. 2016, 27(1): 154-162.

[19] Wu M, Zhang Y D, Hui S, et al. Microwave magnetic properties of Co[sub 50]/(SiO[sub 2])[sub 50] nanoparticles[J]. Applied Physics Letters. 2002, 80(23): 4404.

[20] Wen F, Yi H, Qiao L, et al. Analyses on double resonance behavior in microwave magnetic permeability of multiwalled carbon nanotube composites containing Ni catalyst[J]. Applied Physics Letters. 2008, 92(4): 42507.

[21] Aharoni A. Exchange resonance modes in a ferromagnetic sphere[J]. Journal of Applied Physics. 1991, 69(11): 7762-7764.

IOP Conf. Series: Materials Science and Engineering **729** (2020) 012039 doi:10.1088/1757-899X/729/1/012039

Preparation of Reduced Graphene Oxide/Magnetic Metal Composites and Its Electromagnetic Wave Absorption Properties

Fuli Yang[1], Xingzhe Hou[1], Lirui Wang[2], Youjiang Li[2] and Miao Yu[2*]

[1] Electric power research institute, Chongqing Electric Power Corporation, State Grid, Chongqing 401123, China

[2] College of Optoelectronic Engineering, Chongqing University, Chongqing 400044, China

* Corresponding author's e-mail: yumiao@cqu.edu.cn

Abstract: Reduced graphene oxide (RGO) was successfully prepared by reducing graphene oxide (GO) by zinc powder in this paper. Then the reduced graphene oxide (RGO) was combined with Fe and Ni nanoparticles to form RGO/Fe and RGO/Ni composite absorbing materials. The morphology, chemical composition, magnetic properties and electromagnetic parameters of RGO/Fe and RGO/Ni were characterized by means of scanning electron microscopy(SEM), X-ray diffraction(XRD), vibrating sample magnetometry (VSM) and vector network analysis (VNA) to study absorbing mechanism of the as-prepared RGO/magnetic-metal composites, including the effect of magnetic particles Fe and Ni on the absorbing properties. The results showed that in the frequency range of 2-18 GHz, the effective absorption band of RGO/Fe was wider than that of RGO/Ni, with larger reflection loss.

1. Introduction

Recently, microwave technology is developing rapidly in digital equipment and radar detection. Inevitably, electromagnetic radiation came into being. Therefore, the national security problems caused by electromagnetic radiation and electromagnetic environment are also becoming more and more serious. Therefore, it is a key and urgent task to develop electromagnetic absorbing materials that reliably and effectively reduce electromagnetic radiation[1-3].

Graphene is a new type of functional material with low density, large specific surface area, good electrical conductivity, and high dielectric constant. It can make external electrons easily polarize and relax under electromagnetic field and consume electromagnetic waves. Therefore, it has a very good application prospect in the field of electromagnetic protection and electromagnetic stealth[4, 5]. Sang-Eui Lee et al. combined graphite sheet with a volume fraction of 5% with epoxy resin to prepare an absorbing sample, and achieved a reflection loss (RL) of -22 dB at a thickness of 7 mm[6]. Singh et al. mixed 10% of layered porous graphene with nitrile rubber, and the results showed that the graphene/butyronitrile rubber exhibited a very high dielectric loss tangent[7]. However, according to the theory of electromagnetic complementation, single or excessive dielectric loss and magnetic loss are not conducive to the improvement of absorbing performance. The impedance mismatch of graphene is due to the high complex permittivity, which severely limits the microwave absorption properties of graphene[8]. At the same time, absorbing materials with a single loss mechanism are increasingly difficult to meet the ever-expanding microwave frequency band. Wherefore, it is an

IOP Conf. Series: Materials Science and Engineering **729** (2020) 012039 doi:10.1088/1757-899X/729/1/012039

effective solution to combine graphene with other absorbing mechanism materials to improve impedance matching and enhance electromagnetic wave absorption performance through the synergistic effect of electromagnetic parameters. Soft magnetic metal is an ideal matching material with graphene because of its high magnetic permeability[9]. Due to the adjustable complex permeability and complex permittivity, graphene/magnetic metal has received more and more attention[10].

In this paper, zinc oxide was used to reduce graphene oxide and metal ions by redox method. Graphene oxide reduced by zinc powder was structurally stable, which contained some residual defects and oxygen-containing functional groups. Moreover, it could introduce polarization relaxation and dipole relaxation [11]. In order to reduce the influence of agglomeration on the absorbing properties of composite materials, the thick graphene oxide prepared by Hummer method was used in this paper. Here, two metal elements, Fe and Ni, were selected to study the effects of magnetic metals on graphene composites.

2. Experiment

2.1. Synthesis of graphene@soft magnetic metal nanocomposites

The graphene oxide (GO) was prepared by the modified Hummer method, and the reduced graphene oxide(RGO) suspension was obtained by reducing GO with zinc powder. The average thickness of the RGO was ~19 nm.

The specific steps for preparing RGO/soft magnetic metal nanocomposites were as follows: First, transparent uniform ferric sulfate (FeSO4) solution was prepared by dissolving 5.56 g of FeSO4·7H2O in 400 mL of deionized water. In addition, 60 mg RGO suspension was weighed to added into the above FeSO4 solution, and stirred well until a uniform mixed solution A was obtained. Next, solution B was formed by dissolving 9.07 g of NaBH4 and 0.8 g of NaOH in 240 mL of deionized water, and magnetically stirred for 10 min to evenly mixed. After the above two solutions of A and B were completed, the two mixed solutions A and B were slowly dropped into the reaction tank to carry out reaction. The mixture was allowed to stand for 6 hours to ensure adequate response. After it was fully reacted, the obtained product was vacuum-filtered and dried, then grinded, an RGO/Fe nanocomposite was obtained. Identically, for the goal of obtaining the RGO/Ni nanocomposite, 5.56 g of FeSO4·7H2O was replaced with 5.257 g of NiSO4·6H2O, and the above preparation was repeated.

2.2. Testing and characterization

According to the transmission line theory, for a single-layer absorbing material with a metal backing plate, based on the measured complex permittivity, permeability data and the given thickness, the usual reflection loss (RL) can be calculated by the following formula [1, 12]:

$$Z_{in} = \left(\frac{\mu_0 \mu_r}{\varepsilon_0 \varepsilon_r}\right)^{1/2} \tanh\left(j\left(\frac{2\pi}{c}\right) fd\sqrt{\mu_r \varepsilon_r}\right) \tag{1}$$

$$RL = 20\lg\left|(Z_{in} - Z_0)/(Z_{in} + Z_0)\right| \tag{2}$$

Where Z_0 is the impedance of free space, Z_{in} is the input impedance between free space and the material interface, c is the speed of light, f is the frequency of the microwave, and d is the thickness of the absorbing material.

The complex permittivity and permeability need to be characterized by vector network analyzer using the coaxial reflection/transmission technique in the 2-18 GHz band. Therefore, paraffin-bonded graphene/magnetic metal composite particles were used to press a ring-shaped sample having an outer diameter of 7 mm, an inner diameter of 3.04 mm, and a thickness of 2 mm, wherein the mass fraction of the composite particles was 40%.

3. Results and discussion

3.1. Characterization of the samples

Spherical magnetic metal nanoparticles were successfully deposited on the surface of the reduced graphene oxide (RGO), and no agglomeration occurred. In the high-power diagram, it could be clearly seen that there were some wrinkles on the RGO film, and the nanoparticles were uniformly grown on the RGO film. The spherical nanoparticles were uniform in size, about 100 nm.

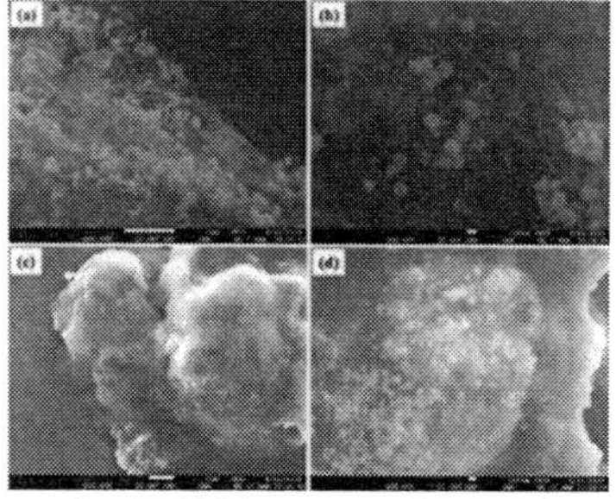

Figure 1. (a), (b) and (c), (d) are the SEM images of RGO/Fe and RGO/Ni nanocomposites.

The XRD pattern of the composite material was shown in figure. 2(a). The diffraction peak of RGO/Fe was located at 2θ=23°and 45°, and diffraction peak at 45°corresponded to the (110) crystal plane of iron. While the diffraction peaks of the RGO/Ni were located at 2θ=23 and 44.8°, and the diffraction peaks at 44.8°matched the nickel (111) crystal plane. The weak diffraction peak at 23°of RGO/Ni and RGO/Fe all corresponded to the RGO (002) crystal plane. Figure. 2(b) showed the hysteresis loop (M-H curve) of RGO/Fe and RGO/Ni at normal temperature. Both materials exhibited typical ferromagnetic properties. The saturation magnetization and coercive force of RGO/Fe were 93.48 emu/g and 76.42 Oe, respectively, while the saturation magnetization and coercive force of RGO/Ni were 3.54 emu/g and 16 Oe, respectively. The magnetic properties of materials were closely related to their morphology, structure, size and crystallinity[13]. The magnetocrystalline anisotropy of the materials can be determined by the value of M_r/M_s[14]. Under the applied electromagnetic field, the magnetocrystalline anisotropy energy of RGO/Fe was 0.039, which was about 3 times that of RGO/Ni. Higher magnetocrystalline anisotropy enabled it to have a larger coercive force *Hc*, and larger *Hc* was often able to make RGO/Fe to obtain a larger imaginary part (μ'') of magnetic permeability, which in turn gived RGO/Fe better absorbing properties.

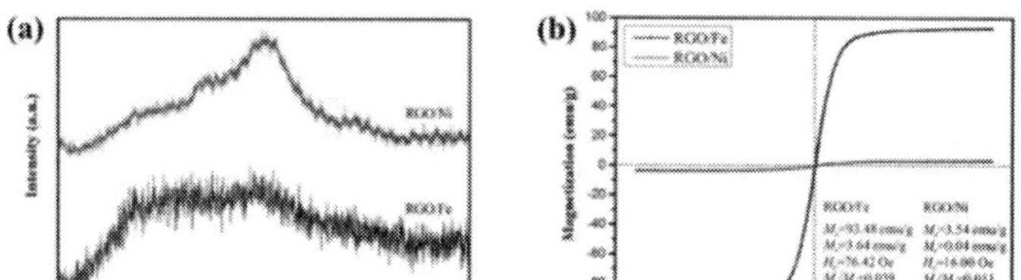

Figure 2. (a) and (b) are the XRD patterns and hysteresis curves of RGO/Fe and RGO/Ni nanocomposites.

3.2. Microwave absorption

It could be obtained from figure. 3 that the real part (ε') and the imaginary part (ε'') of permittivity of the two composite materials had large fluctuations in the frequency range of 2-18 GHz. The ε' of RGO/Fe gradually decreased with increasing frequency, which was caused by the polarization lag of the electric dipole. The overall trend of the complex permittivity of RGO/Ni was gradually decreasing with increasing frequency at 2-13 GHz, while large fluctuations appeared at 14-18 GHz. The ε'' of

RGO/Fe and RGO/Ni slowly increased with frequency in the low frequency band. After 8 GHz, the ε" of RGO/Fe fluctuated around 4, yet RGO/Ni was fluctuating violently. The fluctuations of the two composite materials at higher frequencies were due to that the dipole moment orientation polarization gradually dominated in the composite material as the frequency increased and the electron polarization was caused by the non-coincidence of the equivalent center.

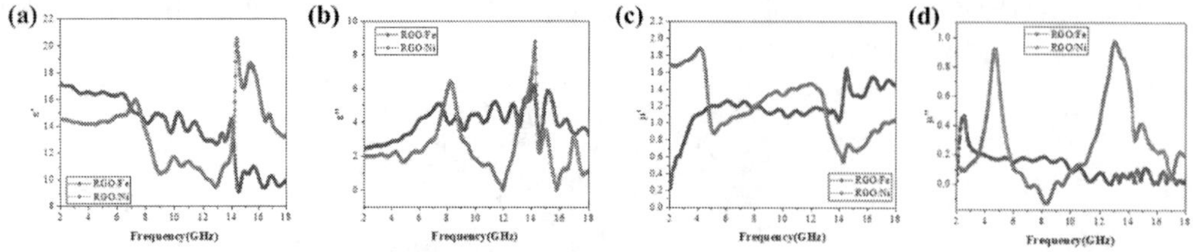

Figure 3. (a) and (b) are the curves of real and imaginary permittivity; (c) and (d) are the curves of the real and imaginary parts of the permeability of RGO/Fe and RGO/Ni with the change of frequency.

The real part (μ') of the complex permeability of RGO/Fe increased first with increasing frequency and then tended to stabilize and then increased, accompanying by slight fluctuations in 5-18 GHz. The μ' of the RGO/Ni magnetic permeability droped sharply at 4-5 GHz and 12.5-14 GHz, and rised slowly in other frequency bands. Whether it was the real part or the imaginary part of the magnetic permeability of RGO/Ni, there were obvious peaks at both the 4 GHz and 13 GHz frequencies. The μ" of RGO/Fe showed a small peak at about 2.5 GHz, which fluctuated slightly in the subsequent bands.

The reflection loss (RL) at different thicknesses was calculated as shown in figure 4. When the thickness of the sample was 2 mm, the maximum reflection loss of RGO/Fe at 9.22 GHz was -25.6 dB, and the effective bandwidth (RL < -10 dB, electromagnetic wave absorption reached 90%) was located at 7.9-11.1 GHz with 3.2 GHz absorption band. As the thickness increased, the absorption peak gradually shifted to the low frequency, movin from 9.22 GHz to 3.6 GHz; and the absorption intensity increased from -25.6 dB at 2 mm to -51.06 dB at 4 mm. The RGO/Ni composite exhibited multiple absorption peaks. When the thickness was 2 mm, it had a broad absorption peak at 8-10 GHz. However, in the 2-18 GHz, the electromagnetic wave absorption was less than -10 dB. When the thickness increased to 3 mm, the maximum reflection loss reached -17.4 dB at 4.44 GHz and the effective bandwidth covered 3.2-7 GHz.

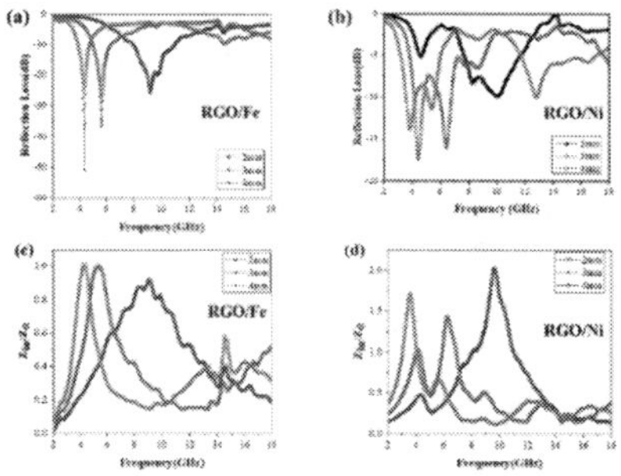

Figure 4. (a) and (b) are respectively the reflection loss curves of RGO/Fe and RGO/Ni with different thickness; (c) (d) are impedance matching curves of RGO/Fe and RGO/Ni with different thickness.

Basic characteristics of electromagnetic absorbing materials: impedance matching and loss capability[15]. The impedance matching (the input impedance of the absorbing material matches the free space impedance) allows the electromagnetic wave to easily enter the material for less reflection. The ideal impedance matching condition is Zin/Z0=1, but it is often not achieved in practice, so often by adjusting Zin/Z0 to be close to 1. It could be seen from figure 4(c) that the best frequency of impedance matching of RGO/Fe at 2-4 mm thickness were 9.22 GHz, 5.9 GHz, and 3.6 GHz, respectively, which corresponded to the frequency of the maximum absorption peak of RGO/Fe.

Meanwhile, it was also found that as the thickness of these samples increases, the frequency with the best impedance matching moved from 9.22 GHz to 3.6 GHz. That was, it moved toward the low frequency, which was consistent with the direction in which the absorption peak moved. From figure 4(d), RGO/Ni had similar results to RGO/Fe.

The dielectric loss tangent （$\tan\delta_\varepsilon$） and magnetic loss tangent （$\tan\delta_\mu$） of the two materials were calculated to further understand the loss mechanism. The RGO/Fe dielectric loss tangent went up with fluctuations in the 2-18 GHz, varying from 0.15 to around 0.5. However, the magnetic loss tangent had a peak at 2.5 GHz and then gradually stabilized around 0.1. It could be inferred that the dielectric loss tangent of RGO/Fe was lower than the magnetic loss tangent in 2-4 GHz, while the dielectric loss in the 4-18 GHz was the main loss mechanism. The RGO/ Ni dielectric loss tangent and the magnetic loss tangent curves had two sharp peaks in the whole test frequency band.

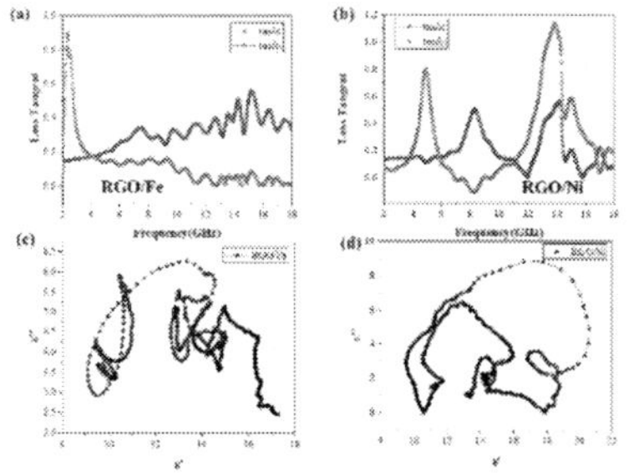

Figure 5. (a) and (b) are tangent curves of dielectric loss and magnetic loss of RGO/Fe and RGO/Ni nanocomposites; (c) and (d) are RGO/Fe and RGO/Ni col-cole semicircle; (e) is the value of C0 of RGO/Fe and RGO/Ni nanocomposites changes with frequency curve.

As is well known, Dielectric loss includes the extremely important loss mechanism of Debye relaxation [16, 17]. From the transmission line theory, the complex permittivity can be expressed by the following formula [18, 19]:

$$\left(\varepsilon'-\frac{\varepsilon_s+\varepsilon_\infty}{2}\right)^2+\left(\varepsilon''\right)^2=\left(\frac{\varepsilon_s-\varepsilon_\infty}{2}\right)^2 \quad (3)$$

Where τ, ε_∞ and ε_s are the polarization relaxation time, the permittivity in optical frequency and the static dielectric constant, respectively. Each Cole-Col semicircle corresponded to a Debye process[20]. These relaxation processes resulted from the migration of carriers between the reduced graphene oxide and the magnetic metal particles, and interfacial polarization, which was due to the accumulation of a large amount of charge between particles and paraffin. Both samples had multiple Cole-Cole semicircles. Compared with RGO/Ni, the Cole-Cole semicircle of RGO/Fe was obviously distorted. These curves showed that in addition to the Debye relaxation process, residual defects in RGO and oxygen-containing functional groups could accumulate electrons which caused electron polarization to form dielectric relaxation.

4. Conclusion

In this paper, RGO/Fe and RGO/Ni nanocomposites were successfully prepared by zinc oxide reduction of graphene oxide and metal ions. The effects of morphology, electromagnetic parameters, impedance matching and Debye relaxation on the absorbing properties were systematically investigated. At a certain thickness, the absorption peaks of RGO/Fe and RGO/Ni are at the best impedance matching frequency. As the thickness increases, absorption peak moved to the low frequency band as well as impedance matching frequency. Meanwhile, due to the large coercive force, good impedance matching and strong loss capability of RGO/Fe, electromagnetic waves could enter the absorbing material as much as possible to induce large attenuation. The visual representation was that the RGO/Fe absorption band was wider with better absorbing performance.

Acknowledgments
The financial support from Electric power research institute of Chongqing Electric Power Corporation of State Grid (52202317001D) is highly acknowledged.

References

[1] Li, X., Feng, J., Du, Y., et al. (2015) One-pot synthesis of CoFe2O4/graphene oxide hybrids and their conversion into FeCo/graphene hybrids for lightweight and highly efficient microwave absorber[J]. Journal of Materials Chemistry A., 3(10): 5535-5546.

[2] Lee, S., Jo, I., Kang, S., et al. (2017) Smart Contact Lenses with Graphene Coating for Electromagnetic Interference Shielding and Dehydration Protection[J]. ACS Nano., 11(6): 5318-5324.

[3] Lin, Y., Dong, J., Zong, H., et al. (2018) Synthesis, Characterization, and Electromagnetic Wave Absorption Properties of Composites of Reduced Graphene Oxide with Porous LiFe5O8 Microspheres[J]. ACS Sustainable Chemistry & Engineering., 6(8): 10011-10020.

[4] Li, Y., Yu, M., Yang, P., et al. (2017) Enhanced Microwave Absorption Property of Fe Nanoparticles Encapsulated within Reduced Graphene Oxide with Different Thicknesses[J]. Industrial & Engineering Chemistry Research., 56(31): 8872-8879.

[5] Wang, G., Zhang, X., Wei, Y., et al. (2013) Polymer composites with enhanced wave absorption properties based on modified graphite and polyvinylidene fluoride[J]. Journal of Materials Chemistry A., 1(24):7031.

[6] Lee, S., Choi, O., Hahn, H., (2008) Microwave properties of graphite nanoplatelet/epoxy composites[J]. Journal of Applied Physics., 104(3): 33705.

[7] Singh, V., Shukla, A., Patra, M., et al. (2012) Microwave absorbing properties of a thermally reduced graphene oxide/nitrile butadiene rubber composite[J]. Carbon., 50(6): 2202-2208.

[8] Gupta, A., Singh, A., Varshney, S., et al. (2014) New insight into the shape-controlled synthesis and microwave shielding properties of iron oxide covered with reduced graphene oxide[J]. RSC Adv., 4(107): 62413-62422.

[9] Luo, P., Liu, Z., Tao, R., et al. (2018) Design of double-layered flaky carbonyl iron/polyurethane wave-absorbing coating and its electromagnetic properties[J]. Electroplating & Finishing, 6: 247-253.

[10] Kong, L., Yin, X., Xu, H., et al. (2019) Powerful absorbing and lightweight electromagnetic shielding CNTs/RGO composite[J]. Carbon., 145: 61-66.

[11] Xu, S., Chen, G., Zhang, H., et al. (2016) Preparation and Electromagnetic Absorption Performance of Ni-RGO[J]. Journal of Inorganic Materials, 6: 567-574.

[12] Saini, L., Patra, M., Dhaka, M., et al. (2018) Ni/graphitic carbon core–shell nanostructure-based light weight elastomeric composites for Ku-band microwave absorption applications[J]. CrystEngComm., 20(32): 4630-4640.

[13] Wang, Z., Bi, H., Wang, P., et al. (2015) Magnetic and microwave absorption properties of self-assemblies composed of core–shell cobalt–cobalt oxide nanocrystals[J]. Phys. Chem. Chem. Phys., 17(5):3796-3801.

[14] Yan, S., (2011). Structure and Electromagnetic Properties of Ferromagnetic Metal Micro/Nanoparticles Covered by Dielectric Layer[D]. Harbin Institute of Technology, Harbin.

[15] Xie, P., Li, H., He, B., et al. (2018) Bio-gel derived nickel/carbon nanocomposites with enhanced microwave absorption[J]. Journal of Materials Chemistry C., 6(32): 8812-8822.

[16] Guo, F., Du, H., Qu, S., et al. (2015) Oxidation resistance and microwave absorption property of core shell urchin-like ZnO\Carbonyl iron powder composite particles[J]. Chinese Journal of Inorganic Chemistry, 4: 755-760.

[17] Zhao, B., Shao, G., Fan, B., et al. (2015) Synthesis of flower-like CuS hollow microspheres based on nanoflakes self-assembly and their microwave absorption properties[J]. JOURNAL OF MATERIALS CHEMISTRY A., 3(19): 10345-10352.

[18] Zhu, W., Wang, L., Zhao, R., et al. (2011) Electromagnetic and microwave-absorbing properties of magnetic nickel ferrite nanocrystals[J]. NANOSCALE., 3(7): 2862-2864.
[19] Zhao, B., Shao, G., Fan, B., et al. (2015) Synthesis of flower-like CuS hollow microspheres based on nanoflakes self-assembly and their microwave absorption properties[J]. J. Mater. Chem. A. 2015, 3(19): 10345-10352.
[20] Du, Y., Liu, W., Qiang, R., et al. (2014) Shell Thickness-Dependent Microwave Absorption of Core–Shell Fe3O4@C Composites[J]. ACS Applied Materials & Interfaces., 6(15): 12997-13006

Research on the change rule of microbial community in the microbial flooding process of Baolige oilfield

Guan Wang[1*], Lisha Duan[1], Feng Han[2], Rui Wang[1], Shasha Liu[1], Xiaoyang Lei[1], Jizhe Jing[1], Yi Gu[1], Yuchao Lei[3]

[1] Engineering Technology Research Institute, Huabei Oilfield Company of PetroChina, Renqiu, He Bei Province, 062552, China

[2] Research Institute of Exploration and Development, Huabei Oilfield Company of PetroChina, Renqiu, He Bei Province, 062552, China

[3] Materials Department of Huabei Oilfield Administration co. LTD, Renqiu, He Bei Province, 062552, China

[*]e-mail: cyy_wangg@petrochina.com.cn

Abstract: In Baolige oilfield, through the research and field implementation of microbial flooding technology, some application effects had been achieved. In order to further improve the implementation effect, the microbial community in the produced liquid during microflooding was tracked, monitored and analyzed by high-throughput sequencing (NGS). The results showed that the microflora diversity was plentiful. The species diversity after micro-flooding was higher than that during micro-flooding. In microbial flooding stage, due to replenish nutrients and exogenous bacteria to reservoir, the advantage bacteria group was mainly Pseudomonas, Acinetobacter, Thauera and Wallingella. After the microbial flooding, due to stop supplying nutrients, the bacteria community changed. The relative abundance of hydrocarbon- degrading bacteria declined and that of anaerobic aerogen increased.

1. Introduction

Microbial enhanced oil recovery (MEOR) is a comprehensive technology that utilizes beneficial activities and metabolites of microorganism to improve oil recovery [1]. According to the microorganism source, MEOR can be divided into exogenous MEOR and endogenous MEOR [2]. The endogenous reservoir microorganisms refer to the microbial community that enters the reservoir with injecting water in the process of water injection development and maintains a relatively stable quantity and species in a certain period [3]. The reservoir microbial metabolism has many types and great variability. It participates in the circulation of matter and energy in the reservoir and plays an important role in the earth ecosystem. According to metabolic types, the reservoir microorganisms mainly include HOB, TGB, FMN, NRB, SRB and MPB [4]. The influence factors of microbial flooding technology efficiency are quantity, type and biological function of the microbial in the reservoir. So it is necessary to understand the structures, properties and changing laws of oil reservoir microbial community roundly and accurately. The recognition of beneficial bacteria and its mechanism action is also vital. They will provide scientific data and methods for the design and optimization of microbial oil recovery technology and then guide the effective application of MEOR technology. The analysis of microbial community structure in reservoir microbial ecology is the key to microbial oil recovery. At present, researchers have conducted a large number of in-depth investigations and studies on the composition and distribution of microbial communities in waterflooding reservoir by pure culture and molecular ecology technology based on 16S r RNA [5].

The average buried depth of Baolige oilfield is 1400m. The formation temperature is 58°C. The salinity is 5726mg/L with $NaHCO_3$ water type. The development pattern is conventional injection. The formation porosity is from 13.4% to 21.7%, the crude oil freezing point is 29°C, the paraffin content is from 10.7% to 10.8% and the colloid asphalt content is from 46.0% to 51.2%. Under the geological conditions, the crude oil viscosity is from 1000 to 2000mPa.s with high oil-water viscosity ratio (34 to 800). The waterflood fingering leads to low efficiency of waterflooding and the watercut rising rate is up to 20%, leading to serious water channeling. Thus some reserves of heavy oil are difficult to be developed effectively with poorer development effect [6]. In Baolige oilfield, overall microbial flooding has been carried out for seven years and reservoir fluid bacteria concentration is up to 10^6/mL. In order to increase microbial oil recovery technology application, in-depth study on flora evolutionary changes and action mechanism are requisite, which can make effective control strategy and give full play to the synergies between the microbial flora.

2. Experiment

2.1. Source of experimental oil and water samples

The oil and water samples were collected from the produced fluid before and after microbial flooding in Baolige oilfield.

2.2. Experimental reagents and instruments

Fast Prep-24 sample rapid processing system (MP of USA).
Mastercycler PCR instrument (Eppendorf of Germany).
High speed refrigerated centrifuge (Beckman Coulter).
Power Soil DNA Isola-tion Kit (MOBIO Laboratories, USA).
Qubit DS-DNA Assay Kit Bradford (Invitrogen, USA).
Hollow Fiber Membrane (MOTIMO, China).
DNA polymerase Ta ka-ra Ex Taq TM (Dalian TaKaRa, China).

2.3. Experimental methods

2.3.1. Pretreatment of reservoir samples and total DNA extraction. The experimental procedures are as follows: (1) Water samples of 10L are taken from oil&water samples after standing for stratification. They are filtered and concentrated by hollow fiber of 0.22μm diameter. (2) The concentrate is transferred to a centrifuge tube, centrifuged for 30min to get the bacteria precipitation. (3) The bacterial precipitation is extracted with the MP Kit FastDNA® SPIN Kit for Soil (MP Biomedicals LLC, Solon, OH, USA) to obtain total genomic DNA which is detected by 1% agarose gel electrophoresis and stored at -20°C environment. The crude oil samples are extracted with PowerSoil® DNA Isolation Kit (Mo Bio Laboratories Inc., Carlsbad, CA, USA) to obtain DNA which is also detected by 1% agarose gel electrophoresis, and then stored at -20°C environment.

2.3.2. Species diversity analysis by Miseq high-throughput sequencing. Experimental procedures and methods refer to GB/7 30989-2014 [8].

3. Results and discussion

3.1. Structure and abundance of reservoir microbial community

Table 1. Identification results of sample microbial community

Taxon	The total number	Predominant bacteria number	Predominant bacteria name (Average abundance %)
Phylum	21	5	Proteobacteria (68.84%), Finnicutes(18.72%), Archaea(7.47%), Bacteroidetes(2.61%)
Class	53	17	Alphaproteobacteria (55.53), α-Proteobacteria(6.26), β-Proteobacteria(4.87), Epsilonproteobacteria(1.64), Deltaproteobacteria (1.49), Class Clostrida(17.6), Thermotogae(0.92), Class Methanobacteria(7.4).
Family	735	63	Pseudomonas(37.17), Acinetobacter(13.93), Alishewanella(2.69), Halomonas(1.32) and Nitrincolar(0.13). Sphingomonas(0.18), Brevundimonas(0.11) and Hyphomonas(0.12). Thauera(1.52), Petrobacter(0.26) and Hydrogenophaga(1.41). Arcobacter(1.76) and Wolinella(0.57). Desulfovibrio(0.38). Thermoanaerobacter(0.62), Caldanaerobacter(0.57), Anaerobranca(0.46), Thermacetogenium(1.63), Pelotomaculum(1.07) and Syntrophomonas(0.24), Desulfotomaculum(0.64) and Fusibacter(0.28). Fervidobacterium(0.69) and Thermotoga(0.24). Methanothermobacter(0.51) and Methanomethylovorans(0.24).

High-throughput sequencing results show that the microbial diversity of reservoir is plentiful. In terms of phylum classification, the reservoir samples belong to 21 groups, but the distribution is very uneven. They mainly distribute in Proteobacteria, Firmicutes, Thermotogae, Bacteroidetes and Euryarchaeota. The dominant bacteria are hydrocarbon oxidation bacteria, Zymophyte, denitrifying bacteria, Syntrophus and Methanogens.

3.2. The variation rules of bacterial species during microflooding

In the early stage of the microbial flooding, species diversity in family is higher. As the continuous replenishment of nutrients and exogenous bacteria liquid to the reservoir, the advantage microflora in reservoir produced fluid is mainly the injected Pseudomonas, Zymophyte, aerobic and anaerobic hydrocarbon-degradation bacteria, while other advantage species decreases rapidly. Pseudomonas and Acinetobacter in Alphaproteobacteria has always been in a major family. Previous studies have shown that these bacteria have a better ability to degrade and emulsify alkanes, aromatic hydrocarbons or their derivatives, indicating that the added nutrients to the reservoir can effectively activate the injected Pseudomonas and beneficial endogenous bacteria in the reservoir, so as to play a role in oil displacement.

Genus	18-42 d70	18-42 d125	18-42 d175	18-42 d205	18-42 d235	18-42 d310	19-18 d70	19-18 d125	19-18 d175	19-18 d205	19-18 d235	19-18 d310	38-63 d125	38-63 d175	38-63 d205	38-63 d235	38-63 d310
Pseudomonas	[illegible]	[illegible]	[illegible]	[illegible]	[illegible]	[illegible]	[illegible]	[illegible]	[illegible]	[illegible]	0.069	0.014	[illegible]	[illegible]	[illegible]	[illegible]	[illegible]
Acinetobacter	[illegible]	[illegible]	[illegible]	0.079	[illegible]	0.027	[illegible]	[illegible]	[illegible]	[illegible]	0.050	0.005	[illegible]	[illegible]	[illegible]	[illegible]	0.001
Thauera	[illegible]	0.080	0.535	0.565	0.002	0.000	[illegible]	0.002	0.051	0.002	0.025	0.000	0.014	0.216	0.068	0.008	0.000
Arcobacter	[illegible]	0.026	0.002	0.010	0.000	0.007	[illegible]	0.002	0.000	0.014	0.000	0.000	0.007	0.000	0.053	0.120	0.000
Wolinella	0.021	0.014	0.008	0.025	0.000	0.006	[illegible]	0.071	0.000	0.440	0.000	0.000	0.076	0.001	0.000	0.000	0.000
Alishewanella	[illegible]	0.072	0.020	0.001	0.000	0.140	[illegible]	0.001	0.011	0.006	0.000	0.000	0.008	0.737	0.372	0.013	0.007
Halomonas	0.002	0.007	0.046	0.871	0.004	0.148	0.066	0.005	0.120	0.034	0.000	0.000	0.032	[illegible]	[illegible]	0.251	0.085
Nitrincola	0.433	0.000	[illegible]	0.000	0.000	0.011	0.128	0.000	0.027	0.007	0.000	0.000	0.000	[illegible]	0.008	0.048	0.001
Desulfovibrio	0.314	0.020	0.001	0.003	0.002	0.000	[illegible]	0.024	0.002	0.006	0.000	0.000	0.330	[illegible]	0.391	0.000	0.000
Fusibacter	[illegible]	0.000	0.002	0.001	0.002	0.000	[illegible]	0.000	0.001	0.000	0.000	0.000	0.000	0.041	0.019	0.000	0.002
Streptococcus	[illegible]	0.001	0.007	0.035	0.073	0.008	0.605	0.006	0.000	0.007	0.028	0.001	0.001	0.000	0.083	0.003	0.081
Hydrogenophaga	0.056	0.077	0.071	[illegible]	0.060	0.073	0.039	0.000	0.007	0.026	0.127	0.009	0.007	0.037	0.417	0.221	0.015
Hyphomonas	0.110	0.004	0.002	0.230	0.000	[illegible]	0.036	0.000	0.145	0.006	0.000	0.000	0.000	0.051	0.175	0.000	0.278
Sphingomonas	0.002	0.001	0.001	0.012	0.000	[illegible]	0.003	0.008	0.004	0.002	0.003	0.001	0.000	0.001	0.003	0.000	0.314
Brevundimonas	0.019	0.001	0.001	0.234	0.002	0.074	0.018	0.006	0.032	0.005	0.000	0.000	0.004	0.766	[illegible]	0.000	0.002
Methanobacteriaceae;g__	0.000	0.014	0.050	0.331	[illegible]	[illegible]	0.000	0.059	0.925	[illegible]	[illegible]	[illegible]	0.003	0.340	[illegible]	0.156	[illegible]
Methanothermobacter	0.010	0.005	0.002	0.614	[illegible]	[illegible]	0.009	0.003	0.007	0.026	0.693	0.068	0.000	0.009	0.080	0.011	[illegible]
Thermacetogenium	0.021	0.012	0.014	0.161	0.185	[illegible]	[illegible]	0.027	[illegible]	[illegible]	[illegible]	[illegible]	0.036	0.333	[illegible]	0.329	[illegible]
Pelotomaculum	0.072	0.004	0.009	0.089	0.148	[illegible]	0.121	0.000	0.275	[illegible]	[illegible]	[illegible]	0.010	0.099	0.872	0.120	[illegible]
Syntrophomonas	0.003	0.000	0.002	0.070	0.027	[illegible]	0.000	0.000	0.225	[illegible]	0.391	0.643	0.000	0.005	0.090	0.010	0.672
Desulfotomaculum	0.005	0.005	0.000	0.058	0.104	0.615	0.022	0.001	0.073	0.524	[illegible]	[illegible]	0.004	0.006	0.313	0.034	0.651
Thermoanaerobacter	0.005	0.000	0.003	0.308	0.129	0.140	0.028	0.000	0.009	0.010	0.333	0.062	0.000	0.000	0.008	0.000	[illegible]
Anaerobaculum	0.021	0.000	0.000	0.006	0.002	0.002	0.018	0.000	0.017	0.027	0.019	0.012	0.000	0.001	0.099	0.005	[illegible]
Fervidobacterium	[illegible]	0.007	0.009	0.152	[illegible]	[illegible]	0.180	0.009	0.246	[illegible]	0.647	[illegible]	0.003	0.006	0.015	0.002	0.258
Thermotoga	0.055	0.021	0.004	0.033	0.029	0.451	0.092	0.009	0.120	0.488	0.171	[illegible]	0.004	0.012	0.218	0.003	0.367
Anaerobranca	0.018	0.000	0.000	0.002	0.202	0.012	0.061	0.000	0.068	0.089	0.960	0.123	0.000	0.002	0.372	0.002	0.187

Figure 1. Structural changes of dominant bacteria at family level in oil wells at different times

In the whole process of microflooding, the hydrocarbon-degradation bacteria is dominated, while the abundance of Syntrophus and Methanogens is very low, which indicates that the hydrocarbon degradation mainly depends on microbial activity at this stage. After the microflooding, the community structure changes rapidly and significantly. With the consumption of nutrients, the abundance of hydrocarbon- degrading bacteria gradually decreases, and the proportion of anaerobic zymophyte (including Thermoanaerobacter, Thermacetogenium, Desulfotomaculum and Syntrophomonas) and Methanogens gradually increases. The methane bacteria number of the well B19-18 increased from 0.9% to 40.6% after 60 days of microbial flooding. The methane bacteria number of well B18-42 increased from 0.05% to 2.4% after 60 days of microbial flooding. The methane bacteria number of well B38-63 increased from 0.3% to 4.1% after 60 days of microbial flooding. All these changes showed that the number of methane produced by bacteria enhanced after microbial flooding. The results are shown in figure 1.

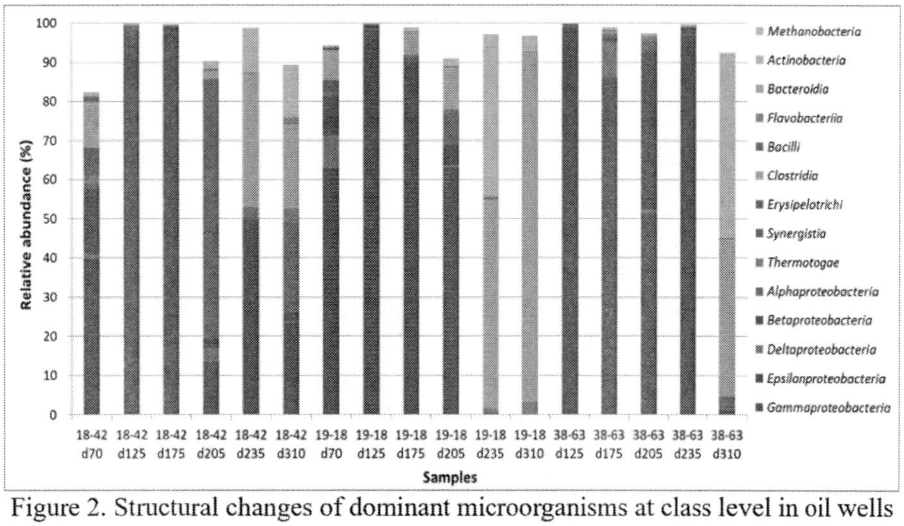

Figure 2. Structural changes of dominant microorganisms at class level in oil wells

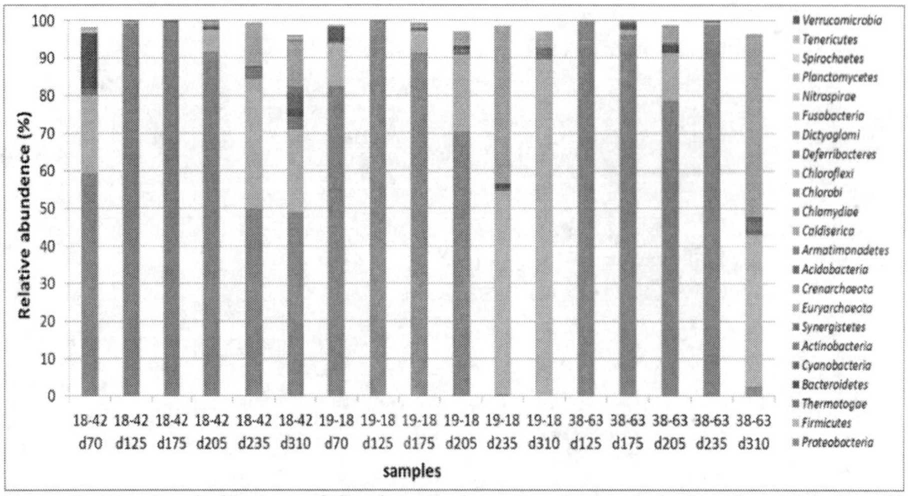

Figure 3. Structural changes of dominant microorganisms at phylum level in oil wells

As shown in figure 2 and figure 3, during the micro-flooding stage, when injecting nutrient and exogenous bacteria solution, the microbial community structure of oil wells is single. The hydrocarbon degradation bacterium in Proteobacteria is dominated. Alphaproteobacteria and Epsilonproteo bacteria are the main class. At the end of microflooding, with the consumption of nutrients, the proportion of Proteobacteria decreases, while the proportion of Firmicutes and Methanogens increases gradually. Microbial community structure begins to complicate and species diversity increases. The proportion of anaerobic bacteria gradually increases, including Thermoanaerobacter, Thermacetogenium and Desulfotomaculum in Methanobacteriaceae. The anaerobic bacteria produce organic acids by using hydrocarbon degradation. Further, interoperated with methanogens, methane is finally produced, which indicates that the methanogenesis metabolism is the main process in this stage.

3.3. analysis of produced fluid

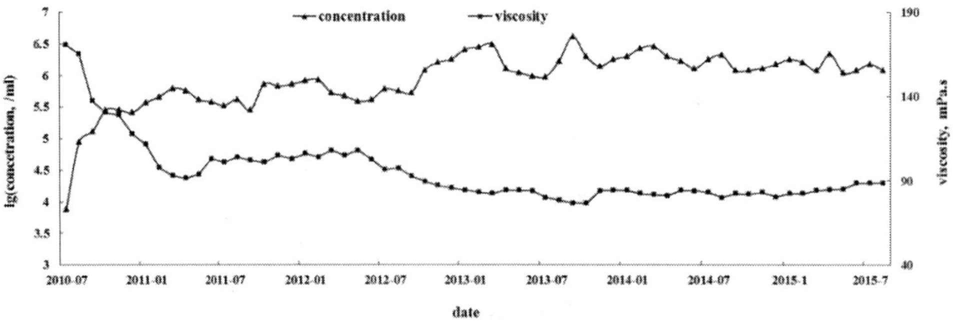

Figure 4. The variation of bacterial concentration and crude oil viscosity of produced fluid in Baolige oilfield

The microbial quantity and crude oil viscosity of the produced fluid from 52 oil wells in Baolige oilfield were continuously monitored. The average bacterial concentration and crude oil viscosity changes are shown in figure 4. The concentration of bacteria in the produced fluid was maintained at 10^6/mL, indicating that a stable microbial field was formed in the reservoir. The average viscosity reduction rate of crude oil was 48.1%. After injection of bacterial fluid and nutrient fluid, the beneficial bacterial community that was easy to drive oil was effectively activated and the number significantly increased. The relationship curve of microbial and oil well production showed that microbial could achieve the effect of increasing and stabilizing production.

Table 2. Surface tension of produced fluid of oil wells (unit: mN/m)

Well	Before microflooding	60 days after microflooding	120 days after microflooding	average reduction rate, %
B18-42	59.12	51.35	50.15	14.16
B19-18	58.35	58.35	51.36	10.87
B38-58	60.18	60.18	52.22	14.04

Table 3. Gas composition of oil wells (unit: %)

	Methane	Ethane	Propane	n-butane	CO_2	N_2
Before microflooding	74.76	4.12	4.02	1.1	1.06	13.46
120 days after microflooding	63.23	5.48	6.92	2.48	1.82	17.58
60 days after microflooding	85.41	2.13	1.57	0.65	1.01	7.48

The monitoring results of production fluid before and after microflooding showed that the content of biological surfactant produced by microorganisms during microflooding in the reservoir was improved and the surface tension decreased by about 12% compared with that before microflooding. After the microflooding, the content of methane increased by 10.65% compared with that before microflooding, indicating that anaerobic zymophyte began to interact with methanogenic archaea to degrade petroleum hydrocarbon and produce methane through oxidation and denitrification. The change of metabolites further indicated that the types and functions of bacteria in the reservoir changed before and after microflooding.

4. Conclusion

1. The biological diversity of reservoir is very rich. In the process of microbial flooding, the microbial community structure in the reservoir is a dynamic changing process. In the process of microbial flooding, the dominant flora is Pseudomonas, Acinetobacter, Thauera and Wallingella.

2. The types and abundance of the dominant bacteria in the reservoir change significantly in the microflooding process. During the period of injecting nutrition agent, the hydrocarbon-degrading bacteria are dominated, while the abundance of Syntrophus and Methanogens is very low. After the microbial flooding, as the consumption of nutrients, the proportion of anaerobic zymophyte (including Thermoanaerobacter, Thermacetogenium, Desulfotomaculum and Syntrophomonas) and Methanogens gradually increases. The anaerobic bacteria produce organic acids by using hydrocarbon degradation. Further, interoperated with methanogens, methane is finally produced. Microbial hydrocarbon degradation and interactional methane production plays an important role in microbial displacement in Baolige oilfield.

References

[1] Chen W.X., Hu R. He Q. (2009) Microbial Enhanced Oil Recovery (MEOR) and Oversea Research Progress in Its Application. Journal of Xi'an Shiyou University (Natural Science Edition), 24(4): 58-61.

[2] Yu D.F., Dong H.P., Yu L., et al. (2012) Laboratory Study of Indigenous Microorganism Activation System in Long Core. Oilfield Chemistry, 29(2): 236-239.

[3] Bao M.T., Kong X.P., Song Y.T., et al. (2004) Selective Activation Condition of Stratal Microflora of S12 Block in Shengli Oilfield. Journal of the University of Petroleum, China, 28(6): 44-48.

[4] Gao P.K., Wang Y.S., Zhang H.Z., et al. (2013) Mechanism of Crude Oil Viscosity Reduction by Two Thermophilic Hydrocarbon-degrading Bacteria. CIESC Journal, 64(11): 4240-4245.

[5] Li H., Mu B.Z. (2008) Recent Advances in Molecular Microbial Ecology of Petroleum Reservoirs. Microbiology, 35(5): 803-808.
[6] Ke C.Y., Wu G., You J. (2013) The Large Scale Application Research on Overall Microbial Flooding Technology in Baolige Oilfield. Oilfield Chemistry, 30(2): 246-250.
[7] Ren F.P., Yu J.L., Chen J.S., et al. (2015) Surface Multiplication Technology of Produced Water after MEOR. Oilfield Chemistry, 32(2): 251-254.
[8] GB/T 30989-2014, (2014) High Throughput Gene Sequencing. Beijing: Standards Press of China.

Research on Investment Distribution Technology of Power Grid Enterprise Overhaul Based on Gini Coefficient Theory

Xiaoyong Yang[1], Ping Xu[1], Jianfeng Shi[2], Zhongxu Li[2] and Chao Chen[3]

[1]Economic and Technological Research Institute, State Grid Zhejiang Power Co.,Ltd., Hangzhou 310000,China;

[2]State Grid Zhejiang Power Co.,Ltd.,Hangzhou 310000,China;

[3]School of Economics and Management, North China Electric Power University, Beijing 102206, China.

Email: hbbdcc@126.com

Abstract. Equipment overhaul is one of the core businesses of power grid enterprises.The scientific and rational allocation of overhaul funds plays an important role in improving the lean management of power grid enterprises.In view of this,this paper is based on the current situation of overhaul funds allocation of 11 municipal power supply companies under the provincial-level power grid companies in China,and based on the relevant theory of Gini coefficient,starting from improving the level of investment efficiency and improving the reliability of equipment operation.The sub-quota degree has been optimized and designed to improve the reasonable level of the over-allocation of funds of Z-provincial power grid companies and improve the level of business operation efficiency.

1. Introduction

With the development of economy and society,the development of power grid enterprises faces a new form,and the investment management method faces new requirements. It is urgent to implement the grid investment strategy and improve the efficiency and efficiency level [1]. Equipment overhaul is one of the core businesses of power grid enterprises. The scientific and rational allocation of overhaul funds plays an important role in improving the lean management of power grid enterprises [2]. However,at present,China's provincial power grid companies have an imbalance in the allocation of equipment overhaul funds,which has an adverse impact on the operation and development of enterprises. Therefore,based on the current situation of the overhaul investment allocation of 11 municipal power supply companies under the Z Power Grid Company of China,combined with the relevant theory of Gini coefficient,this paper studies the improvement of investment efficiency level and the reliability of equipment operation. The technology has improved the reasonable level of investment in the overhaul of the Z provincial power grid company and improved the level of business operation efficiency.

2. The basic principle of Gini coefficient theory

2.1. Lorenz curve

The Lorenz curve is proposed to solve the problem of the distribution of national income. It first queues the population of a country according to the income from low to high. After considering the percentage of income from any percentage of the population with the lowest income,the corresponding relationship between the cumulative percentage of the population and the cumulative percentage of income is plotted on the graph to form the Lorenz curve. The Lorenz curve can be used to compare and analyze the wealth inequality of a country in different eras or in different countries in the same era. It is widely used as a convenient graphical method to summarize the information of income and wealth distribution [3].

The figure below shows the manifestation of the Lorentz curve. In the figure,the horizontal axis OH represents the cumulative percentage of the population (grouped by income from low to high),the vertical axis OM represents the cumulative percentage of income,and the arc OL is the Lorenz curve. As shown below:

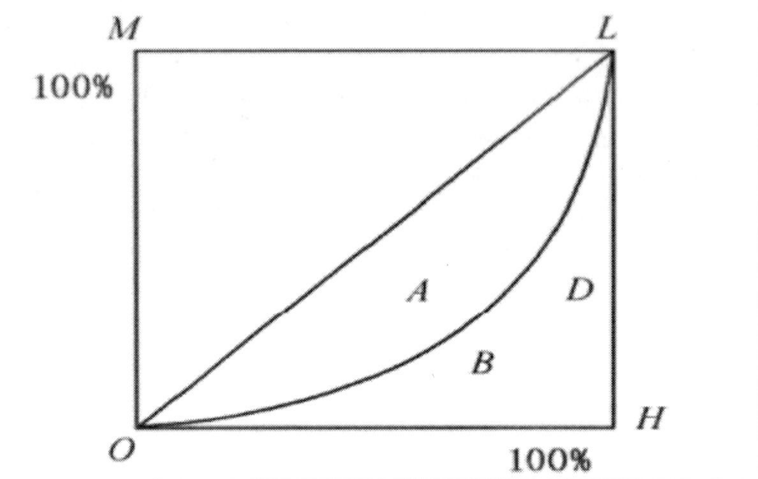

Figure 1. Lorenz curve diagram.

The degree of curvature of the Lorentz curve reveals important information. It reflects the degree of inequality in income distribution,the greater the degree of curvature indicates the more unequal income distribution.Usually a country's income distribution does not reach complete fairness or unfairness,but is somewhere in between. The corresponding Lorentz curve is represented by an arc between the 45° line OC and the broken line OXC,such as the arc LC.

2.2. Gini coefficient

The Gini coefficient is a ratio indicator defined by the Lorenz curve,the absolute equality curve,and the absolute inequality curve to quantitatively analyze the inequality of income distribution [4]. The formula for calculating the Gini coefficient (G) is as follows:

$$G=\frac{S_A}{S_A+S_B}$$

As shown in FIG. 1 above,SA represents the area enclosed between the absolute equality curve OC and the Lorentz curve LC,and SB represents the area enclosed between the Lorentz curve LC and the absolute inequality curve OXC. Among them,when SA=0s,the Gini coefficient is zero,indicating that the income distribution is completely equal,that is,the person who occupies the first percentage of the population receives the same percentage of income; when SB=0,the Gini coefficient is 1,indicating that the income distribution is absolutely unequal. That is,a very small number of people account for the vast majority of income. The more the income distribution tends to be unequal,the greater the curvature of the Lorenz curve,the value range is 0 to 1,and 0.4 is usually used as the warning line for the income distribution gap.

3. Empirical analysis

In this study,combined with the main features and main objectives of the overhaul project,the unit's investment efficiency technology improvement level index (the unit investment increase in electricity and unit investment power supply reliability rate increase rate of 0.2 and 0.8 respectively),for the overhaul investment efficiency,technology upgrade.The situation is calculated comprehensively. As shown in the following table:

Table 1. Table of cumulative percentage of unit investment efficiency technical level improvement indicators.

Serial number	Municipal unit	Cumulative percentage (increased level of unit investment power supply reliability rate)	Cumulative percentage (sold sales)	Unit investment efficiency technology improvement level	Overhaul investment	Cumulative percentage of overhaul investment
1	Z company	24.96%	14.04%	22.78%	2.46%	2.46%
2	Q company	33.94%	16.23%	30.40%	3.30%	5.76%
3	S company	37.41%	40.00%	37.93%	7.40%	13.16%
4	H company	52.18%	41.75%	50.09%	7.60%	20.76%
5	L company	60.04%	46.84%	57.40%	7.85%	28.61%
6	T company	73.61%	66.66%	72.22%	8.81%	37.42%
7	J company	78.72%	71.66%	77.31%	10.85%	48.27%
8	W company	87.24%	74.12%	84.62%	10.89%	59.16%
9	G company	94.35%	75.61%	90.60%	11.42%	70.58%
10	N company	97.59%	95.08%	97.09%	13.34%	83.92%
11	H company	100%	100%%	100.00%	16.09%	100.00%

The cumulative percentage of the investment improvement technology level of each city-level company based on the control index unit is the abscissa,and the cumulative percentage of the newly added investment in the initial 2017 after the adjustment of the power supply enterprises in each city is the Lorenz curve on the ordinate,and the power supply enterprises in each city according to the slope size. Incremental sorting on the Lorentz curve.

The Lorentz plot,based on the data shown in Table 3,is shown in Figure 2:

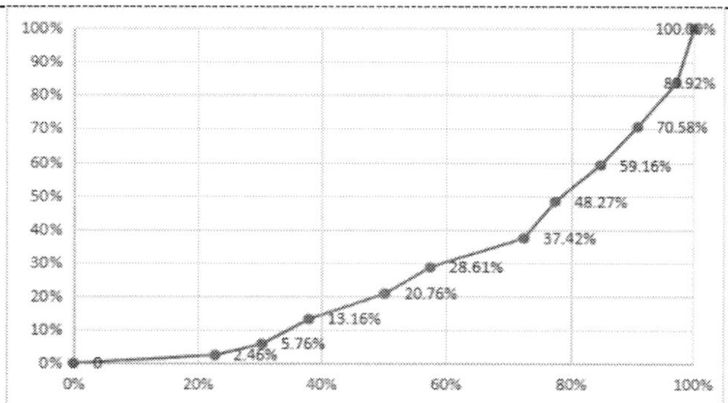

Figure 2. Lorentz curve before optimization.

In order to ensure the fairness and rationality of the allocation of overhaul investment,from the perspective of asset size,municipal units with large asset scales should allocate more funds for overhaul; from the perspective of economic benefits of overhaul investment,municipal units with high unit investment and high power supply should More allocation of overhaul funds; from the perspective of safe and reliable power grids,units with lower power supply reliability rates should allocate more funds for overhaul[5]. Combining various influencing factors,combining the basic process of Gini coefficient optimization analysis,determining the optimal target and constraints for comprehensive optimization planning analysis,the optimized Gini coefficient is 0.3384.

Table 2. Table of cumulative percentage of technical level improvement indicators of unit investment benefit after optimization.

Serial number	Municipal unit	Cumulative percentage (increased level of unit investment power supply reliability rate)	Cumulative percentage (sold sales)	Unit investment efficiency technology improvement level	Overhaul investment
1	Z company	24.96%	14.04%	22.78%	8.46%
2	Q company	33.94%	16.23%	30.40%	15.76%
3	S company	37.41%	40.00%	37.93%	21.16%
4	H company	52.18%	41.75%	50.09%	26.76%
5	L company	60.04%	46.84%	57.40%	38.61%
6	T company	73.61%	66.66%	72.22%	46.42%
7	J company	78.72%	71.66%	77.31%	52.27%
8	W company	87.24%	74.12%	84.62%	59.16%
9	G company	94.35%	75.61%	90.60%	67.58%
10	N company	97.59%	95.08%	97.09%	88.92%
11	H company	100%	100%%	100.00%	100.00%

The Lorentz plot of the optimized control index plotted against the data shown in Table 2 is shown in Figure 3:

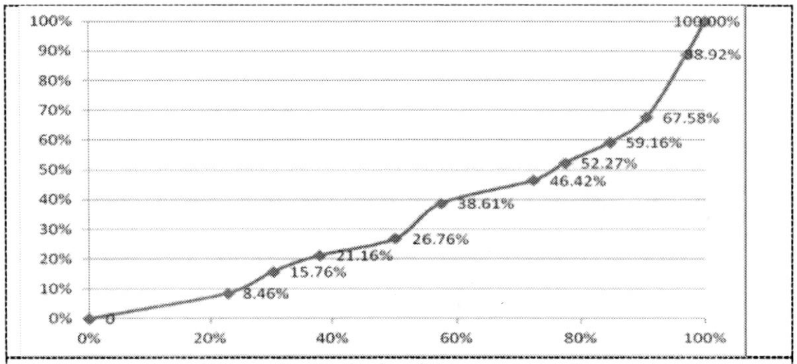

Figure 3. Optimized Lorentz graph.

Combined with the optimized Gini coefficient,it can be concluded that the optimization plan for the overhaul investment of each municipal unit in 2018 is as follows:

Table 3. Table of the 2018 overhaul investment allocation of each municipal unit after optimization.

Serial number	Municipal unit	2018 overhaul plan investment(100 million yuan)	Optimized investment in 2018 after overhaul(100 million yuan)
1	H company	12.34	6.49
2	N company	10.23	5.60

3	W company	8.35	4.14
4	J company	8.76	4.30
5	C company	5.83	9.09
6	S company	5.68	5.99
7	G company	8.32	4.49
8	Q company	2.53	5.29
9	Z company	1.89	6.46
10	T company	6.76	16.37
11	L company	6.02	8.50

4. Conclusion

In this paper,the Gini coefficient theory is used to optimize the overhaul funds of 11 municipal units in Z province of China,and the optimized investment allocation plan is obtained,which not only satisfies the demand for electricity (ie benefit) in economically developed regions,but also guarantees the completion of various regions[6]. The possible balanced development (ie fairness) reflects the optimization concept of "optimizing investment,scientific investment,focusing on efficiency and sustainable development".

References

[1] Zhang Weiqin. Strengthening the management and control of power grid investment and improving the investment efficiency of power grid [J]. China High-tech Enterprise,2016 (02): 176-177.
[2] Yang Wei. Research on optimization model of technical transformation and repair of primary power equipment in power grid [D]. South China University of Technology,2017.
[3] Cuncun Duan,Bin Chen. Analysis of global energy consumption inequality by using Lorenz curve[J]. Energy Procedia,2018,152.
[4] Zehorit Dadon-Golan,Iris BenDavid-Hadar,Joseph Klein. Revisiting educational (in)equity: Measuring educational Gini coefficients for Israeli high schools during the years 2001–2011[J]. International Journal of Educational Development,2019,70.
[5] Gong Yi. On how to improve the management level of technical renovation and overhaul projects [J]. Shandong Industrial Technology, 2014 (24): 227.
[6] Xiao Min. Research on New Ideas of Power System Technical Reform and Overhaul Project Management[J].Shandong Industrial Technology,2018(17):202.

IOP Conf. Series: Materials Science and Engineering **729** (2020) 012042 doi:10.1088/1757-899X/729/1/012042

Research on Forecasting Technology of Single Base Tower Cost in Transmission Line Engineering

Ying Du[1], Jiyuan Zhang[1], Chao Wang[1], Mingyong Wan[1], Ping Zhou[1] and Chao Chen[2]

[1] Economic and Technological Research Institute, State Grid Sichuan Electric Power Company, Chengdu 640041, China;

[2] School of Economics and Management, North China Electric Power University, Beijing 102206, China.

Email: hbbdcc@126.com

Abstract. For a long time, the overhead transmission line project budget, completion settlement and financial final accounts are all valued according to the entire line. The project construction management and asset collection have not yet been refined. The management is not separately calculated according to the actual investment of different towers, and cannot be reflected in each base tower. The actual cost level and the accuracy of the cost need to be further improved. Therefore, this paper combines the actual pain points and difficulties of overhead line engineering cost management, and innovatively builds a pricing model based on a single base tower. When the company strengthens the construction drawing budget management, improves the project settlement level, accurately controls risks, and realizes high-quality management of cost management. The level of leanness is of great practical significance.

1. Introduction

With the development of economy and society, the development of power grid enterprises faces a new form, and investment management faces new requirements. It is urgent to implement a precise investment strategy for power grids to comprehensively improve the efficiency and efficiency level [1]. At the same time, in the context of power system reform, the power grid Enterprises will face cash flow reduction, investment capacity weakened, and grid investment is subject to government control, so investment accuracy needs to be further improved [2]. The overhead transmission line project is a type of engineering with long space continuity. It is affected by factors such as topography, geology, meteorology and transportation conditions. The actual value between tower and tower is very different [3-4]. In the current industry standards, whether it is the "Power System Construction Budgeting and Calculation Regulations (2013 Edition)" or "DL/T5205-2011 Power Construction Engineering Engineering Quantity List Pricing Specification-Transmission Line", it cannot solve the single-base of overhead transmission line engineering[5]. The problem of distinguishing tower unit values. Therefore, how to determine the actual value of the tower unit and calculate the value of each base tower requires an urgent study of the pricing method in the single base tower mode.

2. Analysis of the status quo and problems of overhead line pricing

2.1. Existing pricing model for overhead lines

From the perspective of the formation of engineering cost,the overhead line engineering body is divided into basic engineering,pole tower engineering,grounding engineering,wiring engineering,accessory engineering and auxiliary engineering according to the unit project,based on the cost of the sub-project of each unit project. In combination with quotas,pre-regulations,standards,and policy documents,certain scientific pricing methods are used for pricing,and the total cost of overhead line projects is formed by combining the cumulative stacking of different cost types[6].

Combined with the actual situation of the current overhead line engineering pricing,the main factors that constitute the cost include "quantity","price" and "fee". That is,all the engineering quantities required for the completion of the project,the required quota system,the price of equipment materials,and the rates of fees for different regions and different conditions. Combined with the different stages of project cost,the pricing model is mainly adopted in the design stage (feasibility study estimation,preliminary design budget calculation,construction drawing budget stage),and the list pricing model is mainly adopted in the bidding stage and construction stage.

2.2. Analysis of the valuation of overhead line pricing

At present,the overhead line engineering pricing system relies on the 2013 version of the pre-regulation and quota system. With the passage of time,new technologies,new materials and new processes are continuously applied in the project,and the precision,refinement and precision of the cost control are continuously deepened. The applicability and accuracy of the overhead line engineering pricing system will gradually fail to meet the needs of the construction and management of overhead line projects in the new era.

(1) Applicability of pricing quota

With the continuous development of overhead line engineering construction technology,new technologies,new materials and new processes are continuously applied in overhead line engineering. The construction organization design,construction methods and construction techniques in the existing overhead line engineering pricing system are to some extent. It has been out of touch with the actual construction process,and the relatively lagging pricing system can no longer meet the needs of overhead line engineering cost management.

(2) Accuracy of the pricing system

The process of completion and settlement of overhead line project mainly depends on the bill of quantities calculation. According to the specific requirements of the contract,the unit implementation price is measured and valued by the unit price or total price model. The progress payment is paid according to the contract,and finally the completion settlement is carried out. The basis of the compilation of the engineering quantity list pricing and the construction drawing budget,the terrain ratio of the construction drawing budget,geological conditions,transportation distance and other parameters directly affect the comprehensive unit price of the project quantity list. Therefore,the pricing process in the settlement stage and the quality of the construction drawing budget are directly related. Controlling the basic parameters of the project is the key and key to control the settlement costs.

(3) Real-time performance of the pricing system

The overhead line engineering price mainly includes the fixed base price and the unit price of equipment materials. The current unit price of the device material is based on the information price of the State Grid. The price level has a certain lag with the actual market level. In some projects,the material cost is excessive. The pricing system cannot change in real time in response to market volatility.

3. Construction of single base tower pricing system

3.1. Single base tower pricing system construction framework

In this paper,the transportation parameters and terrain ratio parameters of overhead line engineering are taken as the entry point,and the transportation parameters and terrain ratio are transferred from the perspective of the whole project to the single base tower,the main engineering quantities such as excavation,foundation,group tower,grounding,insulation and metal string. The material fee and the installation fee can be attributed to the single-base tower naturally. The cost of the line and the cost of the fittings in the file can be collected and collected in a distributed manner,and the cost of the single base and the cost of the share are aggregated to form a "oriented "Innovation system for the pricing model of overhead transmission lines".

The innovative system of the overhead line engineering pricing model for "pole tower" can distinguish the cost difference between different towers of the whole project,better realize the collection of asset values,and better clarify the value difference between different towers,which is an overhead line. The innovative model of engineering pricing provides a basis for the precise management of overhead line engineering assets.

3.2. Construction of single base tower pricing system

The cost structure of the overhead line engineering pricing model for "pole tower" is basically the same as the cost structure of the traditional construction budget. The difference is that in the form of expression,the former is represented by a single base tower,which represents the cost of each base tower. Expressed as the cost of a single project.

In the project division,the basic engineering of overhead transmission line engineering,pole tower engineering,grounding engineering,accessory engineering insulator string,jumper string production and installation,heavy hammer,pressure equalizing ring,clamp installation,auxiliary engineering slope protection,retaining wall. The drainage ditch and the basic permanent cofferdam can calculate the body cost according to each base tower; the transportation cost and comprehensive terrain increase fee of these unit projects can also be calculated on a base-to-base basis. In the wire-laying project and the attachment project,the shock-proof hammer and the spacer bar calculate the single-base cost according to the apportionment method.

The evaluation layer of the overhead line engineering pricing model for "rocket tower" mainly specifies the calculation method of the cost. The research group proposes the method according to the cost collection,according to the actual calculation and the number of towers. The equipment material according to the actual calculation can be directly attributed to the single base tower and directly calculated according to the actual value of the equipment materials. The calculation of the number of towers is based on the fact that the equipment materials that cannot be directly attributed to the single base tower and their installation costs are evenly distributed by the number of towers. The weighted average calculation refers to the calculation of the collection of the costs that cannot be directly attributed to the single base tower according to the weight between the different towers in a weighted average manner.

3.3. Single base tower pricing application case

The line project of a certain place is newly built with a total length of 2.7kM (overhead 2.5kM,cable 0.2kM). The terrain ratio is 100% in the hills,and the geological division is 40% for rocks,50% for loose sands,and 10% for ordinary soils.

The construction simulation of the example project,the main parameters of the measured project are shown in the following table. The price of the quota and equipment materials used in the study is the same as that of the existing pricing system.

IOP Conf. Series: Materials Science and Engineering **729** (2020) 012042 doi:10.1088/1757-899X/729/1/012042

Table 1. sample data.

Serial number	Tower type	Human transport（km）	Car transport（km）	terrain
1	1GGD2_24	0.45	3	hills
2	1GGD2_33	0.4		hills
3	1D2-SJC2/23	0.4		hills
4	1D2-SJC2/20	0.5		hills
5	1D2-SJC3/24	0.24		hills
6	1D2-SJC4/18	0.35		hills
7	1D2-SJC4/22	0.3		hills
8	1D2-SJC4/26	0.15		hills

(1) Cost calculations that cannot be directly attributed to a single base tower

1) Wire-laying project: The cost of the overhead line project under the existing pricing system is 61,142.00 yuan.

2) Attachment project: For the cost of the installation of the anti-vibration hammer,refer to the 13 version of the fixed terrain increase factor table. The attachment project hill terrain increase factor is 5% and the mountain increase coefficient is 20%. The proportion of hilly terrain in the example project is 5%*6/(5%*6+2*20%)=43%,and the proportion of mountain topography is 2*20%/(5%*6+2*20%)= 57%.

(2) Cost calculation directly attributable to a single tower

The cost of calculating the 1# tower is shown in the table below.

Table 2. Cost of 1# tower.

Table 2. Cost of 1# tower

Serial number	Project or expense name	Calculation method	Amount of expenses (yuan)
1	Basic engineering	Calculated by calculation	22.0358
2	Tower engineering		14.8145
3	Grounding engineering		1.1725
5	Attachment engineering	Materials and fittings that can be directly attributed to a single base tower are calculated on a real basis	3.0119
		Materials that cannot be directly attributed to a single base tower are calculated according to the terrain of the tower	0.0831
6	Auxiliary engineering	Calculated by calculation	0.1354
	total		41.2532

Similarly,the single base cost of other towers can be calculated,as summarized in the table below.

Table 3. Budget investment under the innovative pricing model.

Serial number	Tower type	Amount of expenses (yuan)
1	1GGD2_24	41.2532
2	1GGD2_33	38.3352
3	1D2-SJC2/23	45.3531
4	1D2-SJC2/20	44.3221
5	1D2-SJC3/24	48.6553
6	1D2-SJC4/18	44.2312
7	1D2-SJC4/22	36.8432
8	1D2-SJC4/26	31.5835
total		330.5768

Compared with the existing pricing system,the accuracy of the ontology investment of the innovative pricing system is increased,and the accuracy ratio is increased by 4.06%.

4. Conclusion

Although the single-base tower pricing method will increase the working time of preparation,review and evaluation to a certain extent,with the improvement of the cost software and the skilled application of relevant personnel,the compilation efficiency can be further improved. Moreover,the depth of the single-base budget is more detailed and closer to the actual situation of the project,which helps to strengthen the cost management of infrastructure projects,standardize the management process,refine the management system,effectively control the project cost,improve the investment efficiency,and promote the scientific management of infrastructure projects. Standardization.

References

[1] Chen Libin. Cost Control of Power Transmission and Transformation Engineering [J]. China New Technology and New Products,2019(11): 122-123.

[2] Fu Meijing,Gu Yanxun,Fan Pengfei,Zeng Lingchi. Exploration of investment impact and investment management of power grid enterprises under the background of new power reform[J]. Times Finance,2016(36): 272-273.

[3] TAN Yongguo,GUAN Qianfeng,WAN Ke. Research on the valuation method of "single base tower" mode for overhead transmission lines [J]. China Power Enterprise Management,2018(15): 62-63.

[4] Wang Xin,Zhai Pengyun,Yuan Jingzhong. Research on Cost Estimation Model of Power Network Overhead Line Engineering[J]. China Power Enterprise Management,2018(24): 76-77.

[5] KONG Jun, CAO Xiaoyu, XIAO Feng.Study on Cost Prediction Model of Transmission and Distribution Engineering Based on Support Vector Machine[J].Modern Electronic Technique,2018,41(04):127-130.

[6] Geng Pengyun, An Lei, Wang Xin.Establishment and Implementation of Transmission Engineering Cost Prediction Model Based on Data Mining Technology[J].Modern Electronic Technique,2018,41(04):157-160.

IOP Conf. Series: Materials Science and Engineering **729** (2020) 012043 doi:10.1088/1757-899X/729/1/012043

Adsorption Properties of Sulfonylcalix[4] arenetetrasulfonate-loaded Resin towards Strontium(II) from Aqueous Solutions

Cong Liu*, Guofeng Su

Institute of Public Safety Research, Tsinghua University, Beijing, 100084, China

*Corresponding author's e-mail: liucong117@mail.tsinghua.edu.cn

Abstract. Sulfonylcalix[4]arenetetrasulfonate-loaded D201 composite (SCAS-D201) was prepared by a simple method and used as an adsorbent for removal of Sr(II) from aqueous solutions. The adsorption of the Sr(II) onto SCAS-D201 composite as a function of contact time, pH, Sr(II), temperature concentration was investigated. Results show that the adsorption capacities of SCAS-D201 composite towards Sr(II) reached a maximum at pH values of ~7. The adsorption has reached equilibrium within 120 min and the kinetics can be well fitted by a pseudo-second-model. The adsorption isotherm was well fitted by the Langmuir model, and the maximum value of adsorption capacity of the adsorbent was 76.43 $mg \cdot g^{-1}$. Possible adsorption mechanisms are also analysed.

1. Introduction

A large amount of radioactive wastes are generated each year from nuclear power plants, nuclear fuel reprocessing and radioisotopes production facilities due to extensive industrial activities. However, these radionuclides have versatile advanced industrial and medical applications if they are purified [1]. Thus, removal of radionuclides from aqueous solutions has become an important topic for sustainable development recently [2].

Various techniques to remove radionuclides are available, such as adsorption, solvent extraction, and ion exchange. Among those techniques, adsorption has been considered to be a favorable technique for its high efficiency, easy operation, and flexibility. It is imperative to have an advanced adsorbent which possesses high adsorption capacity, chemical stability and low cost for radionuclides treatment.

Calixarenes, third-generation supramolecular host compounds, have been regarded as molecular platforms for the construction of metal ion carriers [3, 4]. The thiacalix[4]arenes represent a class of receptor compounds. They were widely studied in the treatment of heavy metals [5, 6]. Hu et al. prepared thiacalix[4]arenetetrasulfonate (TCAS) modified 717 anion exchange resin and pointed out that the selective adsorption property of the adsorbent is attributed to complexation of the TCAS with heavy metals [7]. Also, sulfonylcalix[4]arene, as a thiacalix[4]arenes derivative, has been investigated to remove of radionuclides in recent years [8-9]. Morohashi et al. described that the excellent complexation ability of octylsulfonylcalix[4]arenes towards quite a few metals such as Sr(II) and Eu(III) [10].

The adsorption kinetics and isotherms of radionuclides adsorptions are particularly significant to the practical application. However, the data with respect to the adsorption of sulfonylcalix[4]arenes-loaded resin towards Sr(II) is barely reported. In order to obtain systematic information on the adsorption kinetics and thermodynamics, sulfonylcalix[4]arenetetrasulfonate-loaded D201 resin (SCAS-D201) was prepared by using a simple method and used as adsorbent for Sr(II) removal.

IOP Conf. Series: Materials Science and Engineering **729** (2020) 012043 doi:10.1088/1757-899X/729/1/012043

2. Experimental section

2.1 Reagents

The main materials are as follows: elemental sulfur, diphenyl ether, *p-tert*-butylphenol, sulfuric acid (98%), nitric acid, sodium hydroxide, chloroform, sodium chloride, ethanol, hydrogen peroxide (30%), trifluoroacetic acid, acetone, $Sr(NO_3)_2$. The above reagents used in the experiment were purchased from Klamar and Aladdin. The D201 resin was purchased from the Beijing Mancang technology Ltd.

2.2 Preparation of SCAS-D201 composite.

The TCA, TCAS, and SCAS were synthesized referring to the documents [11-13]. In the general preparation procedure of SCAS-D201, pretreated D201 resin (3 g) was immersed in a 50 mL of SCAS aqueous solution (14.42 $mmol \cdot L^{-1}$) with slowly stir for 24 h. Subsequently, the SCAS-D201 composite was filtered off and washed 3 times with ultrapure water. It dried in an oven at 45 °C for overnight. Fig. 1 shows the schematic of preparation of SCAS-D201.

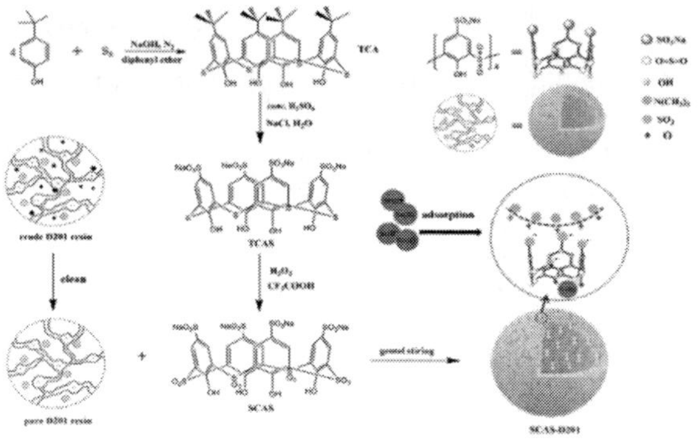

Figure 1. The schematic of preparation of SCAS-D201 composite and its adsorption of Sr(II).

2.3 Characterization of SCAS-D201 composite.

The SEM was observed by a Quanta 250 scanning electron microscopy instrument (FEI, USA) at 25 kV. The FTIR spectra for samples were recorded by a Nicolet iS10 infrared spectrometer (Thermo Fisher, USA) in the frequency range of 4000-400 cm^{-1}. The weighed quantities of materials were subjected to TGA on a TG/DTA6200 instrument (Seiko instruments inc, Japan) in nitrogen atmosphere at a heating rate of 10 $°C \cdot min^{-1}$ from 40 to 800 °C.

2.4 Batch adsorption experiments.

Generally, 20 mL aqueous phase and 10 mg SCAS-D201 composite were mixed in a lot of flasks and horizontally shaken at 150 rpm for 3 h maintained at temperatures between 25 and 40 °C unless otherwise stated. Then the liquid was filtered through 0.22 μm filter membranes. The concentration of the cations was measured using a MP-AES atomic emission spectrometer (Agilent, USA). Adsorption isotherms were determined by mixing 10 mg SCAS-D201 adsorbent with 20 mL of varying concentration solutions (10-150 $mg \cdot L^{-1}$). Adsorption percent (E %) and adsorption capacity (q_e, $mg \cdot g^{-1}$) was calculated according to the equations (1-2), respectively:

$$E = (C_0 - C_e) \times 100 / C_0 \quad (1)$$

$$q_e = (C_0 - C_e) V / m \quad (2)$$

where C_0 represents the initial concentrations of the cations in the flasks, C_e ($mg \cdot L^{-1}$) represents the equilibrium concentrations of the cations, V (L) is the volume of the solution, and m (g) is the mass of adsorbent.

3. Results and discussion

3.1 Characterization.

SEM images, which were shown in Fig. 2, provide the morphologies of D201 resin, SCAS, and SCAS-D201 composite. As demonstrated in Fig. 2a, the spherical D201 resin exhibited rich pores, which can afford abundant space for loading the molecular modifier. From Fig. 2b, SCAS showed irregular shape that can be ascribed to its growth characteristic. It is important to note that the appearance of the SCAS-D201 (Fig. 2c) remained in spherical shape which indicated that the original structure of D201 resin was retained before and after the impregnation synthesis except a few minor cracks appeared, which was due to the external forces effect on the D201 resin in the immobility process. Due to the impregnation of the SCAS molecules, the surface of the SCAS-D201 was covered by the white layer. (Fig. 2d).

The FTIR spectra of pure D201 resin, SCAS and SCAS-D201 composite are presented in Fig. 2e. In the FTIR spectrum of SCAS, the absorption bands at 1198 and 1044 cm^{-1} corresponding to the $-SO_3$ peaks, while the absorption bands at 1326 cm^{-1} corresponding to the asymmetrical sulfonyl stretching vibrations of SCAS. The IR spectrum of SCAS-D201 composite is similar to that of pure D201 resin. The obvious difference between the spectra is that new absorption peaks appeared near 1326, 1198 and 1044 cm^{-1}, which can also be observed in the spectrum of SCAS, confirming the immobilization of SCAS onto D201 resin.

Figure 2f reveals the TGA curves of blank D201 resin, SCAS and SCAS-D201 composite. For SCAS, there was 12 % weight loss within 40~100 °C due to the evaporation of water. From 325 to 800 °C, the main stage on thermal degradation curves was assigned to the decomposion of SCAS molecular skeleton. Over the range of 325 to 450 °C, the weight loss of 18.1 % for SCAS-D201 was greater than that of 16.6 % for D201 resin, which is attributed to the decomposion of the loaded SCAS of the SCAS-D201 composite. The weight loss rate of SCAS-D201 and SCAS was apparently greater than that of D201 resin between 450 and 800 °C, indicating SCAS was introduced into the D201 resin.

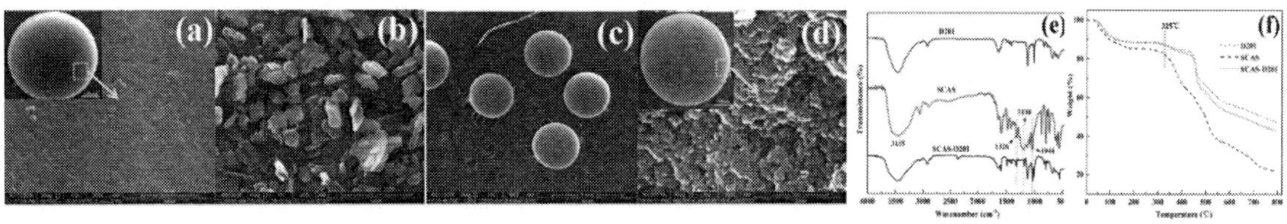

Fig. 2. SEM images of (a) D201 resin, (b) SCAS, (c) SCAS-D201 (×100) and (d) SCAS-D201 (×160000), (e) FTIR and (f) TGA of D201 resin, SCAS, and SCAS-D201 composite

3.2 Effect of pH.

As can be seen in Fig. 3a, the adsorption percentage of Sr(II) (E_{Sr}) via SCAS-D201 composite increased with increasing pH, and reached the maximum at approximately pH values of 7 for Sr(II). However, the E_{Sr} with D201 resin was less than 20% under the experimental conditions, indicating poor adsorption with D201 resin. It can be inferred that the observed adsorption ability is essential contributed to the loaded SCAS of the SCAS-D201 composite. The complex of loaded SCAS with Sr(II) can be explained by the Hard and Soft Acids and Bases (HSAB) theory [14]. The bridging sulfonyl group (O=S=O) of the SCAS categorized as borderline acid which is sensitive to Sr(II), leading to excellent adsorption ability of the SCAS-D201 composite for the selected metals. Since the adsorption of Sr(II) with SCAS-D201 composite is pH dependent, it also appears that the phenolic hydroxyl groups of the loaded SCAS had contribution to the complex of SCAS-D201 composite with Sr(II). It is reported that the complex of SCAS with metal ions is accomplished by coordination of the O=S=O group and phenolic oxygen [15]. It can be confirmed in the FTIR spectra of SCAS-D201 composite before and after the adsorption of Sr(II) (Fig. 3b). The SCAS-D201-Sr(II) complexes were obtained by air drying after the adsorption at pH 7.

IOP Conf. Series: Materials Science and Engineering **729** (2020) 012043 doi:10.1088/1757-899X/729/1/012043

As can be seen in Fig. 3b, the characteristic adsorptions at 3455 cm^{-1} in the FTIR spectrum of SCAS-D201 composite are attributed to OH stretching, while that of the SCAS-D201-Sr(II) complexes were observed at 3421 cm^{-1}. The OH stretching vibrations of the adsorbed species were shifted to lower frequencies, which indicated that -OH were coordinated with tested metals. Previously investigated complexes of *p-tert*-butylcalix[5]arene hexacarboxylic acid ligand with Cu(II) showing that C=O stretching vibration of ligand-Cu(II) changed positively and intensely [16]. In our experiment, the band due to the O=S=O stretching of SCAS-D201 at 1326 cm^{-1} shifted to 1329 cm^{-1} after adsorption of Sr(II). This higher frequencies compared to the free SCAS-D201 denoting the complexation of the O=S=O groups of SCAC with Sr(II). Obviously, the decreased relative intensities (green lines) of the asymmetrical sulfonyl vibration stretching to the $-SO_3$ peaks imply that the bridging O=S=O groups were coordinated with the metal ions as well. However, the sulfonate groups ($-SO_3$) of the SCAS are extensively regarded as poor ligands that can hardly participate in the metals adsorption [17]. Therefore, it is quite obvious that the typical coordination pattern of SCAS-D201 composite to the Sr(II) is mainly depend on the O=S=O groups and hydroxy groups of the loaded SCAS.

Thus, the changes of E_{Sr} with increasing pH can be explained by the fact that the ionization of -OH of SCAS was inhibited at lower pH value, which is adverse to the coordination of O=S=O group and phenolic oxygen, resulting in poor adsorption of the SCAS-D201 composite towards radionuclides in strong acidic solution.

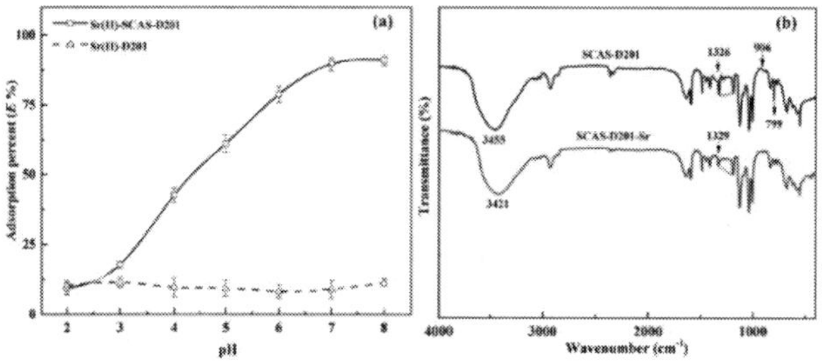

Fig. 3. (a) Effect of pH on the adsorption of the Sr(II) onto SCAS-D201 and D201. (b) FTIR spectra of the SCAS-D201 composite before and after the adsorption of Sr(II) Conditions: T = 298 K, m/V = 0.5 $g \cdot L^{-1}$, [Metal] = 10 $mg \cdot L^{-1}$, contact time = 4 h, shaking speed: 150 rpm.

3.3 Asorption Kinetics.

The adsorption kinetics were studied at different contact time and temperature. As can be seen in Fig. 4, the adsorption capacity increased rapidly in the first 30 min, which can be attributed to the effective complexation of the loaded SCAS with Sr(II). Then, it reached equilibrium at approximately 65 min for. However, it needs less time to reach equilibrium with the increase of temperature, which can be explained by the phenomenon that increasing temperature leads to an improved diffusion rate of ions across the external boundary layer.

In order to clarify the kinetic mechanism on the adsorption of SCAS-D201 composite towards Sr(II), the obtained data was fit to the pseudo-first-order, pseudo-second-order and intraparticle diffusion models to evaluate the time-dependent adsorption process. The pseudo-first-order, pseudo-second-order *and* nd intraparticle diffusion rate expressions are given in equations (3-5), respectively:

$$\ln(q_e \quad q_t) = \ln q_e \quad k_1 \tag{3}$$

$$t \quad q_t = 1 \quad (k_2 q_e^2) \quad t \quad q_e \tag{4}$$

$$q_t = k_i t^{0.5} \quad C \tag{5}$$

where q_e is the adsorption capacities at equilibrium time, q_t ($mg \cdot g^{-1}$) is the adsorption capacities at given time t, k_1 (min^{-1}) is the rate constant of the pseudo-first-order adsorption, k_2 ($g \cdot mg^{-1} \cdot min^{-1}$) is the rate constant of the pseudo-second-order adsorption, k_i ($mg \cdot g^{-1} \cdot min^{0.5}$) is the rate constant of the intraparticle diffusion model adsorption, and C ($mg \cdot g^{-1}$) is the constant proportional to the boundary layer thickness.

Plots of these models for the adsorption of SCAS-D201 composite towards Sr(II) are shown in Fig.5. The relative parameters are summarized in Table 1. The results illustrate that the adsorption kinetics of Sr(II) are fitted better by the pseudo-second-order kinetic model ($R^2 > 0.99$) than the pseudo-first-order model. The pseudo-second-order model is based on the assumption that the chemisorption between binding sites of the adsorbent and metal ions is the rate-limiting step of the adsorption [18]. Thus, it can be concluded that the chemical interaction between SCAS-D201 composite and tested ions is the primary adsorption manner in the adsorption process.

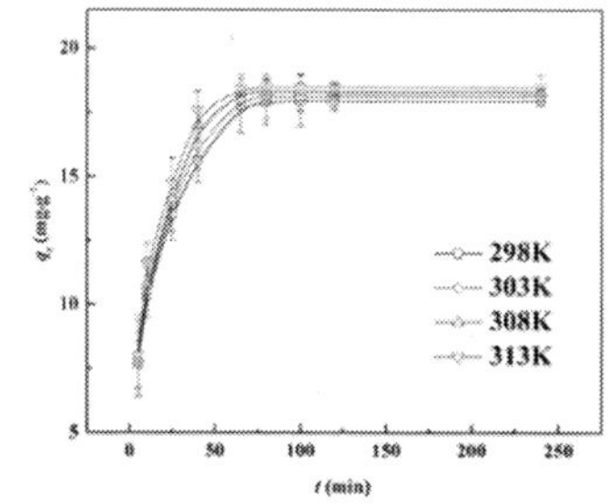

Fig. 4. Adsorption kinetics of Sr(II). Conditions: m/ V = 0.5 $g \cdot L^{-1}$, [Metal] = 10 $mg \cdot L^{-1}$, pH = 7, shaking speed: 150 rpm.

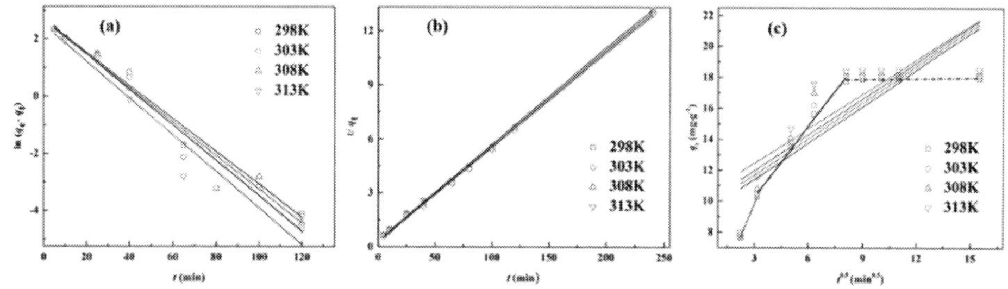

Fig. 5. plots of (a) pseudo-first-order, (b) pseudo-second-order, and (c) intraparticle diffusion kinetics models for Sr(II). Conditions: m/ V = 0.5 $g \cdot L^{-1}$, [Metal] = 10 $mg \cdot L^{-1}$, pH = 7, shaking speed: 150 rpm.

Table 1. Parameters Obtained from Adsorption Kinetics Studies

model	parameter	Absolute temperature (K)			
		298	303	308	313
pseudo-first-order	q_e($mg \cdot g^{-1}$)	15.14	14.17	14.54	12.23
	k_1(min^{-1})	6.19×10^{-2}	5.90×10^{-2}	5.75×10^{-2}	6.07×10^{-2}
	R^2	0.94271	0.96740	0.99449	0.94247
pseudo-second-order	q_e($mg \cdot g^{-1}$)	18.64	18.80	18.93	19.04
	k_2 ($g \cdot mg^{-1} \cdot min^{-1}$)	8.49×10^{-3}	9.03×10^{-3}	9.79×10^{-3}	11.06×10^{-3}
	R^2	0.99862	0.99876	0.99779	0.99889
Intraparticle diffusion	C($mg \cdot g^{-1}$)	9.06	9.36	9.67	10.24
	k_i ($mg \cdot g^{-1} \cdot min^{0.5}$)	7.84×10^{-1}	7.79×10^{-1}	7.72×10^{-1}	7.41×10^{-1}
	R^2	0.66589	0.7065	0.71562	0.64656

Conditions: m/V = 0.5 $g \cdot L^{-1}$, [Metal] = 10 $mg \cdot L^{-1}$, pH = 7, shaking speed: 150 rpm.

3.4 *Adsorption Isotherm Models.*

The adsorption isotherms were studied by changing the initial concentrations of the metal ions at pH 7, for an adsorbent dosage of 0.5 $g \cdot L^{-1}$, a contact time 3 h, and a temperature of 298 K. As presented in Table 2, the adsorption capacity increased with increasing initial ion concentration, which is due to the higher cations concentration, the more active adsorption sites of SCAS-D201 composite involved, and thus the larger the adsorption capacity. Although the adsorption capacity increased with increasing initial cations concentration, the ions removal demonstrated a downward trend. The distribution coefficient, K_d ($K_d = q_e/ C_e$, mL·g^{-1}), is a significant parameter to evaluate the adsorption performance. In this experiment, K_d is greater than 1×10^4 mL·g^{-1} in low initial ions concentration ([Sr(II)] < 20 mg·L^{-1}), indicating that the SCAS-D201 composite is an excellent adsorbent [19].

To understand the mode of interaction between tested cations and SCAS-D201 composite, the adsorption equilibrium was also investigated using Langmuir and Freundlich isotherm models, which can be defined by equations (6-7) in their respective order:

$$C_e \quad q_e = C_e \quad q_m \quad 1 \quad (q_m k_L) \tag{6}$$

$$\ln q_e = n^{\ 1} \ln C_e \quad \ln k_F \tag{7}$$

where C_e (mg·L^{-1}) is the final equilibrium concentration, q_e (mg·g^{-1}) and q_m (mg·g^{-1}) are the adsorption capacity at equilibrium and the maximum adsorption capacity at monolayer coverage, respectively. K_L (L·mg^{-1}) is the Langmuir constant, K_F (mg·g^{-1}) is the Freundlich constant, n is related to energetic heterogeneity.

The fitted curves are presented in Fig.6, and the relative parameters q_m, k_L,and k_F can be determined from the slope and intercept. The values of these relative parameters are listed in Table 3. The correlation coefficients clearly indicate that the Langmuir model (R^2 > 0.99) is appropriate for representing the adsorption process. The adsorption capacity q_m (76.43 mg·g^{-1}) is consistent with the experimental saturation amount q_e (76.36 mg·g^{-1}), indicating that the porous surface of SCAS-D201 composite is homogeneous and that the adsorption can be described as monolayer adsorption.

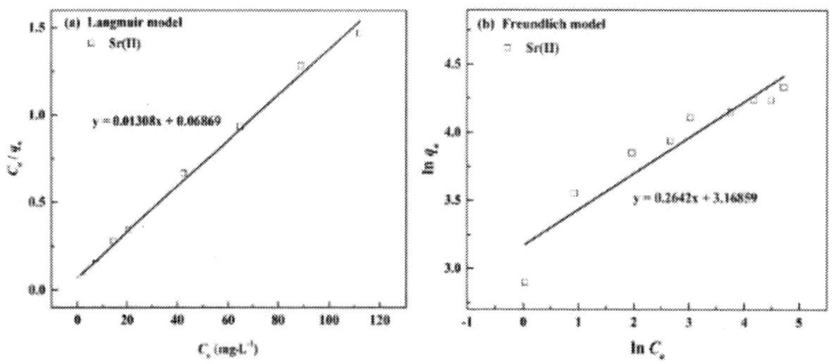

Fig. 6. Adsorption isotherms and plots of (a) Langmuir and (b) Freundlich isotherm models for Sr(II) . Conditions: T = 298 K, m/V = 0.5 g·L^{-1}, contact time = 3 h, pH = 7, shaking speed: 150 rpm.

Table 2. Effect of Initial Ion Concentration on Adsorption of Sr(II) by SCAS-D201

C_0 (mg·L^{-1})	C_e (mg·L^{-1})	q_e (mg·g^{-1})	E (%)	K_d (mL·g^{-1})	RSD (%)
10.11	1.03	18.16	89.81	1.76×10^4	1.16
19.98	2.53	34.9	87.34	1.38×10^4	0.89
30.62	7.12	47	76.75	6.6×10^3	1.6
40.03	14.36	51.34	64.13	3.58×10^3	0.37
51.12	20.65	60.94	59.6	2.95×10^3	1.36
74.25	42.47	63.56	42.8	1.5×10^3	2.04
99.87	65.1	69.54	34.82	1.07×10^3	0.53
123.66	88.96	69.4	28.06	0.78×10^3	1.02
150.37	112.19	76.36	25.39	0.68×10^3	1.43

Conditions: T = 298 K, m/V = 0.5 g·L^{-1}, contact time = 3 h, pH = 7, shaking speed: 150 rpm

Table 3. Langmuir, Freundlich, Temkin, and D-R Model Fitting Parameters for Adsorption on SCAS-D201 composite.

model	parameter		
Langmuir	q_m (mg·g^{-1})	k_L (L·mg^{-1})	R^2
	76.43	0.19	0.99435
Freundlich	k_F (g·mg^{-1})	n	R^2
	23.77	3.79	0.88407

Conditions: T = 298 K, m/V = 0.5 g·L^{-1}, contact time = 3 h, pH = 7, shaking speed: 150 rpm.

4. Conclusions

In this work, SCAS-D201 was prepared by using a simple method. Characterization results confirmed that SCAS was immobilized onto the D201 resin successfully. The SCAS-D201 composite showed excellent Sr(II) removal capability under neutral conditions. The adsorption capacities of SCAS-D201 composite reached maximum at pH values of ~7 for Sr(II). The adsorption capacities increased rapidly with time, and the equilibrium was reached within 2 h. The equilibrium adsorption data fit the Langmuir isotherm model best, and the maximum adsorption capacity of the SCAS-D201 composite for Sr(II) was 76.43 mg·g^{-1}. The adsorption kinetics of Sr(II) are fit to the pseudo-second-order model. This research exhibits potential of the SCAS-D201 composite to be a superior adsorbent for removing radionuclides from aqueous solutions.

Acknowledgments

This work was supported by the National Key Research and Development Plan (2017YFC0704100).

References

[1] G. Naidu, T. Nur, P. Loganathan, J. Kandasamy, S. Vigneswaran, Separation & Purification Technology, **163**, 238-246(2016)
[2] S. Wang, P. Yu, B.A. Purse, M.J. Orta, J. Diwu, W.H. Casey, B.L. Phillips, E.V. Alekseev, W. Depmeier, D.T. Hobbs, T.E. Albrecht-Schmitt, Advanced Functional Materials, **22**, 2241-2250(2016)
[3] D.S. Guo, Y. Liu, Chemical Society Reviews, **41**, 5907-5921(2012)
[4] J.P. Chinta, B. Ramanujam, C.P. Rao, Coordination Chemistry Reviews, **256**, 2762-2794(2012)
[5] N. Morohashi, F. Narumi, N. Iki, T. Hattori, S. Miyano, Chemical Reviews, **106**, 5291-5316(2006)
[6] D.S. Guo, K. Wang, Y. Liu, Journal of Inclusion Phenomena & Macrocyclic Chemistry, **62**, 1-21(2008)
[7] X. Hu, Y. Li, W. Yan, X. Li, H. Li, L. Xiao, P. Zhang, Desalination, **259**, 76-83(2010)
[8] T. Kikuchi, M. Nogami, K. Suzuki, Journal of Alloys & Compounds, **374**, 272-276(2004)
[9] T. Kikuchi, K. Suzuki, Journal of Alloys & Compounds, **408**, 1287-1290(2006)
[10] N. Morohashi, N. Iki, A. Sugawara, S. Miyano, Tetrahedron, **57**, 5557-5563(2001)
[11] H. Kumagai, M. Hasegawa, S. Miyanari, Y. Sugawa, Y. Sato, T. Hori, S. Ueda, H. Kamiyama, S. Miyano, Tetrahedron Letters, **38**, 3971-3972(1997)
[12] N. Iki, T. Fujimoto, S. Miyano, Chemistry Letters, **27**, 625-626(1998)
[13] N. Iki, T. Horiuchi, H. Oka, K. Koyama, N. Morohashi, C. Kabuto, S. Miyano, Journal of the Chemical Society Perkin Transactions, **11**, 2219-2225(2001)
[14] R.G. Pearson, Journal of Chemical Education, **45**, 581(1968)
[15] B.B. Adhikari, M. Gurung, H. Kawakita, K. Ohto, Analyst, **136**, 3758-3769(2011)
[16] A.P. Côté, G.K.H. Shimizu, Coordination Chemistry Reviews, **245**, 49-64(2003)
[17] H.A. Kooijman, Industrial & Engineering Chemistry Research, **41**, 3326-3328(2002)
[18] Y. Ge, D. Xiao, Z. Li, X. Cui, Journal of Materials Chemistry A, **2**, 2136-2145(2014)

IOP Conf. Series: Materials Science and Engineering **729** (2020) 012043 doi:10.1088/1757-899X/729/1/012043

[19] S. Ma, L. Huang, L. Ma, Y. Shim, S.M. Islam, P. Wang, L.-D. Zhao, S. Wang, G. Sun, X. Yang, M.G. Kanatzidis, Journal of the American Chemical Society, **137**, 3670-3677(2015)

IOP Conf. Series: Materials Science and Engineering **729** (2020) 012044 doi:10.1088/1757-899X/729/1/012044

A Review on the Molecular Mechanism, Superiorities, Applications, Limitations and Experimental Workflow of CRISPR/Cas-9 System, and the Future of Gene Engineering

Xiaoke Wang

Supervisor: Lingfang Tang

The Affiliated High School of Peking University

Abstract: Gene engineering has been in development since the 1970s. The appearance of CRISPR/Cas-9 system, a gene engineering technology, even brought the wave of developing to its unprecedented peak. Meanwhile, the drastic growth and maturity of CRISPR made the public, represented by popular presses, to question the integrity and rightfulness of scientists to develop and apply CRISPR system, causing intense worldwide ethical battles of humanity. However, without relative knowledge background, rational and logical judgment can be hard to make. Using qualitative research techniques, this research has thoroughly reviewed previous literatures that introduced the molecular mechanism, superiorities over other technologies, applications and limitations of CRISPR/Cas-9 system in order to give comprehensible introduction for the public. Using quantitative research methods, this research investigated the attitudes came from the public and science community toward the ethical issues around the use of gene engineering tools such as CRISPR in the form of numerical comparisons. More than 50% of people stand neutral in this battle, but there were more supporters of developing gene engineering technologies in the science community than that of in the public presses, which produces a necessity for countries to call for a pause of current developments, and to assemble national leaders to have a serious discussion.

1. Introduction

Have you ever thought about designing your own babies with enhanced intelligence or strength? Gene engineering has the potential to allow us to make it happen someday in the not-distant future. However, gene engineering has been a hot, controversial topic in recent years, gaining supportive and contradictory forces' attention from the globe. Last year, a Chinese scientist named He Jiankui announced that he has successfully edited the genomes of two human embryos using a gene engineering technology - CRISPR/Cas-9 system. These two babies, with genes modified, were given with the ability of preventing HIV infection. World-wide debates around the rightfulness of editing human-embryo genomes and of employing gene engineering technologies were initiated ever since. Mainstream scientists doubt the safety of the clinical use of this technology on human embryos since it is different from editing genes of adults, as the traits being modified will be inheritable, therefore influencing the gene pool globally. While other scientists supported it because of the significant, profound leap of science and ethic it leads to. Regardless the morality of such attempts, the general public are lacking related knowledge about what CRISPR is and what CRISPR can do. Only armed with knowledge, the judgments can be made rationally and logically, therefore leading us to make the right decision on the use and development of gene engineering. This research is divided into mainly two parts; to provide the necessary informative background of CRISPR, the first part of research is a thorough review of previous literature on the molecular mechanism, superiorities, applications, limitations and methodological workflow of the CRISPR/Cas-9 system. To trigger the audiences to ponder over the gene engineering issue, the second half of this research reevaluates the use of gene engineering technologies such as CRISPR by reviewing the articles published by popular presses and the literatures written by scientists in chronological order, so to predict the future of gene engineering.

2. Applications of CRISPR

CRISPR, as a newly-emerged bioengineering technology, has tremendously great potential in many areas of study and extensive applications, which include but not limited to: medical applications, agricultural and environmental applications. This section will present how the world is a better place with CRISPR by listing out various functions derived from its genome engineer nature of CRISPR.

2.1 Medical Applications

2.1.1 Genome and base editing

CRISPR provides scientists the ability to do certain editing on a base or a particular locus on the genome with incredible precision, which could be applied on massive production of a desired protein or hormone, corrections of gene mutations and treatments to genetic diseases.

2.1.2 Genomic-scale screening

High revolutionized screening approached are able to be conducted by using the RNA interference (RNAi) pathway, whereas, interestingly, nowadays some researches have shown that CRISPR system could accomplish the particular same goal with higher adaptability. Traditionally, CRISPR/Cas9 screens have been used to study intracellular phenotypes by combining with positive, negative or marker/reporter gene selection.

2.1.3 Cell therapy

The immune cell therapy and stem cell therapy can be realized by CRISPR/Cas-9 system. For example, in a research that studies the repression and activation of bacterial gene expression, patients with cancer or autoimmune diseases were treated with CRISPR-engineered T-cells with PD-1 gene-knockout. These clinical trials of CRISPR system have been validated for approved treatment of muscle-invasive bladder cancer, castration-resistant prostate cancer, metastatic renal cancer and metastatic non-small cell lung cancer.

2.1.4 Modification of human embryos

Before one is born, using CRISPR technology, it became possible for heritable disease-carrying parents to correct the pathogenic gene mutation and prevent the progression of the disease right from the beginning. In 2017, a Chinese research group consisted of Ma et al. claimed that they have corrected the heterozygous MYBPC3 mutation in the genes of a human embryo before the embryonic implantation happened. Notably, embryonic genome editing with a base editor showed higher efficiency: targeted deep sequencing on injected embryos revealed that 17 out of 17 (100%) or 6 out of 9 (67%) embryos carried the targeted point mutations at the target site in FANCF or DNMT3B gene, respectively. Despite these achievements have raised much more attention on the ethical implications, they did support that CRISPR system is remarkably beneficial for the whole population as long as it is used with a kind initiative.

2.2 Agriculture applications

CRISPR/Cas-9 technology can be applied to the production of trans-genetic rice with higher yield production and higher quality. Recent breakthroughs are around the generation of climate-change-resistant crops and with higher nutrient values.

2.3 Environmental applications

Along with understand CRISPR/Cas-9 technologies deeper and more thorough, our understanding of microbes and its interactions with phages has improved as well. In 2018, a group of researchers comprehensively collected information about CRISPR/Cas systems and anti-CRISPR systems to construct a CRISPRminer web server, seeking to accelerate their understanding of the co-evolutionary relationship between the microbes and phages. This unprecedented work facilitated us to explore more about the unknowns with, unfortunately, limited knowledge, but the return of it is what we are really looking forward to.

3. Experimental workflow of CRISPR

The CRISPR system is capitalized in labs for various purposes but with similar core methodologies. Here I present a generalized step of workflow of employing CRISPR/Cas-9 technology in the labs: 1) Target selection for sgRNA and design of reagents. 2) Construction of reagents. 3) Delivery of sgRNA and Cas Enzyme to the required cells. 4) Design of repair template (optional). 5) Clonal isolation and expansion modified cell lines. 6) Functional testing and validation. 7) Clonal expansion of desired cells.

3.1Method

Reviewing previous literatures, the standpoints of scientific authors on the development of CRISPR technology can be reasonably inferred from the discussion and acknowledgement sections. Similarly, the attitudes from the public, represented by public presses, can also be interpreted by analyzing the published articles. I reviewed 60 literatures investigating CRISPR that were submitted on Nature by scientific authors, and 60 articles published by popular presses in both mainland China and also other countries. Each standpoint was recorded, and all of the reviewed documents were published after 2015 and selected randomly, meaning that the numbers collected are able to represent the latest and most authoritative voices. In doing so, the total number of supporters, the neutral's, and the opponents in these 120 documents were added up, presented in a bar graph, showing a broad statistical difference.

4. Data and data analyzation

4.1 Data

As shown in the figure above, among the 60 scientific authors, there were 18 supporters and 7 opponents. Most researchers (35) stood neutral regarding the issue of developing CRISPR technologies. While in the 60 presses who represented the voice of public, there were only 7 of them who stood supportive. However, the number of denying CRISPR technology development was significantly higher than that of the scientists, which was up to 19. The rest of them (34) took stances at neutral in this specific problem.

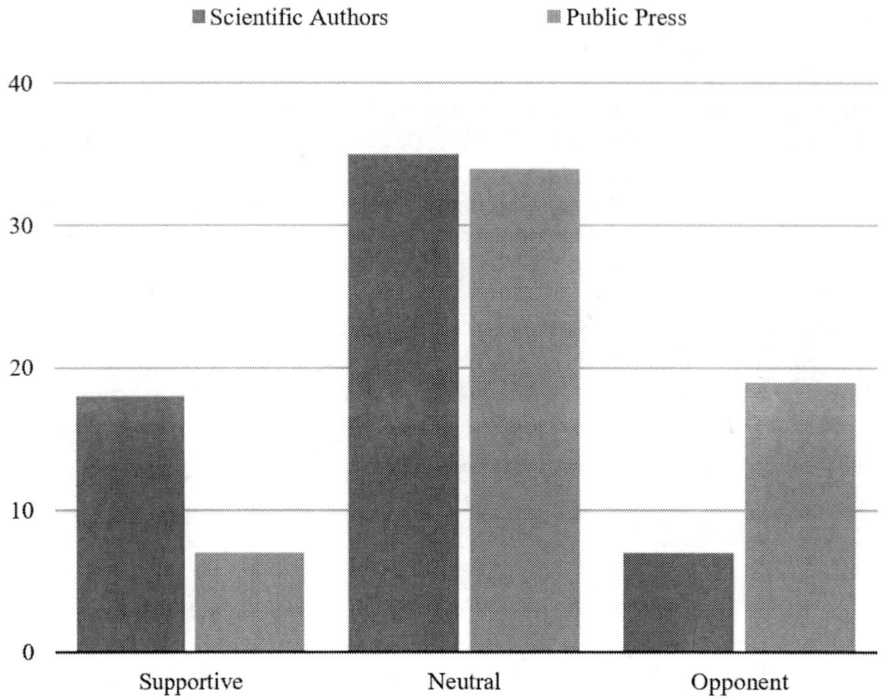

Figure 1. Numbers of supportive, neural and opponent stands in scientific authors and public.

4.2 Data analysis

One of the most probable reasons for why there were more supporters of CRISPR than that in the public is that, compared with the public, scientists have richer knowledge database and deeper understanding of what CRISPR is capable of bringing us in the very non-distant future. For instance, various possible clinical, pharmaceutical uses derived from developing CRISPR technology. For the same reason, the public tends to not fully understand the real power of CRISPR and not sure about its future directions. Such concerns may make them to reject CRISPR, and to desire for a cease in unregulated developments, which explains why there was a higher count of people who against CRISPR. There was a similar share of people who took neutral stands in both communities. That can be caused by the profound understanding of both the advantages and disadvantages of CRISPR, since CRISPR could either lead the world into a better, bigger picture, or a worse, uncontrolled place.

4.3 Discussions

One of the reasons so many scientists are obsessed with CRISPR is that, it not only allows them to correct the mutations that cause genetic diseases, but it also has the potential to develop many beneficial applications for the society, for example, high-yield rice production. Although ethical issues are sometimes raised along with its prevalence, we have to admit that we cannot truly keep a powerful tool like CRISPR under-developed. This research eventually aims to call for a pause in the unregulated developments of gene engineering technologies, so that the public and the science community could have a serious discussion on its future directions. Here are some of the most urgently important topics of the debate around CRISPR: 1) Safety. We need to ensure the safety of gene-editing, especially if it were to be used on human embryos. 2) Prioritization. The use of gene engineering based on technologies like CRISPR should be limited to the cases where there are no alternatives, or the abuse of power in any form could be dangerous. 3) Access. The right to have access to CRISPR should be available for everyone, but not only the wealthy. However, to ensure this could happen really depends on much more complicated issues that yet to be solved around the globe. 4) Engagement. It is incredibly important that citizens everywhere have a voice in deciding the use of powerful technologies like CRISPR, concerning their own welfares where other might not relate, and making a careful, thorough decision. Hopefully, a compromise could be achieved.

One of the founders of CRISPR system, Jennifer Doudna, thinks that it would be a mistake for us to keep these technologies under-developed only because we are uncertain about and fear for what it might bring us. For following researchers, regardless of expertise, should give more serious thinking about the morality of using technologies like CRISPR, and also how could we prevent the possible negative consequences as a result of unregulated development of these life-changing technologies.

5. Conclusion

The CRISPR/Cas-9 technology is based on a bacterial immune system that helps bacterium to fight viral infections. It uses a programmable enzyme called Cas-9 that can be programed with little bits of RNA with particular sequences to locate and cut cleavages on the DNA strand. This double-strand break (DSB) triggers the cell to repair the cut, and, in the process, introduce a change in DNA precisely. Through analytical reviewing, I found that there were more supporters of developing gene engineering technologies in the science community than that of in public presses, and there were less opponents among the scientists than that in the presses. Most people stand neutral in this battle.

Acknowledgments

This research did not display how have the attitudes changed in the science community and the public over time, and did not review the documents in the chronological order of publish dates, which could be one of the topics of future analyzations. The interpretation of standpoints came from each document could be subjective to some levels, as I did not have access to actually communicate with all the authors. But I promise that I was being as objective as I could, and I only reviewed those articles that expressed their opinions relatively clear. I gratefully thank Dr. Tang who provided me professional instructions and advices, and also the technical access offered by Peking University Biology Lab. I give my heartfelt thanks to all the ones who have helped me during the whole research process in any form.

References

[1] Marilynn Marchione (2018), Chinese researcher claims first gene-edited babies. Retrieved From https://www.apnews.com/4997bb7aa36c45449b488e19ac83e86d
[2] Cui Zhang, Renfu Quan and Jinfu Wang. (2018). Development and Application of CRISPR/Cas9 Technologies in Genomic Editing. Human Molecular Genetics, Vol. 27, No. R2, R79–R88.
[3] Bikard, D., Jiang, W., Samai, P., Hochschild, A., Zhang, F. and Marraffini, L.A. (2013) Programmable repression and activation of bacterial gene expression using an engineered CRISPR-Cas system. Nucleic Acids Research, 41, 7429–7437.
[4] Ma, H., Marti-Gutierrez, N., Park, S.W., Wu, J., Lee, Y., Suzuki, K., Koski, A., Ji, D., Hayama, T., Ahmed, R. et al. (2017) Correction of a pathogenic gene mutation in human embryos. Nature, 548, 413–419.
[5] Fan Zhang et al. (2018). CRISPR-miner is a knowledge base for exploring CRISPR-Cas systems in microbe and phage interactions. Nature: Communications Biology. Vol.1, Article number: 180.
[6] Jennifer Doudna (2015), How CRISPR lets us edit our DNA [Video File]. Retrieved from https://www.ted.com/talks/jennifer_doudna_we_can_now_edit_our_dna_but_let_s_do_it_wisely
[7] Amitai G, Sorek R. (2016) CRISPR-Cas Adaptation: Insights into the Mechanism of Action. *Nat. Rev. Microbiol.* 14(2):67–76
[8] Barrangou R, Doudna JA. (2016) Applications of CRISPR Technologies in Research and Beyond. *Nat. Biotechnol.* 34(9):933–41
[9] Charpentier E, Marraffini LA. (2014) Harnessing CRISPR-Cas9 Immunity for Genetic Engineering. *Curr. Opin. Microbiol.* 19:114–19

[10] Doudna JA, Gersbach CA. (2015) Genome Editing: The End of the Beginning. *Genome Biol.* 16(1):292
[11] Esvelt KM, Mali P, Braff JL, Moosburner M, Yaung SJ, Church GM. (2013) Orthogonal Cas9 Proteins for RNA-Guided Gene Regulation and Editing. *Nat. Methods* 10(11):1116–21
[12] Fu Y, Foden JA, Khayter C, Maeder ML, Reyon D, et al. (2013) High-Frequency Off-Target Mutagenesis Induced by CRISPR-Cas Nucleases in Human Cells. *Nat. Biotechnol.* 31(9):822–26
[13] Heidenreich M, Zhang F. (2016) Applications of CRISPR Cas Systems in Neuroscience. *Nat. Rev. Neurosci.* 17(1):36–44
[14] Hou Z, Zhang Y, Propson NE, Howden SE, Chu L-F, et al. (2013) Efficient Genome Engineering in Human Pluripotent Stem Cells using Cas9 from *Neisseria meningitidis. PNAS* 110(39):15644–49
[15] Isaac RS, Jiang F, Doudna JA, Lim WA, Narlikar GJ, Almeida R. (2016) Nucleosome Breathing and Remodeling Constrain CRISPR-Cas9 Function. *eLife* 5: e13450
[16] Jinek M, Chylinski K, Fonfara I, Hauer M, Doudna JA, Charpentier E. (2012) A Programmable Dual-RNA-Guided DNA Endonuclease in Adaptive Bacterial Immunity. *Science* 337(6096):816–21
[17] Kleinstiver BP, Prew MS, Tsai SQ, Topkar VV, Nguyen NT, et al. (2015) Engineered CRISPR-Cas9 Nucleases with Altered PAM Specificities. *Nature* 523(7561):481–85
[18] Leenay RT, Maksimchuk KR, Slotkowski RA, Agrawal RN, Gomaa AA, et al. (2016) Identifying and Visualizing Functional PAM Diversity across CRISPR-Cas Systems. *Mol. Cell* 62(1):137–47
[19] Mulepati S, Héroux A, Bailey S. (2014) Structural biology. Crystal structure of a CRISPR RNA–Guided Surveillance Complex Bound to a ssDNA Target. *Science* 345(6203):1479–84
[20] Ran FA, Cong L, Yan WX, Scott DA, Gootenberg JS, et al. (2015) *In vivo* Genome Editing Using *Staphylococcus aureus* Cas9. *Nature* 520(7546):186–91
[21] Strong A, Musunuru K. (2017) Genome Editing in Cardiovascular Diseases. *Nat. Rev. Cardiol.* 14(1):11-20
[22] Tsai SQ, Joung JK. (2016) Defining and improving the genome-wide specificities of CRISPR–Cas9 nucleases. *Nat. Rev. Genet.* 17(5):300–12
[23] Vassylyev DG, Sekine S-i, Laptenko O, Lee J, Vassylyeva MN, et al. (2002) Crystal Structure of a Bacterial RNA Polymerase Holoenzyme at 2.6 Å Resolution. *Nature* 417(6890):712–19
[24] Wang H, La Russa M, Qi LS. (2016) CRISPR/Cas9 in Genome Editing and Beyond. *Annu. Rev. Biochem.* 85:227–64
[25] Xiong X, Chen M, Lim WA, Zhao D, Qi LS. (2016) CRISPR/Cas9 for Human Genome Engineering and Disease Research. *Annu. Rev. Genom. Hum. Genet.* 17(1):131–54
[26] Zhao H, Sheng G, Wang J, Wang M, Bunkoczi G, et al. (2014) Crystal Structure of the RNA-Guided Immune Surveillance Cascade Complex in *Escherichia coli. Nature* 515(7525):147–50

IOP Conf. Series: Materials Science and Engineering **729** (2020) 012045 doi:10.1088/1757-899X/729/1/012045

Preparation of Rich Hydrocarbon Fuels by Cornus Wisoniana Crude Oil Catalytic Coupling

Pengying Lai[1, 2], Jilie Li[1], Zhihong Xiao[2] and Yifu Miu[2], Aihua Zhang[1, 2*]

[1]Central South University of Forestry and Technology, Changsha, Hunan, 410004, China

[2]Hunan Academy of Forestry, Changsha, Hunan, 410004, China

*Corresponding author' s e-mail: zhangaihua909@163. com

Abstract. Cornus wisonian is a woody energy plant unique to the southern region, the fruit oil can be used as the important raw materials in catalytic cracking. This experiment studied the catalytic cracking decarboxylation reaction of Cornus wisoniana oil. Through experiments, CaO / KF made by ourselves was proved to have good catalytic effect, gas production rate is lower than 12.0%, the by-product of the bottom of the kettle residue is controled under 5.0%, the yield of liquid fuel oil products is at 83.0% or more. Because of high acid value, pyrolysis oil can't be directly burned. In this study, p-toluenesulfonic acid was used as a catalyst for alcohol esterification. Under the Optimization condition, acid value is lower than 1.0 mg KOH/g. Then alcohol esterification was used to reducing acid of cracking bio-fuel products and p-Methylbenzene sulfonic acid as catalyst. After reducing the acid, this kind of fuel is closer than traditional fossil fuels, able to alternative to diesel, and to achieve energy conservation.

1. Introduction

Cornus wisoniana is an important industrial oil energy tree species, mainly distributed in the south of the Yangtze River[1-4]. At present, the Cornus wisoniana oil is mainly converted into fatty acid methyl ester (biodiesel) by acid, alkali or enzyme catalyzed transesterification reaction of alcohol oil to prepare bioenergy. However, the production process of methyl ester biodiesel is complicated and its application range is limited[5]. Post-processing pollution emissions are difficult to solve and other issues restrict their further development.

The catalytic cracking process completely changes the physical and chemical properties from the molecular structure, and the obtained fuel oil has high calorific value, and the properties and composition are closer to the petrochemical oil[6-8]. The high-temperature cracking of the Cornus wisoniana oil to prepare the hydrocarbon-rich fuel is convenient and its requirements are simple. However, the bio-based hydrocarbon-rich fuel has a higher acid value, which reduces the stability of the fuel oil and even causes corrosion of the equipment, so it is necessary to carry out deacidification treatment[9]. In this study, the self-made refined Cornus wisoniana oil was used as raw material to study the catalytic cracking coupling esterification reaction process of woody oil, aiming to improve the quality of fuel oil. The effects of catalytic cracking parameters on the components of pyrolysis oil and the effects of catalytic esterification on the composition of pyrolysis fuel were investigated and the corresponding spectral analysis was carried out.

IOP Conf. Series: Materials Science and Engineering **729** (2020) 012045 doi:10.1088/1757-899X/729/1/012045

2. Materials and methods

2.1. Experimental materials and reagents
Methanol (AR, Tianjin Zhiyuan Chemical Reagent Co., Ltd.); Cornus wisoniana oil (hair oil, homemade); KF (AR, Shanghai Shanpu Chemical Co., Ltd.); KOH (AR, Beijing Tianfuda Industrial Co., Ltd.); p-toluenesulfonic acid (AR, Shanghai Feige Chemical Co., Ltd.); CaO (AR, Beijing Tianfuda Industrial Co., Ltd.)

2.2. Instrument
DF-101S magnetic stirrer; AUY-220 analytical balance; GC-2014 gas chromatograph; IS-5 FT-IR spectrometer; Scion SQ-bruker gas chromatography mass spectrometer.

2.3. Experiment method

2.3.1 Catalytic cracking method of Cornus wisoniana. The raw material light skin oil and the catalyst (the amount of the catalyst added is 1.5%) are introduced into a straight three-port glass reactor, stirred and uniformly mixed, and the reactor is assembled, and the electric furnace is directly preheated to 100 °C ~ 150 °C to remove residual moisture. The temperature is raised to 400 °C ~ 560 °C for the cracking reaction,and the product is a hydrocarbon-rich fuel.

2.3.2 Method for esterification of Cornus wisoniana oil cracking fuel. The Cornus wisoniana oil cracking product catalyzes the esterification reaction. The specific operation steps are as follows: 100 g of the catalytic cracking product is accurately weighed and transferred to a 250 mL three-necked flask at room temperature, and then p-toluenesulfonic acid is added as a catalyst in a flask, which will be called A good amount of methanol was transferred to a 250 mL constant pressure funnel. It is cooled by a serpentine condenser and heated in an oil bath. The temperature was raised to the experimentally required temperature, and the mixture was dropped into a flask in a constant pressure funnel, and the reflux state of the flow was observed to adjust the temperature. The water discharge in the water separator was observed to judge the degree of conversion of the reaction. Until the system has no moisture to stop heating, naturally cool to room temperature. Transfer to a separatory funnel to separate excess methanol and water, first wash with 5% NaCl salt, then add deionized water for 2~ 3 times to obtain an esterified oil.

2.3.3 Analysis of main components of Cornus wisoniana oil. The analysis of the fatty acid composition of the oil and fat was carried out by gas chromatography.
The chromatographic conditions were as follows: FFAP-quartz glass capillary column, column length 60 m, column diameter 0.3 mm; carrier gas N2: 81 kPa; carrier gas split ratio 100:1; H2 flow rate: 30 mL/min; air flow rate: 500 mL/min; Inlet temperature 260 °C; FID temperature 280 °C; set temperature range: 180 °C ~ 240 °C, heating rate: 6 °C / min.

3. Results and discussion

3.1. Analysis of the composition of raw material Cornus wisoniana

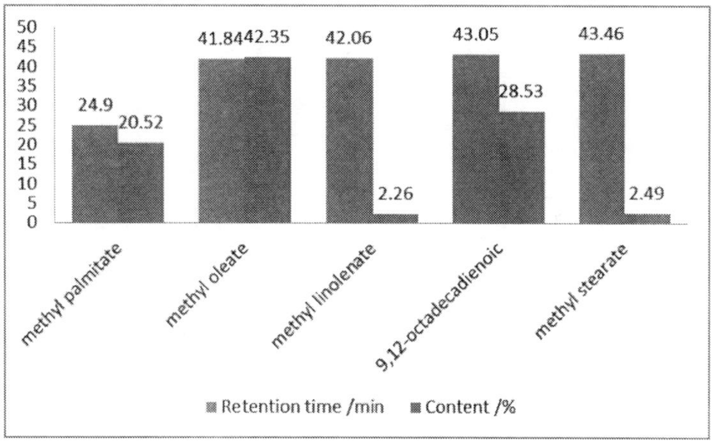

Figure 1. Cornus wisoniana oil major components

By gas chromatography analysis, it was found that the fatty acid above C16 was 96.12%, and the oleic acid was more than 42%. It can be speculated that Cornus wisoniana is a good industrial raw material for the preparation of hydrocarbon-rich fuel.

3.2. *Catalytic cracking reaction of Cornus wisoniana oil*

3.2.1. Effect of temperature on catalytic cracking reaction. Under the condition of reaction time 80 min and catalyst dosage 0.8% (raw material oil mass ratio), the effect of cracking temperature on the catalytic cracking reaction of Cornus wisoniana oil was investigated. The temperature range was from 400 °C to 560 °C.

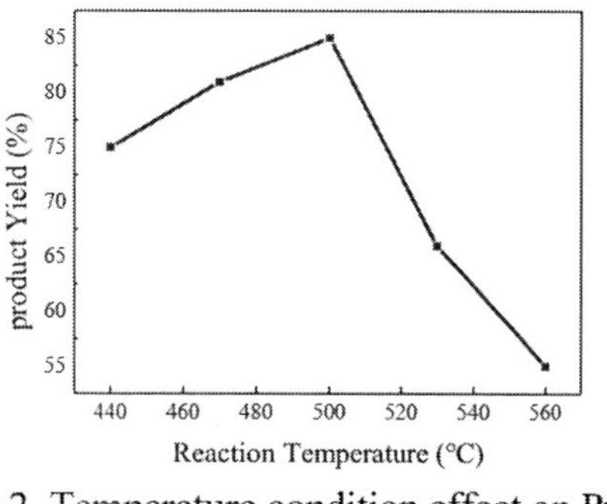

Figure 2. Temperature condition effect on Pyrolysis

It can be seen from Figure 2 that the optimum reaction temperature is 500 °C, and the product yield is the highest. Excessive reaction temperatures result in the production of more small molecular components, which are discharged as non-condensable gases, causing waste of resources; too low reaction temperatures lead to incomplete cracking.

3.2.2. Effect of CaO/KF content on catalytic cracking reaction. Under the condition of reaction time of 80 min and reaction temperature of 500 °C, the effect of the amount of catalyst on the catalytic cracking experiment of Cornus wisoniana oil was investigated.Investigate the range of catalyst dosage: 0.0%~1.6%.

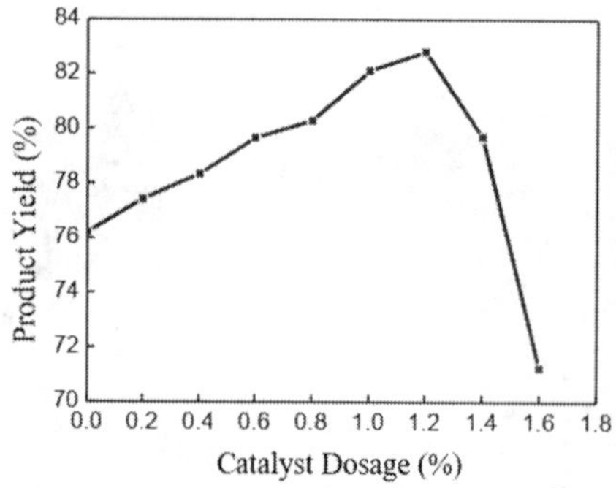

Figure 3. Catalyst content effect on Pyrolysis

When the content of CaO/KF is less than 1.2%, the amount of catalyst in the reaction system is insufficient, resulting in a decrease in reactivity, and the optimum reaction equilibrium is not achieved. The CaO/KF content of 1.2%~1.6% of the catalyst dosage stage results in low product yield due to the excessive number of active sites of the catalyst, and the excessive catalytic performance causes more small molecular components to be discharged, resulting in a reduction in effective products. When the CaO/KF content was 1.2%, the liquid phase yield reached a maximum of 82.11%.The optimum catalyst dosage is 1.2%.

3.2.3. Effect of Time on Catalytic Thermal Cracking reaction .Under the condition of reaction temperature of 500 °C and catalyst content of 1.2%, the influence of reaction time on the catalytic cracking experiment of Cornus wisoniana oil was investigated. The main time range was 30min~100min.

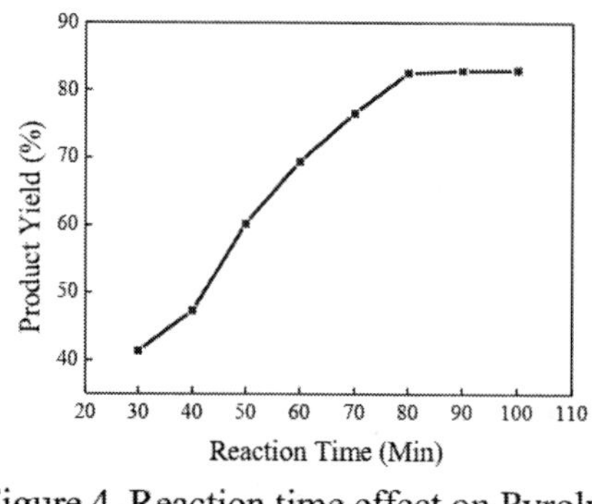

Figure 4. Reaction time effect on Pyrolysis

It can be seen from Fig. 3 that the reaction time of CaO/KF catalytic cracking is 80 min, and the product yield reaches 80% or more. The long or short reaction time will cause an increase in side reactions and even coking, resulting in product loss.

3.3. Product Mass Spectrometry

Reaction conditions: the catalyst KF/CaO content is 1.2%, the reaction time is 80 min, and the reaction temperature is 500 °C. Under this condition, the liquid phase cracking fuel yield is greater than 82%, and the temperament spectrum is as follows.

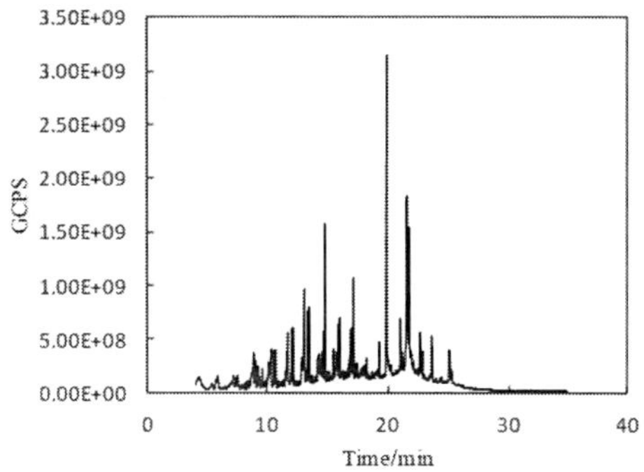

Figure 5. GC - MS chromatogram of catalytic pyrolysis product

According to the gas mass spectrogram results in Fig. 5, the Cornus wisoniana oil cracking products mainly consist of the following components: saturated alkane (C8~C21) and olefin (C8~C21) compounds, and carboxylic acid compounds cycloalkane compounds and aldehyde compounds having corresponding carbon number distribution. Among them, olefin compounds, carboxylic acid compounds and aldehyde compounds are the main reasons for the instability of cracking products. The verification results of GC - MS chromatogram provide a basis for the necessity of subsequent post-esterification treatment

3.4. cleavage products catalyze esterification

Esterification conditions are as follows. The molar mass ratio of methanol to Cornus wisoniana pyrolysis oil is 9:1, using p-toluenesulfonic acid as catalyst, the amount of which is 4% of the mass of the crude oil, the catalytic esterification reaction is 5 h, and the reaction temperature is 85. °C.

Table 1. Pyrolysis of pyrolysis product after esterification

Indicators	Before esterification pyrolysis	After esterification pyrolysis	bio-diesel	0# diesel
Density(20°C)(Kg/m^3)	0.88	0.85	0.87	0.845
Kinematic Viscosity(mm^2/s)	4.46/ 40°C	3.55/ 40°C	4.49/ 20°C	3.52/20 °C
Acid Value(mgKOH/g)	136.44	0.86	0.51	0.56
Calorific value(KJ/g)	36.2	39.2	38.2	43.1

It can be seen from Table 1 that after esterification of the Cornus wisoniana cracking oil, the acid value is reduced by 99.36%, which indicates that the acid component in the cracked oil is significantly reduced after esterification. The density is also reduced by 0.8 from 0.82 and the kinematic viscosity is also decreased by 21.31%, which may be related to the complex reaction occurring inside the oil in the presence of catalyst after methanol addition, and also related to the physical dilution of methanol. The calorific value increased by 7.66%. The above indicates that the physicochemical properties of the cracked oil after the p-toluenesulfonate is significantly improved. Comparing with biodiesel, it is found that the esterified Cornus wisoniana cracking oil is closer to traditional diesel fuel than biodiesel.

Infrared analysis before and after esterification of photocatalytic oil catalytic cracking products:

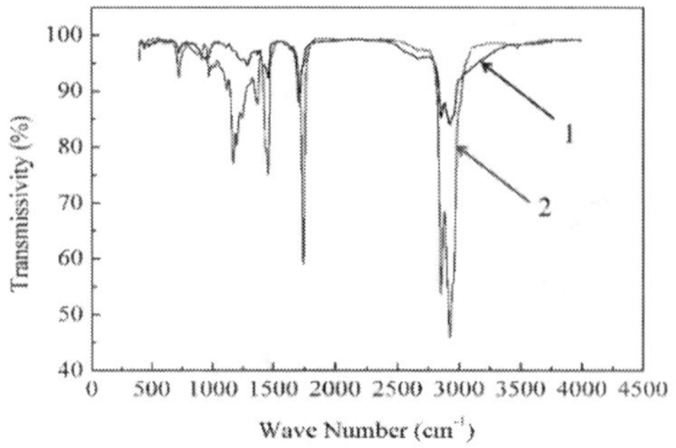

—1: before esterification; —2: After esterification

Figure 6. Pyrolysis oil esterification infrared spectrogram

It can be seen from Fig.6 that the infrared spectrum of the hydrocarbon-rich fuel catalyzed esterification is very different, mainly in the following bands: 2800cm^{-1}~3200cm^{-1}, 1700cm^{-1}~1800 cm^{-1}, 1000cm^{-1}~1200 cm^{-1}. The specific analysis is as follows: 2800 cm-1~3200 cm-1 represents the vibration of saturated CH. The catalytic reaction reacts the unsaturated part of the molecule to increase the saturated CH structure, so the peak shape of the hydrocarbon-rich fuel is deeper after the reaction; 1700cm^{-1}~1760 cm^{-1} represents the ester carbonyl group. After esterification, the peak shape is sharper and stronger than before esterification, indicating that there are a large number of ester-based functional groups in the hydrocarbon-rich fuel after the reaction, which ensures the stability of the fuel; 1000 cm^{-1}~1200 cm^{-1} represents the absorption peak caused by the vibration of the CO bond in the ester. After the esterification, the hydrocarbon-rich fuel has a stronger peak in the wavenumber range, indicating that most of the carboxylic acid in the hydrocarbon-rich fuel after esterification. It has been converted to the corresponding ester.

4. Conclusion

In this experiment, the catalytic cracking reaction of the Cornus wisoniana oil was studied and analyzed, and the following conclusions were drawn.

The optimum conditions for the catalytic cracking of the Cornus wisoniana oil are as follows: the catalyst KF/CaO content is 1.2%, the reaction time is 80 min, and the reaction temperature is 500 °C. Under this condition, the liquid phase cracking fuel yield is greater than 82%; Pyrolysis fuel The finished fuel is obtained by esterification after high esterification. The molecular structure and physical and chemical properties are more similar to 0# diesel. It can replace some fossil energy by mixing to achieve energy saving and emission reduction.

Acknowledgments

The authors gratefully Hunan Province introduces foreign intellectual special projects (2019YZ3026), Hunan Natural Science Foundation (2018JJ2218), Provincial Key Research and Development Plan (2017NK2310) Changzhutang Self-dependent Innovation Demonstration (2018XK2005) and Hunan Provincial Key Laboratory (2018TP1029) for financial support.

References

[1] Xiao, Z. H., Zhang, A. H., Jiang, W.W., Li, Y., Zhang, H. Q., & Li, P. W., et al. (2014). Research on catalytic coupling of cornus wisoniana oil for biological rich-hydrocarbon fuel. J. Applied Mechanics & Materials, 5: 448-453.

[2] Xiao, Z.H., Jiang, W.W., Lin, L., Zhang, A.H., (2014). Catalytic cracking of cornus wisoniana oil to liquid bio-fuel oil using kf/cao as a solid base catalyst. Applied Mechanics & Materials, 477-478, 1446-1451.

[3] Huang, D.M., Jiang,W.W, Lin, L., (2013) Study on the Catalyst Performance on Cornus wisoniana Oil Catalytic Cracking Prepared Biological Fuel Oil. J. Applied Mechanics & Materials, 477-478:1457-1463.

[4] Lin, L., Cui, H., Xiao, Z., Zhang, A., (2015) Synthesis of kf/cao as a catalyst for the production of bio-fuel from cracking of cornus wisoniana oil. European Journal of Lipid Science & Technology, 117(3), 406-410.

[5] Valle, B., Aramburu, B., Benito, P.L., (2018) Biomass to hydrogen-rich gas via steam reforming of raw bio-oil over Ni/La2O3-αAl2O3 catalyst: Effect of space-time and steam-to-carbon ratio[J]. Fuel, 216:445-455.

[6] Yu, J., Biller, P., Mamahkel, A., (2017) Catalytic hydrotreatment of bio-crude produced from the hydrothermal liquefaction of aspen wood: a catalyst screening and parameter optimization study. J. Sustainable Energy & Fuels, 34: 1-19

[7] Yan, W.H., Wang, K., Duan, P.G., (2017) Catalytic hydropyrolysis and co-hydropyrolysis of algae and used engine oil for the production of hydrocarbon-rich fuel. J. Energy, 133: 1153-1162

[8] Liu, Y., An, Z., Yan, H., (2019) A Conceptual Coupled Process for Catalytic Cracking of High Acid Crude Oil. J. Industrial & Engineering Chemistry Research, 59: 4794-4801.

[9] Mcheik, A., Hijazi, A., Diab,H., (2017) Hydroboration of Methyl Esters of Fatty Acids. J. European Scientific Journal, 13: 324-341.

IOP Conf. Series: Materials Science and Engineering **729** (2020) 012046 doi:10.1088/1757-899X/729/1/012046

Effect of microwave radiation on pore structure of SCR denitration catalyst

Honghong Shen[1], Qunyin Gu[2], Daoyuan Wen[2], Gang Liu[1, 3], Weijun Gao[2], and Jianxing Ren[1]

[1]Colleg of Energy and Mechanical Engineering, Shanghai University of Electric Power, Shanghai, China

[2]Department of International Environmental Engineering, `The University of Kitakyushu, Fukuoka, Japan

[3]Guodian Tongling Power Generation Co.Ltd, Anhui, China

e-mail: hh20180606@126.com

Abstract. Based on the background of SCR denitrification technology widely used in pulverized coal utility boilers, the characteristics of denitrification catalysts used in industry were studied. Microwave radiation was used to prepare 8Mn-8Fe/Al_2O_3 denitrification catalyst by using microwave as an efficient and fast heating method. The results show that microwave irradiation can make the catalyst heated from inside to outside without temperature gradient, make the molecule oscillate rapidly, change the specific surface area and pore structure of the catalyst, thus improving the catalytic activity of the catalyst.

1. Introduction

Nitrogen oxides (NO_X) [1]mainly come from flue gas emissions from coal-fired power plants and automobile exhaust emissions. The hazards of nitrogen oxides are mainly manifested in four aspects: Nitrogen oxides entering human respiratory system can cause diseases such as tracheitis, pneumonia, emphysema, etc. Nitrogen oxides and human hemoglobin binding ability is stronger than CO, which makes human body anoxic; Nitrogen oxides are an important factor in acid rain formation. Acidic rain can destroy forest vegetation and pollute water body, and bring incalculable loss to our country; chemical reaction between nitrogen oxides and hydrocarbons under sunlight produces a toxic photochemical smog, which seriously threatens human life; the chain reaction between N_2O in nitrogen oxides and ozone in stratosphere depletes ozone, resulting in ozone hole in the ozone layer. National "Twelfth Five-Year Plan"[2] put forward the most stringent environmental protection standards. By 2020, the plan will reduce NO_X emissions by another 10%. The situation of emission reduction is urgent.

Selective Catalytic Reduction SCR[3] is a widely used flue gas denitrification technology at present. Catalyst is the core of SCR denitrification. Commercial V_2O_5 titanium dioxide catalyst has high cost, high reaction window temperature, and its active component, heavy metal vanadium, is an environmental pollutant. Therefore, under increasingly stringent NO_X emission standards and waste management system, an efficient and environmentally friendly catalyst is developed. Agents have far-reaching significance.

2. Denitrification mechanism

In the SCR reactor, the reactant and the flue gas flow through the gas film on the outer surface of the catalyst, and then diffuse to the outer surface of the catalyst; the reactant then diffuses from the outer surface of the catalyst to the microporous surface of the catalyst wall, and then is adsorbed on the surface of the catalyst; the chemical reaction occurs in the gas phase with the adsorption on the surface of the catalyst, and the products are formed and catalyzed. The surface desorption of the chemical reagent gradually diffuses to the outer surface of the catalyst, and finally diffuses from the outer surface of the catalyst through the gas film to the mainstream gas, which is taken away by the flue gas. The micro-reaction mechanism is shown in Figure 1.and Figure 2.

At present, SCR denitrification catalysts used in industry are mainly is ocatalysts, which have high denitrification efficiency at 250 - 350 °C reaction temperature. However, the problems of catalyst poisoning and severe poisoning have also been paid more attention. Jiang Ye[4] et al. put forward that coal-fired flue gas contains a certain concentration and a small amount of water vapor which reacts with escaping and flue gas, which will not only reduce the activity of catalyst, but also block the downstream pipeline. At present, scholars at home and abroad have carried out different levels of research on selective catalytic reduction catalysts. The research on new catalysts mainly focused on iron-based, copper-based and doped new catalysts. Wang Fang [5]et al. found that the activation energy was low and the activity was high. The denitrification efficiency reached more than 90% at 250 °C. The SCR activity of Cssanova[6] catalyst prepared by precipitation method is directly related to its content at low and medium temperatures. Xu Liting [7]et al found that the maximum denitrification efficiency of the catalyst was 9.1% at 325 °C. Lu Huixia[8] et al. found that the specific surface area of siderite doped with rare earth elements was larger, the crystallization of catalyst was improved, the surface acidity was improved, and the denitrification efficiency was obviously improved. After 7 hours of continuous operation, the denitrification efficiency of siderite catalyst was still more than 80%, and its sulfur resistance and persistence were also improved. Sun Jiaxing [9]et al. found that when the optimum formulation of the catalyst was 6%, the maximum denitrification efficiency could reach 80% below 200 °C in the study of copper-based catalysts modified by transition metal oxides. Catalyst is the core of SCR flue gas denitrification technology. The development of non-toxic, harmless, durable, sulfur-resistant and low-temperature catalysts is of great significance to reduce the operating cost of SCR denitrification system.

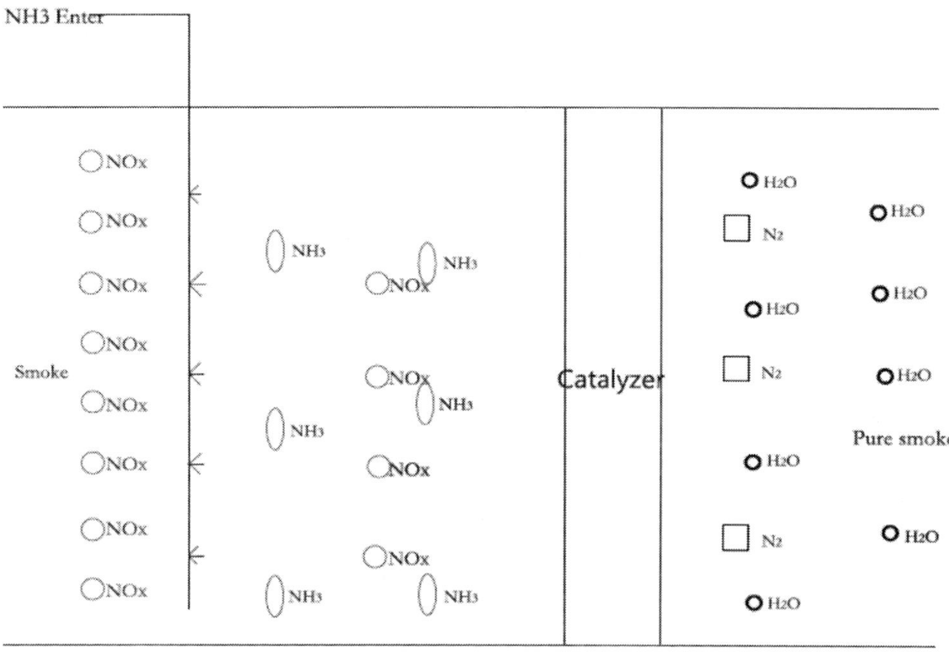

Figure1.Microreaction Mechanism of SCR Flue Gas Denitrification

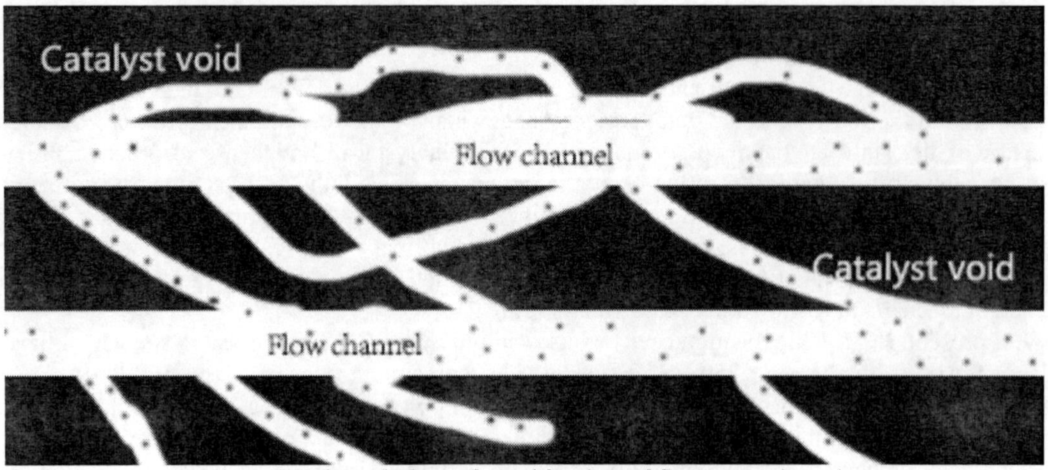

Figure2.Enlarged view of combination of flue gas and catalyst

3. Experiments

The low temperature supported denitrification catalyst 8Mn-8Fe/Al_2O_3 was prepared by co-impregnation method. The effects of common drying method and microwave drying method on the catalyst were compared. The main chemical reagents used are iron nitrate $(NO_3)_3 \cdot 9H_2O$, ammonia (25%) and manganese acetate $(CH_3COO)_2 \cdot 4H_2O$. The main experimental equipments are oven, microwave oven, magnetic stirrer, muffle oven, electronic balance, measuring cylinder, beaker and glass rod, PH test paper, etc. The content of metal elements in the impregnation process is calculated by mass fraction.

Common preparation method: 50 mL deionized water is put into beaker with dosage barrel, then a certain amount of iron nitrate and manganese acetate is weighed with electronic balance and put into beaker. A solution containing 8% of Mn and 8% of Fe is prepared. The solution is stirred with glass rod until it is completely dissolved 230-425 mm of Al_2O_3 particles are put into the solution and dripped. The ammonia water was continuously stirred to pH=8-9, then stirred for 0.5 h in 50 °Cwater bath by magnetic stirrer, and kept for 48 h at room temperature. Then it was dried at 120 °C in oven for 24 h, then calcined in muffle oven at 450 °C for 5 h. The catalyst obtained was O-8Mn-Fe/Al_2O_3.

Microwave preparation method: The operation before the method is the same as that of the ordinary preparation method. After the method is stationary, the microwave oven is used for drying, the drying power is 350 W, the drying time is 15 minutes, and then calcined in a muffle oven at 450 °C for 5 hours. The catalyst obtained is M-8Mn-8Fe/Al_2O_3.

4. Result discussion

The specific surface area and pore size of O-8Mn-8Fe/Al_2O_3, M-8Mn-8Fe/Al_2O_3 [10]and two kinds of samples were characterized and analyzed by means of specific surface area and porosity analyzer. The BET formula was used to calculate the specific surface area and pore size of the samples. The results are shown in Table 1. Compared with ordinary drying, the specific surface area of the catalysts increased after microwave drying. Additionally, the average pore volume increased slightly, mainly because microwave heating was carried out both inside and outside at the same time, and the heating speed was faster. Moisture was gasified instantaneously. The pore structure of the catalyst was improved to a certain extent by leaving the catalyst by blasting. The average pore size of the catalyst was 6.03 nm, belonging to the mesoporous range, which made the volume flow rate of the gas flow larger (high space velocity). In this case, the resistance of gas diffusion within the catalyst is small, which can ensure a higher denitrification efficiency.

Tab. 1 Calculation results by BET method

Sample	Specific surface area ($m^2 \cdot g^{-1}$)	Mean gap ($cm^3 \cdot g^{-1}$)	Average pore size nm
General drying	185.64	0.372 4	8.02
Microwave drying	248.83	0.375 7	6.03

It can be seen from the pore size distribution in Figure 3 [11], the pore size of samples after microwave drying is mostly distributed in 2-4nm, while the pore size of samples after ordinary drying is mostly distributed in 12NM, that is to say, microwave treatment does not significantly change the pore type of catalyst, but changes the pore size distribution, so that the pore structure of catalyst becomes developed, so as to enhance its adsorption capacity and improve its adsorption capacity. Denitration

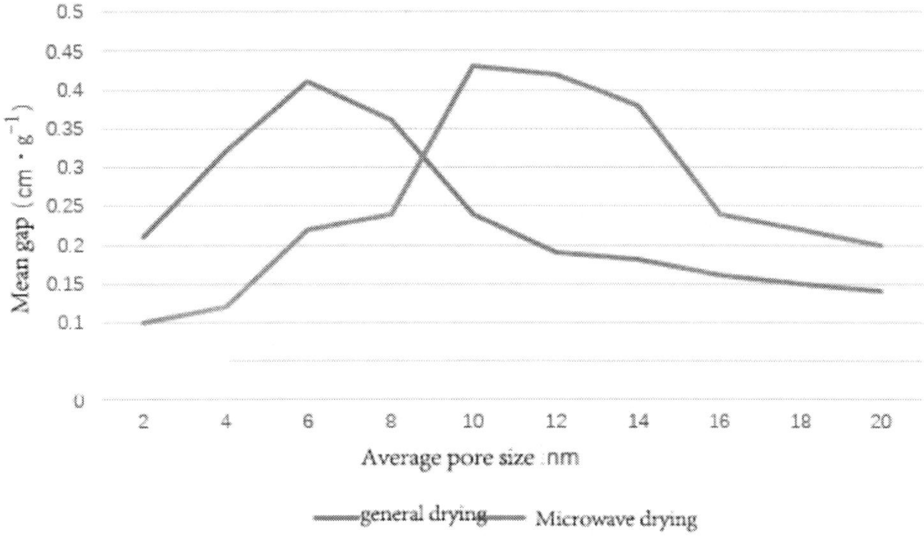

Fig.3 Sample adsorption isotherm

Tab2. Effect of two drying methods on denitration efficiency at different temperatures

Temperature/°C	Denitration rate of microwave drying /%	Denitration rate of common drying /%
80	94	58
120	96	79
160	99	95
200	98	96
240	97	97

The grains on the surface of ordinary dried samples are connected into a large number of particles, and the number of micropores is small. According to bet analysis, the micropores are also small, which will hinder the gas diffusion to the catalyst, and is not conducive to the SCR denitration reaction. However, the samples after microwave drying have a good dispersion and a uniform distribution of grains, which increases the contact area between catalyst and gas, making the gas It is easier to adsorb on the catalyst surface, so as to promote the SCR denitration reaction, and the denitration efficiency is also improved at low temperature, as shown in Table 2 [11]. After microwave drying, the pore structure of the sample is more developed than that of common drying, the agglomeration of catalyst is improved, and its micro performance is improved. This is consistent with bet characterization. Therefore, the

IOP Conf. Series: Materials Science and Engineering **729** (2020) 012046 doi:10.1088/1757-899X/729/1/012046

samples dried by microwave have larger specific surface area, more developed pore structure, better particle dispersion, and the denitrification performance is far better than the catalyst dried by ordinary

References

[1] Kampa.M,Castanas E. Human health effect of air pollution[J].Environment all pollution,2008,151:362-367.

[2] Ministry of Environmental Protection. Fully implement the work plan of ultra-low emission and energy saving transformation of coal-fired power plant[S], Beijing,2015

[3] Xi'an Institute of thermal engineering, SCR Flue Gas Denitration Technology in thermal power plant[M]Beijing, China Electric Power Press,2013

[4] Jiang Y, Summary on deactivation of selective catalytic reduction denitrification catalyst[J], Journal of Electrical Engineering of China,2013, 33 （14） 19-30

[5] Wang F, SCR denitrification technology and mechanism of iron-based catalysts under magnetic field[J]Nanjing, Southeast University,2011

[6] MarziaCasanova, *Jordi*Llorca, AmodSagar,etal. Mixed iron-erbiunvanadate NH_3 -SCR catalysts[J].Catalysis Today.2015,241:159-168

[7] Xu Lt, Denitrification Characteristics of Doped Modified Catalysts[J], Journal of Coal Science, 2017.42(7):1885-1891

[8] Lu HX,Gui KT, Study on the modification of SCR catalyst for denitrification at low temperature[J] ,Journal of power engineering, 2017,37(9):727-731

[9] Sun JX, Study on denitration performance of modified copper（$MnOx-CeO_2$ based catalyst by SCR at low temperature,[J], Journal of power engineering, 2018,38(9):733-739

[10] Wang Xiaobo, Wang Fang, Yao Guihuan, Performance of low temperature De-NO_Xbased on Zr-Mn-Fe/Al_2O_3[J], 2010,40（2）：388-392.

[11] Cai S, The effect of microwave radiation on the denitrification activity of iron-based catalyst SCR at low temperature[J], Journal of power engineering, 2015,35（12）：993-1011

Study on mechanical properties of cemented filling materials using by low-chlorine fly ash

Tengyu Shi [1], Siqi Zhang[1,*], Ni Wen[1],Keqing Li[1], Ke Wang[1] and Haojing Ba[1]

[1]School of Civil and Resource Engineering, University of Science and Technology Beijing, Beijing 100083, China

[*]Corresponding author's e-mail: zsq2017@ustb.edu.cn

Abstract. In this study, the blast furnace slag and desulfurized gypsum were used to make cementing materials, mixed with the tailings and fluidized bed municipal solid waste incineration(MSWI) fly ash to make high performance cementing filling material. The results show that: (1) The best ratio of test is low chlorine fly ash: slag: desulfurization gypsum = 20%: 68%: 12%, the MPC of cementing material is 82%, and the cement-sand ratio is 1:4, after curing for 28 days in a standard curing box with a curing temperature of 40°C and a relative humidity of 90%, the compressive strength reaches 22.81MPa, the flow rate reaches 240 mm. (2) The maximum amount of fly ash can reach 60%, after 28 days of curing, the sample strength is 8.70 MPa, the flow rate is 220 mm, both of which meet the requirements. (3) Comparing the slag system and cement system with 60% fly ash, it was found that the strength reached 10.36 MPa after 28 days of curing, the flow rate of the cement system was 215 mm, which did not reach the self-flow standard. It is possible to provide a low cost cementing filling material containing waste incineration fly ash in future.

1. Introduction

Recent years have witnessed the rapid development of the waste incineration industry in China. In the incineration process, fly ash containing a large amount of hazardous wastes such as heavy metals, soluble salts and dioxins is produced, which must be treated before discharge [1]. Maintenance and stabilization are the main methods for the treatment, while cement is the main material for solidifying the fly ash [2,3]. However, the hazard of high chlorine content in fly ash shall not be ignored because it not only aggravates the internal corrosion of cement kiln equipment, leading to abnormal operation [4], but also directly hinders the hydration process, thus increasing the solubility of heavy metals in the solidified body [5]. Therefore, the chlorine content in the fly ash from waste incineration is an important factor affecting its utilization in the cement industry.

The highly-efficient and low-pollution clean incineration technology-circulating fluidized-bed boiler technology developed in China in the past decade has such advantages as good smoke emission performance and dioxin formation control. It can also ensure that the emission of NO_x, sulfide and acid gas can comply with environmental requirements. In addition, it has high incineration efficiency with coal as the auxiliary fuel, which can significantly reduce operating costs [6]. However, it also has such disadvantage as high self-consumption power consumption, large amount of fly ash discharge and abrasion of furnace refractory [7]. At present, China's incineration plants using fluidized bed technology reach 38% by volume and 40% by scale [8], second only to those using boiler grate technology.

Wang et al. [9] found that the chlorine and sulfur content of fly ash from fluidized bed are lower than that of fly ash from boiler grate. The average chlorine content of the two is 1.71% and 15.41%, respectively, and the average content of SO_3 is 2.875% and 10.67%. Due to coal blending, the heavy metal content is several times lower than that of the fly ash from boiler grate. When the chlorine content in the fly ash is low, the content of fly ash in the co-disposal of the cement kiln can be increased, thus realizing the scale application. The replacement of cement for construction as the basic material is low in disposal cost, low in resource consumption, and low in greenhouse gas emissions [10]. At present, China's research on waste incineration of low-chlorine fly ash is still in the basic stage. For example, Liu [11] found through fly ash effect and microscopic analysis that when the ratio of fly ash from fluidized bed to cement mass is 30%: 70% and 25%: 75%, the strength reaches 38.1MPa and 31.8MPa after 28d, indicating that the fly ash from fluidized bed has better pozzolanic activity. There are also some foreign studies on fly ash from fluidized bed. Wu et al. [12] have shown that fly ash from fluidized bed combustion can promote slag hydration. When the ratio of fly ash to slag is 3:7, after maintenance for 56 days, the strength can reach 72.4% (31.43 MPa) of the strength of ordinary Portland cement mortar test block. Jouni Rissanen [13] found that when the fly ash content is low (10% and 20%), it can reach almost the same strength (50MPa) as ordinary Portland cement system after 90 days. When the fly ash content is 40%, the compressive strength can still reach 88% (44MPa) of cement strength. Although many studies have shown that fly ash from fluidized bed has the potential to replace cement, further research is needed for verification.

The filling mining technology that uses tailings as aggregates to form cemented fillers in combination with cementing agents has been widely used worldwide due to its advantages of energy saving, safety and high efficiency. In recent years, it has been found that slag under the excitation of FGD gypsum can replace ordinary Portland cement as a cementing agent. Therefore, this paper takes the combination of low-chlorine fly ash, slag and FGD gypsum to replace ordinary Portland cement as the binding agent, and adds tailings to prepare full-solid waste high-performance mine filling material. This paper focuses on its mechanical properties and the addition amount of low-chlorine fly ash, with a purpose to realize large-scale utilization and harmless treatment of fly ash from waste incineration, and at the same time, to solve the problem in the low-cost replacement of cement-based binding materials in cement-filled mining.

2. Test material

Slag is from a steel plant in Hebei Province. Its specific surface area is $500m^2/kg$ after grinding [14]. FGD gypsum is from in a power plant in Hebei Province. Its specific surface area is $320m^2/kg$ after drying. Tailings is from an iron ore mine in Hebei Province, and its main mineral component is quartz. Cement used here is ordinary Portland cement P.O42.5 with specific surface area of $355m^2/kg$. Fly ash from waste incineration (fly ash) is from the circulating fluidized bed in an incineration plant in Tangshan City. The chemical composition of each raw material is shown in Table 1. The XRD analysis result of the fly ash is shown in Figure 1. It can be seen that the main minerals of fly ash are quartz, calcite, anhydrite, calcium oxide, aluminum oxide and sodium sulfate.

Table 1. Chemical composition of the raw materials. (wt.%)

Sample	CaO	Cl	SiO_2	Al_2O_3	MgO	SO_3	Fe_2O_3	TiO_2	Na_2O	K_2O	MnO
Slag	42.25	—	26.35	15.23	10.82	1.82	1.21	0.98	0.56	0.39	0.21
FGD Gypsum	48.37	0.42	2.71	0.89	1.35	43.59	0.53	—	—	0.27	—
Tailings	12.29	—	55.63	9.64	5.33	2.01	9.40	0.66	2.23	2.00	0.23
Cement	63.52	—	20.26	5.05	2.85	2.56	3.56	0.42	—	1.00	0.28
Fly ash	27.24	3.38	28.31	15.04	3.82	6.38	5.95	1.15	2.28	2.54	0.15
Sample	P_2O_5	F	ZnO	SrO	BaO	CuO	ZrO_2	PbO	Cr_2O_3	Co_2O_3	Br
Slag	—	—	—	—	—	—	—	—	—	—	—
FGD Gypsum	—	1.27	—	—	—	—	—	—	—	—	—
Tailings	0.27	—	—	—	—	—	—	—	—	—	—
Cement	—	—	—	—	—	—	—	—	—	—	—
Fly ash	2.58	—	0.53	0.20	0.13	0.12	0.06	0.05	0.04	0.03	0.02

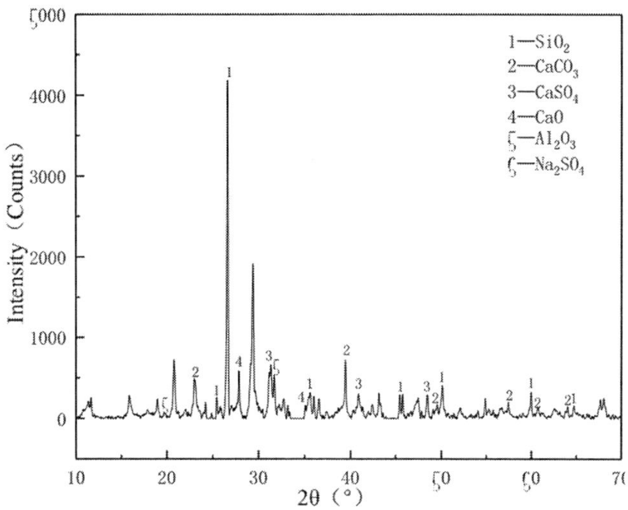

Figure 1. XRD analysis spectrum of MSWI fly ash.

3. Test method

3.1. Sample preparation

The mortar test block is prepared based on the experimental scheme. First, weigh the tailings, slag, FGD gypsum and water. Then, pour the tail sand into the stirring pot, add the binding material, and stir it for 5 minutes. After measuring the fluidity, place the slurry in a 40mm × 40mm × 160mm mold on the vibrating table and ensure that the slurry is filled with the mold. Next, seal it with plastic wrap and place it in a standard maintenance box with a temperature of 40 °C [15] and a relative humidity of 90% or more. After 3 days, remove the mold and test the 3D compressive strength of the test piece. After the mold is removed, place it in a standard maintenance box for 7 days and 28 days and measure the compressive strength.

3.2. Performance test

The compressive strength of the filled test piece is measured using a YES-300 digital display hydraulic pressure tester.

The fluidity of the filling slurry is measured using a NLD-3 type fluidity meter.

4. Test result and discussion

4.1. Strength test of fly ash content

According to the previous experiments [15], the tentative cement-to-sand ratio is 1:4, the slurry concentration is 82%, and the FGD gypsum proportion in cementing material is 10%. The five gradients of fly ash proportion are designed as 0%, 20%, 40%, 60%, and 80%. The test scheme is shown in Table 2, and the strength results are shown in Figure 2.

Table 2. Exploratory test of the amount of fly ash.

Number	Cement-sand ratio	Concent-ration	Tailings (g)	FGD Gypsum (%)	Fly ash (%)	Slag (%)	Water (ml)	Fluidity (mm)
A1					0	90		225
A2					20	70		238
A3	1:4	82%	960	10	40	50	263.5	230
A4					60	30		220
A5					80	10		208

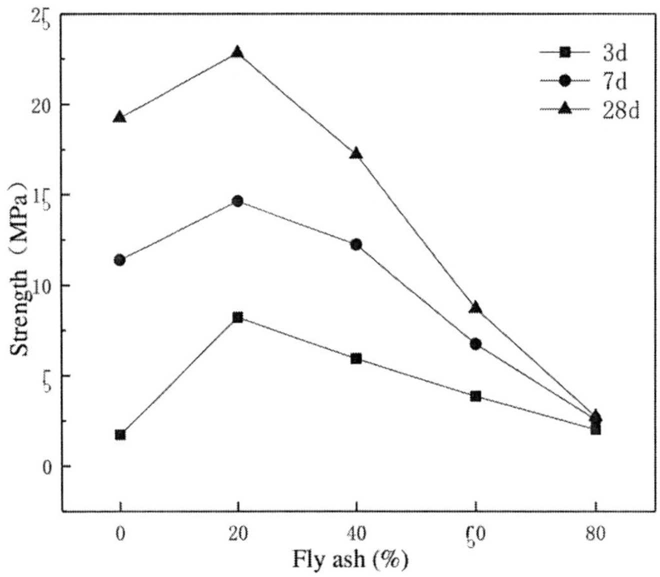

Figure 2. Determination strength results of the amount of fly ash.

It is found from Table 2 that the fluidity of the slurry increases first and then decreases with the increase of the mixing amount of fly ash. When the amount exceeds 60%, the fluidity does not reach the standard of 220 mm at which point the filling slurry can flow by itself. When the amount reaches 20%, the fluidity reaches 238 mm. This is because the fly ash hydration reaction rate is slow and the fly ash will gradually harden in the later stage. The early fly ash is equivalent to the water reducing agent [16], which will increase the slurry fluidity. In addition, the fly ash can serve as the ball in the slurry. Its adsorption effect on the surface leads to the morphological effect of the electric double layer structure, thus significantly improving the fluidity. But the capacity of the particles approaches saturation in this process [16,17].

It is found from Figure 2 that the strength decreases with the increase of the mixing amount of fly ash. The filling test piece made based on Formula A5 has the strength of only 2.74 MPa after 28 days of maintenance, failing to meet the requirement of mine filling, that is, 5 MPa. The blocks made based on other formulas all meet the strength requirement. The fly ash is similar to the cementing material such as slag in terms of component, but it has a different mineral composition and structure and is inactive, so the use of fly ash will reduce the active components in the system thus reducing the hydration product, resulting in a decrease in the strength of the test piece. The fly ash is strong in water absorption, thus reducing the amount of water required for hydration, resulting in incomplete hydration and a decrease in the strength of the test piece. Therefore, when the fly ash content is 20%, the compressive strength reaches the maximum of 22.81 MPa; when the fly ash content is 60%, the strength is 8.70 MPa after 28d maintenance, which basically meets the requirements.

4.2. Strength test of FGD gypsum content

When the fly ash content is 20% and the other test conditions are the same, this paper designs five gradient FGD gypsum content, namely, 6%, 8%, 10%, 12%, and 14%. See Table 3 for the test scheme, and see Figure 3 for the strength test results.

Table 3. Exploratory test of the amount of gypsum.

Number	Cement-sand ratio	Concent-ration	Tailings (g)	FGD Gypsum (%)	Fly ash (%)	Slag (%)	Water (ml)	Fluidity (mm)
B1				6		74		198
B2				8		72		215
B3	1:4	82%	960	10	20	70	263.5	245
B4				12		68		240
B5				14		66		232

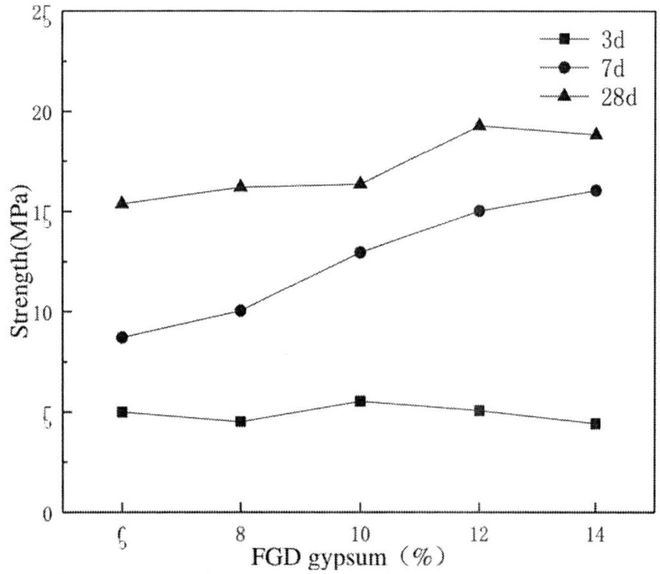

Figure 3. Determination strength results of the amount of gypsum.

It can be seen from Table 3 that the fluidity does not reach the standard at which point the filling slurry can achieve complete self-flow when the gypsum content is 6% or 8%, which is because insufficient gypsum and coagulation time will lead to the phenomenon of instantaneous condensation, meaning that the mortar condenses and hardens too early with low fluidity. When the mixing amount is 10%, the fluidity reaches the optimal value. It can be seen from Figure 3 that the strength increases

IOP Conf. Series: Materials Science and Engineering **729** (2020) 012047 doi:10.1088/1757-899X/729/1/012047

first and then decreases with the increase of gypsum content, and the strength of each test block can reach more than 5 MPa after 28 days of maintenance. Some studies have found that FGD gypsum has a significant excitation effect on fly ash, which can increase the setting time and the strength of the paste. However, excessive FGD gypsum will lead to the existence of a large amount of hemihydrate gypsum in the system. With more active Al_2O_3 dissolved in the fly ash vitreous, ettringite continues to emerge. Both of them are expansive, and the abundant existence of them will lead to a decrease of strength [18], which is consistent with the results. When the mixing amount exceeds 12%, the rate of strength increase begins to decrease. Therefore, with the influence of the setting time on the fluidity and strength taken into consideration, when the gypsum content is 12%, the fluidity reaches 240 mm, and the strength reaches 19.28 MPa.

4.3. Test on the best MPC of cementing material

When the fly ash content is 20%, the gypsum content is 12%, and the other test conditions are the same, this paper designs four concentration gradients, namely, 78%, 80%, 82%, and 84%. The test plan is shown in Table 4, and the strength test results are shown in Figure 4.

Table 4. Exploratory test of the best MPC of cementing material.

Number	Cement-sand ratio	Concent-ration	Tailings (g)	FGD Gypsum (%)	Fly ash (%)	Slag (%)	Water (ml)	Fluidity (mm)
C1		78%					336.5	268
C2	1:4	80%	960	12	20	68	300	252
C3		82%					263.5	235
C4		84%					227	205

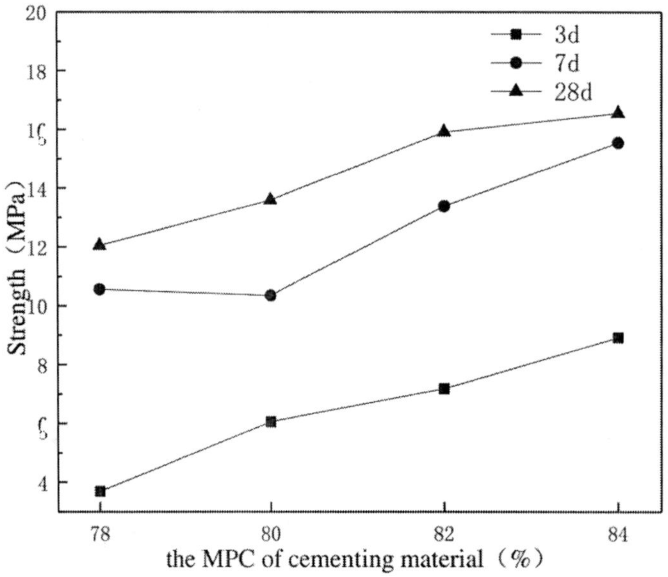

Figure 4. Determination strength results of the best MPC of cementing materia.

It is found from Table 4 that the fluidity decreases with increasing concentration. When the concentration is 84%, the fluidity is only 205 mm. But the excessive amount of water dilutes the slurry, leading to segregation, which has a certain influence on the fluidity performance. As the water content continuously reduces, the amount of water required for the hydration reaction becomes insufficient, leading to lower fluidity.

It is found from Figure 4 that the strength increases with increasing concentration, and that the strength values all meet the requirements after 28 days of maintenance. With lower concentration, when there is more water required for the hydration reaction, the free water will be filled in the voids of the solidified body, resulting in a decrease in strength. When the concentration reaches 84%, the rate of increase in strength slows down, which is because holes caused by thick slurry [19] result in a decrease in strength. Therefore, the optimum concentration is taken as 82%, at which points the slurry is the strongest with good fluidity performance.

4.4. Test on best cement-sand ratio

When the fly ash content is 20%, the concentration is 82%, and the other test conditions are the same, this paper designs eight cement-sand ratio gradients, namely, 1:1, 1:2, 1:3, 1:4, 1: 5, 1:6, 1:7 and 1:8. The test plan is shown in Table 5, and the strength test results are shown in Figure 5.

Table 5. Exploratory test of optimum percentage of cement-sand ratio.

Number	Cement-sand ratio	Concent-ration	Tailings (g)	FGD Gypsum (%)	Fly ash (%)	Slag (%)	Water (ml)	Fluidity (mm)
D1	1:1							150
D2	1:2							160
D3	1:3							212
D4	1:4							240
D5	1:5	82	960	12	20	68	263.5	255
D6	1:6							250
D7	1:7							242
D8	1:8							235

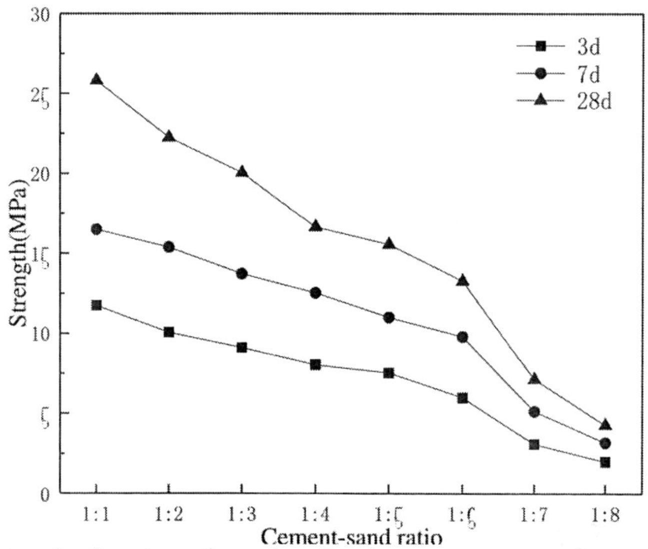

Figure 5. Determination strength results of optimum percentage of cement-sand ratio.

It is found from Table 5 that the fluidity increases first and then decreases with the increase of the sand content. When the cement-sand ratio is 1:4, the fluidity meets the requirements. The iron ore tailings contain a large amount of tiny particles, which can improve the fluidity of the mortar with a small amount, and lead to a decrease in the plasticity with an excessive amount. It can be seen from

IOP Conf. Series: Materials Science and Engineering **729** (2020) 012047 doi:10.1088/1757-899X/729/1/012047

Figure 5 that the strength increases with the increase of the cement-sand ratio. However, when the ratio is 1:8, the strength is reduced to 4.26 MPa after 28 days. The iron ore tailings contain the calcium silicate component similar to cement, which is of a certain activity. The use of an appropriate amount of tailings under the premise of not reducing the fluidity leads to a slight reduction in strength but an increase in the toughness and performance of the mortar [20]. However, the tailings do not have cementing properties, and excessive tailings will lead to the reduction in hydration products. Therefore, considering the results of fluidity and strength, following the principle of using fly ash as much as possible, the cement-sand ratio should be set as 1:4 at which point the flow performance is good, and the strength reaches 16.64 MPa after 28 days.

4.5. Test of ordinary Portland cement system

Based on the above test results, when the concentration, the cement-sand ratio, and the maintenance conditions are the same, this paper replaces the cement with binding materials such as slag and FGD gypsum, and prepares mortar test blocks in the ordinary Portland cement system in which the fly ash content is 0%, 20%, 40%, and 60%. The test plan is shown in Table 6, and the strength test results are shown in Figure 6.

Table 6. Exploratory test of Portland cement system.

Number	Cement-sand ratio	Concentration (%)	Tailings (g)	Cement (%)	Fly ash (%)	Water (ml)	Fluidity (mm)
E1	1:4	82	960	100	0	263.5	220
E2				80	20		245
E3				60	40		225
E4				40	60		215

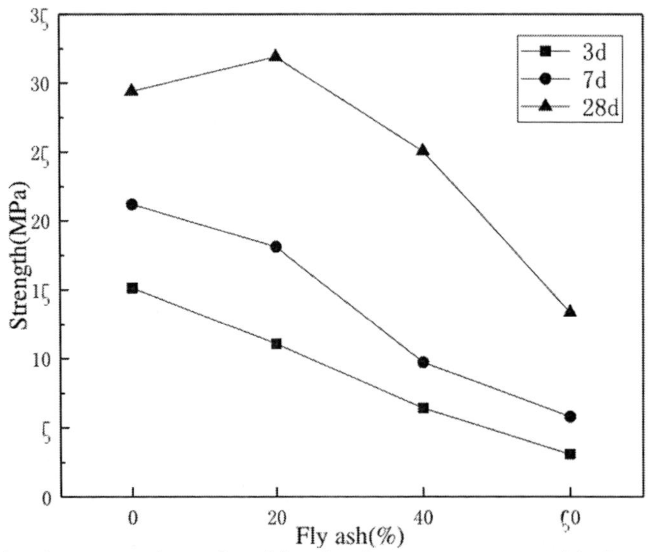

Figure 6. Determination strength results of Portland cement system with the amount of fly ash.

It is found from Table 6 that the amount of fly ash has a certain influence on the fluidity of the cement system. When the fly ash content is 20%, the fluidity performance is the best, reaching 240mm. When the fly ash content reaches 60%, the fluidity only reaches 215mm. Some studies have found that the right amount of fly ash can reduce the plastic viscosity and yield stress of cement mortar, and at the same time, increase its fluidity [20]. It can be seen from Figure 6 that the incorporation of fly ash reduces the strength of the test piece, especially the strength at the early stage. Greater amount of the mixing content leads to more significant decrease in strength. When the fly ash content is 20%, the strength after 7d and 28d of maintenance is close to or exceeds that of the test block without fly ash addition. Therefore, the optimum content of fly ash in the cement system is 20%, at which point the fluidity performance is the strongest, the strength reaches 31.89 MPa; the maximum content is 40%, at which point the fluidity is 225 mm, and the strength is 25.06 MPa.

5. Conclusion

The content of chlorine and sulfur in the fly ash in the circulating fluidized bed is small. The combination of low-chlorine fly ash and the slag in the high-mixing-content fluidized bed can replace the ordinary Portland cement to prepare the full tailings cement filling material under the excitation of FGD gypsum.

When ratio of fly ash: slag: FGD gypsum of the cementing material is 20%: 68%: 12% (the specific surface area of slag is 500 m^2/kg; that of the FGD gypsum is 320 m^2/kg), the concentration is 82%, and the cement-sand ratio is 1:4, the test block's strength will reach 16.33 MPa after 28 days in a standard maintenance tank with a temperature of 40 °C and a relative humidity of over 90%.

Compared with the ordinary Portland cement system, the slag system has a higher strength than the slag system after 28 days. But when the fly ash content reaches 60%, the fluidity of the cement system does not reach the standard of complete self-flow. The low-chlorine fly ash-slag-FGD gypsum system with the fly ash content of 60% and the fluidity of 220mm can have the strength of 7.45MPa after 28d, which can meet the filling strength requirements and the self-flow standard.

Acknowledgement

This work was supported by the key research and development program of Hebei province [Grant Number 18273807D].

References

[1] Pan Y. (2015) Risk assessment of toxicity of domestic waste incineration fly ash and its resource utilization process. http://kns.cnki.net/kns/detail/detail.aspx?FileName=1015991153.nh&DbName=CDFD2016.
[2] Shi, H.S. (2007) Municipal waste incineration fly ash treatment technology and its resource utilization in cement production. Cement, 10: 1-4.
[3] Wang, W., Wu N. (2013) Research status of resource utilization of municipal solid waste incineration fly ash. Northern Environmental, 25: 27-28.
[4] Wu, B.R., Wang, D.Y.,Chai, X.L. (2015) Dechlorination mechanism of municipal solid wastes incineration fly ash by biological process. China Environmental Science, 35:2470-2476.
[5] Li, X.D., Liu, Y.D. (2007) Experimental study on elution of chlorine from fly ash water and its sintering stabilization. Acta Scientiarum Naturalium Universitatis Pekinensis, 6: 752-758.
[6] Chen, Y.C., Gui, B.F. (2005) Discussion of CFB technology used in living waste incineration. Jiangxi Energy, 4: 47-48.
[7] Bie, R.S., Wang, G.Q. (2003) Application of CFB combustion technology to incineration of municipal solid waste. Techniques and Equipment for Environmental Pollution Control, 4: 79-82.
[8] Xu,W.L., Wang, L.H. (2007) Current status and development forecast of waste incineration technology in China. China Environmental Protection Industry, 11: 24-29.
[9] Wang, L., Jin, Y.Y., Li, R.D. (2010) Pollution characteristics of municipal solid waste incineration fly ash. Environmental Science & Technology, 33: 21-26.
[10] Shi, H.S., Cheng, P., Guo, X.L. (2013) Research Status of Resource Utilization of Washed Municipal Solid Waste Incineration Fly Ash in Cement Production. Fly Ash Comprehensive Utilization, 6: 51-56.
[11] Liu, W.S., Zhang, C.H., Zhang D.J. (2008) Effect and Microscopic Analysis of Fluidized Bed

Fly Ash Solidification Incineration Fly Ash. Chinese Journal of Environmental Engineering, 09: 1260-1264.

[12] Wu, Y., Huang, R., Tsai,C. (2015)Utilizing residues of CFB co-combustion of coal, sludge and TDF as an alkali activator in eco-binder. Construction and Building Materials, 80: 69-75.

[13] Jouni, R., Katja Ohenoja, P.D.D.S., Paivo Kinnunen, P.D. (2017) Partial Replacement of Portland-Composite Cement by Fluidized Bed Combustion Fly Ash. J. Mater. Civ. Eng. 29: 04017061.

[14] Wang, K., Ni, W., Zhang, S.Q. (2018) Study on waste incineration fly ash-slag based gelling system and solid cadmium. Nonferrous Metals Engineering, 8: 123-127.

[15] Yang, H., Ni, W., Zhang, S.Q.(2018) Mechanism of solid cadmium in fly ash with metallurgical slag and cement incineration. Chinese Journal of Engineering, 40: 1027-1035.

[16] Zhu, X.L. (2015) Current Status and Technology Selection of Waste Incineration Power Generation Fly Ash Treatment. Electric Safety Technology, 17:59-62.

[17] Yan, X.H. (2018) Effect of Admixture on Mortar Performance and Mechanism.http://kns.cnki.net/kns/detail/detail.aspx?FileName=1018862295.nh&DbName=CMFD2018.

[18] Liu, Y., Fang, R., Cui, B.Q. (2018) Effect of Desulfurized Gypsum on Performance of Coal Powder Ash Filling Paste. Journal of Xianyang Normal University, 33:1-4.

[19] Lu, Q.Y., Du, J.F. (2018) Experimental study on the effect of iron tailings sand on the performance of dry mortar. Beton Chinese Edition —— Ready-mixed Concrete, 9:30-32.

[20] Yang, Q.R., Zhao, Z.Z., Zhang, Q.Z. (2019) Influence of Several Factors on the Rheological Properties of Cement Mortar. Journal of Building Materials, 22:506-515.

IOP Conf. Series: Materials Science and Engineering **729** (2020) 012048 doi:10.1088/1757-899X/729/1/012048

Preparation of Biodiesel Based on Alkaline Ionic Liquid [Bmim]OH Catalyzed Castor Oil

Aihua Zhang[1,3], Qin Wang[1,2],Yidan He[1,3], Pengying Lai[1,3],Yifu Miu[1] and Zhihong Xiao[1*]

[1] Hunan Academy of Forestry, Changsha, Hunan, 410004, China

[2] Changsha University of Science and Technology, Changsha, Hunan, 410114, China

[3] Central South University of Forestry and Technology, Changsha, Hunan, 410004, China

[*]Corresponding author's e-mail: xzhh1015@163.com

Abstract. In this experiment, a new basic ionic liquid [Bmim]OH was synthesized from 1-methylimidazole and sodium hydroxide, which was used to catalyze the preparation of biodiesel from castor oil. The catalyst of [Bmim]OH was compared with KOH and tetrabutylammonium hydroxide during the study. The experimental results showed that the latter two were better than the latter two. The orthogonal experimental design scheme was used to optimize the process conditions. The optimized catalytic conditions of alkaline ionic liquid [Bmim]OH were as follows: the catalyst dosage was 1%, the molar ratio of alcohol to oil was 6:1, the reaction temperature was 40 °C, and the reaction time was 60min. Under the optimized conditions, the crude methyl ester mixture yield is higher than 97%, the castor oil is more than 95% converted to the product methyl ester, and the catalyst of [Bmim]OH is used for six times without significant consumption, and the catalytic performance is stable.

1. Introduction

Castor oil is one of the oils with a high single fatty acid content, mainly hydroxy oils. The main component of fatty acids is ricinoleic acid, which accounts for more than 85% of the total acid. There are also a small amount of palmitic acid, oleic acid, linoleic acid, linolenic acid and stearic acid. The unsaponifiable matter is mainly sterols [1]. Castor oil contains a large amount of hydroxyl groups, and its solubility is very different from other oils. It has poor solubility in hydroxy solvents, but castor oil can be mixed with anhydrous ethanol at any room temperature, and 1 volume of castor oil can be dissolved in 5 volumes of 95% ethanol. Castor oil has unique properties. It does not solidify at -18°Cand deteriorate at 500°C-600°C, and does not burn. The preparation of biodiesel from modified castor oil has great potential in China.

At present, the main production methods of biodiesel are direct mixing, microemulsification, pyrolysis and transesterification[2]. Compared with other methods, transesterification has the advantages of mature technology, low cost and stable product properties, and has become the main method of biodiesel production. According to the type of catalyst used or whether catalyst is used, transesterification can be divided into chemical transesterification, enzymatic transesterification and supercritical fluid method [3]. Supercritical fluid method requires high molar ratio of alcohol to oil and reaction temperature as high as 350°C. The reaction conditions are relatively harsh, and it is difficult to achieve large-scale industrial applications [4]. Biodiesel prepared by enzymatic has the advantages

IOP Conf. Series: Materials Science and Engineering **729** (2020) 012048 doi:10.1088/1757-899X/729/1/012048

of mild reaction conditions, low alcohol consumption, easy collection of products and low environmental pollution. However, it is still in the stage of technological exploration. There is no report on the production practice of biodiesel prepared by enzymatic method. Moreover, the cost of lipase is too high and it is easy to deactivate. The by-product glycerol produced during the reaction is easy to adhere to the lipase surface, which makes the reaction difficult. The development of enzymatic transesterification of biodiesel was limited [5]. Chemical transesterification is the main method of biodiesel production because of its mature technology and low cost. However, both homogeneous and heterogeneous catalysts have their shortcomings, such as the catalyst can't be recycled or the post-treatment method is more complex during in the chemical transesterification.

Ionic liquid catalysts have a series of excellent properties, such as low vapor pressure, wide liquid temperature range, good compatibility with reactants, easy operation and treatment, recyclable use, designable molecules, acid-base catalysis and phase transfer catalysis. In many catalytic reactions, ionic liquids have shown high activity and selectivity. In the field of catalysis and organic synthesis, acidic ionic liquid catalysis has been playing a very important role [6-9].

Inspired by literature [10-13], this study attempted to prepare biodiesel from castor oil catalyzed by basic ionic liquid of [Bmim]OH, studied and optimized the catalytic process conditions and catalyst stability, which laid a foundation for industrial application of the catalytic system.

2. Materials and methods

2.1. Experimental materials

Methanol (AR, Tianjin Kemiou Chemical Reagent Co., Ltd.); NaOH(AR, Tianjin Fengchuan Chemical Reagent Co., Ltd.); KOH(AR, Tianjin Fengchuan Chemical Reagent Co., Ltd.); NaCl(AR, Tianjin Kemiou Chemical Reagent Co., Ltd.); 1-Methylimidazole (AR, Sinopharm Chemical Reagent Co., Ltd); Acetonitrile (AR, Tianjin Kemiou Chemical Reagent Co., Ltd.); Ethyl acetate (AR, Sinopharm Chemical Reagent Co., Ltd); Castor oil (CP, made by Hunan Academy of Forestry); n-butyl bromide (AR, Sinopharm Chemical Reagent Co., Ltd)

2.2. Experimental Methods

2.2.1. Synthesis of [Bmim] OH. 1-Methylimidazole 16.2g was added to a 250mL flask with thermometer and magnet. The flask was slowly filled with nitrogen, heated to 70°C, slowly added 27.4g of equal molar n-butyl bromide, and the pale yellow bromide [Bmim]Br was obtained after 4h of dropping reaction. [Bmim]Br 0.05mol was dissolved in 50mL dry acetonitrile, placed in 250mL round bottom flask, heated to 25°C, added KOH 0.06mol, reacted for 12h, stopped stirring, filtered the reaction materials, removed KBr, rotated evaporation to remove the solvent in the filtrate, and obtained crude ionic liquid [Bmim]OH. The crude ionic liquids were washed three times with ethyl acetate, and then dried in vacuum for 24 hours and obtained a yellowish transparent viscous liquid [Bmim]OH. The chemical reaction formulas are as follows:

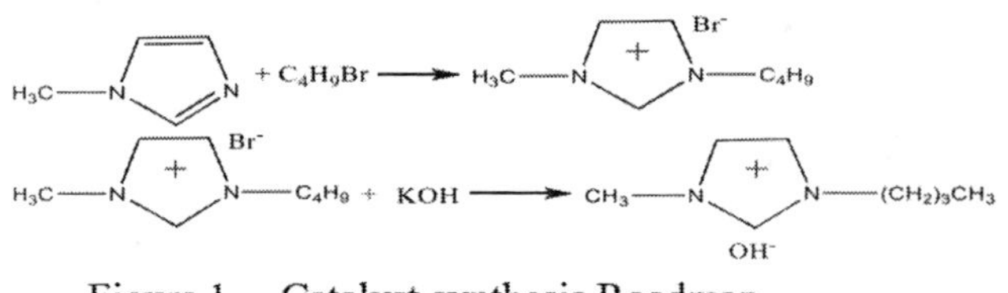

Figure 1. Catalyst synthesis Roadmap

2.2.2. Transesterification reaction. Take a certain amount of castor oil and dry it for 2 hours in a rotary evaporator at 140 °C and 0.095mpa. Accurately weigh 200g of dried castor oil, put it into a 500ml three port flask, preheat to a certain temperature, and add alkaline ionic liquid and methanol into the three port flask together. After the reaction for a period of time, it is transferred to the

IOP Conf. Series: Materials Science and Engineering **729** (2020) 012048 doi:10.1088/1757-899X/729/1/012048

separation funnel for standing and layering. The catalyst phase at the lowest layer is recycled and the glycerine phase is separated. The methyl ester at the upper layer is first distilled under normal pressure, and the excess methanol is recovered, and then washed twice with 10% brine to remove the soap salt. The refined biodiesel was obtained by vacuum distillation at 198°C and 5pa.

3. Results and analysis

3.1. Physical and chemical constants of castor oil.

Table 1 lists the physical and chemical parameters of castor oil. It can be seen that castor oil is an excellent raw material for preparing biodiesel.

Table 1. The physical and chemical constants of castor oil

Analysis item	acid value /(mg.g^{-1})	viscosity (40°C)/(mm^2.s^{-1})	Saponification value /(mg.g^{-1})	Iodine value g/100g	Refractive index (n^{20}_D)
index	1	≥14	176～180	82～86	1.477～1.481
Analysis item	Ricinic acid /%(w)	oleic acid /%(w)	Molar mass (g/mol)	relative density (g/cm^3)	Ignition point /°C
index	82～88	4～7	961.7	0.956	322

3.2. Catalyst selection.

The catalytic performance of different kinds of catalysts for the preparation of biodiesel was investigated. The experimental conditions were as follows: the amount of catalyst was 1% of castor oil mass, the alcohol oil ratio was 6:1, the reaction temperature was 40 °C, and the reaction time was 50 minutes.

Table 2. Comparison of catalysis with different catalyzers

Type of catalyst	productivity %
KOH	92.8
TBAH	78.1
[Bmim]OH	94.4

TBAH can be seen from table 2 that the catalytic performance is poor, and the post-treatment of the product is complex; KOH has good catalytic performance, but it can not be reused after the reaction, and a large number of waste alkali liquor is produced, causing serious environmental pollution. The catalytic performance of [bmim] Oh is good, almost no consumption in the reaction, and the catalytic performance is stable.

3.3. Optimization of synthesis conditions.

In the preparation of biodiesel from castor oil catalyzed by [Bmim]OH, four factors and three levels of orthogonal test were used to optimize the molar ratio of alcohol to oil, the amount of catalyst, the reaction temperature and the reaction time. The level of selected factors is shown in Table 3, and the results of orthogonal test are shown in Table 4.

Table 3. Factors and levels of orthogonal experiment

Factor / Level	Catalyst dosage /%	Alcohol oil ratio	Reaction temperature /°C	Reaction time /min
	A	*B*	*C*	*D*
1	0.9	5:1	30	40
2	1.0	6:1	40	50
3	1.1	7:1	50	60

Table 4. Results of orthogonal experiment

Number	1	2	3	4	Conversion rate %
1	1	1	1	1	77.51
2	1	2	2	2	87.72
3	1	3	3	3	85.67
4	2	1	2	3	91.06
5	2	2	3	1	96.42
6	2	3	1	2	92.37
7	3	1	3	2	81.03
8	3	2	1	3	94.09
9	3	3	2	1	86.10
K_1	250.9	249.6	263.97	260.03	Sum=791.97
K_2	279.85	278.23	264.88	261.12	
K_3	261.22	264.14	263.12	270.82	Average=88.0
R	28.95	28.63	1.76	10.79	

At the level of orthogonal test, the transesterification can be carried out well, and the static mixture has obvious stratification. Theoretically, the mass of methyl ester and glycerin from 200 g castor oil is about 200.9 g and 20.7 g respectively. The highest conversion of the product was 96.42%, which indicated that the transesterification reaction was basically complete under suitable reaction conditions.

According to table 4, the conversion rate of A2 Level transesterification is the highest, and the amount of catalyst is 1.0% of the oil weight. It is found in the experiment that excessive amount of catalyst will aggravate product saponification and make post-treatment difficult. The conversion rate of B2 level transesterification is the highest. From the trend of 1-3 level, it can be seen that the optimal amount of alcohol oil mole ratio is 6:1. The conversion rate of C2 level transesterification is C factor. The highest, but from the 1-3 level trend, there is little difference.

There is an optimal reaction temperature for transesterification. Low temperature will slow down the reaction and reduce the conversion rate. Too high temperature will accelerate the volatilization of methanol and reduce the volume concentration of methanol in the liquid phase, which will also reduce the conversion rate. Therefore, in the actual operation process, the lower reaction temperature can be selected, which will not affect the reaction speed too much, but also effectively avoid the reduction of conversion rate caused by methanol volatilization and reduce energy consumption; among the D factors, the conversion rate of D3 level transesterification reaction is the highest, but the difference is slightly larger than that of C factor from the perspective of 1-3 level trend. It can be seen that in a certain period of time, with the reverse reaction, the With the increase of reaction time, the conversion of biodiesel increased obviously. This is because in the initial stage of the reaction, the reaction has not reached equilibrium. Increasing the reaction time can make the reaction reach equilibrium, thus increasing the conversion of biodiesel.

The range is $R_A > R_B > R_D > R_C$. From the data in Table 4, it can be concluded that the values of R_C and R_D are far less than R_A and R_B, indicating that the amount of catalyst and molar ratio of alcohol to oil have a significant impact on the conversion rate. Therefore, $A_2B_2C_2D_3$ is the best experimental condition for the preparation of biodiesel by catalytic castor oil, that is, the amount of catalyst is 1%, the molar ratio of alcohol to oil is 6:1, the reaction temperature is 40°C, and the reaction time is 60 minutes.

3.4. Stability of [Bmim]OH.

Ionic liquid is a new type of catalyst. The separation of product and ionic liquid can be realized by simple phase separation. The separated ionic liquid can be reused after simple treatment. The stability of ionic liquid catalyst was investigated under the optimized conditions. The results are shown in

Fig.2.

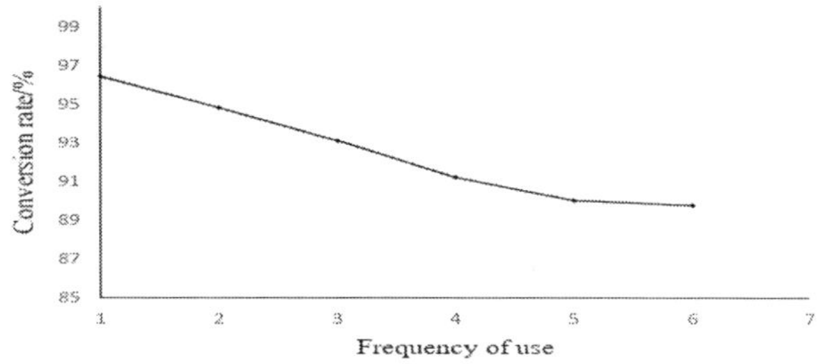

Figure 2. Stability of the ionic liquid

It can be seen from Figure 2 that [Bmim]OH can be stably recycled for more than six times. In the figure, after the second reaction, the product is stratified. The lower [Bmim]OH is directly used for the third reaction without any treatment, and the other several reaction catalysts are recycled for the next experiment. It can be seen that, except for the third cycle, the conversion rate of the product biodiesel slightly decreased, the catalyst activity of the other cycles basically did not change, indicating that [bmim] Oh catalyst has good stability.

3.5. Analysis of physical and chemical constants of biodiesel

Table 5. Main quality indexes of biodiesel

Item	Analysis method	Product sample	*ASTM*6751-03	$0^{\#}$diesel standard
Acid value /(mg.g^{-1})	GB/T 5530-2005	0.4052	≤0.8	-
Closed flash point /°C	GB 267-88	179	≥130	≥55
Density (20°C)/(g.cm^{-3})	GB/T 2540-1981	0.938	0.875～0.90	0.820～0.860
Kinematic viscosity (40°C)/(mm^2.s^{-1})	GB/T 265-1988	15.1	1.9～6.0	2.0～4.5(20°C)
Calorific value /(KJ/g)	GB/T 384-1981	40.1	-	
Water content /%(*w*)	GB/T 260-1977	Trace	≤0.05	Trace
Ash content /%(*w*)	GB/T 508-1965	0.005	≤0.02	-
Freezing point /°C	GB/T 510- 83	-19.2	-	0
Pour point /°C	GB/T 3535- 83	-23.1	-	-2
Cloud point /°C	GB/T 6986-86	-14.6	-	7

It can be seen from table 5 that this experimental product basically meets the American ASTM 6751-03a biodiesel standard, some of which meet the $0^{\#}$ diesel quality index of China, but the kinematic viscosity of castor oil methyl ester exceeds the standard, which may be caused by the following two aspects: first, the average carbon chain length of the biodiesel molecule is longer than that of the mineral diesel molecule; second, the castor oil methyl ester contains hydroxyl group, which can produce molecules The hydrogen bond between. Therefore, diluents, such as ethanol and petrochemical diesel oil, need to be added in use.

4. Conclusion and discussion

In this study, castor oil is used as raw material and [bmim] Oh is used as catalyst to prepare biodiesel. The optimum process conditions are as follows: the amount of catalyst is 1% of the quality of raw oil, the molar ratio of alcohol to oil is 6:1, the reaction temperature is 40°C, the reaction time is 50min, and the conversion rate is over 95%.

Acknowledgments
The author is grateful to the Hunan Provincial Natural Science Foundation (2018JJ2218), the introduction of foreign intellectual special projects in Hunan Province (2019YZ3026), the Changsha, Zhuzhou and Xiangtan National Independent Innovation Demonstration Project (2018XK2005) and the Hunan Provincial Key Laboratory (2018TP1029).

References

[1] Xuebing Xu. Oil chemistry [M].Bei Jing:China Commercial Publishing House,1993:326~327.
[2] Yiping Wang, Yi Zhai, Jinli Zhang, et al. Research Progress of Biodiesel Preparation [J]. Chemical industry and engineering progess, 2003, 22(8~12).
[3] Helin Wang, Wudi Zhang, Shiqing Liu, et al. Study on production of biodiesel from castor oil [J]. Energy engineering, 2007, 3 (24~26).
[4] Freedman B, Butterfield R O, Pryde E H. Transesterifi- cation Kinetics of Soybean Oil. J Am Oil Chem Soc, 1986, 63(10):1 375~1 380.
[5] Boocock D G B, Konar S K, Mao V, et al. Fast One Phase Oil Rich Processes for the Preparation of Vegetable Oil Methyl Esters. Bioresour Technol, 1996, 11(1):43~50.
[6] Ruxiong Li, Jianji Wang. Synthesis and application of ionic liquid. Chemical reagents, 2001, 23(4):211~215.
[7] Zhao H. Current Studies on Some Physical Properties of Ionic Liquids. Physics and Chem- istry of Liquids, 2003, 41(6):545~557.
[8] Cloe A C,Jensen J L,et al. Nnovel BrÖnsted Acidic Ionic Liquids and Their Use as Dual Solvent-Catalysts. J Am Chem Soc, 2002, 124(21): 5 962~5 963.
[9] Yoshizawa M, Hirao M, Ito-Akita K, et al. Ion Conduction in Zwitterionic-Type Molten Salts and Their Polymers. J Mater Chem, 2001, 11 (4):1 057~1 062.
[10] Branco L C, Rosa J N, Moura Ramos J J, et al. Preparation and Characterization of New Room Temperature Ionic Liquids. Chem Eur J. 2002, 8(16):3 671~3 677.
[11] Zhu Huaping, Yang Fan, He Mingyuan. BrÖnsted Acidic Ionic Liquid 1-Methylimi-dazolium Tetrafluor-oborate: A Green Catalyst and Recyclable Medium for Esterification. Green Chem, 2003, 5(1):38~39.
[12] Pernak J, Czepukowicz A. New Ionic Liquid and Their Antielectrostatic Properties. Ind Eng Chem Res, 2001, 40 (11) : 2 379~2 383.
[13] Atef A, Jean P B. Efficient Combination of Recycle Task Specific Ionic and Microwave Dielectric Heating for the Synthesis of Lipophilic Esters. Org Process Ras Dev, 2005, 9(4):743~748.

Removal of Oil from SZ36-1 Simulated Oil Sands Using Sophorolipid Biosurfactant

Junqi Wang[1,2], Juan Wang [3], Sihao Dong[3], Guobin Li[3], Xiangping Kong[1,3*]

[1]The Key Laboratory of Well Stability and Fluid & Rock Mechanics in Oil and Gas Reservoir of Shaanxi province, Xi'an 710065, China

[2]School of Petroleum Engineering of Xi'an Shiyou University, Xi'an 710065, China

[3]College of Chemistry and Pharmacy, Qingdao Agricultural University, Qingdao 266109, China

[*]Corresponding author's e-mail: kxp2004@163.com

Abstract. The sophorolipid biosurfactant was used to remove the oil from the SZ36-1 simulated oil sands. The effects of the sophorolipid concentration, the mass ratio of sophorolipid solution to oil sands (liquid/solid ratio), and the extraction time on the oil extraction efficiency were investigated. Based on the single-factor tests, the optimum results were obtained under the conditions of sophorolipid solution concentration to be 100 $mg \cdot L^{-1}$, the liquid/solid ratio to be 50:1 and the extraction time to be 2 h. The corresponding oil extraction rate from oil sands reached up to 42%, which was comparable to the result from 2.5 $g \cdot L^{-1}$ chemical surfactant of sodium dodecyl sulfate (the value of 2.5 $g \cdot L^{-1}$ was higher than its critical micelle concentration). The oil extraction rate of crude oil with sophorolipid solution was almost not affected by the existence of sodium chloride. The sophorolipid biosurfactant has high surface activity and good oil removing capacity, and is expected to replace the synthetic surfactants to be used in petroleum exploitation and crude oil pollution treatment.

1. Introduction

At present, oil sands mining, tertiary oil recovery and oil pollution treatment have attracted the attentions of researchers in the petroleum industry[1-5]. Surfactants have been widely used in these industrial processes due to their good physicochemical characteristics such as detergency, emulsification, dispersion and solubilization effects[3]. Compared with traditional chemical synthetic surfactants, biosurfactants have excellent characteristics of high surface/interface activity, low toxicity, easy biodegradation, good temperature/salt tolerance[6], and have great application potential in these fields.

Sophorolipid is complex glycolipids comprised of the disaccharide sophorose linked to a hydroxy fatty acyl group by a glycosidic bond, and can be divided into two predominant forms: the macrolactone and the free carboxylic acid[6-8]. As one kind of important biosurfactants, the sophorolipid has been frequently used in medicine, cosmetics, environmental protection as well as petroleum exploitation. In this work, the oil extraction efficiency of sophorolipid solution from oil sands was investigated, and the results were compared with those from the common chemical surfactant of sodium dodecyl sulfate.

2. Experimental

2.1. Materials

Crude oil SZ36-1[9] (high viscosity, low wax, heavy oil) was produced by China National Offshore Oil Corporation. Sophorolipid[6] and 120# solvent oil were supplied by the Key Laboratory of Marine Chemistry Theory and Technology of Ministry of Education and Shanxi Sunshine Chemical Co., Ltd., respectively. The sands were taken from the Aoshanwei Beach of Jimo District, Qingdao. All the other chemicals used in this work were of analytical grade and all solutions were prepared with deionized water.

2.2. Methods

The SZ36-1 simulated oil sands were prepared according to the following method. The crude oil was extracted with chloroform to remove impurities. The lower organic layer was collected, and was evaporated at 40 °C to remove chloroform to obtain the crude oil for analysis. The sands were washed with 0.2 M acetic acid solution and were sterilized subsequently. The obtained crude oil was dissolved in petroleum ether, and poured into the sterilized sands, and then stirred and mixed thoroughly. After the petroleum ether was completely volatilized, it was naturally air-dried for 5 days to obtain simulated oil sands with the oil content of about 70 $mg \cdot g^{-1}$.

In the extraction experiments, effects of the sophorolipid concentration, the ratio of extraction agent to oil sands, and the extraction time were investigated with single-factor experiments. All the experiments were repeated three times, and the mean values were used as results for calculation and analysis. The oil extraction rate (R) from oil sands was calculated according to equation (1):

$$R = \frac{C_0 - C}{C_0} \times 100\% \quad (1)$$

where R is the oil extraction rate, %; C_0 and C are the oil contents in the original oil sands and the treated oil sands by surfactant solution, respectively, $mg \cdot g^{-1}$. C_0 and C were determined with the solvent oil extraction method[10]. The oil extraction rate was close to 100% under the conditions of mass ratio of 120# solvent oil to oil sands to be 10:1, repeated times of 3, extraction time of 40 min and extraction temperature of 75 °C by the magnetic stirring. The oil content of 120# solvent oil solution was measured with UV-2600 UV-Vis spectrophotometer (Shimadzu Scientific Instruments Inc., Japan).

The crude oil for analysis was dissolved in 120# solvent oil to obtain about 50 $mg \cdot L^{-1}$ oil solution. Taking 120# solvent oil as a control, a full scan with a wavelength of 200-600 nm was performed to get the UV-visible absorption curve of the oil solution using the UV-2600 UV-Vis spectrophotometer. The UV-visible absorption curve of 120# solvent oil was similarly obtained using deionized water as a control.

3. Results and discussion

3.1. The UV-visible absorption curve

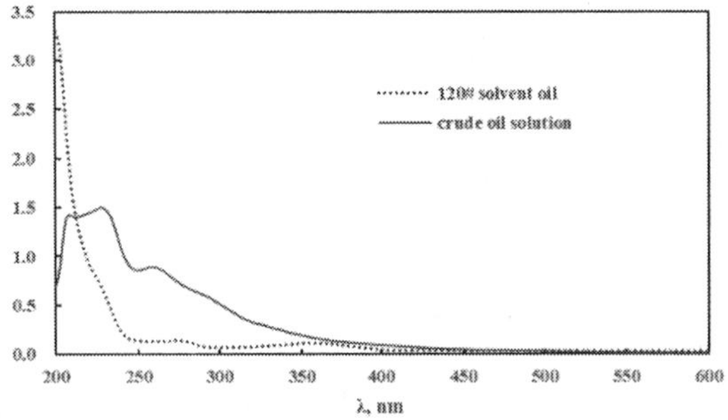

Figure 1. The UV-visible absorption curves of 120# solvent oil and crude oil solution

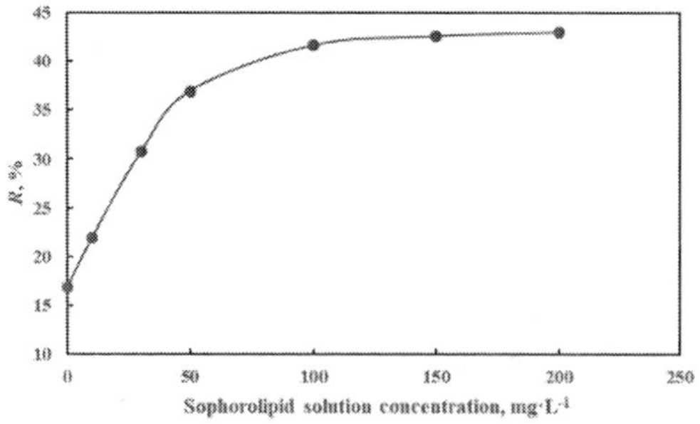

Figure 2. Effect of the sophorolipid solution concentration on extraction rate of crude oil

As shown in Figure 1, both 120# solvent oil and crude oil have no absorption peak in the visible region, suggesting that the visible light region is not suitable for measuring crude oil content in 120# solvent oil solution. Taking 120# solvent oil as a control, the maximum absorption wavelength of crude oil is 228 nm. However, the absorption of 120# solvent oil is strong and the absorption intensity changes greatly near the wavelength of 228 nm, which can lead to large measurement errors. The wavelength of 228 nm is not suitable for determining the crude oil content in 120# solvent oil solution. At a wavelength of 260 nm, the crude oil has an absorption peak and the absorption intensity is almost unchanged near this wavelength. The absorption intensity of 120# solvent oil is weak, and there is almost no change around the wavelength of 260 nm. Thus, the background interference caused by 120# solvent oil is small, and the measurement results of crude oil content in 120# solvent oil solution has high accuracy. Therefore, the wavelength of 260 nm is found to be preferable for determining the crude oil content in 120# solvent oil solution.

3.2. The extraction experiments
The crude oil was dissolved in 120# solvent oil to obtain crude oil standard solutions at different concentrations of 10.00, 20.00, 30.00, 40.00 and 50.00 $mg \cdot L^{-1}$, respectively. The absorbances of these standard solutions were measured at 260 nm using UV-2600 UV-Vis spectrophotometer. A standard curve was drawn based on the absorbances of the standard solutions and the corresponding mass concentrations, and the linear regression equation was $A=0.0183\rho-0.0074$ with R^2 of 0.9990. The crude oil from oil sands and treated oil sands with sophorolipid solution was extracted with 120# solvent oil, respectively. Then the absorbance was measured, and the crude oil content in 120# solvent oil solution was obtained according to the standard curve. Thus, the oil extraction rate could be calculated.

3.2.1. Effect of sophorolipid solution concentration on oil extraction rate. At room temperature, the effect of sophorolipid solution concentration on oil extraction rate was tested with sophorolipid solution to oil sands of 50:1 (m:m) and extraction time of 6 h under the electromagnetic stirring conditions. As shown in Figure 2, the oil extraction rate increased rapidly within the sophorolipid solution concentrations of 0 to 50 $mg \cdot L^{-1}$, and then increased steadily within the concentrations of 50 to100 $mg \cdot L^{-1}$ until it remained almost unchanged. It may be explained by the critical micelle concentration (CMC) of sophorolipid solution. When the sophorolipid solution concentration is lower than the CMC, the surface activity of sophorolipid solution increases rapidly as the concentration of the sophorolipid solution increases. Correspondingly, the oil extraction rate will increase rapidly with the increasing sophorolipid solution concentration. The CMC of sophorolipid solution is about 36.5 $mg \cdot L^{-1}$[6], so the oil extraction rate increases rapidly as the concentration of the sophorolipid solution increases at concentration of lower than 50 $mg \cdot L^{-1}$. When the concentration of the sophorolipid solution is within 50 to 100 $mg \cdot L^{-1}$ (slightly higher than CMC), the surface activity of sophorolipid solution and the oil extraction rate increase steadily with the increasing sophorolipid concentration. When the concentration of the sophorolipid solution is larger than 100 $mg \cdot L^{-1}$ (far higher than the value of CMC), the surface activity of sophorolipid solution and the oil extraction rate almost do not change with the increasing sophorolipid concentration. Therefore, considering the extraction efficiency and economic factors, the concentration of the sophorolipid solution was selected to be 100 $mg \cdot L^{-1}$.

3.2.2. Effect of sophorolipid solution to oil sands ratio on oil extraction rate. The effect of mass ratio of sophorolipid solution to oil sands (liquid/solid ratio) on oil extraction rate was investigated with sophorolipid solution concentration of 100 $mg \cdot L^{-1}$ and extraction time of 6 h. As can be seen from Figure 3, the oil extraction rate of crude oil first increased and then decreased slightly with the increased ratios of liquid/solid. When the ratios were less than 50:1, increasing the values of ratio resulted in the increased contact between the oil sands and the solvent, thereby increased the oil extraction rate. When the liquid/solid ratio was higher than 50:1, the oil sands and the solvent could not be mixed completely under the electromagnetic stirring condition, which lead to the slightly decrease of oil extraction rate. As shown in Figure 3, the maximum oil extraction rate of slightly higher than 42% was obtained when the liquid/solid ratio was 50:1. Thus, 50:1 was chosen as the optimum ratio of sophorolipid solution to oil sands.

3.2.3. Effect of extraction time on oil extraction rate. The effect of extraction time on oil extraction rate was tested with liquid/solid ratio of 50:1 and sophorolipid solution concentration of 100 mg·L^{-1} under the electromagnetic stirring conditions. As shown in Figure 4, the oil extraction rate increased in the first 2 hours, and then remained almost constant with the increase of extraction time. The dissolution of crude oil in the sophorolipid solution needs a certain amount of time. After the sophorolipid solution and oil sands interacted for 2 hours, the crude oil was in a dissolved equilibrium state in the sophorolipid solution, and the maximum oil extraction rate of about 42% was obtained. Therefore, the extraction time was selected to be 2 hours.

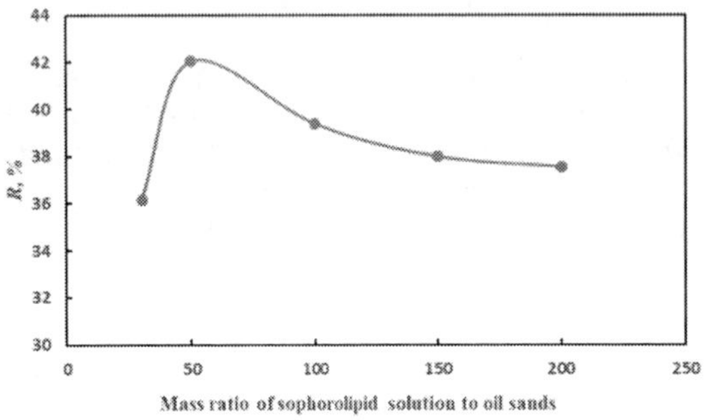

Figure 3. Effect of the mass ratio of sophorolipid solution to oil sands on extraction rate of crude oil

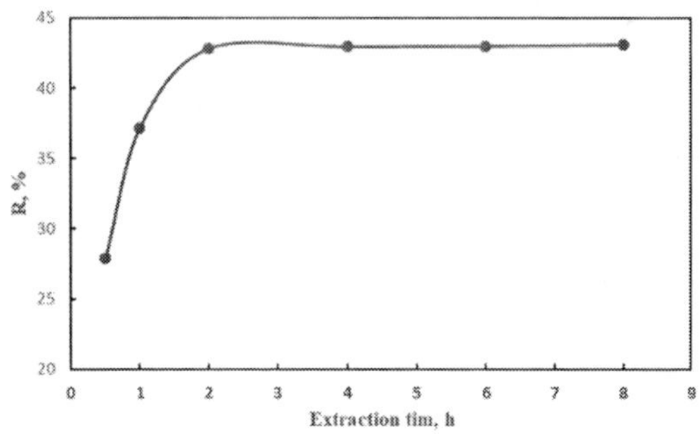

Figure 4. Effect of the extraction time on extraction rate of crude oil

3.3. Effect of NaCl concentration

The effect of NaCl concentration on oil extraction rate was investigated with liquid/solid ratio of 50:1, the sophorolipid solution concentration of 100 mg·L^{-1} and extraction time of 2 h. When the concentrations of sodium chloride were 10, 30, 50, 80 and 100 g·L^{-1}, respectively, the oil extraction rates of crude oil were all slightly higher than 42%, which was consistent with no existence of sodium chloride. The results showed that the oil extraction rate of crude oil with sophorolipid solution was almost unaffected by concentrations of NaCl, which could be explained by the fact that the surface activity of sophorolipid solution was almost not affected by existence of inorganic salts[8].

3.4. Comparison with sodium dodecyl sulfate solution

Sodium dodecyl sulfate (SDS) is one of the widely used chemical surfactants in petroleum exploitation and crude oil pollution treatment. Therefore, the ability of the sophorolipid biosurfactant to extract crude oil from the oil sands can be estimated by comparing the crude oil extraction rates of the sophorolipid solution to those from the SDS solution. The crude oil extraction rates using 100 mg·L^{-1} sophorolipid solution and 2.5 g·L^{-1} SDS (the CMC is about 2.3 g·L^{-1}) were investigated with agent to oil sands ratio of 50:1 and extraction time of 2 h, respectively. The oil extraction rates from oil sands using the sophorolipid solution and SDS were 42.8% and 43.4%, respectively. It suggested that sophorolipid biosurfactant with much lower concentration could achieve comparable oil extraction efficiency to chemical surfactant SDS of higher concentration.

4. Conclusions

The optimum conditions for the oil removal from SZ36-1 simulated oil sands using sophorolipid biosurfactant were obtained with sophorolipid solution concentration of 100 mg·L^{-1}, the liquid/solid ratio of 50:1 and extraction time of 2 h. Under these conditions, the oil extraction rate from oil sands could reach up to 42%, which was comparable to the result from 2.5 $g \cdot L^{-1}$ chemical surfactant of SDS. The oil extraction rate of crude oil with sophorolipid solution was almost not changed with the increasing concentration of sodium chloride. The sophorolipid biosurfactant is expected to replace the synthetic surfactants to be used in the petroleum exploitation and crude oil pollution treatment.

Acknowledgments
This work was subsidized by the Key Research & Development Plan Project of Shaanxi Province of China (2018KW-037), the Open Project of the Key Laboratory of Well Stability and Fluid & Rock Mechanics in Oil and Gas Reservoir of Shaanxi Province of China (WSFRM20180303003) and the Students' Innovation Education Project Funds of Qingdao Agricultural University.

References

[1] Kim, S.-J., Choi, D.H., Sim, D.S., Oh, Y.-S. (2005) Evaluation of bioremediation effectiveness on crude oil-contaminated sand. Chemosphere, 59: 845-852.
[2] Saxena, N., Kumar, S., Mandal, A. (2018) Adsorption characteristics and kinetics of synthesized anionic surfactant and polymeric surfactant on sand surface for application in enhanced oil recovery. Asia-Pac. J. Chem. Eng., 13(4): 1-14.
[3] Bera, A., Kumar, T., Ojha, K., Mandal, A. (2013) Adsorption of surfactants on sand surface in enhanced oil recovery: Isotherms, kinetics and thermodynamic studies. Appl. Surf. Sci., 284: 87-99.
[4] Fu, L.P., Zhang, G.C., Ge, J.J., Liao, K.L., He, Y.F., Wang, X., Li H.T. (2017) Study on dual-frequency ultrasounds assisted surfactant extraction of oil sands. Fuel Process. Technol., 167: 146-152.
[5] Zou, C.J., Zhao, P.W., Ge, T.T., Li, D., Ye, H., Huang, G. (2016) Bitumen recovery from Buton oil sands using a surfactant under the effect of ultrasonic waves. Energ. Source. Part A, 38(2): 270-276.
[6] Song, D.D., Liang, S.K., Wang, J.T. (2011) Structure characterization and physi-chemical properties of sophorolipid biosurfactants. Environ. Chem., 30(8): 1474-1479 (in Chinese).
[7] Sembayeva, A., Berhane, B., Carr, J.A. (2017) Lipase-mediated regioselective modifications of macrolactonic sophorolipids. Tetrahedron, 73(14): 1873-1880.
[8] Wang, J., Kong, X.P. (2015) Interaction of sophorolipids with polyacrylamide. Appl. Chem. Ind., 44(9): 1655-1658 (in Chinese).
[9] Zhang, R.H., Qu, T., Dai, C.Y. (2002) Make the best use of crude oil SZ36-1 to produce premium bitumen. Petrol. Asphalt, 16(3): 46-49 (in Chinese).
[10] Han, D. Y., Cao, Z. B., Xu, X. Q. (2013) Study on solvent extraction of Mongolia outcrop oil sands. Energ. Source. Part A, 35(14): 1368-1374.

IOP Conf. Series: Materials Science and Engineering **729** (2020) 012050 doi:10.1088/1757-899X/729/1/012050

An Investigation of Hydraulic-Fracturing Applied to Marine Gas Hydrate Reservoirs

Yanchang Liu[1*], Minggang Sun[1],Gongtao Wang[2], Qing Wang[3] and Lian Zhao[3]

[1]Electrical Information Engineering Department, Northeast Petroleum University, Daqing, Heilongjiang, 163000, China

[2]Information Engineering Branch of Liaohe Petroleum Exploration Bureau Co., Ltd., Panjin, Liaoning, 124000, China

[3]Test company of PetroChina Dagang Oilfield Tianjin Branch, Tianjin, 300000, China

[*]Corresponding author's e-mail: bq_lyc@sina.com

Abstract. Engineers have been struggling to harvest natural gas from the marine hydrates through drilling wells in the past two decades. The harvesting process has not been successful due to engineering problems such as wellbore collapse and sand production. This study proposes to use frac-packing method to address the wellbore collapse and sand production issues to break through the bottleneck of commercial production of natural gas from marine gas hydrates. An analytical model was developed to describe sequential initiation and simultaneous propagation of multiple vertical fractures in horizontal wells for frac-packing gas hydrate reservoirs. Result of sensitivity analyses with the models indicates that to frac-pack horizontal gas hydrate wells for solid-production control, increasing injection rate and flow behavior index of fracturing fluid will create more friction and thus more resistance to flow in large fractures, promoting the creation of more short fractures, and thus larger treated reservoir volume for higher gas production rate and gas recovery.

1. Introduction

Natural gas hydrates (NGH) are ice-like structures in which gas, most often gas, is trapped inside of water molecules. Unlike the ice gas hydrates are in fact highly flammable, a property that makes these crystalline structures an attractive future energy source. Hydrates are a much more abundant source of natural gas than conventional deposits. According to the U.S. Geological Survey, global stocks of gas hydrates range account for at least 10 times the supply of conventional natural gas deposits, with between 100,000 and 300,000,000 trillion cubic feet of gas yet to be discovered. If these sources of natural gas could be safely, efficiently and cheaply tapped into, gas hydrates could potentially displace coal and oil as the top sources of the world's energy [1].

Since 2002, Canada, the United States, Japan, China and India have carried out well testing and production of NGH[2~7]. Max and Johnson[8] reviewed the current practices in exploration and production of oceanic natural gas hydrate. There are mainly four methods for gas production from Nature Gas Hydrate (NGH) deposit: 1) depressurization [9,10]– this method should decrease the pressure in NGH deposit below the hydrate dissociation pressure; 2) thermal stimulation[11~13]– this approach will use external heat to make the temperature in the NGH deposit above hydrate dissociation temperature with hot water/brine/steam; 3) thermodynamic inhibitor injection[14,15]–this technique is designed to inject chemicals, such as salts and alcohols, to change the hydrate pressure–

temperature equilibrium conditions; 4) a combination of these methods[16]. Guo and Shan[17] presented a feasibility study of harvesting natural gas from seabed hydrates using moving risers.

The energy industry consensus is that commercial-scale production remains years away due to unsolved technical and environmental issues. Engineers have been struggling to harvest natural gas from the marine hydrates through drilling wells in the past two decades. The harvesting process has not been successful due to engineering problems such as wellbore collapse and sand production [18]. This study proposes to use frac-packing method to address the wellbore collapse and sand production issues to break through the bottleneck of commercial production of natural gas from marine gas hydrates. Frac packing is a technology that is widely used in developing unconsolidated high-permeability oil reservoirs to mitigate sand production problems [19]. It is a completion technique that merges two distinct processes – hydraulic fracturing and gravel packing. Instead of using gravel, sand and or silt may be used to fill in the fractures and block the fine particles that are otherwise produced from gas hydrate reservoirs during depressurization. However, it is not clear how fractures are created in the gas hydrate deposits. The fracturing process is investigated through mathematical modeling in this investigation.

Early research work on fracture creation was reported by Roussel and Sharma [20], Manchanda and Sharma[21] and Roussel et al. [22]. Recent investigations were performed by Rucker et al. [23], King and Valencia[24] and Li et al. [25]. They have shown that geological conditions, including the presence of natural fractures, in-situ formation stresses, and stress shadowing of passive hydraulic fractures around the active hydraulic fractures, play important roles in controlling propagation of hydraulic fractures. However, gas hydrate reservoirs are much more homogeneous than conventional oil and gas reservoirs. Mathematical models for homogeneous reservoirs may be used for describing fracture creation and propagation. Conservation of energy and calculus of variation were utilized to form a hierarchical fracture model by Cheng[26] to describe the fracture complexity. Following Cheng et al.'[27] work, Feng et al. [28] defined a fracture complexity indicator as the number of fracture branches. Result of their analyses indicates that the fracture complexity increases near linearly with the fluid injection rate and time. The fracture complexity also increases with the elastic modulus of rock, which is consistent with observations in the practice of volume fracturing of shale gas formations. Nevertheless, Cheng's [26] model and Feng et al's[28] model do not provide quantitative description of the uneven propagation of fracture branches.

This work developed an analytical solution in closed-form to describe formation of vertical fractures. Sequential initiation and simultaneous propagation of multiple vertical fractures with non-Newtonian fluid in horizontal wells were mathematically described. The effects of fluid injection rate and flow behavior on fracture length were identified. An analytical model describing fluid flow in horizontal fractures was also derived for CO2-gas fracturing vertical wells. This model is prepared for designing effective frac-packing of gas hydrate reservoirs.

2. MATHEMATICAL MODEL

Hydraulic fractures open against the minimum stress and propagate in the maximum stress direction. Wherever the formation stress in the vertical direction is greater than the stresses in the horizontal directions, vertical fractures are created [20]. Horizontal wells are drilled to create rectangular hydraulic fractures for improving productivity of oil and gas wells [29]. An analytical model of vertical fractures in horizontal wells for describing sequential initiation and simultaneous propagation of multiple vertical fractures with non-Newtonian fluid in horizontal wells is developed in this work. Due to the limited space of this paper, detailed derivation will be given elsewhere. The resultant equations are presented in this section. The dynamic propagation distance of the first fracture is described by

$$x = \left(\frac{2n+2}{2n+1} C_1 \right)^{\frac{2n+1}{2n+2}} t^{\frac{2n+1}{2n+2}} \quad (1)$$

where x is the fracture propagating distance (length) in ft; n is fluid flow behavior index (dimensionless); t is flow injection time in second, and is a group of constants defined as

$$C_1 = \left[\frac{372.25gq_i(P_p - P_t)}{Kh_f}\right]^{\frac{1}{2n+1}}\left[\frac{0.5q_i}{(3+n^{-1})h_f}\right]^{\frac{n}{2n+1}} \quad (2)$$

where is the gravitational acceleration factor in lbm-ft/lbf-s2; is the injection rate of fracturing fluid into one side of perforation cluster in ft3/s (the total injection rate per cluster is 2); is pressure outside the perforation cluster in lbf/ft2; is the pressure at fracture tip in lbf/ft2; K is fluid consistency index in cp; and is fracture height in ft. The dynamic average fracture width is expressed as

$$w = \frac{\left(\frac{2n+2}{2n+1}\right)\left(\frac{2n+2}{2n+1}C_1\right)^{-\frac{2n+1}{2n+2}}}{h_f} q_i t^{\frac{1}{2n+2}} \quad (3)$$

where w is the average fracture width in ft.

When the width of the fracture reaches a critical value wc, the rock stress at the initiation point of the fracture should cause failure of rock and initiate the second fracture. The critical average fracture width may be calculated by wc = where is the maximum fracture width at the wellbore which can be estimated using Rahman and Rahman's[30] equation:

$$w_f = 9.15^{\frac{1}{2n+2}} 3.98^{\frac{n}{2n+2}} \left[\frac{1+2.14n}{n}\right]^{\frac{n}{2n+2}} K^{\frac{n}{2n+2}} \left[\left(\frac{q_i}{2}\right)\frac{h_f^{1-n} x_f}{E}\right]^{\frac{1}{2n+2}} \quad (4)$$

where E is the Young's modulus of formation rock, xf is the fracture length from wellbore to the fracture tip. Inserting w = wc into Eq. (3) and solving for time t give the critical time for the first fracture to reach the critical length:

$$t_{c1} = \left[\frac{w_c h_f}{q_i\left(\frac{2n+2}{2n+1}\right)\left(\frac{2n+2}{2n+1}C_1\right)^{-\frac{2n+1}{2n+2}}}\right]^{2n+2} \quad (5)$$

Letting t = tc in Eq. (1), the critical length of the first fracture at the critical time is expressed as

$$x_{01} = \left(\frac{2n+2}{2n+1}C_1\right)^{\frac{2n+1}{2n+2}} t_{c1}^{\frac{2n+1}{2n+2}} \quad (6)$$

Assuming the width of the first fracture stabilizes at wc, the additional propagation length of the first fracture after the second fracture is created can be expressed by

$$x_1 = -x_{01} + \left(x_{01}^{\frac{n+1}{n}} + \frac{n+1}{n}C_1' t\right)^{\frac{n}{n+1}} \quad (7)$$

Where

$$C_1^{'} = \left[\frac{372.25 g w_c (P_p - P_t)}{K}\right]^{\frac{1}{n}} \left(\frac{0.5 w_c}{3+n^{-1}}\right) \quad (8)$$

After initiation of the second fracture, the propagating velocity of the first fracture is given by

$$v_1 = C_1^{'}\left(x_{01}^{\frac{n+1}{n}} + \frac{n+1}{n}C_1^{'}t\right)^{-\frac{1}{n+1}} \quad (9)$$

and the fluid flow rate in the first fracture is expressed as

$$q_1 = h_f w_c v_1 = h_f w_1 C_1^{'}(x_{01}^{\frac{n+1}{n}} + \frac{n+1}{n}C_1^{'}t)^{\frac{-1}{n+1}}$$

(10)

The fluid injection rate into the second fracture is therefore

$$q_2 = q_i - q_1 = q_i - h_f w_c C_1^{'}(x_{01}^{\frac{n+1}{n}} + \frac{n+1}{n}C_1^{'}t)^{\frac{-1}{n+1}} \quad (11)$$

where is in ft3/s. The dynamic propagation length of the second fracture is given by

$$x_2 = \left(\frac{2n+2}{2n+1}C_2\right)^{\frac{2n+1}{2n+2}} t^{\frac{2n+1}{2n+2}} \quad (12)$$

where

$$C_2 = \left[\frac{372.25 g q_2 (P_p - P_t)}{K h_f}\right]^{\frac{1}{2n+1}} \left[\frac{0.5 q_2}{(3+n^{-1})h_f}\right]^{\frac{n}{2n+1}} \quad (13)$$

The dynamic average width of the second fracture is expressed as

$$w_2 = \frac{\left(\frac{2n+2}{2n+1}\right)\left(\frac{2n+2}{2n+1}C_1\right)^{-\frac{2n+1}{2n+2}}}{h_f} q_2 t^{\frac{1}{2n+2}} \quad (14)$$

When the width of the second fracture reaches the critical value wc, the rock stress at the initiation point of the fractures should cause rock failure and initiation of the third fracture. The critical time is expressed as

$$t_{c2} = \left[\frac{w_c h_f}{q_2\left(\frac{2n+2}{2n+1}\right)\left(\frac{2n+2}{2n+1}C_2\right)^{-\frac{2n+1}{2n+2}}}\right]^{2n+2} \quad (15)$$

The critical length of the second fracture at the critical time is given by

$$x_{02} = \left(\frac{2n+2}{2n+1} C_2 \right)^{\frac{2n+1}{2n+2}} t_{c2}^{\frac{2n+1}{2n+2}} \quad (16)$$

At the critical time for the second fracture, the length of the first fracture is expressed as

$$X_1 = \left(x_{01}^{\frac{n+1}{n}} + \frac{n+1}{n} C_1' t_{c2} \right)^{\frac{n}{n+1}} \quad (17)$$

In general, for the kth fracture, the full length is expressed as

$$x_{0k} = \left(\frac{2n+2}{2n+1} C_k \right)^{\frac{2n+1}{2n+2}} t_{ck}^{\frac{2n+1}{2n+2}} \quad (18)$$

where

$$C_k = \left[\frac{372.25 g q_k (P_p - P_t)}{K h_f} \right]^{\frac{1}{2n+1}} \left[\frac{0.5 q_k}{(3+n^{-1}) h_f} \right]^{\frac{n}{2n+1}} \quad 19)$$

and

$$t_{ck} = \left[\frac{w_c h_f}{q_k \left(\frac{2n+2}{2n+1} \right) \left(\frac{2n+2}{2n+1} C_k \right)^{-\frac{2n+1}{2n+2}}} \right]^{2n+2} \quad (20)$$

where

$$q_k = \frac{q_i}{k} \quad (21)$$

At time t the number of fractures is

$$k = \frac{q_{it}}{x_{0k} w_c h_f} \quad (22)$$

The procedure for computing the propagation distance x0k and the critical time tck for the kth fracture is to solveEqs. (18) through (22) numerically. Then sum up all the critical times to get the corresponding total injection time.

3. MODEL ANALYSIS

The analytical model for vertical fractures was coded in a computer program to get instant numerical solutions. A sensitivity analysis was run with the computer program to identify engineering factors affecting the sequential initiation and simultaneous propagation of multiple fractures in the SRV. Table 1 presents basic data for the analysis.

Table 1.Data Set Used in Sensitivity Analysis

Flow rate into one side of perforation cluster (qi)	7.5; 10	bpm
Pressure in perforation cluster (Pp)	5500	psi
Pressure at fracture tip (Pt)	5000	psi
Overall density of fracturing fluid (ρL)	165.36	lbm/ft3
Fracture height (hf)	100	ft
Flow consistency (K)	1	cp
Fluid behavior index (n)	1; 1.1	
Critical fracture width (wc)	0.033	inch

Tables 2 summarizes model-calculated number of fractures and fracture length. Figure 1 illustrates fracture sketches plotted with the data in Table 2. The fractures are drawn in scale in the vertical axis (the direction of the maximum horizontal stress) and no-scale in the horizontal axis. These figures imply that the fractures grow and spread out with time not following a linear order. Figure 2 shows a comparison of two data sets with the same value of fluid flow behavior index, indicating that shorter fractures can be created by higher fluid injection rates. Figure 3 presents a comparison of two data sets with the same fluid injection rate, indicating that shorter fractures can be created with higher value of fluid flow behavior index. The effects of injection rate and flow behavior index on the fracture length are due to the high friction pressures caused by the high-values of these two parameters, hindering the growth of large/long fractures. Figure 4 demonstrates the effect of flow rate on the number of fractures, indicating that more fractures can be created by higher fluid injection rates. Figure 5 presents the effect of flow rate on the number of fractures, indicating the more fractures can be created with higher value of fluid flow behavior index. The effects of injection rate and flow behavior index on the number of fractures are caused the limited intake of fluid by the long fractures due to high friction pressures.

Table 2.Model-computed propagation of fractures for q_i=7.5 bpm and n = 1.1

Injection Time (min)	Number of Fractures	Fracture Length (ft)
0.9	1	136
4.9	2	297
13.9	3	455
30.4	4	623
56.3	5	793

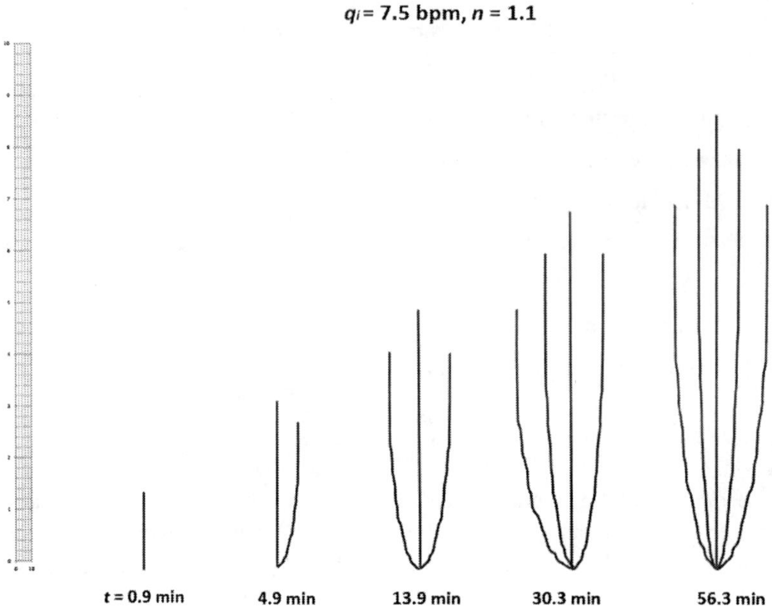

Figure 1. Fracture sketch drawn for qi = 7.5 bpm and n = 1.1.

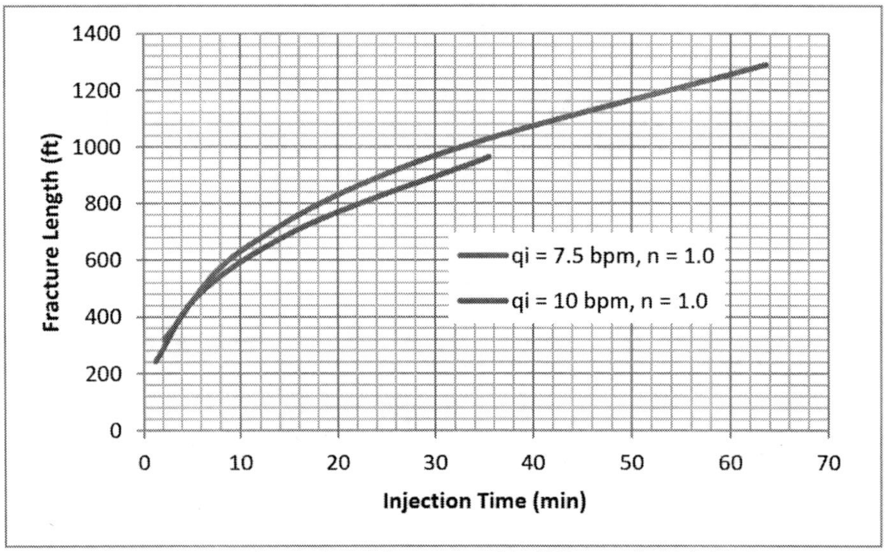

Figure 2. A comparison of two data sets with different fluid injection rates.

IOP Conf. Series: Materials Science and Engineering **729** (2020) 012050 doi:10.1088/1757-899X/729/1/012050

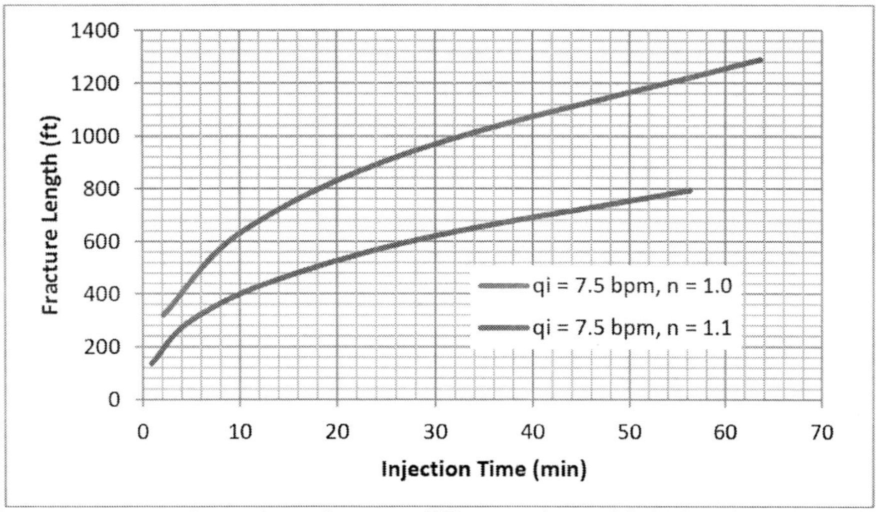

Figure 3. A comparison of two data sets different values of fluid flow behavior index.

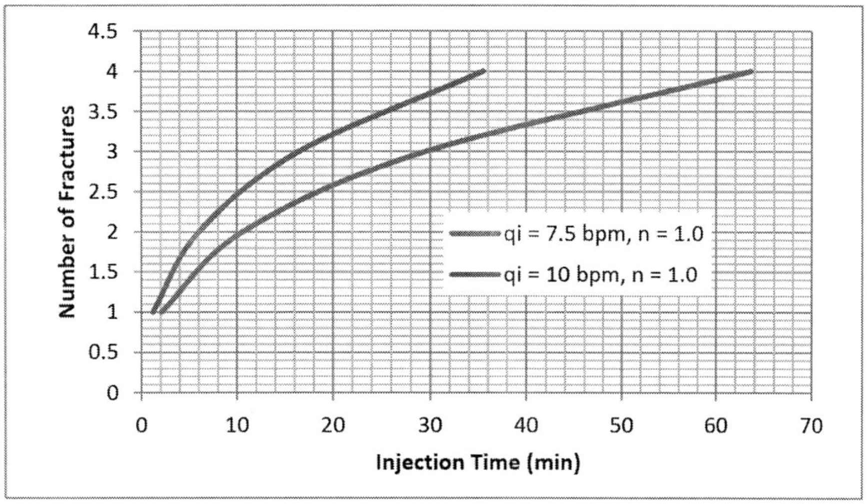

Figure 4. Effect of injection rate on the number of fractures.

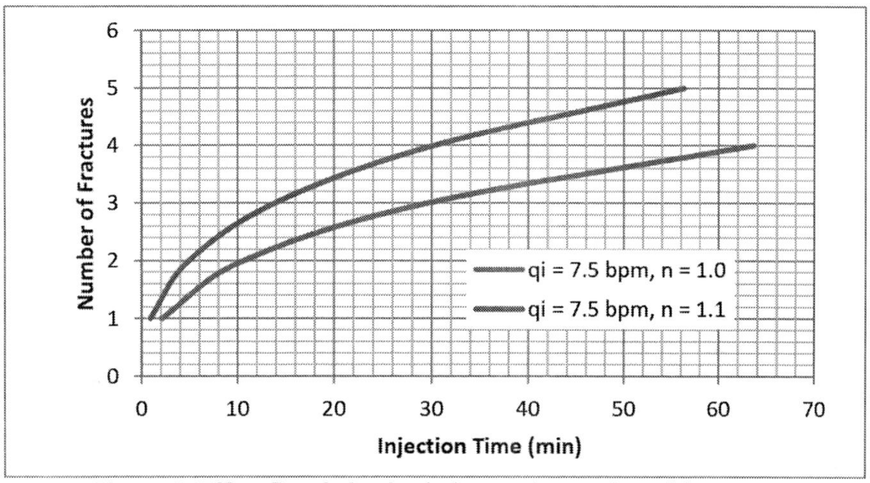

Figure 5. Effect flow behavior index on the number of fractures.

4. CONCLUSIONS

Analytical model was developed in this work to describe sequential initiation and simultaneous propagation of multiple vertical fractures in horizontal wells. This model is prepared designing frac-packing of gas hydrate reservoirs. Result of sensitivity analyses with the model allows for drawing the following conclusions. To frac-pack horizontal gas hydrate wells for solid-production control, increasing injection rate of fracturing fluid will create more friction and thus more resistance to flow in large fractures, promoting the creation of more short fractures, and thus larger treated reservoir volume (TRV) for higher gas production rate and gas recovery. Using dilatant fracturing fluid ($n > 1$) will cause more friction and thus more resistance to flow in large fractures owing to the shear-thickening property of the fluid. This will promote the creation of more short fractures and thus larger treated reservoir volume (TRV) for higher gas production rate and gas recovery.

Acknowledgement
This research was supported by Northeast Petroleum University Foundation Founding No. 2018GP2D-04 and No. 2018QNQ-06.

References

[1] Ruppel, C. 2014. Gas Hydrates and the Future of Natural Gas (PDF), Gas Hydrates Project, Woods Hole, MA: U.S. Geological Survey, archived from the original (PDF) on 6 November 2015, retrieved 28 December 2014.

[2] Li, G, Moridis GJ, Zhang K, Li XS., 2011. The use of huff and puff method in a single horizontal well in gas production from marine gas hydrate deposits in the Shenhu Area of South China Sea. J Petrol Sci Eng 2011;77:49–68.

[3] Hiroko, T. (2013). An Energy Coup for Japan: "Flammable Ice. New York Times. Retrieved March 14, 2013.

[4] Shyi-Min, L. (2015). A global survey of gas hydrate development and reserves: Specifically in themarine field [J]. Renewable and Sustainable Energy Reviews, 41, 884-900.

[5] Zhang Wei, Bai Fenglong, Shao Mingjuan, et al. (2017a). Progress of offshore natural gas hydrate production tests in Japan and implications [J].Marine Geology & Quaternary Geology, 37(5):27-33.

[6] Zhang Wei, Shao Mingjuan, Tian Qianning. (2017b). Technical progress of a Pilot Project to Produce Natural Gas Hydrate in Japanese Waters [J]. Petroleum Drilling Techniques, 45(5):98-102.

[7] Zhou, S., Chen, W., Li, W., Zhou, Q., J., & Shi, H. S. (2017). Research on the solid fluidization well testing and production for shallow non-diagenetic natural gas hydrate in deep water area. China Offshore Oil Gas, 29(4), 1-8.

[8] Max, M.D., Johnson, A.H. 2016. Exploration and Production of Oceanic Natural Gas Hydrate. Springer International Publishing. pp. 39–73.

[9] Ahmadi, G., Ji, C.A., Smith, D.H. 2007. Production of natural gas from methane hydrate by a constant downhole pressure well. Energy Convers. Manage. 48 (7), 2053–2068.

[10] Li, G.; Li, X. S.; Chen, Q.; Chen, Z. Y., 2010. Numerical Studies of Gas Production from Gas hydrate zone in Shenhu Area, South China Sea. Acta Chim. Sin. 2010, 68 (11), 1083–1092.

[11] Kawamura, T., Ohtake, M., Sakamoto, Y., Yamamoto, Y., Haneda, H., Komai, T., Higuchi, S., 2007. Experimental study on steam injection method using methane hydrate core samples. Proceedings of the Seventh (2007) ISOPE Ocean Mining Symposium. Lisbon, PORTUGAL, 1–6 July, pp. 83–86.

[12] Li, G, Moridis GJ, Zhang K, Li XS., 2011. The use of huff and puff method in a single Li, G., Tang, L.G., Huang, C., Feng, Z.P., Fan, S.S., 2006. Thermodynamic evaluation of hot brine stimulation for natural gas hydrate dissociation. J. Chem. Ind. Eng. China 57 (9), 2033–2038.

[13] Li, G., Li, X.S., Tang, L.G., Li, Q.P., 2008. Control mechanisms for methane hydrate production by thermal stimulation. Paper 5783, Proceedings of the 6th International Conference on Gas Hydrates. Vancouver, British Columbia, Canada, 6–10 July 2008.
[14] Kawamura, T., Yamamoto, Y., Ohtake, M., Sakamoto, Y., Komai, T., Haneda, H., 2005. Experimental study on dissociation of hydrate core sample accelerated by thermodynamic inhibitors for gas recovery from natural gas hydrate. Paper 3023, Proceedings of the 5th International Conference on Gas Hydrates. Trondheim, Norway, 13–16 June 2005.
[15] Najibi, H., Chapoy, A., Haghighi, H., Tohidi, B., 2009. Experimental determination and prediction of methane hydrate stability in alcohols and electrolyte solutions. Fluid Phase Equilib. 275 (2), 127–131.
[16] Moridis, G. J., Collett, T. S., Boswell, R., Kurihara, M., Reagan, M. T., Koh, C., Sloan, E. D. 2009. Toward Production from Gas Hydrates: Current Status, Assessment of Resources, and Simulation-Based Evaluation of Technology and Potential. SPE Reservoir Eval. Eng. 2009, 12 (5), 745–771.
[17] Guo, B. and Shan, L. Heat Transfer in Counter-Current Two Phase Flow Applied to Feasibility Study of Harvesting Natural Gas from Seabed Hydrates, International Journal of Heat and Mass Transfer, 126 (2018) 603-612.
[18] Zhang, X., Lu, X., & Li, P. (2018). A comprehensive review in natural gas hydraterecovery methods. SCIENTIA SINICA Physica, Mechanica & Astronomica, 49(3), 034604.
[19] Ghalambor, A., Ali, S.A. and Norman W.D., 2009, "Frac Packing Handbook," Society of Petroleum Engineers Text Book Series, Richardson, Texas.
[20] Roussel, N.P., and Sharma, M.M., 2011, "Optimizing Fracture Spacing and Sequencing in Horizontal-Well Fracturing," SPE Production and Operations 26(2): 173-184.
[21] Manchanda, R., and Sharma, M.M., 2012, "Impact of Completion Design on Fracture Complexity in Horizontal Wells," SPE Annual Technical Conference and Exhibition, San Antonio, Texas, October 8-10, Paper SPE 159899.
[22] Roussel, N.P., Manchanda, R., and Sharma, M.M., 2012, "Implications of Fracturing Pressure Data Recorded during a Horizontal Completion on Stage Spacing Design," SPE Hydraulic Fracturing Technology Conference, Woodlands, Texas, February 6-8, Paper SPE 152631.
[23] Rucker, W. K., Bobich, J., & Wallace, K. (2016, August 1). Low Cost Field Application of Pressure Transient Communication for Rapid Determination of the Upper Limit of Horizontal Well Spacing. Unconventional Resources Technology Conference. doi:10.15530/URTEC-2016-2460806.
[24] King, G. E., & Valencia, R. L. (2016, February 1). Well Integrity for Fracturing and Re-Fracturing: What Is Needed and Why? Society of Petroleum Engineers. doi:10.2118/179120-MS
[25] Li Y, Rui Z, Zhao W, et al. Study on the mechanism of rupture and propagation of T-type fractures in coal fracturing[J]. Journal of Natural gas science and engineering, 2018, 52: 379-389.
[26] Cheng, Y., 2016. "A Mathematical Model to Predict Fracture Complexity Development and Fracture Length," MS thesis. University of Louisiana at Lafayette, Lafayette, Louisiana.
[27] Cheng, Y., Guo, B., and Wei, N., 2015. "Prediction of Fracture Population and Stimulated Reservoir Volume in Shale Gas/Oil Reservoirs". Paper SPE 176833 presented at Asia Pacific Unconventional Resources Conference and Exhibition, November, Brisbane, Australia.
[28] Feng, F., Wang, X, Guo, B. and Ai, C. 2017. Mathematical model of fracture complexity indicator in multistage hydraulic fracturing, Journal of Natural Gas Science and Engineering, 38 (2017) 39-49.
[29] Guo, B., Sun, K. and Ghalambor, A.: Well Productivity Handbook, Gulf Publishing Company, Houston (2008).

[30] Rahman, M.M., and Rahman, M.K. A Review of Hydraulic Fracture Models and Development of an Improved Pseudo-3D Model for Stimulating Tight Oil/Gas Sand. Energy Resources Part A – Recovery Utilization and Environmental Effects, 2010, 32(15):1416-1436.

Study on diffusion of NaOH solution in zinc-rich epoxy coating using EIS

Chao Yang, Zili Li

College of Pipeline and Civil Engineering, China University of Petroleum (East China), Qingdao, Shandong, 266580, China

*Corresponding author's e-mail: zilimenhuzu@163.com

Abstract. In order to explore the diffusion process of NaOH solution in zinc-rich epoxy coating, in this paper, conventional three-electrode system was used to test the open circuit potential and electrochemical impedance spectrum, and equivalent electric circuit was fitted to get the related parameters including capacitance, resistance, phase angle and et al to reveal the diffusion process. The results show that three stages are divided including diffusion in coating, finite layer diffusion and double layer diffusion, and two criterions are put forward.

1. Introduction

Due to the high density and bonding strength between the coating and metal, zinc-rich epoxy coating is widely used in metal anticorrosion. On the other hand, the zinc also provides cathodic protection for the metal substrate. However, with the increase of usage time, the diffusion of water/electrolyte solution in the coating and the electrochemical reaction process on the coating/metal interface will inevitably occur [1]. Therefore, it is of great significance for the study of clarifying the change of coating properties caused by water/electrolyte diffusion in zinc-rich epoxy coating and the electrochemical reaction process at the interface.

For a long time, a large number of scholars have discussed the diffusion process of NaCl solution and deionized water in epoxy coating on the basis of Fick diffusion, and established EIS (electrochemical impedance spectrum) models of water diffusion at different stages [2]. In terms of water/electrolyte diffusion of zinc-rich epoxy coatings, relevant scholars have studied the preparation, appearance and related morphologies of different types of zinc-rich epoxy coatings, but there are few reports on the EIS characteristics of zinc-rich epoxy coatings under alkaline conditions. Therefore, this article adopts the traditional three electrode system to research the diffusion process of zinc-rich epoxy coatings using electrochemical test, including OCP (open circuit potential) and EIS, to establish the models of diffusion and electrochemical reaction.

2. Formatting the title, authors and affiliations

The standard three-electrode system was adopted in the experiment, in which the working electrode was a piece with zinc-rich epoxy coating with a thickness of 30±5μm and an area of $6.25cm^2$, reference electrode was saturated calomel electrode (SCE), Pt electrode worked as auxiliary electrode. Preparation of working electrode: the metal matrix was X80 steel in $25\times25\times2mm^3$. After the process of acetone degreasing - deionized water cleaning - anhydrous ethanol degreasing, the electrode was dried for 24h. Specimens was welded copper wires with epoxy resin encapsulation leaving exposed area of 6.25cm^2 [3], then artificial coated with 70% zinc rich epoxy coating, tested the thickness of the normal temperature cooling after 48h.

Experimental conditions: Experimental solution was prepared by analytical pure NaOH and deionized water with a mass fraction of 3% was prepared, and the all experiment were carried out at a constant temperature of 25°C.

Electrochemical test was conducted using electrochemical workstation PARSTST2273. Firstly, the OCP test was conducted. When the OCP fluctuation did not exceed 3mV within 300s, the whole system was considered to be in equilibrium state for EIS test. In this paper, the average value under the equilibrium state is taken as the OCP in different time periods.

The range of test frequency in EIS was set as 100kHz~10mHz, and the disturbance voltage is 10mV. One set of experimental data was tested every 15 minutes. The data analysis was adopted by the PowerSuite and ZSimpWin software that came with the system. The experiment was repeated three times to ensure the accuracy. However, due to the complexity of the coating system, the data fitting results could not be completely consistent, but their change rules were highly consistent. Therefore, one group of experimental data was selected for analysis [4].

3. Results and discussion

3.1. Open circuit potential

Fig.1 shows the change curve of the OCP of the test sample with time. As can be seen from the figure, the OCP of the bare steel and coated samples was -0.38$V_{vs.SCE}$ and -1.24$V_{vs.SCE}$, respectively, showing a great decrease. From the perspective of thermodynamics, the zinc-rich epoxy coating greatly reduced the driving force of metal corrosion, reflecting the corrosion protection of the coating on the metal. At the same time, along with time of the experiment in 15 min, the OCP decreased to -1.30$V_{vs.SCE}$ with a decrease of 0.6$V_{vs.SCE}$, suggesting that the corrosive medium gradually diffused into the coating. However, the OCP moved positively with time, illustrating a decrease of protecting corrosion with an increasing OCP. When the experiment reached 45min, the rise trend of its OCP became significantly slow, which suggested that the solution might have reached the coating/metal interface and electrochemical reaction occurred.

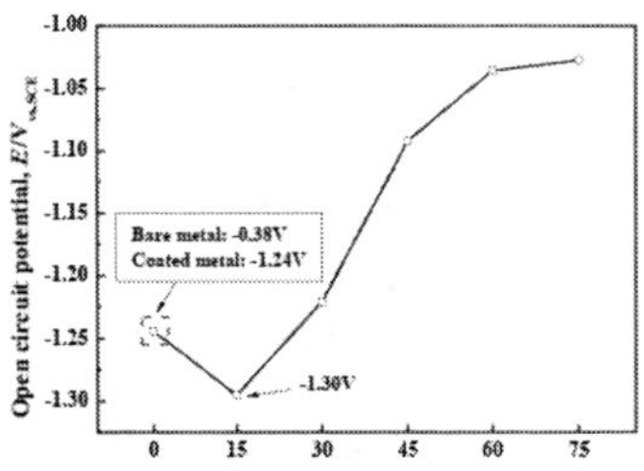

Fig.1 Open circuit potential over time

3.2. Nyquist diagrams

Fig.2 shows the test results of EIS behaviour of coating at different diffusion stages. Tab.1 shows the fitting results [5].The solid line in the figure was the fitting curve, and the discrete point was the test data. As can be seen from the figure, the Nyquist plots were divided into three stages.

At the initial stage of solution diffusion (0~30min), the EIS characteristics of the coating showed a complete capacitive reactance arc, that is to say, its equivalent circuit was solution resistance in series with characteristic parameters, therefore, fitting circuit was chosen as $R_s(C_cR_c)$.Considering the dispersion characteristics of the coating system, CPE(Q) is generally used to replace coating capacitor C_c. As can be seen from the figure, with the development of diffusion, the radius of the capacitive reactance arc decreased gradually, presenting the R_c decreased gradually while the Q increased. Generally speaking, in relevant studies, the resistance of epoxy coating is generally 10^8~10^{10} orders of magnitude, and when it drops to 10^6 orders of magnitude, the coating has failed [6]. However, with the addition of zinc, the electrical conductivity of the coating was significantly enhanced, and the resistance of the coating was also significantly reduced. As can be seen from the data in the table, at the beginning of the spread, the resistance of the coating itself was $3.435\times10^4\Omega\cdot cm^2$, and gradually reduce to $1.161\times10^4\Omega\cdot cm^2$,because of the diffusion of NaOH solution. Meanwhile, it can be seen that the Q for coating capacitance increased from 1.399×10^{-5}F/cm to 4.256×10^{-5}F/cm, while the dispersion coefficient of n=0.51~0.59 remained basically unchanged, indicating that the roughness of the electrode surface was equal to the uniformity of the distribution of corrosion current density in different test times [7], which had no impact on the test results.

When the experiment was carried out for 45min, the EIS showed obvious characteristics of double capacitive reactance arc and there were two time constants, which indicated that the solution had diffused to reach the coating/metal interface for electrochemical reaction and there was an obvious limited layer diffusion process. Therefore, the equivalent circuit was described by $R_s(Q(R_cO(C_{dl}R_{ct})))$: the first time constant of $\tau_1=Q_c\cdot R_c$, representing the nature of the coating system, the second time constant $\tau_2=C_{dl}\cdot R_{ct}$, reflecting the information on the coating/metal interface, and the element O in the fitting circuit represented the diffusion process of finite layers. At this point, it can be seen that the solution had diffused to the coating/metal interface and the electrochemical reaction has begun. However, it can be seen from the limited layer diffusion characteristics of dual capacitive reactance that the electrochemical reaction had just taken place at this time, and there was still no complete double electric layer formed on the metal surface, that is to say, no new concentration gradient layer was formed on the metal surface.

With experiment lasted for 60 min, the finite layer diffusion impedance characteristics disappeared, while there was an obvious Warburg impedance with $n\approx0.5$, illustrating a complete electric double layer on metal surface has formed. At this time, diffusion process on interface was controlled by the concentration gradient of the new layer, presenting that capacitance of electric double layer firstly increased quickly and then basically remained the same, while the charge transfer resistance greatly increased. In fact, when the solution diffuses in the epoxy coating without zinc, once the solution arrived at the coating/metal interface, the electrochemical reaction occurred: in cathode $O_2+2H_2O+4e^-\rightarrow4OH^-$, and in anode $Fe\rightarrow Fe^{2+}+e^{2-}$. When the diffusion in the coating has been completed, the coating capacitance basically remain unchanged, and the coating stripping because of the accumulation of corrosion products to lead to the loss of adhesion between coating/metal. However, based on the fitted results of 45min~75min in this paper, it can be found that when the solution reached the coating interface, the coating capacitance first decreased and then increased. This was because a cathodic protection of Zn-Fe was formed: in anode $Zn\rightarrow Zn^{2+}+2e^-$. On the other hand, the addition of zinc in coating system changed its nature, thus made the coating capacitance reduce, and with the consumption of zinc, the coating capacitance gradually increased [8].

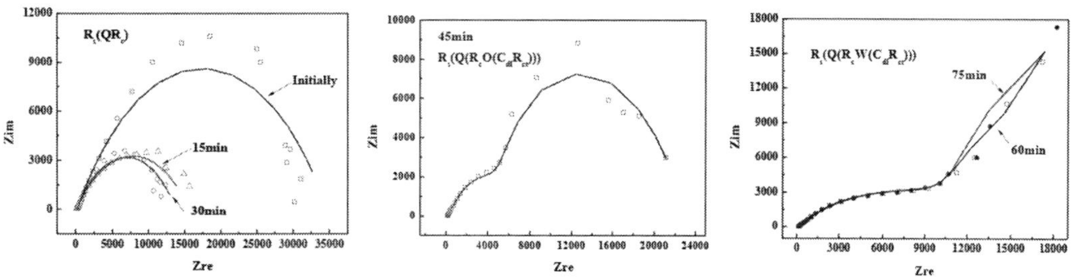

Fig.2 Nyquist diagrams over time

IOP Conf. Series: Materials Science and Engineering **729** (2020) 012051 doi:10.1088/1757-899X/729/1/012051

Tab.1 Fitted results of equivalent electric circuits

Element \ Time	0	15min	30min	45min	60min	75min
R_s	269.9	184.7	157.5	141	129.5	117.7
$Q/10^{-5}$	1.399	3.303	4.256	2.48	3.443	3.966
n	0.5918	0.5587	0.5122	0.5633	0.5457	0.5146
$R_c/10^4$	3.435	1.54	1.161	0.2035	1.273	1.413
$O/10^{-5}$	-	-	-	2.588	-	-
B	-	-	-	0.142	-	-
$W/10^{-3}$	-	-	-	-	1.162	2.98
$C_{dl}/10^{-5}$	-	-	-	8.601	87.48	75.28
$R_{ct}/10^4$	-	-	-	1.554	2.714	5.115

3.3. Bode plots

Fig.3 shows the change curve of phase Angle of coating with time. As can be seen from the figure, the phase Angle change can be divided into three distinct stages, meeting with the analysed results of Nyquist diagrams. In the first stage (0~30 min), the curve was only one peak, showing the characteristics of coating system, and its biggest phase Angle slightly decreased (44.7°→41°→39.9°) with time. Due to the addition of Zn, the phase Angle did not show the obvious characteristics of capacitive reactance (90°). In this stage, the maximal phase Angle frequency increased slightly (7.9Hz→13.7Hz→24Hz). As the experiment progressed to 45min, two significant time constants (two peaks) appeared, which indicated that the solution diffused to the interface and the electrochemical reaction began to occur. As the experiment reached the third stage (60min~75min), it indicated that the "peak" of the double electric layer was significantly shifted to the low frequency, and the maximal phase Angle and frequency basically remained unchanged in the third stage. At the same time, in the second and third stages, the maximal phase Angle and frequency representing the coating characteristics were also basically unchanged. The results were the same as those of Nyquist plots.

Thus, according to the above analysis, process of solution diffusion in zinc-rich epoxy coating are divided into three distinct phases: in the first stage with only one peak in Bode plots, its capacitance increases and the resistance is reduced, therefore time constant ($\tau_1=Q_c \cdot R_c \approx 0.5$) keeps mostly unchanged. The second stage is the finite layer diffusion process on the coating/metal interface, which is manifested as the appearance of diffusion impedance O, and two obvious time constants appear in the Bode figure. The third stage is the electrochemical reaction process on the interface, which shows an obvious Warburg diffusion impedance. In the Bode figure, it shows that the peak value of the dual-electrical measurement features gradually moves to the low frequency, and basically does not change with time. However, due to the cathodic protection effect of zinc, the coating capacitance shows a characteristic of first decreasing and then increasing.

In conclusion, the appearance of limited layer diffusion impedance O can be used as the judgment basis for the solution reaching the coating/interface, while the appearance of Warburg impedance can be used as the judgment basis for the complete formation of new concentration diffusion layer (double electric layer) on the metal interface.

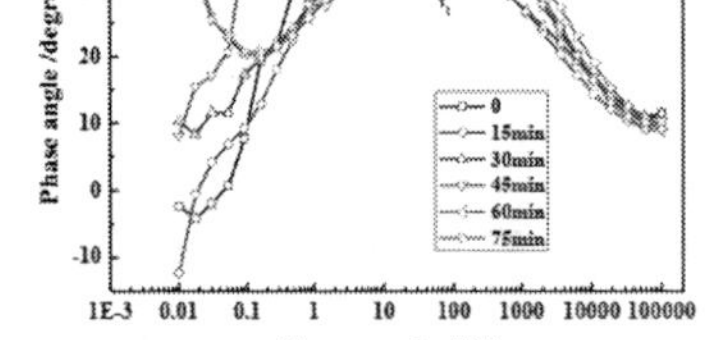

Fig.3 Phase angle over time

4. Conclusion

The diffusion process of NaOH solution in zinc-rich epoxy coating is studied by electrochemical method, and two conclusions are drawn as follows.

(1) The diffusion process of NaOH solution in zinc-rich epoxy coating is divided into three stages: the first stage is the diffusion process of solution in the coating, which shows that the coating capacitance increases, the impedance decreases, and the time constant is basically unchanged. The second process is the finite layer diffusion process on the coating/metal interface, which shows that the Nyquist plot shows two apparent capacitive reactivity arcs and the Bode plot shows two peaks. The third process is the diffusion in the double-layer capacitor, which represents the maximal phase Angle of the double-layer capacitor moves to the low-frequency area. Due to the addition of zinc, the electrochemical reaction process with zinc as the anode mainly occurs.

(2) In this paper, it points out that the appearance of limited layer diffusion impedance O could be used as the judgment basis for the solution reaching the coating/interface, while the appearance of Warburg impedance could be used as the judgment basis for the complete formation of new concentration diffusion layer (double electric layer) on the metal interface.

Reference

[1] X Liu, J Xiong, Y Lv, et al. Study on corrosion electrochemical behavior of several different coating systems by EIS [J]. Progress in Organic Coatings, 2009,64: 497–503.

[2] C G Oliveira, M G S Ferreira. Ranking high-quality paint systems using EIS. Part I: intact coatings [J]. Corrosion Science, 2003,45:123-138.

[3] L Y Xu, X Su, Z X Yin, et al. Development of a real-time AC/DC data acquisition technique for studies of AC corrosion of pipelines [J]. Corrosion Science, 2012,61:215-223.

[4] N J W Revers, H P Huinmink, O C G Adan, et al. Water uptake in thin nylon 6 films as measured by electrochemical impedance spectroscopy and magnetic resonance imaging [J]. Electrochimica Atca, 2013, 94: 219-228.

[5] C M Abreu, M Izquerdo, M Keddam, et al. Electrochemical behavior of zinc-rich epoxy paints in 3% NaCl solution [J]. Electrochimica Acta, 1996, 41(15): 2405-2415.

[6] J D Scantlebury, K Cali. The application of AC impedance to study the performance of lacquered aluminium specimens in acetic acid solution [J]. Progress in Organic Coatings, 1997, 31 (3): 201-214.

[7] H Bryan, E O Mark, T Bemard, et al. Determination of effective capacitance and film thickness from constant-phase-element parameters [J]. Electrochimica Acta, 2010,55:6218-6227.

[8] K Schaefer, A Miszczyk. Improvement of electrochemical action of zinc-rich paints by addition of nanoparticulate zinc [J]. Corrosion Science, 2013, 66: 380-391.

IOP Conf. Series: Materials Science and Engineering **729** (2020) 012052 doi:10.1088/1757-899X/729/1/012052

Study on the optimization of timing for horizontal well re-fracturing in tight oil reservoir -- A case study of Baikouquan Formation in North Mahu Oilfield, Junggar Basin, Western China

Alimujiang Sidike[1], Xie Bin[1], Zeng Bin[2], Wu Guodong[1], Yang Zhaozhong[3*], Chen Jianlin[1], Meng Jigang[1], Huang Bo[1] and Li Xiaogang[3]

[1] PetroChina Xinjiang Oilfield Company, Karamay, Xinjiang, 834099, China

[2] Beijing Gepetto Petroleum Technology Co., LTD, Beijing, 100102, China

[3] State Key Laboratory of Oil and Gas Reservoir Geology and Exploitation, Southwest Petroleum University, Chengdu, Sichuan, 610500, China

Corresponding email: yangzhaozhongswpu@163.com

Abstract. There are abundant tight oil resources in China, and "horizontal well + volume fracturing" technology is the one of the main means to develop such reservoirs. The production of tight oil reservoir declines generally rapidly after the initial fracturing, therefore it is necessary to be conducted re-fracturing operation to restore the oil productivity. The determination of the best time of re-fracturing is an important factor to ensure the re-fracturing effect. In this paper, a mathematical model of oil-water phase seepage was established by applying the principle of finite difference method (FDM). A unified grid system was adopted for the division of the matrix and fractures, and the equations were solved by the IMPES method. The reliability of the proposed seepage model gets verified by the historical matching based on the field production data. The oil production after re-fracturing at different moments of time after initial fracturing was predicted. The simulations show that more cumulative oil production can be obtained if re-fracturing is implemented after the fourth year of initial fracturing. During the period after the 4th year of the initial fracturing, the earlier the re-fracturing is implemented, the more the cumulative oil production will be harvested. Considering the cost of re-fracturing operation, the economic benefits of re-fracturing after the sixth year of initial fracturing were evaluated. The results show that the cost can be recovered about one year after the re-fracturing, the net present value (NPV) after four years of re-fracturing reaches 69.95 million yuan, with an output-input ratio of 5.3.

1. Introduction

There is a great prospect for the exploration and development of tight oil in the unconventional oil and gas field. The large oil-gas basins in China, such as Ordos, Junggar, Songliao, and Qaidam, are rich in tight oil resources [1-3]. In recent years, "horizontal well + volume fracturing" technology has gradually become one of the main means to develop tight oil reservoirs, which greatly improves the production of tight oil [4]. However, due to the poor physical properties of tight oil reservoirs, the formation energy drops sharply along with the oil exploitation. However, an effective injection-production pattern can not be formed for horizontal wells, resulting in insufficient replenishment of energy. What's worse, under the action of fracture closure, the proppants deform, embed and break, resulting in the rapid decline of fracture conductivity. Because of this, some hydraulic fractures in horizontal wells will be invalid, causing rapid decline of oil production and shortened stable production period [5]. In order to recover the productivity of such horizontal wells, the re-fracturing technology is considered as one of the most potential and currently most widely-used measures to stimulate the oil and gas production [6].

In the 1950s, researchers began to carry out the study on vertical well re-fracturing, while the research on horizontal well re-fracturing started relatively later. Since the 21st century, thanks to the large-scale development of shale gas, more and more attention began to be focused on the research on horizontal well re-fracturing [7]. The determination of the timing of re-fracturing is an important aspect in the research of horizontal well re-fracturing. That is, to answer the question of when to implement the re-fracturing to achieve the best oil and gas stimulation effect from the horizontal wells. In the case of the vertical well re-fracturing, it is generally considered that the best timing for re-fracturing is the moment when the change of the ground stress becomes the strongest. Since during this time, the re-fracturing can effectively improve the sweep efficiency and reduce the dead oil area. When it comes to the horizontal well, there are many factors that affect the timing of the re-fracturing. In this case, numerical simulation is usually used to optimize the timing of the horizontal well re-fracturing, with the cumulative oil production or net present value (NPV) being usually taken as the objective function.

Lantz et al. (2008) [8] found that the optimal timing of horizontal well re-fracturing in Bakken shale oil reservoirs changes depending on the completion conditions of the initial fracturing and the difference of the re-fracturing technology. Generally speaking, the effect of re-fracturing is at its best level in 2 to 3.5 years after the initial fracturing. Tavasoli et al. (2013) [9] improved the CMG model and established a numerical model to predict the timing of re-fracturing by the use of the daily gas production and cumulative gas production as the optimization goals. The results indicate that the best effect of re-fracturing is achieved when the gas production falls to 10% - 15% of the initial production. Based on the three different price prediction models of natural gas, Cafaro (2016) [10] did the optimization by taking net present value (NPV) as the objective function and found the optimal frequency and timing of re-fracturing of shale gas horizontal wells corresponding to each price change trend. Based on the finite element method (FEM), Pang et al. (2015) [11] established the mathematical model of oil-water phase seepage, and predicted the daily oil production change of a horizontal well in Changqing Oilfield conducted by re-fracturing after the initial fracturing in the 5th, 6th, 7th and 8th years respectively. By taking the cumulative oil production as the optimization target, 7 years after the initial fracturing was selected as the optimal timing for the re-fracturing operation.

The Baikouquan Formation, locating at North Mahu Oilfield, Junggar basin, China is composed mainly of low permeable glutenite rock, where is rich in tight oil resources. The development mode of "horizontal well + volume fracturing" has been applied widely there. The production of some of those horizontal wells has declined seriously and need to be conducted re-fracturing [12]. However, North Mahu Oilfield does not have the practical experience of re-fracturing before, and there is no previous research on the optimization of the timing for horizontal well re-fracturing aiming at this oilfield. In this paper, taking Well X as a case for study, the oil-water phase seepage model was established, and the equations were discretized based on the finite difference method (FDM) to establish the numerical model, and the equations were solved by Implicit Pressure Explicit Saturation (IMPES) method. This model can be used to calculate the change of oil production after re-fracturing with different timing, and to evaluate the economic benefits of Well X after re-fracturing with the cost of re-fracturing taken into consideration.

2. Establishment of mathematical model

2.1. Model hypothesis

- The reservoir is rectangular, and the flow of fluid is isothermal seepage.

• Two dimensional plane flow (ignoring the influence of fluid gravity).
• There is only oil and water two-phase seepage in the reservoir, and the oil and water do not dissolve each other, and each conforms to Darcy's law.
• Both rock and fluid can be compressed.
• The heterogeneity and anisotropy of rock is considered.
• The capillary force of oil and water is considered.
• The hydraulic fracture extends completely through the reservoir, with the fracture height equal to the reservoir thickness.

2.2. Seepage equations

2.2.1. Reservoir matrix system
The differential equations of oil-water two phase seepage:

$$\frac{\partial}{\partial x}(\rho_o \frac{k_x k_{ro}}{\mu_o}\cdot\frac{\partial p_o}{\partial x})+\frac{\partial}{\partial y}(\rho_o \frac{k_y k_{ro}}{\mu_o}\cdot\frac{\partial p_o}{\partial y})=\frac{\partial}{\partial t}(\varphi\rho_o s_o) \quad (1)$$

$$\frac{\partial}{\partial x}(\rho_w \frac{k_x k_{rw}}{\mu_w}\cdot\frac{\partial p_w}{\partial x})+\frac{\partial}{\partial y}(\rho_w \frac{k_y k_{rw}}{\mu_w}\cdot\frac{\partial p_w}{\partial y})=\frac{\partial}{\partial t}(\varphi\rho_w s_w) \quad (2)$$

where k [μm^2] is the absolute permeability of formation, k_{ro} is the relative permeability of oil phase, k_{rw} is the relative permeability of water phase, μ_o [mPa·s] is the oil viscosity, μ_w [mPa·s] is the water viscosity, s_o is the oil saturation, s_w is the water saturation, p_o [MPa] is the pressure of oil phase, p_w [MPa] is the pressure of water phase, φ is the formation porosity.

The auxiliary equations are:

$$p_c = p_o - p_w \quad (3)$$

$$s_o + s_w = 1 \quad (4)$$

The initial conditions are:

$$p_m(x, y, 0) = p_i \quad (5)$$

$$s_w(x, y, 0) = s_{wi} \quad (6)$$

The closed outer boundary conditions of matrix system are:

$$\left(\frac{\partial p_o}{\partial x}\right)_{x=0} = 0$$
$$\left(\frac{\partial p_o}{\partial x}\right)_{x=L_x} = 0$$
$$\left(\frac{\partial p_o}{\partial y}\right)_{y=0} = 0 \quad (7)$$
$$\left(\frac{\partial p_o}{\partial y}\right)_{y=L_y} = 0$$

The inner boundary condition of matrix system is under the constant flowing bottomhole pressure (FBHP):

$$p_{wf} = C \tag{8}$$

where p_c [MPa] is the capillary pressure, p_i [MPa] is the initial formation pressure, s_{wi} is the initial water saturation, p_{wf} [MPa] is the FBHP.

2.2.2. Fracture system

For the fracture system, the fluid flow is considered as one-dimensional, and the capillary force in the fracture is not considered. Therefore, the oil-water phase seepage equations of the fracture system are expressed as:

$$\frac{\partial}{\partial x}(\rho_o \frac{k_{fx}k_{ro}}{\mu_o} \cdot \frac{\partial p_{fo}}{\partial x}) = \frac{\partial}{\partial t}(\varphi \rho_o s_o) \tag{9}$$

$$\frac{\partial}{\partial x}(\rho_w \frac{k_{fx}k_{rw}}{\mu_w} \cdot \frac{\partial p_{fw}}{\partial x}) = \frac{\partial}{\partial t}(\varphi \rho_w s_w) \tag{10}$$

The initial conditions are:

$$p_f(x,0) = p_i \tag{11}$$

$$s_w(x,0) = s_{wi} \tag{12}$$

The closed outer boundary conditions of fracture system are:

$$\begin{cases} \left(\frac{\partial p_f}{\partial x}\right)_{x=0} = 0 \\ \left(\frac{\partial p_f}{\partial x}\right)_{x=L_f} = 0 \end{cases} \tag{13}$$

The inner boundary condition of fracture system is under the constant FBHP:

$$p_{wf} = C \tag{14}$$

2.3. Discretization and solution of the equations

In order to satisfy the flow continuity, the pressure balance and the calculation of matrix and fracture systems, a unified grid system is adopted for the division of the matrix and fracture. As the width of the fracture differs by several levels of magnitude when compared with the grid size, the matrix and fracture grids are uniformly divided along the fracture direction, which is shown in figure 1. The five diagonal matrix equations are obtained by the central differentiation of the oil and water seepage differential equations. By selecting p_o and s_w as independent variables, and using the IMPES method to solve it, that is, use the implicit iterative method to solve p_o, and use the explicit iterative method through "one-step pressure, multi-step saturation" to solve s_w.

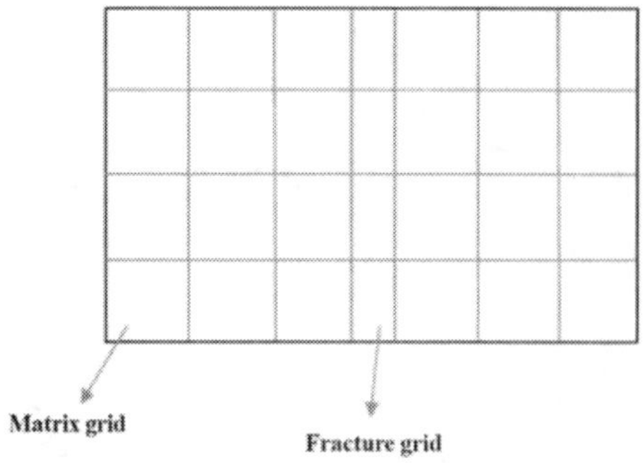

Figure 1. Pattern of grid division

The production of oil flowing from the fracture into the well is:

$$q_{of}(i,j) = \frac{k_f k_{ro} \cdot w_f / 2[p_f(i,j) - p_{wf}]}{\mu_o B_o (\ln(r_e / r_w) + S)} \tag{15}$$

where k_f [μm^2] is the fracture permeability, w_f [m] is the fracture width, S is the skin factor, B_o is the oil volume factor, r_e [m] is the apparent grid radius, r_w [m] is the wellbore radius.

3. Model verification

3.1. Physical model of initial fracturing

Well X is an open-hole horizontal well deployed in the Baikouquan Formation, North Mahu Oilfield. The horizontal section is 800m at length, and 14 hydraulic fractures are generated after the initial fracturing. The production of this well decreases rapidly after the initial fracturing, therefore it is necessary to carry out the re-fracturing. The physical model of production prediction after the initial fracturing of Well X is established, as shown in figure 2, in which the red straight line represents the hydraulic fracture, the thicker black straight line symbolizes the wellbore, with a model area of 1600m × 600m.

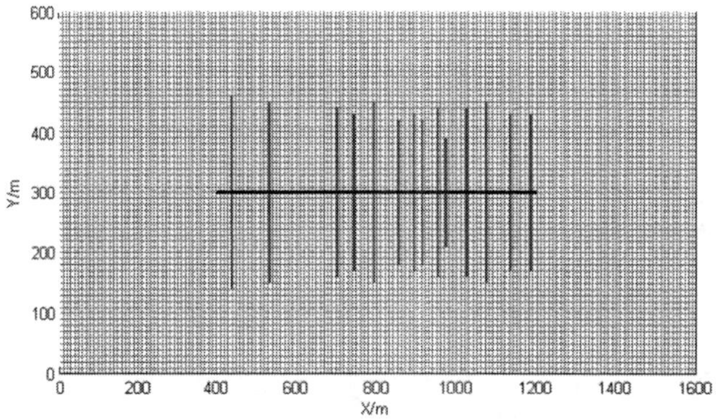

Figure 2. Physical model of production prediction after initial fracturing of Well X

3.2. Input parameters

The input reservoir parameters for the production prediction after the initial fracturing of Well X can be seen in Table 1. According to the fracture simulation and the potentiometry monitoring results, the length, conductivity and the average fracture width of all the fractures created by initial fracturing in Well X can be calculated, with the results shown in table 2. By inputting the above parameters into the model, the oil production after initial fracturing can be calculated.

Table 1. Reservoir parameters for production prediction of Well X after initial fracturing

Parameter name	Unit	Value	Parameter name	Unit	Value
Formation pressure	MPa	44.8	Initial water saturation	%	48.57
Porosity	%	9.42	Rock compression coefficient	MPa^{-1}	3×10^{-4}
Permeability	μm^2	1.16×10^{-3}	Water compression coefficient	MPa^{-1}	4.5×10^{-4}
Temperature	°C	81.8	Oil compression coefficient	MPa^{-1}	8.4×10^{-4}
Effective reservoir thickness	m	12.0	Water volume factor	Dimensionless	1.0
Oil volume factor	Dimensionless	1.1	Oil viscosity	mPa·s	8.04
Oil density	kg/m^3	800	Oil-water relative permeability curve	/	See the fitting curve in figure 3

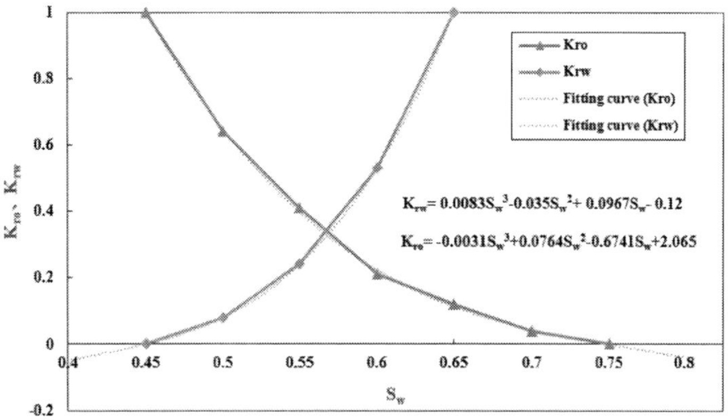

Figure 3. Oil-water relative permeability curve of Well X

Table 2. Fracture parameters of initial fracturing of Well X

No.	Half- length of fracture (m)	Initial average fracture conductivity (D·cm)	Average fracture width (mm)
1	160	59.7	6.8
2	150	68.8	9.1
3	140	60.9	7.5
4	130	68.7	5.1
5	150	58.4	5.5
6	120	73.0	5.1
7	130	58.3	2.9
8	120	59.4	3.2
9	140	61.3	5.1
10	90	56.5	6.1
11	140	58.5	6.8
12	150	68.6	5.5
13	130	59.1	8.3
14	130	69.7	6.5

3.3. Historical fitting of the oil production

Well X was conducted initial fracturing 5.5 years ago, i.e., about 1960 days. By adjusting the parameters such as the oil-water relative permeability relationship, the curve of predicted oil production is consistent with the curve of actual oil production. After adjustment, the curve of predicted daily oil production and actual daily oil production in relation to time after the initial fracturing is obtained and shown in figure 4. Also, the corresponding cumulative oil production can be calculated accordingly, as shown in figure 5. It can be seen that the trend of the curve of predicted accumulative oil production is consistent with that of the actual accumulative production.

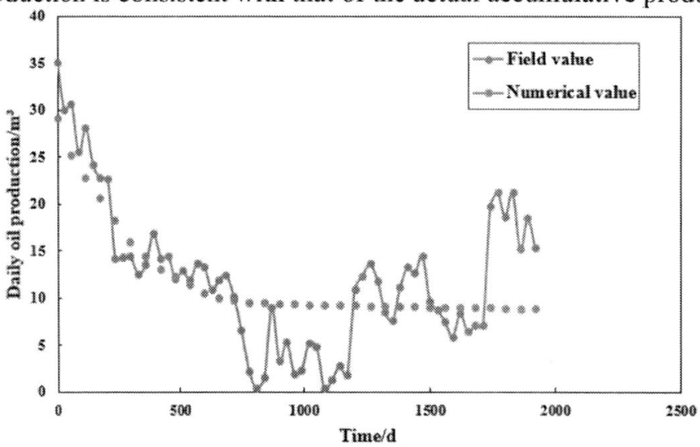

Figure 4. The change curve of daily oil production after initial fracturing of Well X

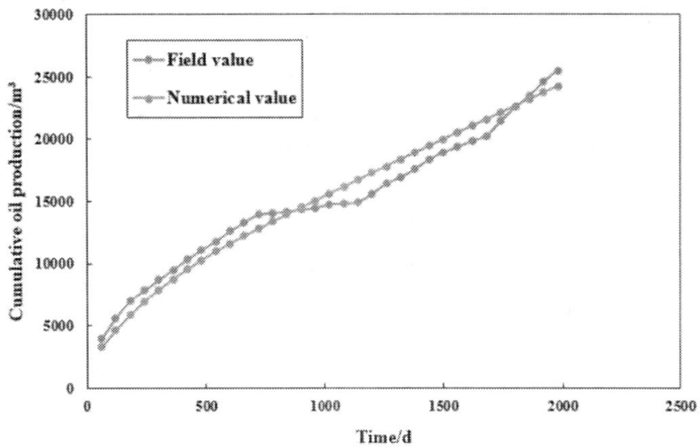

Figure 5. The change curve of cumulative oil production after initial fracturing of Well X

The established mathematical model is used to predict the formation pressure distribution of 1 month and 5.5 years respectively after the initial fracturing of Well X, with the results shown in figure 6 and figure 7. It can be seen that the formation pressure around the intersection of the wellbore and the fracture is about 28.1MPa, while the FBHP measured by downhole pressure gauge in Well X one month after initial fracturing is 29.1MPa (the data is sourced from the daily production report of Well X), which is close to the prediction result of model 28.1MPa, with an error rate of 3.4%. This verifies the accuracy of the established model in this paper. After that, along with the continuous exploitation of crude oil, although there is small amount of change in the formation pressure around the wellbore, the low-pressure area gradually spreads to the area between the fractures and the direction of fracture tips, which eventually forms an elliptical low-pressure area with the long axis parallel to the fracture direction 5.5 years later, with the pressure value of the low-pressure area being about 26 MPa.

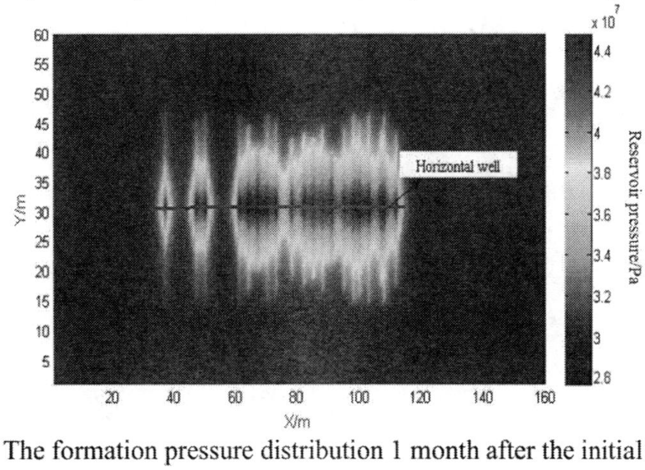

Figure 6. The formation pressure distribution 1 month after the initial fracturing

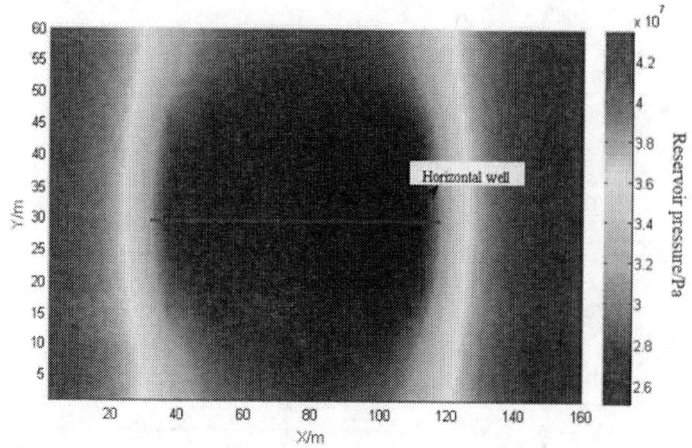

Figure 7. The formation pressure distribution 5.5 years later after the initial fracturing

4. Optimization of re-fracturing timing

4.1. Physical model and input parameters of re-fracturing

By substituting the fracture parameters and the updated boundary conditions of the re-fracturing design into the established model, we can calculate the oil production change after re-fracturing implemented at different moments of time after the initial fracturing. According to the re-fracturing plan of Well X, in addition to the re-fracturing of the original 14 old fractures, based on the logging interpretation results, 6 locations in the horizontal section were selected for perforating new holes, and 6 new fractures were created at these locations. In this paper, the oil production of Well X after re-fracturing in the 3rd, 4th, 5th, 6th and 7th year after the initial fracturing are simulated respectively, and the calculation time is totally 10 years after the initial fracturing.

The physical model of re-fracturing production prediction is shown in figure 8, in which the red line represents the old fracture and the blue line represents the new fracture. According to the software simulations, the average conductivity of the old fracture after re-fracturing is 48.04D cm, with the average fracture width of 4.52mm; the average conductivity of the new fracture is 60.18dD cm, with the average fracture width of 5.26mm.

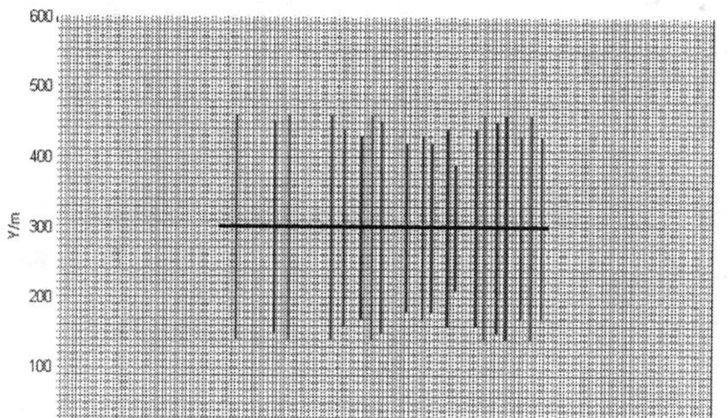

Figure 8. Physical model of re-fracturing production prediction of Well X

4.2. Results of predicted oil production

4.2.1. Daily oil production

Figure 9 shows the prediction curves of daily oil production after re-fracturing at different times after the initial fracturing of Well X. It can be seen that the daily oil production capacity of Well X has been rapidly recovered no matter when the re-fracturing is carried out. The maximum daily oil production can reach 40m³/d at the initial stage, and then the daily oil production is rapidly reduced. Generally, the oil production tends to become stable 2 years after the re-fracturing is implemented. Moreover, the stable daily oil production after re-fracturing is higher than that without re-fracturing.

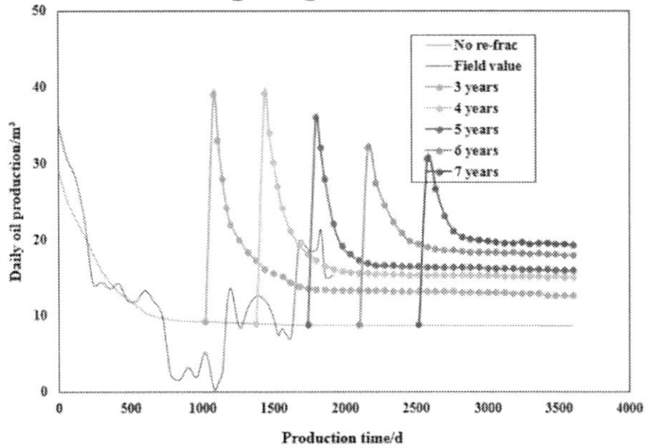

Figure 9. Daily oil production curves of Well X after re-fracturing at different time after initial fracturing

4.2.2. Cumulative oil production

Figure 10 shows the prediction curves of cumulative oil production after re-fracturing at different times after the initial fracturing of Well X. Table 3 indicates the cumulative oil production in 10 years after the initial fracturing. It can be seen that when re-fracturing is implemented after the 4th year of the initial fracturing, the cumulative oil production is higher, implying that it is the best time to implement the re-fracturing after the fourth year. It can be concluded from figure 10 that after the fourth year of initial fracturing, the earlier the re-fracturing is implemented, the higher the cumulative oil production will be obtained. It has been 5.5 years since the initial fracturing of Well X. Although that means Well X has missed the best opportunity for re-fracturing, according to the production prediction results, the earlier the re-fracturing is, the more oil harvest is.

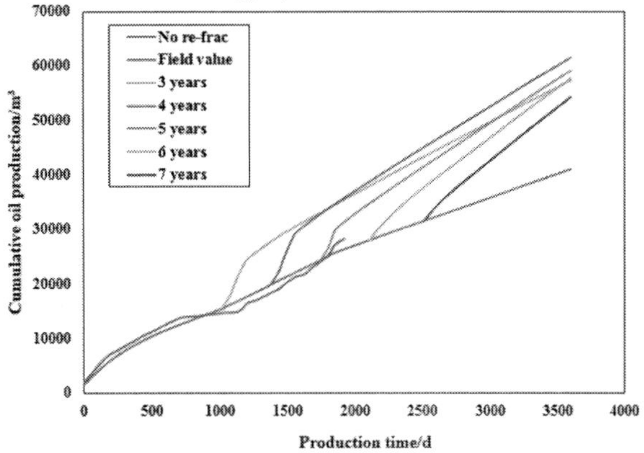

Figure 10. Cumulative oil production curves of re-fracturing at different times after the initial fracturing of Well X

Table 3. 10-year cumulative oil production obtained after re-fracturing at different times after initial fracturing of Well X

Time interval between initial fracturing and re-fracturing (year)	10-year cumulative oil production (m^3)
No re-fracturing	41037
3	57306
4	61427
5	59009
6	57659
7	54168

5. Evaluation of the economic benefits

Although the application of re-fracturing in horizontal wells can increase the oil production notably, the operation of re-fracturing needs to spend a certain cost. According to the design scheme of Well X, the amount of fracturing fluid, proppant, temporary plugging agent and other materials can be known, and the cost of fracturing materials can be estimated based on the market price. By taking the rental cost of equipment and the labor cost into consideration, the total cost of implementing re-fracturing of Well X is estimated to be 16.28 million yuan. For the crude oil produced by re-fracturing, based on the international oil price of 70 US dollars/barrel and the exchange rate at 6.9 US dollars, the price of crude oil can be calculated to be about 2900 yuan/m^3, the income after the sale of crude oil at different times after re-fracturing can be calculated.

The economic benefit of refracturing is evaluated by net present value (NPV) and output/ input ratio (η). The formulas are:

$$\text{Net present value (NPV)=Total present value of future return} - \text{Initial investment} \tag{16}$$

$$\text{Output/input ratio } (\eta) = \text{Crude oil sales income/Cost of re-fracturing} \tag{17}$$

Since Well X has been exploited for 5.5 years, although the best opportunity for re-fracturing has been missed, it is still that the earlier the re-fracturing is implemented, the better the effect will be. Therefore, the NPV curve and the output / input ratio curve of re-fracturing after the 6th year after the initial fracturing are calculated and drawn, as shown in figure 11 and figure 12.

As known from figure 11, the cost can be recovered in about one year after re-fracturing. After 350 days of exploitation, NPV acquired from re-fracturing began to catch up and even surpass the NPV from those without re-fracturing. 4 years after the re-fracturing (that is, 10 years after the initial fracturing), the NPV reaches 69.95 million yuan. If no re-fracturing was carried out, NPV would be only 38.03 million yuan. Therefore, compared with non-refracturing, NPV after re-fracturing increased 31.92 million yuan. It can be seen from figure 12 that 4 years after the re-fracturing (i.e., 10 years after the initial fracturing), the output input ratio reaches 5.3.

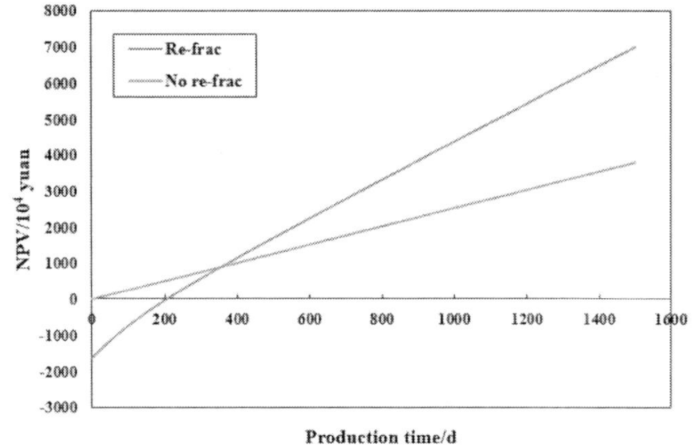

Figure 11. NPV curve of re-fracturing after 6 years since the initial fracturing

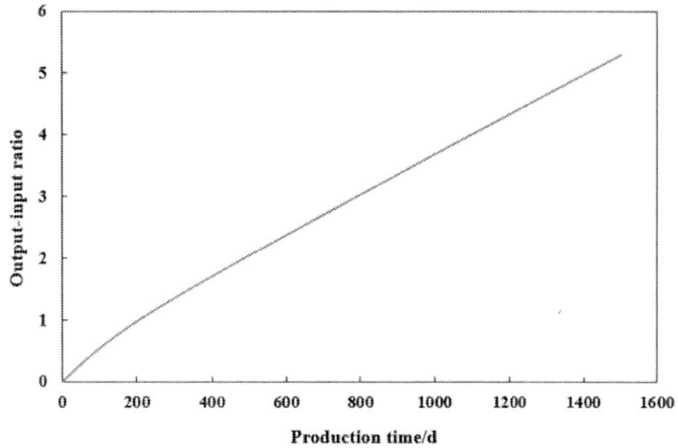

Figure 12. Output-input ratio curve of re-fracturing after 6 years since the initial fracturing

6. Conclusion

- The mathematical model of oil-water phase seepage is established, with the matrix and fracture divided by a unified grid system, and the equations are discretized based on the FDM method and is solved by IMPES method.
- In the study case of Well X, by adjusting the parameters such as the relation of oil-water phase permeability, the rule of change of the predicted oil production curve obtained by fitting is basically consistent with that of the actual oil production curve. The predicted results of formation pressure decline are consistent with the testing results of downhole measurement by pressure gauge, which verifies the accuracy of the established model.
- The numerical simulation results indicate that more cumulative oil production can be realized after the 4th year of initial fracturing of Well X, which reaches 61427m^3. Although the best time of re-fracturing has been missed in Well X, it is recommendable that the re-fracturing should be carried out as early as possible. Considering the cost of re-fracturing operation, the NPV reaches 69.95 million yuan after 4 years since the re-fracturing (that is, 10 years after the initial fracturing), with an output-input ratio reaching 5.3.

Acknowledgments
The research was supported by the Sichuan Science and Technology Project (18YYJC1108) and the cooperative project "Research on temporary plugging volume fracturing for horizontal wells and its field application" between PetroChina Xinjiang Oilfield Company and Beijing Gepetto Petroleum Technology Co., LTD.

References

[1] Du, J.H., He, H.Q., Yang, T., et al. (2014) Progress in China's tight oil exploration and challenges. China Petroleum Exploration, 19: 1-9. (in Chinese)

[2] Hu, S.Y., Zhu, R.K., Wu, S.T., et al. (2018) Profitable exploration and development of continental tight oil in China. Petroleum Exploration and Development, 45: 230-242.

[3] Chen, Q.L., Deng, Y.L., Wei, J., et al. (2018) Types, distribution and play targets of Lower Cretaceous tight oil in Jiuquan Basin, NW China. Petroleum Exploration and Development, 45: 227-238.

[4] Xu, J.W., Li, J.M., Wu, Y.Y., et al. (2019) Exploration and practice of volume fracturing technology in horizontal well of Mahu tight conglomerate reservoirs. China Petroleum Exploration, 24: 241-249. (in Chinese)

[5] Guo, J.C., Tao, L., Zeng, F.H. Optimization of refracturing timing for horizontal wells in tight oil reservoirs: A case study of Cretaceous Qingshankou Formation, Songliao Basin, NE China. Petroleum Exploration and Development, 46: 153-162.

[6] Wang, X.J., Wang, W., Li, B.R., et al. (2018) The application of refracturing technique in horizontal wells of Daqing Oilfield. In: Abu Dhabi International Petroleum Exhibition &Conference. Abu Dhabi. pp. 1-10.

[7] French, S., Rodgerson, J., Feik, C. (2014) Re-fracturing horizontal shale wells: case history of a Woodford Shale pilot project. In: SPE Hydraulic Fracturing Technology Conference. Woodlands. pp. 1-24.

[8] Lantz, T.G., Greene, D.T., Eberhard, M.J., et al. (2007) Refracture treatments proving successful in horizontal Bakken wells; Richland Co, MT. In: Rocky Mountain Oil & Gas Technology Symposium. Denver. pp. 1-8.

[9] Tavassoli, S., Yu, W., Javadpour, F., et al. (2013) Selection of candidate horizontal wells and determination of the optimal time of refracturing in Barnett Shale. In: SPE Unconventional Resources Conference Canada. Calgary. pp. 1-14.

[10] Cafaro, D.C., Drouven, M.G., Grossmann, I.E. (2016) Optimization models for planning shale gas well refracture treatments. AIChE Journal, 62: 4297-4307.

[11] Pang, P, Liu, Z.Y, Wang, H.Z., et al. (2015) Numerical simulation of refracturing opportunity. Petroleum Geology and Oilfield Development in Daqing, 34: 83-87. (in Chinese)

[12] Yang, Z.Z., He, R., Li, X.G., et al. A hybrid model for selecting horizontal candidate wells for re-fracturing of tight oil reservoirs—A case study in MH Oil Field, Junggar Basin, Western China. In: SPE Russian Petroleum Technology Conference. Moscow. pp. 1-22.

IOP Conf. Series: Materials Science and Engineering **729** (2020) 012053 doi:10.1088/1757-899X/729/1/012053

Design and Test of Self-Rescue Safety Belt for Power Line Operation at Heights

Zhao Zhiyong[1*], Zhang Jianzhong[2], Hua Yijun[3] , Fei Chunming[1], Fu Jian1, Wu Jianling[4], Gao Juhua[1], Fang Huayu[1]

[1] Training Center of State Grid Zhejiang Electric Power Co., Ltd. China.

[2] State Grid Zhejiang Tongxiang Power Supply Co., Ltd. China.

[3] State Grid Zhejiang Jinhua Power Supply Co., Ltd. China.

[4] State Grid Zhejiang Electric Power Co., Ltd. China.

*Corresponding author's e-mail: lyjjp@qq.com, 157611870@qq.com

Abstract. The self-rescue ability of electric power workers is of great significance to the safety of workers. However, the current use of safety belts can only ensure that operators do not fall to the ground, but it is not convenient for operators to carry out independent rescue and return to the wire. In this paper, a new type of safety belt design scheme is proposed to improve the self-rescue ability of electric power workers. The biggest difference between this new type of safety belt and the current one is that it integrates a new type of buffer bag, which integrates pedal belt and can help operators change their posture and climb up. Through the test of the safety belt with the new buffer bag, it is found that the new safety belt not only has the basic safety protection ability, but also can effectively assist the operators to carry out self-rescue, which greatly improves the safety protection level of the workers of power lines.

1. Introduction

In some operations high above the ground, because of the relatively narrow working space, high difficulty and greater danger, there are always warnings of safety signs. Another indispensable protective measure is the safety belt, which is worn by operators when working at heights, to prevent workers from falling and to protect workers from or reduce personal protective equipment caused by falling. Therefore, whether the performance of safety belts can meet the need to protect users' safety in use is of vital importance [1]. In China, safety belts are classified into three categories according to the nature of their workplaces: fencing safety belts, regional restricted safety belts and falling suspension safety belts [1]. The main research object of this paper is the safety belt for operation at heights of power system, which belongs to the type of falling suspension safety belt.

As early as 1985, the national standards[3] for safety belts have been formulated and promulgated. The technical performance indicators of the former Soviet Union have been extensively used as references in the contents of the early standards. And it has been used until 2009. Over the past 20 years, the standard has played an important role in quality supervision and market supervision. However, with the improvement of production technology and the emergence of new materials, the standard version 85 no longer meets the needs of safety belts. In the 2009 edition of safety belt standards [4], the standards have changed greatly in view of the differences between domestic standards and foreign standards, as well as the requirements of avoiding restricting technological progress and absorbing foreign standards. Due to the continuous improvement of technology, advanced design abroad is also affecting the design and style of safety belts in China. In the past, safety belts were heavily single-belt or single-belt. However, with the progress of the times, users also have more requirements for security in the use process. Therefore, full-body safety belts are gradually coming into our sight.

At present, in the norms of operation at heights of power system, although it has been generally required to use safety belts to ensure the safety of operators. However, because the operation of power lines is mostly in the field, it is difficult for line operators to carry out rescue work once they fall behind. If rescue is not timely, it will probably lead to casualties. In fact, due to the small drop distance and the protection of cushion packs and all-around safety belts, the operators have the ability of self-rescue in most cases. However, the current all-around safety belts do not have the function of assisting self-rescue, resulting in the passive waiting for rescue of the fallen personnel, which leads to the fact that the fallen personnel can only passively wait for rescue. There are great safety risks. Therefore, the main work of this paper is about how to improve the self-rescue capability of the whole body self-rescue safety belt.

This paper is organized as follows: Section 2 produces the autonomous rescue requirements and characteristics of the electric safety belt, i.e. the design objectives of the new type safety belt; Section 3 introduces two different design ideas and compares their advantages and disadvantages; Section 4 tests the optimal design according to the national standards.

2. Design requirements

In this section, design requirements of the autonomous rescue safety belt for electric operation are discussed in detail.

In the process of operation of electric power, especially in line construction, it is often tens of meters or even hundreds of meters high from the ground. Once the operator falls in line inspection, even if he is not injured, if the rescue is not timely and long-term suspended at heights, it is very likely to lead to casualties. At this time, if the safety belt can assist the falling personnel to rescue independently and let them climb back to the wire, it can completely avoid the occurrence of such accidents. The following two requirements must be met in order for the safety belt to assist the operator in self-rescue:

(1) Safety belt should be able to assist the person falling to change his posture

When the operator falls behind, he will be in a hanging posture. This posture is suitable for long-term hanging and waiting for rescue, but it is not easy to focus on self-rescue. Therefore, there must be ways to help them change their posture; the ideal state is the chest suspension.

(2) Safety belt should be able to assist rescue workers to climb back to the wire independently

If there is no auxiliary device, it is very difficult to climb the safety belt rope at heights by pure arm force after the operator who uses the conventional whole body safety belt falls down the wire. It is necessary to have auxiliary equipment to help him climb back to the wire.

3. Design schemes and comparison

Methods. According to the above design requirements, we can list two kinds of implementation ideas and give the corresponding implementation schemes as shown in Table 1 below. At the same time, we also give the evaluation of its implementation difficulty.

Table 1. Design ideas and comparison

Design requirements	Realization Ways	Design ideas	Difficulty	Dangerous degree
Change body posture	By arm force	Changing Hanging Point by Arm Force	Hard	High
	By arm force and legs	Use pedal belt to assist in changing hanging point	Easy	Low
Auxiliary return line	By Arm force	Use climbing ropes and lifts	Hard	Low
	By arm force and legs	Use pedal straps to assist climbing	Easy	Low

Optimum scheme. From the comparison results in Table 1, it can be seen that whether changing the body posture or assisting return line, simply using arm force and the combination of hands and feet is not only difficult to achieve, but also dangerous. Therefore, the optimal solution to the problem should be the combination of hands and feet.

Under the guidance of this idea, the detail ways of realization are as follows:

(1) Use pedal straps to assist posture changing

Using pedal belt, the operator can use pedal to climb a certain distance, release the back hook and hang it to the front hook, thus changing the suspension mode from the back to the chest, so as to facilitate subsequent climbing.

(2) Use pedal straps to assist climbing

If the pedal belt is long enough and the pedal force is utilized while the arm grabs the safety belt and climbs upward, the traverse can be returned. However, the top of the long enough pedal band needs to be suspended on the line, and the pedal band needs to be folded up under normal conditions, otherwise it will affect the operation. In view of this, the improved scheme draws lessons from the idea of safety belt buffer pack, folds and binds normally, and pulls open and hangs down by human gravity when falling. According to this idea, the upper end of the folding pedal belt is connected with the buckle of the safety belt, and the lower end is connected with the buffer package, which can achieve the above purpose.

Following this idea, the schematic diagram of the improved electric power safety belt structure with self-rescue capability is shown in Fig. 1 below. Figure (a) is in the initial state and figure (b) is in the state after falling with pedal strap unfolded and buffer package ripped.

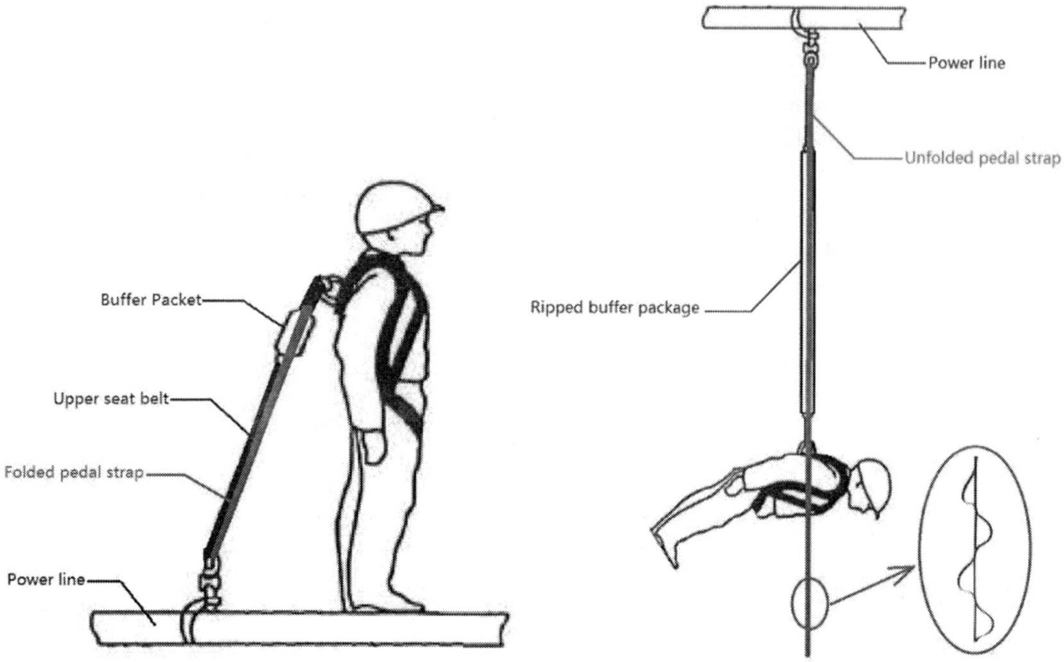

(a) initial state of safety belt (b) state of safety belt after falling
Fig. 1 schematic diagram of the improved electric power safety belt with self-rescue capability

The image of state of posture change after falling is shown in Fig. 2. Figure (a) shows the initial state after falling, figure (b) presents state of posture after change, and figure (c) shows the state of autonomous climbing.

(a) Initial state after falling (b) posture after change (c) autonomous climbing.
Fig.2 The image of state of posture change after falling

4. Experiment results and discussion

Standard adopted. At present, the latest national standards for safety belts are the of "Safety Belt" and "Safety Belt Testing Method" version 2009[4]. In the standards, the overall static and dynamic load testing methods of safety belts and the overall static load testing methods of falling suspension safety belts are required. The standard of the 2009 edition is consistent with the basic requirements of the European Standard [5], that is, the overall static tension is not less than 2kN. For the safety of the falling suspension safety belt, considering the human safety in the process of falling suspension, referring to domestic and foreign statistical data, the peak value of impact force in the dynamic mechanical properties of the standard of the 2009 edition is set at no more than 6kN.

Fortunately, the advantage of the improved design of the safety belt in this paper is that only folding pedal belt is added without changing the existing body of the full-body suspension safety belt, thus avoiding the complex testing process of the performance of the safety belt body, and only need to test whether the folding bag of the pedal belt can be opened normally and its loading capacity.

Experiment design. According to the analysis above, experiments for the new safety belt include two aspects, one is opening capability, and another one is carrying capacity. According to Statistics in State Grid, the weight range of normal line workers is 50kg to100kg. By using safety factor 2, the maximum open weight is 25kg, and minimum load weight is 200kg. By using pedal strap with minimum load weight 220kg and folded by fragile strapping tape, the following experiment results are shown in Table 2 and Table 3 with 10 samples for each test.

Table 2 maximum open weight test results

Serial number	Sample weight(kg)	Results
1	10	failed
2	12	failed
3	14	success
4	16	success
5	18	success
6	20	success
7	21	success
8	22	success
9	23	success
10	24	success

Table 3 minimum load weight test results

Serial number	Sample weight(kg)	Results
1	180	success
2	190	success
3	200	success
4	210	success
5	220	success
6	230	success
7	240	success
8	250	success
9	260	failed
10	270	failed

Results discussion. With the help of pedal strap, new safety power line operation is endowed with self-help ability, but the maximum open weight and minimum load weight must be up to standard. Through the experiments carried out, it can be concluded that the new safety belt meets the requirement of latest national standard version 2009. Moreover, two samples have been used successfully in construction site of Zhejiang electric power company.

Acknowledgements.
This work was sponsored by the State Grid Zhejiang electric power co., ltd. science and technology project "Research and application of trapped rescue for electric power workers" (No. 5211JP170002) .

References

[1] Donna Glassbrenner, Joseph S. Carra, James Nichols. Recent estimates of safety belt use. Journal of Safety Research 2004, (35): 237-244

[2] Tang yimin, Tong suifang, Jiang ruiliang. The function and characteristic of safety belts in falling protection. The 20th Academic Seminar on Occupational Safety and Health across the Taiwan Straits, Hong Kong and Macao, 2012: 291-298

[3] Classification of Casualty Accidents of Enterprise Workers (GB 6441-1986).

[4] National Standard Safety belt Standard GB6095-2009

[5] Arnab Majumdar, Robert B. Noland ∗, Washington Y. Ochieng. A spatial and temporal analysis of safety-belt usage and safety-belt laws. Accident Analysis and Prevention 2004,(36) : 551-560

Impact analysis and risk assessment of supply and demand bilateral randomness on power system

Dunnan Liu[1], Hua Li[1*], Rui Ge[2], Xiaofeng Xu[1], Nan Wang[3]

[1] School of Economics and Management, North China Electric Power University, Beijing , 102206, China

[2] State Power Dispatch Control Center, Beijng, 100031, China

[3] State Gid Electric Pwer co., LTD. In Hebei Province, Shijiazhuang, Hebei, 050021, China

*Corresponding author's e-mail: lihua_ncepu@163.com

Abstract. With the high proportion of new energy access, the randomness of the supply side increases.The electrical power system system is transformed from a unilateral stochastic system on the demand side to a bilateral stochastic system on the supply and demand side.Firstly, this paper studies the potential risks of power system under the randomness of supply and demand sides;Then, according to the principles of integrity, relevance, operability, quantitative and qualitative combination, the evaluation index system of bilateral randomness for the safety and stability risk of power system is constructed.Finally, a risk assessment model is constructed based on the static subjective weights of FAHP and similarity clustering analysis and the dynamic objective weights of the improved CRITIC weighting method.

1. Introduction

The power system is widely embedded and deeply affects the economic and social development. Any changes in its operational characteristics and its corresponding technological upgrades will affect many aspects of the country and society. The traditional supply-side thermal power generation has a high proportion and strong regulation ability. The output of the thermal power unit can respond faster with the change of the demand side power load.

The large-scale intermittent renewable energy grid connection makes the supply side controllability gradually decrease, the randomness gradually increases, and the basic operational characteristics of the power system change accordingly. From the demand side single-sided stochastic system to the supply and demand two-sided stochastic system [1].

2. Analysis of the Influence of Bilateral Randomness on Power System Security and Stability.

Renewable energy power generation has strong random volatility. With the development of clean energy, the scale of renewable energy integration is expanding, and the proportion of renewable energy generation in the power supply side is constantly increasing, requiring a large number of conventional power sources to complement each other to stabilize the volatility [2]. There is no coordination between current renewable energy construction and grid planning. With the further deepening of the bilateral randomness characteristics, the mutual influence between the source networks will gradually become obvious. The specific performance is that the energy consumption of renewable energy is insufficient, and a large number of abandoned water, abandoned wind and abandoned light appear, resulting in serious waste of resources and the controllability of the output of the electric side is poor, which increase the influence of the bilateral random volatility on the operation of the system [3].

3. Bilateral randomness to power system security and stability risk assessment index system

3.1. Index system construction

This section mainly constructs the risk assessment index system of power system security stability from the aspects of power supply complementation and source network coordination [4]. As shown in Table 1.

1) Power supply complementary risk R_1

a. Capacity validity R11: Describe the effective utilization of the capacity of the machine assembly machine, reflecting the abundance of the unit's power generation

b. Power generation intensity R12: Describe the continuity of the unit's power generation, which is the ratio of the equivalent full power generation hours of the unit to the statistical time of 8760 hours.

c. Operating safety R13: Describe the degree of safe operation of the unit, reflecting the degree of impact of unit outage and grid failure on unit reliability.

d. Relative benefit coefficient and R14: Describe the superiority of the renewable energy unit relative to the thermal power unit.

2) Coordination risk between power and grid R_2

a. Peaking ability R21: Reflecting the indicators of renewable energy consumption, the stronger the peaking ability of the system, the larger the range of power fluctuations allowed by the system.

b. Transmission capacity R22: Reflecting the indicators of renewable energy consumption capacity, the stronger the system's transmission capacity, the more energy that renewable energy delivers, and the greater the capacity of renewable energy.

c. Grid-connected technology R23: Indicators reflecting the ability of renewable energy to absorb, the more mature the grid-connected technology, the less volatility caused by the integration of renewable energy, and the more renewable energy that can be absorbed.

d. Power load matching degree R24: Reflecting the degree of matching between power supply and load, the higher the matching degree between power supply and load, the lower the frequency of abandoned power generation and power failure.

e. System security R25: Indicators that reflect the operational level of the power system, including static and dynamic security. Static security mainly refers to the ability of the system to maintain the initial stable operating state due to line failure or load fluctuation.

f. Power quality R26: Indicators that reflect the operating level of the power system, mainly related to voltage, voltage sag, voltage fluctuations and flicker, harmonics, and frequency.

g. Abandon power loss R27: The indicator reflecting the degree of incompatibility between the power and the grid is the sum of the waste of resources caused by the abandonment of water wind or light.

h. Loss of power failure R28: The indicator reflecting the degree of uncoordinated power and grid is the sum of direct and indirect losses caused by power failure.

4. Potential risk assessment model under the bilateral randomness of supply and demand

4.1. Static subjective weights based on FAHP and similarity clustering analysis

FAHP is run by the maker's subjective judgment [5]. Subjective authority is a feature of the method. Therefore, this chapter establishes the index sorting vector matrix L of the expert group based on FAHP, applies the similarity clustering analysis to establish the authoritative factor matrix γ of the expert group, and combines the two matrices to determine the static subjective weight of the potential risk assessment index system under the bilateral randomness of supply and demand.

The steps for calculating the authoritative factor matrix γ are:

Step 1: Use L to establish a similarity correlation matrix $S=\left[s_{pq}\right]_{m\times n}$

The following formula is used to specifically define the similarity of the row vectors lp and lq, and the similarity is represented by the vector angle cosine s_{pq} (p≠q; p,q=1,2,...,n).

$$s_{pq}=\frac{l_p l_q^T}{\|l_p\|\|l_q\|} \tag{4-1}$$

Where l_q^T is the transpose of vector l_q, and $\|l_p\|$ and $\|l_q\|$ are the norms of the corresponding vector.

Step 2: Expert Cluster Analysis

(a) Find the largest off-diagonal element Sab in matrix S, and put experts e_a and e_b into set E; (b) Let $S_{ab}=S_{bd}=0$, determine the largest off-diagonal element Sac(S_{bc}), and classify e_c into set E; (c) Let $S_{ac}=S_{ca}=0$(Let $S_{bc}=S_{cb}=0$), determine the off-diagonal element S_{cd}, and classify ed into set E; (d) And so on, until all experts are included in the set E; (e) Determine a threshold T. If $S_{lk}\le T$, the experts el, e_k are classified into one class, and if not, el and ek are different categories.

Step 3: Authoritative factor α between expert categories

If the clustering is successful, n experts total r categories, where the expert e i belongs to the h-th (l ≤ h ≤ r) class, and the class includes φ_h experts, then the authoritative factor α_h of the class is:

$$\begin{cases} N_i=\sum_{j=1,j\neq i}^{n}\|l_i-l_j\|^2 \\ N_h'=\left(\sum_{i=1}^{n}N_i\right)/\varphi_h \\ \alpha_h=\left(\varphi_h^2/N_h'\right)/\sum_{t=1}^{r}\left(\varphi_t^2/N_h'\right) \end{cases} \tag{4-2}$$

Among them, Ni is the comprehensive consistency difference value between the index sorting vectors established by experts e $_i$ and other experts; N_h' is the comprehensive consistency difference value between the category h expert and other expert categories.

Step 4: Calculate the authority coefficient β in the expert categories

The authoritative coefficient β_i of the expert e_i in the h-th category is:

$$\beta_i=\frac{1}{1+cC_{Ri}}/\sum_{j=1}^{\varphi_h}\frac{1}{1+cC_{Rj}} \tag{4-3}$$

Where c is the adjustment scale factor, let c=10; C_{Ri} and C_{Rj} are the consistency ratios of the fuzzy judgment matrices established by the experts e_i and e_j in the h-th category, respectively.

Step 5: Calculate γ

The authoritative factor of expert e $_i$ is:

$$\gamma_i=\alpha_h\beta_i \tag{4-4}$$

4.2. Dynamic objective weight based on improved CRITIC Empowerment Method

The model based on the CRITIC Empowerment Method to obtain objective empowerment is [6]:

$$\begin{cases} \sigma_i=\left[\sum_{k=1}^{u}(x_{i,k}-x_{i,ave})^2/u\right]^{\frac{1}{2}} \\ r_{ij}=\dfrac{\sum_{k=1}^{u}(x_{i,k}-x_{i,ave})(x_{j,k}-x_{j,ave})}{\sum_{k=1}^{u}(x_{i,k}-x_{i,ave})^2\sum_{k=1}^{u}(x_{j,k}-x_{j,ave})^2} \\ \omega_2(i)=\left[\sigma_i\sum_{j=1,j\neq i}^{v}(1-r_{ij})\right]/\left[\sum_{i=1}^{v}\sigma_i\sum_{j=1,j\neq i}^{v}(1-r_{ij})\right] \end{cases} \tag{4-5}$$

Where σi is the standard deviation of the index i; u is the number of decision schemes; xi,k and xj,k are the normalized parameter values of the indicators i and j in the k-th decision scheme, respectively ; rij is the correlation coefficient for indicators i and j ; w2(i) is the dynamic objective weight, and v is the number of indicators. The parameters of indicators i and j are normalized.

Because CRITIC lacks the discreteness of the metrics, this chapter uses the formula of information entropy to consider the weights between the indicators to compare strength, conflict, and discreteness.

$$\omega_2(i)=\frac{(\sigma_i+s_i)\sum_{j=1,j\neq i}^{v}(1-r_{ij})}{\sum_{i=1}^{v}(\sigma_i+s_i)\sum_{j=1,j\neq i}^{v}(1-r_{ij})} \tag{4-6}$$

(3) Obtaining optimal variable weight

For the index i, the formula for calculating the optimal variable weight is as follows:

$$\omega(i)=\frac{\omega_1(i)\omega_2(i)}{\sum_{i=1}^{t}\omega_1(i)\omega_2(i)} \tag{4-7}$$

Where w1(i) and w2(i) are the static subjective weights and dynamic objective weights of indicator i, respectively; t is the total number of indicators of the same category as indicator i.

5. Case analysis

The power system safety and stability risk assessment system proposed in the previous paper is used to obtain the safety and stability risk value of the power system.

Step 1: Calculate the risk value of each micro indicator

Calculate the risk value of each micro-indicator of the power system in the recent development stage according to the previous steps, as shown in Table 2.

Table 1. Risk values of different development stages of micro indicators

Macro risk	Micro indicator	Recent	Medium-term	Long-term
Power supply complementary risk R_1	Capacity validity R_{11}	12.047	13.681	10.354
	Power generation intensity R_{12}	15.368	12.976	10.252
	Operational safety R_{13}	16.324	15.024	10.687
	Relative benefit coefficient R_{14}	11.578	12.649	10.208
Coordination risk between power and grid R_2	Peaking ability R_{21}	9.427	8.514	7.138
	Transmission capacity R_{22}	10.088	8.289	7.651
	Grid-connected technology R_{23}	11.235	9.802	7.064
	Power load matching degree R_{24}	13.864	16.334	11.240
	System security R_{25}	14.392	11.446	8.083
	Power quality R_{26}	15.027	10.066	5.9863
	Abandon power loss R_{27}	14.239	10.872	9.624
	Loss of power failure R_{28}	11.947	10.368	6.248

Step 2: Calculate the optimal variable weight

According to the static subjective weight solving process, the calculated static subjective weights are:

R1=[0.2867,0.3302,0.1846,01985],

R2=[0.1045,0.1372,0.1672,0.0967,0.1824,0.0726,0.1285,0.1106], R= [0.4026,0.5974]。

Combined with the contents of Table 4-1, the dynamic objective weights of the micro-indicators are determined according to the improved CRITIC Empowerment Method:

R_1'=[0.2525,0.4302,0.1066,0.2107],

$R_2^{'}$ =[0.1503,0.1244,0.0987,0.1234, 0.1406,0.1720,0.1047,0.0859]。

R_1^{*} =[0.2634,0.5147.0.07131,0.1515], R_2^{*} =[0.1284, 0.1395,0.1349,0.0975, 0.2096,0.1021.0.1100,0.0776] was calculated according to the optimal variable weight formula.

Step 3: Calculation of overall risk value

Based on the above calculation results, the weighted average method is used to obtain the safety value of the two-sided randomness to the power system security stability. As shown in Table 3.

Step 4: Risk rating

In the case study, the safety and stability risk value of the power system is 12.601. Compared with the risk feature vector, the risk level is close to the third level, and the risk is medium risk.

Table 2. Power system security stability risk value

assess target	Risk value	Risk description
R1	10.028	medium
R2	14.336	medium
R3	12.601	medium

The safety and stability risk level of the power system after the renewable energy is connected to the grid is general. Therefore, based on the risk assessment model established in this section, the safety and stability risk situation of the power system is consistent with the actual situation, and the validity of the model is verified.

6. Conclusion

This paper studies the impact of bilateral randomness on the safety and stability of power system, analyzes the risk of bilateral randomness on the safety and stability of power system, constructs a risk assessment index system, and establishes a risk assessment model.

Acknowledgements

Science and Technology Project of State Grid Corporation of China, Integrated Operation Control Strategy of Cross District Whole Network Adapting to the Development of New Energy, Number: SGTYHT/16-JS-198.

References

[1] He Ping. Research on the influence of large-scale wind power access on power system stability and control measures [D]. South China University of Technology, 2014.

[2] Mu Yunfei. Research on power system safety assessment and control of wind farms and electric vehicles [D]. Tianjin University, 2012.

[3] YE Shengyong. Research on transient stability assessment of power system based on machine learning [D]. Southwest Jiaotong University, 2010.

[4] Li Shiqian, Shen Yuqiang, Hu Minda, Zhou Lifang. Hidden risk assessment system for relay protection of intelligent substation[J].Electrical Engineering, 2019(14):97-99.

[5] Wu Xu, Yuan Yao, Wang Yang. FAHP-based QoS-aware trusted cloud service selection model[J].Computer and Digital Engineering,2019,47(09):2233-2241.

[6] CHEN Wei, ZHAO Min, ZHOU Hua, YI Peng. Fuzzy comprehensive water quality assessment method and application of Jiangyan District based on game theory combined weighting[J]. Hydropower Science, 2019, 37(09): 32-35 +91.

IOP Conf. Series: Materials Science and Engineering **729** (2020) 012055 doi:10.1088/1757-899X/729/1/012055

The research progress on micro-arc oxidation of aluminum alloy

Dong Yuting, Liu Zhiyang, Ma Guofeng

Liaoning Province Key Laboratory for Advanced Materials Preparation Technology, Shenyang University, Shenyang 110044

Abstract. The micro-arc oxidation of aluminum alloy is a kind of surface treatment technology for in-situ growth of a hard ceramic film with Al2O3 as the main effect of instantaneous high temperature and high pressure generated by arc discharge on the surface of aluminum alloy. This technology can significantly improve the wear resistance, corrosion resistance, hardness and other properties of the aluminum alloy matrix. This paper introduces the characteristics, influencing factors, research status of domestic and foreign micro-arc oxidation technology, and prospects for future research directions.

1. Introduction

Micro arc oxidation (MAO), also known as plasma electrolytic oxidation (PEO), micro plasma oxidation (MPO). It produces instantaneous high temperature and high pressure on the surface of aluminum and its alloys, through the combination of electrolyte and corresponding electrical parameters and arc discharge. And the ceramic film layer mainly composed of the base metal oxide is grown in situ. The micro-arc oxidation technology of aluminum alloy is mature and easy to operate. After a series of reactions, a dense oxide film with Al2O3 as the main component is formed. The reaction equation is as follows:

Cathodic reaction:

$$2H++2e-\rightarrow H2\uparrow \quad (1)$$

Anode reaction:

$$4OH-4e-\rightarrow 2H2O+2O2\uparrow \quad (2)$$

$$Al-3e-\rightarrow Al3+ \quad (3)$$

$$Al3++3OH-\rightarrow Al(OH)3 \quad (4)$$

$$2Al(OH)3\rightarrow Al2O3+3H2O \quad (5)$$

$$2Al3++3SiO32-\rightarrow 3SiO2+Al2O3 \quad (6)$$

$$4Al+3O2\rightarrow 2Al2O3 \quad (7)$$

$$2Al3++3O2-\rightarrow Al2O3 \quad (8)$$

$$\gamma\text{-}Al2O3\rightarrow \alpha\text{-}Al2O3 \quad (9)$$

The oxide film formed by the micro-arc oxidation process has good wear resistance, high hardness, strong corrosion resistance, less defects on the surface of the film layer, and the structure tends to be dense.

2. Principle of micro-arc oxidation technology

The principle of micro-arc oxidation technology is as follows [1]: the metal such as aluminum, magnesium, titanium or its alloy is placed in the electrolyte under the electric field environment as the anode, the electrolytic cell is the cathode, and the voltage is applied (up to 1000V) and larger current, as shown in Fig. 1.

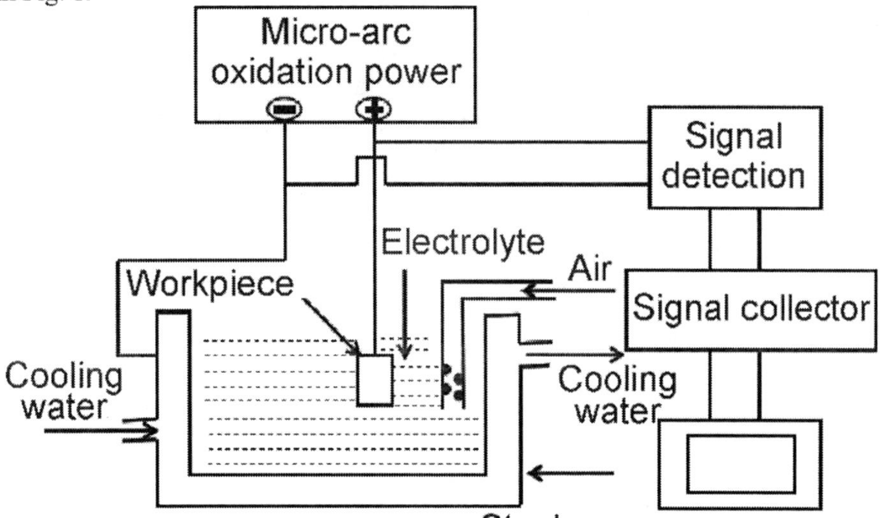

Fig 1. Schematic diagram of principle of MAO technology

Immediately after energization, a thin metal oxide insulating film is formed on the metal surface. The formation of a complete insulating film is a necessary condition for micro-arc oxidation treatment. On this basis, the applied voltage of the workpiece rises steadily. When a certain critical value is reached, some weak links on the insulating film are broken first. It will occur in the micro-area arc discharge phenomenon, and instantly form an ultra-high temperature region (103-104K) [2]. This causes the oxide and the base metal to be melted or even vaporized. After the molten material is in contact with the electrolytic solution, a ceramic film layer is formed by chilling. Since the breakdown always occurs at a relatively weak portion of the oxide film, and a new oxide film is formed in the original portion after the breakdown. The breakdown point then shifts to other relatively weak areas. Repeatedly, a uniform oxide film is formed on the metal surface.

3. Development of micro-arc oxidation technology

The development of micro-arc oxidation technology can be traced back to the early 1930s. German scientists Gunterschulze and Betz first reported that the metal immersed in the solution will have spark discharge on the surface under the action of high-voltage electric field [3]. Through further observation and research, it was found that the ceramic oxide film can be successfully prepared on the metal surface by using this spark discharge phenomenon. Due to the protective effect of the oxide film, the magnesium alloy anticorrosion was originally applied by this technology [4]. So far, magnesium alloy spark discharge technology is still under study. In the 1970s, the United States, Germany, the Soviet Union and other countries saw the application potential of micro-arc oxidation technology, and joined the research of this technology. The University of Illinois, Karl Marx University of Germany and other units used DC or single-phase pulsed power supply mode to comprehensively analyze aluminum, titanium and other valve metal spark discharge deposition layers, and named them as anode spark deposition [5-6] and spark discharge anodization [7-8]. Compared with the traditional anodizing technology, the biggest advancement of this technology is that the microplasma generated on the surface of the metal during oxidation can be sintered into a crystalline state in a high temperature and high pressure region. The film layer prepared on the aluminum alloy, α-Al2O3 and γ-Al2O3 occupy its main film component [9-10]. In 1977, the institute of inorganic chemistry of the Soviet Union began research on this technology. They used an alternating voltage mode and obtained a similar film layer, in which α-Al2O3 and γ-Al2O3 still occupy the main components of the oxide film. This technique is named micro-arc oxidation. So far, the micro-arc oxidation technology has been greatly developed [11].

At present, the countries that are engaged in the research of micro-arc oxidation technology in the world mainly include Russia, the United States, Germany, Japan, etc.. Among them, Russia is more advanced. Russia has proposed a relatively complete theory in the research of micro-arc oxidation mechanism, and successfully applied to aviation, petroleum, textile and other industrial fields [12]. Since the 1990s, China has begun to pay attention to the technology of micro-arc oxidation. At present, it has begun to take the form of wear-resistant, corrosion-resistant and decorative film layers, but it is still in its infancy. Generally speaking, the research level of micro-arc oxidation technology abroad is slightly ahead of the domestic. But at present, domestic and foreign countries have not entered the stage of large-scale industrial application. Therefore, in order to understand and master the technology in depth, it is still necessary to carry out a large number of research experiments to prepare a ceramic film layer with better performance, so that it can be practically applied in more fields.

4. Factors Affecting Microarc Oxidation of Aluminum Alloy

4.1. Operating Voltage

The arcing voltage at the beginning of the micro-arc oxidation reaction is closely related to the composition of the electrolyte and the composition of the aluminum alloy. When the working voltage is too low, the micro-arc oxidation reaction is slow and the quality of the oxide film is not good. Ablation occurs when the operating voltage is too high. Moreover, the operating voltage can also determine the thickness of the final oxide film. Therefore, the working voltage has a very important influence on the micro-arc oxidation of aluminum alloy.

4.2. Current density

The current density has an important influence on the growth process, microstructure, element composition, phase composition and corrosion resistance of the aluminum alloy micro-arc oxidation film. As the current density increases, the thickness of the film layer increases and becomes more dense. However, when the current density is too large, the micro-arc oxidation reaction becomes severe, causing the film layer to become rough or even crack.

4.3. Electrolyte composition

The electrolyte used in micro-arc oxidation technology is generally divided into two categories: acidic electrolyte and alkaline electrolyte. However, due to the environmental pollution caused by acidic electrolyte, most of the micro-arc oxidation technology used is alkaline electrolyte. Because the alkaline electrolyte has a dissolution effect on the aluminum alloy, the alkaline concentration of the electrolyte cannot be too high, and sodium hydroxide or potassium hydroxide is usually used to adjust the pH of the electrolyte. Studies have shown that anions in the electrolyte composition participate in the formation of an oxide film, thereby changing the composition, structure, phase structure and properties of the micro-arc oxidation coating. The ceramic layer is selective for the adsorption of particles in the electrolyte, and the order of adsorption from strong to weak is SiO32-, PO43-, VO43-, MoO42-, WO42-, B4O72-, CrO42- [13]. The electrolytes commonly used in micro-arc oxidation are formulated with acid salt systems, phosphate systems, aluminate systems, etc. In addition, complex salts are used in many studies [14].

4.4. Additive

Zhang Xinyu et al [15] found that the addition of DTA and SDBS in the electrolyte can improve the stability and film formation rate of the electrolyte, but has little effect on the hardness and corrosion resistance of the film. Luo Haihe et al [16] studied the effect of adding different concentrations of (NaPO3)6 on the properties of magnesium alloy micro-arc oxidation ceramic coatings. They found that adding appropriate concentration of (NaPO3)6 can effectively increase the thickness of the coating. It can also improve the chemical composition and structure of ceramic coatings to improve their corrosion resistance.

4.5. Temperature
When the temperature is low, the film grows faster because the large amount of heat released during the reaction at a lower temperature is easily lost, so that the sintering is denser and the performance is better. On the contrary, it will accelerate the loss of active ingredients in the solution, resulting in the failure of the electrolytic solution [17].

5. Application of Microarc Oxidation of Aluminum Alloy
The micro-arc oxidation technology is environmentally friendly and highly efficient. The ceramic coating prepared has the advantages of high strength, wear resistance and corrosion resistance, and is widely used in various fields such as aerospace and petrochemical industries. By using this technology, a porous ceramic film layer having excellent heat resistance, wear resistance and corrosion resistance is grown in situ on a metal surface such as an aluminum alloy, thereby changing the defects of the metal itself. It has a better development space in the military field such as aerospace and has expanded the application range of aluminum alloy.

6. Development trend of micro-arc oxidation

6.1. Development of process database system
The parameters involved in the micro-arc oxidation process are complex and variable, which is not conducive to automated operation. To solve this problem, a process database can be established for the device. The process database contains all the parameters that need to be set for the micro-arc oxidation process. When you use it, you only need to call a micro-arc oxidation process to specify the set value for each parameter of the process, which will greatly improve the efficiency of use.

6.2. Development of low energy micro-arc oxidation process
The conventional micro-arc oxidation process processes ultra-large complex light metal components with high equipment requirements, long working hours and high energy consumption. By designing a scanning mobile cathode and controlling the micro-arc oxidation process, energy consumption can be effectively reduced. In addition, according to Chinalco.com, Beiteng (Shanghai) Technology Co., Ltd. has successfully developed low-voltage micro-arc oxidation technology. Compared with traditional micro-arc oxidation technology and anodizing technology, it not only has the advantages of low voltage, low energy consumption, low processing cost, high efficiency, high environmental protection, and recyclability, but also does not cause any damage to the surface of the base metal.

7. Conclusion
This paper introduces the principle and development of aluminum alloy micro-arc oxidation technology. The factors affecting the micro-arc oxidation technology of aluminum alloy are Operating voltage, current density, electrolyte composition, additives and temperature. We can control these influencing factors to produce a better quality ceramic film. The aluminum alloy adopts micro-arc oxidation technology to change its own defects and expand the application field. There are still some shortcomings in micro-arc oxidation technology. I think that in the future, we should focus on developing process database systems and low-energy micro-arc oxidation processes.

Acknowledgements
The authors gratefully acknowledge the China Postdoctoral Science Foundation (No. 2016M601333), Liaoning Provincial Natural Science Foundation of China (No.2019-MS-236), the Shenyang Science and Technology Project (No. 2018-013-0-14).

References

[1] ZUO H B, KONG Q S, SHANG J Q. Plasma enhanced chemical surface ceramic transfor mation technology [J]. Materials Protection, 1995,28(7): 117-120.
[2] XUE W, WANG C, LI Y. Effect of micro-arc discharge surface treatment on the tensile properties of Al-Cu-Mg alloy [J]. Materials Letters, 2002,56: 737-743.
[3] GUNTERSCHULZE A, BETZ H. Neue untersuchungen uber die elektrolytische ventilwirkung: II die oxydschicht von Sb, Bi, W, Zr, Al, Zn, Mg[J]. Zeitschrift Für Physik, 1932,78(3/4):196-210.
[4] GÜINTERSCHULZ A, BETZ H. Die elektronenstromung in isolatoren bei extremen feldstarken[J]. Zeitschrift Für Physik,1934,91(1/2):70-96.
[5] VAN T B, WIRTZ G P, BROWN S D. Anodic spark deposited coatings[J]. America Ceramic Society Bulletin, 1976, 55(4):412.
[6] WIRTZ G P, BROWN S D, KRIVEN W M. Ceramics coatings by anodic spark deposition[J]. Materials. and Manufacturing Processes,1991,6(1):87-115.
[7] KRYSMANN W, KURZE P, DITTRICH K H, et al. Process characteristics and parameters of anodic oxidation by spark discharge (ANOF) [J]. Crystal Research and Technology,1984,19(7):973-979.
[8] KURZE P, KRYSMANN W, SCHREKENBACH J, et al. Colored ANOF layers on aluminum[J]. Crystal Research and Technology, 1987,22(1):53-58.
[9] VAN T B, BROWN S D, WIRTZ G P. Mechanism of anodic spark deposition[J].American Ceramic Society Bulletin,1997,56:105-107.
[10] DITTRICK K H, GLEOAZD L. Micro-arc oxidation of aluminum alloy components [J]. Crystal Research & Technology,1984,19(1):93-96.
[11] LI S H, YIN Y J, CHENG J S. Technology of microarc oxidation and ceramics of material surface[J]. Special Casting & Nonferrous Alloys,2001,35(1):35-36.
[12] ZHANG W H, HU Z Q, MA J. The development of research on microarc oxidation technology in Russia[J]. Light Alloy Fabrication Technology,2004,32(1):25-29.
[13] LI H X, LI W J, SONG R G, et al. Effects of different current densities on properties of MAO coatings embedded with and without α-Al2O3 nano additives [J]. Materials Science and Technology,2013,28(5):565-568.
[14] WHEELER J M, CURRAN J A, SHRESTHA S, et al, Microstructure and multi-scale mechanical behavior of hard anodized and plasma electrolytic oxidation(PEO) coatings on aluminum alloy 5052[J]. Surface and Coatings Technology,2012,207:480-488.
[15] ZHANG X Y, SHI Y L. Technique of plasma microarc oxidation and its application[J]. Journal of Qingdao Institute of Chemical Technology,2002,23(1):69-73.
[16] LUO H H, CAI Q Z, WEI B K, et al. Effects of additive concentration on microstructure and corrosion resistance of ceramic coatings formed by micro-arc oxidation on AZ91D Mg alloy[J]. The Chinese Journal of Nonferrous Metals,2008,18(6):1082-1088.
[17] Curran J A, Clyne T W. The thermal conductivity of plasma electrolytic oxide coatings on aluminium and magnesium[J]. Surface Coating Technology,2005,199:177-183.

Alkali treatment of bamboo powder mixed polypropylene and its effect on properties

Chenxin Li[1*], Weilin Guo, Chen Liu, Taoyi Shen, Tong Chen, Lijun Wang and Linxue Kong[2*]

[1]Department of Environmental and Material Engineering, Shanghai Polytechnic University, Shanghai, 201209, China.

[2]Frontier Materials Research Institute, Deakin University, Geelong Victoria 3220 Australia.

[*]Corresponding author's e-mail: 13456606409@163.com

Abstract. Bamboo powder (BP) was treated with different concentrations of alkali solution, and modified polypropylene material of bamboo powder/polypropylene blend was prepared. Using the scanning electron microscopy (SEM) and X-ray diffractometry (XRD) to study the effects of alkali treatment and various alkali concentrations on the micro-structure and mechanical properties of the modified polypropylene. The test results illustrate that the alkali treatment can remove the amorphous cement between lignocellulose, form more fine mesh structure, and increase the surface area, which is beneficial to improve the mechanical interlock between the fiber and the polymer matrix. Thus, improving the mechanics performance of composites by this method. As the increase of alkali solution concentration, the impact strength of bamboo powder/polypropylene composites show the trend of rising up and then descending. When the concentration of sodium hydroxide is 3%, the bamboo powder/polypropylene composite material has better normal temperature impact strength. When the sodium hydroxide concentration is 9%, the bamboo powder/polypropylene composite has the optimal yield strength.

1. Introduction

With the acceleration of industrialization, the demand for materials has deeply increased. Polypropylene (PP) is a kind of thermoplastic synthetic resin with excellent performance. It is the shortest and fastest growing variety among the five general-purpose thermoplastic resins[1], which has excellent comprehensive properties, such as good processing performance, high tensile strength and modulus of elasticity, low relative density, excellent stress resistance and chemical resistance; raw materials are abundant, non-toxic, easy to recycle, low energy consumption, and less pollution. Therefore, it is widely used in the automotive industry, electronics, household appliances, furniture, packaging and building materials[2]. However, polypropylene also has various disadvantages. For instance, large molding shrinkage, low temperature brittle fracture, poor abrasion resistance, bad notched impact strength, and low heat distortion temperature[3].

As technology advances, high performance materials are required to the variety of industries. At the same time, environmental protection and eco-friendly living space development have always been the focus of attention over all countries of the world. The environmental pollution caused by the soaring economy and industrialization cannot be underestimated, and sustainable development is imminent, exactly. Single and environmentally hazardous materials are no longer sufficient to meet market needs, and there is increasingly number of researchers are turning their attention to environmentally friendly composites.

Researchers have attached more importance to natural resources, which are numeric in storage, widely sourced, and low in cost[4-6]. The natural plant fiber is easy to take, simple to make, less loss to processing equipment during experiments, and low energy consumption. The comprehensive performance of most natural fibers is remarkably prominent. The density of plant fibers is lower than that of inorganic fibers. In addition to, compared to the general metal materials, the tensile strength can reach 500 MPa approximately[7-10]. At the same time, the strength is much higher than that of metallic materials. Last but not least, natural plant fibers release less carbon dioxide when they are incinerated or land-filled in natural plant fiber/thermoplastic resin composites. Repeated melting and cooling processes can maintain plasticity and facilitate recycling[11,12]. When it comes to the composites, there are the first generation composite materials of glass fiber/thermoplastic resin composite materials and the second generation composite materials such as carbon fiber and synthetic fiber/thermoplastic resin composite materials. Moreover, the third generation composite material--natural plant fiber/ thermoplastic resin composite material have the outstanding characters like eco-friendly and good mechanical properties.

Among natural plants, bamboo is one of the fastest growing plants with a short maturity[13]. It is worth mentioning that China is one of the centers of origin and distribution of bamboo plants and the main bamboo producing country in the world. However, to some extent, the utilization rate of bamboo resources in China is poor. Generally, there are bamboo handicrafts, household item, etc. There are few applications in engineering, and more functions of bamboo fiber have yet to be explored. From the perspective of bamboo fiber morphology, it can be divided into bamboo fiber bundles, bamboo pulp fibers and bamboo powder particles[12]. What surprises the researchers is that bamboo fiber is comparable in stiffness and strength to glass fiber. Renewability and suitable mechanical properties make bamboo a superior reinforcement for natural fiber reinforced thermoplastic composites. At the same time, bamboo fiber has many uncontrollable factors due to its complex composition, and the technical problems faced by the composite of bamboo fiber and thermoplastic resin need to be solved urgently. The high polarity of the bamboo fiber surface makes it poorly compatible with the non-polar polymer, which seriously affects the mechanical properties of the composite. The main components of bamboo fiber, cellulose, hemicellulose and lignin, as well as a small amount of pectin, fatty waxy gray matter, etc., are all soluble in NaOH solution. Therefore, with the help of the alkali treatment of bamboo powder, the blending ability between the polymer and the polymer is improved, so that the common BP/polypropylene composite material achieves better mechanical properties[14,15].

The main purpose of this experiment is to explore the effect of alkali-treated bamboo powder on the micro-structure and mechanical properties of bamboo powder/polypropylene composites. At the same time, choosing the optimum alkali solution concentration of BP under a certain condition and using SEM and XRD to analyze bamboo. Impact strength and tensile properties of the prepared samples were measured simultaneously, and comprehensive analysis was carried out in order to draw conclusions.

2. Materials and Methods

2.1. A Materials

In this study, the following materials were used as follows: Polypropylene (injection molding-grade; Zhejiang Ningbo Taisu industry Co., Ltd., China); the BP(60-80 mesh; Jiangsu Lianyungang Straw processing Co., Ltd., China) were extracted from moso bamboo. Chemicals such as sodium hydroxide (NaOH with minimum assay 99.00%; Sinopharm Chemical Reagent Co.; Ltd., China) were used for the surface modification of the BP.

2.2. A Sample Preparation

2.2.1. Chemical Modification. The BP were treated with NaOH solution at 1%, 3%, 5%, 7%, 9%, 10% (NaOH to water mass ratio) .The BP were soaked in NaOH solution for 24 h at room temperature. The BP were then washed to neutral with distilled water to allow absorbed alkali to leach from the powder and then to pump filter to a certain extent. Then put the bamboo powder to the oven dried at 140°C for another 6h. The dried powder were broken in a high-speed mill and put aside.

2.2.2. Preparation of the BP-Reinforced PP Composites. Untreated or treated BP, PP were blended at a mass ratio of 10:90. Then material master batch was prepared in a co-rotating twin-screw extruder and then pelletized for injection mold after water cooling.Setting temperature of the co-rotating twin-screw extruder were 190, 185, 185, 185, 185, 190°C, respectively, and rotor speed of the main machine and feeder was 5.0 and 2.0 rpm, respectively. The obtained master batch was dried at 80°C for 4h in the oven. Then using the injection molding machine to make the samples. The temperature of the injection molding machine were 180, 180, 180, 180, 180, respectively.

2.3. A Sample Preparation

2.3.1. Mechanical Properties. Tensile test was carried out with a tensile speed of 50 mm min-1. Notch impact test was carried out and the bottom radius of notch is 2.0mm and the notch angle is -150.43°.

2.3.2. X-Ray Diffraction. Bamboo Power were characterized by XRD. The continuous scanning angle range use with a step interval of 0.05° and the machine was operation at 40kV and 40mA.

2.3.3. Scanning Electron Microscopy. The surface structures of raw BP and BP treated by NaOH were evaluated by SEM.

3. Results

Through the bamboo powder modified polypropylene, the impact properties and tensile properties of the obtained bamboo powder/polypropylene composite material are improved. The mechanical properties of the composites obtained by NaOH-treated bamboo powder are better than those of untreated bamboo powder composite polypropylene. The experimental results show that when the concentration of NaOH is 3%, the impact performance of the composite is up to 78.205KJ/m^2, which is 18.5% higher than that of untreated BP, and 67.7% higher than that of pure PP without BP. When the NaOH concentration is 9%, the composite has the best yield strength, which can be about 22 MPa, which is better than the BP/PP composite without alkali treatment.

4. Disscussion

4.1. The Effect of Alkali Treatment

The mechanical properties of the untreated BP-reinforced PP composites and ABP-reinforced PP composites in different concentration. And the changes of impact strength are shown in the figure 1.

As shown in the figure 1 below, the impact properties and yield strength of bamboo powder/polypropylene composites blended with bamboo powder are significantly improved, which indicate that bamboo fibers are indeed a good natural material for modified thermoplastic resins. At the same time, it can be concluded that with the increase of alkali concentration, the impact performance of the BP/PP composites firstly rise up and then go down. To sum up, when NaOH concentration is 3%, the properties of impact reach the best situation. From the figure 2, when the alkali concentration is 9%, the yield strength of the composite material is superior, and the general trend is the same as the impact performance.

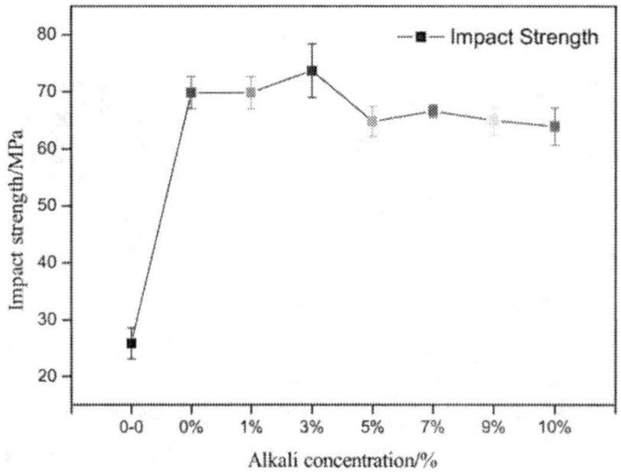

Figure 1. The effect of alkali concentration on composite impact strength.

The addition of bamboo powder increases the toughness of the composite material. The alkali-treated bamboo powder, hemicellulose, pectin and other impurities are totally reduced, the cellulose content is increased, and the fiber function is fully exerted. However, as the concentration of the alkali solution increases, the toughness decreases. One of the reason is that a large amount of hydroxyl groups (-OH) exist on the surface of the bamboo fiber, and strong hydrogen bonds are formed between the molecular chains of the bamboo cellulose, which makes the surface strong. Water absorption and chemical polarity, while polypropylene is non-polar and hydrophobic, so that the interface wet-ability and adhesion between the two materials deteriorate, resulting in a decrease in the mechanical properties of the composite. Considering the comprehensive consideration, the choice of 3% alkali concentration is a better parameter for the subsequent exploration of the effect of suitable dispersants on the properties of the composite.

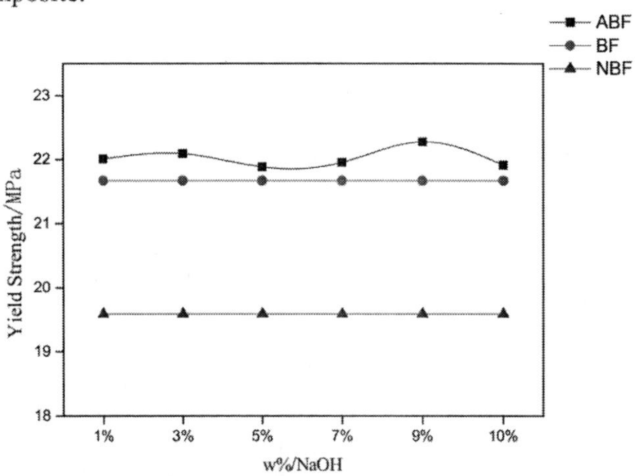

Figure 2. The effect of alkali concentration on composite yield strength.

4.2. Morphology

It is acknowledged that the interface strength between the reinforcements and matrix has a significant influence on the properties of the composites since the interface is regarded as one of the most essential factors to improve the composite characteristics.

The SEM images for the surface of untreated BP and ABF are presented in Figure. As shown in the figure 3(a) and figure 3(b), compared with the smooth surface structure of the BP without alkali treatment, the surface of the alkali-treated bamboo powder is rough and presents pores along the fiber direction. The main reason is that alkali treatment can remove lignocellulose and other impurities, enhance the interfacial adhesion between bamboo powder and polymer, and increase the surface active point of the fiber, form more fine mesh structure and increase the surface area. It will help to improve the mechanical interlock between the fiber and the polymer matrix and the mechanical properties of the composite.

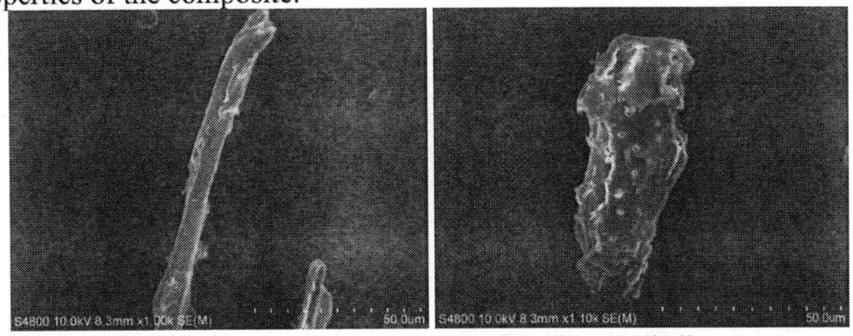

Figure 3(a). SEM images of bamboo powder without alkali treatment.

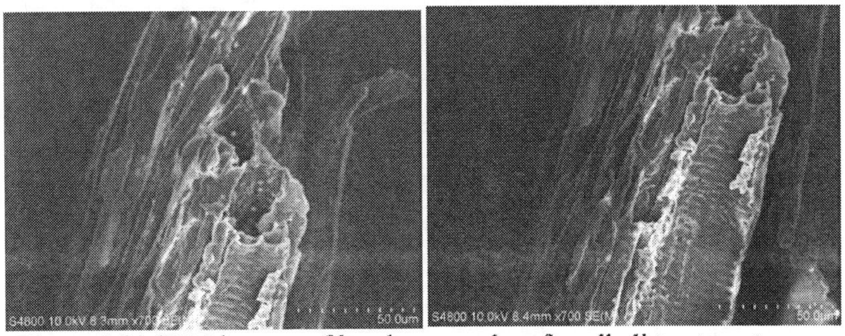

Figure 3(b).SEM images of bamboo powder after alkali treatment.

4.3. X-ray diffraction analysis

As is depicts in the figure 4, after the alkali-treated bamboo powder, the amorphous region of cellulose is destroyed, and the regular crystallinity of the structure is not damaged, and the crystallinity is remarkably increased that indicates that the alkali treatment removes other impurities and the surface of the fiber becomes rough, which is beneficial to the mechanical properties of the bamboo powder/polypropylene composite.

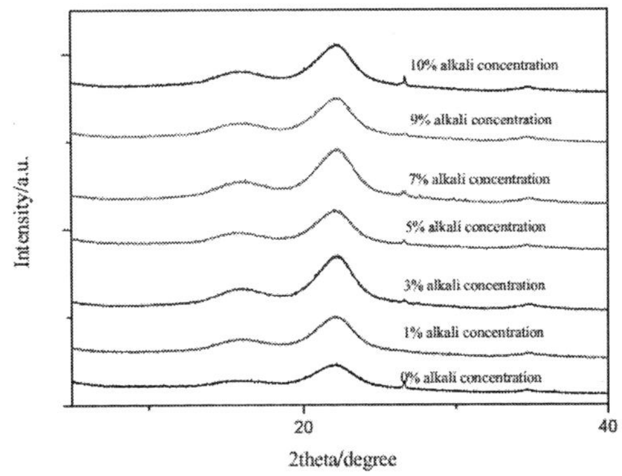

Figure 4. XRD images of different NaOH concentrations to treat bamboo powder.

5. Conclusions

Alkali treatment of bamboo powder will effectively improve the mechanical properties of BP/PP composites. In the range of 1%-10%, there is an optimum alkali treatment concentration to obtain the best impact strength and yield strength. However, as the alkali concentration increases, the mechanical properties of the composites tend to decrease, which is closely related to the interfacial compatibility between the bamboo powder and the polymer. How to improve the dispersion of bamboo powder in polymers is also one of the focuses of future research.

References

[1] Zehui Jiang, Zhengjun Sun, Haiqing Ren. (2006) Application of advanced bio-composites in wind blades. J. Library Theory and Practice. Commun., 123: 126–130 (in Chinese).

[2] M.R. Sanjay, P.Madhu, Mohammad Jawaid, P.Senthamaraikannan, S.Senthil, S. Pradeep. (2018) Characterization and properties of natural fiber polymer composites: A comprehensive review. J. Journal of Cleaner Production. Commun., 172: 566–581.

[3] Rui Wang, Chun-hong Wang, Si Zhao, Hai-jiang Tong, Li-yan Liu, Fei Yu. (2006) Raw Bamboo Fiber Reinforced Composites. J. Plastic. Commun., 35: 38–41(in Chinese).

[4] Xiuying Hong, Zutai Huang, Wenbin Yang.(2005) Application and development of plant fiber in composite materials. J. Cellul Sci Technol. Commun., 13: 55–59 (in Chinese).

[5] Wenjing Guo, Zheng Wang, Fucheng Bao. (2008) The Status and Trend of Natural Fiber Biodegradable Plastic Bio-Composites. J. Scientia Silvae Sinicae. Commun., 44: 157–162 (in Chinese).

[6] Anon. (2002) The competitiveness of natural fiber based composites in the automotive sector. J. The sisal Agribusiness in Brazili. Commun., 684: 113–139.

[7] Lulin Zhang, (2009) Bamboo Fiber Composites Microstruture Design, Preparation and Proerties Research. https://www.cnki.net/.

[8] Ranan A K, Mandal A, Mitra B C, et al. (1998) Short fiber-reinforced polypropylene composites. J. Journal of applied polymer science. Commun., 69: 329–338.

[9] Mitra B C, Basak R K, Sarkar M. (1998) Studies on jute-reinforced composites, its limitations, and some solutions through chemical modifications of fibers. J. Journal of applied polymer science. Commun., 67: 1093–1100.

[10] Bledzki A K, Gassan J. (1999) Composite reinforced with cellulose based fibers. J. Progress in polymer science. Commun., 24: 221–274.

[11] H Farmington, Michigan. (1996) State of the art report on fiber reinforced concrete. J. American Concrete Institute. Commun., 96: 544.

[12] Xiaolong Zhu, (2015) Study on Structure and Properties of Bamboo Powder/Polypropylene Composites. https://www.cnki.net/.

[13] Fu Gu, Yitao Zheng, Wujie Zhang, Xing Yao, Deng Pan, Alecia Sze Mun Wong, Jianfeng Guo, Philip Hall, Nusrat Sharmin.(2018) Can bamboo fibers be an alternative to flax fibers as materials for plastic reinforcement? A comparative life cycle study on polypropylene/flax/bamboo laminates. J. Industrial Crops & Products. Commun., 121: 372–387.

[14] Chunhong Wang, Shengkai Liu. (2015) Effect of alkali treatment on properties of bamboo fiber and bamboo fiber reinforced polypropylene composites. J. Acta Materiae Compositae Sinic. Commun., 32: 683–690.

[15] Zhang L, Feng S H. (2006) Study on improving compatibility of thermoplastics/plant fibers. J. Engineering Plastics Application. Commun., 34:75–78 (in Chinese).

IOP Conf. Series: Materials Science and Engineering **729** (2020) 012057 doi:10.1088/1757-899X/729/1/012057

Application of F^- modified Fe-SAPO-34 as a New Photo-catalyst in Printing and Dyeing Wastewater Treatment

Ying Cao[1,2], Taoyi Shen[1], Weilin Guo[1], Xianzhen Song[1], Lingxue Kong[1] and Lijun Wang[1,2*]

[1] Faculty of Environmental and Materials Engineering, Shanghai Polytechnic University, Shanghai 201209, P.R. China

[2] Research Center of Resource Recycling Science and Engineering, Shanghai Polytechnic University, Shanghai 201209, P.R. China

Email: ljwang@sspu.edu.cn

Abstract. The development of new photocatalysts with a wide range of applications is one of the important challenges in the field of photocatalysis research. In this thesis, ferric phosphate was used as a transition metal salt, and HF was used as a mineralizer to prepare Fe-SAPO-34 photocatalyst, which was applied to the degradation of organic dye wastewater. The materials were characterized by XRD, SEM and other characterization methods. After the photocatalyst is supported on the molecular sieve, it usually exhibits excellent performance in catalyzing the degradation of different dyes, and generally higher than the catalytic efficiency of the unmodified catalyst under the same conditions.

1. Introduction

With the rapid development of industry, the problem of water pollution has become more and more serious in recent years. Traditional technologies of sewage treatment, such as adsorption and coagulation, are costly and easily cause secondary pollution. Nowadays, one of the most promising processing technologies is the photo-Fenton oxidation technology[1], which utilizes transition metals like iron, manganese, etc. as a catalyst, hydrogen peroxide as an oxidant, and strong oxidizing under ultraviolet or visible light. It produces strong oxidizing hydroxyl radicals under ultraviolet or visible light, which can completely oxidize the dye molecules to water and carbon dioxide. It has the characteristics of high processing efficiency, low cost and no secondary pollution[2]. However, its ability to use solar energy is not strong, and only ultraviolet light of $\lambda < 300$ nm can be utilized, and most of the visible light area in solar energy is not fully utilized. The porous structure of the molecular sieve gives it a large specific surface area, and this characteristic also makes it an ideal molecular sieve carrier. The contact area between the contaminant and the molecular sieve can be effectively increased if the photo-catalyst can be supported on it. Moreover, it can effectively solve the problem of the decrease in the intensity of the optical radiation caused by the suspension of the photocatalyst, thereby improving the catalytic efficiency.

This thesis mainly deals with the preparation of molecular sieve supported photocatalyst and its application in the treatment of printing and dyeing wastewater. Iron-based catalysts have a wide range of applications in the field of photocatalysis[3-5]. The phosphating slag is rich in iron salts. The synthesis of Fe-SAPO-34 molecular sieve catalyst with iron salt in phosphating slag has strong research value for photocatalytic degradation of dye wastewater. In this study, Fe-SAPO-34 was synthesized by hydrothermal synthesis using iron phosphate as a metal source and HF as a mineralizer. The photocatalytic performance under UV light was studied by degrading rhodamine B.

2. Materials and Methods

2.1. Preparation of Samples

The hydrothermal method was used to synthesize Fe-SAPO-34 molecular sieve. The HF, silicon source, phosphorus source, aluminum source, $FePO_4$ and appropriate amount of water were mixed uniformly in a certain proportion, then the template reagent diethylamine was added, mixed uniformly and transferred to the reaction kettle. After sealing, it was heated to 180°C, washed and dried to obtain a molecular sieve sample.

2.2. Medicine and Characterization

The morphology of the samples was studied by scanning electron microscopy (SEM) (HITACHI-S4800, Hitachi Corp-oration, Japan) with the accelerating voltage at 10 kV and current at 10 mA. X-ray diffraction (XRD) (D8-Advance, Bruker Corporation, Germany) measurements were performed to record the phase composition and crystalline structure of the powder samples, operating at 40 kV and 40 mA with a monochromatized Cu/Kα radiation (λ = 0.15418 nm) source. Phosphate residue was supplied by an auto components company in Zhejiang, China., hydrofluoric acid (HF, ≥40%). Other chemicals used in the present work were of analytical grade and used without further purification.

2.3. Photocatalytic Experiment

In this experiment, 50mg/L rhodamine B was used to simulate dye wastewater, 10mmol/L hydrogen peroxide was used as an oxidizing agent. After adding 0.5 g/L of the catalyst, a 500W mercury lamp was used to simulate ultraviolet light. After adding rhodamine B solution and a certain amount of catalyst and H_2O_2 to the reaction vessel, the reaction vessel was placed in a photocatalytic reactor (XPA-7, Xujiang Power Plant, China) to start photocatalytic reaction. Samples were taken every 20 minutes. The sample was filtered through a 0.45 um filter needle, and its absorbance at the maximum absorption wavelength λ = 544 nm was measured by an ultraviolet-visible spectrophotometer (UV-752, Shanghai Yoke Instrument. Co., Ltd., China)[6-8]. The concentration of rhodamine B at different times was calculated according to Lambert Beer's law. The reaction conforms to the first-order degradation kinetics model, and the reaction rate constant $K=In(C_0/C_t)$. Where C_0 is the initial concentration of rhodamine B and Ct is the concentration of rhodamine B solution at different times. The reaction time T is plotted on the abscissa and the reaction rate constant K is plotted on the ordinate. Then a straight line is fitted, and the slope K of the straight line is the degradation rate of rhodamine B in different systems.

3. Results and Discussions

The XRD diffraction angular position and intensity are used to identify the phase composition of the unknown sample. Fig.1 shows five peaks at 9.48°, 16.2°, 20.7°, 25° and 31°, corresponding to (101), (211), (104), (401) of SAPO-34 (JCPDS) respectively[9]. The diffraction intensity of Fe-SAPO-34(I) is lower than that of the Fe-SAPO-34(II) sample treated with HF. Except for the addition of HF, the molar ratio of the other components was the same as that of Fe-SAPO-34(I). As a result, the presence of surface HF can significantly promote the growth and dispersion of molecular sieve crystallites on the surface of the support, and improve the structural regularity of the molecular sieve grains[10]. A large number of literatures have shown that the relative crystallinity of Fe-SAPO-34 molecular sieve can be significantly improved under the F ion system, and it also has a certain influence on the introduction of its component Fe.

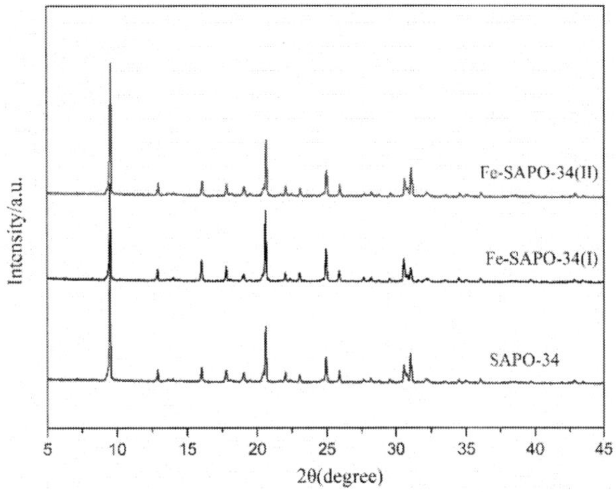

Figure 1. XRD patterns of SAPO-34, Fe-SAPO-34(I) and Fe-SAPO-34(II)

It can be seen from the Fig.2 that the morphology of Fe-SAPO-34(I) and Fe-SAPO-34(II) molecular sieves are all cubic. Fe-SAPO-34(I) has an average particle size of about 5 microns. The Fe-SAPO-34(II) synthesized by the addition of HF has a particle size of about 1 um. Therefore, the SAPO-34 molecular sieve synthesized by using HF as a mineralizer not only has a regular shape, but also synthesizes SAPO-34 with a small grain size[11].

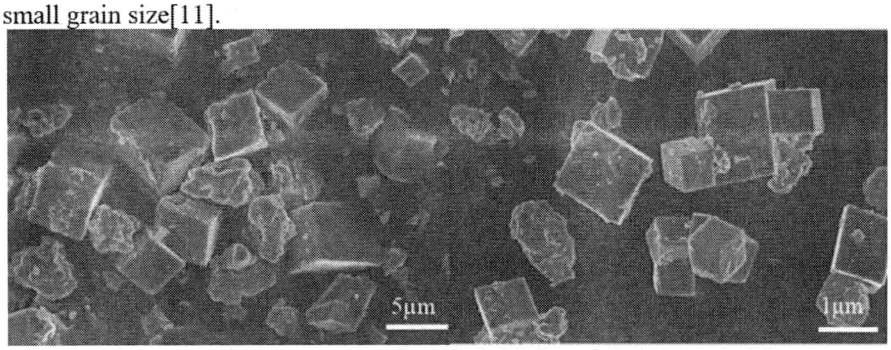

Figure 2. SEM image of Fe-SAPO-34(I) and Fe-SAPO-34(II)

The photocatalytic performance of the catalyst in different systems was investigated by using rhodamine B simulated dye wastewater and 500W mercury lamp to simulate ultraviolet light. In the UV system, the degradation efficiency of rhodamine B was 7.4% after 20 minutes of mercury lamp irradiation, indicating that UV light can promote the degradation of rhodamine B[12-13]. In the UV/H_2O_2 system (10.2%), H_2O_2 decomposed under ultraviolet light to produce a strong oxidizing hydroxyl radical ($E°(\cdot OH/H_2O_2) = 1.9\text{-}2.7$ V), which can effectively degrade dye molecules. When Fe-SAPO-34(II) was used together with H_2O_2, the degradation efficiency of rhodamine B increased to 30.2%. This indicated that Fe-SAPO-34(II) molecular sieve can catalyze the decomposition of hydrogen peroxide to produce hydroxyl radicals, thereby improving the degradation efficiency of dye molecules. The catalyst captures photons through the surface, which in turn produces photogenerated electrons and holes[14]. The photocatalytic activity of the catalyst is increased, so the degradation rate of rhodamine B is increased. In order to compare the degradation effects of the catalysts, the same quality of Fe-SAPO-34(I) and the modified catalyst Fe-SAPO-34(II) were added to the reaction system under the same conditions. The modified catalyst can completely degrade the dye molecules within 60 min, and its degradation effect is better than that of the unmodified catalyst. Because Fe-SAPO-34(II) has a more complete crystal form and a larger specific surface area, it can effectively capture dye molecules, which has more excellent catalytic degradation performance than Fe-SAPO-34(I). In addition, the kinetics of

IOP Conf. Series: Materials Science and Engineering **729** (2020) 012057 doi:10.1088/1757-899X/729/1/012057

the degradation reaction of rhodamine B, which Fe-SAPO-34 molecular sieve catalyst, was also studied. It was found that the reaction was pseudo-first order reaction model, in line with LH kinetic model. In order to quantitatively analyze the degradation effect of Fe-SAPO-34 under ultraviolet light, the degradation rate constant K was fitted.

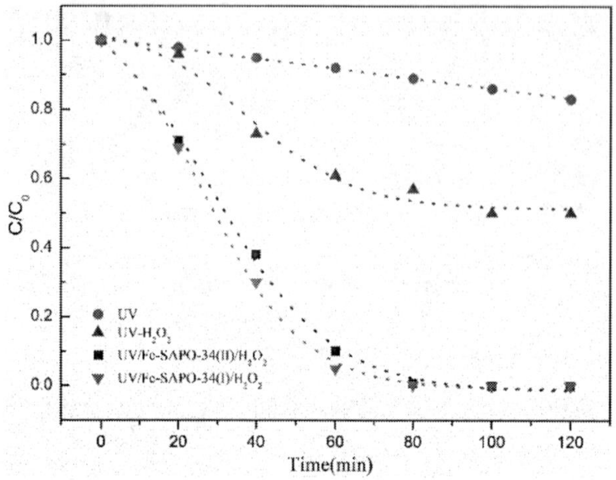

Figure 3. Degradation rate of rhodamine B at different system

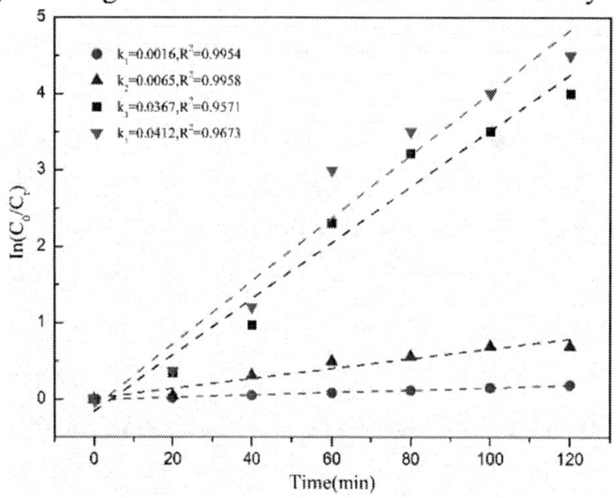

Figure 4. Kinetic equations under different systems

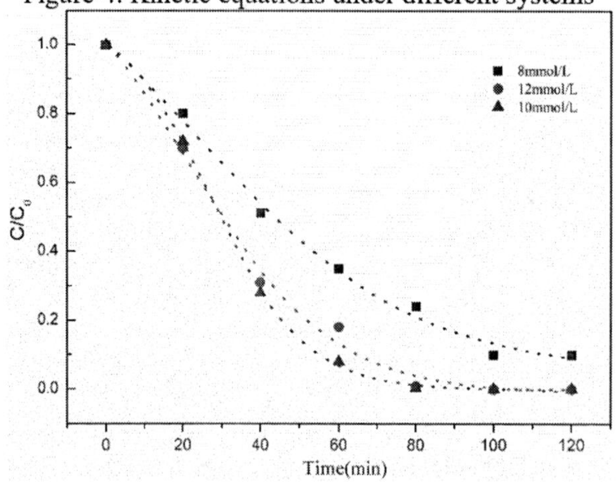

Figure 5. Effects of H_2O_2 dosage on the catalytic degradation of rhodamine B

Furthermore, in order to study the relationship between the reaction conditions and the catalytic degradation effect, the effect of the dosage of H_2O_2 on the degradation effect was investigated, as shown in Fig.5. It illustrated the effect of the amount of hydrogen peroxide added on the degradation rate of rhodamine B. Under the catalysis of a sufficient amount of catalyst, as the amount of hydrogen peroxide added increases, the amount of hydroxyl radicals generated increases as well. However, when the amount of hydrogen peroxide added exceeds 12 mmol/L, due to insufficient active sites provided by the catalyst, the system had a considerable amount of hydrogen peroxide which is not reacted with the catalyst exists. This has a certain quenching effect on the hydroxyl radicals that have been produced[15], so the degradation rate of rhodamine B will decrease.

4. Summary and Conclusion

In summary, Fe-SAPO-34 molecular sieves were prepared by using iron phosphate as a transition metal. And HF was used as a mineralizer. The research shows that in the synthesis of molecular sieves, HF can not only play a role in mineralization, but also function as a template to stabilize the molecular sieve structure. After the F^- is added to the initial gel, the synthesized molecular sieve has a higher crystallinity and a smaller average particle diameter. In this paper, the prepared catalyst is used to degrade rhodamine B dye molecules in a heterogeneous Fenton system. As a result, the surface-modified Fe-SAPO-34 catalyst exhibited excellent catalytic activity under ultraviolet light.

Acknowledgement

This work was supported by Shanghai Polytechnic University Innovation Project (No. 2019-yjscx-013), and Material Science and Engineering Key Subject of Shanghai Polytechnic University (No. XXKZD1601).

References

[1] Yan GY, Wang XX, Fu XZ. (2004) A Primary Study on Photocatalytic Properties of ZSM-5 Zeolite. Chem Res Chinese U, 5:942-944.
[2] Zhang YF, Teng J, Zhang XY, et al. (2003) Research Progress in Printing and Dyeing Wastewater Treatment Technology. Industrial Water Treatment, 23: 23-27.
[3] Yan L, De ZS. (2006) Study progress of dye wastewater treatment by advanced oxidation processes. Industrial Water Treatment, 26:1-5.
[4] Sui G, Liu T, Li J, et al.(2014) Preparation and photocatalytic degradability of Pr-TiO_2 nanocomposite catalyst. T Auxiliaries, 12: 4-5.
[5] Wei L, Zhang Y, Chen S, et al. (2019) Synthesis of nitrogen-doped carbon nanotubes-$FePO_4$ composite from phosphate residue and its application as effective Fenton-like catalyst for dye degradation. J. Environ. Sci, 76:191-201.
[6] Tian XZ, Liao S, Kun WU, et al. (2008) Synthesis of Ammonium Ferric Phosphate Hydrate by Solid State Reaction and Its Catalysis Effect to Butyl Acetate. Guangxi Sci, 2:41-43.
[7] Chen YN, Zhang SG, Zhang ML, et al. (2016) Research progress on the influence factors and control of SAPO-34 molecular sieve grain size and morphology. Guangdong Chemical Industry, 43: 139-141.
[8] Xu J, Guan XA, Sun JZ. (2016) Synthesis of SAPO-17 Molecular Sieves. Tianjin Chemical Industry, 30: 17-19.
[9] Liu HX, Xie ZK, Zhang CF, et al. (2003) Synthesis of SAPO-34 molecular sieves with hydrogen fluoride-triethylamine complex template. Chinese J of Catalysis, 24: 279-283.
[10] Wang YP, Xu J, Zhang XX, et al. (2018) Rapid and stable synthesis and characterization of SAPO-17 molecular sieves in HF and KCl systems. Modern Chemicals, 379(05):165-169.
[11] Liu C, Wang F, Liu G, et al. (2016) Preparation of nanometer mesoporous iron hydroxy phosphate supported gold catalyst and its catalytic epoxidation performance of styrene. Modern Chemical Industry, 10: 87-91.
[12] Guo S , Zhang G , Yu JC. (2015) Enhanced photo-Fenton degradation of rhodamine B using

graphene oxide-amorphous $FePO_4$ as effective and stable heterogeneous catalyst. Journal of Colloid & Interface Science, 448:460-466.
[13] Zhao H , Li H , Li X , et al. (2017) The promotion effect of Fe to Cu-SAPO-34 for selective catalytic reduction of NOx, with NH_3. Catalysis Today,297:84-91.
[14] Wei Y , Zhang D , Xu L , et al. (2008) Synthesis, characterization and catalytic performance of metal-incorporated SAPO-34 for chloromethane transformation to light olefins. Catalysis Today, 131:262-269.
[15] Andonova S , Tamm S , Montreuil C , et al. (2016) The effect of iron loading and hydrothermal aging on one-pot synthesized Fe/SAPO-34 for ammonia SCR. Applied Catalysis B Environmental, 180:775-787.

Preparation of porous iron hydroxy phosphate by phosphate slag and its application for adsorbing heavy metal Pb^{2+}

Weilin Guo *[1], Chenxin Li[1], Ying Cao[1], Xianzhen Song[1], Lijun Wang[1], Lingxue Kong[2]

[1]Department of Environmental and Material Engineering, Shanghai Polytechnic University, No. 2360 Jinhai Road, Pudong New Area in Shanghai, China.

[2]Frontier Materials Research Institute, Deakin University, Geelong Victoria 3220 Australia.

E-mail: weilin-guo@foxmail.com

Abstract: Porous iron hydroxy phosphate was prepared by hydrothermal purification, ball milling modification, hydrothermal crystallization and high temperature roasting of phosphating slag, and was used to treat lead-containing wastewater. The adsorbent was characterized by XRD, SEM, FTIR and TG, and the effect of the amount of adsorbent and the pH value of the system on the adsorption performance of PIHP was investigated. The results showed that when the amount of PIHP added was 0.5g, the removal rate of pb^{2+} could reach 91% after 120min. When the amount of PIHP added was 0.2 g and the pH of the system was 2.15, the removal rate of pb^{2+} after 120min could reach 95%. This research has opened up a new direction for the resource utilization of phosphating slag.

1. Introduction

With the rapid development of the metal industry, phosphating treatment technology is more and more widely used, but a large amount of by-product phosphating slag is generated in the phosphating process. The phosphating slag contains some heavy metal ions and a large amount of phosphate ions, which easily cause acidification of the soil and eutrophication of water bodies [1]. According to statistics, nearly 900 million tons of phosphating slag is produced every year in China. Most of the phosphating slag treatment methods are piled up or landfilled, which seriously pollutes the water body and also occupies a large amount of land resources [2]. The main component of phosphating slag is $FePO_4$, which also contains some trace metal elements. The extraction of these useful substances from phosphating slag has a very broad application prospect [3, 4].

On the other hand, with the development of industry, heavy metal wastewater has become one of the most serious environmental pollution [5], pb^{2+} are harmful to the human and cause great damage to the human kidney and immune system [6]. Iron hydroxy phosphate is a derivative of iron phosphate, and significant progress has been made in the study of the removal of heavy metals by the inorganic adsorbent iron hydroxy phosphate in recent years [7,8]. Due to its special crystal chemical characteristics, iron hydroxy phosphate has wide acceptability and adsorption fixation effects on various metal cations, and has good ion exchange and surface adsorption capacity, thus it becomes a new type of environmental mineral material and has a good application prospect in the field of removing heavy metal cations in wastewater and soil pollution [9,10]. In this study, the phosphating slag was used as raw material to synthesize porous iron hydroxy phosphate by precursor thermal decomposition method [11], and the adsorption performance of heavy metal ions (pb^{2+}) was studied.

2. Materials and Method

2.1. Materials

Phosphating slag (provided by an automobile parts company in Zhejiang), phosphoric acid (AR, ≥85%), absolute ethanol (AR, ≥99.7%), cetyltrimethylammonium bromide (AR, ≥99.0%) , oxalic acid (AR, ≥99.5%), lead chloride (AR, ≥99.5%), hydrochloric acid (AR, ≥99.5%), sodium hydroxide (AR, ≥99.5%), deionized water from ultrapure water system (UPT-I-60L, Super Pure Technology China)).

2.2. Adsorbent preparation

The catalyst preparation process mainly included three processes which included phosphating slag purification, iron phosphate ball milling, and preparation of porous iron hydroxy phosphate. First, the phosphating slag, deionized water and phosphoric acid were stirred at a mass ratio of 3:3:1 in a beaker for 4 hours, and the heating temperature was 80°C, and then 30 g phosphating slag, 30 g deionized water and 1 g phosphoric acid were hydrothermally purified in a polytetrafluoroethylene reactor at a hydrothermal temperature of 150°C for 4 hours. Then 20 g purified iron phosphate were ball milled with 20 g absolute ethanol and 120 g agate ball for 5 hour. The ball-milled iron phosphate, deionized water, oxalic acid and hexadecyl trimethyl ammonium bromide (CTAB) were hydrothermally reacted at 200°C for 8 hours in a polytetrafluoroethylene reactor at a mass ratio of 5:40:2:2 to prepare iron hydroxy phosphate precursor. The iron hydroxy phosphate precursor was calcined in a nitrogen atmosphere at a calcination temperature of 600°C for 4 hours to form a porous structure.

2.3. Heavy metal ion removal experiment

The lead-containing wastewater was simulated with 100 mg/L $PbCl_2$ solution, different masses of PIHP were added to 200 mL lead-containing wastewater, and the pH value of the system was adjusted by HCl and NaOH. 10 mL solution was filtered through a needle and the residual Pb^{2+} concentration was determined by inductively coupled plasma atomic emission spectrometry (ICP-AES) (ICAP 6000, Thermo, USA), then the lead ion removal rate of the lead-containing wastewater and the amount of adsorption of the adsorbent were calculated. All adsorption experiments were stirred at 350 r/min. The calculation formula of the lead ion removal efficiency η and the adsorption amount q of the adsorbent are as follows:

$$\eta = \frac{C_0 - C_t}{C_0} \times 100\% \qquad (1)$$

$$q = \frac{V(C_0 - C_t)}{m} \qquad (2)$$

η is the removal efficiency of lead ions and q is the adsorption amount. C_0 is the initial concentration of lead ions in the solution and C_t is the concentration of lead ions in the solution at a certain moment, and **V** is the volume of the solution.

2.4. Characterization method

The phase composition and crystal structure of the powder sample were measured by X-ray diffraction (XRD) (D8-Advance, Bruker Corporation, Germany), and operation was carried out at 40 kV and 40 mA with aminobrominated Cu Kα radiation (λ = 0.15418 nm) source. The composition of the sample was analyzed by a synchronous thermogravimetric analyzer (TG) (STA 449C, NETZSCH instrument Corporation, Germany) under a nitrogen atmosphere at a heating rate of 10°C/min. The surface morphology of the sample was analyzed by a scanning electron microscope (SEM) (S-4800, Hitachi Corporation, Japan) with an acceleration voltage of 10 kV and a current of 10 mA. The functional groups and the chemical bonds in the samples were analyzed by fourier transform infrared spectroscopy (FTIR) (Vertex 70, Bruker, Germany),and the KBr was chosen as a medium.

3. Results and discussion

3.1. XRD characterization

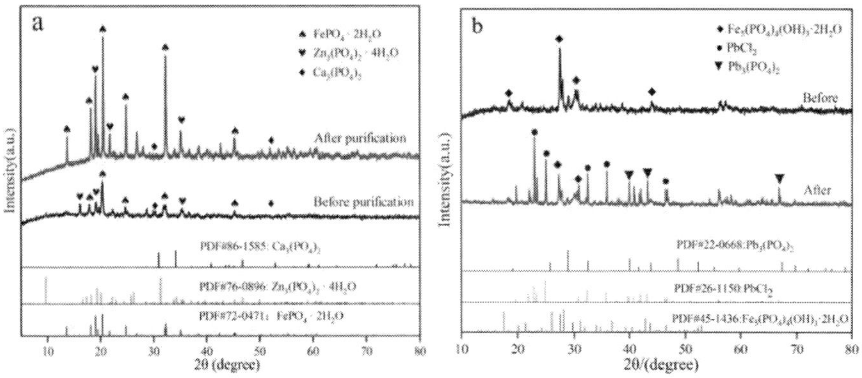

Figure 1. (a) the xrd diagram before and after phosphating slag purification and (b) the xrd diagram before and after PIHP adsorption of Pb^{2+}

The results of XRD analysis before and after purification of phosphating slag were shown in Fig a, the diffraction peaks of the three crystals which appeared in the phosphating slag before purification were matched with the PDF#72-0471, PDF#76-0896 and PDF#86-1585 standard cards respectively, indicating that the phosphating slag contained $FePO_4{\cdot}2H_2O$, $Zn_3(PO_4)_2{\cdot}4H_2O$ and $Ca_3(PO_4)_2$. The XRD diffraction peak became sharper after the phosphating slag was hydrothermally purified, indicating that the content and crystallinity of $FePO_4{\cdot}2H_2O$ was greatly improved after hydrothermal purification. As shown in Figure b, the main component of PIHP was iron hydroxy phosphate which was matched with the standard card of PDF#45-1436 before the adsorption of Pb^{2+}. $PbCl_2$ and $Pb_3(PO_4)_2$ appeared after the adsorption of Pb^{2+} which matched with the standard cards of PDF#26-1150 and PDF#22-0668. It was indicated that the absorption of $PbCl_2$ by PIHP had two effects of adsorption and ion exchange.

3.2. SEM characterization

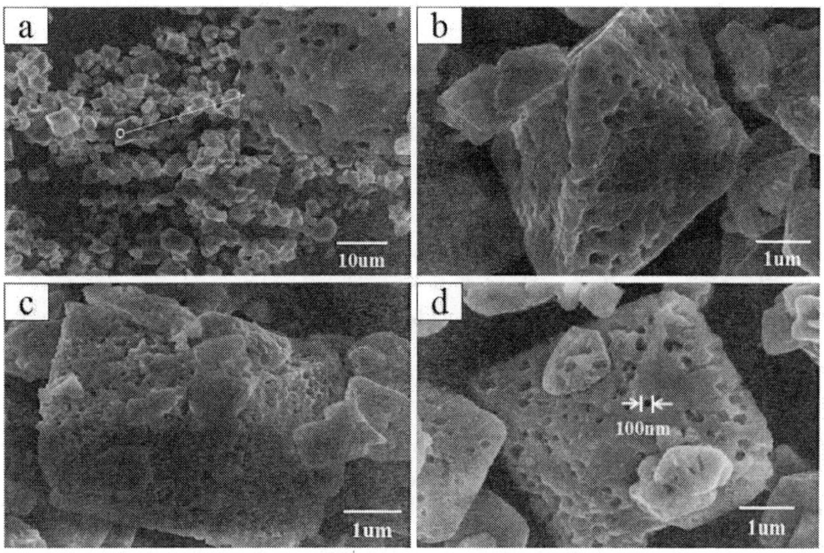

Figure 2. SEM image of porous iron hydroxy phosphate

Figure 2 was the SEM image of porous iron hydroxy phosphate. The iron hydroxy phosphate had a regular octahedral structure, but the surface structure changed to form a porous structure after high-temperature calcination, this was mainly because some iron ions and oxalates form iron oxalate complexes during the preparation of iron hydroxy phosphate, and the iron oxalate decomposed and released CO and CO_2 during the high temperature roasting process, and formed a porous structure on the surface of iron hydroxy phosphate [12], the pore size was about 100 nm.

3.3. FTIR and TG characterization

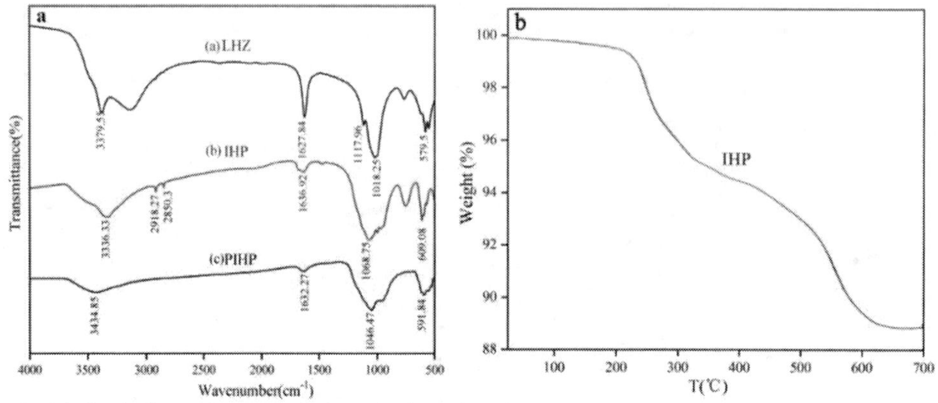

Figure 3. (a) the infrared image of phosphating slag, iron hydroxy phosphate, porous iron hydroxy phosphate and (b) the thermograviogram of iron hydroxy phosphate

Figure a was the infrared spectrum analysis diagram of iron phosphate, iron hydroxy phosphate and porous iron hydroxy phosphate. As shown in (a), the iron phosphate had vibration peaks at 33789.55, 1627.84, 1117.66, 1018.25, 579.5 cm^{-1}. 3375.55 and 1627.84 cm^{-1} were corresponded to the stretching vibration and bending vibration of O-H in crystallization water, and 1117.96 and 579.5cm^{-1} were correspond to the anti-expansion vibration and asymmetric bending vibration of P-O bond, and 1018.25 cm^{-1} were corresponded to the stretching vibration of Fe-O-P. As shown in (b), the iron hydroxy phosphate had vibrational peaks at 3336, 2918, 2850, 1636, 1068, 751, and 609 cm^{-1}, 3336 and 1636 cm^{-1} were corresponded to the stretching vibration and bending vibration of O-H in the crystallization water, 2918 and 2850cm^{-1} were corresponded to the stretching vibration of $-CH_3$ and $-CH_2$, and 1068cm^{-1} and 609cm^{-1} were corresponded to the anti-expansion vibration and asymmetric bending vibration of P-O bond. The analysis results showed that iron hydroxy phosphate contained crystal water, hydrocarbon compound on the surface was presumed to be CTAB remaining on the surface in the process of preparing iron hydroxy phosphate. As shown in (c), the porous iron hydroxy phosphate showed weak vibration peaks at 3134 and 1632 cm^{-1}, which wre corresponded to the stretching vibration and bending vibration of O-H in crystal water, and the 1046 and 591 cm^{-1} were corresponded to the anti-expansion vibration and asymmetric bending vibration of the P-O bond. The result showed that the porous iron hydroxy phosphate contained a small amount of crystal water, and its main component was iron hydroxy phosphate.

Fig b was the TG analysis diagram of adsorbent PIHP precursor, the mass was reduced by 5.4% between 250°C and 400°C, in which the crystallization water was lost, and the mass was reduced by 5.8% between 400°C and 620°C, in which the iron oxalate decomposed to form Fe_2O_3, CO and CO_2.

3.4. Pb^{2+} adsorption experiment

3.4.1. Effect of adsorbent dose on adsorption effect

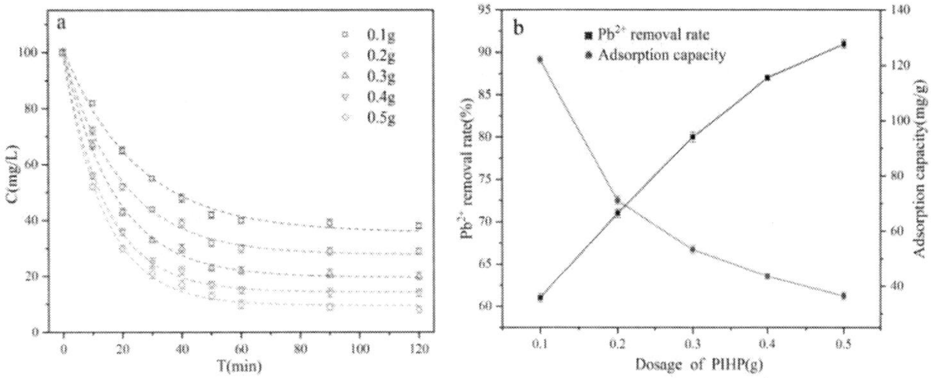

Figure 4. Effect of the amount of adsorbent added on the removal rate of Pb^{2+} and the amount of adsorbent adsorbed

The concentration of Pb^{2+} gradually decreased with the increase of adsorption time and reached the adsorption equilibrium at 60 min. The removal rate of Pb^{2+} after 120 min was 61% when the amount of PIHP added was 0.1 g, the removal rate of Pb^{2+} increased to 91% when the amount of PIHP added increased to 0.5 g. The removal rate of Pb^{2+} in the solution increased with the increase of PIHP addition, but the unit adsorption amount of the adsorbent decreased, which may be due to the increase of the active sites of adsorption as the adsorbent increased, the excess active sites could not exert its adsorption effect when the amount of the adsorbent was too much, resulting in a decrease in the unit adsorption capacity of the adsorbent [13].

3.4.2. Effect of PH on adsorption effect

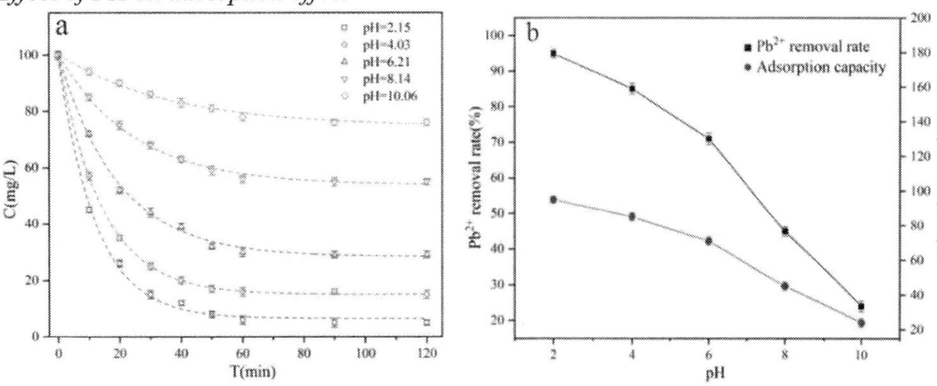

Figure 5. Effect of pH of the system on Pb^{2+} removal rate and adsorbent adsorption capacity

0.2 g PIHP was added to 200 mL of $PbCl_2$ solution and the pH value of the solution was changed by adding HCl or NaOH to study the effect of pH value on the adsorption of pb^{2+}. As shown in Figure a, the effect of pH on the adsorption of pb^{2+} was very obvious, and the adsorption amount increased with the decreased of pH value. The decrease of Pb^{2+} concentration in the system was mainly due to the three main functions of dissolution-precipitation, ion exchange and surface physical adsorption [14]. The adsorption of Pb^{2+} by PIHP depended on the concentration of phosphate dissolved, the phosphate dissolved by PIHP gradually decreased with the increased of pH, the ability to remove Pb^{2+} by dissolution-precipitation was weakened and some hydroxides also appearred, which changed the adsorption mechanism and inhibited the adsorption of Pb^{2+} by PIHP [15].

4. Conclusion

This study used industrial solid waste phosphating slag as raw material to prepare porous iron hydroxy phosphate for treating lead-containing wastewater, the porous iron hydroxy phosphate was prepared by hydrothermal purification, ball milling modification, hydrothermal crystallization and calcination treatment, the particle size was about 3 μm and the pore size was about 100 nm. The adsorption results showed that the removal rate of Pb^{2+} increased gradually with the increased of PIHP, but the unit adsorption amount of PIHP decreased, the removal rate of Pb^{2+} could reach 91% when the addition amount of PIHP was 0.5 g. The adsorption capacity of PIHP for pb^{2+} increased with the decreased of pH value of the solution, the removal rate of pb^{2+} after 120 min could reach 95% when the amount of PIHP added was 0.2 g and the pH value of the system was 2.15. The results showed that the heavy metal ion adsorbent which prepared by the solid waste phosphating slag exhibited excellent performance in the treatment of Pb^{2+}-containing wastewater, and achieved the purpose of waste treatment.

Acknowledgements
This work was supported by Shanghai Polytechnic University Graduate Program Foundation (EGD18YJ0057), Material Science and Engineering Key Subject of Shanghai Polytechnic University (No. XXKZD1601), and Gaoyuan Discipline of Shanghai-Environmental Science and Engineering (Resource Recycling Science and Engineering).

References

[1] Fan H Q, Li S Y. (2007) Research progress on the status and resource utilization of solid waste phosphating slag. Modern Chemicals, 27(1): 97-99.
[2] Liu D M, Fang K, Wu F Y. (2005) Study on the development and utilization of phosphorus slag, Mining News, 429 (3): 21–25.
[3] Kuo Y M. (2012) An alternative approach to recovering valuable metals from zinc phosphating sludge. Journal of Hazardous Materials, 201(1): 265-272.
[4] Wu X Y, Huang C C, Lang Q C, et al. (2015) Research progress and prospects of phosphating slag resource utilization. Renewable Resources and Circular Economy, 227(8): 38-41.
[5] Liu J Y, Liu L H, Xue J R, Lü C Q, Li T, Hu B Q. (2018) Research progress in adsorption treatment of heavy metal wastewater. Environmental Chemistry, 37(09):2016-2024.
[6] Li H L, Li T, Yang G, et al. (2017) Synthesis of thiol-functionalized mesoporous calcium silicate and its adsorption characteristics for heavy metal ions. Journal of Environmental Chemical Engineering, 5(6): 6201- 6215.
[7] Yan Y N, Chai L Y, Yang Z H, et al. (2017) Immobilization of Cd and Pb in soils by polymeric hydroxyl ferric phosphate. Chinese Journal of Nonferrous Metals, 216(5):32-35.
[8] Wu R P. (2014) Preparation of polyhydroxyphosphate and its application in soil remediation of lead and cadmium. Central South University, 15(2):45-47.
[9] Fulazzaky MA, Salim NAA, Abdullah NH, et al. (2014) Precipitation of iron hydroxyl phosphate of added ferric iron from domestic wastewater by an alternating aerobic anoxic process. Chemical Engineering Journal, 253:291- 297.
[10] Zhu M X, Ding K Y, Xu S H, et al. (2009) Adsorption of phosphate on hydroxyaluminum- and hydroxyiron-montmorillonite complexes. Journal of Hazardous Materials, 165(1): 645-651.
[11] Huang Y H, Xie Z G, He Y H, et al. (2007) Thermal decomposition process of oxalate coprecipitation precursors. Powder Metallurgy Materials Science and Engineering, 12(1):45-47.
[12] Yao Y L, Zhang C F, Fan Y Q, et al. (2015) Preparation of porous FeNi alloy powder by thermal decomposition of oxalate and its thermal decomposition kinetics. Chinese Journal of Process Engineering, 15(1): 84-88.
[13] Fan Z, Jing L I , Jianhua T, et al. (2013) Advance of the treatment of heavy metal wastewater by adsorption. Chemical Industry and Engineering Progress, 251:67-69.
[14] Wang Y Y, Liu Y X, Lu HH, et al. (2018) Competitive adsorption of Pb(II), Cu(II), and Zn(II) ions onto hydroxyapatite-biochar nanocomposite in aqueous solutions. Journal of Solid State Chemistry, 261:53-61.

IOP Conf. Series: Materials Science and Engineering **729** (2020) 012058 doi:10.1088/1757-899X/729/1/012058

[15] Sun Y X, Zhou S, Yang H. (2010) Preparation of hydroxyapatite nanoparticles and their adsorption properties for Pb^{2+} ions. Environmental Chemistry, 29(6): 1059-1062.

Quantitative determination of Lisinopril Dihydrate in pharmaceutical preparations by direct potentiometriy

S Bozhanov[1], M Smerikarova[1] and V Maslarska[1*]

[1]Department of Chemistry, Faculty of Pharmacy, Medical University-Sofia, Sofia, 1000, Bulgaria

*Corresponding author's e-mail: vmaslarska@mail.bg

Abstract. A direct potentiometric titration method was applied for determination of Lisinopril Dihydrate. The method was based on the treatment of the primary data with nonlinear regression procedure. An accessible and easy-to-use algorithm for the evaluation (quantification) of a system with acid-base properties has been developed to provide a quick and unambiguous answer for the quantitative content of Lisinopril Dihydrate. The acid-base constants of Lisinopril Dihydrate were determined in aqueous solutions (at constant ionic strength 0.2 mol/l KCl and temperature 25°C). The potentiometric method developed was validated according to ICH and shows very good accuracy and precision. The present approach can be successfully used in routine analysis of the drug in quality control laboratories.

1. Introduction

Combination therapy with Lisinopril – a potent ACE inhibitor (chemically known as (2S)-1-[(2S)-6-amino-2-[[(1S)-1-carboxy-3-phenylpropyl]amino]hexanoyl]pyrrolidine-2-carboxylic acid) and Hydrochlorothiazide – a diuretic from the group of thiazides (chemically known as 6-chloro-1,1-dioxo-3,4-dihydro-2H-1λ6,2,4-benzothiadiazine-7-sulfonamide) has been shown to be effective in treating high blood pressure and heart failure [1]. Chemical structures of both drugs are shown in figure. 1.

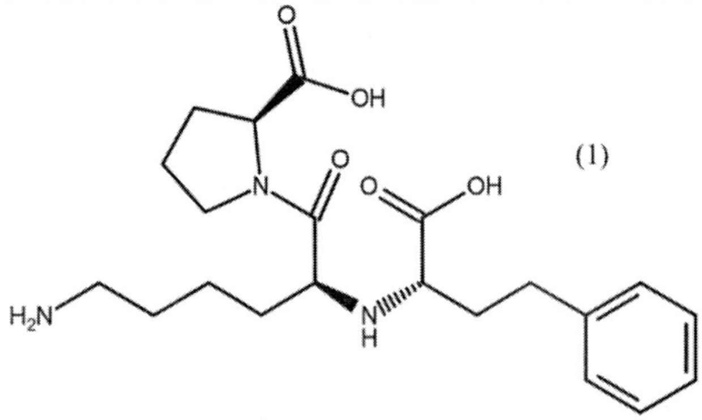

(1)

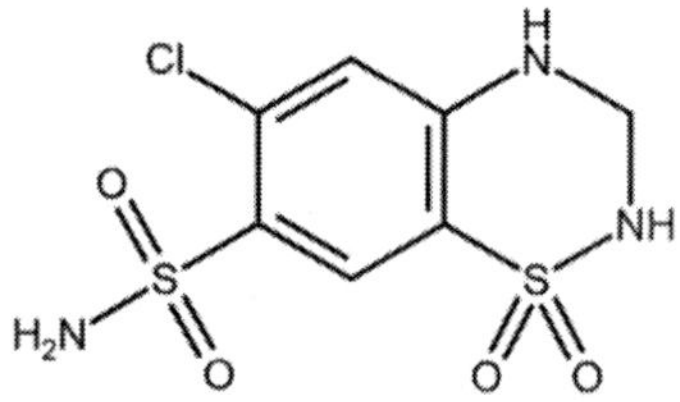

(2)

Figure 1. Chemical structures of Lisinopril (1) and Hydrochlorothiazide (2).

Wide use of such drugs requires reliable methods for quality control of bulk substances and pharmaceutical formulations. Various methods such as titrimetry [2], voltammetry [3, 4], spectrophotometry [5, 6] and spectrofluorimetry [7, 8] have been reported for determination of the two drugs. Analysis based on HPTLC [9] and HPLC [10-13] methods are also available. Despite the variety of techniques, direct potentiometry probably remains the simplest method in terms of materials used, equipment availability and labor intensity.

Complex interactions were established in each acid-base system, that is why the acid-base equilibrium depends on many factors. Such complicated relations can be described by regression methods. Most often linear or nonlinear regression was applied to equations of the type V = f (h) or V = f (E) for pH / ml data and V / E data respectively. This work was based on the earlier developed approach for data processing of potentiometric titration of substances with acid - base properties [14-16]. Using the mass balance equations and proton stability of the species, a well-defined formula valid for each titration of mono- and polyfunctional protoliths was derived - Equation (1)

$$V = V_o * \left(h - K_w/h + \sum_{m=1}^{M} \sum_{n=1}^{N} B_{o,m} * \bar{n}_{H,m}\right)/(C_t - h + K_w/h) \quad (1)$$

where, V - is volume of the added titrant, ml; V_o - initial volume of the titrated solution, ml; h - concentration of protons, mol/l; K_w - autoprotolysis constant of water; M - number of components; N - maximum number of protons that a protolyte can accept or release; B_o - concentration of the analyzed base (resp. A_o for acid), mol/l; C_t - total concentration of the titrant, mol/l.

As a result, the proton stability acid-basic constants β_n (resp. pK), autoprtolysis constant (K_w) of the water and the concentration of the analysed substance (B_o) were obtained.

The theoretical foundations of the proposed method were based on the fact that acid - base interactions were represented as complex formation process between the base (B) or the acid (A) and the ligand H (protons). Steady-state equilibrium represents interaction in a mononuclear system. The aim of this study is to develop and validate an analytical method for routine analysis of Lisinopril Dihydrate in pharmaceutical preparations using direct potentiometry.

2. Material and methods

2.1. Measurements

2.2. The measurements were carried at constant ionic strenght (KCl, 0.2 mol/l) in thermostated vessel (25°C). A 713 Metrohm pH-meter, equipped with combined electrode (ref. 6.0228.000 Pt1000) with temperature sensor and auto burette "Radiometer" ABU 80 were used.Reagents and solutions

Sodium hydroxide – p.a. and Potassium chloride – p.a. (Merck, Darmstadt, Germany) were used without purification. Lisinopril Dihydrate standard was obtained from Sigma Aldrich. Tablet formulation containing Lisinopril Dihydrate 20 mg and Lisinopril 20 mg + Hydrochlorothiazide 12.5 mg in VITOPRIL plus tablets STADA were obtained commercially. All chemicals investigated, corresponded to p.a. purity and were used without purification.

Sodium hydroxide (0.01 mol/l) in water was prepared by dilution of certified volumetric solutions with carbon-dioxide free redistilled water. The solution of sodium hydroxide was standardized with standard solution of hydrochloric acid.

2.3. Preparation of the standard solution

Aqueous solution (0.01 mol/l) of Lisinopril Dihydrate standard was prepared. The titrant used was a standard solution of sodium hydroxide (0.01 mol/l).

2.4. Sample Preparation

2.4.1. Procedure for Lisinopril Dihydrate. A solution of the analyzed substance at concentration of 0.01 mol/l and a constant ionic strength 0.2 mol/l was prepared. Aliquot samples of 10.0 ml and 90.0 ml 0.2 mol/l KCl were titrated at 25±0.1°C with a standard0.01mol/l NaOH.

IOP Conf. Series: Materials Science and Engineering **729** (2020) 012059 doi:10.1088/1757-899X/729/1/012059

2.4.2. Procedure for Lisinopril Dihydrate in Tablets. Twenty tablets were weighed accurately and ground into a fine powder. One portion of the powder equivalent to 300 mg of Lisinopril Dihydrate (0.01 mol/l) was accurately weighed into a 100 ml volumetric flask. The extraction was done by adding 40 ml of water and shaking for 20 minutes. After that the volume was adjusted to the mark with distilled water and the solution was filtrated using a filter paper. Aliquot samples of 10.0 ml were mixed with 90.0 ml 0.2 mol/l KCl and titrated according to the above described scheme.

3. Results and discussion

The Equation 1 was applied to the analysis of Lisinopril Dihydrate substance. With the help of nonlinear regression pK_a values were calculated, than were set as constants for further data processing and calculating of the analytical results. Analysis of Lisinopril Dihydrate according to European Pharmacopoeia 8.0 was also performed. The results obtained by the two methods were in very good agreement (table 1). Further, in order to validate the method, the procedure was applied for the analysis of synthetic mixtures of the pharmaceutical form of Lisinopril Dihydrate as well as placebo sample.

Table 1. Determination of Lisinopril Dihydrate substance by pharmacopoeial method and withdirect potentiometry.

Substance	pK_{a1}	pK_{a2}	Found, %	Ph.Eur.8
Lisinopril Dihydrate	3.49	4.40	99.53	99.80

All of the analytical validation parameters for this proposed method were determined according to ICH guidelines [17, 18] as follows:

3.1. Selectivity

A placebo blank containing calcium hydrogen phosphate dehydrate, magnesium stearate, cornstarch, mannitol and colloidal anhydrous silica was prepared, extracted and solution made as described under "procedure for tablets". Aliquot samples were titrated according to the recommended procedure. It was found that there was no interference between the analyte and placebo.

3.2. Accuracy

Accuracy studies were performed by titration of samples at three different concentration levels. The accuracy was evaluated as percentage relative standard deviation (%) between the measured and taken amounts/concentrations. Relative standard deviation was found to be less than 2% which was within the limits according to ICH. The results of this study are compiled in table 2.

Table 2. Accuracy for determination of Lisinopril Dihydrate.

Lisinopril Dihydrate, taken, mg	Intra-day accuracy		Inter-day accuracy	
	Lisinopril Dihydrate found*, mg	%, RSD	Lisinopril Dihydrate found*, mg	%, RSD
10.0	10.21	0.684	10.12	1.135
20.0	19.78	0.345	20.03	0.432
30.0	29.93	0.231	29.56	0.324

* Average of three independent procedures.

3.3. Precision

Precision of the proposed method was established by six-time analysis of Lisinopril Dihydrate in substance and tablets. The results obtained are presented in table 3.

Table 3. Results of the Precision analysis of Lisinopril Dihydrate and tablets 20 mg.

№	Substance, %	Tablets, mg
1	99.95	20.13
2	100.05	19.89
3	99.87	20.34
4	99.29	20.16
5	99.86	20.11
6	100.5	19.56
Parameter		
Mean	99.92	20.03
Stand.Deviation	0.389	0.272
Rel.SD,%	0.389	1.358
Conf. Interval	0.408	0.285
% Error	0.408	1.424

3.4. Application to combined dosage form

The proposed method was applied to combined dosage form. Results of the statistical processing of the obtained method (table 4) repeat data for the analysis of Lisinopril 20 mg and Hydrochlorothiazide 12.5 mg in VITOPRIL plus tablets STADA.

Table 4. Results for combined dosage form – VITOPRIL plus, STADA

№	Lisinopril, mg	Hydrochlorothiazide, mg
1	19.90	12.51
2	20.10	12.48
3	19.96	12.42
4	20.11	12.55
5	20.03	12.48
6	20.06	12.53
Parameter		
Mean	20.03	12.49
Stand.Deviation	0.082	0.046
Rel.SD,%	0.411	0.368
Conf. Interval	0.086	0.048
% Error	0.431	0.386

4. Conclusions

The applied approach provides an accessible procedure for processing the experimental data in order to quantify substances with acid-base properties. A quantitative method suitable for routine laboratory analysis of Lisinopril Dihydrate substance, Lisinopril Dihydrate with Hydrochlorothiazide in tablets was developed. The method was validated and has been proven to have very good reproducibility and accuracy.

Acknowledgements

The authors are thankful to Faculty of Pharmacy, Medical University-Sofia, Bulgaria for providing lab facilities.

References

[1] Sukalo, A., Deljo, D., Krupalija, A., Zjajo, N., Kos, S., Curic, A., Divkovic, G., Hubjar, S., Smailagic, M., Hodzic, E., Marjanovic, D., Medjedovic, S. (2016) Treatment of hypertension with combination of lisinopril/hydrochlorothiazide. Med. Arch., 70(4): 299-302.

[2] Olalowo, A.W., Adegbolagun, O.M., Bamiro, O.A. (2015) Simple potentiometry and phenolphthalein-based titrimetric methods of analysis for Lisinopril tablets. Afr. J. Pharm. Pharmacol., 9(6): 165-172.

[3] González-Vargas, C., Serrano, N, Ariño, C., Salazar, R., Esteban, M., Díaz-Cruz, J.M. (2017) Voltammetric determination of anti-hypertensive drug hydrochlorothiazide using screen-printed electrodes modified with l-glutamic acid. Chemosensors, 5: 25-36.

[4] Valezi, C.F., Eisele, A.P.P., Sartori, E.R. (2017) Versatility of a carbon paste electrode coupled to differential pulse voltammetry for determination of lisinopril with its associations (hydrochlorothiazide and amlodipine). Anal. Methods, 9: 4599-4608.

[5] Shraitah, M., Okdeh, M.M.S. (2016) New method for spectrophotometric determination of lisinopril in pure form and in pharmaceutical formulations. Mod. Chem. Appl., 4(1): 172-175.

[6] Patel, N., Prabhu, P., Walia, S. (2015) Spectrophotometric quantitative estimation of hydrocholorothiazide and lisinopril in bulk drugs and pharmaceutical dosage form. World J. Pharm. Res., 4(9): 1684-1701.

[7] Derayea, S.M., Badr El-din, K.M., Mohammed, F.F. (2018) An innovative validated spectrofluorimetric method for determination of lisinopril in presence of hydrochlorothiazide; application to content uniformity testing. Spectrochim. Acta A, 188: 318-323.

[8] Derayea, S.M., Badr El-din K.M., Mohammed, F.F. (2017) Selective spectrofluorimetric method for determination of Lisinopril in pharmaceutical preparations and in presence of hydrochlorothiazide: Application to content uniformity testing. Luminescence, 32(8): 1482-1487.

[9] Pandya, J.J., Sanyal, M., Shrivastav, P.S. (2017) Simultaneous densitometric analysis of amlodipine, hydrochlorothiazide, lisinopril, and valsartan by HPTLC in pharmaceutical formulations and human plasma. J. Liq. Chromatogr. Relat. Technol., 40(9): 467-478.

[10] Mohammed, N.S., Mohammed, A.J. (2016) Development and validation of RP-HPLC method for the determination of hydrochlorothiazide in bulk drug and pharmaceutical dosage form. Chromatogr. Res. Int., 2016, Article ID 1693024, pp. 1-7.

[11] Shah, J.V., Shah, P.A., Shah, P.V., Sanyalc, M., Shrivastava, P.S. (2017) Fast and sensitive LC–MS/MS method for the simultaneous determination of lisinopril and hydrochlorothiazide in human plasma. J. Pharm. Anal., 7:163–169.

[12] Eid, M., El - Shabrawy, Y., El - Shaheny, R. (2017) Green micellar HPLC analysis of three angiotensin - converting enzyme inhibitors in their mixtures with hydrochlorothiazide and modeling of their retention behavior by fitting to Foley's model. J. Sep. Sci., 40(18): 3646-3654.

[13] Dawud, E.R., Shakya, A.K. (2019) HPLC-PDA analysis of ACE-inhibitors, hydrochlorothiazide and indapamide utilizing design of experiments. Arab. J. Chem., 12: 718–728.

[14] Maslarska, V., Tencheva, J., Budevsky, O. (2003) New approach in the treatment of data from an acid–base potentiometric titration I. Monocomponent systems of monofunctional acids and bases. Anal. Bioanal. Chem., 375(2): 217–222.

[15] Maslarska, V., Tencheva, J., Budevsky, O. (2005) A new approach to data treatment in acid-base potentiometric titration II. Determination of polyprotic acids and bases. Chem. Anal. (Warsaw), 50(5): 815-823.

[16] Maslarska, V., Tencheva, J., Budevsky, O. (2011) Potentiometric analysis of mixtures of acids or bases. Der Chemica Sinica, 2(6): 325-330.

[17] International Conference on Harmonization (2005) ICH harmonized tripartite guideline Validation of analytical procedures: text and methodology Q2 (R1) ICH, Geneva, Nov; 2005.

[18] European Pharmacopoeia 8th Edition, Vol. 2, 2014.

IOP Conf. Series: Materials Science and Engineering **729** (2020) 012060 doi:10.1088/1757-899X/729/1/012060

The establishment of a newly modified numerical global springback compensation method

Wei Xiong [1], Wenfeng Zhang [1, 2*] Yong Lv[1] Chuan Li [1], Xiaohui Song [1] and YushanXia [3]

[1]School of Mechanical Engineering, Gulin University of Aerospace Technology, Guilin 541004 China

[2]Guangxi Colleges and Universities Key Laboratory of Robot & Welding, Guilin University of Aerospace Technology, Guilin 541004, China

[3]Key Laboratory of Contemporary Design and Integrated Manufacturing Technology, Ministry of Education, Northwestern Polytechnical University, Xi'an 710072 China

*Corresponding author's e-mail: zwf@guat.edu.cn

Abstract. Part shape deviation caused by spring back is a quality problem in sheet metal forming. There are two major approaches, adjusting tooling shapes and adopting process control. This paper focused on tooling contour design. A new die design approach, called global compensation method (GC), has been proposed and it possesses following advantages. Compensation factor in displacement adjustment method (DA) is extended to a compensation matrix, thus every node has its own compensation magnitude and compensation direction, which are related to the global forming error.

1. Introduction

Springback during the unloading phase of sheet metal forming leads to deviation from design part shape. It affects products' geometrical acceptability. There are several means to reduce the dimensional error caused by springback. One of them is trying to eliminate springback by modifying forming processes. Another instance is increasing blank holder pressure or using inverse approach in deep drawing process. These methods possess the benefits of convenience in die design, but their application may be restricted by parts' materials and shapes. Stretch-warp bending is not quite suitable for parts with internal ribs like integral panels. As blank holder pressure increases the chance of excessive thinning augments too.

Although iterative die design can be carried out with actual dies and parts, trail-and-error method will not be included in this paper. Since it consumes too much time and money, interest is attracted by new measures based on numerical simulation. Initially, Karafillisand Boyce[1][2][3]suggested what is called "spring forward" method to design the tooling and the blinder shape with a finite element(FE) grogram. Opposite section moments or tractions are applied on the desired part shape, which was mistaken for the initial unbending part in Ref.[4], to generate new trial for tooling shape. Besides arguments of lacking convergence in asymmetric cases[5][6], a few more works have to be done in applying loads to the desired shape. Webb and Hardt[7] viewed the iterative die design process as a close loop control system and introduced deformation transfer function to relate changes of die shapes to that of part shapes. Attempts have also been made using optimization algorithm[8][9][10][11]. These methods involve considerable complexity in formulation and implemen-tation as part of a special-purpose finite element program[5]. GAN and WAGONER [5][6] proposed displacement adjustment(DA) method, whose concept is to move the die surface nodes in the direction opposite to the springback error. To remedy this issue, a natural solution is to include abscissa and/or z direction in compensation, such as the three dimensional smooth displacement field in Ref.[12]. Although the RD approach reflects the tendency of rotation during springback to some extent, it deviates a lot from the right direction in V-bending [12]. However, far more work is needed to put it into practice. Because not only calculating distinction of normal directions but also an angle factor for an optimal compensation direction are necessary in the approach.

In this paper, endeavor has been made to improve DA method by extending the compensation factor to a compensation matrix. By doing so, not only both compensation magnitude and direction, but the global issue of spring back is taken into account.

2. The establishment of the modified global springback compensation method

2.1. The compensation matrix

The compensation matrix can be introduced from the traditional DA method. The principle of DA can be described as follows:

$$C = D - \alpha(S - D) \tag{1}$$

where, C, D and S are topological identical sets of points in the compensated geometry, desired geometry and springback part geometry, respectively. They represent shape of the three geometries. α is the compensation factor. When the DA is performed iteratively, it can be represented by Eq.(2).

$$C^{j+1} = C^j - \alpha(S^j - D^j) \tag{2}$$

where, j represents the j-th iteration. In Eq.(2) and Eq.(1) α is identical, which means every point in the die employs the same compensation magnitude. However, optimal compensation factors vary from one point to another. So, an intuitive solution is to utilize different compensation factor for different points. This can be represented by a matrix equation:

$$\begin{bmatrix} \vec{c}_1^{\,j+1} \\ \vec{c}_2^{\,j+1} \\ \vec{c}_3^{\,j+1} \\ \vdots \\ \vec{c}_n^{\,j+1} \end{bmatrix} = \begin{bmatrix} \vec{c}_1^{\,j} \\ \vec{c}_2^{\,j} \\ \vec{c}_3^{\,j} \\ \vdots \\ \vec{c}_n^{\,j} \end{bmatrix} - \begin{bmatrix} \alpha_1 & & & & \\ & \alpha_2 & & & \\ & & \alpha_3 & & \\ & & & \ddots & \\ & & & & \alpha_n \end{bmatrix} \left(\begin{bmatrix} \vec{s}_1^{\,j} \\ \vec{s}_2^{\,j} \\ \vec{s}_3^{\,j} \\ \vdots \\ \vec{s}_n^{\,j} \end{bmatrix} - \begin{bmatrix} \vec{d}_1 \\ \vec{d}_2 \\ \vec{d}_3 \\ \vdots \\ \vec{d}_n \end{bmatrix} \right) \tag{3}$$

where, $\vec{c}_i^{\,j}$, $\vec{s}_i^{\,j}$ and $\vec{d}_i^{\,j}$ are points in *C*, *S* and *D*. Here, only elements in the diagonal have non-zero value. So it still considers springback compensation in a local manner. However, if non-zero values are assigned to elements in upper and/or lower triangle as Eq.(4), a vector in the compensated die geometry is related to a set of forming errors in the entire part, thus springback compensation can be treated in a global approach.

$$A = \begin{bmatrix} \alpha_{11} & \alpha_{12} & \alpha_{13} & & \alpha_{1n} \\ \alpha_{21} & \alpha_{22} & \alpha_{23} & \cdots & \alpha_{2n} \\ \alpha_{31} & \alpha_{32} & \alpha_{33} & & \alpha_{3n} \\ & \vdots & & \ddots & \\ \alpha_{n1} & \alpha_{n2} & \alpha_{n3} & & \alpha_{nn} \end{bmatrix} \tag{4}$$

In the Eq.(4), A represents compensation matrix. Except its advantages in providing different compensation magnitudes and directions for different nodes in the die geometry, its evaluation is still needed to be solved when it is applied in compensation practice. This will be discussed in the following section.

2.2. Evaluation of the compensation matrix

Unlike most references mentioned above, where compensation factors are evaluated in a static manner, here the compensation matrix is related to the forming deviation. The evaluation of its elements is self-adapting. To achieve this, the problem of die design is studied from an unusual, macroscopic view to introduce the concept of forming function. And the details such as compensation magnitude and direction will be discussed later.

The process of employing finite element method (FEM) to calculate the forming deviation of a certain die can be summarized like this: First, a set of mesh node coordinates, which represents the contour of the die, is imputed. Then after computation a set of forming errors, which describes deviations between the part's shape after stamping and that of designation, is outputted. In the case of die design, forming errors are usually converted to the form of deviations in every node of the die mesh. Therefore, the number of deviations is equal to that of die nodes, or coordinates of die nodes if deviations are decomposed to components in three coordinate's directions. It can be represented by a set of equations:

$$\begin{cases} f_1(x_1,x_2,\cdots,x_m,y_1,y_2,\cdots,y_m,z_1,z_2,\cdots,z_m)=\delta_1 \\ f_2(x_1,x_2,\cdots,x_m,y_1,y_2,\cdots,y_m,z_1,z_2,\cdots,z_m)=\delta_2 \\ \vdots \\ f_m(x_1,x_2,\cdots,x_m,y_1,y_2,\cdots,y_m,z_1,z_2,\cdots,z_m)=\delta_m \end{cases} \tag{5}$$

where,$x_i, y_i, z_i, (i = 1,2, \dots m)$are mesh nodes' coordinates in the die,$\delta_i, (i = 1,2, \dots m)$are forming errors. Die design is equal to search a group of coordinates that make $\delta_i = 0$. This is the non-linear equation solving concept for die design presented in Ref.[4]. Eq.(5) is called forming function and its zero value, or zero values if multiple solutions exit, is the desired die contour.

By converting die contour design question to solving nonlinear equations, algorithm borrowing from numerical analysis can be used. However, not all numerical methods are suitable to solving these equations due to several characters of them. First, these equations cannot be written down analytically, so calculation of derivatives is not possible. Any method, which needs derivatives, cannot be employed. Second, these equations can only be work out as a whole, therefore, some methods requiring computation certain equations in them can be eliminated from consideration. In this paper, the following quasi Newton (QN) algorithm is used to evaluate the compensation matrix:

(1) Set initial value $y^{(0)}$, permitted error ε and initial Hessian matrix $H_0 = I$, $k = 0$;

(2) Calculate $F\left(\{x,y,z\}^{(k)}\right)$. if $\left\|F\left(\{x,y,z\}^{(k)}\right)\right\| < \varepsilon$,then let mold contour be $y = y^{(k)}$, terminate program;

(3) Calculate $y^{(k+1)} = y^{(k)} - H_k F\left(\{x,y,z\}^{(k)}\right)$;

(4) Calculate $F\left(\{x,y,z\}^{(k+1)}\right)$, $s^{(k)} = y^{(k+1)} - y^{(k)}$, $\delta^{(k)} = F\left(\{x,y,z\}^{(k+1)}\right) - F\left(\{x,y,z\}^{(k)}\right)$, $H_{k+1} = H_k + \frac{1}{\delta^{(k)T}\delta^{(k)}}\left(s^{(k)} - H_k\delta^{(k)}\right)\delta^{(k)T}$;

(5) $k+1 \Rightarrow k$, go to (2).

where, $\{x,y,z\}^{(k)}$aredie's mesh node coordinates in k th FEM calculation, $F\left(\{x,y,z\}^{(k)}\right)$refers Eq.(5) for short. And the Hessian matrix in the algorithm is an evaluation of the compensation matrix A.

2.3. Algorithm of springback compensation

The algorithm mentioned above can be used directly to design die contour. In certain cases, it converges faster than DA. However, the algorithm is sensitive to the initial guess of the die contour. To avoid its limit of local convergence, while to keep its advantages in regulating compensation magnitude for individual node based on the forming error on it and employing global compensation matrix, a homotopy algorithm is introduced to springback compensation.

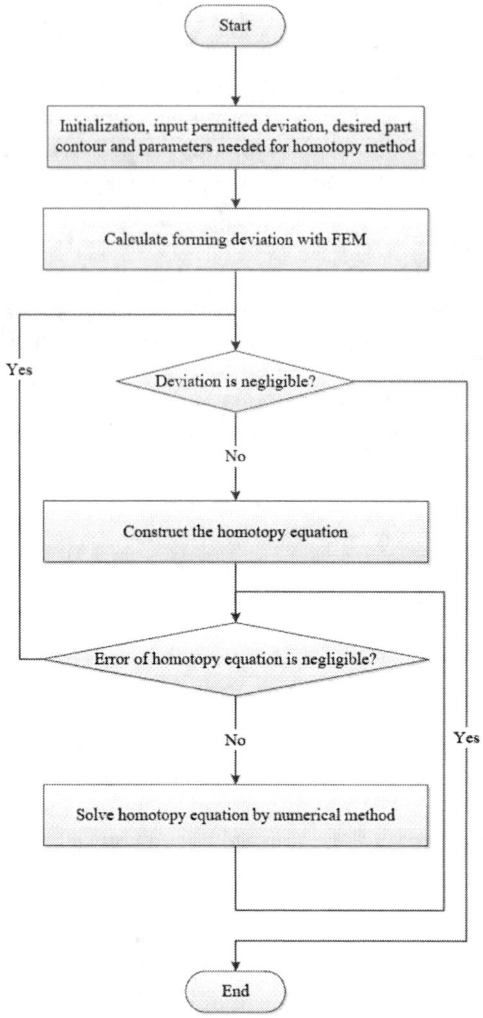

Figure 1.Global compensation algorithm for MATLAB

The basic idea of homotopy method is creating serial equations and employing the solution of the last equation as the initial value for the current equation. Here Eq.(5) is referred as $F(x) = \delta$ for short, the equation needs to be work out is $F(x) = 0$. A serial homotopy equation can be constructed by embedding parameter t_i as follows:

$$H(x,t_i)=0 \tag{6}$$

where, $0 = t_1 < t_2 < \cdots < t_N = 1$. Eq.(6) satisfies following requests:

$$H(x,t_N)=F(x) \tag{7}$$

$$H(x,t_0)=F_0(x) \tag{8}$$

where, $F_0(x)$is an equation that its solution$x_0$is already known. Assuming x^i is the solution of the ith homotopy equation in Eq.(6), if $t_i - t_{i-1}$ is small enough, it can be expected that x^{i-1} is enough close to x^i and is capable of serving the initial value to solve the ith homotopy equation with a local convergence algorithm.

One form of homotopy equation that tally with the requests from Eq.(7)and Eq.(8) is

$$H(x,t)=tF(x)+(1-t)A(x-x_0) \tag{9}$$

where, $F_0(x) = x - x_0$ and x_0 can take the design contour(D_0) of the part needs to be formed. Eq.(9) can be solved one after another through local convergence methods.

In this paper, the quasi Newton method described above is used to solve homotopy equation (Eq.(9)) and to evaluate the compensation matrix. However, another problem should be taken into consideration. The determinant of $\delta^{(k)T}\delta^{(k)}$ in the fourth step of quasi Newton method might be close to zero, if so, it will be added with a damping parameterλI to avoid unfavorable effects.

The springback compensation method can be described with N-S flowchart as Fig. 1.This flowchart is design for realization with MATLAB, which does not support "until" type loop, therefore only "when" type loop type is used.

2.4. Realization of the method

There are several details needs to be coped with in the realization of GC for springback compensation. In the first place, deviations in the part's mesh nodes must be defined and calculated. Because the part's contour is described by a set of mesh nodes after FEM simulation of its forming. It is convenience to compare these points with the nodes that represent the part's design shape, which is employed as the initial value for iterative springback compensations. And it is also expedient to adopt the mesh nodes, which is in the very moment that the part has completely contact with the die in the first FEM calculation, to represent the part's design contour. Therefore, deviations in three dimensions in each and every part nodes can be work out. However not all distances of corresponding nodes represent forming deviation, they may be caused by global movements of node sets. Therefore, before these deviations actually are calculated in computer program, registration job must be done to minimize the effects of global motion. The registration scheme is similar to that in Ref.[4], except the initial registration. First, it calculates the maximum and minimum distances of the two set of nodes only in stamping direction. Then it moves the nodes in the contour after forming the average of the two extremums towards that represents the design contour. It is not only the program is simpler but its effect is better. In accurate registration, the scheme still use genetic algorithm and its parameters are same to that in Ref.[4].

Second, forming deviations in the part's mesh nodes must be transferred to that in the die's mesh nodes. Though the die's contour can be represented by the part's mesh nodes in the moment that the part makes fully contact with the die. The two groups of nodes have different configuration, which means there is no one to one relation between them. A remedy is numerical interpolation. Since the deviations are computed by comparing the design contour and the part's contour after forming, and the part's design contour has the same configuration of the shape in the contact moment in forming, every node's deviation can be directly moved to the later node set, which has the same contour with the die but different meshes. Then deviations in three coordinate direction in a certain mesh node in the die can be interpolated with Shepard method like this:

$$\delta\left(x^{i}\right)=\frac{\sum_{k=1}^{M}W_{k}\left(x^{1},x^{2},x^{3}\right)f_{k}^{i}}{\sum_{k=1}^{M}W_{k}\left(x^{1},x^{2},x^{3}\right)},i=1,2,3 \tag{10}$$

where,x^i, $(i = 1,2,3)$are threeCartesian coordinates for the certain mesh node in the die, $\delta(x^i)$ are deviation in the same direction. And the number of M nodes that are close to the certain mesh node are chosen from the set of the part's mesh that taking on the die's contour. $f_k^i$is the deviation of the kth node in x^idirection. $W_k(x^1, x^2, x^3)$are weights. They can be computed as follows:

$$W_{k}\left(x^{1},x^{2},x^{3}\right)=\left[\sum_{i=1}^{3}(x^{i}-x_{k}^{i})^{2}\right]^{-1} \tag{11}$$

Through using interpolation method, the deviations are expediently considering in all three dimensions. And the programming work is less than B Spline fitting employed in Ref.[4].

Last but not least, the FEM model's robustness should be taken into consideration. Since iterative springback compensation methods are carried out by a computer program, which is realized with MATLAB here, stamping and springback simulation must be able to complete with a range of different die contours and part shapes. Unlike single simulation, boundary conditions, analysis steps, mesh size and types and so on can be manually modified according to specific needs of the certain forming simulation. The FEM model used in the compensation program should be robust enough to accomplish forming simulation with different die contours that might appears in springback compensation. When ABAQUS is employed, as stamping commonly carried out with ABAQUS/Explicit, it has sufficient robustness. The problem usually occurs in springback simulation with Standard module in ABAQUS. Therefore, special treatment has to be down to ensure the FEM model's robustness. Scripts that describe the springback FEM model are generated semi-automatically. First, a template of the springback script is created. It contains statements that import stamping results, material property design, and assembly and so on, but leaves any statement related to the part's mesh, such as assigning material section to elements in the mesh, creating analysis steps and boundary conditions, empty. Then the MATLAB program reads stamping file and analyze its mesh. After that these mesh related statements are filled in the template in a way that allows the unloading processes from one set of nodes to anther according to their distances to the final boundary condition.

3. Numerical Simulations

In order to test the effectiveness of the proposed compensation method, a series of cylinder, sphere and saddle parts were subjected to die design using GC, QN and DA.Three methods all compensate along three coordinate directions. The compensation factor for DA method is 1.0. For GC, $t_i = 0.05(i-1), i = 1,2,\cdots 21$, $A = 0.01$ and $\lambda = 0.1$.

In all these shape cases, the modified methods successfully converge to ideal part shapes due to their simplicity.

4. Conclusion

In this paper, a new global method based on compensation matrix is proposed. In the method, homotopy algorithm is used to calculate die shapes iteratively and a damping parameter is also introduced into the method. The method converges faster and is not sensitive to the initial die contour. The new method has been compared with the displacement adjustment method and the quasi Newton method.

Compared with displacement adjustment method, in which compensation magnitude and compensation direction are important aspects, the compensation factor has been replaced by a compensation matrix. It enables different compensation magnitudes and directions for different nodes. And the compensation is not only based on the information of the very node but also information from other nodes in the entire part. This fits the global nature of the springback problem.

The new method also use a numerical approach to evaluate the compensation matrix based on forming shape deviations. Theoretical computation of the matrix will be included in the future work. And the method will be used in practical parts, such as creep age formed aircraft integral panels and automobile panels.

Acknowledgments

This paper is supported by This work was financially supported by the National Science Foundation for Young Scientists of China (Grant No. 51601044), Guangxi Natural Science Foundation for Young Scientists of China (NO. 2015GXNSFBA139225, 2017GXNSFBA198213), Guangxi college and university young teacher fundamental ability promotion project (KY2016YB528), Guangxi Natural Science Foundation 2018JJA120628, National Science Foundation 51705094, the Opening Project of Guangxi Colleges and Universities Key Laboratory of robot & welding: Re-search of microstructure and mechanical properties on CLAM steel after EBW and the Doctor's Research Initial Fund of Gulin University of Aerospace Technology: Researches on Rapid Die Design methods for Sheet metal, and Study on improving the microstructure stability of the heat resistant steel by modifying the precipitates.

References

[1] KARAFILLIS AP, BOYCE MC. (1995) Tooling and binder design for sheet metal forming processes compen-sating springback error [J]. Journal of Machine Tools and Manufacture,,36:503-526.

[2] KARAFILLIS A P, BOYCE M C. (1992) Tooling design in sheet metal forming using springback calcula-tions[J].International Journal of Mechanical Sciences, 34:113-131.

[3] KARAFILLIS A P, BOYCE M C. (1992) Tooling design accommodating springback errors[J].Journal of Mate-rials Processing Technology,32(1-2):499-508.

[4] Xiong Wei, Gan Zhong, Xiong Shipeng, et al. (2014) Rapid Springback Compensation for Age Forming Based on Quasi Newton Method[J]. Chinese Journal of Mechanical Engineering, 27(3):551-557.

[5] GAN WEI,WAGONER R H. (2004) Die design method forsheetspringback[J].International Journal of Me-chanical Sciences, 46:1097-1113.

[6] GAN WEI, WAGONER R H,MAO KUNMING, et al. (2004) Practical methods for the designof sheet formed components[J].Journal of Engineering Materials and Technology, 126:360-367.

[7] R.D. Webb, D.E. Hardt. (1991) A transfer function description of sheet metal forming for process con-trol[J].Transactions of the ASME, 113:44-52.

[8] Nielen Stander, Mike Burger, Xinhai Zhu, et al. (2002) An Optimization Procedure for Springback Compensa-tion using LS-OPT[C].7th international LSDYNA users conference, Dearborn, MI, 19–21 May 15/1–10.

[9] Nielen Stander, Mike Burger, Xinhai Zhu, et al. (2002) Springback compensation in sheet metal forming using a successive response surface method. Proceedings of the 9th AIAA/ISSMO Symposium on Multidis-ciplinary Analysis and Optimization. 2002-2039.

[10] Cimolin F, Vadori R, Canuto C. (2008) Springback Compensation In Deep Drawing Applications A New Me-thodology For Automatic Die Compensation Through A Suitable Optimization[J]. Meccanica,43: 101–113.

[11] Naceur H, Guo Y Q, Ben S. (2006) Response surface methodology for design of sheet forming parameters to control springback effects[J]. Computers & Structures, 84:1651–1663.

[12] Weiherm J, Rietman B, Kose K, Ohminus S, Petzoldt M. (2004) Controlling springback with compensation strategies[C]. NUMIFORM 2004, Ohio State University. p. 1011–1015.

The application of a newly modified numerical global springback compensation method

Wenfeng Zhang[1, 2], Wei Xiong[1*] Chuan Li[1], Yong Lv[1] Xiaohui Song[1] and YushanXia[3]

[1]School of Mechanical Engineering, Gulin University of Aerospace Technology, Guilin 541004 China

[2]Guangxi Colleges and Universities Key Laboratory of Robot & Welding, Guilin University of Aerospace Technology, Guilin 541004, China

[3]Key Laboratory of Contemporary Design and Integrated Manufacturing Technology, Ministry of Education, Northwestern Polytechnical University, Xi'an 710072 China

[*]Corresponding author's e-mail:zwf@guat.edu.cn

Abstract. A newly modified numerical global springback compensation methodwas used to simulate the die process in this paper. The numerical approach is given to evaluate the compensation matrix, and the matrix changes according to forming deviations.And GCremedies quasi Newton (QN) method's sensitivity to initial die contour. To verify its effectiveness, GC, DA and QN have been used to design dies for a series sheet metal parts with cylinder, sphere and saddle shapes. Summarizing simulation and experiment results, the method in general converges faster and more stable.

1. Introduction

There are several means to reduce the dimensional error caused by springback. Some of them tried to eliminate springback by modifying forming processes[1-2]. Some of them attempted to increase the blank holder pressure or use inverse approach in deep drawing process to limit the error[3-4]. However, their application may be restricted by parts' materials and shapes,although these methods possess the benefits of convenience in die design.

GAN and WAGONER [5-6] proposed displacement adjustment(DA) method, whose concept is to move the die surface nodes in the direction opposite to the springback error. The distances and directions that die nodes move along were not specified in the concept. And according to different compensation magnitudes and direc-tions, a series version of DA algorithm can be derived. In the first version, compensation is only made in the y direction, which is parallel to the punch travel direction[7]. And the shape error is added directly to the current die shape nodal positions. This method converges faster than "spring forward" algorithm and exhibits stabilization for cases involves large springback shape changes and non-symmetric parts[8-10].It deviates a lot from the right direction in V-bending [10]. The third type of compensation direction relates to normal orientation. This method is still feasible when there is no mesh nodes available, such as cases that tools' shape are described by point clouds in reverse engineering or CAD geometries[9].Contrary to the RD approach, it works well in V-bending cases, where rotation is the main phenomenon during springback, but would have a poor performance when displacement direction deviates from the normal orientation[7].Besides this, calculation normal direction needs extra work, so it is not a readily measure as the former two.

Another way of utilizing normal directions to compensate springback is the comprehensive compensation (CC) presented in Ref.[11]. Being different from moving nodes in normal direc-tions, this method takes the difference of normal orientations of die contours and that of part shapes into consideration. As figures in the paper illustrated, its performance was better than several current methods. However, far more work is needed to put it into practice. Because not only calculating distinction of normal directions but also an angle factor for an optimal compensation direction are necessary in the approach.

The compensation factor was extended to acompensation matrix in the DA method in this paper. By doing so, not only both compensation magnitude and direction, but the global issue of spring back is taken into account.

2. Numerical Simulations

In order to test the effectiveness of the proposed compensation method, a series of cylinder, sphere and saddle parts were subjected to die design using GC, QN and DA.Three methods all compensate along three coordinate directions. The compensation factor for DA method is 1.0. For GC, $t_i = 0.05(i-1), i = 1,2,\cdots 21$, $A = 0.01$ and$\lambda = 0.1$.

2.1. Cylinder cases

In the simulation, FEM simulations were performed with ABAQUS. Compensation methods, DA,QN and GC, were implemented through a MATLAB program. The sheets were loaded by a 20MPa pressure to ensure fully contact with the single die. A single sheet was meshed with 300 S4R shell elements and 5 integration points in its thickness direction. The material used was aluminum alloy 2124T851, whose properties came from Ref.[19]. The Young's modulus is 69251MPa, Poisson's ratio is 0.33. The dimension of sheets was 150×50×2mm.Compensation iteration stopped when the absolute shape deviation was lower than 0.8mm.

As shown in Fig. 1, GC needs less FEM calculations than DA method in all five radius and converges faster than QN method in most cases. Fig. 2 illustrates the result of GC in R=800mm cylinder case. As shown in Fig. 2, the die shape changes small in the first compensation, large in the second one and small again in the third one. The changes satisfy the expectation that die shape changes large when it is away from the desired one, and it changes small when close to the desired one.

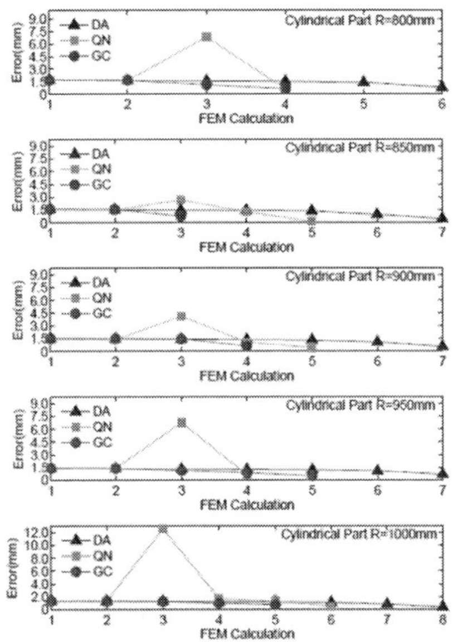

Figure1. Comparisonsofshapeerrorbetween three methodsfor five cylinders

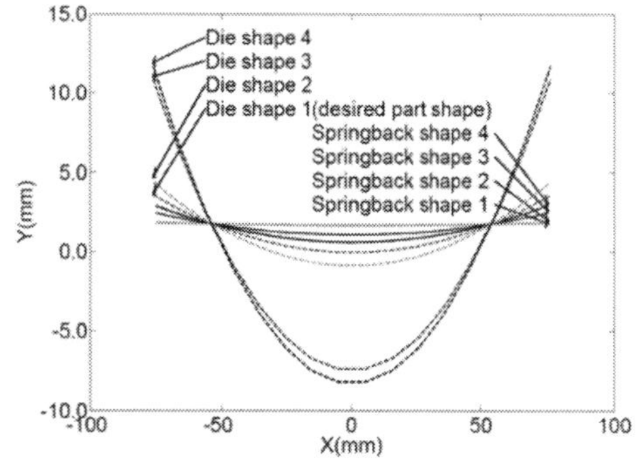

Figure2.Results from GC, cylinder R=800mm.

IOP Conf. Series: Materials Science and Engineering **729** (2020) 012061 doi:10.1088/1757-899X/729/1/012061

Cylindrical parts with radius 800mm, 850mm, 900mm and 1000mm were subjected to die design with a to-lerance ±0.8mm through the three methods. Multiple radius were chosen to test methods' sensitivity for initial die contours. And desired part shapes were employed as initial die contours. Fig. 1 shows the maximum absolute shape error for each FEM calculation. There are two loops in Homotopy method. Here, times of FEM calculation rather than iteration are taken into account, since it is much more time-consuming than other computation.

2.2. Sphere cases

In the spherical simulation cases, the settings were nearly the same to the cylindrical cases, except parts using here were spheres with radius 800mm, 850mm, 900mm and 1000mm. And if iteration consumed more than 7 FEM calculations, computation would be stopped. The results are shown in Fig. 3 and Fig. 4. Again the GC demonstrates good performance. Generally speaking it converges fast among the three approaches. The DA method needs one or two more FEM calculation than that of the GC. The main difference is in the QN method. Except the case with radius 900mm, the QN method consumes more than 7 FEM computations and does not show trances of converges. This is attributed to its nature of local convergence.Fig. 4 shows shape changes of die and part during the GC iteration with radius 900mm sphere. Since part shapes are three dimensional in this section, only center contours along direction of parts' length are shown in Fig. 4. In the first three FEM calculations, the die contour changes small. However, it converts dramatically in the last FEM calculation. Referring Fig. 3, shape error begins to change large in the third calculation. This indicates that small changes in die shape can leads to big changes in part's shape error. There is non-linearity between shape error and die shape. Therefore, the DA method is an attempt to copy a nonlinear problem with a linear approach.

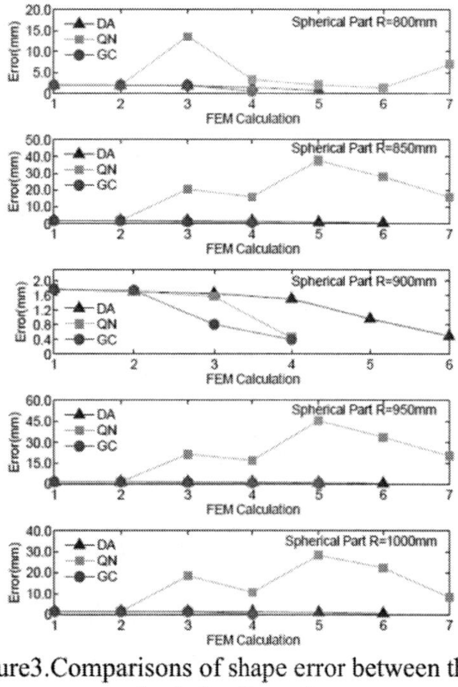

Figure3.Comparisons of shape error between three methods for five spheres

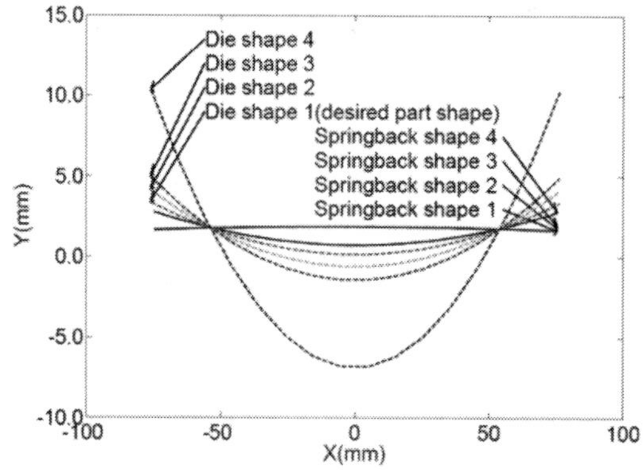

Figure4. Results from GC, sphere R=900mm.

2.3. Saddle cases

The last set of cases employed in numerical simulations was parts with saddle shape. Here, saddle represented another type of double curve parts who have curvatures in two opposite directions. Identical radius was selected for the two reverse orientations. Therefore, utilized parts could be referred by saddle parts with ra-dius 800mm, 850mm, 900mm and 1000mm.The results are shown in

IOP Conf. Series: Materials Science and Engineering **729** (2020) 012061 doi:10.1088/1757-899X/729/1/012061

Fig. 5 and Fig. 6. Through several itera-tions have cost 7 FEM calculations; they all reached the desired shape error. Contrary to its poor performance in sphere cases, the QN method resumes converges faster than the DA method in saddle cases. And the GC is the fast approach in three fifth of all saddle cases. Since the GC method uses a modified quasi Newton method to solve homotopy equation, it over-predicts compensation magnitudes in some cases as what the QN method does. These cases are shown by saddle part with radius 800mm and 1000mm in Fig. 5.

Fig. 6 shows shape changes of die and part during the GC iteration with radius 850mm saddle. It re-flects an ideal situation for the idea that attempts to accelerate compensation by modifying compensation factor (matrix) according to shape error. That is to evaluate the compensation matrix with the initial two FEM calculation results and the third FEM calculation meets the required shape tolerance.

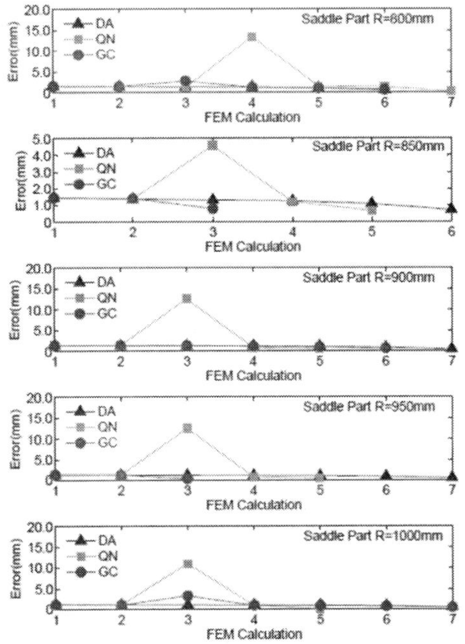

Figure5. Comparisons of shape error between three methods for five saddles.

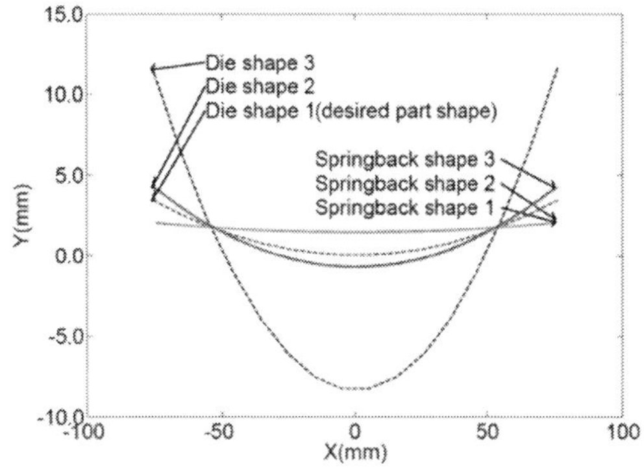

Figure6. Results from GC, saddle R=850mm

3. Application of the method

To verify the accuracy of the modified numerical global spring back compensation method, the forming process was carried on a hydraulic forging press. The die was made of low carbon steel and several layers of rubber covered on the sheet metal to simulate pressure loading condition, as shown in Fig. 7. Based on the mod-ified method, the spring backed sheets were processed sequentially in the order shown in section 2, and the result was shown in Fig. 8. Comparing experimental results from sheets with approximate size in Ref.[20] and Ref.[13], the result is content.

There are several causes for the deviations in Fig. 8. First, due to limits of equipment, the Young's modulus is not precisely measured. Second, parameters of 304 stainless steel used in experiments is not perfect consistent with each other. Third, the FEM model employed in the springback compensation program also has errors. Through to improve FEM model's accuracy is beyond this paper, it does have influence on the experimental results.

Figure7. Sheet metal forming experiment

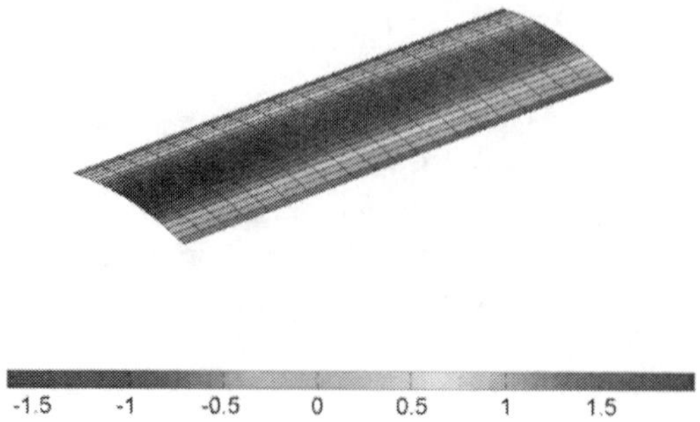

Figure8. Shape error of forming metal sheet(mm).

4. Results and discussion

Contrast to other classes of compensation methods, which tend to predict the tooling shape based on local stress/strain information of the last forming increment[20],the DA, QN and GC methods employ last forming shape deviations to do that. Compared with DA method, in GC and QN method, compensation coefficient has been extended to a compensation matrix that is called Hessian matrix. Through doing so, compensation magni-tudes are not unified in the die's mesh nodes and are adjusted according to the deviations in every iteration. Furthermore, the compensation magnitude in a node is not restricted to a multiple of deviation in the very node, but a linear combination of deviations in all die nodes. The latter is more reasonable than the former. Since springback is a global issue of the entire sheet[20]. The forming result on a certain die node is at least related to nodes around it.Hence, with above factors combined together and in certain type of cases, GC and QN method converges faster than DA method.

Besides the algorithm to solve forming equations, another difference comparing the approach used in Ref. [4] is the compensation direction. In Ref. [4], the nodes in die's contour mesh was only adjusted vertically. So a group of y coordinates represented the die shape. It was suitable when parts' curvature is small as age forming cases studied in Ref.[4]. To expand its application, all three coordinate directions, x, y and z, are utilized to de-scribe die shape and they are arranged to an array as showed in Eq.(5).Since elements in compensation matrix varies, the actual compensation direction of a certain node is a linear combination of x, y and z. It changes with elements in compensation matrix. So compensation direction adjustment can also be achieved by utilizing the concept of compensation matrix.

To compare performance of the three methods when compensation is only made in stamping direction (y di-rection), three parts with radius 900mm cylinder, sphere and saddle parts were subjected to die design. Other setting was the same as in section 3. The results are shown in Fig. 9.The only method whose results changed significantly is the QN method. In cylindrical and spherical cases, QN yield divergence results, which converged using three coordinates. While GC and DA method remains stable.

Besides the instability, there is a key defect in QN method when it is applied in springback compensation. Quasi Newton method is a local convergence algorithm in numerical analysis, that is to say, the initial contour used in the iterative program should be close enough to that of the die producing parts satisfying design. Otherwise, the die contour sequence generated by the program will not converge, as shown in Fig. 2. It is a harsh request, because a close-enough die contour is difficult to design when designers do not know the contour that it needs to be close to. For the same reason, these iterative springback compensation methods usually employ parts' design shapes in initial tryouts. The GC, however, is a global convergence algorithm. With same initial die contours, it can converge to the desired die shapes when QN cannot.

Although Quasi Newton algorithm is employed in the GC to solve homotopy equations, there is an improvement compared to the algorithm in Ref. [4]. A damping parameterλI is introduced into the method. It can suppress, though not eliminate, that shape error drastically increases, as those in the third FEM calculation of QN method. The damping parameter improves GC's performance. However, it is not the only cause of its advantages over the QN method. To illustrate this, diverged cases in section 3 were recalculated with QN, which had damping parameter. All four cases, spherical parts with radius 800mm, 850mm, 950mm and 1000mm, still did not converge as shown in Fig. 10.

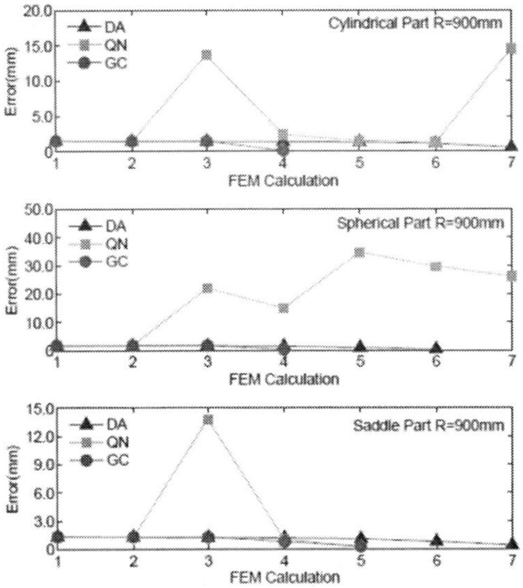

Figure9. Results of three methods compensating only in stamping direction.

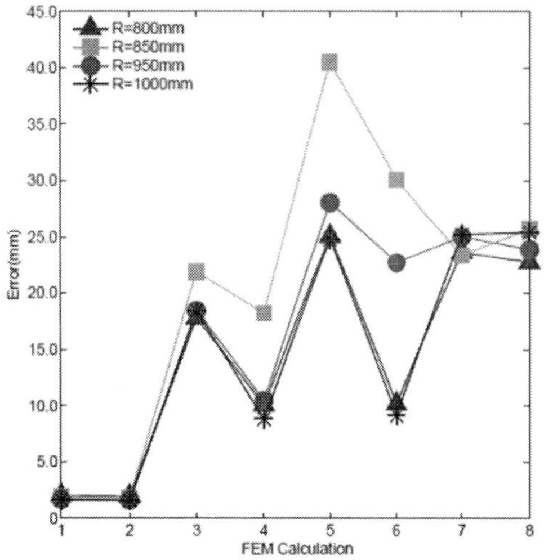

Figure10. Effects of ramping factor on the QN method

5. Conclusion

In this paper, a newglobal method based on compensation matrix was proposed. In the method, homotopy algorithm was used to calculate die shapes iteratively and a damping parameter was introduced into the method. The method converges faster and is not sensitive to the initial die contour.Thenew method was compared with the displacement adjustment method and the quasi Newton method.The following conclusion could be made from the results:

(1) Comparedwith displacement adjustment method, in which compensation magnitude and compensation direction were important aspects, the compensation factor had been replaced by a compensation matrix. It enabled different compensation magnitudes and directions for different nodes.

(2) The compensation was not only based on the information of the very node but also from other nodes in the entire part. This fit the global nature of the springback problem.

(3) The new method also used a numericalapproach to evaluate the compensation matrix based on forming shape deviations. Theoretical computation of the matrix would be included in the future work. And the method would be used in practical parts, such as creep age formed aircraft integral panels and automobile panels.

Acknowledgments

This paper is supported by This work was financially supported by the National Science Foundation for Young Scientists of China (Grant No. 51601044), Guangxi Natural Science Foundation for Young Scientists of China (NO. 2015GXNSFBA139225, 2017GXNSFBA198213), Guangxi college and university young teacher fundamental ability promotion project (KY2016YB528), Guangxi Natural Science Foundation 2018JJA120628, National Science Foundation 51705094, the Opening Project of Guangxi Colleges and Universities Key Laboratory of robot & welding: Re-search of microstructure and mechanical properties on CLAM steel after EBW and the Doctor's Research Initial Fund of Gulin

University of Aerospace Technology: Researches on Rapid Die Design methods for Sheet metal, and Study on improving the microstructure stability of the heat resistant steel by modifying the precipitates.

References

[1] KARAFILLIS AP, BOYCE MC. (1995) Tooling and binder design for sheet metal forming processes compen-sating springback error [J]. Journal of Machine Tools and Manufacture,,36:503-526.

[2] Xiang AnYang, FengRuan. A die design method for springback compensation based on displacement adjustment[J]. International Journal of Mechanical Sciences,2011,(53):399-406.

[3] LINGBEEK R, HUETINK J, OHNIMUS S, et al. The development of a finite elements based springback compensation tool for sheet metal products[J]. Journal of Materials Processing Technology, 2005, 169(1):115-125.

[4] R. A. Lingbeek, W. Gan, R. H. Wagoner, et al. Theoretical verification of the displacement adjustment and springforward algorithms for springback compensation[J]. Int J Mater Form (2008) 1:159–168.

[5] Meinders T, Burchitz I A, Bonte M H A, et al. Numerical product design: Springback prediction, compensation and optimization[J]. International Journal of Machine Tools & Manufacture, 2007, 48(5):499-514.

[6] Lingbeek R, Meinders T, Ohnimus S, et al. Springback Compensation: Fundamental Topics and Practical Application[J]. International Journal of Forming Processes, 2006. 403–406.

[7] Lingbeek RA, Hue´ tink J, OhnimusS,Weiher J. Iterative springback compensation of numisheet benchmark #1. In:Smith L,etal.,editors. Proceedings of the 6th NUMISHEET conference, vol. A,Detroit; 2005.p.328–333.

[8] ZHANG LEI. Age Forming Simulation and Springback Prediction of Aluminum alloy 2124[D]. Xian: NorthwesternPolytechnical University, 2011. (in Chinese)

[9] Hang Shawn Cheng, Jian Cao, Z. Cedric Xia. An accelerated springback compensation method[J]. International Journal of Mechanical Sciences, 2007, (49) :267-279.

[10] Naceur H, Guo Y Q, Ben S. (2006) Response surface methodology for design of sheet forming parameters to control springbackeffects[J]. Computers & Structures, 84:1651–1663.

[11] Weiherm J, Rietman B, Kose K, Ohminus S, Petzoldt M. (2004) Controlling springback with compensation strategies[C]. NUMIFORM 2004, Ohio State University. p. 1011–1015.

IOP Conf. Series: Materials Science and Engineering **729** (2020) 012062 doi:10.1088/1757-899X/729/1/012062

Fabrication of Ag-incorporated TiO_2 film on TiNb substrate and evaluation on its antibacterial activity against *E. coli*

Liao Xiaogang[1], Li Gang[1,*], Xiang Xiaoqian[1], Fu Tao[2] and Tang Guo[1]

[1]School of Chemistry and Chemical Engineering, Chongqing University of Technology, Chongqing, 400054, P.R. China

[2]School of Life Science and Technology, Xi'an Jiaotong University, Xi'an, Shaanxi, 710049, P.R. China

[*]Corresponding author's e-mail: 47922138@163.com

Abstract. For the purpose of endowing TiNb alloy with antibacterial function, a two-step strategy was proposed to fabricate Ag-incorporated TiO_2 film on TiNb substrate in this study. X-ray diffraction (XRD), Field-emission scanning electron microscopy (FESEM) and X-ray photoelectron microscopy (XPS) were used to characterize the crystal phases, morphologies and chemical compositions of the samples. The results showed that a layer of anatase TiO_2 film with two kinds of morphologies was formed by hydrothermal treatment of TiNb alloy in NH_4F-H_2O_2 solution. When the as-obtained TiNb-based TiO_2 film was subjected to a further ultraviolet light irradiation in $AgNO_3$ solution, metallic Ag granules with size range from 60 nm to 1.3 μm were uniformly deposited on the surface of TiO_2 particles. *Escherichia coli* (*E. coli*) was selected as a model to estimate the bacteriostatic effect of the Ag-TiO_2 film. The experimental result indicated that Ag-incorporated TiO_2 exhibited excellent antibacterial activity against *E. coli*. It suggests that this two-step method for preparation of Ag-TiO_2 composite film may provide an alternative strategy to reduce bacterial infection for Ti-implants in clinical application.

1. Introduction

Titanium and its alloys not only can be used as surgical instruments and tools, but also can be served as load-bearing materials for replacement and repair of human hard tissue, because of their advantages, such as excellent comprehensive mechanical properties, strong corrosion resistance capability, good biocompatibility, easy processing, and so forth [1, 2]. However, with the increase of clinical use of titanium and its alloys in human body, medical workers become increasingly aware of the seriousness of implant failure caused by bacteria infection after surgery [3, 4]. Thus, it is of great importance to take positive and effective measures to avoid the possible infections originated from biomedical Ti-based implants. Introducing of elements with antibacterial function, such as Ag [5], Cu [6], Zn [7], etc, into the surface of Ti and its alloys is an effective way to prevent them from infection after implantation in body. Among these antibacterial elements, Ag has been regarded as the preferred material for antibacterial treatment of Ti-based implants in recent years, due to its merits of broad-spectrum antimicrobial activity, less cytotoxic to human cells, and less possibility to cause drug resistance, and so on [3, 8].

TiNb, a new kind of near β-type titanium alloy, has much lower and better matching modulus of elasticity with human bone tissue than that of pure titanium. And also it does not contain any toxic and side-effect elements. These merits make TiNb alloy have more attractive application prospects in the biomedical field. Nevertheless, like other titanium alloys, the surface of TiNb was needed to be modified with antimicrobial agents. Hence, to endow the TiNb alloy with antibacterial function, a two-step strategy was employed to fabrication of Ag-incorporated TiO_2 composite film on TiNb substrate in this study. Furthermore, *E. coli* was used to evaluate the bacteriostatic effect of the as-prepared TiNb-based composite film.

2. Experimental section

2.1. Material preparation

Ti20Nb slices with a size of 10 mm×10 mm×1 mm were used as the substrates. Prior to the experiments, the slices were first polished to a mirror image using silicon carbide abrasive paper. Then, the polished slices were ultrasonic cleaned with distilled water and ethanol, followed by drying in ambient air. A two-step strategy was performed to prepare Ag-TiO_2 composite film on TiNb substrate. Namely, TiO_2 film was produced on the surface of TiNb alloy by hydrothermal method and then Ag was introduced into the film through photo-deposition.

For the preparation of TiNb-based TiO_2 film, a typical process was as followed. First, 1.554 g of NH_4F was dissolved into a mixture solution composed of 420 mL of de-ionized water and 280 mL of H_2O_2 (30%, mass percent). After that, a piece of the pre-treated TiNb alloy slice was transferred into a Teflon-lined autoclave with a capacity of 100 mL, in which 60 mL of the above mixed solution was pre-filled. Then, the autoclave was sealed and reacted with a period of 72 h at 180 °C. When the reaction finished, the TiNb alloy slice was taken out and soaked in NaOH aqueous solution with a concentration of 1 mol/L for 36 h. Finally, the slice was rinsed repeatedly with de-ionized water until a neutral solution was achieved.

For the fabrication of Ag-incorporated TiO_2 composite film, a photo-deposition method was employed. That is, five pieces of TiNb alloys experienced the hydrothermal treatment were placed into a crystallizing dish, into which 40 mL of $AgNO_3$ solution with a concentration of 12.5 mmol/L was added. When irradiated for 10 min under a high pressure mercury lamp with a power of 300 W, the slices were taken out, rinsed with de-ionized water and dried in a vacuum oven.

2.2. Characterization and analysis

Crystal phases of the samples were analyzed by Grazing incidence X-ray diffraction (GIXRD) with Cu Kα radiation in the 2θ range of 15-75° (Philip X' pert PRO). A Quanta 600F field emission scanning electron microscope (FESEM, FEI, United States) was utilized to observe the morphologies of the samples. X-ray photoelectron spectroscopy (XPS) was performed using a Kratos XSAM-800 instrument with Al Kα (1486.6 eV) X-ray source. All binding energies were referenced to the C 1*s* located at 284.8 eV.

2.3. Antibacterial activity evaluation

Escherichia coli (*E. Coli*), which is gram-negative, was selected as a model to estimate the bacteriostatic effect of the sample. Prior to the experiments, culture dishes and culture tubes were sterilized with high temperature steam, while the sample and filter membrane was disinfected with 75 wt% of medical alcohol. The initial concentration of *E. Coli* bacteria was first adjusted to 1.5×10^5 CFU/mL by dilution. Next, 10 μL of bacterial solution was dropped onto the surface of the specimen, and then the surface was covered soon by a piece of sterile filter membrane. When the specimen was placed in the dark at room temperature for 3 h, it was then transferred to a 5 mL centrifuge tube and mixed homogeneously with 3 mL of sterile water under shaking. After that, 150 μL diluted solution was dropped onto LB agar solid culture medium and then coated uniformly on the whole surface using a glass spreader. After incubation at 37 °C for 24h, the active bacteria were photographed and counted.

3. Results and discussion

3.1. Crystal phase analysis

XRD was used to analyze the crystalline phases of the samples. For the sample obtained by hydrothermal treatment of TiNb alloy in NH_4F-H_2O_2, several diffraction peaks located at approximately 2θ=25.17°, 36.81°, 37.77°, 38.40°, 47.79°, 53.80°, 54.77°, 62.48°, 68.61°and 69.85° were observed in the XRD pattern (Figure 1a), which could be indentified as the (101), (103), (004), (112), (200), (105), (211), (204), (116) and (220) planes of anatase TiO_2. XRD result demonstrates that anatase TiO_2 film was formed on the surface of TiNb alloy through a hydrothermal treatment. Interestingly, compared with the standard diffraction data of anatase TiO_2 (JCPDS Card 21-1272), all the diffraction peaks shifted slightly toward the direction of low diffraction angle. This may be due to the doping effect of Nb element. As widely known, the ion radii of Nb^{5+} and Ti^{4+} are 0.070 nm and 0.068 nm, respectively. Therefore, Ti^{4+} in the lattice of TiO_2 is easily replaced by Nb^{5+}, which finally resulted in the lattice distortion. While for the sample obtained by hydrothermal treatment and photo-deposition in sequence, except for the original diffraction peaks, another three new diffraction peaks were found in the XRD pattern (Figure 1b). The peaks situated at 2θ=38.12°, 44.28°, and 64.43° could be well indexed as the (111), (200) and (220) planes of metallic Ag (JCPDS Card 04-0783). This suggests that elemental Ag was successfully incorporated into the TiO_2 film.

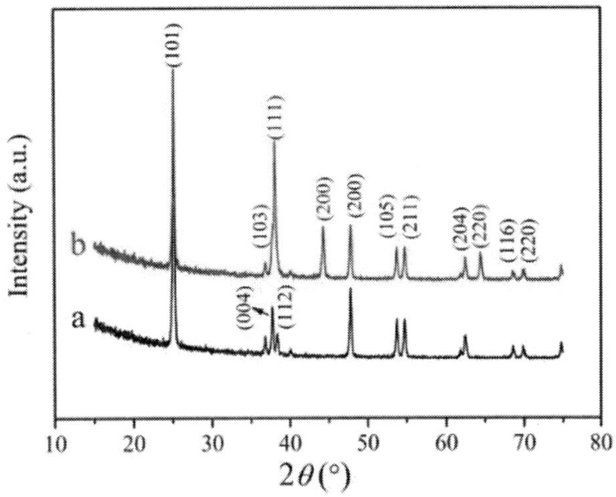

Figure 1. XRD patterns of the samples endured with hydrothermal treatment (a), hydrothermal treatment and photo-deposition in sequence (b).

3.2. Morphology analysis

The surface FESEM image of the TiNb alloy underwent a hydrothermal treatment in NH_4F-H_2O_2 solution was shown Figure 2a. Two kinds of morphologies could be observed from the image. One was the well-defined micron-sized crystal, and the other was the agglomeration-like structure with irregular shape. From the magnified SEM image (Figure 2b), we known that the well-defined micron-sized crystal displayed a truncated octahedron structure, while the agglomeration-like structure was composed of many sub-micron sized crystalline grains with sharp edges. According to the cell structure of the anatase single crystal, it was easy to identify that the one square surface was {001} facet and the four isosceles trapezoidal surfaces were {101} facets for these well-defined TiO_2 micron-sized single crystals. However, for those agglomerations assembled by sub-micron sized crystalline grains, it was hard to determine the accurate crystal plane structure. When the hydrothermal treated sample was subjected to a further photo-deposition in $AgNO_3$ solution, lots of granules with a diameter of 60 nm to 1.3 μm were found to deposit on the surfaces of these two different kinds of TiO_2 particles (Figure 2c). By comparing image of the sample before photo-deposition, these newly formed granules were considered to be Ag. Electron back-scatter diffraction (EBSD) image was carried out to further confirm this. As expected, there was a strong contrast difference between the deposited granules and TiO_2 particles (Figure 2d). Based on the principle of EBSD technique, the particles with brighter colour were none other than Ag.

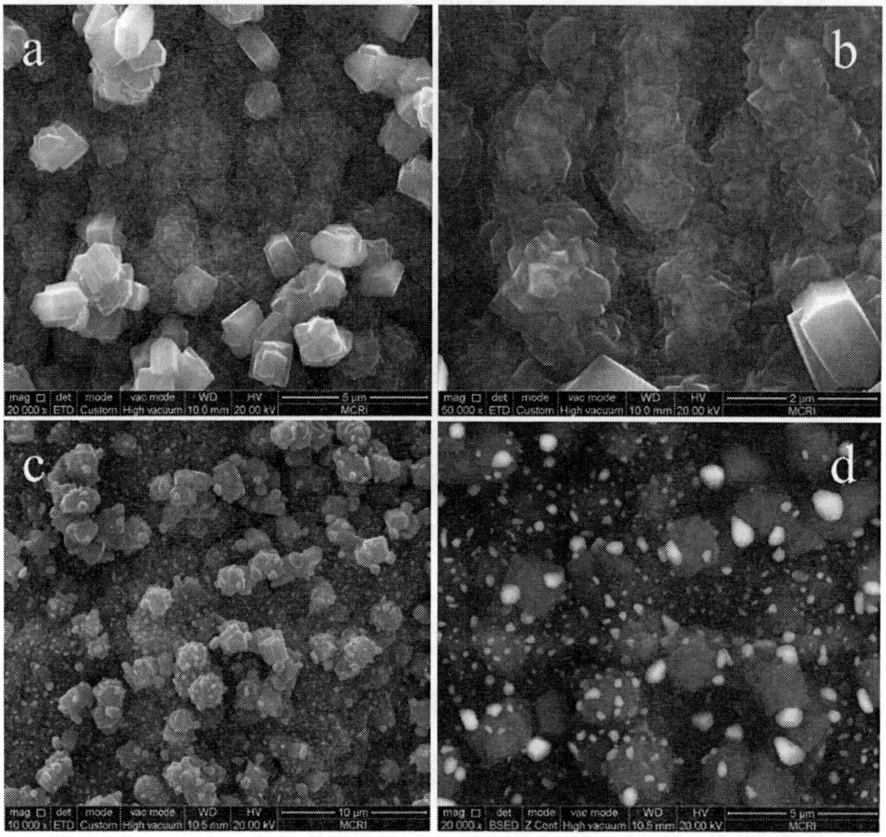

Figure 2. FESEM images of TiNb alloy endured with hydrothermal treatment (a~b) and sequent photo-deposition (c~d).

3.3. Chemical composition analysis

Figure 3a shows the wide-scan XPS spectrum of the TiNb alloy sample which was experienced two consecutive steps of hydrothermal reaction and photo-deposition. As presented, signals from Ti, Nb, O, Ag and C elements were found in the spectrum. The signal of C 1*s* located at 284.8 eV was ascribed to the adventitious hydrocarbon from instrument itself. High resolution XPS spectrum of Ti 2*p* is shown in Figure 3b. Two peaks at binding energies of 459.2 eV and 464.8 eV was attributed to Ti $2p_{3/2}$ and Ti $2p_{1/2}$, respectively [9]. This indicates that Ti element was in the form of Ti^{4+} in the film. High resolution XPS spectrum of O 1*s* shown in Figure 3c was asymmetrical, which could be fitted as two oxygen peaks. The peak at 530.4 eV was due to Ti-O bond of TiO_2, while the peak located at 532.1 eV was attributed to hydroxyl species on the surface [10]. For the high resolution XPS spectrum of Nb 3*d* (Figure 3d), two peaks with binding energy of 207.4 eV and 210.2 eV were observed, which were attributed to Nb $3d_{5/2}$ and Nb $3d_{3/2}$, respectively. This reveals that Nb existed as Nb^{5+} in the sample [11]. For the high resolution XPS spectrum of Ag 3*d* (Figure 3e), two peaks appeared at 368.1 eV and 374.1 eV were ascribed to Ag $3d_{5/2}$ and Ag $3d_{3/2}$, respectively. Additionally, the difference of binding energies of the two peaks was 6.0 eV. This was a criterion for confirming that Ag existed as the metallic state [12]. According to the results of XRD, SEM and XPS, it enabled us to reasonably conclude that Ag-TiO_2 was formed on the TiNb alloy after hydrothermal treatment and photo-deposition in order.

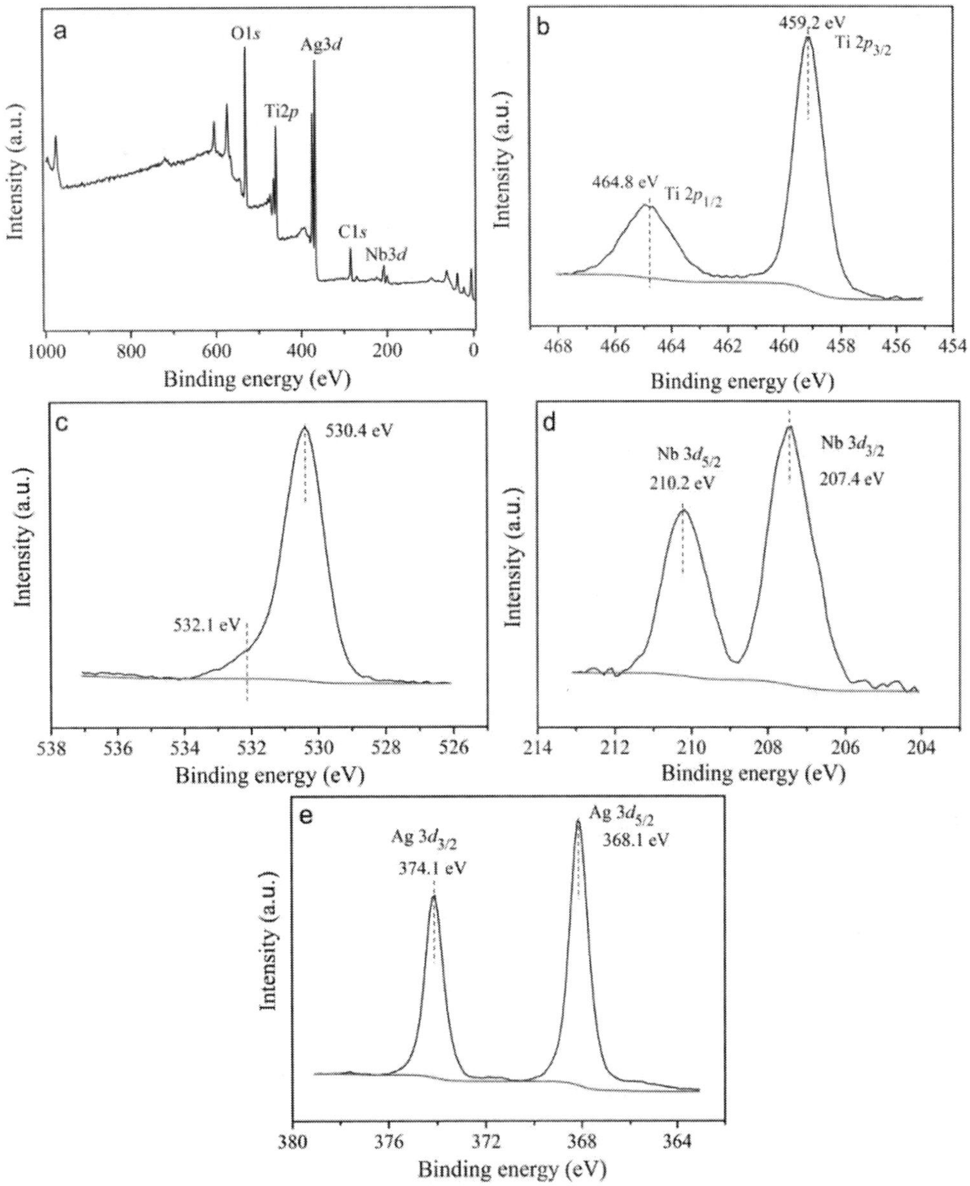

Figure 3. Wide-scan XPS spectrum of the sample (a), and high resolution spectra of Ti 2*p* (b), O 1*s* (c), Nb 3*d* (d) and Ag 3*d* (e).

3.4. Antibacterial activity

Using *E. coli* as a model, the antibacterial activity of the Ag-containing TiO_2 film, obtained by hydrothermal treatment of TiNb alloy and subsequent photo-deposition, was evaluated by coating-plate counting method. Figure 4 shows the antibacterial activity of the samples. For the blank sample (Figure 4a), lots of bacterial colonies were observed in the culture dish. While for the Ag-incorporated sample (Figure 4b), no bacterial colonies were found, suggesting that Ag-TiO_2 composite film had good antibacterial properties in dark condition. The excellent bacterial activity of Ag-TiO_2 was attributed to the release of Ag^+ in the solution [13]. Our study reveals that TiNb-based Ag-TiO_2

composite film prepared by this two-step method may be a potential candidate in clinical application as implants.

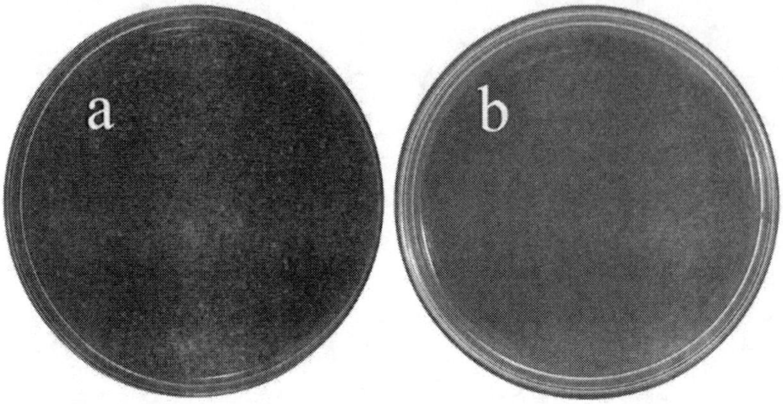

Figure 4. Antibacterial experimental results of the samples: blank control (a) and Ag-TiO_2 (b).

4. Conclusions

In this study, Ag-incorporated anatase TiO_2 film was successfully fabricated on TiNb substrate through a facile two-step strategy. That is, anatase TiO_2 film was first formed by hydrothermal treatment of TiNb alloy in NH_4F-H_2O_2 solution. After irradiation the as-obtained TiO_2 film in $AgNO_3$ solution with ultraviolet light, elemental Ag was then uniformly deposited on the surface of TiO_2 particles. *E. Coli* was used to evaluate antibacterial activity of the Ag-TiO_2 composite film. The result indicated that the incorporation of Ag was effective to restrain the growth of the testing bacteria.

Acknowledgments

This work was financially supported by Scientific and Technological Research Program of Chongqing Municipal Education Commission (KJ1600926), Chongqing Research Program of Basic Research and Frontier Technology (cstc2016jcyjA0438), China Postdoctoral Science Foundation (2016M592637, 2017T100676, Xm2016072) and National Innovation and Entrepreneurship Training Program for College Students (2019CX054).

References

[1] Correa, D.R.N., Kuroda, P.A.B., Lourenço, M.L., Buzalaf, M.A.R., Mendoza, M.E., Archanjo, B.S., Achete, C.A., Rocha, L.A., Grandini, C.R. (2018) Microstructure and selected mechanical properties of aged Ti-15Zr-based alloys for biomedical applications. Materials Science and Engineering: C, 91:762-771.

[2] Mutlu, I. (2016) Synthsis and characterization of Ti-Co alloy foam for biomedical applications. Transactions of Nonferrous Metals Society of China, 26:126-137.

[3] Hou, X.G., Mao, D., Ma, H.Y., Ai, Y.K., Zhao, X.L., Deng, J.H., Li, D.J., Liao, B. (2015) Antibacterial ability of Ag-TiO_2 nanotubes prepared by ion implantation and anodic oxidation. Materials Letters, 161:309-312.

[4] Guo, Z.J., Chen, C., Gao, Q., Li, Y.B., Zhang, L. (2014) Fabrication of silver-incorporated TiO_2 nanotubes and evaluation on its antibacterial activity. Materials Letters, 137: 464-467.

[5] Boutinguiza, M., Fernández-Arias, M., del Val, J., Buxadera-Palomero, J., Rodríguez, D., Lusquiños, F., Gil, F.J., Pou, J. (2018) Synthesis and deposition of silver nanoparticles on cp Ti by laser ablation in open air for antibacterial effect in dental implants. Materials Letters, 231:126-129.

[6] Yao, X.H., Zhang, X.Y., Wu, H.B., Tian, L.H., Ma, Y., Tang, B. (2014) Microstructure and antibacterial properties of Cu-doped TiO_2 coating on titanium by micro-arc oxidation. Applied Surface Science, 292:944-947.

[7] Zhang, X.Y., Wang, H.Z., Li, J.F., He, X.J., Huang, R.Q., Huang, X.B., Tian, L.H., Tang, B. (2016) Corrosion behavior of Zn-incorporated antibacterial TiO_2 porous coating on titanium. Ceramics International, 42:17095-17100.

[8] Wu, C.J., Zhang, G.X., Xia, T., Li, Z.N., Zhao, K., Deng, Z.W., Guo, D.Z., Peng, B. (2015) Bioinspired synthesis of polydopamine/Ag nanocomposite particles with antibacterial activities. Materials Science and Engineering: C, 55:155-165.

[9] Ma, L.J., Zhang, T., Song, R., Guo, L.J. (2018) In-situ Raman study of relation between microstructure and photoactivity of CdS@TiO_2 core-shell nanostructures. International Journal of Hydrogen Energy, 43:13778-13787.

[10] Cai, L., Yao, J.W., Li, J.B., Zhang, Y.F., Wei, Y. (2019) Sonochemical synthesis of BiOI-TiO_2 heterojunction with enhanced visible-light-driven photocatalytic activity. Journal of Alloys and Compounds, 783: 300-309.

[11] Wan, J., Yang, P., Guo, X.L., Zhou, R.X. (2019) Investigation on the structure-activity relationship of Nb_2O_5 promoting CeO_2-CrOx-Nb_2O_5 catalysts for 1,2-dichloroethane elimination. Molecular Catalysis, 470: 75-81.

[12] Lu, J.L., Yang, J.J., Singh, S.C., Zhan, Z.B., Yu, Z., Xin, W., Huang, T., Guo, C.L. (2019) Hierarchical micro/nanostructured TiO_2/Ag substrates based on femtosecond laser structuring: A facile route for enhanced SERS performance and location predictability. Applied Surface Science, 478: 737-743.

[13] Yuan, Z., Liu, P., Hao, Y.S., Ding, Y., Cai, K.Y. (2018) Construction of Ag-incorporated coating on Ti substrates for inhibited bacterial growth and enhanced osteoblast response. Colloids and Surfaces B: Biointerfaces, 171:597-605.

IOP Conf. Series: Materials Science and Engineering **729** (2020) 012063 doi:10.1088/1757-899X/729/1/012063

Label-free electrochemical aptasensor for GPC3 detection based on RGO-CS-Fc/Pt NPs

Yuqiu Li, Wenzhan Li and Guiyin Li*

School of Life and Environmental Sciences, Guilin University of Electronic Technology, Guilin, Guangxi 541004, China

*Corresponding author: liguiyin01@163.com (G.Y. Li)

Abstract: Glypican-3 (GPC3) is a promising biomarker of early detection of hepatocellular carcinoma (HCC). Herein, an label-free electrochemical aptasensor for the detection of GPC3 was developed based on reduced graphene oxide-chitosan-ferrocene /platinum nanoparticles (RGO-CS-Fc/Pt NPs). GPC3 aptamer (GPC3-Apt) was immobilized onto the surface of the modified screen-printed carbon electrode (SPE) by RGO-CS-Fc/Pt NPs. When the GPC3 was added to the platforms of aptasensor, the specific recognition reaction between the GPC3-Apt and GPC3 produced aptamer-antigen complex and arranged on the electrode surface, which made the change of electrochemical signals of Fc obtained by differential pulse voltammetry (DPV). Hence, the calibration graph (peak current vs GPC3 concentration) can be used for quantification of GPC3. The measured peak current and GPC3 concentration have a good linearity in the range of 0.05-10.0 μg/mL, a correlation coefficient of 0.9805 and a low detection limit of 0.018 μg/mL (S/N=3). Moreover, the designed aptasensor showed good performance with excellent selectivity, acceptable reproducibility and stability. Thus, this label-free electrochemical aptasensor has promising applications for the early detection of HCC.

1. Introduction

Hepatocellular carcinoma (HCC) is one of the most common cancers with a high mortality rate, especially in China. Therefore, potential biomarkers are urgently required for the early detection of HCC and the development of novel therapeutic methods [1].Glypican-3 (GPC3) is a membrane-bound heparin sulfate proteoglycan and highly expressed in HCC tissues, while GPC3 is negative in normal liver tissue or benign hepatocellular nodules. Hence, GPC3 has been used as a biomarker for molecular imaging detection and therapeutic intervention in HCC [2, 3]. At present, an enzyme-linked immunosorbent assay (ELISA) or a radioimmunoassay is commonly used to detect the level of GPC3 in serum and tissues of HCC [4, 5]. These methods rely on anti-GPC3 antibodies. However, the clinical application of the GPC3 antibodies is still limited because of its high immunogenicity, high production cost, and low stability [6]. Therefore, it is urgent to find a new type of recognition molecule with high specificity to bind GPC3 and to develop methods for detection of serum GPC3.

Aptamers are low molecular weight (6-30 kDa) single-stranded RNA or DNA oligonucleotides, and are bind to the target molecule with specificity and efficiency. Aptamers can recognizes and binds to the related target in a specific and high binding affinity via an adaptive recognition manner [7]. Therefore, aptamers that specifically binds to HCC cells in tissue samples or cell lines may facilitate the establishing novel approaches for HCC early diagnosis. Zhang et al successfully screened a batch of GPC3 aptamers and established a fluorescence analysis for the diagnosis of HCC [8]. To date, the aptamer-based electrochemical biosensor (aptasensor) had been considered to be the simple and selective tool for the detection of target molecule [9, 10]. Recently, we constructed a label-free electrochemical aptasensor for the detection of AFP using AFP-aptamer as the recognition molecule and thionin/reduced graphene oxide/gold nanoparticles (TH/RGO/Au NPs) as the sensor platform. The aptasensor provided a promising application of aptamer with low cost, high selectivity in clinic [11].

At present, nanomaterials and nanostructures have been widely used in electrochemical biosensing systems to achieve better sensitivity [12]. Many different nanomaterials such as graphene, gold nanoparticles, platinum nanoparticles, carbon nanotubes have been used to improve the performance of aptasensors [13]. Herein, we have prepared the well-dispersed reduced graphene oxide-chitosan-ferrocene (RGO-CS-Fc) nanocomposites and used them to modify the surface of screen-printed carbon electrode (SPE). Pt NPs with good electronic conductivity were electrodeposited onto the surface of modified SPE. Then, GPC3 aptamer (GPC3-Apt) was immobilized onto the surface of the RGO-CS-Fc/Pt NPs/SPE. Thus, a highly sensitive electrochemical aptasensor has been constructed for detection GPC3 protein based on RGO-CS-Fc/Pt NPs. The analytical performance in terms of working curve, linear range, specificity, reproducibility and stability of the proposed GPC3 aptasensor were discussed.

2. Experimental

2.1 Chemicals and reagents

Graphene oxide (GO) was obtained from Timenano Company (Chengdu, China). Ferrocene formic acid, glutaraldehyde, chloroplatinic acid (H_2PtCl_6) and hydrazine hydrate were supported by Sinopharm chemical Reagent Co., Ltd (Beijing, China). N-hydroxysulfosuccinimide sodium salt (Sulfo-NHS), N-(3-dimethylaminopropyl)-N-ethyl-carbodiimide hydrochloride (EDC), chitosan and ascorbic acid were purchased from Sigma (St. Louis, MO, USA). All solutions were prepared with ultrapure water of 18 MΩ·cm purified from Milli-Q purification system (Milli-Pore, Bedford, MA, USA). The GPC3 aptamer (GPC3-Apt) [8] with the following sequences (5'-NH_2-TAACGCTGACCTTAGCTGCATGGCTTTACATGTTCCA-3') was purchased from Sangon biotech Co., Ltd (Shanghai, China).

Electrochemical measurements were carried out on a CHI-660E electrochemical workstation (Shanghai Chenhua Instruments Co., China) using a conventional three screen-printed carbon electrode system. In this system, one carbon paste electrode or modified carbon paste electrode is used as the working electrode, another carbon electrode is used as the auxiliary electrode, and pseudo-reference Ag/AgCl electrode is used as the reference electrode.

2.2 Preparation of the RGO-CS-Fc nanocomposite

RGO-CS-Fc nanocomposite has been prepared using hydrazine hydrate as the reducing agent following our previous method [14]. Briefly, 1.0 mg GO was dispersed to 10 mL ultrapure water and ultrasonic agitation for 2 h to obtain GO solution, then 10 mg ascorbic acid was slowly added and stirred for 12 h to attain RGO. Then, 0.2 mg chitosan was dispersed into 10.0 mL acetic acid solution (1.0%) to form chitosan solution. Next, 2.0 mg ferrocene formic acid, 1.0 mL 10 mM EDC and 4.0 mL 10 mM sulfo-NHS solution were added to the chitosan solution and stirred for 4 h. Thirdly, 10.0 mL 1.0 mg/mL RGO solution was added the above solution and stirred for another 12 h. Lastly, the mixed solution was centrifuged for 30 min at 20000 rpm to remove the supernatant. The precipitate (RGO-CS-Fc nanocomposite) was washed twice with ultrapure water, and re-dispersed in 10.0 mL ultrapure water to form a 1.0 mg/mL solution and stored in refrigerator (4 °C) for future use.

2.3 Fabrication of the electrochemical GPC3 aptasensor

Firstly, the SPE was immersed in 0.5 M H_2SO_4 solution and subjected to electrochemical cyclic voltammetry (CV) scan for 10 cycles. The voltage was set from -0.2 V to 1.0 V and the speed was 100 mV/s. Secondly, Pt NPs were electrodeposited onto the activated SPE in 0.01% H_2PtCl_6 solution under the potential of -0.2 V for 120 s while the magnetic stirring. Thirdly, 8.0 μL 1.0 mg/mL RGO-CS-Fc solution was added dropwise onto the surface of SPE/ Pt NPs, and incubated for 30 minutes at 25 °C.

Fourthly, SPE/ Pt NPs /RGO-CS-Fc was immersed in 2.5% glutaraldehyde solution for 15 minutes and cleaned with PBS solution three times. After drying, 2 μL 0.1μM GPC3-Apt was added dropwise to the surface of SPE/ Pt NPs /RGO-CS-Fc and incubated at 25 °C for 2 h. Finally, 0.5% BSA solution was added to the surface of SPE/ Pt NPs /RGO-CS-Fc/GPC3-Apt to block nonspecific site. Thus, the label-free electrochemical aptasensor (SPE/ Pt NPs /RGO-CS-Fc/GPC3-Apt) has been come into being and stored in refrigerator (4 °C) for future use.

2.4 Detection of GPC3 with the electrochemical aptasensor

The detection of GPC3 was performed by differential pulse voltammetry (DPV) method with SPE/ Pt NPs /RGO-CS-Fc/GPC3-Apt as the working electrode and PBS as the supporting electrolyte. In a typical process, 2 μL of GPC3 (different concentration) standard solution were added dropwise to the surface of SPE/ Pt NPs /RGO-CS-Fc/GPC3-Apt, and incubated for 30 min at 25 °C. Subsequently, the modified electrode (SPE/ Pt NPs /RGO-CS-Fc/GPC3-Apt/GPC3) was immersed vertically in the center of the glass bottle filled with 5.0 mL PBS solution (0.2 M, pH=6.0), and record the electrochemical signals using the DPV method for quantitative analysis. Three times was detected for each sample, and the results were calculated as mean±RSD.

3. Results and discussion

3.1 Analytical principle of electrochemical detection of GPC3 based on Pt NPs /RGO-CS-Fc modified SPE

The analytical principle of the electrochemical detection of GPC3 based on Pt NPs /RGO-CS-Fc modified SPE was shown in Figure 1. In this assay, the bare SPE was activated by H_2SO_4, and electrodeposited of Pt NPs onto the surface of SPE through the isostatic potentiometric deposition. RGO-CS-Fc nanocomposite was immobilized onto the surface of SPE/ Pt NPs by electrostatic adsorption. After that, the GPC3-Apt was captured by means of physical and chemical adsorption on the surface of SPE/ Pt NPs /RGO-CS-Fc. When the GPC3 solution was dropped onto the surface of SPE/ Pt NPs /RGO-CS-Fc/ GPC3-Apt, the specific recognition reaction between the GPC3-Apt and GPC3 produced aptamer-antigen complex and arranged on the electrode surface, which made the change of electrochemical signals of Fc. The value of electrochemical signal obtained by DPV decreased with the increase of the GPC3 concentration. Thus, through measuring the change of electrochemical signals of Fc, GPC3 can be detected with high sensitivity. Herein, RGO-CS-Fc was employed as the nanocarrier to load GPC3-Apt. The decorated Fc acted not only as a bridging molecule to effectively capture GPC3-Apt, but also as the electron transfer mediator to provide the electrochemical signal. Under the synergistic effect of Pt NPs with good electroconductibility and RGO-CS-Fc with high specific surface area and good biocompatibility, the electrochemical signal of Fc could be amplified to indicate the GPC3 level.

3.2 The feasibility of the electrochemical aptasensor for the detection of GPC3

The feasibility of the developed aptasensor for GPC3 detection was illustrated in Figure 2A with the DPV method. Seen from Figure 2A, there was a remarkable current response (103.60 μA) at -0.096 mV without GPC3 (curve a). While a mild current response (73.26 μA) appeared at -0.088 mV with 5.0 μg/mL GPC3 (curve b) and a small of current response (56.17 μA) revealed at -0.093 mV in with 10.0 μg/mL GPC3 (curve c). Hence, the aptasensor showed the analyte-dependent current, which provided a measurement for determining the GPC3 protein.

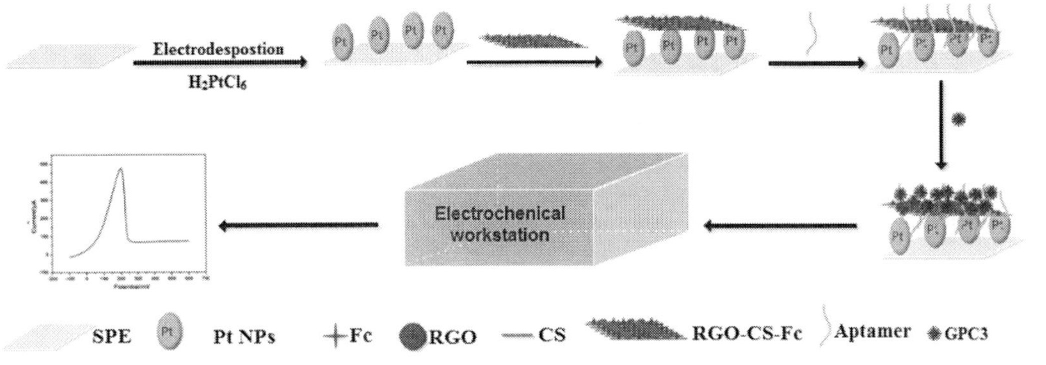

Figure 1 Analytical principle of electrochemical detection of GPC3 based on Pt NPs /RGO-CS-Fc modified SPE

3.3 Analytical performance of the electrochemical aptasensor for the detection of GPC3

Furthermore, the standard GPC3 solutions with different concentrations were tested using the proposed aptasensor with the DPV method, and the results were shown in Figure 2B. The current response of the aptasensor decreased along with the increasing concentration of GPC3. As the concentration of GPC3 increasing, more amount of GPC3 was specifically recognized by GPC3-aptamer and more aptamer-antigen complexs formed, which brought about the gradually decreasing of electrochemical signal response. In the range of 0.05-10.0 μg/mL, the concentration of GPC3 has a good linear relationship with the corresponding current of the aptasensor. The linear equation was Y= -3.4478X+90.0089 with a correlation coefficient of 0.9805. The limit of detection (LOD) was calculated to be 0.018 μg/mL at a signal-to-noise ratio of 3 (S/N = 3).

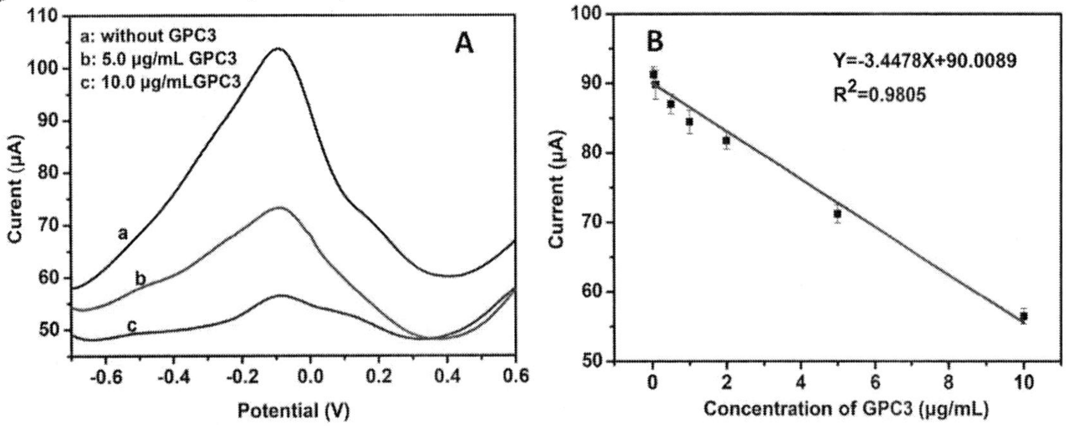

Figure 2 (A) the feasibility of the developed aptasensor for GPC3 detection using the DPV method. (B) Working curve of GPC3 by the developed aptasensor

3.4 Specificity, reproducibility, and stability of the GPC3 aptasensor

Investigating the specificity of the electrochemical aptamer was very important. The aptasensor was incubated with 10.0 μg/mL interfering substances, including alpha-fetoprotein (AFP), bovine serum albumin (BSA), immunoglobulin G (IgG), human serum albumin (HSA), immunoglobulin E (IgE) and recorded the current response by DPV method. The aptasensor's current response value were 95.79 μA for AFP, 92.78 μA for BSA, 91.89 μA for IgG, 94.68 μA for HSA, 90.93 μA for IgE, respectively, while the current response value for CPC3 was 55.72 μA, which revealed the designed aptasensor had favorable specificity.

Moreover, the reproducibility of the aptasensor was evaluated using five freshly prepared modified electrodes to detect the same concentration of GPC3 (10.0 μg/mL). All five electrodes had similar current response value (54.96 μA, 55.72 μA, 53.93 μA, 56.55 μA, 59.86 μA) with the relative standard deviation (RSD) of 4.02%, indicating that the proposed aptasensor had well reproducibility for GPC3 detection.

Additionally, the stability of the aptasensor was investigated by keeping the same electrode at 4 °C and being testing the current value per 4 days from 1 to 16 days. The maximum and the minimum current response were maintained at 94.38% and 86.05% of the original response, which demonstrated the aptasensor had excellent operational stability.

4. Conclusions

In summary, a label-free electrochemical aptasensor based on Pt NPs /RGO-CS-Fc modified SPE has been successfully developed for the measurement of GPC3. The GPC3-Apt was immobilized on the surface of SPE/ Pt NPs /RGO-CS-Fc. In the presence of GPC3, the specific recognition reaction between the GPC3-Apt and GPC3 produced aptamer-antigen complex and arranged on the electrode surface, which made the change of electrochemical signals of Fc record by DPV method. The linear relationship between the current response and the GPC3 concentration was very well over the range from 0.05 to 10.0 μg/mL with a readily achievable LOD of 0.018 μg/mL (S/N = 3). Furthermore, the developed label-free electrochemical aptasensor had favorable specificity, well reproducibility and acceptable stability. Therefore, this ultrasensitive electrochemical aptasensor offers great application promises in detection of plasma GPC3 for HCC early diagnosis.

Acknowledgments

This work was supported by the National Nature Science Foundation of China (No. 81760534), and Fund of the College students' innovative projects of Guangxi province (No. 201810595006).

References

[1] M.S. Grandhi, A.K. Kim, S.M. Ronnekleiv-Kelly, I.R. Kamel, M.A. Ghasebeh, T.M. Pawlik, Hepatocellular carcinoma: From diagnosis to treatment, Surg Oncol 25(2) (2016) 74-85.
[2] F.B. Zhou, W.T. Shang, X.L. Yu, J. Tian, Glypican-3: A promising biomarker for hepatocellular carcinoma diagnosis and treatment, Med Res Rev 38(2) (2018) 741-767.
[3] M. Montalbano, C. Rastellini, J.T. McGuire, J. Prajapati, A. Shirafkan, R. Vento, L. Cicalese, Role of Glypican-3 in the growth, migration and invasion of primary hepatocytes isolated from patients with hepatocellular carcinoma, Cell Oncol (Dordr) 41(2) (2018) 169-184.
[4] H.H. Han, Y.J. Qiu, Y.Y. Shi, W. Wen, X.P. He, L.W. Dong, Y.X. Tan, Y.T. Long, H. Tian, H.Y. Wang, Glypican-3-targeted precision diagnosis of hepatocellular carcinoma on clinical sections with a supramolecular 2D imaging probe, Theranostics 8(12) (2018) 3268-3274.
[5] W. Yuanyuan, Z. Chenjie, L. Jing, Z. Ling, L. Mingsong, X. Bing, Value of detection of serum glypican-3 level in diagnosis and therapeutic effect evaluation of primary hepatocellular carcinoma, J South Med Univ 37(8) (2017) 1060-1065.
[6] Y. Fu, D.J. Urban, R.R. Nani, Y.F. Zhang, N. Li, H. Fu, H. Shah, A.P. Gorka, R. Guha, L. Chen, M.D. Hall, M.J. Schnermann, M. Ho, Glypican-3 Specific Antibody Drug Conjugates Targeting Hepatocellular Carcinoma, Hepatology (2018).
[7] R.B. Ladju, D. Pascut, M.N. Massi, C. Tiribelli, C.H.C. Sukowati, Aptamer: a potential oligonucleotide nanomedicine in the diagnosis and treatment of hepatocellular carcinoma, Oncotarget 9(2) (2018) 2951-2961.
[8] M. Zhao, L. Dong, Z. Liu, S. Yang, W. Wu, J. Lin, In vivo fluorescence imaging of hepatocellular carcinoma using a novel GPC3-specific aptamer probe, Quant Imaging Med Surg 8(2) (2018) 151-160.
[9] M. Yousefi, S. Dehghani, R. Nosrati, H. Zare, M. Evazalipour, J. Mosafer, B.S. Tehrani, A. Pasdar, A. Mokhtarzadeh, M. Ramezani, Aptasensors as a new sensing technology developed for the

detection of MUC1 mucin: A review, Biosensors and Bioelectronics 130 (2019) 1-19.

[10] S. Dehghani, R. Nosrati, M. Yousefi, A. Nezami, F. Soltani, S.M. Taghdisi, K. Abnous, M. Alibolandi, M. Ramezani, Aptamer-based biosensors and nanosensors for the detection of vascular endothelial growth factor (VEGF): A review, Biosensors and Bioelectronics 110 (2018) 23-37.

[11] G. Li, S. Li, Z. Wang, Y. Xue, C. Dong, J. Zeng, Y. Huang, J. Liang, Z. Zhou, Label-free electrochemical aptasensor for detection of alpha-fetoprotein based on AFP-aptamer and thionin/reduced graphene oxide/gold nanoparticles, Analytical biochemistry 547 (2018) 37-44.

[12] S. Yang, F. Zhang, Z. Wang, Q. Liang, A graphene oxide-based label-free electrochemical aptasensor for the detection of alpha-fetoprotein, Biosens Bioelectron 112 (2018) 186-192.

[13] Y. Zhang, J. Xia, F. Zhang, Z. Wang, Q. Liu, A dual-channel homogeneous aptasensor combining colorimetric with electrochemical strategy for thrombin, Biosens Bioelectron 120 (2018) 15-21.

[14] G. Li, J. Zeng, L. Zhao, Z. Wang, C. Dong, J. Liang, Z. Zhou, Y. Huang, Amperometric cholesterol biosensor based on reduction graphene oxide-chitosan-ferrocene/platinum nanoparticles modified screen-printed electrode, J Nanopart Res 21(7) (2019) 162-177.

Influence of Carbothermal Reduction Process on the Microscopic Properties of Lime-Free Roasting Chromite Ore Processing Residue (COPR)

Geng CHENG[1], Xuejun QUAN[1], Huazheng LUO[1], Weiyang BAI[1], Cunfang LU[1]*

[1] School of Chemistry and Chemical Engineering, Chongqing University of Technology, Chongqing, 400050, China

*Corresponding author's e-mail: lucunfang@cqut.edu.cn

Abstract: The lime-free roasting chromite ore processing residue (COPR) was more environmentally friendly than the lime-based COPR. And its components in lime-free COPR should be reused as secondary resource. The process of carbothermal reduction associated with magnetic separation was adopted in this study to reclaim the iron oxide in lime free roasting COPR. The microscopic properties of COPR samples treated by carbonthermal reduction were characterized by scanning electron microscopy (SEM), X-ray diffraction (XRD), etc. The effects of process parameters on iron recovery rate were studied by single-factor experiments. The results indicated that the carbothermal reduction roasting process was beneficial to the destruction of the iron-rich spinel in the original lime-free roasting COPR and the enrichment of iron metal in lime-free roasting COPR. The optimal conditions were obtained as follows: reduction calcination temperature of 1100 °C, reduction time of 2 hours, coke ration of 3.5 times of theoretical content. The main components of the separated fraction were iron particles and some magnesium aluminum spinel.

1. Introduction

The heavy metal contamination is a serious problem to the environment, because the anthropogenic activities from mining, processing and applications of these metals have increased enormously during the past few decades and have become a challenge for life on earth [1-2]. Chromium was widely used in many industrial processes such as electroplating, leather tanning and mineral extraction, and as a common ingredient used in protective coatings, pigments, and stainless steel [3-7].

Chromite ore processing residue (COPR) is a by-product generated from the extraction process of chromium (Cr) from chromite ore. The total Cr concentration in COPR was up to 5%, including the hexavalent chromium Cr(VI) concentration up to 2% on dry weight basis [8–10].

According to the flux used in the roasting process, the chromium salt manufacturing process could be divided into the lime-based roasting process and the lime-free roasting process [11-12]. The lime-based COPR were more toxic than the lime-free COPR, for the hexavalent chromium Cr (VI) in COPR was mainly existed in the form of calcium chromate, $CaCrO_4$. The leaching of Cr(VI) in $CaCrO_4$ was a slow dissolution process for the solubility of Cr(VI) was poor [13-15].

So far there are still millions of metric tons of lime based chromite ore processing residue (COPR) were deposited in urban areas around the world in Europe and the US such as Glasgow, U.K., and Jersey City, as well as at chromite ore processing plants operating in Japan, India, Pakistan and the former Soviet Union [16–17].

Comparing with the lime-based COPR, the lime-free COPR were much more environmentally friendly. The main components of the lime-free roasting COPR were iron oxide, chromite oxide and etc. For lime-free COPR, the iron oxide content was about 40%, which is similar to the iron content of some lean iron ore mines [18]. Magnetized roasting is one of the most effective methods for the treatment of the lean iron ore [19-20]. Therefore, the magnetization roasting method was adopted to in this study to investigate the iron separation of iron metal from lime-free COPR.

2. Materials and methods

2.1. Materials

The lime-free roasting COPR samples used in the experiments were supplied by a chromium salt manufacturer in the southwest of China. Reduction agents used in the carbonthermal experimental were industrial coke. The components of COPR were shown in Table 1.

Table 1. Composition of COPR (%)

Al_2O_3	CaO	Cr_2O_3	Fe_2O_3	MgO	SiO_2
20.90	0.15	14.76	41.87	12.62	1.18

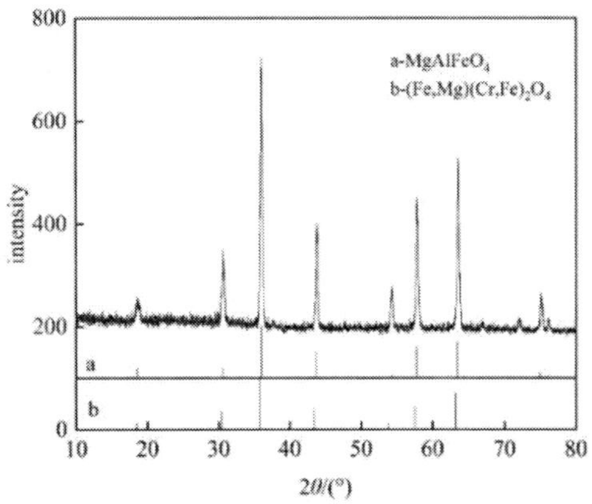

Figure1. XRD results of the original COPR

2.2. Methods and procedure

COPR and the coke powder were uniformly grounded and mixed to place in a crucible after they were dried at 105 °C for 24 hours. The carbon content required for calcination was calculated according to the stoichiometry of equation (1).

$$MgO \cdot Fe2O3 + 3C = MgO + 2Fe + 3CO \quad (1)$$

Mixtures of COPR and coke were roasted in muffle furnace at certain temperatures. When the calcination was completed, the samples were taken out and immersed into water to avoid the oxidation of by oxygen in air during the cooling time. The magnetic separation process was carried out by Ø50 magnetic tube. The magnetic fraction ratios were calculated by the magnetic weight divided by the weight of the original COPR.

3. Results and discussion

3.1. Morphology and microscopic characteristic of lime free COPR

IOP Conf. Series: Materials Science and Engineering **729** (2020) 012064 doi:10.1088/1757-899X/729/1/012064

a) Original SEM of lime-free COPR

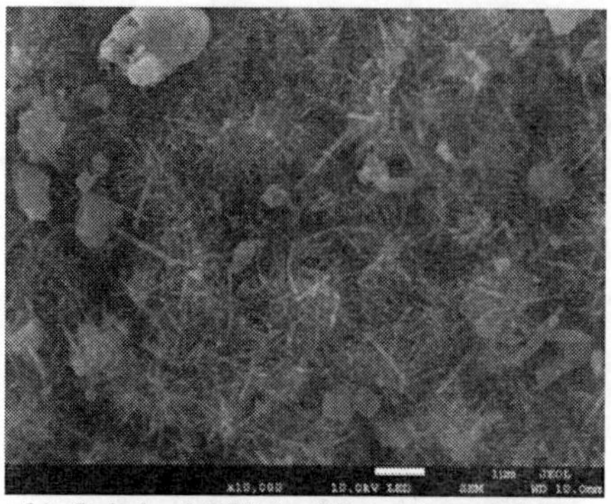

b) Original SEM of lime-free COPR

c) Surface morphology of COPR after calcination

d) Fe distribution on the surface of COPR after calcination

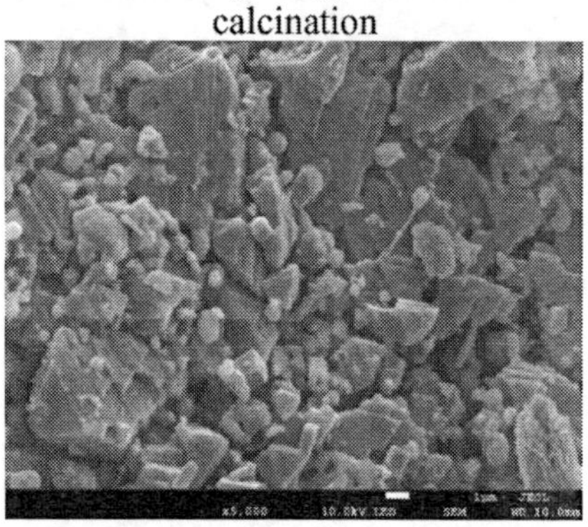

e) Surface morphology of COPR after calcination and magnetic separation

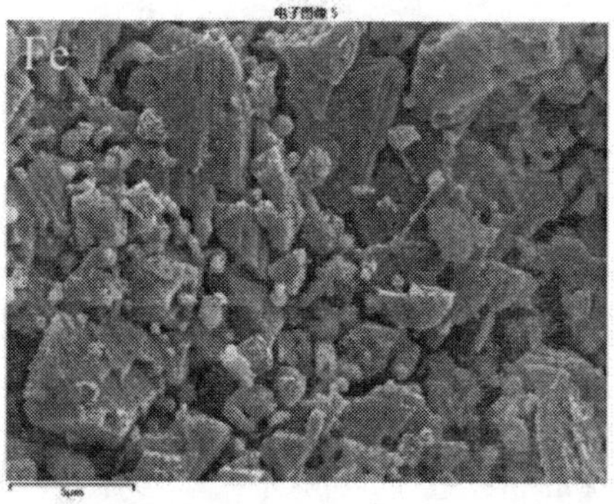

f) Fe distribution on the surface of COPR after calcination and magnetic separation

Figure 2. SEM photographs and elements distributions examination of the COPR

The surface morphology and Fe distribution of COPR was shown in Fig 2. It can be seen from Figure 2 a) and b) that the original lime free COPR had a rough surface. There was a disorderly grass like layer on the top of the residue, and encapsulated are some little tone particles. Figure 2 c) showed that the COPR surface morphology of after carbonthermal reduction became compact and smooth. It seemed that the rough surface of the original COPR was burn-off during the calcination process. Little iron metals could be observed through EDS in with an average particle size less than 1 μm, also crystals of regular shape could be observed Figre 2 d). It may be the crystals of magnesium aluminate spinel according to the XRD results. Figure 2 e) and f) were the magnetic fraction of COPR. There were plenty of magnesium aluminum spinel crystals still in the magnetic part. In addition, the particles sizes of the enrichment iron metal were significantly increased with an average size at about 2-3μm.

3.2. Carbonthermal reduction process results

Single-factor experiments were conducted to study the effects of carbonthermal reduction parameters on the magnetic separation process, including carbon addition amount, reduction temperature and reduction time. During the experiments, the magnetic field intensity of the separation process changed from zero to 500 mT (micro Tesla).

3.2.1. Effect of carbon content on the magnetic separation results

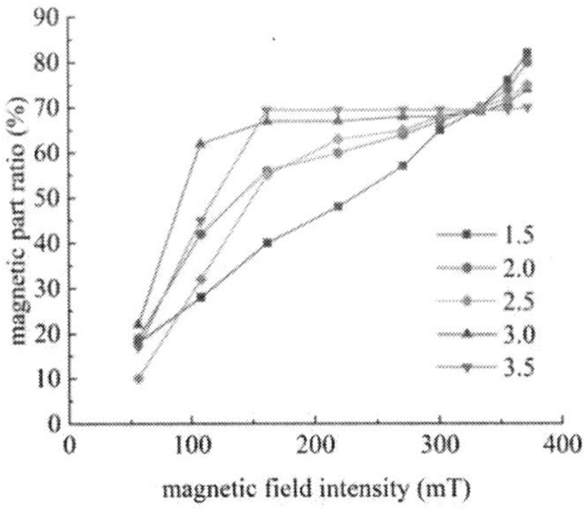

(a) magnetic mass fraction under various coke amounts

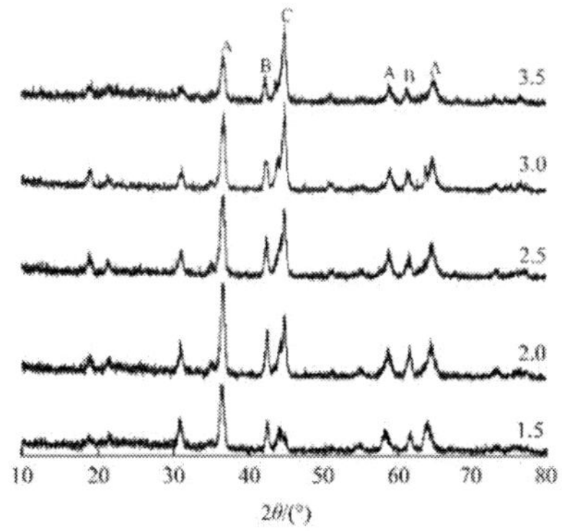

(b) XRD results for roasted COPR, A - $MgFeAlO_4$, B - FeO, C - Fe

Figure3. Effects of coke addition amount on the magnetic part ratio,;

To investigate the effect of coke amount addition on the iron oxide reduction reactions, different times of stoichiometric ratio amount of coke ranging from 1.5 to 3.5 was added roasted at 1100 °C for 2 hours. The results were shown in Fig. 3.

It can be seen that with the increase of the coke addition, the magnetic fraction ratio increased from 10 % to 70 % when the magnetic intensity was about 100-150mT. For samples times of stoichiometric ratio amount of coke were 3.0 and 3.5, the magnetic mass separation reached 70 % under the condition that the magnetic intensity was 150 mT. And the corresponding magnetic intensity was about 350 mT when the times of stoichiometric ratio amount of coke were 1.5, 2.0 and 2.5.

Besides, the increase of coke amount was helpful to enhance the reduction reaction of the iron oxides in COPR. From Fig. 3 b), there were obvious iron metal peak in the XRD pattern. The peak corresponding to iron metal was become higher and sharper with the increase of coke amount.

3.2.2. Effect of reduction temperature on the magnetic separation results

Influence of roasting temperatures from 800 °C to 1200 °C was investigated under condition that times of stoichiometric ratio amount of coke was 3.5 and roasting time was 2 hours. Fig.4 (a) and (b) show the results of magnetic mass fraction and XRD. It can be observed that high temperature was benefit to the improvement of the magnetic mass fraction, with an average ratio of 30 %, when the roasting temperature increased from 800 °C to 1100 °C. However, the magnetic mass fraction was only 40% at the temperature of 1200 °C. It may be caused by the burning of coke at high temperature or the formation of iron compounds, such as carbide.

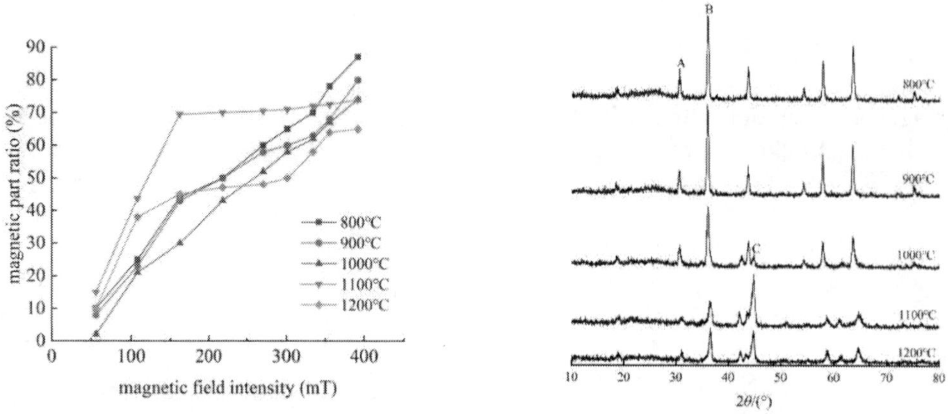

(a) magnetic mass fraction under various magnetic field intensities

(b) XRD results for roasted COPR, A - $MgFeAlO_4$, B – FeO, C - Fe

Figure 4. Effects of temperature on the magnetic part ratio

3.2.3. Effect of reduction time on the magnetic separation results

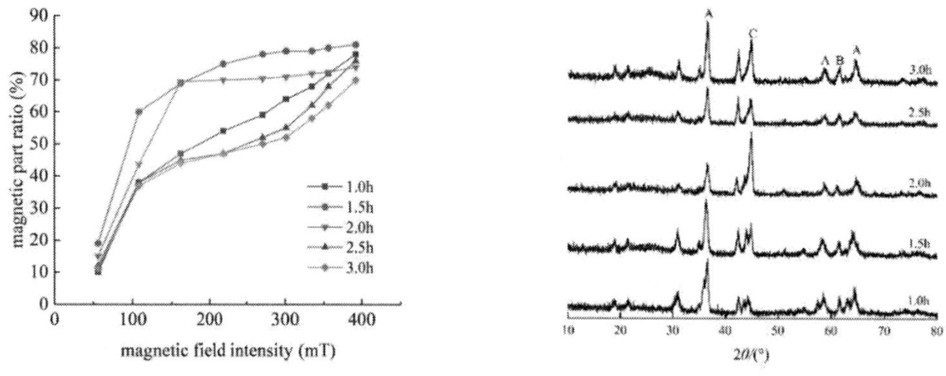

(a) magnetic mass fraction under various magnetic field intensities

(b) XRD results for roasted COPR from 1.0 h to 3.0 h; A - $MgFeAlO_4$, B - FeO, C - Fe

Figure5. Effects of roasting time on the magnetic part ratio

Influence of roasting time ranging from 1.0 hour to 3.0 hours were investigated under conditions that times of stoichiometric ratio amount of coke was 3.5 and roasting temperature was 1100 °C. With the magnetic field intensity increased from 50 mT to 400 mT, the magnetic fraction mass ratio reached about 70-80 % for all the samples. However, it is about 150 mT for samples roasted for 2 hours. And the metal iron of the corresponding sample had the comparable peak in Fig. 3b) with samples roasted for longer times. Thus, the optimal roasting time was 2 hours.

4. Conclusions

Carbothermal reduction method was adopted to treat the lime-free COPR in this study. The experiment results indicated that the carbothermal reduction and magnetic separation process was effective to reclaim the iron metals in lime-free COPR. Under the conditions that of the times of theoretical carbon content, calcination temperature and calcination times were 3.5, 1100 °C and 2 hours, particles of iron metal with a particle size of 1 μm about were enriched. After the magnetic separation treatment, the particle sizes of the enrichment iron were significantly increased with an average size at of about 2-3μm. Due to the size of the iron particles has a large influence on the magnetic properties, the iron particles of the smaller particles not existed in the magnetic part. The magnetic magnesium aluminate spinel in the COPR was not destroyed during the carbothermal reduction process.

Acknowledgement
Supported by the Science and Technology Research Program of Chongqing Municipal Education Commission (Grant No. KJQN201801140).

References

[1] A. Benedetti, Defining soil quality: introduction to round table, in: S. deBertoldi, F. Pinzari(Eds.), COST Actions 831, JointWCs Meeting, Biotechnology of soil: monitoring conservation and remediation, 1998, pp 29–33.
[2] M. Wazne, S.C. Jagupilla, D.H. Moon, S.C. Jagupilla, C. Christodoulatos, M.G. Kim, Assessment of calcium polysulfide for the remediation of hexavalent chromium in chromite ore processing residue (COPR), J. Hazard. Mater. 143 (2007) 620–628.
[3] D.H. Moon, M. Wazne, D. Dermatas, C. Christodoulatos, A.M. Sanchez, D.G. Grubb, M. Chrysochoou, M.G. Kim, Long-term treatment issues with chromite ore processing residue (COPR): Cr6+ reduction and heave, J. Hazard. Mater. 143 (2007) 629–635.
[4] Weng CH, Huang CP, Allen HE, Cheng AHD, Sanders PF. Chromium leaching behaviour in soil derived from chromite ore processing waste. Sci Total Environ 1994; 154:71–86.
[5] James BR. The challenge of remediating chromium-contaminated soil. Environ Sci Technol 1996;30:248A–51A.
[6] Chrysochoou M, Moon DH, Dermatas D, Wazne M, Christodoulatos C, Meng X, et al. Mineralogical analysis of chromite ore processing residue by X-ray powder diffraction. In: Allenman BC, Kelley ME, editors. 8th international in situ and on-site bioremediation symposium, 6–9 June. Baltimore, Maryland: Battelle Press; 2005.
[7] Dermatas D, Chrysochoou M, Moon DH, Pardali S, ChristodoulatosC, Lazarte CA, et al. Mineralogical characterization of chromite ore processing residue at Dundalk Marine Terminal Area 1800. In: Allenman BC, Kelley ME, editors. 8th international in situ and on-site bioremediation symposium, 6–9 June. Baltimore, Maryland: Battelle Press; 2005.
[8] Chrysochoou M, Dermatas D, Moon DH, Christodoulatos C, Wazne M, French C, et al. Investigation of barium treatment of chromite ore processing residue. J ASTM Int 2006;3(6). doi:10.1520/JAI13314.
[9] Moon DH, Chrysochoou M, Dermatas D, Christodoulatos C, Kaouris M, Morris J. Investigation of ettringite formation in chromite ore processing residue. In: Allenman BC, Kelley ME, editors. 8th international in situ and on-site bioremediation symposium, 6–9 June. Baltimore, Maryland: Battelle Press; 2005.
[10] Moon DH, Dermatas D, Chrysochoou M, Shen G. An investigation of the heaving mechanism related to chromite ore processing residue. J ASTM Int 2006;3(6). doi:10.1520/JAI13309.
[11] T. Burke, J. Fagliano, M. Goldoft, R.E. Hazen, R. Iglewicz, T. McKee, Chromite ore processing residue in Hudson County, New Jersey, Environ. Health Perspect. 92 (1991) 131–137.
[12] G. Darrie, Commercial extraction technology and process waste disposal in the manufacture of chromium chemicals from ore, Environ. Geochem. Health 23 (2001) 187–193.
[13] J.G. Farmer, R.P. Thomas, M.C. Graham, J.S. Geelhoed, D.G. Lumsdon, E. Paterson, Chromium speciation and fractionation in ground and surface waters in the vicinity of chromite ore processing residue disposal sites, J. Environ. Monit. 4 (2002) 235–243.
[14] M. Chrysochoou, D. Dermatas, M. Asce, D.G. Grubb, M. Asce, D.H. Moon, C. Christodoulatos, Importance of mineralogy in the environmental characterization and treatment of chromite ore processing residue, J. Geotech. Geoenviron. Eng. 36 (2010) 510–521.
[15] Arndt, U., Batz, M., Bellinghausen, R., Hans-Dieter, B., Heinrich, H., Norbert, L., HansHeinrich, M., Hans-Georg, N., Hans, R., Bernhard, S., 1996. Method for manufacturing alkali chromates from chromite ore. US Patent Application, US 5505918.

[16] YANG Yang. It is very serious of chromium salts waste slag pollution in China [J]. Chemical Industry Management, 2003(1): 4–6. (in Chinese)
[17] ZHU Ji. The two key factors for disposal of chrome residue [J]. Inorganic Chemicals Industry, 2004, 36(5): 1–4. (in Chinese)
[18] Zhang Min. Experimental study on ore dressing in an iron ore mine in Panxi [J]. Mining and Metallurgical Engineering, 2017, 37(1): 57-59.
[19] Zhu Deqing, Zhao Qiang, Qiu Guanzhou. Magnetization roasting-magnetic separation process of Anhui limonite [J]. Journal of University of Science and Technology Beijing 2010,
[20] Chen G,Wang, J J,Wang X H, et al. An investigation on the kinetics of chromium dissolution from Philippine chromite ore at high oxygen pressure in KOH sub-molten salt solution[J]. Hydrometallurgy, 2013, 139(3): 46-53.

IOP Conf. Series: Materials Science and Engineering **729** (2020) 012065 doi:10.1088/1757-899X/729/1/012065

DFT Study on the Hydrogen Evolution Reaction on CoP(110) Surface

Chen Cheng[1], Xingyu Peng, [1]Pingbo Li[1], Peng Ye[1] and Xueqiang Qi[1]*

[1] College of Chemistry and Chemical Engineering, Chongqing University of Technology, Chongqing, 400054, China.

*Corresponding author's e-mail: xqqi@cqut.edu.cn

Abstract. Hydrogen evolution reaction (HER) through water splitting depends on the development of efficient non-noble catalysts, among which the cobalt phosphide has been considered as one of the promising catalyst. Herein, the Gibbs free energy towards HER on CoP(110) surface was studied with density functional theory (DFT) calculation. It can be found that the crystal parameters calculated with both GGA-PBE and GGA-PW91 functionals consist well with the experimental result. The CoP(110) may possess good HER activity since it has good conductivity and appropriate Gibbs free energy (ΔG_H) for hydrogen adsorption.

1. Introduction

The generation of hydrogen fuel from water as an alternative to fossil oils has been considered as a promising green technology with the ongoing energy crisis and climate change[1]. Hydrogen evolution reaction (HER) through electrolytic water splitting has been received increasing worldwide concern. Platinum is considered as state-of-the-art electrocatalysts for HER. However, the scarcity and high price hamper their large-scale applications compared with the traditional natural gas reforming method. Thus it is of high priority to research earth-abundant and cost-effective catalysts as possible alternatives to noble metals[2]. Recently, transition metals (Mo, Co, Ni, Fe, etc.) sulfides and phosphides have been testified that they own good catalytic activity toward HER[3]. Transition metal phosphides (TMPs) which own similar structure with hydrogenase are considered as promising electrocatalysts for the HER, since they possess both good corrosion resistance and good electrical conductivity as an electrode[4]. It has been established that CoP has presented good catalytic activity towards HER, however, most of studies were focused on the experimental researches, but the deep and comprehensive understanding of the relationship between the catalytic activity and the electronic and geometric structures of CoP was rarely reported.

Herein, DFT calculations were performed to study the properties of the CoP(110) surface. It has been found that the surface formation energy of CoP(110) was a little bit positive. However, the Gibbs free energy for H adsorption on CoP(110) was comparative with that on Pt. Thus the CoP(110) may perform good catalytic activity towards HER.

2. Methodology

A preliminary screening of different functionals suggested that both PW91 and PBE functional are suitable to describe accurately the structural and electronic properties of CoP bulk. Thus the DFT calculations were performed with the Perdew-Burke-Ernzerhof (PBE) generalized gradient approximation (GGA) exchange-correlation functional for the properties of CoP(110)[5]. The (110) facet was modeled through cutting from the optimized bulk structure. The k-point meshes were set to 2×2×1 and the core electrons were treated with ultrosoft pseudopotentials[6]. A cutoff energy of 340 eV was used for all calculations. The convergence criterion for the structural optimizations was a maximum force of 0.05 eV/Å and a maximum displacement of 0.002 Å. A vacuum layer of 15 Å thickness was used along the z-direction to eliminate totally the interaction among different surfaces. The surface formation energy (E_s) was calculated according to equation (1):

$$E_s = (E_{slab} - N^* E_{bulk})/2A \quad (1)$$

where E_{slab} is the energy of the optimized CoP surface and E_{bulk} is the energy of CoP bulk. N is the number of CoP bulk contained in CoP surface and A is the surface area of CoP(110).

The Gibbsfree energy (ΔG_H) of H adsorption on CoP(110) surface was calculated according to the method proposed by Nørskov et al[7].

$$\Delta G_H = E_{total} - (E_{surface} + \frac{1}{2} E_{H2}) + \Delta E_{ZPE} - T\Delta S \quad (2)$$

where E_{total} is the total energy for the adsorption state, $E_{surface}$ is the energy of the pure surface, E_{H2} is the energy of hydrogen molecule, ΔE_{ZPE} is the zero-point energy change and ΔS is the difference in entropy between H that is adsorbed and in the gas phase, at 101325 Pa.

3. Results and discussion

3.1. The calculation of bulk CoP

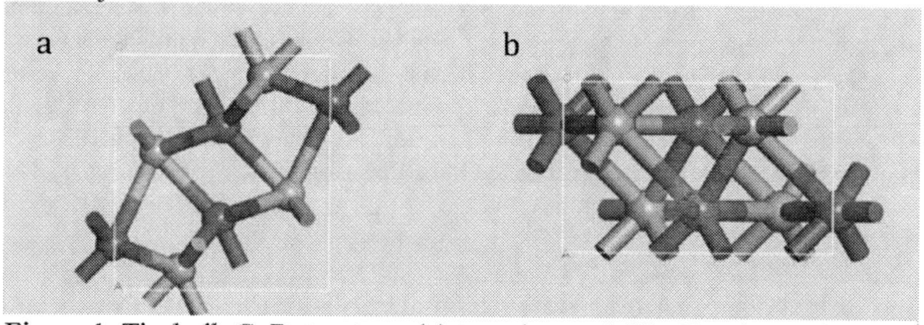

Figure 1. The bulk CoP structure, (a) top-view and (b) side-view. Blue and pink balls represent Co and P atoms, respectively.

Table 1. The computed bond lengths between Co and its six coordinated P atoms with different exchange-correlation functionals, respectively.

Functionals	Bond length of Co-P (Å)					
LDA-CAPZ	2.335	2.316	2.316	2.196	2.195	2.150
GGA-PW91	2.392	2.361	2.359	2.246	2.243	2.187
GGA-PBE	2.387	2.373	2.371	2.234	2.233	2.183
GGA-WC	2.361	2.346	2.344	2.217	2.217	2.171

Table 2. Computed lattice parameters and volumes of CoP with different exchange-correlation functionals

Functionals	Parameters (Å)			Volume (Å^3)
	a	b	c	
LDA-CAPZ	3.261	4.926	5.366	86.200
GGA-PW91	3.327	5.024	5.484	91.664
GGA-PBE	3.325	5.024	5.482	91.580
GGA-WC	3.299	4.980	5.423	89.093
Ref [8]	3.281	5.077	5.587	93.100

As shown in figure 1, the computationally optimized lattice structure of bulk CoP has a monoclinic structure where Co atoms are 6-fold coordinated to P atoms, and the P atoms are also 6-fold coordinated to Co atoms. In order to identify the best method for the description of CoP, four different exchange-correlation functionals have been screened in this paper. As listed in table 1 and table 2, the bond lengths of Co-P vary not too much with different functionals. All the functionals give lattice parameters within about 8% deviation with respect to the available experimental data[8]. However, the GGA-PBE and GGA-PW91[9] functionals perform a lower deviation compared to GGA-WC[10] and LDA-CZAP[11]. Thus the GGA-PBE functional was chosen for the following calculations.

3.2. The calculation of CoP(110)

As shown in figure 2, it can be found that one 3-fold coordinated and two 5-fold coordinated P atoms are exposed to the vacuum, while two 4-fold coordinated and one 5-fold coordinated Co atoms are exposed to the vacuum. What's more, the optimized surface area of CoP(110) is 33.52 Å^2 and the surface formation energy (Es) is only 0.124 eV/Å^2.

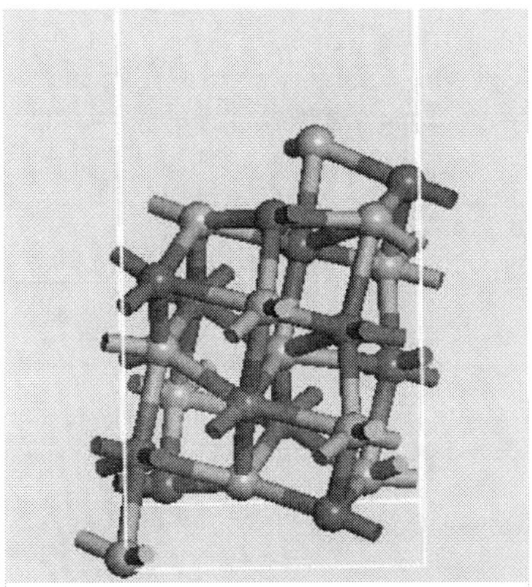

Figure 2. The optimized CoP(110) surface. Blue and pink balls represent Co and P atoms, respectively.

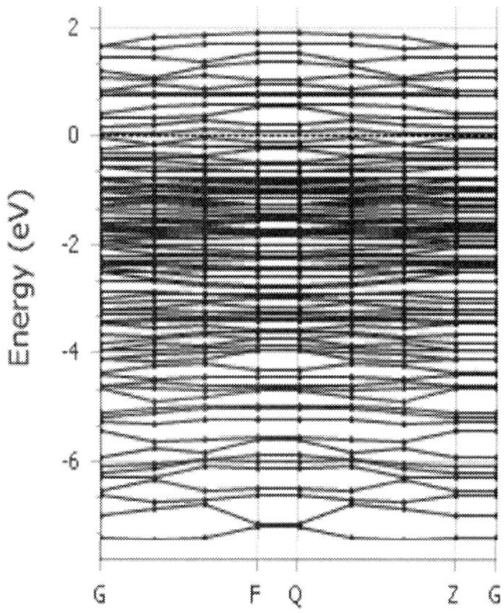

Figure 3. Band structure of CoP(110)

The band structure of CoP(110) was shown in figure 3, there is no significant band gap around the Fermi level, which represents that the CoP(110) surface possess good conductivity.

3.3. Catalytic Activity of CoP(110) toward HER

The Gibbs free energy of hydrogen reaction (ΔG_H) is a reasonable descriptor for HER catalysts. A good HER catalyst should possess a value of ΔG_H close to 0 eV, which keeps a balance between the transfer of hydrogen and the removal of adsorbed hydrogen. As shown in figure 4, the ΔG_H of HER on CoP(110) surface is found to be -0.068 eV, which is a little bit closer to thermoneutral than that of Pt (-0.09 eV), which means that the CoP(110) may perform even better catalytic activity toward HER than that of Pt.

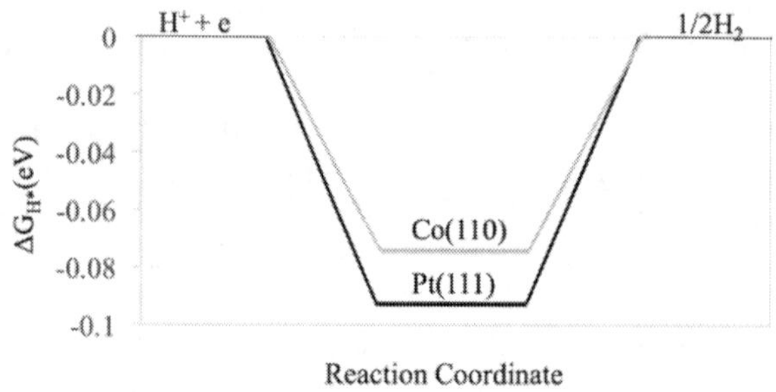

Figure 4. HER free energy diagrams for the three CoP(110) surface.

4. Conclusions

Recent experimental works have identified CoP as a promising catalyst for HER. In order to gain insights into the surface stability and the catalytic activity toward HER, DFT study on CoP(110) surface was performed. The results suggest that the surface formation energy of CoP(110) surface is 0.124 eV/Å^2. The CoP(110) may possess good catalytic activity towards HER, since the CoP(110) performs good conductivity and the Gibbs free energy of HER on CoP(110) is closer to thermoneutral.

Acknowledgements

This work was supported by the Foundation and Frontier Research Project of Chongqing of China (cstc2018jcyjAX0513), Science and Technology Research Program of Chongqing Municipal Education Commission (KJQN201801125).

References

[1] Jiao, Y., Zheng, Y., Jaroniec, M., Qiao, S.Z. (2015) Design of electrocatalysts for oxygen and hydrogen involving energy conversion reactions. Chem. Soc. Rev., 44: 2060-2086.

[2] Wang, J.H., Cui, W., Liu, Q., Xing, Z.C., Asiri, A.M., Sun, X.P. (2016) Recent progress in cobalt-based heterogeneous catalysts for electrochemical water splitting. Adv. Mater., 28: 215-230.

[3] Shi, Y., Zhang, B. (2016) Recent advances in transition metal phosphide nanomaterials: synthesis and applications in hydrogen evolution reaction. Chem. Soc. Rev., 45: 1529-1541.

[4] Pu, Z., Liu, Q., Jiang, P., Asiri, A.M., Obaid, A.Y., Sun, X. (2014) CoP nanosheet arrays supported on a Ti plate: an efficient cathode for electrochemical hydrogen evolution. Chem. Mater., 26: 4326-4329.

[5] Perdew, J.P., Burke, K., Ernzerhof, M. (1996) Generalized gradient approximation made simple. Phys. Rev. Lett., 77: 3865.

[6] Kresse, G., Hafner, J. (1994) Norm-conserving and ultrasoft pseudopotentials for first-row and tran-sition elements. Journal of Physics: Condensed Matter, 6: 8245.

[7] Nørskov, J.K., Rossmeisl, J., Logadottir, A., Lindqvist, L., Kitchin, J.R., Bligaard, T. Origin of the overpotential for oxygen reduction at a fuel-cell cathode. J. Phys. Chem. B, 2004, 108: 17886.

[8] Rundqvist, S. (1962) Phosphides of the B31(MnP) structure type. Acta. Chem. Scand., 16: 287-292.

[9] Perdew, J.P., Wang, Y. (1992) Accurate and simple analytic representation of the electron-gas correlation energy. Phys. Rev. B, 45: 13244-13249.

[10] Wu, Z., Cohen, R.E. (2006) More accurate generalized gradient approximation for solids. Phys. Rev. B, 73: 235116.

[11] Ceperley, D.M., Alder, B.J. (1980) Ground state of the electron gas by a stochastic method. Phys. Rev. Lett., 45: 566-569.

IOP Conf. Series: Materials Science and Engineering **729** (2020) 012066 doi:10.1088/1757-899X/729/1/012066

Quantitative Evaluation Method of Contract Execution Deviation Risk in Power Spot Market

Lin Zhang[1,2*], Yujian Wu[1,2], Jian Liu[1,2], Pengfei Li [3]

[1]NARI Group Co., Ltd. (State Grid Electric Power Research Institute Co., Ltd.), 210003, *China*

[2]Beijing Kedong Electric Power Control System Co., Ltd, *100194, China*

[3]School of economics and management, *North China Electric Power University, 102206, China*

*Corresponding author's e-mail:huadianlipengfei@163.com

Abstract: In the power spot market, the price risk is the biggest among the power purchase schemes under multiple time scales. Equipment failure, scheduling affairs, external policy adjustment and market fluctuation may cause the deviation of contract execution. This paper analyzes the uncertain risk degree of random variables by identifying risk factors. The probability density of the random variable is analyzed, and the conditional risk-value function of the random variable is established. Finally, based on monte carlo stochastic simulation, risk indicators are used to quantitatively evaluate the risk of power purchase and sale.

1. Introduction

Quantitative risk assessment is the numerical calculation of risk indicators. At present, risk indicators mainly include standard deviation, standard deviation coefficient, value at risk (VaR), conditional value at risk (CVaR), etc. Among them, the market subject pays more attention to the risk that the revenue from electricity purchase and sale is less than the expected revenue. The standard deviation and standard deviation coefficient cannot describe the direction of the risk, and the value at risk does not meet the consistency and additivity.

In the current research on the risk of spot power market, in order to promote the optimization of resource allocation, it is necessary to form the electricity price signal that reflects the physical operation characteristics of the power system. Some experts and scholars adopt credit evaluation system for risk early warning management [1]. For the user side, one kind of research focuses on the analysis of over-sensitive influence of prediction accuracy, resource characteristics of the user side and cost on the control of real-time deviating electric quantity of the selling company [2]. In the other category, the user similarity calculation method is adopted to automatically identify the required parameters and achieve the optimal clustering effect [3]. However, the evaluation results of these two methods are highly influenced by subjective factors. Therefore, conditional risk value based on objective probability data is used as risk index. It is more helpful for the selling companies to avoid the risk of contract execution deviation in the electricity spot market and improve their profitability [4-5] to evaluate the electricity purchase and sale risks of the main players in the electricity retail market with the highest profit and the lowest risk [6].

Because the mean and standard deviation of day-ahead clearance price have a strong correlation with day-ahead market load, it can be expressed by a unitary linear regression equation with day-ahead clearance price as the dependent variable and day-ahead market load as the independent variable. Therefore, in this paper, the risk condition function of income is constructed to minimize the risk of electricity purchase while striving for the maximum income.

2. Analysis of Contract Deviation Factors in Power Spot Market

For electricity sales revenue, the capacity electricity price, electricity price and peak, flat and valley time periods of the price package shall be determined by the power selling company and power users through bilateral independent negotiation； 电 The real-time load of power users is a random variable, which can be divided into two parts: planned load and random load error. Among them, the planned load is caused by the change of electricity users' production and living behaviors, which is manifested as the change of the overall trend of the daily load curve, with a long cycle and a large range of changes； random errors of load are caused by unpredictable daily changes, which are shown as small fluctuations under a certain daily load curve with short period and small range. The planned load corresponds to the annual and monthly medium and long-term risks of the load, and the random error of the load corresponds to the day-ahead and real-time short-term risks of the load.

For the electricity purchase cost, the contracted power is decomposed according to the standardized contract curve, and the contract price is independently negotiated by the selling company and the user, both of which are determined variables； day-ahead bid winning load, day-ahead clearance price and real-time clearance price are all random variables. In the day-ago and real-time market, the number of electricity selling companies and electricity users is large and the market share is scattered. Generally, there is no market monopoly. It is approximately believed that a single electricity selling company and electricity users cannot affect the clearance price, that is, the electricity selling company and electricity users are price receivers. Therefore, it is believed that the uncertainty of day-ahead bid-winning load, day-ahead clearance price and real-time clearance price is caused by the non-monopoly position of the selling companies and users in the market and the unpredictability of other market players' trading strategies.

Most customers of the power retail package are enterprises, which have the characteristics of large power consumption, complex electrical equipment, small power demand elasticity, and sensitive to the influence of market policies. Therefore, the deviation of the user's contract execution mainly comes from two aspects: the lack of internal control and the change of the external environment.

Equipment failure may cause deviation in contract execution. After signing the power consumption contract with the selling company, the time period and load of power consumption may be restricted by the contract. In the case of failure or even aging of electrical equipment, the load is likely to change uncontrollably. When the equipment is out of service or the power consumption increases sharply, it is possible to break through the agreed scope of the contract and cause deviation in the execution of the contract.

Scheduling transactions may cause deviations in contract execution. The contract performance of power users is closely related to the production scheduling plan. Scientific and reasonable scheduling plan can effectively balance the power demand curve and make the demand of electricity characteristic load contract. If the plan for electricity use is not reasonable, and the electrical appliances run over schedule, resulting in a large amount of load demand within a short period of time or the total power exceeds the limit of the contract, the contract implementation deviation will be caused.

External policy adjustment may cause deviation of contract execution. As an uncontrollable factor in contract execution, policy adjustment often affects power consumption characteristics of large industrial enterprises. In the period of environmental rectification, production inspection, policy shutdown, etc., the factory cannot carry out production according to the established plan, and most of the load stops working, and the scheduling cannot foresee such events, so it is difficult to make timely adjustments to meet the requirements of the contract.

Market fluctuation may cause deviation of contract. The market fluctuation is the most significant factor affecting the electricity contract. In order to cope with the possible increase in market demand, the general industrial production users will reserve inventory in advance. However, when the market is saturated, work stoppages will occur, and the electricity load will fall to a lower level, which will bring a big deviation to the monthly contract execution.

3. Quantitative Risk Assessment

Conditional value-at-risk is the expectation that the loss is greater than the value-at-risk under a certain degree of confidence. The conditional value-at-risk considers all kinds of situations in which the loss is greater than the value-at-risk. The conditional value at risk calculated by profit is opposite to that calculated by loss, that is, the greater the conditional value at risk, the smaller the risk. The calculation method is as follows.

$$R_{\mathrm{V}} = \left\{ x \mid \int_{-\infty}^{x} f(r)\mathrm{d}r = \alpha \right\} \quad (1)$$

$$R_{\mathrm{CV}} = \frac{1}{1-\alpha}\int_{-\infty}^{R_{\mathrm{V}}} rf(r)\mathrm{d}r \quad (2)$$

Where,

R_{V}—Risk Mechanism (VaR)

x—Quantile of Probability Density Function

r—Revenue

$f(r)$—Probability Density Function of Revenue

α —Confidence Coefficient

R_{CV} —Conditional value-at-risk (CVaR)

Due to the different number of agent users and electricity of different market players, the income level of electricity purchase and sale is also different. In order to compare the purchase and sale risks of different market players, it is necessary to normalize the conditional risk value, that is, the conditional risk value divided by the expected return.

$$R'_{\mathrm{CV}} = \frac{R_{\mathrm{CV}}}{E(R)} = \frac{R_{\mathrm{CV}}}{\int_{-\infty}^{+\infty} rf(r)\mathrm{d}r} \quad (3)$$

Where,

R'_{CV} —Normalized Conditional Value-at-risk

$E(R)$—Expected Revenue

4. Risk Warning

Risk early warning can be carried out in the following ways:

(1) Maximum Load

According to the maximum normal load value of the industry, the maximum power load is agreed according to the current contract agreement and the actual situation of users. When the user appears or exceeds the maximum load, the user shall be informed in time to restrict the power consumption.

(2) Cumulative Power Consumption at Time Point

By balancing the ideal value with the accumulated power consumption at the normal time point in history, the base value of the accumulated power consumption at each time point for a specific user is obtained. When there is a big deviation between the electricity consumption of the user at this point and the base value, the user shall be informed in time to restrict the electricity consumption.

(3) Time Point Warning

Carry out calculation and analysis on the above indexes at the planned time point, and form the routine inspection values, which can be used as an important basis for analyzing the law of user electricity consumption and formulating the next phase contract.

(4) Extreme Data Warning

When a certain index of the user is abnormal, the early-warning mechanism is triggered to calculate the deviation of contract execution for the user immediately, which is an effective means to ensure the safe and stable operation of the system and the execution of the contract.

5. Contract Execution Deviation Control Strategy

According to the above analysis, in order to realize the effective avoidance of risks in the spot power market, the following Suggestions are put forward.

(1) Optimize Contract Decomposition Method, Lock in the Income Level

The core of day-ahead market risk aversion is to reduce the deviation between the contract decomposition curve and the day-ahead bid winning load curve. Generally, service providers are the passive recipients of the market price in the spot electricity market. Therefore, the main measures to avoid the day-ahead market risk are to optimize the contract decomposition mode, refine the contract decomposition curve, and basically lock in the medium and long-term income level of the service provider with a reasonable contract for difference.

For the main selling electricity, the principle of contract decomposition is to use the contract to lock in earnings in advance and reduce the market risk. In the period when the clearance price is small and the fluctuation is small, the contract decomposition quantity is small and the electricity is mainly purchased through the spot market. In the period of high clearance price and high fluctuation, the contract decomposition power is large. Contract decomposition mainly includes the decomposition and combination of generating sets according to the standard decomposition curve provided by the power trading center, as well as the two ways of generating sets, selling companies and users to negotiate the decomposition independently.

1) if the generator set standard decomposition curve provided by the power trading center is used for decomposition and combination, the power seller shall take into account the type of generator set signed the contract, typical output curve, clearing price and other factors at the node. Taking the maximum total market return and the minimum risk of spot market as the objective function, the optimal contract decomposition curve combination is obtained by using the optimization model.

2) If the generator set and the selling company, the customer independently negotiate decomposition plan. Then both parties need to negotiate bilaterally to form a contract decomposition mode acceptable to both parties.

(2) Improve Load Prediction Accuracy, Smooth Energy Storage Load Fluctuation

The core of real-time market risk aversion is to reduce the deviation between real-time load curve and day-ahead winning load curve. Therefore, the main measures for real-time market risk avoidance are as follows: first, to improve the accuracy of service providers' short-term and ultra-short-term load prediction of agent users; second, to collect real-time power consumption information of agent users based on smart electricity meters, Internet of things and other technologies, and to use energy storage technology to smooth load fluctuation and reduce load deviation.

6. Conclusion

According to the power market settlement method, the main risk factors are different in the day-ahead and real-time markets. For day-ahead market, day-ahead clearance price is the main risk factor. A normal distribution probability density function model associated with day-ahead market load space can be used to approximate the day-ahead clearance price. For the real-time market, real-time load and real-time clearing price are the main risk factors. The normal distribution probability density function model associated with the real-time market load space can be used to approximate the real-time clearing price.

Generally, compared with day-ahead market, the load space of real-time market is smaller and the corresponding risk is smaller. Therefore, compared with the accuracy of load lifting prediction and energy storage smoothing load fluctuation, optimizing contract decomposition and reasonably determining contract for difference have more advantages in terms of technology and economy, which will be the main measures to avoid risks in the spot power market.

Acknowledgements
This research was financially supported by research and development of the key technology of electric energy trading for spot market and green energy trading, Number: 52467M180053.

References

[1] Tian Lin, Gan Beiyu, Sun Qian, Ji Tianyao, Sheng Jiansheng, Xu Lixin, Xie Yuting, Jing Zhaoxia, Tan Jie, Hu Xiuzhen, Luo Jinqing. Credit risk management of guangdong electric power spot market [J]. China southern power grid technology,2019,13(06):50-56.
[2] Gao Rui, Guo Hongxia, Yang Ping, Guo Minhua, Ren Zhijun. Real-time market deviation electric quantity control strategy of power selling companies based on user side adjustable resources [J]. Power construction,2019,40(06):114-122.
[3] Jia Chen. Analysis and decision of power purchase behavior of power selling companies under the power market environment [D]. Taiyuan university of technology,2019.
[4] Gao Yi. Study on electricity purchase and sale strategies of e-commerce sellers considering the quota system of renewable energy [D]. Huazhong university of science and technology,2019.
[5] Ma Hui, Chen Hugo, Chen Ye, Liu Wentao, Lin Shaohua, Zhang Xuan, Bai Yang, Luo Gang, Lai Xiaowen, Wang Yang. Mechanism design of spot market of electric power in southern China (starting from guangdong province) [J]. China southern power grid technology,2018,12(12):42-48.
[6] Lu Bo. A study on the operation strategy of power selling companies considering risk and demand response [D]. Southeast university,2018.

Prediction of Miscibility of HTPB Curing System and Solvents by Molecular Dynamics Simulations

Liu Bo, Bu Xiaoyu, Wang Lin

Xi' an Institute of High Technology and Science, Xi' an 710025, P. R. China

liubopaper@163.com

Abstract. In order to study on HTPB curing system and solvent's miscibility, molecular dynamic simulations were performed to analyse curing system's miscibility with water, ethyl alcohol, methylbenzene, acetone and tetrahydrofuran from several aspects like solubility parameter, molecular configuration and density distribution. The result shows that the computed results of the solubility parameter have high consistency with experimental values and the maximum relative error is less than 3%. The solubility parameter differences of acetone, tetrahydrofuran and curing system are relatively small with a homogeneous distribution and a relatively high miscibility. Methylbenzene's miscibility is general while the ethyl alcohol's and water's miscibility are relatively poor.

1. Introduction

HTPB propellants' curing system is mainly comprising of adhesives (hydroxyl-terminated polybutadiene, HTPB), curing agent (toluene diisocyanate, TDI), and crosslinking agent (tris-1-(2-methylaziridynyl) phosphinoxide, MAPO). These three components will have the cure reaction and conjugated reaction, and then form a three-dimensional reticular formation of curing system, which is the major skeleton of HTPB propellant, having an important effect on propellants properties. The curing system of propellant is unstable and easy to occur the oxidation reaction, conjugated reaction, chain scission and other chemical reaction, finally leading to an aging and invalid propellant.

In order to recover the high-energy components of aging propellant, suitable solvent must be selected to process the curing system. There are so many solvents with complicated properties; therefore, it is unrealistic to study simply by doing experiments. With the development of computational chemistry, molecular dynamics simulations have ever become a vital method to study the miscibility of blends [1-3]. The objectives of this study were to develop a computational model based on molecular dynamics method to predict the miscibility of curing system and solvent from three aspects, including solubility parameter, molecular configuration and density distribution, which will provide the important theoretical support for follow-up experiments and craft.

2. Model and Method

2.1. Model construction

In order to look for a suitable solvent of compatible curing system, according to the structural characteristics of HTPB curing system, five common solvents (water, ethyl alcohol, methylbenzene, acetone and tetrahydrofuran) were selected. The molecular models can be set up in accordance with the molecular structures of solvents. The results can be seen as Figure 1.

IOP Conf. Series: Materials Science and Engineering **729** (2020) 012067 doi:10.1088/1757-899X/729/1/012067

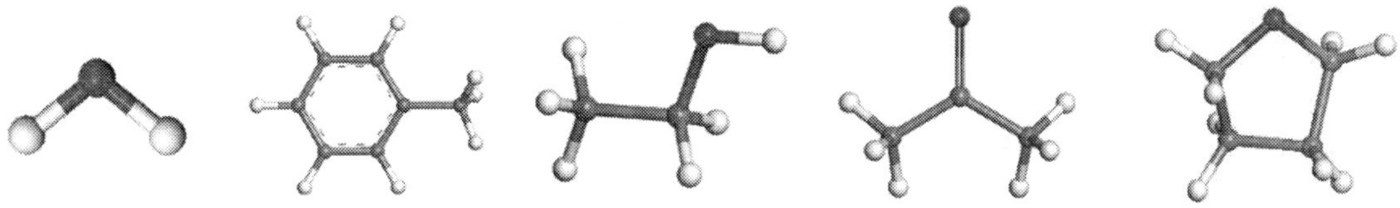

a. Water b. Methylbenzene c. Ethyl alcohol d. Acetone e. Tetrahydrofuran

Figure 1. Molecular structure of five solvents

On the condition of 293K, 1.01×10^5Pa, we firstly optimized the single molecular model and calculated the single point energy. Take the conformation with the lowest energy as the next initial one for simulation. Amorphous models of solvents were established according to different solvents' molecular models and densities. For the purpose of assuring the accuracy and calculated quantity of the calculation, the "size effect" will be minimized and the atomic number of different solvent modules should be less than 1500.The amorphous model construction details of 5 solvents listed above can be observed in the Fig.1.Take H_2O as an example, its amorphous molecular model can be seen from the Fig.2. Models of other solvents will not be listed one by one.

Table 1 Model construction details of five solvents

Solvents	Formula	Numbers of molecular	Numbers of atoms	Density/(g·cm^{-3})
Water	H_2O	400	1200	1.00
Methylbenzene	C_7H_8	100	1500	0.87
Ethyl alcohol	C_2H_6O	150	1350	0.79
Acetone	C_3H_6O	150	1500	0.79
Tetrahydrofuran	C_4H_8O	100	1300	0.89

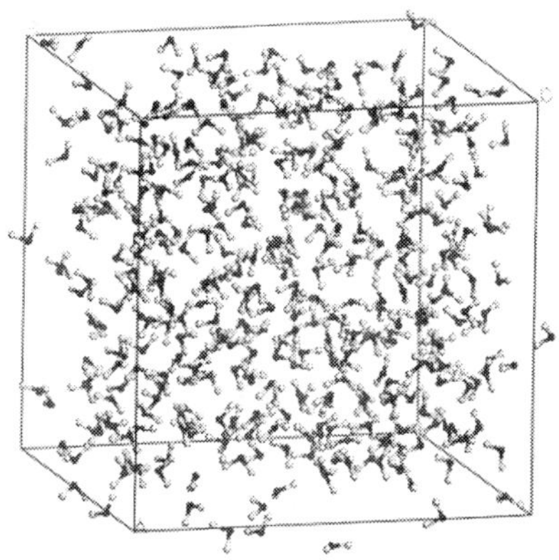

Figure 2. Simulated amorphous molecular model of water

2.2. *Molecular Dynamics Simulation details*

All the calculations were performed using the Materials Studio. In order to minimize unreasonable local structures inside the models, the molecular models were separately operated by annealing treatment. There are five circulations in total, firstly, improving the temperature from 300K to 600K, then to 300K. In every 60k, the molecular dynamics (MD) was operated in isothermal-isobaric (NPT) system, which will provide a reasonable balanced geometrical conformation for the next MD simulations to make the system reach equilibrium as fast as possible. Finally, the MD simulation was operated on NPT system at 350ps, 250ps for structural improvement and 100ps for result analyzing. Analyzing and calculating based on using MD to simulate tracks, and therefore many properties like solubility parameter and density distribution could be obtained.

Select the COMPASS force field in all simulation. The initial velocity of each molecule will be sampled according to Maxwell-Boltzmann random distribution of adopting the Velocity Verlet calculation. The electrostatic interaction and van der Waals force were separately measured with Ewald and Atom Based method. The Anderson thermostat and barostat were used to maintain the temperature and pressure, respectively.

3. Results and Discussion

3.1. Conclusion and Discussion

The judgment of MD system equilibrium mainly depends on temperature and energy, which means the fluctuation changes of temperature and energy are small and the deviation is less than 5%[4,5].Taking the HTPB curing system as an example, we will analyze the temperature and energy changes of 350ps-MD simulations at final 100ps.The changing situation of temperature and energy of MD at final 100ps is illustrated in Figure 3. The fluctuation changes of temperature and energy are both less than 5%, indicating the system has reached its equilibrium and can be used for functional analysis. Similarly, this method was performed to determine the solvents system.

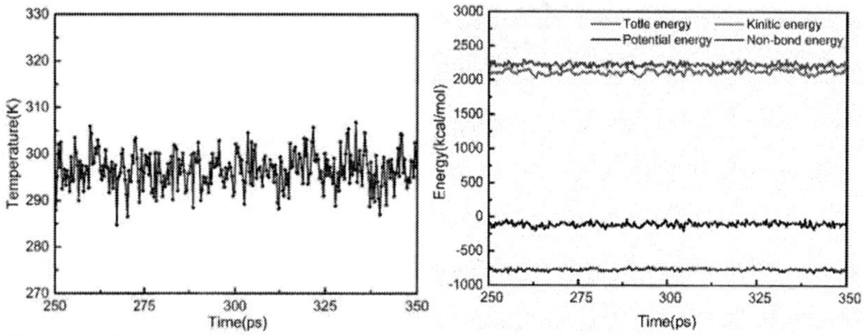

Figure 3. The temperature and energy as a function of MD simulation time

3.2. Solubility Parameter

According to the principle of solubility parameter to determine the solvents' miscibility[6], when the solubility parameter of HTPB curing system is similar to the solvents', the swelling process could be continue The solubility parameter of HTPB curing system at 293K is 19.58$J^{1/2}\cdot cm^{-3/2}$with molecular dynamics method. The results of the solvents' solubility parameter could be seen as Table.2. The solvents' δ_{MD} ,solvents' experiment $\delta_{Ref.}$ [7] as well as the difference $|\Delta\delta_{MD}|$ of HTPB curing system's and solvents', were separately listed.

From Table 2, the 5 solvents' calculated value remains higher agreement with experimental values and the maximum relative error are all less than 3%, indicating the reliability of solvents' solubility parameter calculated by MD. There was great difference between different solvent's and curing system's solubility parameter $|\Delta\delta_{MD}|$,among which acetone's and tetrahydrofuran's $|\Delta\delta_{MD}|$ are relatively small (0.11 $J^{1/2}\cdot cm^{-3/2}$ and 0.22 $J^{1/2}\cdot cm^{-3/2}$ separately), indicating that acetone and tetrahydrofuran have higher miscibility with HTPB curing system; the value of water's $|\Delta\delta_{MD}|$ is maximum, which is 27.81 $J^{1/2}\cdot cm^{-3/2}$, indicating that water has smaller miscibility with HTPB curing system.

Table 2. Solubility Parameter of five solvents （$J^{1/2}cm^{-3/2}$）

Solvents	δ_{MD}	$\delta_{Ref.}$ [7]	Relative error	$\lvert\Delta\delta_{MD}\rvert=\lvert\delta_{MD}-\delta_{\text{固化体系}}\rvert$
Water	47.39	47.8	0.86%	27.81
Methylbenzene	17.87	18.3	2.35%	1.71
Ethyl alcohol	26.31	26.5	0.72%	6.73

Acetone	19.69	20	1.55%	0.11
Tetrahydrofuran	19.36	19.2	0.83%	0.22

3.3. Molecular Configuration

In order to have a further study on different solvents' miscibility with HTPB curing system, five solvents and HTPB curing system were formed to the blend module. The molecular model parameters of 5 blends are shown as Fig.3. In order to make sure the accuracy and calculating amount, the molecular number of HTPB, TDI and MAPO in 5 blends remains the same, the number of atoms in the blends was about 4400 at total.

Table 3. Molecular models parameters of HTPB curing system and solvents

Blends	Number of HTPB	Number of TDI	Number of MAPO	Number of solvents	Number of atoms
HTPB/acetone	5	5	2	90	4424
HTPB/tetrahydrofuran	5	5	2	60	4424
HTPB/methylbenzene	5	5	2	60	4424
HTPB/ethyl alcohol	5	5	2	70	4434
HTPB/water	5	5	2	300	4424

The calculated molecular configuration of 5 blends could be seen as Figure 3. The red regions was in the presence of solvent molecules. After the MD -calculation, the molecular models of 5 blends become different: (1) there are more acetone molecules inside the HTPB curing system with a homogeneous distribution, indicating acetone's high miscibility with HTPB curing system. Meanwhile, the distribution of tetrahydrofuran molecule is homogeneous, indicating tetrahydrofuran's high miscibility with HTPB curing system. (2) Solvent molecules of water, ethyl alcohol and methylbenzene inside the blend model were obviously separated from the curing system, especially most of water molecules mainly distributed out of the curing system and there were no water molecule inside the curing system, indicating that water's worst miscibility HTPB with curing system. As above explanation shown, the order of 5 solvents' miscibility with curing system is Acetone>tetrahydrofuran>methylbenzene>ethyl alcohol>water, which was in accordance with conclusion of solubility parameter.

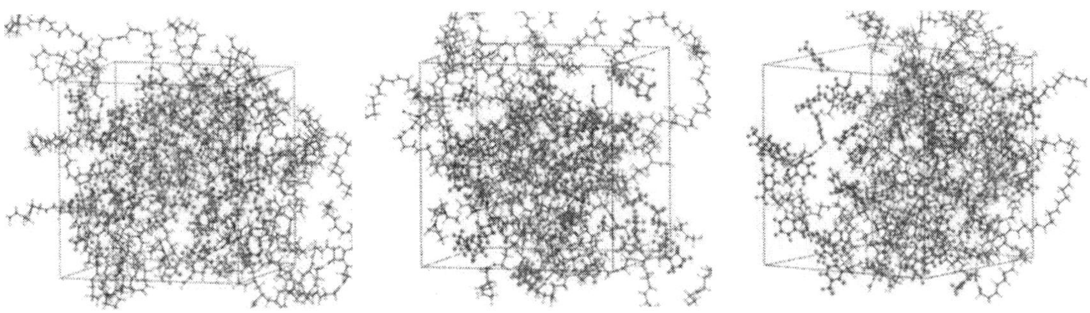

a. HTPB/acetone b. HTPB/tetrahydrofuran c. HTPB/methylbenzene

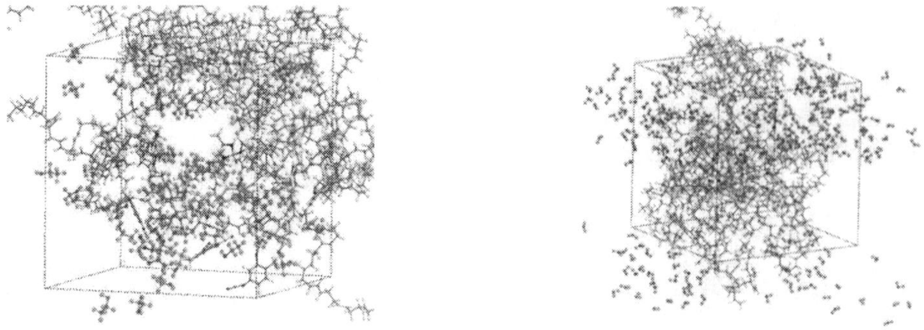

d. HTPB/ethyl alcohol e. HTPB/water

Figure 4. Molecular Configuration of five blends

3.4. Density Distribution

Fig.5 shows the density distribution of five blends. The differences between different blend's density distribution were obviously find, leading to conclusions that: (1) among 5 solvents, density distribution of acetone and tetrahydrofuran was relatively homogeneous with lower density difference, indicating acetone's and tetrahydrofuran's good miscibility with HTPB curing system; (2) other solvents' density distribution in the figure4 showed obvious density distribution with great density differences, indicating these three solvents were not mixed with curing system, which indicated their poor miscibility with curing system. These conclusions were in accordance with the structure of the solubility parameter and molecular configuration mentioned above, providing a further explanation for different solvent's miscibility with curing system.

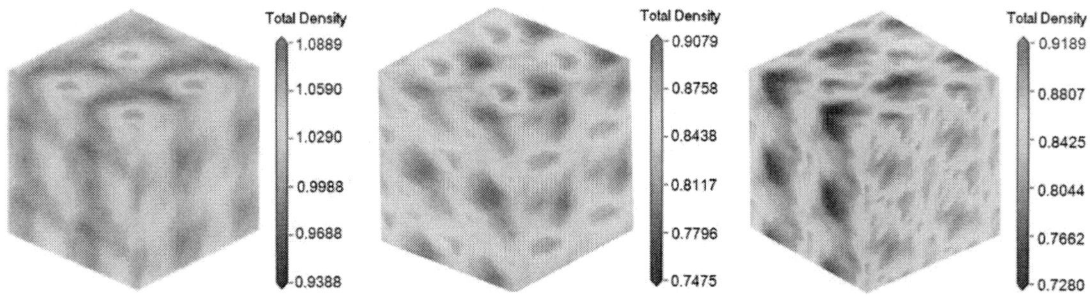

f. HTPB/acetone g. HTPB/tetrahydrofuran h. HTPB/methylbenzene

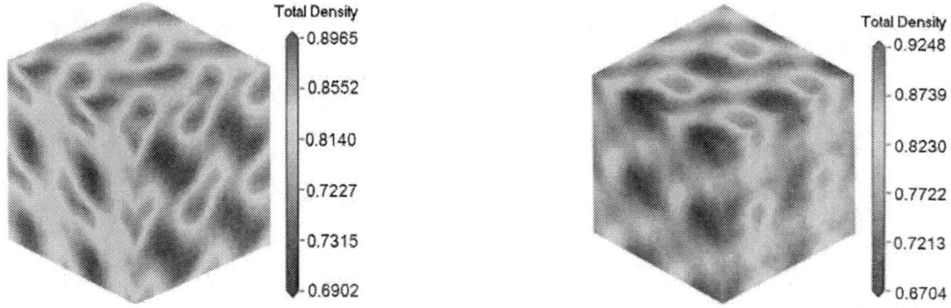

i. HTPB/ethyl alcohol j. HTPB/water

Figure 5. Density distribution of five blends

4. Conclusion

The molecular models of five solvents and HTPB curing system were established with molecular dynamics method, providing important theoretical supports for solvents' miscibility with curing

IOP Conf. Series: Materials Science and Engineering **729** (2020) 012067 doi:10.1088/1757-899X/729/1/012067

system by calculating the solubility parameter, molecular configuration and density distribution of curing system and solvents. The solubility parameter is a reference to determine solvent's miscibility with curing system; after MD calculation on 5 solvents, the solubility parameter differences of acetone, tetrahydrofuran and curing system are small, with a homogeneous distribution and high miscibility. Methylbenzene's miscibility remains general while the miscibility of ethyl alcohol and water is poor.

References

[1] Fu, X. L., Fan, X. Z., Ju, X. H., Qi, X. F., Li, J. Z., & Yu, H. J. 2015 *RSC Advances* **5** 52844
[2] Gupta, J., Nunes, C., Vyas, S., & Jonnalagadda, S. 2011 *The Journal of Physical Chemistry B* **115** 2014
[3] Zhao, Y., Zhang, X., Zhang, W., Xu, H., Xie, W., Du, J., & Liu, Y. 2016. *The Journal of Physical Chemistry A* **120** 765
[4] Taylor, D. E., Strawhecker, K. E., Shanholtz, E. R., Sorescu, D. C., & Sausa, R. C. 2014 *The Journal of Physical Chemistry A* **118** 5083
[5] Yang, H., Ze-Sheng, L., Qian, H. J., Yang, Y. B., Zhang, X. B., & Sun, C. C. 2004. Polymer, **45** 453
[6] Belmares, M., Blanco, M., Goddard, W. A., Ross, R. B., Caldwell, G., Chou, S. H., &Thomas, C. 2004 *Journal of computational chemistry* **25** 1814
[7] Burke, J. 1984. Solubility parameters: theory and application.

IOP Conf. Series: Materials Science and Engineering **729** (2020) 012068 doi:10.1088/1757-899X/729/1/012068

Study on Microwave Method for Jet Fuel's Silver Corrosion Test

Liping TONG*, Jun Ma, Yiwei FEI, Kun Wu

Air Force Logistics College, Xuzhou, Jiangsu, China, 221000

Phone: +86 516-82371636 Fax: +86 516-82371636

Telephone: 13814436290

* Corresponding author: E-Mail: 364511840@qq.com

Abstract: In this paper, the mechanism of microwave-catalyzed jet fuel's silver strip corrosion reaction is analyzed from the perspective of microwave chemistry principle, and the influence of microwave conditions on silver strip corrosion reaction is discussed from the aspects of test time, microwave power, load, etc. The optimal reaction conditions for silver strip corrosion are obtained through orthogonal analysis.

1. Analysis of microwave influence on crystalline materials' surface polarization

The effect of microwave on crystalline materials' surface properties is mainly the effect on crystalline dielectrics' surface properties. Dielectric is a substance with high resistivity and poor conductivity. Charged particles carried by dielectric are confined in molecules of dielectric and cannot move macroscopically [1]. Under the action of an external electric field, charged particles in the dielectric will undergo microscopic displacement, causing molecules to polarize. According to the types of microscopic particles participating in polarization, dielectric molecular polarization[2] can be roughly divided into electron displacement polarization, ion displacement polarization, dipole steering polarization (orientation polarization) and interface polarization. The relaxation time of the first two polarizations is between 10-15-10-16 seconds and 10-12-10-13 seconds, while the time of microwave alternating electric field vibration for one week is between 10-9-10-12 seconds. Therefore, microwave will not cause electron polarization and ion polarization, but only dipole steering polarization and interface polarization. Considering that the influence of microwave on the surface properties of crystalline materials is mainly reflected in the polarization of dipoles and interface polarization of crystalline materials, i.e. the polarization of dipoles and interface polarization of dielectrics play a major role in the energy absorption of materials under microwave.

2. Catalytic mechanism of microwave on corrosion reaction of silver sheet

The microwave accelerates the corrosion of jet fuel's silver flakes mainly from the local accumulation of space charges on the silver flakes surface, resonance absorption of lattice defects on the silver flakes surface and microwave, and reaction diffusion characteristics of covalent bond structure of the corrosion product silver sulfide, etc.

1. The thermal effect of microwave-catalyzed chemical reaction is manifested in that microwave penetrates into the interior of the object at the speed of light and is absorbed by the movement of electrons and ions or the polarization of defect dipoles, which does work on charged particles in the reaction system, causing the temperature of the system to rise, forming an "integral" heating effect

inside and outside the reaction substance and accelerating the reaction.

2. The thermal effect [3] of microwave-catalyzed chemical reactions is manifested in the electromagnetic field effect and microwave characteristics. It simultaneously affects the overall movement of the colliding "molecular pair" and the movement of the two molecules relative to their common center of mass, causing charged particles such as polar molecules, polar atoms and polar ions in the reaction system to undergo transitions between rotational energy levels, changing their activation states, thus causing the internal energy and potential energy of the initial reaction molecules to be increased and finally accelerating the speed of chemical reactions.

3. The interfacial polarization effect of the metallic silver sheet under the action of microwave will cause the surface free charge to move, causing it to generate macro dipole moment, forming local accumulation of surface space charge, making the metallic silver sheet easier to absorb microwave energy and improving the reactivity of the silver sheet with sulfide.

4. In the microwave field, weak bonds and defective parts on the surface of the silver sheet generate local resonance coupling energy transfer with the microwave, exposing the fresh surface and increasing the reaction interface, resulting in uneven energy on the surface of the silver sheet and forming "microwave hot spots" and corrosion reaction active centers. Sulfide corrosion occurs in these parts. Polar molecules in the system have "mutual friction effect" and "molecular stirring effect". The activity of sulfide ion reaction with the silver sheet is improved by destroying the physical adsorption film on the surface of the silver sheet. 5. Silver sulfide generated by the action of active sulfur in silver flakes and jet fuel has covalent bond components, low electrical conductivity, small dielectric loss factor, shallow depth of action with microwave, difficult occurrence of deep corrosion reaction, and easy occurrence of spread corrosion reaction.

3. Analysis of influencing factors on corrosion reaction of silver sheet catalyzed by microwave

3.1. Analysis of influence of microwave test time on corrosion reaction results

Oil samples with corrosion grade 2 are selected for corrosion test by microwave rapid determination method, and the results of different test times are observed, as shown in Table 1.

Table1. Corrosion of Silver Sheets at Different Test Times

Time	corrosion level	phenomenon
1 min	level 1	light brown
2 min	level 1	light brown
3 min	level 2	The light brown color increases and the corrosion area expands.
4 min	level 2	Brown becomes heavier and corrosion area expands.
5 min	level 2	Light brown contact lines and boundary lines begin to appear
6 min	level 2	The pale brown color of the boundary line has increased.
8 min	level 2	The markings on the border turned pale purple.
10 min	level 2	Purple is getting heavier and spreading to one side.
12 min	level 2	Purple becomes heavier and its area expands.
15 min	level 2	The boundary line turns blue, the purple area on one side
17 min	level 2	The boundary line turns blue, the purple area on one side expands, and the grade 2 corrosion feature is obvious.
20 min	level 2	Area of the blue boundary line is enlarged, and purple area is enlarged to the edge of silver sheet, causing serious corrosion.

Judging from the above test results, the sample had little effect on the silver sheet 2 minutes before the test, and began to have effect on the silver sheet about 3 minutes before the test, and the corrosion to the silver sheet gradually increased with the time. The test results of 10 ~ 15 minutes are comparable with those of SH/T0023-1990(2006) standard method.

3.2. Analysis of influence of microwave power on corrosion reaction results

For oil samples with corrosion grade 2, the corrosion test shall be conducted by microwave rapid determination method. The microwave power shall be adjusted to 50mA, 100mA, 150mA and 200mA respectively. The test results are shown in Table 2.

Table 2. Corrosion of Silver Sheet under Different Microwave Power

Power	time	phenomenon
50mA	15min	corrosion is relatively light, and the corrosion phenomenon can be grade 1.
100mA	15min	The corrosion is aggravated, and the corrosion level 2 phenomenon is relatively weak.
150mA	15min	The corrosion is serious, and the corrosion level 2 phenomenon is obvious.
200mA	15min	Increased corrosion

It can be seen that when the microwave power is 50mA, the corrosion phenomenon is relatively light and can be classified as grade 1 corrosion. With the increase of microwave power, the corrosion phenomenon also increases. When the microwave power is 150mA, the grade 2 corrosion phenomenon is more obvious. It can be seen that when the microwave power is close to 150mA, the results of the silver strip corrosion microwave rapid test method are better comparable with the test results of SH/T0023-1990(2006) standard method.

3.3. Analysis of Influence of Microwave Loading on Corrosion Reaction Results

Take 8 groups of oil samples, including 6 Class 2 oil samples, 1 Class 3 oil sample and 1 Class 4 oil sample. The determination results are shown in Table 3.

Table 3. Simultaneous Tests of Multiple Groups of Oil Samples

Oil sample	Phenomenon	Corrosion level
Level 2	Blue, Purple, Black Plaque	Level 2
	Blue, Purple, Black Plaque	Level 2
	Blue, Purple, Black Plaque	Level 2
	Blue, Purple, Black Plaque	Level 2
	Blue, Purple, Black Plaque	Level 2
	Blue, Purple, Black Plaque	Level 2
Level 3	Uniform Black Film	Level 3
Level 4	Uniform depth blackening	Level 4

The results show that it is basically consistent with 2 ~ 3 loads, slightly relieved but not obvious, and can still be classified as grade 2 corrosion. It shows that increasing microwave load has no obvious effect on corrosion results.

3.4. study on the addition of microwave absorbent for jet fuel silver plate corrosion

The following experiments were conducted to determine the amount of microwave absorbent.

Microwave absorbent with mass percentages of 8%, 10%, 12%, 14%, 20% and 30% (corresponding volumes of 4 mL, 5 mL, 6mL, 7 mL, 10ml, 15ml) were taken, respectively, and poured into jet fuel (46 mL, 45 mL, 44 mL, 43 mL, 40ml, 35ml), which were corroded by silver plates to form 50mL mixed solution. After treating the silver in accordance with SH/ t0023-1990 (2000) standard method, it is suspended in the oil sample of the corrosion detection bottle. Put the corrosion test bottle containing the oil sample into a special microwave container and place it in the center of the microwave rapid tester (in order to minimize the error, the special microwave container for each test is placed in the same position), and adjust the microwave power of 150 mA. The reaction time was 10min. After the reaction, the surface corrosion of the silver sheet was observed. The experimental results were shown in table4.

Table4. influence of microwave absorbent on silver plate corrosion

The amount of microwave absorbent added	Heating time	Power	Phenomenon
8%	10min	150mA	pale yellow
10%	10min	150mA	mauve
12%	10min	150mA	purple and wheat yellow
14%	10min	150mA	corner is blue
20%	10min	150mA	one side Angle is blue, the corrosion is serious
30%	10min	150mA	without obvious markings, the whole silver piece is wheat yellow

After the reaction, special microwave container with a large number of solution, the corrosion detection of the volume of a solution with the increase of the amount of microwave absorber is reduced, that with the addition of more weibo absorber, reaction system explosive boiling phenomenon is serious, the evaporation of mixture in the glass container wall rapid condensation, flow to the bottom of the container. It can be seen from the experimental phenomenon that the addition amount of microwave absorbent is not the more the better, but has an optimal value. According to the reaction results, when the addition amount of microwave absorbent was 10%, 12% and 14% respectively, the reaction effect of silver sheet was better. When the addition amount of microwave absorbent was 12%, the corrosion phenomenon of silver sheet was blue, purple and wheat yellow, and the reaction effect was the best.

Then microwave absorbents with mass percentages of 10%, 12% and 14% (corresponding volumes of 5 mL, 6mL and 7 mL) were taken and poured into jet fuel (45 mL, 44 mL and 43 mL) with silver plate corrosion level 2 to form 50mL mixed solution. The silver plate was treated as SH/ t0023-1990 (2000) standard and suspended in the oil sample of the corrosion detection bottle. Put the corrosion test bottle containing the oil sample into a special microwave container and place it in the center of the microwave rapid tester. Adjust the microwave power of 150 mA. Take out the flask and measure the temperature at 4, 6, 7, 8, 10, 12, 14, 16, 18 and 20 min after heating.

Analysis of experimental results: when the addition of microwave absorbent was 10%, 12% and 14%, the temperature rise rate of the reaction system increased with the increase of the addition of microwave absorbent. However, when the addition of microwave absorbent was 12% and 14%, the difference of temperature rise rate was relatively small, and from the reaction of silver plate, when the addition of microwave absorbent was 12%, the corrosion reaction effect of silver plate was better, which was consistent with the color of SH/ t0023-1990 (2000) stipulated silver plate corrosion test method for jet fuel. Therefore, the best value of microwave absorbent is 12%. It was found that the reaction rate of active sulfur in silver plate and jet fuel could be effectively improved by adding microwave absorbent, but the more the amount of microwave absorbent, the better. With the increase of the addition amount of sensitizer, the active sulfur was also diluted, thus reducing the effective collision probability between the active sulfur and the silver sheet, and the corrosion rate of the silver sheet showed a downward trend. Therefore, the addition amount of sensitizer had an optimal value.

4. Study on optimum experimental conditions for silver strip corrosion

Through the research results of influencing factors of microwave-catalyzed silver flake corrosion reaction of jet fuel, it is shown that microwave power, test time and dosage of sensitizer have great influence on the corrosion degree of silver flake.

The experimental results show that the test results of microwave-catalyzed silver strip corrosion at the time of 10 ~ 15 minutes and the microwave power of 150mA are quite comparable with the test results of SH/T0023-1990(2006) standard method, and the increase of microwave load has little effect on the corrosion results.

The effects of microwave power, test time, dosage of sensitizer and volume of reaction system on the corrosion rate of silver flakes are investigated by orthogonal test method. According to each factor, three levels are selected to form a four-factor three-level orthogonal table, and the optimal combination of four factors is obtained, namely, the optimal experimental conditions: the dosage of sensitizer is 5ml, the reaction time is 15min, the microwave power is 150mA, and the volume of the reaction system is 50ml.

5.Conclusion

In this paper, the mechanism of microwave catalytic accelerated silver strip corrosion was analyzed. By investigating the influence factors of microwave test time, microwave power and microwave load, it is concluded that the best test conditions of microwave catalytic silver strip corrosion reaction were 5ml of sensitizer, 15min of reaction time, 150mA of microwave power and 50ml of reaction system volume.

References

[1] He G.Q., Gan W, Qian J.H. Discussion on improving the silver strip corrosion test method for jet fuel. military oil, 2002, 5:36-37.
[2] Tong L.p., Fei Y.W., Liang G.H. Study on corrosion of sulfide in jet fuel and its analytical detection technology. Journal of Air Force Logistics College, 2007, 18 (2): 69- 73.
[3] Zhang Zh.T, Zhong R.Q. Fundamentals of microwave heating technology. Beijing: Electronics Industry Press, 2003. 78.

Low Temperature Catalytic Performance of Manganese and Cerium Complex Oxide Catalyst towards Toluene

Keliang Wang[1], Xiuzhen Liu[2], Shunheng Tu[1], Le Zhang[1], Wen Li[1], Chan Jiang[1], Daiqi Ye[3*]

[1] Institute of Environmental Sciences, Guangdong Polytechnic of Environmental Protection Engineering, Foshan, Guangdong, 528216, China

[2] Department of Economics & Trade, Xinhua College of Sun Yat-sen University, Dongguan, Guangdong, 523133, China

[3] School of Environment and Energy, South China University of Technology, Guangzhou 510006, China

* Corresponding author's e-mail: cedqye@scut.edu.cn

Abstract: A series of $MnO_X(X)–CeO_2$ complex oxide catalysts with various Mn/Ce mole ratios were prepared successfully by sol-gel method and were characterized by BET, XRD, SEM, H_2-TPR and XPS techniques, and the catalytic activity towards toluene were tested. The results showed that $MnO_X(X)–CeO_2$ complex oxide catalysts presented superior low temperature catalytic activity than pure CeO_2 and MnOx, MnOx(0.6)-CeO_2 exhibited the highest catalytic activity with T_{50}=125℃ and T_{99}=197℃ under the condition of GHSV of 10000h^{-1} and toluene concentration of 100ppm. The characterization results also indicated that the catalytic activity was relavant closely with specific surface area, catalyst morphology and valence of manganese oxides. Also, $MnO_X(X)–CeO_2$ formed solid solution with fluorite structure after the introduction of manganese, and produced quantities of manganese species, atom adsorbed oxygen and crystal oxygen, which promoted the adsorption and activation of gaseous oxygen and the mobility and transfer of active oxygen.

1. Introduction

Currently, Volatile Organic Compounds (VOCs) have become the main atmospheric pollutants, and catalysis is one of the most efficient methods to degrade VOCs [1]. As the widely use in the industry, the noble metal catalysts have the deficiencies including the shortage, valuableness and instability, though they have the advantage of high activity [2]. However, because of the high thermostability, inexpensive, and high catalytic activity, the oxide catalyts of transition metal have obtained abundant attentions. Relative to the single component transition metal oxide catalyst, the two or more components of the transition metal composite oxide might express higher catalytic activity and stability. Under certain conditions, the complex oxide catylsts might put up the same activity as the noble metal catalysts [3]. Using the Cu-Mn-Ce-Zr composite oxide as the catalytic combustion towards toluene, the reaction temperature can be lower than 200℃ [4]. With the reason of low-temperature activity and excellent oxygen storage capacity, the Mn-Ce- compound catalysts also gain amount of attractions. Yuan *et al* [5] have discovered that Ce-doped MnO_X-CeO_2/TiO_2 has lower catalytic temperature than the MnO_X/TiO_2, with the removal rate higher than 90% in the range of 300-

400℃. Liu *et al* [6] have expressed that the Ce/Mn compounds have higher oxidation state than the MnOx by the activation of MnOx's lattice oxygen, which might show up a more superior catalytic activity towards formaldehyde. As is known to all, series of factors might increase the low-temperature activity. However, the method to efficiently increase the activity and the way to control the differences of catalyst surface and the composition of species are still unclear. Herein, we apply the sol-gel method to investigate a series of $MnO_X(X)–CeO_2$, exploring the effect of the Mn-/Ce- doped catalysts with different molar ratio towards toluene, characterized by BET, XRD, H_2-TPR and XPS techniques, and the catalytic activity towards toluene were tested.

2. Materials and Methods

2.1. Catalysts
The sol-gel method was used to investigate a series of $MnO_X(X)–CeO_2$ catalysts. In typically, adding a certain amount of Cerium nitrate ($Ce(NO_3)_3 \cdot 6H_2O$) and Manganese acetate ($Mn(CH_3COO)_2 \cdot 4H_2O$) to the ethanol, then adding some Citric acid to form a homogeneous solution , with stiring and heating in water bath at 80℃. After reaction, drying for 12h and then roasting at 550℃ for 5h, different composition of MnO_X-CeO_2 catalysts were obtained. Additionally, the pure MnO_2 catalyst and CeO_2 catalyst were prepared using the same method.

2.2. Catalyst Characterization
XRD measurements were taken on a D8 Advance diffractometer with Ni-filtered Cu Kα radiation (k = 0.15418 nm, 40 kV, 40 mA) in the 2θ range of 5-90°. The XRD phases were identified by comparison with the reference data from Joint Committee on Power Diffraction Standards (JCPDS) data files. The specific surface area, pore volume and pore size of the samples were determined by nitrogen adsorption on an ASAP 2020 M, and the BET method was utilized to calculate the specific surface area (S_{BET}). H_2-TPR experiments were performed with an Auto Chem II. The samples were pretreated in Ar at 300℃ for 30min. After that, the experiments were carried out from 50℃ to 900℃ at a heating rate of 10℃/min in 10% H_2 /Ar (30 mL/min). The H_2 consumption was determined by a TCD detector. XPS was used to analyze the chemical state and surface composition of samples. The binding energy calibration was performed using the C1s peak in the background as the reference energy (284.6 eV).The peak-differentation-imitating was obtained through the XPS Peak 4 program.

2.3. Catalyst activity test
The catalyst test apparatus are composed of the valve system, the reactor and the detector, with using the toluene as target reactant. An air stream bubbling through a saturator filled with liquid toluene carried toluene vapor. The toluene destruction efficiency as a function of reaction temperature was measured in the case of space velocities of 10,000 h^{-1}. All gases were controlled by mass flow meters. Reactants and products were analyzed by gas chromatograph (GC-7890II) with FID detector.

3. Results and Discussion

3.1 Catalyst activity evaluation
The catalytic activities of the as-prepared catalysts towards toluene are shown in Fig.1. The catalyst of single oxide has only low activity towards toluene, the 50% conversion temperature (T_{50}) of CeO_2 and MnO_X are 185℃ and 155℃, with the 99% conversion temperature (T_{99}) are over 300℃ and 225℃, respectively. Compared with the single oxide, the manganese and cerium complex oxide catalysts have more excellent low-temperature activity, with the toluene conversion rate is above 90% when the temperature more than 200℃. As shown in Fig.1, the MnO_X (0.6)-CeO_2 might have the best activity, the T_{50} and T_{99} are 125℃ and 197℃. The catalytic activity orders of MnOx(X)-CeO_2 are from MnOx(0.6)-CeO_2, MnOx(0.4)-CeO_2, MnOx(0.8)-CeO_2, MnOx(0.2)-CeO_2, MnOx to CeO_2 in sequence.

The above results might be attributed to the high oxygen storage capacity of CeO_2, increasing the dispersity of Mn. What's more, CeO_2 and MnOx exist the synergistic effect, which might enhance the catalytic activity towards toluene.

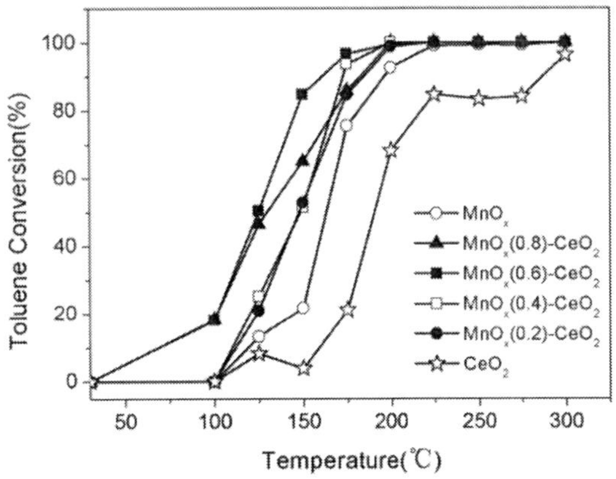

Figure 1. Comparison of catalytic activity towards toluene on MnOx(X)-CeO_2 catalysts

3.2 BET analysis

The specific surface area of the MnOx(X)-CeO_2 are listed in Table.1. The results express that the specific surface area of MnOx and CeO_2 are 11.96m^2/g and 15.12m^2/g respectively, indicating that both the two single oxide have relatively low specific surface area. Compared with the above two samples, the CeO_2 doped manganites (MnOx(X)-CeO_2) get much larger specific surface area, which can be attributed to the lattice imperfection ,and the lattice imperfection causing by the replacement of Mn^{n+} can make the crystalline and the average crystallite size decrease, leading the specific surface area enlarge[7,8]. With the combination of the above 3.1, we can obviously find that moderate Ce-doped and the large specific surface area can develop the catalytic activity towards toluene, revealing that the Ce-doped and the specific surface area are the important factors to the catalytic performance.

Table 1. BET of MnOx(X)-CeO_2 catalysts

MnOx(X)-CeO_2	CeO_2	MnOx(0.2)-CeO_2	MnOx(0.4)-CeO_2	MnOx(0.6)-CeO_2	MnOx(0.8)-CeO_2	MnOx
Surface area (m^2/g)	15.12	54.87	75.27	67.44	73.40	11.96

3.3 XRD

Fig.2 appears the x-ray diffraction patterns of MnOx(X)–CeO_2 catalysts. As shown in Fig.2, the peaks of the obtained spectrograms are fitting with the characteristic peaks of cubic fluorite structure CeO_2 [9]. Meanwhile, the spectrograms of MnOx appear the characteristic peaks of Mn_2O_3 at 23.1°, 33.9° and 55.2°, with appearing the characteristic peaks of MnO_2 at 45.1°and 65.6°, but there is only one characteristic peak of Mn_3O_4 at 38.1°[10],which suggest that most existence form of the MnOx catalysts are present in high-valent manganese oxides through the sol-gel method. Studies [11] have shown that the activity order of the non-load type manganese oxide was $Mn_2O_3 > MnO_2 > MnO$, suggested that numerous of Mn_2O_3 and MnO_2 was benefit to the enhancement of the low-temperature activity.

The XRD patterns of MnOx(0.2)-CeO_2 to MnOx (0.8)-CeO_2 show only the characteristic diffraction peaks of CeO_2 cubic fluorite structure, and there is no characteristic peak of MnOx. It is presumably in the reason of CeO_2 and MnOx forming a fluorspar structure of the Mn-Ce solid solution or the amorphous manganese oxide. In addition, it can be seen from Fig.2 that the main diffraction peak of $MnO_X(X)$-CeO_2 catalyst is similar to CeO_2 with the increase of Mn introduction, indicating that the doping of Mn does not change the main structure of CeO_2, but the diffraction peaks of the composite oxide are broadened and passivated, and the position of the CeO_2 characteristic diffraction peaks is shifted in the high angle direction, indicating that more manganese species enter into the CeO_2 lattice and form a manganese-cerium solid solution, causing the CeO_2 cell shrinkage[12]. This is because smaller manganese ions inset into the CeO_2 lattice instead of the cerium ion, resulting in smaller interplanar spacing and unit cell parameters [13]. Therefore, the addition of Mn is beneficial to the formation of manganese solid solution.

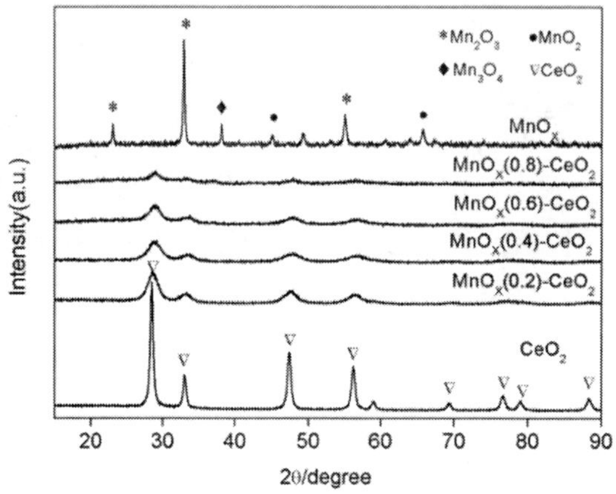

Figure 2. XRD patterns of the $MnO_X(X)$–CeO_2 complex oxide catalysts

3.4 H_2-TPR

Fig.3 expresses the H_2-TPR graph of the $MnO_X(X)$-CeO_2. From the Fig.3, there are two big reduction peaks, and the relative temperatures of reduction peaks are 338℃ and 446℃. According to the literatures [14], the reduction peak of lower temperature means MnO_2 and Mn_2O_3 are reduced to Mn_3O_4, the other peak means Mn_3O_4 are reduced to MnO, which means MnO_X finally changed into MnO. Pure CeO_2 shows two weak reduction peaks at about 380℃ and about 500℃,which are relative with the reduction of Ce^{4+}[15]. It can be seen from the peak area that CeO_2 consumes less H_2 than MnO_X. Compared with the reducibility of MnO_X, the $MnO_X(X)$–CeO_2 also appears two peaks of reduction, and reduction temperature is shifted to the low temperature, indicating that the Mn-doped

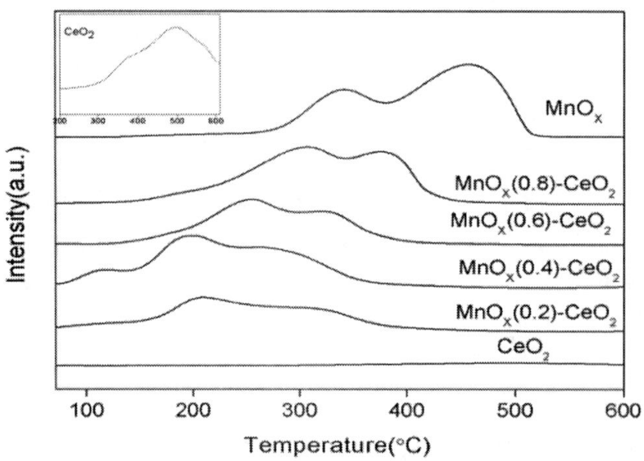

Figure 3. H_2-TPR profiles of $MnO_X(X)$-CeO_2 catalysts

can improve the redox performance of the catalyst. Based on XRD results (3.3), importing moderate amount of Mn into the $MnO_X(X)-CeO_2$ can generate more lattice defects and oxygen vacancies developing the ability of oxygen migration, transformation and release capacity, with enhancing the redox capacity of the catalyst. However, with excessive amount of Mn introduction, some Mn ions accumulated on the catalyst surface rather than entering into the solid solution structure, which might weaken the flow of oxygen performance.

3.5 XPS

Fig.4 expresses the XPS graph of the O1s peak fitting, and the results are shown in Table 2. As seen in Fig.4, pure CeO_2 appears three different binding energy O1s peak, the binding energies appearing in 533eV, 531eV and 528eV are assigned to O_2^-, O^- and O^{2-}, respectively[16]. The O1s peak of lattice oxygen (O^{2-}) appears to migrate toward high energy with being attributed to the process of "O→Mn" electronic transmission [17]. In addition, compared with pure CeO_2, $MnOx(X)-CeO_2$ have stronger ability of oxygen adsorption. The results show that the adsorbed oxygen atom (O^-) is a reactive oxygen species of catalytic combustion activity [18]. From the Table.2, MnO_X (0.6)-CeO_2 has gotten more lattice oxygen leading the strength of the catalytic activity at low temperature, and the result is consistent with the results of H_2-TPR (3.4).

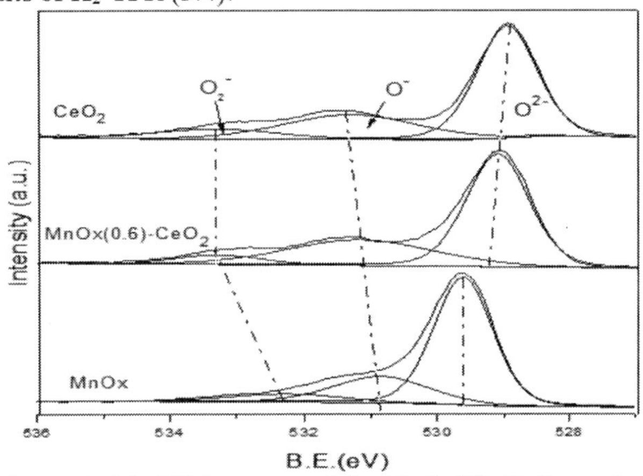

Figure 4. O1s XPS spectra of the $MnO_x(X)-CeO_2$ catalysts

Table 2. The distribution of Mn and oxygen species of Catalysts

Catalysts	Mn distribution (%)		Oxygen distribution (%)		
	Mn^{4+}	Mn^{3+}/ Mn^{2+}	O^{2-}	O^-	O^-_2
CeO_2	--	--	60.1	28.9	11
$MnO_X(0.6)-CeO_2$	75.6	24.4	60.9	32.0	7.1
MnO_X	56.8	43.2	68.2	23.3	8.5

Additionally, Fig.5 expresses the XPS graph of the Mn2p peak fitting. As shown in Fig.5, pure MnO_X appears three different binding energy Mn2p peaks, the binding energies appearing in 641-643eV and 640eV are assigned to $Mn^{4+}$and Mn^{3+}/ Mn^{2+}, respectively. Compared with pure MnO_X, $MnOx(0.6)-CeO_2$ catalyst has more Mn^{4+},and the range of energy level become wider, meaning that high-valence manganese species are more conducive to improve the catalytic reaction of low temperature activity, and the result is consistent with the results of XRD(3.3).

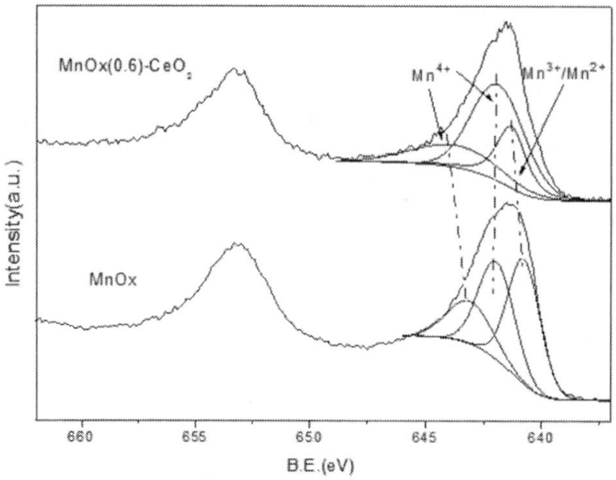

Figure 5. Mn 2p XPS spectra of the $MnO_x(0.6)$-CeO_2 catalysts

3.6 Morphology of MnO_X (0.6)-CeO_2

The morphologies of the as-synthesized samples were examined by SEM. As observed in Fig. 6A and Fig.6B, the pictures express the MnO_X and CeO_2 are mainly composed of nanorods, combined with the results of XRD (Fig.2). What's interesting, from the Fig6C, the morphology of MnO_X (0.6)-CeO_2 was a structure of nano-cluster, because of the surface energy reduction. In order to develop the Ce-doped formation, the high-resolution image of MnO_X (0.6)-CeO_2 (Fig.6D) show that the nano-blocks of Cerium oxide equably distributed on the MnOx, providing a large number of catalytic active sites. In addition, the Mn-Ce solid solution can be further improved the co-catalyst of CeO_2, leading an excellent catalytic activity to low temperature catalytic combustion towards toluene [15].

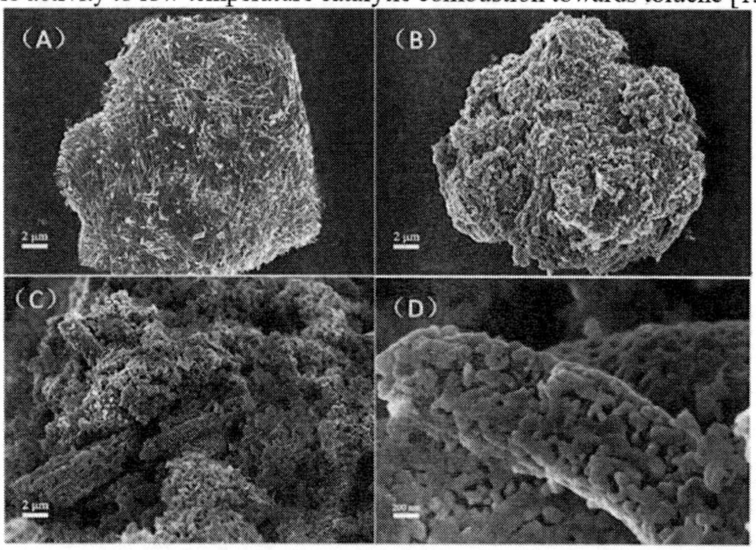

Figure 6. SEM images of the as-synthesized samples MnO_X (A), CeO_2 (B), MnO_X (0.6)-CeO_2 (C) and the high-resolution image of MnO_X (0.6)-CeO_2 (D)

3.7 Stability test of MnO_X (0.6)-CeO_2

In commercial applications, catalytic stability as well as activity, is a very important factor. The stability performance of the MnO_X (0.6)-CeO_2 catalyst is tested under the condition of GHSV of $10000h^{-1}$ and toluene concentration of 100ppm is shown in Fig.7. We can easily find from the figure that the toluene conversion still approach to 90% after 50 hours reaction period at 197℃. It means that

IOP Conf. Series: Materials Science and Engineering **729** (2020) 012069 doi:10.1088/1757-899X/729/1/012069

the as-prepared catalyst has an excellent stability in the catalytic combustion towards toluene.

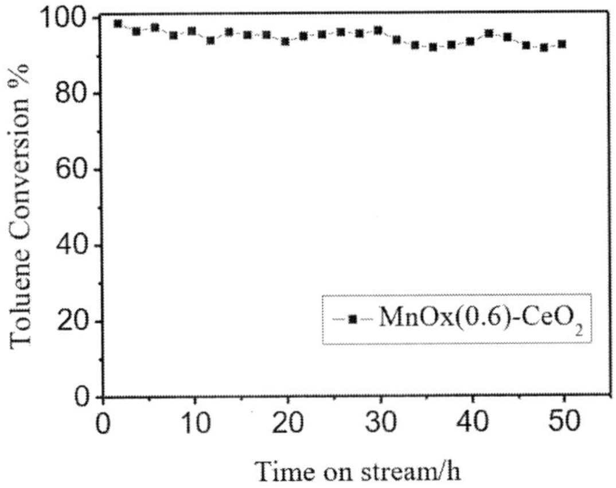

Figure 7. Stability test of MnOx(0.6)-CeO_2 catalyst

4. Conclusions

Among the MnO_X(X)-CeO_2 catalysts prepared, the MnOx(0.6)-CeO_2 exhibits the highest catalytic activity with T_{50}=125℃ and T_{99}=197℃ under the condition of GHSV of 10000h^{-1} and toluene concentration of 100ppm. The low-temperature catalytic activity of MnO_X(X)–CeO_2 towards toluene is assuredly relative to the specific surface area, manganese oxide morphology and manganese oxide valence state. The interaction between MnOx and CeO_2 form the fluorite structure solid solution, a large number of high valence manganese species,O^- and O^{2-}, promoting the gaseous oxygen adsorption on the catalyst surface, migration, activation and conversion of reactive oxygen species. Additionally, the MnOx(0.6)-CeO_2 catalyst has an excellent stability.

References

[1] Li W.B., Chu W.B., Zhuang M., et al, (2004) Catalytic oxidation of toluene on Mn-containing mixed oxides prepared in reverse microemulsions. Catalysis Today, 93:205-209.
[2] Takeguchi T., Aoyama S., Ueda J., et al, (2003) Catalytic combustion of volatile organic compounds on supported precious metal catalysts. Topics in Catalysis, 23:159-162.
[3] Dimitrios D., Theophilos I. (2009) VOCs oxidation over CuO-CeO_2 catalysts prepared by a combustion method. Applied Catalysis B: Environmental, 89:295–302.
[4] Yu F.J., Zhang L.D. (2001) Catalytic oxidation of benzene on Cu-Mn-Ce-Zr-O catalyst. Journal of Beijing University of Chemical Technology, 28(4):67-72.
[5] Qi L.L., Yao J., You H. (2010) Structured catalyst MnOx-CeO_2/TiO_2 for toluene removal. Journal of Harbin Institute of Technology, 42(6): 990-994.
[6] Liu X., Lu J., Qian K., et al. (2009) A comparative study of formaldehyde and carbon monoxide complete oxidation on MnO_X-CeO_2 catalysts. Journal of Rare Earths, 27(3): 418-424.
[7] Azalim S., Franco M., Brahmi R., et al. (2011) Removal of oxygenated volatile organic compounds by catalytic oxidation over Zr-Ce-Mn catalysts. Journal of Hazardous Materials, 188(3): 422-427.
[8] Picasso G., Gutierrez M., Pina M.P., et al. (2007) Preparation and characterization of Ce-Zr and Ce-Mn based oxides for n-hexane combustion: Application to catalytic membrane reactors. Chemical Engineering Journal, 126(3): 119-130.
[9] Dai Y., Wang X.Y., Dai Q.G., et al. (2012) Effect of Ce and La on the structure and activity of MnOx catalyst in catalytic combustion of chlorobenzene. Applied Catalysis B: Environmental, 112(5): 141-149.
[10] Liao Y.N., Fu M.L., Chen L.M., et al. (2013) Catalytic oxidation of toluene over nanorod-

structured Mn-Ce mixed oxides. Catalysis Today, 216(6): 220-228.

[11] Ramesh K., Chen L., Chen F., et al. (2008) Re-investigating the CO oxidation mechanism over unsupported MnO, Mn_2O_3 and MnO_2 catalysts. Catal Today, 131(4): 477-482.

[12] Machida M., Uto M., Kurogi D., Kijima T. (2000) MnOx-CeO_2 binary oxides for catalytic NOx sorption at low temperatures. Sorptive removal of NOx. Chemistry of materials, 12(10): 3158-3164

[13] Wu X.D., Liu S., Weng D., *et al.* (2011) MnOx-CeO_2-Al_2O_3 mixed oxides for soot oxidation: Activity and thermal stability. Journal of Hazardous Materials, 187(3): 283-290.

[14] Arena F., Torre T., Raimondo C., Parmaliana A. (2001) Structure and redox properties of bulk and supported manganese oxide catalysts. Phys. Chem. Chem. Phys, 3(10) 1911-1917.

[15] Arena F., Trunfio G., Negro J., Fazio B., Spadaro L. (2007) Basic Evidence of the Molecular Dispersion of MnCeOx Catalysts Synthesized via a Novel "Redox-Precipitation" Route. Chem. Mater, 19 (9):2269–2276.

[16] Li H.F., Lu G.Z., Dai Q.G., *et al.* (2011) Efficient low-temperature catalytic combustion of trichloroethylene over flower-like mesoporous Mn-doped CeO_2 microspheres. Applied Catalysis B: Environmental, 102(4): 475-483.

[17] Zhang M., Fu M.L., Ye D.Q. (2011) Characteristic of Surface Oxygen Species and Catalytic Property on MnOx-CeO_2 for Soot Combustion. Journal of the Chinese Society of Rare Earths, 29(3): 303-309.

[18] Dai Q.G., Huang H., Zhu Y., *et al.* (2012) Catalysis oxidation of 1, 2-dichloroethane and ethyl acetate over ceria nanocrystals with well-defined crystal planes. Applied Catalysis B: Environmental, 117-118(1): 360-368.

IOP Conf. Series: Materials Science and Engineering **729** (2020) 012070 doi:10.1088/1757-899X/729/1/012070

Production and characterization of modified biochar by corn cob and its ability to absorb phenol

Guanghua Wang[1,2], Zheng Zhang[1,2*], Wenbing Li[2], Changjun Du[2] and Tie Chen[2]

[1]Research Center for Green and Intelligent Coal Chemical Engineering, Wuhan University of science and technology, Wuhan, Hubei Province, 430081, China

[2]Institute of Chemistry and Chemical, Wuhan University of science and technology, Wuhan, Hubei Province, 430081, China

[*]Corresponding author's e-mail: zhangzheng@wust.edu.cn

Abstract. The adsorption performance of KOH modified biochar as adsorbent for phenol in aqueous solution was investigated. The effects of initial pH of the solution, initial concentration of phenol, contact time and temperature on the adsorption capacity of phenol were investigated. Adsorption isotherms, adsorption kinetics and thermodynamic analysis were examined. The results showed that the specific surface area of modified biochar was 487.64$m^2 \cdot g^{-1}$, the content of C is 72.06 wt.% and the value of H/C is 0.05, the aromatization degree of modified biochar is very high. The results showed that the maximum adsorption capacity of the modified biochar is 217.06 $mg \cdot g^{-1}$ at 298 K, which is higher than that of biochar. The adsorption process accords with the Freundlich adsorption isotherm model and pseudo-second-order kinetic equation. Thermodynamic analysis shows that adsorption is spontaneous and exothermic.

1. Introduction

Biochar is a kind of solid carbon material obtained by pyrolysis and carbonization of plant biomass animal manure and municipal sludge under the condition of hypoxia or oxygen restriction[1]. It is a kind of refractory solid material with high carbon content and high aromatization, which has superior thermal stability, porosity and high specific surface area. Biomass materials come from a wide range of sources, agricultural and forestry waste, livestock and poultry waste, municipal waste and so can serve as raw materials for the preparation of biochar[2]. As a further resource, agricultural and forestry waste has been widely studied by scholars all over the world. It is full of resources and low in utilization. The preparation of biochar can provide a useful channel for the recycling and utilization of agricultural wastes. In recent years, the majority of scholars have used organic waste such as animal manure, animal bones, plant roots, sawdust and wheat straw to absorb harmful heavy metal ions and organic pollutants in the soil. Other researchers use biochar to prepare electrode materials for lithium ion batteries[3-4].

Phenolic organic wastewater is highly toxic and difficult to degrade, which directly endangers aquatic organisms in the process of oxidation and degradation, which results in the decrease of oxygen content and deterioration of water quality, which leads to the death of aquatic organisms. Phenolic untreated wastewater is harmful to human health[5-7]. Toxic substances in the wastewater are enriched in water, vegetables, fruit, grain and other crops, and finally enter into the human body, which is a serious threat to human health[8-9]. It is pointed out that 90% of human cancer is induced by chemical carcinogens in the surrounding environment, and organic compounds account for more than 80% of these carcinogens. The clean treatment and recycling of phenolic organic wastewater have profound and positive effects on the sustainable development of water resources. In this paper, the preparation process, microstructure and adsorption properties for phenol of corncob biochar were studied. The kinetics, adsorption isotherm and thermodynamic characteristics of the adsorption process are studied. It serves as a theoretical basis for the application of corncob biochar in the field of organic pollution wastewater treatment[10].

2. Materials and methods

2.1. Preparation of biochar

The raw material of biomass was obtained by drying corn cob in natural air, washing it with oxygen displacement water and anhydrous ethanol alternately for three times, and drying in a vacuum drying box at 105°C for 12 hours. The dried corncob was crushed and ground through a 100 mesh sieve. The biomass powder was gathered and set aside. Under the protection of N_2, the biomass powder was heated up to 600°C at 5°C·min^{-1}, keepping heating for 2 h. And then cooled to 150°C, the product was taken out. After cooling to room temperature, the product was washed to neutral with deionized water, and the product was dried at 105°C for 8 hours to obtain the initial biochar, named BC, and put it in a sealed bag and put it in a dehumidifying storage tank[11].

2.2. Preparation of modified biochar

Using ethanol as solvent, the biochar was soaked in KOH solution, and the impregnation ratio was 1: 3(wt.%). The mixture of biochar and KOH was obtained by vacuum drying at 75°C for 12 h. The mixture was heated up to 600°C at 5°C·min^{-1} and activated for 120 min under N_2 protection. After natural cooling to 150°C, the product was cooled to room temperature, washed with 1mol·L^{-1} hydrochloric acid for 1 h, then washed to neutral with distilled water. The modified biochar was obtained by drying the washed product in 105°C vacuum drying box for 8 hours. The modified biochar was named MBC and was dried and preserved.

2.3. Characterization of biochar and modified biochar

The content of C, H, N, S in biochar was determined by organic element analyzer, the oxygen contents were calculated by difference. The adsorption amount of nitrogen on the samples at 77K was determined by using specific surface area and pore meter, and the specific surface area of each sample was calculated according to BET equation. The infrared spectra were measured in the range of 4000-400cm^{-1} by the infrared spectrometer model 6700. The absorbance of phenol was measured by 752N UV-vis spectrophotometer.

3. Results and discussion

3.1. Properties of modified Biochar

The composition of elements, atomic ratio, ash content, specific surface area and charring yield of biochar and modified biochar are shown in Table 1. The carbon content in the modified biochar increased obviously, and the hydrogen, nitrogen and oxygen contents also decreased. As H/C value decreases, the degree of carbonization increases gradually. Porosity has a strong effect on the adsorption performance of activated carbon. The results showed that the activation of biochar provided high specific surface area and micropore volume in the presence of potassium hydroxide. The specific surface area of modified biochar is 487.64 m^2·g^{-1}, most of which are composed of micropores.

Table 1. Characteristic of the biochar and modified biochar.

	Biochar	Modified biochar
Elemental analysis(wt.%)		
Carbon	62.45	72.06

Hydrogen	2.87	3.78
Nitrogen	0.22	0.15
Sulfur	-	-
Oxygen (by difference)	34.46	24.01
Surface properties		
BET surface area ($m^2 \cdot g^{-1}$)	15.32	487.64

Although it is not easy to identify all chemical species on the surface of modified carbon with chemical modification, information about the chemical properties of carbon surface can be obtained by using FTIR technique. Figure 1 shows the FTIR analysis of biochar and activated carbon. Stretching vibration of hydroxyl group(—OH、C—H) with wide peak near 3409、874 cm^{-1}. The absorption peak at 2929 cm^{-1} is adipose -CH_2-asymmetric stretching vibration. The peak at 1754 cm^{-1} refers to the stretching vibration peak of C=O in the aldehydes. The peak at 1704 cm^{-1} is generally considered to be the stretching vibration peak of C=O in carboxyl or ketones. The absorption peak at 1589 cm^{-1} is the expansion vibration of the aromatic ring C=O、C=C. Stretching vibration absorption peak of phenolic hydroxyl at 1278 cm^{-1}. At 1118 cm^{-1}, the stretching vibration absorption peak of C-O in carbohydrates is observed. FTIR analysis showed that the surfaces of BC and MBC contained abundant oxygen-containing functional groups such as —OH、—COH and —COOH.

In order to observe its surface morphology, scanning electron microscopy (SEM) was used to observe the biochar and modified carbon. The SEM images of biochar and modified biochar are shown in figure 2. The surface morphology of biochar and modified biochar is significantly different. The SEM images show that there are almost no micropores in biochar, which was verified from microporous BET results (15.32 $m^2 \cdot g^{-1}$). According to the diagram, a complete pore structure was observed in the modified biochar activated by KOH, which can clearly display the pore line of the pore. However, the modified biochar showed almost uniform pore structure in the process of activation, indicating the formation of the new matrix. In addition, the activated carbon also showed uniform, highly porous and obvious honey comb structure, indicating that phenol is likely to be captured and adsorbed.

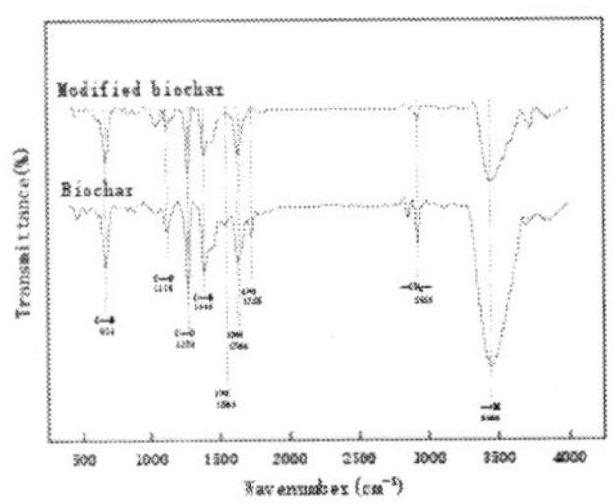

Figure 1. FTIR spectra of biochar and modified biochar.

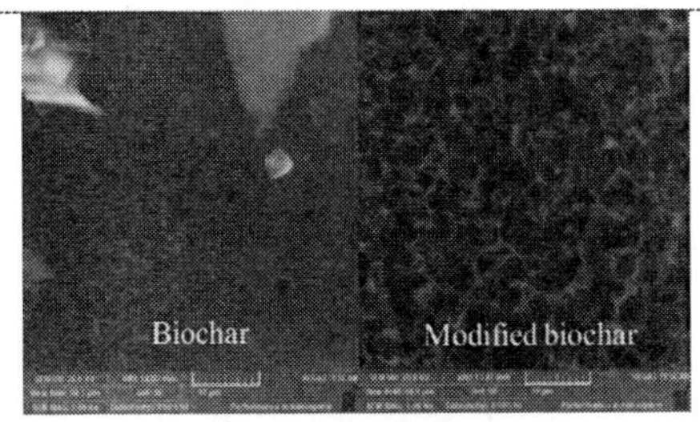

Figure 2. SEM images of biochar and modified biochar.

3.2. Effect of temperature on adsorption process

The adsorption isotherm of phenol by MBC is shown in figure 3. At the same equilibrium concentration, the maximum adsorption amount of phenol increased slightly with the increase of temperature, which indicated that the adsorption of phenol by MBC was an endothermic reaction, but temperature had little effect on it. The Gibbs free energy changes (ΔG^0), entropy change (ΔS^0) and enthalpy change (ΔH^0) are calculated by the following equation (1) and van't Hoff equation (2)[12]:

$$\Delta G^0 = -RTlnK_L \tag{1}$$

$$K_L = \frac{q_e}{C_e} \tag{2}$$

In the formula R [8.314 J·(mol·kg)$^{-1}$] is a gas constant, T(K) is the Kjeldahl temperature, K_L is the parameter of Langmuir isothermal model.

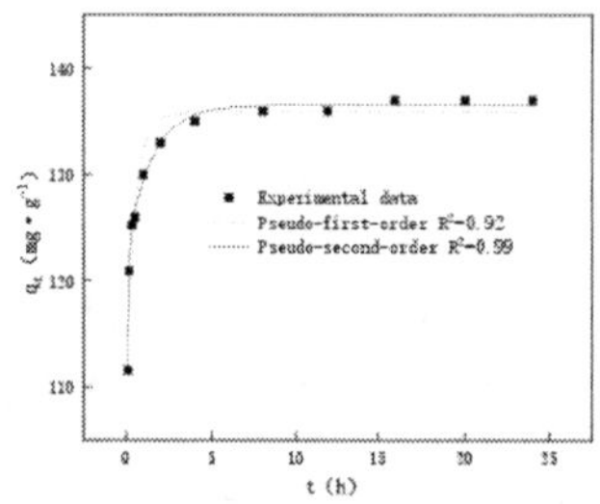

Figure 3. Effect of adsorption time on adsorption of Phenol by modified biochar.

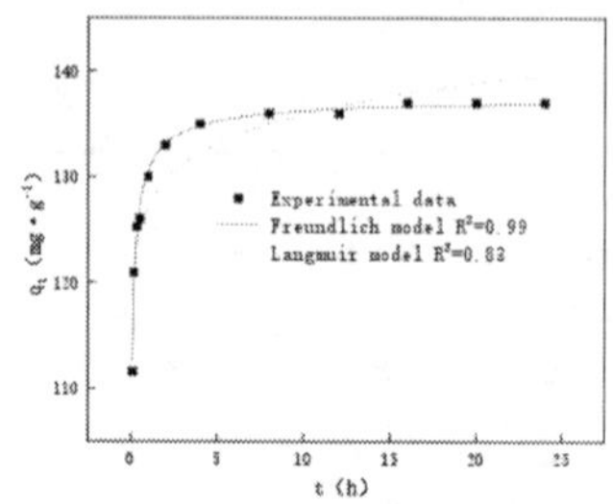

Figure 4. Kinetics of phenol adsorption onto modified biochar.

Table 2 shows that Δ $G^0 < 0$ indicates that the adsorption process is spontaneous in the range of experimental temperature, and the value of Δ G^0 decreases with the decrease of temperature, which indicates that the decrease of temperature can improve the spontaneous characteristics of phenol adsorption and removal by MBC. It is helpful to promote the adsorption process. In addition, all Δ G^0 values are between -20~0 kJ mol^{-1}, indicating that the adsorption of phenol by MBC is mainly a physical adsorption process. Δ $H^0 < 0$ indicates that the adsorption of phenol by MBC is an exothermic process, which is consistent with the experimental results. The negative value of Δ S^0 indicates that the adsorption process is accompanied by the change of structure between MBC and phenol, which indicates that phenol increases from the free state of dissolution in aqueous solution to the state of adsorption.

Table 2. Thermodynamic parameters for the adsorption of phenol onto modified biochar.

T(K)	KL	ΔG^0(kJ·mol^{-1})	ΔH^0(kJ·mol^{-1})	ΔS^0(kJ·mol^{-1})
298	3.08	-2.94	-7.99	-16.94
308	2.63	-2.48		
318	2.33	-2.24		

3.3. Adsorption kinetics

Biochar adsorption is a dynamic process, and the study of adsorption kinetics is to better understand the adsorption process and efficiency. A large number of adsorption processes can be described by two commonly used models, which are the pseudo-first order reaction kinetics equation of Lagergren and pseudo-second stage reaction kinetics equation. When the initial concentration of phenol solution is at 200mg·L-1, adsorption ability and Kinetics of modified Biochar for Phenol, as shown in figure 4.

A simple pseudo-first order equation (3) is given by Lagergren equation[13]:

$$log(q_e\text{-}q_t) = logq_e\text{-}\frac{k_1 t}{2.303} \tag{3}$$

The pseudo-second order equation based on the adsorption equilibrium capacity is studied. The equation is expressed in the following equation (4):

$$\frac{t}{q_t} = \frac{1}{k_2 q_e^2} + \frac{t}{q_e} \tag{4}$$

Table 3. Kinetic parameters for adsorption rate expression.

q_e(exp)	pseudo-first order kinetic model			pseudo-second order kinetic model		
($mg \cdot g^{-1}$)	$k_1(min^{-1})$	q_e(cal)($mg \cdot g^{-1}$)	R^2	k_2 ($g \cdot mg^{-1} \cdot min^{-1}$)	q_e(cal)($mg \cdot g^{-1}$)	R^2
140.31	0.29	17.89	0.89	0.10	140.85	1

All the constants and regression coefficients of the model are given in the table 3. On the basis of regression coefficient, pseudo-second order model is more suitable to describe adsorption kinetic data. The calculated value of q_e (140.85 $mg \cdot g^{-1}$) is close to the experimental value of 140.31 $mg \cdot g^{-1}$. It is shown that the pseudo-second-order kinetic model can prove that the adsorption mechanism is correct.

4. Conclusion

Pyrolytic studies have shown that corncobs can produce more valuable functional products (biochar), which are currently used only as agricultural and forestry waste. In this study, biochar prepared by pyrolysis of corncob was activated into modified biochar by chemical activation, which was used to remove phenol in aqueous solution. The parameters of ΔG^0 and ΔH^0 showed that the adsorption of phenol on MBC was spontaneous and exothermic. Kinetic studies show that the quasi-second-order model is suitable to describe the adsorption process of phenol. The isotherm study shows that the Freundlich isotherm model is in good agreement with the experimental equilibrium data. The results show that MBC activated by biochar can effectively remove phenol from wastewater.

References

[1] Vu T M, Trinh V T, Doan D P, Van H T, Nguyen T V, Vigneswaran S, Ngo H H. (2017) Removing Ammonium From Water Using Modified Corncob-Biochar. Science of The Total Environment, 579: 612-619.

[2] Lee J, Kim K, Kwon E E. (2017) Biochar as a Catalyst. Renewable and Sustainable Energy Reviews, 77: 70-79.

[3] Chen J, Wang C, Pan Y, Farzana S S, Tam N F. (2018) Biochar Accelerates Microbial Reductive Debromination of 2,2′ ,4,4′ -Tetrabromodiphenyl Ether (Bde-47) in Anaerobic Mangrove Sediments. Journal of Hazardous Materials, 341: 177-186.

[4] Zhu L, Lei H, Wang L, Yadavalli G, Zhang X, Wei Y, Liu Y, Yan D, Chen S, Ahring B. (2015) Biochar of Corn Stover: Microwave-Assisted Pyrolysis Condition Induced Changes in Surface Functional Groups and Characteristics. Journal of Analytical and Applied Pyrolysis, 115: 149-156.

[5] Yang J, Pan B, Li H, Liao S, Zhang D, Wu M, Xing B. (2015) Degradation Ofp - Nitrophenol On Biochars: Role of Persistent Free Radicals. Environmental Science & Technology, 50: 694-700.

[6] Ahmed M B, Zhou J L, Ngo H H, Johir M A H, Sornalingam K. (2018) Sorptive Removal of Phenolic Endocrine Disruptors by Functionalized Biochar: Competitive Interaction Mechanism, Removal Efficacy and Application in Wastewater. Chemical Engineering Journal, 335: 801-811.

[7] Yoshida S, Iwamura S, Ogino I, Mukai S R. (2016) Adsorption of Phenol in Flow Systems by a Monolithic Carbon Cryogel with a Microhoneycomb Structure. Adsorption, 22: 1051-1058.

[8] Kruszka B, Wiśniewski M, Terzyk A P. (2016) Phenol Adsorption On Different Nano-Sized Carbon Materials: First Comparative Study. Adsorption, 22: 437-444.

[9] Lorenc-Grabowska E. (2016) Effect of Micropore Size Distribution On Phenol Adsorption On Steam Activated Carbons. Adsorption, 22: 599-607.

[10] Carvajal-Bernal A M, Gómez-Granados F, Giraldo L, Moreno-Piraján J C. (2016) Calorimetric Evaluation of Activated Carbons Modified for Phenol and 2,4-Dinitrophenol Adsorption. Adsorption, 22: 13-21.

[11] Angın D, Köse T E, Selengil U. (2013) Production and Characterization of Activated Carbon

Prepared From Safflower Seed Cake Biochar and its Ability to Absorb Reactive Dyestuff. Applied Surface Science, 280: 705-710.
[12] Shin W. (2017) Adsorption Characteristics of Phenol and Heavy Metals On Biochar From Hizikia Fusiformis. Environmental Earth Sciences, 76: 1-9.
[13] Wang C, Wang T, Li W, Yan J, Li Z, Ahmad R, Herath S K, Zhu N. (2014) Adsorption of Deoxyribonucleic Acid (Dna) by Willow Wood Biochars Produced at Different Pyrolysis Temperatures. Biology and Fertility of Soils, 50: 87-94.

IOP Conf. Series: Materials Science and Engineering **729** (2020) 012071 doi:10.1088/1757-899X/729/1/012071

Micellar and Interfacial Behavior of Mixed Systems Containing Anionic-nonionic Gemini Surfactant

Hong Chen[1*] , Baowei Zhu[1] and Haiyan Liu[1]

[1]Department of Chemical Engineering, Yingkou Institute of Technology, Yingkou, 115014, China

[*]Corresponding author's e-mail: hong_chen1967@126.com

Abstract. The Anionic-nonionic Gemini Surfactant oxalamide polyoxyethylene lauryl ether(9)succinate sodium sulfonate(OPLES-9) was synthesized. It was mixed with the anionic surfactant sodium lauryl sulfate(SDS) and the nonionic surfactant dodecanol polyoxyethylene(9) ether (AEO_9) respectively in different ratios to enhance application performance. The experimental results showed that the surface performance of the mixed systems was more superior to the relevant single surfactants. The mixed critical micelle concentration (CMC) of the binary surfactant systems was determined by surface tension measurement. Critical micelle concentration of mixed system reached the minimum 0.146mmol˙L^{-1} when the molar fraction of OPLES-9 to SDS was 1:1. Critical micelle concentration of OPLES-9/AEO_9 mixed system reached the minimum 0.259mmol˙L^{-1} when the molar fraction of OPLES-9 to AEO_9 was 0.9:1.

1. Introduction

Most of surfactants are used in the form of mixtures in practical application, therefore, surfactant mixtures became an important subject in industrial production. A mixed surfactant system often can show a unique performance, for example, lower surfactant tension and better wettability. The surfactant mixtures can achieve the excellent properties [1-7].

The anionic-nonionic Gemini surfactant—oxalamide polyoxyethylene lauryl ether(9)succinate sodium sulfonate(OPLES-9, for short) has shown interesting applications in detergents, cosmetics, pesticides, and medicine because of its good dermatological compatibility and biodegradability. The mixed system properties of OPLES-9 and sodium lauryl sulfate (SDS, for short), OPLES-9 and dodecanol polyoxyethylene(9) ether (AEO_9,for short)were studied in this paper. The aim is to provide a theoretical basis for the practical application of the mixed system of OPLES-9/SDS and OPLES-9/AEO_9.

2. Experiments

2.1Materials and Apparatus

Materials: The anionic-nonionic Gemini surfactant—OPLES-9 (lab-synthesized surfactant) is analyzed by FT-IR, mass spectrometry, elementary analysis, and is consistent with the target product. SDS(reagent grade ,Shenyang Xinhua Reagent Factory), AEO_9(reagent grade, Shenyang Xinhua Reagent Factory), the water of the experiment is ultrapure water(lab-synthesized water).

Apparatus: K100 Surface tensiometer(KRUSS,Germany),BP2MKII Dynamic surface tensiometer(KRUSS,Germany), Ultrapure water machine(Sichuan youpu super pure technology co. LTD) ,etc..

2.2 Synthesis of oxalamide polyoxyethylene lauryl ether(9)succinate sodium sulfonate

OPLES-9 was synthesized through the first step amidation, then through the second step esterification, followed by the third step sulfonation according to the Scheme 1. Fig.1 and Fig.2 are IR and mass spectrum of Oxalamide lauryl polyoxyethylene ether(9) succinate sodium sulfonate.

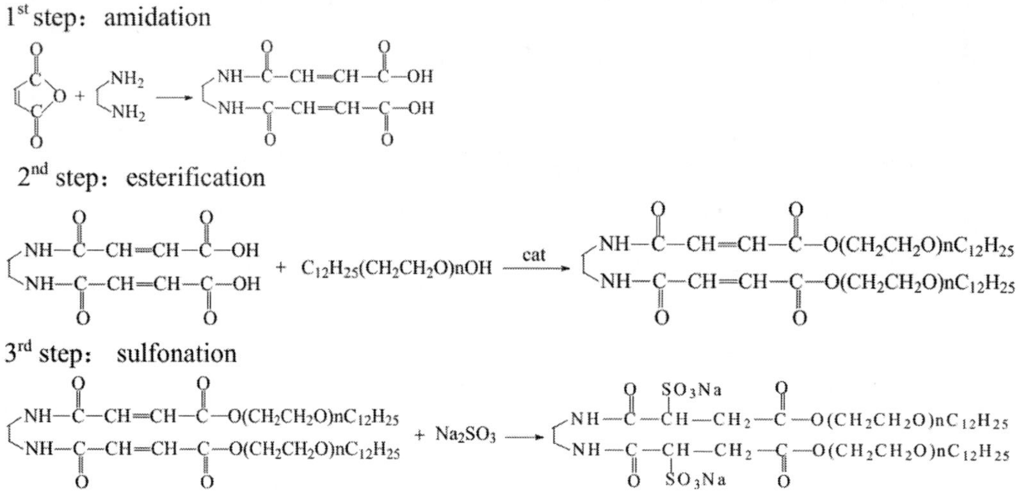

Scheme 1: Synthesis of oxalamide lauryl alcohol polyoxyethylene ether succinate sodium sulfonate

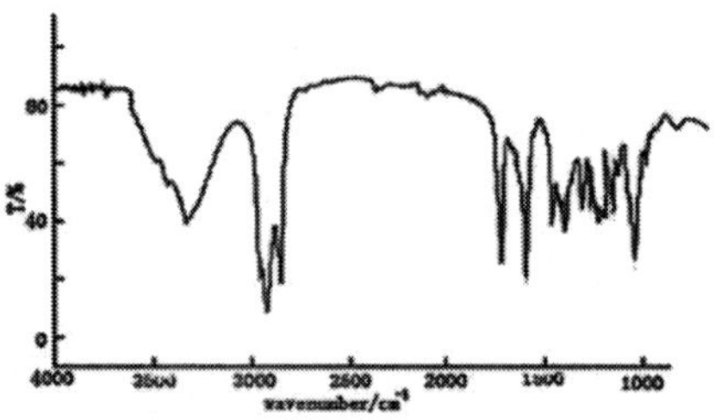

Fig. 1 IR of Oxalamide lauryl polyoxyethylene ether(9) succinate sodium sulfonate

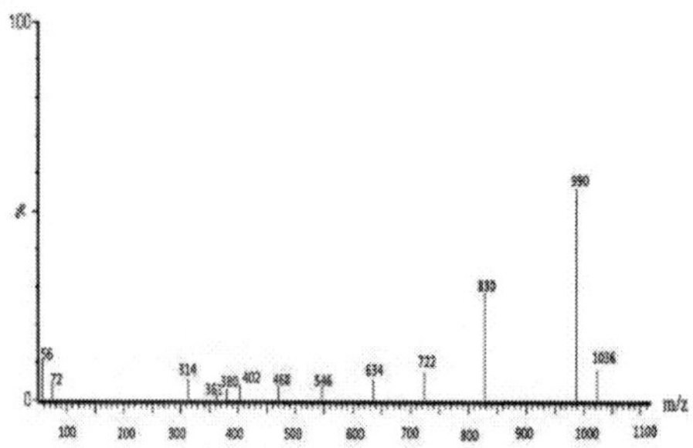

Fig. 2 Mass spectrum of Oxalamide lauryl polyoxyethylene ether(9) succinate sodium sulfonate

Fig. 1 and Fig. 2 show that the synthesized product is the target product.

2.3 Determining mixed proportion

The anionic-nonionic Gemini surfactant OPLES-9 was selected for compounding with SDS and AEO_9, separately. Mixed proportion was selected 0:1, 1:10, 3:10. 5:10, 7:10, 9:10, 1:0 (mole:mole) respectively.

2.4 Physicochemical Characterization

Equilibrium surface tension(γ)of mixed systems of OPLES-9/SDS, OPLES-9/AEO_9 was measured by K100 automatic surface tensiometer under 25°C, respectively. The equilibrium surface tensions were taken after the change in surface tension was less than 0.1 m Nm^{-1} every 5 min. The CMC and γ_{cmc} was obtained from the breakpoint of the curve of surface tension versus the logarithm of molar concentration.

3. Results and Discussion

3.1 The surface chemistry of the mixed systems of OPLES-9/SDS and OPLES-9/AEO₉

Surface tension(γ) of mixed systems of OPLES-9/SDS and OPLES-9/AEO_9 was measured, respectively. Fig. 3 shows the curve of surface tension(γ) to the total concentration (lgC) of the mixed systems. As shown in figure 3, the surface tension of the mixed system of OPLES-9 and SDS is lower than that of OPLES-9. The surface tension of the mixed system of OPLES-9 and AEO_9 is also lower than that of OPLES-9.

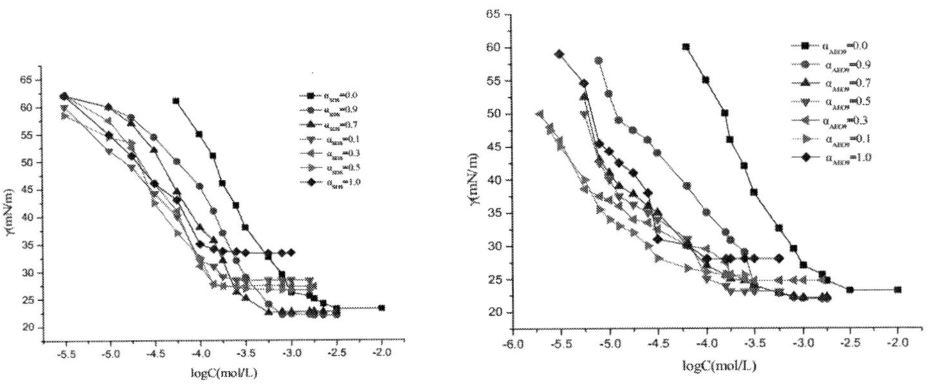

Fig. 3 Representative plot of surface tension curves lg surfactant concentration for different mole fraction of the OPLES-9 and other surfactants

The calculation shows that γ_{cmc} of OPLES-9/SDS system is slightly greater than that of OPLES-9/AEO_9 system. The reason may be that sulfuric acid group in SDS has stronger hydration than that of hydroxyl group in AEO_9 to form stronger hydration layer and more adsorption of water molecules on the surface. SDS molecules are repelled by more adsorption of water molecules when they continue to spread into the surface area, which results in the γ_{cmc} of OPLES-9/SDS system is slightly greater than that of OPLES-9/AEO_9 system. The regular solution theory(RST) is a medium to understand the interaction between surfactant molecules in mixed system. The regular solution theory is used to solve the surfactant molecules interacting with each other in the mixed system since the regular solution theory is applied to the ideal mixed system successfully by Clint [8] and to the nonideal mixed system by Rubingh [9].

The interaction parameters β (β^{σ} and β^{m}) of the mixed system and the properties of the single surfactant can be used to determine whether there is synergistic effect in the mixed system.

(1) Judgment criteria of ability of reducing surface tension:

$$\beta^{\sigma}<0 \qquad \beta^{\sigma}-\beta^{m}<0 \;; \quad \left|\beta^{\sigma}-\beta^{m}\right|>\left|\ln \frac{C_1^0 C_2^m}{C_2^0 C_1^m}\right|$$

(2) Judgment criteria of efficiency of reducing surface tension:

$$\beta^{\sigma}<0 \;; \quad \left|\ln \frac{C_1^0}{C_2^0}\right|<\left|\beta^{\sigma}\right|$$

(3) Judgment criteria of forming micellar power:

$$\beta^{m}<0 \;; \quad \left|\ln \frac{C_1^m}{C_2^m}\right|<\left|\beta^{m}\right|$$

The calculated data is shown in table 1.

IOP Conf. Series: Materials Science and Engineering **729** (2020) 012071 doi:10.1088/1757-899X/729/1/012071

Table 1 Composition and interaction parameters of the OPLES-9/SDS and OPLES-9/AEO_9 mixed system at 298K

	x_1^m	β^m	x_1^σ	β^σ	β^σ-β^m
α_{SDS}					
0.1	0.185	-4.549	0.239	-4.868	-0.319
0.3	0.279	-4.191	0.298	-3.452	0.739
0.5	0.340	-4.083	0.369	-3.299	0.784
0.7	0.373	-2.282	0.441	-2.971	-0.689
0.9	0.537	-1.343	0.583	-2.509	-1.166
α_{AEO9}		-3.284[a]		-3.421[a]	-0.137[a]
0.1	0.170	-6.621	0.292	-10.235	-3.614
0.3	0.213	-5.746	0.331	-9.438	-3.692
0.5	0.261	-5.670	0.341	-7.389	-1.719
0.7	0.286	-4.464	0.386	-7.469	-3.005
0.9	0.300	-1.749	0.435	-4.357	-2.608
		-4.848[a]		-7.779[a]	2.922[a]

Synergy parameter βm and βσ are an important index of the synergistic effect of mixed surfactant system. β<0 means that two components of mixed system are attracted each other. β=0 expresses that mixed system is the ideal mixture solution in the micelle or surface layer. β>0 means that two components of mixed system are repelled each other.

Table 1 shows that both β^m and β^σ value are negative, and β^σ is $|\beta^\sigma| > \left|\ln C_1^0 / C_2^0\right| = 1.31$ in the OPLES-9/SDS system and β^σ is $|\beta^\sigma| > \left|\ln C_1^0 / C_2^0\right| = 2.86$ in the OPLES-9/AEO_9 system, which indicates that OPLES-9/SDS and OPLES-9/AEO_9 system all meet the conditions of reducing surface tension efficiency and synergistic effect.

α_1 is 0.3-0.5, β^σ-β^m> 0 and $|\beta^\sigma - \beta^m| > \left|\ln C_1^0 C_2^m / C_2^0 C_1^m\right| = 0.55$ in OPLES-9/SDS system, which indicates that synergistic effect shows decreasing effect when the surface tension decreases to a maximum. α_1 is 0.7-0.9, β^σ-β^m<0 in OPLES-9/SDS system, which indicates that surface tension reduction efficiency shows increasing effect. Synergy parameters are $|\beta^\sigma - \beta^m| > \left|\ln C_1^0 C_2^m / C_2^0 C_1^m\right| = 0.70$ in OPLES-9/AEO_9 system, which indicates that surface tension reduction efficiency always shows increasing effect.

The reasons why synergistic effect in OPLES-9/AEO_9 system is better than that of OPLES-9/SDS system are as follows. Firstly, electrostatic stability of OPLES-9/AEO_9 system is increased by AEO_9 molecule weakening the repelling effect between OPLES-9 molecules when OPLES-9 and AEO_9 form mixed micelle. Second, because oxygen molecules in AEO_9 and water molecules can form hydrogen bonding and AEO_9 shows the characteristics of weak cationic surfactants in aqueous. Therefore, there is stronger interaction force between the OPLES-9 and AEO_9 molecules than that of OPLES-9 and SDS molecules. Micelles of OPLES-9 and AEO_9 are more likely to form and surface adsorption layer is close [10-12].

3.2 Thermodynamic properties of micelles of mixed systems

The mixed thermodynamic function is calculated using the hypothetical data given by the regular solution theory(RST) in order to obtain additional information about the mixed system. The mixed excess entropy (S^E) is considered to be zero and the entropy of mixed system is equal to the entropy of the ideal case (S^M=S^{ideal}) by RST theory. Therefore, the relationship between excess free energy (G^E), excess enthalpy (H^E) and mixed enthalpy (H^M) must be considered.

$$G^E = H^E = \Delta H_M = RT\Sigma x_1 \ln f_1 \quad (1)$$

Excess free energy (G^E) represents the deviation degree of the actual state and ideal state ($G^E = G^M - G_M^{ideal}$).The value of G^E is related to the interaction parameter β_{12} according to the RST theory. For the two component system, G^E is as follow.

$$G^E = RT\beta_{12}x_1(1-x_1) \quad (2)$$

Expression of the ideal mixed free energy ΔG_M^{ideal} is $\Delta G_M^{ideal} = RT\sum_i x_i \ln x_i$ (3)

The RST proposed by Rubingh [7] is more commonly used in calculation of micelle. According to the RST, the relationship between the activity coefficient f and the interaction parameter β_{12} is as follow.

$$f_1 = \exp \beta_{12}(1-x_1)^2 \quad (4)$$

$$f_2 = \exp \beta_{12}x_1^2 \quad (5)$$

x_1 represents the mole fraction of component 1(OPLES-9) in the mixed micelle. β_{12} is the interaction parameter of OPLES-9 and SDS, OPLES-9 and AEO_9, which illustrates the deviation degree of the real solution to the ideal solution in the two components system.

Expression of real mixed free energy is $\Delta G_M = RT\sum_i x_i \ln f_i x_i$ (6)

The entropy value can be calculated by the following formula.

$$T\Delta S_M = \Delta H_M - \Delta G_M \quad (7)$$

The calculated data is shown in table 2.

Table 2 variation of the mixing thermodynamic functions, micelle and activity coefficients for the mixed system at 298K

	x_1^m	f_1	f_2	ΔG_M /kJ·mol^{-1}	ΔH_M /kJ·mol^{-1}	$T\Delta S_M$ /kJ·mol^{-1}
α_{SDS}						
0.1	0.198	0.051	0.832	-3.043	-1.801	1.242
0.3	0.280	0.113	0.715	-3.570	-2.100	1.470
0.5	0.342	0.170	0.614	-3.878	-2.281	1.597
0.7	0.374	0.410	0.722	-2.968	-1.368	1.640
0.9	0.539	0.751	0.672	-2.533	-0.823	1.710
α_{AEO9}						
0.1	0.169	0.012	0.825	-3.460	-2.322	1.138
0.3	0.212	0.030	0.770	-3.686	-2.395	1.291
0.5	0.260	0.046	0.679	-4.142	-2.713	1.429
0.7	0.284	0.105	0.694	-3.753	-2.264	1.489
0.9	0.299	0.426	0.855	-2.429	-0.909	1.520

From Table 2, it can be seen that ΔG_M is all negative value, which indicates that the micellation of the mixed system is a spontaneous process. ΔG_M value of mixed system of OPLES-9/SDS and OPLES-9/AEO_9 both reach the minimum when the mole ratio of OPLES-9 and SDS, OPLES-9 and AEO_9 is equal. The reason for the increase in entropy value is explained as follow. On the one hand, the regular water structure is destroyed around the carbon chain by the hydrophobic groups escaping water phase. On the other hand, the degree of freedom is increased by hydrophobic group transferring to inside the micelles of liquid hydrocarbons.

4. Conclusion

Critical micelle concentration of mixed system reaches the minimum 0.146 $mmol \cdot L^{-1}$when the molar fraction of OPLES-9 to SDS is 1:1. The performance of reducing the surface tension of OPLES-9/SDS mixed system exhibits the decreasing effect as molar fraction of OPLES-9 to SDS changes from 0.3 to 0.5. Performance of reducing the surface tension of OPLES-9/SDS mixed system exhibits increasing effect as molar fraction of OPLES-9 to SDS changes from 0.7 to 0.9.

Critical micelle concentration of OPLES-9/AEO_9 mixed system reaches the minimum 0.259 $mmol \cdot L^{-1}$when the molar fraction of OPLES-9 to AEO_9 is 0.9:1. The micellation of either OPLES-9/SDS or OPLES-9/AEO_9 mixed system is a spontaneous process. The synergistic effect of the OPLES-9/SDS and OPLES-9/AEO_9 mixed system can make their performance better, so there is a very wide application prospect.

Acknowledgements
Thanks for supporting by Foundation of Liaoning Province Educational Commission of China （Project L2014592 ） and Yingkou Institute of Technology.

References

[1] Alami E O, Holmberg K. (2003) Heterogemini surfactants. Advances in colloid and interface science, 100-102: 13-46
[2] Sood AK, Singh K, Kaur J et al. (2012) Mixed micellization behavior of the m-2-m gemini surfactants with some conventional surfactants at different temperatures. J Surf Deterg., 15: 327-338
[3] Sheikh MS, Kabir-ud-Din DAA (2011) Synergistic interaction of gemini surfactant pentanediyl-1,5-bis(dimethyl cetyl ammonium bromide) with conventional (ionic and non-ionic) surfactants and its impact on the solubilization. Colloids Surf A., 378: 60-66
[4] He F, Li R, Wu D (2010) Monolayers of mixture of alkyl amino methyl rutin and lecithin at the air/water interface. J Colloid Interface Sci., 349: 215-223
[5] Yan L, Lu L, Ping M, et al. (2018) Equilibrium and Dynamic Surface Properties of Cationic/Anionic Surfactant Mixtures Based on Carboxylate Gemini Surfactant. J Surf Deterg., 21: 845-858
[6] Din K, Sheikh MS, Dar AA (2009) Interaction of a cationic gemini surfactant with conventional surfactants in the mixed micelle and monolayer formation in aqueous medium. J Colloid Interface Sci, 333: 605–612
[7] Zhou M, ChenYP, Zou JX, et al. (2018) Recent Advances in the Synthesis of Sulfonate Gemini Surfactants. J Surf Deterg., 21: 443-453
[8] Clint J.H. (1974) Micellization of Mixed Nonionic Surface Active Agents. Journal of the Chemical Society, Faraday Transactions, 71: 1327-1334.
[9] Rubingh D.N., Mittal K.L. (1979) Solution Chemistry of Surfactants, Plenum Press, New York.
[10] Luo SQ, Wang YX, Wang MN, et al. (2018) Synergistic Interaction and Aggregation Behavior in a Mixture of a Tripolar Zwitterionic Surfactant and an Anionic Surfactant. J Surf Deterg., 21: 899- 908
[11] Savaş Ö, Şemsettin U (2019) Interfacial Behavior of Anionic/Cationic Flotation Collectors in Mixed Aqueous Solutions and Their Effect on Flotation Recovery of Quartz.doi. J Surf Deterg., 22: 61-71
[12] Jumpertz T, Tschapek B, Infed N, et al. (2011) High-Throughput Evaluation of the Critical Micelle Concentration of Detergents. A.nalytical Biochemistry, 408: 64-70

IOP Conf. Series: Materials Science and Engineering **729** (2020) 012072 doi:10.1088/1757-899X/729/1/012072

Study on Thermal Oxidation and Viscosity Degradation for Synthetic Aviation Lubricant Fluids

Fei Yi-Wei*, **Ruan Shao-Jun, Ma Jun, Guo Feng**

Air Force Logistics College, Xuzhou, China, 221000

*Corresponding author: yiweifei50 @163.com

Abstract: The thermal oxidation and viscosity degradation for poly-α-olefins (PAO) and di-esters (DOS) was studied under the GC/MS investigation. Through the oxidative pyrolysis conditions, these two synthetic base fluids were characterized to get their simulated "areo-engine" conditions' for thermal stability, such as oxidative and kinematic viscosity variation. The experimental results suggests that the oxidation products generated from oxidation reactions of DOS is mainly single esters, di-esters, ketone, ether, organic acids and alcohol, which results much bigger total acid values than that of PAO treated in similar conditions. The substantial increment of the kinematic viscosity of DOS reveals its limited thermal oxidation stability compared to PAO.

1. Introduction

The synthetic lubricating fluids, poly-α-olefins (PAO) and synthetic esters (DOS) are formulated to suit the aviation applications. Compared to the mineral fluids, PAO and DOS have unique properties of high viscosity, low corrosive and volatility and toxicity, etc. The excellent viscosity index property of PAO and DOS allows the synthetic lubricants to be used at extended operation temperature ranges. However, the PAO and DOS base oils are gradually hard to meet the increasing standards of next generation aviation lubrication oils, due to the limited application properties and promoting demands of high-performance aero-engine at conditions of ultra temperature, revolving speed and catalyst of transition metal ions [1]. In particular, viscosity thinning and colour deepening of in-use oil embarrass normal usage of the aero-engine, which influences lubricants' comprehensive properties severely, such as lubricity, viscosity and volatility, so as to understand the molecular composition through the oxidative pyrolytic process.

Sophisticated detection strategies have been developed to identify the molecular information of the lubricants. The gas chromatography/mass spectrometer (GC/MS), has been widely accepted for evaluating the molecular change of lubrication oils, is effective for analysing the most volatile species, through which we could monitor the thermal-oxidative stability of lubricants. Nowadays, the GC/MS method has been successfully applied in deep analysis of medicine molecules[2], special steroids[3], common contaminants[4], mainly species of crude oils[5] and synthetic polymers used in petroleum products[6].

Both the PAO and DOS base oils are important synthetic lubricating fluids[7-9], which is need further investigations to understand the chemical reactions occurs under extreme condition that simulated the aviation engine operated at harsh condition. In this paper, the molecular structure analysis of the PAO, DOS and their oxidation products were characterized through viscosity testing experiments and GC/MS.

IOP Conf. Series: Materials Science and Engineering **729** (2020) 012072 doi:10.1088/1757-899X/729/1/012072

2. Experimental

2.1 Fluid Sample and Reagents
The PAO and DOS base oil are were purchased from Yiping Lubrication Company of SINOPEC. Other reagents used in this paper like petroleum ether, acetone, n-hexane and so on were analytical pure degree, and the reagents applied while they are received without any further treatment.

2.2 High temperature oxidation of the oil samples
All of the received oil samples including PAO and DOS are oxidized at 170°C, 200°C and 300°C, respectively, in a 300mL stainless-steel, mechanically stirred autoclave with the stirring rate of 800r/min. The oxidation time lasted for 2 hours, then the products are allowed to cool down to room temperature and collected for the kinematic viscosity examination according to ASTM D445 standard to evaluating the deterioration of the viscosity properties of the oxidized oil samples. For convenience in description, the PAO, DOS and the oxidized oil samples treated at 170°C, 200°C and 300°C are named as P_0, P_1, P_2, P_3 and E_0, E_1, E_2, E_3, respectively.

2.3 Analytical Methods
PAO, DOS and their degraded samples were analyzed with a Frontier Fourier transform infrared (FTIR, Perkin Elmer Company) spectrometer by collecting 64 scans at a resolution of 8 cm^{-1} in reflectance with measuring region from 4000 to 400 cm^{-1}.

Afterwards, fluid samples were analyzed with a Perkin Elmer GC/MS (GC Clarus 680 and MS Clarus SQ 8 T), which was equipped with a capillary column coated with MS-5 and a quadrupole analyzer with a m/z range from 45 to 550 and operated in EI mode. The capillary column was heated from 50°C to 280°C at a rate of 10°C/min and held at 280°C for 8 min.

3. Results and Discussion

3.1 FTIR Analysis
The FT-IR method is applied to study the functional groups variation in the base oils used in this paper after oxidation reactions. As studied before, the synthetic lubricant oils are used to be changed both in chemical composition and physical properties. Their oxidation products, which are mainly harmful to the physicochemical performances of these lubrication oils, should be formed during the oxidative thermal cracking process [10].

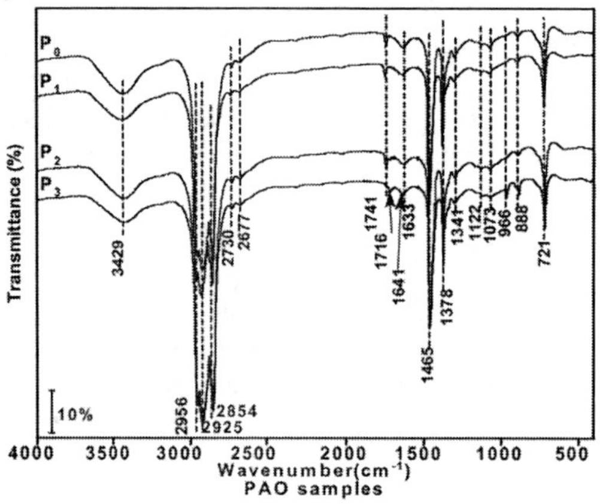

Fig.1 Infrared spectra of PAO and their reacted samples

As shown in Figure 1 and 2, the functional groups of PAO and DOS are changed during oxidation reactions based on the FTIR analysis of PAO and DOS oil samples according to the experimental conditions. The IR absorbance peaks centered at around 2844-2963 cm^{-1} and 1461 cm^{-1} are assigned

IOP Conf. Series: Materials Science and Engineering **729** (2020) 012072 doi:10.1088/1757-899X/729/1/012072

the energetic absorption of aliphatic moieties (AMs) from Pn species (n=0-3), which is much strong than those of E_n species, indicating that AM-rich species in P_n are predominant species. The AM-rich species in P_n is mainly from the oxidation reaction of the special structure of PAO, which is the multi-aligned side-chains comb-likely and straight chain alkynes skeleton bones molecular structure. The strong absorption peak at ~3431 cm^{-1} and 3462 cm^{-1} could be attributed to stretching vibration of –OH group. The stretching vibration of –OH groups in E_n is also stronger than that in P_n. As shown in Figure 1, the relative weak absorption peak at ~1640, 1642, 955 and 891 cm^{-1} could be attributed to the absorbance effect of C=C group that might be β-elimination reactions occur in the P_n, indicating the existence of olefins in P_3.

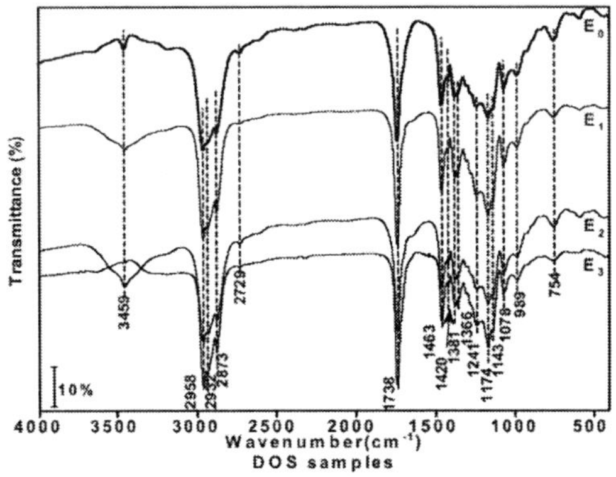

Fig.2 Infrared spectra of DOS and their reacted samples

Relatively absorption peak at ~1735 cm^{-1} could be assigned to the –C=O bond of carboxyl groups in En, which is nearly weak in P_n, which is also proved that the -C=O is the dominate group in the esters of DOS base oils. The characteristic absorption peak of the carbonyl group of the aldehydes, ketones and carboxylic acids appear at ~1718 cm^{-1}. Such absorption peaks of PAO are very weak, suggesting that fractional substances are oxygenized. The absorbance of more -(CH_2)n- at 754 cm^{-1} of but the DOS oil samples are much stronger, which indicates that the DOS has been thermal oxidized and generated many small oxidation products.

3.2 GC/MS Analysis

The GC/MS analysis is applied to investigate the molecular structure of the oxidized oil samples to further research the thermal cracking effect of high temperature oxidation atmosphere. As shown in Figure 3, as much as 94 micro chemicals (MCs) were found in the GC and MS results. Both the PAO and DOS oil samples are tested. The chemicals include 13 n-alkanes (NAs), 26 iso-paraffins (IPs), 51 olefins (OFs) and 4 other compounds (OCs) shown in the left graph in Figure 3 PAO reacted oil samples. For the DOS oil samples, 53 MCs were identified, includes 24 small esters (SEs), 13 micro-di-esters (MDEs), 9 ketones, 3 alcohol, 1 organic acids (OAs), 1 alkane and 2 ethers. The results indicate that the chemical composition of oxidation products in oxidation reactions of PAO is much more complicated than DOS, which also generates more oxidation products during oxidation process of the base oils.

As shown in figure 3a, the main products from chemical pyrolysis reactions of PAO are IPs, NAs and OFs. The carbon number of the backbone of oxidation products of NAs is from 12 to 24, and the C20 chemical species are dominant compounds. The relative content of RC and OFs are equivalent to that of NAs, which the carbon number the backbone of oxidation products of DOS is from 12 to 23, indicates that oxidative molecular recombination reactions happened with increasing oxidation temperature. This result is quite similar to the oxidative molecular composition of the former PAO products. The results give the higher content of long-chain macromolecular species and quite less oxygen-containing organic compounds (OCOCs) in the oxidation products. The results also show that the existence of OCOCs is coincident with the results from the FTIR analysis in figure 1 and 2. However, the RC of OCOCs is nearly 1.03 % and only show in the oxidative products in PAO. The existence of RC of OCOCs indicates the higher oxidation degree of PAO base oils which is depicted in Scheme 1.

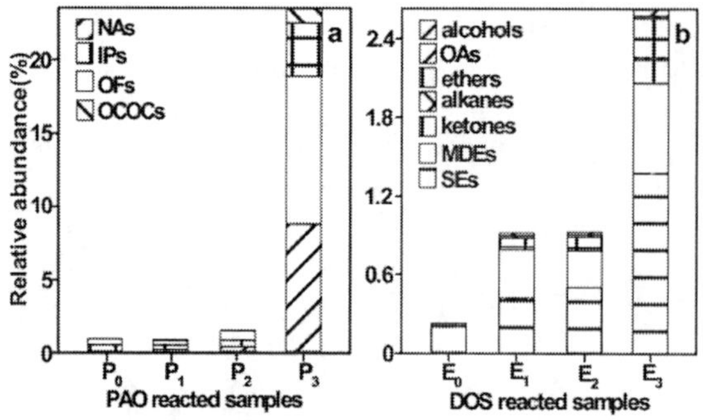

Fig. 3 Relative abundance of PAO (a), DOS (b) and their reacted samples

The first oxidation cracking of PAO is such as –CH_2- bonds in the backbone in the comb-like molecular to produce large amounts of olefins in the process. Thus some of the olefins are oxidized and rearranged by further reactions. According to oxidation mechanism, the oxidation of NAs or IPs could be described to the cleavage of weak covalent bonds at the tertiary carbon atom.

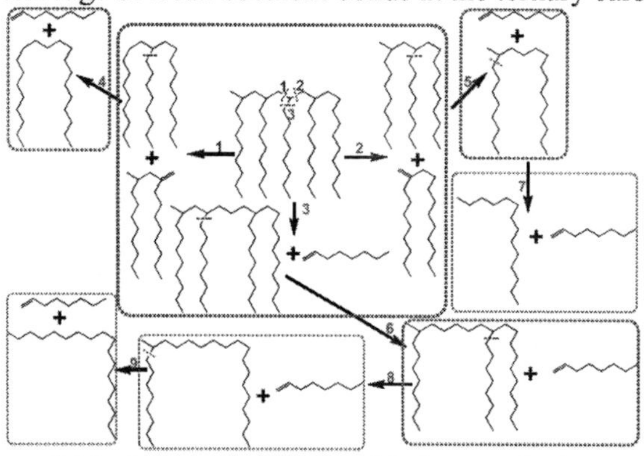

Scheme 1 Mechanism for PAO oxidative pyrolysis reactions

The formation of di-alkenes or vinyl radicals is the part of vial dehydrogenation of the further decomposition of OFs and OCOCs. However, the oxidation of dialkenes and vinyl radicals could efficiently destroy the C-C bonds to produce very small aromatic products at t higher emperatures exceed 300°C, the materials could results the darken oils colour.

As depicted by figure 2b, the order: $E_3 >> E_2 > E_1 > E_0$ is found in the RC of SEs decreases order. With carbon number of the backbone from 13 to 16, the SEs in E_1, E_2 and E_3 are mainly short-chain species, gives the RC of SEs from E_3 (1.36%) is triple more than that from E_2 (0.39%). The yields of short-chain SEs indicates that the thermal oxidation dramatically increased. Much lower than that from E_3 is founded, few significant differences could be described in the results of the RC of MDEs (0.27%), ketone (0.11%), OAs (0.019%), ether (0.004%) and alcohol (0.04%) between E_1 and E_2. Special chemical, such as 6-(2-ethylhexyloxy)-6- oxohexanoic acid are found in the oxidation products, however, the detection of the species might be rarely accurate. Moreover, the products in OCOCs from E_3 is related to the fact that the acid value of E_3 is 12.598 mgKOH/g, with 0.02% of RC, which is much higher PAO, chemicals likes 2-cyclopentenylethenylpentanone is very small in ketone from E_1, E_2 and E_3, might contributing to the darkened colour in oxidized oil samples.

Seven OCOCs were confirmed in the oxidation products from the oxidation reactions of DOS oil samples, which are demonstrated in figure 4. Few reports have been reported on the detection of OCOCs in the products from DOS oxidation reactions which produce various compounds were identified from fluid oxidation.

The results show the positive oxidation effect due to the oxidation atmosphere, which leading to the formation of oxygenated species. The oxidation reactions of benzyl and alkyl radicals indicate the observed increase in OCOCs formation at higher oxidation temperatures. The intermediate oxidation products formed from the reactions between the ketone through the dialkenes and vinyl radical and oxygen containing species.

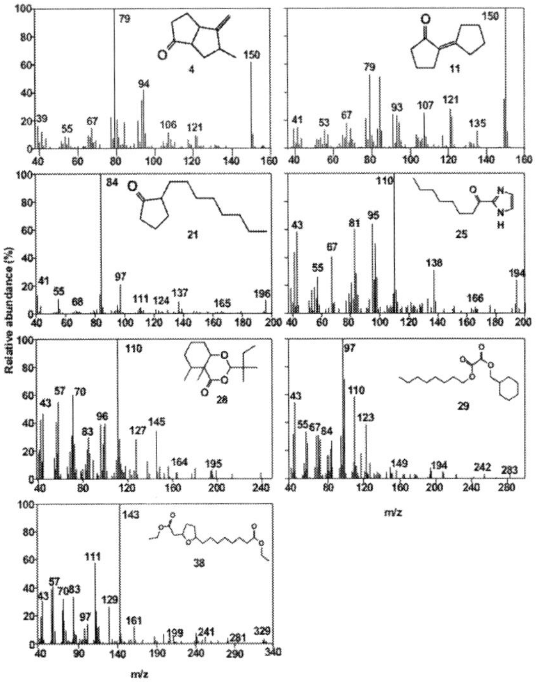

Fig. 4 Mass spectrums of some chromophore detected from DOS reacted samples

Then, some highly condensed groups (solids phase) are produced into the black deposit, which may further polymerize with the same free radicals to generate aromatic radicals, to performed as building blocks to synthases the heavy species by consecutive reactions and highly condense.

After oxidized at high temperature, the compounds are much more than the detected species in the GC/MS results. The evidence of the light black solid deposit stay in the oil samples has very high molecular weight and contain some smaller molecules in the pore structures to release at certain conditions like the higher temperature and better solvent were applied. It is impossible to analysis the entire chemical compounds in the solid deposit since the GC/MS analysis can not perform while the analytes can not be vaporized.

3.3 Viscosity Analysis.

The certain sulfur and nitrogen containing species would substantially deteriorate the viscosity properties of the oxidized oil samples. According to the standards, lubricating oils should relative clean and has proper viscosity. In this paper, kinematic viscosity is measured according to the ASTM D445 standard to predict the actual service life of lubricant while served at high temperature. And less tendency it has to viscosity increases during the usage results more resistant a lubricant in thermal-oxidation resistance[10].

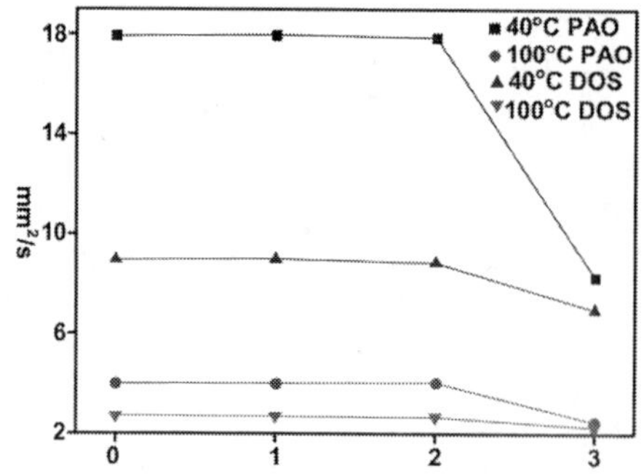

Fig. 5 Trend in viscosity of PAO and DOS reacted samples

Each point shown in figure 5 represents the viscosity variation of the oxidized base oil samples, which shows the thermal stability properties of the PAO and DOS, respectively. The higher content of long-chain AMs than DOS could be found in the viscosity of the two synthetic lubricants including PAO and DOS, which is follows the PAO > DOS rules as shown in figure 5. However, DOS shows better oxidative thermal stability at temperatures lower than 200°C, which only has 23.8 % of RC in MCs. Species such as methyl and ethyl di-ester, which is more thermal stable than PAO oxidation products with a higher NAs content than DOS.

4. Conclusion

Oxidation thermal treatments is an effective method to evaluating the base oils of PAO and DOS base on the GC/MS analysis and kinematic viscosity testing. The results also provide solid evidences to understand oxidative mechanism of the PAO and DOS base oil. The experimental results suggests that the oxidation products generated from oxidation reactions of DOS is mainly single esters, di-esters, ketone, ether, organic acids and alcohol, which results much bigger total acid values than that of PAO treated in similar conditions. The substantial increment of the kinematic viscosity of DOS reveals its limited thermal oxidation stability compared to PAO. The significant decrease of viscosity in PAO reveals its lower thermal stability compared to DOS.

References

[1] Brajendra K. Sharma, Arthur J. Stipanovic. Development of a new oxidation stability test method for lubricating oils using high-pressure differential scanning calorimetry. Thermo- chimica Acta, 2003, (402): 1-18.
[2] Zhang Hong-bo, Zhao Lu. Quantitative analysis of methcathinone by gas chromatography/mass spectrometry. J Shanxi Med Univ., 2013, 40(12): 939-940.
[3] Xing Fan, Wei Xian-yong, Zong Zhi-min. Application of gas chromatography/msaa spectrometry in studies on separa- tion and identification of organic species in coals. Fuel, 2013, (109): 28-32.
[4] Liang Cun-feng. Detection of polycyclic aromatic hydrocarbons and phthalate esters in drinking waters with GC-MS. Journal of Beijing Institute of Petro-Chemical Technology, 2013, (21): 14-17.
[5] Shang Ji-wu, Zhang Yi-he. Analysis of hazardous organic residues from sodium hydrosulfite industry and utilization as raw materials in a novel solid lubricant production. Journal of Hazardous Materials, 2011, (198): 65-69.
[6] Frank Cheng-Yu Wang, Walter C. Buzanowski. Polymer additive analysis by pyrolysis–gas chromatography III. Lubricants. Journal of Chromatography A, 2000, (891): 313-324.
[7] Dlubek G., Bamford D., Henschke O.,et al. The local free volume in metallocene-catalysed poly(a-olefin)s: a positron lifetime study. Polymer, 2001, (42): 5381-5388.

[8] Martin Kohler, Norbert V. Heeb. Characterization of ageing products of ester-based synthetic lubricants by liquid chromatography with electrospray ionization mass spectrometry and by electrospray ionization mass spectrometry. Journal of Chromatography A, 2001, (926): 161-165.
[9] Yanjarappa M. J., Sivaram S. Recent developments in the synthesis of functional poly(olefins). Progress in polymer science, 2002, (27): 1347-1398.
[10] Brajendra K. Sharma, Arthur J. Stipanovic. Development of a new oxidation stability test method for lubricating oils using high-pressure differential scanning calorimetry. www.elsevier.com/locate/tca, 2003, (402): 1-18.

IOP Conf. Series: Materials Science and Engineering **729** (2020) 012073 doi:10.1088/1757-899X/729/1/012073

Study on microwave pyrolysis of Mengdong lignite by Fe/Ca@Al_2O_3 catalyst

Guanghua Wang[1,2], Xiaoyuan Li[1,2*], Qingdong Wang[1,2], Zhiyong Ma[2] and Xiaodan Li[2]

[1]Research Center for Green and Intelligent Coal Chemical Engineering, Wuhan University of science and technology, Wuhan, Hubei Province, 430081, China

[2]Institute of Chemistry and Chemical, Wuhan University of science and technology, Wuhan, Hubei Province, 430081, China

[*]Corresponding author's e-mail: lixiaoy2001@163.com

Abstract: Microwave pyrolysis coal has the characteristics of fast heating, short time and high content of active components. In order to further improve the production and relative content of hydrogen(H_2) and carbon monoxide(CO) during microwave pyrolysis of lignite, Fe/Ca@Al_2O_3 catalyst was synthesized by co-precipitation with Al_2O_3 as the carrier, and the effect of Fe/Ca ratio on the production and content of active components (CO+H_2) in pyrolysis gas were studied. With the addition of catalysts, the pyrolysis process was advanced, and the final temperature rise was high. 12Ca/Fe catalyst has the most obvious effect on improving the production of H_2 and CO, with an increase rate of 36.6% and 54.8%, respectively.

1. Introduction

Microwave pyrolysis coal has the characteristics of fast heating, short time and high content of active components. However, lignite has poor microwave absorption ability. By adding semi-coke and other wave-absorbing materials, the heating state of lignite can be improved and the pyrolysis efficiency can be further improved[1,2]. The study[3,4] shows that the salt or oxide of Fe, Ca and other metal elements in the ash content of lignite can improve the production and quality of pyrolysis gas, among which Fe has an obvious catalytic effect on the formation of H_2. Therefore, it is very important to study the basic characteristics of microwave pyrolysis of lignite and its gas production.

2. Experimental

2.1. Coal samples

This experiment takes Mongdong lignite as the research object. Before the experiment, the coal sample was firstly dried and crushed, and the coal with particle size less than 0.15mm was screened out and put into an airtight bag for storage. Weigh 20.00±0.02 g coal sample, put it in a vacuum drying oven, dry it at 65 ℃ for 12 h, and get 1# coal sample. The industrial analysis of coal samples is shown in table 1.

Table 1 Industrial analysis of coal samples

Coal sample	M_{ad}/%	V_{ad}/%	A_{ad}/%	FC_{ad}/%
1#	14.3	28.9	4.8	52.0

2.2. Catalyst materials and preparation method

In this experiment, coprecipitation method was selected to prepare the catalyst. The specific steps were as follows: According to Ca/Fe ratio of 3:1, 6:1, 9:1 (the ratio of the amount of material), as seen in table 2, take corresponding quality $Ca(NO_3)_2{\bullet}4H_2O$ and $Fe(NO_3)_3{\bullet}9H_2O$, then take a certain amount of $Al(NO_3)_3{\bullet}6H_2O$ and $CO(NH)_2$, and add 100 ml deionized water. All of them were fully stirred to dissolve. The conical flask containing mixture was put into the oil bath pot and stirred vigorously by electromagnetic stirring way from ambient temperature to 120 ℃. And take it out after 4 h to filter when it was cooled. Subsequently, the solid should be drying 8h in a vacuum drying oven where the temperature was 80 ℃. Then it was placed in a muffle oven and burned at 875 ℃ for 4 h. After cooling, take it out and put into a sealed bag for later using.

Table 2. Name and number of catalyst

catalyst	Fe/Ca	Sample
$3Ca/Fe@Al_2O_3$	3:1	5#
$6Ca/Fe@Al_2O_3$	6:1	6#
$9Ca/Fe@Al_2O_3$	9:1	7#
$12Ca/Fe@Al_2O_3$	12:1	8#

2.3 Experimental equipment and methods

The experiment was carried out in a modified Galanz P70D20TL-D4 microwave oven with a resonant cavity size of 220×200×180 mm, magnetron model of M24FB-610A and rated output power of 700 W. Weigh a certain amount of 1# coal sample and catalyst and grind them evenly in a mortar according to the mass ratio of 5:2. The mixed samples were numbered according to table 2. Accurately weigh 3.5 g sample in the quartz reactor for each pyrolysis experiment. The upper part of the samples was covered with quartz cotton with a diameter of 40 mm and a thickness of 20 mm. The purpose was to prevent the coal powder from entering the subsequent pipelines along with the pyrolysis airflow and causing the pipeline blockage. Before the experiment control N_2 flow rate of 250 ml/min, scavenging for 10 minutes, then the flow control for 25 ml/min, start the microwave oven, microwave output power setting of 700 W, pyrolysis gas through condensing unit, in turn tar filter unit, gas drying device, the gas sampling by aluminum foil bag collection, every 2 minutes to replace an air pocket, collect altogether seven times, the pyrolysis time about 14 min, the gases in the air pocket product first collected by drainage method measuring different times gas volume(V_j). Then, Gasboard-3100 gas component analyzer (Wuhan Sifang Photoelectric) was used to measure gas volume fraction(x_i). CO_2, CO and CH_4 were detected by infrared sensor, H_2 was detected by thermal conductivity sensor, and the detection accuracy was ±1%. The output of corresponding components was calculated according to equation (1).

$$y_i = \frac{\sum V_j \times x_{ji}}{m_{daf}} \tag{1}$$

Average gas generation rate in different time periods was calculated according to equation (2), where m_{daf} was the dry ash-free mass of coal sample:

$$v_i = \frac{y_i}{t} \tag{2}$$

3. Results and discussion

3.1 Pyrolysis gas generation characteristics

Average generation rates of CO, CO_2, CH_4 and H_2 in different time periods were shown in figure 1. Different proportions of Ca/Fe catalysts have a significant impact on the release process of H_2. H_2 was generated in 5# and 8# coal samples within 0-2 min, among which 5# coal sample reached the maximum

H_2 production rate in the pyrolysis process, which was 50.3 ml•g^{-1}•min^{-1}, and 8# coal sample reached the maximum H_2 production rate within 2-4 min, which was 59.9 ml•g^{-1}•min^{-1}, and then gradually decreased. The peak shape of the change curve of H_2 production rate of 6# coal sample was relatively sharp, and the peak rate was the highest of the five coal samples, 66.9 ml•g^{-1}•min^{-1}, and then gradually decreased, which was an explosive H_2 production process. The change trend of H_2 production rate of the 1# and 7# coal samples was similar. Compared with the 6# coal samples, the peak shape was relatively mild. The H_2 production rate first gradually increased to the peak value of 39.8 ml•g^{-1}•min^{-1} and 53.3 ml•g^{-1}•min^{-1}, respectively, and then gradually decreased. By analyzing the H_2 generation rate curve, it can be seen that 3Ca/Fe@Al_2O_3 and 12Ca/Fe@Al_2O_3 make the H_2 process advance, and the addition of these three catalysts was improved because the temperature changes in the material within 0-2 min. The dielectric constant and dielectric loss factor of the material enable the material to be well coupled with the microwave field, reduce the ineffective reflection of the microwave, efficiently convert the electromagnetic energy into internal energy, and quickly reach the temperature required for pyrolysis, which improved the energy use efficiency. The rate of CO production was similar to that of H_2. In the process of CO_2 release, except for 5# and 8# coal samples, the CO_2 release of all other coal samples was concentrated in 2-4 min. Among them, the CO_2 generation rate of 6# coal sample was the highest among 5# coal samples, which was 11.2 ml•g^{-1}•min^{-1}. In addition, in the middle and late stage of pyrolysis (4-8 min), 1# coal sample still has an obvious CO_2 release process. The 5# and 8# coal samples reached the maximum CO_2 release rate within 0-2 min, which were 5.8 ml•g^{-1}•min^{-1} and 4.8ml•g^{-1}•min^{-1}, respectively. After 4 min, there was no obvious CO_2 release sign, and the CO_2 generation process basically ended. By comparing the CO_2 release rate curves of 1# and 7# coal samples, it can be seen that they have almost the same CO_2 release rate between 0-4 min. However, after 4 min, the material enters the constant temperature range, and the CO_2 generation rate of 7# coal sample slowed down significantly, which may be caused by the reaction of CO_2 generated by pyrolysis and CaO in the catalyst, leading to the reduction of CO_2 collected. The release law of CH_4 was similar to that of CO_2.

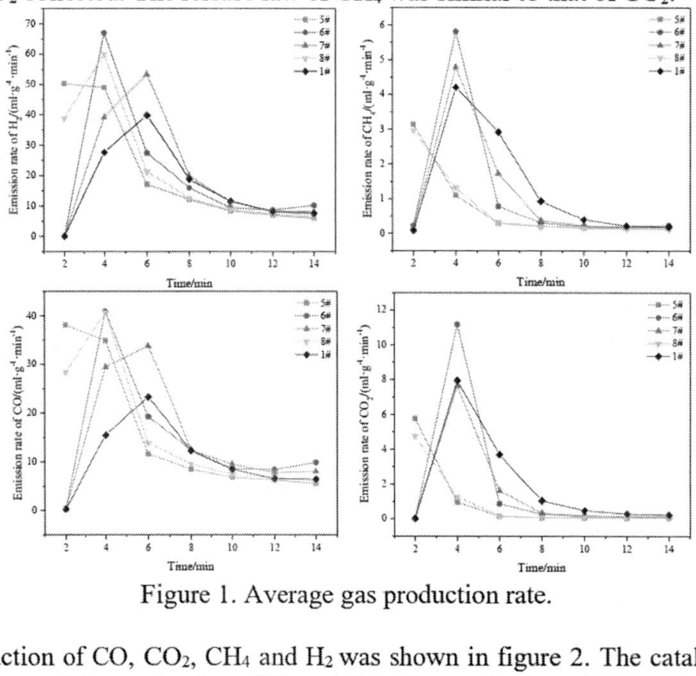

Figure 1. Average gas production rate.

The total production of CO, CO_2, CH_4 and H_2 was shown in figure 2. The catalysts with different Ca/Fe ratios all improved the production of H_2 and CO. The production of H_2 was in order from low to high: 8#>5#> 7#>6#>1#. And the yield was respectively 310.7 ml•g^{-1}, 300.2 ml•g^{-1}, 281.6 ml•g^{-1}, 277.9 ml•g^{-1} and 227.5 ml•g^{-1}. The production of CO was in order from low to high:8#>5#>7#>6#>1#, and the yield was respectively 225.38 ml•g^{-1}, 223.8 ml•g^{-1}, 202.6 ml•g^{-1}, 199.5 ml•g^{-1}, and 145.5 ml•g^{-1}. The variation trend of H_2 yield with different catalysts was very similar to that of CO. The CO_2 yield of 8#

IOP Conf. Series: Materials Science and Engineering **729** (2020) 012073 doi:10.1088/1757-899X/729/1/012073

coal sample decreased by 52.7% compared with that of 1# coal sample, and the corresponding catalyst was 12Ca/Fe. As Ca/Fe increased from 3:1 to 12:1, CO_2 production showed a trend of first increasing and then decreasing, among which CO_2 production of coal sample 6# was the highest, which was 25.5 ml $\cdot g^{-1}$. The highest CH_4 yield was 17.8 ml$\cdot g^{-1}$ for 1# coal sample, followed by 15.5 ml$\cdot g^{-1}$ and 15.3 ml$\cdot g^{-1}$ for 6# and 7# coal sample, respectively. The lowest CH_4 yield was 10.4 ml$\cdot g^{-1}$ and 10.3 ml$\cdot g^{-1}$ for 5# and 8# coal sample, respectively. The total gas production and the content of each component were shown in Table 3. It can be seen from the figure that the addition of catalyst increased the total gas production, and the total output of 8# coal sample gas was the highest, 559.2 ml$\cdot g^{-1}$. Compared with the 1# coal sample, the CO content in the pyrolysis gas composition of the other coal samples was significantly improved, and the CO_2 content was significantly decreased, indicating that the addition of the catalyst increased the content of the effective components in the gas and improved the quality of the pyrolysis gas.

Table 3. Total gas production and relative content of each component

Sample	CO/vol.%	CO_2/vol.%	CH_4/vol.%	H_2/vol.%	Production/(ml/g)
1#	34.8	6.5	4.3	54.4	418.1
5#	40.8	2.0	1.9	54.7	548.6
6#	38.5	4.9	3.0	53.6	518.4
7#	39.0	3.9	2.9	54.2	519.7
8#	40.3	2.3	1.8	55.6	559.2

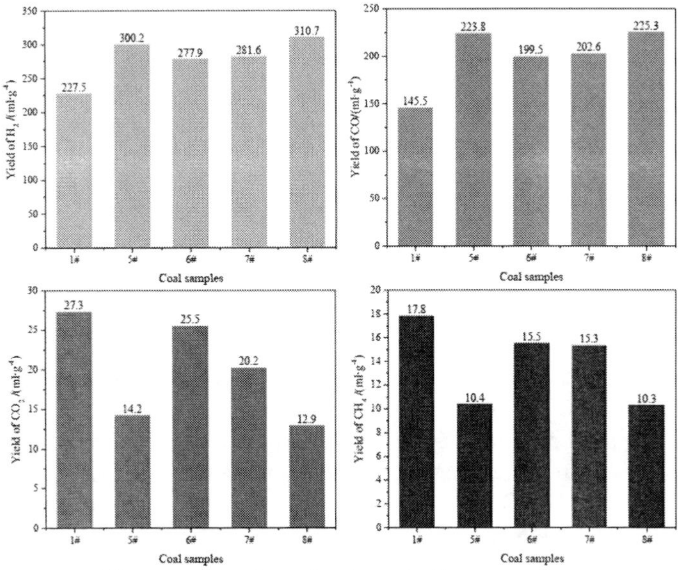

Figure 2. Total production of all components

Table 4 showed the weight loss of samples before and after pyrolysis and the temperature at the end of pyrolysis experiment. It can be seen from the table that after adding the catalyst, the final temperature of the sample was improved, indicating that the addition of the catalyst increases the apparent dielectric loss factor of the sample, the material absorbing ability was enhanced, and the efficiency of converting electromagnetic energy into internal energy was improved. The temperature was the highest after adding 3Ca/Fe@Al_2O_3, and further increasing the content of Ca in the catalyst caused the temperature of the material to decrease. Compared with the 1# coal sample, the weight loss rate of the sample also increased after adding catalyst, which indicated that the volatile products increased. This was because the catalyst not only reduced the reaction activation energy, but also promoted the mutual coupling between the material and electromagnetic wave, thus promoting the generation of gas.

Table 4. Loss weight of coal sample and final temperature of pyrolysis

	1#	5#	6#	7#	8#
T/°C	608.0°C	657.6°C	619.6°C	616.1°C	629.4°C
Loss/%	48.8%	56.6%	54.0%	54.3%	55.5%

3.2 Semi-coke FTIR analysis

Infrared spectra of raw coal and semi-coke after pyrolysis were shown in figure 3. As can be seen from the figure, within the range of wave Numbers of 2820 cm^{-1}-2950 cm^{-1} and 1100 cm^{-1}-1780 cm^{-1}, the absorption peak strength of the coal sample after pyrolysis was weakened. Including 2820 cm^{-1} - 2850 cm^{-1} within the scope of the corresponding absorption peak of aliphatic hydrocarbon vibration absorption peak of -CH_2 and -CH_3, their strength showed that the fat in the process of pyrolysis hydrocarbon side chain fracture, fall off from the brown coal macromolecular, become volatile and escape from the system, notice that after adding catalyst, absorption peak weakened further, show that the preparation of the catalyst was catalyst for this process. 1100 cm^{-1}-1780 cm^{-1} was mainly the absorption peak of oxygen-containing functional groups such as ether bond, ester bond and carbonyl group. Compared with raw coal pyrolysis semi-coke, the absorption peak in this range was obviously weak. It was speculated that the reason was that the catalyst promoted the removal of functional groups such as carbonyl group and the reaction with CH_4, increasing the content of CO and H_2 in the gas products, which was consistent with the gas production data.

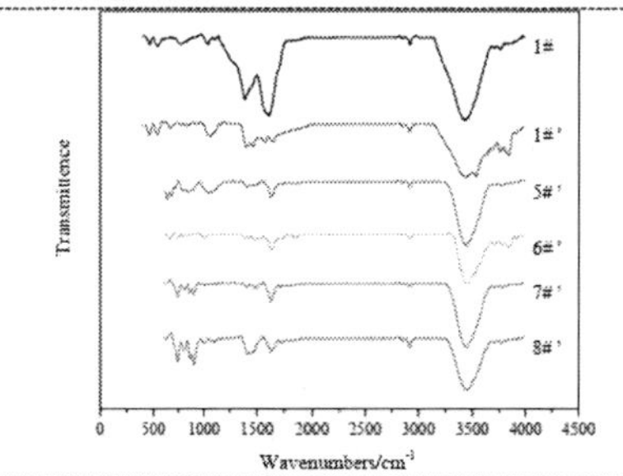

Figure 3. FTIR map of raw coal and semi-coke after Pyrolysis.

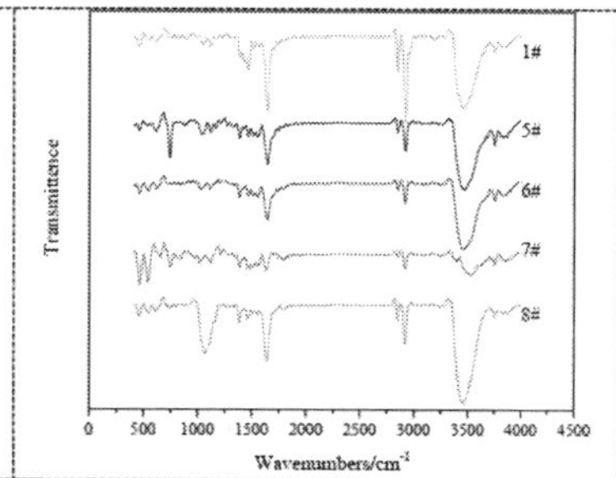

Figure 4. FTIR spectrum of tar.

3.3 Tar FTIR analysis

Infrared spectra of pyrolysis tar of different coal samples were shown in figure 4. As can be seen from the figure, the positions of the peaks were basically similar, indicated that the types of chemicals in the tar were basically the same. Compared with 1# coal sample, the absorption peak intensity in the infrared spectrum of the remaining coal sample pyrolysis tar in the range of 2825 cm^{-1}-2950 cm^{-1} and 1300 cm^{-1}-1500 cm^{-1} was weakened, indicated that the fat in the tar was -CH_2, -CH_3, etc. Hydrocarbon side chain content was reduced [5,6]. This aspect was due to the adsorption of Fe element in the rich position of the lignite macromolecule, which leaded to the weakening of the chemical bond between the side chain and the host, which promoted the shedding of small molecular groups. On the one hand, the catalyst formed a hot spot under the microwave field, and the volatiles undergo secondary decomposition during the precipitation process, resulting in a decrease in the aliphatic hydrocarbon side chain and an increase in gas production in the tar. The 5# coal sample pyrolysis tar has an obvious absorption peak at 750 cm^{-1}, which corresponded to the out-of-plane bending vibration of four adjacent H in the aromatic ring, indicating that 0.03Ca/Fe@Al_2O_3 catalyst promoted the transfer from H to tar. The degree of polycondensation of the aromatic compound in the tar was lowered, and the number of substituents was decreased.

4. Conclusions

In the experiment of microwave pyrolysis, the addition of catalysts made the pyrolysis process advance and the final temperature rise was high, indicating that the addition of catalysts increased the apparent dielectric loss factor of materials and enhanced the coupling ability with the microwave field. All the four catalysts inhibited the release of CO_2 in the microwave pyrolysis experiment, of which 12Ca/Fe had the most obvious effect. Compared with 1# coal sample, CO_2 production decreased by 52.7%. As the release of CO_2 was inhibited, the total content of CO and H_2 in the gas composition was increased. In conventional pyrolysis experiments, only 12Ca/Fe catalyst reduced CO_2 production. In the experiment of microwave pyrolysis, 12Ca/Fe catalyst has the most obvious effect on improving the production of H_2 and CO, with the increased rates of 36.6% and 54.8%, respectively. Compared with microwave pyrolysis, although the catalytic effect on the release of H_2 and CO was more obvious under conventional pyrolysis, the total production was far behind, indicating that microwave catalytic pyrolysis can improve the production of CO and H_2 more effectively.

Acknowledgements

This research was supported by Open Foundation of State Key Laboratory of High-efficiency Coal Utilization and Green Chemical Engineering (2018-K19), the Excellent Young Talents Training Program of WUST (2017xz009) and the Defense Pre-research Program of WUST (YY1050003).

References

[1] Huang Y F, Peite Chiueh, Wen H K, Lo S L.(2013) Microwave pyrolysis of rice straw: products, mechanism, and kinetics. Bioresource Technology, 142(7): 620-624.
[2] Wang Q D, Wang G H, Li W B, Chen B. (2016)Production of Hydrogen - Rich Syngas from Lignite using Different Pyrolysis Methods. Energy Technology, 4(6): 751-757.
[3] Zhao X Q, Guo B W, Wang W L,Chen G F,Mao Y P,Sun J,Song Z L.(2017) Experimental study on microwave pyrolysis of three Chinese lignite. Journal of Analytical and Applied Pyrolysis, 2017, 124: 303-309.
[4] Haque Kazi E.(1999) Microwave energy for mineral treatment processes—a brief review. International Journal of Mineral Processing, 57(1): 1-24.
[5] Yao Q X,Li Y b,Tang X,Gao J W,Wang R C,Zhang Y J,Sun M,Ma X X.(2019)Separation of petroleum ether extracted residue of low temperature coal tar by chromatography column and structural feature of fractions by TG-FTIR and PY-GC/MS. Fuel, 245: 122-130.
[6] Ye C P,Yang Z J,Li W Y,Rong H L,Feng J.(2017) Effect of adjusting coal properties on HulunBuir lignite pyrolysis [J]. Fuel Processing Technology, 2017, 156: 415-420.

IOP Conf. Series: Materials Science and Engineering **729** (2020) 012074 doi:10.1088/1757-899X/729/1/012074

Solvent swelling behaviour of Shenmu-Fugu and Shengli coals

Yan-Bin Wei[1], Xian-Yong Wei[2,3,*], Ting-Ting Sun[1] and Zhi-Min Zong[2]

[1] School of Chemical Engineering, Xuzhou College of Industrial Technology, Xuzhou, Jiangsu, 221140, China

[2] Key Laboratory of Coal Processing and Efficient Utilization, Ministry of Education, China University of Mining & Technology, Xuzhou, Jiangsu, 221116, China

[3] State Key Laboratory of High-efficiency Coal Utilization and Green Chemical Engineering, Ningxia University, Yinchuan, Ningxia, 750021, China

*Corresponding author's e-mail: wei_xianyong@163.com

Abstract: Swelling behaviors of Shenmu-Fugu bituminous coal (SFBC) and Shengli lignite (SLL) in various organic solvents were investigated at temperatures ranging from 0 °C to 55 °C. The results show that tetrahydrofuran (THF) has a relatively higher swelling ratio in comparison to other solvents, suggesting that THF likely disrupts coal-coal cross-links between macromolecular network structures in coal. SLL swelled more highly than SFBC in methanol (MA), indicating that MA interacts strongly with specific sites on SLL. Swelling ratios of SFBC in carbon disulfide (CDS), recycle liquefaction solvent (RLS) and initial liquefaction solvent (ILS) are appreciably higher than those of SLL. Coal swelling increases with raising temperature within the temperature range studied (0-55 °C), especially in polar solvent.

1. Introduction

Solvent swelling is a widely used, convenient and low-cost method for understanding the nature of fossil fuels with cross-linked macromolecular network structures. Solvent swelling has been extensively used to determine solubility parameters [1], to characterize solvent-macromolecule interactions [2], for calculating the average molecular weight between cross-links, the cross-link density, and the number of repeat units per cross-link of macromolecular network structures [3-5]. The covalent cross-links and the noncovalent interactions of coal-solvent and coal-coal are two important factors in determining the extent of swelling [6-8]. In addition, swelling ratio is closely related to coal rank [8,9], swelling time [10], solvent type [11-13], and temperature [14]. Understanding the relationships is helpful for optimizing coal conversion processes, especially coal hydroliquefaction (CHL).

Solvent swelling of coal is a complex process. Numerous previous attempts have been made to better understand the swelling behaviour of coal [8,14-18]. A set of nonpolar solvents (cyclohexane, *n*-hexane, benzene, ethylenediamine, and carbon disulfide), polar solvents (pyridine, *N*-methyl-2-pyrrolidone (NMP), tetrahydrofuran (THF), butylamine, and hexylamine), and binary mixtures (pyridine-water, pyridine-toluene, pyridine-cyclohexane) have been used in these studies.

Hu et al. [19] investigated the effect of solvent swelling on the extraction of Xinglong coal. The results show that swelling pretreatment of coal with THF or pyridine can significantly enhance conversion and light-fraction yield of supercritical extraction process. Recently, direct coal liquefaction in the world, both on research and industrialization, has become a hot issue, and has been pushed to the peak of interest, especially for China which is a rapid developing country with limited petroleum reserves [20]. Although a large amount of work has been done, CHL is not yet competitive economically due to its harsh reaction condition and huge equipment investment. A variety of attempts, e.g., pre-swelling of coal [21] and the use of highly active catalysts [22,23], have been made to better develop economically feasible processes for CHL. Swelling is considered to be a critical process to determine coal reactivity for a liquid-solid mixture or a slurry system [11], such as CHL. Shenmu-Fugu bituminous coal (SFBC) and Shengli lignite (SLL) are of high quality being low in ash content, and extremely low in sulphur and suitable for direct liquefaction. Understanding swelling behavior of SFBC and SLL under different conditions is a subject for optimization of mild and directional CHL process. The aim of this investigation was to compare the solvent swelling behavior of SFBC and SLL in various organic solvents at different temperatures in order to provide a scientific basis for future sustainable development and industrialization of CHL.

2. Experimental

2.1. Coal sample and solvents

Recycle liquefaction solvent (RLS), initial liquefaction solvent (ILS), SFBC and SLL were provided by Beijing Research Institute of Coal Chemistry, China Coal Research Institute. Coal samples were pulverized to 200 mesh sieve and then dried in a vacuum at 80 °C for 24 h. Table 1 shows the proximate and ultimate analyses of the dried samples. Solvents carbon disulfide (CDS), methanol (MA), acetone (AT) and THF are commercially purchased analytical-pure reagents and were distilled prior to use.

Table 1. Proximate and ultimate analyses (*wt*%) of SFBC and SLL.

Sample	Proximate analysis			Ultimate analysis (*wt* %, daf)				
	M_{ad}	A_d	V_{daf}	C	H	N	S	O_{diff}
SFBC	5.33	6.32	30.74	79.82	4.73	1.05	0.50	13.90
SLL	20.41	19.00	45.99	72.20	4.95	0.30	1.10	21.45

2.2. Swelling procedure

About 3 g of the coal samples were placed in constant diameter glass tubes (5 mm i.d., ca. 120 mm length) and centrifuged in a Hitachi CR 22E centrifuge at 7500 r/min for 5 min and then the initial sample layer height (h_i) was measured. Solvent was then added to the tube followed by ultrasonic irradiation in a JY92-IID ultrasonic tub for a prescribed period of time at indicated temperature. Then, the tube was centrifuged again as above, and the final sample layer height (h_f) was measured. The volumetric swelling ratio is equal to the ratio of h_f to h_i. The amount of solvent added was determined to ensure that the height of the solvent surface is a little higher than that of the coal surface.

3. Results and discussion

3.1. Effects of time and solvent

As Table 2 shows, the swelling of SFBC reaches equilibrium within 2 min in MA, 3 min in THF and CDS and 4 min in RLS and ILS. Equilibrium swelling ratio in different solvents decrease in the order: THF (2.00) >> CDS (1.33) > MA(1.24) > RLS (1.19) > ILS (1.14). The equilibrium swelling of SLL needs 3 min in AT, 4 min in MA, CDS and THF, 5 min in RLS and ILS, as exhibited in Table 2 and Figure 1. We obtained swelling ratio values of 1.65, 1.33, 1.24, 1.09, 1.04, and 1.03 in THF, MA, AT, CDS, RLS, and ILS, respectively.

Table 2. Equilibrium swelling time and swelling ratios of SFBC and SLL in various solvents at 35 °C.

Sample	Equilibrium swelling time (min)						swelling ratio					
	THF	CDS	MA	AT	RLS	ILS	THF	CDS	MA	AT	RLS	ILS
SFBC	3	3	2	/	4	4	2.00	1.33	1.24	/	1.19	1.14
SLL	4	4	4	3	5	5	1.65	1.09	1.33	1.24	1.04	1.03

THHF and pyridine are commonly quite good swelling solvents for bituminous coals. Swelling ratio of SFBC in THF is drastically higher than that in other solvents and also higher than that of SLL, as exhibited in Figure 1. THF is specifically interacting solvent, meaning that it is able to dissociate specific interactions (particularly hydrogen-bonding) within the coal. In comparison to other solvents, the much greater efficacy of THF may be partly due to the greater electron donor strengths of THF, but may also be due to the greater affinity of THF for coal [24]. Wang et al. [25] suggested according to the type of THF adsorption isotherm that THF may break the original hydrogen-bond cross-links between macromolecular chains in coal.

Swelling of coal changes with coal rank and solvent type. As shown in Figure 1, swelling ratios of SFBC in CDS, RLS and ILS are appreciably higher than those of SLL. Swelling ratios of SLL in polar solvents are significantly higher than those in nonpolar solvents. Takanohashi et al. [12] investigated swelling of low-rank coals with various solvents at room temperature. The results show that low-rank coals swelled more highly than higher-rank (subbituminous coals and bituminous coals) coals in polar solvents. As Figure 1 exhibits, swelling ratio of SLL in MA is appreciably higher than that in CDS, whereas for SFBC, swelling ratio in MA is slightly lower than that in CDS. According to the results reported by Takanohashi et al. [12], swelling ratio of low-rank coals in MA is appreciably higher than those in CDS, which are consistent with our results. In general, low-rank coals contain larger amounts of polar groups, e.g., carboxyl and hydroxyl, to interact with solvent by forming hydrogen bonds, which play a very important role in determining the swelling of coal in polar solvents [8]. This may be the dominate reason that SFBC has lower swelling ratio values than SLL in MA. In addition, higher-rank coals except anthracites comprise of more aromatic rings (ARs), especially more condensed ARs. The strong interactions between CDS and the ARs may be an important reason for higher swelling ratio of SFBC in CDS than in MA by weakening π–π interactions between ARs in SFBC. Wang et al. [25] also expected that nonpolar swelling solvent adsorb on coals by π–π interactions with ARs. The low swelling of SLL in CDS can be attributed to the strong ionic forces. Chen et al. believe that coal swelling in nonpolar solvents is mainly controlled by coal-coal interactions and that ionic forces in lignite restrains swelling of coal [8].

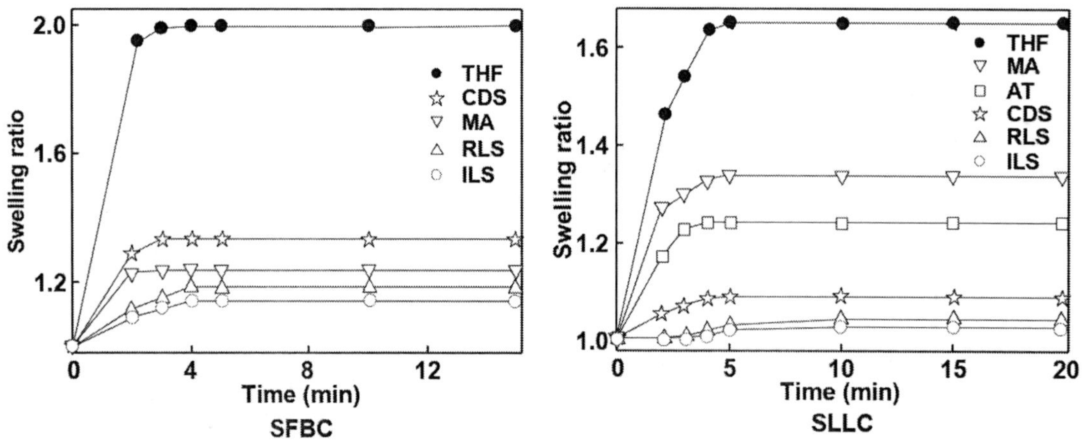

Figure 1. Time profiles of SFBC and SLL swelling in different solvents at 35 °C.

Although RLS and ILS are factually used in CHL process, both of them are poor solvents for SFBC and SLL swelling under the experimental conditions. RLS and ILS mainly consist of aliphatic and aromatic hydrocarbons with much higher molecular weights than THF, CDS and MA, correspondingly making them have much higher viscosity (16.84 cp at 40 °C for ILS, 15.87 cp at 40 °C for RLS). The high viscosity may retard the solvent access into micropores in coal, so the swelling of coals in them is very negligible, especially for SLL. These facts indicate that the high viscosity of ILS and RLS have significantly adverse effects on the higher conversion yield of CHL. Therefore how to improve swelling ratio of liquefaction coal in ILS and RLS is advantageous for the coal liquefaction process.

3.2. Effect of temperature

As Figure 2 exhibits, swelling ratio of SFBC appreciably increased with raising temperature and reach maximums at 35 °C in RLS and at 45 °C in other solvents. Noteworthily, swelling ratio of SFBC in both MA and ILS remarkably increased with raising temperature from 35 to 45 °C so that swelling ratio of SFBC in ILS becomes the same as that in RLS at 45 °C, but swelling ratio of SFBC in MA is still lower than that in CDS.

Swelling ratio of SLL in THF remarkably increased to ca. 1.71 from ca. 1.27 with raising temperature from 0 to 55 °C, only slightly increased in MA and AT and almost no change in CDS, ILS and RLS, as illustrated in Figure 2.

Several researchers examined the swelling behavior of different coals in various organic solvents at various temperatures [26,27]. Their results show that the ultimate swelling ratios are independent of temperature within the temperature range studied and that the rates of swelling are increased. Otake et al. [14] also confirmed that there is no significant effect of temperature on the extent of swelling in the range from 10 to 60 °C. But in the present experiments, we obtained the different results and noted swelling ratios of SFBC and SLL appreciably increased with raising temperature, especially in polar solvent. It is speculated that the rates of solvent diffusion increase with raising temperature. Increasing the temperature favors uptake of solvent by coal to strengthen the interactions of coal-solvent, especially for coal-polar solvent.

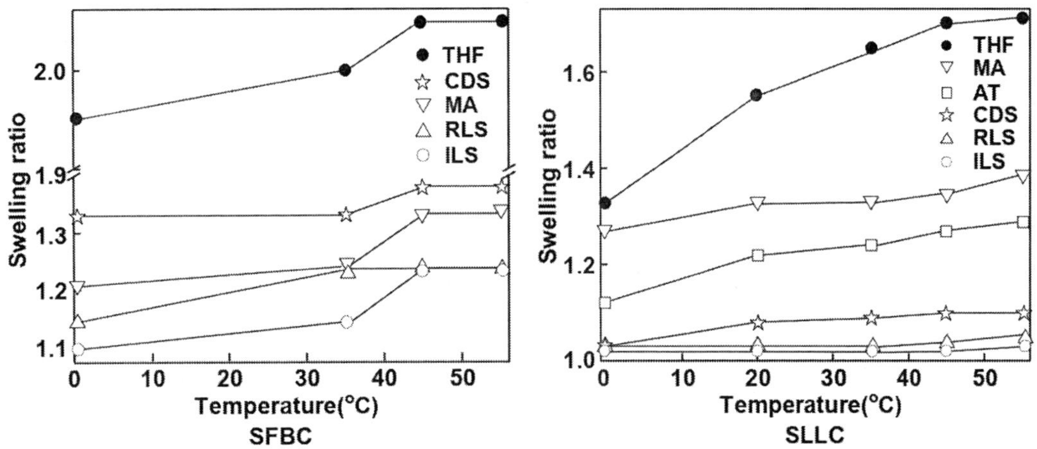

Figure 2. Temperature profiles of SFBC and SLL swelling in different solvents for 15 min.

4. Conclusions

The swelling of SFBC and SLL reach equilibrium within 5 min in various solvents. THF is quite good swelling solvents and RLS and ILS are very poor swelling solvents for SFBC and SLL. Swelling ratios of SFBC in CDS, RLS and ILS are appreciably higher than those of SLL. SLL has substantially higher swelling ratios in MA in comparison to SFBC. Swelling ratios of SFBC and SLL appreciably increased with raising temperature, especially in polar solvent.

Acknowledgments

This work was supported by the Key Research and Development Program (Social Development) of XuZhou (Grant KC18152) and a Project Funded by the Priority Academic Program Development of Jiangsu Higher Education Institutions.

References

[1] Painter, P.C., Graf, J., Coleman, M.M. (1990) Coal solubility and swelling. 1. Solubility parameters for coal and the Flory χ parameter. Energy Fuels, 4: 379-384.

[2] Barr-Howell, B.D., Peppas, N.A. (1985) Importance of junction functionality in highly crosslinked polymers. Polymer Bull, 13: 91-96.

[3] Painter, P.C., Park, Y., Sobkowiak, M., Coleman, M.M. (1990) Coal solubility and swelling. 2. Effect of hydrogen bonding on calculations of molecular weight from swelling measurements. Energy Fuels, 4: 384-393.

[4] Painter, P.C., Graf, J., Coleman, M.M. (1990) Coal solubility and swelling. 3. A model for coal swelling. Energy Fuels, 4: 393-397.

[5] Hruljova, J., Savest, N., Oja, V., Suuberg, E.M. (2013) Kukersite oil shale kerogen solvent swelling in binary mixtures. Fuel, 105:77-82.

[6] Green, T.K., Larsen, J.W. (1984) Coal swelling in binary solvent mixtures: Pyridine-chlorobenzene and N, N-dimethylaniline-alcohol. Fuel, 63: 1538-1543.

[7] Hall, P.J., Marsh, H., Thomas, K.M. (1988) Solvent induced swelling of coals to study macromolecular structure. Fuel, 67: 863-866.

[8] Chen, C., Gao, J.S., Yan, Y.J. (1998) Role of noncovalent bonding in swelling of coal. Energy Fuels, 1998, 12: 1328-1334.

[9] Sanada, Y., Honda, H. (1966) Swelling equilibrium of coal by pyridine at 25 °C. Fuel, 45: 295-300.

[10] Nishioka, M. (2001) Dependence of solvent swelling on coal concentration: A theoretical investigation. Energy Fuels, 15: 1270-1275.

[11] Nishioka, M. (2002) Role of solvation for coal swelling in slurry. Energy Fuels, 16: 1109-1115.

[12] Takanohashi, T., Yanagida, T., Iino, M. (1996) Extraction and swelling of low-rank coals with various solvents at room temperature. Energy Fuels, 10: 1128-1132.

[13] Ndaji, F. E., Thomas, K. M. (1995) Effects of solvent steric properties on the equilibrium swelling and kinetics of solvent swelling of coal. Fuel, 74: 842-845.

[14] Otake, Y., Suuberg, E. M. (1997) Temperature dependence of solvent swelling and diffusion processes in coals. Energy Fuels, 11: 1155-1164.

[15] Niekerk, D.V., Halleck, P.M., Mathews, J.P. (2010) Solvent swelling behavior of Permian-aged South African vitrinite-rich and inertinite-rich coals. Fuel, 89: 19-25.

[16] Chen, H.K., Miura, K., Li, B.Q., Li, W. (2003) Adsorption and swelling behaviors of coal in pyridine containing binary solvent mixtures. Journal of Fuel Chemistry and Technology, 31: 6-11.

[17] Pande, S., Sharma, D.K. (2001) Studies of kinetics of diffusion of *N*-methyl-2-pyrrolidone (NMP), ethylenediamine(EDA), and NMP+EDA (1:1, vol/vol) mixed solvent system in Chinakuri coal by solvent swelling techniques. Energy Fuels, 15: 1063-1068.

[18] Milligan, J.B., Thomas, K.M., Crelling, J.C. (1997) Solvent swelling of maceral concentrates. Energy Fuels, 11: 364-371.

[19] Hu, H.Q., Quan, H., Guo, S.C. (1997) Solvent swelling of coal for improved supercritical extraction. Journal of Fuel Chemistry and Technology, 25: 23-226.

[20] Liu, Z.Y., Shi, S.D., Li, Y.W. (2010) Coal liquefaction technologies-development in China and challenges in chemical reaction engineering. Chem Eng Sci., 65: 12-17.

[21] Shui, H.F., Liu, J.L., Wang, Z.C., Cao, M.X, Wei, X.Y. (2009) Effect of pre-swelling of coal at mild temperatures on its hydro-liquefaction properties. Fuel, 90: 1047-1051.

[22] Wang, Z.C., Shui, H.F., Pei, Z.N., Gao, J.S. (2008) Acidity and catalytic property of SO_4^{2-}/ZrO_2 on the hydro-liquefaction of coal. Journal of Fuel Chemistry and Technology, 36: 10-14.

[23] Şimşek, E.H., Karaduman, A., Olcay, A. (2001) Liquefaction of Turkish coals in tetralin with microwaves. Fuel Processing Technology, 73: 111-125.

[24] Yun, Y., Suuberg, E.M. (1998) Cooperative effects in solvent swelling of a bituminous coal. Energy Fuels, 12: 798-800.

[25] Wang, N., Sasaki, M., Yoshida, T., Kotanigawa, T. (1998) Characterization of different possible

solvent-coal interaction mechanisms by the relationship between the volumetric swelling of coals and heat release in swelling solvent. Energy Fuels, 12: 531-535.
[26] Ndaji, F.E., Thomas, K.M. (1993) The kinetics of coal solvent swelling using pyridine as solvent. Fuel, 72: 1525-1530.
[27] Otake, Y., Suuberg, E.M. (1998) Solvent swelling rates of low rank coals and implications regarding their structure. Fuel, 77: 901-904.

IOP Conf. Series: Materials Science and Engineering **729** (2020) 012075 doi:10.1088/1757-899X/729/1/012075

Low NOx combustion mechanism of coke oven gas with excess air coefficient and flue gas recirculation

Guanghua Wang[1,2], Zihao Yan[1,2*], Qingdong Wang[1] and Guocheng Wang[2]

[1]Research Center for Green and Intelligent Coal Chemical Engineering, Wuhan University of science and technology, Wuhan, Hubei Province, 430081, China

[2]Institute of Chemistry and Chemical, Wuhan University of science and technology, Wuhan, Hubei Province, 430081, China

*Corresponding author's e-mail: zhy_wust@163.com

Abstract: With the more stringent NOx emission limit from the coking chemical industry, the coking production with only one type of low NOx combustion technologies cannot meet the emission requirement. Thus a new low NOx experimental system was developed with the help of the external flue gas recirculation (FGR) and reasonable excess air coefficient, where the goal is to reduce NOx emission effectively. And an experiment was carried out on the 18 KW coke oven gas (COG) combustion device, and the effects of excess air coefficient, the ratio of flue gas recirculation and air preheating temperature on the combustion characteristics and NOx formation were extensively studied. The results show that there is an important effect of excess air coefficient, ratio of FGR and air preheating temperature on the temperature distribution inside the furnace and the formation of NOx. When the air excess coefficient of 1.2 and an external flue gas recirculation rate of 20% to 25%, the emission of thermal nitrogen oxides can be significantly reduced

1. Introduction

The ministry of environmental protection promulgated the emission standards for coking chemical industry pollutants in 2012[1]: From January 1, 2015, The NOx emission limit for existing companies is 500 mg/m^3. The emission limits imposed by some special areas are 150 mg/m^3. Through investigations of most coking chemical companies, it is found that the nitrogen oxide content of flue gas in tamping coke ovens above 5.5 m is generally between 1200 and 1400 mg/m^3, which seriously exceeds the emission limits, Therefore, the control of nitrogen oxides in the coking industry is imminent[2]. The low-nitrogen combustion technology of COG mainly focuses on how to avoid the formation of local high-temperature zone, so that the furnace temperature is evenly distributed along the high and horizontal directions, and the generation of thermal nitrogen oxides is reduced. Its main technologies include low nitrogen burners, flue gas recirculation and reasonable control of the air excess coefficient, etc. However, with low controllable precision, the industrial coke oven is bulky and complex, which takes a lot of manpower and material resources to directly test on the coke oven scale, and even causes environmental pollution.

Studies have shown that[3]when the temperature reaches 1800 K, the thermal NOx formation rate is significantly accelerated, and the growth rate will increase by 6 to 7 times for every 100 K increase in temperature. Baltasar et al.[4]believe that flue gas recirculation can significantly reduce NOx emissions without a significant impact on CO emissions by numerical simulations. Song Shao Peng et al.[5]believe that flue gas recirculation mainly inhibits the formation of thermal NOx, and the effect on temperature is greater than the influence of oxygen partial pressure, simultaneously it has a certain inhibitory effect on rapid NOx formation. Wang Zhi Ning et al.[6]studied the influence of internal/external flue gas recirculation on NOx formation in natural gas combustion process on a 0.8 MW pilot test bench. It was found that when the external flue gas recirculation rate increased to more than 20%, the rapid NOx mechanism played a leading role in overall NOx formation. Through numerical simulation, Stanislaw Gamrat et al.[7]found the flue gas external circulation can produce a more uniform temperature field in the coke oven combustion chamber, and the amount of NO_X produced linearly decreases with the strengthening of the flue gas recirculation through numerical simulation.

In order to meet the increasingly strict emission standards of the coking chemical industry, we design a laboratory-scale COG combustion test platform using a combination of flue gas recirculation technology and a reasonable control of the air excess coefficient. The effects of flue gas recirculation ratio, air excess coefficient and air preheating temperature on coke oven gas combustion stability, burnout rate, furnace temperature distribution and NO_X formation characteristics were studied.

2. Test system

2.1. experimental device

The test device is mainly composed of a burner with a nozzle, a pre-mixing chamber for flue gas, a mixed air heater, a flue gas cooler and a flue gas recirculation system. The burner nozzle is a non-premixed burner with a power of 18 kW. The structure parameters of the furnace body are reduced by 5 times in proportion to the 6.3 m coke oven vertical fire channel. The system is shown in Figure 1.

Seven temperature-measuring thermocouples of different heights are installed vertically in the furnace. The thermocouples T_1 to T_7 are installed vertically from the bottom of the furnace to the top of the furnace. The temperature measurement point mentioned in the later question represents the different height positions in the furnace of the furnace. The external flue gas recirculation system consists of the return pipe, a flue gas return regulating valve, a mixed air flow regulating valve, a wind-flue mixing chamber, an air preheater heater. The test data such as combustion air flow, flue gas return flow, air preheating temperature and flue gas analyzer are all fed back to the console in real time to accurately adjust various test parameters.

2.2. Fuel characteristics and experimental conditions

The test was carried out using the prepared COG, and its composition characteristics are shown in Table 1. The combustion furnace is first heated by liquefied petroleum gas(LPG). When the temperature exceeds 1600 K, the fuel is replaced with the COG. When the temperature of the combustion chamber is stable, and the instruments such as the flue gas analyzer and the gas analyzer are working normally, the experiment operation and data collecting begins.

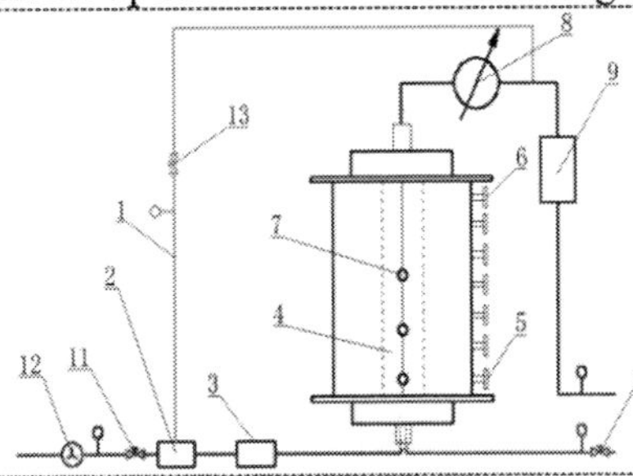

Figure 1. Low NO_X experimental system
1-Flue gas recirculation pipe; 2-Pre-mixingchamber; 3-Mixed air heater; 4-Furnace; 5-T_1; 6-T_7;7-Fire hole; 8-Flue gas heat exchanger; 9-Flue gas analyzer; 10-Gas valve; 11-Air valve; 12-Blower; 13-Reflux gas valve.

IOP Conf. Series: Materials Science and Engineering **729** (2020) 012075 doi:10.1088/1757-899X/729/1/012075

Table 1. Fuel characteristics

The volume fraction of Gas component/%				
H_2	CH_4	CO	CO_2	N_2
58	26	7	2	7

Under test conditions without FGR, firstly, the preheating temperature of the combustion air is 600 ° C, the gas flow rate is 2.84 m^3/h, and the air excess coefficient is 1.05, 1.10, 1.15, 1.20, 1.25. The test of the influence of the air excess coefficient on the NOx emission characteristics of COG combustion was carried out. Then, the COG combustion test was carried out under the condition that the preheating temperature of the combustion air was 500 °C, the flow rate was 2.84 m^3/h, the air excess coefficient was 1.20, and the flue gas circulation rate was 0, 15%, 20%, 25%. The impact of flue gas recirculation rate on NOx emissions is also analyzed.

3. Experimental results and analysis

3.1. Influence of air excess coefficient on NO_X generation

The air excess coefficient has an important influence on the furnace combustion temperature. The furnace temperature distribution can reflect the combustion state of the COG, and the maximum temperature in the furnace plays an important role in controlling the generation of thermal NO_X.

Figure 2 shows the temperature distribution of the furnace after the flue gas recirculation ratio is 0, the preheating temperature of the air is 650 °C under different air excess coefficients. With the increases of the air excess coefficients, the temperature in the furnace generally rises briefly and then decreases. According to the analysis, when the air excess coefficient is less than 1.10, the COG is not sufficiently contacted with oxygen, resulting in a lower overall temperature in the furnace. The increase of the air excess coefficient causes the air inlet flow rate to increase, which lengthens the flame height of the Combustion, resulting in a rapid drop in the temperature of the T1 thermocouple located at the bottom of the furnace. Figure 3 shows the variation of NOx production and maximum furnace temperature with air excess coefficient. In the low air excess coefficient, the amount of NO_X generated is large, and the amount of NO_X is mainly controlled by the fast type NO_X generation mechanism. When the air excess coefficients is higher than 1.1, the maximum temperature in the furnace gradually decreases, and the amount of NO_X is gradually decreased. This is due to the relatively sufficient combustion of COG and a reduction in local high temperature zones. It is worth noting that when air excess coefficient is between 1.10 and 1.25, the maximum temperature and the rate of NOx production show a high degree of consistency. With the rapid decrease in the maximum temperature and the slow recovery, the amount of NO_X produced rapidly drops, and then becomes a slow decline. This is because when the temperature rises above 1700 K, the generation of NOx is controlled by the fast type NOx turn to the thermal type NOx control. When air excess coefficients is greater than 1.15, the volume of the furnace having a temperature higher than 1700 K is reduced, and the amount of NOx generated is decreased. Therefore, when the air excess coefficient is controlled to 1.2, the maximum temperature in the combustion furnace can be lowered, and the generation of NOx can be effectively controlled.

IOP Conf. Series: Materials Science and Engineering **729** (2020) 012075 doi:10.1088/1757-899X/729/1/012075

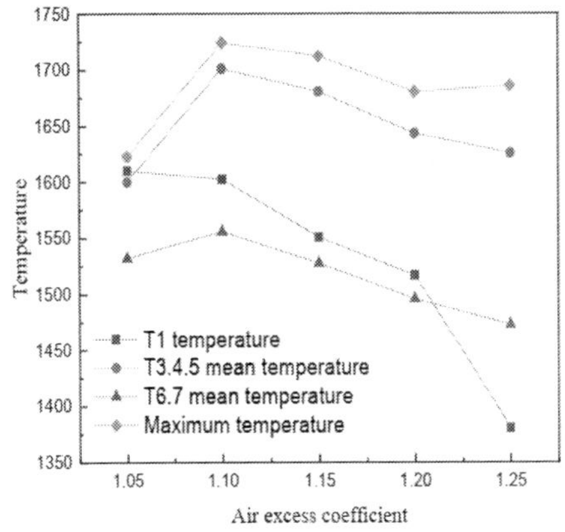

Figure 2. Influence of air excess coefficient on temperature distribution in furnace.

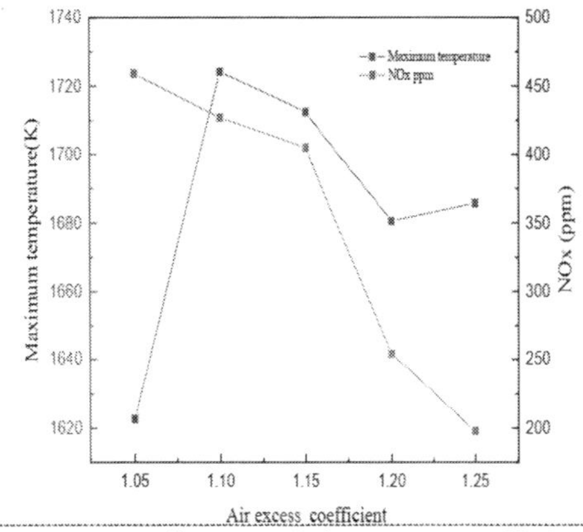

Figure 3. Influence of air excess coefficient on Max-temperature and NOx production in furnace.

3.2. Influence of flue gas recirculation on NO_X generation

The flue gas recirculation ratio has an important impact on NO_X emissions. Figure 5 shows the variation of NO_X and the maximum temperature in the furnace with the flue gas recirculation ratio when the coke oven gas flow rate is 2.84 m^3/h, the mixed air preheating temperature is 550 °C, and the air excess coefficient is 1.2. In the absence of flue gas recirculation, the temperature difference in the furnace is about 100 K. When the flue gas circulation is introduced, the temperature difference is reduced to less than 40 K. It can be seen that the introduction of flue gas recirculation can produce a more uniform temperature field in the combustion chamber. The numerical simulation experiment of COG combustion by Stanislaw Gamrat et al.[7]was verified. It can be seen from Figure 6 that when the flue gas circulation rate is increased from 0 to 20%, the amount of NO_X produced is reduced by 48%, and then flue gas recirculation ratio is increased, and the NO_X content changes slowly.

In summary, it can be concluded that when the air excess coefficient is 1.2 and the flue gas recirculation ratio is 20%, the generation of NO_X in the combustion furnace can be significantly reduced, but excessively high circulation amount also brings various degrees of side effects. Therefore, it is recommended that the maximum smoke circulation rate should not exceed 25%.

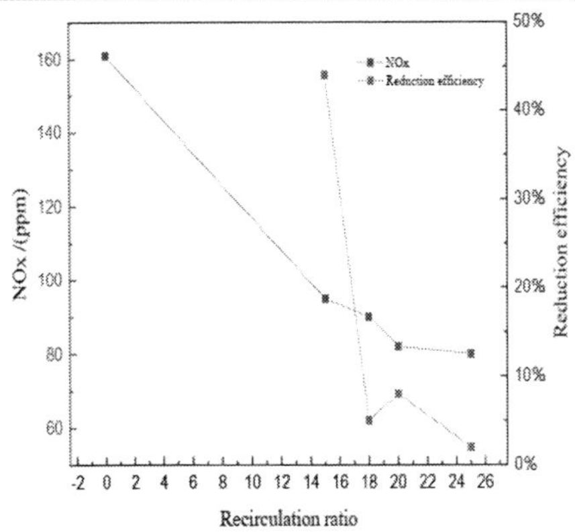

Figure 4. Influence of flue gas recirculation on NO_X production.

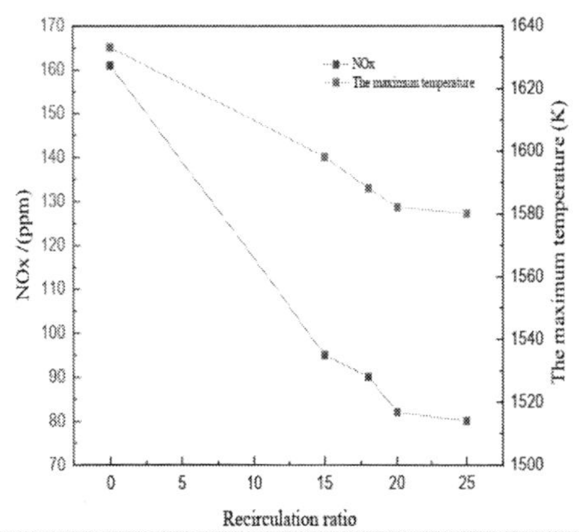

Figure 5. Influence of flue gas recirculation on Max-temperature and NO_X production.

3.3. Effect of air preheating temperature on NO_X formation and COG combustion characteristics

When the COG is burned, the flame profile is completely invisible, and the furnace is bright yellow, and the combustion is relatively stable. During the entire test, the smoke analyzer almost failed to detect the formation of CO, and no black carbon was formed on the fire hole.

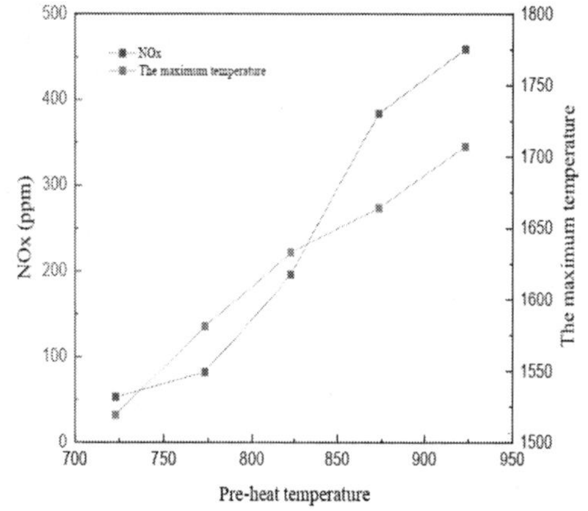

Figure 6. Influence of air preheating temperature on Max-temperature and NOx production.

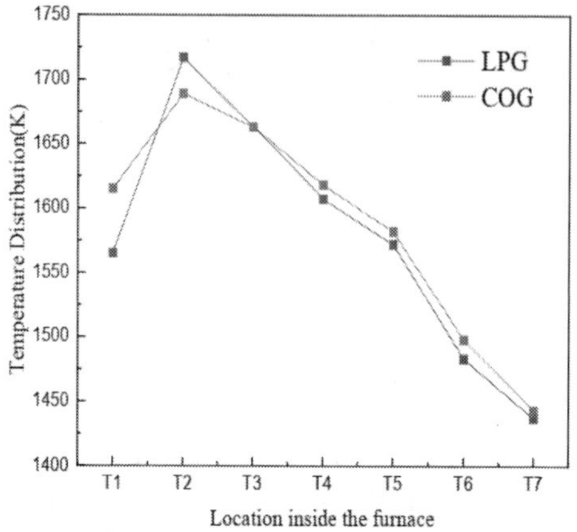

Figure 7. Influence of fuel species on temperature distribution inside the furnace

Comparing Figure 3 and Figure 4, when the preheating temperature of the combustion air is lowered by 100 K, the content of nitrogen oxides is decreased by 36% at this time. It can be seen that the generation of NOx is mainly controlled by the mechanism of the generation of thermal NOx. Figure 6 shows that the increase in air preheating temperature can directly increase the average temperature of COG combustion, the average temperature in the furnace rises by 47 K for every 50 K increase in air preheating temperature. Figure 7 is a comparison of COG and LPG combustion, the temperature distribution at different locations within the furnace when the average temperature in the furnace is the same. COG and LPG show the same trend of temperature change. However, the maximum temperature of LPG combustion is higher than COG, and the temperature at the inlet is lower than COG.

4. Conclusion and discussion

In this paper, a new low NOx experimental system for reducing NOx emissions from COG was designed and developed. The parameters of air excess coefficient, flue gas recirculation ratio and air preheating temperature were studied, the conclusions are as follows:

As the air excess coefficient increases, the amount of NOx decreases rapidly, and as it continues to increase, the rate of decline becomes slower. Controlling the air excess coefficient to 1.2 has an important effect on lowering the maximum temperature in the furnace and reducing the amount of NOx produced. Flue gas recirculation not only makes the temperature distribution in the furnace more uniform, but also reduces local high temperature areas and significantly reduces NOx emissions. The increase of air preheating temperature can directly increase the average temperature of COG combustion, the average temperature in the furnace rises by 47 K for every 50 K increase in air preheating temperature. When the air excess coefficient is 1.2 and the flue recirculation ratio is 20%, the NOx production amount can be reduced by 40%.

Acknowledgements

The authors gratefully acknowledge the Supported by National Key R&D Program of China (2017YFB0304300&2017YFB0304303).

References

[1] Ministry of Ecology and Environment of the people's Republic of China, (2012)Emission Standard of Pollutants For Coking Chemical Industry: China Environmental Science Press
[2]Dong X Z. (2017)The present situation of coking denitrification. Shandong Chemical industry,

46(19): 189-190.
[3] Turns Stephen R. (1995)Understanding NOx formation in nonpremixed flames: Experiments and modeling. Progress in Energy & Combustion Science, 21(5): 361-385.
[4] Baltasar J, Maria G, et al. (1997)Flue gas recirculation in a gas-fired laboratory furnace : measurements and modelling. Fuel, 76(10): 919-929.
[5] Song S P, Zhuo J K, Li N, et al. (2016)Low NOx Combustion Mechanism of a Natural Gas Burner With Fuel-staged and Flue Gas Recirculation. Proceedings of the Csee, 36(24): 6849-6858
[6] Wang Z N, Yang X H, Zhang Y, et al. (2019)I-/e-FGR effect on NO_X emission of natural gas combustion. Chemical Industry And Engineering Progress, 36(9): 4327-4334
[7] Stanislaw G, Jakub P, Jakub B, et al. (2016)Influence of external flue gas recirculation on gas combustion in a coke oven heating system. Fuel Processing Technology, 152: 430-437.

IOP Conf. Series: Materials Science and Engineering **729** (2020) 012076 doi:10.1088/1757-899X/729/1/012076

Study on adsorption performance of pine needle biochar on rhodamine B

Wenbing Li[1,2], Tie chen[2*], Zheng Zhang[1,2], Changjun Du[2] and Guanghua Wang[1,2]

[1]Research Center for Green and Intelligent Coal Chemical Engineering, Wuhan University of science and technology, Wuhan, Hubei Province, 430081, China

[2]Institute of Chemistry and Chemical, Wuhan University of science and technology, Wuhan, Hubei Province, 430081, China

[*]Corresponding author's e-mail: 1070789057@qq.com

Abstract. Biochar is regarded as a kind of high-efficiency, environmentally friendly and economical new adsorbent because of its low preparation cost, large specific surface area, developed pore structure and rich surface functional groups. It has broad application prospects in the field of sewage treatment. Using pine massoniana as raw material, the pine needle charcoal was obtained by pyrolysis at 300 °C, and the material was characterized by Fourier transform infrared spectroscopy (FTIR) and scanning electron microscopy (SEM). The dyeing wastewater was simulated with rhodamine B to study the pine needle. The adsorption performance of biochar showed that the adsorption kinetics of rhodamine B and the pseudo-second-order kinetic equation were higher, indicating that the adsorption rate was controlled by liquid membrane diffusion and intraparticle diffusion. The liquid membrane diffusion was main control step.

1. Introduction

Dyes are widely used in textile, leather, printing and dyeing, cosmetics and other industries. The environmental impact caused by dye wastewater generated in the production process is also increasingly serious. The dye wastewater has a large amount of COD, high chroma, deep color, high toxicity and complex composition. It has great harm to the water environment and human health. Therefore, there is an urgent need to find an efficient and low-cost dye wastewater treatment method[1]. The adsorption method is widely used in dye wastewater treatment research because of its simple operation, high efficiency, abundant source of adsorbed materials and low cost[2]. Biochar is a solid carbon-rich product produced by high temperature pyrolysis of bioorganic materials under anoxic conditions. The raw materials for the preparation of biochar are mainly biomass waste, and the utilization of waste resources can effectively alleviate environmental pollution[3]. The preparation methods of biochar mainly include thermal cracking, hydrothermal carbonization, gasification, etc. Different bioorganic materials, pyrolysis temperature, modification methods and other factors will have an impact on the performance of biochar. Existing biochar research has focused on areas such as soil performance regulation, global climate change, and renewable biofuels. In recent years, the application of biochar in the field of sewage treatment has gradually become a research hotspot[4]. At present, biochar adsorption technology is gradually applied in the field of sewage treatment. It has been shown that biochar has a good adsorption effect on typical pollutants such as heavy metals, nitrogen and phosphorus, which are widely present in aquaculture wastewater[5]. After being treated by magnetization, the biochar has good magnetic properties and is easily separated from the aqueous phase[6]. Compared with commercial activated carbon products, it is easier to apply to sewage treatment. At the same time, because biochar has good adsorption capacity for nitrogen and phosphorus, it can be used as a slow-release fertilizer after adsorption, and has agricultural environmental friendliness[7]. In this paper, the biochar was made from Pine massoniana and the dye wastewater was simulated with rhodamine B. The adsorption process of pine needle biochar on rhodamine B was studied.

2. Materials and methods

2.1. Materials and instruments

Materials: Wild horsetail pine needles in Jingmen, Hubei; Rhodamine B; Analysis of pure KOH. Instruments: SK2-4-10 tube type resistance furnace; Ultra-violet and visible spectrophotometer; Drying oven.

2.2. Preparation of pine needle biochar

Selection of Hubei jingmen wild pine massoniana, water to wash away the dust on the pine needles, dry and cut into small pieces, put in the oven baked 2 h under 80 ℃ atmosphere, remove the pine needles in the crucible, put in tube furnace, the crucible modulation tube furnace at 300 ℃ and 10 ℃ / min up to the insulation 2 h, cooling them, remove the biochar milling grinding into powder, washed with distilled water, into the dryer after drying, sieving, weighing and set aside[8].

2.3. Adsorption test of rhodamine B

The Put 0.1g biochar in a 250ml conical flask, add 100ml diluted 100mg /L rhodamine B solution into it, put it in a constant temperature water bath, stir it with magnetic force at the rotating speed of 400r/min, take samples at intervals, and take the supernatant to determine the concentration of rhodamine B. According to equation (1), the adsorption amount of biochar was calculated, and the mean value of all experiments was taken three times in parallel.

$$q = (C_0 - C_t)V/M \tag{1}$$

In the equation q- adsorbing capacity, mg/g;
C_0-Initial concentration of rhodamine B, mg/L;
C_t-Residual concentration in solution, mg/L;
V- The volume of rhodamine B solution, L;
M-The quality of the adsorbent, g.

3. Results and discussion

3.1. Characteristics of pine needle biochar

The chemical composition of biochar is shown in table 1. Compared with traditional peat, pine needle charcoal has higher carbon and hydrogen content and lower ash content. The strong adsorption ability of pine needle carbon is mainly due to its larger surface area and micropore volume. Although its polarity is better than that of traditional peat, which is not conducive to its adsorption of non-polar compounds, its surface area and micropore still play a dominant role[9].

The organic structure of biochar consists of two parts: a stacked graphene layer and an aromatic structure randomly dotted with a graphene layer, which promotes the biochar with a large specific surface area and a rich void structure. The larger specific surface area enhances the physical adsorption capacity of the biochar, and the rich void structure helps to adsorb organic matter of comparable molecular weight. The pine needle charcoal prepared by Pinus massoniana has a very rich void structure, and the mesoporous structure accounts for nearly 90% of the total void structure. The adsorption of rhodamine B mainly occurs in the mesoporous structure.

Table 1. Properties of pine needle charcoal and peat.

	Pine needle carbon	Peat
Elemental analysis (wt.%)		
Carbon	65.1	60.5
Hydrogen	4.3	3.9
Nitrogen	1.8	3.3
Ash content	9.8	18.7
Oxygen (by difference)	34.46	24.01
Surface properties		
BET surface area ($m^2 \cdot g^{-1}$)	389.4	285.6

Peat after time is greater than 0.5 h, adsorption quantity increases slowly, gradually into balance, in order to assure the so equilibrium time of adsorption can be fully for 5 h, homemade pine needles biochar adsorption quantity increases with the increase of the time, when the time is more than 5 h, the trend of increase gradually slow, gradually into balance, in order to assure the so equilibrium time of adsorption can be fully for 70 h.

Through comparison, it can be seen that the adsorption time of purchased peat to reach equilibrium is shorter than that of homemade pine needle biochar. The reason for this phenomenon may be that the pore diameter and surface area of peat are smaller, and part of the physical impurities inherent in the pine needle adsorbed on the fired activated carbon during the firing process[10].

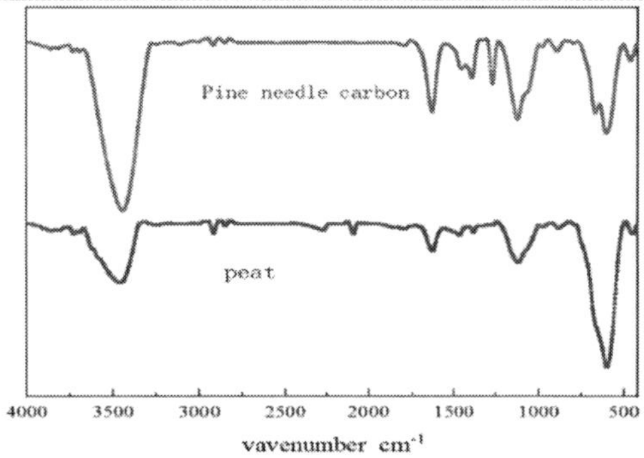

Figure 1. FTIR spectra of Pine needle charcoal and peat.

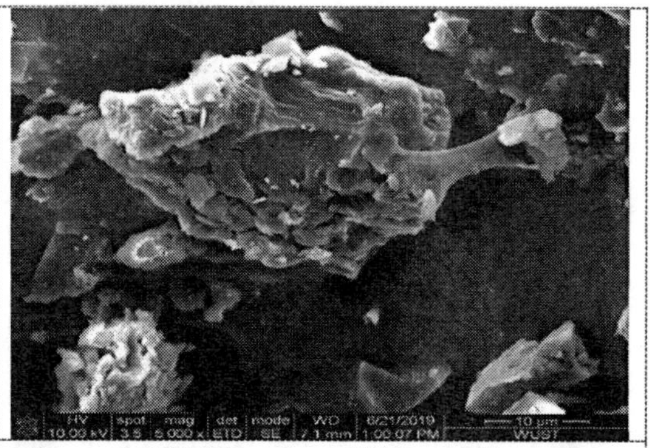

Figure 2. SEM images of Pine needle carbon.

3.2. Adsorption kinetics

When the initial concentration of rhodamine B solution was 100 mg/L, the amount of adsorbent was 50 mg/L, the pH was 7.0, and the temperature was 25 ℃, the effects of different adsorption times on rhodamine B were studied. The results were shown in the figure 3 below. The reaction was rapid within 2.0 h, and slow after 2.0 h. The initial rapid reaction was due to the high affinity sites on the surface of the adsorbent, and rhodamine B could occupy these sites quickly[11]. Later, the sites were reduced, and the internal diffusion of the particles was dominant, with a slow adsorption rate, which reached the adsorption equilibrium at 4.0h. The experimental data were fitted with the quasi-first-order kinetic equation, quasi-second-order kinetic equation and particle diffusion equation.

$$ln(Q_e - Q_t) = ln\, Q_e - k_1 \quad (2)$$

$$\frac{t}{Q_t} = \frac{1}{k_2 {Q_e}^2} + \frac{1}{Q_e} t \quad (3)$$

$$Q_t = k_3 t^{1/2} + C \quad (4)$$

Kinetic parameters obtained by linear fitting, including correlation coefficients (R2), Q_e, k_1, k_2, k_3, and C. It can be concluded from the table that the linearity of the quasi-secondary fit is higher than the quasi-first-order fit. It indicated that the adsorption mechanism of pine needle charcoal on rhodamine B may be chemical adsorption. In the intraparticle diffusion equation, if C = 0, it indicates that internal diffusion is the only factor affecting the adsorption process. However, as can be seen from the table, C>0. Therefore, internal diffusion is not the only control step affecting the adsorption of pine needle carbon on rhodamine B, and its adsorption process is affected by many factors[12].

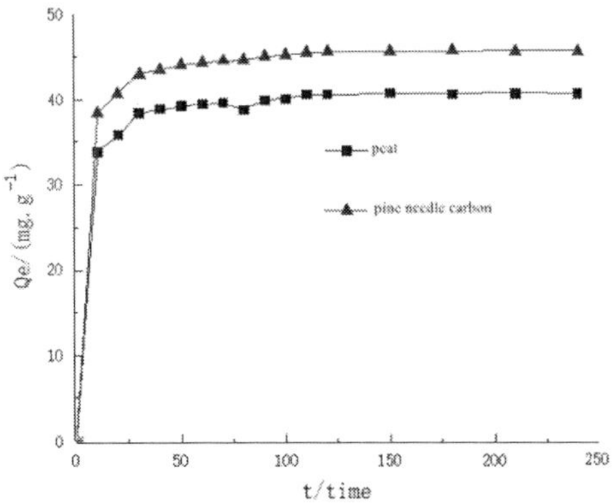

Fig 3.Effect of adsorption time on adsorption of rhodamine B by pine needle charcoal

4. Conclusion

The adsorption performance of biological carbon of pine needles is better than that of grass carbon on the market, and the self-made biological carbon of pine needles is a good adsorbent of dye, which has the characteristics of easy access to materials, simple production, low cost, fast adsorption efficiency, environmentally friendly, etc. It is a new adsorption material that can be vigorously promoted. The adsorption rate of pine needle charcoal to high and low concentration dye solution was faster. The dye adsorption process is mainly concentrated in the first 1 h, low concentration of dye solution. The adsorption can reach the equilibrium faster, the stabilization time is short, the diffusion speed of high concentration dye solution dye to the pores inside the adsorbent is slow, the duration is longer, the adsorption rate of the diffusion phase within the particles has an important influence on the adsorption rate of the whole process.

References

[1] Duan S, Xu X, Liu X, et al. (2018) Highly enhanced adsorption performance of U(VI) by non-thermal plasma modified magnetic Fe3O4 nanoparticles. Journal of Colloid and Interface Science, 513: 92 - 103.

[2] Karthick K, Namasivayam C, Pragasan L A. (2017) Kinetics and isotherm studies on acid dye adsorption using thermal and chemical activated Jatropha husk carbons . Environmental Progress & Sustainable Energy, 37(2):719 - 732.

[3] Reddy D H K, Yun Y-S. (2016) Spinel ferrite magnetic adsorbents: Alternative future materials for water purification? Coordination Chemistry Reviews, 315: 90 - 111.

[4] Warner C L, Chouyyok W, Mackie K E, et al. (2016) Manganese Doping of Magnetic Iron Oxide Nanoparticles: Tailoring Surface Reactivity for a Regenerable Heavy Metal Sorbent . Langmuir, 28(8): 3931 - 3937.

[5] Gao S, Zhang W-W, Zhou H-P, et al. (2018) Magnetic composite Fe3O4/CeO2 for adsorption of azo dye. Journal of Rare Earths, 986-993.

[6] Li X-Y, Wang X, Song S, et al. (2012) Selectively Deposited Noble Metal Nanoparticles on Fe3O4/Graphene Composites: Stable, Recyclable, and Magnetically Separable Catalysts. Chemistry - A European Journal, 18(24): 7601–7607.

[7] Cheng S, Zhang L, Ma A, et al. (2018) Comparison of activated carbon and iron/cerium modified

IOP Conf. Series: Materials Science and Engineering **729** (2020) 012076 doi:10.1088/1757-899X/729/1/012076

activated carbon to remove methylene blue from wastewater. Journal of Environmental Sciences, 65:92 - 102.

[8] Roberts K, Gloy B, and Joseph S, et al.（2010） Life cycle assessment of biochar systems: Estimating the energetic, economic and climate change potential. Environ Sci Technol, 44（2）: 827-833.

[9] Xiao X, Chen B, Zhu L.（2014） Transformation, morphology, and dissolution of silicon and carbon in rice straw-derived biochars under different pyrolytic temperatures. Environmental Science & Technology, 48(6):3411-3419.

[10] Fidel R B, Laird D A, Thompson M L, et al. （2017）Characterization and quantification of biochar alkalinity. Chemosphere, 167:367-373.

[11] Wang X J, Wang Y, Wang X, et al.（2011）Microwave-assisted preparation of bamboo charcoalbased iron-containing adsorbents for Cr(VI) removal. Chemical Engineering Journal, 174(1):326-332.

[12] Gan C, Liu Y, Tan X, et al.（2015） Effect of porous zinc - biochar nanocomposites on Cr(VI) adsorption from aqueous solution. Rsc Advances, 5(44):35107-35115.

IOP Conf. Series: Materials Science and Engineering **729** (2020) 012077 doi:10.1088/1757-899X/729/1/012077

Effect of Fe-modified ZSM-5 catalyst during microwave assisted pyrolysis of Mengdong lignite for the oil production

Guanghua Wang[1,2], Xiaodan Li[1,2*], Qingdong Wang[1] and Wenbing Li[2]

[1]Research Center for Green and Intelligent Coal Chemical Engineering, Wuhan University of science and technology, Wuhan, Hubei Province, 430081, China

[2]Institute of Chemistry and Chemical, Wuhan University of science and technology, Wuhan, Hubei Province, 430081, China

[*]Corresponding author's e-mail: 1062090946@qq.com

Abstract. Microwave catalytic pyrolysis of Mengdong lignite by Fe-modified ZSM-5 catalysts was studied. The effects of pyrolysis method and the ZSM-5 molecular sieves loading different amounts of Fe_2O_3 on the total tar yield were investigated. The results show that the microwave heating technology could effectively improve the yield of lignite pyrolysis tar; In the conventional pyrolysis, the total yield of raw coal pyrolysis oil was 2.19%. When the Fe_2O_3 loading was 5%, the catalyst had the best effect with the yield of 3.56%, and the increase rate was 62.56%; Under microwave condition, the pyrolysis oil yield of raw coal is 6.43%. When the Fe_2O_3 loading increased from 0 to 10%, the total yield of tar increased from 7.02% to 9.47%. When the Fe_2O_3 loading was 10%, the catalyst had the best effect with the increase rate of 62.56%.

1. Introduction

Microwave heating is widely used in various industries due to its advantages of fast heating speed, energy saving, selective heating, and environmental protection. In recent years, many researchers have been keen to use microwave as a heating source to study the pyrolysis of biomass, solid waste, fossil fuels, etc[1-3]. Uslu et al.[4] studied the temperature rise of coal in the microwave field, and found that the absorbing materials such as Fe_3O_4 and SiC are beneficial to the rapid heating of coal in the microwave field, and the pyrolysis temperature can be raised to above 1000°C in 3 minutes. Xinzhe Lan et al. investigated the effect of particle size on the quality and yield of low-metamorphic coal microwave pyrolysis products. The study showed that under microwave heating conditions, the temperature of low-metamorphism coal could rise to 750°C after heating for about 10 minutes; Compared with conventional pyrolysis, the yield of pyrolysis tar obtained by microwave pyrolysis was 4% higher, which is 12%; And the content of CO, H_2 and CH_4 in pyrolysis gas also increased.

Microwave and catalyst have the promotion effect on lignite pyrolysis, and the total yield of tar and the light degree of tar are higher than that of ordinary pyrolysis[5-7]. However, there are few studies on the catalytic pyrolysis of lignite in the microwave field. According to the characteristics of microwave heating, the catalyst for microwave field can be designed to pyrolyze lignite, in order to obtain more tar and lighter components. Zeolite molecular sieve is a mesoporous material with uniform pore size distribution, which can allow organic macromolecules to enter and undergo a series of reactions such as cracking, cycleization and isomerization under the action of acidic sites[8]. Considering its large specific surface area, it is conducive to the thermal cracking reaction. In this paper, ZSM-5 molecular sieve is selected as the carrier with the chemical formula of $0.7CaO \cdot 0.3Na_2O \cdot Al_2O_3 \cdot 2.0SiO_2 \cdot 4.5H_2O$ and the ratio of silicon to aluminum is 2. And Fe_2O_3 was loaded by sedimentation method, trying to catalytically crack the primary pyrolysis product entering its pore structure, the effects of the different amounts load of Fe_2O_3 on the total tar yield was studied.

2. Materials and methods

2.1. Preparation of coal sample

In this paper, Mengdong brown coal was taken as the research object, and the raw coal was pulverized and sieved to select coal samples with a diameter of 40 mesh to 60 mesh. Then the selected coal samples were placed in a vacuum drying oven, after drying for 12h at 80°C, they were sealed for storage.

2.2. Preparation of catalyst

The Fe_2O_3@ZSM-5 catalyst was prepared by deposition and sedimentation method. The specific operation steps were as follows: firstly, the ZSM-5 molecular sieve was calcinated at 400°C for 4 h under the air atmosphere, in order to remove impurities in the hole; and then using a sample pulverizer grounded the molecular sieve to a particle size of <0.15 mm. Weighing 10.00g molecular sieve in the conical flask, and weigh the corresponding mass of $Fe(NO_3)_3 \cdot 9H_2O$ according to the Fe_2O_3 loading of 1%, 5%, 10% and 15% adding into the conical flask, while adding a certain amount of $CO(NH_2)_2$ as a precipitant, and finally adding 100 ml of deionized water, after fully stirring for 30 minutes under 40°C water bath, it was heated to 90°C and kept for 6 h, when the turbid liquid was cooled to room temperature, it was suction filtered and washed. The obtained solid was dried in a vacuum oven for 8 h at a temperature of 80 °C and then calcinated at 875°C for 4 h, after cooling, it was taken out and placed in a sealed bag for storage.

2.3. Microwave catalytic pyrolysis experiment

This experiment was carried out in a modified Galanz P70D20TL-D4 microwave oven with the resonant cavity size of 220×200×180mm, the magnetron model of M24FB-610A, and the rated output power of 700W. A certain amount of coal sample and catalyst were weighed and mixed mechanically at a mass ratio of 5:2. The mixed samples were numbered according to Table 1. In each pyrolysis experiment, 35g samples were accurately weighed and placed in a quartz reactor. The upper part of the sample was covered with quartz wool with a diameter of 40mm and a thickness of 20mm. The purpose was to prevent the coal powder from entering the subsequent pipeline along with the pyrolysis airflow and causing the pipeline to be blocked. Control the N_2 flow rate to 250ml/min before the experiment, after purifying for 10 minutes, adjust the flow rate to 25ml/min, start the microwave oven, set the microwave output power to 700w. Successively condensed in three stages by ice water, the pyrolysis pyrolysis product entered the acetone washing system, and then entered the drying tower. The tar attached to the outlet end of the reactor and the condenser was washed with acetone into the round bottom flask, and the water in the tar was absorbed by anhydrous sodium sulfate, and then the acetone was distilled off by a rotary evaporation method at 65°C in a water bath. After that, the tar was dried at 30°C in the dry box to constant weight and weighed to obtain the quality of tar. The total pyrolysis time was 15 minutes.

Table 1. catalyst name and sample number.

catalyst	Fe_2O_3 load (mass fraction)	Sample number
0%Fe_2O_3@ZSM-5	0	1#
1%Fe_2O_3@ZSM-5	1%	2#
5%Fe_2O_3@ZSM-5	5%	3#
10%Fe_2O_3@ZSM-5	10%	4#
15%Fe_2O_3@ZSM-5	15%	5#

2.4. Conventional catalytic pyrolysis experiment

In order to compare with microwave catalytic pyrolysis, catalytic pyrolysis experiments under conventional heating conditions were carried out in a tube furnace. 35g coal samples were weighed into a quartz reactor, then placed in a vertical tube furnace heated to 600°C in advance, and the rest of the operations were the same as microwave catalytic pyrolysis.

3. Results and discussion

3.1. Characterization of the catalysts

3.1.1. SEM analysis

Figure 1 is a scanning electron micrograph of the catalyst. Most ZSM-5 molecular sieves after milling were spherical with different sizes. After loading Fe_2O_3, due to the interaction between the metal oxide and the carrier, adhesion between the spherical particles occurs.

3.1.2. XRD analysis

Figure 2 shows the XRD patterns of Fe_2O_3@ZSM-5 catalysts with different loadings. As can be seen from the figure, the characteristic peak of ZSM-5 molecular sieve gradually weakens as the load increases from 1% to 15%. And a similar phenomenon is observed in the experiment of Shuping Zhang et al.[9, 10], which may be due to the interaction between Fe_2O_3 and ZSM-5 molecular sieve carrier, leading to the decrease of the latter's crystallinity. At the same time, it was also observed that 10% Fe_2O_3@ZSM-5 and 15% Fe_2O_3@ZSM-5 have a relatively broad peak in the range of 25°-30°, which is also one of the evidences of the decrease in crystallinity. It should be noted that after loading Fe_2O_3, no corresponding characteristic peaks were observed in the XRD patterns, indicating that Fe_2O_3 was uniformly loaded on the catalyst surface and did not aggregate to form detectable crystals.

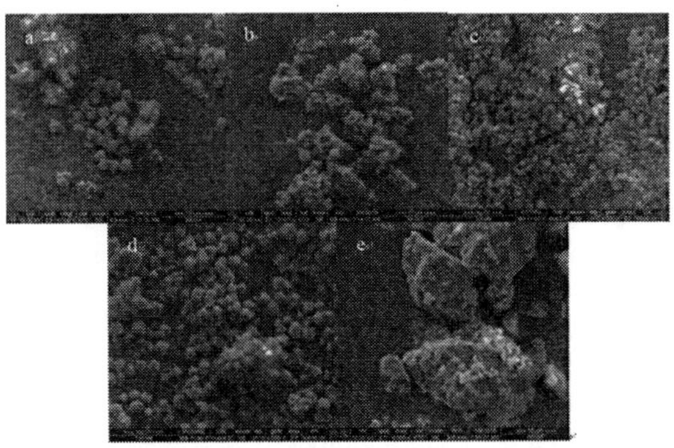

Figure 1. SEM images of the catalyst (a-ZSM5; b-1%; c-5%; d-10%; e-15%)

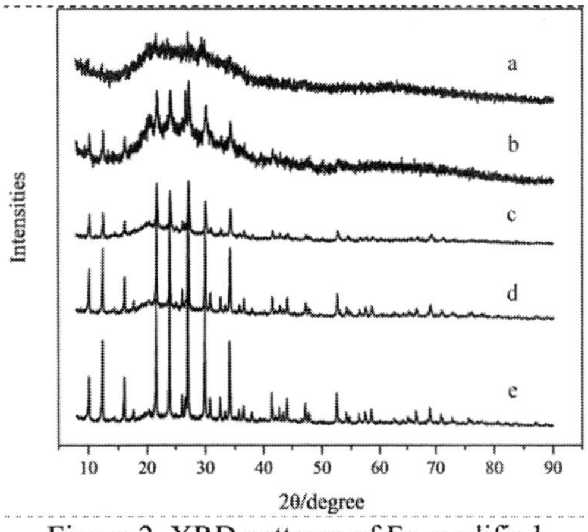

Figure 2. XRD patterns of Fe-modified ZSM-5(a-ZSM5; b-1%; c-5%; d-10%; e-15%)

3.2. analysis of microwave catalytic pyrolysis oil

3.2.1. Tar yield analysis

Figure 3 shows the effect of ZSM-5 molecular sieve catalyst with different mass of Fe_2O_3 on the microwave pyrolysis results of lignite. The pyrolysis process of Fe_2O_3 with load of 0, 1wt%, 5wt%, 10wt% and 15wt% was studied respectively in the experiment. The results show that the tar yield reaches the maximum 9.47wt% when the ZSM-5 molecular sieve catalyst with a loading of 10wt% was added. while under the same conditions, the yield of microwave pyrolysis tar was only 6.43wt% without the addition of catalyst, the relative increase of 47.29%. Therefore, the composite catalyst can effectively improve the yield of lignite microwave pyrolysis tar.

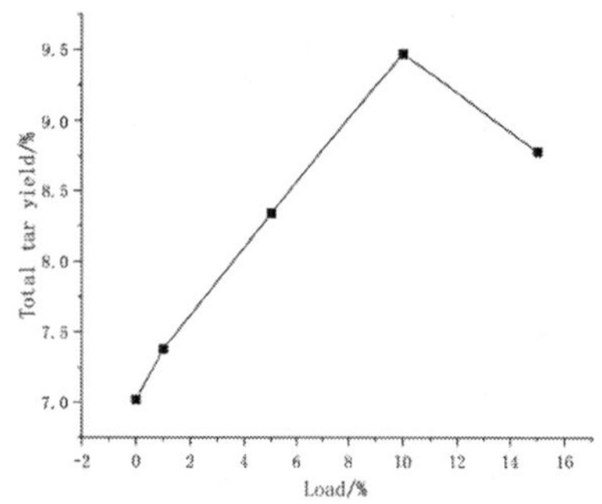

Figure 3. Effect of molecular sieve catalysts with different loadings on the yield of lignite microwave pyrolysis tar.

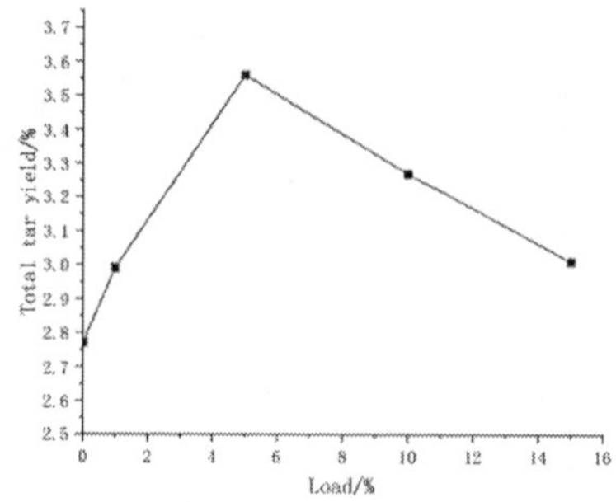

Figure 4. Effect of molecular sieve catalysts with different loadings on the yield of lignite conventional pyrolysis tar.

3.2.2. Coal sample weight loss analysis

Table 2 shows the weight loss of the samples before and after pyrolysis and the temperature at the end of the pyrolysis experiment. As can be seen from the table, the final temperature of samples increased after adding catalysts, indicating that the addition of the catalyst improved the apparent dielectric loss factor of samples, enhanced the material's wave absorption ability, and improved the efficiency of converting electromagnetic energy into internal energy. When the Fe_2O_3 load was 15%, the pyrolysis temperature could reach 695.9°C. The weight loss rate of the sample was up to 59.16%, and the corresponding catalyst was 10% Fe_2O_3@ZSM5. Meanwhile, the tar yield is up to 9.47%.

Table 2. weight loss of coal sample and final temperature of pyrolysis.

Coal sample	final temperature /°C	weight loss ratio/%
1#	644.9	55.7
2#	673.0	51.5
3#	672.0	57.9
4#	694.9	59.2
5#	695.9	55.8

3.3. Analysis of conventional catalytic pyrolysis tar yield

Figure 4 shows the effect of ZSM-5 molecular sieve catalyst with different Fe_2O_3 mass on the conventional pyrolysis of lignite. The pyrolysis process of Fe_2O_3 with load of 0, 1wt%, 5wt%, 10wt% and 15wt% was studied respectively in the experiment. The results showed that when the ZSM-5 molecular sieve catalyst with a loading of 5wt% was added, the tar yield reached the maximum of 3.56wt%, while under the same conditions, the yield of conventional pyrolysis tar was only 2.19wt% without the addition of catalyst, the relative increase of 62.56%. Therefore, the composite catalyst could effectively improve the yield of lignite pyrolysis tar.

3.4. Comparison between conventional catalytic pyrolysis and microwave catalytic pyrolysis

Fe_2O_3@ZSM-5 series catalysts have different effects on pyrolysis oil production under two heating modes. Under the conventional heating condition, all the catalysts increased the yield of tar to some extent, and the effect was best when the Fe_2O_3 loading was 5%. Under the condition of microwave heating, all the catalysts also increased the yield of tar to some extent, but when the Fe_2O_3 loading was 10%, the tar yield was the highest. Speculated that the cause of this difference is that under conventional heating, the catalytic effect mainly depends on the mesoporous structure of the zeolite molecular sieve and the active site, and catalytic cracking of the lignite primary depolymerization product, however too strong acidity can lead to rapid carbon catalyst deactivation, the Fe_2O_3 loading can effectively improve the molecular sieve acidity [11], but it also reduces the specific surface area of the catalyst. If the load exceeds a certain limit, the mesoporous structure of the zeolite molecular sieve is blocked and the catalytic effect is reduced. In the process of microwave heating, in addition to the above catalytic effects, the catalyst also plays the role of a microwave absorber, producing special "active sites" such as "hot spots" and "microplasma", which are related to the loading of Fe_2O_3, and the experiment shows that the effect is best when the load amount is 10%.

4. Conclusion

This paper mainly discusses the effects of heating method and different amounts of Fe_2O_3 loaded on ZSM-5 molecular sieve on microwave assisted pyrolysis of Mengdong lignite for the oil production. The yield of microwave pyrolysis tar of lignite was about 6.43wt%. Compared with the conventional pyrolysis results under the same conditions, the yield of lignite pyrolysis tar was relatively increased by 193.61%, which proved that microwave heating technology could effectively improve the yield of lignite pyrolysis tar. According to the XRD characterization of the catalyst, Fe_2O_3 does not form crystals which can be detected on the surface of the catalyst, and the load was relatively uniform, with the increase of Fe_2O_3 load, the crystallinity of the catalyst decreased. Moreover, the addition of the catalyst increased the dielectric loss factor of the material, accelerated the microwave pyrolysis process, and improved the yield of tar, which had the best effect when the Fe_2O_3 loading is 10%, the total yield of tar is up to 9.47%, and the tar yield is relatively increased by 47.29%, therefore, Fe/ZSM-5 molecular sieve catalyst can promote the effective improvement of lignite microwave pyrolysis tar yield.

Acknowledgements

This research was supported by Open Foundation of State Key Laboratory of High-efficiency Coal Utilization and Green Chemical Engineering (2018-K19), the Excellent Young Talents Training Program of WUST (2017xz009) and the Defense Pre-research Program of WUST (YY1050003).

References

[1] Zhang Y., Fan L., Liu S., et al. Microwave-assisted co-pyrolysis of brown coal and corn stover for oil production [J]. Bioresour Technol, 2018, 259.

[2] Lester Ed, Kingman Sam, Dodds Chris, et al. The potential for rapid coke making using microwave energy [J]. Fuel, 2006, 85(14-15): 2057-2063.

[3] Monsefmirzai Parisa, Ravindran Mythili, Mcwhinnie William R., et al. Rapid microwave pyrolysis of coal - Methodology and examination of the residual and volatile phases [J]. Fuel, 1995, 74(1): 20-27. [4] Huang Y. F., Chiueh P. T., Kuan W. H., et al. Microwave pyrolysis of rice straw: products, mechanism, and kinetics [J]. Bioresour Technol, 2013, 142(7): 620-624.

[4] Uslu T., Atalay Ü. Microwave heating of coal for enhanced magnetic removal of pyrite [J]. Fuel Processing Technology, 2004, 85(1): 21-29.

[5] Zhang Yaning, Fan Liangliang, Liu Shiyu, et al. Microwave-assisted co-pyrolysis of brown coal and corn stover for oil production [J]. Bioresour Technol, 2018, 259: S0960852418304292.

[6] Liu Shuqin, Zhang Yanjun, Tuo Kaiyong, et al. Structure, electrical conductivity, and dielectric properties of semi-coke derived from microwave-pyrolyzed low-rank coal [J]. Fuel Processing Technology, 2018, 178: 139-147.

[7] Zhao Xiqiang, Guo Bingwen, Wang Wenlong, et al. Experimental Study on Microwave Pyrolysis of Three Chinese Lignite [J]. Journal of Analytical & Applied Pyrolysis, 2017, 124: 303-309.

[8] Veses A., Puértolas B., Callén M. S., et al. Catalytic upgrading of biomass derived pyrolysis vapors over metal-loaded ZSM-5 zeolites: Effect of different metal cations onthebio-oil final properties [J]. Microporous & Mesoporous Materials, 2015, 209: 189-196.

[9] Zhang Shuping, Zhang Houlei, Liu Xinzhi, et al. Upgrading of bio-oil from catalytic pyrolysis of

pretreated rice husk over Fe-modified ZSM-5 zeolite catalyst [J]. Fuel Processing Technology, 2018, 175: 17-25.

[10] Sun Laizhi, Zhang Xiaodong, Lei Chen, et al. Comparision of catalytic fast pyrolysis of biomass to aromatic hydrocarbons over ZSM-5 and Fe/ZSM-5 catalysts [J]. Journal of Analytical & Applied Pyrolysis, 2016, 121: 342-346.

[11] Tan Pinglian. Active phase, catalytic activity, and induction period of Fe/zeolite material in nonoxidative aromatization of methane [J]. Journal of Catalysis, 2016, 338: 21-29.

Optimization and control of NOx in coke oven vertical flue under flue gas recirculation

Guanghua Wang[1,2], Guocheng Wang[1,2*], Qingdong Wang[1] and Zihao Yan

[1]Research Center for Green and Intelligent Coal Chemical Engineering, Wuhan

[2]Institute of Chemistry and Chemical, Wuhan University of science and technology, Wuhan, Hubei Province, 430081, China

[*]Corresponding author's e-mail: 1972973091@qq.com

Abstract. The orthogonal experimental design with CFD simulation was carried out to simulate combustion in the coke oven vertical flue at external flue gas recirculation. The control of nitrogen oxides (NOx) generation was analysed by different factors such as air excess coefficient, oxidant preheating temperature, flue gas recirculation and oxidizer inlet cross-sectional area ratio. The simulation results showed that the the orthogonal experimental design with CFD had a significant effect on the combustion process. The flue gas recirculation ratio effectively improved the temperature distribution uniformity in the vertical flue, and had a significant effect on NOx production in the vertical flue. Within the parameters of the orthogonal design, when the air excess coefficient was 1.1, the oxidant preheating temperature was 1200 K, the flue gas recirculation ratio was 30%, and the oxidant inlet cross-sectional area ratio was 0.5, the NOx production in the coke oven vertical flue was the smallest, lower than the industry emission standard.

1. Introduction

Coking industry is one of the most important supporting industries of steel industry. Coke acts as the reducing agent and heat source in the reduction process of iron ore. However, the production process of coke also generates a series of pollution problems, such as soot, NOx, SO_2, etc. Among them, NOx is mainly produced by the combustion process of coke oven gas (blast furnace gas) in the coke oven combustion chamber. Currently, reducing NOx in flue gas is mainly divided into the two methods: source control and terminal control. The source control is to inhibit the formation of NOx from the root. It mainly includes staging combustion technology, thick and thin combustion method, flue gas recirculation technology and so on. Poraj et al. [1]found that air staging combustion technology had an active role in coke oven heating system. Smolka et al. [2]Conducted two-dimensional simulation of coke oven vertical flue under the condition of smoke external circulation by controlling the mass flow of oxidant inlet constant, and concluded that when smoke external circulation ratio was 0.2-0.25, the production amount of NOx was the smallest. Terminal treatment of reducing NOx in flue gas includes biological method, physical method and chemical method, etc[3]. Mature technologies on the market mainly includes SCR, SNCR and dual ammonia solution[4].

In this paper, numerical simulation was carried out for the single opposing vertical flue in the combustion chamber of JN60 coke oven which are widely applied in coking industry, and the external flue gas circulation and the combustion process in the coke oven vertical flue are considered as a whole, which is regarded as a strengthening effect of flue gas internal circulation. Through the orthogonal design method, the simulation calculation of the combustion process under the external circulation of flue gas in the vertical flue of coke oven was carried out.

2. Model and calculation method

2.1. physical model

Figure 1 shows the three-dimensional structure diagram and the two-dimensional top view of the coke oven vertical flue. The coke oven gas occurs in the combustion process in the upward vertical flue, and then flue gas produced passes through the bridge window into the downward vertical flue. most of the flue gas discharge through outlet while the remaining through recirculation window participates in combustion process with air and coke oven gas [3]. The model of coke oven vertical flue is divided into structured hexahedral grids with a total number of 157,780 grids which shows good grid quality.

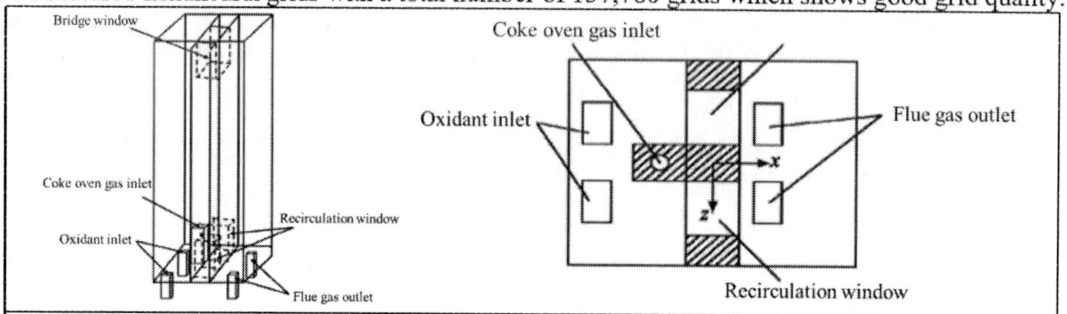

Figure 1. The three-dimensional structure diagram (left) and the two-dimensional top view (right) of the coke oven vertical flue.

2.2. Mathematical model

The turbulence model was the standard k-ε model. The radiation model was the P-1 model. The combustion model was modelled by material transport and eddy-dissipation model was analysed the turbulence-chemistry interaction. The NOx model was the thermodynamic NOx model.

2.3. Boundary conditions of model

Coke oven vertical flue is mainly composed of solid furnace wall and gases which join combustion reaction. The furnace wall of coke oven vertical flue is composed of silicon bricks whose density is 2323 kg/m^3 and specific heat capacity is 858 J/(kg·K). The average thermal conductivity of that is 1.90 W/(m·K). The density of coke oven gas is approximately 0.45 kg/m^3. As a secondary energy source, coke oven gas is complex in composition. In this study, a quantitative standard was adopted to conduct numerical simulation. The coke oven gas is composed of 12% H_2, 41% CH_4, 19% CO, 19% N_2 and 9% CO_2 (mass fraction). and air is composed of 23% oxygen and 77% nitrogen (mass fraction).

2.4. Circulation model of flue gas

In order to simulate the flue gas recirculation combustion process in the coke oven vertical flue, the combustion in the coke oven vertical flue and the flue gas circulation are regarded as a system. In this case, the flue gas external circulation is actually an intensification method for the flue gas internal circulation process, as shown in the figure 2:

Formulas involved in the circulation of flue gas in coke oven vertical flue:

$$m_{OX} = m_A + R \cdot m_{EX} \tag{1}$$

$$R = m_R / m_{EX} \tag{2}$$

$$y_{OXi} = (y_{EXi} \cdot m_R + y_{Ai} \cdot m_A) / m_{OX} \tag{3}$$

IOP Conf. Series: Materials Science and Engineering **729** (2020) 012078 doi:10.1088/1757-899X/729/1/012078

Where m_{OX}, m_{EX} and m_F respectively represent the mass flow rate of oxidant, mass flow rate of air and the mass flow rate of coke oven gas. m_R is the flue gas recirculation mass flow rate. y_{OXi}, y_{EXi} and y_{Ai} respectively are the mass fraction of component i.

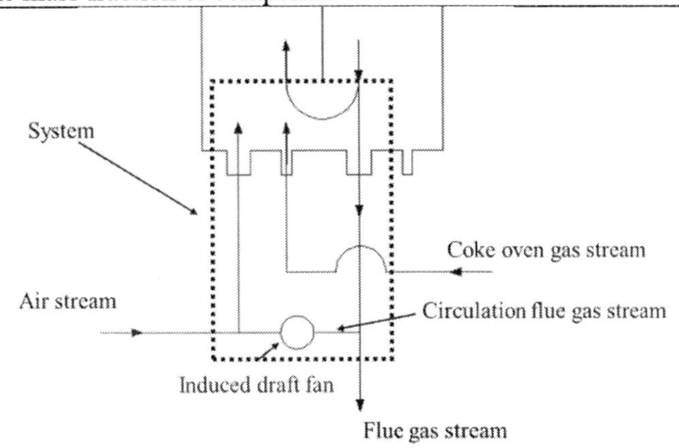

Figure 2. The combustion system of coke oven vertical flue under the flue gas circulation.

3. Orthogonal experimental design

The air excess coefficient (α), the cross-sectional area ratio of the oxidant inlet (S_r), the pre-heating temperature of the oxidant inlet (T_{OX}) and the flue gas recirculation ratio (R) of the coke oven vertical flue were taken as the main influencing factors on the combustion process in the coke oven vertical flue. The table 1 was designed for numerical calculation of the vertical combustion process. And each factor was divided into four levels. And the $L_{16}(4^5)$ orthogonal table of the vertical flue combustion process was designed.

Table 1. Design factors and levels of the vertical flue combustion process

Factors		levels			
		1	2	3	4
α	A	1.05	1.10	1.15	1.2
T_{OX} (K)	B	1000	1300	1100	1200
R (%)	C	10	30	20	0
S_r	D	1	0.25	0.5	0.75

4. Results and discussion

4.1. Orthogonal design-direct analysis

Orthogonal design-direct analysis of the experimental results was shown in table 2 and 3. K_i (i =1,2,3,4) represents the sum of experimental results with the same level number in any column. And k_i is the arithmetic mean value of the experimental results with the same level number in any column. R represents the range and R= Max $\{k_1, k_2, k_3, k_4\}$-min $\{k_1, k_2, k_3, k_4\}$ in any column. The range is positively correlated with the experimental results. Therefore, according to the intuitive analysis of experimental results, the priority order of factors affecting experimental results was CDBA. The range of factor C was the most influential, and in other word, the main factor affecting the generation of NOx was the flue gas recirculation ratio. At the same time, in order to reduce the NOx generation in the combustion process of coke oven vertical flue, the minimum value of any column of K_i should be selected as the evaluation index of the optimal scheme. After comparison, the optimal scheme was $A_2B_4C_2D_3$. As shown in table 4, The experimental results of the optimal scheme and the orthogonal experimental group were in comparison. It could be found that both number 8 and optimal theme can effectively reduce NOx generation in coke oven vertical flue. The optimal scheme had the best effect and belongs to the optimal scheme.

Table 2. Orthogonal design-direct analysis of the experimental results

number	A	B	Error	C	D	y(NOx)	NOx(mg/m^3)	T_{mean}(K)	T_{max}(K)
1	1	1	1	1	1	0.00369	826	1781	2430
2	1	2	2	2	2	0.00216	523	1749	2402
3	1	3	3	3	3	0.00084	215	1676	2255
4	1	4	4	4	4	0.00554	1304	1824	2501
5	2	1	2	3	4	0.00181	443	1734	2355
6	2	2	1	4	3	0.00491	1151	1759	2466
7	2	3	4	1	2	0.00492	1155	1820	2517
8	2	4	3	2	1	0.00018	50	1555	2170
9	3	1	3	4	2	0.00562	1298	1846	2562
10	3	2	4	3	1	0.00557	1190	1772	2559
11	3	3	1	2	4	0.00106	269	1663	2337
12	3	4	2	1	3	0.00022	60	1604	2155
13	4	1	4	2	3	0.00099	226	1663	2360
14	4	2	3	1	4	0.00525	1098	1839	2573
15	4	3	2	4	1	0.00526	1162	1756	2547
16	4	4	1	3	2	0.00436	898	1840	2584

Table 3. Orthogonal design-direct analysis of the NOx generation

factor		A	B	Error	C	D
NOx(mg/m^3)	K_1	2869	2793	3143	3138	3227
	K_2	2798	3962	2189	1067	3874
	K_3	2816	2801	2661	2746	1652
	K_4	3383	2312	3874	4915	3114
	k_1	717	698	786	785	807
	k_2	700	990	547	267	968
	k_3	704	700	665	686	413
	k_4	846	578	968	1229	779
	R	146	413	421	962	555
		A_2	B_4		C_2	D_3

According to the simulation results of orthogonal design experiment, the horizontal trend chart of each factor was drawn. As shown in figure 3 (a), (b), (c), (d), with the increase of air excess coefficient and oxidant pre-heating temperature, the formation of NOx did not change significantly, while the ratio of flue gas external circulation ratio and the cross-sectional area of oxidant inlet had an obvious influence on the formation of NOx. From the orthogonal test table, it could be found that when the ratio of flue gas external circulation was 30% and the cross-sectional area ratio of oxidant inlet was 0.5, the production amount of NOx was the smallest.

4.2. Analysis of variance

The intuitive analysis method of orthogonal design is simple and intuitive. However, the intuitive analysis method cannot estimate the magnitude of the error, and cannot accurately estimate the importance of each factor to the experiment results. Therefore, analysis of variance is used to further analyse the orthogonal test group.

IOP Conf. Series: Materials Science and Engineering **729** (2020) 012078 doi:10.1088/1757-899X/729/1/012078

As shown in table 5, after calculating the deviation sum of squares (SS), degrees of freedom (df) and mean square of the experimental results of the orthogonal design group, it could be found that the factor $MS_A<MS_E$, which indicated that factor A was a secondary factor to the generation of NOx in the vertical flue. Therefore, factor A could be put into the Error. Then, the deviation sum of squares, degrees of freedom and mean square of Error all changed.

Table 4. The experimental results of the optimal scheme and the orthogonal experiment number 8

Factors	Optimal scheme	Number 8
α	1.1	1.1
T_{ox}(K)	1200	1200
R (%)	30	30
S_r	0.5	1
NOx(mg/m^3)	44	50

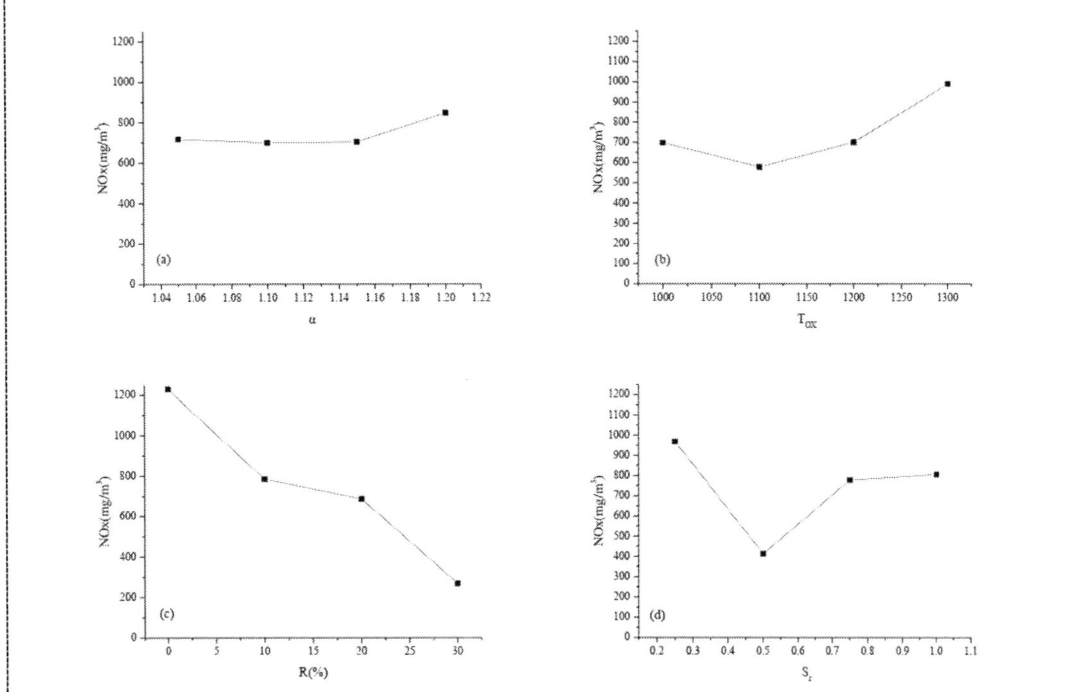

Figure 3. Each factor horizontal trend chart of the orthogonal experimental results circulation.

According to the F distribution table, $F_{0.05}(3,6)$ =4.76, in which $F_C>F_{0.05}(3,6)$, so factor C had significant influence on the experimental results. While $F_B<F_{0.05}(3,6)$ and $F_D<F_{0.05}(3,6)$, which indicated that factor B and D had no significant influence on the experimental results. Therefore, it could be found that the generation of NOx in the vertical flue of coke oven was mainly affected by the recirculation ratio of flue gas. It was because the flue gas recycling effectively reduced the partial pressure of O_2 in the air under the condition of a certain oxygen content, the maximum temperature in the vertical flue decreased significantly.

Table 5. Sum of squares of the experimental results

factors	SS	df	MS
A	2.60E-06	3	8.68E-07
B	8.20E-06	3	2.73E-06

C	3.62E-05	3	1.21E-05
D	1.41E-05	3	4.69E-06
Error	7.75E-06	3	2.58E-06
SUM	6.11E-05	15	4.07E-06

Table 6. New sum of squares of the experimental results

factors	SS	df	MS	F	Significance
B	8.20E-06	3	2.73E-06	1.578	
C	3.62E-05	3	1.21E-05	6.994	*
D	1.41E-05	3	4.69E-06	2.711	
e^{Δ}	1.04E-05	6	1.73E-06		
SUM	6.11E-05	15	4.07E-06		

5. conclusions

In order to better analyse the generation of NOx in the vertical flue of coke oven under flue gas recirculation, the orthogonal design and numerical simulation method were adopted, and remarkable results were obtained. In the case that the heat flux density was guaranteed, the orthogonal test results showed that when the air excess coefficient was 1.1, the oxidant preheating temperature was 1200K, the flue gas recirculation ratio was 30%, and the oxidant inlet cross-sectional area ratio was 0.5, the NOx generation in the coke oven vertical flue was the minimum, which was 44mg/m^3. There was no doubt that the environmental protection requirements of coking industry production. The flue gas recirculation had a significant effect on the formation of NOx in the vertical flue. While the air excess coefficient, the preheating temperature of the oxidant and the proportion of the inlet cross-sectional area of the oxidant have no obvious influence on the formation of NOx.

Acknowledgments

The authors gratefully acknowledge the Supported by National Key R&D Program of China (2017YFB0304300&2017YFB0304303).

References

[1] Poraj, J., Gamrat, S., Bodys, J., Smolka, J., Adamczyk, W. (2016) Numerical study of air staging in a coke oven heating system. Clean Technologies and Environmental Policy, 18(6): 1815-1825.

[2] Gamrat, S., Poraj, J., Bodys, J., Smolka, J., Adamczyk, W. (2016) Influence of external flue gas recirculation on gas combustion in a coke oven heating system. Fuel Processing Technology, 152: 430-437.

[3] Li, H., Liu, S., Liew, C., Gatts, T., Wayne, S., Clark, N., Nuszkowski, J. (2018) An Investigation of the Combustion Process of a Heavy-Duty Natural Gas-Diesel Dual Fuel Engine. JOURNAL OF ENGINEERING FOR GAS TURBINES AND POWER-TRANSACTIONS OF THE ASME, 140(0915029).

[4] Ouyang, Z., Liu, W., Man, C., Zhu, J., Liu, J. (2018) Experimental study on combustion, flame and NOx emission of pulverized coal preheated by a preheating burner. FUEL PROCESSING TECHNOLOGY, 179: 197-202.

IOP Conf. Series: Materials Science and Engineering **729** (2020) 012079 doi:10.1088/1757-899X/729/1/012079

Dry surface modification of heavy calcium carbonate powder with two modifiers

Min Li[1], Xiaohong Yang[1], Jianbing Chen[1]

[1]School of materials and envionmental engineering, Chizhou college, Chizhou, Anhui 247000, China

*Corresponding author's e-mail: 714821655@qq.com

Abstract. In this work, The modification of heavy calcium carbonate powder was studied by dry surface modification with stearic acid and anhydrous citric acid. we have developed a facile and inexpensive dry surface method to fabricate a lower oil absorption value of calcium carbonate surface. The modification process was analyzed by modifying temperature, modifying time, modifiers percentage. Infrared spectral and TG showed that stearic acid modifier was mainly reducing oil absorption value of $CaCO_3$ powder from 33.9 ml/100 g to 19.0 ml/100 g.

1. Introduction

Fillers are used in the industries of plastics, rubber, paint, fibers, and so on. One of the most challenging issues for the use of calcium carbonate|($CaCO_3$) is the hydrophilic property.

The successful surface modification of $CaCO_3$ with a hydrophobic species would widen the range of potential applications.Dry surface modification method is cheaper than wet surface modification method. In general, stearic acid[1],KR-TTS titanate coupling agents[2],aluminate coupling agent (ACA)[3], oleic acid [4-5]are often present as modifiers. The hydrophobic characters of modified calcium carbonate powder contain water contact angle,activity index,oil absorption value. Oil absorption value is seldom evaluated the modified calcium carbonate powder.

In this paper,stearic acid and anhydrous citric acid were used to modify heavy calcium carbonate. Oil absorption value is evaluated the modified calcium carbonate powder.The surface modification processes are carefully studied , at the same time, the mechanism of surface modification was analyzed.

2. Experimental

1500 mesh calcium carbonate ($CaCO_3$), stearic acid and anhydrous citric acid were used as starting materials. In typical procedure, firstly, $CaCO_3$ and certain content anhydrous citric acid were put into mental jar by use of QM-2L ball mill equipment at 100r/min about 1h .After,stearic acid was put into the same jar milled 1h at the condition.Then, the 50g of mixture $CaCO_3$ was placed in culture dish ,and it was heated at certain certain period of time at some constant temperature in drying oven to modification .After the reaction was completed, as a primary criterion the oil absorption value of sample was employed to evaluate the surface modification effect. The oil absorption value is quantity of Dioctyl-Phthalate/5.00.The quantity of Dioctyl-Phthalate is determined by titrating the modified power(5g) until it becoming a group.The dispersity of modified calcium carbonate powder was also evaluated by sedimentation. Firstly, modified $CaCO_3$ powder(5g) and absolute ethyl alcohol

IOP Conf. Series: Materials Science and Engineering **729** (2020) 012079 doi:10.1088/1757-899X/729/1/012079

solvent(50mL) are added into beaker. Then the beaker was place on the 78-1 magnetic stirring apparatus and stirred 30 min.At last, suspension with modified $CaCO_3$ powder was put into graduated cylinder(50mL), and the sedimentation height of suspension changes according to sedimentation time.Fourier transform infrared (FTIR) spectra were recorded with a Nicolet Is10 spectrometer at room temperature. TG and DTA was tested by use of Netzsch 214.

3. Results and discussion

3.1 The analysis of modification condition
Oil absorption value of unmodified $CaCO_3$ powder is 33.9 ml/100 g through testing. Many factors can impact the modification effect. These parameters are modification temperature, anhydrous citric acid percentage, stearic acid percentage ,and modification time. The oil absorption value was employed as reference standard.

3.1.1 The effect of modification temperature
Fig.1 shows the oil absorption value of the modified powder with different modification temperature. The anhydrous citric acid and stearic acid is respectively at 5wt%.It can be seen that with the increasing modification temperature, the oil absorption value first goes down, then it stays flat, at last it goes up.It is thought that increasing temperature enhances the reaction activity between the modifier and $CaCO_3$ surface[2].While the high temperature decomposes the modifiers.This is because that the anhydrous citric acid is resolved according to its nature.A kind of burnt smelled up during modification when the temperature above 175℃. Thus, the optimum modification temperature is at 125~150°C.

3.1.2 The effect of anhydrous citric acid percentage
Fig.2 is the oil absorption value of the modified powder with different anhydrous citric acid percentage,when the stearic acid percentage is at 5wt%. It is contributed to physical or chemical reaction between $CaCO_3$ powder and the modifier. It can be seen that anhydrous citric percentage has an obvious influence on modification effects. The oil absorption value increases along with the increase of anhydrous citric acid percentage.The reason is that more anhydrous citric acid first coated $CaCO_3$ surface,which hinders stearic acid modification effects.

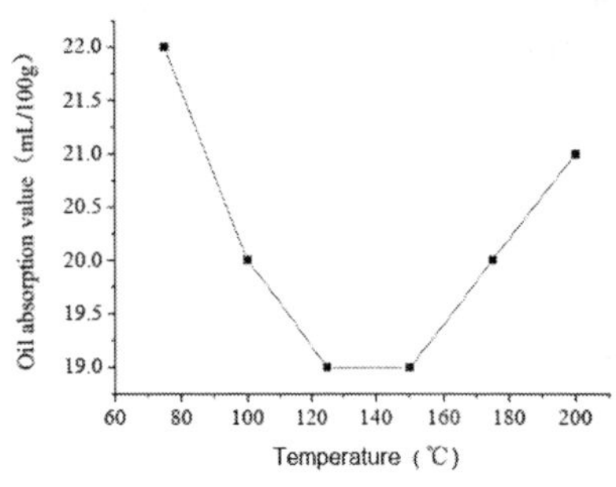

Fig.1. The oil absorption value of the modified powder with different modification temperature.

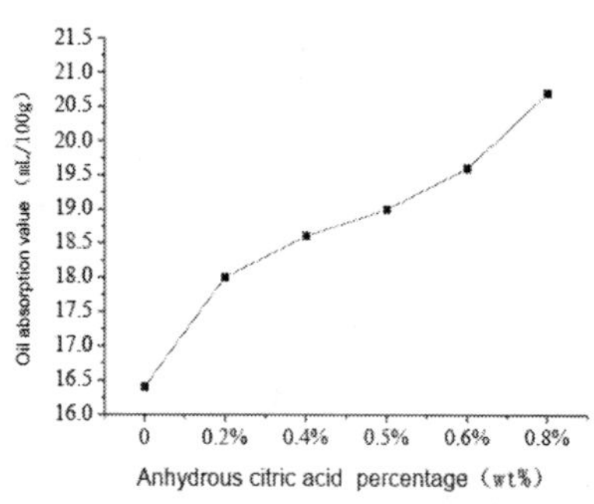

Fig.2. The oil absorption value of the modified powder with different anhydrous citric acid percentage.

3.1.3 The effect of stearic acid percentage
Fig.3 is the oil absorption value of the modified powder with different stearic acid percentage. It can be seen that with the increase stearic acid percentage from 0.3wt% to 0.4wt%, the oil absorption value reduces in a directly and then it keep at 19.0 mL/100 g when the stearic acid percentage is from 0.4wt% to 0.7wt%. We considered that optimum 0.4 wt% content of stearic acid may lead to full

surface coverage. In order to ensure modification stability, 0.5wt% content of stearic acid could be used.

3.1.4 The effect of modification time

Sufficient modification time must be met in order to reach good modification performances. Fig.4 is the oil absorption value of the modified powder at different modification time.When modification time rises from 1h to 2h , the oil absorption value just lowers from 20.0 mL/100 g to 19.0 mL/100g.The modification time is determined as 2h in this test.

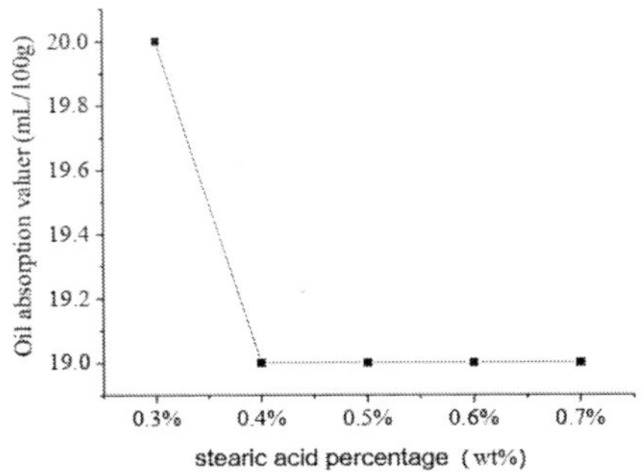

Fig.3. The oil absorption value of the modified powder with different stearic acid percentage.

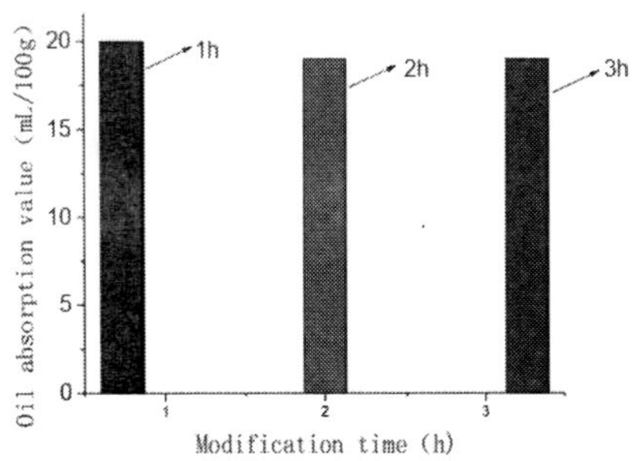

Fig.4. The oil absorption value of the modified powder at different modification time.

3.2 The analysis dispersity of modified powder

The results show that the stearic acid percentage is the major influencing factor to reduce oil absorption value. At the same time, considering the dispersity of modified powder filled in the industries of plastics, rubber and so on, a further experiment was developed. Fig.5 shows sedimentation height of power in suspension with sedimentation time. It can be seen that the settling velocity of stearic acid modified powder in absolute ethyl alcohol solvent is faster than the anhydrous citric acid and stearic acid modified powder. We thought that surface of two agents modified powder maybe covered a part of anhydrous citric acid, which created a bigger repulsive force hindering particle agglomeration. And small particles can be dispersed, which caused powder slowing settlement. The phenomenon shows that dispersity of stearic acid modified powder is inferior than two agents.

3.3 TG and DTA analysis

Fig.6 is TG and DTA of 0.5wt% anhydrous citric acid and 0.5% stearic acid modified $CaCO_3$ powder. It can be seen that decalescence point appears in the DSC curve of modified powder is about 160.0°C, which is due to that the decomposition of anhydrous citric acid is at 175℃. Then exothermic process with temperature rising above 200℃, which is because stearic acid chains were ruptured releasing heat. When the temperature up to about 350℃, all modifiers are decomposed and volatiled. The process proves anhydrous citric acid and stearic acid are all covered on surface of calcium carbonate power.

IOP Conf. Series: Materials Science and Engineering **729** (2020) 012079 doi:10.1088/1757-899X/729/1/012079

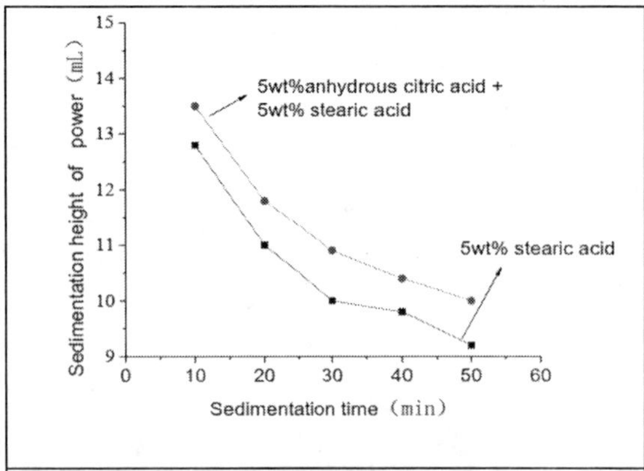

Fig.5. Sedimentation height of suspension with sedimentation time.

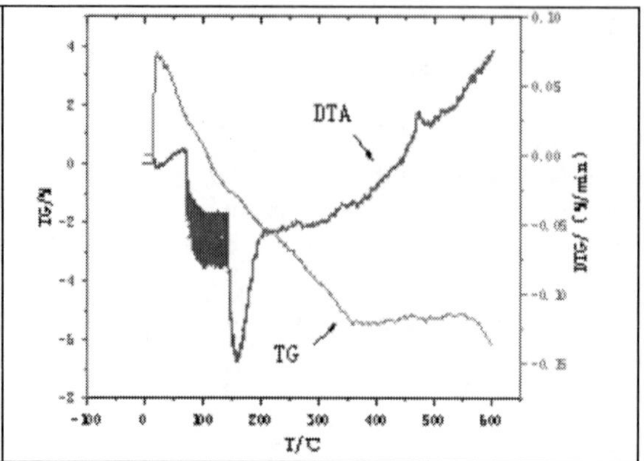

Fig.6. TG and DTA of modified powder.

3.4 Infrared spectral analysis

The formation mechanism between $CaCO_3$ powder and the substrate can be identified with FTIR. FTIR spectra (Fig. 7.) of anhydrous citric acid (line a), stearic acid (line b), and the modified anhydrous citric acid and stearic acid modified $CaCO_3$ powder (line c). Compared to the spectrum of line b, the asymmetric [$\nu as(CH_2)$] and symmetric [$\nu s(CH_2)$] methylene vibration peaks appear clearly at 2918 and 2846 cm^{-1} of line c, which indicates the existence of the stearic acid on the surface. The modified $CaCO_3$ powder was washed with n-propyl alcohol before FTIR test. So it can be considered that stearic acid has bound onto the surface of the $CaCO_3$ powder to achieve modification, while anhydrous citric acid was washed away owing to weak action between it and $CaCO_3$ powder.

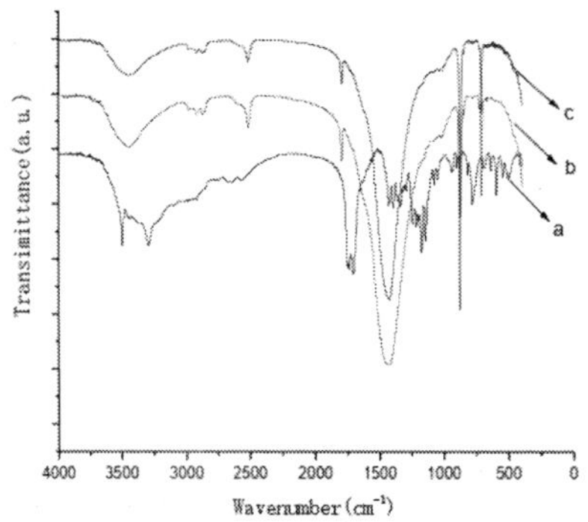

Fig.7. FTIR spectra of (a) anhydrous citric acid, (b) stearic acid, (c) 0.5wt% anhydrous citric acid and 0.5wt% stearic acid modified $CaCO_3$ powder.

4. Conclusion

In this work, we have developed a facile modification method. The optimum technological parameters were the following: modification temperature was at 125-150℃, anhydrous citric acid percentage was 0.2wt% and stearic acid percent was 0.4wt%, modification time was 2h.

The oil absorption value of modified $CaCO_3$ have been reduced greatly from 33.9 ml/100 g to 19.0 ml/100 g, the reason was that stearic acid modifier was bound onto the surface of $CaCO_3$ powder. The input of anhydrous citric acid can improve dispersity of modified $CaCO_3$ powder in absolute ethyl alcohol solvent.

Acknowledgments

This work was financially supported by Key Project of Natural Science Research of Universities in Anhui Province (KJ2018A0577), Key Laboratory of Micro-Nano Powder and Advanced Energy Materials of Anhui Highter Education Institutes, Chi zhou University, and Chi Zhou Research Academy of Non-metallic Mining Industry Development.

References

[1] Hua Zhang, Xiaofei Zeng, Yongfeng Gao, et al. (2011)A Facile Method To Prepare Superhydrophobic Coatings by Calcium Carbonate. American Chemical Society,Industrial & Engineering Chemistry Research, 50, 3089–3094.

[2] Zehong Wang, Jinghua Chen, Wanzhong Yin, et al. (2010)Study on the Surface Modification of Nanometer Calcium Carbonate. Advanced Materials Research, 92,229-234.

[3] Langchen Li, Yue Zhang. (2009)Wet surface modification of light calcium carbonate powder by aluminate coupling agent. Advanced Materials Research,79-82 ,1967-1970.

[4] Lina Zhao, Xu Zhao, Suxia Ren, et al. (2009)In Situ Synthesis and Surface Modification of Calcium Carbonate. Acta Phys.Chim.Sin,25(1):47-52.

[5] Chengyu Wang, Cheng Piao, Xianglin Zhai, et al. (2010)Synthesis and character of super-hydrophobic $CaCO_3$ powder in situ. Powder Technology,84-86.

IOP Conf. Series: Materials Science and Engineering **729** (2020) 012080 doi:10.1088/1757-899X/729/1/012080

Exploration on Spent FCC Catalyst for Decolorization of Vegetable Oil

Qiang Liu [1], Hui Li [1], Sheng Zeng [1], Lixin Liu[2] and Dulin Yin [1*]

[1]National & Local Joint Engineering Laboratory for New Petro-chemical Materials and Fine Utilization of Resources, College of Chemistry and Chemical Engineering, Hunan Normal University, Changsha, Hunan, 410012, P. R. China

[2]Hunan Jinhai Technology Co., Ltd., Changsha, Hunan, 414100, P. R. China

[*] Corresponding author's e-mail: dulinyin@126.com

Abstract. In the present work, an FCC equilibrium catalyst (E-cat) was used as a decolorizer in decolorization of vegetable oil. E-cat is an excellent and stable decolorizer, which can even be directly applied to the decolorization of vegetable oil after being discharged. The process conditions were investigated and optimized, mainly the bleaching temperature, time, and E-cat amounts. The spectrophotometric method was used to calculate the decolorization rate after detecting the absorbance of vegetable oil. As the bleaching time, temperature and E-cat amounts increase, the effects on the decolorization of rapeseed oil by E-cat were all improved to varying degrees. The best decolorization utility of E-cat was performed at 115°C and an E-cat ratio of 5wt% for 40min, the decolorization rate of the rapeseed oil was 93.3%. Compared with the fresh FCC catalyst with high catalytic activity, the decolorization capability of E-cat was not much reduced. It indicates that the structural characteristics of E-cat are more desirable in the decolorization of vegetable oil, as evidenced by the alkali treatment test.

1. Introduction

Globally, vegetable oil has become an indispensable part of daily life, such as palm oil[1], rapeseed oil[2], peanut oil, olive oil[3] and so on. In China, rapeseed oil is the main edible vegetable oil, which has been shown to contribute to high absorptivity in the human body and certain functions of softening blood vessels and delaying aging. The annual production of rapeseed has reached 14 million tons in recent years in China. Production has been steadily rising since the crop entered the food market in the 1970s and is predicted to continue in this upward trend as global oil consumption increases[4]. The crude rapeseed oil derived from pressing or leaching is disadvantageous for storage and hydrogenation to prepare downstream products such as margarine and shortening because of colloids, free carboxylic acid and colorants[5]. In the refining of rapeseed oil, the processes of degumming, deacidification, deodorization and decolorization are generally carried out, and the decolorization process is an important operation[6]. By removing the colorants from the crude rapeseed oil, it not only gives the consumer a sense of comfort, but also enriches the way the product is used and extends the shelf life of the product[7].

At present, the main process for decolorization of oil is adsorption. Commonly used decolorizer include activated kaolin[8, 9], activated bentonite[10] and activated clay based on aluminosilicate mineral processing. Activated carbon[11] and synthetic porous silica-based products[12] are also being studied as decolorizer. These decolorization materials require a large amount of resources to be consumed. The catalyst used in the fluid refining cracking of the petroleum refining industry is commonly referred to as FCC catalyst, with an annual global supply of approximately 840,000 tons[13]. The catalyst is mainly made up of natural and synthetic aluminosilicates such as kaolin, faujasite and aluminum sol[14]. In the process of use, the FCC catalyst is deactivated due to the refinement of particle size as a result of catalyst attrition, sintering during calcination regeneration at around 700 °C and amorphization[15]. It is therefore necessary to continuously add fresh FCC catalyst and continuously discharge the deactivated catalyst (commonly referred to as FCC equilibrium catalyst, E-cat) in such process. The progress in the resource utilization of such the largest tonnage of spent catalysts in the world is often reviewed, and its application potential is expected to be explored. Obviously, E-cat is a stable porous aluminosilicate material with acidic active sites on the surface, exhibiting adsorption and catalysis, but so far there has been relatively little research attention on adsorption properties. In this work, the influencing factors on the decolorization performance of the E-cat for rapeseed oil were investigated.

2. Experimental

2.1. Materials

All reagents are analytically pure by purchase. The alkali-refined rapeseed oil was taken from the rapeseed oil alkali refining device of Daodaoquan Grain and Oil Yueyang Co., Ltd., The E-cat was discharged from the catalytic cracking unit of Sinopec Changling Branch, of which the BET surface area and pore volumn are 94.7 m^2/g and 0.13 cm^3/g respectively from as measured using NH_3-TPD.

2.2. Decolorization experiment

The experiment was carried out in a round bottom flask with a stirring speed of 200 r/min during the decolorization. First, we put 50ml of alkali-refined rapeseed oil into the flask. Secondly, E-cat was added after reaching the bleaching temperature. The decolorization was allowed to progress for a period of time, during which oil samples were taken at regular intervals and tested after E-cat was removed from the oil by centrifugation, the process had been carried out at a bleaching temperature of 105°C and an E-cat amount of 4wt% for 40min.

3. Calculation of decolorization rate

In previous studies, the spectrophotometric method has been used to detect the absorbance of vegetable oil and calculate the decolorization rate (R) of the alkali-refined rapeseed oil[16, 17]. In this various solution system, a spectrum was obtained by detection, and a wavelength of 670 nm was selected as an object of investigation. The absorbance of vegetable oil was detected by a TechcompUV-1102 II spectrophotometer (Techcomp Corporation, Shanghai, China). The standard curve was plotted at the wavelength of 670 nm and the correlation coefficients were 0.999. In the appropriate range, it can be used to conveniently calculate the decolorization rate (R) of the alkali-refined rapeseed oil.

$$R(\%) = (A_0 - A_t)/A_0 \times 100\% \qquad (1)$$

where R is the decolorization rate (%) of the oil, A_0 is the absorbance of the original oil, and A_t is the absorbance of the decolorized oil.

4. Results and Discussion

Due to the dealumination of the zeolite catalyst component, new discharged E-cat has different particle sizes and external surface areas, which will affect its decolorization performance. During the use of the FCC catalyst, the process temperature is generally above 700 °C. The discharged E-cat might absorb water in a natural air atmosphere during the storage period. The effects of these factors were investigated on the decolorization of vegetable oil by E-cat. It can be seen from Table 1 that the decolorization abilities of E-cat having particle sizes in the range of 60-300 mesh did not show much variation, for the decolorization of the alkali refining rapeseed oil. This suggested that the application of E-cat to the decolorization process was not limited by the need for screening. The decolorization performance of E-cat was slightly increased after calcination (Table 1). It was found that the correlation between the water content in E-cat and the decolorization effect was not very significant. It also suggested that the water molecules physically block or produce a surface-interface effect on E-cat, thereby inhibiting the decolorization of E-cat on the alkali-refined rapeseed oil. The calcined E-cat absorbed water and reached saturation, then its decolorization performance decreased less. It was therefore theoretically possible to apply the just discharged E-cat directly to the decolorization process of the vegetable oil refining process, which will reduce the number of required operations and energy consumption.

Table 1. Effect of physical treatment of E-cat on decolorization

E-cat	Origin	Screened (mesh)				Calcined	Reabsorb
		60-100	100-150	150-200	200-300		
R (%)	57.7	52.5	45.8	47.8	54.8	61.3	58.9

In the decolorization process of rapeseed oil refining, the bleaching temperature is generally 105 °C[7]. If the temperature is too high, some oxidation reactions will occur in the oil[18, 19]. The decolorization ability of E-cat was investigated at a suitable temperature range: as the bleaching temperature increases, the colour of oil was significantly lighter. The results showed that the increase in temperature greatly enhanced the decolorization effect of E-cat on rapeseed oil (Figure. 1. a). Within the appropriate range, the maximum temperature was therefore the optimum temperature for decolorization of vegetable oil. Increasing the temperature will improve the application effect of E-cat, so it was obvious that the influence of bleaching time should also be investigated: this was carried out at the low temperature of 75 °C. As was evident from Figure. 1. b, as the bleaching time increased, colorants in the alkali-refined rapeseed oil were continuously adsorbed. During the whole decolorization process, the decolorization rate of rapeseed oil by E-cat slowly eased. Bleaching vegetable oil more than forty minutes, E-cat can fully exert its adsorption performance. Not only the decolorization temperature and time should be considered, but the amount of decolorizer was also a key operating parameter in the decolorization process of the rapeseed oil refining. The effect of E-cat amounts on decolorization of the oil was investigated under the more favourable conditions (115°C, 40min). As the amount of E-cat increases, the colour of the alkali-refined rapeseed oil became lighter

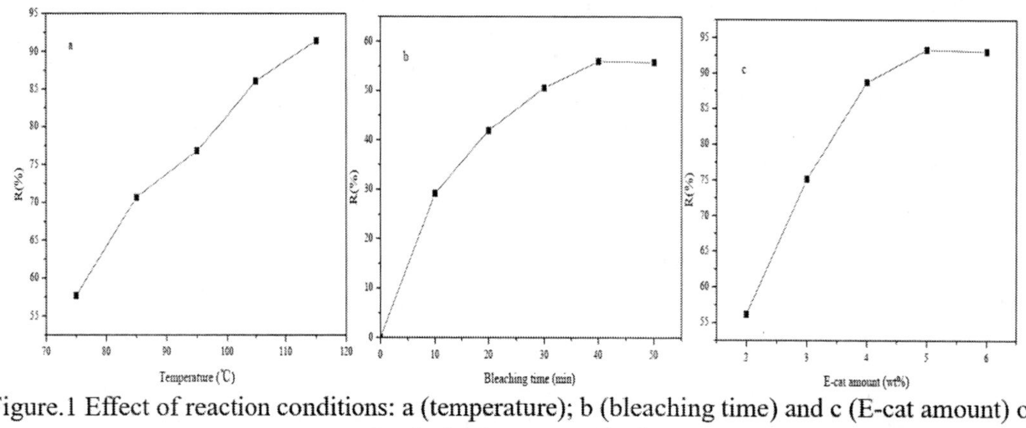

Figure.1 Effect of reaction conditions: a (temperature); b (bleaching time) and c (E-cat amount) on decolorization, respectively.

(Figure. 1. c). When the amount of E-cat reached 5wt%, the decolorization rate of rapeseed oil was 93.7%. A further increase in the amount of E-cat did not significantly enhance the decolorization effect.

Table 2. Effect of poison tests on the decolorization of FCC catalysts

FCC catalysts	F-cat	E-cat	
		untreated	Base-treatment
R (%)	94.7	93.7	16.8

As seen from Table 2, compared with E-cat, the fresh FCC catalyst (F-cat) did not have much higher decolorization performance on rapeseed oil. This suggested that higher catalytic activity was not required for rapeseed oil bleaching. E-cat was treated by impregnation with NaOH solution. By alkali treatment, the adsorption ability of E-cat declined greatly. It indicated that the acid sites on E-cat were neutralized by alkali, and the lye also destroyed the silica-alumina structure of the molecular sieve. In summary, E-cat and F-cat are molecular sieve catalysts with excellent structure, which was the suitable decolorizer for the decolorization of rapeseed oil.

5. Conclusions

E-cat was applied for the decolorization process of vegetable oil refining, which was a supplement and optimization of the structure of resource use to make full use of the FCC catalyst. Its decolorization performance was investigated for the decolorization process of the oil. The results showed that E-cat was an excelled decolorizer for the decolorization of vegetable oil. The increase in bleaching temperature, time, and E-cat amounts promoted the decolorization of rapeseed oil. The decolorization rate of the alkali refined rapeseed oil by E-cat was 93.7% under optimized decolorization conditions (bleaching temperature, 115°C; bleaching time, 40min; E-cat amounts, 5wt%). Although these process conditions were important, the structural characteristics of E-cat were more important for decolorization of vegetable oil.

Acknowledgements

This work was supported by the National Natural Science Foundation of China (No. 21776068 and 21606082), Collaborative Innovation Center of New Chemical Technologies for Environmental Benignity and Efficient Resource Utilization and Department of Science and Technology of Hunan Province.

References

[1] Mat Yasin, M.H. (2017) Potentials of palm oil as new feedstock oil for a global alternative fuel: A review. Renew. Sust. Energ. Rev. , 79:1034-1049.
[2] Å. H. Dahlén, J. (1973) Chlorophyll content monitoring of Swedish rapeseed and its significance in oil quality. J. Am. Oil Chem. Soc. , 50:312A-312A.
[3] Criado, M.N. (2007) Comparative study of the effect of the maturation process of the olive fruit on the chlorophyll and carotenoid fractions of drupes and virgin oils from Arbequina and Farga cultivars. Food Chem. ,100(2): 748-755.
[4] Woodfield, H.K. and J.L. Harwood (2017) Oilseed Crops: Linseed, Rapeseed, Soybean, and Sunflower, in Encyclopedia of Applied Plant Sciences (Second Edition), B. Thomas, B.G. Murray, and D.J. Murphy, Editors. Academic Press: Oxford. 34-38.
[5] Abraham, V. and J.M. de Man (1986) Hydrogenation of canola oil as affected by chlorophyll. J. Am. Oil Chem. Soc., 63(9): 1185-1188.
[6] Chapman, D.M., E.A. Pfannkock, and R.J. Kupper (1994) Separation and characterization of pigments from bleached and deodorized canola oil. J. Am. Oil Chem. Soc., 71(4): 401-407.
[7] Sherwin, E.R. (1978) Oxidation and antioxidants in fat and oil processing. J. Am. Oil Chem. Soc., 55(11): 809-814.

[8] L. Aung, L. (2014) The capacity of activated kaolins to remove colour pigments from rice bran oil: The effects of acid concentration and pre-heating prior to activation. Clay Miner. , 49.
[9] Nandi, B.K., A. Goswami, and M.K. Purkait (2009) Adsorption characteristics of brilliant green dye on kaolin. J. Hazard. Mater. , 161(1): 387-395.
[10] Didi, M.A. (2009) Colza oil bleaching through optimized acid activation of bentonite. A comparative study. Appl. Clay Sci. , 42(3): 336-344.
[11] Madu, C.P. and L. Lajide (2013) Physicochemical characteristics of activated charcoal derived from melon seed husk. J. Chem. Pharm. Res. , 5: 94-98.
[12] Ahmad, A.L. (2009) Adsorption kinetics and thermodynamics of β-carotene on silica-based adsorbent. Chem. Eng. J. , 148(2): 378-384.
[13] Ferella, F., V. Innocenzi, and F. Maggiore (2016) Oil refining spent catalysts: A review of possible recycling technologies. Resour. Conserv. Recy. , 108:10-20.
[14] Liu, H. (2007) A novel FCC catalyst synthesized via in situ overgrowth of NaY zeolite on kaolin microspheres for maximizing propylene yield. Catal. Today, 125(3): 163-168.
[15] Cerqueira, H.S. (2008) Deactivation of FCC catalysts. J. Mol. Catal. A-Chem. , 292(1): 1-13.
[16] Kıvan (2010) A novel method for color determination of edible oils in L*a*b* format. Eur. J. Lipid Sci. Tech. , 109(2): 157-164.
[17] Endo, Y., C.T. Thorsteinson, and J.K. Daun (1992) Characterization of chlorophyll pigments present in canola seed, meal and oil. Spectrochim. Acta A , 69(6): 564-568.
[18] Melendez-Martinez, A.J. (2007) Relationship between the colour and the chemical structure of carotenoid pigments. Food Chem. , 101(3): 1145-1150.
[19] Niewiadomski, H., I. Bratkowska, and E. Mossakowska (1965) Content of chlorophylls and carotenes in rapeseed oil. J. Am. Oil Chem. Soc., 42(8): 731-734.

Efficient adsorption removal of Cd^{2+} from aqueous solutions by HNO_3 modified bamboo-derived biochar

Wenxiao Tang[1#], Ningning Cai[1#], Hongxia Xie[1#], Yachun Liu [1*], Zhifeng Wang[2], Youhong Liao[1], Taotao Wei[1], Chao Zhang[1], Zaihui Fu[1], and Dulin Yin[1]

[1] National & Local Joint Engineering Laboratory for New Petro-chemical Materials and Fine Utilization of Resources, Key Laboratory of Resource Fine-Processing and advanced materials of Hunan Province and Key Laboratory of Chemical Biology and Traditional Chinese Medicine Research (Ministry of Education of China), College of Chemistry and Chemical Engineering, Hunan Normal University, Changsha, Hunan, 410081, China

[2] Hunan Kangshou Pharmaceutical Co., Ltd., Changsha, Hunan, 410200, China

[*]Corresponding author's e-mail: yacl315@163.com

[#]These authors contributed equally to this work

Abstract. In this work porous biochar was prepared from bamboo powder by $ZnCl_2$ chemical activation. The effects of preparation, modification and adsorption conditions on its removal performance of Cd^{2+} in simulated wastewater were investigated by static adsorption. The adsorption kinetics and isotherms of Cd^{2+} were further studied. The results showed that the good adsorption property was obtained over the biochar prepared under optimized preparation and modification conditions. The biochar reached the highest removal efficiency of 95.53% under optimum adsorption conditions. The adsorption process of Cd^{2+} was well in accordance with the pseudo-second adsorption kinetics equation, and the equilibrium adsorption capacity was up to 16.62 $mg \cdot g^{-1}$. The Freundlich isothermal adsorption model better fitted compared with the Langmuir model for this adsorption system and the maximum adsorption capacity was as high as 44.54$mg \cdot g^{-1}$. After four adsorption-desorption recycling, the adsorption removal percentage decreased a little, indicating that biochar had good regeneration performance. In short, this inexpensive porous biochar has the potential to become a practical adsorbent of cadmium.

1. Introduction

With the continuous development of economy and society, the harm of heavy metal ions containing industrial wastewater to the ecological environment and human health is becoming an increasingly serious problem in many developing countries[1, 2]. For example, cadmium ions in industrial wastewater have mutagenic, carcinogenic, teratogenic effects and thus cause serious damage to the respiratory system, kidney and other organs[3-6]. At present, there are a lot of reports on the removal methods of cadmium ions in wastewater, such as chemical precipitation, ion exchange and adsorption[1-6]. Among these methods, the adsorption using commercial activated carbon as an sorbent is a popular and effective means due to its high surface area, well-developed internal microporosity, and plentiful surface functional groups[7-9].

Conventional activated carbon sorbents, however, is to a certain extent limited owing to the existing problems originated from the raw material source including non-renewable resources such as coal tar and petroleum coke pitch derived one-time fossil resources and renewable resources like wood. A large-scale consumption of these natural resources can cause great environmental pressures. On the other hand, bamboo is an abundant and renewable biomass with its advantages of fast growth, extensive planting and high yield[10, 11], and accordingly leading to the rapid development of the bamboo products industry and the simultaneous production of a large number of bamboo processing residues. The residues can be easily converted to bamboo-derived porous biochar. The material possesses a high specific surface area, a high micropore ratio and abundant surface acidic groups by chemical activation. This preparation method of activated carbon has emerged as a very attractive approach due to its low cost, an easy process and mild conditions ascompared to physical activation[12, 13]. In recent years, some biochars have been reported to adsorb organic and inorganic contaminants with the high adsorption capacity[14, 15]. However, there are a few studies on the treatment of cadmium ions containing wastewater by bamboo-derived porous biochar[5, 16] especially little those on the influences of preparation, modification of the biochar on its removal performance of Cd^{2+} in simulated wastewater. Therefore, the systematical study of the effects of preparation and modification of the biochar on its adsorption removal performance, the adsorption kinetics and equilibrium of Cd^{2+} is of very high significance in the treatment of Cd^{2+}containing wastewater.

In this work, bamboo-derived porous biochar was used as the adsorbent for removing the Cd^{2+} in the simulated wastewater. The optimum preparation conditions of biochar and the effects of pH, adsorption temperature, adsorption time and the amount of adsorbent on the removal of cadmium ions in simulated wastewater were in details examined. The adsorption kinetics and equilibrium were studied, and the regenerative adsorption performance was investigated. This work will provide the theoretical and practical basis for biochar in the treatment of cadmium in industrial wastewater.

2. Experimental

2.1. Reagents and raw materials

The cadmium chloride, concentrated hydrochloric acid, concentrated nitric acid, concentrated sulfuric acid, sodium hydroxide, zinc chloride and sodium bicarbonate were analytical grade reagents and purchased from Sinopharm Group Chemical Reagent Co. Ltd. Bamboo powder (40-60 mesh) was prepared from bamboo processing residues from a local bamboo products factory by drying, grinding, and sieving. Wood commercial activated carbon was purchased from a local supplier. The chemicals were all used as received.

2.2. Preparation and modification of materials

Weigh 20g of bamboo powder and 10g of zinc chloride, respectively. $ZnCl_2$ was dissolved with a certain volume of distilled water to form the solution clarified by adding proper volume of concentrated hydrochloric acid. The bamboo powder was impregnated with the solution by the incipient wetness method for 12h at room temperature and then placed in oven at 60°C for 12h. The dried precursor was placed into the tube furnace and activated at 500°C or 700°C in N_2 for 3h. The carbonized precursor was washed with hot distilled water containing a proper amount of concentrated hydrochloric acid, and then repeatedly washed with distilled water until the filtrate became neutral. Finally, the washed material was dried at 110°C for 12h to yield the bamboo-derived porous biochar (denoted as BC-50-500 or BC-50-700, respectively, where BC is the abbreviation of the bamboo-derived porous biochar, 50 represents the amount of $ZnCl_2$ and 500 or 700 is the activation temperature).

1.0 g of BC-50-500 or 1.0 g of BC-50-700 was added to a 25 mL round-bottomed flask. 10 mL of concentrated sulfuric acid was slowly dropped to the flask under strong stirring with a glass rod, and was then treated at 180°C in oil bath for 5h under magnetic stirring. Next the filtrate was repeatedly washed with distilled water until the pH value of the filtrate was neutral, followed by final drying in the oven at 110 °C for 12 h to yield the modified material (designated as BC-50-500-S or BC-50-700-S, respectively, where S refers to the sulfonation modification by concentrated sulfuric acid).

1.5 g of BC-50-500 or BC-50-700 or AC was added to a 25 mL round-bottomed flask. 9 mL of 65% concentrated nitric acid was slowly added to the flask under strong stirring with a glass rod, and was then treated at different temperature (83°C or 100°C or 120°C) for 3 h under magnetic stirring. Next the filtrate was repeatedly washed with distilled water until the pH value of the filtrate was neutral, followed by final drying in the oven at 110°C for 12 h to yield the modified material (designated as BC-50-500-N-83 or BC-50-700-N-83 or BC-50-700-N-100 or BC-50-700-N-120 or AC-N-100, where N refers to the oxidation modification by concentrated nitric acid).

2.3. Static adsorption experiment

A simulated wastewater containing Cd^{2+} with mass concentration of (25-200 mg/L) was prepared by $CdCl_2 \cdot 2.5H_2O$ dissolved in a certain volume of distilled water. 25mL of simulated Cd^{2+} wastewater and a certain amount of biochar (from 0.025g-0.25g) was placed in 50mL round bottom flask. The pH (range from 3-7) of the system was adjusted with 0.1mol/L NaOH solution or 0.1mol/L HCl solution. The system was placed in thermostatic water bath (temperature's range from 25°C-75°C) under magnetic stirring in a certain period of time (30-120min). The Cd^{2+} concentration in the filtrate was determined by atomic absorption spectrophotometer [TAS-990 (Beijing analysis Instrument Co.,Ltd.)] and the removal percentage of Cd^{2+} was calculated according to (1) [17].

$$\textit{Removal percentage (\%)} = (C_0 - C_e) / C_0 \qquad (1)$$

Where C_0 refers to the initial concentration of Cd^{2+} in the simulated wastewater, and C_e refers to the concentration of Cd^{2+} in the filtrate after saturated adsorption. Each experiment was conducted in triplicate, and the observed data deviation between three parallel experiments was lower than ca. 5%.

2.4. Recycling test

After centrifugal separation of the biochar with the solution, the concentration of Cd^{2+} in the supernatant was determined by AAS. The spent biochar and 20mL hydrochloric acid with 0.5mol/L concentration were added to a 50 mL round-bottomed flask, refluxed in a thermostatic water bath at 50°C for 30min, and then filtrated. The filter residue was repeated 3 times according to the preceding procedure, washed with distilled water to neutral, dried, and then reused for adsorption of Cd^{2+}.

3. Results and Discussion

3.1. Cd^{2+} adsorption properties of different porous carbon materials

Table 1. Comparision of the adsorption capacity and removal percentage of various materials[a]

Sample	Adsorption capacity/($mg \cdot g^{-1}$)	Removal percentage (%)
BC-50-500	10.78	43.10
BC-50-700	9.42	37.66
BC-50-500-S	10.57	42.26
BC-50-700-S	6.35	25.41
BC-50-500-N-83	11.93	47.70
BC-50-700-N-83	13.59	54.34
BC-50-700-N-100	17.29	69.16
BC-50-700-N-120	13.25	53.01

[a]Adsorption conditions: 25 mL of simulated cadmium-containing wastewater (100 mg/L), 0.1000 g of adsorbent dosage, adsorption time of 90 min and adsorption temperature of 35°C.

Table 1 lists the adsorption capacity and the removal percentage of cadmium ions in simulated wastewater over some representative samples. As shown in Table 1, Cd^{2+} adsorption capacity of BC-50-700 is slightly less than that of BC-50-500. It may be that BC-50-700 possesses less acid functional groups and slightly lower pore structure parameters due to its higher preparation temperature [9, 15].

The result implies that the surface acid groups and pore structure of biochar are the two main factors affecting the adsorption of cadmium ions. The Cd^{2+} adsorption properties of materials after HNO_3 oxidation were improved to varying extent, which indicate that the increase of acid groups is beneficial to improve the Cd^{2+}adsorption performance of materials[7-9, 15]. However, the improved Cd^{2+} adsorption capacity of BC-50-700-N-83 was stronger than that of BC-50-500-N-83, which may be due to BC-50-700's better structural stability as compared to BC-50-500 [7-9, 15]. Furthermore, BC-50-700-N-100 exhibited the best Cd^{2+} removal percentage for the simulated cadmium-containing wastewater. This result may be due to BC-50-700's better structural stability leading its higher acidic functional groups via HNO_3 oxidation modification [7-9, 15], which benefit to the adsorption of Cd^{2+}. BC-50 -700-N-120 had a lower adsorption capacity similar to that of BC-50-500-N-83, which may be due to a further increase in HNO_3 oxidation temperature leading to the greater damage to its pore structure [7-9, 15]. The adsorption capacity of BC-50-500-S and BC-50-700-S were not enhanced, suggesting that sulfonation did not seem to improve its adsorption performance.

3.2. Effect of adsorption conditions on Cd^{2+} removal performance

3.2.1. Effect of pH on Cd^{2+} removal percentage

The effect of pH on Cd^{2+} removal percentage over BC-50-700-N-100 (the optimum bamboo-derived biochar obtained by HNO_3 oxidation modification) is shown in Figure 1 under the same conditions of 25 mL of cadmium-containing simulated wastewater, 0.20 g of biochar, 90 min and 35°C. It can be seen from Figure 1 that the removal percentage of Cd^{2+} was lower at pH 3. When pH was enhanced from 3 to 4, the removal percentage of Cd^{2+} was increased from 75.36% to 86.96%, and then the value increased gradually[5]. The lower removal efficiency in strongly acidic media is due to the competition over the available adsorption sites between Cd^{2+} and H^+ with the high H^+ concentration under the low pH value, and thus suppresses Cd^{2+} adsorption[6, 18, 19]. With the increase of the pH value, the adsorption capacity of H^+ decreased, accordingly, Cd^{2+} adsorbed gradually increased[5, 20, 21]. When the pH is greater than 7, Cd^{2+} begins to precipitate[6, 18]. Therefore, pH 7.0 in the simulated wastewater is the reasonable value for Cd^{2+} adsorption.

3.2.2. Effect of temperature on Cd^{2+}removal percentage

The effect of temperature on Cd^{2+} removal percentage over BC-50-700-N-100 is shown in Figure 2 under the same other conditions. As shown in Figure 2, the temperature increased from 25°C to 35°C, the removal percentage was enhanced from 90.23% to 94.65%. With further increasing temperature, the value decreased gradually. The result is consistent with the result of adsorption versus temperature in the literature[22]. The increase in adsorption reflects chemical interactions between the ions and the biochar, indicating that the adsorption process is endothermic in nature, whereas the decrease in adsorption with the further increase of temperature may be due to the weakening of adsorptive forces between $Cd^{2+}$and adsorption sites of the adsorbent[22]. Thus the proper adsorption temperature is about 35°C.

3.2.3. Effect of adsorbent amount on removal percentage of Cd^{2+}

The effect of the adsorbent amount on Cd^{2+} removal percentage over BC-50-700-N-100 is shown in Figure 3 under constant other conditions. It can be observed from Figure 3 that the Cd^{2+} removal percentage increased with increasing the biochar's amount from 0.025 to 0.200 g. When the biochar dosage was more than 0.200g, the removal percentage of Cd^{2+} remained basically unchanged. At the beginning of adsorption, the increased biochar's dosage provided a large number of the unoccupied adsorption sitesto Cd^{2+}[17, 23],accordingly, the Cd^{2+}removal percentage was rapidly increase[19]. At the later stage of adsorption, the mass concentration of Cd^{2+} decreased gradually and thus the adsorption driving force of Cd^{2+} to the surface of biochar weakened. Although the adsorption removal percentage of Cd^{2+} was enhanced, the unit adsorption capacity of biochar was reduced. As a result, the adsorbent amount of 0.2000 g / 25mL is appropriate.

IOP Conf. Series: Materials Science and Engineering **729** (2020) 012081 doi:10.1088/1757-899X/729/1/012081

3.2.4. Effect of contact Time

The effect of contact time on the Cd^{2+} removal percentage over BC-50-700-N-100is shown in Figure 4 under constant other conditions. As shown in Figure 4, the Cd^{2+} removal percentage increased with prolonging the contact time. The value reached 95.37% when the time was 90 min. As the time went on, the Cd^{2+} removal percentage remained essentially unchanged. At the initial stage, the high Cd^{2+}concentration and adsorption sites contributed to the fast adsorption rate. With the progress of the adsorption, the Cd^{2+}concentration and adsorption sites decreased and thus the adsorption rate slowed down until the adsorption equilibrium was reached[22-24]. Accordingly, the adsorption time of 90 min is suitable. In summary, when the Cd^{2+} concentration is 100 mg/L, the optimum conditions for Cd^{2+} adsorption are pH 7, adsorption temperature of 35°C, adsorbent dosage of 0.2000g/25mL and adsorption time of 90 min.

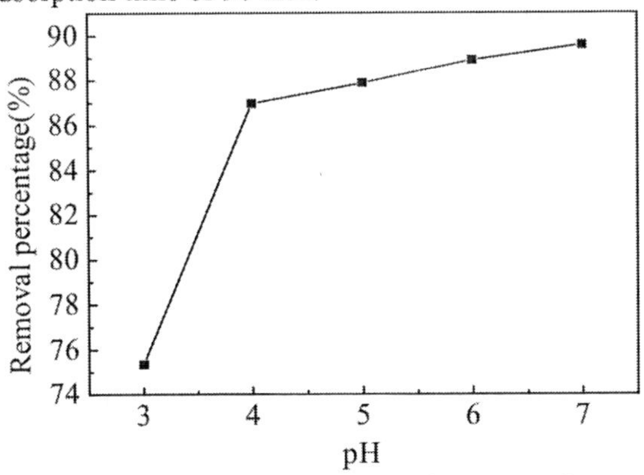

Figure 1. Effects of pH value on the removal percentage of Cd^{2+}.

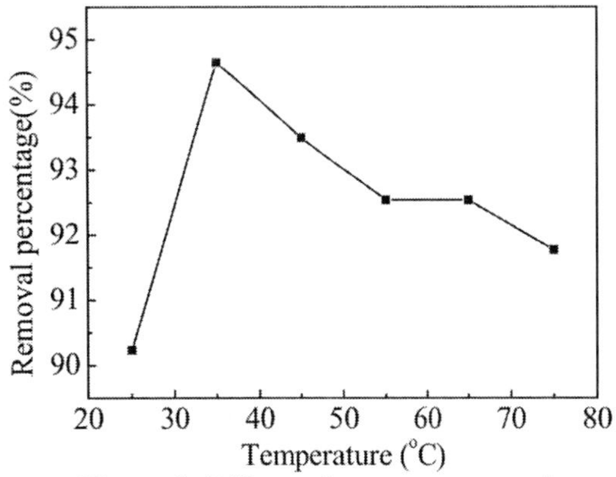

Figure 2. Effects of temperature on the removal percentage of Cd^{2+}.

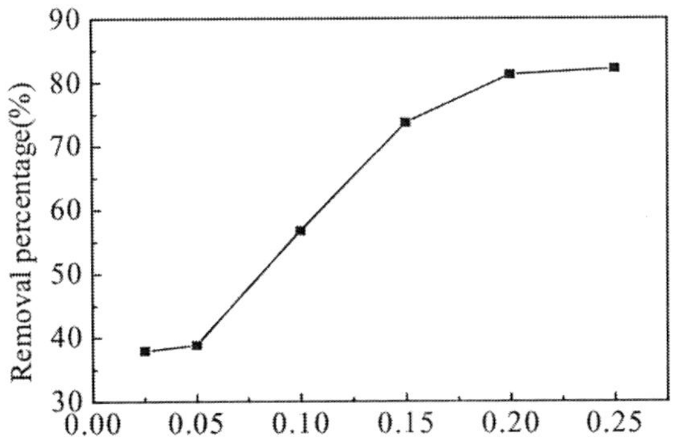

Figure 3. Effects of adsorbent dosage on the removal percentage of Cd^{2+}.

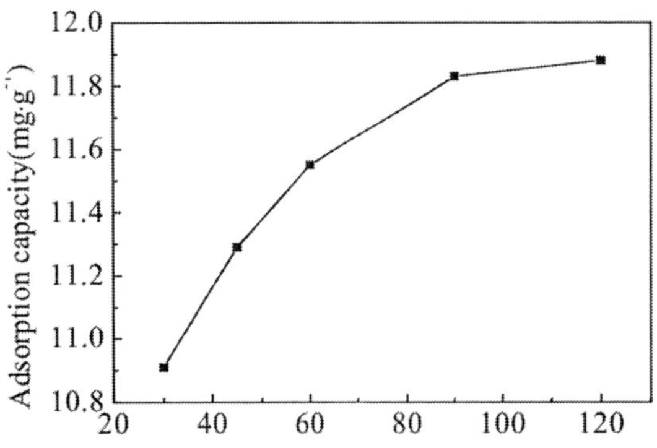

Figure 4. Effects of contact time on the removal percentage of Cd^{2+}.

3.2.5. Effect of temperature on Cd^{2+}removal percentage

The effect of temperature on Cd^{2+} removal perc

3.3. Cd^{2+} adsorption kinetics and isotherm

3.3.1. Adsorption kinetics

The kinetic parameters are usually measured by fitting the kinetic data based on the Lagergren quasi-second order kinetic equation, contributing to study the mechanism of the adsorption process [25]. It can be seen from Fig.4 that the adsorption rate of Cd^{2+} over BC-50-700-N-100 was increased sharply in the initial 45 min owing to biochar possessing a large number of the unoccupied adsorption sites.

With the prolongation of adsorption time, the active sites for adsorbing Cd^{2+} were relatively reduced, and thus Cd^{2+}adsorption rate became slow. After 90 min, the adsorption of Cd^{2+}achieves equilibrium, and the adsorption percentage reached the maximum. The fitting results based on the equation (2) was shown in Figure 5.

$$t/q_t=1/k_2q_e^2+t/q_e \quad (2)$$

where q_t is the amount of Cd^{2+} adsorbed by biochar at time t ($mg \cdot g^{-1}$); q_e is equilibrium adsorption capacity ($mg \cdot g^{-1}$); k_2 is the rate constant in the pseudo-second order kinetic equation ($g\ mg^{-1} \cdot min^{-1}$) [2, 26-30]. It can be calculated from Fig. 5 that q_e is 16.62 $mg \cdot g^{-1}$ and the linear correlation coefficient R^2=0.9999, which indicates that the pseudo-second order kinetic model can well describe the adsorption kinetics of Cd^{2+}.

3.3.2. Adsorption isotherm

The adsorption isotherm of Cd^{2+} on over BC-50-700-N-100 at 35°C is shown in Fig.6. With the increase of Cd^{2+} concentration, the equilibrium adsorption capacity of Cd^{2+} also increased. When the equilibrium concentration of Cd^{2+} reached 20mg/L, saturated adsorption was achieved. This result is due to the limited amount of adsorption sites in the case of the constant amount of biochar. For the low initial Cd^{2+} concentration, sufficient active sites can be provided for the adsorption, so that the equilibrium adsorption capacity of biochar increases gradually. When the initial Cd^{2+} concentration is high, the active sites are relatively insufficient, leading to rapid saturation of the adsorption sites.

Langmuir and Freundlich adsorption isotherm models are commonly used to describe the adsorption equilibrium of liquid-solid systems. The Langmuir model is based on this assumption that the adsorption of components is a monomolecular layer that occurs on a uniform surface, while the Freundlich model is built on the hypothesis that the adsorption of metal ions occurs on non-uniform surfaces. The corresponding mathematical expression is as follows, respectively[2, 24, 31-33]:

$$1/q_e= (1/QK_1)\times(1/C_e)+1/Q \quad (3)$$

$$\lg q_e=n\lg C_e+\lg K_2 \quad (4)$$

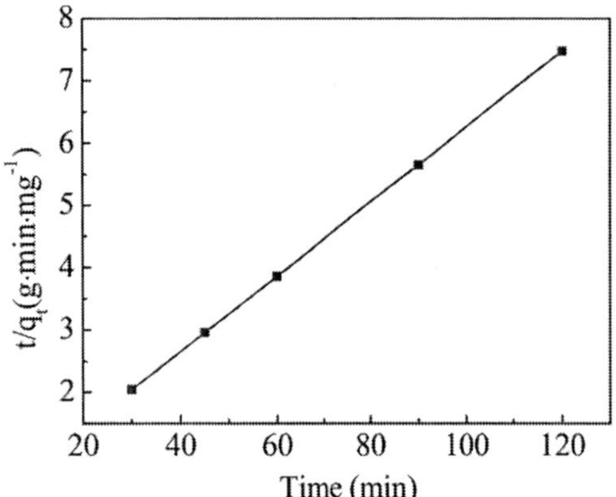

Figure 5. The adsorption kinetics of Cd^{2+} on biochar.

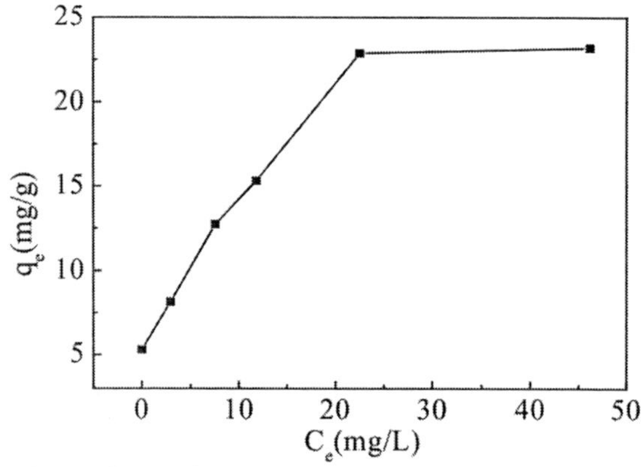

Figure 6. Adsorption isotherm of biochar for Cd^{2+} at 35°C.

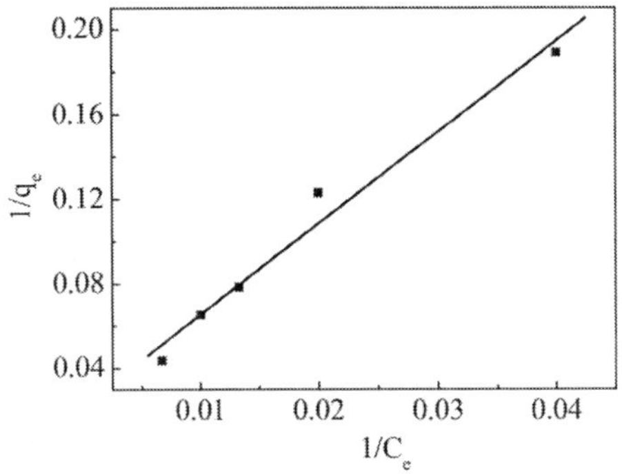

Figure 7. Langmuir adsorption isotherm of Cd^{2+} on biochar at 35°C.

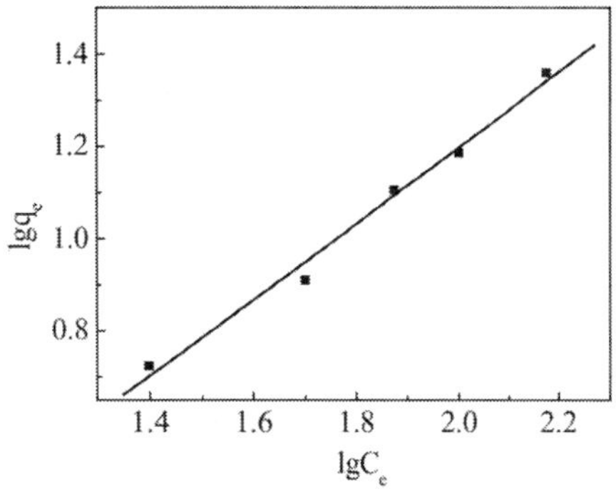

Figure 8. Freundlich adsorption isotherm of Cd^2+ on biochar at 35°C.

where q_e is the equilibrium adsorption capacity ($mg \cdot g^{-1}$); C_e is the adsorption equilibrium concentration of Cd^{2+}($mg \cdot L^{-1}$); Q is the maximum adsorption capacity ($mg \cdot g^{-1}$); K_1 is Langmuir adsorption equilibrium constant ($L \cdot mg^{-1}$); K_2 and n are Freundlich adsorption constants.

The adsorption experiments data were fitted with the Langmuir and Freundlich models, respectively. As shown in Fig.7, Langmuir equation was fitted to $1/q_e=4.3014/C_e+0.02245$. The maximum adsorption capacity of Cd^{2+} was calculated as $44.54 mg \cdot g^{-1}$, and the correlation coefficient R^2 was 0.9702, which indicated that the adsorption behavior of biochar to Cd^{2+} accorded with Langmuir adsorption isotherm. Meanwhile, the adsorption data of biochar was fitted by Freundlich adsorption isotherm(see Fig.8), and the equation was obtained as $lgq_e= -0.449+0.823lgC_e$ where the correlation coefficient R^2 is 0.9859, and 1/n=0.823. The large value 1/n indicates that the adsorption capacity of the adsorbent is strong and that the biochar by HNO_3 oxidation modification is a highly adsorbable adsorbent to Cd^{2+}[32]. The above results show that the Freundlich isothermal adsorption model fitted better than the Langmuir isothermal adsorption model for this adsorption system, indicating that the surface of the porous biochar is not uniform.

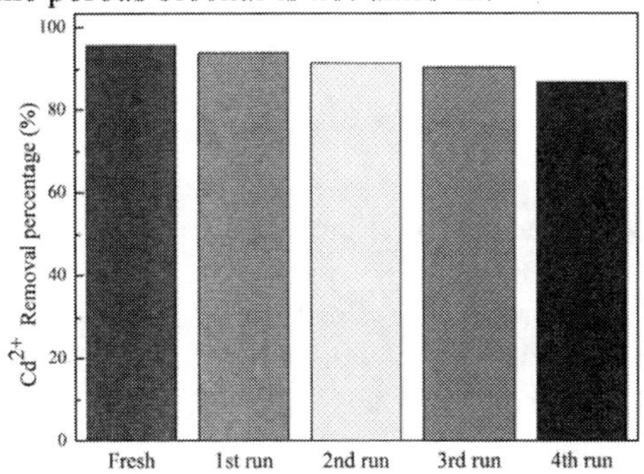

Figure 9. Effects of regeneration times on the removal percentage of Cd^{2+}.

3.4. Regeneration and recycling performance

The regeneration of over BC-50-700-N-100 was carried out by reflux in dilute hydrochloric acid[25]. The adsorption performance after 4 runs regeneration is shown in Figure 9. It can be seen from Fig. 9 that the Cd^{2+} removal percentage of biochar was reduced from 95.53% to 90.36% after 3 runs regeneration. For the 4th run regeneration, the removal percentage is reduced to 86.88%, which indicates that the biochar exhibits good regeneration performance.

4. Conclusions

In summary, in this work a bamboo-derived porous biochar with a good adsorption removal capacity to cadmium ions in the simulated wastewater was prepared by 50% $ZnCl_2$ chemical activation of bamboo processing residues at 700°C and further modification with the concentrated HNO_3 oxidation at 100°C. At the Cd^{2+} concentration of 100mg/L, the biochar achieved the adsorption removal efficiency of 95.53% under the optimum adsorption conditions (at pH 7, at adsorption temperature of 35°C, at adsorbent dosage of 0.2000g/25mL and in adsorption time of 90min). The adsorption process of Cd^{2+} on biochar was in accordance with the pseudo-second-order adsorption kinetics equation. Meanwhile, the Freundlich isothermal adsorption model fitted better than Langmuir isothermal adsorption model for this Cd^{2+} adsorption system. After four adsorption-desorption recycling, the adsorption capacity of Cd^{2+} over biochar has no significant decline, which indicates that biochar has high adsorption capacity and good regeneration performance. Such cheap and readily available biochar may therefore become a promising adsorbent for removal of cadmium ions in the industrial wastewater.

Acknowledgments

We acknowledge the financial support for this work by the National Natural Science Foundation of China (21073057, 21676079), the Natural Science Foundation of Hunan Province (11JJ6008, 2018JJ3335), the Innovation Platform Open Foundation of Hunan College (14K059, 18K016), and Hunan Normal University graduates innovative experiment project (201610542074, 201810542141).

References

[1] Sud, D., Mahajan, G., Kaur, M. P. (2008) Agricultural waste material as potential adsorbent for sequestering heavy metal ions from aqueous solutions – A review. Bioresource Technol., 99 (14): 6017–6027.

[2] Febrianto, J., Kosasih, A. N., Sunarso, J., Ju, Y.-H., Indraswati, N., Ismadji, S. (2009) Equilibrium and kinetic studies in adsorption of heavy metals using biosorbent: A summary of recent studies. J. Hazard. Mater., 162 (2): 616–645.

[3] Molino, A., Erto, A., Natale, F. D., Donatelli, A., Iovane, P., Musmarra, D. (2013) Gasification of Granulated Scrap Tires for the Production of Syngas and a Low-Cost Adsorbent for Cd(II) Removal from Wastewaters. Ind. Eng. Chem. Res., 52 : 12154–12160.

[4] Lo, S.-F., Wang, S.-Y., Tsai, M.-J., Lin, L.-D. (2012) Adsorption capacity and removal efficiency of heavy metal ions by Moso and Ma bamboo activated carbons. Chem. Eng. Res. Des., 90 (9): 1397–1406.

[5] Aziz, H. A., Adlan, M. N., Ariffin, K. S. (2008) Heavy metals (Cd, Pb, Zn, Ni, Cu and Cr(III)) removal from water in Malaysia: Post treatment by high quality limestone. Bioresource Technol., 99 (6): 1578–1583.

[6] Argun, M. E., Dursun, S., Karatas, M., Gürü, M. (2008) Activation of pine cone using Fenton oxidation for Cd(II) and Pb(II) removal. Bioresource Technol., 99 (18): 8691–8698.

[7] Qiu, J., Wang, G., Bao, Y., Zeng, D., Chen, Y. (2015) Effect of oxidative modification of coal tar pitch-based mesoporous activated carbon on the adsorption of benzothiophene and dibenzothiophene. Fuel Process Technol., 129: 85–90.

[8] Zhou, A., Ma, X., Song, C. (2009) Effects of oxidative modification of carbon surface on the adsorption of sulfur compounds in diesel fuel. Appl. Catal. B: Environ., 87 (3): 190–199.

[9] Bhatnagar, A., Hogland, W., Marques, M., Sillanpää, M. An overview of the modification methods of activated carbon for its water treatment applications. Chem. Eng. J., 219: 499–511.

[10] Tan, Z., Qiu, J., Zeng, H., Liu, H., Xiang, J. (2011) Removal of elemental mercury by bamboo charcoal impregnated with H_2O_2. Fuel, 90 (4): 1471–1475.

[11] Yang, W., Wang, H., Zhang, M., Zhu, J., Zhou, J., Wu, S. (2016) Fuel properties and combustion kinetics of hydrochar prepared by hydrothermal carbonization of bamboo. Bioresource

Technol., 205: 199–204.
[12] Liu, Q.-S., Zheng, T., Wang, P., Guo, L. (2010) Preparation and characterization of activated carbon from bamboo by microwave-induced phosphoric acid activation. Ind. Crop Prod., 31 (2): 233–238.
[13] Ahmad, A. A., Hameed, B. H. (2009) Reduction of COD and color of dyeing effluent from a cotton textile mill by adsorption onto bamboo-based activated carbon. J. Hazard. Mater., 172 (2): 1538–1543.
[14] Mohan, D., Sarswat, A., Ok, Y. S., Pittman, C. U. (2014) Organic and inorganic contaminants removal from water with biochar, a renewable, low cost and sustainable adsorbent – A critical review. Bioresource Technol., 160: 191–202.
[15] Yang, E., Yao, C., Liu, Y., Zhang, C. Jia, L., Li, D., Fu, Z., Sun, D., Robert Kirk, S., Yin, D. (2018) Bamboo-derived porous biochar for efficient adsorption removal of dibenzothiophene from model fuel. Fuel, 211: 121–129.
[16] Lu, K., Yang, X., Shen, J., Robinson, B., Huang, H., Liu, D., Bolan, N., Pei, J., Wang, H. (2014) Effect of bamboo and rice straw biochars on the bioavailability of Cd, Cu, Pb and Zn to Sedum plumbizincicola. Agr. Ecosyst. Environ., 191: 124–132.
[17] Yong, S. K., Bolan, N., Lombi, E., Skinner, W. (2013) Synthesis and Characterization of Thiolated Chitosan Beads for Removal of Cu(II) and Cd(II) from Wastewater. Water Air Soil Poll., 224: 1720–1732.
[18] Zeng, G., Liu, Y., Tang, L., Yang, G., Pang, Y., Zhang, Y., Zhou, Y., Li, Z., Li, M., Lai, M., He, X., He, Y. (2015) Enhancement of Cd(II) adsorption by polyacrylic acid modified magnetic mesoporous carbon. Chem. Eng. J., 259: 153–160.
[19] Tan, I. A. W., Chan, J. C., Hameed, B. H., Lim, L. L. P. (2016) Adsorption behavior of cadmium ions onto phosphoric acid-impregnated microwave-induced mesoporous activated carbon. J. Water Process Eng., 14: 60–70.
[20] Zheng, W., Li, X.-m., Wang, F., Yang, Q., Deng, P., Zeng, G.-m. (2008) Adsorption removal of cadmium and copper from aqueous solution by areca—A food waste. J. Hazard. Mater., 157 (2): 490–495.
[21] Li, X., Wang, S., Liu, Y., Jiang, L., Song, B., Li, M., Zeng, G., Tan, X., Cai, X., Ding, Y. (2017) Adsorption of Cu(II), Pb(II), and Cd(II) Ions from Acidic Aqueous Solutions by Diethylenetriaminepentaacetic Acid-Modified Magnetic Graphene Oxide. J. Chem. Eng. Data, 62 (1): 407–416.
[22] Wang, K., Gu, J., Yin, N. (2017) Efficient Removal of Pb(II) and Cd(II) Using NH_2-Functionalized Zr-MOFs via Rapid Microwave-Promoted Synthesis. Ind. Eng. Chem. Res., 56 (7): 1880–1887.
[23] Garg, U., Kaur, M. P., Jawa, G. K., Sud, D., Garg, V. K. (2008) Removal of cadmium (II) from aqueous solutions by adsorption on agricultural waste biomass. J. Hazard. Mater., 154 (1): 1149–1157.
[24] Zhao, F., Repo, E., Yin, D., Sillanpää, M. E. T. (2013) Adsorption of Cd(II) and Pb(II) by a novel EGTA-modified chitosan material: Kinetics and isotherms. J. Colloid Interf. Sci., 409: 174–182.
[25] Debnath, S., Nandi, D., Ghosh, U. C. (2011) Adsorption–Desorption Behavior of Cadmium(II) and Copper(II) on the Surface of Nanoparticle Agglomerates of Hydrous Titanium(IV) Oxide. J. Chem. Eng. Data, 56 (7): 3021–3028.
[26] Yan, L., Zhao, Q., Jiang, T., Liu, X., Li, Y., Fang, W., Yin, H. (2015) Adsorption characteristics and behavior of a graphene oxide–Al13 composite for cadmium ion removal from aqueous solutions. RSC Adv., 5 (83): 67372–67379.
[27] Shin, E. W., Karthikeyan, K. G., Tshabalala, M. A. (2007) Adsorption mechanism of cadmium on juniper bark and wood. Bioresource Technol., 98 (3): 588–594.
[28] Wang, H., Zhao, Y., Ma, L., Fan, P., Xu, C., Jiao, Ch., Lin, A. (2016) Preparation of Composite Modified Expanded Graphite and Its Adsorption on Acid Brilliant Blue Dye. Chem. J.

Chinese U., 37 (2): 335--341.
[29] Zhao, F., Repo, E., Sillanpää, M., Meng, Y., Yin, D., Tang, W. Z. (2015) Green Synthesis of Magnetic EDTA- and/or DTPA-Cross-Linked Chitosan Adsorbents for Highly Efficient Removal of Metals. Ind. Eng. Chem. Res., 54 (4): 1271–1281.
[30] Iqbal, M., Iqbal, N., Bhatti, I. A., Ahmad, N., Zahid, M. (2016) Response surface methodology application in optimization of cadmium adsorption by shoe waste: A good option of waste mitigation by waste. Ecol. Eng., 88: 265–275.
[31] Chen, Y., Gao., L., He, M., Wei, Y. (2014) High-capacity Polyvinyltetrazole-grafted Chelating Resin for Adsorption of Heavy Metal Ions. Chem. J. Chinese U., 35, (7): 1596–1602.
[32] Fouladi Tajar, A., Kaghazchi, T., Soleimani, M. (2009) Adsorption of cadmium from aqueous solutions on sulfurized activated carbon prepared from nut shells. J. Hazard. Mater., 165 (1): 1159–1164.
[33] Zheng, W., Li, X.-m., Yang, Q., Zeng, G.-m., Shen, X.-x., Zhang, Y., Liu, J.-j. (2007) Adsorption of Cd(II) and Cu(II) from aqueous solution by carbonate hydroxylapatite derived from eggshell waste. J. Hazard. Mater., 147 (1): 534–539.

IOP Conf. Series: Materials Science and Engineering **729** (2020) 012082 doi:10.1088/1757-899X/729/1/012082

Electrochemical Decontamination of Platinum Nanocrystals and its Applications for Methanol Electrooxidation

Leilei Lu*, Mi Zhu, Yaoping Zheng, Rong Yang

School of science, Xi'an University of Technology, Xi'an, Shaanxi, 710054, China

*E-mail: lullxaut@xaut.edu.cn

Abstract: Polyvinylpyrrolidone (PVP) is an effective surface-capping agent for engineering metal surface structure but is hardly to be thoroughly removed. In this study, the electrochemical treatment was applied for the decontamination of polyvinylpyrrolidone (PVP) on platinum (Pt) nanocrystals. Spherical, cubic, octahedral and cuboctahedral Pt nanocrystals were prepared by the colloidal method with PVP K30 as the capping agent. Electrochemical decontamination of Pt nanocrystals was performed by using square-wave potential steps and methanol electrooxidation was used as a probe reaction. High-frequency potential steps were set up between a high voltage and a low voltage to avoid surface distortion. Cyclic voltammetry (CV) and chronoamperometry measurements were used to evaluate the catalytic activity of the cleaned Pt nanocrystals for methanol. The results showed that electrochemical decontamination was an effective way to clean Pt nanocrystals and the cuboctahedral Pt nanocrystals displayed the best electrocatalytic activity for methanol electrooxidation.

1. Introduction

Shape-controlled synthesis of platinum-group metals is of fundamental and practical importance in energy catalysis field. Both anode oxidation reaction and cathode reduction reaction are very sensitive to the surface structure and the exposed surface sites [1-3]. PVP is one of the most effective surface-active capping agents for the morphology control of Pt-based nanocrystal. The main reason is that the addition of PVP can simply obtain uniform nanocrystals loaded on various carbons with good reproducibility [4-5].

Because PVP can combine strongly with the surface of platinum through chemical adsorption, it can prevent the direct interaction between the surface active sites and the reagent molecules, thus reducing the activity or selectivity of catalytic reaction [6]. Even if Pt nanocrystals were carefully cleaned by extensive washing with organic solvents, the distinctive hydrogen adsorption/desorption peaks of Pt CVs in blank sulfuric acid were still inconspicuous. This is mainly ascribed to the high-molecular-weight PVP which can strongly block the Pt active sites [7-8]. Therefore, it is necessary to apply a thorough cleaning process for Pt nanocrystals before electrochemical measurements.

Various methods have been reported in the literature for the removal of surface-capping PVP from the nanoparticle surface. The most effective method should be the heat-treatment of PVP-capped nanoparticles at different temperatures. However, the particle shapes experienced a dramatic transformation into spherical-like shapes when the temperature is raised above ~500 °C At higher temperatures, surface melting becomes obvious leading to coalescence of the surfaces of neighboring nanocrystals and a decrease in the volume occupied by the assembled nanocrystals and hence, careful control of temperatures is necessary for the heat-treatment method [9]. Another technique for removal of stabilizer from nanoparticles is by using UV–ozone exposure in order to eliminate the traces of PVP still presenting on the surface [10-11]. However, it is difficult to generalize this technique and the results depend on the chemical nature of the specimen and the capping agents. The removal of PVP has also been reported by applying CV up to 1.5 V [12-13], CO-adsorption-induced removal [14]. However, it has been generally accepted that too many cycles up to high potentials may degrade the surface structure [15-16]. Especially, Pt(100) surface could only maintain its structure under 0.9 V. While, CV cycles below 0.9 V couldn't clean PVP adsorbed on Pt nanocrystals. Applying a high voltage to the Pt surface will introduce defects to the primary surface. That is mainly because oxygen could adsorb on the Pt(100) surface in the potential region higher than 0.9 V and oxygen atom could crowd out some Pt atoms from its equilibrium position [17]. This process is irreversible. Therefore, it is of great significance to develop a simple and efficient cleaning method for Pt nanocrystals to avoid using too many organic solvents. In this work, a facile electrochemical decontamination method was applied to the PVP-capped Pt nanocrystals by square-wave potential steps. High-frequency potential steps were set up between a high voltage and a low voltage to remove the chemically adsorbed PVP molecules without disturbing the surface structure and its performances. Methanol electrooxidation was then adopted as a probe reaction to evaluate the electrocatalytic activity and stability of treated Pt nanocrystals.

2. Experimental Section

2.1 Synthesis of Pt nanocrystals

Pt nanocrystals were synthesized by the colloidal method described previously [15]. The precursor Pt salt is $H_2PtCl_6{\cdot}6H_2O$, and the stabilizer is PVP K30. Briefly, 3 mL 0.375 M PVP and 1.5 mL 0.0625 M $H_2PtCl_6{\cdot}6H_2O$ solutions in ethylene glycol (EG) were added gradually to the boiling EG over a 16-min period. The resulting mixture was refluxed for an additional 5 min. The product was then centrifuged and washed by a large amount of acetone, ethanol, and hexane before the electrochemical test. Different amounts of AgNO3 solution in EG were added to the reaction mixture prior to the addition of PVP and Pt precursors to adjust the morphology of Pt nanocrystals. For Pt cube, cuboctahedron and octahedron synthesis, Ag salt/Pt salt molar ratio were adjusted to 1.1 mol %, 11 mol %, and 32 mol %, respectively.

2.2 Characterization and electrochemical measurements

Transmission electron microscopy (TEM) and high-resolution transmission microscopy (HRTEM) experiments were made on a JEM-3010 microscope operated at 300 kV.

Before the electrochemical measurements, Pt nanocrystals with different shape were dispersed in ultra-pure water by sonicating for at least 20 min and then 6 μL suspension was transferred to the surface of a glassy carbon (GC) electrode with a diameter of 4 mm with a micropipette. The GC electrode was then dried in air at room temperature. A saturated Hg/Hg_2SO_4 electrode and Pt foil were used as the reference and counter electrode, respectively. Conventional three-compartment electrochemical cells were used for all the electrochemical tests. The electrolyte solution was deaerated with high-purity Ar (99.99%) for at least 20 min before each measurement. Cyclic voltammograms (CVs) were taken in the potential region of 0.05 ~ 1.2 V. Chronoamperometric curves were performed at 0.80 V for 600 s after a potential step from 0.05 V in 0.5 mol·L-1 H_2SO_4 + 0.5 mol·L^{-1} CH_3OH. All electrochemical measurements were performed with a CHI electrochemical workstation (CH Instruments, model CHI750d) at room temperature (ca. 20 °C). All the potentials in this paper are calibrated to a reversible hydrogen electrode (RHE).

3. Results and Discussion

3.1 Morphology characterization of platinum nanocrystals

Figure 1 shows TEM and HRTEM images of cubic (a, d), octahedral (b, e) and cuboctahedral (c, f) Pt nanocrystals synthesized in this reaction. These Pt nanocrystals are uniform in shape with a narrow size distribution (cubic Pt: 7.0 nm, octahedral Pt: 10.0 nm, cuboctahedral Pt: 9.5 nm). The HRTEM image (Figure 1d) of a Pt cube demonstrates the exposed (100) facets with an interplanar spacing of 0.20 nm. While HRTEM image of Pt octahedra demonstrates an interplanar spacing of 0.23 nm, in which all side faces are covered with (111) facets. The cuboctahedral Pt was covered by both (100) and (111) facets.

According to the reaction mechanism proposed by Yang's group, the Ag+ ion plays a crucial role in controlling the shape and surface structure of the Pt nanocrystals. As the concentration of Ag ion increases in the reaction mixture, the majority of the Pt particles changes from the cubes to the cuboctahedra, and eventually to the octahedra [15].

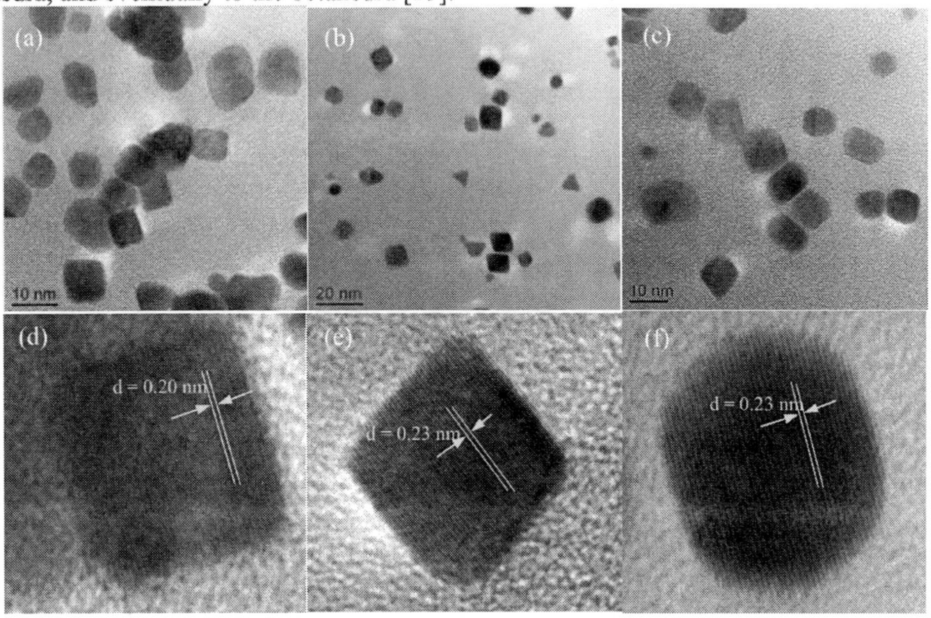

Figure 1. TEM and HRTEM images of cubic (a, d), octahedral (b, e) and cuboctahedral (c, f) Pt nanocrystals.

3.2 Electrochemical testing

In our previous tests, the as-prepared and carefully washed platinum nanocrystals were used for the electrocatalytic oxidation of methanol. However, no electrooxidation current was detected. Therefore, a repeated potential step sequence between 0.05 V and 1.4 V with a pulse width of 0.5 s was applied to the PVP-capped Pt nanocrystals. The short pulse width was selected to avoid long-time oxygen adsorption and its influence on surface structure. Figure 2a shows the CV curves of spherical platinum nanocrystals before and after electrochemical treatment in 0.5 mol·L^{-1} H_2SO_4 solution with a scanning rate of 0.05 V·s^{-1}. As can be seen in Figure 2a, the adsorption and desorption peaks of hydrogen and oxygen are very small before electrochemical cleaning since the active sites were blocked by chemically adsorbed PVP. After electrochemical treatment for 100 steps, the featured hydrogen and oxygen peaks become clear and tend to stabilize. This indicates that the potential step treatment can effectively remove PVP species strongly adsorbed on platinum surfaces which were hard to be removed by solvent washing method. At the same time, this method could protect the surface structure from reconfiguration so that the catalytic activity of platinum nanocrystals with different morphologies could be distinguished in the subsequent methanol electrooxidation experiments.

It has been widely accepted that the hydrogen adsorption/desorption peaks of CVs for Pt in H_2SO_4 solution is highly sensitive to the surface morphology of Pt, which could be used to characterize the predominant surface of Pt particles [18-20]. Figure 2b shows stable CV curves platinum nanocrystals with different morphologies after electrochemical treatment in 0.5 mol·L^{-1} H_2SO_4 solution. As shown in this figure, the CVs display the typical shape for the blank voltammograms of Pt. The curves show the presence of adsorption states associated with (110) and (100) sites at 0.12 and 0.27 V, respectively [21]. The relative peak height for the (110) and (100) Pt sites is an important information for the shape

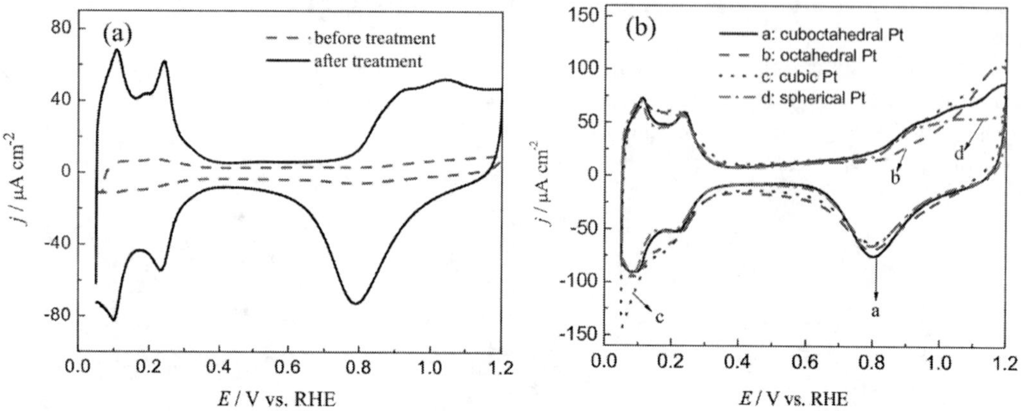

Figure 2. (a) CVs of spherical platinum nanocrystals before and after potential step treatment; (b) CV curves of Pt nanocrystals of different morphologies in 0.5 mol·L^{-1} H_2SO_4 solution.

and dominant facet yield of the Pt nanoparticles [16]. Spherical Pt nanocrystals show clear representative polycrystalline Pt features, which has a higher (110) peak height than (100) peak [22]. For the octahedral and cubic shape Pt nanoparticles, the current in the potential range between 0.11 ~ 0.23 V is a bit higher than that of spherical Pt in the CV curve, which is an indication of the Pt nanoparticles containing mainly (111) facets and (100) step sites. These electrochemical results are in good agreement with the TEM images. It has been known that (100) surface with long-range ordered domains will be destroyed under the potential higher than 0.9 V while (111) facets could maintain the stable surface structure until 1.2 V [23]. Therefore, the shoulder around 0.35 V which indicates the existence of long-range ordered (100) terrace sites was not well identified.

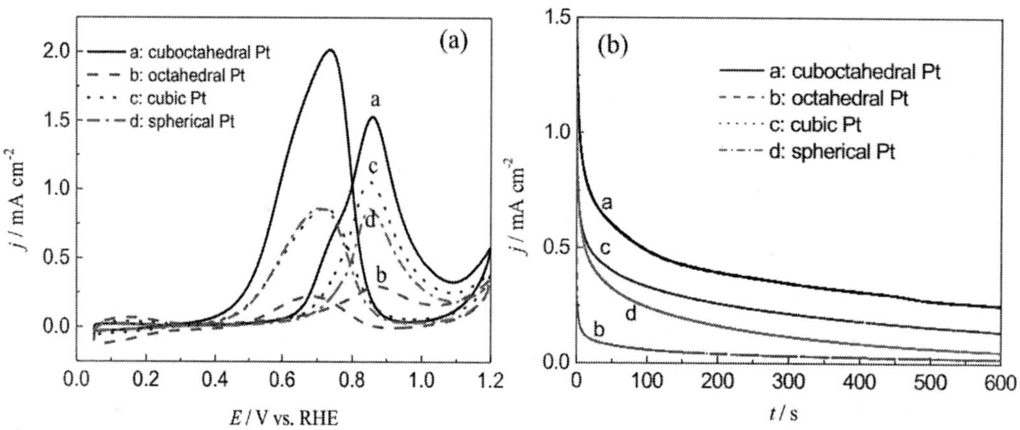

Figure 3. (a) CVs and (b) Chronoamperometric curves of methanol electrooxidation on Pt nanocrystals of different morphologies in 0.5 mol·L^{-1} H_2SO_4 + 0.5 mol·L^{-1} CH_3OH.

Figure 3 shows (a) the CVs and (b) chronoamperometric (CA) curves of methanol electrooxidation on Pt nanocrystals of different morphologies in 0.5 mol·L^{-1} H_2SO_4 + 0.5 mol·L^{-1} CH_3OH. It should be clearly seen in Figure 3a that hydrogen adsorption was depressed significantly in the solution containing methanol. This was due to the dissociative adsorption of methanol. It was known that methanol started to decompose on Pt from 0.05 V and the resulting intermediates (CO_{ad}, etc) adsorb on the Pt surfaces, which will poison the surface sites for methanol oxidation. This was regarded as the indirect oxidation of methanol. In the higher potential region, these intermediates were oxidized and the direct oxidation of methanol will occur. For the cubic Pt nanocrystals as an example, during positive scanning process, the onset potential of methanol electrooxidation was 0.45 V. The peak potential located at 0.85 V with a peak current density of 1.06 $mA \cdot cm^{-2}$, which included both the oxidation of methanol dissociative adsorption species (i.e. indirect oxidation of methanol) and the direct oxidation of methanol from aqueous solution. In the higher potential region, the electrode surface will be covered by adsorbed oxygen species. In the negative scanning process, along with the desorption of oxygen species, methanol oxidation appeared again and resulted in another oxidation peak with a maximum current density of 0.88 $mA \cdot cm^{-2}$ at 0.72V. As the potential sweep negatively, the dissociative adsorption of methanol occurred and poisonous intermediates formed and occupied Pt surface sites again, which would retard the direct oxidation of methanol and cause the oxidation current decrease. As can be seen from Figure 3a, the platinum nanocrystals after electrochemical cleaning show quite different catalytic activity for methanol electrooxidation. The positive oxidation peak currents decrease in the order of cuboctahedral Pt > cubic Pt > spherical Pt > octahedral Pt. The peak current for cuboctahedral Pt was 5 times higher as that of octahedral Pt. On the other hand, the residual current in the hydrogen region of octahedral Pt was much higher than others, indicating the octahedral Pt has better resistance to methanol dissociative adsorption. It has been proved that methanol proceeds via both direct and indirect pathways which are governed by the initial C–H and O–H bond activation, respectively [24]. The primary path requires an ensemble size of between 3~4 Pt atoms and is much more structure sensitive than the secondary path. That's the intrinsic reason for the morphology-dependent electrooxidation activity of methanol. Furthermore, Oezaslan and co-workers found that PVP-based impurities have more pronounced effects on the Pt hydr(oxide) region than on the hydrogen ad/desorption region [25], thus the electrochemical cleaning of PVP will show more important impact on the direct-oxidation of methanol in higher potential region.

It has been previously reported by Wieckowski that methanol electrooxidation activity at Pt low index crystal planes increased in the order of Pt(111) < Pt(100) < Pt(110) [22]. Furthermore, the stepped surfaces, which were considered as models for surface defects, could also give some practical instructions for shape-dependent electrocatalysis. For example, Housmans and co-workers found that methanol oxidation activity on Pt[n(111)×(110)] high index planes increased with the step density, suggesting that the presence of steps with a (110) orientation catalyzes methanol decomposition, CO oxidation and also the direct methanol oxidation [26; 27]. In addition, the catalytic activity of nanoparticles of well-defined edges and large surface boundaries are said to be greatly dependent on the bounding planes where catalytic activity follows the (111) < (100) < (110) order, as in the case of hydrogen-related reactions [28-30]. In a previous paper reported by Feliu and co-workers [31], different surface sites on platinum samples have been quantitatively analyzed by specific site-probe reactions. It has been shown clearly that for the surfaces of cubic and octahedral Pt nanoparticles, the ordered domains were mainly (100) and (111) sites. While the ordered domains for spherical Pt represented a much smaller fraction of the total surfaces. In their cover picture, they suggested that the current for (110) sites was mainly attributed to the edge and corner atoms of a cuboctahedral Pt particle. Therefore, it is reasonable to suppose that the much higher activity of cuboctahedral Pt originated from its higher fraction of edge and corner Pt atoms.

The stability of the catalysts with different morphologies has also been investigated by CA curves recorded at 0.80 V, which was shown in Figure 3b. A large anodic current is observed on cuboctahedral Pt with only a small drop. A large current is also found on cubic Pt and spherical Pt, whereas a much lower current is observed on octahedral Pt. The final methanol electrooxidation current density on cuboctahedral Pt after 600 s is nearly 13 times higher than that on octahedral Pt. The electrooxidation activity and stability for methanol increase in the order cuboctahedral Pt > cubic

Pt > spherical Pt > octahedral Pt, which is in accordance with the results of CV curves. These results provide a reasonable guidance for the structure and shape design of Pt nanocrystals, especially for the electrocatalysts with high index facets, which have been considered as promising candidate for practical catalysis applications [32].

4. Conclusions

In this study, we developed a facile electrochemical decontamination method of PVP on Pt nanocrystals by electrochemical treatment and investigated its influences on the electrooxidation of methanol. Pt nanocrystals with different morphologies was synthesized by the colloidal method. The as-prepared Pt nanocrystals were electrochemically cleaned by square-wave potential steps before electrocatalytic performance measurements. It was found that the high-frequency potential step method was an effective method for the thorough decontamination of PVP on platinum nanocrystals and protection of preferential-oriented surfaces. More and more surface sites were exposed as the potential steps proceeding, which was clearly indicated from the increasing hydrogen and oxygen adsorption/desorption peaks of the CV curves. The electrochemically cleaned Pt nanocrystals were then applied to methanol electrooxidation. The results showed that the cuboctahedral Pt nanocrystals displayed the best catalytic activity and stability for methanol electrooxidation in acidic solution. The electrocatalytic activity of Pt nanocrystals for methanol increases in the order of cuboctahedral Pt > cubic Pt > spherical Pt > octahedral Pt. This work puts forward a simple and effective method for the thorough decontamination of PVP-capped Pt surfaces and is helpful in understanding the relationship between the electrochemical performances and the surface structure of the catalysts.

Acknowledgments

This work was funded by the National Natural Science Foundation of China (No. 21503166), the International Science Technology Cooperation Program of China (No. 2015DFR50350) and Innovation Capability Support Program of Shaanxi (No. 2019-TD019).

References

[1] Xia Y., Xiong Y., Lim B. and Skrabalak S. E. (2009) Shape-Controlled Synthesis of Metal Nanocrystals: Simple Chemistry Meets Complex Physics? Angewandte Chemie-International Edition, 48: 60-103.
[2] Tian N., Zhou Z.-Y., Sun S.-G., Ding Y. and Wang Z. L. (2007) Synthesis of Tetrahexahedral Platinum Nanocrystals with High-Index Facets and High Electro-Oxidation Activity. Science, 316: 732-5.
[3] Tao A. R., Habas S. and Yang P. (2008) Shape control of colloidal metal nanocrystals. Small, 4: 310-25.
[4] Fernandez P. S., Ferreira D. S., Martins C. A., Troiani H. E., Camara G. A. and Martins M. E. (2013) Platinum nanoparticles produced by EG/PVP method: The effect of cleaning on the electro-oxidation of glycerol. Electrochimica Acta, 98: 25-31.
[5] Hei H., Wang R., Liu X., He L. and Zhang G. (2012) Controlled synthesis and characterization of nobel metal nanoparticles. Soft Nanoscience Letter, 2: 34-40.
[6] Xie S. F., Choi S. I., Xia X. H. and Xia Y. N. (2013) Catalysis on faceted noble-metal nanocrystals: both shape and size matter. Current Opinion in Chemical Engineering, 2: 142-50.
[7] Lopez-Sanchez J. A., Dimitratos N., Hammond C., Brett G. L., Kesavan L., White S., Miedziak P., Tiruvalam R., Jenkins R. L., Carley A. F., Knight D., Kiely C. J. and Hutchings G. J. (2011) Facile removal of stabilizer-ligands from supported gold nanoparticles. Nature Chemistry, 3: 551-6.
[8] Niu Z. and Li Y. (2014) Removal and Utilization of Capping Agents in Nanocatalysis. Chemistry of Materials, 26: 72-83.
[9] Wang Z. L., Petroski J. M., Green T. C. and El-Sayed M. A. (1998) Shape Transformation and

Surface Melting of Cubic and Tetrahedral Platinum Nanocrystals. The Journal of Physical Chemistry B, 102: 6145-51.

[10] Crespo-Quesada M., Andanson J. M., Yarulin A., Lim B., Xia Y. N. and Kiwi-Minsker L. (2011) UV-Ozone Cleaning of Supported Poly(vinylpyrrolidone)-Stabilized Palladium Nanocubes: Effect of Stabilizer Removal on Morphology and Catalytic Behavior. Langmuir, 27: 7909-16.

[11] Baker L. R., Kennedy G., Krier J. M., Van Spronsen M., Onorato R. M. and Somorjai G. A. (2012) The Role of an Organic Cap in Nanoparticle Catalysis: Reversible Restructuring of Carbonaceous Material Controls Catalytic Activity of Platinum Nanoparticles for Ethylene Hydrogenation and Methanol Oxidation. Catalysis Letters, 142: 1286-94.

[12] Solla-Gullón J., Montiel V., Aldaz A. and Clavilier J. (2000) Electrochemical characterisation of platinum nanoparticles prepared by microemulsion: how to clean them without loss of crystalline surface structure. Journal of Electroanalytical Chemistry, 491: 69-77.

[13] Inaba M., Ando M., Hatanaka A., Nomoto A., Matsuzawa K., Tasaka A., Kinumoto T., Iriyama Y. and Ogumi Z. (2006) Controlled growth and shape formation of platinum nanoparticles and their electrochemical properties. Electrochimica Acta, 52: 1632-8.

[14] Gong K., Vukmirovic M. B., Ma C., Zhu Y. and Adzic R. R. (2011) Synthesis and catalytic activity of Pt monolayer on Pd tetrahedral nanocrystals with CO-adsorption-induced removal of surfactants. Journal of Electroanalytical Chemistry, 662: 213-8.

[15] Song H., Kim F., Connor S., Somorjai G. A. and Yang P. (2005) Pt Nanocrystals: Shape Control and Langmuir−Blodgett Monolayer Formation. The Journal of Physical Chemistry B, 109: 188-93.

[16] Han S.-B., Song Y.-J., Lee J.-M., Kim J.-Y. and Park K.-W. (2008) Platinum nanocube catalysts for methanol and ethanol electrooxidation. Electrochemistry Communications, 10: 1044-7.

[17] Rodes A., Zamakhchari M. A., El Achi K. and Clavilier J. (1991) Electrochemical behaviour of Pt(100) in various acidic media: Part I. On a new voltammetric profile of Pt(100) in perchloric acid and effects of surface defects. Journal of Electroanalytical Chemistry and Interfacial Electrochemistry, 305: 115-29.

[18] Martinez-Rodriguez R. A., Vidal-Iglesias F. J., Solla-Gullon J., Cabrera C. R. and Feliu J. M. (2014) Synthesis of Pt Nanoparticles in Water-in-Oil Microemulsion: Effect of HCl on Their Surface Structure. Journal of the American Chemical Society, 136: 1280-3.

[19] Berlin E., Garbarino S., Guay D., Solla-Gullon J., Vidal-Iglesias F. J. and Feliu J. M. (2013) Electrodeposited platinum thin films with preferential (100) orientation: Characterization and electrocatalytic properties for ammonia and formic acid oxidation. Journal of Power Sources, 225: 323-9.

[20] Liu J., Fan X., Liu X., Song Z., Deng Y., Han X., Hu W. and Zhong C. (2017) Synthesis of Cubic-Shaped Pt Particles with (100) Preferential Orientation by a Quick, One-Step and Clean Electrochemical Method. Acs Applied Materials & Interfaces, 9: 18856-64.

[21] Serrano-Ruiz J. C., Lopez-Cudero A., Solla-Gullon J., Sepulveda-Escribano A., Aldaz A. and Rodriguez-Reinoso F. (2008) Hydrogenation of alpha, beta unsaturated aldehydes over polycrystalline, (111) and (100) preferentially oriented Pt nanoparticles supported on carbon. Journal of Catalysis, 253: 159-66.

[22] Herrero E., Franaszczuk K. and Wieckowski A. (1994) Electrochemistry of Methanol at Low Index Crystal Planes of Platinum: An Integrated Voltammetric and Chronoamperometric Study. The Journal of Physical Chemistry, 98: 5074-83.

[23] Lu L. L., Yin G. P., Wang Z. B. and Gao Y. Z. (2009) Electro-oxidation of dimethyl ether on platinum nanocubes with preferential {100} surfaces. Electrochemistry Communications, 11: 1596-8.

[24] Neurock M., Janik M. and Wieckowski A. (2009) A first principles comparison of the mechanism and site requirements for the electrocatalytic oxidation of methanol and formic acid over Pt. Faraday Discussions, 140: 363-78.

[25] Safo I. A. and Oezaslan M. (2017) Electrochemical Cleaning of Polyvinylpyrrolidone-capped Pt

Nanocubes for the Oxygen Reduction Reaction. Electrochimica Acta, 241: 544-52.
[26] Housmans T. H. M. and Koper M. T. M. (2003) Methanol Oxidation on Stepped Pt[n(111) × (110)] Electrodes: A Chronoamperometric Study. The Journal of Physical Chemistry B, 107: 8557-67.
[27] Solla-Gullon J., Vidal-Iglesias F. J., Lopez-Cudero A., Garnier E., Feliu J. M. and Aldaza A. (2008) Shape-dependent electrocatalysis: methanol and formic acid electrooxidation on preferentially oriented Pt nanoparticles. Physical Chemistry Chemical Physics, 10: 3689-98.
[28] Markarian M. Z., El Harakeh M. and Halaoui L. I. (2005) Adsorption of Atomic Hydrogen at a Nanostructured Electrode of Polyacrylate-Capped Pt Nanoparticles in Polyelectrolyte. The Journal of Physical Chemistry B, 109: 11616-21.
[29] Narayanan R. and El-Sayed M. A. (2005) Catalysis with Transition Metal Nanoparticles in Colloidal Solution: Nanoparticle Shape Dependence and Stability. The Journal of Physical Chemistry B, 109: 12663-76.
[30] Nguyen Viet L., Ohtaki M., Nogami M. and Tong Duy H. (2011) Effects of heat treatment and poly(vinylpyrrolidone) (PVP) polymer on electrocatalytic activity of polyhedral Pt nanoparticles towards their methanol oxidation. Colloid and Polymer Science, 289: 1373-86.
[31] Solla-Gullon J., Rodriguez P., Herrero E., Aldaz A. and Feliu J. M. (2008) Surface characterization of platinum electrodes. Physical Chemistry Chemical Physics, 10: 1359-73.
[32] Sheng T., Tian N., Zhou Z.-Y., Lin W.-F. and Sun S.-G. (2017) Designing Pt-Based Electrocatalysts with High Surface Energy. Acs Energy Letters, 2: 1892-900.

Effect of Different Reduction Methods on the Structure of Graphene Oxide

Ting Feng[1], Leilei Zheng[2], Jin Huang[3], Zhen Chu[4], Dengrong Xue[5] and Zhongxin Liu[1]*

[1]Key Laboratory of the Ministry of Education, Hainan Key Laboratory of Fine Chemical Engineering, Hainan University, Haikou, Hainan, 570228, China.

*Corresponding author's e-mail: 491602387@qq.com(Ting Feng), 1209877460@qq.com(Leilei Zheng), 1585426720@qq.com(Jin Huang), 923254969@qq.com(Zhen Chu), 1683438502@qq.com(Dengrong Xue), liuzhongxin@hainu.edu.cn(Zhongxin Liu).

Abstract. The samples of graphene oxide were prepared by Hummers method, and then they were reduced and dried by different methods. The effects of different treatment methods on the graphene oxide's morphology, size and oxygen content were analysed by IR, Roman and TEM. The results show that the reduction method of increasing temperature can effectively reduce the oxygen content of graphene oxide but increase the defect density. However, the ultrasonic cell pulveriser can effectively reduce the oxygen content of graphene oxide in the reduction process and has little effect on the defect density. Therefore, low-defect and low-oxygen reduced graphene oxide can be quickly obtained by an ultrasonic cell pulveriser.

1. Introduction

In 2004, British scientist professor A. Geim and his team used the method of mechanical stripping to successfully prepare graphene in experiments for the first time [1]. Since then, graphene has attracted extensive attention because of its excellent properties. Graphene is a type of carbon elemental substance. Carbon atoms of graphene form two-dimensional materials with hexagonal structural units through sp^2 hybridization. Due to its special two-dimensional structure, graphene has excellent electrical, optical, thermodynamic and mechanical properties, which make it widely used in many fields. Reduced graphene oxide has properties between graphene and graphene oxide [2]. Reduced graphene oxide can be combined with a variety of materials due to its surface containing a variety of oxygen-containing functional groups, making it applicable for biomedical applications [3-4]. Graphene oxide can be reduced by reducing agents, heating and electrochemical reduction methods. During the reduction process, graphene oxide will lose a large number of oxygen-containing functional groups on the surface of graphene oxide, making its C/O ratio increased [5-8].

In this paper, different treatment methods such as ultrasonic, microwave, drying and freeze-drying were used respectively or together in the reduction process of graphene oxide. Through comparing their effects on the reduction degree of graphene oxide, effective methods are proposed to prepare certain type of graphene oxide to prepare for subsequent application.

2. Experimental section

2.1. Materials

Graphite powder (C, AR, >98.0%) were supplied by Tianjin Dengke Chemical Reagent Co., Ltd. Potassium Permanganate ($KMnO_4$, AR, >99.5%), Concentrated Sulfuric Acid (H_2SO_4, AR, 95.0% - 98.0%), Hydrochloric Acid (HCl), Hydrogen Peroxide (H_2O_2, GR, 30%), all above chemicals were used directly without further purification.

Table 1. Different treatment conditions of graphene oxide.

Graphene oxide	Microwave (1 min | 3 times)	Ultrasonic (30min)	Untreated	Untreated	Ultrasonic cell disrupter (30 s | 3 times)
Freeze drying	7-1	7-2		7-3	7-4
Drying at 60℃			7		

2.2. Preparation of Graphene oxide and reduced graphene oxide

Graphene oxide was prepared from graphite by a modified Hummers method [9]. Take 2.5 g of graphite powder and slowly add it into the beaker with 80mL of concentrated sulfuric acid under the condition of ice water bath, then add 7.5 g of potassium permanganate into the beaker and stir for 10 minutes, the solution turns dark green. Add the magnetic rotor to the beaker and stir it for 6 hours. After 6 hours, the solution turns to dark green viscous substance, slowly add 110 mL distilled water to the beaker, the solution starts to turn yellow, then add 15 mL of 10% hydrogen peroxide solution to oxidize the incomplete reaction substance, and the solution turns to golden yellow after complete oxidation. Wash the golden yellow liquid with 5% hydrochloric acid solution for 3 times, and then wash it with distilled water until there is no sulfate ion in the washing solution to get wet graphene oxide sample.

Table 1 shows the symbolisms of samples. No.7 sample was dried directly at 60 ℃. No.7-1 sample was freeze-dried after microwave treatment for 3 times. No.7-2 sample was freeze-dried after ultrasonic treatment for 30 minutes. No.7-3 sample was freeze-dried in original solution. No.7-4 sample was freeze-dried after ultrasonic cell pulverizer treatment for three times. No.7-5 sample was 99% high-purity graphene as control.

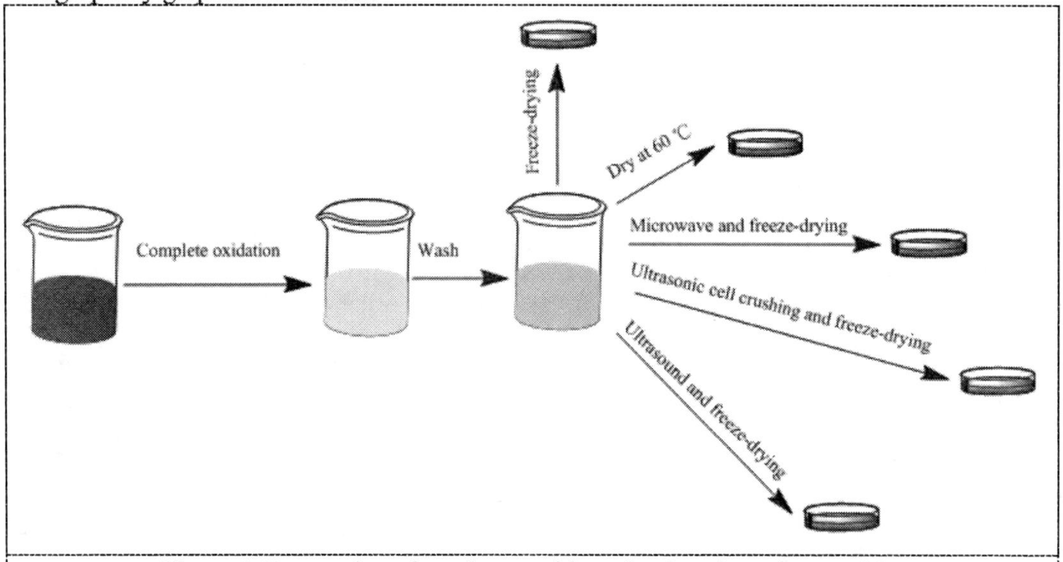

Figure 1. Preparation of graphene oxide and reduced graphene oxide.

2.3. Characterization

The Transmission Electron Microscopy (TEM) were characterized by JEOL JEM 2100, with TEM running at an accelerating voltage of 200 kV. The Raman spectra were obtained by a microscopic confocal laser Raman spectrometer (in Via Reflex), with a Spectra-Physics argon ion laser and excited by the 514.5 nm line. Fourier-transform Infrared Spectroscopy (FT-IR) was recorded on a Bruker

TENSOR27 IR spectrometry. The structural analysis of the samples was characterized by power X-ray Diffraction (XRD) using a Bruker D8-advance with Cu Kα (λ = 1.5418 Å). The size and morphology were obtained by a scanning electron microscopy (SEM, Hitachi S-4800).

3. Results and discussion

3.1. Infrared test results of different samples

In the infrared spectrum, the broad bands at around 3440 and 1384 cm^{-1} are assigned to the stretching vibration and deformation vibration of hydroxyl (-OH) from oxygen-containing functional groups and water molecules. The asymmetric and symmetric stretching bands of -CH_2- are performed with two bands at 2926 and 2855 cm^{-1}. The peak of 1717 cm^{-1}, 1633 cm^{-1}, 1053 cm^{-1} are caused by the stretching vibration of carboxylic (C=O), the aromatic ring vibrations from graphene skeletal (C=C) and the symmetrical stretching vibration of epoxy groups (C-O-C) respectively [10]. It can be seen from Figure 2 that the peaks of high-purity graphene (sample 7-5) in 3440 and 1384cm-1 (- OH), 1717cm-1 (C = O) and 1053cm-1 (C-O-C) are not obvious with respect to the freeze-dried original sample (7-3), and the peaks of the samples after different treatments are all changed, but the changes of sample 7-1 and 7-2 are not obvious, but the changes of sample 7 and 7-4 are obvious, indicating the increased temperature and ultrasonic cell pulverizer can promote the removal of oxygen-containing groups more easily, and the ultrasonic cell pulverizer has a very good effect on the removal of oxygen-containing groups in a relatively short period of time.

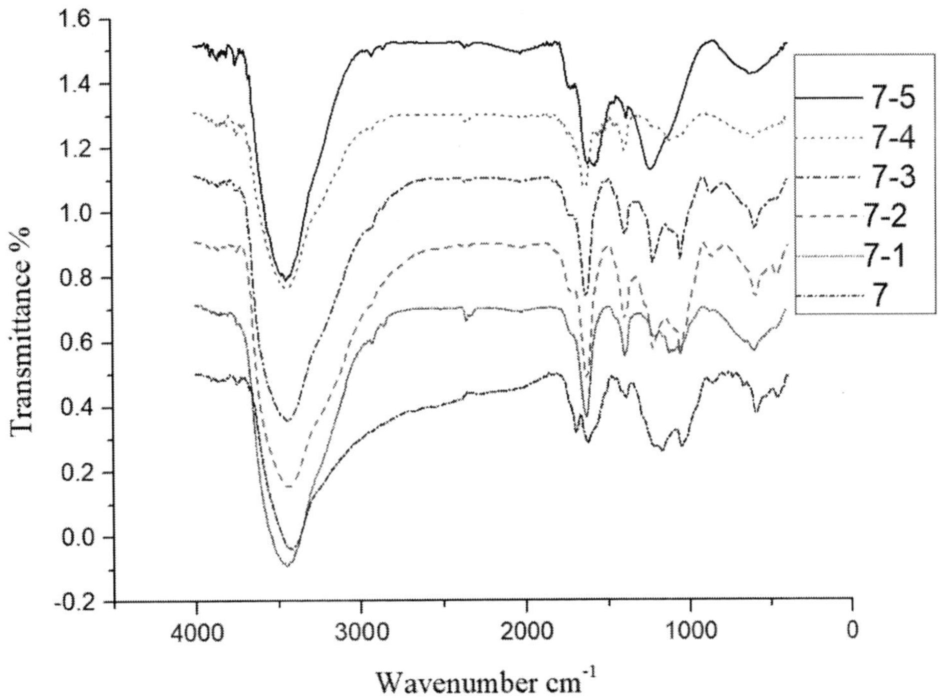

Figure 2. IR spectra of different samples.

3.2. Roman test results of different samples

In Raman spectrum, G- peak is the main characteristic peak of graphene, which is caused by the in-plane vibration of sp2 carbon atom. It appears near 1580 cm-1. The peak reflects the number of graphene layers. The intensity of G- peak increases approximately linearly with the number of layers. The D- peak is the disordered vibration peak of graphene, which can be used to characterize the structure defect or the edge defect of graphene. Generally, the defect density in graphene is characterized by the intensity ratio ID / IG of D- peak and G- peak and the half width of G- peak [11]. It can be seen from figure 3 that the ID/IG ratio of sample 7-5 high purity graphene is the smallest, and the lines of sample 7-2, 7-3 and 7-4 almost coincide, while the ratio of sample 7 is the largest. The comparison showed that the defect density of go increased during heating and microwave reduction, but did not increase during ultrasonic reduction.

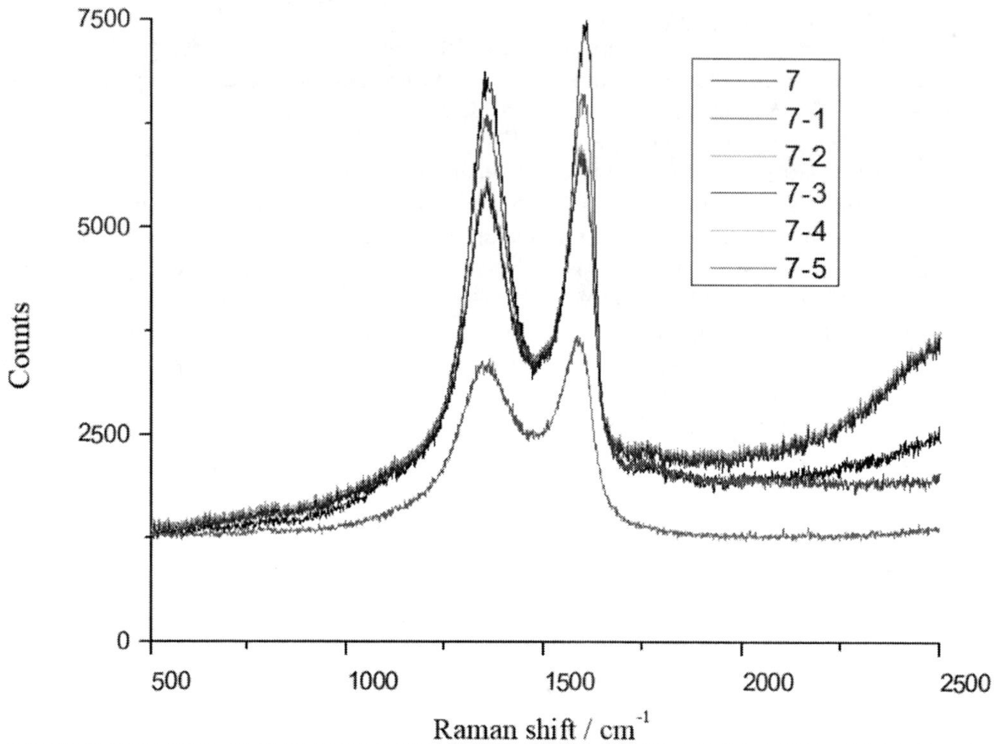

Figure 3. Raman spectra of different samples.

3.3. TEM test results of different samples

The SEM view of sample 7 (Figure 4) shows that the size of the stripped go is about 10 microns. Dried and reduced flakes of go look like wrinkled-paper, which correlates to the imperfect C=C bonds and thermal instability influenced by the oxygen functional groups in the sp^3 hybridization model due to incomplete reduction during the hydrothermal route [12].

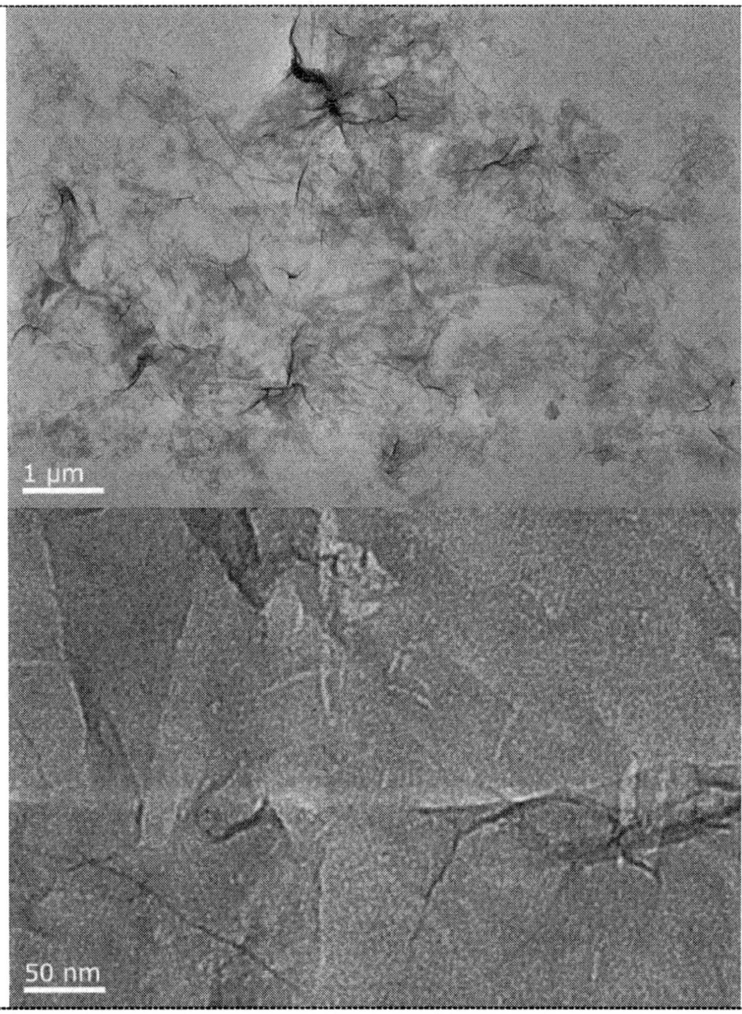

Figure 4. SEM images of the sample 7.

4. Conclusions

In this paper, the effect of the reduction graphene oxide prepared by hummers under different conditions is reported. It can be concluded from the experimental comparison that the reduction method with elevated temperature can effectively reduce the oxygen content of graphene oxide but increase the defect density. The ultrasonic cell pulverizer can effectively reduce the oxygen content of graphene oxide while reducing the defect density. Therefore, low-defect and low-oxygen reduced graphene oxide can be quickly obtained by an ultrasonic cell pulverizer.

Acknowledgment

This work was supported by Innovation and Scientific Research Project for Postgraduates of General Institutions of Higher Learning in Hainan Province (Hys2018-54)

References

[1] Novoselov, K., S. (2004) Electric field effect in atomically thin carbon films. J. Science, 306(5696): 666-669.
[2] Eda, G., Chhowalla, M. (2010) Chemically derived graphene oxide: towards large-area thin-film electronics and optoelectronics. J. Advanced Materials, 22(22): 2392-2415.

[3] Georgakilas, V., Otyepka, M., Bourlinos, A.B., Chandra, V., Kim, N., Kemp, K.C., et al. (2012). Functionalization of graphene: covalent and non-covalent approaches, derivatives and applications. J. Chemical Reviews, 112(11): 6156-6214.

[4] Shim, G., Kim, M.G., Park, J.Y., Oh, Y.K. (2016) Graphene-based nanosheets for delivery of chemotherapeutics and biological drugs. J. Advanced Drug Delivery Reviews, 105(Part B): 205-227.

[5] Stankovich, S., Dikin, D.A., Piner, R.D., Kohlhaas, K.A., Kleinhammes, A., Jia, Y., et al. (2007) Synthesis of graphene-based nanosheets via chemical reduction of exfoliated graphite oxide. J. Carbon, 45(7): 1558-1565.

[6] Chen, C.M., Zhang, Q., Yang, M.G., Huang, C.H., Yang, Y.G., Wang, M.Z. (2012) Structural evolution during annealing of thermally reduced graphene nanosheets for application in supercapacitors. J. Carbon, 50(10): 3572-3584.

[7] Sungjin Park, Jinho An, Jeffrey R. Potts, Aruna Velamakanni, Shanthi Murali, Rodney S. Ruoff. (2011) Hydrazine-reduction of graphite- and graphene oxide. J. Carbon, 49(9): 3019-3023.

[8] Li, C., Shi, G. (2014) Functional gels based on chemically modified graphenes. J. Advanced Materials, 26(24): 3992-4012.

[9] W.S. Hummers, R.E. Offeman. (1958) Preparation of graphitic oxide. J. American Chemical Society, 208: 1334-1339.

[10] Yang, C., Liu, J., Ma, L., Li, A., Liu, Z. (2016) Synthesis and structure change of graphene oxide/GdF3: Yb, Er nanocomposites with improved upconversion luminescence. J. Materials Research Bulletin, 84: 283-287.

[11] Neelgund, G.M., Oki, A. (2016) Influence of carbon nanotubes and graphene nanosheets on photothermal effect of hydroxyapatite. J. Journal of Colloid and Interface Science, 484: 135-145.

[12] Meyer, J.C., Geim, A.K., Katsnelson, M.I., Novoselov, K.S., Booth, T.J., Roth, S. (2007) The structure of suspended graphene sheets. J. Nature, 446(7131): 60-63.

IOP Conf. Series: Materials Science and Engineering **729** (2020) 012084 doi:10.1088/1757-899X/729/1/012084

Preparation of porous composite materials with semi-coke based activated carbon doped with graphene oxide

Zhiyuan Yang[1,2*], HaiLong Ning[1], Hong Jia[1], Yinyan Li[1], Zhuoyue Meng[1] and Zhiping Chen[1]

[1]College of Chemistry and Chemical Engineering, Xi'an University of Science and Technology, Xi'an, Shaanxi, 710054, China

[2]Key Laboratory of Coal Resources Exploration and Comprehensive Utilization, Ministry of Land and Resources. Xi'an, Shaanxi, 710021, China

*Corresponding author's e-mail: zhiyuanyang@126.com.

Abstract: In this paper, the waste semi-coke is used as the carbon precursor, doped with a certain amount of graphene oxide (GO), and activated by KOH to obtain graphene-doped activated carbon. The changes of morphology and structure were compared by SEM, the properties of specific surface area and pore structure were tested by nitrogen adsorption at low temperature, and the changes of iodine adsorption capacity of samples with different ratio were tested. Finally, the graphene-doped activated carbon with specific surface area of 1072.08 m^2/g and iodine adsorption capacity of 1223.99 mg/g was obtained. The optimal ratio of semi-coke to graphene oxide was 300:1, KOH and the semi-coke ratio is 4:1, which provides an excellent adsorbent for the adsorption of methane and carbon dioxide.

1. Introduction

The demand for natural gas in national production and life is increasing, and it is still in a state of short supply. As a main component of coal bed methane, CH_4 mainly exists in the coal voids in free state. However, due to technical reasons, the utilization rate of methane is not high, so people pay more and more attention to its enrichment technology[1-2]. With the rapid development of modern industry, the global greenhouse effect is becoming more and more serious. The concentration of harmful gases in the atmosphere of CO_2 and methane, which are six kinds of greenhouse gases, is increasing, resulting in the formation of a large number of aerosols (PM > 2.5 micro particles), the harm to the ozone layer is more serious, and the global temperature is also increasing year by year. Therefore, carbon capture technology has extremely vital significance [3-5].

Semi-coke is obtained from the pyrolysis of low cohesive or non cohesive high volatile bituminous coal at medium and low temperature. It has the characteristics of high chemical activity, high fixed carbon content, developed pore structure, low ash content and low volatile content. Semi-coke is a new type of carbon material, which has a block structure and a particle size of about 10-60 mm and a light black appearance. However, semi-coke is often used as waste material to discharge directly, which causes environmental pollution[6-8]. Therefore, the reasonable use of semi-coke powder to prepare porous carbon for adsorption and separation is not only conducive to the development of carbon capture technology and concentration technology of low concentration coal bed methane, but also can effectively improve the classified utilization of coal, reduce environmental pollution, provide basic adsorption data for gas separation, promote the development of preparation technology, promote the rapid development of waste gas treatment and other related fields When semi-coke is used as carbon source, the cost of raw materials is reduced to a certain extent, which is conducive to promoting the transformation of the semi-coke industry to the direction of comprehensive utilization, low energy consumption, low pollution and low carbon, expanding the application field of semi-coke resources, extending the semi-coke industry chain, promoting the development of semi-coke industry, generating huge economic benefits, and realizing sustainable economic development and environmental protection significance. However, the effect of the activated carbon prepared by semi-coke on the enrichment of methane and the adsorption of carbon dioxide is not obvious.

The microstructure of graphene oxide (GO) is a kind of honeycomb lattice two-dimensional structure composed of carbon atoms and SP^2 hybrid orbital. Its surface has a large number of functional groups, excellent mechanical properties and chemical stability[9]. Therefore, it is interesting that the porous carbon materials based on graphene will not bring by-products to the reaction, but also can be used as carbon materials to adjust the pore structure, specific surface area and other properties of porous carbon, making it develop in a favorable direction[11].

Herein, semi-coke is used as carbon source and graphene oxide is doped to form a new type of composite. Porous structure is produced by alkali activation. The influence of different ratio of graphene oxide and semi-coke on the composite is explored. The addition of graphene oxide can not only improve the diffusion rate of the adsorbate, but also regulate the pore structure and specific surface area, and increase the utilization rate of the adsorption active site. What's exciting is that compared with the pure bulk activated carbon, the activated carbon doped with graphene oxide has a better structure, and there is no need to worry about the production of any intermediate products in the reaction process, which provides favorable conditions for the adsorption of methane, carbon dioxide and other substances.

2. Experiment

2.1. Preparation of graphene oxide (GO)

Graphene oxide was synthesized by graphite powder oxidation method with improved Hummers method. Concentrated H_2SO_4 (130 ml) was added into the mixture of graphite powder (5.0 g) and $NaNO_3$ (2.5 g), and the mixture was cooled to 0°C. Slowly add $KMnO_4$ (20 g) in batches to keep the reaction temperature below 5°C for 30 minutes. The reaction was heated to 40 °C and stirred for 1.5 h. at this time, water was added slowly (230 ml), resulting in a large exothermic temperature of 98 °C. Introduce external heating to maintain the reaction temperature at 98 °C for 20 minutes, then remove the heat and cool the reaction with a water bath for 10 minutes. Add additional water (700 ml) and 30% hydrogen peroxide (50 mL) to the suspension. Under the action of ultrasound, the suspension of go was repeatedly filtered with 5% HCl aqueous solution and water until the pH value of the solution became neutral without SO_4^{2-}. Powder solid GO obtained by freeze-drying[10].

2.2. Preparation of graphene oxide based semi-coke activated carbon composite

Firstly, KOH, semi-coke and graphene oxide are weighed in proportion, and a small amount of deionized water is added to graphene oxide to make it completely dissolved, and ultrasound is used to make it evenly dispersed in deionized water for two hours. At the same time, add a certain amount of deionized water to KOH, stir it fully to make it just completely dissolved, then add a certain mass ratio of semi coke powder and dissolved graphite oxide, stir it for 2 h under 70 °C magnetic force to make it viscous, then pour the viscous liquid into the porcelain boat, put it into the tubular furnace, and raise the temperature at the temperature rate of 5 °C/min at the starting temperature of 25 °C with the protective gas of N_2, After heating to 800 °C and keeping for 60 min, cooling to room temperature in nitrogen atmosphere, filter the sample repeatedly with 5% HCl aqueous solution and water to make it neutral, and dry in vacuum at 105 °C for 2 h to obtain the composite.

IOP Conf. Series: Materials Science and Engineering **729** (2020) 012084 doi:10.1088/1757-899X/729/1/012084

2.3. Low temperature nitrogen adsorption test

The specific surface area and pore structure of the test samples were tested by ASAP2020 physical adsorption instrument of Micromeritics Company. The adsorption-desorption isotherm curve of nitrogen at 77 K was measured by static capacity method. The total pore volume was calculated by the adsorption capacity of adsorption isotherm at 0.99 relative pressures. S_{BET} was calculated by BET method from the adsorption capacity data. The pore size distribution was calculated by t-plot method, and then the pore size distribution was calculated by density pan function theory.

2.4. Scanning electron microscopy

The scanning electron microscope (SEM) of phenom Pro type is used to detect the micro morphology of the test sample: the magnification is 80-150000, the resolution is 8 nm, and the acceleration voltage is 5-15 kv, which is mainly used to characterize the microstructure characteristics of the sample, such as particle size, morphology, etc.

2.5. Determination of iodine adsorption value

Iodine adsorption value is mainly used as the adsorption test of coal based activated carbon. Its main principle is to determine the residual iodine concentration by titration with sodium thiosulfate standard solution after the quantitative sample and iodine standard solution are fully shaken under specific conditions, so as to determine the iodine adsorption amount.

Iodine adsorption value can be used as a qualitative indicator of microporous structure of activated carbon. Iodine adsorption value is determined according to GB /T 7702.7-2008.

3. Results and discussion

3.1. Surface topography

The micro surfaces of semi-coke and activated carbon were observed by SEM. Figure 1 shows semi-coke and graphene doped semi-coke porous carbon (semi-coke: GO = 300:1). Figure 1(a) shows that the surface of semi-coke is not completely smooth and the whole is relatively dense, with very few pores and very few pore structures.

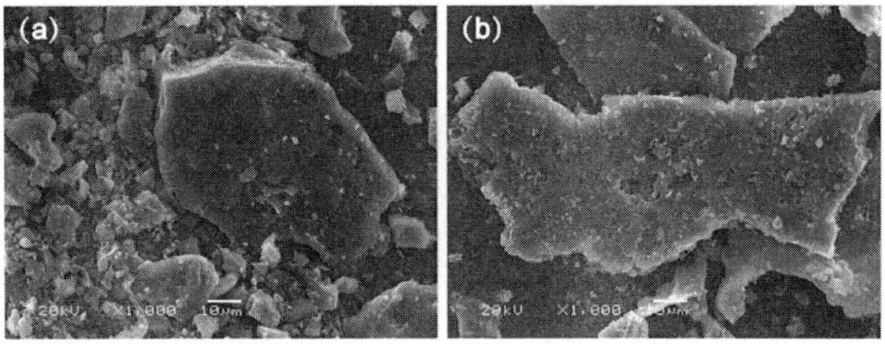

Figure 1. (a) Scanning electron micrograph of semi-coke sample, (b) scanning electron micrograph of composite carbon with ratio of semi-coke to graphene oxide of 300:1.

In Figure 1(b), it can be seen that the activated semi-coke compared with semi-coke, the activated semi-coke has obviously generated small new pores, and there are some small particles, the pores are relatively developed, the structure is loose and porous. This should be because the disordered carbon structure in the semi-coke is activated by KOH at high temperature and etched into pores. These abundant pore structures also promote the increase of the specific surface area of the activated carbon and provide a favorable environment for the separation and adsorption of the activated carbon.

3.2. Texture characteristics

Figure 2 shows the adsorption-desorption isotherms of 100:1, 200:1, 300:1, 400:1 and pore size distribution of graphene based semi-coke porous carbon.

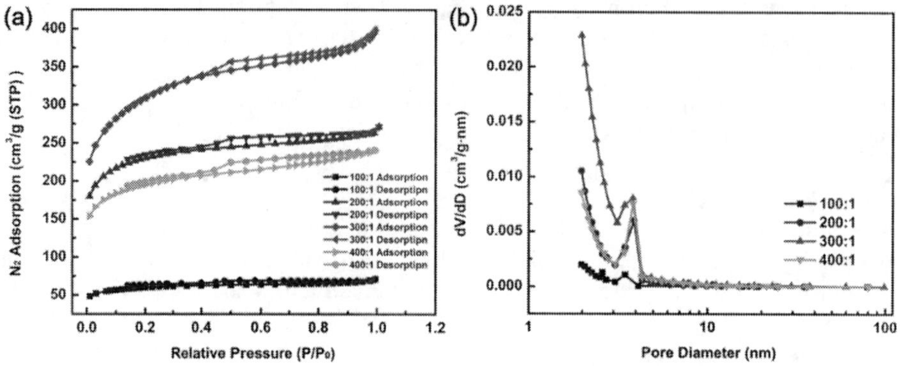

Figure 2. (a) Nitrogen adsorption-desorption curve of composite porous carbon with different proportion of graphene oxide and (b) pore size distribution diagram.

Through the nitrogen adsorption test at low temperature, it can be seen from the adsorption and desorption curve in Figure 2(a) and the pore structure distribution in Figure 2(b), the adsorption and desorption curves of 200:1 and 400:1 are partially inconsistent under the condition of high pressure. Compared with the changes of pore structure of semi-coke based composite porous carbon under different ratios, the pore development degree of graphene based composite porous carbon varies with the amount of go added The increasing trend of the composite porous carbon is first increasing and then decreasing, which is consistent with its pore size distribution. With the addition of go, the adsorption isotherm of the composite porous carbon gradually presents a hysteresis loop, and the hysteresis loop type is H4, indicating that the composite material has a large number of mesoporous structures, which is the typical micro mesoporous structure with narrow fracture pores. This is related to the pore size distribution and the average pore size.

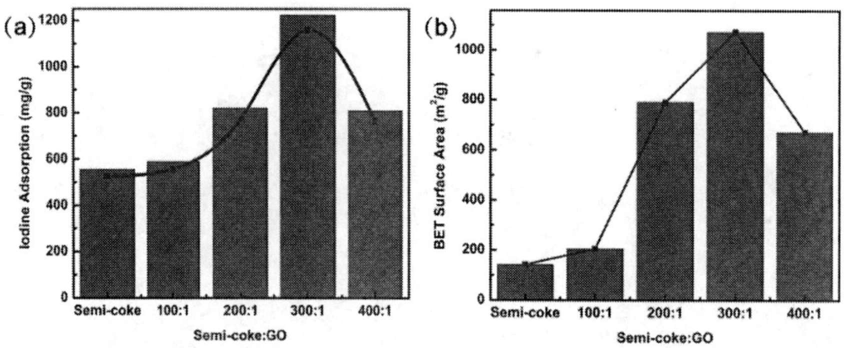

Figure 3. (a) Iodine adsorption capacity and (b) specific surface area of composite porous carbon with different proportion of semi-coke graphene oxide.

Figure 3 shows the iodine adsorption value and BET surface area of graphene based semi-coke porous carbon with different proportions of semi-coke and go. Through the test of iodine adsorption, it is found that the laws of iodine adsorption capacity and specific surface area of semi-coke and graphite oxide with mass ratio of 100:1, 200:1, 300:1 and 400:1 are consistent. When the mass ratio of semi-coke to graphite oxide is 300:1, the specific surface area and iodine adsorption amount reach the maximum.

IOP Conf. Series: Materials Science and Engineering **729** (2020) 012084 doi:10.1088/1757-899X/729/1/012084

4. Conclusion

The composite carbon material was prepared by tube furnace chemical activation method with semi-coke as carbon precursor and graphite oxide prepared by graphite powder in different mass ratio. The activation reagent was KOH, the activation temperature was 800 °C, and the activation time was 60 minutes. The results show that when the mass ratio of semi-coke to graphite oxide is 300:1, the ratio of alkali to carbon is 4:1, the iodine adsorption value is 1223.99 mg/g, the specific surface area is 1072.08 m^2/g, the iodine adsorption value and the specific surface area are the best, the pore structure is also greatly improved, and the prediction has a good effect on the gas adsorption.

Acknowledgments

This study is supported by the Natural Science Foundation of China (No.41772166, 21706206), Shaanxi Key Industry Chain Innovation Project (2017ZDCXL-GY-10-01-02, 2018GY-076), Major Project from Key Laboratory of Coal Resources Exploration and Comprehensive Utilization, Ministry of Land and Resources (SMDZ-2019ZD-2) and Xi'an Science and Technology Innovation Guidance Project (201805036YD14CG20(6)).

References

[1] Yao, Y., Liu, D., Tang, D., Tang, S., Huang, W. (2008) Fractal characterization of adsorption-pores of coals from North China: an investigation on CH_4 adsorption capacity of coals. Int. J. Coal. Geol., 73: 27-42.

[2] Yang, Z., Wang, D., Meng, Z., Li, Y. (2019). Adsorption separation of CH_4/N_2 on modified coal-based carbon molecular sieve. Sep. Purif. Technol., 218: 130-137.

[3] Gómez-Pozuelo, G., Sanz-Pérez, E.S., Arencibia, A., Pizarro, P., Sanz, R., Serrano, D.P. (2019) CO_2 adsorption on amine-functionalized clays. Micropor. Mesopor. Mat., 282: 38-47.

[4] Dong, F., Yu, B., Pan, Y. (2019) Examining the synergistic effect of CO_2 emissions on $PM_{2.5}$ emissions reduction: Evidence from China. J. Clean. Prod., 223: 759-771.

[5] Yang, Z., Wang, D., Meng, Z., Li, Y. (2019) Adsorption separation of CH4/N2 on modified coal-based carbon molecular sieve. Sep. Purif. Tech., 218:130-137.

[6] Li, X.H., Ma, J.S., Li, L.L., Li, B.F., Feng, J., Turmel, W., Li, W.Y. (2016) Semi-coke as solid heat carrier for low-temperature coal tar upgrading. Fuel. Process. Technol., 143: 79-85.

[7] Yao, H., He, B., Ding, G., Tong, W., Kuang, Y. (2019) Thermogravimetric analyses of oxy-fuel co-combustion of semi-coke and bituminous coal. Appl. Therm. Eng., 156: 708-721.

[8] Meng, Z., Yang, Z., Yin, Z., Li, Y., Song, X., Zhao, J., Wu, W. (2020). Effects of coal slime on the slurry ability of a semi-coke water slurry. Powder. Technol., 359: 261-267.

[9] Yuan, K., Xu, Y., Uihlein, J., Brunklaus, G., Shi, L., Heiderhoff, R., Que, M., Forster, M., Chassé, T., Pichler, T., Riedl, T., Chen, Y., Scherf, U. (2015) Straightforward generation of pillared, microporous graphene frameworks for use in supercapacitors. Adv. Mater., 27: 6714-6721.

[10] Xie, A., Dai, J., Cui, J., Lang, J., Wei, M., Dai, X., Li, C., Yan, Y. (2017) Novel graphene oxide–confined nanospace directed synthesis of glucose-based porous carbon nanosheets with enhanced adsorption performance. Acs. Sustain. Chem. Eng., 5: 11566-11576.

[11] Wang, T., Zhao, L., Shen, J.N., Wu, L.G., Van der Bruggen, B. (2015) Enhanced performance of polyurethane hybrid membranes for CO_2 separation by incorporating graphene oxide: the relationship between membrane performance and morphology of graphene oxide. Environ. Sci. Technol., 49: 8004-8011.

IOP Conf. Series: Materials Science and Engineering **729** (2020) 012085 doi:10.1088/1757-899X/729/1/012085

Preparation of composite catalyst and its catalytic cracking properties of coal tar

Zhiyuan Yang[1,2*], Xiaoqian Ju[1], Zhuanzhuan Ma[1], Jingjing Zhao[1], Zhuoyue Meng[1], Zhiping Chen[1]

[1]College of Chemistry and Chemical Engineering, Xi'an University of Science and Technology, Xi'an, Shaanxi, 710054, China

[2]Key Laboratory of Coal Resources Exploration and Comprehensive Utilization, Ministry of Land and Resources. Xi'an, Shaanxi, 710021, China

[*]Corresponding Author: Zhiyuan Yang. E-mail: zhiyuanyang@126.com.

Abstract. China is rich in coal resources. Coal tar produced by coal combustion is a large amount of surplus. Naphtha and diesel oil with similar properties to kerosene and diesel oil can be produced by catalytic cracking of coal tar oil. In this article, metal-loaded UiO-66 and Y-type zeolite were synthesized by solution thermal method and microwave assisted method. And a high temperature resistant active component catalyst UiO-66-Y was prepared by compounding the two. According to the GC-MS characterization of the original coal tar and the products after cracking with UiO-66-Y, the results show that part of the aromatic compounds in the coal tar added with catalyst will be decomposed and a large number of light component esters will be synthesized at the same time. It provides a possible way to solve the current energy dilemma.

1. Introduction

China's resources are short and single[1-2]. As the main energy in China, most of the coal is used by combustion, which leads to a large amount of coal tar surplus[3]. However, the dependence on oil imports is increasing in China year by year. It had been found that naphtha and diesel oil which was very similar to those of diesel oil and kerosene used in our daily life can be prepared from coal tar by adding catalyst at high temperature[4]. If we can use these fuel oils instead of kerosene or diesel, we can save our country from this energy dilemma to a great extent. Therefore, the ideal way to solve the current dilemma is to design high-efficiency catalysts, which can make the heavy component in coal tars into light component with high cost performance through catalytic cracking.

Metal organic frameworks (MOFs) are formed by chemical bonds between metal ions and organic ligands[5-6]. UiO-66 is a MOF material, which is mainly composed of $C_8H_6O_4$ as organic ligand and Zr^{4+} as metal ion[7]. Compared with traditional porous materials, it has many advantages such as high porosity and high specific surface area. Microporous Y-type zeolite has large pore diameter and high temperature resistance, so it is not easy to decompose under heating[8]. The ratio of surface area to mass of the material is high, and the density is low, which provides a favourable way for the diffusion of macromolecular particles.

In this paper, composite materials were synthesized by using Y-type zeolite and UiO-66 by microwave-assisted method and solution thermal method. Taking nickel and cobalt as active metals with high cost performance, a kind of UiO-66-Y metal supported catalyst materials with high temperature resistance and active component activity was obtained to improve the catalytic activity and stability of coal tar cracking. By means of various characterization methods, the interaction between MOFs and active components is revealed, the internal reasons for the better performance of metal organic frameworks materials than other catalysts are found, and the role of MOFs catalytic materials in the reaction.

2. Experimental

2.1. Preparation of UiO-66

2.1.1. Preparation of UiO-66 by solution thermal method. Weighed 4.66 g $ZrCl_4$ and 3.32 g terephthalic acid, dissolved them in 120 mL DMF, stirred them at 25 ℃ for 30 min, make them dissolved, transferred them to the lined PTFE reactor, reacted at 120 ℃ for 24 h, cooled after removal, washed them twice with DMF and CH_3OH, centrifuged them at 7000 r/min for 10 min, then dry the products at 90 ℃ for 8 h, grind them to obtain white solid powder samples, recorded as UiO-66-s. Add 3 mL hydrochloric acid to the above raw materials and repeat the remaining steps of the above operation. The sample is recorded as UiO-66-sa.

2.1.2. Preparation of UiO-66 by microwave assisted method. Weigh 4.66 g $ZrCl_4$ and 3.32 g white terephthalic acid powder, dissolve them in 120 mL DMF, stir them at 25 ℃ for 30 min, make them dissolved, transfer them to the microwave hydrothermal synthesis instrument, set the final temperature to 120 °C for 4 h, then take them out. Cooling, centrifugation, filtration and drying are the same as solution thermal method, grind the white powder sample and record it as UiO-66-m. Add 3 mL hydrochloric acid to the raw materials and repeat the remaining steps of the above operation to get the sample recorded as UiO-66-ma.

2.2. Preparation of metal loaded catalyst

Take 1.9729 g, 3.9458 g and 5.9186 g $Ni(NO_3)_2 \cdot 6H_2O$ respectively and dissolve them in three beakers, add 8g Y-type zeolite (UiO-66-ma) into the beaker, and add a proper amount of distilled water to dissolve them evenly, conduct ultrasound in the ultrasonic cleaner for 20 min, and stir at room temperature for 30 min. Take it out and put it in a sealed place for 24 h at room temperature. Dry the sample in a drying oven. Put in nitrogen, calcine it at 300 ℃ for 3 h to get a green powder, corresponding to the loading of 5%, 10% and 15% Ni/Y (Ni/UiO-66).

Replace the raw materials with 1.9729 g, 3.9458 g and 5.9186 g $Co(NO_3)_2 \cdot 6H_2O$ and repeat the remaining steps of the above operation to obtain dark red powder, corresponding to the loading of 5%, 10% and 15% Co/Y (Co/UiO-66).

2.3. Preparation of composite materials

Weigh 3 g of non-calcined UiO-66 (5% Ni/UiO-66) and Y-type zeolite (10% Ni/Y). Mixed by mechanical stirring evenly, placed in a muffle furnace, and heated at 300 °C for 3 h under a nitrogen atmosphere to obtain a sample of UiO-66-Y.

3. Results and discussion

3.1. Characterization of UiO-66

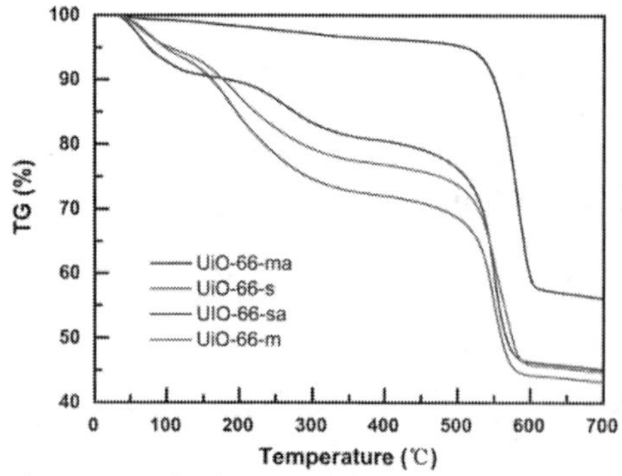

Figure 1. Thermogravimetric curve of UiO-66. (UiO-66-s; UiO-66-sa; UiO-66-m; UiO-66-ma)

It can be seen from the Figure 1 that the reason for the weight loss of the four catalysts at about 100 °C is the loss of water molecules on the surface of UiO-66 due to the high temperature. The reason of weight loss in the temperature range from 100 °C to 600 ℃ is due to the collapse and decomposition of the internal skeleton structure of UiO-66 materials.

Figure 2. Scanning electron micrograph of four catalysts.
(a: UiO-66-s; b: UiO-66-sa; c: UiO-66-m; d: UiO-66-ma)

Figure 2 is the SEM of four catalysts. It can be seen from the figure that the crystal particles of UiO-66-ma (d) are more uniform, have more pore structure and appearance structure, and the proportion of crystal area in the polymer is higher.

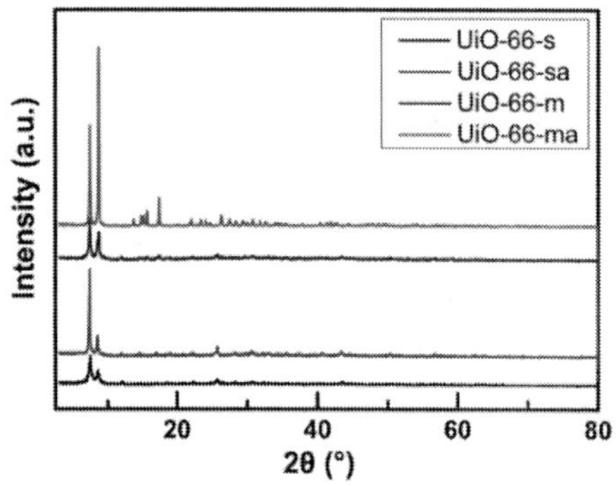

Figure 3. XRD patterns of UiO-66-s, UiO-66-sa, UiO-66-m and UiO-66-ma.

The XRD figures of the four catalysts are shown in Figure 3. The samples prepared under different conditions show the same curve as the reported UiO-66, which shows that the purity of the materials we synthesized is higher, and the crystal morphology of the sample UiO-66-ma is better, and the characteristic peak is more consistent with the original literature.

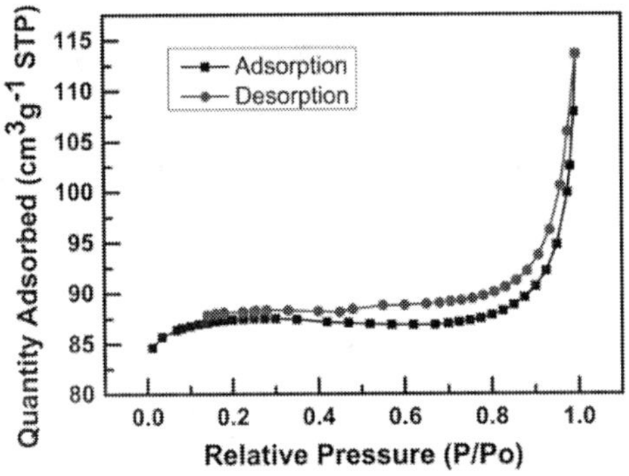

Figure 4. The adsorption-desorption isotherms of UiO-66-ma.

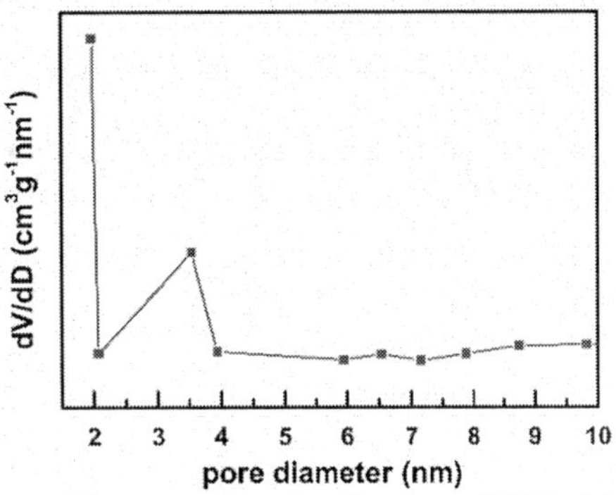

Figure 5. The pore size distribution of UiO-66-ma.

We have tested the nitrogen adsorption curve of UiO-66-ma in Figure 4 and Figure 5, the best crystal type sample, and obtained that its N_2 isotherm is type I, indicating that it is mainly microporous structure, while the adsorption desorption curve is not completely overlapped, and the part that is not overlapped forms a ring, indicating that the material contains pore structure.

3.2. Characterization of metal supported UiO-66 and Y-type zeolite

As shown in Figure 6 and Figure 7, the XRD patterns of Y-type zeolite and UiO-66 loaded with different amounts of Ni and Co are shown. It can be seen that the characteristic peaks of 5% Ni/UiO-66, 15% Co/UiO-66, 10% Ni/Y and 5% Co/Y are the most obvious, the diffraction peaks are the same as those in the original literature, and the crystal forms are better.

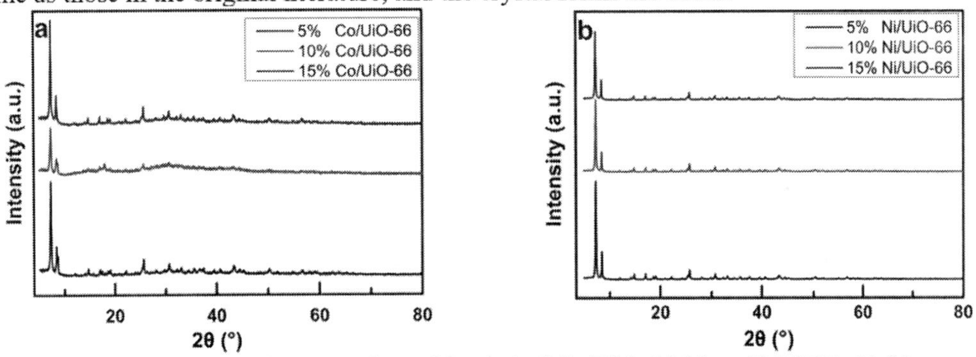

Figure 6. The XRD images of metal loaded of Co/UiO-66 (a) and Ni/UiO-66 (b).

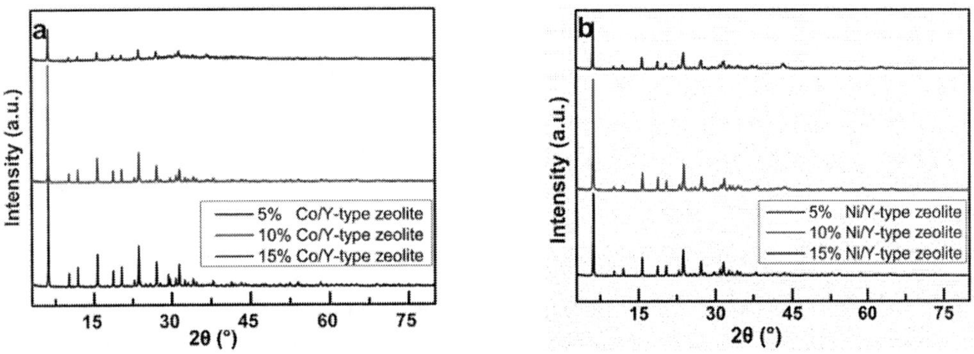

Figure 7. The XRD images of metal loaded of Co/Y (a) and Ni/Y (b).

3.3. Performance evaluation of composite catalyst

10 g coal samples were respectively put into a low-temperature distillation by aluminium retort. The pyrolysis products were collected during the experiment. The collected gas was analyzed by gas chromatograph.

It can be seen from Figure 8 that, most substances are detected at about 5-25 min, and less substances are detected after 40 min. A total of 158 substances have been detected, including 59 kinds of fat, 57 kinds of aromatic compounds, 13 kinds of alcohol compounds and 15 kinds of phenolic compounds containing benzene ring.

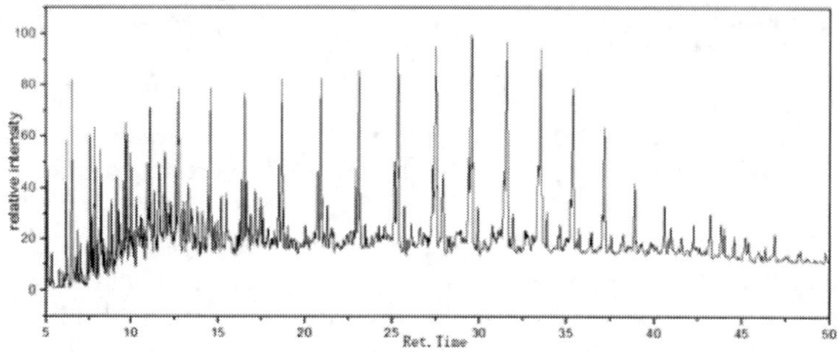

Figure 8. The GC-MS total ion flow chromatogram of coal tar

As can be seen from Figure 9, most substances are detected within 5-40min. According to the identification results of all substances detected by GC-MS, 237 substances were found in coal tar and catalyst after pyrolysis, including 12 aromatics, 218 fats and 12 acids. Compared with the detection types of coal tar without catalyst, the number of aromatic compounds decreased significantly, and the types of ester compounds increased significantly.

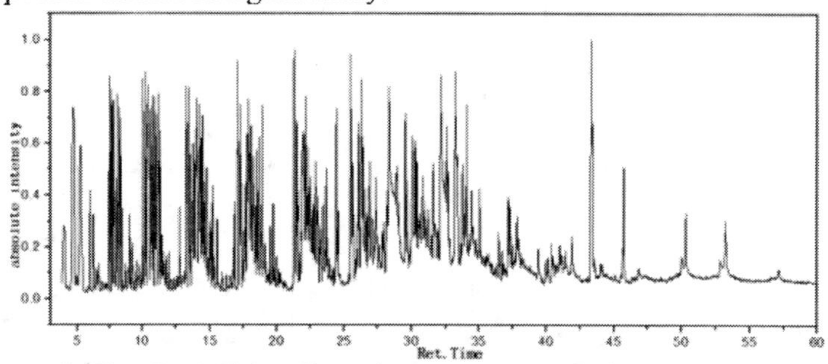

Figure 9. The GC-MS ion flow chromatogram of UiO-66-Y and coal tar

4. Conclusion

UiO-66-Y metal-loaded catalytic materials were prepared by combining UiO-66 and Y-type zeolite by microwave assisted method and solution thermal method. The composite materials not only have the advantages of MOFs materials: high area per unit mass and low density, but also has the advantages of molecular sieve materials: high temperature resistance and high maximum load per unit area. Through GC-MS analysis of raw coal tar and its products after cracking on UiO-66-Y catalyst, it is found that UiO-66-Y can promote the decomposition of some aromatic compounds in coal tar and synthesize a large number of ester compounds at the same time.

Acknowledgments

This study is supported by Shaanxi Key Industry Chain Innovation Project (2017ZDCXL-GY-10-01-02, 2018GY-076), the Natural Science Foundation of China (No.41772166, 21706206), Major Project from Key Laboratory of Coal Resources Exploration and Comprehensive Utilization, Ministry of Land and Resources (SMDZ-2019ZD-2) and Xi'an Science and Technology Innovation Guidance Project (201805036YD14CG20(6)).

IOP Conf. Series: Materials Science and Engineering **729** (2020) 012085 doi:10.1088/1757-899X/729/1/012085

References

[1] Yang, X., Teng, F. (2018) The air quality co-benefit of coal control strategy in China. Resour. Conserv. Recy., 129: 373-382.

[2] Meng, Z., Yang, Z., Yin, Z., Li, Y., Song, X., Zhao, J., Wu, W. (2020). Effects of coal slime on the slurry ability of a semi-coke water slurry. Powder. Technol., 359: 261-267.

[3] Cui, D., Xu, C., Chung, K.H., Shi, Q. (2015) Potential of using coal tar as a quenching agent for coal gasification. Energ. Fuel., 29: 6964-6969.

[4] Kusy, J., Andel, L., Safarova, M., Vales, J., Ciahotny, K. (2012) Hydrogenation process of the tar obtained from the pyrolisis of brown coal. Fuel, 101: 38-44.

[5] Yaghi, O., Davis, C.E., Li, G., Li, H. (1997) Selective guest binding by tailored channels in a 3-D porous zinc (II)−benzenetricarboxylate network. J. Am. Chem. Soc., 119: 2861-2868.

[6] Yaghi, O., Davis, C., Groy, T. (1997) Construction of a new open-framework solid from 1, 3, 5-cyclohexanetricarboxylate and zinc (II) building blocks. J. Chem. Soc. Dalton, 2383-2384.

[7] andiah, M., Nilsen, M.H., Usseglio, S., Jakobsen, S., Olsbye, U., Tilset, M., Lillerud, K. (2010) Synthesis and stability of tagged UiO-66 Zr-MOFs. Chem. Mater., 22: 6632-6640.

[8] Le, T., Wang, Q., Pan, B., Ravindra, A.V., Ju, S., Peng, J. (2019) Process regulation of microwave intensified synthesis of Y-type zeolite. Micropor. Mesopor. Mat., 284: 476-485.

IOP Conf. Series: Materials Science and Engineering **729** (2020) 012086 doi:10.1088/1757-899X/729/1/012086

Application of gold nanoparticles in photoacoustic imaging

Wen He[1], Xiaoxia Wang[1] ,Xing Gao[1*], Zaijun Lu[2], and Jibing Song[3*]

[1]School of Chemistry and Bioengineering, Qilu Institute of Technology, Jinan, Shandong, 250200, China

[2]School of Chemistry and Chemical Engineering, Shandong University, Jinan, Shandong, 250100, China

[3]College of Chemistry, Fuzhou University, Fujian, Fuzhou, 350108, China

[*]Corresponding author's e-mail: gaoxing-620@163.com and jibinsong@fzu.edu.cn

Abstract. Photoacoustic (PA) imaging is a noninvasive imaging mode that be rapidly developed in recent years for medical diagnosis and treatment by using contrast agents. Among them, gold nanomaterials (GNPs) offer a suitable platform for PA imaging, owing to their advantageous physical/chemical/biological properties and high-intensity absorption in the near-infrared (NIR) region. In this review, we firstly concentrate on the recent development of GNPs and their assemblies for PA imaging. Then, the future development in the research of GNPs as PA imaging contrast agents are discussed.

1. Introduction

PA imaging is a noninvasive medical imaging mode which combines the advantages of the excellent contrast of traditional optical imaging with high penetrability (up to 5-6 cm) of ultrasound imaging and has rapidly developed in recent years for biomedical diagnosis and treatment. The imaging system explotis PA effect that was primary reported in 1880 by Alexander Graham Bell [1]. Concisely, as the contrast agent absorbs the pulsed laser, the thermoelastic expansion of biological tissues will generate broadband sound waves at megahertz frequencies.

These broadband ultrasonic waves can be found by ultrasound transducers on the tissue surface, and changed to electric signals which processed appropriately to obtain PA image [2]. Since there were no strong light sources and sensitive detectors at that time, until the late 1990s, PA imaging developed rapidly and was widely used in biomedical fields.

Contrast agents for PA imaging can be classified into endogenous and exogenous contrast agents depending on the source. Endogenous contrast agents, such as melanin and hemoglobin, offer functional and structural message for PA imaging. However, exogenous contrast agents, e.g. small molecule dyes, fluorophores, nanomaterials, which can be imaged at the cellular and molecular levels, are able to greatly enhance these imaging modalities. For the moment, owing to the specific size and shape dependent physiochemical properties, remarkable bio/chemical inertness, and the optical characteristics originated from the giant electromagnetic field near the surface of GNPs, GNPs have become one of the most prospective exogenous contrast agents for PA imaging[3-4].

Here, the recent development of GNPs and their assmelies in PA imaging are outlined. Then, the perspectives of GNPs for PA imaging, especially about their clinical translation, are discussed.

2. Classification of plasmonic gold nanoparticles(GNPs) and their assemblies

GNPs based on their Localized Surface Plasmon Resonance (LSPR) effect adjusted by different sizes, shapes, and couplings of the nanoparticles [5], have been actively used as contrast agents for PA imaging[6].Besides the original gold nanospheres, in recent years, researchers have synthesized many complicated nanostructures[7]including gold nanorods[8-9], gold nanocages[10], gold nanostars[11], gold nanoshells[12], and so on (Figure 1).

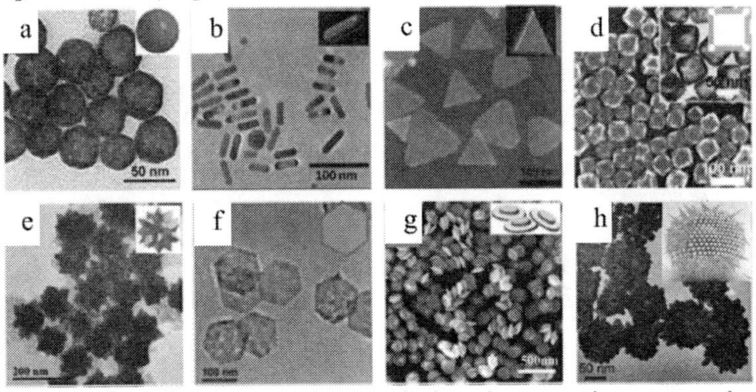

Figure 1. Scanning electron microscope and transmission electron microscope micrographs of gold nanospheres (a) , gold nanorods (b) , gold nanoprisms (c) , gold nanocages (d) , gold nanostars (e), gold nanoplates (f), gold nanodisks (g) and gold nanoshells (h) [4].

By collecting the literature, it is found that there is not much research on original gold nanospheres (GNSPs) for PA imaging. Because the optical absorption spectrum of GNSPs, typically at approximately 520 nm, is not fall into the biological window (650-1100 nm), in which the least blood and tissue attenuation occurs[2]. However, by modification with polymers or magnetic materials, GNSPs have red-shifted absorption spectra and thus could offer stronger PA signals in vivo. Lu[13] synthesized poly(ethylene glycol)-coated GNSPs (40-50 nm). They displayed intensive resonance absorption adjusted to the NIR region, with a maximum absorption peak at 800 nm, and its PA efficiency is remarkably higher than that of blood. Bai[14] developed a contrast agent based on GNSPs and Fe_3O_4. Utilizing the outstanding magnetic properties of Fe_3O_4 and strong NIR absorption performance of GNSPs, such nanocomposites had been used for targeted PA imaging of cancer cells, with a red-shift to 830 nm.

Gold nanorods (GNRs) equipped with tunable NIR absorption behavior by increasing in particle aspect (length to width) ratio that has been proved by Jokerst[15-16], so they are the most potential PA imaging contrast agents. However, without changing the aspect ratio, Chen[17]reported the synthesis of miniaturized GNRs absorbing at 1,064 nm, a wavelength within the NIR-II window, which ensures low background noise from endogenous absorbers in tissue, that are 5-11 times smaller than regular-sized GNRs with a similar aspect ratio. Miniaturized GNRs generated 3.5 times stronger PA signals under nanosecond pulsed laser illumination and displayed obviously improvement in overall tumour PA signals by 4.5-fold compared to the large counterparts (Figure 2) [17]. These findings provide another strategy for the design of future PA contrast agents and have the potential to stimulate more PA molecular imaging applications in the NIR-II window.

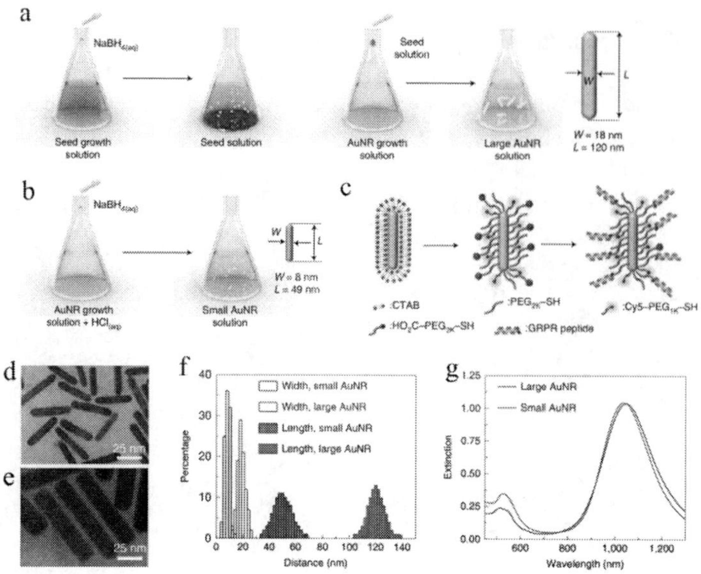

Figure 2. Synthesis procedure (a-c) and characterization (d-g) of small and large GNRs[17].

GNRs synthesized by the laboratory method usually have a film coated with a surfactant on the surface, which inhibits the activity of biological cell because of active surface electrons. Moon[18] synthesized the reduced graphene oxide coated GNRs (r-GO-GNRs). Theoretical research shows that compared with naked GNR or GO-GNR, the enhanced electromagnetic field around r-GO-GNR is 4 times stronger. In addition, due to the extremely high PA amplitude in the 4-11 MHz operating frequency of ultrasonic transducers, r-GO-GNRs are expected to be promising deep tissue imaging probes. Furthermore, Wu[19] used traditional Chinese medicine acupuncture as an auxiliary method, and developed an effective method to use a combination of acupuncture and (polyethylene glycol)-coated GNRs (PEG-GNRs) as a composite contrast agent to enhance the contrast of PA imaging. Yan[20]designed a comprehensive platform, that combines extremely sensitive nucleic acid probes with polyethylenimine-modified GNRs, named as GNRs-PEI/FIRE. GNRs-PEI/FIRE, which has potential applications in the diagnosis and treatment of tumors, displayed higher target miRNA in vitro /in vivo detection sensitivity and enhanced PA imaging signals compared to controls.

For application of PA imaging to diagnosis, GNRs have been suggested as molecular probes because they show strong PA signals. Umehara[9]synthesized GNRs (G-NIs) with NI (2-nitroimidazole derivative) units on the surface, and characterized its characteristics, which can be used as an indicator of hypoxia in PA imaging. GNRs showed a powerful PA signal (Figure 3)[9]. Due to the reducing properties of intracellular reductase to NI units, the binding of NI units and nanoparticles causes them to accumulate in hypoxic cells and tissues. Therefore, G-NIs are allowed to accumulate in hypoxic cells to display their PA signals in hypoxic tumor cells and tissues. G-NIs is a promising PA nanoprobe that can be used to track tumor hypoxia.

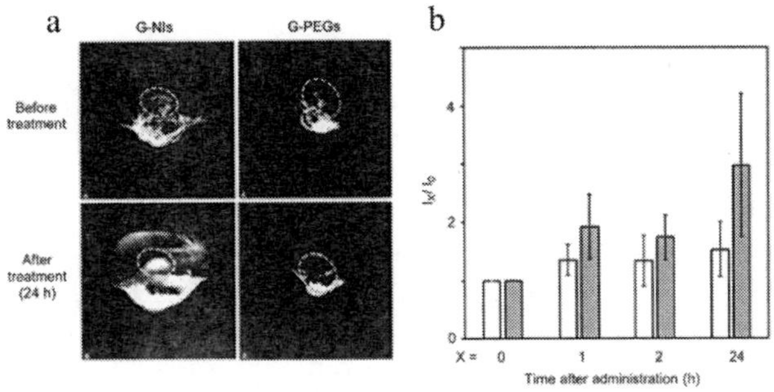

Figure 3. PA images of tumor-bearing mice (a) and ratio of PA signals intensity after injection of G-NIs (right) or G-PEGs (left) (b) [9].

GNPs tied to polymer brushes were widely used as building blocks for building functional assemblies with various structures (Figure 4a) [21] as a result of the unique collective characteristics of GNPs assemblies that have multiple advantages comparing to the single GNPs. Due to the strong plasmon coupling effect between adjacent GNPs of the module, the LSPR spectrum often appears red-shifted, and the PA signal is greatly enhanced.

Among GNPs assemblies, gold nanovesicles (GNVs) have been widely used in biomedical applications[22]. As shown in Figure 4b, they can be formed by self-assembly They can be formed by self-assembly of amphiphilic polymer grafted with GNPs[23]. As mentioned earlier, due to the plasmon wavelength of GNRs can enshroud the visible and NIR areas by adjusting aspect ratio, GNRs is an ideal candidate for GNVs[24-26]. For instance, Song [25] prepared a small GRVs (about 60 nm in size), which is assembled from small GNRs (about 8×2 nm). These GNRs use PEG and polylactic acid-glycolic acid (PLGA) coating and synthetized by emulsion method. Their Maximum absorption wavelengths are from 800 to 1050 nm. At the same optical density (OD) value, the PA intensity of the GNVs is about ten times higher than that of GNRs illuminated with 808 nm laser. Subsequently, based on Au@PEG-Fe_3O_4@PS, Au@PS-Fe_3O_4@PEG and Au-Fe_3O_4@PEG/PS, Song[27-28]prepared Double-Layered vesicles (DL-Ve 1and DL-Ve 2) and Mono-Layered vesicles (ML-Ve 3), separately. Under the same OD_{785}, PA amplitude of DL-Ve 2 was 2.2 times and 5 times stronger than that of DL-Ve 1 and ML-Ve 3, respectively. Interestingly, on account of the janus amphiphilic gold–iron (II,III) Oxide Nanoparticles as a core, they indicated the capacity to be used as a dual modal imaging (optical and magnetic resonance imaging)agent in vivo. They adopted the DL-Ve 2 as dual modal imaging agents to determine vesicle accumulation in the tumor area after intravenous injection of vesicles into PBS. The sustained improvement of the PA signals and strength over time in the tumor region confirmed the accumulation and distribution of the vesicle in the tumor[22]. Using high-resolution 3D PA images of the tumor, they obtained a lot of information about the tumor, such as location, size, and morphology.

Except the dual PA and MR imaging, in recent years, researchers explored various strategies to make GNVs have more imaging capabilities, rather than giving it its own physical and chemical characteristics, achieving multimodal imaging of tumors based on GNVs, such as PET/PA[25-26]dual modal imaging, and even FL/PT/PA trimodal imaging[29].

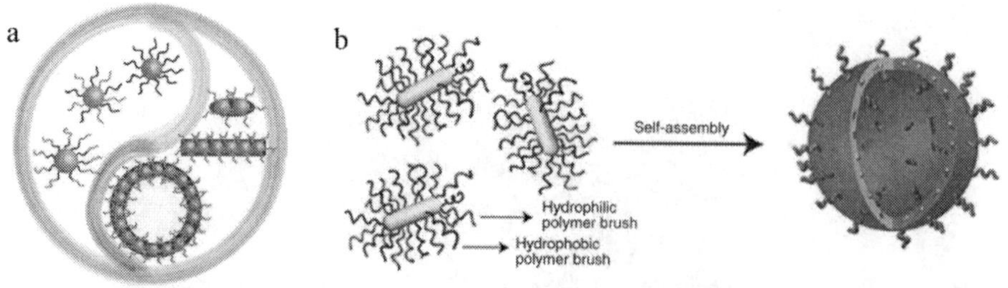

Figure 4. a) Schematic illustration of polymer coated gold nanoparticles and their assemblies[21]. b) Schematic illustration of the self-assembly of amphiphilic gold nanorods with mixed polymer brushes into vesicular structures [23].

3. Conclusions and perspectives

In this review, we concentrated on the latest development of GNPs for PA imaging, due to their chemical/biological inertness in various cell and animal models, adjustable optical properties and low cytotoxicity. However, in order to make GNP have better capability in the field of biomedicine, and even turn to clinical translation, more efforts are still needed[22]. The first, researchers should take measures to improve the photostability of GNPs, during PA imaging some plasmonic GNPs can change their shapes[9]. Second, apply measures to produce strong PA imaging signals at relatively low doses of GNPs and multi-modality biomedical imaging by janus GNPs. Third, since a few endogenous biomolecules can absorb or emit NIR-II light, it is necessary to develop suitable contrast agents used for PA imaging in the NIR-II window[4].

Acknowledgments

The authors are thankful to Research Project Foundation of Qilu Institute of Technology (No. QL19K046) for financial support of this work.

References

[1] Bell, A.G. (1880) On the production and reproduction of sound by light. Am. J. Sci., 20: 305–324.
[2] Li, W.W., Chen, X.Y. (2015) Gold nanoparticles for photoacoustic imaging. Nanomedicine, 10(2): 299–320.
[3] Madkour, L.H. (2018) Applications of gold nanoparticles in medicine and therapy. Pharm. Pharmacol. Int. J. 6(3):157-174.
[4] Lyu, Y., Li, J.C., Pu, K.Y. (2019) Second Near-Infrared Absorbing Agents for Photoacoustic Imaging and Photothermal Therapy. Small method. 3(11): 1900553.
[5]Park, J.E., Kim, M., Hwang, J.H., Nam. J.M. (2017) Golden opportunities: plasmonic gold nanostructures for biomedical applications based on the second near-infrared window. Small methods. 1(3):1600032.,
[6] Liu Y.J., Bhattarai P., Dai Z.F, Chen, X.Y. (2018) Photothermal therapy and photoacoustic imaging via nanotheranostics in fighting cancer. Chem. Soc. Rev., 48: 2053–2108.
[7] Fu, Q.R., Zhu, R., Song, J.B., Yang, H.H., Chen, X.Y. (2019) Photoacoustic imaging: contrast agents and their biomedical applications. Adv. Mater. 31(6): 1805875.
[8] Knights, O.B., Ye, S.J., Ingram, N., Freear, S., Mclaughlan, J.R. (2019) Optimising gold nanorods for photoacoustic imaging in vitro. Nanoscale Adv., 1: 1472–1481.
[9] Umehara, Y., Kageyama, T., Son, A., Kimura, Y., Kondo, T., Tanabe, K. (2019) Biological reduction of nitroimidazole-functionalized gold nanorods for photoacoustic imaging of tumor hypoxia. RSC Adv., 9: 16863–16868.
[10] Zhan, C.Y., Huang, Y., Lin, G.F., Huang, S.L., Zeng F., Wu, S.Z. (2019) A gold nanocage/cluster hybrid structure for whole-body multispectral optoacoustic tomography imaging, EGFR inhibitor delivery, and photothermal therapy. Small. 15(33): 1900309.

[11] Bhowmik, D., Culver, K.S.B., Liu,T.T., Odom, T.W. (2019) Resolving single-nanoconstruct dynamics during targeting and nontargeting live-cell membrane interactions. ACS Nano, DOI: 10.1021/acsnano.9b03144.
[12] Culver, H.R., Wechsler, M.E., Peppas. N.A. (2018) Label-free detection of tear biomarkers using hydrogel coated gold nanoshells in a localized surface plasmon resonance-based biosensor. ACS Nano, 12(9): 9342-9354.
[13] Lu, W., Huang, Q., Ku, G., Wen, X.X., Zhou, M., Guzatov, D., Brecht, P., Su R., Oraevskyc A., Wang L.H.V., Li C. (2010) Photoacoustic imaging of living mouse brain vasculature using hollow gold nanospheres. Biomaterials, 31(9): 2617–2626.
[14] Bai, L.Y., Yang, X.Q., An, J., Zhang, L., Zhao, K., Qin, M.Y., Fang B.Y., Li C., Xuan Y., Zhang X.S. Zhao, Y.D., and Ma, Z.Y. (2015) Multifunctional magnetic-hollow gold nanospheres for bimodal cancer cell imaging and photothermal therapy. Nanotechnology, 26(31): 315701.
[15] Jokerst, J.V., Cole, A.J., Sompel, D.V.D., Gambhir, S.S. (2012) Gold nanorods for ovarian cancer detection with photoacoustic imaging and resection guidance via raman imaging in living mice. ACS Nano, 6(11): 10366–10377.
[16] Jokerst, J.V., Thangaraj, M., Kempen, P.J., Sinclair, R., Gambhir, S.S. (2012) Photoacoustic imaging of mesenchymal stem cells in living mice via silica-coated gold nanorods. ACS Nano, 6(7): 5920–5930.
[17] Chen, Y.S., Zhao,Y., Yoon, S.J., Gambhir, S.S., Emelianov, S. (2019) Miniature gold nanorods for photoacoustic molecular imaging in the second near-infrared optical window. Nat. Nanotech., 14: 465–472.
[18] Moon, H., Kumar, D., Kim, H., Sim, C., Chang, J.H. , Kim, J.M., Kim, H., Lim D.K. (2015) Amplified photoacoustic performance and enhanced photothermal stability of reduced graphene oxide coated gold nanorods for sensitive photoacoustic imaging. ACS Nano, 9(3): 2711–2719.
[19] Wu, D., Zhang, X.X., Rong, J., Jiang, H.B. (2019). Photoacoustic molecular imaging using combined acupuncture and gold nanorods as a composite contrast agent. Innov. Opt. Health Sci., 12(3): 1941004.
[20] Yan, N., Wang, X.J., Lin, L., Song, T.J., Sun, P.J., Tian, H.Y., Liang, H.J., Chen, X.S. (2018) Gold nanorods electrostatically binding nucleic acid probe for in vivo microRNA amplified detection and photoacoustic imaging-guided photothermal therapy. Adv. Funct. Mater. 28(22): 1800490.
[21] Song, J.B., Niu, G., Chen, X.Y. (2017). Amphiphilic polymer-guided plasmonic assemblies and their biomedical applications. Bioconjugate Chem., 28(1): 105–114.
[22] Huang, Y., Huang, P., Lin, J. (2019) Plasmonic gold nanovesicles for biomedical applications. Small Methods, 3(3): 1800394.
[23] Song, J.B., Huang, P., Chen, X.Y. (2016) Preparation of plasmonic vesicles from amphiphilic gold nanocrystals grafted with polymer brushes. Nat. Protoc., 11(11): 2287–2299.
[24] Song, J.B., Pu, L. , Zhou, J.J., Duan, B., Duan, H.W. (2013) Biodegradable theranostic plasmonic vesicles of amphiphilic gold nanorods. ACS Nano, 7(11): 9947–9960.
[25] Song, J.B., Yang, X.Y, Jacobson, O., Huang, P., Sun, X.L., Lin, L.S., Yan, X.F., Niu, G., Ma, Q.J., Chen, X.Y. (2015). Ultrasmall gold nanorod vesicles with enhanced tumor accumulation and fast excretion from the body for cancer therapy. Adv. Mater. 27(33): 4910–4917.
[26] Song, J.B., Yang, X.Y, Jacobson, O., Lin, L.S., Huang, P., Niu, G., Ma, Q.J., Chen, X.Y. (2015) Sequential drug release and enhanced photothermal effect of hybrid reduced graphene oxide-loaded ultrasmall gold nanorod vesicles for cancer therapy. ACS Nano, 9: 9199–9209.
[27] Song, J.B., Wu, B.H., Zhou, Z.J., Zhu, G.Z., Liu, Y.J., Yang, Z., Lin, L.S., Yu, G.C., Zhang, F.W., Zhang, G.F., Duan, H.W., Stucky, G.D., Chen, X.Y. (2017). Double-layered plasmonic-magnetic vesicles by self-assembly of janus amphiphilic Au-Fe_3O_4 nanoparticles. Angew. Chem. Int. Ed., 56: 8110–8114.
[28] Song, J.B. Wu, B.H., Zhou, Z,J. Zhu G.Z., et al. (2017) Double-layered plasmonic-magnetic

vesicles by self-assembly of janus amphiphilic gold-iron (II,III) oxide nanoparticles. Angewandte Chemie, 129(28): 8222-8226.

[29] Lin, J., Wang, S.J. Huang, P., Wang, Z., Chen, S.H., Niu, G., Li, W.W., He, J., Cui, D.X., Lu, G.M., Chen, X.Y., Nie, Z.H. (2013) Photosensitizer-loaded gold vesicles with strong plasmonic coupling effect for imaging-guided photothermal/photodynamic therapy. ACS Nano, 7(6): 5320–5329.

IOP Conf. Series: Materials Science and Engineering **729** (2020) 012087 doi:10.1088/1757-899X/729/1/012087

Halloysite-TiO_2-Ag composites: Preparation,characterization and photodegradation

Yuanyuan Du[1*], Ying Fu[1], Xing Gao[1],Wen He[1] and Pengwu Zheng[2*]

[1]School of Chemical and Biological Engineering, Qilu Institute of Technology, Jinan, Shandong, 250200, China

[2]School of Pharmacy, Jiangxi Science and Technology Normal University, Nanchang, Jiangxi, 330013, China

[*]Corresponding author's e-mail:duyuan01@126.com zhengpw@126.com

Abstract. Halloysite (HNTs)-TiO_2-Ag composites were fabricated by depositing anatase TiO_2 and nano Ag particles on the HNTs surfaces. The composite was characterized with corresponding instrumental analysis methods. Compared with HNTs-TiO_2 composites, the HNTs-TiO_2-Ag composites exhibited better photocatalytic activity in the decomposition of MB and the persistent organic pollutant 4-nitrophenol(4-NP) under UV irradiation. After 2.5 hours of UV light irradiation, HNTs-TiO_2-Ag composites removed about 97.63% and 51% of MB and 4-NP, respectively.

1. Introduction

Dyes are widely used in environmental, chemical, textile, cosmetics and other fields. [1]. However, dyes are very difficult to handle because they are difficult to be degraded into non-toxic small molecules, and they are very stable to light and heat. [2]. Dyes can cause serious damage to ecological balance and human health when discharged into nature. Adsorption technology is widely used to remove some dyes, especially those that are not easy to biodegrade. Various adsorbents such as carbon nanotubes, graphene oxide, zeolites, clays, biomass and polymeric materials have attracted the attention of researchers[3].

Halloysite (HNTs) is a natural alumino silicate clay mineral with large specific surface area, unique hollow microtubular structure and good biocompatibility and thermal stability. HNTs is generally composed of multiple layers, its external diameter is 10-20 nm, and the inner diameter is about 40-100 nm. It has a chemical formula of $Al_2Si_2O_5(OH)_4{\cdot}nH_2O$, which is similar to kaolinite except for the presence of an additional water monolayer between the adjacent clay layers[4]. HNTs occurs mainly in two polymorphs: the anhydrous form, with and interlayer of 7 Å, and the hydrated form, with expanded interlayer spacing of 10 Å, as a result of the incorporation of water in the inter lamellar space[5]. HNTs tubes consist of gibbsite octahedral sheet (Al–OH) groups on the internal surface and siloxane groups (Si–O–Si) on the external surface[6]. Due to the different chemical groups, the HNTs has a negative charge on the outer surface and a positive charge on the inner surface in the pH range 2-8. HNTs has a highly negatively charged outer surface, which makes it have superior adsorption, pore filtration, and ion exchange properties. These excellent properties make HNTs possible as an adsorbent, a drug carrier, and a fixed catalytically active molecule.

TiO_2 is the most excellent photocatalytic material due to its non-toxicity, high photocatalytic activity and reusability.

However, TiO_2 nanoparticles are prone to aggregate, resulting in an adverse effect on the photocatalytic activity[7]. Consequently, great efforts have been made to block the aggregation of TiO_2 nanoparticles, such as coating method. The Ag nanoparticles are deposited on the TiO_2 nanoparticles, the fabrication of metal-semiconductor oxide composites can improve the photocatalytic activities of semiconductor oxide photocatalysts. The electron transfers from semiconductor oxides to metal nanoparticles can occur, which can decrease the number of electron/hole recombination events and improve the photocatalytic performance of semiconductor oxide-based catalysts[8].

In the present work, the HNTs-TiO_2-Ag composites are prepared by sol-gel method and the structure of the composite was characterized by various testing methods. In addition, in order to explore the photocatalytic performance of the material, the work studied the photocatalytic degradation of MB and 4-NP systematically.

2. Materials and methods

2.1. Materials

The HNTs were obtained from Hebei Province, China. MB dye and 4-NP were provided by Tianjin Damao Chemical Reagent Co., Ltd. All other reagents are of analytical grade .

2.2. The Preparation of HNTs-TiO_2-Ag composites

Glacial acetic acid (4 mL) and deionized water(10 mL)were added in 30 mL absolute ethanol, followed by the addition of 8.5 g silver nitrate. And then, the concentrated nitric acid was added to the solution to make the pH within 2-3. The solution was protected from light and stirred for 10 min at the room temperature.

15 g of Tetrabutyltitanate was mixed with a certain amount of absolute ethanol. The mixture was mechanically stirred for 20 min at 25℃. HNTs (6 g) were slowly added in the mixture. The obtained suspension was continuously stirred for 1 h. The solution was added dropwise to the suspension and the sol can be obtained after stirring for 1h. The sol was stored at 25℃ to form better nanoparticles . Firstly, the products were centrifuged with ethanol and oven-dried. And then, HNTs-TiO_2-Ag composites were calcined in Muffle furnace at 300℃.

2.3. Fourier transform infrared spectroscopy

The spectra of composites were obtained on a BIO-RAD FTS 3000 IR Spectrum Scanner. The resolution of the instrument was 4 cm^{-1}, the analysis range was 300-4000 cm^{-1}, and the wave number accuracy was 0.01 cm^{-1}. The wavelength range of each absorption band was observed and analyzed and the functional groups in the molecule were inferred.

2.4. X-ray diffraction

HNTs, HNTs-TiO_2 and HNTs-TiO_2-Ag powders were placed in a sample reservoir to hold the sample reservoir in the diffractor. Scanning was operated using a Bruker D8-S4 X-ray diffractometer. The test parameters were: Cu target, scanning speed was 4 ° / min, the range of diffraction angle is 2θ = 10-70 °, Cu Ka wavelength was 1.542 Å.

2.5. Transmission electron microscope

Drops of HNTs-TiO_2 and HNTs-TiO_2-Ag aqueous suspensions were delivered onto the copper grid and air dried. The samples were analysed using a TEM JEM-1200EX.

2.6. Photodegradation Experiments

The glass bottles that contain adsorbents and dyes (MB) were exposed to ultraviolet light. The concentration of MB was measured by an ultraviolet spectrophotometer at regular intervals. The degradation rate is represented by $(C_0-C_t)/C_0$, C_0 is the initial concentration of methylene blue and C_t

is the concentration measured after photocatalytic degradation for a certain period of time[9]. MB dye was replaced with 4-NP (0.01 g/L) and the degradation rate of 4-NP could be measured.

3. Results and discussion

3.1. Morphology

The morphologies of the composites are obtained by TEM. In Fig. 1(a-b), TiO_2 particles or clusters are deposited on the HNTs. The size of the TiO_2 particles or clusters is about 10-40 nm. In Fig. 1(c-d), Ag and TiO_2 particles The size of the particles is about 10-20 nm, and they are evenly distributed on the surface of the HNTs tube with little agglomeration. It can be seen from the figure that the addition of nano Ag weakens the aggregation of TiO_2.

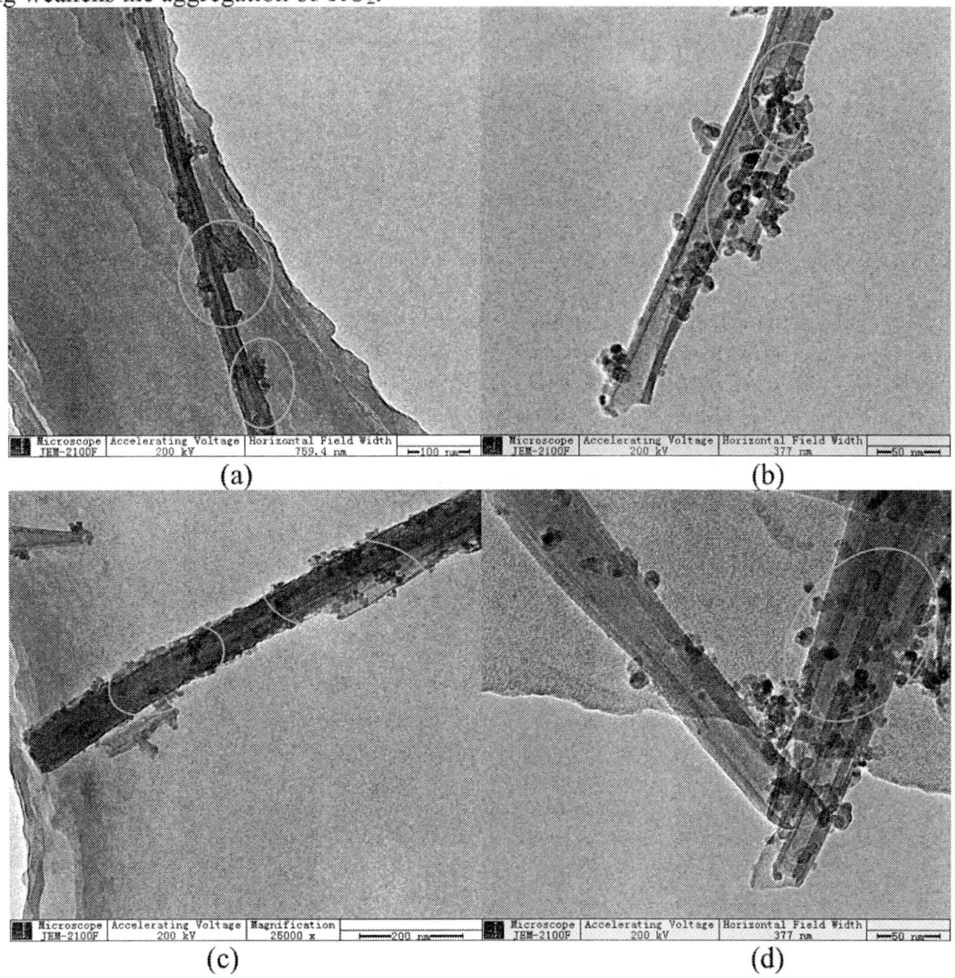

Fig 1. TEM micrograph of the HNTs-TiO_2(a-b) and HNTs-TiO_2-Ag(c-d).

3.2. FTIR

Fig. 2 exhibits the FTIR spectra of HNTs. HNTs-TiO_2 and HNTs-TiO_2-Ag. The double peaks at 3696 and 3623 cm^{-1} are due to the stretching vibrations of Al-OH groups at the surface of HNTs[10,11]. The hydroxy bending vibration absorption peak of HNTs appear at 916.02 and 1028.084 cm^{-1}[12]. The above mentioned peaks all appear in FTIR of HNTs-TiO_2 except for HNTs-TiO_2-Ag. Due to the loading of Ag particles, the absorption peaks at 3696 and 3623 cm^{-1} are blue-shifted and moved to 3660 and 3704cm^{-1}, respectively.

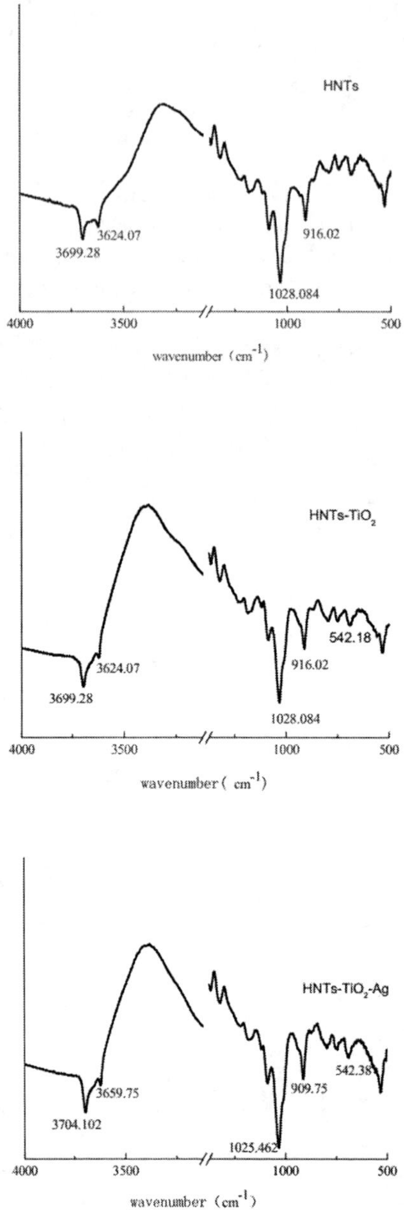

Fig 2. FTIR spectra of HNTs.HNTs-TiO_2 and HNTs-TiO_2-Ag.

3.3. XRD

The X-ray diffraction patterns of the HNTs. HNTs-TiO_2 and HNTs-TiO_2-Ag are obtained. As you can see from the figure, the powder of HNTs-TiO_2-Ag display distinct peaks at 2θ values of about 25.3, 37.9, 47.6, 54.8 and 62.7[13]. The diffraction peak of the nanosilver is at a 2θ value of 38.1[14]. The above data indicates that Ag and TiO_2 particles are attached to the halloysite nanotubes, which is consistent with the TEM and FTIR results. The diffraction peak of Ag is not very obvious. This may be caused by: (i) The diffraction peaks of Ag are not obvious due to the low Ag content; (ii) The diffraction peak of Ag overlaps the diffraction peak of TiO_2[8].

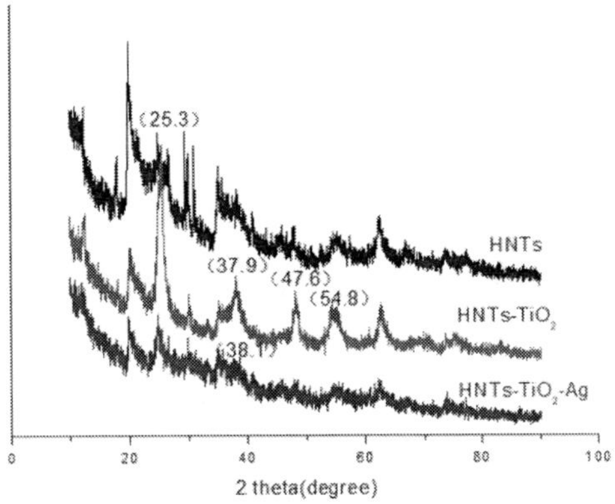

Fig 3. X-ray diffraction patterns of HNTs.HNTs-$TiO_2$and HNTs-TiO_2-Ag.

3.4. Photodegradation

As shown in Fig. 4 , after photocatalysis for 2.5 h, the absorbance decreased from 2.318 to 0.055 for HNTs-TiO_2-Ag composite. The degradation rate of HNTs-TiO_2-Ag composite to methylene blue was 97.63%, while the degradation rate of HNTs-TiO_2 composite was 51.08%. The removal of MB is the result of both adsorption and photocatalysis, but HNTs-TiO_2-Ag exhibits better dye removal performance[15]. As shown in the Fig. 5, after photocatalysis for 2.5 h, the degradation rate of 4-NP by HNTs-TiO_2-Ag composites was 51%, while the degradation rate of HNTs-TiO_2 composite was 20.75%. The results show that the addition of Ag particles can improve the photocatalytic performance of the composites.

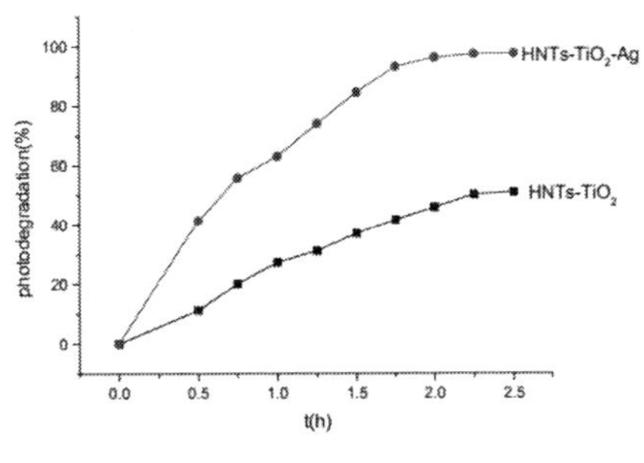

Fig 4. Photocatalytic degradation of MB on HNTs-TiO_2 and HNTs-TiO_2-Ag.

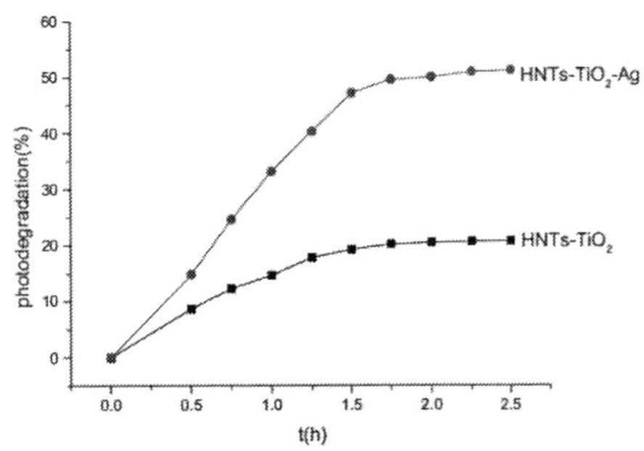

Fig 5. Photocatalytic degradation of 4-NP on HNTs-TiO_2 and HNTs-TiO_2-Ag.

4. Conclusions

In this paper, TiO_2 and Ag particles were successfully deposited on the surface of HNTs by sol-gel method. The doping of the Ag particles reduces the agglomeration of the TiO_2 particles, which improves the photocatalytic performance of the photocatalyst to some extent. The prepared HNTs-TiO_2-Ag composite has excellent adsorption performance and photocatalytic performance and can improve the removal of dyes. In terms of sewage treatment, HNTs-TiO_2-Ag composite show good application value.

Acknowledgments
This research was supported by the National College Students Innovation and Entrepreneurship Training Program(201713998027) and Research Fund Project(QL19K040).

References

[1] Belkassa, K., Bessaha, F., Marouf-Khelifa, K. (2013) Purification of halloysite by magnetic separation. Colloids Surf. A, 421: 26.
[2] Crini, G. (2006) Non-conventional low-cost adsorbents for dye removal: A review. Bioresource Technology, 97: 1061-1085.
[3] Du, Y.Y., Zheng, P.W. (2014) Adsorption and photodegradation of methylene blue onTiO_2-halloysite adsorbents. Korean J. Chem. Eng., 31: 2051–2056.
[4] Abdullayev, E., Joshi, A., Wei, W.B. (2012) Enlargement of Halloysite Clay Nanotube Lumen by Selective Etching of Aluminum Oxide. Nano, 6: 7216-7226.
[5] Yah, W.O., Takahara, A., Lvov, Y.M. (2012) Selective Modification of Halloysite Lumen with Octadecylphosphonic Acid: New Inorganic Tubular Micelle. J. Am. Chem. Soc., 134: 1853−1859.
[6] Xie, Y.F., Qian, D.Y. , Wu, D.L. (2011) Magnetic halloysite nanotubes/iron oxide composites for the adsorption of dyes. Chem. Eng. J., 168: 959.
[7] Wang, R.J., Jiang, G.h., Ding. Y.W. (2011) Photocatalytic Activity of Heterostructures Based on TiO_2 and Halloysite Nanotubes. Appl. Mater. Interfaces, 3: 4154–4158.
[8] Yu, D.H., Yu, X.D., Wang, C.H. (2012) Synthesis of Natural Cellulose-Templated TiO_2/Ag Nanosponge Composites and Photocatalytic Properties. Appl. Mater. Interfaces, 4: 2781−2787.
[9] Zheng, P.W., Du, Y.Y., M, X.F. (2015) Amylose–halloysite–TiO_2 composites: Preparation, characterization and photodegradation. Applied Surface Science, 329: 256-261.
[10] Luo, P., Zhao, Y.F., Zhang, B. (2010) Study on the adsorption of Neutral Red from aqueous solution onto halloysite nanotubes. Water Res., 44: 1489.
[11] Zhang, Y., Fu, L.J., Yang, H.M. (2012) Insights into the physicochemical aspects from natural halloysite to silica nanotubes. Colloids Surf. A, 414: 115.
[12] Tierrablanca, E., Romero-García, J., Roman, P. (2010) Biomimetic polymerization of aniline using hematin supported on halloysite nanotubes. Appl. Catal. A, 381: 267.
[13] Papoulis, D., Komarneni, S., Panagiotaras, D. (2013) Halloysite-TiO_2 nanocomposites: Synthesis, characterization and photocatalytic activity. Appl. Catal. B, 132-133: 416-422.
[14] Wilke, C. M., Tong, T., Gaillard, J. F., Gray, K. A. (2016) Attenuation of Microbial 417 Stress Due to Nano-Ag and Nano-TiO_2 Interactions under Dark Conditions. Environ. Sci. 418 Technol., 50: 11302-11310.
[15] Liu, L., Wan, Y.Z., Xie, Y.D. (2012) The removal of dye from aqueous solution using alginate–halloysite nanotube beads. Chem. Eng. J., 187: 210–216.

Preparation and properties of polydiacetylene based temperature sensors

Hui Wang[a*], Jiaxin Yu[a], Xing Gao and Hailing Ding

Department of chemistry, Qilu Institute of Technology, Jinan City, Shandong Province, 250000, China

[a]These authors contributed equally to this work.

[*]Corresponding author's e-mail: 924334478@qq.com

Abstract: Polydiacetylene vesicles with different concentrations were prepared, which have a good temperature-induced discoloration effect. As the temperature rises, the appearance colors appear blue-violet, purple, rose-red, watermelon-red and orange-red. The UV-Vis spectrophotometer was used to study the thermochromism process of the corresponding polydiacetylene molecules. As the temperature increased, the maximum absorption peak at 650 nm gradually decreased, the maximum absorption peak at 550 nm gradually increased, consistent with the appearance of color change. In addition, the blue-phase polydiacetylene molecule is not fluorescent, while the orange-red polydiacetylene molecule is fluorescent, and the fluorescence emission peak is at 560 nm. Fourier transform infrared spectroscopy was used to investigate the temperature-induced discoloration properties of polydiacetylene molecules. The possible reason was that with the increase of temperature, the perturbation of polydiacetylene molecules caused the distortion of the side chain structure, resulting the conformation transition of ene-yne conjugates. As the bone structure changes, the effective conjugation length becomes shorter, which in turn changes the color of the polydiacetylene molecule.

1. Introduction

Diacetylene compound is a conjugated compound that does not exist as a single substance in nature and needs to be separated and purified from some natural substances[1]. In addition, diacetylene monomers undergo topochemical polymerization under ultraviolet light or gamma rays to form polydiacetylene[2]. The polymerization process requires no catalyst or initiator, which ensures the purity of polydiacetylene to a large extent. The premise of topological chemical polymerization of diacetylene molecules is accurate assembly, and this kind of assembly method is diverse[3-4]. (2) assembly of polydiacetylene liposome[5]; (3) assembly of polydiacetylene nanofibers[6]; (4) assembly of polydiacetylene composite material[7]. The special properties of diacetylene determine the research value in the field of chemistry and biology.

Polydiacetylene is a typical conjugated polymers with the π conjugated main chain structure of alkene alkyne alternation. When affected by external environment stimulation, the delocalized electron motion state of polydiacetylene changes, which will cause a strong color change of polydiacetylene molecules. The obvious color change can be seen by the naked eyes during the experiment, which is reflected in the transition of the system color from blue to orange red. Because this color change is easy to be observed by the naked eyes, it has superior optical and electrical properties, and is widely used in the sensing field: (1) polydiacetylene chemical sensor, which can be roughly divided into polydiacetylene temperature sensor, polydiacetylene pH sensor, polydiacetylene ion identification sensor, polydiacetylene gas sensor and polydiacetylene organic molecular sensor[8-10]; (2) polydiacetylene biosensor, which can be roughly divided into polydiacetylene pathogen microbial sensor and polydiacetylene biomolecular sensor[11]. The outstanding characteristics of these sensors are simplicity, sensitivity, high economic benefit, visible to the naked eyes and easy to distinguish, which can be applied to the field of food safety rapid detection and immune analysis[12]. Cao Chen et al. prepared a chemical sensor based on molecular recognition of melamine. This sensor is sensitive to pH, and changes in color and state will occur within a certain pH range. As pH increases, the color changes from blue to red, and the solution state changes from liquid to solid[13]. Kim et al. prepared the temperature sensor, which could produce intuitive color change under the condition of 20°C-130°C, and the color change range changed from red to yellow, which could realize the detection of a variety of substances at a higher temperature[14].

Using 10,12-pentacosadiynoic acid as raw material and polydiacetylene temperature sensor as the starting point, polydiacetylene vesicles were prepared by different assembly processes. Through the study of the temperature sensors, it is found that they can achieve a variety of discoloration under the condition of 80°C and short time. The discoloration mechanism of polydiacetylene was studied by Fourier transform infrared spectrometer. The discoloration mechanism may be that with the increase of temperature, the conformation of ene-alkyne conjugated skeleton changed, the effective conjugation length became shorter, leading the color change and fluorescence emission. The advantages of this kind of sensor are high sensitivity, good discrimination, strong stability and low price.

2. Reagents and instruments

Anhydrous ethanol; Polyvinyl alcohol; 10,12-pentacosadiynoic acid. Fluorescence spectrophotometer (F-2700, hitachi high-tech company); UV-Vis spectrophotometer (UV-1750, shimazu company, Japan); Fourier transform infrared spectrometer (Nicolet-iS5, semel technologies co., LTD.).

3. Experimental methods

3.1 Preparation of polydiacetylene vesicles

10,12-pentacosadiynoic acid ethanol solutions of different concentrations (0.001 mol/L, 0.005 mol/L, 0.01 mol/L) were prepared, and appropriate amount of deionized water was added to the samples after evaporation to obtain the corresponding prepolymer of the vesicle solution.

The resulting propolymer solution was sonicated for 15 min at 60°C, Following sonication, the solution was cooled at 4°C for 12 hours. Polydiacetylene vesicles are prepared by radiation under UV light.

4. Results and discussion

4.1 The effect of ultraviolet radiation time on the polymerization degree of polydiacetylene *vesicle*

The prepared diacetylene vesicles were placed in the UV analyzer and treated with the UV lamp for 1 minute, 2 minutes and 3 minutes respectively. As the irradiation time increases, the degree of polymerization increases (Figure 1).

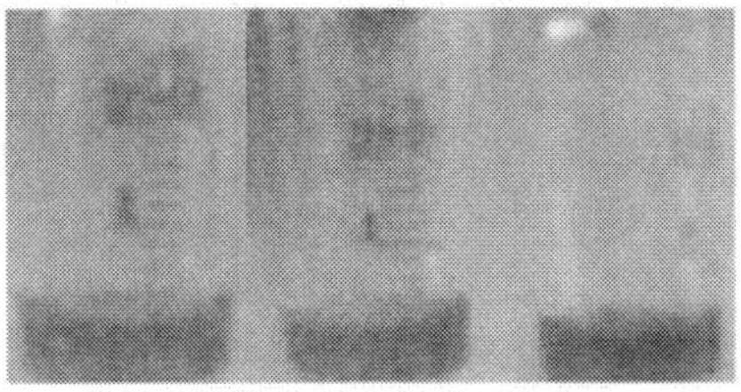

Figure 1. 0.001 mol/L polydiacetylene vesicles were radiated by UV lamp for 1 min, 2 min and 3 min.

4.2 Effects of temperature on polydiacetylene vesicles

4.2.1 Chromogenic state of polydiacetylene vesicles at different temperatures. The obtained polydiacetylene vesicles with three different concentrations (0.001 mol/L, 0.005 mol/L and 0.01 mol/L) were put into an electric thermostatic blast drying oven for heating. The thermochromic reaction of polydiacetylene vesicles was investigated at different temperatures (40°C, 50°C, 60°C, 70°C and >80°C). It can be observed that with the increase of temperature, the color of the polydiacetylene vesicle changes from blue to orange-red, and goes through dark blue, blue-purple, purple, rose-red and watermelon red in the middle. When the temperature is >80°C, the color remains orange-red, and the discoloration process of each concentration is shown in Table 1.

Table 1. Color development of polydiacetylene vesicles with three different concentrations at different temperatures.

	Room temperature	40°C	50°C	60°C	70°C	80°C
0.001 mol/L	Blue	Purple	Mei red	Mei red	Mei red	Orange red
0.005 mol/L	Blue	Dark blue	Dark blue	Dark blue	Watermelon red	Orange red
0.01 mol/L	Blue	Dark blue	Purple	Mei red	Mei red	Orange red

4.2.2 Visible spectral analysis of Polydiacetylene vesicles at different temperatures. The ordered polyacetylene monomers undergo topological chemical polymerization under the guidance of ultraviolet light, forming hyperconjugates, which greatly reduces the π-π* electronic transition ability, making polyacetylene molecules have strong absorption in the visible light area, and the maximum absorption peak is about 650nm. With the temperature rising, the corresponding maximum absorption peak of polyacetylene molecules gradually moves to 550nm[15]. Polydiacetylene molecules with different concentrations have the same change trend with temperature. Figure 2 shows the visible spectral analysis of 0.001 mol/L polydiacetylene vesicles under different temperature stimulation.

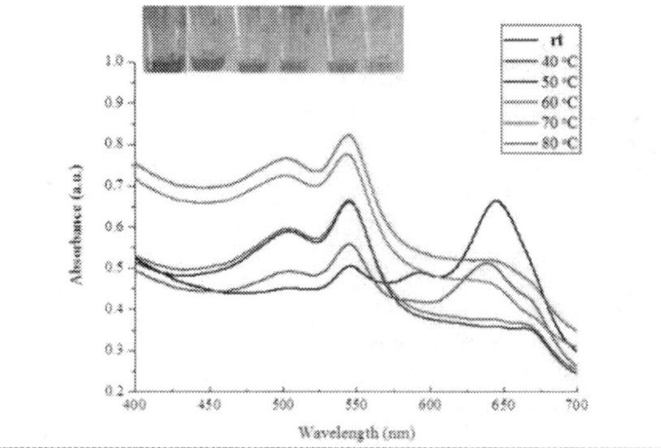

Figure 2. Visible spectra of 0.001 mol/L polydiacetylene vesicles at different temperatures during the blue-purple-rose-orange transition.

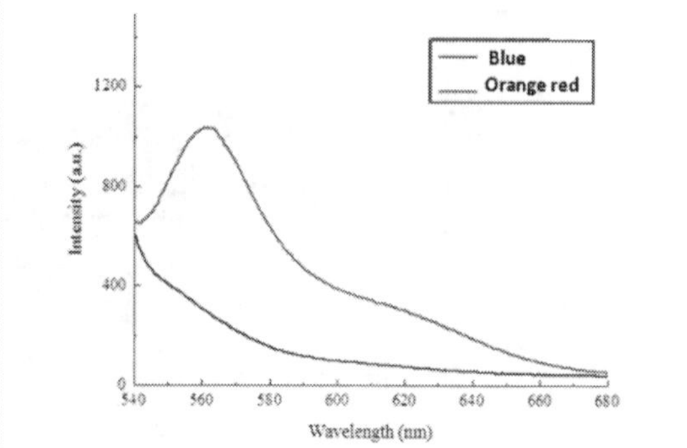

Figure 3. Fluorescence spectra of blue phase and orange phase in the polydiacetylene vesicle solution of 0.005 mol/L.

4.2.3 Colorimetric response analysis of Polydiacetylene vesicles. In order to quantitatively describe the color response of polydiacetylene that changes from blue to orange, CR value of colorimetric response was defined through its absorption spectrum data[16]. The value represents the degree to which the color of polydiacetylene changes from blue to red in the process of color respose. The larger CR value is, the color of poly acetylene tends to be red. The calculation formula is referenced from the literature[17-18].

Table 2. Colorimetric response values of Polydiacetylene at different temperatures (room temperature, 40°C, 50°C, 60°C, 70°C, 80°C) of 0.005 mol/L.

	CR_{rt}	$CR_{40°C}$	$CR_{50°C}$	$CR_{60°C}$	$CR_{70°C}$	$CR_{80°C}$
0.005 mol/L	0	1.7%	11.5%	22.5%	42.5%	49.3%

4.3 Investigation of temperature response time of Polydiacetylene vesicles

The obtained three polydiacetylene vesicles with different concentrations were put into an electric thermostatic blast drying box for heating. The response time of thermochromic reaction of polydiacetylene vesicles was explored at different temperatures (40°C, 50°C, 60°C, 70°C and >80°C). Can be observed when at a certain temperature, explore different times (2 min, 6 min, 10 min) polydiacetylene vesicle color from blue to other color, dark blue, violet, purple, mei red, watermelon red, when the temperature > 80°C color orange red unchanged, the concentration process as shown in Table 3.

Table 3. 0.001 mol/L polydiacetylene vesicles responded to color changes at 40°C, 50°C, 60°C, 70°C and 80°C for 2 min, 6 min and 10 min, respectively.

	40°C	50°C	60°C	70°C	80°C
2 min	Blue	Blue	Purple	Mei red	Orange red
6 min	Blue	Purple	Purple	Mei red	Orange red
10 min	Blue	Purple	Mei red	Mei red	Orange red

4.4 Fluorescence analysis of vesicles

In the process of stimulus response of polydiacetylene, fluorescent signal would also appear. 520nm was chosen as the excitation wavelength to analyze the fluorescence of polyacetylene. There was a fluorescence emission peak at 560nm for orange red polyacetylene, while there was no fluorescence emission peak for blue red polyacetylene, which indicated that orange red polyacetylene had fluorescence, while blue phase had no fluorescence[19]. This is mainly due to the fact that the atomic orbital of the lowest excited state of the blue phase molecule is intermediate symmetry (g- symmetry), and the atomic orbital of the ground state of the polydiacetylene molecule is also g symmetry, thus forming a dipole forbidden transition. The atomic orbital of the molecule in the lowest excited state of the polydiacetylene red phase is central antisymmetry (u-symmetry), which is allowed to transition in contrast to the blue phase. Therefore, the polydiacetylene red phase can produce fluorescence[20].

Due to the advantages of high sensitivity and good stability of fluorescence signal detection, the fluorescence detection of polydiacetylene molecule can make up for the inherent shortcomings of colorimetric response in analysis and detection to some extent, so that it has the advantages of natural dual signal detection in biological and chemical sensors and greatly improves the sensitivity of sensors[21].

4.5 Analysis of mechanism of polydiacetylene discoloration

Infrared spectra of PDAs of different colors are shown in Figure 4. When the temperature rises from 20°C to 80°C, the appearance color gradually changes from blue to orange-red, and the corresponding side chain hydrocarbon stretching vibration signal moves towards high wave number, from blue phase 2860 cm^{-1} and 2928 cm^{-1} to orange phase 2865 cm^{-1} and 2938 cm^{-1}. Although the change in displacement is small, the reason may be that the conformation may be distorted to some extent during the color transition.

Here, the possible mechanism of thermochromic properties is proposed as follows: with the increase of temperature, the side chain disturbance of polydiacetylene leads to the change of the conformation of the ene-acetylene conjugate skeleton, and the effective conjugate length becomes shorter, thus leading to the color change of poly acetylene molecules.

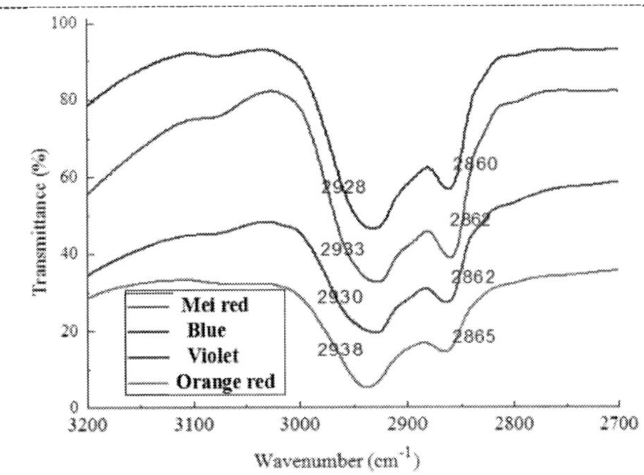

Figure 4. Infrared spectra of polydiacetylene molecules with different colors.

5. Conclusion

The resulted blue-phase polydiacetylene changed into blue-purple, purple, rose-red, watermelon red and orange-red with the increase of temperature in the range of room temperature to more than 80°C. The fluorescence properties were studied by fluorescence spectrophotometer, and the results showed that the orange red copolymerized acetylene molecules had fluorescence emission peak (560 nm), while the blue copolymerized acetylene molecules had no fluorescence emission peak, indicating that the orange red copolymerized acetylene molecules had fluorescence, while the blue phase had no fluorescence. The mechanism of diacetylene discoloration is studied by Fourier transform infrared spectrometer, which may be as follows: the increase of temperature promotes the disturbance of its side chain, leading to the color transformation of the conformation distortion of the main chain of diacetylene.

Polydiacetylene vesicle has special optical color-changing properties and better stability compared with the vesicle system prepared by phospholipid, which provide a possibility for their application in the fields of food safety rapid detection and biological chemical sensors.

Reference

[1] Tang, J., Lu, S.G., Duan, X.J., Luo, f., Han, Y.W., et al. (2015) Color detection of cationic surfactant by polydiacetylene vesicles modified with aminobenzene boric acid. J. Science and technology of food industry, 36(15): 78-82.

[2] Niu, L.P., Li, H., Dou, X.Y., (2014) Progress in Beta molecular sieve research. J. Shandong chemical industry, 43(15): 52-54.

[3] Li, Y.J., Zhou, P., She, L.S., Lin, Z.Y., Guo, L.G., Qiu, B., Chen, G.N., et al. (2013) Research on lysozyme biosensor based on poly acetylene vesicles. In: Photochemistry.Xi an.pp. 1.

[4] Fuzhou university. (2013) Biosensor construction and application research based on poly acetylene self-assembly bionic vesicle technology. http://www.cnki.net.

[5] Chen, Z.X., Cao, C., Deng, S.P., (2012) Color development mechanism and thermodynamics of a Bola type amphiphilic molecule loaded with polydiacetylene vesicles during the recognition of melamine. J. Acta physicochemical sinica, 28(06): 1320-1328.

[6] Duan, X.J., Lu, S.G., Han, Y.W., (2012) Study on vesicles self-assembly based on zinc oxide (polydiacetylene/phospholipid) nanocomposites. In: Construction and application of new functional self-assembly. Cheng du.pp.1

[7] Deng, J.L., Guo, C.X., Lu, W.S., Liu, T., Jiang, L., et al. (2006) Biolecular recognition of diacetylene vesicles based on molecular assembly. J.Advances in chemistry, (11): 1397-1408.

[8] Zhejiang gongshang university. (2011) Basic research on rapid detection technology based on molecular recognition principle. http://www.cnki.net.

[9] Li, X.Q., Chen, Z.X., Guo, G.M., Deng, S.P., et al. (2010) Construction of poly acetylene color-changing vesicle technology and its application in rapid detection of food safety. J Food science and technology, 35(07): 303-307.

[10] Liu, T., Deng, J.L., Guo, C.X., Zhang, L.Z., Jiang, L., et al. (2008) Research progress of poly acetylene vesicle biosensor. In: Colloidal and interface chemistry. Tian jin.pp. 1.

[11] Liu, T., Deng, J.L., Guo, C.X., Zhang, L.Z., Jiang, L., et al. (2007) Research progress of poly acetylene vesicle biosensor. In: Colloidal and interface chemistry. Fu zhou.pp. 1.

[12] Guo, C. X., Liu, T., Jiang, L., (2006) Effect of size of glycolipid-functionalized polydiacetylene vesicles on discoloration detection ability. In: Chemistry and society. Chang chun.pp. 1.

[13] Zhejiang gongshang university (2012). Basic research on visual detection of melamine based on color-changing vesicles. http://www.cnki.net.

[14] Kim, J., Lee, Y., Yang, D., Lee, J., Lee, G., et al. (2005) Polydiacetylene-based fluorescent sensor chip. J. Am. Chem. Soc, (127): 17580-17581.

[15] Guo, C., Boullanger, P., Liu, T., Size effect of polydiacetylene vesicles functionalized with glycolipids on their colorimetric detection ability. J. Phys. Chem.

[16] Yin, S., Song, B., Liu, G., Wang, Z., Zhang, X., (2007) Self-organization of the polymerizable bolaamphiphiles bearing diacetylene mesogenic group. J. Langmuir, (23): 5936-5941.

[17] Zhang, Y.X., Gao, X.Q., (1985) Application of surfactants in food industry. (2): 29-34.

[18] China tissue engineering research. (2011) Immunoassay of polybutylene nanoparticles against escherichia coli. http://www.cnki.net.

[19] Li, Y.J. (2013) Research on lysozyme biosensor based on polydiacetylene vesicles. In: Photochemistry. Xi an.pp.

[20] Su, A.M., Li, F., Determination of melamine in protein powder by HPLC. J. Hubei agricultural sciences, 2011: 378-380.
[21] Xi, H.L., Liu, K.Y., (2011) Determination of melamine in eggs by imprinted solid phase extraction gas chromatography-mass spectrometry.J. Chinese journal of veterinary medicine, (1): 2.

IOP Conf. Series: Materials Science and Engineering **729** (2020) 012089 doi:10.1088/1757-899X/729/1/012089

Synthesis of triple-doped core/shell/shell upconversion nanoparticles

Ying Fu[1*], Yuanyuan Du[1], Qixing Hu[1], Xing Gao[1], Wen He[1] and Zhihua Li[2*]

[1]School of Chemical and Biological Engineering, Qilu Institute of Technology, Jinan, Shandong, 250200, China

[2]College of Chemistry, Chemical Engineering and Materials Science, Shandong Normal University, Jinan, Shandong, 250014, China

[*]Corresponding authors: 15753190890@163.com
lizhihua2006@126.com

Abstract. A novel triple-doped core/shell/shell $NaGdF_4/NaGdF_4$:Yb,Tm,Nd/$NaGdF_4$ upconversion nanoparticles(UCNPs) has been designed and synthesized by thermal decomposition. Yb^{3+} and Nd^{3+} were used as sensitizers and Tm^{3+} with multi-level was activators. Three-layout nanoparticles exhibits 1.5 times strong emission than two-layout structure. The UC spectrum exhibits two bands centered at 525–535 and 578–585 nm from Nd^{3+}, corresponding to $^4G_{7/2}$–$^4I_{9/2}$ and $^4G_{7/2}$–$^4I_{11/2}$ transitions after excitation at 800 nm. These nanocrystals are promising an attractive luminescent biomarkers for biological and magnetic resonance imaging.

1. Introduction

Over the past few years, rare earth doped upconversion nanocrystals have been extensively studied[1-5]. In photoluminescence, upconversion luminescence is an anti-stokes phenomenon that long-wave radiation with low energy absorption is converted into short-wave radiation with high energy[6-7]. Rare earth ions are widely used in upconversion luminescent materials due to their abundant energy level structure, usually composed of activators, sensitizers and matrix materials. Activators provide luminescent centers with abundant energy levels. The commonly used rare earth ions are Er^{3+}, Ho^{3+} and Tm^{3+} which have multi-level. Sensitizers absorb energy and transfer it to the activator ions, sensitize the luminescence of other rare earth ions, and improve the luminescence properties. Commonly used sensitizers are Yb^{3+} and Nd^{3+}[8]. The rare earth-converted nanocrystals have the advantages of high sensitivity, low toxicity, high chemical stability, etc., but the fluorescence intensity is low, which greatly limits its development[9-10]. Core/shell upconversion nanomaterials can suppress the fluorescence quenching on the surface of the materials, fill the lattice defects of incomplete coordination of rare earth ions, and thus improve the fluorescence intensity of rare earth upconversion nanomaterials[11-12].

In the present study, we successfully prepared the triple-doped core/shell/shell structure of $NaGdF_4/NaGdF_4$:Yb,Tm,Nd/$NaGdF_4$ nanocrystals by thermal decomposition in the presence of oleic acid and 1-octadecene[13]. We chose $NaGdF_4$ as the host material because of its high upconversion efficiency. In the middle layer, Yb^{3+} ions can play the part of energy-transfer bridging ions between sensitizer Nd^{3+} ions and energy acceptor Tm^{3+} ions[11]. Nd^{3+} has two NIR excitation bands at 525 and 590 nm, and transitions from $^4G_{7/2}$ to $^4I_{9/2}$ and $^4G_{7/2}$ to $^4I_{11/2}$, respectively[14-15]. It was worth noting that we found that UCNPs of the three-layer core/shell/shell structure had 1.5 times the fluorescence intensity of the two-layer structure.

2. Experiment

2.1 The experimental materials:

Gd_2O_3、Yb_2O_3、Tm_2O_3、Nd_2O_3 (99.9%, Sigma), oleic acid (OA; 90%, Sigma), 1-octadecene (ODE; >90%, Sigma), cyclohexane (>90%, Sigma), Anhydrous ethanol (>99.5%, Sigma), Anhydrous methanol (>99.5%, Sigma) ,Hydrochloric acid (AR), rare-earth chlorides are prepared by our group.

2.2 Synthesis of rare-earth chlorides:

At first, the molar ratio of Gd_2O_3、Yb_2O_3、Tm_2O_3、Nd_2O_3 is 80.5%:18%:0.5%:1%, which are stirred in hydrochloric acid solution from turbid to a clear solution. The solution is evaporated and crystallized under 100 degrees Celsius to obtain dried rare-earth chlorides.

2.3 Synthesis of $NaGdF_4$:

At room temperature, 1mmol $GdCl_3$, 6mL OA, 15mL ODE were put into a 100mL three-necked flask. The flask was put in the heating mantle and the solution was heated for 30 min at 160 °C. Close the heating button and allow the solution to cool down to room temperature. Then pipette 10mL NaOH(2.5mmol, 0.1g)-NH_4F(4mmol, 0.148g)-methanol solution into the three-necked flask, and keep the reaction at room temperature for 30 min. Increase the temperature to 110 °C for 15min under the condition of vacuum, then raise the temperature to 330 °C for 60min under nitrogen atmosphere. Shut off the heating button and allow the solution to cool down to room temperature while stirring.

2.4 Synthesis of $NaGdF_4$/$NaGdF_4$:Yb,Tm,Nd:

Transfer the solution of the flask into a beaker, settle of the solution for one night, pour out the supernatant, and then transfer the contents to the centrifuge tube. Spin down the product at 4,000 r.p.m., disperse $NaGdF_4$ nanoparticle with cyclohexane.

Repeat step of Synthesis of $NaGdF_4$, weigh 0.8mmol $GdCl_3$:Yb,Tm,Nd instead of $GdCl_3$, and adjust the dosage of NaOH-NH_4F-methanol solution.

2.5 Synthesis of $NaGdF_4$/$NaGdF_4$:Yb,Tm,Nd/$NaGdF_4$:

Handling the sample of $NaGdF_4$/$NaGdF_4$:Yb,Tm,Nd nanoparticle in the same way with $NaGdF_4$.

Repeat step of Synthesis of $NaGdF_4$, weigh 0.5mmol $GdCl_3$ and 10mL NaOH(1.25mmol, 0.05g)-NH_4F(2mmol, 0.074g) -methanol solution instead of 1mmol $GdCl_3$ and 10mL NaOH(2.5mmol, 0.1g)-NH_4F(4mmol, 0.148g) -methanol solution.

3. Results and discussion

3.1 The high-resolution images

As shown in the high-resolution (HRTEM) images from Figure 1a, the $NaGdF_4$ core nanoparticles are nearly uniform nanosphere morphology with an average diameter of about 12×15 nm without aggregation. We successively compounded two shells of $NaGdF_4$/$NaGdF_4$:Yb,Tm,Nd with an average diameter of 16×40 nm through an epitaxial growth process. The HRTEM images of Figure 1b reveal that the nanospheres have a rod-like structure. The core/shell/shell UCNPs have an average diameter of 22×45 nm and figure 1c shows the HRTEM images.

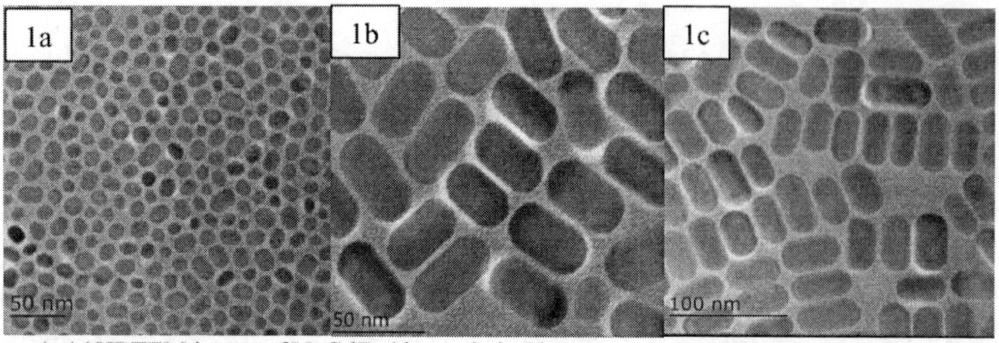

Figure 1. (a)HRTEM image of $NaGdF_4$ (the scale is 50 nanometers); (b) HRTEM image of $NaGdF_4$: Yb^{3+} (18 mol%), Tm^{3+} (0.5 mol%), Nd^{3+} (1 mol%)(the scale is 50 nanometers); (c)HRTEM image of $NaGdF_4$/$NaGdF_4$: Yb,Tm,Nd/$NaGdF_4$ (the scale is 100 nanometers).

3.2 The X-ray diffraction patterns

The crystallinity and the composition of the synthesized nanoparticles patterns of the samples were obtained. The X-ray diffraction (XRD) patterns of $NaGdF_4$ confirm the presence of a pure β phase structure, the XRD image in Fig.2a. Figure 2b shows the XRD pattern of $NaGdF_4$/$NaGdF_4$:Yb,Tm,Nd. The core/shell UCNPs show slightly narrower peak bandwidths. The XRD patterns of core/shell/shell UCNPs are similar to that of core UCNPs. The diffraction peak positions and intensities are in good agreement with the data for the reference (JCPDS card No. 27-0699)[14].

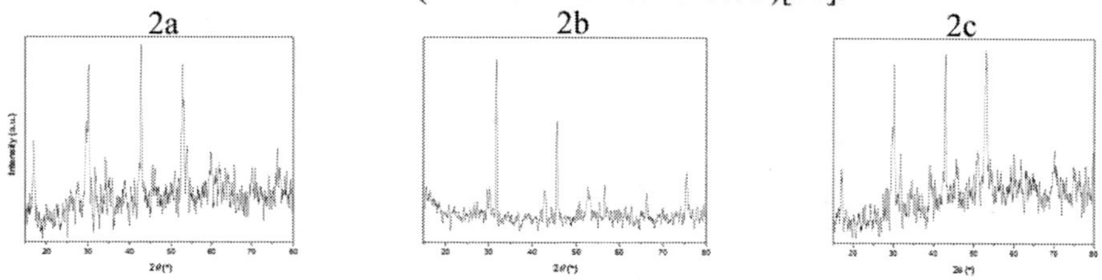

Figure 2. XRD pattern of (a)$NaGdF_4$;(b)$NaGdF_4$/$NaGdF_4$:Yb,Tm,Nd; (c)$NaGdF_4$/$NaGdF_4$: Yb,Tm,Nd/$NaGdF_4$.

3.3 Fluorescence analysis

Upconversion spectra of UCNPs are characterized under 800nm irradiation with the nearly same concentrations. By comparison, the experiment shows that fluorescence intensity of core/shell/shell nanoparticles exhibits 1.5 times strong emission than two-layout structure as shown in Figure 3. The UC spectrum exhibits two bands centered at 525–535 and 578–585 nm from Nd^{3+}, corresponding to $^4G_{7/2}$–$^4I_{9/2}$ and $^4G_{7/2}$–$^4I_{11/2}$ transitions[14-15]. These data demonstrate that core/shell/shell UCNPs can effectively improve the luminous efficiency.

IOP Conf. Series: Materials Science and Engineering **729** (2020) 012089 doi:10.1088/1757-899X/729/1/012089

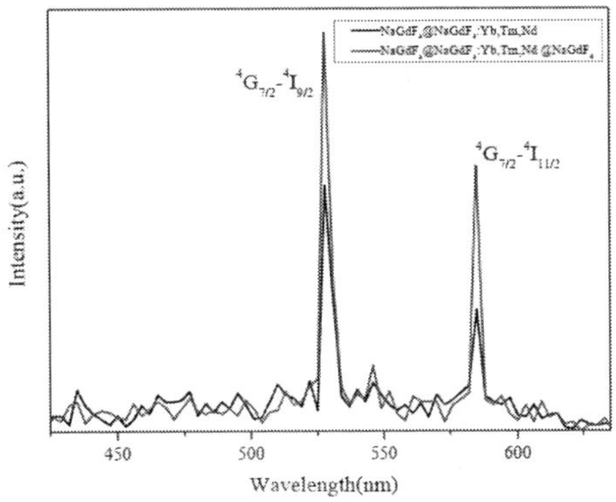

Figure 3. UC photoluminescence emission spectra of $NaGdF_4$/$NaGdF_4$:Yb,Tm,Nd and $NaGdF_4$/$NaGdF_4$: Yb,Tm,Nd/ $NaGdF_4$ under 800 nm NIR excitation.

4. Conclusion

In short, we have developed a new class of triple-doped core/shell/shell UCNPs. $NaGdF_4$, the host matrix was doped with Yb^{3+}、$Nd^{3+}$and Tm^{3+}as sensitizers and activator. HRTEM and XRD were employed to probe the morphology and structures of nanoparticles. It has been proved that the synthesized UCNPs were rod-like structures having high crystallinity. Under near 800nm excitation, the three-layer core/shell/shell structure exhibited more efficient activator emission. The application in biotechnology is limited by overexposure and tissue penetration depth because of the influence of 980 nm irradiation as well as high absorption and scattering that occurs in biological tissues[16]. To reduce the impacts of penetration depth, absorption and scatter in bio-tissues, we can choose 800nm laser excitation. This study represents an important advance in the development of luminescent markers suitable for biolabeling applications, enabling in situ imaging of biological and chemical events without overheating issues.

Acknowledgment

This work is supported by the National Natural Science Foundation of China (Grant No. 21445001).

References

[1] Zhang, J.J., Zhao, S.L., Xu, Z., Zhang, L.G., Zuo, P.F., Wu, Q.X., (2019). Near-infrared light-driven photocatalytic $NaYF_4$:Yb,Tm@ZnO core/shell nanomaterials and their performance. RSC Advances, 9(7):3688–3692.

[2] Liu, K., Liu, X.M., Zeng, Q.H., Zhang, Y.L., Tu, L.P., Liu, T., Kong, X.G., Wang, Y.H., Cao, F., Lambrechts, S. G., Zhang, H. (2012) Covalently assembled NIR nanoplatform for simultaneous fluorescence imaging and photodynamic therapy of cancer cells. ACS Nano., 6: 4054–4062.

[3] Kumar, P., Singh, S., Gupta, B. K. (2016) Future prospects of luminescent nanomaterial based security inks: from synthesis to anti-counterfeiting applications. Nanoscale, 8(30):14297–14340.

[4] Bettinelli, M., Carlos, L., Liu, X. (2015) Lanthanide-doped upconversion nanoparticles, Physics Today, 68: 38-44.

[5] Zhou, J., Liu,Q., Feng, W., Sun, Y., Li, F.Y. (2015) Upconversion Luminescent Materials: Advances and Applications. Chem. Rev., 115: 395−465.

[6] Yang, Y.M., Shao, Q., Deng, R.R., Wang, C., Teng, X., Cheng, K., Cheng, Z., Huang, L., Liu, Z., Liu, X.G., Xing, B.G. (2012) In Vitro and In Vivo Uncaging and Bioluminescence Imaging by Using Photocaged Upconversion Nanoparticles. Angew. Chem., Int. Ed., 51: 3125–3129.

[7] Wang, Y., Zheng, K.Z., Song, S.Y., Fan, D.Y., Zhang H.J., Liu, X.G. (2018) Remote manipulation of upconversion luminescence. Chem. Soc. Rev., 47:6473-6485.
[8] Shen, J., Chen, G., Vu, A., Fan, W., Bilsel, S., Chang, C., Han, G.(2013) Engineering the Upconversion Nanoparticle Excitation Wavelength: Cascade Sensitization of Tri-doped Upconversion Colloidal Nanoparticles at 800 nm. Adv. Optical Mater., 1:644–650.
[9] Feng, Z.W., Hu, D.J., Liang, L.L., Xu, J., Cao, Y.Y., Zhan, Q.Q., Guan, B.O., Liu, X.G., Li, X.P. (2019) Laser-Splashed Plasmonic Nanocrater for Ratiometric Upconversion Regulation and Encryption. Adv. Optical Mater., 1900610.
[10] Liu, Y.S., Tu, D.T., Zhu, H.M., Li, R.F., Luo, W.Q., Chen, X.Y. (2010) A Strategy to Achieve Efficient Dual-Mode Luminescence of Eu^{3+} in Lanthanides Doped Multifunctional $NaGdF_4$ Nanocrystals. Adv. Mater., 22:3266–3271.
[11] Xie, X.J., Gao, N.Y., Deng, R.R., Sun, Q., Xu, Q.H., Liu X.G. (2013) Mechanistic Investigation of Photon Upconversion in Nd^{3+}-Sensitized Core−Shell Nanoparticles. J. Am. Chem. Soc., 135:12608−12611.
[12] Chen, G.Y., Ohulchanskyy, T., Liu, S., Law, W.C., Wu, F., Swihart, M, Agren, H., Prasa, P. (2012) Core/Shell $NaGdF_4$:Nd^{3+}/$NaGdF_4$ Nanocrystals with Efficient Near-Infrared to Near-Infrared Downconversion Photoluminescence for Bioimaging Applications. ACS Nano., 6: 2969–2977.
[13] Liu, D., Jing, Y., Wang, K., Wang, Y.D., Luo, G.S. (2019) Reaction study of α-phase $NaYF_4$:Yb,Er generation via a tubular microreactor: Discovery of an efficient synthesis strategy. Nanoscale, DOI: 10.1039/C8NR09957J.
[14] Zhang, X.W., Zhao, Z., Zhang, X., Cordes, D., Weeks, B., Qiu, B., Madanan, K., Sardar, D., Chaudhuri, J. (2015) Magnetic and optical properties of $NaGdF_4$:Nd^{3+}, Yb^{3+}, Tm^{3+} nanocrystals with upconversion/downconversion luminescence from visible to the near-infrared second window. Nano Research, 8: 636–648.
[15] Wang, Y.F., Sun, L.D., Xiao, J.W., Feng, W., Zhou, J.C., Shen, J., Yan, C.H. (2012) Rare-Earth Nanoparticles with Enhanced Upconversion Emission and Suppressed Rare-Earth-Ion Leakage. Chem. Eur. J., 18, 5558–5564.
[16] Cai, W.B., Shin, D.W., Chen, K., Gheysens, O., Cao, Q.Z., Wang, S., Gambhir, S., Chen, X.Y. (2006) Peptide-labeled near-infrared quantum dots for imaging tumor vasculature in living subjects. Nano Lett., 6: 669–676.

IOP Conf. Series: Materials Science and Engineering **729** (2020) 012090 doi:10.1088/1757-899X/729/1/012090

A label free colorimetric assay for detection of sulfite root based on Cu^{2+}

Yefei Ma*, Xiaofen Fu, Wen He, Xing Gao*

School of Chemistry and Bioengineering, Qilu Institute of Technology, Jinan, Shandong, 250200, China

*Corresponding author's e-mail: myf2009090181@163.com and gaoxing-620@163.com

Abstract. The reaction between sulfite and hydrogen peroxide causes the absorbance decrease, which provides the basis to determine the sulfite content that is sensitive to the absorbance change. Cu^{2+} can catalyze the chemical reaction between hydrogen peroxide (H_2O_2) and the 2,2'-Azinobis-(3-ethylbenzothiazoline-6-sulphonate) (ABTS) effectively, which will cause the change of absorbance. In this paper, we presented an experiment system to study the influence of pH value, H_2O_2 content, Cu^{2+} content and reaction time on the reaction system. The comparison of experiment system results indicated that the optimal condition was pH value of 7.0, H_2O_2 of 500 μM, Cu^{2+} of 250 μM and 5 minutes reaction time. Under the optimal conditions, the absorbance (Y) and SO_3^{2-} content (C) have a linear relationship Y=-861.35C+0.8923. The coefficient of determination (R^2) in the linear relationship is 0.9983, which indicates a strong correspondence between C and Y. Then, the established method was used to detect the sulfite in wine. The recovery rate was between 96.01% and 106.90%, which proves this method is efficient for sulfite determination.

1. Introduction

Sulfites have been widely used as a food additive in the food industry to maintain color, prevent bacteria and retain freshness[1-3]. Most of wines contain sulfites that are produced naturally during the fermentation process[4]. During the wine production, sulfites (or sulfur dioxide) are also added as preservative and inhibitor to arrest fermentation and prevent oxidation[5-6]. Although sulfites are safe for most people, they are a threat for 1% of the nonasthmatic population and 5% of asthmatics population who are sulfite sensitive[7-8]. Thus, it is necessary to detect the concentration of sulfites in the wine with a high precision and label them on the package prominently.

There are usually two methods for quantifying total sulfites in foods and beverages. One method is direct iodometry, which uses the color change to judge whether the chemical reaction is ended or not. Although the iodine direct titration method is relatively simple, there is a subjective error, and an iodine complex may be formed during the titration to cause interference. This method is easily interfered by compounds formed by iodine in the process of titration[9]. The accuracy of this method is questionable. Another method is European Union usually adopt more formal (AOAC recognized) Monier-Williams (OMW) based on optimization Detection method, which needs a long-time sample preparation and cannot be used in the rapid visual inspection on-site[10]. To address the challenge, great efforts have been made to detect the sulfite assays with better performance, such as Capillary electrophoresis[11], electrochemical analysis[12], fluorescence analysis[13-14], Chemiluminescence[15], colorimetry [16]. Although these methods have made some progresses in sulfite detection, they still have some shortcomings. For instance, Capillary electrophoresis with the wide, application range has a poor stability and is difficult to achieve good repeatability. The fluorescence analysis is fast methods, which requires large instruments and a large one-time investment. Although the electrochemical analysis method has high sensitivity and wide measurement range, it requires cumbersome modification and washing steps. It is precisely because of these shortcomings that these methods cannot be used for on-site inspection of sulfite, which cannot meet the needs of relevant law enforcement agencies for spot sampling. Therefore, it is of great practical significance to establish a research system that can be used for rapid visualization determination of sulfite on site.

Spectrophotometry has the advantages of being simple and fast and does not need large and expensive instruments. However, there are not many reports on the application of spectrophotometry to the detection of sulfite. Therefore, it is still a great challenge to build a determination of sulfite based on spectrophotometry. The chromogenic reaction of 2,2-azinobis(3-ethylbenzothiazoline)-6-sulfonic acid (ABTS) and hydrogen peroxide (H_2O_2) is widely used in spectrophotometric studies. ABTS will be oxidized to green cation under the action of appropriate oxidants. Relevant studies have shown that ABTS can react quickly with hydroxyl radicals To form colored $ABTS^{\cdot+}$[17-20]. Cu^{2+} reacts with H_2O_2 to form hydroxyl radical, which can be used to catalyze the oxidation of ABTS[21-23].

In this study, we found that Cu^{2+} has the ability to catalyze the oxidation of ABTS by H_2O_2 to generate colored $ABTS^{\cdot+}$ and bring the change of absorbance. When SO_3^{2-} are added to the system, the absorption of $ABTS^{\cdot+}$ is reduced (Scheme 1). Because the H_2O_2 is consumed in the reaction between SO_3^{2-} and H_2O_2, it reduces the absorbance of the system. Therefore, the new sensing method established based on the principle that sulfite consumes hydrogen peroxide and reduces the absorbance of reaction system can be applied in the detection of SO_3^{2-}. The proposed method does not need any chemical labeling, making it convenient, fast and low-cost.

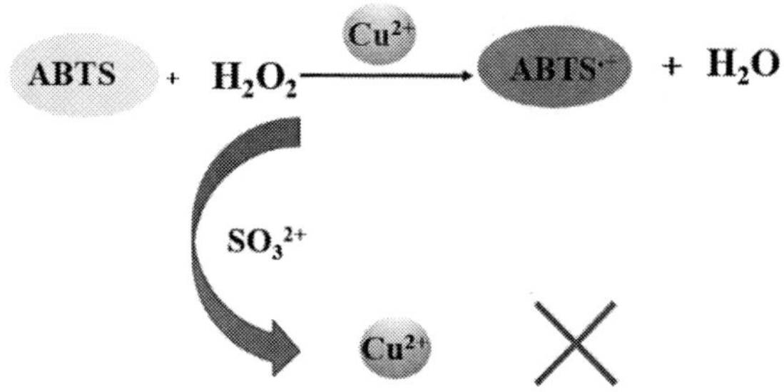

(1) $Cu^{2+} + H_2O_2 \longrightarrow Cu^{+} + O_2^{-} + 2H^{+}$

(2) $Cu^{+} + H_2O_2 \longrightarrow Cu^{2+} + OH^{-} + {}^{\cdot}OH$

(3) ${}^{\cdot}OH + ABTS \longrightarrow ABTS^{\cdot+} + OH^{-}$

Scheme 1. Schematic illustration of the colorimetric sensing system for SO_3^{2-} detection.

2. Material and Methods

2.1. Reagent and chemicals

2,2-azinobis(3-ethylbenzothiazoline)-6-sulfonic acid (ABTS) were obtained from Sigma-Aldrich Chemical Co. Copper chloride. H_2O_2 and Anhydrous sodium sulfite (Na_2SO_3) were purchased from Sinopharm Chemical Reagent Co., Ltd. (Shanghai, China). All the other reagents in this work were of analytical grade. The ultraviolet-visible (UV-vis) absorption spectra were recorded using a Cary UV-1750 spectrophotometer.

2.2. Detection of SO_3^{2-}

The sample solution of 1500 μL mixture containing 250 μM Cu^{2+}, 10 mM tirs-HCl (pH 7.0), 500 μM H_2O_2 and various concentrations of SO_3^{2-} was incubated for 5 min at room temperature. After that, ABTS

(2 mM) was added and after adding ABTS reaction for 5 min. Finally, the UV-vis spectra were recorded at room temperature 5 min later by using a UV-1750 UV-visible spectrophotometer (Shimadzu, Japan).

3. Results and discussion

3.1. Cu^{2+}-catalyzed H_2O_2+ ABTS reaction

The catalytic ability of Cu^{2+} was first verified in the study (Figure 1a). In the absence of Cu^{2+}, there is no significant difference in terms of the absorption spectrum and color between ABTS solution (vial 1) and ABTS + H_2O_2 mixture (vial 2). When Cu^{2+} was added into the ABTS solution (vial 3), absorption spectrum and color change were not obvious. However, great changes of absorption spectrum and color were observed after the addition of Cu^{2+} to ABTS + H_2O_2 mixture. The color of the Cu^{2+} +ABTS + H_2O_2 mixture becomes greener as the concentration of Cu^{2+} increases(from vial 4 to 6). These results indicate that Cu^{2+} can catalyze the oxidation of ABTS by H_2O_2 to form colored $ABTS^{\bullet+}$, which increases the absorbance of the sensing system.

3.2. Cu^{2+}-catalyzed H_2O_2+ ABTS reaction

When SO_3^{2-} is added to the ABTS + H_2O_2 + Cu^{2+} system, SO_3^{2-} reacts with H_2O_2 to generate hydroxyl radicals, which results in the H_2O_2 consumption and corresponding reduction of $ABTS^{\bullet+}$. SO_3^{2-} reacts with H_2O_2 to generate hydroxyl radicals, consuming H_2O_2 and resulting in the reduction of. It can be inferred that SO_3^{2-} can reduce the absorbance of ABTS + H_2O_2 + Cu^{2+} system. In order to explore whether the sensing system can be used for the detection of SO_3^{2-}, the feasibility of reducing ABTS + H_2O_2 + Cu^{2+} reaction by SO_3^{2-} is investigated. As shown in Figure 1b, it can be known that when sulfite was added into the ABTS + H_2O_2 + Cu^{2+} system, the absorbance of the system decreases at 415 nm and the color fades. Furthermore, with the increase of SO_3^{2-} concentration, the absorbance gradually decreases and the color of the sample solution becomes lighter (vial 2,3,4). However, in presence of SO_3^{2-}, absorption spectrum and color change of the ABTS + Cu^{2+} system were not obvious (vial 5). The results confirmed that the absorbance of the reaction system with SO_3^{2-}decreases because of the reaction of SO_3^{2-} with H_2O_2 consumption. These results show that the principle designed in this research is feasible, and the sensor system can be used for the quantitative detection of SO_3^{2-}.

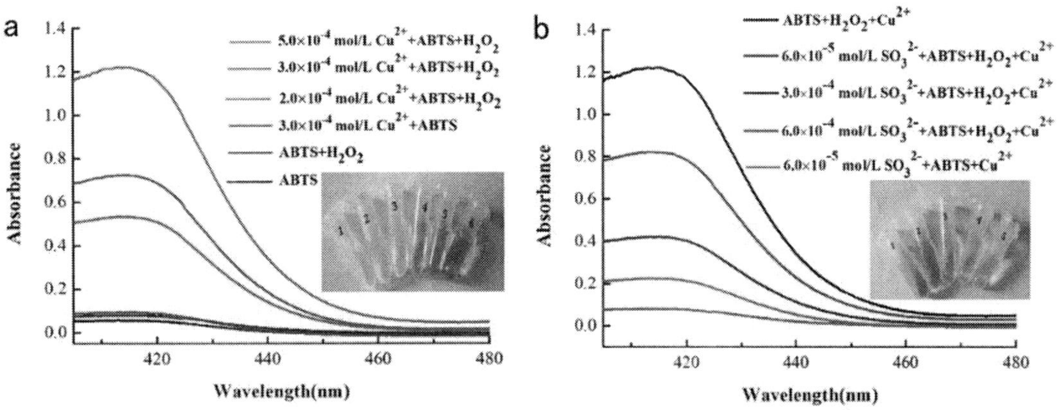

Figure 1. UV-visible absorption spectrum of different sample solutions. (a) Inset : color change diagram of different sample solutions.1: ABTS; 2: ABTS+H_2O_2; 3: 300 μM Cu^{2+}+ABTS; 4: 200 μM Cu^{2+}+ ABTS+H_2O_2; 5: 300 μM Cu^{2+}+ABTS+H_2O_2; 6: 500 μM Cu^{2+}+ABTS+H_2O_2. (b) Inset: color change diagram of different sample solutions. 1: ABTS+H_2O_2+Cu^{2+}; 2: 60 μM SO_3^{2-} +ABTS+Cu^{2+}+H_2O_2; 3: 30 μM SO_3^{2-}+ABTS+Cu^{2+}+H_2O_2; 4: 600 μM SO_3^{2}+ABTS+Cu^{2+}+H_2O_2; 5: 60 μM SO_3^{2-}+ABTS+Cu^{2+}. The concentrations of H_2O_2, ABTS and Cu^{2+} were 2 mM, 2 mM and 500 μM respectively.

IOP Conf. Series: Materials Science and Engineering **729** (2020) 012090 doi:10.1088/1757-899X/729/1/012090

3.3. Optimization of experimental conditions

In order to improve the sensitivity of the ABTS+H_2O_2+Cu^{2+} system for detecting SO_3^{2-}, many experimental conditions were investigated and optimized.

The pH value of the buffer solution has a certain influence on the absorbance of the ABTS+H_2O_2+Cu^{2+} system. The A/A_0 under different pH were explored (where A is absorbance in the absence of SO_3^{2-} and A_0 is absorbance in the presence of SO_3^{2-}). As shown in Figure 2a, the A/A_0 increases with the increase of pH, and then A/A_0 reached the maximum at pH 7.0. Thus, the optimal pH value is determined to be 7.0, which will be used in subsequent research.

As the reactant, the concentration of H_2O_2 has a great influence on the absorbance of the reaction system. Therefore, the concentration of H_2O_2 is optimized in this experiment. From the Figure 2b, it can be seen that the absorbance of the system increases gradually when the final concentration of H_2O_2 reaches 500 μM. It can be seen that the absorbance of the system is gradually stable, so the optimal concentration of H_2O_2 is 500 μM.

Cu^{2+} as catalyst has a certain influence on the colorimetric sensing system, so the Cu^{2+} concentration was optimized in this experiment. It can be seen from Figure 2c that with the increase of Cu^{2+} concentration, the absorbance also increases. When it increases to 250 μM, it can be seen that the absorbance gradually tends to be stable. Therefore, the optimal Cu^{2+} concentration determined in this experiment is 250 μM, and this concentration will be used for subsequent research.

The reaction time of SO_3^{2-} with H_2O_2 was also optimized. It can be seen from Figure 2d the absorbance of the system after the reaction time of SO_3^{2-} and H_2O_2 exceeds 5 min. The change is not obvious, so the optimal reaction time of SO_3^{2-} and H_2O_2 in this experiment is 5 min.

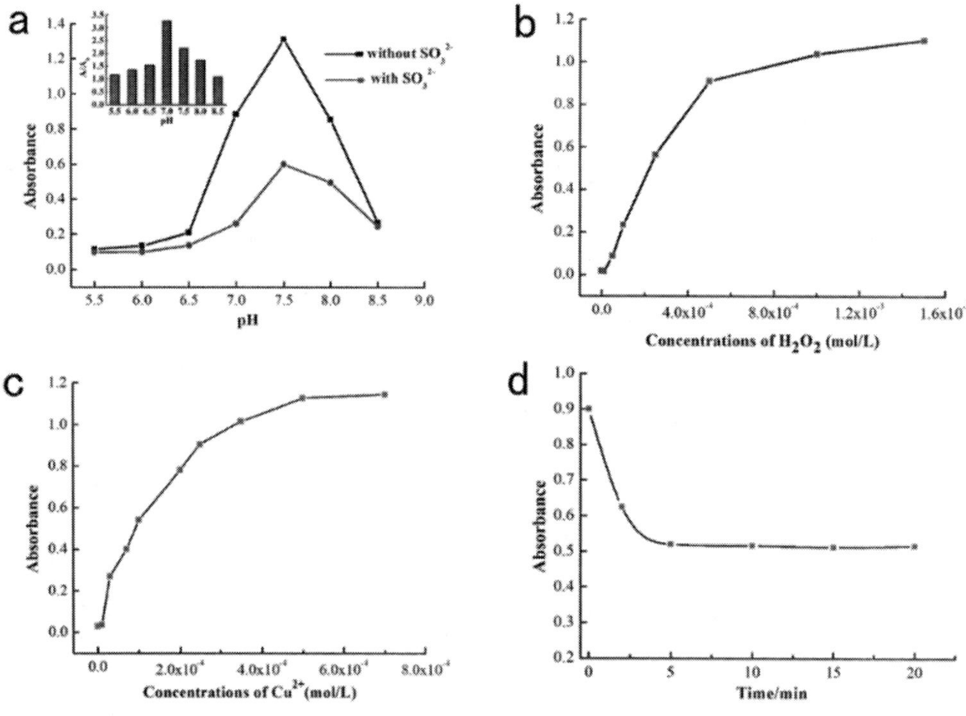

Figure 2. (a) The absorbance of the systems in the absence or in the presence of SO_3^{2-} under different pH. (b) The effect of the amount of H_2O_2 on the assay performance. (c) The effect of the amount of Cu^{2+} on the assay performance. (d) The optimization of the of SO_3^{2-} with H_2O_2 reaction time. The concentrations of ABTS and SO_3^{2-} were 2 mM and 400 μM respectively.

3.4. Assay performance of the present method for SO_3^{2-} detection

Under the optimal conditions, the absorption spectrum of SO_3^{2-} with different concentrations was scanned at the wavelength of 405 ~ 480 nm, the color photos of sample solution and the drawing of standard curve were taken. It can be seen from Figure 3a that when the concentration of SO_3^{2-} increases gradually, the absorbance value of its absorption spectrum decreases gradually at 415 nm. In Figure. 4c we can see that when the concentration of SO_3^{2-} was in the range of 20 μM ~ 1 mM, there is a good linear relationship with the absorbance. The linear equation is Y = -861.35 C+0.8923 (R^2=0.9983) (Y is the peak fluorescence intensity and C is the SO_3^{2-} concentration).With the increase of SO_3^{2-} concentration, the color of the solution gradually changed from dark green to light. In addition, the relative standard deviations (RSDs) were 2.8%, 3.1% and 2.5% in three repetitive assays of 20 μM, 400 μM and 1 mM SO_3^{2-}, which shows that the method has good stability.

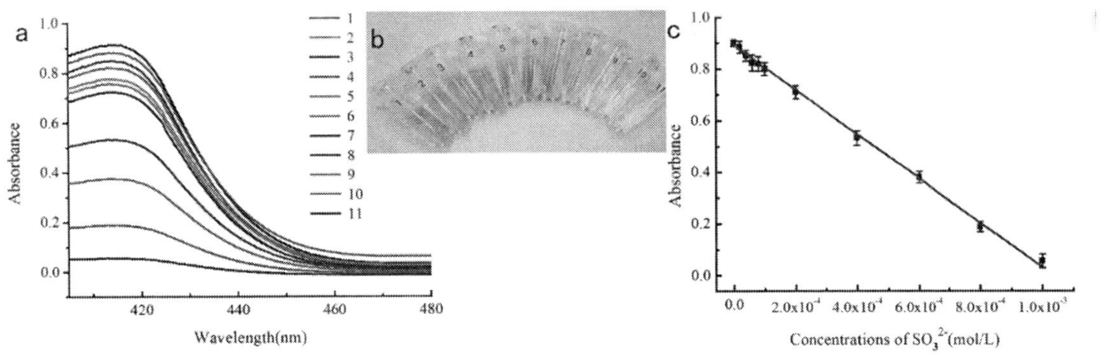

Figure 3. (a) UV-visible absorption spectra in the presence of SO_3^{2-} at different concentrations from 0 to 1 mM. (b) Color change in the presence of different concentrations of SO_3^{2-}. (c) SO_3^{2-} standard curve. The error bars represent the standard deviation of the three measurements.

3.5. Selectivity of the present method

In order to further verify the feasibility of this sensing system, six different ions and SO_3^{2-} were selected for comparison under the same reaction conditions, so as to verify the selectivity of ABTS+H_2O_2+Cu^{2+} system to SO_3^{2-}. It can be seen from Figure4 that when Na^+, Mn^{2+}, K^+, Cl^-, SO_4^{2-}, NO_3^- with higher concentration than SO_3^{2-} are used, it is found that these ions have little effect on the absorbance of the system, while the absorbance of the system decreases rapidly when SO_3^{2-} with smaller concentration is added, which shows that the system is highly selective for SO_3^{2-}.

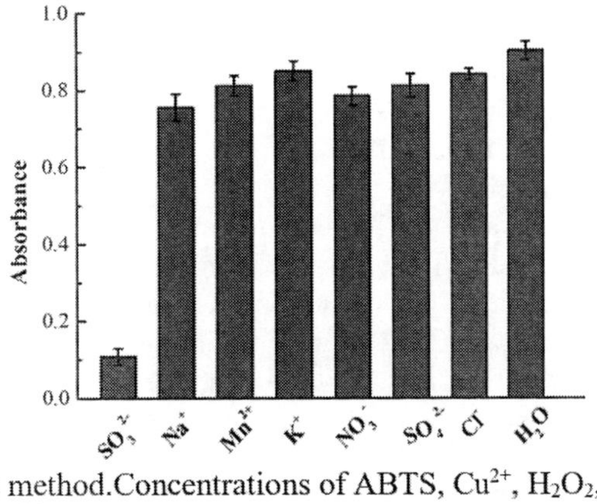

Figure 4. Selectivity of this method.Concentrations of ABTS, Cu^{2+}, H_2O_2, and SO_3^{2-} were 2 mM, 250 μM, 500 μM and1 mM, respectively. Concentration of other metal ions were 2 mM. Error bars were estimated from three replicate measurements.

3.6. Determination of actual samples

In order to use the sensor system established in this experiment for the detection of actual samples, this experiment has carried out the analysis and determination of wine. In the actual sample wine, different concentrations of sulfite standard solution have been added as samples to be determined. The results show that the absorbance value of the sample decrease with the increasement of sulfite concentration added in wine. In addition, the standard addition recovery experiment on wine samples was carried out, the standard addition recovery rate of sulfite in wine is between 96.01% and 106.90%(Table 1). The standard addition recovery experiment shows that the method established in this experiment has potential application for the detection of sulfite.6.5. Positioning figures

Table 1 Experimental results of determination of sulfite in wine

Sample	Added(μM)	Found(μM)	Recovery(%)	RSD (n=3, %)
1	0	0	-	-
2	100	106.90	106.90	4.9
3	300	294.99	98.33	2.6
4	500	480.05	96.01	3.4

4. Conclusion

Based on the principle that SO_3^{2-} can reduce the absorbance of Cu^{2+}+ABTS+H_2O_2 reaction system, a simple and rapid spectrophotometric method was developed for the detection of sulfite and the sensitive analysis of sulfite was realized. This method has some obvious advantages. It does not need any chemical synthesis and enzyme participation. It has low cost and high stability. Because the whole detection process does not need to go through the complex steps such as washing, it is very convenient. In addition, the method has high selectivity. The sensor system established in this paper is applied to the detection of real sample wine. From the recovery of standard addition, it can be concluded that the method is feasible, and the result is reliable. Therefore, the sensor system established in this paper provides a new way for the detection of sulfite.

Acknowledgments

The authors are thankful to Research Project Foundation of Qilu Institute of Technology (No. QL19K044) for financial support of this work.

References

[1] Yang, W.H., Purchase, E.C. (1985) Adverse reactions to sulfites. Cmaj, 133(9), 865-867.
[2] Omsoliu, G., Rojasgraü, M.A., Alandes González, L., Varela, P., Solivafortuny, R., Hernando Hernando, M. I. (2010) Recent approaches using chemical treatments to preserve quality of fresh-cut fruit: a review. Postharvest Biology Technology, 57(3), 139-148.
[3] Rahman M.S. (2007) Food Preservation: Overview. In: Rahman M.S.(Ed), Handbook of Food Preservation, Second Edition. CRC press. pp.3-18.
[4] Eschenbruch, R. (1974) Sulfite and sulfide formation during winemaking-a review. Am. J. Enol. Viticult., 25(3), 157-161.
[5] Santos, M.C., Nunes, C., Saraiva, J. A., Coimbra, M.A. (2012) Chemical and physical methodologies for the replacement/reduction of sulfur dioxide use during winemaking: review of their potentialities and limitations. European Food Research Technology, 234(1), 1-12.
[6] Divol, B., Du, T.M., Duckitt, E. (2012) Surviving in the presence of sulphur dioxide: strategies developed by wine yeasts. Applied Microbiology Biotechnology, 95(3), 601-613.
[7] Grotheer, P., Marshall, M., Simonne, A. (2005) Sulfites: separating fact from fiction. Department of Family, Youth, and Community Sciences, University of Florida. http://citeseerx.ist.psu.edu/viewdoc/download?doi=10.1.1.322.297&rep=rep1&type=pdf
[8] Knodel, L.C. (1997) Current issues in drug toxicity: Potential health hazards of sulfites. Toxic Subst.

Mech., 16(3), 309-311.
[9] Gonçalves, L.M., Pacheco, J.G., Magalhães, P.J., Rodrigues, J.A., Barros, A.A. (2010) Determination of free and total sulfites in wine using an automatic flow injection analysis system with voltammetric detection. Food Additives Contaminants, 27(2), 175-180.
[10] Koch, M., Köppen, R., Siegel, D., Witt, A., Nehls, I. (2010). Determination of total sulfite in wine by ion chromatography after in-sample oxidation. Journal of agricultural and food chemistry, 58(17), 9463-9467.
[11] Masár, M., Danková, M., Olvecká, E., Stachurová, A., Kaniansky, D., Stanislawski, B. (2005) Determination of total sulfite in wine. zone electrophoresis-isotachophoresis quantitation of sulfate on a chip after an in-sample oxidation of total sulfite. Journal of Chromatography A, 1084(1), 101-107.
[12] Devadas, B., Sivakumar, M., Shen, M.C., Cheemalapati, S. (2015) An electrochemical approach: switching structures of rare earth metal praseodymium hexacyanoferrate and its application to sulfite sensor in red wine. Electrochimica Acta, 176, 350-358.
[13] Li, H., Zhu, H., Sun, M., Yan, Y., Zhang, K., Huang, D., et al. (2015) Manipulating the surface chemistry of quantum dots for sensitive ratiometric fluorescence detection of sulfur dioxide. Langmuir, 150728141434003.
[14] Wang, J., Hao, Y., Wang, H., Yang, S., Tian, H.Y., Sun, B., et al. (2017) Rapidly responsive and highly selective fluorescent probe for bisulfite detection in food. Journal of Agricultural Food Chemistry, 65(13), 2883-2887.
[15] Rawal, R., Pundir, C.S. (2012) Development of an amperometric sulfite biosensor based on so x /pbnps/ppy modified ito electrode. International Journal of Biological Macromolecules, 51(4), 449-455.
[16] Zhang, J., Xu, X., Yang, X. (2012) Role of tris on the colorimetric recognition of anions with melamine-modified gold nanoparticle probe and the visual detection of sulfite and hypochlorite. Analyst, 137(15), 3437-3440.
[17] Ding, N., Yan, N., Ren, C., Chen, X., Chem, A. (2010) Colorimetric determination of melamine in dairy products by Fe_3O4 magnetic nanoparticles-H_2O_2-ABTS detection system. Analytical Chemistry, 82(13), 5897-9.
[18] Persano, S., Valentini, P., Kim, J.H., Pompa, P.P. (2013) Colorimetric detection of human papilloma virus by double isothermal amplification. Chemical Communications, 49(90), 10605-10607.
[19] Qin, Y., Zhang, L., Ye,G., Zhao, S. (2014) Homogeneous label-free colorimetric strategy for convenient bleomycin detection based on bleomycin enhanced Fe(II)-H_2O_2-ABTS reaction. Analytical Methods, 6(19), 7973-7977.
[20] Aliaga, C., Lissi, E.A. (2015) Reaction of 2,2′-azinobis (3-ethylbenzothiazoline-6-sulfonic acid (ABTS) derived radicals with hydroperoxides. kinetics and mechanism. International Journal of Chemical Kinetics, 30(8), 565-570.
[21] Zhang, L., Li, M. , Qin, Y., Chu, Z., Zhao, S. (2014) A convenient label free colorimetric assay for pyrophosphatase activity based on a pyrophosphate-inhibited Cu^{2+}-ABTS-H_2O_2 reaction. Analyst, 139(23):6298.
[22] Wang, H.B., Li, Y., Dong, G.L., Gan, T., Liu, Y. (2017) A convenient and label-free colorimetric assay for dopamine detection based on the inhibition of Cu (II)-catalyzed oxidation of 3, 3′, 5, 5′-tetramethylbenzidine- H_2O_2 system. New Journal of Chemistry, 41(23), 14364-14369.
[23] Han, X., Qin, Z.X., Zhao, M.Y., Song, J.J., Qu, F., Qu, F.L., Kong, R.M. (2018) Convenient and sensitive colorimetric detection of melamine in dairy products based on Cu(II)-H_2O_2-3, 3′, 5, 5′-tetramethylbenzidine syste. RSC Advances 8(61), 34877-34882.

Preparation and Drug Release Properties of Fe_3O_4/Chitosan Magnetic Microspheres

Mingming Li[1*], Xuyan Zhang[2], Shiping Zhan[1] and Weimin Hou[1]

[1] Department of Environment and Chemical Engineering, Dalian University, Dalian, Liaoning province, 116622, China

[2] Gastroenterology, The 967 Hospital of the People's Liberation Army, Dalian, Liaoning province, 116000, China

[*] Corresponding author's e-mail: magenness@163.com

Abstract. Fe_3O_4/chitosan(CS) magnetic microspheres were prepared by emulsion crosslinking method, in which 5-fluorouracil (5-Fu) was loaded as model drug. The preparation conditions were optimized by orthogonal experiment with the the package rate as an index. The structure and morphology of the magnetic microspheres were characterized by scanning electron microscopy(SEM). The drug release properties of the microspheres in vitro were investigated. The optimal preparation conditions of drug loaded magnetic microspheres were as follows: 0.3 g 5-Fu, 5 mL glutaraldehyde, 3 h crosslinking time and 0.2 g/ml acetic acid concentration. Under these conditions, the drug loading efficiency was up to 6.35% as well as the package rate was up to 35.2%. The Fe_3O_4/chitosan magnetic microspheres also displayed an excellent drug release behavior under the experimental conditions of simulated intestinal fluid. The releasing content reached a peak in 10 h.

1. Introduction

The clinical applications of some anti-cancer drug is limited because of its toxicity, such as 5-fluorouracil(5-Fu)[1]. Drug and an appropriate magnetic material can be performed into controlled drug-released magnetic microspheres. Under the external magnetic field, it can carry the drug to a special target site and the drug can be controlled released on that spot, so that the dosage and toxic effects of drugs are both significantly reduced, and the treatment efficiency can be enhanced[2-3]. However, most of magnetic microsphere treatments used currently are not satisfied, due to their small magnetic flux, less carrying capacity and short drug time[4]. The preparation of drug-released magnetic microspheres with high magnetic flux, large drug loading as well as small size has become the hot spot in current studies.

Chitosan(CS), a kind of natural polysaccharide, is widely used as drug carrier, due to its good biocompatibility and biodegradability. Paramagnetic or super magnetic Fe_3O_4 nanoparticles under the effect of the external magnetic field can kill tumor cells when the temperature rises to 40 ℃-45 ℃[5]. This paper was attempted to prepare a drug carrier system, in which chitosan was used as enveloping material, 5-Fu as model drug, Fe_3O_4 as kernel, and glutaraldehyde as crosslinking agent. The preparation conditions were optimized by drug loading efficiency and package rate. In addition, the morphology and drug release properties were determined.

2. Materials and methods

2.1. Materials and Instrument

5-Fu (purity>98%, Zhuorui Chem. Shanghai, China), chitosan (purified, Haiyun Chem. Dalian, China), Fe_3O_4 nanoparticles (self-made), glutaricaldehyde (AR, Hangzhou Luer Tech Co., Ltd. Hangzhou, China), acetic acid (AR, Tianjin Kemiou Chemical Reagent Co., Tianjin, China), Phosphate buffer solution (PBS, pH 6.86, Tianjin Damao Chemical reagent Factory, Tianjin, China), dialysis bag (Beijing JingKeHongDa, Beijing, China), and distilled water were used in this experiment.

Ultraviolet spectrometer (UV2100, Unico, USA), scanning electron microscopy (SU3500, Hitachi, Ltd, Japan).

2.2. Preparation of Fe_3O_4 nanoparticles

According to Ref.[6-7], Fe_3O_4 nanoparticles were prepared. The preliminary studies showed that they had good magnetic properties.

2.3. Preparation of 5-Fu loaded Fe_3O_4/chitosan magnetic microspheres

0.1 g (dry mass) Fe_3O_4 nanoparticles mud paste, 5-Fu and 0.6 g chitosan were mixed in 30 mL acetic acid solution for 1 h with intense stirring, and then the mixture was installed to remove the foam generated by the agitation.

In a 250 mL three-neck flask, 50 mL Liquid paraffin was mixed with 1.5 mL span80, which was used as dispersing agent; then the above mixture was slowly dripped under severe agitation; at last 1 mL 50% glutaraldehyde was added and reacted at 35 °C for a period of time. After holding for 12 h, the precipitates of nanoparticles were washed by petroleum ether, ethanol and distilled water for three times, respectively. Then 5-Fu loaded Fe_3O_4/chitosan magnetic microspheres were obtained after centrifugation and drying in a vacuum oven.

2.4. Morphology of 5-Fu loaded Fe_3O_4/Chitosan magnetic microspheres

Scanning electron microscopy(SEM)was used for observation of the size and morphology of the magnetic microspheres. The obtained microspheres were vibrated with ultrasound for 30 min, and treated by spray-gold for SEM observation.

2.5. Determination of drug loading efficiency and package rate

5-Fu (10 mg) was dissolved in phosphate buffer solution (pH 6.68) and diluted to 100 mL. 1.0, 2.0, 3.0, 4.0, and 5.0 mL solution were used and diluted to 50 mL, respectively. After 1 h lasting ultrasonic treatment, UV-vis spectrometer were used to measure 265 nm absorbance frequency for the standard curve.

5-Fu loaded Fe_3O_4/chitosan magnetic microspheres were measured, added with phosphate buffer solution (pH 6.68). After the microspheres were dissolved, the solution was diluted for UV-vis spectrometer measuring, and compared with the standard content curve. The formulas of the drug loading and package rate are as follows:

$$Drug\ loading\left(\%\ w/w\right)=\frac{mass\ of\ 5-Fu\ in\ microspheres}{mass\ of\ microspheres\ recovered}\times 100\% \quad (1)$$

$$Package\ rate\left(\%\ w/w\right)=\frac{mass\ of\ 5-Fu\ in\ microspheres}{total\ mass\ of\ 5-Fu}\times 100\% \quad (2)$$

2.6. In Vitro experiment of drug release

10 mg 5-Fu loaded Fe_3O_4/chitosan magnetic microspheres was put into a dialysis bag, added with 10 mL phosphate buffer solution (pH 6.68), then was kept at temperature of 37 °C. 5 mL solution was regularly measured for absorbance and at the same time 5 mL fresh buffer solution was added. The release amount of 5-Fu via time curve was measured as mentioned above.

3. Results and discussion

3.1. The standard curve of 5-Fu

The relationship of 5-Fu concentration with absorbance was analyzed by linear regression With the correlation coefficient R^2=0.9999 as follows:

$$A = 0.0521C + 0.0229 \tag{3}$$

A is absorbance value, whereas *C* is mass concentration of 5-Fu (μg/mL).

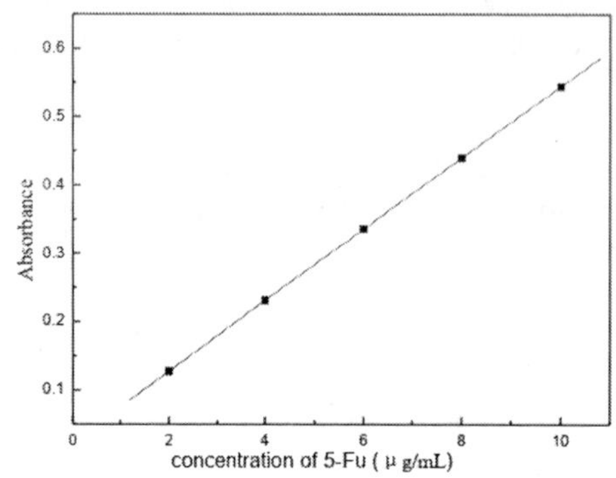

Figure 1. The standard curve of 5-Fu,

3.2. Optimization of the preparation conditions

The optimal preparation conditions were selected by orthogonal experiment with the package rate as an index. The 5-Fu dosage, glutaraldehyde dosage, crosslinking time and acetic acid concentration were estimated as four major factors due to the pre-experiments. The experimental design is shown in Table 1.

The results of orthogonal experiment is shown in Table 2. According to these data, sorted in order of size, the Ranges of the four factors were R_D>R_A>R_B>R_C, which indicated that 5-Fu dosage was the most important factor affecting the encapsulation rate of the preparation of magnetic microspheres, whereas other influencing factors were acetic acid concentration, glutaraldehyde dosage, and crosslinking time in order. No. 3 experiment $A_1B_3C_3D_3$ displayed the best performance. The optimal preparation conditions were as follows: 0.2 g/ml acetic acid concentration, 5 mL glutaraldehyde, 3 h crosslinking time and 0.3 g 5-Fu. Under these conditions, the package rate was up to 35.2%, as well as the drug loading efficiency was up to 6.35%.

Table 1. Factor design of the orthogonal experiment

	Factor A: acetic acid concentration (g/mL)	Factor B: glutaraldehyde dosage (mL)	Factor C: crosslinking time (h)	Factor D: 5-Fu dosage (g)
1	0.2	2	2	0.15
2	0.4	4	2.5	0.2
3	0.6	5	3	0.3

Table 2. Analysis of the orthogonal experiment

No.	Factor A	Factor B	Factor C	Factor D	the package rate (%)
1	1	1	1	1	10.3
2	1	2	2	2	23.6

IOP Conf. Series: Materials Science and Engineering **729** (2020) 012091 doi:10.1088/1757-899X/729/1/012091

3	1	3	3	3	35.2
4	2	1	2	3	35.0
5	2	2	3	1	23.4
6	2	3	1	2	32.2
7	3	1	3	2	30.6
8	3	2	1	3	30.8
9	3	3	2	1	25.9
K_1	23	25.3	24.4	19.9	
K_2	30.2	25.9	28.2	28.8	
K_3	29.1	31.1	29.7	33.7	
R	7.2	5.8	5.3	13.8	

3.3. Morphology of 5-Fu loaded Fe_3O_4/chitosan magnetic microspheres

The morphology of 5-Fu loaded Fe_3O_4/chitosan magnetic microspheres(experiment No.3) is shown in Fig.2.

As seen in Fig.2, the magnetic microspheres were regularly spherical, and surface was smooth. They were homogeneous in size, and the range of microspheres size is from 5 to 10 μm.

3.4. Drug releasing property

The releasing property of 5-Fu loaded Fe_3O_4/chitosan magnetic microspheres(experiment No.3) is shown in Fig.3.

As seen in Fig.3, 5-Fu loaded Fe_3O_4/chitosan magnetic microspheres hold a quick release within 6 h and the whole releasing quantity was achieved to be about 74.2%. Then the release rate became more slowly, and the release rate is about 85.6% in 24 h. Thus, the release curve, which had a quicker release in the beginning and a slower release in the late, showed that 5-Fu loaded Fe_3O_4/chitosan magnetic microspheres had a good releasing performance.

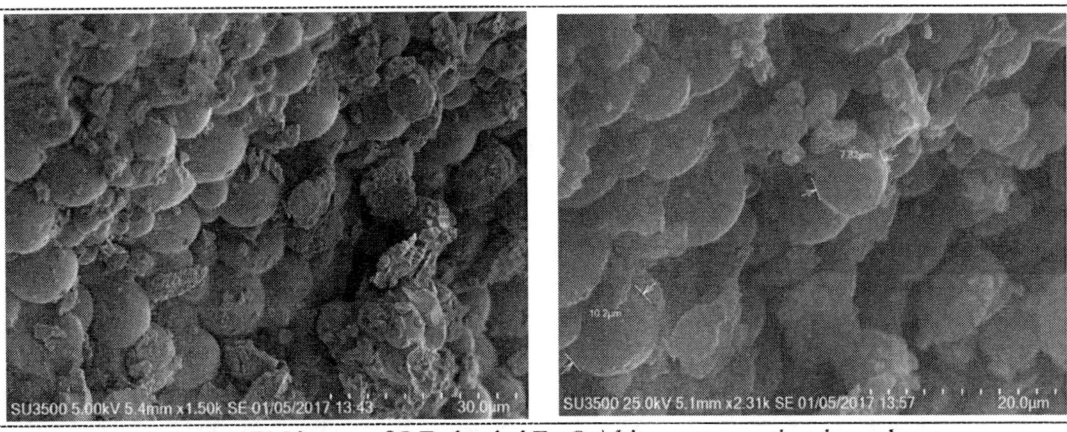

Figure 2. SEM image of 5-Fu loaded Fe_3O_4/chitosan magnetic microspheres

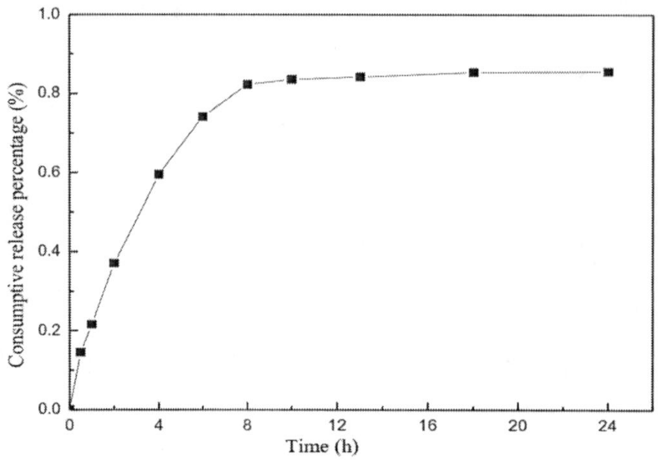

Figure 3. The releasing property of 5-Fu loaded Fe_3O_4/chitosan magnetic microspheres in phosphate buffer solution (pH 6.68)

4. Conclusion

In this work, the 5-Fu loaded Fe_3O_4/chitosan magnetic microspheres were successfully prepared. The preparation conditions were optimized by orthogonal experiment, and the experiment No.3 possessed the best performance in both of package rate and drug loading efficiency, which was 35.2% and 6.35%, respectively. In the phosphate buffer solution (pH 6.68), the accumulation 5-Fu release rate of magnetic microspheres in 24 h is 85.6%, showing a obvious sustained release effect.

Acknowledgments

This research is supported by the Project(No.20170540036) of Liaoning Natural Science Foundation, China.

References

[1] Wang, D.S., Li, J.G., Li, H.P., et al. (2009) Preparation and drug releasing property of magnetic chitosan-5-fluorouracil nano-particles. Transactions of Nonferrous Metals Society of China, 19: 1232-1236.

[2] Katrine, D.M., Camilla, S., Stefania, B., et al. (2013) Development of an ex vivo retention model simulating bioadhesion in the oral cavity using human saliva and physiologically relevant irrigation media. International Journal of Pharmaceutics, 448: 373-381.

[3] Mennini, N., Furlanetto, S., Cirri, M., et al. (2012) Quality by design approach ofr developing chitosan-Ca-alginate microspheres for colon delivery of celecoxib-hydroxypropyl-β-cyclodextrin-PVP complex. European Journal of Pharmaceutics and Biopharmaceutics, 80: 321-331.

[4] Zhu, L.Z., Ma, J.W., Jia, N.Q., et al. (2009) Chitosan-coated magnetic nanoparticles as carriers of 5-fluorouracil: Preparation, characterization and cytotoxicity studies. Colloids and Surfaces B: Biointerfaces, 68: 1-6.

[5] Halbreich, A., Groman, E.V., Raison, D., et al. (2002) Damage to the protein synthesizing apparatus in mouse liver in vivo by magnetocytolysis in the presence of hepatospecific magnetic nanoparticles. Journal of Magnetism and Magnetic Material, 248: 276-285.

[6] Wang, X.D., Wu, P., Jin, S.P., et al. (2014) Preparation of Fe_3O_4/Chitosan magnetic composite particle loading drug by co-precipitation in situ and its release behavior in vitro, Acta Materiae Compositae Sinica, 31: 158-165.

[7] Chu, M.H., Chen, F., Li, X.Q., et al. (2015) Preparation of Fe_3O_4/carboxyl chitosan magnetic composite particle loading drug, Journal of Functional Materials, 19: 19144-19147.

IOP Conf. Series: Materials Science and Engineering **729** (2020) 012092 doi:10.1088/1757-899X/729/1/012092

Direct Synthesis of Glycol Ethyl Ethers from Glycol with Modified HZSM-5 as Catalyst

Gangsen Li[1], Xing Wen[1], Junwen Wang[1] and Yadong Zhang[2]

[1] College of Chemistry and Chemical Engineering, Taiyuan University of Technology, Taiyuan, Shanxi, China;

[2] School of Chemical Engineering and Energy, Zhengzhou University, Zhengzhou 450001, China

*Corresponding author's e-mail: lgs_chem@163.com

Abstract. Glycol ethyl ethers possess many excellent properties and are widely used in paint industry, chemical industry and so on. In this paper, glycol ethyl ethers were synthesized from glycol with the modified HZSM-5 as catalyst. The modified HZSM-5 was characterized by XRD, SEM, EDS, BET and NH_3-TPD. The results showed that the modification did not change the structure and morphology of HZSM-5, sulfur was adsorbed on the surface of catalyst and strong acid sites appeared. The influence of ratio of the reactants, reaction temperature, reaction pressure, and time on the yield and selectivity was studied. The optimal reaction condition was performed with the molar ratio of glycol and ethanol 1:4 under 7MPa pressure at 453K for 4 h. Under these conditions, the conversion of glycol was 68.12%, and the selectivity of glycol ethers could be up to 98.61%.

1.Introduction

Glycol ethyl ethers, including glycol monoethyl ether and glycol diethyl ether, were important chemical material. It can be used as solvent, jet fuel, deicing agent and chemical intermediate etc. [1, 2] Traditionally, glycol ethers were synthesized through addition reaction of epoxy ethane [3-10]. There are limitations to this method owing to flammability and explosibility of epoxy ethane. Syngas method was also not ideal for the high reaction pressure, low yield and low selectivity of glycol ethers [11, 12].

Although ethylene glycol (EG) is a safe row material for synthesis of glycol ethyl ethers, it is also derived from petroleum in traditional method. With the industrialization of preparation of EG from coal [13-17], production of EG will increase a lot in the future and synthesis of glycol ethers from EG will have advantages in price [18] and more safety than epoxy ethane method. However there are few literatures about synthesized of glycol ethers from EG. Tomoharu Oku synthesized glycol monomethyl ether from glycol with Cs-P-Si mixed oxide as catalyst, but the conversion of glycol and selectivity of glycol monomethyl ether were low [19]. Yadong Zhang also prepared glycol ether from EG which was catalyzed by aluminium chloride and diammonium hydrogenphosphate, but it is also less desirable for the short life and difficult recovery of the catalysts [20]. Junfeng Liu synthesized diethylene glycol ethyl ether from diethylene glycol and ethanol catalized by heteropoly acid, the conversion of diethylene glycol was 92% and the selectivity for diethylene glycol ethyl ether was 73.5% [21]. Owing to unique pore structure controlled acidity and high hydrothermal stability, ZSM-5 has became one of the most widely studied and used catalysts. Furthermore, the intrinsic micropore size of ZSM-5 is similar to that of small organic molecules, which offer the benefits shape to selectivity of glycol ethers in the micropore space. Those characters make ZSM-5 become one of the most important shape-selective catalysts, which are widely used in alkylation, isomerization, cracking, disproportionation and other processes [22]. In this paper, modified HZSM-5 catalysts were firstly prepared, characterized and then employed for the direct synthesis of glycol monoethyl ether from EG and ethanol. In order to obtain optimal reaction conditions, single factor experiment was conducted. Finally the reusability of the modified catalyst was tested.

2. Results and discussion

2.1 Optimization of reaction conditions

With the desired catalysts in hand, the effects of reaction temperature, feed ratio, pressure, time and the concentration of modification were investigated respectively by means of single factor experiment method. The results were shown in Table 1-5.

2.1.1 Effect of concentration of modification on the performance of catalysts. In order to explore the influence of concentration of modification on the performance of catalysts, HZSM-5 was firstly modified by ammonium sulfate aqueous solution from 0.1 mol•L^{-1} to 0.7 mol•L^{-1}. The catalytic performances were measured under the following reaction conditions: reaction pressure was 7 MPa, reaction temperature was 473K, reaction time was 4 hours and the molar ratio of EG to ethanol was 1:3. The result was shown in Table 1.

Table 1. Effect of modifier concentration on catalytic ability

Concentration of modifier /mol·L^{-1}	$Conv_1$/%	Sel_2/%	Sel_3/%	$Sel_{(2+3)}$/%
0.1	47.2	69.0	10.4	79.4
0.3	53.8	71.5	17.5	89.0
0.5	69.1	61.5	26.6	88.1
0.7	66.5	62.2	24.9	87.1

Reaction conditions, n (EG): n (ETOH)=1:3, 473K, 4h, 7MPa, the loading of catalyst 0.6%.

As shown in Table 1, when the concentration less than 0.5 mol•L^{-1}, the conversion of EG and the selectivity of ethylene glycol diethyl ether increased with the increasing of modification concentration, but the selectivity of ethylene glycol monoethyl ether slightly decreased. The total selectivity of ethylene glycol ethers almost kept unchanged when the concentration is larger than 0.3 mol•L^{-1}. When the modification concentration reached 0.5 mol•L^{-1}, the yield of glycol ethyl ethers reached the top. Based on these results, the optimum concentration of ammonium sulfate was 0.5 mol•L^{-1}.

2.1.2 Effect of reaction temperature on the reaction results. The effect of reaction temperature on the conversion of EG, selectivity and yield of product 2 and 3 were investigated and the results were summarized in Table 2.

Table 2. Effect of reaction temperature on the results

Reaction temperature/ K	$Conv_1$/%	Sel_2/%	Sel_3/%	$Y_{(2+3)}$ /%
473	75.74	55.16	35.54	68.70
453	68.77	64.20	28.24	63.57
433	59.91	67.63	27.84	57.20
413	53.39	72.43	20.17	49.44
393	6.45	92.17	00.00	5.94

Reaction conditions: n (EG): n (ETOH) =1:3, 3h, 7MPa, the loading of catalyst 0.6%.

Table 2 showed that the conversion of EG and the yield of glycol ethyl ethers increased with the increasing of reaction temperature, but the selectivity of glycol ethyl ethers decreased. Although the yield of glycol ethyl ethers could reach the highest value (68.70%) at 473 K, the selectivity of glycol

IOP Conf. Series: Materials Science and Engineering **729** (2020) 012092 doi:10.1088/1757-899X/729/1/012092

ethyl ethers was the lowest, which means the appearance of more byproducts. It also could be seen that the yield of glycol ethyl ethers was 63.57% at 453 K, which was only lower than at 473 K, but the conversion of EG was much higher than at 433 K, so that 453 K was the optimum reaction temperature.

2.1.3 Effect of reaction time on the reaction results. The dependence of EG conversion, products selectivity and yield on reaction time were investigated, the results were shown in Table 3.

Table 3. Effect of reaction time on the results

Reaction time/h	$Conv_1$/%	Sel_2/%	Sel_3/%	$Y_{(2+3)}$/%
8	75.63	48.57	38.12	65.56
6	75.66	49.26	39.29	66.99
4	74.92	55.43	37.50	69.62
2	66.72	67.19	29.05	64.21
1	67.63	64.64	29.35	63.56

Reaction conditions: n (EG): n (ETOH)=1:4, 473K, 7MPa, the loading of catalyst 0.6%.

It could be seen from Table 3 that reaction time had great influence on the conversion of glycol and selectivity of glycol monoethyl ether and glycol diethyl ether. When the reaction time was 4 h, the conversion of glycol reached the peak value. Meanwhile, the selectivity of glycol ethyl ethers increased firstly, but then decreased with the reaction time prolonged. The optimum reaction time was 4 h.

2.1.4 Effect of reaction pressure on the reaction results. The effect of reaction pressure was also studied by single factor experiment method, and the results were listed in Table 4.

Table 4. Effect of reaction pressure on the results

Reaction pressure (MPa)	$Conv_1$/%	Sel_2/%	Sel_3/%	$Y_{(2+3)}$/%
2	20.29	90.92	2.49	18.95
3	39.57	65.41	30.54	37.97
4	46.38	64.92	32.65	45.26
5	51.07	65.81	31.96	49.93
6	54.82	64.39	31.57	52.6
7	60.37	63.99	32.59	58.31

Reaction conditions: n (EG): n (ETOH)=1:3, 3 h, 413 K, the loading of catalyst 0.6%.

It was evident from Table 4, better conversion of EG and higher yield of glycol ethyl ethers could be obtained with the increasing of reaction pressure, but selectivity of glycol monoethyl ether and glycol diethyl ether almost kept unchanged when the reaction pressure was higher than 3 MPa. At the same reaction temperature, the lower pressure and the less liquid ethanol in the autoclave, there were not enough effective contact of ethanol with the catalyst. Less contact will lead to less collisions of activated ethanol molecule with glycol, resulting in low conversion of EG. For the same reason, less collisions of activated ethanol molecule with glycol monoethyl ether, less glycol monoethyl ether converted to glycol diethyl ether, resulting in lower glycol diethyl ether yield and higher selectivity of glycol monoethyl ether. On the contrary, the higher pressure the more liquid ethanol in the autoclave, the conversion of EG and yield of glycol diethyl ether increased with the increasing of reaction pressure. The pressure of 7 MPa was suitable for this transformation.

2.1.5 Effect of feed ratio on the reaction results. In order to investigate the effect of feed ratio of EG and ETOH, the values of n(EG):n(ETOH) were varied from 1:2 to 1:6 in the following experiments, keeping the other reaction conditions unchanged. The results were summarized in Table 5.

Table 5. Effect of feed ratio on the results

n(EG):n(ETOH)	$Conv_1$/%	Sel_2/%	Sel_3/%	$Y_{(2+3)}$ /%
1:2	56.42	65.77	25.69	51.60
1:3	61.58	64.93	30.23	58.59
1:4	65.37	63.99	32.59	63.13
1:5	67.89	54.53	42.32	65.75
1:6	68.06	56.55	40.41	65.99

Reaction conditions: 433 K, 4 h, 7 MPa, the loading of catalyst 0.6%.

As indicated in Table 5, the higher feed ratio of ethanol to EG could afford better conversion of EG and higher selectivity of glycol diethyl ether, but lower selectivity of glycol monoethyl ether. Increasing the amount of ethanol will bring more activated ethanol, which could react with EG better, resulting in higher EG conversion and more glycol monoethyl ether. Then the glycol monoethyl ether further reacted with activated ethanol to generate more glycol diethyl ether causing higher selectivity of glycol diethyl ether and lower selectivity of glycol monoethyl ether. When the feed ratio of ethanol to EG was higher than 3, the total selectivity of glycol ethyl ethers kept stable. However, too much ethanol would cause extra difficulty in the separation of products. Based on these observation, 1:4 was chosen as the optimum feed ratio of EG to ethanol.

2.1.6 Reusability of the catalyst. The reusability of the catalyst is important for the large scale operation from industrial point of view. The reusability was evaluated by operation under the optimized conditions (1:4, 4 h, 453 K, 7 MPa, the quantity of catalyst 0.6%). After every run, the catalyst was recovered by filtration from crude reaction mixture, and calcined at 773 K for 2 h. The results were shown in Figure 1.

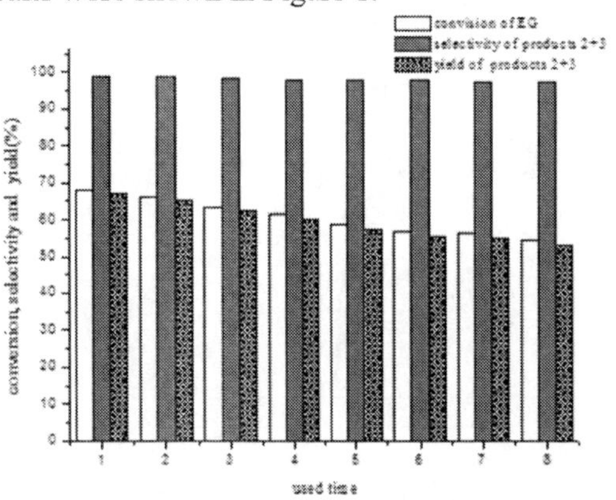

Fig. 1 Reused times of the catalyst affected the conversion of EG, selectivity and yield of glycol ethers

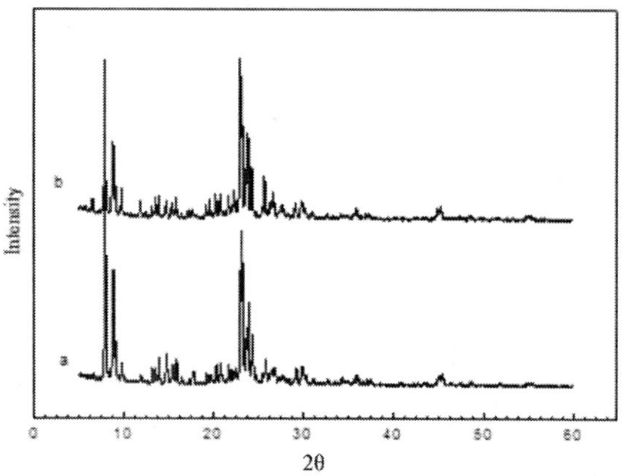

Fig.2 XRD patterns of (a) the unmodified HZSM-5, (b) the modified HZSM-5

Figure 1 showed that the conversion of glycol and the yield of glycol ethyl ethers decreased slightly with the reused time increasing, but the selectivity of glycol ethyl ethers was almost unchanged. This suggested that the catalyst has high stability and excellent reusability.

2.2 Characterization of catalysts

2.2.1 XRD of catalyst. The XRD patterns of unmodified and modified HZSM-5 were shown in Figure 2.The XRD patterns of unmodified HZSM-5 (a) and modified HZSM-5 (b) showed intense and sharp lines values at 7.9°, 8.8°, 23.1°, 23.3°, 23.9°. These signals are characteristic of ZSM5-type zeolites, and they are agreement well with the XRD patterns of ZSM-5 in the Ref. [23]. That means the modification does not affect the structure of HZSM-5.

2.2.2 SEM and EDS. It is well known that the size and the relative content of surface elements of the catalyst have strong effect on their catalytic performance. Figure 3 is the image and EDS of the catalyst.

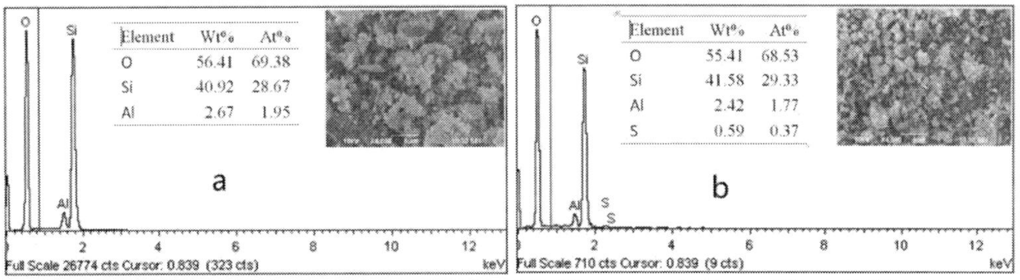

Fig.3 SEM and EDS of (a) unmodified HZSM-5, and (b) modified HZSM-5

As indicated in Figure 3, the morphology of catalysts had not changed after modification. The particle size was still between 1~5μm. The EDS showed that there were only O, Si, Al elements on the surface of unmodified HZSM-5, and the S element was introduced into the surface of catalyst after modification, which was the reason for the appearance of strong acid sites as shown in Figure 4.

2.2.3 Specific surface area and pore structure. The quantitative comparison of the BET surface area and pore size of catalysts were given in Table 6.

Table 6. BET surface pore volume and pore size

Catalysts	$S_{BET}/m^2\cdot g^{-1}$	pore volume/$cm^3\cdot g^{-1}$	pore size/nm
Unmodified HZSM-5	340.36	0.2705	3.1790
Modified HZSM-5	292.87	0.2259	3.0859

As shown in Table 6, the BET surface area, the pore volume and the pore size decreased obviously after modification, which means the modification was not only stick to the surface of catalyst but also get into the inner surface of the pores, resulting in the decrease of the above three parameters.

2.2.4 NH_3-TPD. The acid property of unmodified and modified catalysts were measured by NH_3-TPD experiments, and was shown in Figure 4.

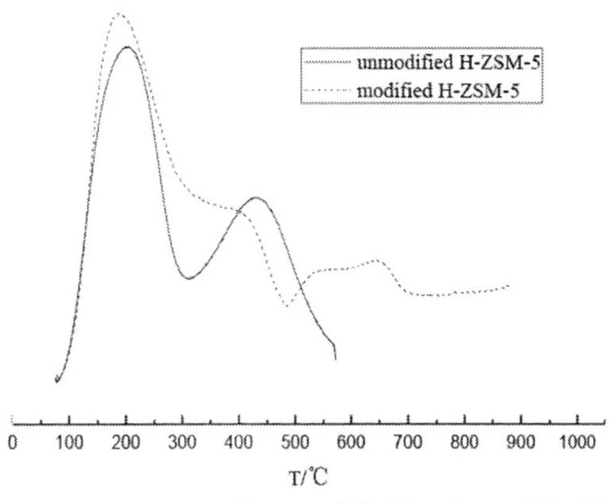

Fig. 4 NH_3-TPD patterns of unmodified and modified HZSM-5

The NH_3-TPD profiles showed that unmodified HZSM-5 had 2 broad desorption feature while the modified had 3 broad desorption feature, it could be assigned to the desorption of physisorbed NH_3 from acid sites. After NH_3 was adsorbed 1 h on the samples at 353 K, the desorption temperatures of unmodified HZSM-5 at 373 K~573 K and 583 K~853 K, and the modified one were monitored at 373 K~573 K, 573 K~753 K and 753 K~963 K respectively, which means that modified HZSM-5 had higher acidity than the unmodified. According to our previous research [21], higher conversion of EG could be obtained with higher acidic catalyst, which means that the catalytic activity could improve with the enhancement of acidity.

3. Experimental

3.1 Material and reagents

All chemicals used in this study were commercially available. They were used without further purification unless otherwise mentioned. HZSM-5 was obtained from the Catalyst Plant of Nankai University, ammonium sulfate was supplied by Sinopharm Chemical Reagent Co., Ltd., EG and ethanol were purchased from Tianjin Fengchuan Chemical Reagent Co., Ltd.

3.2 Catalysts preparation

The catalysts were prepared by excessive impregnation. Firstly, ammonium sulfate were dissolved in deionized water, and the concentrations were 0.1 to 0.9 $mol \cdot L^{-1}$, HZSM-5 with the mass of 1 g was dipped into different aqueous, then stirred for 8 h at 353 K, filtered the suspension, finally dried for 8 h at 373 K, and calcined at 773 K for 6 h.

3.3 Characterization of the catalysts

X-ray diffraction (XRD) was obtained by using a BDX-3200 with Ni-filtered Cu Kα radiation. The X-ray tube was operated at 40 kV, 20 mA, and 2θ was ranged from 5° to 60°. BET surface areas and pore structure were measured by N_2 adsorption at the temperature of liquid nitrogen using a Micromeritics Chemisorb 2720 (U.S.A) with 0.10 g catalyst and support sampled respectively. Before measurement, every sample was heated under ultrapure He atmosphere at a flow rate of 30 $mL \cdot min^{-1}$ at 773 k for 1 h to remove surface adsorbate, cooled to 353 K and saturated with 5% NH_3-He before flushed with He to remove physisorbed NH_3. The effluent of NH_3-TPD was quantified by a flame ionization detector (FID) under a heating rate of 10 $K \cdot min^{-1}$. The morphology and size of the samples were identified by scanning electronic microscopy (SEM), which was conducted on a JEM-100CXII instrument combined with energy dispersive spectrometer (EDS).

3.4 General procedure of synthesis of glycol ethyl ethers

Glycol ethyl ethers were synthesized in a 250 mL high-pressure autoclave (made in Weihai, China) equipped with a magnetically driven impeller. In a typical run, the reactor were charged with 100 g mixture of EG and ethanol, 0.5 g catalyst (modified HZSM-5). The reactor was pressurized with N_2 to 3 MPa and then was heated to 433 K and stirred for a designated period of time. Then the reactor was cooled quickly to a low temperature by placing in an ice/water bath, and the pressure was released slowly. The reaction mixture was filtered and the catalyst was washed by ethanol for reusing. The remained liquid mixture was weighted and analyzed by gas chromatography (GC). The possible reaction pathways of EG were shown in Scheme 1. The conversion of EG and selectivity of product 2 and 3 were calculated by the Formulas.

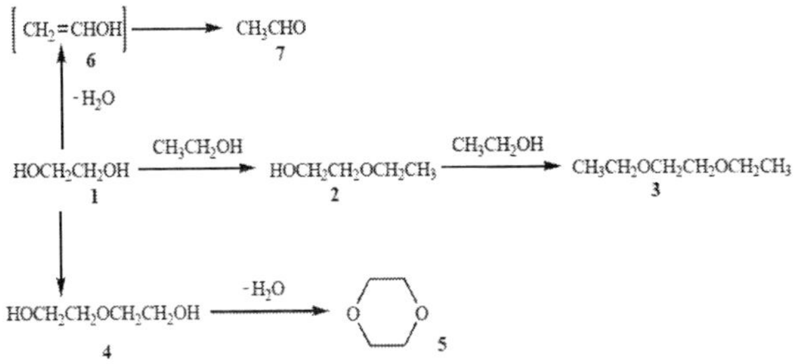

$$Conv_1 = \frac{M_{0EG} - M_{1EG}}{M_{0EG}} \times 100\%$$

$$Sel_2 = \frac{M_2}{M_{0EG} - M_{1EG}} \times 100\%$$

$$Sel_3 = \frac{M_3}{M_{0EG} - M_{1EG}} \times 100\%$$

M_{0EG}, mole percent of EG before reaction (mol•kg^{-1}); M_{1EG}, mole percent of EG after reaction (mol•kg^{-1}); M_2, mole percent of product 2 after reaction (mol•kg^{-1}); M_3, mole percent of product 3 after reaction (mol•kg^{-1}).

4. Conclusion

It can be seen from the experiment, the optimal reaction condition was performed with the molar ratio of glycol and ethanol 1:4 under 7MPa pressure at 453K for 4 h by modified HZSM-5 as catalyst. Under these conditions, the conversion of glycol was 68.12%, and the selectivity of glycol ethers could be up to 98.61%.

Acknowledgments

We thank the National Natural Science Foundation of China (81101748);
Natural Science Foundation of Shanxi Province (201801D121058).

References

[1] Chen Y.F., Qin Y.S., Market and production technology advances of ethylene glycol ether. Chin. J. Adv. Fin. Petrochem., 2002, 3, 28-34.
[2] Liu J.F., New progress on technology for synthesis of glycol ethers, Chin. J. Special. Petrochem., 1997, 1, 53-56.
[3] Wang Y. J., Liu J. L., Li W. Z., Etherification of ethanol with ethylene oxide by heteropolyacid catalysts, Chin. J. Petrochem. Tech., 1999, 28, 95-99..
[4] Wang Y.J., Liu J.L., Li W.Z., Synthesis of 2-butoxy ethanol with narrow-range distribubution catalysed by supported heteroplyacid, J. Mol. Catal. A: Chem, 2000, 159, 71-75.
[5] Du X.L., Li S.Z., Zhang D.S., Improvement of DH-1 montmorillonite solid acid catalysts, Chin. J. Special. Petrochem., 2013, 30, 23-24.
[6] Dong W.Y., Li H.H., Gao F., Liu S.Q., Guo J.M., Liu J.F., Guo Z.J., Solid acid catalyst suitable for producing glycol ethers, P. CN1005133B. 1985.
[7] Dong W.Y., Gao T.N., Wang J.Z., Liu S.Q., Li H.H., Solid strengthened catalyst for preparation of glycol ether, P. CN1033742C, 1997
[8] Laemmle G.J., Rightmer K.C., Anhydrous catalyst for manufacture of glycol ethers, P.US3972948, 1976
[9] Calcagno B., Cocuzza G., Torreggiani G., Process for the production of glycol ethers, P.US3935279, 1976.
[10] King S.W., Heterogeneous alkoxylation using anion-bound metal oxides, P. US4727199A, 1988.
[11] Roger G.D., John F. K., Process for low pressure synthesis of ethylene glycol from synthesis gas, P. US4565896 A, 1986.
[12] John F.K., Process for preparing ethylene glycol monoalkyl ethers, P. US4357477 A, 1982.
[13] Tahara S., Fujii K., Nishihira K., Matsuda M., Mizutare K., Process for continuously preparing ethylene glycol, P. US4453026 A, 1984.
[14] Bartley W.J., Process for the preparation of ethylene glycol, P. US4677234, 1987.
[15] Thakur D.S., Roberts B.L., Sullivan T.J., Vichek A.L., Hydrogenation catalyst, process for preparing and process of using said catalyst, P. US5345005, 1994.

[16] Liao X.Z., Lu L., Ning C.L., Dai C.Y., Gao Z., Zhang C.L., Recent progress in the catalysts for methyl glycolate synthesis via dimethyl oxalate hydrogenation, Chin. J. Chem. Ind. Eng. Prog., 2011, 30, 2349-2356.
[17] Chen Q.L., Yang W.M., Teng J.W., Recent advances in coal to chemicals technology developed by SINOPEC, Chin. J. Catal., 2013, 34,217-224.
[18] Chen Y.D., Li G.F., The technological progress of synthesizing glycol using coal to replace petro-ethylene, J. Univ. Sci. Technol. Chin., 2009, 39, 1-10.
[19] Tomoharu O., Yoshitaka A., Hideaki T., Takao I., Continuous chemoselective methylation of functionalized amines and diols with supercritical methanol over solid acid and acid-base bifunctional catalysts, J. Am. Chem. Soc., 2004, 126, 7368-7377.
[20] Zhang D., Li G.S., Zhang Y. D., Synthesizing Ethylene Glycol Monoethyl Ether Directly from Ethanol and Ethylene Glycol, J. Chem. Eng. Chin. Univ., 2014, 28, 1091-1096.
[21] Liu J. F., Liu Y., Yi P. G., Synthesis of Diethylene Glycol Ethyl Ether with Diethylene Glycol and Ethanol by the Catalysis of Heteropoly Acid, Appl. Catal. A: Gen. 2004, 277, 167- 171.
[22] Sazama P., Sobalik Z., Dedecek J., Jakubec I., Parvulescu V., Bastl Z., Rathousky J., Jirglova H., Enhancement of activity and selectivity in acid catalyzed reactions by dealuminated hierarchical zeolites. Angew. Chem. Int. Ed.., 2013, 52, 2038-2041.
[23] Marta M., Husley M., Luis M., Angela B. S., Alberto A., Aura L., Delfin M., Roger S., Joaquín L. B. Microwave-assisted polyol synthesis of Pt/H-ZSM5 catalysts, Micropor. Mesopor. Mat., 2010, 131, 342–349.

Preparation of polyacrylonitrile based carbon nanofibers by electrospinning with different additives from direct coal liquefaction residue

Tongxin Qiao, Haoxing Wang, Yingjie Yang, Xiangna Lu, Guixia Fan, and Peng Li*

School of Chemical Engineering, Zhengzhou University, Zhengzhou 450001, China

*Corresponding author's e-mail: lipeng1987040@126.com

Abstract. Direct coal liquefaction residue is of valuable materials for preparation of carbon materials due to its high carbon content and aromatic structures. In this work, extract, extraction residue, and oxidized extraction residue from direct coal liquefaction residue were employed as additives for preparation of polyacrylonitrile based carbon nanofibers by electrospinning, followed by air stabilization and carbonization processes. The results show that the additives were evenly dispersed in the carbon nanofibers, and the diameters of nanofibers were smaller than that pure polyacrylonitrile based carbon nanofibers. After the extraction residue was pre-treated with hydrogen peroxide, the morphology of nanofibers became more smooth and uniform.

1. Introduction

Direct coal liquefaction residue (DCLR), a coal by-product but also a valuable carbon resource from direct coal liquefaction process [1], is of high carbon content and rich in condensed and heteroatom-containing aromatic structures. Direct utilization of CLR was investigated by gasification [2] and combustion [3]. However, high ash and sulphur content in CLR is considered as the precursors of environmental contaminator when CLR was used directly as fuels. For example, the nitrogen and sulfur are emitted as relevant oxides (NOx and SOx) during coal combustion, causing serious environmental pollutions [4]. The mineral materials in CLR would be transferred into the atmosphere as the toxic substance or hazardous particles during combustion, such as fine particular matter (PM2.5), causing serious health problem [5, 6]. Therefore, efficiently and environmental friendly utilization of the direct coal liquefaction residue from direct coal liquefaction process is an important research topic. The DCLR can be used as raw materials for acquiring valuable polycyclic aromatic hydrocarbons (PAHs) [7, 8], and it was also a good precursor for preparation of carbon materials, such as carbon microfiber [9], mesoporous carbons [10], carbon nanofibers/carbon foam monolithic composite [11], hierarchical porous carbons [12].

Electrospinning is one of the efficient approaches to prepare nanofiber materials, due to the wide source of raw materials, simple operation process and low cost [13]. Mooste [14] *et al.* reported that non-precious metal cathode catalysts prepared from polyacrylonitrile (PAN) based composite fibers have great advantages in low-temperature fuel cell applications. Jayawickramage [15] investigated the mixture of PAN and lignin for preparation of carbon nanofiber by electrospinning. It was concluded that heat and activation treatments can exhibit strong specific surface area, electrical conductivity and mesoporosity of nanofibers. Moreover, good performance of the nanofibers was also detected when it was used as the supercapacitor electrode.

IOP Conf. Series: Materials Science and Engineering **729** (2020) 012093 doi:10.1088/1757-899X/729/1/012093

In this work, we prepared the spinning solution by adding the modified direct coal liquefaction residue as the additives to PAN for electrospinning, to achieve the efficient utilization of coal liquefaction residue, and also provide a reference for decreasing the cost of PAN based carbon nanofibers.

2. Experiment

2.1 Materials

The DCLR was obtained by liquefying Shenmu-Fugu subbituminous coal at 455 °C under 19 MPa of hydrogen in a pilot-scale direct coal liquefaction unit operated by Shenhua Group Corporation Ltd., China. It was pulverized to < 74 μm followed by desiccation in a vacuum at 80 °C for 24 h prior to use. The DLCR was conducted into a Soxhlet extractor, followed by solvent extraction with isometric acetone/CS_2 mixed solvent, to afford the extract (E) and extraction residue (ER). The ER was oxidized by using hydrogen peroxide at 40 °C for 6 h, followed by filter and dried, to prepare oxidized ER (OER). All the reagents used in the experiments are analytical grade, and all the organic solvents were distilled with a Büchi R-300 rotary evaporator prior to use.

2.2 Preparation of carbon nanofibers

The PAN (2 g) with none additive was ultrasonically dissolved in 20 mL of *N,N*-dimethylformamide (DMF) to afford the spinning solution labelled as P-N. Then PAN (1.8 g) with additive of E (0.2 g), ER (0.2 g), and OER (0.2 g) from DCLR were also separately dissolved in DMF (20 mL), to afford the spinning solutions labelled as P-E, P-ER, and P-OER, respectively. The mass ratio of additive to PAN is 1/9. Then, 5 mL of the solution was transformed into a syringe with a needle diameter of 0.5 mm. The solution was spun into nanofibers through a positively charged needle by using an electrospinning apparatus at 18 kV and a negatively charged aluminum foil at -3.00 kV to collect the fibers with a spinning distance of 15 cm, injection speed of 0.1 mm/min, and room temperature. The carbon nanofibers precursors were collected from the aluminum foil, then placed in an air atmosphere at the heating rate of 2 °C/min up to 250 °C, holding for 1 h. Finally, carbon nanofibers were obtained by carbonizing at 450 °C for 2 h under Ar.

The surface of the carbon nanofibers and their precursors were characterized by a Nova Nano SEM450 Ultrasonic Field Emission Scanning Electron Microscope (SEM) provided by FEI.

3. Results and discussion

3.1 SEM analysis of carbon nanofibers and their precursors

Table 1. Distributions on the diameter (nm) of carbon nanofibers and their precursors

	P-N	P-E	P-ER	P-OER
Carbon nanofibers precursors	344.2-395.2	251.8-298.5	414.1-426.9	255.9-302.4
Carbon nanofibers	287.2-303.4	213.8-224.4	234.3-259.0	170.9-226.4

As shown in Table 1 and SEM images of P-N, P-E, P-ER, and P-OER in Figure 1, it is clear that the pristine carbon nanofibers show regular fibrous morphology with diameters mostly in the range of 251.8−426.9 nm, by adding E, ER, and OER from DCLR as additives in to PAN. Carbon nanofibers precursors with smaller diameters were obtained by adding E and OER, but larger diameters by adding ER. That is because both of the E and OER were well dissolved in DMF as well as PAN to form a homogeneous solution, thereby having good spinnability during electrospinning. The ER from DCLR has larger average molecular mass and poor solubility in DMF. In contrast to the P-ER, the diameters of the nanofibers of P-OER were more uniform, indicating that oxidation can improve the spinnability of ER with PAN.

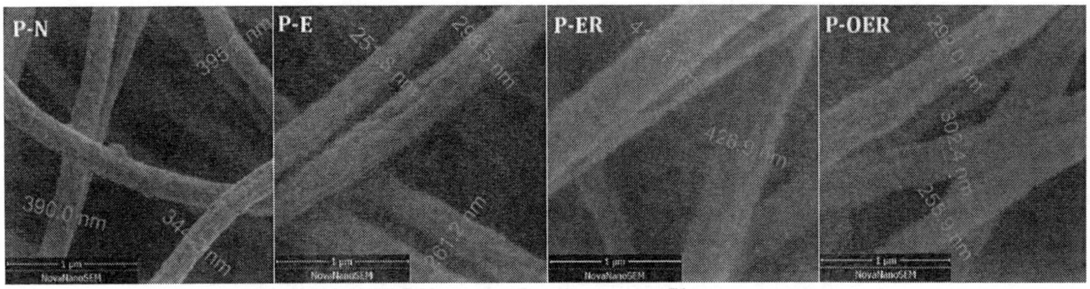

Figure 1. SEM images of carbon nanofibers precursors

As displayed in Table 1 and Figure 2, the surface of the nanofibers became more smooth after carbonization treatment, and the diameters of the carbon nanofibers were also reduced. The diameter of carbon nanofibers has smaller diameters with additives than that pure PAN. Carbon nanofibers with smaller diameter and more smooth surface were prepared with additive of E than that of ER. The P-ER had the smallest average diameter and most regular fibrous morphology than the others, indicating the oxidation treatment shows positive effect on the morphology and diameter of carbon nanofibers.

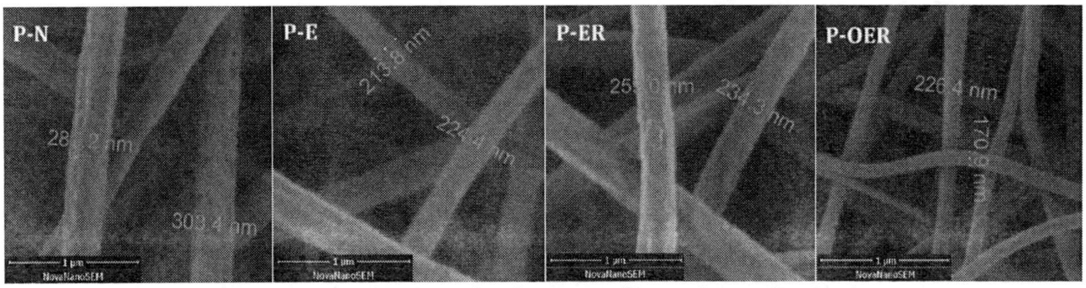

Figure 2. SEM images of carbon nanofibers after carbonization

3.2 Raman spectroscopy analysis of carbon nanofibers

Table 2. Parameters of carbon nanofibers by Raman analysis

Sample	*D* peek (cm^{-1})	*G* peek (cm^{-1})	*2D* peek (cm^{-1})	I_D/I_G
PA	1360	1581	2725	1.1227
PC-R	1359	1574	2720	1.0963
PC-O	1356	1578	2718	1.0715
PC-E	1360	1584	2719	1.0486

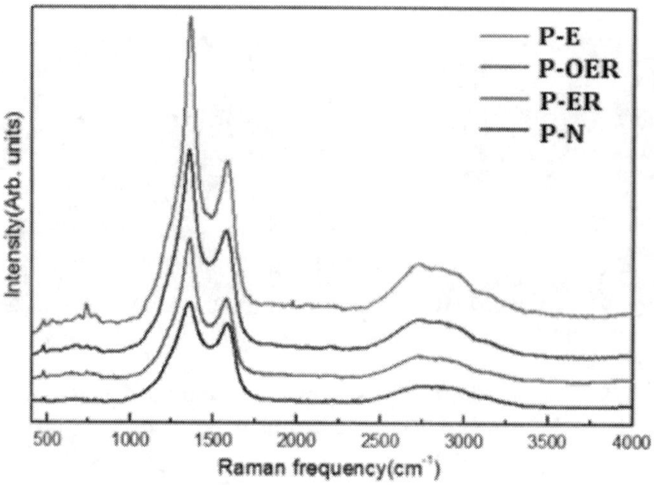

Figure 3. Raman spectrograms of carbon nanofibers

According to the Raman spectroscopy fitting analysis, the Raman spectrograms and I_D/I_G values of P-N, P-E, P-ER, and P-OER were illustrated in Table 2 and Figure 3. After incorporating the modified raw materials, the graphitized degree of carbon nanofibers is significantly improved. There are two possible reasons: one is that participation of modified raw materials reduces the proportion of carbon atoms in the disordered structure. The other is that some components in the modified raw materials promoted the high-temperature carbonization process. In addition, the appearance of 2*D* peak indicates that graphene exists in the nanofibers after carbonization.

4. Conclusions

The E, ER, and OER from DCLR were successfully added into the PAN, thereby forming the uniform nanofibers. Adding DCLR's components into PAN as additives can obviously improve the morphology of carbon nanofibers, and the oxidation treatment provided positive effect on the morphology and diameter of carbon nanofibers.

Acknowledgements

This work was subsidized by the National Natural Science Foundation of China (Grant 21606210 and 51704263), Postdoctoral Foundation of Henan Province, and Training Program of Innovation and Entrepreneurship for Undergraduates of Henan Province (Grant 201910459142).

References

[1] Hirano, K. (2000). Outline of NEDOL coal liquefaction process development (pilot plant program). *Fuel processing technology*, *62*(2-3), 109-118.

[2] Chu, X., Li, W., Li, B., & Chen, H. (2008). Sulfur transfers from pyrolysis and gasification of direct liquefaction residue of Shenhua coal. *Fuel*, *87*(2), 211-215.

[3] Zhou, J. H., Fang, L., Chen, J., Liu, J. Z., Zhou, Z. J., & Cen, K. F. (2005). Dynamic sulfur emission from combustion of blends of coal liquefaction residue and biomass. *Journal of Fuel Chemistry and Technology*, *33*(5), 626.

[4] Kambara, S., Takarada, T., Toyoshima, M., & Kato, K. (1995). Relation between functional forms of coal nitrogen and NOx emissions from pulverized coal combustion. *Fuel*, *74*(9), 1247-1253.

[5] Liu, J., Mauzerall, D. L., Chen, Q., Zhang, Q., Song, Y., Peng, W., ... & Lin, W. (2016). Air pollutant emissions from Chinese households: A major and underappreciated ambient pollution source. *Proceedings of the National Academy of Sciences*, *113*(28), 7756-7761.

[6] Wong, C. M., Tsang, H., Lai, H. K., Thomas, G. N., Lam, K. B., Chan, K. P., ... & Thach, T. Q. (2016). Cancer mortality risks from long-term exposure to ambient fine particle. *Cancer Epidemiology and Prevention Biomarkers*, *25*(5), 839-845.

[7] Li, P., Wei, X. Y., Sun, X. H., Lu, Y., Zong, Z. M., Mukasa, R., ... & Fan, X. (2013). The isolation of condensed arenes from Shenmu-Fugu coal liquefaction residue. *Energy Sources, Part A: Recovery, Utilization, and Environmental Effects*, *35*(23), 2250-2256.

[8] Li, P., Zhao, Y. Q., & Fan, G. X. (2018). Quantitative analysis of alkanoate and condensed arenes in the extracts from direct coal liquefaction residue by ultrasonication-assisted solvent extraction using alcohols. *Energy Sources, Part A: Recovery, Utilization, and Environmental Effects*, *40*(10), 1266-1272.

[9] Zhou, Y., Xiao, N., Qiu, J., Sun, Y., Sun, T., Zhao, Z., ... & Tsubaki, N. (2008). Preparation of carbon microfibers from coal liquefaction residue. *Fuel*, *87*(15-16), 3474-3476.

[10] Zhang, J., Jin, L., Zhu, S., & Hu, H. (2012). Preparation of mesoporous activated carbons from coal liquefaction residue for methane decomposition. *Journal of Natural Gas Chemistry*, *21*(6), 759-766.

[11] Xiao, N., Zhou, Y., Qiu, J., & Wang, Z. (2010). Preparation of carbon nanofibers/carbon foam monolithic composite from coal liquefaction residue. *Fuel*, *89*(5), 1169-1171.

[12] Zhang, J., Jin, L., Cheng, J., & Hu, H. (2013). Hierarchical porous carbons prepared from direct coal liquefaction residue and coal for supercapacitor electrodes. *Carbon*, *55*, 221-232.

[13] Liu, Q., Zhu, J., Zhang, L., & Qiu, Y. (2018). Recent advances in energy materials by electrospinning. *Renewable and Sustainable Energy Reviews*, *81*, 1825-1858.

[14] Mooste, M., Kibena-Põldsepp, E., Vassiljeva, V., Merisalu, M., Kook, M., Treshchalov, A., Kisand, V., Uibu, M., Krumme, A., Sammelselg, V., & Tammeveski, K. (2019). Electrocatalysts for oxygen reduction reaction based on electrospun polyacrylonitrile, styrene–acrylonitrile copolymer and carbon nanotube composite fibres. *Journal of Materials Science*, *54*(17), 11618-11634.

[15] Jayawickramage, R. A. P., Balkus Jr, K. J., & Ferraris, J. P. (2019). Binder free carbon nanofiber electrodes derived from polyacrylonitrile-lignin blends for high performance supercapacitors. *Nanotechnology*, *30*(35), 355402.

IOP Conf. Series: Materials Science and Engineering **729** (2020) 012094 doi:10.1088/1757-899X/729/1/012094

The role of DNA methyltransferases in smooth muscle cell phenotypic switch

Jinhua Ding[1], Hongmin Zhang[1], Ying Yuan[1], Liwen Huang[1], Nan Wang[1*]

[1] College of Biotechnology, Tianjin University of Science and Technology, P.R. China,300457

[*]Corresponding author's e-mail: wn929@tust.edu.cn

Abstract: Phenotypic switching of vascular smooth muscle cells (VSMCs) is a common pathological process of multiple cardiovascular diseases. DNA methylation plays a key role in atherosclerosis, but its mechanism is not clear. DNA methyltransferases (DNMT3a and DNMT1) could be involved in the development of phenotypic switching process of VSMCs. Therefore, the purpose of this research is to probe the action of DNA methyltransferases on phenotypic switching process of VSMCs. First, Treatment with platelet-derived growth factor-BB (PDGF-BB) induced human aorta VSMC (HA-VSMC) morphological changes from contractile to synthetic phenotype. EdU staining confirmed that PDGF-BB could induce HA-VSMC proliferation. PDGF-BB remarkably augmented the mRNA and protein level of proliferation marker gene CyclinD1 and osteopontin (OPN) in HA-VSMCs. The mRNA and protein level of differentiation marker SM22 and SM-α-actin were obviously upregulated in HA-VSMCs after PDGF-BB treatment. Next, the level of DNMT1 and DNMT3a were examined in PDGF-BB-treated HA-VSMCs by RT-PCR and western blotting. The high level of DNMT1 and DNMT3a was observed in PDGF-BB-induced proliferation model. Therefore, DNMT3a and DNMT1 could be involved in phenotypic switch of vascular smooth muscle cells.

1. Introduction

Phenotypic switch of vascular smooth muscle cells (VSMCs) plays an important role in the initial stage of vascular proliferative disease. In adult tissue, VSMCs presents a physiological contractile phenotype, which is characterized by high expression of contractile markers and decreased proliferation and migration of cells[1]. After the injury to the vessel, VSMCs undergo phenotypic modulation from a contractile phenotype back to a proliferative phenotype, which is related to the significant decreased selective marker gene expression and the promoted proliferation and migration.

Platelet-derived growth factor-BB (PDGF-BB) is considered to be the key regulator of VSMC phenotypic switching, which stimulates phenotype regulation from differentiated phenotype to proliferative phenotype[2]. PDGF-BB is one of the most effective mitogens and chemokines in VSMCs. PDGF-BB has also been showed to activate a variety of signal transduction pathways, including extracellular signal-regulated kinase (ERK), phosphatidylinositol 3-kinase (PI3K)/AKT and p38 mitogen-activated protein kinase (MAPK) pathways by binding with PDGF receptor [3].

Epigenetic modification, such as DNA methylation, plays an essential role in the regulation of gene expression. De novo methyltransferases DNMT3A and DNMT3B mainly catalyzs the methylation of unmodified DNA, and DNMT1 is essential to maintain CpG DNA methylation [4]. Hypomethylation of CpG islands in the promoter region leads to gene expression, while hypermethylation results in gene

IOP Conf. Series: Materials Science and Engineering **729** (2020) 012094 doi:10.1088/1757-899X/729/1/012094

silence. In this study, we investigated the role of DNA methyltransferase in phenotypic transformation of VSMCs[5].

2. Materials and methods

2.1. Cell culture
Human aortic VSMCs (HA-VSMCs) were preserved by our laboratory. HA-VSMCs were incubated in dulbecco's minimum essential medium-high glucose (DMEM-HG) containing 10% FBS and 1% penicillin-streptomycin, in humified air with 5% CO_2.

2.2. EDU assay
To explore the pro-proliferation role of PDGF-BB on HA-VSMCs, 1×10^5 cells were cultured with 40 ng/mL PDGF-BB for 24h, and then EdU (5-ethynyl-2'-deoxyuridine) incorporation test was carried out with Cell-Light EdU imaging detection kit. Mainly, the cells were incubated with EdU, followed by fixation and permeabilization. EdU was detected by click reaction using Alexa Fluor 488 azide. The nuclear DNA was visualized by DAPI. In the four random high power fields each well, the incorporation rate of EdU is expressed as the percentage of EdU positive nucleus in the total nucleus.

2.3. Real-time RT-PCR
Total RNA was extracted from VMSCs using Trizol reagent (Solarbio) according to the manufacturer's instructions. RNA was reverse-transcribed into cDNA with random primers using M-MLV reverse transcriptase. Thermal cycling conditions were as follows: 1 cycle at 95°C for 2 min, 40 cycles at 95°C for 10 s, 60°C for 34s, and 95°C for 1min. The PCR primers used are as follows: CyclinD1: F-5'GCTGTGCATCTACACCGACAACTC', R-5'TTGCGGATGATCTGTTTGTTCTCCT'; OPN: F-5'AGTACCCTGATGCTACAGACGAG3', R-5'CGTTTCATAACTGTCCTTCCCAC3', DNMT1: F-5'TATCCGAGGAGGGCTACCTGGC3', R-5'TGGGGCTAGGTGAAGGTTCAGGC3'; DNMT3a: F-5'TATTGATGAGCGCACAAGAGAGC3', R-5'GGGTGTTCCAGGGTAACATTGAG3'; GAPDH: F-5'ATGGAAATCCCATCACCATCTT-3', R-5'CGCCCCACTTGATTTTGG3'.

2.4. Western blotting
The total protein was extracted from cells using extraction buffer (0.5% Triton X-100, EDTA, phenylmethyl sulfonylfluoride, and complete protease inhibitors). Cell extracts were separated by SDS-PAGE and then proteins were transferred onto PVDF membranes. The membranes were blocked with 5% skim milk powder, probed with anti-β-actin (1:200 dilution, Santa Cruz), anti-Cyclin-D1 (1:600 dilution, Santa Cruz), anti-OPN (1:600 dilution, Santa Cruz), anti-DNMT1 (1:500 dilution, Santa Cruz) and anti-DNMT3a (1:500 dilution, Santa Cruz) antibodies overnight at 4°C, followed by incubation with IRDyeTM-800 conjugated secondary antibodies for 30 min at RT. After washing, protein bands were visualized using an Odyssey infrared imaging system (Gene Company, Lincoln, NE, USA).

2.5 Statistical analysis
Data were expressed as the means ±standard deviation, and subsequently analyzed by t-test. The difference was statistically significant ($P < 0.05$).

3. Results

3.1. PDGF-BB promoted the proliferation of HA-VSMCs confirmed by EdU staining

To explore the pro-proliferative role of PDGF-BB in HA-VSMCs, 40 ng/mL PDGF-BB was used to treat HA-VSMCs for 24 hours, and EdU staining was carried out. As shown in Figure. 1, PDGF-BB-treated group had more EdU positive cells compared to control group, suggesting that smooth muscle cells stimulated with PDGF-BB are in a proliferative state. The results showed that PDGF-BB could induce the proliferation of HA-VSMCs.

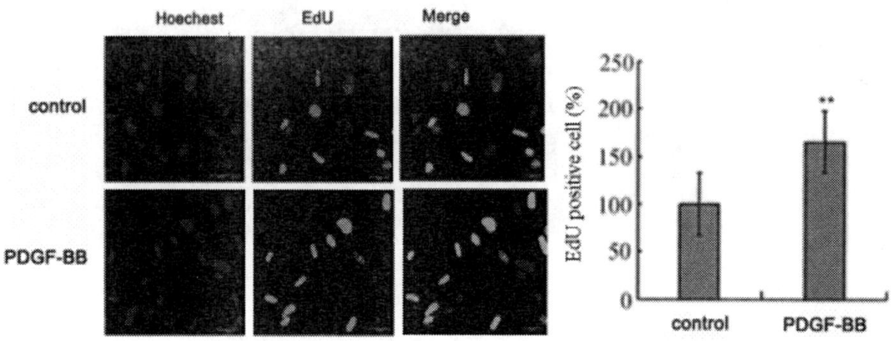

Figure 1. PDGF-BB promoted the proliferation of HA-VSMCs confirmed by EdU staining

3.2. PDGF-BB induced the upregulation of proliferative markers CyclinD1 and OPN in HA-VSMCs

In the early stages of atheromatosis, vascular smooth muscle cells undergo the loss of contract markers and the gain of uncontrolled proliferative activity. Cyclin D1 is a cell cycle regulator necessary for G1 phase progression and is also a marker of SMC proliferation[6]. OPN is a pleiotropic cytokine involved in SMC proliferation, which is upregulated in phenotypic switch of SMCs from a contract phenotype to a pathological synthetic phenotype[7]. To study the effect of PDGF-BB on the cyclin D1 and OPN, HA-VSMCs were incubated with 40ng/mL PDGF-BB for 48h, and then the RNA and protein levels were tested by RT-PCR and western blotting. The increased mRNA and protein level of cyclin D1 and OPN were observed in PDGF-BB-stimulated HA-VSMCs (Figure. 2). These data further illustrated that PDGF-BB could stimulate the proliferation of SMCs.

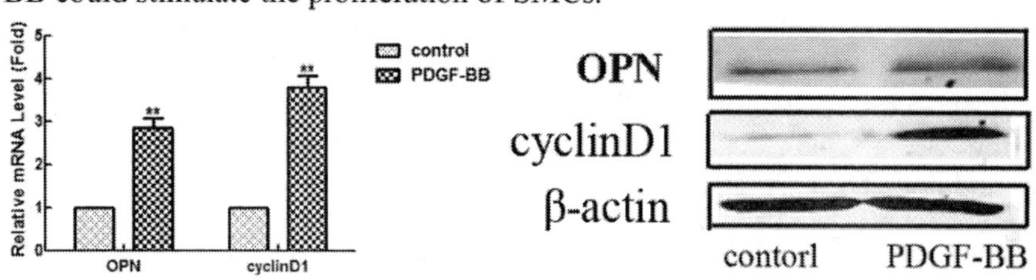

(a) (b)

Figure 2. PDGF-BB could induce the upregulation of proliferative markers CyclinD1 and OPN in HA-VSMCs. (a) The mRNA level of CyclinD1 and OPN were tested by RT-PCR. (b) The protein level of CyclinD1 and OPN were detected by western blotting.

3.3. PDGF-BB inhibited the expression of differentiated markers SM-α-actin and SM22 in HA-VSMCs

To investigate the effect of PDGF-BB on the expression of differentiated markers SM22 and SM-α-actin, HA-VSMCs were incubation with 40 ng/mL PDGF-BB for 48 h, and then the mRNA and protein level of SM22 and SM-α-actin were detected by RT-PCR and western blotting. As shown in

Figure. 3, the decreased level of SM22 and SM-α-actin were observed in HA-VSMCs stimulated by PDGF-BB, suggesting the loss of differentiation phenotype.

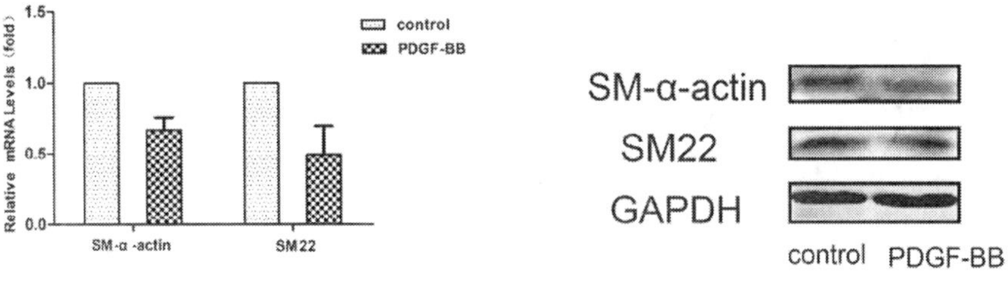

(a) (b)

Figure 3. PDGF-BB inhibited the expression of differentiated markers SM22 and SM-α-actin in HA-VSMCs. (a) The mRNA level of SM22 and SM-α-actin were tested by RT-PCR. (b) The protein level of SM22 and SM-α-actin were detected by western blotting.

3.4. The expression of DNMT3a and DNMT1 were upregulated in PDGF-BB-processed HA-VSMCs

To investigate the effect of PDGF-BB on the expression of DNMT1 and DNMT3a, the mRNA level of DNMT1 and DNMT3a were tested in PDGF-BB-treated HA-VSMCs by RT-PCR and the protein expression were detected by western blotting. The data were illustrated in Figure. 4. Treatment with PDGF-BB could lead to the added expression of DNMT3a and DNMT1 in HA-VSMCs. These result illustrated that DNMT1 and DNMT3a could play important roles in phenotypic switch of smooth muscle cells.

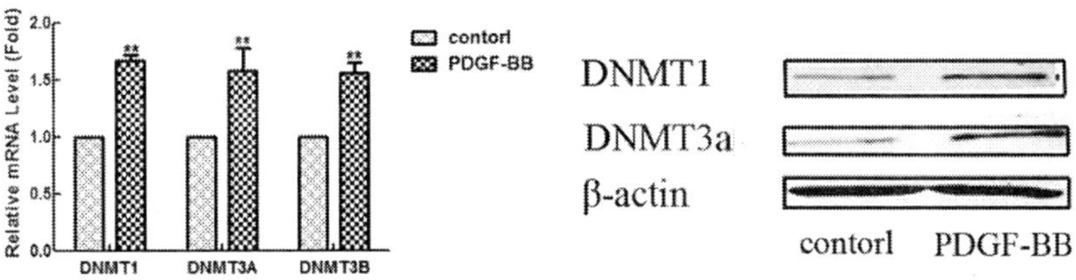

(a) (b)

Figure 4. The level of DNMT3a and DNMT1 were upregulated in PDGF-BB-treated HA-VSMCs. (a) The mRNA level of DNMT3a and DNMT1 were tested by RT-PCR. (b) The protein level of DNMT3a and DNMT1 were detected by western blotting.

4. Conclusion

PDGF-BB could promote the proliferation of SMCs and inhibit its differentiation, confirmed by markedly upregulating the expression of SMC proliferative markers, such as cyclin D1, OPN and PCNA and downregulating the expression of SMC contract markers, including calponin, αSMA and SM22 in HA-VSMCs [8]. Similar to other studies, we found PDGF-BB could induce the proliferation of SMCs. DNA methylation happens on the fifth carbon of cytosine through DNA methyltransferase enzymes (DNMT1, several DNMT3B isoforms, and two isoforms of DNMT3A) [9]. CpG methylation of key regions can regulate gene expression. Our results showed that DNMT1 and DNMT3a could play essential roles in phenotypic switch of smooth muscle cells. The increased expression of DNMT1 and DNMT3a in PDGF-BB-treated HA-VSMCs could be associated with the decrease of differentiation marker express.

5. Discussion

The abnormal proliferation and migration of VSMCs is a common pathological feature of the occurrence and development of various cardiovascular diseases, such as pulmonary hypertension, hypertension, atherosclerosis, et al, and VSMCs phenotype transformation plays a key role in the process of VSMCs proliferation and migration. Therefore, understanding the mechanism of VSMC

IOP Conf. Series: Materials Science and Engineering **729** (2020) 012094 doi:10.1088/1757-899X/729/1/012094

phenotype transformation is of great significance for the prevention and treatment of cardiovascular diseases. The influencing factors of phenotype transformation of VSMCs include the surrounding environment, including mechanical force, extracellular matrix, neuroregulation, cytokines and their inhibitors. Cytokines can also activate different signaling pathways to induce the phenotype transformation of VSMCs , but the specific mechanism needs further study.

Acknowledgments
This work was financially supported by the Tianjin Natural Science Foundation (17JCZDJC33600) and the Innovative Research Team of Tianjin Municipal Education Commission (TD13-5015).

References
[1] Schwartz SM. (1997)Smooth muscle migration in atherosclerosis and restenosis. The Journal of clinical investigation,100(11):S87-9.
[2] Ferns GA, Raines EW, Sprugel KH, Motani AS, Reidy MA, Ross R.(1991) Inhibition of neointimal smooth muscle accumulation after angioplasty by an antibody to PDGF. Science (New York, NY), 253 (5024):1129 - 32.
[3] Claesson-Welsh L. (1994) Platelet-derived growth factor receptor signals. The Journal of biological chemistry,269(51):32023 - 6.
[4] Schumacher A, Petronis A.(2006)Epigenetics of complex diseases: from general theory to laboratory experiments. Curr Top Microbiol Immunol,310: 81 - 115.
[5] Bradley J. Toghill1*, Athanasios Saratzis1 , Peter J. Freeman2 , Nicolas Sylvius2 , UKAGS collaborators and Matthew J. Bown1.(2018)SMYD2 promoter DNA methylation is associated with abdominal aortic aneurysm (AAA) and SMYD2 expression in vascular smooth muscle cells.Clin Epigenetics,10(29):2-15.
[6] Kyuho Jeong,* Jung-Hyun Kim,* James M. Murphy,* Hyeonsoo Park, Su-Jeong Kim, Yelitza A.R. Rodriguez .(2019)Nuclear Focal Adhesion Kinase Controls Vascular Smooth Muscle Cell Proliferation and Neointimal Hyperplasia Through GATA4-Mediated Cyclin D1 Transcription.Circulation Research,.125(2):152-166.
[7] Choi KH, Kim JE, Song NR, Son JE, Hwang MK, Byun S, et al. (2010)Phosphoinositide 3-kinase is a novel target of piceatannol for inhibiting PDGF-BB-induced proliferation and migration in human aortic smooth muscle cells. Cardiovascular research, 85(4):836 - 44.
[8] Santos KF, Mazzola TN, Carvalho HF. (2005)The prima donna of epigenetics: the regulation of gene expression by DNA methylation. Braz J Med Biol Res, (38): 1531 - 1541.
[9] Fang Y1, Xia W2, Cai W3, Zhang X4, Zhang J4, Fu X5, Li S3, Fang X3, Sun S6, Wang Z6, Zhang X2, Zhu S7,Li J8.(2019)Effects of TLR4 overexpression on sperm quality, seminal plasma biomarkers, sperm DNA methylation and pregnancy rate in sheep.

IOP Conf. Series: Materials Science and Engineering **729** (2020) 012095 doi:10.1088/1757-899X/729/1/012095

Detailed description of the mathematical modeling of the catalytic naphtha reforming process dynamics

Xiaoyang Yi [1], Peng Zhang[2]* and Changlu Hu[2]

[1]Department of chemical engineering and environment University of Petroleum, Beijing, 102249, China

[2]Reforming and Aromatics Research Laboratory, PetroChina Co Ltd. Petrochemical Research Institute, Beijing, 102200, China

*Corresponding author's e-mail: zhangpengmail@petrochina.com

Abstract. This work is dedicated to the construction of a kinetic model for the naphtha reforming process. By assembling the components into reactivity based lumps, a new reaction network model rooted on 33 lumps and 101 reactions was studied. The tabu search algorithm was used to determine the global optimal solution. The experimental data from the laboratory scale isothermal reactor was used to investigate the kinetics. By comparing with the test data, the average deviation of the reformed products calculated from the proposed model reached 2.5%.

1.Introduction

Catalytic naphtha reforming is practiced extensively in the petroleum-refining industry to convert petroleum naphtha for use as high-performance gasoline fuel. Naphtha is a low octane cut from atmospheric distillation of crude oil from the topping plant usually in the range between 30°C and 200°C, and constitutes typically from 15% to 30% by weight of the crude oil.[1] This cut contains some impurities such as sulphur, oxygen, and nitrogen compounds. As a high-octane gasoline blending component, reformed gasoline has reached 1/3 of the gasoline pools of developed countries such as Europe and the United States. Zhou Xiang developed a detailed kinetic model of the CCR process through KMT software,[2] depicting the paraffin and naphthene isomers up to the C9 component, while the aromatic isomers are fully described up to C10. The coking kinetics and the corresponding deactivation of the catalyst are integrated into the model. Wei designed an economic model predictive control (EMPC) and production mode conversion strategy through economic model predictive control (EMPC) and established a dynamic process model to simulate the naphtha reforming process.[3]

2.Chemistry of the Process

The main reactions during the reforming process are shown in Figure 1. The key to a successful simulated catalytic reforming process is to establish a well-developed description of the naphtha reforming kinetics.[4] The reforming process follows pathbelow. (1) Dehydrogenation of a cycloalkane to an aromatic hydrocarbon. (2) Paraffin in naphtha is cyclized to a cycloalkane or converted to an isoparaffin. (3) Dehydrogenation of aromatic hydrocarbons or paraffin. Under typical naphtha reforming conditions, these reactions realize the conversion to aromatics.

IOP Conf. Series: Materials Science and Engineering **729** (2020) 012095 doi:10.1088/1757-899X/729/1/012095

3.Experiment

3.1. Research reform response

The commercial platinum stannum bimetallic catalyst Pt-Sn/γ-Al2O3 was used as the reforming catalyst.

3.2. Study detailed kinetics using laboratory-scale reactor experiments

The kinetic study instrument generally includes the section of the preparation and supply of the raw materials, the reactor, and the detection unit. The preparation of the raw materials is to highly purify the raw materials and control the content of harmful substances. The reactor is the key part of the kinetics assay.

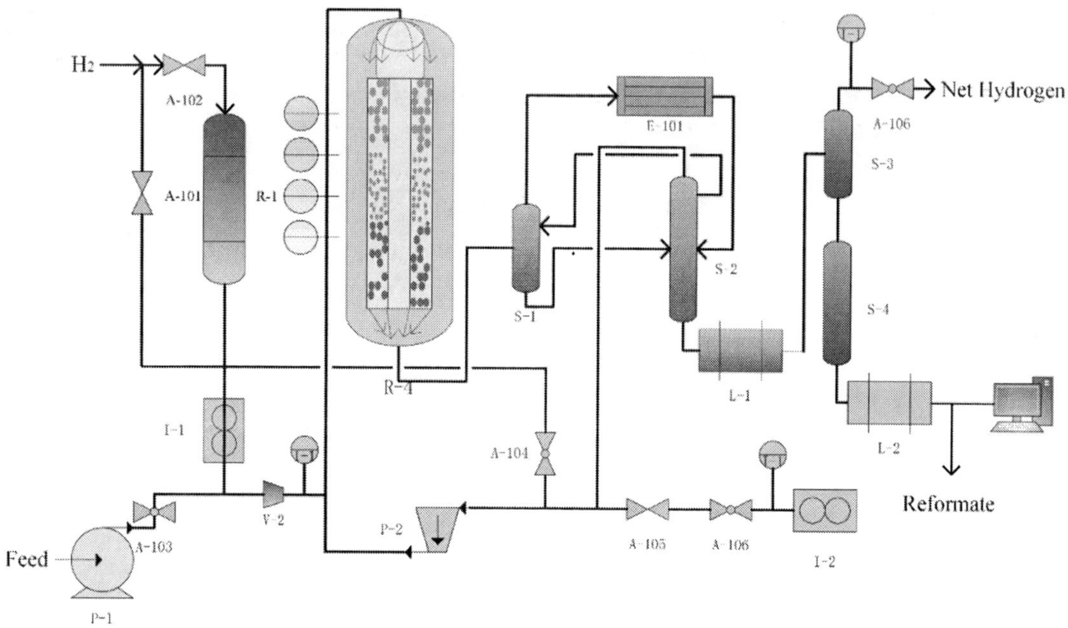

Figure 1. Process flow scheme for the reforming process.

According to the method provided above, the process includes two main stages: (1) reaction stage, and (2) regeneration stage. The reaction stage can be determined as the core of the catalytic reforming process.[5]

3.3. Raw material and product distribution analysis

The operating conditions during the investigation and the basic properties of raw material were summarized in Table 1, The raw materials were analyzed on a PIONA column using a gas chromatograph (GC) to determine their detailed composition.

Table 1. Specifications of catalyst, feed, reactors and of plant for fresh catalyst

	n-paraffins	i-paraffins	olefins	naphthenes	aromatics
C4	0.007	0.000	0.003	0.000	0.000
C5	1.167	0.422	0.000	0.325	0.000
C6	3.792	3.492	0.000	4.396	0.228
C7	5.236	4.155	0.000	11.884	1.371
C8	5.824	5.508	0.000	12.312	2.976

C9	4.361	6.493	0.000	9.182	3.109
C10	2.936	5.924	0.000	1.801	1.736
C11	0.203	0.861	0.000	0.204	0.051
C12	0.000	0.027	0.000	0.005	0.010

4. Kinetics Scheme Construction

4.1. Dynamics of the reforming process

4.1.1 Method

This study built mathematical model for kinetic reactions, using laboratory data. The main purpose is to predict output parameters and detect contours to improve process performance.A tabu search algorithm is implemented to simulate existing processes in the study,[6, 7] while MATLAB program task is used to accomplish the task. The kinetic model developed was an extension of Badiea'swork,[8] which utilized the aggregated representation of the reaction.

4.1.2 Dynamic model of the reform process

From our view, there are three major drawbacks that may exist in the existing kinetic models.[9] (a) The lack of detail of the naphthenic reaction network leads to the inability to correctly predict the five-membered naphthenes and six-membered naphthenes, which may lead to erroneous calculations of process energy balance. (B) Excessive simplification of the list of components, in particular, Tri-methyl-benzene and methyl-ethylbenzene, which may lead to the incorrect predictions of the aromatics value of the reformate. (C) Incomplete reactor model caused by different reactor flow patterns.

In our study, we tried to overcome the above shortcomings. In order to consider the predictability of better naphtha composition, the following modifications were made to the original model.

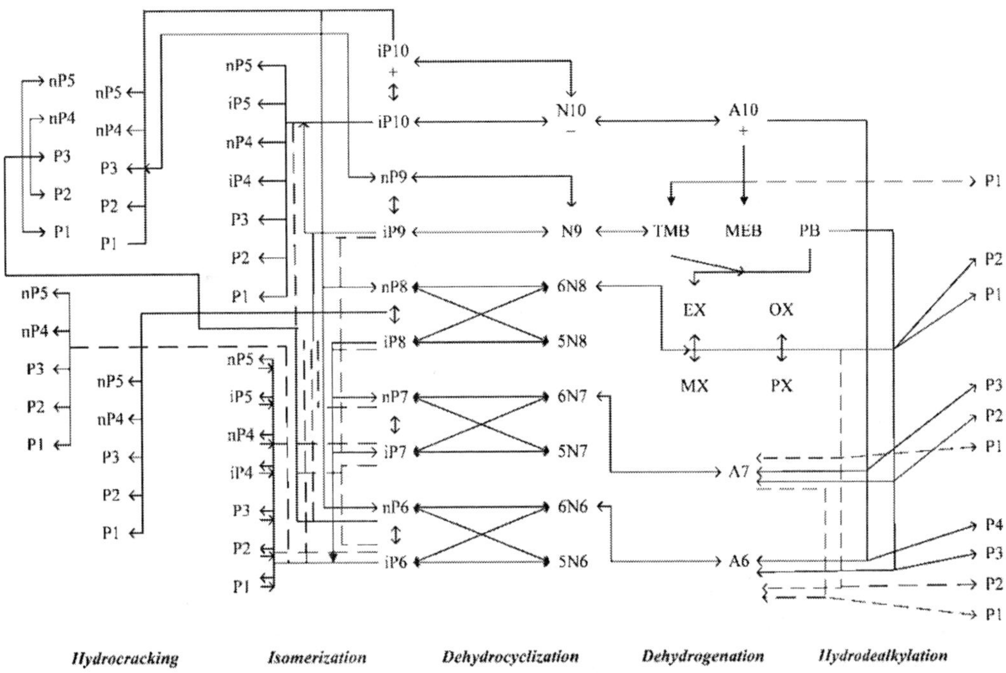

Figure 2. Reaction network scheme for the naphtha reforming process.

The rate of chemical reaction of naphtha depends on several factors, such as activation energy, temperature, pressure, and the factors that influence the effects of pressure, temperature and pressure on the kinetic parameters can be expressed as follows

$$k_j = k_{0j}exp\left(\frac{-E_{0j}}{RT}\right)P^{b_{0j}}$$

Where k_{0j} represents the frequency factor, E0J represents the activation energy (KJ/Kmol), b0j represents the activation energy, R represents the gas constant (KJ/Kmol k), and T represents the temperature (K).

4.1.3.Model Parameter Estimation Based on Tabu Search Algorithm

The tabu search algorithm is used to optimize the estimated kinetic parameters to determine the minimum error of the model output and laboratory data. Considering the same conditions as the operating conditions of the laboratory in the model, estimating the frequency factor k_0 for each reaction in the reaction network is a very important step.[10]

According to Arrhenius's law, there are usually three kinetic parameters. To verify the proposed kinetic model, the predicted data of the kinetic model needs to be compared with the real data, and the kinetics of each reaction is estimated in the kinetic reaction network. The estimation of parameters is very important.[11] According to the Arrhenius's law, a total of 300 kinetic parameters, including 100 reaction activation energy factors (E1,...., E100), and 100 correction factors (k1, ..., k100) need to be determined. These parameters can be used to minimize the difference of results between the predictions and the experiments by using the methods of the objective function. And by applying the optimization algorithm to get the best convergence function, the dynamic parameter objective function (KPOF) is as follows

$$KPOF = \left(\frac{F_{i,out,cal} - F_{i,out,act}}{F_{i,out,act}}\right)^2$$

$F_{i,out,cal}$ represents the molar flow rate of the component (i) calculated by the kinetic model and $F_{i,out,act}$ represents the molar flow rate of the actual evaluation data component (i) obtained by the laboratory evaluation device. The algorithm source of the kinetic parameters is determined by the tabu search method, and the convergence speed is accelerated by the tabu search optimization algorithm (TS).

5.Results and discussion

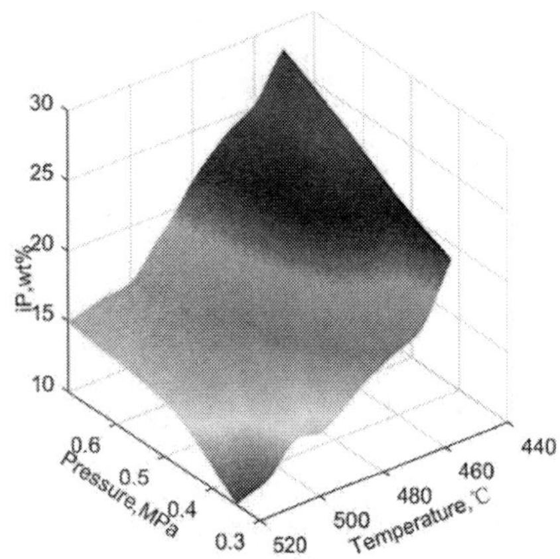

Figure 3. The combined effect of temperature and pressure on product distribution

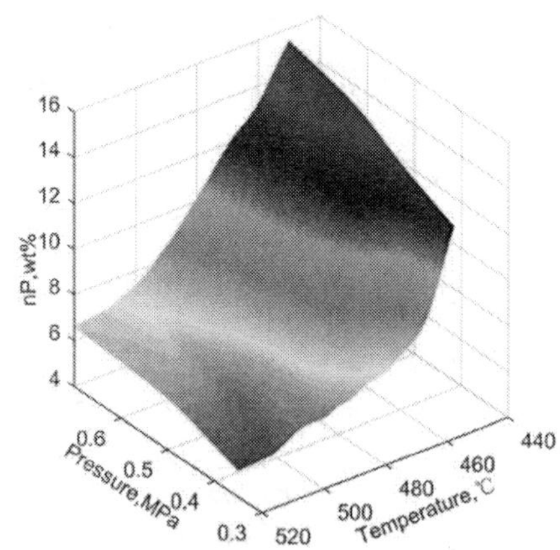

Figure 4. The combined effect of temperature and pressure on product distribution

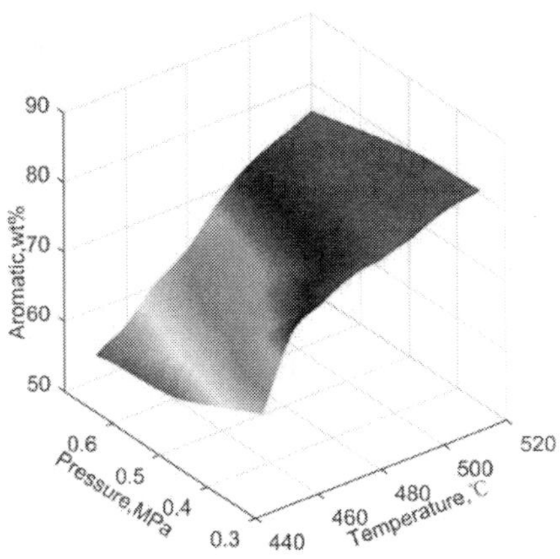

Figure 5. The combined effect of temperature and pressure on product distribution

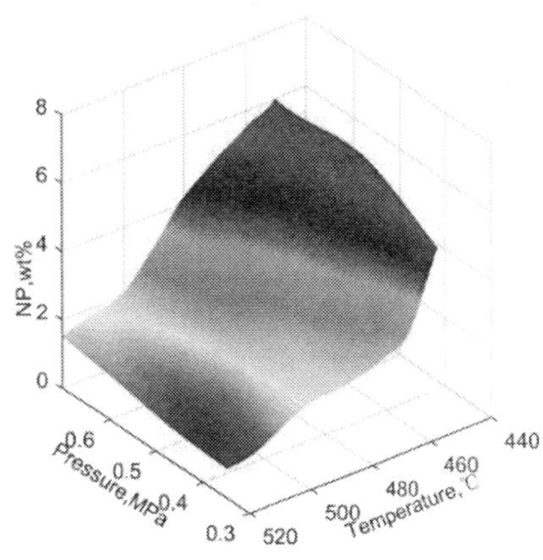

Figure 6. The combined effect of temperature and pressure on product distribution

It can be seen from Figures 3 to 6 that since the pressure is about 0.5 MPa at a low temperature, the hydrocracking reaction is not fast enough to compete with the isomerization reaction. Similarly, it can be seen that the normal paraffin reduction rate is low before 470°C, At lower pressures, the higher the carbon number, the faster the rate of hydrocracking and cyclization, However, as the carbon number of paraffin gets lower, the content is unchanged. The pressure reduction contributes to the dehydrogenation of naphthenes to aromatics. It conforms to the law of catalytic reforming reaction.

6. Conclusion

A process model for the naphtha reforming reaction was established, which was developed to predict the behavior of the CCR and SR process reactions. The modeling process mainly took those aspects into consideration: (1) changes in operating conditions; (2) changes in the molar flow rate of the feedstock; (3) changes in the molar composition of the feedstock. The proposed model is a new network of various reactions, consisting of 33 blocks and 101 reactions. A tabu search algorithm with multiple initial features was used to determine the global optimal solution, and the estimated parameters obtained from the model were tested for different experimental data. The reformate composition and temperature profile were obtained to provide a degree of conversion for each of the reactors. The simulation results are in good agreement with the plant data, with an average deviation of less than 2.5%. Also, the distribution and characteristics of the reformate are described based on material mass balance.

7. Appendices

KPOF dynamic parameter objective function. TS tabu search optimization algorithm. Y_N Naphthenic component content. Y_A Aromatic hydrocarbon component content. K_{epj} Reversible reaction constant. Y_{np} Normal paraffin content. $Y_{ip\ Isoparaffin\ content}$. $F_{i,\ out,\ cal}$ molar flow rate of the component (i) calculated by the kinetic model. $F_{i,\ out,\ act}$ molar flow rate of the actual evaluation data component (i) obtained by the laboratory evaluation device. k_{0j} frequency factor. E_{0j} activation energy, kj/kmol. b_{0j} activation energy. R gas constant, kj/kmol•k. T temperature, K

Acknowledgments

The author would like to thank PetroChina's major science and technology project (2017E-08), the continuous reforming technology development and application, and thank Peng Zhang for providing guidance for this work.

IOP Conf. Series: Materials Science and Engineering **729** (2020) 012095 doi:10.1088/1757-899X/729/1/012095

References

[1] GARY J H, HANDWERK G E, KAISER M J. Petroleum refining: technology and economics [M]. CRC press, 2007.

[2] ZHOU X, HOU Z, WANG J, et al. Molecular-Level Kinetic Model for C12 Continuous Catalytic Reforming [J]. Energy & fuels, 2018, 32(6): 7078-85.

[3] WEI M, YANG M, QIAN F, et al. Dynamic Modeling and Economic Model Predictive Control with Production Mode Switching for an Industrial Catalytic Naphtha Reforming Process [J]. Industrial & Engineering Chemistry Research, 2017, 56(31): 8961-71.

[4] BOYAS R, FROMENT G. Fundamental kinetic modeling of catalytic reformer [J]. Ind Eng Chem Res, 2009, 48(1107-19.

[5] ABGHARI S Z, ALIZADEHDAKHEL A, MOHADDECY R S, et al. Experimental and modeling study of a catalytic reforming unit [J]. Journal of the Taiwan Institute of Chemical Engineers, 2014, 45(4): 1411-20.

[6] STORN R, PRICE K. Differential evolution–a simple and efficient heuristic for global optimization over continuous spaces [J]. Journal of global optimization, 1997, 11(4): 341-59.

[7] GLOVER F. Tabu search—part II [J]. ORSA Journal on computing, 1990, 2(1): 4-32.

[8] BABAQI B S, TAKRIFF M S, KAMARUDIN S K, et al. Mathematical modeling, simulation, and analysis for predicting improvement opportunities in the continuous catalytic regeneration reforming process [J]. Chemical Engineering Research and Design, 2018, 132(235-51.

[9] BARE W H, MULHOLLAND R J, SOFER S S. Design of a self-tuning rule based controller for a gasoline refinery catalytic reformer [J]. IEEE transactions on automatic control, 1990, 35(2): 156-64.

[10] GRAJCIAR L, HEARD C J, BONDARENKO A A, et al. Towards operando computational modeling in heterogeneous catalysis [J]. Chemical Society Reviews, 2018, 47(22): 8307-48.

[11] STEINFELD J I, FRANCISCO J S, HASE W L. Chemical kinetics and dynamics [M]. Prentice Hall Englewood Cliffs (New Jersey), 1989.

IOP Conf. Series: Materials Science and Engineering **729** (2020) 012096 doi:10.1088/1757-899X/729/1/012096

Preparation of nanocomposite CNTs/PbO and CNTs/CuO via microemulsion process

Xiaofen Fu[1], Yefei Ma[1], Xing Gao, Chunli Zhu*

School of Chemistry and Biological Engineering, Qilu Institute of Technology, Jinan, 250200, P. R. China.

*Corresponding author: Chunli Zhu; e-mail: chunli0125@163.com

[1]These two authors contributer equally to this work.

Abstract. The nanocomposite CNTs/PbO and CNTs/CuO were prepared via W/O microemulsion process. The properties of the microemulsion system for preparing the nanocomposite were studied by drawing ternary phase diagrams of microemulsion. The effects of the ratio of surfactant/cosurfactant, temperature, and salt concentration on the phase region of W/O microemulsion were investigated. A suitable microemulsion system is composed that the ratio of CTAB/cosurfactant was 4:6, the ratio of surfactant/cosurfactant was 2:3, and the reaction temperature was 25℃. The microscopic morphology of the nanocomposite CNTs/PbO and CNTs/CuO was characterized by transmission electron microscopy TEM. The specific surface area decreased obviously after nanocomposite PbO/CNTs and CuO/CNTs mixed in a 1:1 ratio by grinding.

1. Introduction

Solid propellant is the power source material of solid rocket motor, and plays an important role in the development of missile and space technology. Solid propellants often use the "lead-copper-carbon" ternary catalyst which is constituted of lead oxide, copper oxide and carbon black. Carbon plays a particular role in the double-base propellant. Powling etc first came up with the theory of lead-carbon catalysis[1]. It is proved that carbon accelerates the reduction of NO, and the existence of lead salt catalyst activates carbon. The study of Lengelle [2] and Denisyuk [3-4] found that the small amount change of carbon black concentration can cause a great change of PbO catalysis. The discovery of carbon nanotubes can further improve the performance of solid propellants. Ramaswamy [5] prepared carbon nanotubes and organics composites as the propellant Matrix, which was significantly enhanced in the ignition, total energy, safety and mechanical properties of the propellant. Therefore, the preparation of nanocomposites with the carbon nanotubes as the carrier has become a research hotspot.

The common synthesis methods of nanocomposites are: sol-gel method, precipitation method, microemulsion method and so on. The sol-gel method has the disadvantages of high cost and long synthesis period. In addition, some metals such as alkali metals that are not easily polymerized by hydrolysis are more difficult to bind firmly into the gel network, which makes the type of Nano oxide obtained by this method limited. The reaction process of precipitation method is simple, but the size of the resulting material is not easy to control, and the size of the particles is not uniform. The microemulsion method is a thermodynamically stable isotropic, transparent, or translucent dispersion system formed by two insoluble liquids [6-8]. It usually consists of active emulsifiers, auxiliary emulsifiers, oil phases, and water phases. Nanoparticles prepared with microemulsion have the

IOP Conf. Series: Materials Science and Engineering **729** (2020) 012096 doi:10.1088/1757-899X/729/1/012096

advantages of small particle size, good dispersion, and no impurities, and the equipment and process of this method are simple. In recent years, more and more attention has been paid to it.

At present, there is relatively little literature on the preparation of carbon nanotubes loaded with metal oxides by microemulsion method. Our group [9] prepared PbO·CuO/CNTs nanoparticles by microemulsion method, and the morphology of the samples was characterized by SEM, TEM, XRD, etc. The nanoparticles prepared greatly improved the activity and selectivity of the catalyst. However, the ratio of PbO and CuO in the prepared PbO·CuO/CNTs namocomposite could not be controlled. Therefore, in this paper, the carbon nanotubes are individually loaded with PbO and CuO, and the CuO/CNTs and PbO/CNTs are fully mixed in a 1:1 ratio by grinding. It is hoped to get the nanocomposite particles with a good load and controllable ratio.

2. Experiment

2.1 Materials

CNTs and CTAB were obtained from Sigma-Aldrich Chemical Co., analysis of pure; Cyclohexane, Cu $(NO_3)_2$ and Pb $(NO_3)_2$ were purchased from Sinopharm Chemical Reagent Co., Ltd.

2.2 Synthesis

Composite of carbon nanotubes and metal oxidizer was synthesized with microemulsion method. The procedures of were CuO/CNTs as follows: First added hexadecyl trimethyl ammonium bromide (CTAB) and n-pentanol into hexahydrobenzene at room temperature, stirred for 30 min, and dropped water slowly by microburet until the solution became transparent and W/O (water-in-oil) microemulsion is got. Added ammonia water, copper nitrate into the solution and stirred for 1 h. At the same time, added purified carbon nanotubes to react for about 24 h. After that centrifuged the mixture, cleaned it by mixture of water and alcohol 5 times, filtered and heated the product in a muffle at 300℃ and got the CuO/CNTs sample. In the same way, the PbO/CNTs sample was obtained. Then mixed CuO/CNTs and PbO/CNTs at a mass ratio of 1:1 together and got the final product after grinding.

2.3 Method

The steps of drawing the ternary phase diagram of microemulsion were as follows: Surfactant CTAB and cosurfactant n-amyl alcohol as a point in the triangulation (As), the mass ratio of the two is 1:2, 2:3, 3:4. The mixture of surfactant and cyclohexane with a mass ratio of 0:10, 1:9, 2:8, ..., 8:2, 9:1, 10:0 is the second point. Distilled water is the third point of the phase diagram. Put them into glass test tubes separately, then put glass test tubes into ultrasonic cleaners, and slowly drip distilled water at a certain temperature. The amount of distilled water is adjusted by the conductivity which determined by the conductivity meter. The maximum area of microemulsion is found out from the ternary phase diagram to determine the best ratio. Change the temperature, the salt concentration of $Cu(NO_3)_2$ or $Pb(NO_3)_2$, repeat the above experiment, and draw a new ternary phase diagram.

3. Results and discussion

3.1 Microemulsion system

During the process of synthesis, the size and shape of the nano-oxides depend on the structure of water nuclei in W/O microemulsion to a great extent. Hence the research on constitution and boundary of phase, which balanced coexistance in the microemulsion, is very important. In present work, ternary diagram was adopted to analyze the essential factor, which affects microemulsion, including temperature, surfactant or cosurfactant and concentration of solution, in order to find the best microemulsion system and thus to control the core formation and growth of oxides on the nanotemplate. Fig. 1 shows the ternary phase diagram of microemulsion.

Figure 1(a) reveals that dimension of W/O microemulsion will be the largest if the proportion of CTAB and n-pentanol is 3:4, that is to say the Hydrophile-Lipophile Balance (HLB) is 10.5. The reason is that as the HLB of the composite surfactant increases, the hydrophilic property of the two molecules goes up, the force of solubilization of polar molecules is reinforced, the content of solubilize water rises and dimension of the microemulsion increases.

It is essential to select the appropriate temperature of preparation with microemulsion. As can be seen from Figure 1(b), the structure of microemulsion changes little with temperature, and after repeating the experiment we confirm that 25℃ is the optimum temperature.

Figure 1(c) and Figure 1(d) shows the effects of the salt concentration of $Cu(NO_3)_2$ and $Pb(NO_3)_2$ on microemulsion. Ions of water nuclei in high concentration could reduce its solubilization and affect the deposit capacity. Ions reacted with polar molecules in surfactant so as to decrease the repulsion between molecules in surfactant, strengthen the side attractive force and increase the rigidity of the water nuclei film. The results show that the appropriate concentration of $Cu(NO_3)_2$ and $Pb(NO_3)_2$ are 0.48 and 0.6 mol/L, respectively.

(a)

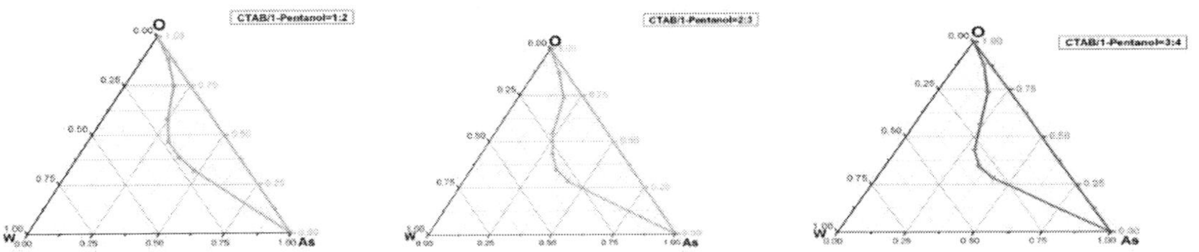

(b)

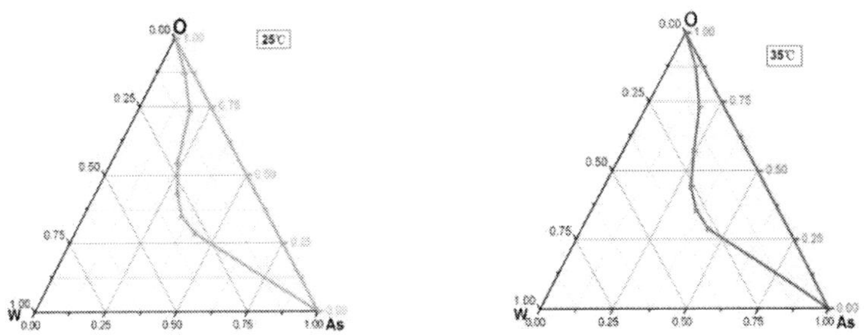

(c)

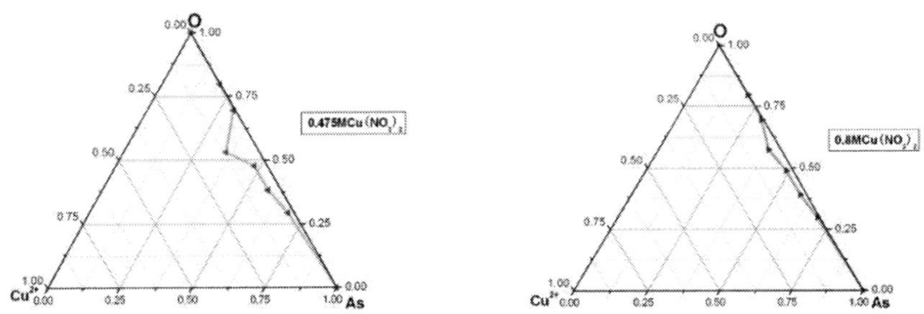

(d)

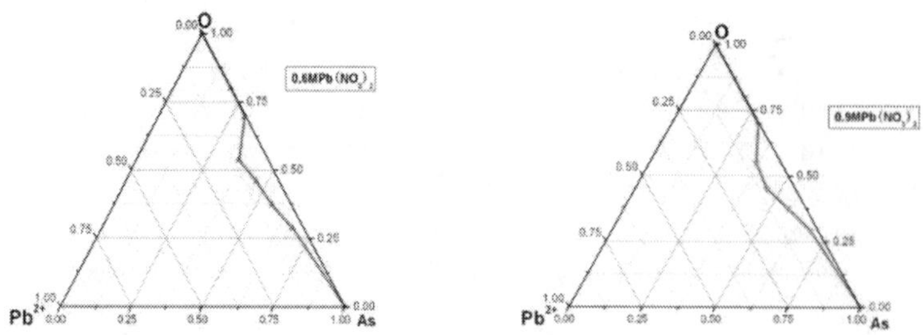

Figure 1. Ternary phase diagram of microemulsion. (a) ratio of surfactant/cosurfactant; (b) temperature; (c) concentration of cupric nitrate and (d) concentration of lead nitrate.

3.2 Microstructure of nanocomposite

The microstructure of nanocomposite PbO/CNTs and CuO/CNTs was characterized by transmission electron microscopy (TEM type: JEM-2011). As shown in the figure 2, many particles are deposited on the surface of carbon nanotubes. The sizes of particles are between 20 nm and 50 nm and evenly deposit in spite of agglomeration of oxide particles in parts. It is clear from the right figure, which demonstrates the amplifying nanotubes that oxide particles firmly combine with the surface of carbon nanotubes and evenly loaded on carbon nanotubes.

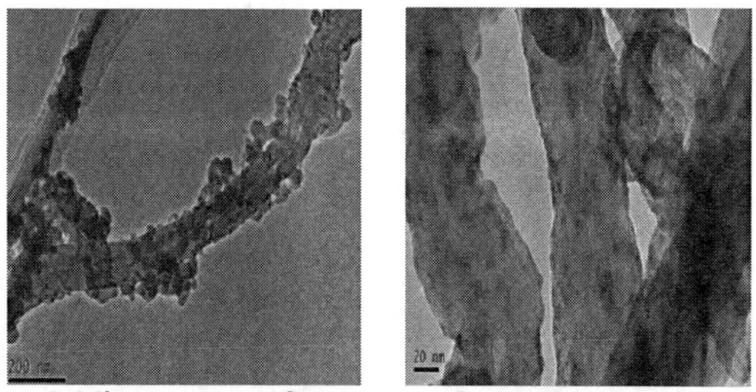

Figure 2. Microstructure of nanocomposite PbO/CNTs and CuO/CNTs

3.3 Specific surface area of nanocomposite

The specific surface area of nanocomposite PbO/CNTs and CuO/CNTs was measured by specific surface area meter (SSAM type: SA3100) with BET method.

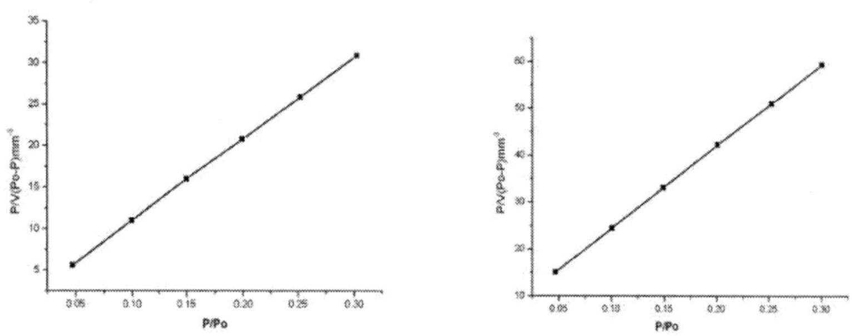

(a) (b)

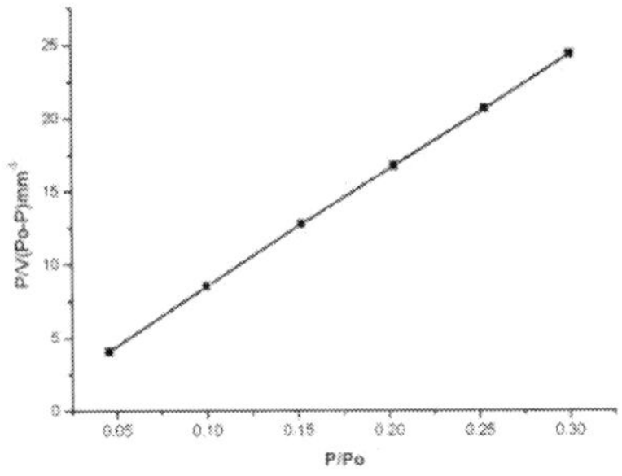

(c)

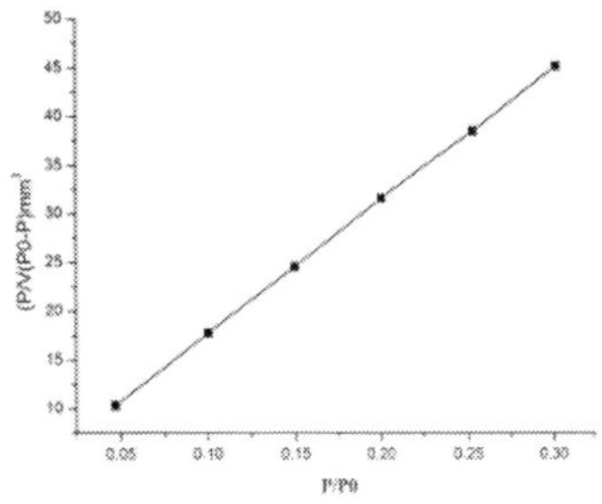

(d)

Figure 3. Absorption of nitrogen on nanomaterials (a) purified CNTs (b) CNTs/CuO (c) CNTs/PbO (d) CNTs/PbO and CNTs/CuO.

From Figure 3, we can find the specific surface area of different samples: D_{CNTs}=43.5 m^2/g, D_{CuO}=35.0 m^2/g, D_{PbO}=18.0 m^2/g, $D_{PbO/CuO}$=25.0 m^2/g. From these data, it can be seen that the specific surface area of CNTs loaded with CuO and PbO respectively decreases compared to the specific surface area of purified CNTs, and the specific surface area of CNTs loaded with PbO decreases more. After mixing both, the surface area is reduced, indicating that the load effect is good, which is consistent with the TEM observation.

4. Conclusion

Microemulsion method was adopted to synthesize nano-metal oxide deposited on carbon nanotubes. Ternary diagram was used to study influencing factors of W/O microemulsion system. The optimum technical parameters of the process are the proportion of surfactant and cosurfactant (3:4), temperature 25℃), and concentrations of $Cu(NO_3)_2$ and $Pb(NO_3)_2$ (0.48 and 0.6 mol/L, respectively). Transmission electron microscopy (TEM) was applied to test the microstructure of the product. The specific surface area decreased obviously after nanocomposite PbO/CNTs and CuO/CNTs mixed in a 1:1 ratio by grinding indicating that the load effect is good.

Acknowledgments

The authors are thankful to Research Project Foundation of Qilu Institute of Technology (No. QL19K057) for financial support of this work.

References

[1] Hewkin, D.J., Hicks, J.A., Powling, J., et al. (1971) The combustion of nitricester-based propellants: ballistic modification by lead compounds. Combust. Sci. Technol., 2: 307-327.

[2] Lengelle, G., Bizot, A., Ferreira, J. (1985) Steady-state burning of homogeneous propellants. In: kuo K K, Summerfild Med. Fundamental of solid propellant combustion, Summerfild M. Vol. 90 progress in astronautics and aeronautics, AIAA, Washington D C.

[3] Denisyuk, A. P., Margolin, A. D., Khubaev, G. V., et al. (1977) The role of soot in the bombusion of ballistic propellants with lead-containing catalysts. Combust. Explo. Shock+, 13(4), 490-496.

[4] Denisyuk, A. P., Demidobal, L. A. (2004) Effect of some catalysts on combustion of double-base propellants. Combust. Explo. Shock+, 40(3): 311 -318.

[5] Bratcher, M., Pesce, R. R., Kaste, P., Ramaswamy, A. L. (2002) Nanotube modification of energetic materials. Proceedings of the 38th Meeting of the JANNAF Combustion Subcommittee. Destin: FL.

[6] Loang, Y. J., Huang, W. J., Tian, Z. Q. (2007) Progress in preparation of nanomaterials by microemulsion. Technol. Nat. Sci., 21 (9): 87-91.

[7] Cheng, Y., Zhao, J., Wang, G. (2013) Synthesis of polypyrrole nanoparticles by microemulsion

polymerization for photocatalysis. Technol. Mater. Sci. Ed., 28 (4): 1312-1319.
[8] Shao, Q. H., Gu, G. B., Zhang, L. J. (2002) By reverse micelles are microemulsion system in the application of nanometer materials. Chem. Eng. Prog., 21 (2): 136-139.
[9] Hui, R., Liu, Y. Y., Jiao, Q. J., Fu, X. F., Yang, T. T. (2010) Preparation of nanocomposite PbO·CuO/CNTs via microemulsion process and its catalysis on thermal decomposition of RDX. J. Phys. Chem. Solids., 71: 149-152.

Research progress of alginate in biomedicine

Zongmei Wu[1], Weixi Fan, Xiaoming Cai, Chunli Zhu*

School of Chemistry and Biological Engineering, Qilu Institute of Technology, Jinan, 250200, P. R. China.

*Corresponding author: Chunli Zhu; e-mail: chunli0125@163.com

Abstract. Sodium alginate was widely used in biomedicine because of its biocompatibility, biodegradability, safety, non-toxicity and easy processing. The development of the application of micelles, microspheres, nanoparticles and hydrogels prepared by sodium alginate polymers in drug release is described in this paper.

1. Introduction

Sodium alginate is a water-soluble natural ionic polysaccharide with excellent properties such as biocompatibility, biodegradability and safety. It is widely used in biomedical applications because of its stability, solubility, viscosity and other properties required for pharmaceutical preparations.

In recent years, natural biodegradable polymers have attracted widespread attention in controlling drug release. The traditional drug release has the disadvantages of too fast release rate, short duration of drug action, unstable drug concentration in the drug receiving system, and short drug half-life, which reduces the therapeutic effect of the drug. The biodegradable sodium alginate polymer has physical and chemical properties stability and easy processing, and can be used as a carrier for a drug release system, also which can be classified into a system of micelles, microspheres, nanoparticles and hydrogels.

2. Alginate micellar

Macromolecules containing both hydrophilic and hydrophobic groups are often referred to as polymeric surfactants and are also described in the relevant literature as amphiphilic polymers, hydrophobically modified water-soluble polymers, micellar polymers, and associated polymers and other related terms. Due to the presence of hydrophilic groups and hydrophobic groups, polymeric surfactants exhibit different association behaviors in different solvents, eventually forming spherical, rod-like, layered association-micelles. Polymer micelles having a core-shell structure can be used as a pharmaceutical carrier to increase the solubility of poorly soluble drugs in aqueous/oil solutions. The drug is then delivered to the lesion site and slowly released for targeted therapy. As a good drug-loading system, polymer micelles can reduce the body's adverse reactions to drugs and improve the bioavailability of drugs.

Sodium alginate is a strongly hydrophilic natural ionic polysaccharide whose macromolecular chain has a coiled conformation in an aqueous solution. In order to expand the application of sodium alginate in biomedicine, in the previous work, the sodium alginate was hydrophobically modified to obtain an amphiphilic sodium alginate derivative-hydrophobically modified sodium alginate (HM-alginate) [1-2]. As shown in Figure 1, the micelle image of hydrophobically modified sodium alginate was confirmed by confocal laser scanning microscope (CLSM). By using fat-soluble Sudan IV as a mimetic drug, it was found that the amphiphilic sodium alginate derivative formed a core-shell micelle in an aqueous solution, and the Sudan IV was encapsulated in its hydrophobic core, thereby increasing the solubility of the fat-soluble drug in an aqueous solution (As shown in Figure 2). Meng[3] studied the release properties of hydrophobically modified sodium alginate, and the results showed that the amphiphilic sodium alginate derivative (0.8 mg/mL) had a slower release rate to clofazimine: after 17 hours of release, it basically reached the release balance, and the cumulative release amount was about 72%. Sodium alginate derivatives make it possible to release the drug due to its unique configuration and adhesion. The application of polymer micelles in drug development has been extensively studied and plays a potential carrier role in the research of drug carriers.

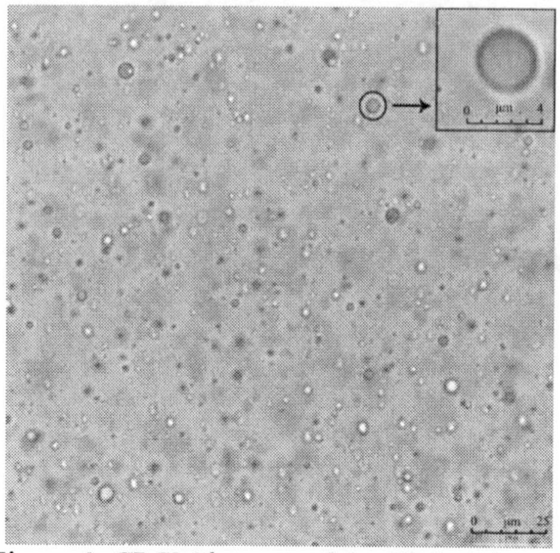

Figure 1. CLSM images of the HM-alginate

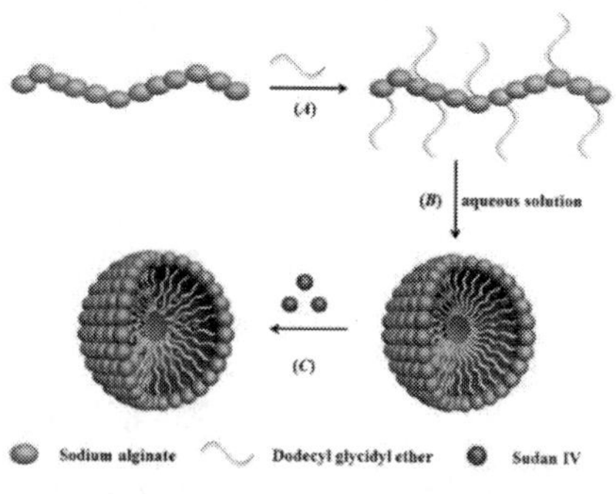

Figure 2. Schematic of the HM-alginate micelle. (A) Synthesis of HM-alginate (B) HM-alginate formed spherical micelles in aqueous solution (C) Sudan IV was loaded in HM-alginate micelles

3. Alginate microspheres

Microsphere is a microparticle polymer framework with a size of about 1-1000 μm, which refers to the microsphere formed by drug dispersion or adsorption in the polymer matrix. The matrix materials for microspheres are usually natural or synthetic polymers, while polysaccharides have the possibility of preparing microspheres to encapsulate biological and drug molecules due to their biodegradability and non-toxic side effects[4]. For example, sodium alginate[5], chitosan[6,7], cellulose[8], okra mucus[9] and other biodegradable polymers are widely used in the design and development of different drug delivery systems.

When sodium alginate is used as carrier to wrap or absorb polymer or target drug, spherical or quasi spherical particles are called sodium alginate microsphere system. The preparation of microspheres is the key to the selection of drug delivery routes and the control of drug release. Sodium alginate microspheres[10] can be prepared by emulsion ion crosslinking, microemulsion and complex coacervation, which can be divided into drug loaded microspheres and material microspheres.

3.1 Drug loaded microspheres
Ghumman[9] used sodium alginate as the dispersion medium and added okra gum to prepare the mixed microspheres under the condition of agitation. After that, oxcarbazepine, a neurogenic drug, was added to prepare alginate/okra pod gum microspheres loaded with oxcarbazepine by Ion Gel Technology. The encapsulation efficiency of the microspheres was 76%-90%, and the average particle size was 496-692 μm. The release rate of the drug was 84.47% at 24 hours.

3.2 Material microspheres
Shi[11] reported the preparation of polymeric microspheres by combining magnetic Fe_3O_4 nanoparticles with sodium alginate: glucose oxidase was immobilized using glutaraldehyde as a crosslinking agent, and the immobilization conditions of the material for glucose oxidase and the enzymatic properties of the magnetic immobilized enzyme were investigated.

Fan[8] reported the preparation of a polyethylenimine-functionalized sodium alginate/cellulose nanocrystal/polyvinyl alcohol core-shell microsphere (pva/sa/cnc@pei) using sodium alginate as a matrix. Fan uses the non-steroidal anti-inflammatory drug diclofenac sodium as a typical drug. Due to the high specific surface area and affinity of pva/sa/cnc@pei, the maximum adsorption capacity of diclofenac sodium is much higher than that of other adsorbent materials. It demonstrates the applicability and potential of pva/sa/cnc@pei microspheres for drug adsorption.

4. Alginate nanocarrier
Targeted drugs refer to drugs with targeting ability, which is a research hotspot of current researchers. The targeted drug can be enriched in the lesion, so that the drug forms a relatively high concentration in the target, and the targeted drug has high selectivity, and the toxic side effect is suppressed while improving the drug effect. Nano carriers are all kinds of nanoparticles with sizes between 1-1000 nm that can dissolve or disperse drugs.

Wan[12] designed and constructed amphiphilic 6-thioguanine sodium alginate prodrug with 6-thioguanine as hydrophobic drug and biodegradable dialdehyde sodium alginate as drug carrier. The prodrug nanoparticles with pH response were prepared by ultrasonic dispersion. The core-shell particle size of 6-thioguanine sodium alginate is about 98 nm, which can significantly inhibit the cytotoxicity of 6-thioguanine and protect L-O2 cells. It can respond to the weak acid pH stimulation decomposition in cancer cells, release the bound drug molecule 6-thioguanine, so as to achieve targeted treatment.

5. Alginate gel
Hydrogel is a kind of polymer with high hydrophilicity and three-dimensional network structure. It has high water locking property and has good permeability to nutrient inflow and excretion of metabolites. When the gel is used as a carrier to coat the drug, the absorbent will expand and form gel diffusion layer in the medium because of the high water absorption of the gel, which will delay the release of the drug.

Huang reported a method of preparing soluble self-healing alginate polymethacrylic acid (SAMA) hydrogel based on natural polysaccharide alginate and functional monomer of sodium methacrylate, and explored the sustained release property of SAMA hydrogel by using 6 g Luo Danming as a drug template. The results showed that the SAMA hydrogel without crosslinking agent could completely release the loaded drug molecules and ensure the full utilization of the drug. The release time of Luo Danming's release from hydrogel containing 0.6 wt% crosslinking agent epichlorohydrin was extended from 5 h to 24 h, and the complete release of the drug was ensured.

The good biodegradability, mechanical strength and multiple biological functions of the hydrogel can effectively prevent the immune response and severe inflammation during drug delivery and tissue engineering. Therefore, hydrogels have potential application properties in the biomedical field.

6. Conclusion and Outlook

Sodium alginate is a natural polymer with adjuvant properties. It has excellent properties such as stability, biodegradability, bioadhesiveness and non-toxicity required for pharmaceutical preparations. It can improve the convenience and compliance of patients in biomedicine. As a pharmaceutical preparation, the sodium alginate has a high added value and a broad market prospect. As the research progresses further and the applied clinical practice will bring huge economic benefits.

Acknowledgments

The authors are thankful to Research Project Foundation of Qilu Institute of Technology (No. QL19K055) for financial support of this work.

References

[1] Wu, Z., Wu, J., Zhang, R., Yuan, S., Lu, Q. & Yu, Y. (2018) Colloid properties of hydrophobic modified alginate: Surface tension, zeta-potential, viscosity and emulsification. Carbohydr. Polym., 181: 56-62.

[2] Wu, J., Wu, Z., Zhang, R., Yuan, S., Lu, Q. & Yu, Y. (2017) Synthesis and micelle properties of the hydrophobic modified alginate. Int. J. Polymer. Mater., 66: 742-47.

[3] Meng, Y., Wu, C., Zhang, J., Cao, Q., Liu, Q. & Yu, Y. (2015) Amphiphilic alginate as a drug release vehicle for water-insoluble drugs. Colloid J., 77: 754-60.

[4] Faidi, A., Lassoued, M. A., Becheikh, M. E. H., Touati, M., Stumbe, J. F. & Farhat, F. (2019) Application of sodium alginate extracted from a Tunisian brown algae Padina pavonica for essential oil encapsulation: Microspheres preparation, characterization and in vitro release study. Int. J. Biol. Macromol., 136: 386-94.

[5] Hussein, N., Omer, H., Ismael, A., Albed Alhnan, M., Elhissi, A. & Ahmed, W. (2019) Spray-Dried Alginate Microparticles for Potential Intranasal Delivery of Ropinirole Hydrochloride: Development, Characterization and Histopathological Evaluatio Pharm.Dev.Technol., pp: 1-36.

[6] Kucukoglu, V., Uzuner, H., Kenar, H. & Karadenizli, A. (2019) In vitro antibacterial activity of ciprofloxacin loaded chitosan microparticles and their effects on human lung epithelial cells. Int. J. Pharm., 569:118578.

[7] Ibe, M. I., Odimegwu, D. C. & Onuigbo, E. B. (2019) Alginate-coated chitosan microparticles encapsulating an oral plasmid-cured live Salmonella enterica serovar Gallinarum vaccine cause a higher expression of interferon-gamma in chickens compared to the parenteral live vaccine. Avian. pathol., 48: 423-28.

[8] Fan, L., Lu, Y., Yang, L. Y., Huang, F. & Ouyang, X. K. (2019) Fabrication of polyethylenimine-functionalized sodium alginate/cellulose nanocrystal/polyvinyl alcohol core-shell microspheres ((PVA/SA/CNC)@PEI) for diclofenac sodium adsorption. J. Colloid Interface Sci., 554: 48-58.

[9] Ghumman, S. A., Bashir, S., Noreen, S., Khan, A. M., Riffat, S. & Abbas, M. (2018) Polymeric microspheres of okra mucilage and alginate for the controlled release of oxcarbazepine: In vitro &in vivo evaluation. Int. J. Biol. Macromol., 111: 1156-1165.

[10] Ding, S. C., Hu, X. L., Guan, P., Du, C. B. (2016). Preparation of surface-imprinted microspheres effectively controlled by orientated template immobilization using highly cross-linked raspberry-like microspheres for the selective recognition of an immunostimulating peptide. J. Mater. Chem. B, 4(8): 1510-1519.

[11] Shi, F., Li, P. W., Dong, R. Z. (2018) Immobilization of glucose oxidase with magnetic Fe_3O_4 and sodium alginate. Food Sci. Technol., 43(01): 262-266.

[12] Wan, Y. Y., Bu, J., Liu, J., Yang, W., Cai, Y., Yin, W., Xu, P., Xu, J., He, M. (2018). PH and reduction-activated polymeric prodrug nanoparticles based on a 6-thioguanine-dialdehyde sodium alginate conjugate for enhanced intracellular drug release in leukemia. Polym. Chem., 9(24): 3415-3424.

IOP Conf. Series: Materials Science and Engineering **729** (2020) 012098 doi:10.1088/1757-899X/729/1/012098

Efficiency Enhancement of Perovskite $CsPbBr_3$ Quantum Dot Light-emitting Diodes by Doped Hole Transport Layer

Mengke Liu, Weiling Luan*, Yihui Huang, Shulv Zhang

Key laboratory of pressure system science and safety, East China University of Science and Technology, Shanghai, 200237, China

*Corresponding author's e-mail: luan@ecust.edu.cn

Abstract: Balanced charge injection is essential to high-performance Perovskite $CsPbBr_3$ quantum dot-based light-emitting diodes (QLEDs). However, low mobility of hole-transport materials (HTMs) severely restrict improving performance of QLEDs. Herein, we provide a novel HTMs to improve the highest occupied molecular orbital (HOMO) energy level structure and carrier mobility by doping poly (9-vinlycarbazole) (PVK) and poly [N, N′-bis(4-butylphenyl)-N, N′-bis(phenyl) benzi-dine] (poly-TPD). We also introduce poly (methyl methacrylate) (PMMA) as electron block layer to further achieve charge injection balance. Finally, an enhanced external quantum efficiency (EQE) of 0.53% and 414.83 cd/m^2 was obtained. Compared with the untreated QLED, this result has been 8-fold enhanced, provides a new approach to attain better performance.

1. Introduction

In the field of optoelectronics, all-inorganic perovskite ($CsPbX_3$, X=Cl, Br, I) light-emitting diodes (LEDs) based on quantum dots (QDs) have become a central issue for their solution processability, low cost, high luminous efficiency and narrow emission spectrum. [1, 2] The LEDs based on $CsPbBr_3$ QDs was first reported to have an external quantum efficiency (EQE) of 0.035% [3], opening a new path for their application in optoelectronic devices.

To improve the property of QLEDs, their emission mechanism should be better understood, and the influencing factors of EQE is as below: [4]

$$EQE = \eta_{out} \cdot r \cdot q \cdot \gamma \qquad (1)$$

Where η_{out} represents the optical coupling factor, which is generally less than 0.2. r represents the part of the radiation attenuation and q represents the PLQY of the emitter. γ reflects the charge balance, and generally a large γ attributes to more balanced charge injection. In recent studies, PLQY has been upgraded to over 90% [5], which is difficult to make further improvements. The effective optical coupling could improve the EQE of perovskite QLEDs, [6] but it is complicated to deal with the refractive index of each layer. The balance of carrier injection has been confirmed to improve the efficiency of LED due to the γ approaching to 1. However, there has been less study carrying on this method to balance charge injection. [7]

Generally, the architecture of perovskite $CsPbBr_3$ QLEDs is Cathode / Hole Injection Layer (HIL) / Hole Transport Layer (HTL) / Emitting Layer / Electron Transport Layer (ETL) / Electron Injection Layer (EIL) / Anode [8]. The perovskite $CsPbBr_3$ QLEDs can be roughly divided into four types according to the transport layer materials: QLEDs based on polymer charge transport layers, QLEDs exploiting small molecule charge transport layers, QLEDs with inorganic charge transport layers and QLEDs using hybrid transport layer.[9] In early researches, the commonly used charge transport

IOP Conf. Series: Materials Science and Engineering **729** (2020) 012098 doi:10.1088/1757-899X/729/1/012098

material was PVK[2], but the mobility of PVK was only 10^{-7}~10^{-6} $cm^2V^{-1}s^{-1}$[10]. However, the mobility of the electron transport materials was much higher than PVK, making it difficult to balance the charge injection. Later, poly-TPD began to be used as hole transport material due to its higher carrier mobility [11], but the carrier mobility of poly-TPD was still lower than that of electron transport layer materials. Therefore, it has been a major challenge to develop high-efficiency all-inorganic perovskite $CsPbBr_3$ QLEDs via employing a better material as hole transport layer.

In addition, the traditional perovskite $CsPbBr_3$ QDs using oleic acid-oleylamine as a ligand has terrible conductivity because of the insulating property of oleic acid and oleylamine. Zeng reported a method to address the issue via ligand density control. [5] Wu replaced the ligand of oleic acid-oleylamine with DDAB by ligand exchange. [12]

The purpose of this study is to enhance the device performance of all-inorganic perovskite $CsPbBr_3$ QLEDs with a hybrid hole transport material doping PVK and poly-TPD, which can improve hole mobility and balance charge injection. An electron blocking layer employing poly (methyl methacrylate) (PMMA) is also introduced to limit the charge injection of electron transport layer. Compared with the LED using poly-TPD as hole transport layer, an 8-fold enhanced EQE of 0.53% is successfully achieved.

2. Experimental

2.1 Chemicals

Oleic acid (OA,90%), 1-ocetadecene (ODE, 90%), oleylamine (OLA, 80%-90%), Lead (*II*) bromide powder ($PbBr_2$) and Cesium carbonate (Cs_2CO_3) were purchased from Sigma Aldrin. Didodecyldimethylammonium Bromide (DDAB), poly (3,4-ethylenedioxythiophene) polystyrene) solution (PEDOT:PSS, Clevis P VP Al 4083), Poly-TPD, PVK and PMMA were purchased from Xian Polymer Light Technology.

2.2 Synthesis of $CsPbBr_3$ QDs

The $CsPbBr_3$ QDs were synthesized following the reported procedures. [12] 15 mL of 1-ocetadecene (ODE), 2 mL of Oleic acid (OA) and 0.4 g of Cesium carbonate (Cs_2CO_3) were mixed in a flask. After 10 min under flowing N_2, the flask was heated to 120 *°C* as Cs-oleate precursor. Meanwhile, 10 mL of ODE and 0.138 g Lead (*II*) bromide powder ($PbBr_2$) were loaded to another flask, heated to 120 *°C* under N_2 flow. After 30 min, 1 mL OLA with 0.046 g of DDAB dissolved and 1 mL OA were added into the flask with $PbBr_2$. When the solute was dissolved, the flask was heated to 160 *°C* for 10 min and 0.8 mL of Cs-oleate precursor was quickly injected into it. After 5 s, the flask was dipped in ice-water to cool down. For purification, firstly, the crude solution was centrifuged at 10000 rpm for 6 min, and then the precipitate was dissolved in a mixture of toluene and ethyl acetate (1:2 v/v). The solution was centrifuged at 10000 rpm for 6 min again. The final product was diluted by n-hexane to 20 mg/mL.

2.3 Device fabrication

Firstly, ITO-coated glass was cleaned in water, acetone, isopropyl alcohol for 10 min respectively and treated with UV-ozone for 15 min. PEDOT:PSS (filtered through a 0.22 μm filter) was spin-coated onto the cleaned ITO-coated glass substrates at 3000 rpm for 40 s and annealed in air for 15 min at 120 *°C*. Hole injection material film was carried out in a glove-box, via spin-coating the material solutions with concentration of 8 mg mL^{-1} at 3000 rpm for 40 s, and annealed for 15 min at 80 *°C*. The $CsPbBr_3$ QDs were deposited by spin coating at 2000 rpm for 40 s and PMMA solution were deposited by spin coating at 4500 rpm for 30 s. Tmpypb and Al/LiF were deposited in a vacuum deposition clamber.

2.4 Characterization

PL characteristics and absorption spectra were measured via Cary Eclipse spectrometer and Cary 50 UV-visible spectrophotometer respectively. Gemini SEM 500 system was used to record SEM images. UPS date were measured by a PREVAC system. JEOL JEM 2100 was utilized to record TEM images and HRTEM images.

3. Results and discussion

$CsPbBr_3$ QDs was synthesis following the method described above. Figure 1 (a) shows exciton absorption and photoluminescence (PL) emission of $CsPbBr_3$. Its absorption peak was located at 504 nm and the excitation peak was at 512 nm with the full width at half maximum (FWHM) of 18 nm. Low FWHM means uniform size of the nanocrystals. From TEM and HRTEM images in Figure 1 (b), cubic phase was confirmed with $CsPbBr_3$ QDs account for the lattice constants of 5.8 Å, 5.8 Å and 5.8 Å. Figure 1 (c) shows the SEM image of the $CsPbBr_3$ QDs film deposited by spin-coating displaying a the smooth surface, indicating that the size of $CsPbBr_3$ nanocrystal is uniform, which is beneficial to improve the film formation of the upper layer.

QLEDs with conventional HTMs obtain worse performance because of unbalanced charge injection, which limit EQE of QLEDs account for a great deal of nonradiative recombination. PVK and poly-TPD were doped in different proportions and then applied in LEDs as HTL. As shown in Figure 2 (a), the optimal doping ratio of poly-TPD and PVK is 1:4, achieving the maximum EQE of 0.33%. Moreover, the performance of QLEDs with PMMA of different concentration were also investigated. From Figure 2 (b) and Figure 2 (c), in contrast to the untreated QLEDs with the EQE of 0.07%, the QLEDs with PMMA of 0.16 mg/mL attain a maximum EQE of 0.53% and a brightness of 414.86 cd/m^2. The complete architecture used for the current study is shown schematically in Figure 3 (a), QLEDs use architecture of ITO (150nm) / PEDOT: PSS (40nm) / HTL (30nm) / $CsPbBr_3$ (30nm) / PMMA / Tmpypb (40nm) / LiF / Al (80nm). And the cross-sectional SEM image in Figure 3 (b) shows multiple layers of the LED, in which the PMMA layer and the LiF layer are indistinguishable, because of their low thickness.

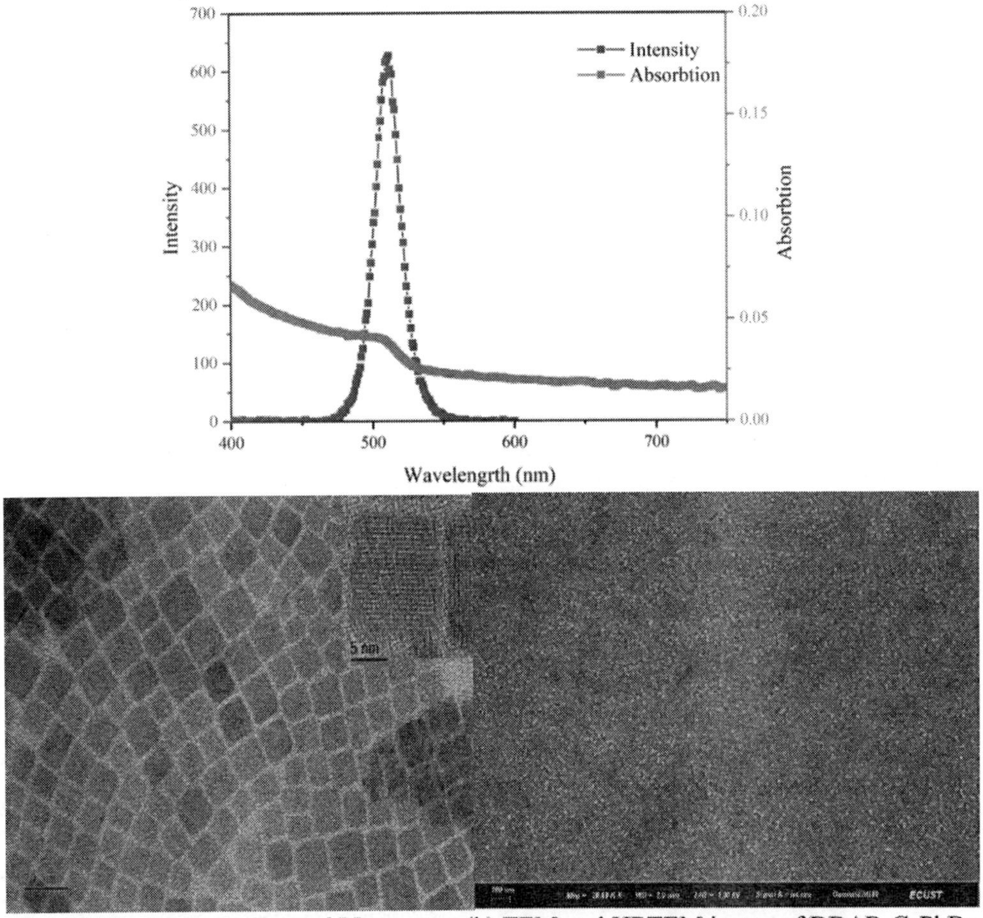

Figure 1. (a) UV–vis absorption and PL spectra, (b) TEM and HRTEM image of DDAB-$CsPbBr_3$ QDs, (c) SEM image of the surface of DDAB-$CsPbBr_3$ QDs film.

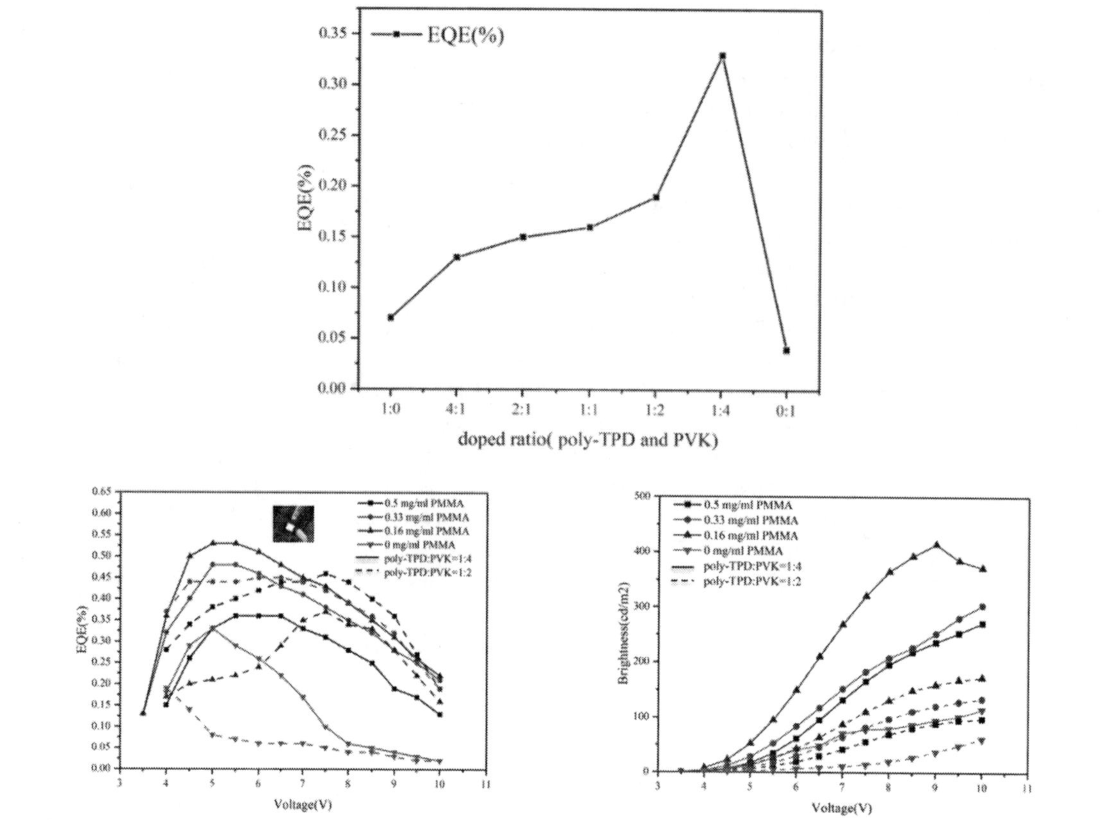

Figure 2. $CsPbBr_3$ QLED performance: (a) The highest EQE of different doping ratios (poly-TPD: PVK), (b) EQE of device with different PMMA concentrations, (c) Brightness of device with different PMMA concentrations.

The conventional HTMs are not conducive to hole transport due to the improper HOMO. From Figure 4 (a), UPS of different HTMs were tested to explain the reason for efficiency improvement. The results show that the HOMO energy level of poly-TPD and PVK is respectively 5.4 eV and 5.8 eV, while the HOMO energy level of hybrid HTMs is 5.51 eV, between the energy levels of PEDOT: PSS and $CsPbBr_3$, indicating an easier hole migration process. In addition, the carrier mobility of HTMs were measured and shown in Table 1. The carrier mobility of poly-TPD: PVK=1:4 is -3.39E-03 $m^2/v.s$, which is higher than others, because doped semiconductors can produce more majority carriers to accelerate recombination of majority carriers and minority carriers. Therefore, QLEDs with treated HTL shows better optical performance account for higher hole mobility approaching electron mobility could increase the radiation recombination of hole and electron in the emitting layer to improve EQE. Then, the current density of QLEDs with electron blocking layers added to the structure was measured. Figure 4 (b) shows that QLEDs utilizing PMMA of 0.16 mg/mL obtain significantly reduced current density, demonstrating the blocking effect of PMMA on electron transport.

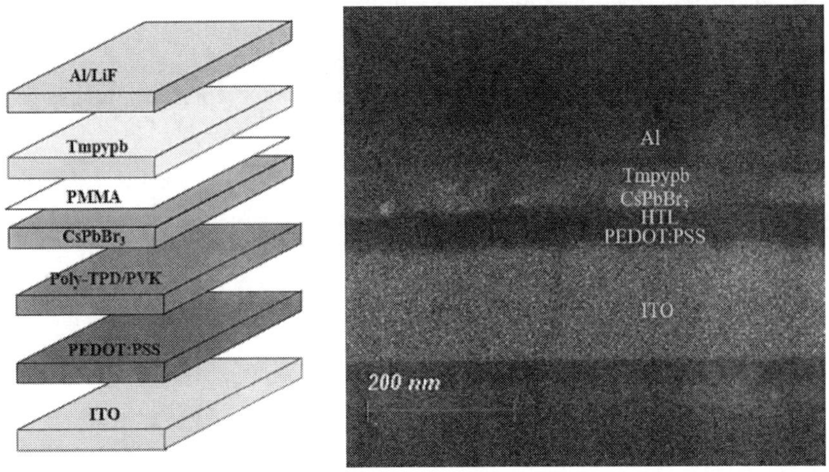

Figure 3. (a) Structure diagram of the device, (b) SEM image of cross section of QLED.

We believe that the blocking effect of PMMA on the current leads to the balanced injection of hole and electron, improved the radiation recombination of carriers in the emitting layer. Because EQE of QLEDs first increases and then decreases as the concentration of PMMA increases, which indicates that the mobilities of hole and electron reach equilibrium in this process. The reduction of EQE could be attributed to the broken carriers balance due to further enhanced blocking effect of PMMA.

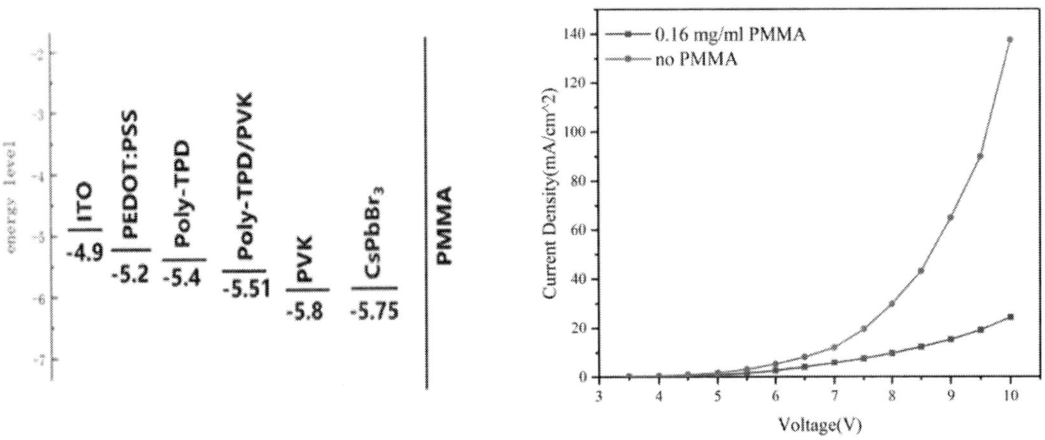

Figure 4. (a) the HOMO of materials measured by UPS. (b) Current density comparison of devices with or without PMMA.

Table 1. Carrier mobility of materials with different doping ratios obtained by Hall Effect Measurement System

Sample No.	Poly-TPD: PVK	carrier mobility[m^2/v.s]
1	∞	-1.72E-05
2	1:1	-4.17E-04
3	1:2	-5.45E-04
4	1:4	-3.39E-03
5	0	-3.50E-06

4. Conclusion

In this study, we adopt a novel method to enhance efficiency of $CsPbBr_3$ QLEDs by using hybrid HTMs as HTL and PMMA to block electron. HTL doped with PVK and poly-TPD, under the synergistic action of PMMA, balances the carrier mobility in the emitting layer, furtherly improves the radiation recombination of exciton and the EQE of QLEDs. The mixture of PVK and poly-TPD shows more suitable mobility properties for LEDs with the doping ratio of 1:4. Moreover, a small amount of PMMA blocks the injection of electrons and balance charge injection. A high brightness of 414.83 cd/m^2 and EQE of 0.53%, which is increased by 8-fold compared to the untreated QLED is achieved.

Reference

[1] Zhang C., Luan, W., Huang, Y., Yang, F. (2019) Growth of perovskite nanocrystals in poly-tetra fluoroethylene based microsystem: on-line and off-line measurements. J. Nanotechnology, 30: 145602-145619.

[2] Liu, B., Wang, L., Gu, H., Sun, H., Demir, H. V. (2018) Highly Efficient Green Light-Emitting Diodes from All-Inorganic Perovskite Nanocrystals Enabled by a New Electron Transport Layer. J. Adv. Optical Mater., 6(11): 1800220- 1800226.

[3] Yantara, N., Bhaumik, S., Yan, F., Sabba, D., Dewi, H. A., Mathews, N., Mhaisalkar, S. (2015) Inorganic Halide Perovskites for Efficient Light-Emitting Diodes. J. J. Phys. Chem. Lett., 6(21): 4360-4364.

[4] Higuchi, T., Nakanotani, H., Adachi, C. (2015) High-Efficiency White Organic Light-Emitting Diodes Based on a Blue Thermally Activated Delayed Fluorescent Emitter Combined with Green and Red Fluorescent Emitters. J. Adv. Mater., 27:2019–2023.

[5] Li, J., Xu, L., Wang, T., Song, J., Chen, J., Xue, J., Zeng, H. (2017) 50-Fold EQE Improvement up to 6.27% of Solution-Processed All-Inorganic Perovskite CsPbBr3 QLEDs via Surface Ligand Density Control. J. Adv. Mater., 29(5): 1603885- 1603893.

[6] Zhang, Q., Tavakoli, M. M., Gu, L., Zhang, D., Tang, L., Gao, Y., Fan, Z. (2019) Efficient metal halide perovskite light-emitting diodes with significantly improved light extraction on nanophotonic substrates. J. Nat. Commun., 10(1): 1-9.

[7] Supran, G.J., Shirasaki, Y., Song, K.W., Caruge, J.-M., Kazlas, P.T., Coe-Sullivan, S., Andrew, T.L., Bawendi, M.G., Bulovic, V. (2013) QLEDs for displays and solid-state lighting. J. MRS. Bull., 38: 703–711.

[8] Coe, S., Woo,W.-K., Bawendi, M.G., Bulovic, V. (2002) Electro-luminescence from Single Monolayers of Nanocrystals in Molecular Organic Devices. J. Nat., 420: 800–803.

[9] Shirasaki, Y., Supran, G.J., Bawendi, M.G., Bulovic, V. (2013) Emergence of colloidal quantum-dot light-emitting technologies. J. Nat. Photonics, 7: 13–23.

[10] Xiao, P., Huang, J., Yan, D., Luo, D., Yuan, J., Liu, B., & Liang, D. (2018) Emergence of nanoplatelet light-emitting diodes. J. Mater., 11(8): 1376-1397.

[11] Subramanian, A., Pan, Z., Zhang, Z., Ahmad, I., Chen, J., Liu, M., Zhang, Y. (2018) Interfacial Energy-Level Alignment for High-Performance All-Inorganic Perovskite CsPbBr3 Quantum Dot-Based Inverted Light-Emitting Diodes. J. ACS Appl. Mater. Inter., 10(15): 13236-13243.

[12] Wu, H., Zhang, Y., Lu, M., Zhang, X., Sun C., Y, W. (2018) Surface ligand modification of cesium lead bromide nanocrystals for improved light-emitting performance. J. Nano., 10 (9): 4173-4178.

IOP Conf. Series: Materials Science and Engineering **729** (2020) 012099 doi:10.1088/1757-899X/729/1/012099

A stable FHD display device based on BCE IGZO TFTs

G M Wan[1*], S M Ge[1], C Gong[2], S Li[2] and X N Lin[1]

[1] School of Electronic and Computer Engineering, Peking University, Shenzhen, Guangdong, 518055, China

[2] China Star Optoelectronics Semiconductor Display Technology Co., Ltd, Shenzhen, Guangdong, 518132, China

[*]Corresponding author's e-mail: wanguangm@126.com

Abstract. In this work, the impact of the deposition process of the SiOx passivation layer on the electrical properties of the BCE IGZO TFTs has been studied. The ΔVth of the TFTs are 2.52 and -1.67 V under PBTS (60°C, 30 V) and NBITS (60°C, -30 V, 4500 nit) after 1 hour, respectively. The stability of these TFTs is verified in 32in FHD display devices, which still could display the picture properly after the 500-hour aging test at 60°C and 90% humidity. A stable FHD display device based on BCE IGZO TFTs was achieved.

1. Introduction

In recent years, the need for thin-film transistors (TFTs) with high performance has been increasing on account of the demand for ultra high definition display, low energy consumption and interactive functionality [1, 2]. A variety of technical solutions have been tried to meet these requirements. Amorphous metal oxide semiconductors (AOSs), such as Indium-Gallium-Zinc-Oxide (IGZO), Indium-Hafnium-Zinc-Oxide (IHZO) and Indium-Gallium-Tin-Oxide (IGTO), have attained great attention due to their excellent performance in cost-effective manufacturing process, good uniformity and reasonable low off-current [3-5]. Among them, IGZO has become an industry-standard channel layer because of its desirable mobility, high Ion/Ioff ratio and good long-time stability.

Many studies were performed to improve the performance of the IGZO TFTs. The back-channel etch (BCE) IGZO TFTs were prepared to reduce parasitic capacitance and production costs [6-8]. The effect of hot carriers on the IGZO lattice under high field strength has been studied [9]. The source/drain/active layer was split to improve TFT robustness under mechanical bending strain [10]. However, there are few reports on the influence of the deposition process of passivation layers on performance of IGZO TFTs.

In this paper, the deposition process of the SiO_x passivation layer was studied to improve the performance of BCE IGZO TFTs. The TFTs were used in 32in full high definition (FHD) display devices. The stability of these devices was verified by an aging test in a high temperature and high humidity environment.

2. Experimental

The cross-sectional schematic of a BCE IGZO TFT is shown in Figure 1. At first, the Mo/Cu layer was deposited by sputtering and patterned as the gate material, followed by the deposition of SiN_x and SiO_x as the gate insulator (GI) equipped by a plasma enhanced chemical vapor deposition (PECVD). Then the IGZO layer was sputtered by megnetron sputtering system from the IGZO targets and further patterned. Afterwards, the double layered source/drain (S/D) electrodes consisted of Mo/Cu were sputtered and patterned with a wet etchant. The channel width (W) and length (L) of the oxide TFTs were 10 μm and 5 μm, respectively. Subsequently, stacked SiO_x/SiN_x passivation layers and a PFA layer were deposited by PECVD and patterned. Finally, the indium tin oxide (ITO) layer was deposited by sputtering and patterned as the pixel electrodes.

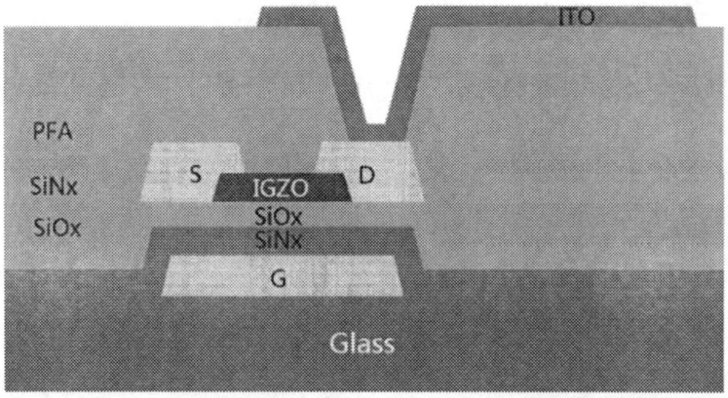

Figure 1 Cross-sectional schematic of the BCE IGZO TFTs.

All electrical measurements were monitored by a Keithley 4200 source meter. The transfer characteristic curves were plotted measuring drain current (I_d) by sweeping gate voltage (V_g) from -25 V to 25 V with a drain voltage (V_d) of 15 V. Electrical properties such as threshold voltage (V_{th}), field effect mobility (μ_{fe}) and Sub-threshold Swing (SS) are extracted from the transfer characteristic curves. The V_{th} is calculated by extracting V_g when I_d reached 1 nA under the condition of V_d = 15 V. The output characteristic curves were plotted measuring I_d by sweeping V_d from 0 V to 30 V with various V_g of 5, 10, 15, 20 and 25 V, respectively. For positive bias temperature stress (PBTS) measurements, a constant V_g of 30 V was applied at 60°C for 1 hour under dark conditions. For negative bias temperature stress under illumination (NBITS) measurements, a constant V_g of -30 V was applied at 60°C for 1 hour under illumination (4500nit). The TFTs were used in 32in FHD display devices to verify their stability and practicality in an aging test at 60°C and 90% humidity.

3. Results and discussion

The influence of the deposition process of the SiO_x passivation layer on the electrical properties of the BCE IGZO TFTs has been studied. Figure 2 represents the transfer characteristics of the TFTs with a SiO_x passivation layer deposited by different process parameters.

As shown in Figure 2(a), the curves illustrate a positive drift as the sputtering power of the SiO_x passivation layer increasing from 5 KW to 10 KW. The V_{th} of the TFTs are -0.6, 1.31, 2.77 and 3.86 V when the power are 5, 6, 8 and 10 KW, respectively. The increasing of V_{th} is due to the increase in the degree of oxidation of the IGZO active layer. As the sputtering power increases, the deposition rate of the SiO_x layer becomes slower (data not shown here), and the oxygen replenishment capacity of the SiO_x passivation layer is strengthened. When the degree of oxidation of the active layer is increased, free carriers in the film are reduced. Therefore, a higher voltage is required to turn the TFTs on.

A negative drift of the curves could be observed in Figure 2(b), as the SiH_4 flow rate increasing from 1200 sccm to 1500 sccm. The V_{th} of the TFTs are 1.77, 1.31, -8.25 and -23.05 V while the SiH_4 flow rate are 1200, 1300, 1400 and 1500 sccm, respectively. As the flow rate of SiH_4 increases, the amount of H^+ entering the IGZO layer increases, which improves the degree of conductivity of the IGZO layer. Thus, the V_{th} of the TFTs drifts to the negative direction.

Figure 2(c) shows the impact of the flow ratio of N_2O to SiH_4 on the electrical properties of the TFTs. As can be seen from the figure that the electrical properties of the TFTs are less sensitive to the ratio. The V_{th} of the TFTs are 1.31, 1.39, 1.85 and 2.17 V, respectively, as the ratio are 30, 45, 60 and 75.

The SS could be calculated from the linear part of the $\log(I_d)$ vs. V_g plot [11]:

$$SS = \frac{dV_g}{dlogI_d} \tag{1}$$

And the μ_{fe} could be estimated from the linear transfer characteristic curves by the equation [12]:

$$\mu_{fe} = \frac{dI_d}{dV_g}\frac{L}{WC_{GI}V_d} \tag{2}$$

where C_{GI} is the GI capacitance per unit area. Table 1 presents the electrical parameters of the BCE IGZO TFTs with a sputtering power of 6 KW, a SiH_4 flow rate at 1300 sccm and a flow ratio of N_2O to SiH_4 at 30. A V_{th} of 1.31 V, SS of 0.99 V/decade, I_{on}/I_{off} of 4.60×10^7 and μ_{fe} of 7.54 cm^2/Vs are observed.

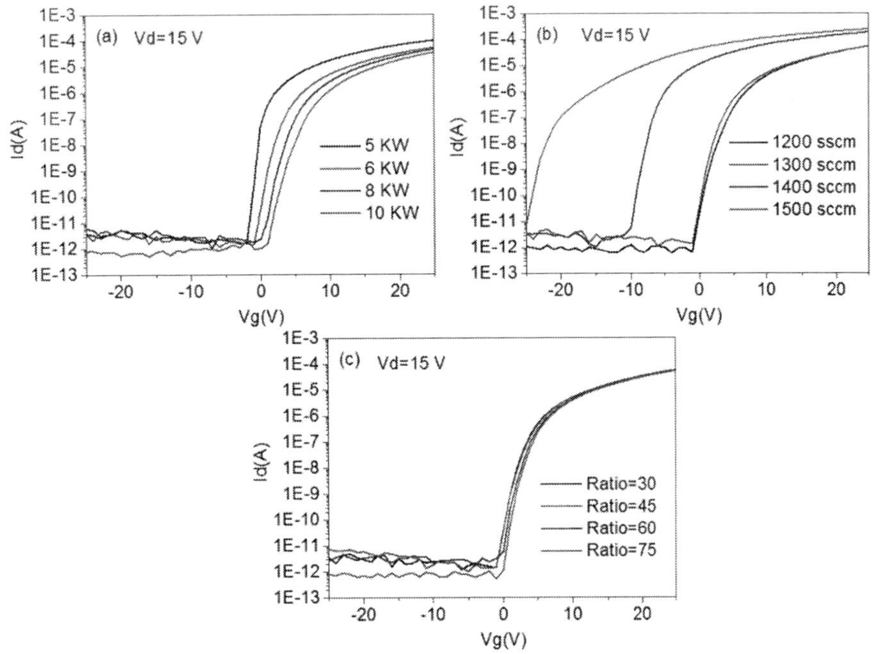

Figure 2 Transfer characteristic curves of the BCE IGZO TFTs with a SiO_x passivation layer deposited by different (a) process power (b) SiH_4 flow rate and (c) flow ratio of N_2O to SiH_4.

Table 1 Electrical parameters of the BCE IGZO TFTs with a sputtering power of 6 KW, a SiH_4 flow rate at 1300 sccm and a flow ratio of N_2O to SiH_4 at 30.

Parameter	Unit	Data
V_{th}	V	1.31
SS	V/decade	0.99
I_{on}/I_{off}		4.60×10^7
μ_{fe}	cm^2/Vs	7.54

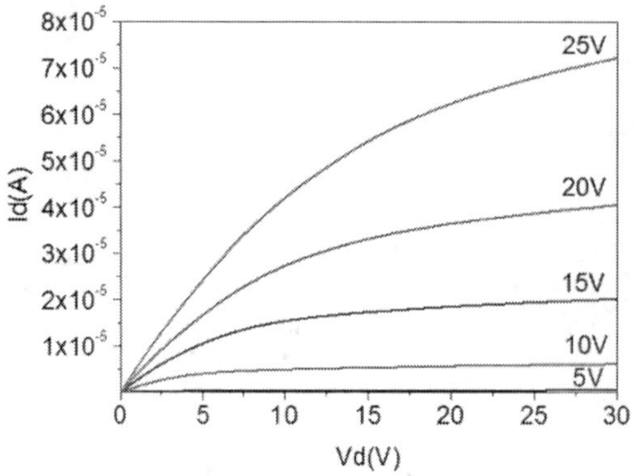

Figure 3 Output characteristic curves of the BCE IGZO TFT with a SiO_x passivation layer deposited with a sputtering power at 6 KW, a SiH_4 flow at 1300 sccm and a flow ratio of N_2O to SiH_4 at 30.

Figure 3 shows the output characteristics of the BCE IGZO TFTs with a SiO_x passivation layer deposited with a sputtering power at 6 KW, a SiH_4 flow at 1300 sccm and a flow ratio of N_2O to SiH_4 at 30. The curves show clear pinch-off, which indicates that the electron transport in the TFTs is controlled by the V_g and V_d [13].

PBTS and NBITS were carried out, to further examine the long-time stability of the TFTs with a sputtering power at 6 KW, a SiH_4 flow at 1300 sccm and a flow ratio of N_2O to SiH_4 at 30. As shown in Figure 4(a), under the high temperature of 60°C and bias voltage of 30 V, the V_{th} of the TFTs shifted from 1.36 V to 3.88 V after 1 hour. Figure 4(b) indicates that the V_{th} drifts from 1.53 V to -0.14 V under NBITS (60°C, -30 V, 4500 nit) after 1 hour.

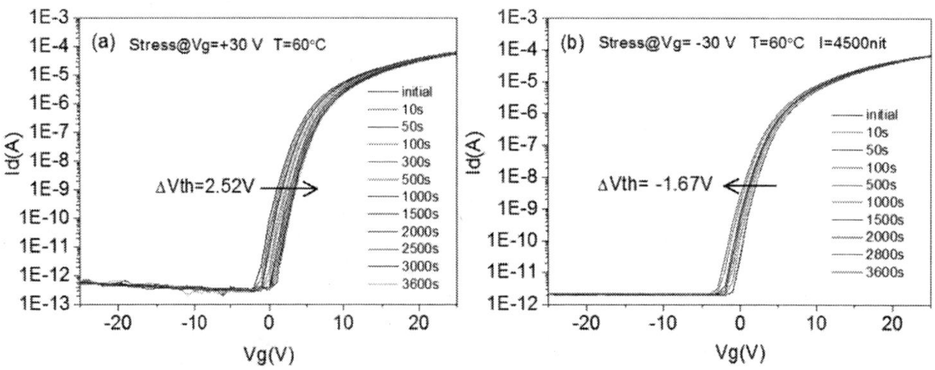

Figure 4 (a) PBTS @ Vg=+30 V, T=60°C and (b) NBITS @ Vg=-30 V, T=60°C, I=4500 nit of the BCE IGZO TFT with a SiO_x passivation layer deposited with a sputtering power at 6 KW, a SiH_4 flow at 1300 sccm and a flow ratio of N_2O to SiH_4 at 30.

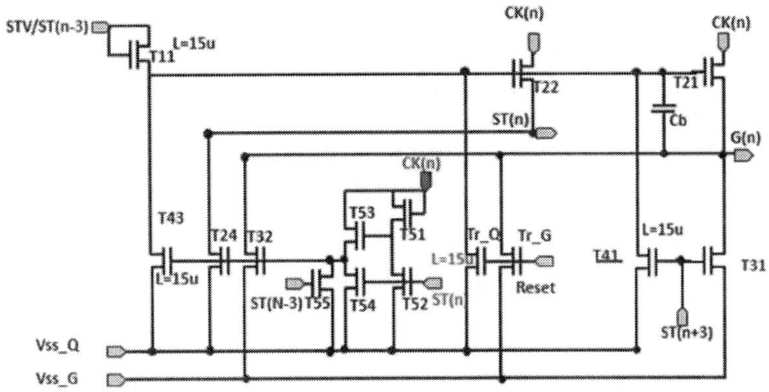

Figure 5 The circuit schematic of the gate-driver on array substrates.

The BCE IGZO TFTs using the above deposition process were applied in 32in FHD display devices. Figure 5 shows the circuit schematic of the gate on array (GOA), which integrates the gate drive circuit on the array substrate to reduce the size of the device frame. The GOA circuit is responsible for outputting the gate signal to the display area, so the stability of the TFTs in the circuit will determine whether the display is normal. In order to verify the stability of the device, an aging test was carried out. Figure 6 shows the photograph of the display device passed the 500-hour aging test at 60°C and 90% humidity. The device still displayed the picture properly, which proves that the TFTs used in the device have good stability and the deposition process of the SiO_x passivation layer does improve the performance of BCE IGZO TFTs.

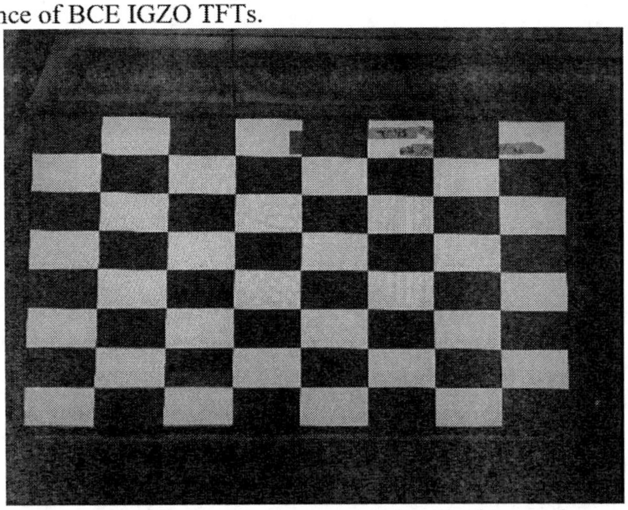

Figure 6 The photograph of the display device passed the 500-hour aging test at 60°C and 90% humidity.

4. Conclusions

The BCE IGZO TFTs with a SiO_x passivation layer were prepared on glass substrates. The influence of the deposition process of the SiO_x passivation layer on the electrical properties of the BCE IGZO TFTs has been studied. The sputtering power and the flow ratio of N_2O to SiH_4 are positively correlated to the V_{th} of the TFTs, while the SiH_4 flow rate and the V_{th} are negatively correlated. A V_{th} of 1.31 V, SS of 0.99 V/decade, I_{on}/I_{off} of 4.60×10^7 and μ_{fe} of 7.54 cm^2/Vs are observed when the SiO_x passivation layer has a sputtering power of 6 KW, a SiH_4 flow rate at 1300 sccm and a flow ratio of N_2O to SiH_4 at 30. The electron transport in the TFTs is controlled by the V_g and V_d as the output characteristic curves show clear pinch-off. In addition, the V_{th} of the TFTs shifted from 1.36 V to 3.88 V after 1 hour under PBTS (60°C, +30 V), while the V_{th} drifts from 1.53 V to -0.14 V under NBITS

(60°C, -30 V, 4500 nit). The TFTs were applied in 32in FHD display devices. The device still displayed the picture properly after the 500-hour aging test at 60°C and 90% humidity. A stable FHD display device based on the BCE IGZO TFTs was achieved.

Acknowledgements

Thanks Shenzhen China Star Optoelectronics Semiconductor Display Technology Co., Ltd for providing sample preparation and measurement platform. It was also supported by the School of Electronic and Computer Engineering, Peking University.

References

[1] Hirao, T., Furuta, M., Furuta, H., Matsuda, T., Hiramatsu, T., Hokari, H., Yoshida, M., Ishii, H., Kakegawa, M. (2007) Novel top-gate zinc oxide thin-film transistors (ZnO TFTs) for AMLCDs. Journal of the Society for Information Display, 15: 17-22.

[2] Barquinha, P., Pimentel, A., Marques, A., Pereira, L., Martins, R., Fortunato, E. (2006) Influence of the semiconductor thickness on the electrical properties of transparent TFTs based on indium zinc oxide. Journal of non-crystalline solids, 352: 1749-1752.

[3] Park, S.J., Jeon, D.Y., Ahn, S.E., Jeon, S., Montès, L., Kim, G.T., Ghibaudo, G. (2013) Static electrical characterization and low frequency noise of a-InHfZnO thin film transistors. Thin Solid Films, 548: 560-565.

[4] Ding, X., Qin, C., Song, J., Zhang, J., Jiang, X., Zhang, Z. (2017) The influence of hafnium doping on density of states in zinc oxide thin-film transistors deposited via atomic layer deposition. Nanoscale research letters, 12: 63.

[5] Kamiya, T., Nomura, K., Hosono, H. (2009) Origins of high mobility and low operation voltage of amorphous oxide TFTs: Electronic structure, electron transport, defects and doping. Journal of display Technology, 5: 468-483.

[6] Ge, S.M., Li, S., Chen, S.J., Kong, X.Y., Meng, Y.H., Shi, W., Shi, L.Q., Wu, W., Liu, X., Gan, Q.M., Zhao, Y., Zhang, C., Chiu, C.Y., Lee, C.Y. (2017) 42-1: Development of Cu BCE-Structure IGZO TFT for a High-ppi 31-in. 8K × 4K GOA LCD. SID Symposium Digest of Technical Papers, 48: 592-595.

[7] Nag, M., Rockele, M., Steudel, S., Chasin, A., Myny, K., Bhoolokam, A., Willegems, M., Smout, S., Vicca, P., Ameys, M. (2013) Novel back-channel-etch process flow based a-IGZO TFTs for circuit and display applications on PEN foil. Journal of the Society for Information Display, 21: 369-375.

[8] Ohara, H., Sasaki, T., Noda, K., Ito, S., Sasaki, M., Endo, Y., Yoshitomi, S., Sakata, J., Serikawa, T., Yamazaki, S. (2010) 4.0-inch active-matrix organic light-emitting diode display integrated with driver circuits using amorphous In–Ga–Zn-Oxide thin-film transistors with suppressed variation. Japanese Journal of Applied Physics, 49: 03CD02.

[9] Park, H.W., Bae, J., Kang, H., Kim, D.H., Jung, P., Park, H., Lee, S., Bae, J.U., Yoon, S.Y., Kang, I. (2019) P-3: A Study on the Hot Carrier Effect in InGaZnO Thin Film Transistors. SID Symposium Digest of Technical Papers, 50: 1222-1225.

[10] Lee, S., Chen, Y., Kim, H., Kim, J., Jang, J. (2019) P-14: Highly Robust Oxide TFT with Bulk Accumulation and Source/Drain/Active Layer Splitting. SID Symposium Digest of Technical Papers, 50: 1263-1266.

[11] Dewey, G., Chu, K.B., Boardman, J., Fastenau, J., Kavalieros, J., Kotlyar, R., Liu, W., Lubyshev, D., Metz, M., Mukherjee, N. (2011) Fabrication, characterization, and physics of III–V heterojunction tunneling field effect transistors (H-TFET) for steep sub-threshold swing. In: 2011 International electron devices meeting, IEEE. Washington. pp. 33.36.31-33.36.34.

[12] Park, J.C., Ahn, S.E., Lee, H.N. (2013) High-performance low-cost back-channel-etch amorphous gallium–indium–zinc oxide thin-film transistors by curing and passivation of the damaged back channel. ACS applied materials & interfaces, 5: 12262-12267.

[13] Cho, M.H., Seol, H., On, N., Kim, T.K., Yun, P.S., Bae, J.U., Park, K.S., Jeong, J.K. (2019) P-13: High Performance a-IGZO Thin-Film Transistors Grown by Atomic Layer Deposition: Cation Combinatorial Approach. SID Symposium Digest of Technical Papers, 50: 1259-1262.

IOP Conf. Series: Materials Science and Engineering **729** (2020) 012100 doi:10.1088/1757-899X/729/1/012100

Welding performance of the dissimilar joints of Fe-Mn-Si alloy and 304 stainless steel

Linlin Liu[1], Jinqiao Du[1], Chaoyu Zhou[2]

[1] School of Mechanical Engineering, Dalian Jiaotong University, Dalian, Liaoning, 116028, China

[2] Department of Mechanical and electrical, Weihai, Shandong, 264210, China

[*]Corresponding author's e-mail: good_linlin@163.com

Abstract. The dissimilar joints of Fe-17Mn-5Si-10Cr-5Ni memory alloy and 304 stainless steel were welded by 5 kw transverse flow CO_2 laser. The microstructure, morphology and composition of the welding of the dissimilar joints were analyzed by metallographic microscope, field emission scanning electron microscopy. The results showed that the crystals of dissimilar joints of Fe-17Mn-5Si-10Cr-5Ni memory alloy and 304 stainless steel were symmetrically distributed on both sides of the joint weld center. At Fe-17Mn-5Si-10Cr-5Ni memory alloy close to the side of the weld, ε martensite phase was also found, which was caused by the $\gamma \rightarrow \varepsilon$ martensite transformation induced by the residual stress.The distribution of dissimilar joint weld elements was in zones obviously, the distribution mode was related to the unique keyhole effect of laser welding, and the weld center metal was still Fe-Mn-Si memory alloy, which had the characteristic of the inverse phase transition of the stress induced martensite.

1. Introduction

Among the numerous shape memory alloys (SMAS), the Fe-Mn-Si based alloys have attracted much attention because of their good shape memory effect, mechanical properties, low cost and latent use [1-4]. So far, a lot of research has been done on the composition design, phase change mechanism and solid solution treatment of Fe-Mn-Si shape memory alloy[5-7]. Fe-Mn-Si alloy inevitably involves welding problems or connection with other dissimilar metal materials in the practical application process, however, at present,there are few reports on the welding of Fe-Mn-Si alloy and dissimilar metals[8]. Therefore, 304 stainless steel and stainless Fe-Mn-Si alloy were selected for CO_2 laser welding test. The mechanical behavior of dissimilar welding joints was evaluated through the analysis of welding area morphology and fracture, and the welding composition, microstructure and change rule of dissimilar welding joints were revealed through the observation of micro-area composition and microstructure.

2. Experiment Material and Method

In this study, the Fe-16.86Mn-4.50Si-10.30Cr-5.29Ni-0.08C(wt%) (shortened form Fe-Mn-Si alloy) and 304 stainless steel were prepared, and Fe-Mn-Si alloy was prepared by vacuum median frequency furnace, the cast ingot was first homogenized at 1200℃, reheated to 1100℃ for 1h. Welding specimens with efficient dimensions of 4mm×1mm×28mm were wire cut by electrical discharge machine. The solution treatment was carried out by austenitizing at 1000℃ for 1h in the presence of argon atmosphere.

A continuous wave (CW) CO_2 laser of 10 μm wavelength (5 kW maximum power) was utilized for material processing, and the optimal welding process parameters were obtained through orthogonal test: P=2000 W, V =250 mm/min, z=-0.5 mm.

Metallographical specimens were observed by OLYMPUS G×51microscope. The sample was polished with No. 1200 water sandpaper. In order to further analyze the microstructure of the weld,etched with a solution of 3g$CuSO_4$+10mLHCl+30mLH_2O and followed by etching with a solution of volume ratio of 3:1 HCL+HNO_3. Microstructure morphology of weld was analyzed by scanning electron microscope (SEM).

3. Result and Discussion

Figure 1 shows cross section of laser welding seam of 304 stainless steel and Fe-Mn-Si alloy. It can be seen that the welding joint of Fe-Mn-Si alloy and 304 stainless steel is well formed with good symmetry, no obvious pores or cracks, presenting an X shape, which is a typical laser penetration welding. The minimum welding width is located in the position above the middle of the welding seam, and the welding width ratio is close to 1:1, which indicates that the laser heat is evenly distributed along the thickness of the plate during welding. There are significant differences only at the bottom of the weld in figure 1, which indicates that metal melting and the inhomogeneity of heat conductivity present slightly larger at the bottom of the weld in the condition of laser welding.

Figure 1. Cross section of laser welding seam of 304 stainless steel and Fe-Mn-Si alloy

Figure 2 shows microstructure of the fusion zone of dissimilar steel joint. It can be seen that the fusion zone of 304 stainless steel side is obviously smaller than that of Fe-Mn-Si alloy side. From the perspective of crystal morphology, the distribution of 304 stainless steel base material to the weld center is in order of plane crystal, cellular crystal and dendrite cellular crystal [9], as shown in figure 2 (a), while the same distribution of microstructure also exists between Fe-Mn-Si alloy base material and the weld center. The difference is that the crystals size of Fe-Mn-Si alloy side is larger than that of 304 stainless steel, and the fusion zone is wider, as shown in figure2(b).

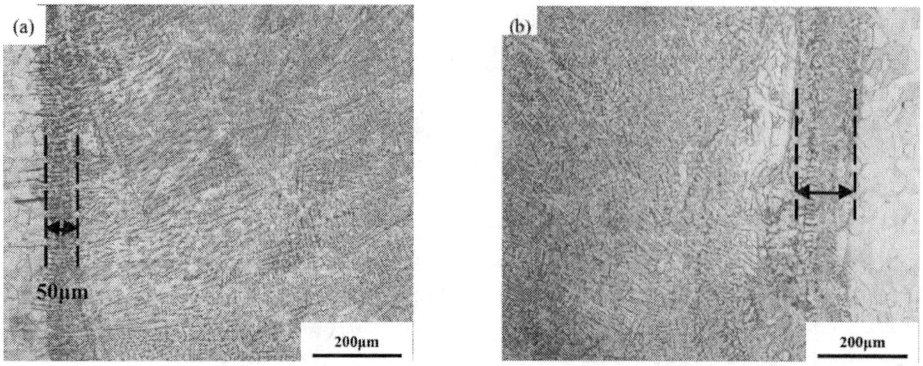

Figure 2. Microstructure of the fusion zone of dissimilar steel joint (a) Fusion zone close to the side of 304 stainless steel; (b) Fusion zone close to the side of Fe-Mn-Si alloy

It is worth noting that there are a large number of fine cellular crystal compositions in the planar crystal on the side of 304 stainless steel (figure 3). The interleaving phenomenon of grain boundary is more serious than that of Fe-Mn-Si alloy. This microstructure morphology can improve the tensile strength of the weld, but too much boundary interleaving means that the resistance to deformation increases. In the bending fatigue test, due to repeated alternating stress, excessive stress concentration will occur near the area, which is easy to germinate tiny micro-cracks.

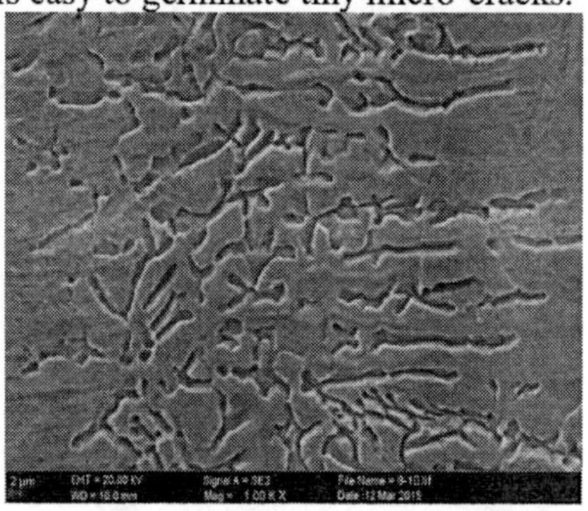

Figure 3. SEM of dissimilar joint fusion zone near the side of 304 stainless steel

Figure 4 shows the microstructure of the base material on both sides of the weld seam of the dissimilar joint. It can be seen that the structure of 304 stainless steel base material is still pure austenite (figure 4(a)), and the structure has not changed, which has the black pock markings are still austenite after welding. The formation of pock markings is because the arrangement of orientation is difference with different austenitic grain. The corrosion rate is different under the same etching condition, and the black one corrodes faster, forming pock markings. In addition to parent austenite, ε martensite is also observed in Fe-Mn-Si alloy base material (figure 4(b)). This is because the rapid heating and cooling of the laser in the welding process will lead to residual stress at the welding seam, which drives $\gamma \rightarrow \varepsilon$ martensite transformation of Fe-Mn-Si alloy. The behavior of stress induced ε martensite transformation of the dissimilar joint of Fe - Mn - Si alloy can release residual stress and improve the mechanical properties of the welding joint.

Figure 5 shows the scanning electron microscope photos of the dissimilar joint and 304 stainless steel joint with the instantaneous fault area. It can be seen that there are a large number of dimples in the instantaneous fault area of both of them, which shows the characteristics of plastic fracture. The difference is that the dimple size (figure 5(a)) and depth of the dissimilar joint are significantly larger than that of the 304 stainless steel joint (figure (b)). The size and depth of the dimple is greater, the toughness of the joint is better, which indicates that the toughness of dissimilar joints welding seam is better than that of 304 stainless steel joints.

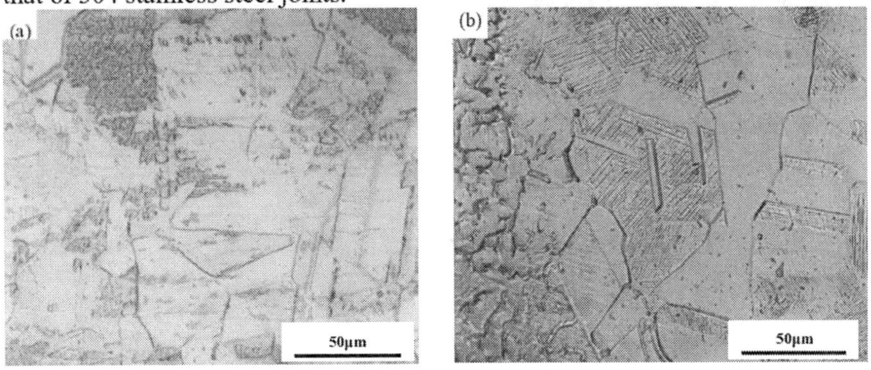

Fig. 4 Microstructure of the base material on both sides of the weld seam of the dissimilar joint (a) 304 stainless steel base material; (b) Fe-Mn-Si alloy

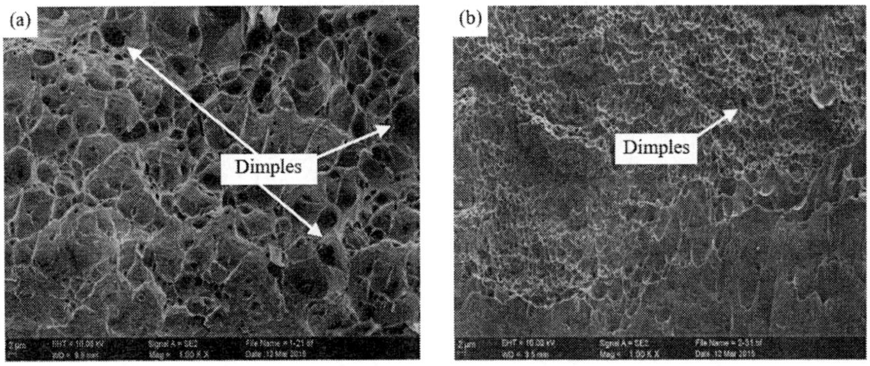

Fig. 5 SEM of the dissimilar joint and 304 stainless steel joint with the instantaneous fault area (a) Dissimilar steel welding joint; (b) 304 stainless steel welding joint

Figure 6 shows the composition test location of dissimilar welding joint. Figure 7 is the element content of each region according to the test position in figure 6. It can be seen that the change of weld element follows the principle that solute diffuses from high concentration to low concentration. From both sides of the welding base material to the center of the weld, the content of each element is continuously "neutralized", and the balance is reached at the center of the weld at 4 positions, which indicates that the unique keyhole effect of laser welding makes the butt joint approximate "paste" together which plays a positive role in maintaining the characteristics of the dissimilar joint base material. In terms of composition, there are differences in composition between 3# and 7# regions, but the weld metals belong to Fe - Mn - Si alloys, which have the characteristics of stress induced martensite positive and inverse phase transformation. Therefore, most of the weld positions of the dissimilar joints have the feature of crack stop, while the weld fusion area (2# area) near the base material of 304 stainless steel has the characteristic of the crack stop disappears, which is exactly consistent with the location of the fracture of the dissimilar joints in the bending fatigue test.

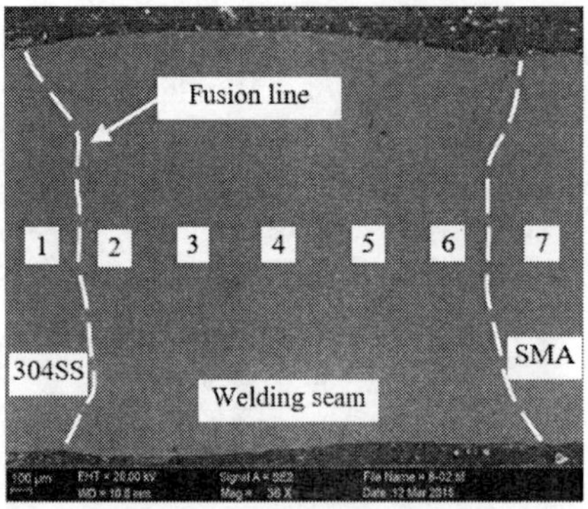

Figure 6. The composition test location of dissimilar welding joint

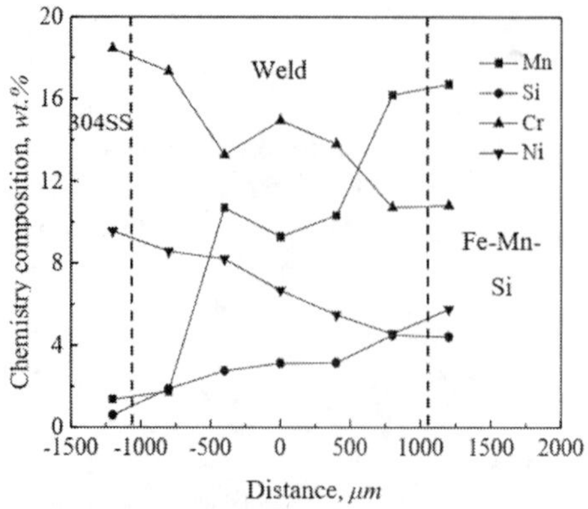

Figure 7. Chemistry composition of dissimilar welding joint (wt.%)

4. Conclusion

(1) The welding joint of Fe-17Mn-5Si-10Cr-5Ni alloy and 304 stainless steel was well formed and presented an X shape. The crystallization of the dissimilar joint started from the fusion zone on both sides of the welding base material and spreaded to the weld center at the same time. The distribution of crystal morphology was symmetrical on both sides of the weld center, followed by plane crystal, cellular crystal and dendrite crystal. E martensite was found in Fe-17Mn-5Si-10Cr-5Ni alloy near the weld, which was caused by residual stress induced $\gamma \rightarrow \varepsilon$ martensite phase transformation.

(2) The size and depth of the dimple of the dissimilar joint between Fe-17Mn-5Si-10Cr-5Ni alloy and 304 stainless steel were obviously larger than that of the dimple of the instantaneous fracture zone in the weld fracture of 304 stainless steel joint, which indicated that the toughness of the dissimilar joint weld was obviously better than that of 304 stainless steel joint.

(3) The composition distribution of laser weld elements in Fe-17Mn-5Si-10Cr-5Ni alloy and 304 stainless steel dissimilar joints was obviously regional, and the distribution mode was related to the keyhole effect of CO_2 laser welding. The weld metal near the side of Fe-17Mn-5Si-10Cr-5Ni alloy still belonged to Fe-Mn-Si memory alloy, which had the characteristics of stress induced martensite positive and reverse phase transformation.

References

[1] Yang, S. Z., Li, N., Li, H. (2016) Effects of electric pulse treatment on shape memory effect and microstructure in a pre-deformation Fe13Mn6Si13Cr4Ni0.1C alloy. Rare metal materials and engineering, 45(1): 0051-0055.

[2] Song, F., Zhang, C. Y., Wang, S. L. (2015) Effect of deformation temperature on shape memory effect of solution treated Fe-Mn-Si based alloys. Transactions of materials and heat treatment, 36(9):1-6.

[3] Liu, B., Zheng, Y. F., Ruan, L. Q. (2011) In vitro investigation of Fe30Mn6Si shape memory alloy as potential biodegradable metallic material. Materials letters, 65: 540-543.

[4] Zhang, Y. C., Song, F., Wang, S. L. (2015) Effect mechanism of Mn contents on shape memory of Fe-Mn-Si-Cr-Ni alloys. Acta metallurgica sinica, 51(2):201-208.

[5] Arghavani, J. F., Auricchio, R. N. (2011) An improved fully symmetric finite-strain phenomenological constitutive model for shape memory alloys. Finite elements in analysis and design, 47:166-174.

[6] Lu, Q., He, G. Q., Chen, S. J. (2015) Microstructure evolution of Fe-Mn-Si shape memory alloy during thermal mechanical training process. Journal of Materials Engineering, 43(4): 8-12.

[7] Xu, D. Q., Peng, H. B., Liu, G. (2011) Influence of thermo-mechanical treatment with different annealing temperature on shape memory effect of a Fe-Mn-Si-Cr-Ni alloy. Transactions of Materials and Heat Treatment, 32(3):30-34.

[8] Zhou, C. Y., Lin, C. X., Xu, P. (2014) In situ synthesis of Fe-Mn-Si memory alloy on the surface of stainless by laser cladding. Rare Metal Materials and Engineering, 43(12):3042-3046.

[9] Qiao, Z. X., Li, Z. M., Liu, Y. C. (2007) Study on microstructure and shape memory effect in weld joints of an Fe-Mn-Si-Cr-Ni alloy. Functional materials, 38(10):1645-1647.

IOP Conf. Series: Materials Science and Engineering **729** (2020) 012101 doi:10.1088/1757-899X/729/1/012101

Fatigue Crack Initiation and Propagation Mechanism of FGH96 PM Superalloy

Zhao Kai[1,2,3,4*], He Yu-huai[1,2,3,4]

[1] AECC Beijing Institute of Aeronautical Materials, Beijing 100095, China

[2] Beijing Key Laboratory of Aeronautical Materials Testing and Evaluation, Beijing 100095, China

[3] Key Laboratory of Science and Technology on Aeronautical Materials Testing and Evaluation, Aero engine corporation of China, Beijing 100095, China

[4] Aviation Key Laboratory of Science and Technology on Materials Testing and Evaluation, Beijing 100095, China

*Corresponding author's e-mail:372537284@qq.com

Abstract. FGH96 is regarded as a promising powder metallurgy superalloy for high-temperature / pressure turbine disc in aerospace industries due to its high resistance / defect tolerance and high working temperature up to 750℃. The initiation and propagation mechanism of FGH96 PM superalloy is researched by in-situ fatigue experiments. The result shows that the crack of gap samples initiates and propagates at the same stress level at 25°C and 300°C. At 300°C, there are no creep characteristics such as the intergranular characteristics found in the sample fracture. The crack at 300°C has a faster initiation and propagation process compared to the crack at 25°C, especially in the late stage of fatigue. The increase of the crack propagation rate of the sample is unrelated to the creep, mainly due to oxidation. Oxidation reduces the fatigue life of materials because it accelerates the initiation and propagation of transgranular cracks.

1.Introduction

In-situ fatigue test under scanning electron microscope (SEM) is a technique which the sample table in SEM is replaced in-suit fatigue assembly. The external loading is applied and the movement of the scanning electron beam is controlled by an external magnetic field to conform to the deformation of the sample[1-3]. The process of crack initiation and propagation can be intuitively understood by tracking and observing the in-situ tensile and fatigue processes of SEM, especially the changes of the sample surface during the loading process, which provides abundant information for the study of crack initiation and propagation[4-6].

In the first half of the 20th century, cyclic slip was considered to be the main cause of microcrack formation. In the 1930s, Gough suggested that fatigue cracks were formed because cyclic slippage exceeded the limit of local strain hardening[7-8]. 1939, Orowan adopted this viewpoint who believed that the exhausting of local ductile led to the increase of the local stress and crack eventually[9]. 1953, Head presented the first crack propagation rate model according to the concept. Forsyth found that materials smooth surfaces could form peaks and valleys due to different slip surfaces being invaded and extruded when materials suffered cyclic strain or stress and the slip band or stayed slip band was generated[10]. Hunsche and Ma[11] found that the interface between stayed slip band and adjacent matrices had a large dislocation density gradient. It was easy to generate the microcrack with the discordance of strain.

2.Experimental

FGH96 P/M superalloy was used to machining samples (Figure 1). The alloy in this investigation was an advanced nickel-base PM superalloy. The chemical composition of P/M nickel-based superalloy was listed in table 1. In-situ fatigue test was performed in the SEM-SERVO 550 machine and the test conditions were listed in table 2.

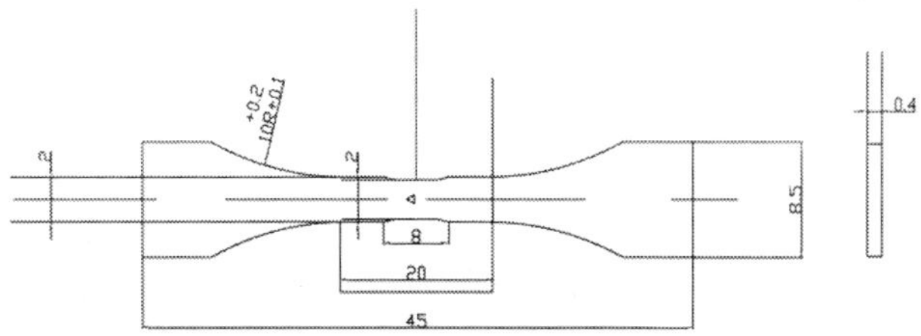

Figure 1. In-situ fatigue test sample

Table 1. Main chemical composition of FGH96 P/M superalloy in wt/%.

Cr	Co	W	Mo	Nb	Al	Ti	Zr	Fe
15.5～16.5	12.5～13.5	3.8～4.2	3.8～4.2	0.60～0.80	1.95～2.3	3.55～3.90	0.03～0.06	≤0.50
Mn	N	O	H	C	S	P	B	Si
≤0.02	≤0.006	≤0.015	≤0.001	0.045～0.060	≤0.0012	≤0.010	0.012～0.020	≤0.10

Table 2. FGH96 PM superalloy in-situ fatigue test conditions.

Sample style	**Temperature**	**Gap depth/mm**	**Thickness/mm**
Unilateral gap	25°C	0.5	0.40
Unilateral gap	300℃	0.5	0.42

3.Results and discussion

3.1.In-situ fatigue test (25°C)

In-situ fatigue test process (25°C) was listed in table 3. At the beginning, the frequency and the stress were low. After that, the frequency was stable at 3Hz and the stress increased. The purpose of this process was to avoid the thin sample warping.

Table 3. In-situ fatigue test process (25°C).

Stress/MPa	500	600	700	750	800	850	850	850	850	850	850
Fatigue cycles/Cycle	306	1000	1240	2097	2003	1044	972	974	1122	1070	270
Frequency/Hz	1	3	3	3	3	3	3	3	3	3	3

The grain appearance near the gap of the sample before loading was shown in Figure 2a. When the stress increased to 850MPa and the sample was operated under the stress for 1044 cycles, 16.4μm long crack was observed. The crack started at the gap edge A and extended into the sample through the grain boundary. There was also a microcrack B near the the crack A (Figure 2b). After 972 cycles of

850MPa, the crack A appearance was shown in Figure 2c. It could be seen that the crack A expanded obviously and its length was 42.8μm. But some microcracks in the middle were not connected. The crack B was not expanded clearly. After 974 cycles of 850MPa, the crack A expanded obviously and its length was 84.1μm. Microcracks in the middle were connected. Also, the crack B was not expanded clearly (Figure 2d). After 1,122 cycles of 850MPa, the crack A expanded to 154μm, and cracks in the middle broadened. Also, the crack B was not expanded clearly (Figure 2e). After 1,070 cycles of 850MPa, the crack A expanded to 320.8μm (Figure 2f). After 270 cycles of 850MPa, the sample broke. The total fatigue life of the sample was 12,098 cycles and the initiation life was 6646 cycles, so the initiation life accounted for 55% of the total life.

During the initiation and propagation of the whole fatigue crack, the crack slipped under the action of the cyclic load and the stress concentration. The microcrack appeared at the edge of the gap and expanded through the grain. When the crack encountered the grain boundary, deflection would occur. At the stress concentration in front of the main crack, new microcracks may form and connect with the main crack to expand. The main crack expanded with the increase of cycle, however, the microcrack next to it did not develop obviously.

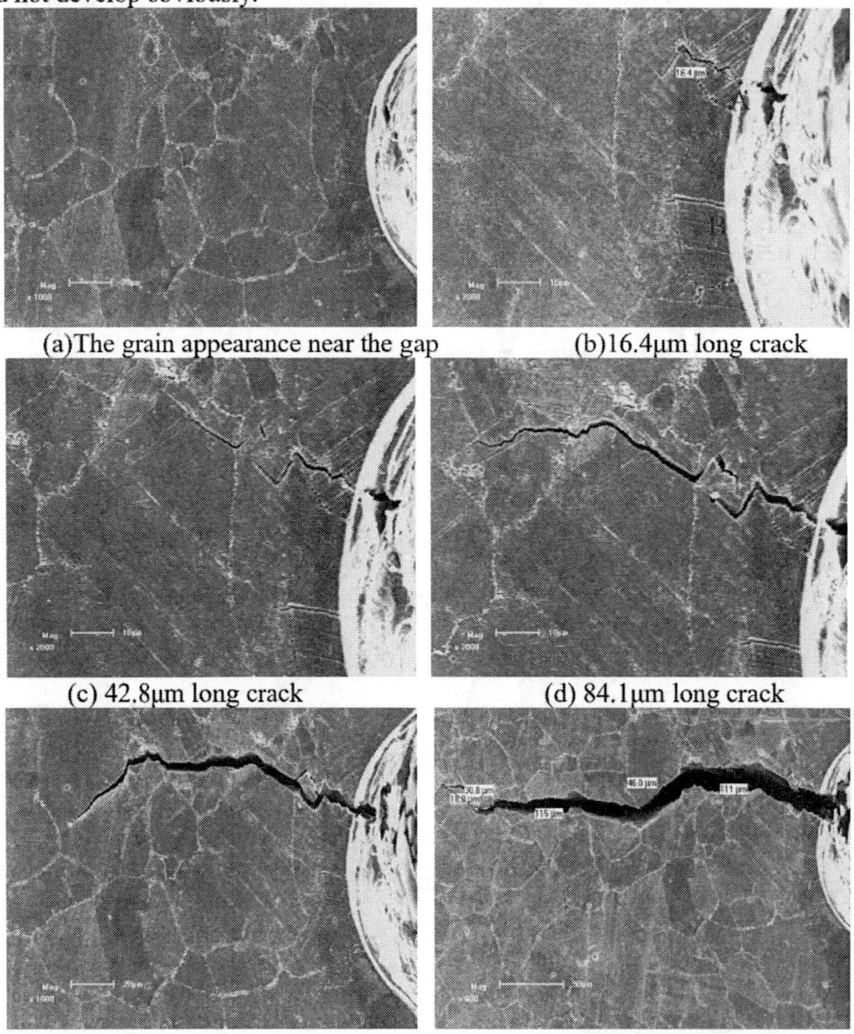

(a)The grain appearance near the gap (b)16.4μm long crack

(c) 42.8μm long crack (d) 84.1μm long crack

(e) 154μm long crack (f) 320.8μm long crack

Figure 2. In-situ fatigue crack expanding process (25°C)

The sample fracture was observed in the SEM. The fracture characteristic of the fatigue propagation in the first stage was class cleavage (Figure 3a). The length of the crack was 65μm and the life in the first stage was 2,503 cycles, so the crack propagation rate in the first stage was 2.5×10^{-5}mm/cycle. The fracture characteristic of the fatigue propagation in the second stage was fatigue stripe (Figure 3b). The length of the crack was 266μm and the life in the first stage was 2,949 cycles, so the crack propagation rate in the second stage was 9×10^{-5}mm/cycle. The fracture characteristic of the fatigue propagation in the late stage was dimple (Figure 3c).

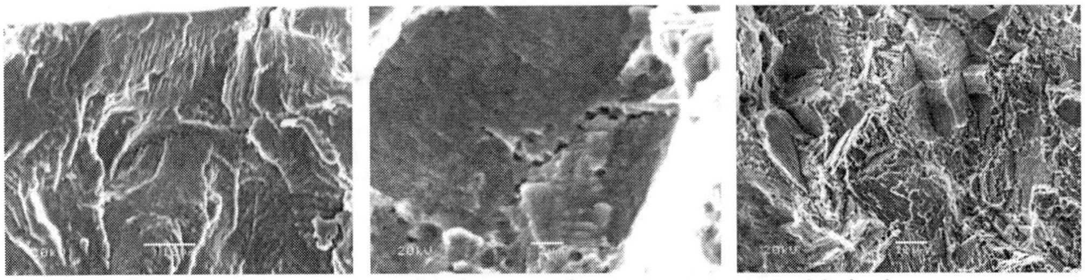

(a) the first stage (b) the second stage (c) the late stage

Figure 3. The sample fracture characteristic

3.2. In-situ fatigue test (300°C)

In-situ fatigue test process (300°C) was listed in table 4. The grain appearance near the gap of the sample before loading was shown in Figure 4a. When the temperature increased to 300°C without load, the material properties were unchanged. When the stress increased to 500MPa, there was no crack appeared. When the stress increased to 850MPa and the sample was operated under the stress for 1174 cycles, 50.67μm long crack appeared (Figure 4b). After 300 cycles of 850MPa, the crack propagated to 87.84μm (Figure 4c). After 302 cycles of 850MPa, the crack propagated to 176.9μm (Figure 4d). After 290 cycles of 850MPa, the sample broke. The microcrack appeared at the edge of the gap and expanded through the grain. When the crack encountered the grain boundary, deflection would occur.

The sample fracture was observed in the SEM. The length of the fatigue propagation in the first stage was short. There was little fracture characteristic so that it was too hard to distinguish the first stage and the second stage. The fatigue stripe could be observed near the source region (Figure 5a). The total length of the first stage and the second stage was 337μm and the total life was 2066 cycles, so the crack propagation rate of the total stages was 1.63×10^{-4}mm/cycle (Figure 5b).

Table 4. In-situ fatigue test process (300°C).

Stress/MPa	500	600	700	750	800	850	850	850	850
Fatigue cycles/Cycle	504	498	1062	1040	1278	1174	300	302	290
Frequency/Hz	1	3	3	3	3	3	3	3	3

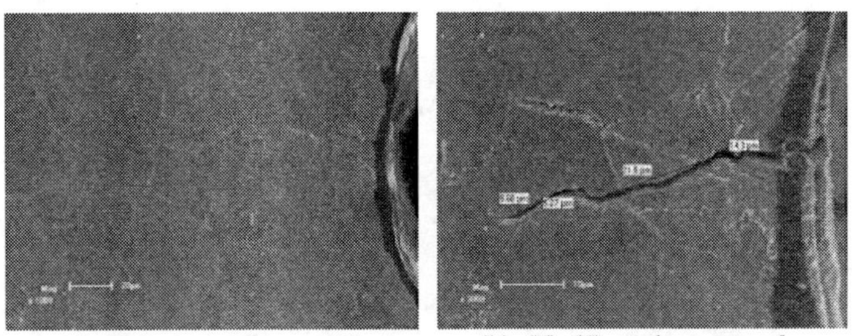

(a) The grain appearance near the gap (b) 50.67μm long crack

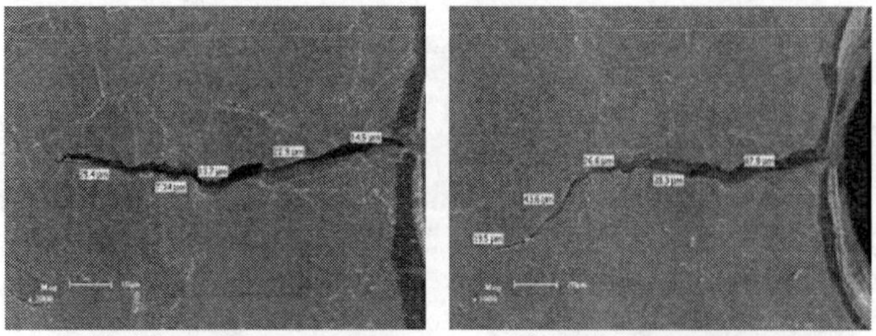

(c) 87.84μm long crack (d) 176.9μm long crack

Figure 4. In-situ fatigue crack expanding process (300°C)

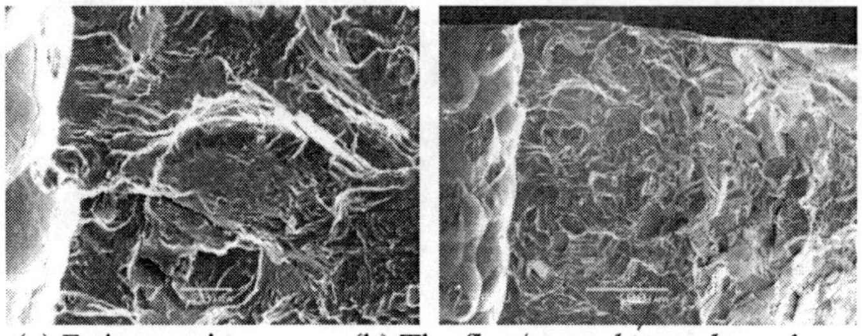

(a) Fatigue stripe (b) The first/second stage boundary

Figure 5. The sample fracture characteristic

4. Discussion

The in-situ fatigue test process at 300°C was as same as the test at 25°C. The total fatigue life of the sample at 300°C was 6448 cycles. The crack appeared when the stress increased to 850MPa. The sample broke after 2,066 cycles of 850MPa. The crack at 300°C had a faster initiation and propagation process compared to the crack at 25°C. The total fatigue life of the sample at 25°C was 12,098 cycles. Also, the crack appeared when the stress increased to 850MPa. The sample broke after 5,452 cycles of 850MPa.

At 300°C, the crack expanded so fast in the last 290 cycles that the length of the crack could not be confirmed. However, the length of the fatigue zone was 337μm measured by the SEM and the total life was 6,448 cycles, so the nominal crack propagation rate was 0.052μm/cycle, which included the fatigue first stage and the fatigue second stage. At 25°C, the length of the fatigue zone was 320.8μm and the total life was 12,098 cycles, so the nominal crack propagation rate was 0.027μm/cycle.

The temperature of 300°C caused the crack to initiate and expand rapidly, especially in the late stage of fatigue. The behavior that the fatigue life of materials decreased with the increased of temperature was usually caused by time-related damage. Generally, there were two kinds of time-related damage at high temperature, namely, creep and oxidation[12]. At 300°C, there were no creep characteristics such as the intergranular characteristics found in the sample fracture. Therefore, the increase of the crack propagation rate of the sample was unrelated to the creep, mainly due to oxidation. Oxidation reduced the fatigue life of materials because it accelerated the initiation and propagation of transgranular cracks.

5. Conclusions

1) At 25°C and 300°C, the crack of gap samples initiates and propagates at the same stress level;

2) At 300°C, there are no creep characteristics such as the intergranular characteristics found in the sample fracture;

3) The crack at 300°C has a faster initiation and propagation process compared to the crack at 25°C, especially in the late stage of fatigue;

4) The increase of the crack propagation rate of the sample is unrelated to the creep, mainly due to oxidation. Oxidation reduces the fatigue life of materials because it accelerates the initiation and propagation of transgranular cracks.

Acknowledgments
Authors wishing to acknowledge assistance or encouragement from colleagues, special work by technical staff or financial support from the National Science and Technology Major Project (2017-IV-0004-0041).

References
[1] Jablonski D A. (1981) The effect of ceramic inclusions on the low circle fatigue life of low carbon Astroloy subjected to hot isostatic pressing. Mater Sci Eng, 48: 189.
[2] Bretheau T, et al. (1988) Inclusion/matrix mechanical interaction an in situ study by tensile and fatigue tests in the scanning electron microscope. Strength of Metals and Alloys. Finland. pp. 87-91.
[3] Xie Xishan, Zhang Lina, Zhang Maicang et al. (2002) Micro-mechanical Behavior Study of Non-metallic Inclusions in Nickel-base P/M Superalloy. Acta Metallurgica Sinica, 6: 635-642.
[4] Ambroise W H, et al. (1987) Crack initiation from the interface of superficial inclusions. Conference: Mechanical Behavior of Materials - V, Beijing, pp. 169-174.
[5] Mander J M, Kortovich C S. (1981) Development of Materials and Process Technology for Dual Alloy Disks. NASA-CR-165224.
[6] Harf F H. (1985) Properties and Microstructurals for Dual Alloy Combinations of Three Superalloys with Alloy 901. NASA-TM-86987.
[7] Schijve J. (2003) Fatigue of structures and materials in the 20th century and the state of the art. International Journal of Fatigue, 25(8): 679-702.
[8] Barter S, Molent L, Goldsmith N, et al. (2005) An experimental evaluation of fatigue crack growth. Engineering Failure Analysis, 12(1): 99-128.
[9] Hunsche A, Neumann P. (1986) Quantitative measurement of persistent slip band profiles and crack initiation. Acta Metallurgica, 34(2): 207-217.
[10] Zou Jinwen, Wang Wuxiang. (1999) Characteristic and quality control of inclusions in P/M superalloy. Powder Metallurgy Technology,7: 201-206.
[11] Guo Wei-min, Feng di, Wu Jian-tao. (2002) Research and development of P/M superalloy metallurgic technigues. Journal of Materials Engineering, 3: 44-48.
[12] Ma BT, Laird C. (1989) Overview of fatigue behavior in copper single crystals-I. Surface morphology and stage I crack initiation sites for tests at constant strain amplitude. Acta Metallurgica, 37(2): 325-336.

IOP Conf. Series: Materials Science and Engineering **729** (2020) 012102 doi:10.1088/1757-899X/729/1/012102

Investigation of all solid state end-pumped Nd:YAG Q-switched laser

LIU Juan, ZHANG Yang, WANG San-zhao, SHI Yu and WANG Cai-li ,

China Building Materials Academy, Beijing 100024, China

*Corresponding author, E-mail: liujuanliuna@126.com

Abstract. High energy 1064 nm Q-switched laser output is obtained by LD array end pumping Nd:YAG. For different divergence angles of fast axis and slow axis of LD array, cylindrical lens and spherical lens are used for beam shaping of LD array. The laser properties of high energy 1064 nm Q-switched laser with pump energy is investigated with different Nd3+ ion doping concentration and different LD temperature. The results indicate that the extremely high gain formed at the end of the laser is easy to generate self-excited oscillation with the increasing of Nd3+ ion doping concentration and the increasing of LD temperature, which eventually results in the saturation of Q-switched laser pulse energy and makes it difficult to improve further. In view of saturation phenomenon of Q-switched laser output, Nd3+ ion doping concentration and LD temperature adjustment is proposed to reduce the absorption rate of laser medium to the pump light, which provides an effective technical means for obtaining high energy end pumped Q-switched pulsed laser. A 77.8mJ and 10 ns 1064 nm laser is generated with the LD temperature of 20 °C and the Nd3+ ion doping concentration of 0.8%, and the corresponding optical conversion efficiency is 13.9%.

1. INTRODUCTION

High-energy Q-switched lasers with low repetition frequency have been widely used in the fields of mass spectrometry, laser lidars and laser indication for the characteristics of high peak power and narrow pulse width [1-5]. The high energy Q-switched 1064 nm laser and its frequency conversion generated 532,355 nm laser are widely used in atmospheric monitoring of the lidar [6,7]. All solid state lasers pumped by semiconductor laser diodes (DPL) have the advantages of small size, long lifetime, high efficiency and good beam quality. The LD spectral line can be well matched with the absorption spectral line of the gain medium, because the LD spectral line of the pump source is narrow and the central wavelength is controllable. So the energy conversion efficiency is high, and the waste heat in the process of laser generation is less, so the wave front distortion caused by heat is reduced, which is conducive to the output of laser with high beam quality. In addition, the heat can be derived by thermoelectric cooler (TEC), which greatly improves the reliability and environmental adaptability of the system, so DPL has become a research hotspot for obtaining high energy Q-switched lasers.

The LD side-pumped Nd:YAG crystal lasers are primarily studied for obtaining high energy 1064 nm Q-switched laser due to their absorption equality and low cost. E. Armandillo et al. designed a Z-shaped geometric optical path based on slab and a Gauss unstable resonator to reduce wave-front distortion and improve laser beam quality. Under the condition of LD pumping energy 750 mJ, a 1064-nm pulsed laser with 100 mJ and 20 ns was produced, and the corresponding light-to-light conversion efficiency was 13.3%[3]. The Chen Weibiao research group of Shanghai Institute of Optics and Machinery of Chinese Academy of Sciences used Z-shaped Nd:YAG slab cut by Brewster angle to conduct the laser beam. LD and laser crystal waste heat are derived by cooling method, and Q-switched by KD*P is used to obtain 1064 nm laser output of 100 mJ and 10 ns. The corresponding light-to-light conversion efficiency is 13%[8]. In 2011, the project group used unstable resonators and two Z-shaped Nd:YAG slabs for laser amplification, and obtained 450 mJ, 10 ns laser output [9]. In 2017, Li Chaoyang presented a 2.36 J, 50 Hz, 9.4 ns laser with a diode side-pumped Nd:YAG MOPA system[10].

Compared with side-pumping form, end-pumping structure owns the superiority of high efficiency, good mode matching, and good wavelength matching, and so on. Meantime, the laser structure in end-pumped is more compact, which is conducive to reducing the volume and weight of the laser. Liu Xiaojuan et al. obtained 1064 nm laser output with pulse width of 7 ns and peak power of 3.4 kW in the Pulsed laser diode array pumped Nd:YVO4/GaAs laser[11]. L. Goldberg et al. generated 60 mJ of 1064 nm Q-switched laser, corresponding to optical conversion efficiency of 17%[3]. Using LD end-pumped Z-shaped Nd:YAG slabs, Paul R. Stysley et al. obtained 107 mJ laser output with an optical efficiency of 24% [12]. It can be seen that the optical conversion efficiency of end pump is better than that of side pump. However, the extremely high gain formed at the end of the laser is easy to generate self-excited oscillation, which eventually results in the saturation of Q-switched laser pulse energy and makes it difficult to improve further.

In this paper, the laser properties of high energy 1064 nm pulse laser with pump energy is investigated with different Nd3+ ion doping concentration and different LD temperature. High energy Q-switched laser is obtained by adjusting Nd3+ ion doping concentration and LD temperature, which can reduce the absorption rate of the laser medium to the pump light. When the temperature of LD is 20 ℃ and the doping concentration of the crystal rod is 0.8%, a 77.8 mJ and 10 ns 1064 nm laser is generated and the corresponding optical conversion efficiency is 13.9%.

2. EXPERIMENT AND ANALYSIS

2.1. Experimental setup

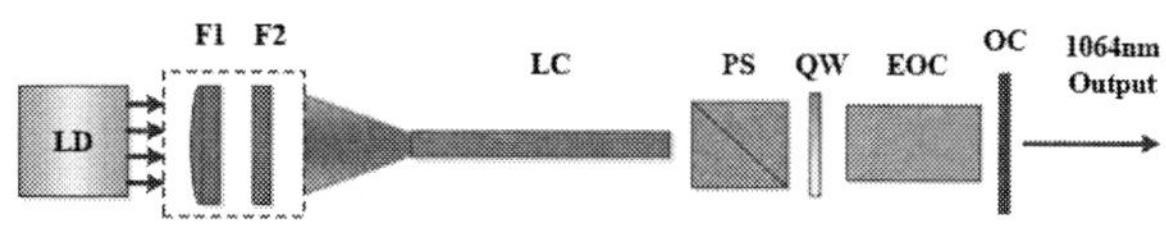

Fig. 1 Experimental device of end-pumped 1064 nm Q-switched laser

The experimental setup of the end-pumped Nd:YAG Q-switching is shown in Fig. 1, which includes the pump source LD, LD shaping lens F1, F2, laser crystal LC, polarizer PS, quarter-wave plate QW, electro-optic crystal EOC and output coupled mirror OC. LD is a single row vertical array consisting of 10 bars, with maximum output energy of 600 mJ. The bar spacing is 0.43 mm, and the divergence angles of the fast and slow shafts are 5 and 10 degrees, respectively. Due to the different divergence angles, the LD fast and slow axis is shaped by cylindrical lens and spherical lens with focal length 80 mm and 30 mm respectively. The distance between cylindrical lens and LD is 6.8 mm; the distance between cylindrical lens and spherical lens is 25.5 mm; the distance between crystal rod and spherical lens is 17 mm. The Nd:YAG rod was 5 mm in diameter and a 40mm long low-doped region with different Nd3+ ion doping concentration. Pump end of Nd:YAG is coated with 1064 nm high reflectivity film and 800 nm high transmittance film, and the other end is plated with 1064 nm anti reflection coating. PS is a polarizing beam splitter, which makes the vertically polarized s light reflect and horizontally polarized P light transmission. The Q-switched component modulates the Q value of the resonant cavity to produce a pulsed laser output, which concludes polarization beam splitter, quarter-wave plate and electro-optic crystal. The output coupling mirror OC with mirror curvature of 1100 mm and pump end face of Nd:YAG with mirror curvature of -2000 mm form the resonant cavity to generate a 1064 nm laser.

IOP Conf. Series: Materials Science and Engineering **729** (2020) 012102 doi:10.1088/1757-899X/729/1/012102

2.2. Experimental results

In the experiment, the output energy of Q-switched 1064 nm laser is measured by energy meter (PE50BF-DIFH-C, Ophir-Spiricon, Inc.) The output energy respect to the pump energy for different Nd3+ ion doping concentration with the LD temperature of 25 ℃ is shown in Fig. 2. It is found that the 1064 nm laser output energy is saturated when the pump energy is greater than 405 mJ with 0.8% Nd3+ ion doping concentration, and the output energy is 59.6 mJ at the pump energy of 498 mJ. The 1064 nm laser output energy does not appear saturation in the pump energy range 467 mJ with 0.6% Nd3+ ion doping concentration, and the output energy is 54.1 mJ at the pump energy of 498 mJ. The output energy of 1064 nm pulsed laser increases continuously with pumping energy with 0.4% Nd3+ ion doping concentration, and the output energy is 57.1 mJ at the pump energy of 498 mJ, and no saturation occurs. When the laser medium doping concentration is high, such as 0.6% or 0.8%, the extremely high gain formed in the front of the Nd:YAG, which can easily generate self-excited oscillation. This leads to the saturation of Q-switched laser pulse energy, which is difficult to further improve. However, the maximum output energy of 1064 nm pulsed laser with 0.8% Nd3+ ion doping concentration is larger than that with 0.4% and 0.6% Nd3+ ion doping concentration in the pump energy range of 498 mJ.

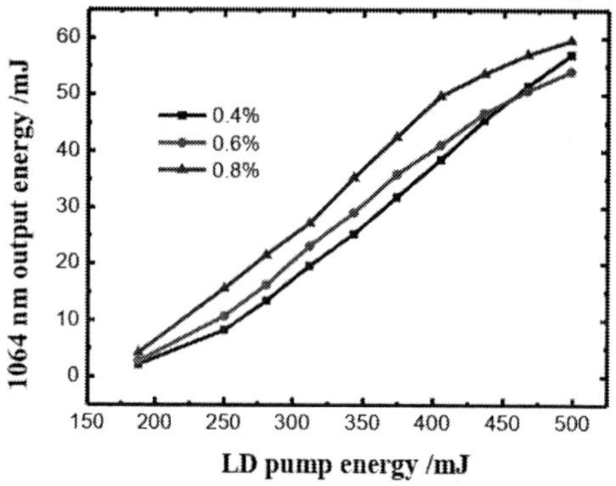

Fig. 2 Variation curves of 1064 nm laser output energy with the pump energy at different Nd3+ doping concentration

The output energy respect to the pump energy for different LD temperature with 0.8% Nd3+ ion doping concentration is shown in Fig. 3. When the LD temperature is 25 ℃, it is found that the 1064 nm laser output energy is saturated when the pump energy is greater than 405mJ. Subsequently, the output energy of 1064 nm pulsed laser at different temperatures is studied. It is found that pumping energy corresponding to the output laser saturates increases continuously as the LD temperature decreases. When the LD temperature is 23 ℃, 1064 nm laser output does not appear saturation in the pump energy range of 498 mJ. When the LD temperature is 20 ℃, the output energy of 1064 nm pulsed laser increases continuously with the increasing of the pump energy, and a 77.8mJ 1064 nm laser is obtained at the pump energy of 560 mJ, and the corresponding optical conversion efficiency is 13.9%. The laser pulse waveform of the maximum output energy is measured by a fast response silicon photo detector (DET10A, Thorlabs) and an oscilloscope (DPO 4104, Tektronix), as shown in Figure 4, and laser pulse width is about 10 ns.

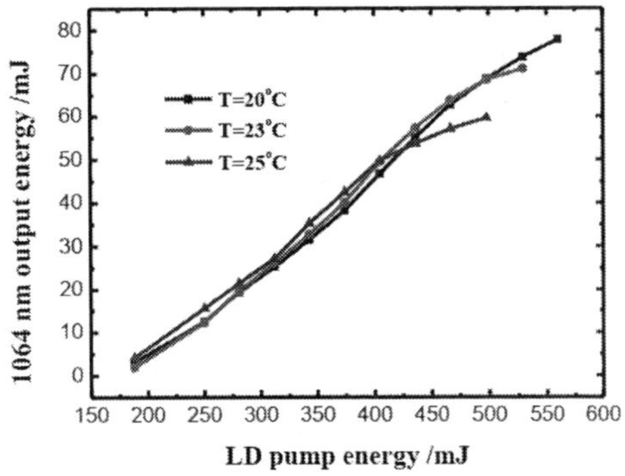

Fig. 3 Variation curves of 1064 nm laser output energy with the pump energy at different LD temperatures

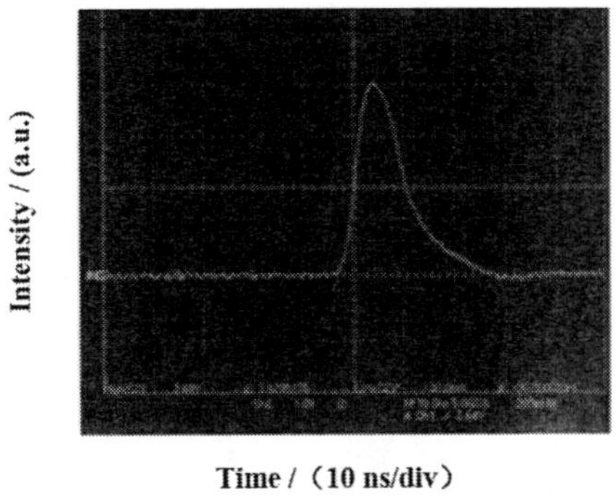

Fig. 4 Pulse waveform of 1064 nm laser

The LD spectra were measured by spectral analyzer (HR2000+CG, Ocean Optics) in the process of temperature changes for studying the saturation phenomenon of laser output energy. The LD wavelength decreases with the decrease of temperature, as shown in Fig. 5. The peak wavelength of LD changes from 796.1 nm to 799.2 nm when the LD temperature increases from 17.5 ℃ to 30 ℃, corresponding to wavelength variation coefficient of 0.25 nm/℃, and the LD wavelength is shifted away from the absorption peak of laser medium by adjusting the LD temperature. So the absorption rate of the gain medium to pump light decreases with the decreasing of LD temperature. The very high gain formed by end-to-end will easily produce self-excited oscillation with the increase of temperature. As a result, the energy of 064 nm pulsed laser pulse saturates, which makes it difficult to improve further.

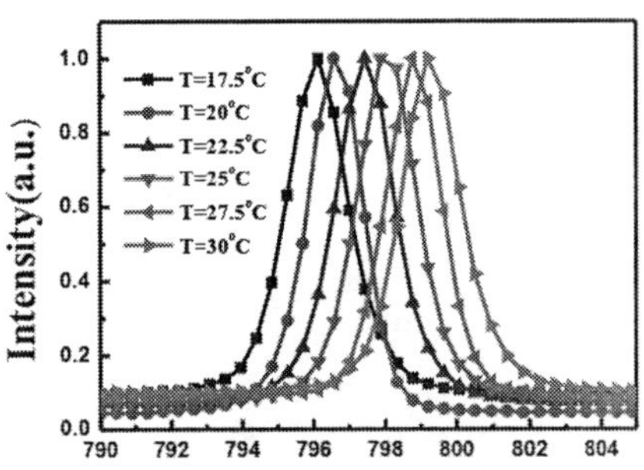

Fig. 5 The wavelength of LD at different temperature

IOP Conf. Series: Materials Science and Engineering **729** (2020) 012102 doi:10.1088/1757-899X/729/1/012102

3. CONCLUSIONS

The laser properties of 1064 nm pulsed laser with different Nd3+ ion doping concentration and different LD temperature are investigated in this study. For different divergence angles of fast axis and slow axis of LD array, cylindrical lens and spherical lens are used for beam shaping. The laser properties of 1064 nm pulsed laser with pump energy is investigated with different Nd3+ ion doping concentration and different LD temperature. The results indicate that the extremely high gain formed at the end of the laser is easy to generate self-excited oscillation with the increasing of Nd3+ ion doping concentration and the increasing of LD temperature, and finally leads to the saturation of Q-switched laser pulse energy.For the saturation phenomenon of 1064 nm pulsed laser energy output, LD temperature adjustment is carried out to make the pump wavelength to deviate from the absorption peak of Nd:YAG crystal, and reduce the end-face gain of laser medium pumping, which effectively suppress the self-excited oscillation, and provide an effective technical means for obtaining high-energy end-pumped Q-switched laser output. A 77.8mJ and 10 ns 1064 nm laser is obtained at the pump energy of 560 mJ with the LD temperature of 20 ℃ and the Nd3+ ion doping concentration of 0.8%, and the corresponding optical conversion efficiency is 13.9%.

References

[1] Goldberg, L., Mcintosh, C., Cole, B. (2011) VCSEL end-pumped passively Q-switched Nd:YAG laser with adjustable pulse energy [J]. Optics Express, 19(5): 4261-4267.
[2] Goldberg, L., Nettleton, J., Schilling, B. et al. (2007) Compact laser sources for laser designation, ranging and active imaging [J]. Proceedings of SPIE, 6552: 65520G-1-8.
[3] Armandillo, E., Norrie, C., Cosentino, A. et al. (1997) Diode-pumped high-efficiency high-brightness Q-switched ND:YAG slab laser [J]. Optics Letters, 22(15): 1168-1170.
[4] Pappalardo, G., Amodeo, A., Pandolfi, M. et al. (2004) Aerosol Lidar Intercomparison in the Framework of the EARLINET Project. 3. Raman Lidar Algorithm for Aerosol Extinction, Backscatter, and Lidar Ratio[J]. Applied Optics, 43(28): 5370-5385
[5] Schilling, B.W., Chinn, S.R., Hays, A.D., Goldberg, L., Trussell, C.W. (2006) End-pumped 1.5 μm monoblock laser for broad temperature operation, Appl. Opt. 45, 6607-6615,
[6] Zhang, W. (2013) Design and calibration of polarization channel in water vapor–cloud-aerosol lidar [D]. Qingdao: Ocean University of China. (in Chinese)
[7] Zhou, H.J. (2013) Studies on remote sensing of aerosol and trace gas vertical distribution by the MAX-DOAS [D]. Hefei: University of Science and Technology of China. (in Chinese)
[8] Ma, X.H., Bi, J.Z., Hou, X., et al. (2008) Conductively cooled all-solid-state zigzag slab laser [J]. Chinese Optics Letters, 6(5): 366-368.
[9] Ma, X.H., Bi, J.Z., Hou, X., et al. (2011) High efficiency diode-pumped slab oscillator and amplifier forspace-based application [J]. Optics & Laser Technology, 43:559–562.
[10] Li, C.Y., Lu, C.Q.*, Li, C., Yang, N. (2017) 2.36 J, 50 Hz nanosecond pulses from a diode side-pumped Nd:YAG MOPA system, Opt.394 :1–5
[11] Liu, X.J., Fu, R.L., et al. (2006) Pulse-LDA-pumped Passively Q-switched Nd:YVO4 Laser with GaAs as Saturable Absorber[J]. Journal of Optoelectronics · Laser, 12（17）:1443- 1447.
[12] Paul, R.S., Demetrios, P., Donald, B.C., et al. (2008) Highly efficient dual head 100 mJ TEM00 Nd:YAG Oscillator,Optics & Laser Technology, 40: 435–440.

IOP Conf. Series: Materials Science and Engineering **729** (2020) 012103 doi:10.1088/1757-899X/729/1/012103

Unique hierarchical Sn_3O_4 nanoflowers decorated with ZnO nanoparticles and their formaldehyde gas sensing properties

Lei Wang[1], Yang Li[1], Song Gao[1] and Wenjing Yue[1*]

[1] School of Information Science and Engineering, University of Jinan, Jinan, Shandong, 250022, China

[*]Corresponding author's e-mail: ise_yuewj@ujn.edu.cn

Abstract. Gas sensors with high response are highly desirable for the detection of harmful gases in real-time. Herein, the ZnO nanoparticles decorated hierarchical Sn_3O_4 nanoflowers are synthesized by a hydrothermal method. The gas sensor based on the ZnO/Sn_3O_4 composites exhibits excellent sensing properties to formaldehyde with a low operating temperature of 180°C. Specifically, the sensor has a response value of 44 for 100 ppm formaldehyde and a fast response time of 5 s. Moreover, it is noteworthy that the proposed gas sensor shows a wide detection range and a low detection limit of 1 ppm. Besides, the ZnO/Sn_3O_4 sensor exhibits good selectivity for formaldehyde and perfect reproducibility. The sensing mechanism of the gas sensor is discussed, and the heterojunction between ZnO and Sn_3O_4 enhances the sensing performance of the proposed gas sensor.

1. Introduction

Formaldehyde, as a common toxic volatile organic compound, is widely found in various plastic products, artificial floor and furniture, and it is the main pollutant after the renovation of new homes. Even at very low concentrations, it can cause headaches, fatigue, difficult breathing and other health threats, and formaldehyde is classified as a primary carcinogen by the World Health Organization [1].

Gas sensors based on metal oxide semiconductor materials have attracted extensive attention in detecting toxic and harmful gases, due to their simple operation and low cost. Designing and preparing gas sensors with high response, fast response and recovery time, low detection limit, and good repeatability has attracted a lot of research. Since the performance of gas sensors is closely related to the morphology of sensing materials, gas sensors based on metal oxide semiconductor materials with different morphologies have been deeply studied, such as nanowires [2], nanorods [3], and flower-like nanostructure [4]. In particular, gas sensors based on metal oxide semiconductors with three-dimensional (3D) hierarchical architecture have attracted strong interest and have been extensively researched, due to the good surface permeability and large specific surface area [5]. In recent years, Sn_3O_4 as an n-type semiconductor with the mixed valence of tin has been synthesized and applied in the field of gas sensors, due to the excellent physical and chemical properties. Liu et al. have synthesized flower-like Sn_3O_4 for ethanol sensor, and the sensor exhibits a low detection limit and good repeatability to ethanol [6]. Ma et al. have prepared the sensor based on Ce-doped Sn_3O_4 and the sensor shows enhanced sensitivity to formaldehyde [7]. However, these gas sensors generally have a low response to the target gas and high operating temperature.

To further improve the properties of gas sensors, heterojunction has been extensively studied as a popular method. In this work, the ZnO nanoparticles decorated hierarchical Sn_3O_4 nanoflowers have been synthesized, and the proposed ZnO/Sn_3O_4 gas sensor exhibits an enhanced response to

IOP Conf. Series: Materials Science and Engineering **729** (2020) 012103 doi:10.1088/1757-899X/729/1/012103

formaldehyde taking the advantage of the heterostructure of ZnO/Sn_3O_4 composites. Excellent gas sensing properties are obtained at the low operating temperature of 180°C, including high response, a low detection limit of 1 ppm, and good repeatability. The results demonstrate the potential application of the synthesized ZnO nanoparticles decorated hierarchical Sn_3O_4 nanoflowers, and the mechanism of the enhanced sensing properties of heterojunction is discussed.

2. Experimental section

2.1. Synthesis of ZnO nanoparticles decorated hierarchical Sn_3O_4 nanoflowers

All the chemical reagents are analytical grade and without any further purification. The synthesis of ZnO nanoparticles decorated hierarchical Sn_3O_4 nanoflowers is carried out by the hydrothermal method. First, 1.128 g $SnCl_2 \cdot 2H_2O$ and 3.675 g $C_6H_5Na_3O_7 \cdot 2H_2O$ are dissolved into 25 mL distilled water under magnetic stirring to form a uniform solution, and then, 0.1 g NaOH is added into the uniform solution and magnetic stirring for 2 h to form the precursor solution. After that, the precursor solution is transferred into a 50 mL Teflon-line stainless steel autoclave and heated at 180°C for 12 h to form Sn_3O_4 suspension, and then cooled down to room temperature naturally. Thereafter, 10 mL of a solution containing 60 mg of ZnO nanoparticles is added to the Sn_3O_4 suspension, stirring for 30 min, followed by hydrothermally heated at 180°C for 6 h. The obtained precipitates are washed several times with distilled water and ethanol. At last, the ZnO nanoparticles decorated hierarchical Sn_3O_4 nanoflowers are obtained after drying for 12 h at 80°C. Besides, pure Sn_3O_4 nanoflowers are synthesized without the addition of the ZnO nanoparticle solution.

2.2. Characterization and sensor fabrication

The crystal of the synthesized ZnO/Sn_3O_4 composites and pure Sn_3O_4 are investigated using X-ray diffractometer (XRD, Bruker D8 Advance) with Cu $K_{\alpha 1}$ radiation. The morphologies of the synthesized samples are examined by field-emission scanning electronic microscope (FESEM, Regulus-8100).

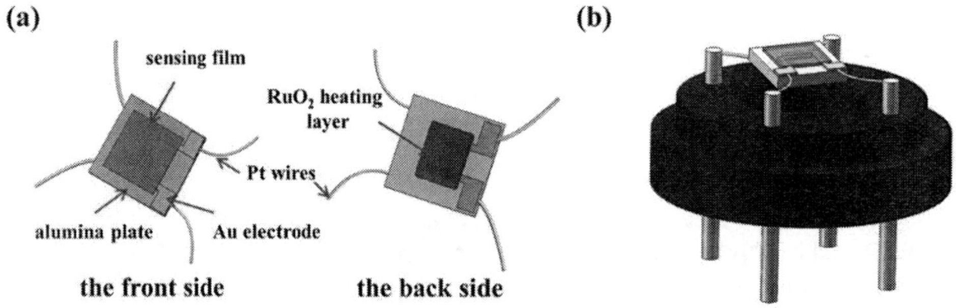

Figure 1. The structure diagrams of (a) the alumina plate and (b) the gas sensor.

The fabrication progress of the gas sensor is as follows: the obtained products are mixed with distilled water to form a homogeneous paste in an agate mortar and then smeared onto the front side of the ceramic plate to form a sensing film to cover the pair of Au electrode. There is a heating layer of ruthenium oxides on the back side, and the operating temperature of the sensor is provided by adjusting the heating voltage. Finally, the Pt wires connecting the Au electrodes are welded onto the pedestal to obtain a gas sensor. The structure diagrams of the gas sensor are shown in figure 1. The gas sensing properties are measured by a static test system. The response (S) of the sensor is defined as $S = R_a/R_g$, where R_a and R_g are the resistance of the gas sensor in air and the test gas, respectively. The response time (T_{res}) and recovery time (T_{rec}) are defined as the time required for reaching a 90% variation of the resistance change during the gas adsorption and desorption process.

3. Results and discussion

3.1. Structural characteristics and gas sensor performance

The crystal structure of the synthesized pure Sn_3O_4 and the ZnO/Sn_3O_4 samples are analysed by XRD. Figure 2(a) exhibits the crystal structure of the pure Sn_3O_4 product, and the main diffraction peaks match well with the triclinic Sn_3O_4 structure standard card (JCPDS: 16-0737). For the XRD pattern of ZnO/Sn_3O_4 sample shown in figure. 2(b), one can see that, besides the diffraction peaks of the triclinic Sn_3O_4, the diffraction peaks belong to ZnO also match well with the hexagonal wurtzite structure standard card (JCPDS 36–1451). No other diffraction peaks are found here, demonstrating the ZnO/Sn_3O_4 composites with a better crystallinity.

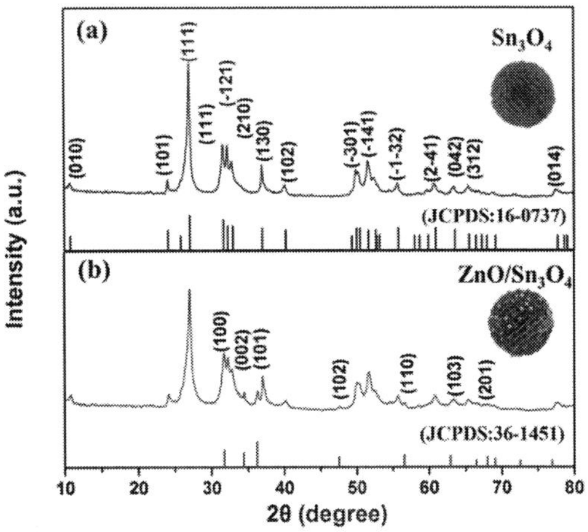

Figure 2. XRD patterns of the (a) pure Sn_3O_4 and (b) ZnO/Sn_3O_4 composites.

The morphologies of the prepared pure Sn_3O_4 and ZnO/Sn_3O_4 samples are characterized by the FESEM. From the low-magnification and high-magnification FESEM images of the pure Sn_3O_4 sample in figure 3(a) and (b), it can be seen that the pure Sn_3O_4 product exhibits the hierarchical nanoflower structure which is consisted of thin nanosheets. For the FESEM images of the ZnO/Sn_3O_4 composites shown in figure 3(c), the sample still maintains the nanoflower structure. From the high-magnification FESEM image of the ZnO/Sn_3O_4 sample in figure 3(d), it can be clearly seen that ZnO nanoparticles are attached to the nanosheets of Sn_3O_4.

The gas sensing properties of the proposed sensors based on the pure Sn_3O_4 and ZnO/Sn_3O_4 are completely studied. Since the response of the gas sensor is affected by temperature, the sensor exhibits different sensing properties at different working temperatures. To determine the optimal operating temperature of the sensor, the responses of the two sensors to 100 ppm formaldehyde at different operating temperatures are tested. As shown in figure 4(a), both sensors have a higher sensing response to formaldehyde at a lower operating temperature, and as the temperature increases, the response of both sensors decreases. At the same time, it can be seen that the response of the ZnO/Sn_3O_4 gas sensor is higher than the pure Sn_3O_4 gas sensor. Although the ZnO/Sn_3O_4 sensor has a higher response to formaldehyde at lower operating temperatures, the recovery time of the sensor at low temperatures is long, as shown in figure 4(b), which is not conducive to the practical application. To detect the gas concentration in real-time, 180°C is selected as the optimal working temperature. The dynamic response of ZnO/Sn_3O_4 gas sensor to 100 ppm formaldehyde at the working temperature of 180°C is shown in figure 4(c), and the response and recovery times are 5 s and 86 s, respectively.

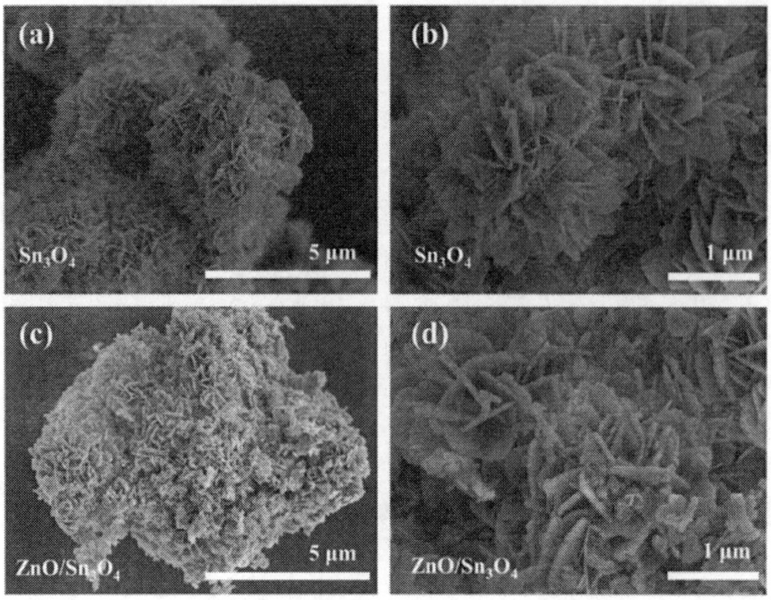

Figure 3. Typical FESEM images of (a-b) pure Sn_3O_4 nanoflowers and (c-d) ZnO decorated Sn_3O_4 nanoflowers.

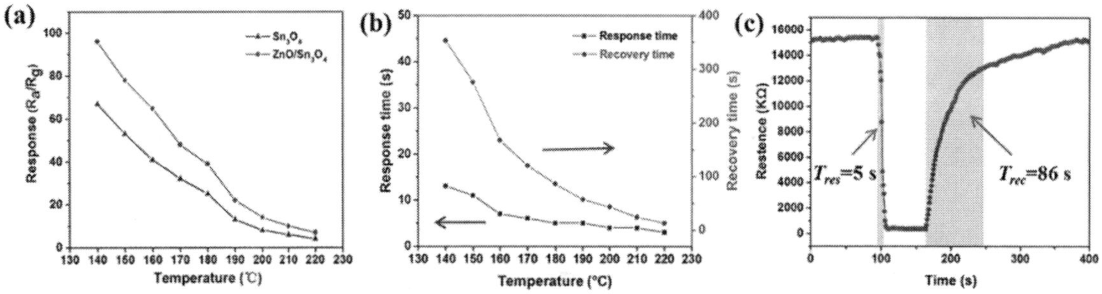

Figure 4. (a) The responses of sensors to 100 ppm formaldehyde at different operating temperatures, and (b) the response and recovery time of the ZnO/Sn_3O_4 sensor at different operating temperatures. (c)The response and recovery curve of the ZnO/Sn_3O_4 sensor to 100 ppm formaldehyde at 180°C.

To further study the sensing properties of the ZnO/Sn_3O_4 gas sensor, the responses of the sensor to different concentrations of formaldehyde are tested, as shown in figure 5(a). It can be observed that the response increases significantly with increasing formaldehyde concentration, and the response to 500 ppm formaldehyde reaches 96, demonstrating that the ZnO/Sn_3O_4 gas sensor has a wide detection range. Besides, it should be noted that the sensor has a certain response (3.5) to 1 ppm formaldehyde, indicating that the ZnO/Sn_3O_4 sensor has a low detection limit. Selectivity is very important in the practical application of gas sensors. Figure 5(b) exhibits the response of the proposed ZnO/Sn_3O_4 sensor to 100 ppm of various testing gases at the operating temperature of 180°C. The response to 100 ppm formaldehyde is 44, which is significantly higher than other test gases, including acetone, ethanol, benzene, ammonia, and xylene. Consequently, the proposed ZnO/Sn_3O_4 sensor could be used to detect formaldehyde. To characterize the repeatability of the ZnO/Sn_3O_4 sensor, the five reversible cycle tests of 100 ppm formaldehyde are performed, as exhibited in figure 5(c). It is observed that the ZnO/Sn_3O_4 sensor shows a reversible response and recovery characteristic, which demonstrates that the sensor has perfect repeatability.

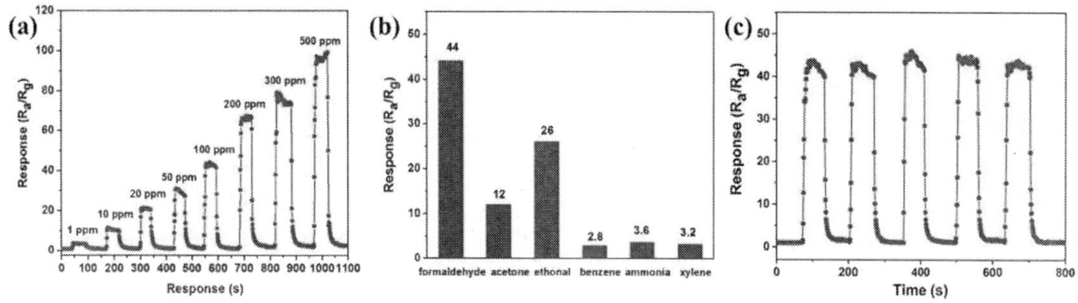

Figure 5. (a)The response curves of the ZnO/Sn_3O_4 sensor to different concentrations of formaldehyde. The selectivity (b) and repeatability (c) of the ZnO/Sn_3O_4 sensor.

3.2. Gas sensing mechanism

As it is known, Sn_3O_4 and ZnO are n-type metal oxide semiconductor, and the electrons are the main carriers. The gas sensing mechanism is related to the change of resistance when the oxygen species on the surface of the sensing materials react with the target gas [8-9]. When the sensor is in the air atmosphere, oxygen molecules will be adsorbed on the surface of the gas sensor and capture free electrons in the sensing materials to form O_2^-, O^-or O^{2-}, increasing the effective barrier height Φ_{eff} and resistance of the sensor [10-11]. When the sensor is exposed to the test gas, formaldehyde will react with the adsorbed oxygen species and release electrons, reducing the resistance of the sensor.

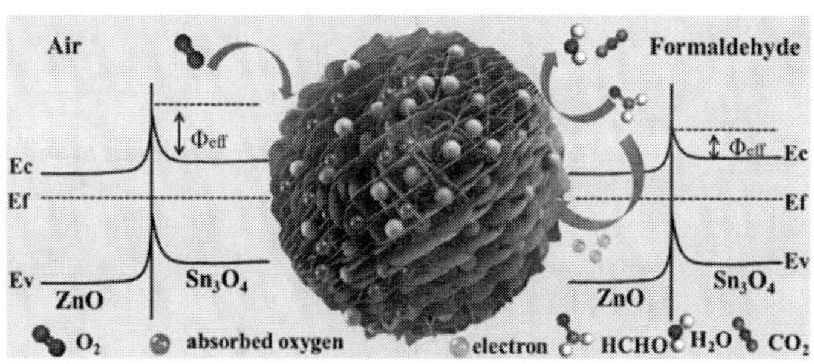

Figure 6. Schematic diagrams of the ZnO/Sn_3O_4 heterojunction energy band structure and the formaldehyde sensing mechanism.

For the ZnO/Sn_3O_4 gas sensor, the n-n heterojunction between ZnO and Sn_3O_4 has an important impact on improving sensor response. Due to the different work function of ZnO (4.45 eV) and Sn_3O_4 (3.9 eV) [12-13], the electrons will transfer from Sn_3O_4 to ZnO, and the electron accumulation layer will be formed on the surface of ZnO. The diagram of the ZnO/Sn_3O_4 heterojunction energy band structure is proposed as shown in figure 6. The increasing electrons on the surface of the sensing material will increase the amount of adsorbed oxygen, further increasing the effective barrier height Φ_{eff} and increasing the resistance of the sensor [14]. When the sensor is exposed to the test gas formaldehyde, these gas molecules could react with the adsorbed oxygen species, releasing electrons back into the conduction band of the sensing material. This process will increase the electrons in the conduction band, and lower the effective barrier height Φ_{eff}, reducing the resistance of the sensor [15]. Therefore, the large resistance change of the sensor results in a high response.

4. Conclusions

In summary, the pure Sn_3O_4 and ZnO nanoparticles decorated hierarchical Sn_3O_4 nanoflowers are successfully synthesized by the hydrothermal method. The crystal structure and morphology of the materials are characterized by the X-ray diffractometer and field-emission scanning electron microscope. The results show that the synthesized Sn_3O_4 exhibits a flower-like structure composed of nanosheets, and the ZnO nanoparticles are attached to the Sn_3O_4 nanosheets. The proposed sensor based on the ZnO/Sn_3O_4 composite exhibits excellent sensing properties to formaldehyde at a low operating temperature of 180°C, including high response, fast response and recovery time, low detection limits and good repeatability. In addition, the sensing mechanism of the gas sensor is discussed. As a result, the proposed sensor based on ZnO/Sn_3O_4 composites exhibits perfect sensing performance, and the sensor can be used to accurately and in real-time detect formaldehyde.

References

[1] Park, H.J., Hong, S.Y., Chun, D.H., Kang, S.W., Park, J.C., Lee, D.S. (2019) A highly susceptive mesoporous hematite microcube architecture for sustainable P-type formaldehyde gas sensors. Sens. Actuators B, 287:437-444.

[2] Li, Z.W. (2017) Supersensitive and superselective formaldehyde gas sensor based on NiO nanowires. Vacuum, 143:50-53.

[3] Choi, S., Bonyani, M., Sun, G.J., Lee, J.K., Hyun, S.K., Lee, C. (2018) Cr_2O_3 nanoparticle-functionalized WO_3 nanorods for ethanol gas sensors. Appl. Surf. Sci., 432:241-249.

[4] Deng, H., Li, H.R., Wang, F., Yuan, C.X., Liu, S., Wang, P., Xie, L.Z., Sun, Y.Z., Chang, F.Z. (2016) A high sensitive and low detection limit of formaldehyde gas sensor based on hierarchical flower-like CuO nanostructure fabricated by sol–gel method. J. Mater. Sci. Mater. Electron., 27:6766-6772.

[5] Zhu, K.M., Ma, S.Y., Tie, Y., Zhang, Q.X., Wang, W.Q., Pei, S.T., Xu, X.L. (2019) Highly sensitive formaldehyde gas sensors based on Y-doped SnO_2 hierarchical flower-shaped nanostructures. J. Alloys Compd., 792:938-944.

[6] Liu, J.Y, Wang, C., Yang, Q.Y, Gao, Y., Zhou, X., Liang, X.S., Sun, P., Lu, G.Y. (2016) Hydrothermal synthesis and gas-sensing properties of flower-like Sn_3O_4. Sens. Actuators B, 224:128-133.

[7] Ma, X.H, Shen, J.L., Hu, D.X., Sun, L., Chen, Y., Liu, M., Li, C.N., Ruan, S.P. (2017) Preparation of three-dimensional Ce-doped Sn_3O_4 hierarchical microsphere and its application on formaldehyde gas sensor. J. Alloys Compd., 726: 1092-1100.

[8] Jiang, W., Meng, L., Zhang, S., Chuai, X., Sun, P., Gao, Y., Liang, X., Lu G. (2019) Enhanced resistive acetone sensing by using hollow spherical composites prepared from MoO_3 and In_2O_3. Microchim. Acta, 186:6.

[9] Xue, D.P., Zhang, S.S., Zhang, Z.Y., (2019) Hydrothermal synthesis of methane sensitive porous In_2O_3 nanosheets. Mater. Lett., 252:169-172.

[10] Wang, T.S., Kou, X.Y., Zhao, L.P., Sun, P., Liu C., Wang, Y., Shimanoe, K., Yarnazoe, N., Lu, G.Y. (2017) Flower-like ZnO hollow microspheres loaded with CdO nanoparticles as high performance sensing material for gas sensors. Sens. Actuators B, 250:692-702.

[11] Du, L.T., Li, H.Y., Li, S., Liu, L., Li, Y., Xu, S.Y., Gong, Y.M., Chen, Y.L., Zeng, X.G., Lian, Q.C., (2018) A gas sensor based on Ga-doped SnO_2 porous microflowers for detecting formaldehyde at low temperature. Chem. Phys. Lett., 713:235-241.

[12] Cao, P., Yang, Z., Navale, S.T., Han, S., Liu, X., Liu, W., Lu, Y., Stadler, F.J., Zhu, D. (2019) Ethanol sensing behavior of Pd-nanoparticles decorated ZnO-nanorod based chemiresistive gas sensors. Sens. Actuators B, 298:126850.

[13] Zeng, W.W., Liu, Y.Z., Mei, J., Tang, C.Y., Luo, K., Li, S.M, Zhan, H.R., He, Z.K. (2019) Hierarchical SnO_2-Sn_3O_4 heterostructural gas sensor with high sensitivity and selectivity to NO_2. Sens. Actuators B, 301:127010.

[14] Liu, J.Y., Wang, T.S., Wang, B.Q., Sun, P., Yang, Q.Y., Liang, X.S., Song, H.W., Lu, G.Y. (2017) Highly sensitive and low detection limit of ethanol gas sensor based on hollow ZnO/SnO_2 spheres composite material. Sens. Actuators B, 245: 551-559.

[15] Arafat, M.M., Haseeb, A.S.M.A., Akbar, S.A., Quadir, M.Z. (2017) In-situ fabricated gas sensors based on one dimensional core-shell TiO_2-Al_2O_3 nanostructures. Sens. Actuators B, 238:972-984.

IOP Conf. Series: Materials Science and Engineering **729** (2020) 012104 doi:10.1088/1757-899X/729/1/012104

Study on Optical Properties and Waveguide Structure Optimization of Semiconductor Lasers with InGaAs Quantum Well

Lei Gu[1, 2], **Lin Li**[1,a]

[1] National Key Lab of High Power Semiconductor Lasers, Changchun University of Science and technology, No. 7186, Weixing Road, Chaoyang District, Changchun 130022, Jilin, China

[2] College of Electrical Engineering, Jilin Engineering Normal University, Changchun 130052, Jilin, China

[a] email: licust@126.com

Abstract. InGaAs quantum wells (QWs) are grown on GaAs substrates by metal-organic chemieal vapor deposition (MOCVD). Sample grown on substrates (100) oriented 2° off towards <111> exhibits relatively high photoluminescence (PL) intensity and narrow full-width at half-maximum (FWHM). Increasing growth rate leads to the enhancement of PL intensity and decrease of FWHM. Asymmetric waveguide layers are applied to reduce the confinement factor of base mode and increase the loss of higher-order mode.

1. Introduction

In the application of optoelectronic devices, InGaAs quantum well structure is one of the research hotspots, and it has been widely concerned. However, a large amount of In needs to be incorporated into the InGaAs active layer in the long-wavelength band, and the mismatch between In and GaAs is large, which makes the strain layer form a mismatch dislocation. In general, by studying substrates with off angles, growth rate and other methods, we can inhibit the growth mode from two-dimensional to three-dimensional transformation, and optimize the growth quality of quantum wells [1]. With the development of semiconductor optoelectronic technology, the performance of semiconductor laser has reached a relatively high level. Long-wavelength semiconductor laser has very important applications in civil and military fields, such as fiber communication, laser welding, laser guidance and so on. In these applications, high output power of laser is often needed. At present, in order to improve the output power of semiconductor lasers, the research institutions at home and abroad mostly adopt large cavity structure or super large cavity structure. Although this structure can improve the cavity catastrophic damage (COD) threshold of laser, the excessive thickness of waveguide layer leads to serious carrier leakage and high-order mode generation, which limits the further improvement of device power [2-3].Use *italic* for emphasizing a word or phrase. Do not use boldface typing or capital letters except for section headings (cf. remarks on section headings, below). Use a laser printer, not a matrix dot printer.

In this paper, MOCVD is used to study substrate bias angle and growth rate to improve the luminescent properties of quantum well materials. At the same time, the asymmetric waveguide layer is used to optimize the waveguide structure of the semiconductor laser, which reduces the output of high-order mode and improves the output power of the laser.

2. Experiment

1) Substrates with off angles

In epitaxial growth, the substrates with off angles have a great influence on the quality of epitaxial materials. Cheng Tien Wan et al. [4] have grown GaAssb/GaAs quantum wells on GaAs (100) bias < 011 > 0 °, 2 °, 6 ° and 15 ° substrates. The experimental results show that with the increase of substrates with off angles, the interface between GaAssb and GaAs is steeper, the crystal quality is improved faster, and the luminescent properties are better. In this experiment, we compared the PL spectra of InGaAs / GaAs single quantum wells grown on substrates with GaAs (100) surface deflection < 111 > 0 °, 2 °, 4 ° and 10 ° and analyzed the influence of substrates with off angles on the luminescence characteristics of the quantum well.

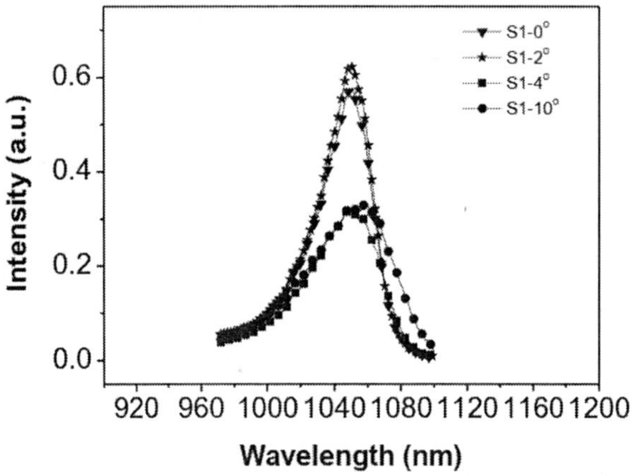

Figure 1. PL spectra of S1 samples at different offcut angles

As shown in Figure 1, the emission peaks of four groups of samples S1-0 °, S1-2 °, s1-4 °, s1-10 ° are 0.56mV, 0.63mV, 0.32mV and 0.33mV, respectively. The emission peaks of S1-0 ° samples are higher than those of S1-2 ° samples, and the emission peaks of S1-2 ° samples are twice of s1-4 °, while the emission peaks of s1-10 ° samples are only slightly larger than those of s1-4 ° samples, still about half of S1-2 ° samples. The FWHM of the four groups of samples were 38.2nm, 37.6nm, 51.5nm and 65.8nm, respectively, indicating that the FWHM of the samples increased with the increase of substrate deflection angle.

2) Growth rate

The growth rate has an important influence on the growth mode of quantum wells [5]. By optimizing the experimental conditions, the gas ratio of In was set to be 33.71%, and two samples of S2 and S3 were grown. Similar to the first group of experiments, the growth rates of the samples were changed to 1.15 μm / h and 0.95 μm / h respectively. As shown in Figure 2, it is the normal temperature PL of the two samples.

It can be seen from Figure 2 that the FWHM of the sample decreases, the luminescence wavelength appears blue shift with the increase of growth rate, and the crystal quality is improved. At this time, the incorporation of in is affected by the growth rate [6].

The results show that increasing the growth rate can increase the luminescent intensity, decrease the FWHM and improve the growth quality. The incorporation of in decreased with the increase of growth rate.

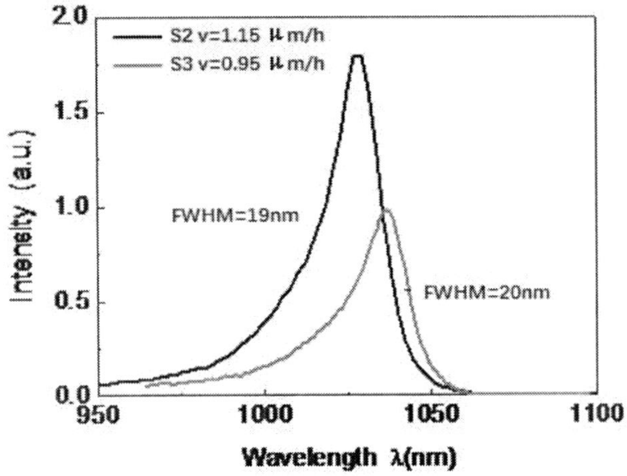

Figure 2. PL spectra of S2、S3 samples at different growth rates

3) Optimization of waveguide structure of laser

Because the thickness of the quantum well directly affects the COD phenomenon of the device, and then affects the output characteristics. The greater the thickness, the higher the maximum output power. However, when the equivalent quantum well component is fixed, the lasing wavelength increases with the increase of the quantum well thickness. According to Matthews Blakeslee's critical thickness theory, the thickness of quantum wells cannot be increased indefinitely, otherwise stress release will occur and defects will occur in quantum wells. The thickness of InGaAs quantum well is 7nm, and the In component is about 0.29.

In symmetric waveguide structure, when the thickness of waveguide layer increases to a certain value, the corresponding first-order and second-order modes will appear, and the corresponding transverse mode limiting factors can also be compared with the limiting factors of the basic mode gradually, but the three are gradually reduced with the increase of the thickness of waveguide layer. However, the limiting factors of the second-order mode and the fundamental mode are close. In the actual operation of the device, the threshold currents of the two modes are similar, and there is mode competition between them, which is not conducive to the high power output of the device and the improvement of the beam quality [7-8]. In order to solve this problem, the epitaxial structure of asymmetric waveguide can be used. As shown in Fig. 3, in this structure, the position of the quantum well is no longer in the center of the whole waveguide, but as mentioned before, because the distribution of the optical field is mainly determined by the waveguide layer, the change of the optical field is very small when the thickness of the waveguide layer is unchanged. This causes the quantum well to deviate from the center of the light field, thus changing its limiting factor for a certain mode [9-10].

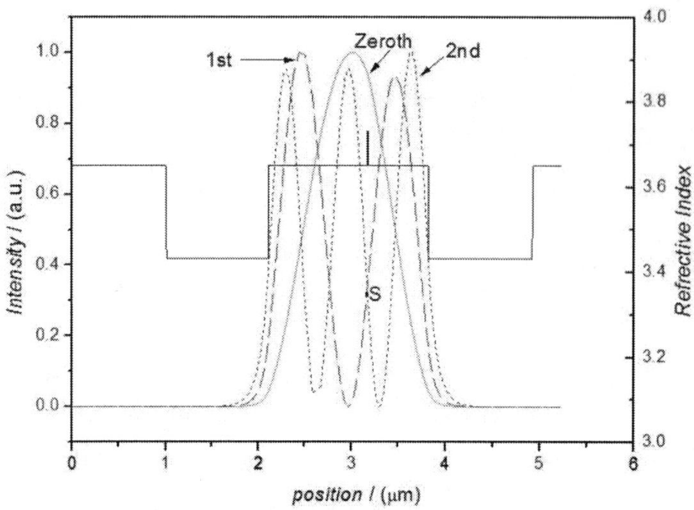

Figure 3.The refractive index of wafer and distribution of the intensity of optical filed with different modes

Set the waveguide thickness to 1.6 μm, we calculated the limiting factors of three modes in symmetric waveguide structure and asymmetric waveguide structure, as shown in the table below:

Table 1.The optical confinement factor of various modes

	Fundamental mode limiting factor (10^{-2})	1 order mode limiting factor (10^{-2})	2 order mode limiting factor (10^{-2})
Symmetrical waveguide structure	0.9365	0.00876	0.8092
Asymmetric waveguide structure	0.8378	0.3083	0.1693

It can be seen from the data in the table that in the symmetric waveguide structure, the difference between the limiting factors of the basic mode and the second-order mode is very small. In the optimized asymmetric waveguide structure [11], although the limiting factors of the first-order mode increase significantly, their values are still about 2.6 times smaller than the limiting factors of the basic mode. The limiting factor of the second-order module is reduced to about 1 / 5 of the limiting factor of the basic module. Therefore, in the asymmetric waveguide structure, the loss of the first mode and the second mode will increase, the threshold current will also be greatly increased, and the mode characteristics of the device will be significantly improved.

3. Summary

The InGaAs quantum well structure was grown on GaAs (100) < 111 > substrate by MOCVD. The effects of substrates with off angles and growth rate on the luminescence characteristics of the variable InGaAs quantum well were studied. When the substrate bias angle is small, the PL intensity of the sample is large, and the FWHM is narrow. The incorporation of in decreased with the increase of growth rate, and the wavelength shifted blue. With the increase of growth rate, the luminescent intensity of the sample increases, FWHM decreases, and the growth quality is improved. The results show that the optimized asymmetric waveguide structure can reduce the limiting factor of the fundamental mode and increase the loss of the higher-order mode.

References

[1] D Schlenker, T Miyamoto, Z Chen, et al .. Growth of highly strained GaInAs/ GaAs quantum wells for 1.2 um wavelength lasers[J] . *Cryst . Growth* , 2000, 209, 27-36.

[2] Prahova P, Moritz I B, Riordan T. High brightness 810 nm long cavity diode lasers with high d /Γ ratio in asymmetric low confinement epitaxial structure [C] . *Proceedings of the 2001 IEEE/LEOS Annual Meeting* , 2001.Vol.1: 135-136.

[3] Ryvkin B S, Avrutin E A. Free carrier absorption and active layer heating in large optical cavity high power diode laser. [J] . *Appl. Phys.*, 2006, 100(10) : 023104-023112.

[4] Wang Chengtien, Su Yankuin, Ricky W.Chuang, *et al.* Improving photoluminescence of highly strained 1.32 μm GaAsSb/GaAs multiple quantum wells grown on misorientation substrate [J]. *Journal of Crystal Growth*, 2008, 310, 4854-4857

[5] A Jasik , A Wnuk , J Gaca , *et al.*. The influence of the growth rate and V/III ratio on the crystal quality of InGaAs/GaAs QW structures grown by MBE and MOCVD methods[J]. *Journal of Crystal Growth* , 2009 ,311(19) : 4423-4432.

[6] F.Bugge, U.Zeimerd,*et al.* . MOVPE growth of highly strained InGaAs/GaAs quantum wells[J] . *Cryst . Growth.* 1998,18,511-5188

[7] Prahova P, Moritz I B, Riordan T . High brightness 810 nm long cavity diode lasers with high d /Γ ratio in asymmetric low confinement epitaxial structure [C] . *Proceedings of the 2001 IEEE/LEOS Annual Meeting*, 2001, Vol . 1: 135-136 .

[8] Ryvkin B S, Avrutin E A . Free carrier absorption and active layer heating in large optical cavity high power diode laser [J] . *Appl . Phys .* , 2006, 100(10) : 023104-023112 .

[9] Wade J K, Mawst L J, Botez D, et al . . 8 W CW Power from Broad-Waveguide Al-Free Active-Region(λ = 0 . 805μm) Diode Lasers [J] . *Electron Lett* . 1998, 34: 1100-1102 .

[10] Botez D . Simple Design Rules for Single-Lobe Operation of (Evanescently Coupled) Index-Guided Phase-Locked Arrays of Diode Lasers [J] . *IEEE . Quantum Electron* . 1988, 24: 2034-2038 .

[11] Hasler K H, Sumpf B, Adamiec P, etal . 5 W DBR Tapered Lasers Emitting at 1060 nm With a Narrow Spectral Linewidth and a Nearly Diffraction-Limited Beam Quality , IEEE *Photon . Technol . Lett* . , 2008, 20(19) : 1648-1650 .

IOP Conf. Series: Materials Science and Engineering **729** (2020) 012105 doi:10.1088/1757-899X/729/1/012105

Study on the preparation of suspension stabilizer for liquid accelerator

Mengqin Guan, Xianghe Li, Xinqi Guo*, Zhuojun Jiang

KZJ New Materials Group Co., Ltd. Xiamen 361101, Fujian, China

*Corresponding author, E-mail: gxq00008@163.com

Abstract. A suspension stabilizer (P-XF) was prepared by copolymerization of acrylamide (AM), acrylic acid (AA), ammonium persulfate (APS), thioglycolic acid (TGA) and potassium hydrogen tartrate (KHT). The suspension stabilizer (P-XF), aluminum sulfate ($Al_2(SO_4)_3$), p-toluenesulfonic acid (PTSA) and triethanolamine (TEA) were used to prepare a liquid accelerator (P-SN_2). The test results show that compared with the liquid accelerator (P-SN1) without P-XF, P-SN_2 not only can effectively shorten the setting time, but also has high stability, with a stable period of more than one year.

1. Introduction

Accelerator is an additive that can make concrete or cement mortar set and harden rapidly. Its main function is to accelerate the setting and hardening speed of shotcrete, reduce rebound loss, prevent shotcrete from falling off due to gravity, increase the thickness of one shot and shorten the interval time between shot layers[1]. China began to study powder accelerator around 1990s[2]. 8604 type accelerator successfully developed by Architectural Research Institute of Ministry of metallurgy in 1989, AC type accelerator developed by Changsha mine Research Institute in 1990, MJ-2000 accelerator developed by Coal General Institute in 1995, and the raw materials for this kind of accelerator are bauxite, lime, gypsum, etc. The 28d strength ratio of concrete mixed with these accelerators is increased to about 90%, but the dosage and production cost of these powder accelerators are high, so the application of them is not ideal. With the development and application of wet jet technology,the research direction of accelerator also turns to liquid accelerator.

The research of liquid accelerator began in the late 1980s, mainly in the United States, Japan, Germany, Switzerland and other countries[3-6]. At the end of the 20th century, the research on liquid accelerators began in China. At present, there are some problems in the market, such as short product storage period and long setting time[7]. The short storage period is mainly reflected in the normal construction temperature range, and the product appears layered precipitation or crystallization about 1 month after production. The long setting time means that the product needs to increase the content of accelerator in the use process, thus increasing the construction cost, and at the same time, it will have different degrees of impact on the durability of concrete. Therefore, it is of great significance to develop a liquid accelerator with good stability, short setting time, low dosage, good adaptability to cement and no adverse effect on durability. A suspension stabilizer (P-XF) was prepared by copolymerization of acrylamide (AM), acrylic acid (AA), ammonium persulfate (APS), thioglycolic acid (TGA) and potassium hydrogen tartrate (KHT), When the suspension stabilizer (P-XF) is used in the synthesis of accelerator, it can solve the adverse effects of liquid accelerator, especially in the aspects of stability and coagulation.

2. Experimental

2.1.Materials

2.1.1.The main synthetic experimental raw materials. acrylamide (AM) , industrial grade; acrylic acid (AA) , industrial grade; ammonium persulfate (APS), industrial grade; thioglycolic acid (TGA), industrial grade; potassium hydrogen tartrate (KHT), industrial grade; aluminum sulfate ($Al_2(SO_4)_3$), industrial grade; p-toluenesulfonic acid (PTSA), industrial grade; triethanolamine (TEA), industrial grade;

2.1.2.Main performance test raw materials for experiment. Cement: Min Fu P.O 42.5 Cement; Sand: standard sand (GB/ T17671-1999), products of Xiamen ISO standard sand co. Ltd; P-SN_1: Liquid accelerator, Self made without P-XF, other synthesis conditions are the same as P-SN_2.

2.2. Synthesis of suspension stabilizer P-XF
Metered water and AM were added to the reactor, heating up to 70°C, APS, AA, and TGA were added to the reactor respectively. The dropping was controlled within 3 h. Add KHT after dripping, and then P-DMG was obtained.

2.3. Synthesis of liquid accelerator P-SN_2
Metered water, P-XF, $Al_2(SO_4)_3$, PTSA and TEA were added to the reactor. After the system temperature reached 40°C, and reaction 1 h, and then P-SN_2 was obtained.

2.4.Performance test method

2.4.1. Setting time of cement pastes. The tests of the setting time of the cement pastes are conducted in accordance with GB/T35159-2017 "Flash setting admixrutes for Shotcrete ".

2.4.2. Determination of mortar strength. The tests of the mortar strength are conducted in accordance with GB/T35159-2017 "Flash setting admixrutes for Shotcrete ".

2.4.3. Gel chromatography test. Waters 1515 was used. The column was connected in series by UltrahydragelTM 250 and UltrahydragelTM 500. The mobile phase was 0.1 mol/L sodium nitrate aqueous solution at a flow rate of 0.8 mL/min. The standard calibration curve was sodium polyacrylate and the running time was 60 min.

3. Experimental results and discussion

3.1. Effect of molar ratio of AA to AM on conversion
The effect of the molar ratio of AA to AM on the conversion was investigated under the same conditions, the experimental results are shown in Figure 1. The results showed that with the increase of the molar ratio of AA to AM, the conversion increased first and then decreased. When the molar ratio of AA to AM is 0.8:1, the conversion is the highest, which is 81.5%.

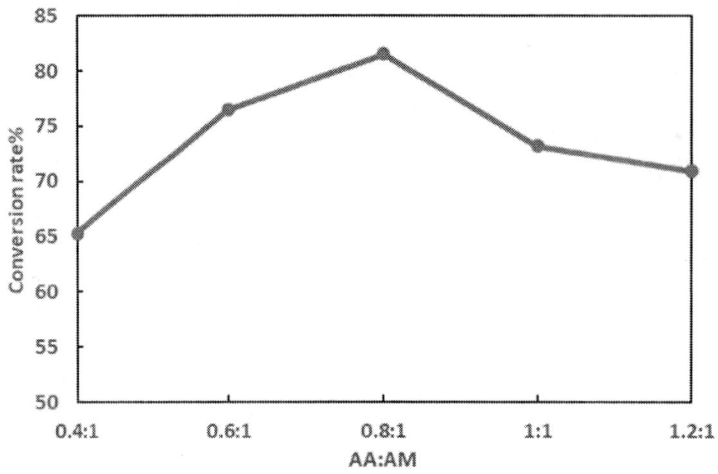

Figure 1. Effect of molar ratio of AA to AM on conversion

3.2. Orthogonal test and analysis

According to the results of previous tests, the dosage of AA and AM was fixed. APS, TGA and KTH were used as three factors to carry out the uniform test design, and two levels of each factor were selected. Taking the conversion rate of polymerization as the investigation index, the orthogonal experiment was designed and arranged. The dosage of three factors is the percentage of total mass of AA and AM. The optimal material ratio is determined through the test, and the test results and analysis are shown in Table 1.

Table 1. Orthogonal test results and analysis

Number	APS/%	TGA/%	KTH/%	conversion tate/%
1#	1.5	0.8	0.5	76.8
2#	1.5	1.6	1.0	80.5
3#	3.0	0.8	1.0	92.3
4#	3.0	1.6	0.5	81.4
Value K1j		78.65	84.55	79.1
Value K2j		86.85	80.95	86.4
Range R		8.2	3.6	7.3

Table 1 shows that APS, TGA and KTH have the following effects on the conversion of polymerization: APS> KTH> TGA. The results of orthogonal test show that the optimal process is APS, TGA and KTH, which are 3.0%, 0.8% and 1.0% of the total mass of AA and am respectively. The P-XF prepared by this process has the highest conversion.

3.3. Effect of reaction temperature on conversion

On the basis of Orthogonal Optimum Technology, the effect of the reaction temperature on the conversion was investigated under the same conditions, the experimental results are shown in Figure 2. The results showed that with the increase of reaction time, the conversion increased first and then stabilized. When the reaction time is 3.0 h, the conversion rate reaches 93.8%. When the reaction time is extended, the conversion rate shows a stable trend. Therefore, the best reaction time is 3.0 h.

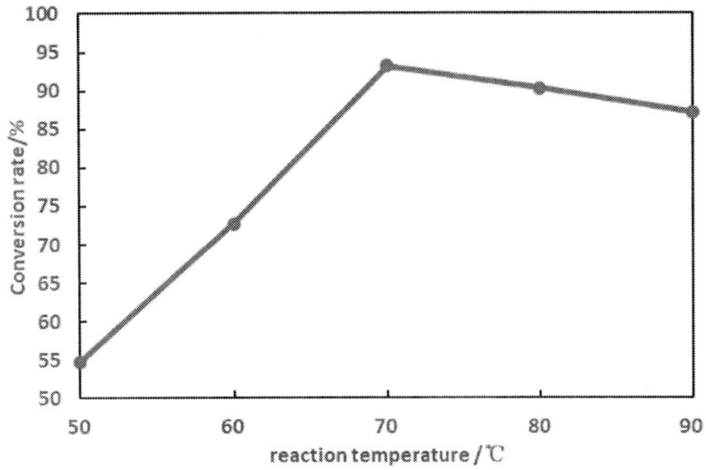

Figure 2. Effect of reaction temperature on conversion

3.4. Effect of reaction time on conversion

The effect of the reaction time on the conversion was investigated under the same conditions, the experimental results are shown in Figure 3. The results showed that the conversion increased first and then stabilized with the increase of reaction time. When the reaction time is 3.0 h, the conversion rate reaches 93.8%. When the reaction time is extended, the conversion rate shows a stable trend. Therefore, the best reaction time is 3.0 h.

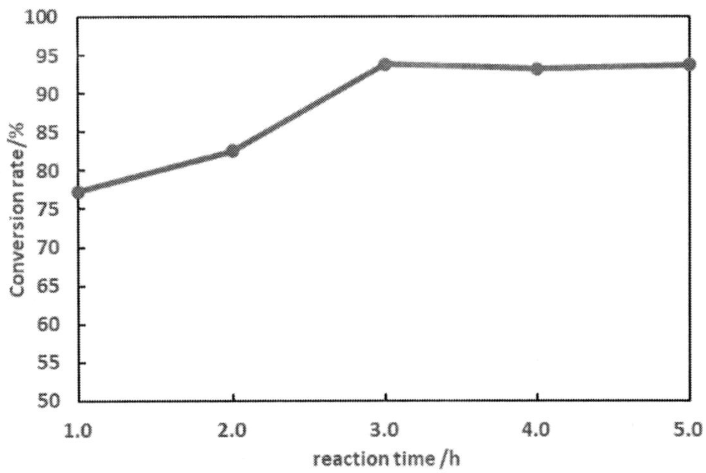

Figure 3. Effect of reaction time on conversion

3.5. Performance test of liquid accelerator synthesized by P-XF

The preparation conditions of P-XF were as follows: the molar ratio of AA to AM is 0.8:1, the dosage of APS, TGA and kth are 3.0%, 0.8% and 1.0% of the total mass of AA and am respectively. the reaction temperature was 70°C, the reaction time was 3 h. Then P-XF, $Al_2(SO_4)_3$, PTSA and tea were used as raw materials to prepare liquid accelerator P-SN_2. P-SN_1 was prepared from $Al_2(SO_4)_3$, PTSA and TEA under the same conditions as P-SN_2.

The setting time, mortar strength and stability of p-sn1 and p-sn2 are tested in accordance with GB / T35159-2017 "Flash setting admixrutes for Shotcrete ", the test results are shown in Table 2.

IOP Conf. Series: Materials Science and Engineering **729** (2020) 012105 doi:10.1088/1757-899X/729/1/012105

Table 2. Performance test results

Accelerator	Dosage /%	setting time /min:s		1 day compressive strength /MPa	Compressive strength ratio /%		Stability/day
		Initial	Final		28d	90d	
P-SN1	8.00	4：30	11：15	7.8	95	100	60
P-SN2	7.00	4：35	10：30	8.9	100	105	450

The test results show that the initial setting time is the same when P-SN2 is 0.88 times of P-SN1, the final setting time is shorter, the strength is higher in one day, the compressive strength ratio is higher in 28 days and 90 days, and the stability of the product is better. It can be seen that the introduction of p-xf into the accelerator can not only effectively shorten the setting time, but also have obvious advantages in the stability of the product, with the stability up to 15 months.

4. Conclusions

（1）According to the molar ratio of AA to AM of 0.8:1, the dosage of APS, TGA and kth are 3.0%, 0.8% and 1.0% of the total mass of AA and am, The reaction temperature is 70 °C, and the reaction time is 3.0h. The suspension stabilizer P-XF is prepared.

（2）The suspension stabilizer made in this paper is used in the synthesis of liquid accelerator, which improves the cohesiveness of liquid accelerator, has short setting time, improves the strength of concrete, does not appear stratification or crystallization, and has a stability of up to 15 months.

References:

[1] Yang L Y, Tian J T, Yang Y B, et al. Research status of liquid accelerator for shotcrete[J]. Tunnel Construction, 2017,5(37):543-550.
[2] Wang Z M, Current status and trends of research and development of shotcrete and accelerators[A]. Research and application progress of concrete admixture in China, 2018:7.
[3] Snyder Raymond C, Snyder Pul F.Liquid concrete accelerator mixtures and methods for use thereof. US4046584, Sep6, 1977.
[4] Nitto chemical industry Co.Ltd.,Japan.Cement settingaccelerator[P].JP59169957A2, Sep26, 1984.
[5] Terashima Isao, Hirano Kenkichi.Setting accelerator and quick-setting blowing cement concrete having high strength[P]. JP11199287A2, Jul27, 1999.
[6] Sommer Marcel,wombacher Franz， Buerge Theodor A.Alkali-free setting and hardening accelerator based on aluminum salts for Cementitious mixes[P] EP1167317A1, Jan2 2002.
[7] Li X H. The research and application of liquid set accelerator[J]. New buliding materials, 2018,45(12):65-68.

IOP Conf. Series: Materials Science and Engineering **729** (2020) 012106 doi:10.1088/1757-899X/729/1/012106

Simulation and application of a new type of slideway beam based on MIDAS jacking construction method

LI Qin[1, a], Jian Gao[1,b] and Qiuyue Zhang[1,c]

[1]College of Civil Engineering and Architecture, Shandong University of Science and Technology, Qingdao, Shandong, 266590, China

[a]q15615610522923@163.com, [b]422252270@qq.com, [c]13615426280@163.com

Abstract. In order to ensure the continuous box girder jacking method in the process of construction safety in Jingzhou City, Mongolia and China Railway Bridge jacking construction method of process engineering background, research on continuous box girder pushing process in the first two layer 8 laminated I-beam spell 56B security control problem model in the process of construction, the slide beam. Based on the control principle of adaptive method, a finite element model of slideway beam during incremental launching is built by MIDAS, and the range of deformation and internal force of the whole structure of slipway beam is put forward during construction. Through monitoring and measuring the deformation of the slideway beam during the incremental launching process, we get the test data and compare with the theoretical value, so as to guide the safety control of the slipway beam during the pushing construction of the continuous box girder. The research results show that during the incremental launching of continuous box girder, the slideway beam can meet the requirements of deformation and internal force. The safety control technology adopted can effectively guide the engineering construction, and the method is feasible.

1. Introduction

Since its successful application in ager bridge in Austria in 1959, the jacking method has been developed rapidly in the history of bridge construction in the world. In 1977, China first used the jacking method to build the dijiahe bridge [1]. Shao H et al [2] first carried out the research of jacking construction in China. Liu S[3] carried out the research of internal force analysis in the stage of jacking construction earlier, established the simplified model of internal force calculation of continuous beam in combination with the engineering practice, and carried out the detailed calculation and analysis. Liang A et al[4]used MATLAB tools to study the dynamic analysis of the construction internal force and displacement of the continuous beam bridge with constant cross section constructed by one-way (two-way) single point (multi-point) jacking method.

However, from many research results, it can be seen that: at present, there is little research and Analysis on the slipway beam in the process of bridge jacking construction. A large number of engineering examples show that the slipway surface of the concrete jacking beam is in contact with each other, and the concrete beam is not tensile under compression. The deformation of the slipway beam will make the slide at the bottom of the beam hollow, resulting in the uneven stress distribution of the beam, resulting in the cracking of the concrete beam. Therefore, we need to study one A new type of slideway beam can ensure the smoothness of the slideway surface during the beam concrete pouring and jacking construction, and effectively solve the cracking problem of the jacking concrete beam caused by the former slideway irregularity.

IOP Conf. Series: Materials Science and Engineering **729** (2020) 012106 doi:10.1088/1757-899X/729/1/012106

2. Finite element solid model of slideway beam

2.1 introduction to finite element model. (1) unit division. The model of slideway beam is built by hexahedral solid element, and the material is Q235 steel. The calculation model is 3 × 6.5m beam section. The longitudinal length of the contact between the slideway beam and each transverse distribution beam in the middle is about 700mm, and the longitudinal length of the contact part is 14 × 50mm; the longitudinal length of the contact part with both ends is 350mm, and the longitudinal length is 7 × 50mm; the beam section without the contact transverse distribution beam is divided into 30 × 200mm units. In the transverse direction, the connecting steel plate and I-beam flange plate are divided into 28 units, of which the transverse length of 4 units is the same as the thickness of I-beam web, and only one unit is divided in the transverse direction of I-beam web. The connection steel plate is divided into one unit vertically, I-beam flange plate is divided into two units and web plate is divided into five units. The model of slideway beam is shown in Fig. 1.

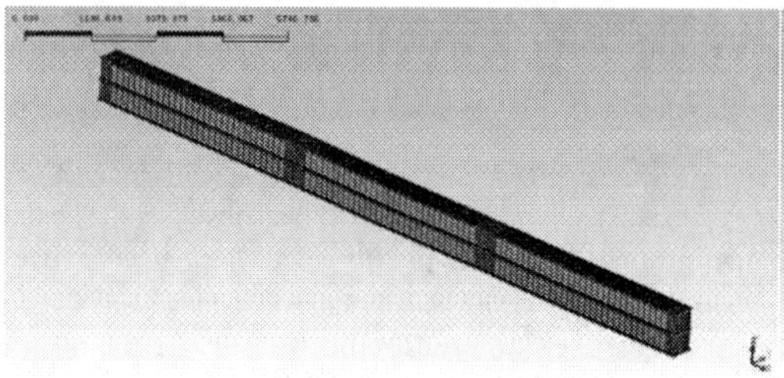

(a) Model axonometric drawing

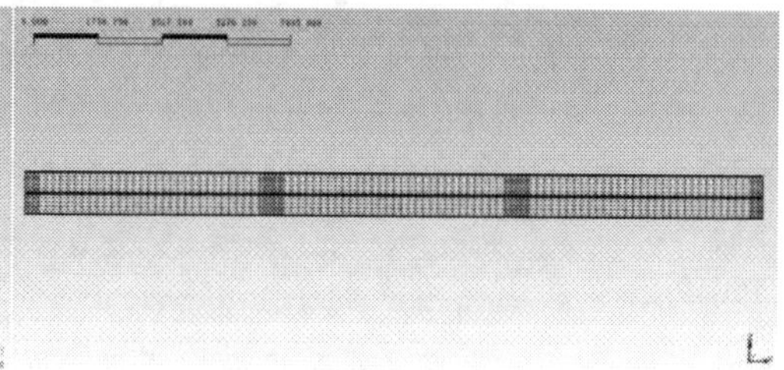

(b) Model elevation

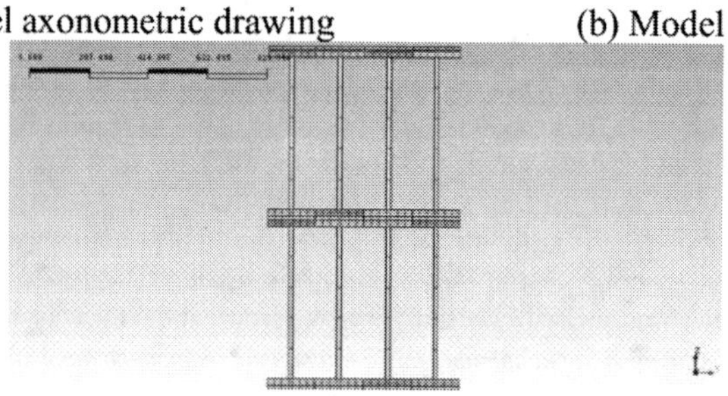

(c) Model side view

Fig.1 Finite element model of slideway beam

(2) calculation load

In the process of launching, the box girder web is always located above the slideway girder, and the two transverse slideway girders are symmetrically arranged. It is considered that the two girders are equally divided into the box girder weight and evenly distributed in the vertical and horizontal directions. Surface pressure is proposed to simulate the dead weight of box girder, and it is calculated that the applied surface pressure is 260*kN/m*. The pressure load model is shown in Fig. 2.

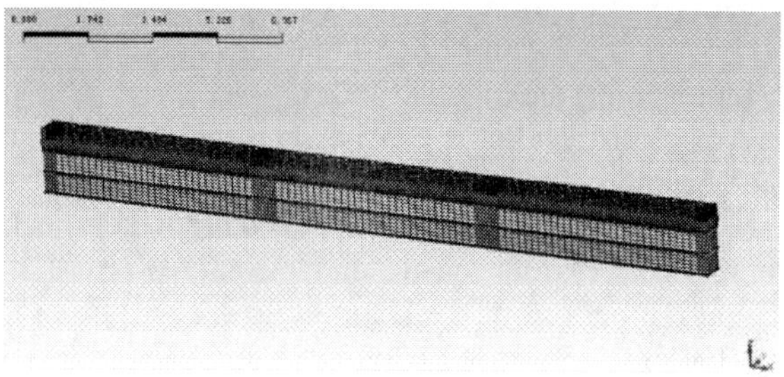

Fig. 2 Calculation load

(3) boundary conditions

The calculation model takes one section of the slideway beam, and the boundary condition is simplified as the boundary condition of three span continuous beam. Considering that there is no horizontal sliding between the slideway beam and the transverse distribution beam, the horizontal direction is also constrained at the constraint of each span. The contact length between the slideway beam and the transverse distribution beam in the longitudinal direction is about 0.7m, and the boundary conditions at the two ends are 0.35m and 0.7m in the middle restrict the X, y and Z directions The boundary conditions of the model are shown in Fig. 3.

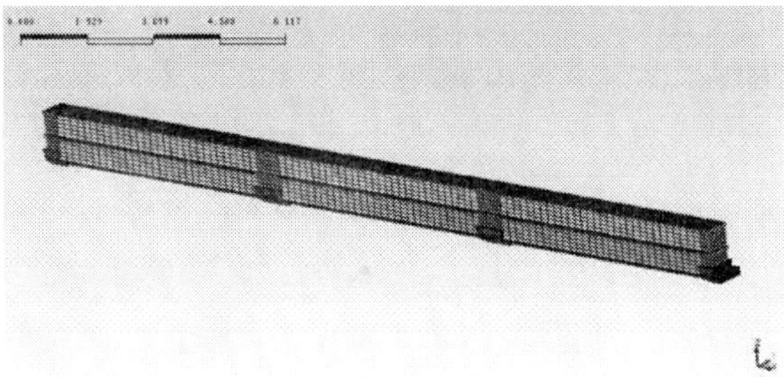

Fig. 3 Boundary conditions

(4) connection simulation between I-beam and connecting plate

In the actual situation of the slideway beam, the I-beam and the connecting plate are only welded at the outermost edge of the outer I-beam. The rigid connection is used for the connection of the model, and the rest of the joints connecting the steel plate and the I-beam contact part are connected by spring element. The connection between connecting plate and I-beam is simulated by spring in horizontal direction.

(5) calculation conditions

According to the possible position of the box girder during the jacking process and considering the unfavorable stress condition of the slideway girder, this model can be divided into three situations according to the longitudinal loading position: three span full load, two span load in the middle and one span load in the side span. In the actual construction process, most cases are three span full load, so this paper simulates one of the working conditions of the finite element model, and carries out the numerical simulation analysis.

2.2 analysis of calculation results

Table 1 calculation results

Working condition	deformation	Result
	Maximum vertical (mm)（mm）	-1.09
Integral deformation	Maximum longitudinal (mm)	0.13
	Relative slip (mm)	0.15
	Maximum vertical (mm)	-1.09
Connecting plate	Maximum value of the first principal stress (MPa)	40.94
	Maximum value of the third main stress (MPa)	-52.14
	Maximum vertical (mm)	-1.06
I-beam	Maximum value of the first principal stress (MPa)	34.69
	Maximum value of the third principal stress(MPa)	-57.9

IOP Conf. Series: Materials Science and Engineering **729** (2020) 012106 doi:10.1088/1757-899X/729/1/012106

The overall deformation cloud diagram of the slideway beam is shown in Fig. 4-5.

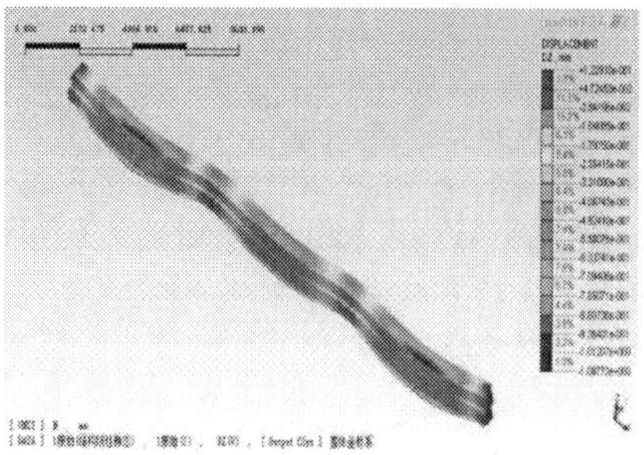

Fig. 4 Cloud chart of overall vertical deformation of slideway beam (mm)

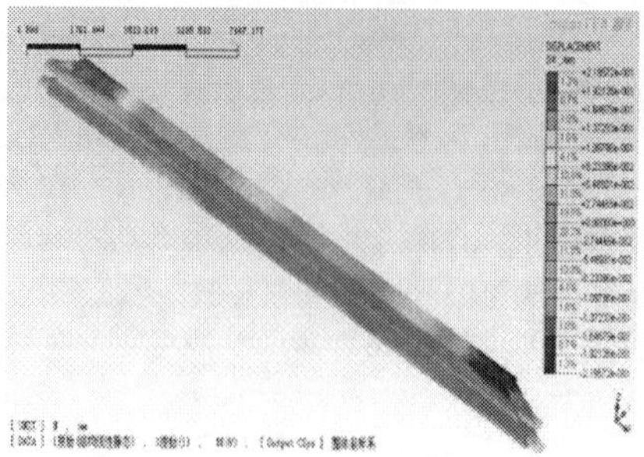

Fig.5 Cloud chart of overall longitudinal deformation of slideway beam (mm)

The cloud diagram of top connecting steel plate is shown in Fig. 6-8.

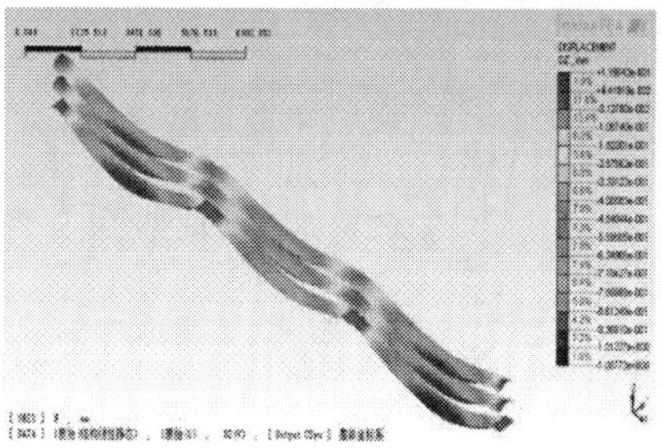

Fig.6 Cloud chart of vertical deformation of connecting steel plate (mm)

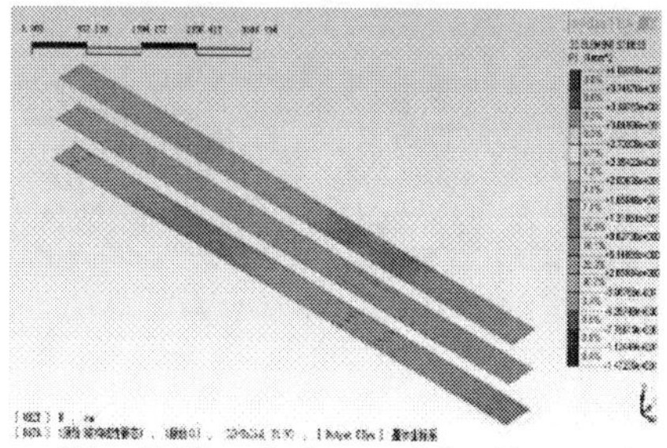

Fig. 7 Cloud chart of the first main stress of connecting steel plate (MPA)

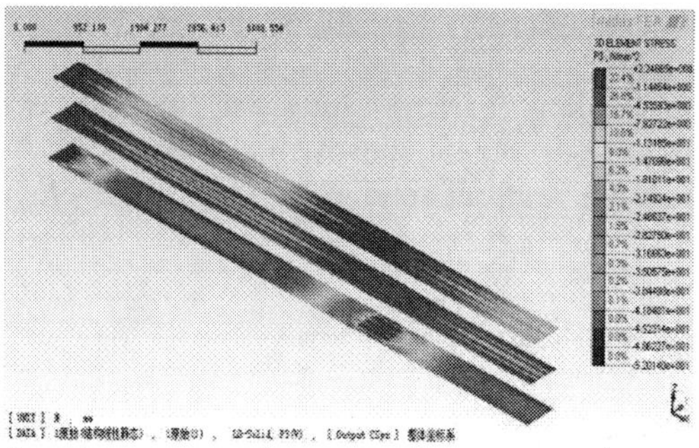

Fig.8 Third main stress nephogram of top connecting steel plate (MPA)

The cloud chart of I-beam is shown in Fig. 9-11.

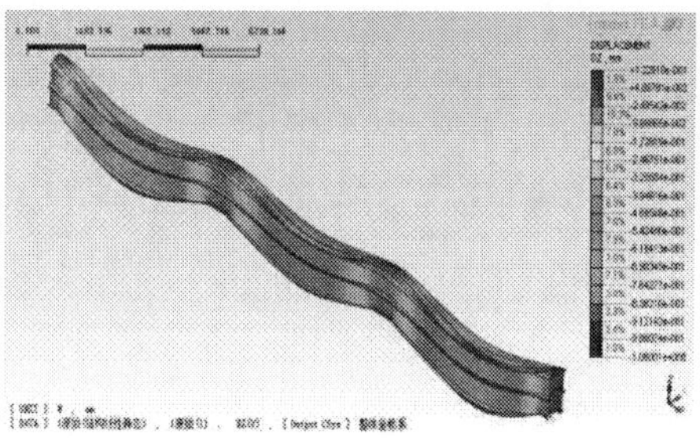

Fig.9 Vertical deformation nephogram of I-beam (mm)

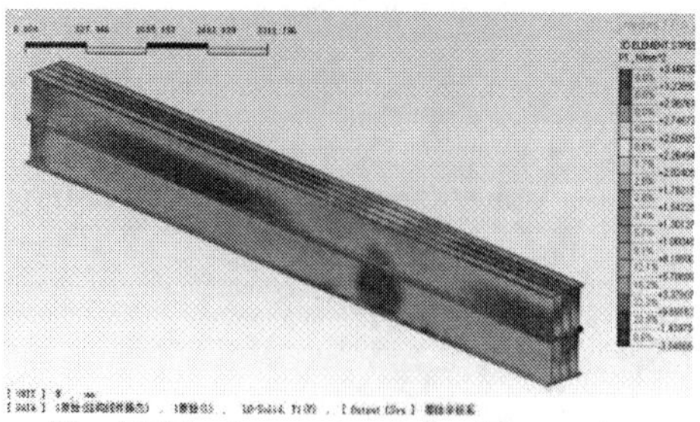

Fig.10 Nephogram of the first main stress of I-beam (MPA)

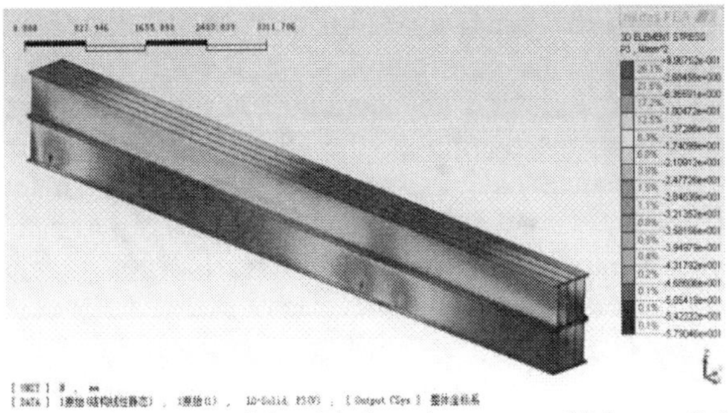

Fig.11 Nephogram of the third main stress of I-beam (MPA)

3. Field test of bearing capacity of slideway beam

3.1 preloading scheme of slideway beam. The standard section of 6.7m long slideway beam is selected for preloading above the erected slideway beam. The maximum construction load of the jacking section beam (45.7m) is 1623.5t, the maximum load weight is about 1.2 times of the maximum construction load, and the maximum load weight per linear meter is 40.96t. Set up the preloading platform on the already installed slideway beam for 40%, 60%, 100%, 110% and 120% loading, arrange the dial indicator at the bottom of 1 / 4 span, 1 / 2 span and 3 / 4 span of the slideway beam to observe the deflection of the slideway beam, and measure the compression of the sand box before and after preloading.

Fig.12 Field diagram of Preloading for deflection test of slideway beam

Fig.13 Site layout of dial indicator

3.2 pre compression test results of slideway beam. When the slideway beam is preloaded to 120% of the maximum construction load, there is no crack in the processing weld of the slideway beam. The measured deflection on site (the reading of dial indicator minus the compression of the sand box) is as follows: the measured deflection of the East 1 / 4 span is 2.13mm, the measured deflection of the East 3.1mm, the measured deflection of the east 3 / 4 span is 2.2mm, the measured deflection of the West 1 / 4 span is 2.08mm, the measured deflection of the west 3 / 4 span is 2.9mm, and the measured deflection of the west 3 / 4 span is 2.04mm.

The fulcrum under the slideway beam is the sand box, which is welded and reinforced with the slideway beam. The dial indicator is fixed on the longitudinal distribution beam, and the longitudinal distribution beam is connected with the transverse distribution beam. Considering the settlement of the sand box in the preloading process, the measured deflection value will be larger than the actual deflection value.

4. Summary

It can be seen from the test results that the slideway beam is still deformed after unloading, which indicates that the measured deflection includes elastic deflection and inelastic deflection when the

IOP Conf. Series: Materials Science and Engineering **729** (2020) 012106 doi:10.1088/1757-899X/729/1/012106

slideway beam is preloaded. In addition, the fulcrum under the slideway beam is the sand box, which is welded and reinforced with the slideway beam, the dial indicator is fixed on the longitudinal distribution beam, and the longitudinal distribution beam is connected with the transverse distribution beam. In consideration of the settlement of the sand box during the preloading process, the theoretical calculation value is the elastic deformation of the slideway beam, so the measured value will be slightly greater than the theoretical calculation value, and the test results are consistent with the analysis. In the test results, the elastic deformation of the slideway beam is close to the theoretical calculation value. When the maximum construction load is 120%, the structure has no abnormal phenomenon. It can be considered that the slideway beam is in a safe state.

References

[1] Zhao R, Zhang S. Research status and development trend of bridge jacking construction. China highway journal, 2016
[2] Shao H, Zhou Y, Luo X. Dijiahe by pushing method . Journal of railway, 1979
[3] Liu S. Analysis of internal force and design of temporary structure in jacking construction stage [J]. Journal of Shijiazhuang Railway University, 1998
[4] Liang A, Wang C, Xu Y. Using MATLAB language to develop structural analysis program for launching construction of continuous girder bridge . Transportation and computer, 2000

IOP Conf. Series: Materials Science and Engineering **729** (2020) 012107 doi:10.1088/1757-899X/729/1/012107

Study on chemical looping reforming reaction of methane based on PSO-BP neural network

Yongbin Liu[1], Xiaoxun Ma[1], and Long Xu[1*]

[1]School of Chemical Engineering, Northwest University, International Science & Technology Cooperation Base of MOST for Clean Utilization of Hydrocarbon Resources, Chemical Engineering Research Center of the Ministry of Education for Advanced Use Technology of Shanbei Energy, Shaanxi Research Center of Engineering Technology for Clean Coal Conversion, Collaborative Innovation Center for Development of Energy and Chemical Industry in Northern Shaanxi, Xi' an, 710069, People' s Republic of China.

*Corresponding author's e-mail: longxuxulong@163.com.

Abstract. Through the study of methane chemical looping reforming reaction, the BP neural network model optimized by the PSO algorithm was established, and the network structure of the PSO-BP model was determined to be 9-11-3. After training the model, the mean square error of the network was finally stabilized at 0.013509, and the learning rate was finally fixed at 0.083453, while the fitting degree of the training sample and the test sample were both above 0.979, indicating that the network of the PSO-BP model had strong learning ability and generalization capability, and was a simulation prediction model with good performance. The PSO-BP model was used to simulate the full set of experimental condition data, and the most experimental condition was the cerium iron composite oxygen carrier prepared by co-precipitation method (No.2) with a molar ratio of 0.7/0.3, a calcination temperature of 800℃, a calcination time of 6h, a reaction temperature of 850℃, a reaction time of 13min, and a circulation number of 0, while it was consistent with the actual experimental.

1.Introduction

Natural gas is a clean and environmentally friendly resource that will play a key role in energy transformation[1]. Natural gas is mainly methane, so in the conversion and utilization of natural gas, the indirect technology of producing synthetic gas from methane and then producing high-value-added chemicals or liquid hydrocarbons has been a research hotspot[2]. Methane chemical looping reforming technology is a new synthetic gas preparation technology based on the concept of chemical looping combustion[3-4]. In this technology, methane catalytic partial oxidation reaction is exothermic and mild. The synthetic gas produced has a hydrocarbon ratio of 2, which can be directly used for methanol and Fischer-Tropsch synthesis. Moreover, it has the advantages of small reactor volume, high efficiency and low energy consumption, and has been widely valued at home and abroad[5].

The artificial neural network is a mathematical model which has been widely applied in recent years, with a strong adaptive and self-learning and nonlinear mapping ability. At present, the artificial neural network is mainly applied in fault diagnosis[6-7], process control and optimization[8-9] and physical property estimation[10]. Wang AL et al.[9] used particle swarm optimization (PSO) algorithm to optimize BP neural network, established foaming quality control model of asphalt and a parameter optimization model of asphalt, and improved the prediction accuracy and generalization capability.

GAO YM et al.[11] used the BP neural network model optimized by PSO and the normal BP neural network model to predict the diesel oil production, and concluded that the PSO-BP model has a higher prediction accuracy than the normal BP model.

The PSO algorithm has the advantages of easy realization, fast convergence speed, and high efficiency, and can make up for the disadvantages of BP neural network such as slow learning process speed and easy falling into local minimum points[9,11]. Therefore, BP neural network optimized by PSO algorithm was proposed in this paper to simulate methane chemical looping reforming reaction, and to use the model to predict the experimental result data under specific experimental conditions and to study the optimal oxygen carrier system preparation and reaction conditions, providing valuable reference for the prediction of results, system analysis, process optimization and other chemical processes.

2. PSO-BP neural network model

2.1. BP neural network model

In this paper, the relevant research data of methane chemical chain reforming reaction in literature[12] is selected. The experiment based on $Ce(NO_3)_2.6H_2O$, with $Fe(NO_3)_3.9H_2O$, $Mn(NO_3)_3.6H_2O$, $Cu(NO_3)_2.3H_2O$ as doping phase, using two different co-precipitation preparation a series of cerium based composite metal oxides. 419 groups of experimental data were extracted from this experiment. After normalization, 80% of the data groups were randomly selected as training samples and the remaining 20% as test samples.

According to the extracted data, Selection of Compound metal oxide type, molar ratio of ceria oxide, molar ratio of composite oxide, preparation method, calcination temperature, calcination time, reaction temperature, reaction time and cycle numbers as input variables, CH_4 conversion rate, CO selectivity and H_2 selectivity were selected as the output variables. Therefore, the establishment of the BP neural network model of the input layer node number is 9, output layer node number is 3.

2.1.1 Selection of the number of hidden layer nodes The number of nodes in the hidden layer of the BP neural network determines the network structure and has a great influence on fault tolerance, information processing ability and generalization ability of the network. In this paper, the approximate range of the number of nodes in the hidden layer is determined by the formula, and the number of nodes with the best comprehensive performance of the BP model is found out through the trial-and-error method, which is the optimal number of nodes in the hidden layer.

$$k = \sqrt{m+n} + a$$

Where k is the number of hidden layer nodes, m is the number of input layer nodes, n is the number of output layer nodes, and a is a constant between 0 and 10.

2.1.2 Selection of activation function In this paper, the tansig function that can accelerate the convergence rate is chosen as the activation function from the input layer to the hidden layer and from the hidden layer to the output layer of the BP neural network. Its function expression is

$$f = \frac{2}{(1+e^{-2n}) - 1}$$

2.1.3 Selection of learning algorithm Adaptive learning rate refers to the timely adjustment of learning rate according to the objective function of the network. In the process of solving the network, the moving step size more in line with the current situation is adopted according to the actual situation, so as to make the convergence of the network more intelligent, which will improve the convergence speed and convergence ability of the network.

2.2. PSO algorithm optimization

The PSO algorithm optimizes the initial weight threshold. Based on the BP neural network model, the weight threshold is treated as a particle. The PSO algorithm is used to globally optimize and find the relative optimal initial weight threshold. The neural network model is called For the PSO-BP model. In the PSO-BP model, the population size is 60, the number of evolution is 30, the learning factors c1 and c2 are both 1.49445, the position change of particles in the search space is set within $[X_{min}, X_{max}]=[-5, 5]$, and the velocity change is set within $[V_{min}, V_{max}]=[-1, 1]$.

3. Results and discussion

Figure 1 shows that when the number of hidden layer nodes is 11, the training error of the network is the smallest, and the network of correlation of training samples and testing samples was higher than other numbers of hidden layer nodes. Therefore, the optimal number of hidden layer nodes in this model is 11. Finally, the PSO-BP neural network structure was selected to be 9-11-3.

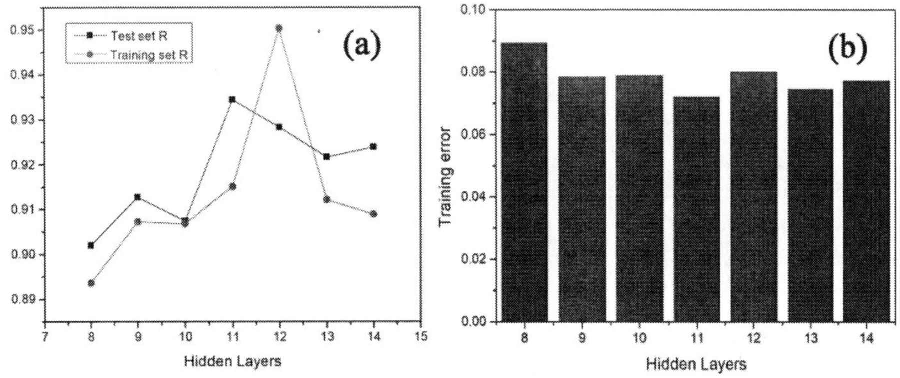

Figure 1. Network performance of different hidden layer nodes

Figure 2 shows that the fitting degree of the PSO-BP model to the training sample data is 0.9847, and the fitting degree of the test sample data can reach 0.97926, indicating that the PSO-BP model has a good training effect.

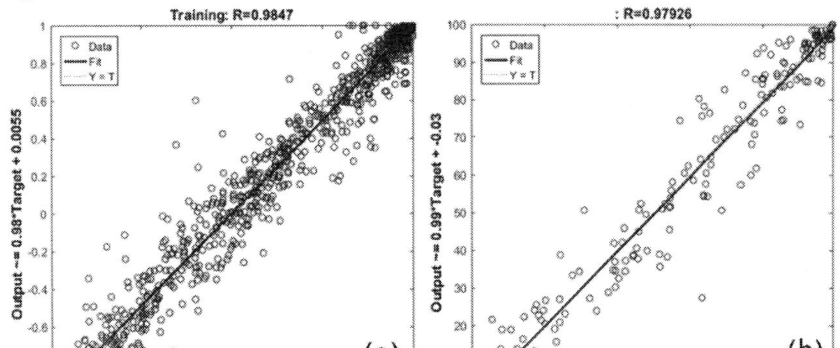

Figure 2. Fitting degree of PSO-BP model to sample data (a is training data, b is prediction data)

It can be concluded from figure 3 and figure 4 that in the first 8000 times of network training, the learning rate is between 1-2, the particles move in the solution space to search for solutions with big step size, and the mean square error of the network decreases rapidly. After about 8000 times of training, the learning rate of the network fluctuates between 0-1, the moving step size of particles in the solution space shrinks and global optimization continues to approach the optimal solution, the decrease rate of the mean square error of the network gradually decreases and eventually tends to be stable, the final training mean square error is 0.013509, and the learning rate is finally fixed at

0.083453. It shows that the PSO-BP neural network model has good convergence ability.

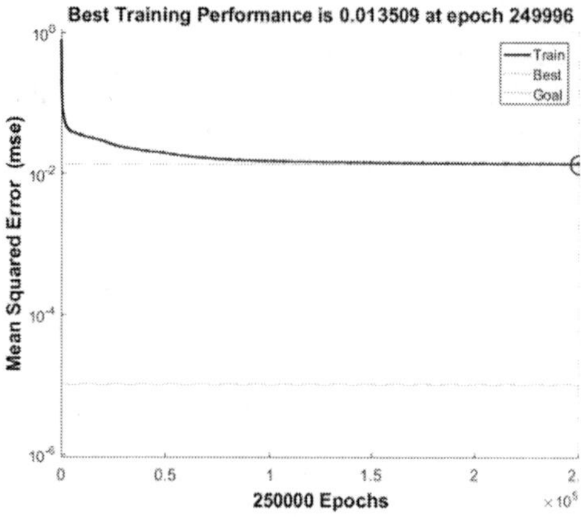

Figure 3. The training mean square error curves of PSO-BP model

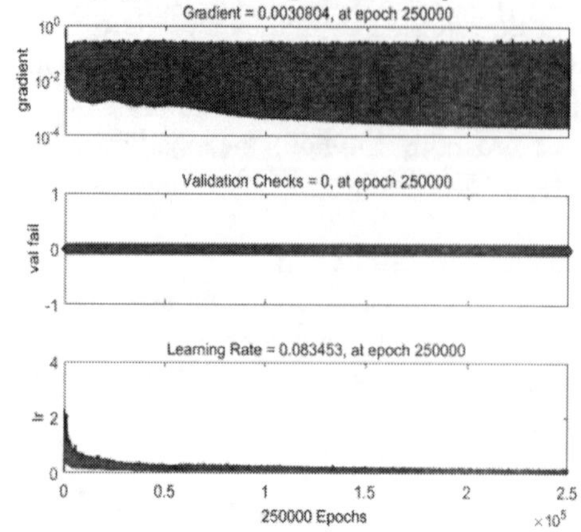

Figure 4. Training gradient curve of PSO-BP model

After optimizing the initial weight threshold, the fitting degree of the PSO-BP model for training samples and test samples was above 0.979, and the learning ability and generalization capability of the network were strong. Therefore, PSO-BP model is a good simulation and prediction model for methane chemical looping reforming reaction.

3.1. Prediction of optimal reaction conditions for methane chemical looping reforming reaction

We extracted 419 groups of experimental conditions and corresponding experimental results from the literature. The experimental conditions worthy of exploration in this system were listed and sorted out, and 123,552 groups of experimental conditions were obtained. By using PSO-BP model to predict these experimental conditions, all the corresponding experimental results can be obtained. All the experimental results were optimized to obtain the optimal preparation conditions and fixed bed reaction conditions for the composite oxygen carrier system when the CH_4 conversion rate, CO selectivity, and H_2 selectivity were optimized. The relevant data are shown in table 1.

Table 1 Best optimal oxygen carrier preparation conditions and reaction conditions.

Experimental condition	Parameters
Compound metal oxide type	$Fe(NO_3)_3.9H_2O$
Molar ratio of ceria oxide	0.7
Molar ratio of composite oxide	0.3
Preparation method	co-precipitation method (No.2)
Calcination temperature	800°C
Calcination time	6h
Reaction temperature	850°C
Reaction time	13min
Cycle numbers	0

PSO-BP neural network simulation to find the best experimental conditions, compared with the experimental data and the related conditions set of data 419 is extracted methane chemical looping reforming reaction results are in agreement. From the 419 sets of actual data, 123,552 sets of highly predictive data were obtained, and the data expansion rate was 293.87 times. It can be seen that the PSO-BP model can provide a valuable reference for finding the optimal oxygen carrier preparation conditions and reactor reaction conditions.

4. Conclusions

Based on the experimental data of methane chemical looping reforming reaction, the PSO-BP model was established with compound metal oxide type, molar ratio of ceria oxide, molar ratio of composite oxide, preparation method, calcination temperature, calcination time, reaction temperature, reaction time and cycle numbers as input variables and CH_4 conversion rate, CO selectivity and H_2 selectivity as output variables. PSO algorithm which solves the instability of BP neural network initial weights and thresholds randomly selected causes problems predictions, improved stability and prediction accuracy of the network. In addition, the PSO-BP model was used to make a complete set prediction of the experimental data and select the optimal reaction conditions, and the results were consistent with the facts. It is shown that the model has good prediction ability and can provide a valuable reference for the chemical process such as result prediction, system analysis, and process optimization.

Acknowledgements

The authors gratefully acknowledge the financial support from the Joint Funds of National Key R&D Program of China (2018YFB0604603), the National Natural Science Foundation of China (21536009), Key R & D Program of Shaanxi Province (2018ZDXM-GY-167), and Shaanxi provincial education department serves local special project (17JF029).

References

[1] Li ZG, Cheng H, Gu TY. (2019) Research on dynamic relationship between natural gas consumption and economic growth in China[J]. Structural Change and Economic Dynamics, 49: 334-339.

[2] Shen Y, Zhao K, He F, Li HB. (2019) The structure-reactivity relationships of using three-dimensionally ordered macroporous $LaFe_{1-x}Ni_xO3$ perovskites for chemical-looping steam methane reforming[J]. Journal of the Energy Institute, 92(2): 239-246.

[3] Zhao K, He F, Huang Z, et al. (2016) CaO/MgO modified perovskite type oxides for chemical-looping steam reforming of methane[J]. Journal of Fuel Chemistry and Technology, 44(6): 680-688.

[4] Di ZC, Cao Y, Yang FL, Zhang K, Cheng FQ. (2019) Thermodynamic analysis on the parametric optimization of a novel chemical looping methane reforming in the separated productions of H2 and CO[J]. Energy Conversion and Management, 192: 171-179.

[5] Huang Z, He F, Zhao K, Zheng AQ, Li HB, Zhao ZL. (2012) Synthesis Gas Production by Chemical-Looping Reforming of Methane Using Lattice Oxygen[J]. Progress in Chemistry,24(8): 1599-1609.

[6] Lu XY, Lin PJ, Cheng SY, Lin YH, Chen ZC, Wu LJ, Zheng QY. (2019) Fault diagnosis for photovoltaic array based on convolutional neural network and electrical time series graph[J]. Energy Conversion and Management, 196: 950-965.

[7] Ma DY, Liang YC, Zhao XS, Guan RC, Shi XH. (2013) Multi-BP expert system for fault diagnosis of powersystem. Engineering Applications of Artificial Intelligence, 26(3): 937-944.

[8] Zhang J, Feng YT, Al-Mahrouqi M H. (2011) Reliable optimal control of a fed-batch fermentation process using ant colony optimisation and bootstrap aggregated neural network models[J]. Computer Aided Chemical Engineering, 29: 663-667.

[9] Wang AL, Fu ZS, Liu FM.(2018) Asphalt foaming quality control model using neural network and parameters optimization[J], International Journal of Pavement Research and Technology, 11, 4: 401-407.

[10] Wang XP, Zhang F, Ding JL, Kung H, Latif A, Verner C. Johnson. (2018) Estimation of soil salt content (SSC) in the Ebinur Lake Wetland National Nature Reserve (ELWNNR), Northwest China, based on a Bootstrap-BP neural network model and optimal spectral indices[J]. Science of The Total Environment, 615: 918-930.

[11] Gao YM, Xing YF, Fu J, Zhang W, Zhao JH. (2017) The application of BP neural network based on a particle swarm optimization to the catalytic cracking reaction-regeneration process[J]. Computers and Applied Chemistry, 11:899-903.
[12] Li KZ. (2008) Preparation of cerium-based composite oxide and its partial oxidation of methane to syngas[D]. Kunming: Kunming University of Science and Technology.

IOP Conf. Series: Materials Science and Engineering **729** (2020) 012108 doi:10.1088/1757-899X/729/1/012108

CeO_2 promoted the performance of Ni/Al_2O_3 catalyst for CRM reaction

Penggang Lv[1], Xiaoxun Ma[1] and Long Xu[1,*]

[1]School of Chemical Engineering, Northwest University, International Science & Technology Cooperation Base of MOST for Clean Utilization of Hydrocarbon Resources, Chemical Engineering Research Center of the Ministry of Education for Advanced Use Technology of Shanbei Energy, Shaanxi Research Center of Engineering Technology for Clean Coal Conversion, Collaborative Innovation Center for Development of Energy and Chemical Industry in Northern Shaanxi, Xi'an, 710069, PR China.

[*]Corresponding author's e-mail: longxuxulong@163.com

Abstract: The promoted effect of CeO_2 on the reaction of Ni/Al_2O_3 catalyst in the syngas of CO_2 reforming of methane (CRM) was studied. The results showed that the catalyst prepared with CeO_2 addition exhibited high activity and stability, and the conversions of CH_4 and CO_2 could reach 96% at 800 °C when the amount of Ce was 5 wt%. SEM, H_2-TPR, CH_4-TPSR and CO_2/O_2-TPO characterization indicated that addition CeO_2 could enhance the interaction between support and Ni species in preparation process, enabling the capacity of catalyst to resist carbon deposition and metal sintering.

1. Introduction

The CO_2 reforming of methane(CRM) reaction (CH_4+CO_2=2CO+2H_2 ΔH°_{298}= +247.3 kJ/mol) can not only consume CH_4 and CO_2, but also produce synthesis gas with low H_2/CO ratio, which is an ideal raw material for production of hydrocarbons via Fischer-Tropsch synthesis and oxygenated chemicals[1].

Because CRM is an extremely endothermic and complex reaction process, the development of catalysts with high activity is the focus of the study[2]. In the process of CRM reaction to generate syngas, it will be affected by competitive side reactions, which will increase the probability of carbon deposition, such as CH_4 decomposition reaction (CH_4=C+2H_2 ΔH°_{298}= +74.9 kJ/mol), RWGS reaction (CO_2+H_2⇋CO+H_2O ΔH°_{298}= +41.2 kJ/mol), and the boudouard reaction (2CO⇋CO_2+C ΔH°_{298}= -172.4 kJ/mol). Therefore, the preparation of catalyst with high activity, long-term stability and low cost is the focus of current research[3]. At present, the catalysts widely used in CRM reaction are mainly noble and non-noble metals, such as Ru, Rh, Pt, Ni, CO, Fe, perovskite oxide, and core-shell catalysts[4-5]. Although noble metal catalysts have higher activity and carbon deposition resistance than non-noble metal catalysts, the high cost and sacristy restrict its application in industry.

Several factors, such as active metals, supports, promoters, synthesis method, calcination temperatures, and particle size, affected the activity and carbon deposition of catalysts[6]. Different active metals such as Ni and supports such as Al_2O_3, MgO, and CeO_2 have been investigated as catalyst in CRM reaction[7]. It is necessary to select appropriate metal as catalyst promoter to enhance the physical and chemical properties of Ni based catalyst and to improve the inhibition of carbon deposition. In this respect, considering that CeO_2 is a special oxide with good oxidation potential (Ce^{4+}/Ce^{3+}), high oxygen storage and high temperature resistance, this property can improve the dispersion of Ni species on the catalyst surface and the gasification performance of carbon deposition[8-9]. Daza et al.[10] investigated the performance of Ni-Ce/clay catalyst in CRM reaction, and found that the addition of Ce not only improved the catalytic activity, but also reduced the formation of carbon deposition[11]. Wang et al.[12] also found the higher reducibility and metal dispersion of Ce promoted Ni/Al_2O_3 catalysts than Ni/Al_2O_3 catalyst.

In this paper, Ni&Ce/Al_2O_3 catalysts were synthesized by impregnation method. The effect of different Ce contents on the performance of catalyst was investigated. The performance of catalysts at different reaction temperatures was compared, and the promotion effect of Ce on the Ni/Al_2O_3 catalyst was analyzed.

2. Experimental

2.1. Catalyst preparation

Al_2O_3 was obtained by calcination $Al_2O_3{\cdot}nH_2O$ (AR, 98%, Chalco Shandong Branch, China) at 550 °C for 2 h, after cooling, grinding and sieving to the desired particle size of between 250 ~ 550 μm Al_2O_3 supports. The catalyst 5NixCe/Al_2O_3 was prepared by impregnation method, $Ni(NO_3)_2{\cdot}6H_2O$ (AR, 98%, Sichuan Xilong Chemical Co., Ltd., China) and $Ce(NO_3)_2{\cdot}6H_2O$ (AR, 98%, Zhengzhou Paini Chemical Reagent Factory, China) as precursors. 1g Al_2O_3 was added to the prepared Ni&Ce metal mixed solution (Ni: 5 wt%; Ce: 0, 1, 3, 5, 10 wt%) and treated by ultrasonic (40 kHz) for 20 min, then laying 18 h and dried overnight at 110 °C. Subsequently, the samples were calcined at 500 °C for 2 h.

2.2. Catalyst characterization

The surface morphology of catalysts were observed via the high resolution scanning electron microscope (HRSEM) using Zeiss EVO18, The operating parameters were 15.0 kV, 20 mbar of low vacuum and 3 nm of resolution, respectively. The temperature programmed reduction（TPR）analysis of the catalysts were carried out on Xianquan Industrial and Trading TP-5080 adsorption instrument. 50 mg catalyst sample was pretreated at 400 °C for 1 h under the condition of 30 mL/min N_2 flow. The experiment was performed from 100 to 900 °C under H_2/N_2 (5/30 mL/min) mixture gas. The contents of CH_4 and H_2 in the gas at the outlet of the reaction were monitored on-line by mass spectrometry. 0.1g catalyst was pretreated at 400 °C for 1h in Ar atmosphere. The experiment was performed from 100 to 900 °C (10 °C/min) under CH_4/Ar (5/30 mL/min) mixture gas. The types of carbon deposition were studied, after CH_4-TPSR analysis, the catalyst cooling down to room temperature in Ar, 5 mL/min CO_2/O_2 was introduced, The experiment was performed from 100 to 900 °C (10 °C/min), and the amount of CO/CO_2 in the reaction tail gas was monitored by mass spectrometry.

2.3. Activity evaluation

The catalyst activity in CRM process was evaluated in a quartz tube vertical fixed-bed reactor of 6 mm ID at atmospheric pressure. Before each run, 0.1 g of the catalyst was loaded in the quartz tube and 20 mL/min of Ar was introduced to replace the air in the reaction tube. The reaction temperature was controlled in the range of 600 ~ 900 °C (10 °C/min) under a flow of pure Ar. Before the reaction, the catalyst was reduced with flowing 5% H_2 diluted with Ar at 800 °C for 2 h. The CO_2 and CH_4 mixture with volume ratio of 1 and total flow rate of 40 mL/min were fed into the reactor once the temperature reached the desirable temperature, and the GHSV was 24 L/g/h. The reaction lasted for 4 h; the outlet gases were cooled and detected by an on-line mass spectrometer (Thermostar™ GSD320).

3. Results and discussion

Figure 1 (a) shows many pores and gullies on the surface of Al_2O_3 support prepared by $Al_2O_3{\cdot}nH_2O$, so it is suitable to be used as catalyst support. In figure 1(b), it is found that the pore structures of Al_2O_3 are filled by NiO after the loading of active metal. However, the dispersion effect of active component

Ni species on the catalyst surface is not ideal, and the particles formed are large. Figure 1(d) shows the $5Ni5Ce/Al_2O_3$ catalyst before reaction, the dispersion of Ni species on the surface of the catalyst containing Ce is relatively uniform. In addition, the catalysts after the reaction of Ni/Al_2O_3 and $5Ni5Ce/Al_2O_3$ are shown in figure 1(c) and (e), respectively. Compared with that before the reaction, the surface of Ni/Al_2O_3-4h catalyst appears sintering phenomenon, while $5Ni5Ce/Al_2O_3$-4h catalyst does not find obvious sintering. Therefore, this indicates that the dispersion of Ni in the catalyst can be improved by adding CeO_2, which can also improve sintering resistance of catalyst.

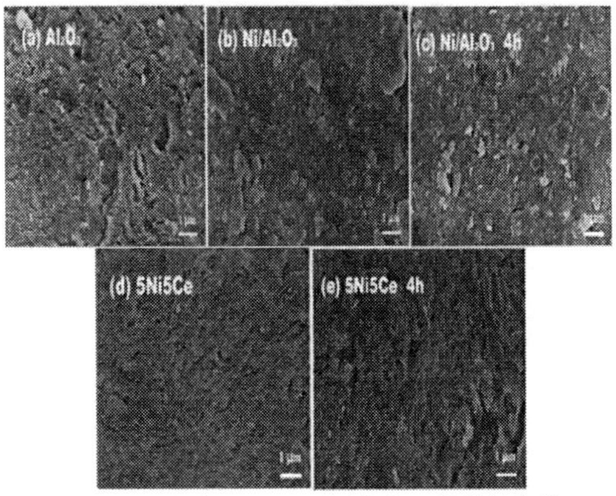

Figure 1. SEM images of Al_2O_3, $5Ni/Al_2O_3$ and $5Ni5Ce/Al_2O_3$ before and after reaction.

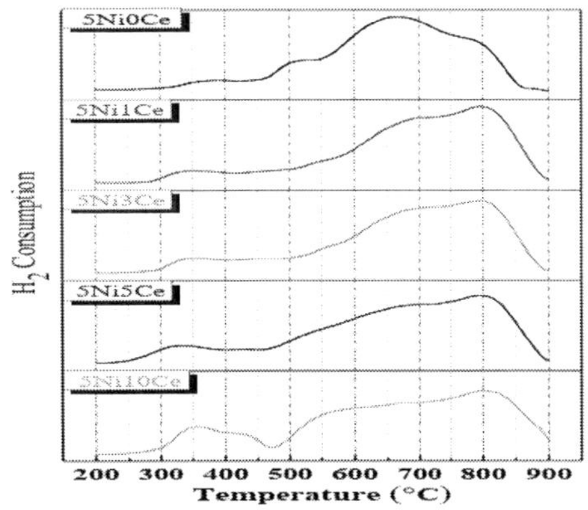

Figure 2. H_2-TPR profile of $5NixCe/Al_2O_3$ catalysts (x=0, 1, 3, 5, and 10 wt%).

The H_2-TPR profiles of catalysts with different Ce contents are shown in figure 2. The TPR profiles could be divided three forms of Ni species on the surface of metal support[13], the temperature zone between 300 °C and 550 °C is related the free-state on the support surface. The temperature zone between 550 °C and 750 °C is attributed to the dispersed-state. The temperature zone between 750 °C and 900 °C is corresponding to the fixed-state. The main reduction peak of 5Ni0Ce catalyst is around 650 °C, which belongs to NiO in dispersed state. However, the catalysts with CeO_2 promoter all showed a large H_2 consumption peak around 800 °C, which should be attributed to NiO in a fixed-state. This indicated a stronger association between NiO and the support, which can inhabit the sintering and loss of Ni particles during the reaction.

In order to study the catalytic effect of CeO_2 on methane decomposition and the mechanism of CRM reaction, figure 3 shows the curves of CH_4 and H_2 current values detected by mass spectrometry during methane temperature programmed surface reaction (TPSR). It can be observed from figure 3 that the current value of H_2 detected on 5Ni0Ce catalyst began to increase around 450 °C, while the current value of CH_4 began to decrease, and the production of H_2 and the consumption of CH_4 reached the maximum at 585 °C. Compared with Ce-containing catalyst, the values of H_2 and consumption of methane increase, and the values are the largest when the amount of Ce is 3%. In figure 3(b), there are three generation peaks of H_2, the main H_2 generation peak is at 600 °C, and the amount of H_2 generation tends to increase again at 850 °C. According to the reaction mechanism of CRM, the three peaks are attributed to H_2 released by CH_4 the stepwise decomposition to CH_X(x= 1, 2, 3). This is because the higher the degree of dissociation, the more demanding the reaction conditions. In addition, from the H_2 generation peaks and the CH_4 consumption peak positions of 5NixCe catalysts, the temperature corresponding to the maximum peaks of H_2 generation and CH_4 consumption moved to the low temperature zone as the amount of CeO_2 added increased. Therefore, the CeO_2 helps keep methane decomposition required temperature down. But the amount of H_2 and the consumption of methane begin to decrease when the amount of CeO_2 is more than 3 wt%. This may be due to the excessive CeO_2 covering the active metal Ni, reducing the number of active sites.

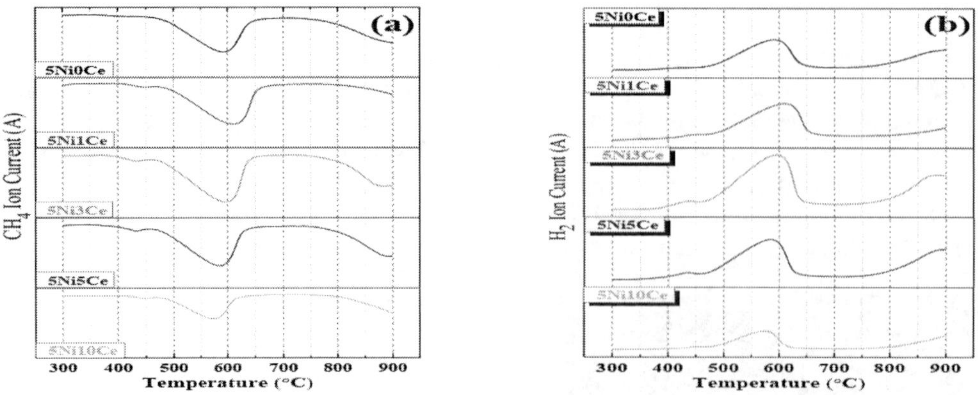

Figure 3. CH_4-TPSR profiles of $5NixCe/Al_2O_3$ catalysts (x=0, 1, 3, 5, and 10 wt%).

The carbon deposition in the CRM can be divided into amorphous carbon and graphite carbon. The amorphous carbon can react with CO_2 to produce CO; graphite carbon is relatively stable and has strong interaction with Ni, which makes it difficult to be gasified. In order to study the type of carbon deposition on the catalyst surface, figure 4(a) shows the CO_2-TPO test of the catalysts after the analysis of CH_4-TPRS. The gasification temperature of carbon on 5Ni0Ce surface area is about 617 °C, the CO formation temperature corresponding to 5Ni1Ce, 5Ni3Ce, 5Ni5Ce, and 5Ni10Ce is 625 °C, 615 °C, 610 °C, 570 °C, respectively. The results show that the gasification temperature of carbon on the surface of the catalyst decreases with the increase of Ce contents. During the CRM reaction, the oxygen vacancies contained in CeO_2 release a large amount of free electrons and transfer to the center of Ni atom, increasing the density of Ni atom electron cloud, so that the catalyst can absorb more CO_2 and promote the CO_2 to eliminate carbon deposition[14].

Figure 4(b) shows the O_2-TPO test of the catalyst after the analysis of CH_4-TPRS, 5Ni0Ce has a CO_2 formation peak at about 500 °C, which is attributed to carbon with weak binding to Ni, and the second CO_2 formation peak appears at 675 °C, which should be graphite carbon with stable catalyst surface. With the increase of CeO_2 contents, the formation peak of CO_2 gradually moves to the low temperature zone, which indicates that the gasification temperature of graphite carbon decreases and the addition of CeO_2 can weaken the interaction between graphite carbon and Ni.

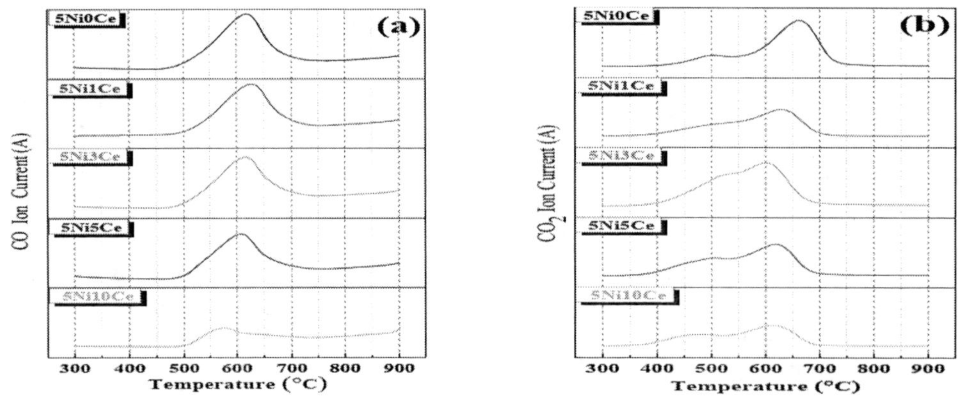

Figure 4. CO_2-TPO (a) and O_2-TPO (b) profiles of $5NixCe/Al_2O_3$ (x=0, 1, 3, 5, 10 wt%).

Figure 5 shows the performance of $5NixCe/Al_2O_3$ catalysts in CRM reaction at different temperatures. The conversions of CH_4 and CO_2 could reach 96% at 800 °C when the amount of Ce is 5 wt%. It can be seen from the conversions that the catalytic performance of catalyst with CeO_2 is significantly improved, especially under 800 °C. This indicates that CeO_2 can significantly improve the catalytic activity of nickel based catalysts. The H_2/CO ratio decreases with the increase of Ce contents, but the H_2/CO ratio corresponding to the catalyst increases when the Ce addition amount is 10%. It is indicated that the optimal addition amount of Ce is 3~5 wt%.

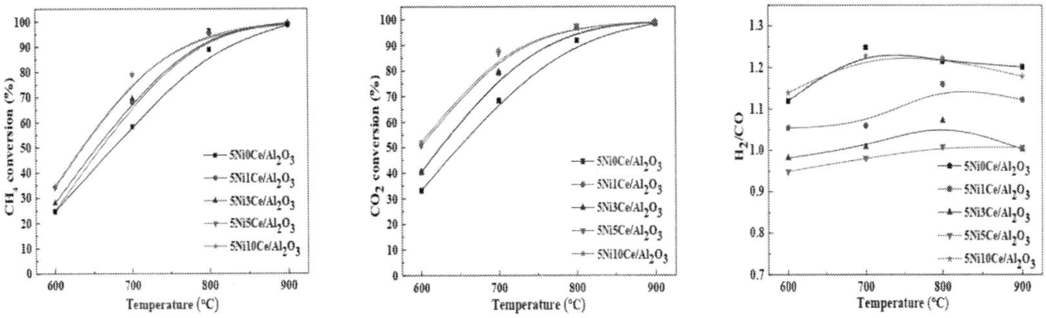

Figure 5. The catalytic performance of $5NixCe/Al_2O_3$: GHSV=24 $L{\cdot}g^{-1}{\cdot}h^{-1}$, CH_4/CO_2=1.

4. Conclusions

The effect of different amounts of CeO_2 on the catalytic performance of $5NixCe/Al_2O_3$ in CRM reforming was investigated. The addition of CeO_2 significantly enhanced the catalytic performance of the Ni/Al_2O_3 catalyst. TPR profiles showed Ce strengthen the interaction between Ni species and support, and improved the sintering resistance of the catalyst. The analysis of CH_4-TPRS and TPO showed that the addition of Ce not only reduced the energy required for CH_4 cracking, but also promoted the carbon deposition on the catalyst to be easily eliminated, which improved the carbon deposition resistance of the catalyst.

Acknowledgments

The authors gratefully acknowledge the financial support from the Joint Funds of National Key R&D Program of China (2018YFB0604603), the National Natural Science Foundation of China (21536009), Key R & D Program of Shaanxi Province (2018ZDXM-GY-167), and Shaanxi provincial education department serves local special project (17JF029).

References

[1] F. Zhang, Z. Song, J. Zhu, J. Sun, X. Zhao, Y. Mao, L. Liu, W. Wang. (2018) Factors influencing CH_4-CO_2 reforming reaction over Fe catalyst supported on foam ceramics under microwave irradiation, International Journal of Hydrogen Energy. 43: 9495–9502.

[2] G. Zhang, Y. Dong, M. Feng, Y. Zhang, W. Zhao, H. Cao. (2010) CO_2 reforming of CH_4 in coke oven gas to syngas over coal char catalyst, Chemical Engineer journal, 156: 519–523.

[3] Y. Zhang, G. Zhang, B. Zhang, F. Guo, Y. Sun, (2011) Effects of pressure on CO_2 reforming of CH_4 over carbonaceous catalyst, Chemical Engineer journal. 173: 592–597.

[4] K. Sutthiumporn, T. Maneerung, Y. Kathiraser, S. Kawi. (2012) CO_2 dry-reforming of methane over $La_{0.8}Sr_{0.2}Ni_{0.8}M_{0.2}O_3$ perovskite (M = Bi, Co, Cr, Cu, Fe): Roles of lattice oxygen on CH activation and carbon suppression, International Journal of Hydrogen Energy. 37: 11195–11207.

[5] Li Z, Li M, Bian Z, Yasotha K, Sibudjing K. (2016) Design of Highly Stable and Selective Core/Yolk-Shell Nanocatalysts-A Review, Journal of CO_2 Utilization. 188: 324-341.

[6] Akbari E, Alavi S M, Rezaei M. (2018) CeO_2, Promoted Ni-MgO-Al_2O_3, nanocatalysts for carbon dioxide reforming of methane. Journal of CO_2 Utilization, 24: 128-138.

[7] Usman M, Wan Daud W M A, Abbas H F. (2015) Dry reforming of methane: Influence of process parameters-A review, Renewable and Sustainable Energy Reviews. 45: 710-744.

[8] S. Zeng, X. Fu, T. Zhou, X. Wang, H Su. (2013) Influence of pore distribution on catalytic performance over inverse CeO_2/Co_3O_4 catalysts for CH_4/CO_2 reforming, Fuel Processing

Technology. 114: 69-74.
[9] Rezaei M , Alavi S M , Sahebdelfar S. (2008) Effects of K_2O Promoter on the Activity and Stability of Nickel Catalysts Supported on Mesoporous Nanocrystalline Zirconia in CH_4 Reforming with CO_2. Journal of Natural Gas Chemistry, 17(3): 278-282.
[10] Daza C E, Kiennemann A, Moreno S, Molina R. (2009) Dry reforming of methane using Ni-Ce catalysts supported on a modified mineral clay, Applied Catalysis A General, , 364: 65-74.
[11] Laosiripojana N, Sutthisripok W, Assabumrungrat S. (2005) Synthesis gas production from dry reforming of methane over CeO_2 doped Ni/Al_2O_3: Influence of the doping ceria on the resistance toward carbon formation, Chemical Engineering Journal, 112: 13-22.
[12] Wang S, Lu G Q. (1998) Role of CeO_2 in Ni/CeO_2-Al_2O_3 catalysts for carbon dioxide reforming of methane, Applied Catalysis B: Environmental,19: 267-277.
[13] Wang S, Wang Y, Hu C. (2018) The effect of $NH_3 \cdot H_2O$ addition in Ni/SBA-15 catalyst preparation on its performance for carbon dioxide reforming of methane to produce H_2, International Journal of Hydrogen Energy, 43: 13921-13930.
[14] Yang Y, Li W, Xu H. (2002) A new explanation for the carbon deposition and elimination over supported Ni; Ni-Ce and Ni-Co catalysts for CO_2-reforming of methane. Reaction Kinetics and Catalysis Letters, 2002, 77(1):155-162.

IOP Conf. Series: Materials Science and Engineering **729** (2020) 012109 doi:10.1088/1757-899X/729/1/012109

Efficient synthesis of γ-valerolactone from ethyl levulinate over Ni/V_2O_5

Xi Cao[1], Heda Dong[2], Han Chen[1], Qiong Xu[1]*, Dulin Yin[1]

[1]National and Local Joint Engineering Laboratory for New Petro-Chemical Materials and Fine Utilization of Resources, College of Chemistry and Chemical Engineering, Hunan Normal University, 410081, Changsha, China

[2]College of Chemistry and Chemical Engineering, Central South University, 410081, Changsha, China

* Corresponding Author: Qiong Xu; email: xuqiong@hunnu.edu.cn; phone: +86-731-88872576; fax: +86-731-88872530.

Email of all the authors: 473166430@qq.com (Xi Cao), 1197587147@qq.com (Heda Dong), 2543164815@qq.com(Han Chen), xuqiong@hunnu.edu.cn(Qiong Xu), dulin-yin@126.com(Dulin Yin)

Abstract: γ-valerolactone (GVL) has been recognized as one of the most value-added chemicals derived from biomass. In this work, an efficient Ni/V_2O_5 catalyst was prepared via solid phase grinding and in-situ reduction. The Ni/V_2O_5 catalyst was employed for catalytic transfer hydrogenation (CTH) of ethyl levulinate (EL) to GVL using 2-propanol as a hydrogen source. An excellent GVL selectivity of 93.04 % at EL conversion of 97.40 % could be achieved at 180 °C in 4 h using 30% Ni/V_2O_5 as the catalyst. This catalyst was able to be reused for five times with the GVL selectivity decreased a little. A possible adsorption mechanism of generate active H species in CTH of EL over Ni/V_2O_5 was proposed.

1. Introduction

As demand for chemicals and fuels increases and fossil resources are about to be depleted, it is imperative to find sustainable alternatives for petroleum-based products. γ-valerolactone (GVL) is an important platform compound derived from renewable biomass [1-5], which may find use as a green solvent, a fuel additive, precursor to alkene based transportation fuels or an intermediate compound to prepare functionlized materials because of its excellent physical and chemical properties such as high boiling point, low volatility, easy degradation, and low toxicity. Thus, the synthesis of GVL is one of the pivotal steps in the transformation of biomass resources into liquid fuels and high-value chemicals.

The biomass derivative levulinic acid (LA) and its esters are direct starting materials for synthesis of GVL. Extensive catalytic routes for the synthesis of GVL from LA/levulinate esters have been continuously improved in recent years [6, 7]. As well-known, metal catalysts have good catalytic activities for the hydrogenation reaction. The Pd [8, 9], Pt [10], Ru [11-13], Ir [14], Rh [15] are the most common precious metals used for the hydrogenation of LA/levulinate esters to GVL, and they exhibited fairly satisfactory catalytic performance under an optimal reaction condition. Even though the precious metal catalysts show good catalytic effects, the high economic cost cannot be ignored. So, much attention has been focused to the nonprecious metal catalysts.

Among the nonprecious metal catalysts, Cu [16], Co [17], Ni [18] catalysts exhibited considerable catalytic performances in this reaction, especially the Ni supported catalysts. Ni supported [19] by Al_2O_3, SiO_2, ZnO, ZrO_2, TiO_2 and MgO were prepared to investigate the synthetic efficiency of GVL from LA. 30 % Ni/SiO_2 can achieve the highest GVL yield of nearly 90 % at 250 °C in the vapor phase at atmospheric pressure. Sakakibara et al. [20] reported Ni/ZrO_2 for the transfer hydrogenation of methyl levulinate (ML) and LA to GVL using 2-propanol as a hydrogen donor to show the yield of 94 % and 92 %, respectively. Fang et al. [21] reported in situ synthesis of biomass-derived Ni/C catalyst for the hydrogenation of LA to GVL, and a high yield of 94.5 % can be obtained. The CTH of ethyl levulinate (EL) to GVL over Cu-Ni bimetallic catalyst with 2-butanol as the hydrogen donor was carried out and provided a 97% yield of GVL product in 12 h at 150 °C [22]. Activated carbon-supported bimetallic nickel-iron catalyst (Ni-Fe/AC) was employed to catalyse the EL converting into GVL in the liquid phase. At 100 °C, 4 MPa H_2, the reaction was carried out for 6 h, and the EL was completely converted, and the obtained GVL yield was 99% [23]. Most of the reported metal based catalysts prepare by impregnation or coprecipitation followed by reduction in H_2 gas. To the best of our knowledge, the Ni loaded V_2O_5 catalyst prepared via a solid phase grinding and in-situ reduction method hasn't been reported in transfer hydrogenation of EL into GVL yet.

With the aim of developing new active, stable catalytic systems through cost effective system, in this work, this green and simple preparation method for metal supported catalyst was used to produce Ni/V_2O_5 catalysts. This method needs no solvent or reductive H_2 gas. No extra reduction is needed before using the metal catalysts. High catalytic activity of the Ni/V_2O_5 catalysts was obtained in synthesis of GVL from ethyl levulinate.

2. Material and Methods

2.1. Materials

Ethyl levulinate (EL, 99%), γ-valerolactone (GVL, 98%), and n-dodecane (98%) were purchased from Aladdin industrial corporation (Shanghai, China). 2-propanol (A.R.) was purchased from Hunan hui-hong reagent co., Ltd (Hunan, China). Citric acid monohydrate (A.R.), nickel nitrate hexahydrate (A.R.), aluminum nitrate nonahydrate (A.R.) and nitrate ferric nitrate (A.R.) were purchased from Sinopharm Chemical Reagent Co., Ltd. Vanadium pentoxide (A.R.) was purchased from Xiangzhong Chemical Reagent Development Center. All chemicals were used as received without further purification.

2.2. Catalysts preparation

Ni/V_2O_5 was prepared by solid phase grinding and in situ reduction method [24]. In a typical procedure, $Ni(NO_3)_2 \cdot 6H_2O$ (1.455 g) and citric acid monohydrate (1.26 g), were mixed and grinded at room temperature for 0.5 h to form the nickel citrate complex. Then V_2O_5 (0.685 g) was added and grinded for another 0.5 h. The paste precursor dried in a drying oven for 12 h and calcined for 3 h at 350 °C with ramp rate of 3 °C /min under nitrogen atmosphere. The catalyst was described as 30% Ni/V_2O_5, represents the loading of Ni in wt% is 30%. Ni/V_2O_5 catalyst with different loading of 5%, 10%, 20%, and 40% were prepared by changing the V_2O_5 feeding amount.

The preparation procedure of Ni/Al_2O_3 and Ni/Fe_2O_3 were very similar to that of Ni/V_2O_5 except that the corresponding nitrates of Al_2O_3 and Fe_2O_3 were added to the nickel citrate complex.

All catalysts were stored in vacuum desiccator and used directly.

2.3. Catalytic activity

Catalytic conversion of EL to GVL was carried out in a 100 mL autoclave with Teflon lined reactor under magnetic stirring. In a typical experiment, the batch was loaded with ethyl levulinate (EL, 5 mmol), 2-propanol (19 mL) as a hydrogen donor and solvent, n-dodecane (1 mmol) as an internal standard and the catalyst (216 mg), and then purged five times with N_2 at atmospheric conditions. The reaction mixture was heated to the prescribed temperature for a desired reaction time with stirring at

500 rpm. After reaction, the reactor was cooled down to room temperature, and the reaction mixture was separated from the catalyst magnetically. The reaction mixture was analyzed by a Flame Ionization Detector (FID) equipped gas chromatograph, GC 2014 (Shimadzu Instruments, Japan) with HP-5 column (30.0 m×0.50 mm×0.32 μm) and the products were confirmed by GCMS-QP2010) (Shimadzu Instruments, Japan).

Generally, EL conversion and selectivity of GVL, ethyl 4-hydroxypentanoate (4-HPE), and isopropyl levulinate (IPL) are calculated using the following formula:

$$\text{EL conversion (\%)} = \left(1 - \frac{\text{Mole of EL after reaction}}{\text{Initial mole of EL}}\right) \times 100 \quad (1)$$

$$\text{GVL selectivity (\%)} = \frac{\text{Mole of GVL produced}}{\text{Mole of EL converted}} \times 100 \quad (2)$$

$$\text{4-HPE selectivity (\%)} = \frac{\text{Mole of 4-HPE produced}}{\text{Mole of EL converted}} \times 100 \quad (3)$$

$$\text{IPL selectivity (\%)} = \frac{\text{Mole of IPL produced}}{\text{Mole of EL converted}} \times 100 \quad (4)$$

2.4. Recycle experiments

The stability of the 30%Ni/V_2O_5 was tested in conversion of EL to GVL under the optimal condition. Autoclave containing the separated catalyst was evacuated and again charged with fresh EL solution for a subsequent catalytic cycle.

2.5. X-Ray Diffraction analysis

XRD patterns of Ni/V_2O_5 catalysts were recorded with an Ultima-IVX-ray powder diffractometer (Rigaku Corporation, Japan) using Ni filtered Cu Ka radiation (λ= 1.5406 Å). Samples were scanned in the 2θ range of 5~80° at 40 kV and 40 mA with a scan speed of 20°/min.

3. Results and discussion

3.1. XRD characterization of Ni/V_2O_5

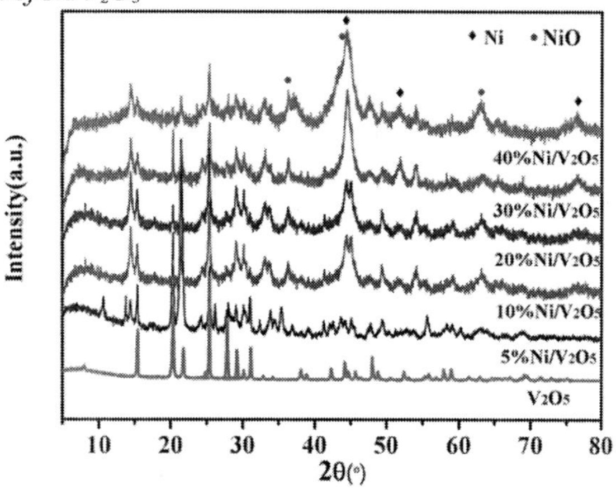

Figure 1. XRD patterns of the Ni/V_2O_5 catalysts with different Ni content

The Crystal structures of the Ni/V_2O_5 catalysts with different Ni loading as well as V_2O_5 support were detected by X-ray powder diffractometer, and results were shown in *Fig. 1*. Compared with the XRD pattern of V_2O_5, the XRD patterns of all the supported catalysts showed metallic Ni and NiO phases. XRD reflections at 44.49° (111), 51.85° (200) and 76.38° (220) correspond to metallic Ni with a face-centered cubic geometry (ICDD. No. 04-002-3695) and the diffraction peaks at 37.24° (111), 43.28°(200) and 62.85° (220) corresponding to NiO phase (ICDD. No. 97-002-8834)[25, 26]. The intensity of these peaks increased as the Ni loading in the catalysts increased.

3.2. Catalytic performance of Ni supported catalysts in hydrogenation of EL to GVL

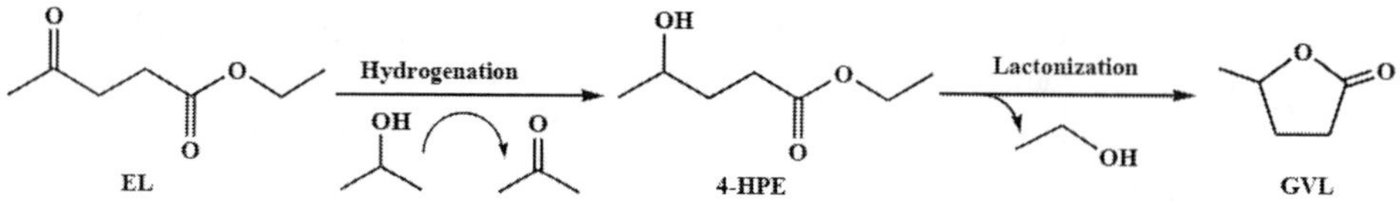

Scheme 1. The pathway of CTH of EL to GVL using 2-propanol as the hydrogen donor.

Table 1. Catalytic transfer hydrogenation of EL to GVL by Ni supported catalysts.

Entry	Catalysts	Conversion /%	Selectivity/%		
			GVL	4-HPE	IPL
1	Ni/Al_2O_3	29.70	88.57	4.39	7.04
2	Ni/Fe_2O_3	81.24	89.12	8.79	2.10
3	Ni/V_2O_5	97.40	94.50	2.72	2.78

Reaction conditions: ethyl levulinate 5 mmol (0.721 g), 19 mL 2-propanol, 216 mg catalyst, reaction temperature 180 °C, reaction time 4 h.

Three different metal oxides Al_2O_3, Fe_2O_3, and V_2O_5 were chosen as the supports to prepare Ni based catalysts. Their catalytic activities were tested in transfer hydrogenation of EL to GVL using 2-propanol as a hydrogen source and solvent. Results were shown in *Table 1*. All the catalysts showed good selectivity of main product GVL while had a significantly difference on conversion of EL. The high selectivity of GVL in all reactions may benefit to the highly selective reduction of carbonyl groups to alcoholic hydroxyl groups in the presence of 2-propanol as hydrogen donors [1]. Ni/V_2O_5 prepared from an acidic oxide support gave the highest EL conversion of 97.40%. In addition to the main product GVL, ethyl 4-hydroxypentanoate (4-HPE) as well as isopropyl levulinate (IPL) were detected by GC and identified by GC-MS. No other intermediate or by-products were detected. The intermediate product 4-HPE is produced by the hydrogenation of EL, which could be esterified and cyclized to synthesize GVL [27]. The pathway of catalytic transfer hydrogenation of EL to GVL using 2-propanol as the hydrogen donor was shown in *Scheme 1*. By-product IPL generated from the trans-esterification of EL and 2-propanol under the action of acidic active sites provided by NiO, V_2O_5 or Al_2O_3. However, the superior EL conversion in Ni/V_2O_5 catalytic system may also due to the appropriate amount acidic active sites in Ni/V_2O_5 to promote the lactonization of 4-HPE. More results discussed in the following part further proved this point. Furthermore, literatures [28, 29] also proved that modulating the Lewis acid or base contents of the catalysts is an effective measure to promote the lactonization efficiency of the catalysts in the CTH of alkyl levulinate to GVL.

In a subsequent experiment, Ni/V_2O_5 was selected as a catalyst to explore the reaction conditions required for its optimum catalytic activity.

3.3. Effect of Ni loading on the hydrogenation of EL

Table 2. Effect of Ni/V_2O_5 with different Ni loading on hydrogenation of EL to GVL

Entry	Load/%	Conversion/%	Selectivity/%		
			GVL	4-HPE	IPL
1	5	25.26	2.56	-	97.45
2	10	16.86	5.67	-	94.43
3	20	16.42	14.26	-	85.74

4	30	97.40	94.50	2.72	2.78
5	40	97.45	95.04	2.98	1.98

Reaction conditions: ethyl levulinate 5 mmol (0.721 g), 19 mL 2-propanolt, 216 mg catalyst, reaction temperature 180 °C, reaction time 4 h.

As can be seen from *Table 2*, when Ni loading was lower than 30%, the EL conversion rate decreased and GVL selectivity increased slightly with the increase of the Ni loading. When the load increased from 20% to 30%, both EL conversion rate and GVL selectivity showed a sudden increase. Oppositely, IPL selectivity decreased dramatically. This may because the loading amount of 30% in Ni/V_2O_5 can give enough Ni metal active sites for the hydrogenation reaction. When the metal catalytic sites is not enough, the acid sites in Ni/V_2O_5 played a dominate role in the catalytic reaction to produce by-product IPL. On the contrary, as the Ni content continued to increase with the loading amount, there were enough metal active sites to reduce EL to the intermediate 4-HPE, which further generated GVL. So the EL conversion rate and selectivity of GVL increased. Obviously, $Ni/V_2O_5$with 30% Ni loading has an appropriately ratio of metal active sites to acid active sites. 30% was chosen as the optimal loading amount.

3.4. *Effect of catalyst dosage on the hydrogenation of EL*

Table 3. Effect of Ni/V_2O_5 dosage on hydrogenation of EL to GVL

Entry	Dosage/%	Conversion/%	Selectivity/%		
			GVL	4-HPE	IPL
1	5	36.39	36.13	35.16	28.71
2	10	70.35	72.58	14.25	13.59
3	20	93.12	68.15	18.05	13.14
4	30	97.40	94.50	2.72	2.78
5	40	96.14	80.75	17.57	1.68

Reaction conditions: ethyl levulinate 5 mmol (0.721 g), 19 mL 2-propanol, catalyst 30% Ni/V_2O_5, reaction temperature 180 °C, reaction time 4 h.

Catalyst dosages from 5% to 30% based on the mass of the substrate were investigated and results were seen from *Table 3*. The EL conversion and GVL selectivity gradually increased, while the intermediate product 4-HPE gradually decreased. Correspondingly, selective of the acid catalysed product IPL decreased with the Ni active centres increased. These results indicate the good catalytic activity of Ni active centres in transfer hydrogenation reaction. However, with the continuous increase of the catalyst dose to 40%, GVL selectivity declined and 4-HPE increased. The possible reason may as follows: the catalytic action of acidic sites on the Ni/V_2O_5 may be suppressed in the presence of too much Ni active centres in the reaction system, so that it delayed transformation of the intermediate into GVL. This, on the contrary, proved the catalytic action of the acid sites on the lactonization of 4-HPE, which discussed in section 3.2. Within the range examined, the optimum catalyst dosage was 30%.

3.5. *Effect of reaction time on the hydrogenation of EL*

Reaction time has a great influence on the EL conversion and products' selectivities, the trends were shown in *Table 4*. The catalytic reaction trend along with the reaction time in CTH of EL to GVL over Ni/V_2O_5 is very similar to that of catalyst dosage. When prolonging reaction time to 4 h, the EL can be transferred into GVL nearly completely, with very little 4-HPE and IPL existed.

Table 4. Effect of reaction time on hydrogenation of EL to GVL over Ni/V_2O_5.

Entry	Time/h	Conversion/%	Selectivity/%		
			GVL	4-HPE	IPL
1	1	21.22	36.325	46.73	16.95
2	2	86.74	85.35	10.11	4.53
3	4	97.40	94.50	2.72	2.78
4	6	98.63	92.07	6.53	1.40

Reaction conditions: ethyl levulinate 5 mmol (0.721 g), 19 mL 2-propanol as solvent, catalyst 30% Ni/V_2O_5, reaction temperature 180 °C.

3.6. Effect of reaction temperature on the hydrogenation of EL

Table 5. Effect of reaction temperatures on hydrogenation of EL to GVL over Ni/V_2O_5

Entry	temperature/°C	Conversion/%	Selectivity/%		
			GVL	4-HPE	IPL
1	150	7.00	2.11	-	97.89
2	160	79.78	67.76	26.35	5.89
3	170	93.48	79.08	19.07	1.85
4	180	97.40	94.50	2.72	2.78
5	190	97.87	86.83	11.52	1.65

Reaction conditions: ethyl levulinate 5 mmol (0.721 g), 19 mL 2-propanol, catalyst 30% Ni/V_2O_5, reaction time 4 h.

Reaction temperature is also an important condition effect the catalytic activity of the catalyst. Catalytic active of Ni/V_2O_5were investigated in different reaction temperature in the range of 150 to 190 °C. The results were shown in *Table 5*. EL conversion was only 7.00% when the reaction was carried out at 150 °C, and the product was mainly IPL, which means the Ni/V_2O_5 catalyst almost had no activity under the reaction temperature of 150 °C. Interestingly, when increasing the reaction temperature to 160 °C, conversion rate of EL and selectivity of GVL were significantly increased to 79.8% and 67.76%, and continuing increasing the reaction temperature to 180 °C, both values kept increasing. In the meantime, the intermediate 4-HPE decreased gradually with the reaction temperature increasing. Based on the above results, only the reaction temperature reached 160 °C can provide sufficient energy to over the catalytic energy barrier in this reaction, and high reaction temperature promoted the catalytic reaction rate. Thus, the suitable reaction temperature is 180 °C.

3.7. Recycling experiments

Table 6. Results of recycling experiments

Entry	Usage times	Conversion/%	Selectivity/%		
			GVL	4-HPE	IPL
1	1	97.04	94.50	2.72	2.78
2	2	92.11	88.40	7.77	3.72
3	3	90.87	81.03	12.03	5.57
4	4	86.34	76.08	16.63	5.94
5	5	88.92	72.29	19.39	5.64

Reaction conditions: substrate 5 mmol (0.721g) ethyl levulinate, 19 mL 2-propanol, 216 mg Ni/V_2O_5, reaction temperature 453 K, reaction time 4 h.

To investigate the recyclability of the 30%Ni/V_2O_5 catalyst, recycle tests for the CTH of EL to GVL were performed under the optimized reaction condition. After each run, the catalyst was separated by magnet, and washed with 2-propanol for 3 times. Then it subjected to the next run. The results are presented in table 6. The catalyst was reused for four catalytic runs with only a slight decrease in the EL conversion. But GVL selectivity decreased from 94.50 % to 72.29 %, and the selectivity of intermediate 4-HPE increased from 2.72 % to 19.39 % in the 5 runs, namely the lactonization of intermediate product 4-HPE to synthesize GVL was slowing down or be blocked. These results indicated that the metal active sites kept well while the acidic active sites may be covered or reduced in the recycle experiments.

3.8. The adsorption mechanism in CTH of EL over Ni/V_2O_5

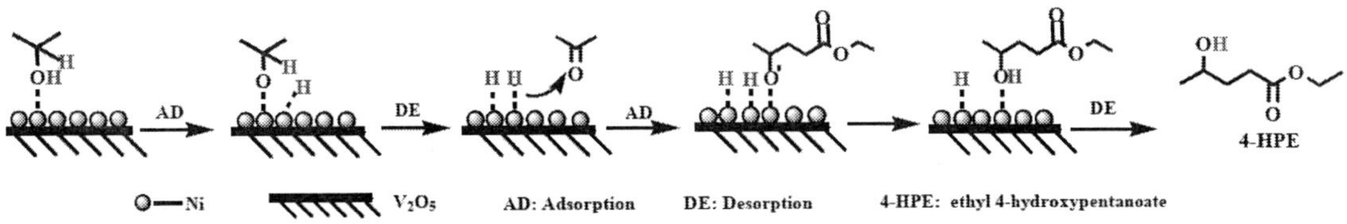

Scheme 2. The adsorption mechanism of generating H active species in CTH of EL over Ni/V_2O_5

Based on the above results, the synthesis of GVL from EL catalysed by Ni/V_2O_5 using 2-propanol as a hydrogen source has two step reactions. Firstly, ethyl levulinate was hydrogenated under the action of Ni catalytic active sites and then the intermediate 4-HPE esterified and cyclized to generate GVL catalysed by the acidic active sites, as shown in *scheme 1*. The use of secondary alcohols as hydrogen sources in CTH is different from the traditional hydrogenation. Thus, the possible adsorption mechanism in CTH of EL over Ni/V_2O_5 to generate the reductive H species was proposed in *scheme 2*. Firstly, the hydroxyl group of 2-propanol was adsorbed on the Ni metal active sites to make a cleavage of O-H bond and generate a surface alkoxy group on the Ni surface. At the same time, 2-propanol is easily undergoing the β-H elimination step after the generation of an alkoxy group on the Ni surface. During the elimination process, a ketone molecule was released and two H-atoms remained on the Ni surface which was the active species for transfer hydrogenation. Then, this was helpful in the transfer of the adsorbed atomic hydrogen to the keto group of EL which was present on the same surface/adjacent surface sites, resulting in the formation of the intermediate 4-HPE on the Ni surface.

4. Conclusions

In this work, the efficient Ni/V_2O_5 catalysts were prepared via a solid phase grinding and in-situ reduction method. An excellent GVL selectivity of 93.04 % at EL conversion of 97.40 % could be achieved at 180°C in 4 h using Ni/V_2O_5 with 30 % Ni loading as a catalyst in CTH of EL to GVL. The catalyst can be reused for 5 times with the GVL selectivity decreased. A possible adsorption mechanism of generating H active species in CTH of EL over Ni/V_2O_5 was proposed. This work provided a green and efficient catalytic route for the synthesis of GVL from bio-derived ethyl levulinate. The solid phase grinding and in-situ reduction can be used as a green and efficient method for preparation of metal supported catalysts in numerous of hydrogenation reactions.

Acknowledgements

This research was financially supported by the National Natural Science Foundation of China (Grant No. 21975070) and Provincial College Students Innovation Experiment Program (S201910542066)

References

[1] Yu, Z.H.; Lu, X.B.; Liu, C.; Han, Y.W.; Ji, N. (2019). Synthesis of γ-valerolactone from different biomass-derived feedstocks: Recent advances on reaction mechanisms and catalytic systems. J. Renewable and Sustainable Energy Reviews, 112: 140-157.

[2] Osatiashtiani, A.; Lee, A.F.; Wilson, K. (2017). Recent advances in the production of γ-valerolactone from biomass-derived feedstocks via heterogeneous catalytic transfer hydrogenation. J. Chem Technol Biotechnol, 92: 1125-1135.

[3] Yan, K.; Yang, Y.Y.; Chai, J.J. (2015). Catalytic reactions of γ-valerolactone: A platform to fuels and value-added chemicals. J. Appl. Catal. B, 179: 292-304.

[4] Alonso, M. D.; Wettstein, S. G.; Dumesic, J. A. (2013). Gamma-valerolactone, a sustainable platform molecule derived from lignocellulosic biomass. Green Chem., 15: 584-595

[5] Bai, Y.Y.; Su, S.H.; Wang, S.Z. (2018). Catalytic conversion of carbohydrates into 5-ethoxymethylfurfural by a magnetic solid acid using gamma-valerolactone as a co-solvent. J.Energy Tech., 6: 1951-1958.

[6] Yan, K.; Jarvis, C.; Gu, J.; Yan, Y. (2015). Production and catalytic transformation of levulinic acid: a platform for speciality chemicals and fuels. Renew Sustain Energy Rev, 51: 986-997.

[7] Du, X.; Liu, Y.; Wang, J.; Cao, Y.; Fan, K. (2013). Catalytic conversion of biomass-derived levulinic acid into γ-valerolactone using iridium nanoparticles supported on carbon nanotubes. J. Chin J Catal, 34: 993–1001.

[8] Yan, K.; Jarvis, C.; Lafleur, T.; Qiao, Y.; Xie, X. (2013). Novel synthesis of Pd nanoparticles for hydrogenation of biomass-derived platform chemicals showing enhanced catalytic performance. J. RSC Advances, 3: 25865-25871.

[9] Yan, K.; Lafleur, T.; Jarvis, C.; Wu, G.S. (2014). Clean and selective production of γ-valerolactone from biomass-derived levulinic acid catalyzed by recyclable Pd nanoparticle catalyst. J. Cleaner Product., 72: 230–232.

[10] Upare, P.P.; Lee, J.M.; Hwang, D.W.; Halligudi, S.B; Hwang, Y.K.; Chang, J.S. (2011). Selective hydrogenation of levulinic acid to γ-valerolactone over carbon-supported noble metal catalysts. J. Ind. Eng. Chem., 17: 287-292.

[11] Selva, M.; Gottardo, M.; Perosa, A. (2012). Upgrade of Biomass-derived levulinic acid via Ru/C-catalyzed hydrogenation to γ-valerolactone in aqueous-organic-ionic liquids multiphase systems. J. ACS Sustain. Chem. Eng., 1: 180-189.

[12] Sudhakar, M.; Kantam, M.L.; Jaya, V.S.; Kishore, R.; Ramanujachary, K.V.; Venugopal, A. (2014). Hydroxyapatite as a novel support for Ru in the hydrogenation of levulinic acid to γ-valerolactone, Catal. Commun., 50: 101-104.

[13] Hengne, A.M.; Biradar, N.S.; Rode, C.V. (2012). Surface species of supported Ruthenium catalysts in selective hydrogenation of levulinic esters for bio-refinery application. Catal. Lett., 142: 779-787.

[14] Li, W.; Xie, J.H.; Lin, H.; Zhou, Q.L. (2012). Highly efficient hydrogenation of biomass-derived levulinic acid to γ-valerolactone catalyzed by iridium pincer complexes. Green Chem., 14: 2388-2390.

[15] Manzer, L.E. (2004). Catalytic synthesis of α-methylene-γ-valerolactone: a biomass-derived acrylic monomer. J. Appl. Catal. A, 272: 249-256.

[16] Hengne A.M.; Rode C.V. (2012). Cu-ZrO_2 nanocomposite catalyst for selective hydrogenation of levulinic acid and its ester to γ-valerolactone. J. Green Chem, 14: 1064-72.

[17] Long, X.; Sun,P.; Li, Z.; Lang, R.; Xia, C.; Li,F. (2015). Magnetic Co/Al_2O_3 catalyst derived from hydrotalcite for hydrogenation of levulinic acid to γ-valerolactone. Chin. J. Catal., 36: 1512–1518.

[18] Rong, Z.; Sun, Z.; Wang, L.; Lv, J.; Wang, Y.; Wang, Y. (2014). Efficient conversion of levulinic acid into γ-valerolactone over Raney Ni catalyst prepared from melt-quenching alloy. Catal. Lett., 144: 1766-1771.

[19] Baithy, M.; Putla, S.; Bellala, V.S.R. (2018). Nanostructured Nickel/Silica catalysts for continuous flow conversion of levulinic acid to γ-valerolacton. J. ACS Omega, 3: 16839-16849.

[20] Sakakibara, K.; Endo, K.; Osawa, T. (2019). Facile synthesis of γ-valerolactone by transfer hydrogenation of methyl levulinate and levulinic acid over Ni/ZrO_2. Catal. Commun., 125, 52-55.

[21] Fang, S.Q.; Cui, Z.B.; Zhu,Y.T.; Wang, C.G.; Bai, J.; Zhang, X.H.; Xu,Y.; Liu, Q.Y.; Chen, L.G.; Zhang, Q.; Ma, L.L. (2019). In situ synthesis of biomass-derived Ni/C catalyst by self-reduction for the hydrogenation of levulinic acid to γ-valerolactone, Energy Chem., 37: 204-214.

[22] Cai, B.; Zhou, X.C.; Miao, Y.C.; Luo, J.Y.; Pan, H.; Huang, Y.B. (2016). Enhanced catalytic transfer hydrogenation of ethyl levulinate to γ-valerolactone over a robust Cu–Ni bimetallic catalyst. ACS Sustain. Chem. Eng., 5: 1322-1331.

[23] Li, C.; Xu, G.Y.; Zhai, Y.X. (2017). Hydrogenation of biomass-derived ethyl levulinate into gamma-valerolactone by activated carbon supported bimetallic Ni and Fe catalysts. Fuel, 203: 23-31.

[24] Venkatesha, N.J.; Ramesh, S. (2018). Citric acid-assisted synthesis of nanoparticle copper catalyst supported on an oxide system for the reduction of furfural to furfuryl alcohol in the vapor phase. Ind. Eng. Chem. Res., 57: 1506-1515.

[25] Mohan, V.; Raghavendra, C.; Pramod, C.V.; Raju, B.D.; Rama Rao, K.S. (2014). Ni/H-ZSM-5 as a promising catalyst for vapour phase hydrogenation of levulinic acid at atmospheric pressure. RSC Advances, 4: 9660-9668.

[26] Kumar, V.V.; Naresh, G.; Sudhakar, M.; Anjaneyulu, C.; Bhargava, S.K.; Tardio, J.; Reddy,V.K.; Padmasri, A.H.; Venugopal, A. (2016). An investigation on the influence of support type for Ni catalysed vapour phase hydrogenation of aqueous levulinic acid to γ-valerolactone. RSC Advances, 6: 9872-9879.

[27] Gürbüz, E.I.; Alonso, D.M.; Bond, J.Q.; Dumesic, J.A. (2011). Reactive extraction of levulinate esters and conversion to γ-valerolactone for production of liquid fuels. ChemSusChem, 4: 357–361.

[28] Kuwahara, Y.; Magatani, Y.; Yamashita, H. (2015). Ru nanoparticles confined in Zr-containing spherical mesoporous silica containers for hydrogenation of levulinic acid and its esters into γ-valerolactone at ambient conditions. J. Catal. Today, 258: 262–269.

[29] He, J.; Li, H.; Lu, Y.M.; Liu, Y.X.; Wu, Z.B.; Hu, D.Y. (2016). Cascade catalytic transfer hydrogenation-cyclization of ethyl levulinate to γ-valerolactone with Al-Zr mixed oxides. J. Appl Catal A Gen, 510: 11–19.

IOP Conf. Series: Materials Science and Engineering **729** (2020) 012110 doi:10.1088/1757-899X/729/1/012110

Preparation of nanofluids and their applications in enhanced CO_2 absorption

Huiyu Han[1], Xiaoxun Ma[1] and Long Xu[1*]

[1]School of Chemical Engineering, Northwest University, International Science & Technology Cooperation Base of MOST for Clean Utilization of Hydrocarbon Resources, Chemical Engineering Research Center of the Ministry of Education for Advanced Use Technology of Shanbei Energy, Shaanxi Research Center of Engineering Technology for Clean Coal Conversion, Collaborative Innovation Center for Development of Energy and Chemical Industry in Northern Shaanxi, Xi'an, 710069, PR China.

[*]Corresponding author's e-mail: longxuxulong@163.com

Abstract: Nanofluids opened up research in the field of thermal energy and mass transfer engineering when it was first proposed by American scientist Choi. In this paper, a two-step method was used to prepare stable nanofluids. Enhancement factor was introduced to study the enhancement of CO_2 absorption by nanofluids under different experimental conditions. The results show that the enhancement of the three nanoparticles in the same solid content is $TiO_2 > Al_2O_3 > SiO_2$; The optimum solid content of SiO_2 and Al_2O_3 is 0.8 g/L, and the optimum solid content of TiO_2 is 0.6 g/L; Within the scope of this study, the larger the particle size, the better the absorption effect. As the organic amine concentration increases, absorption of solution improves.

1. Introduction

In 1969, Ramachandran and Sharma first theoretically inferred that for slurry systems with fast chemical reactions and sparingly soluble particles, the particles would significantly accelerate the rate of gas absorption[1]. Later, Kars et al. used the stirred tank to study the effect of activated carbon particles on the gas-liquid mass transfer in the Na_2SO_3-O_2 absorption system, and confirmed the existence of the strengthening effect[2]. This discovery motivates that we will expand our research into nanoscale particles[3-5].

In 1995, the concept of nanofluids was first proposed by Choi from the National Laboratory of Argonne, USA that Nanofluids are stable and uniform two-phase suspensions are obtained by dispersing nanoparticles in an organic or inorganic liquid medium in a certain proportion and manner[6]. In theory, nanofluids belong to a colloidal dispersion system, and its properties are related to the properties, content, scale, shape and physical and chemical properties of the nanoparticles[7-10].

The concept of nanofluids was first applied to the field of heat transfer. These studies have basically found that the addition of nanoparticles can increase the thermal conductivity of the base fluid, providing a new idea for enhancing heat transfer[11]. According to the similarity between heat transfer and mass transfer, nanoparticles should also affect the mass transfer process inside the fluid. Therefore, research on mass transfer in nanofluids has also begun.

At present, the research on gas-liquid mass transfer of nano-particles is mainly focused on the physical absorption of insoluble gases. There are few studies on the absorption process of chemical reactions. Therefore, the effect of nanoparticles on gas-liquid mass transfer in the presence of chemical reactions was studied in the context of organic amines absorbing CO_2[12]. In this paper, 20nm, 40nm and 60nm SiO_2, TiO_2 and Al_2O_3 nanoparticles were used, and MEA and MDEA with large difference in absorption rate were selected to prepare nanofluids with good stability. The effects of particle parameters such as nanoparticle type, particle size and solid content on absorption were studied.

2. Experimental section

2.1. Preparation of nanofluids

In this experiment, nanofluids was prepared by a two-step method. Firstly, Monoethanolamine (AR, Tianjin Komio Chemical Reagent Co., Ltd.), N-Methyldiethanola Mine (99%, Tianjin Komio Chemical Reagent Co., Ltd.) having a mass fraction of 30% was separately disposed, and the nano-SiO_2, TiO_2, Al_2O_3 (Hebei Gegui Welding Material Co., Ltd.) was weighed according to the solid content of 0.4-1.6 g/L with an electronic balance (Shenyang Longteng Electronics Co., Ltd.), and added to the solution. The mixture was stirred for 20 min with a DF-101S collector-type constant temperature heating magnetic stirrer (Zhengzhou Kefeng Instrument Equipment Co., Ltd.), and ultrasonically dispersed for 40 min using a ultrasonic cell pulverizer (Ningbo Xinyi ultrasonic equipment Co., Ltd.).

2.2. Stability of nanofluids

The advantage of the "two-step method" is that it is easy to operate and is suitable for the preparation of most nanofluids. The disadvantage is that the prepared nanofluids have poor stability. In order to reduce this effect, Generally, we will add appropriate dispersing agents or surface chemical modification of the nanoparticles during the preparation process[13].

In this experiment, four different types of dispersing agents, sodium dodecyl benzene sulfonate(AR,Tianjin Tianli Chemical Reagent Co., Ltd.), sodium polyacrylate(AR, Tianjin Damao Chemical Reagent Factory), Triton X-100(AR, Xibao Biotechnology Co., Ltd.), N, N, N-trimethyl-1-dodecanaminium bromide(99%, Tianjin Dengfeng Chemical Reagent Factory), were selected. Figure 1 shows the morphology of the nanofluids with four dispersants added in sequence after standing for 24 h. It can be seen from the figure that the nanofluid with sodium dodecyl benzene sulfonate has no obvious precipitates, indicating that the dispersant basically meets the experimental requirements.

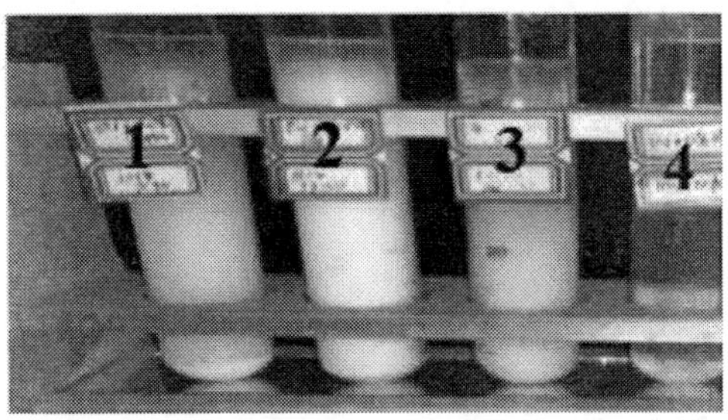

1-Nanofluid prepared by adding sodium dodecyl benzene sulfonate; 2-Nanofluid prepared by adding sodium polyacrylate; 3-Nanofluid prepared by adding Triton X-100; 4-Nanofluid prepared by adding N, N, N-trimethyl-1-dodecanaminium bromide.

Figure1. Effect of dispersant on the stability of nanofluids.

Chemical modification refers to the use of a chemical reagent as an intermediate medium to bind nanoparticles to a base liquid molecule. 1.5 g of nano-SiO_2 was ultrasonically dispersed in 14 mL of toluene, 0.6 g of triethylamine was added and stirred uniformly, by the addition of 0.8 g of trimethyl chlorosilane, and condensed and refluxed at 120 °C for 4 h. The obtained suspension was centrifuged, and the precipitate was washed five times with toluene, after vacuum drying at 60 °C, the surface-grafted modified nano-SiO_2 spheres were fabricated.

2.3. Nanofluid enhanced mass transfer

2.3.1. Introduction to experimental equipment and principles. As shown in figure 2, N_2 and CO_2 are depressurized by a pressure reducing valve and then enter the mass flow device to control the total mixed gas flow rate of 67 mL/min, and the initial volume fraction of CO_2 is 10.45%. The mixed gas enters the absorption liquid from the nozzle at the bottom of the absorber. During the ascending process, a part of CO_2 is absorbed, and the other part is discharged with the N_2 from the upper outlet of the absorber. After the exhaust gas is dried, it enters the infrared flue gas analyzer, and the CO_2 absorption and absorption rate are calculated by measuring the CO_2 volume fraction in the exhaust gas for characterization, To characterize the absorption capacity of the absorption liquid for CO_2.

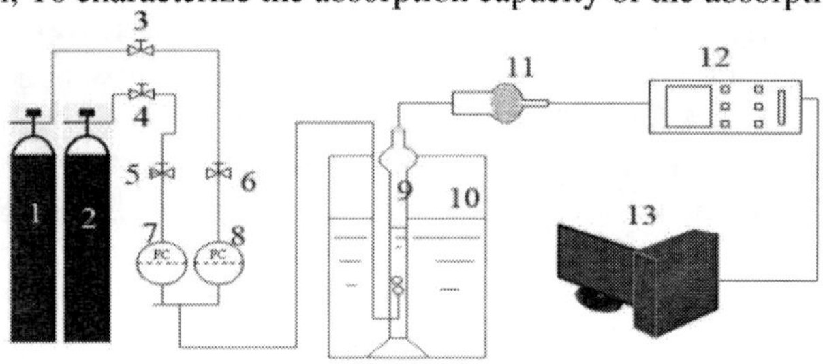

1-N_2; 2-CO_2; 3, 4-pressure reducing valve; 5, 6-normal valve; 7-8-mass flow controller; 9-absorber; 10-constant water tank; 11-drying tube; 12-infrared flue gas analyzer; 13-computer

Figure 2. absorption experiment device.

2.3.2. Data analysis method. In order to reflect the difference in absorption capacity between the two-component nanofluid and the organic amine solution, we introduce an absorption enhancement factor:

$$E = \left(\frac{V_{CO_2-nanofluid}}{V_{CO_2-pure-solution}} \right)_{\text{Have the same fluiddynamic conditions}} \quad (2\text{-}1)$$

3. Experimental results and discussion

3.1 Nanoparticle-MDEA Nanofluid Absorption Experiment

Figure 3 shows the breakthrough curves of 40nm, 0.4-1.6 g/L TiO_2-MDEA nanofluid and MDEA solution. The figure shows that the volume fraction of CO_2 in the exhaust gas of the nanofluid experiment is lower than that of the blank experiment, indicating that more CO_2 is absorbed, which proves that the addition of TiO_2 nanoparticles enhances the absorption of CO_2.

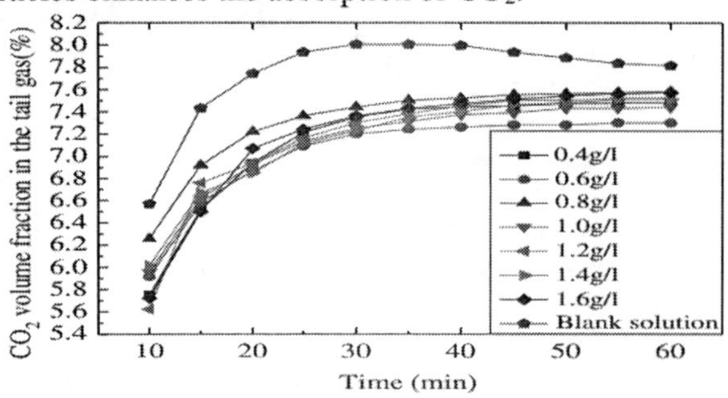

Figure 3. Comparison of MDEA blank solution and TiO_2 nanofluid absorption experiment.

Figure 4 compares the variation of the enhancement factors of TiO_2, SiO_2 and Al_2O_3 with the solid content, the particle size is the same as 40nm. The enhancement effect of the four nanoparticles at the same solid content is TiO_2>Al_2O_3>SiO_2. SiO_2 , even appears to hinder absorption. Because the SiO_2 nanoparticles are highly hydrophilic, causing an increase in the viscous resistance of the particle motion and a decrease in the disturbance to the mass transfer boundary layer; TiO_2 particles have a certain adsorption effect on CO_2, so their enhancement factor is larger than other particles.

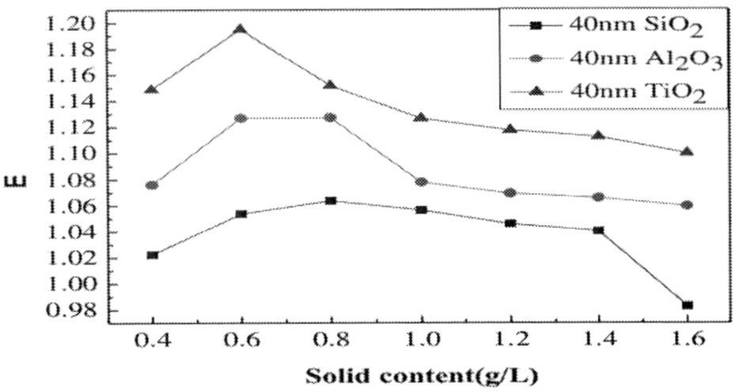

Figure 4. Effect of nanoparticle types on absorption enhanced ticks.

For the same type of nanoparticles, increasing the solid content means an increase in the number of particles, and the possibility of agglomeration of the nanoparticles increases. In addition, the surface tension and dynamic viscosity of the nanofluids also change slightly, which respectively affect the detachment radius of the bubble at the nozzle and the residence time in the liquid phase[14]. Figure 5 shows the variation of the enhancement factor of several nanoparticles used in the experiment with the solid content. The optimum solid contents of SiO_2, Al_2O_3, and TiO_2 were 0.8 g/L, 0.8 g/L, and 0.6 g/L, respectively, because the TiO_2 particles were easily agglomerated. The TiO_2 and Al_2O_3 nanoparticles have a high density, once agglomeration occurs, the enhancement factor will decrease rapidly.

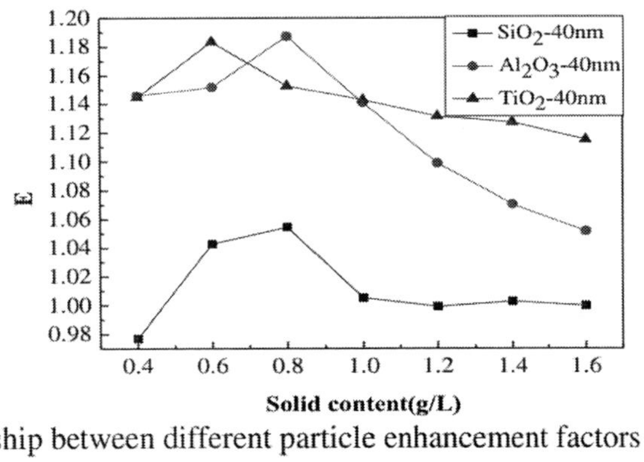

Figure 5. Relationship between different particle enhancement factors and solid content.

3.2. Nanoparticle-MEA nanofluid absorption experiment

Figure 6 compares the enhancement factors of TiO_2-MEA nanofluids at three particle sizes. The enhancement factor of the nanofluid decreases as the particle size decreases, because the particle size decreases, the mass of the particles decreases, and the motion inertia decreases; In addition, the relative surface area of the particles increases, and the viscous resistance increases. Both of which cause the particle's ability to disturb the mass transfer boundary layer to be weakened.

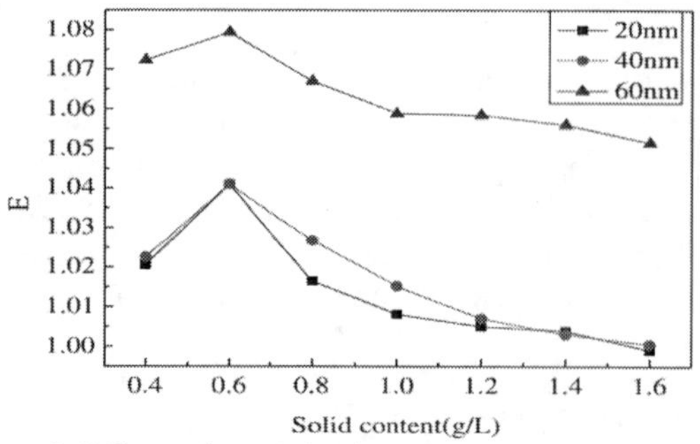

Figure 6. Effect of particle size on enhancement factor.

Comparing figure 7, the absorption curves of the two amine solutions vary greatly. This is because the nitrogen atom in the MDEA structure is not directly connected to the hydrogen atom, and the zwitterion cannot be formed during the reaction with CO_2. Therefore, MDEA does not react directly with CO_2 like MEA. The higher the concentration of the organic amine, the greater the absorption of CO_2.

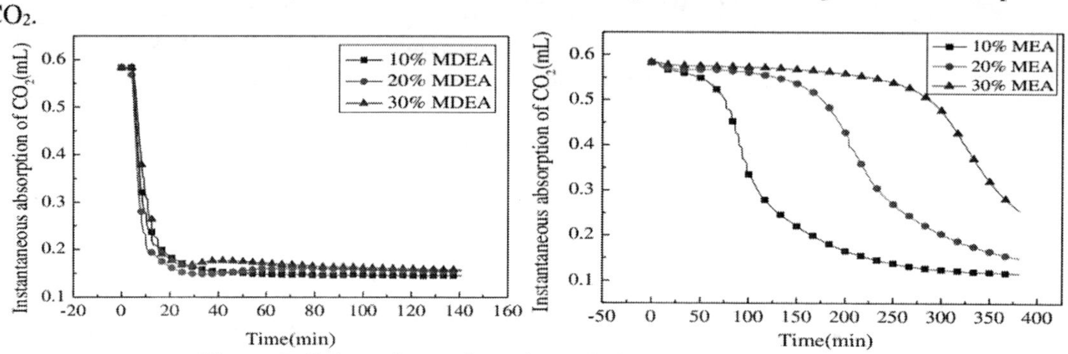

Figure 7. Effect of organic amine solution on CO_2 absorption.

4. Conclusions

In this paper, a two-step method was used to prepare nanofluids. At the same time, enhancement factors were introduced to study the effect of nanofluids on mass transfer efficiency. The following conclusions were obtained: the nanofluid prepared by adding sodium polyacrylate has good stability and meets the experimental requirements; the enhancement effect of nanoparticles on MDEA/CO_2 absorption system is more obvious than MEA/CO_2 system; under the same solid content, TiO_2 have the largest enhancement factor, and SiO_2 has the smallest, even inhibiting effect; there is an optimum solid content, the value varies depending on the type of particle; in the particle size range used herein, the smaller the particle size, the smaller the enhancement factor; within the concentration range of the experimental study, the higher the concentration of the organic amine, the greater the absorption of CO_2.

Acknowledgments

The authors gratefully acknowledge the financial support from the Joint Funds of National Key R&D Program of China (2018YFB0604603), the National Natural Science Foundation of China (21536009), Key R & D Program of Shaanxi Province (2018ZDXM-GY-167), and Shaanxi provincial education department serves local special project (17JF029).

References

[1] Ramchandran, P.A., Sharma, M.M. (1969) Absorption with fast reaction in a slurry containing sparingly soluble fine particles. J. Chemical Engineering Science, 24(11):1681-1686.

[2] Kars, R.L., Best, R.J., Drinkenburg, A. A. H. (1979) The sorption of propane in slurries of active carbon in water. J. The Chemical Engineering Journal, 17(2): 201-210.
[3] Daungthongsuk, W., Wongwises, S. (2007) A critical review of convective heat transfer of nanofluids. J. Renewable & Sustainable Energy Reviews, 11(5): 797-817.
[4] Wang, X.Q., Mujumdar, A.S. (2007) Heat transfer characteristics of nanofluids: a review. J. International Journal of Thermal Sciences, 46(1): 1-19.
[5] Ma, H.B., Wilson, C., Borgmeyer, B., et al. (2006) Effect of nanofluid on the heat transport capability in an oscillating heat pipe. J. Applied Physics Letters, 88(14): 14-16.
[6] Choi, S.U.S., Eastman, J.A. (1995) Enhancing thermal conductivity of fluids with nanoparticles. J. Asme Fed: 231(1), 99-105.
[7] Gurav, P., Naik, S.S., Ansari, K., et al. (2014) Stable colloidal copper nanoparticles for a nanofluid: production and application. J. Colloids & Surfaces A Physicochemical & Engineering Aspects, 441(3): 589-597.
[8] Nguyen, C.T., Desgranges, F., Roy, G., et al. (2015) Temperature and particle-size dependent viscosity data for water-based nanofluids – hysteresis phenomenon. J. International Journal of Heat & Fluid Flow, 28(6): 1492-1506.
[9] Mojarrad, M.S., Keshavarz, A., Ziabasharhagh, M., et al. (2014) Experimental investigation on heat transfer enhancement of alumina/water and alumina/water–ethylene glycol nanofluids in thermally developing laminar flow. J. Experimental Thermal & Fluid Science, 53(2): 111-118.
[10] Safaei, M.R., Togun, H., Vafai, K., et al. (2014) Investigation of heat transfer enhancement in a forward-facing contracting channel using fmwcnt nanofluids. J. Numerical Heat Transfer Part A Applications, 66(12), 1321-1340.
[11] Dhuria, R., Dalia, V., Sunthar, P. (2017) Diffusiophoretic enhancement of mass transfer by nanofluids. J. Chemical Engineering Science, 176: 632-640.
[12] Kim, J.H., Jung, C.W., Yong, T.K. (2014) Mass transfer enhancement during CO_2 absorption process in methanol/Al_2O_3 nanofluids. J. International Journal of Heat & Mass Transfer, 76(6): 484-491
[13] Wei, B., Zou, C., Li, X. (2017) Experimental investigation on stability and thermal conductivity of diathermic oil based TiO_2 nanofluids. J. International Journal of Heat & Mass Transfer, 104: 537-543.
[14] Yuan, Y., Li, X., Tu, J. (2017) The effects of nanoparticles on the lift force and drag force on bubbles in nanofluids: a two-fluid model study. J. International Journal of Thermal Sciences, 119: 1-8.
[15] Versteeg, G.F., Duck, L.A.J.V., Swaaij, W.P.M.V. (2010) Cheminform abstract: kinetics between CO_2 and alkanolamines both in aqueous and nonaqueous solutions. an overview. J. Cheminform, 28(6): no-no.

AUTHOR INDEX